Organosilicon Chemistry VI

Edited by N. Auner and J. Weis

Further Reading from Wiley-VCH

Jutzi, P., Schubert, U. (Eds.)
Silicon Chemistry
2003. 3-527-30647-1

Auner, N., Weis, J. (Eds.)
Organosilicon Chemistry V
From Molecules to Materials
2003. 3-527-30670-6

Meyer, G., Naumann, D., Wesemann, L., (Eds.)
Inorganic Chemistry in Focus II
2005. 3-527-30811-3

Schubert, U., Hüsing, N.
Synthesis of Inorganic Materials, 2nd Ed.
2005. 3-527-31037-1

Organosilicon Chemistry VI

From Molecules to Materials
Volume 1

Edited by Norbert Auner and
Johann Weis

WILEY-VCH Verlag GmbH & Co. KGaA

Prof. Dr. N. Auner
Department of Inorganic Chemistry
University of Frankfurt
Marie-Curie-Straße 11
60439 Frankfurt am Main
Germany

Prof. Dr. J. Weis
Consortium of Electrochemical Industry GmbH
Zielstattstraße 20
81379 Munich
Germany

Library of Congress Card-No.: applied for

A catalogue record for this book is available from the British Library.

Bibliographic information published by Die Deutsche Bibliothek
Die Deutsche Bibliothek lists this publication in the Deutsche Nationalbibliografie; detailed bibliographic data is available in the Internet at http://dnb.ddb.de

Printed on acid-free paper.

Printing: Strauss GmbH, Mörlenbach
Bookbinding: Litges & Dopf Buchbinderei GmbH, Heppenheim.
Printed in the Federal Republic of Germany

ISBN-13: 978- 3-527-31214-6
ISBN-10: 3-527-31214-5

Preface

In about 1990, the idea of establishing a national conference on organosilicon chemistry was born, bringing together researchers from academia and industry for scientific discussion and exchange of experience and knowledge in order to strengthen the organosilicon-based community, particularly in Germany. When we started with the first Munich Silicon Days in 1992, jointly organized by the *Gesellschaft Deutscher Chemiker* and *Wacker-Chemie GmbH,* we were overwhelmed by the exceptionally large number of participants. The impressive number of participating students and young scientists was especially convincing evidence of the high level of interest in this meeting. The date for the first Munich Silicon Days was not chosen by accident: it marked a very exciting anniversary, the 50th birthday of the *Direct Synthesis*. Because of this celebration, the two pioneers, Prof. Eugene G. Rochow and Prof. Richard Müller, participated in this symposium and were honored with the "Wacker Silicone Award".

Impressed by the scientific quality of the contributions presented and stimulated by the large number of requests for access to information about the symposium from those scientists who could not attend it, we decided to edit the first Volume of *Organosilicon Chemistry — From Molecules to Materials*.

After four successful Munich Silicon Days, in 1992, 1994, 1996, and 1998, the time was ripe to expand the organization from a national to a European level. This happened between 1998 and 2001, and Munich Silicon Days was transformed into European Silicon Days. A European Advisory Board was established, represented equally by academia and industry, which took responsibility to organize future conferences at different places and countries. Meanwhile, the Wacker Silicone Award became a highly regarded honor for outstanding scientists from all over the world. European Silicon Days remained an appropriate venue for the presentation of this prestigious award. At the 2nd European Silicon Days in 2003, the recipient of the Wacker Silicone Award was Prof. Dr. T. Don Tilley of the University of California, Berkeley, who presented highlights of his research in transition metal organosilicon chemistry in a fascinating plenary lecture at this meeting. He continues an impressive list of former awardees.

2003 Prof. Dr. T. Don Tilley
(II. European Silicon Days, September 2003)
2001 Prof. Dr. M. Weidenbruch
(I. European Silicon Days, September 2001)
1998 Prof. Dr. R. Corriu
(IV. Munich Silicon Days, April 1998)
1996 Prof. Dr. H. Schmidbaur
(III. Munich Silicon Days, April 1996)
1994 Prof. Dr. E. Hengge
(II. Munich Silicon Days, August 1994)

1992	Prof. Dr. R. Müller and Prof. Dr. E. G. Rochow (I. Munich Silicon Days, August 1992)
1990	Prof. Dr. H. Sakurai
1989	Prof. Dr. R. West
1988	Prof. Dr. N. Wiberg, Prof. Dr. R. Tacke (junior award)
1987	Prof. Dr. R. Jutzi, Prof. Dr. N. Auner (junior award)

Despite the above-mentioned many changes over the years, we have remained committed to our task of editing the scientific contributions presented during the Symposia. The present edition of *Organosilicon Chemistry VI — From Molecules to Materials* covers the diversity of silicon chemistry as well as the scientific fascination of dealing with this element. This two-volume book is divided into six chapters reflecting the wide reach of organosilicon research. The first volume of *Organosilicon Chemistry VI* contains Chapters I–IV. Chapter I deals with the chemistry of organosilicon-based reactive intermediates, while Chapter II covers different aspects of molecular inorganic silicon chemistry. Chapter III involves the basic chemistry of silicon in the coordination sphere of transition metals, including such topics as silylene complexes, catalysis, and silicon-metal bonds. Finally, Chapter IV contains contributions on the role of silicon in organic and bioorganic chemistry. The second volume of *Organosilicon Chemistry VI* contains Chapters V and VI: Chapter V switches to the often more complex systems of organosilicon compounds in and for industrial applications, whereas Chapter VI covers the contributions on solid silicon-based materials.

The present edition of *Organosilicon Chemistry VI — From Molecules to Materials* will — like all its forerunners — essentially continue to stimulate young researchers to focus on basic silicon science and its transfer into consumer-oriented applications which provide innovative solutions to current problems. In addition, it should serve as a solid basis for learning more about modern developments in a fascinating area of inorganic chemistry which can hardly be surpassed for diversity.

March 2005 Norbert Auner and Johann Weis

Acknowledgments

First of all we thank the authors for their contributions and intense cooperation, which made this overview of current organosilicon chemistry possible. The tremendous work to achieve the attractive layout of this volume was performed by Dr. Yu Yang, and Mrs. Hannelore Bovermann helped to organize the editorial work.

Furthermore we are very grateful to Dr. Sven Holl for his very active assistance to read, compare and correct.

We thank all of them for their admirable engagement!

Prof. Dr. Norbert Auner
Johann Wolfgang Goethe-Universität
Frankfurt

Prof. Dr. Johann Weis
Wacker Chemie GmbH
München

Contents

Chapter I

Organosilicon-Based Reactive Intermediates

Chapter II

Molecular Inorganic Silicon Chemistry

Chapter III

Transition Metals in Organosilicon-Based Chemistry

Chapter IV

Silicon in Organic and Bioorganic Chemistry

Chapter V

Organosilicon Compounds for Industrial Applications

Chapter VI

Silicon-Based Materials

Introduction: Organosilicon Chemistry — Recent Highlights in a Fascinating Research Area

In Volume V of "*Organosilicon Chemistry — From Molecules to Materials*", the editors tried to stimulate discussion about "future trends in organosilicon chemistry", an important subject for the very start of the 21st Century. At the 2nd European Silicon Days in 2003, numerous presentations of outstanding quality from academia and industry convincingly demonstrated that these trends are being pursued in world-wide research, providing an increasingly broader platform for the design of compounds and materials with extraordinary chemical, physical, and biological properties, which promise to open up new fields in different applications. The scientific contributions presented at the symposium are briefly summarized in this introduction to provide an "appetizer", hopefully stimulating the reader to embark on an in-depth study of all the papers.

Chapter I covers different aspects of *organosilicon-based reactive intermediates.* Among these are molecules containing divalent silicon atoms — silylenes R_2Si, which have been in the limelight for four decades. As structural analogs of carbenes, R_2C, the promising aspects of silylenes for synthetic and mechanistic studies were recognized immediately. The hope for great synthetic value has been richly fulfilled, and, as Peter Gaspar pointed out in his lecture *Learning from Silylenes and Supersilylenes*, "There are many examples of molecules that we owe to silylene chemistry. Learning about structural and chemical similarities and differences between carbenes and silylenes, it is now time to broaden the scope of our interests and look at the distinctive features of the chemistry of higher divalent group 14 compounds". Another research field of unbroken interest is that of doubly bonded silicon — and the recent results are really fascinating, as shown by M. Kira's report on *New Molecular Systems with Silicon–Silicon Multiple Bonds,* covering the synthesis and structural investigations of compounds such as tetrakis(trialkylsilyl)disilenes, persilylated cyclotrisilenes, cyclotetrasilenes, spiropentasiladienes, and even a trisilaallene, the first stable silicon compound with a formal sp-hybridized silicon atom. Mechanistic studies of intramolecular rearrangements of "small" silenes, as well as the continuous search for stable compounds with Si=C-π-bonds are of ongoing interest, as demonstrated by contributions of L. Gusel'nikov and H. Öhme. Exciting results were reported by R. West, namely that the reaction of a tetraphenylsilole with 1,1-dichloro-2,3-diphenylcyclopropene gives a highly stable silicon diradical which does not react with water and alcohols and which transforms into a 1,4-disilapentalene upon heating. Furthermore, the Haifa group of Y. Apeloig was successful in the kinetic stabilization of polysilyl radicals — an excellent contribution combining both experimental and theoretical investigations. In addition to detailed investigations on silicon cations $R_3Si^{\oplus}$, P. Jutzi's group studied those of the type $RSi:^{\oplus}$ as described in his contribution *Spectroscopic Evidence for the Formation of the Pentamethylcyclopentadienylsilicon Cation*: the synthesis, structure and bonding in $Me_5C_5Si:^{\oplus}$.

Chapter II of the proceedings deals with basic aspects of *molecular inorganic silicon chemistry.* Starting with matrix investigations on the reactivity of silicon atoms, G. Maier convincingly demonstrated that product formation in the Müller-Rochow Process is mainly dependent on the silicon particle size: silicon atoms in the gas phase do not need any catalyst or promoter to react with haloalkyls to yield haloalkylsilanes. The formation of, e.g., dimethyldichlorosilane occurs via silylene intermediates. This remarkable result may influence further investigations on the important Direct Process for the production of silicones. Another very important technical process is the synthesis of trichlorosilane, the starting material for the production of polycrystalline silicon for, e.g., semiconductor and photovoltaic applications. Three contributions of H. Lieske's group deal with the basic understanding and optimization of this process and thus demonstrate the strong link between academia and industry in this area of research.

Three contributions from the groups of G. Roewer, H. Grützmacher, and R. Nesper cover the formation of various novel BCCSi-substituted borazines and synthetic approaches to the functionalization of silicon-based non-oxidic ceramics, and papers from W. Mitzel, U. Klingebiel, H. Karsch, M. Driess, and A. Bauer deal with the chemistry of aminosilanes and a possible transfer into application, e.g., the formation of Si–N-containing polymers and ceramics.

In his lecture *Silanols as Precursors to Cyclo- and Polysiloxanes*, M. Veith described the syntheses of six- and eight-membered cyclosiloxanes $(X_2SiO)(Y_2SiO)_2$ and $(X_2SiO)_2(Y_2SiO)_2$ with different organic and inorganic substituents at the silicon. These substituents have a quite significant impact on the three-dimensional structures of the compounds. The incorporation of Al and Ga into polysiloxanes is achieved using $(t\text{BuOMH}_2)_2$ (M = Al, Ga) as synthetic building blocks and opens up new entries for the formation of novel heteroatom-substituted polysiloxanes. Another entry was shown by K. Jurkschat, who presented recent results on *The Origin of Ring Strain and Conformational Flexibility in Tri- and Tetrasiloxane Rings and their Heavier Group 14 Congeners.* A kinetically-controlled ring-opening polymerization of strained trisiloxane rings is the most powerful process for the preparation of high-molecular-weight silicone polymers. By contrast, the thermodynamically controlled opening of $(R_2SiO)_3$ rings or of (virtually) strain-free tetrasiloxane rings provides mixtures of low-molecular-weight polymers and substantial amounts of redundant cyclic oligomeres. Finally, U. Herzog reported on a novel telluro-analog of the hexaphenylcyclotrisiloxane $(Ph_2SiTe)_3$. G. Tekautz presented his research on silicon-phosphorus and silicon-arsenic cages, T. Schollmeier reported on Si–Sn four- and six-membered rings, and M. Mehring talked about his work on homo- and heterometallic bismuth silanolates.

Over a couple of decades, the synthesis, structure, and reactivity of hyper- (mainly hexa-) coordinate silicon has remained exciting: contributions from the groups of G. Roewer, R. Tacke, and D. Kost reported on their latest results in this fascinating research area. Very similar to that chemistry, the synthesis and properties of polysilanes are in the focus of quite a number of research groups. For the controlled formation of polysilanes, oligo- and polysilylanions could be the key. Contributions from the groups of C. Marschner, M. Niemeyer, W. Uhlig, and C. Krempner deal with this topic and cover the field from the synthetic point of view as well as the related aspects of photophysical properties (R. Crespo, H. Stüger). Finally, M. Unno summarized his research on *Reactions of Octasilacubanes*, which are intriguing polyhedral polysilanes because of their unique electronic properties arising from their highly strained Si–Si σ-bonded framework.

Chapter III basically collects contributions on the *role of transition metals in organosilicon-based chemistry*. The review of the Wacker-Silicone Awardee T. Don Tilley on *Transition Metals in Organosilicon Chemistry* fundamentally describes two different strategies toward the exploitation of reactive metal-silicon bonds. One type involves the early transition metals, since d^0 metal–silicon single bonds are unsupported by metal-to-silicon π-backbonding and are therefore weaker than other transition metal–silicon bonds: these M–Si bonds activate various unsaturated compounds via insertion and participate in the activation of single bonds (e.g., C–H and Si–C), possibly being responsible for the action as catalysts for the dehydropolymerization of hydrosilanes to polysilanes. Most recently, even the catalytic silylation of methane has been observed.

A second type of reactive metal–silicon bond involves multiple bonding, as might exist in a silylene complex, $L_nM{=}SiR_2$. The synthesis of isolable silylene complexes has led to the observation of new silicon-based reactivity patterns: redistribution at silicon occurs via bi-molecular reactions of silylene complexes; with osmium silylene complexes, reactions have been observed that mimic proposed transformations in the Direct Process. And, very recently, ruthenium silylene complexes have been reported to be catalytically active in hydrosilylation reactions.

This industrially and chemically very important process was perfectly brought into the audience's focus by the lecture of one of the pioneering researchers in that field, John F. Harrod, entitled *Hydrosilylation: a Personal Perspective on a Scientifically and Technologically Fascinating Chemical Methodology*. Since its discovery, homogeneous catalytic hydrosilylation has become a tool for the development of many new areas of chemistry and technology, including surface functionalization, exotic polymer architectures (e.g., dendrimers), and organic synthesis (particularly enantioselective catalysis).

A remarkable series of papers by U. Schubert, Z. V. Belyakova, H. Maciejewski, B. Marciniec, B. Weyershausen, and F. Baumann deal with specific chemical aspects of this reaction in fundamental research as well as in application and thus provides a broad overview of this important topic.

F. Hoffmann und G. Roewer reported on reactions of cyclohexasilylpotassium with transition metal compounds, and C. Marschner described a Cp-free hafnium silyl-substituted complex. A series of four papers from the group of W. Malisch deal with the description of silanol functionalities in organometallic iron and tungsten complexes and the synthesis of perchlorinated metallo siloxanes.

The *role of silicon in organic and bioorganic chemistry* is the key focus in Chapter IV. The use of norbornyl groups attached at silicon as protecting groups is introduced by D. Heldmann from Wacker-Chemie. Brilliant reseach is presented by C. Strohmann and his group, reporting the preparation of enantiomerically pure *N*-methyl-2-silyl-substituted pyrrolidine, diastereomerically enriched α-lithiated benzylsilanes, and silyllithium compounds and their stereochemical transformations. A paper by M. Otto describes the reagent disilylmercury and a new approach to amino carbenes. Impressive results are reported from a cooperation between the Technical University of Graz and Wacker-Chemie on the topic of the *Electrochemical Synthesis of Functional Organosilanes* starting from chlorosilanes and organic halides. The electrochemical reduction is a very simple and elegant method to produce silanes X-SiR_2-RY′.

Organosilicon-based compounds could be promising tools for optoelectronic applications, e.g., in LED technology. Three papers report on this topic. Another area which is becoming increasingly important deals with the biological activity of organosilicon compounds, and this is convincingly demonstrated by seven contributions including investigations on silicon derivatives of 2-trifluoro-acetylfuran and their oximes, silicondiols, 1,4′-silaspiro[tetralin-1,4′-piperidine] type compounds, silocanes, and silatranes.

The second main part of Organosilicon Chemistry VI focuses on *organosilicon compounds for industrial applications* (Chapter V) and on *silicon-based materials* (Chapter VI). Chapter V begins with a very perspective- and future-oriented contribution by J. W. White of the Dow Corning Corp. entitled *Silicon Science and Technology — An Industrialist's View of the Future*. This paper reflects on changes in Industrial Silicon Chemistry over the last decade and focuses on the outlook for the next 10 years. The topics include a variety of selected industries such as Transportation, Electronics, Biotechnology, and Energy. Future needs are for high-quality products offering complete application and service solutions. As White claims, the "fusion of scientific and commercial disciplines is essential, and it will be at the boundaries between the physical, biological, and engineering sciences that we will find the most productive areas".

From that overview M. De Poortere went into more detail, giving a summary on *The Markets for Silicones*: during the last decade the average global growth rate was about 6%, with most of the growth coming from developing Asian countries. In 2002, the global market for silicones totaled about €8 billion based on a production volume slightly over 2 000 000 tonnes. Sales have been evenly distributed between America, Europe, and Asia. De Poortere's paper clearly focuses on the different types of silicones produced, their shares in the global market, and their applications resulting from very specific chemical and physical properties. J. Chojnowski provided an excellent overview of the *Synthesis of Organofunctional Polysiloxanes of Various Topologies*, including the anionic ring-opening polymerization of cyclotrisiloxanes and the functionalization via organic radicals pendant to the polysiloxane chain resulting in different influences on the properties of the silicone. He also focused on organic siloxane block and graft copolymers, on star-branched polymers, and finally on the formation of polysiloxane–silica hybrids.

P. C. Hupfield (Dow Corning Ltd.) described a very efficient phosphazene-based catalyst and its use in siloxane synthesis. The following papers deal with hydrolysis and condensation reactions of α,ω-diols $OHSiMe_2O(SiPh_2)_nSiMe_2OH$ ($n = 1 - 4$), of carbofunctional amino-alkoxysilanes, and of octyltriethoxysilane. Remarkable results were obtained with silacyclobutene-PDMS-copolymers, exhibiting unusual thermal behavior. The more application-oriented papers include topics such as *Monofunctional Silicone Fluids and Silicone Organic Polymers* (W. Keller), the tailoring of silicone oil for aluminum production (S. Breuning), silicone antifoams with improved persistence (J. Huggins), silicone mist supressors for fast paper coating processes (H. Delis), and silicone polymers for coating, cosmetic, textile, and fabric care (K. Stark). R. Wagner and researchers from GE Bayer Silicones reported on *Siloxane-Based Copolymers for Textile Finishing Purposes — General Structure Concepts, Application Aspects, and Behavior on Fiber Surfaces*: combined

XPS/FTIR experiments suggest that the new polysiloxane softeners show superior substantivity to fiber surfaces. Other topics included the synthesis of new UV-curable alkoxysiloxanes modified with tris(trimethylsilyl)methyl derivatives (A. Kowalewska) and of methacrylic silanes as versatile polymer building blocks and surface modifiers (J. Pfeiffer, Wacker-Chemie). A contribution by A. Lejeune (OSi Specialities) dealt with *Hydrolysis Studies of Silane Crosslinkers in Latexes.* The study introduces methods of following the rate of hydrolysis and subsequent crosslinking by silanol condensation of post-added as well as directly copolymerized silanes in fully formulated latexes. J.-M. Pujol of Rhodia Silicones presented a talk about *Mastering Crosslinking in Silicone Sealants*, and he pointed out that silicone sealants can be tailored to specific requirements today. The cure rate and mechanical properties can be adjusted by the selection of appropriate curing agents.

In a similar context, H. Mack (Degussa AG) reflected on *Selecting the Right Aminosilane Adhesion Promoter for Hybrid Sealants* and provided an overview of recent developments with novel and nonsensitizing aminosilanes used as adhesion promoters to make high-performance and low-modulus construction sealants. In his paper, R. Bärsch discussed the *Influence of Different Stresses on the Hydrophobicity and the Electrical Behavior of Silicone Rubber Surfaces.* Other topics in Chapter V can only be mentioned briefly: G. V. Stepanov described *Silicone Magnetoelastic Composite*, S. N. Kholod reported on *Modifiers for Compounded Rubbers Based on Fluoro- and Phenylsiloxane Rubbers*, and N. Semenkova focused on *Polyphenylsilsesquioxane–polydiorganosiloxane Block-Copolymers.* O. Schäfer from Wacker-Chemie gave a basic overview of *Thermoplastic Silicone Elastomers*: PDMS-rubbers require extremely high-weight-average molecular weights to develop useful properties. To achieve this they must be chemically crosslinked. This additional crosslinking step can be avoided with thermoplastic silicone rubbers: the PDMS backbone was successfully modified with urea segments, which tend to crystallize at room temperature. Silicone-urea copolymers exhibit a combination of very interesting properties, such as very low glass transition temperatures, high oxidative stability, low surface energy, hydrophobicity, high gas permeability, good electrical properties, and biocompatibility. Additional topics include *Silicone Hybrid Copolymers* (W. Hiller) and the *Hydrolytic Stability of Silicone Polyether Copolymers* (M. G. Pigeon).

Chapter V closes with a remarkable contribution by A. Lork of Wacker-Chemie covering the field of *Nanoscale Networks for Masonry Protection.* Porous surface structures of hydrophilic mineral construction materials require protection against water, pollutants, and chemical and microbial attack. Since most damaging effects are caused by water uptake, the facades should be efficiently impregnated and must also feature good vapor permeability so that damp walls can dry out quickly. Lork's paper answers the basic question how silicone resin networks attach to mineral substrates, and she points out that the excellent long-term stability of the silicone resin system is the result of the strong chemical interaction/physical adhesion of the network to the mineral substrate, the unique molecular structure with its inorganic and organic hybrid character, and the hardness of the network.

Chapter VI of the proceedings deals with *silicon-based materials.* F. Schüth gave an impressive overview on *Mesostructured Silica and Organically Functionalized Silica — Status and*

Perspectives. Ordered mesostructured and mesoporous silica has been known for little more than ten years. Tremendous progress has been made with respect to precise control of the structure, texture, and chemical functionality of the surface of these materials. His lecture surveyed the synthesis of such materials, with a focus on organically ordered mesoporous materials. Quite a number of contributions dealt with amorphous fumed silica, its *Physical-Chemical Features and Related Hazard Risk Assessment* (M. Heinemann), the description of fractal aggregates (C. Batz-Sohn), and the *Characterization of Size and Structure of Fumed Silica Particles in Suspension* (F. Babick). E. Brendle reported on *Adsorption of Water on Fumed Silica*, and in a second paper he summarized research on *Methylene Chloride Adsorption on Pyrogenic Silica Surfaces*.

Some contributions cover the development of specific materials and analytical methods to measure the characteristic properties of solid particles, such as particle sizes, surfaces areas, mechanical strengths, or solid-matrix interactions. Thus, papers from M. Heinemann and S. Hild deal with the characterization of silica-polymer interactions using Scanning Force Microscopy, while C. Panz uses the combination of special basic silica, fitting silanes, and adequate hydrophobization conditions to generate high-performance silica with new properties.

Zeolites and zeosils (as zeolite-like substances) are of great academic and industrial importance, as they could be produced naturally by biomineralization processes as well as by standard sol–gel procedures. These materials possess a pure silica framework, and therefore interactions between these microporous hosts and guest species are very weak. R. Nechifor summarized his investigations on *Chemical Vapor Deposition of Iodine into Pure Silica Hosts* and pointed out that the degree of interaction can be estimated qualitatively by the color of the compounds and more quantitatively by spectroscopic methods, especially UV-Vis and Raman spectroscopy. The synthesis of the insertion compounds is a multi-step procedure involving first the preparation of the host material using organic compounds as structure-directing agents. This is followed by the removal of the organic molecule included in the pores (channels) and finally the insertion of the guest material. This quite common preparation process could be generally used for the design of tailor-made materials for e.g., catalysis, separation, reinforcing materials, and fillers for plastics and ceramics. In a similar context, metal-doped silica gels exhibit a wide range of optical properties for optoelectronic applications. M. Milbradt presented his results aiming to combine the outstanding properties of nano/microtubes and metal-doped silica gel. There are many kinds of dopants, e.g., transition metals and rare-earth ions. An advantage of these materials is the high heat resistance of the tubular structures, up to ~250 °C with organic and up to 800 °C with only metal functionalization.

Under the topic *Branched Functionalized Polysilane–Silica Hybrids for Immobilization of Catalysts*, K. Rozga-Wijas gave a summary of a range of hybrids of functionalized polysiloxanes of well-defined structures having various topologies (linear, comb-branched, and dendritic-branched). These hybrids were generated by grafting polysiloxane to the prefunctionalized surfaces of porous silica micro-particles. Subsequent functionalization of the polysiloxane-silica hybrid with tertiary phosphine groups, which are excellent ligands for transition metal moieties, accomplished the synthesis of a new class of catalysts.

The importance of biomineralization as a natural process for the production of siliceous skeletons has just been mentioned. The formation of this biomineral requires bio-organic compounds such as polysaccharides, proteins (silicatein and silaffin), or polyamines. P. Bärnreuther of the Behrens group reported on investigations into the interaction of such polyamines with silicic acid or silica in aqueous solution. Thus, linear polyamines with a defined degree of polymerization were prepared, which accelerate the decrease of concentration of soluble silicic acid and influence the time of gelling. The results indicate that polypropyleneimine has a catalytic effect on the condensation and that it is able to postpone gelling, allowing for a longer time for structural changes. This is a fact that might be important for biomineralization processes.

Hydrolysis and condensation reactions of alkoxysilanes are the "master key" reactions in the synthesis of silica-based materials. In this case, sol–gel processing is limited, especially for the synthesis of monolithic mesostructured materials, because of the obligatory presence of a compatibilizing agent, typically an alcohol, which is often detrimental to liquid crystal-like phases. N. Hüsing, in her talk on *Polyol-Modified Silanes as Precursors for Mesostructured Silica Monoliths*, presented the preparation of large mesostructured silica monoliths from ethylene glycol-modified silanes in the presence of block copolymer surfactants; the material is directly prepared from liquid crystal-like phases of the surfactants in water. The resulting monoliths show very high densities, high porosities, and extraordinarily high surface areas.

In this expanding research field, M. Wong Chi Man introduced a general method for the preparation of *Self-Organized Bridged Silsesquioxanes*: the introduction of urea groups into the bridging organic unit during the hydrolysis of a corresponding trialkoxysilane creates a self-assembly of the molecular precursors to afford supramolecular architectures. According to the hydrolysis conditions, helical, tubular, and spherical solids were obtained.

The increasing importance of the synthesis of ionic conductors via the sol–gel route is shown by B. Orel's contribution *Carbamatosil Nanocomposites with Ionic Liquid: Redox Electrolytes for Electrooptic Devices*: organic-inorganic hybrid materials were produced from an organically modified alkoxysilane with oxalic acid as a gelation promoter, while inorganic or organic iodides and iodide were introduced as redox species and triethylene glycol was used as co-solvent. Finally, J. Ebenhoch (Wacker-Chemie) introduced *Silicone Nanospheres for Polymer and Coating Applications*.

Two papers from R. Nesper and M. Weinmann cover various approaches to the preparation of functional silicon-based non-oxidic ceramics, especially those consisting of the Si–B–C–N moieties which exhibit an extraordinary high-temperature stability. Chapter VI ends by describing carbosilane elastomers as promising membrane materials (N. N. Ushakov) and investigations into the *Chemical Functionalization of Titanium Surfaces* with 1-trichlorosilylalkanes.

As mentioned in the "Preface", the main object of this monograph is to stimulate especially young students and scientists to focus on the broad field of organosilicon research. To this end, the final contribution to Organosilicon Chemistry VI, provided by M. W. Tausch in cooperation with Wacker-Chemie, deals with a comprehensive *Documentation of Silicones for Chemistry Education and Public Understanding*.

In his presentation, J. White put the question: "Is there a future for Silicon Chemistry?", and at the end of his talk he answered with an emphatic *yes*! The two editors wholeheartedly endorse his opinion and believe that the impressive variety of topics in this volume afford a convincing proof of this assessment!

We sincerely hope that *Organosilicon Chemistry — From Molecules to Materials* Vol. VI will encourage many young scientists all over the world to continue the work in this exciting field of chemistry.

March 2005 Prof. Dr. Norbert Auner, Prof. Dr. Johann Weis

Chapter I

Organosilicon-Based Reactive Intermediates

Learning from Silylenes and Supersilylenes

Peter P. Gaspar*

Department of Chemistry, Washington University, Saint Louis, MO 63130-4899, USA
Tel.: +1 314 935 6568 — Fax: +1 314 935 4481
e-mail: gaspar@wuchem.wustl.edu

Keywords: silylenes, supersilylenes, silanetriyl cations

Summary: Answers are suggested to the questions: what have we learned from silylene chemistry and what do we look forward to learning in the future? Included are initial observations of the chemistry of "supersilylenes", four-valence electron monovalent silicon cations R–Si:$^+$ whose electronic structure endows them with the potential for forming as many as three new bonds in a single reactive collision.

Introduction

As someone who has been involved in the study of silylenes for nearly 40 years, I might be expected to indulge in a series of reminiscences outlining the history of this field through its participants and their accomplishments. Instead I have set myself a more difficult task of drawing general lessons about chemical reactivity from the "extended organic chemistry" of silylenes. There will inevitably be historical passages in this story.

Interest in molecules containing divalent silicon atoms, silylenes R_2Si:, was awakened in the 1960s by the dawning recognition that divalent carbon compounds, carbenes R_2C:, participate in novel and synthetically useful π-bond addition and σ-bond insertion processes [1]. It was natural to wonder whether divalent compounds of silicon, germanium, and tin would undergo reactions analogous to those of carbenes [2].

It was clear from the beginning that new routes would be required to silylenes, since silicon analogs of the common carbene precursors, diazoalkanes, diazirines, and ketenes, are unstable or unknown. Our first reports on silylene reactions utilized a few million $^{31}SiH_2$ molecules produced from high-energy radioactive silicon atoms by poorly understood hydrogen atom acquisition processes. Using hot-atom chemistry techniques, insertion of $^{31}SiH_2$ into H–Si bonds and addition to C=C π-bonds were found in the 1960s, as well as an interesting difference between silylenes and carbenes to which I shall return below [3]. An early indication of the synthetic potential of silylene chemistry was the formation of unsubstituted silole [4].

Lesson 1 — The Reversibility of Silylene Reactions

During the 1960s and 1970s, clean methods were developed for the generation of silylenes [5]. It required many years, however, for workers in the field to recognize the connection between the most useful reactions for the production of silylenes, extrusion from disilanes and siliranes, and reactions that efficiently consume them, insertions into σ-bonds and addition to π-bonds. Yet staring us in the face was a phenomenon quite different from carbene chemistry: nearly all silylene reactions are *reversible* [6]. This provides a compact framework within which to present a selection of the reactions that lead to silylenes and those reactions that consume them, as shown in Scheme 1.

$$:SiR^1R^2 + SiWXYZ \underset{\text{extrusion}}{\overset{\text{insertion}}{\rightleftharpoons}} Z{-}SiR^1R^2{-}SiWXY \qquad (1)$$

$$:SiR^1R^2 + R^3C{\equiv}CR^4 \underset{\text{retro-addition}}{\overset{\text{addition}}{\rightleftharpoons}} \text{silirene } (R^3C{=}CR^4)SiR^1R^2 \qquad (2)$$

$$:SiR^1R^2 + WXC{=}CYZ \underset{\text{retro-addition}}{\overset{\text{addition}}{\rightleftharpoons}} \text{silirane } (WXC{-}CYZ)SiR^1R^2 \qquad (3)$$

$$:SiR^1R^2 + \text{butadiene} \underset{\text{retro-addition}}{\overset{\text{addition}}{\rightleftharpoons}} \text{silacyclopent-3-ene } SiR^1R^2 \qquad (4)$$

Scheme 1. Reversible reactions producing silylenes in one direction and consuming them in the other.

The extrusion of SiH_2 from disilane (Reaction 1, R^1, R^2, W, X, Y, Z = H) was proposed by Stokland in 1946 [7] and placed on a firm foundation in the 1960s by the work of Purnell and Walsh [8], and by Ring, O'Neal and coworkers [9]. The pyrolysis of disilane became the standard source of SiH_2 for gas phase studies. A few years later Stokland was also the first to propose the forward reaction, insertion of SiH_2 into an H–Si bond of silane [10], and Skell found H–Si insertion by $SiMe_2$ into $HSiMe_3$ [11]. The breakthrough that ushered in the use of silylenes as synthetic reagents and led to the discovery of many new silylene reactions both in solution and the gas phase was the 1966 discovery by Atwell and Weyenberg that extrusion of $SiMe_2$ from $(MeOSiMe_2)_2$ (Reaction 1, R^1, R^2, W, X = Me, Y, Z = OMe) occurs under mild conditions (< 200 °C) [12].

Reaction 1 accommodates a variety of groups, the only limitation being that group Z migrates more easily than R^1 and R^2.

Addition of silylenes to alkynes (Reaction 2) and alkenes (Reaction 3) had been suggested on the basis of secondary products, but the first three-membered ring adduct of a silylene directly detected was tetramethylsilirene (Reaction 2, R^1, R^2, R^3, R^4 = Me) in 1976 [13], the same year that Seyferth discovered that the extrusion of $SiMe_2$ from hexamethylsilirane (Reaction 3, R^1, R^2, W, X, Y, Z = Me) at low temperatures (ca. 75 °C) is an efficient and mild source of silylenes [14]. Boudjouk discovered that photoextrusion of silylenes from siliranes requires no chromophore beyond the silirane ring [15]. Stereospecific addition of several silylenes to olefins was demonstrated in both photochemical and mild pyrolysis experiments [16]. Formal 1,4-addition of silylenes to 1,3-dienes yielding 1-silacyclopent-3-enes was discovered in the 1960s and was first thought to distinguish silylenes from carbenes, whose additions to 1,3-dienes generally yield vinylcyclopropanes. Extensive mechanistic studies have demonstrated that 1,2-addition is also the primary step in silylene-diene additions, but the instability of alkenylsiliranes leads to a variety of products, whose formation is in most cases reversible [17]. One example is shown in Scheme 2.

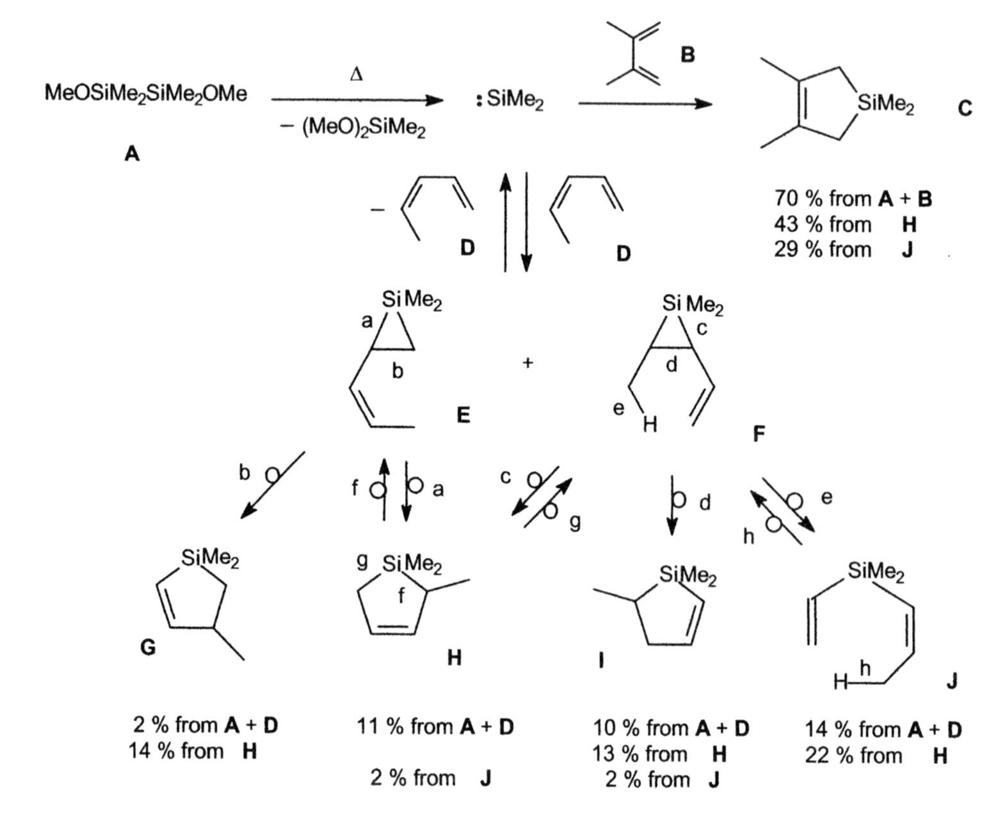

Scheme 2. The reversible addition of $SiMe_2$ to *cis*-piperylene.

That the primary step in the reaction of $SiMe_2$ to *cis*-piperylene is a 1,2-addition forming two isomeric alkenylsiliranes was revealed by their rearrangement products [18], including that of a retro-home-ene reaction **J** [19], formed in higher yield than the product of formal 1,4-addition. Formation of **J** is seen to be reversible, with a 29 % yield of $SiMe_2$ trapping product formed upon its pyrolysis in the presence of 2,3-dimethylbutadiene [20]. A 37 % yield of piperylenes arises from pyrolysis of **J**, suggesting that silylene retro-additions can form carbon-carbon bonds (here an unsymmetrical C=C) and thus possess synthetic utility. $SiMe_2$ is trapped in 43 % yield upon pyrolysis of **H**, and yields as high as 72 % have been found upon extrusion from other 1,1-dimethylsilacyclopent-3-enes [21].

The reversibility of silylene reactions has taught the chemical community how to make a number of reactive intermediates, including germylenes GeR_2 [22] and stannylenes SnR_2 [23] by extrusion from germa- and stannacyclopent-3-enes, and phosphinidenes R–P [24], and phosphinidene chalcogenides R–P=Z (Z = O, S) [25], by extrusion from phosphiranes and phosphirane chalcogenides. But the chief lesson is that reversibility itself appears to be a general phenomenon in main group chemistry, to whose discovery the study of silylenes has made an important contribution.

Lesson 2 — The Behavior of Electron-Deficient First-Row Species can be Used to Predict Heavier-Element Chemistry

In the early days of silylene chemistry it was written that: "The real danger is that the distinctive features of the chemistry of the divalent compounds of silicon, germanium, and tin will be overlooked because the interpretation of their reactions is colored by the expectation that their behavior will resemble that of carbenes" [2]. I was urging colleagues and warning my own coworkers to let silylenes be silylenes and thus find their non carbene-like behavior, but the typical silylene insertion and addition reactions displayed in Scheme 1 are striking in their similarity to those of carbenes. Even the rearrangements of carbenes find mechanistic parallels in silylene chemistry, but some differences emerge as seen in Scheme 3.

Silylenes only undergo rearrangements at elevated temperatures, and the high barriers for their intramolecular reactions have made possible the convenient study of their intermolecular chemistry. Alkylcarbenes undergo rapid H-shifts, converting them to olefins even at room temperature, and thus the intermolecular insertions and additions of alkyl carbenes have remained rather obscure.

Silylenes, like carbenes, can rearrange by a 1,2-shift of a group to the divalent atom. Silylsilylene SiH_3SiH: rearranges to disilene $H_2Si{=}SiH_2$ sufficiently slowly that both species can be trapped by butadiene, and an activation barrier for the H-shift of 10 ± 2 kcal/mol was estimated from the temperature dependence of the product ratio [26]. The H-shift that converts CH_3CH: to $CH_2{=}CH_2$ is so rapid that intermolecular trapping of the carbene is inefficient [27]. An H-shift from an alkyl group to a silylene center is sufficiently slow that the formation of silenes is not competitive with other reactions of alkylsilylenes [28].

A methyl-shift is much slower than an H-shift for both silylenes and carbenes. In the case of

$Me_3Si–SiH$:, generated in the gas-phase at 570 °C in the presence of butadiene, the yields of products from an Me-shift forming $Me_2Si=SiHMe$ totaled 10 %, while silylene insertion into a C–H bond forming a disilirane intermediate led to products in 12 % total yield, and trapping of unrearranged silylene led to products totaling 25 % yield [29].

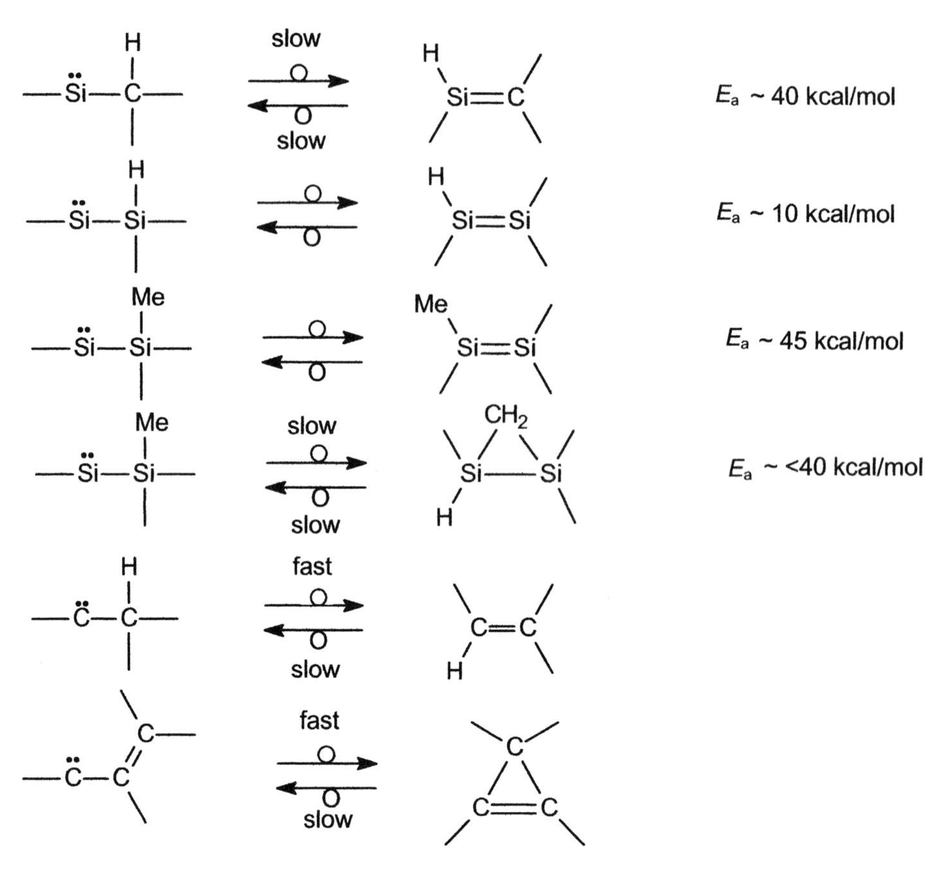

Scheme 3. Some primary steps in the rearrangements of silylenes and carbenes.

More extensive rearrangements involving silylene-to-silylene or carbene-to-carbene interconversions require elevated temperatures and often follow different mechanisms. Organosilicon chemistry was revolutionized by the discovery by Wulff, Goure and Barton that $Me_3Si–SiMe$: undergoes a series of silylene-to-silylene rearrangements by intramolecular C–H insertion forming a disilirane intermediate, followed by ring-opening via competing H- and Me-shifts as shown in Scheme 4 [30]. These are, of course, extrusions (retro-insertions).

In contrast, carbene-to-carbene rearrangements are often reversible vinylcarbene-to-cyclopropene interconversions. High temperatures are required for silylene rearrangements because ring-closure via C–H insertion is mildly exothermic and has a significant barrier, higher than that for ring-opening [31]. In the carbene case, ring-closure via intramolecular addition is highly exothermic with a very low barrier, but ring-opening is highly endothermic.

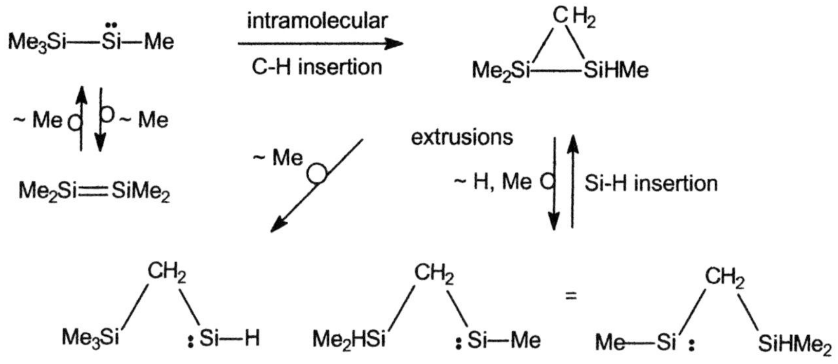

Scheme 4. The mechanism of the Barton rearrangement of α- to β-silylsilylenes [30].

It is well known that phenyl carbenes undergo carbene-to-carbene rearrangements as shown in Scheme 5, the first-step being ring-closure to a bicycloheptatriene which disrupts the aromatic π-system [32]. The analogous phenylsilylene-to-silaphenylcarbene rearrangement does not seem thermodynamically or kinetically feasible, frustrated by the lower energy of a silylene compared to a carbene and by the destabilization of silabenzene relative to benzene [33]. Singlet SiH_3–CH: is predicted to be 69.0 kcal/mol higher in energy than SiH_2=CH_2 [34], which in turn is predicted to be 4.0 kcal/mol lower than singlet CH_3–SiH: [35].

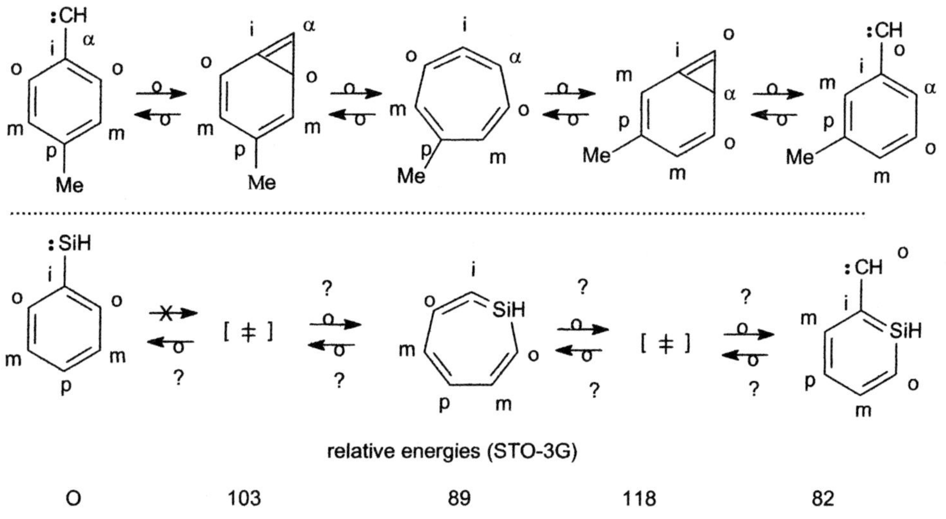

Scheme 5. Initial steps in the rearrangement of *p*-tolylmethylene [32] and the predicted energetics of the rearrangement of phenylsilylene to 2-silaphenylmethylene [33].

That most reactions of silylenes could be predicted from the chemistry of carbenes teaches the valuable lesson that one can predict many seemingly unlikely reactions of main-group reactive intermediates from the behavior of their first-row analogs.

Lesson 3 — Slowing Down the Reactions of Carbene-Like Species Allows Processes to Occur that Cannot be Observed for Carbenes

The lower heats of formation of silylenes compared with carbenes translate into somewhat lower, albeit similar, reactivity. There is a general lesson here that slowing down the reactions of carbene-like species makes possible processes like dimerization that are more or less impossible for carbenes. We have already seen that slowing down their rearrangements allows the observation of intermolecular reactions of alkylsilylenes not observed for alkylcarbenes. The dimerization of silylenes to disilenes, discovered in the gas phase [36], was employed to brilliant effect by West and coworkers in the condensed phase synthesis of isolable disilenes [37]. While these were not the first disilenes, silylene dimerization opened the way for the preparation of many new molecules, including digermenes from the dimerization of germylenes [38].

Triplet Ground-State Silylenes

One major difference between silylenes and carbenes is that, until recently, all known silylenes had singlet electronic ground states [17], while many carbenes have triplet ground states [39]. Kutzelnigg, in his seminal paper on bonding in higher main-group elements, explained this by suggesting that the promotion energy for the hybridization required for the formation of the lowest triplet state is compensated by the formation of stronger bonds in the case of carbenes, but *not* in the case of silylenes [40]. Since 1984, attention has focused on the energy difference between the frontier orbitals and the resulting dependence on bond angle and substituent electronegativity of the promotion energy from the σ^2 electronic configuration of the lowest singlet state to the σ_{p} configuration of the lowest triplet state [41]. These factors are displayed in Scheme 6.

ZSiZ bond angle increases with size of Z

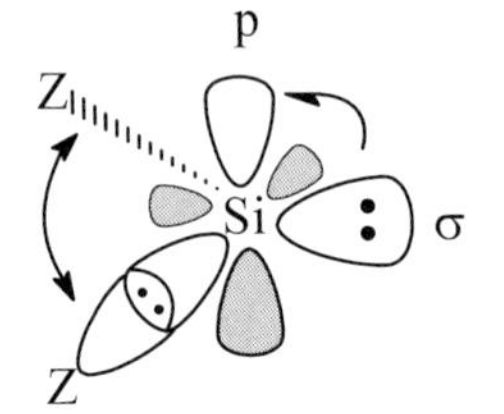

The σ-to-p promotion energy depends on the p character of σ, which increases with increasing ZSiZ bond angle *and* with decreasing electronegativity of Z

Scheme 6. Effects of bond angle and substituent electronegativity on the σ-to-p promotion energy for $:SiZ_2$.

For a silylene to have a triplet ground state, the one-electron promotion energy attendant upon moving one of the lone-pair electrons of the σ^2 singlet to the p-orbital must be offset by the decrease in the repulsion upon separation of the nonbonding electrons. For the majority of carbenes, the energy lowering upon separating the lone-pair electrons outweighs the one-electron energy increase due to moving an electron from an s-weighted to a pure p orbital. For silylenes, with their larger valence orbitals, the decrease in the repulsion between the nonbonding electrons of the σ^2 singlet upon conversion to a σ_p triplet is much smaller. That is why silylenes tend to have singlet ground states [41]. Apeloig has recently provided a quantitative basis for these qualitative notions [42].

It is difficult to control the size of a frontier orbital, but the promotion energy can be decreased by increasing the p-character of the in-plane σ-nonbonding orbital. Increasing the bond angle does this by increasing the s-character of the orbitals that the divalent silicon atom contributes to the substituent-to-silicon bonding orbital. Decreasing the electronegativity of the substituent Z shifts the distribution of the shared electron-pair in the Z–Si bond toward the silicon atom, and this also increases the s-character of the silicon bonding orbital and thus must increase the p-character of the nonbonding σ-orbital. Hence, at any Z-Si-Z angle, the promotion energy decreases with a decrease in the electronegativity of Z.

We began our attempts to generate a triplet ground-state silylene based on the prediction by Gordon that at angles larger than 130°, triplet SiH_2 would be lower in energy than the singlet [43].

Our first "designer" silylene Ad_2Si was a failure, undergoing stereospecific addition to 2-butenes [44]. Both the premise that a $\angle$C-Si-C > 130° would suffice for a dialkylsilylene to possess a triplet ground state *and* the estimate of that bond angle for Ad_2Si were incorrect. Calculations by Grev and Shaefer revealed that the crossover angle beyond which the ground state of silylene becomes a triplet depends strongly on the electronegativity of the substiuents [45]. The revelation that for disilylsilylene $(H_3Si)_2Si$: the predicted crossover angle was only ca. 115° led to the strategy that has ultimately produced triplet silylenes – the use of trialkylsilyl R_3Si substituents with alkyl groups large enough to open the central Si-Si-Si angle sufficiently to raise the singlet electronic state above the triplet.

Semi-empirical calculations suggested that the tri*iso*propylsilyl group ^{i}Pr$_3$Si might do the job [46], and calculations by Apeloig predicted that $(^iPr_3Si)_2Si$: would have a triplet ground state, but by an uncomfortably small 1.4 to 1.7 kcal/mol energy difference $\Delta E = (E_S - E_T)$ [47]. While the observed stereospecific addition of $(^iPr_3)_2Si$: to *cis*-2-butene could be explained by a breakdown of the "Skell rule" of carbene chemistry that predicts nonstereospecific addition by triplet carbenes [39], addition from the σ^2 singlet state had to be considered a strong possibility [48]. A low-lying σ^2 singlet state could be thermally populated and react more rapidly than the σ_p triplet even if the latter were the ground state [41, 48]. Recent calculations by Yoshida predict that $(^iPr_3)_2Si$: has a singlet ground state with the σ_p triplet 1.3 to 1.7 kcal/mole higher in energy, with an estimated error of at most 1 kcal/mol [49]. Thus the lowest S and T states of $(^iPr_3)_2Si$: seem to be nearly degenerate.

Both Apeloig [47] and Yoshida [49] made the unequivocal prediction that $(^tBu_3Si)_2Si$: would possess a triplet ground state $\Delta E = (E_S - E_T) = 7.1$ [47] or 4.5 to 4.9 kcal/mol [49], and Wiberg had reported the possible generation of $(^tBu_3Si)_2Si$: and reactions that included the abstraction of hydrogen atoms, which suggested that it might be reacting from a triplet state [50]. The organometallic route employed by Wiberg left the possibility of intervention by "silylenoids", but in retrospect it does appear that Wiberg generated triplet $(^tBu_3Si)_2Si$:.

Wiberg's experiments, however, suggested that two $^tBu_3Si$ groups provided such extreme steric shielding that intermolecular trapping of the reactive intermediate, whether silylene or silylenoid, would fail. Our interest in the *reactions* of triplet silylenes therefore led us to study a slightly less congested silylene $(^tBu)_3Si$–Si–$Si(^iPr)_3$, for which model compounds suggested that the Si-Si-Si angle would be intermediate between $(^iPr_3Si)_2Si$: and $(^tBu_3Si)_2Si$: [41, 51]. This was later supported by Yoshida's calculations, which predicted that the σ_p triplet state of $(^tBu)_3Si$–Si–$Si(^iPr)_3$ and the σ^2 singlet are nearly degenerate [52]. When generated by photochemical extrusion in the presence of suitable trapping agents, this silylene can be captured by intermolecular reactions. When generated in the absence of trapping agents, a four-membered ring is formed by formal C–H insertion [41, 51] analogous to the major product found by Wiberg [50] from $(^tBu_3Si)_2Si$:. Reaction from a triplet state was proposed for $(^tBu)_3Si$–Si–$Si(^iPr)_3$ via the mechanism shown in Scheme 7.

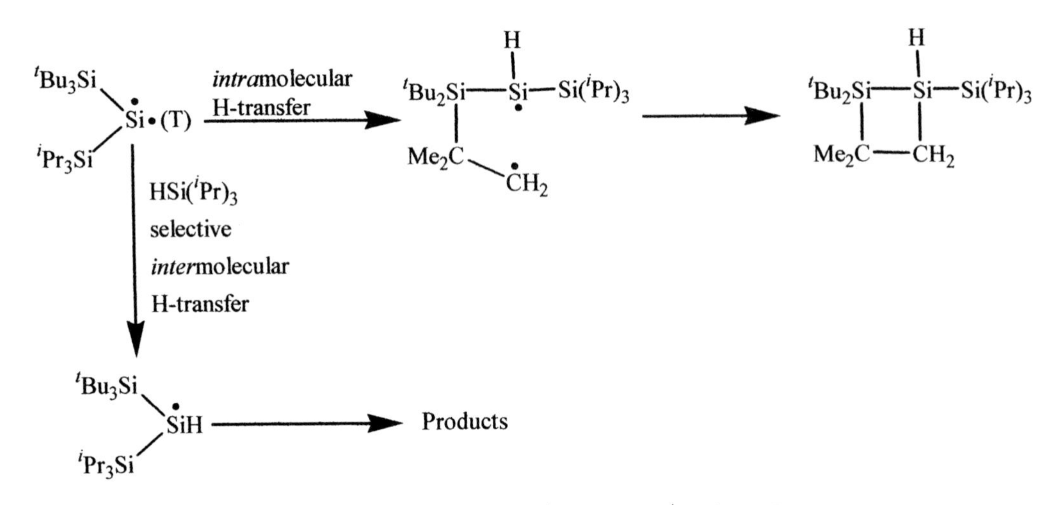

Scheme 7. Proposed mechanism for reactions of triplet $(^tBu)_3Si$–Si–$Si(^iPr)_3$ [41, 51].

A recent kinetic study of the extrusion of $(^tBu)_3Si$–Si–$Si(^iPr)_3$ from a silirane suggested that the silylene was formed in a singlet state by a non-least motion, concerted, stereospecific process, but that it reacted from a triplet state via the mechanism of Scheme 7 [53].

Of course, finding reactions from a triplet state does not guarantee that the ground state of $(^tBu)_3Si$–Si–$Si(^iPr)_3$ is a triplet. As Yoshida has pointed out [49], if the ground state is a singlet, with the lowest triplet only a kcal/mol higher in energy, singlet-triplet equilibration is likely, and triplet reactions would be observable under conditions that the singlet is not efficiently trapped.

Very recently Sekiguchi has unambiguously generated $(^tBu_3Si)_2Si$: and recorded an ESR spectrum that appears to demonstrate that it possesses a triplet ground state [54]. The quest for a triplet ground-state silylene seems to be over, but since Sekiguchi has, like Wiberg, not found any intermolecular reactions beyond H-atom abstraction also detected for $(^tBu)_3Si$–Si–$Si(^iPr)_3$, the search for triplet silylene *chemistry* has just begun.

Monovalent Silicon Cations — "Supersilylenes" with Four Valence Electrons

Most of the known (singlet) silylene reactions, additions, insertions, and rearrangements, involve the concerted formation of two bonds, made possible by the acceptance of an electron pair by the silylene LUMO and the donation of the electron pair in the singlet silylene HOMO. What reactions would we expect for a species that differs from a silylene by the presence of *two* LUMOs and a positive charge, as is the case for four-valence electron monovalent silyl cations Z-Si:$^+$? Some possibilities are presented in Scheme 8. These are all of interest. The first is a Lewis acid – Lewis base reaction producing a six valence electron silicon atom, a silylene, with a charged substituent. The silylene-like addition and insertion reactions forming two bonds are new routes to difficultly accessible trivalent silicon cations. The most novel feature of these four valence electron R–Si:$^+$ species is the possibility that they might form *three* bonds in a single step. If they do so, they will certainly deserve the nickname of "supersilylenes".

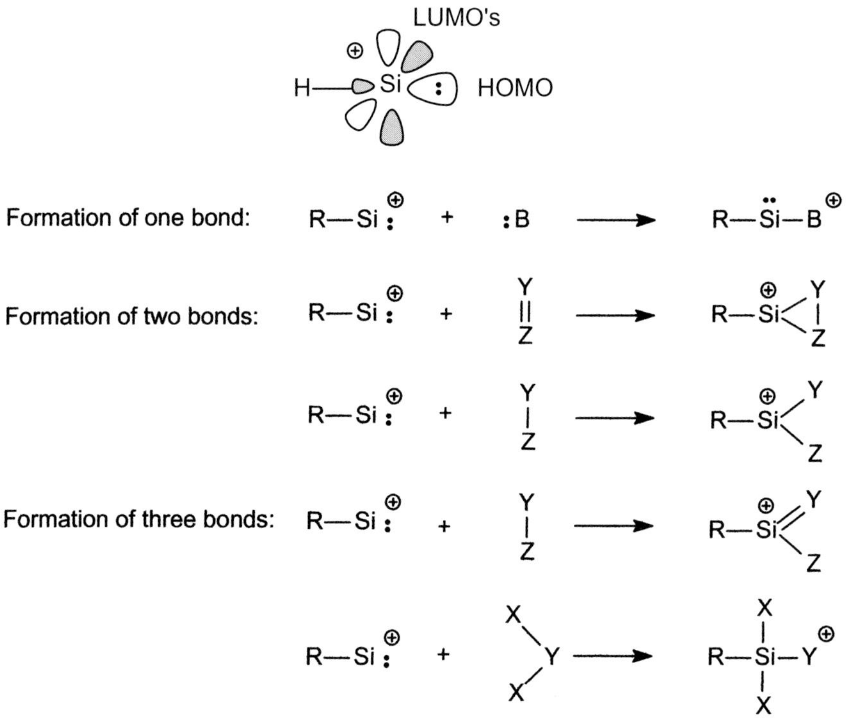

LUMO's

H—Si: HOMO

Formation of one bond: $R\text{—}Si:^{\oplus} + :B \longrightarrow R\text{—}\ddot{Si}\text{—}B^{\oplus}$

Formation of two bonds: $R\text{—}Si:^{\oplus} + Y{=}Z \longrightarrow$ R—Si$^{\oplus}$(Y=Z) (three-membered ring)

$R\text{—}Si:^{\oplus} + Y\text{—}Z \longrightarrow$ R—Si$^{\oplus}$(Y)(Z)

Formation of three bonds: $R\text{—}Si:^{\oplus} + Y\text{—}Z \longrightarrow$ R—Si$^{\oplus}$(=Y)(Z)

$R\text{—}Si:^{\oplus} + X_2Y \longrightarrow$ R—Si(X)(X)—Y$^{\oplus}$

Scheme 8. Frontier orbitals of H–Si:$^+$ and potential reactions for four-valence electron monovalent silicon cations.

Study of several monovalent silicon cations has been initiated [55] employing a quadrupole ion trap mass spectrometer equipped with a gas chromatograph to introduce neutral precursor molecules from which ions are produced by electron impact [56]. Neutral substrate molecules with which ions of interest are reacted are introduced as minor (< 1 %) dopants in the helium buffer gas constantly leaked into the ion cell in order to maintain a background pressure of 10^{-3} torr [55]. Two provocative early observations on R–Si:$^+$ ions are shown in Scheme 9.

Stepwise addition of two pyridine molecules to Ph–Si:$^+$, whose reversibility was established in collision-induced dissociation (CID) experiments, seems to be due to the formation of one bond at a time, the monovalent silicon cation reacting as a Lewis acid. That two, but no more than two, pyridine molecules are accepted by Ph–Si:$^+$ points to the silicon atom as the site of addition. In this scenario, addition of the first pyridine forms a distonic silylene. That this is a plausible process is indicated by the reaction of the parent silanetriyl cation H–Si:$^+$ with diethylamine $HNEt_2$. CID of the product ion established its structure as a four-membered ring whose most likely source is a two-step process: formation of a silylene intermediate by a Lewis acid–Lewis base reaction followed by intramolecular insertion of the silylene into a methyl C–H bond. Three bonds are formed in a single reactive encounter, but the stepwise process is much more likely than the more interesting concerted reaction.

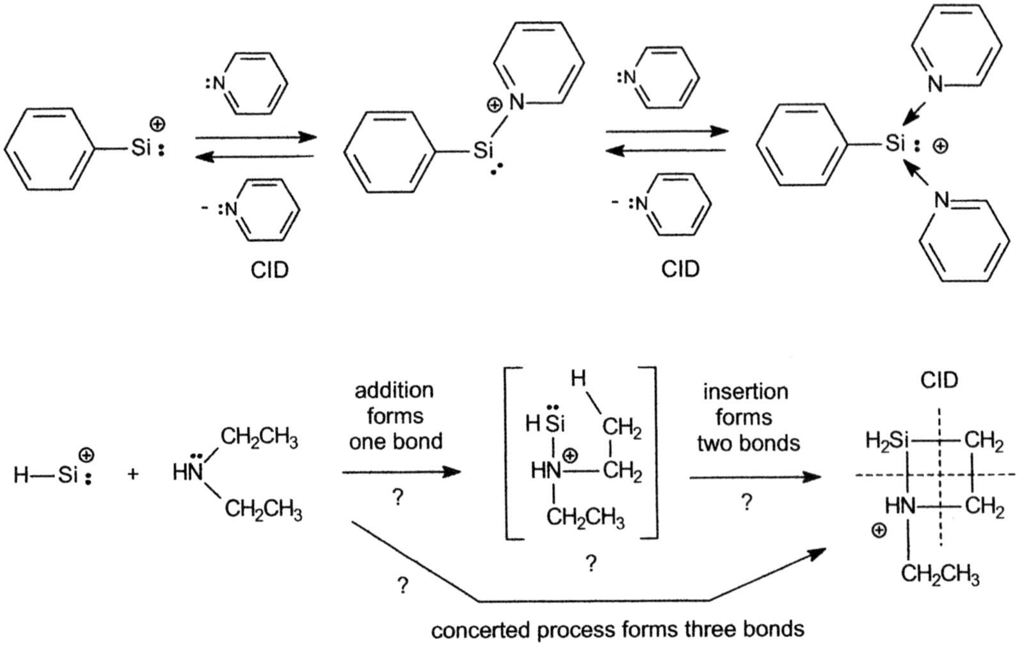

Scheme 9. Reactions of Ph–Si:$^+$ with pyridine and H–Si:$^+$ with diethylamine.

Experiments with R–Si:$^+$ are continuing to explore their reactivity, particularly with substrates capable of revealing silylene-like concerted formation of two bonds and the holy grail: concerted formation of three bonds.

What do we Look Forward to Learning in the Future From Silylenes and Supersilylenes?

Much time has been spent in learning how to make ground-state triplet silylenes, but the strategy of employing bulky, electropositive substituents such as trialkylsilyl groups [45] has proved to be self-defeating in the exploration of the chemistry of triplet silylenes. The steric shielding provided by groups large enough to force open the central bond angle to the point where the ground state is a triplet, in effect shuts down intermolecular reactions. Triplet silylenes are needed whose divalent silicon atoms are sterically accessible. One approach has been known for more than two decades: the use of such extremely positive substituents as lithium atoms. Gordon [57] and Scheafer [58] have predicted that SiHLi and $SiLi_2$ would have triplet ground states. We failed in an attempt to generate Me_3Si–Si–Li photochemically [59], but the generation of lithiosilylenes from 1-lithio siliranes or silirenes seems feasible. Another approach to triplet silylenes with sterically accessible divalent centers involves the construction of rigid scaffolds that can impose large central bond angles without hindering reaction [60].

Our questions regarding "supersilylenes" are more ambitious. Of course we would like to know what kinds of reactions they undergo, and especially if they participate in processes in which concerted formation of more than one bond occurs. While charged species like $R\text{–}Si:^+$ and $R\text{–}C:^+$ are convenient to generate and study in the gas-phase, neutral molecules such as R–B: and R–Al: should also be accessible and would allow study under conditions that permit larger scale reactions in solution.

If, as seems possible, four-valence electron cations prefer Lewis acid – Lewis base reactions forming just one bond at a time, opportunity will be provided to answer a very basic question about six-valence electron species like carbenes, silylenes, nitrenes, phosphinidenes, etc.: how can their tendency toward concerted formation of two bonds be explained? The presence of HOMOs and LUMOs of accessible energies may be insufficient. So the last question I will leave with you is: how can we modify the structures of electron-deficient species that are reluctant to participate in concerted reactions forming more than one bond in order to encourage them to do so?

It may seem as if I am asking why do carbenes and silylenes do two-bond forming chemistry rather than behaving like carbenium ions and silyl cations? This is equivalent to asking when is a lone pair active and when is it inert? This may not be interesting when comparing a second-row with a third-row element, but surely it becomes more so when comparing different species based on the same element, like R_2Si: and $R\text{–}Si:^+$, both with an unshared electron pair.

Acknowledgments: Coauthors of this paper are the coworkers who have elucidated much of silylene chemistry and have begun to explore the chemistry of "supersilylenes": W. C. Eckelman, S. A. Bock, C. A. Levy, P. Markusch, J. D. Holten, J. J. Frost, R.-J. Hwang, R. T. Conlin, B. J. Cornett, Y.-S. Chen, B. H. Cohen, A. P. Helfer, S. Konieczny, E. C.-L. Ma, S. H. Mo, M. Xiao, D. Lei, B. H. Boo, D. L. Svoboda, J. Frueh, M. E. Lee, M. North, F. Gonzalez, J. B. Wilking, D. H. Pae, D. J. Berger, T. Haile, K. L. Bobbitt, D. J. Berger, T. Chen, W. R. Winchester, P. Jiang, D.

Zhou, D. Read, X. Liu, J. S. Prell, J. L. Cichon.

Financial support of this work by the National Science Foundation under grant No. CHE-0316124 is gratefully acknowledged.

References

[1] W. Kirmse, *Carbene Chemistry* 1st edn., Academic Press, New York, **1964**.

[2] P. P. Gaspar, B. J. Herold, Silicon, germanium, and tin structural analogs of carbenes, in *Carbene Chemistry* 2nd edn. (Ed.: W. Kirmse), Academic Press, New York, **1971**, p. 504.

[3] P. P. Gaspar, B. D. Pate, W. C. Eckelman, *J. Am. Chem. Soc.* **1966**, *88*, 3878; P. P. Gaspar, S. A. Bock, W. C. Eckelman, *J. Am. Chem. Soc.* **1968**, *90*, 6914.

[4] P. P. Gaspar, R.-J. Hwang, W. C. Eckelman, *J. Chem. Soc., Chem. Commun.* **1974**, 242; D. Lei, Y.-S. Chen, B. H. Boo, J. Frueh, D. L. Svoboda, P. P. Gaspar, *Organometallics* **1992**, 559; V. N. Khabashesku, V. Balaji, S. Boganov, S. A. Bashkirova, P. M. Matveichev, E. A. Chernyshev, O. M. Nefedov, J. Michl, *Mendeleev Commun.* **1992**, 38.

[5] P. P. Gaspar, Recent advances in silylene chemistry, in *Reactive Intermediates* Vol 1 (Eds.: M. Jones, Jr., R. A. Moss), Wiley, New York, **1978**, p. 229.

[6] For the first connection between the energy barrier for a silylene insertion and the tendency for the insertion product to undergo the reverse reaction see: I. M. T. Davidson, *J. Organomet. Chem.* **1970**, *24*, 97.

[7] K. Stokland, *Trans. Faraday Soc.* **1948**, *44*, 545.

[8] J. H. Purnell, R. Walsh, *Proc. Roy. Soc., Ser. A* **1966**, *293*, 543.

[9] E. M. Tebben, M. A. Ring, *Inorg. Chem.* **1969**, *8*, 1787; M. A. Ring, H. E. O'Neal, *J. Phys. Chem.* **1992**, *96*, 10848.

[10] K. Stokland, *Kgl. Nor. Vidensk. Selsk. Skr.* **1950**, [N.S.] No. 3, p. 1.

[11] P. S. Skell, E. J. Goldstein, *J. Am. Chem. Soc.* **1964**, *86*, 1442 (two papers).

[12] W. H. Atwell, D. R. Weyenberg, *J. Organomet. Chem.* **1966**, *5*, 594; *Angew. Chem. Int. Ed. Engl.* **1969**, *8*, 469.

[13] R. T. Conlin, P. P. Gaspar, *J. Am. Chem. Soc.* **1976**, *98*, 3715.

[14] D. Seyferth, D. C. Annarelli, S. C. Vick, *J. Am. Chem. Soc.* **1976**, *98*, 6382; D. Seyferth, S. C. Vick, *J. Organomet. Chem.* **1977**, *125*, C11.

[15] P. Boudjouk, V. Samaraweera, R. Sooriyakumaran, J. Chrusciel, K. R. Anderson, *Angew. Chem. Int. Ed. Engl.* **1988**, *27*, 1355.

[16] V. J. Tortorelli, M. Jones, Jr., *J. Am. Chem. Soc.* **1980**, *102*, 1425; M. Ishikawa, K.-I. Nakagawa, M. Kumada, *J. Organomet. Chem.* **1979**, *178*, 105; V. J. Tortorelli, M. Jones, Jr., S.-H. Wu, Z.-H. Li, *Organometallics* **1983**, *2*, 759; D. Seyferth, D. C. Annarelli, D. P. Duncan, *Organometallics* **1982**, *1*, 1288.

[17] P. P. Gaspar, R. West, Silylenes, in *The Chemistry of Organosilicon Compounds*, Vol. 2 (Eds: Z. Rappoport, Y. Apeloig), Wiley, Chichester, **1998**, p. 2463.

[18] D. Lei, R.-J. Hwang, P. P. Gaspar, *J. Organomet. Chem.* **1984**, *271*, 1.

[19] D. Lei, P. P. Gaspar, *J. Chem. Soc., Chem. Commun.* **1985**, 1149.

[20] D. Lei, P. P. Gaspar, *Organometallics* **1986**, *5*, 1276.

[21] D. Lei, P. P. Gaspar, *Organometallics*, **1985**, *4*, 1471.

[22] D. Lei, P. P. Gaspar, *Polyhedron* **1991**, *10*, 1221.

[23] D. Zhou, P. P. Gaspar, Dimethylstannylene: generation and trapping reactions, in *Abstracts of Papers, 225th ACS National Meeting, New Orleans LA, March 23 – 27, 2003* **2003**, ORGN-566.

[24] X. Li, D. Lei, M. Chiang, P. P. Gaspar, *J. Am. Chem. Soc.* **1992**, *114*, 8526; *P, S, Si & Related Elements* **1993**, *76*, 74.

[25] P. P. Gaspar, A. M. Beatty, X. Li, H. Qian, N. P. Rath, J. C. Watt, *P, S, Si & Related Elements* **1999**, *144-146*, 277; P. P. Gaspar, H. Qian, A. M. Beatty, A. D'Avignon, J. L.-F. Kao, J. C. Watt, N. P. Rath, *Tetrahedron* **2000**, *56*, 105.

[26] P. P. Gaspar, B. H. Boo, D. L. Svoboda, *J. Phys. Chem.* **1987**, *91*, 5011.

[27] H. M. Frey, *J. Chem. Soc.* **1962**, 2293.

[28] M. E. Lee, M. A. North, P. P. Gaspar, *P, S, & Si* **1991**, *56*, 203.

[29] B. H. Boo, P. P. Gaspar, *Organometallics* **1986**, *5*, 698.

[30] W. D. Wulff, W. W. F. Goure, T. J. Barton, *J. Am. Chem. Soc.* **1978**, *100*, 6236.

[31] I. M. T. Davidson, R. J. Scampton, *J. Organomet. Chem.* **1984**, *271*, 249.

[32] R. C. Joines, A. B. Turner, W. M. Jones, *J. Am. Chem. Soc.* **1969**, *91*, 7754; W. J. Baron, M. Jones, Jr., P. P. Gaspar, *J. Am. Chem. Soc.* **1970**, *92*, 4739; M. J. S. Dewar, D. Landman, *J. Am. Chem. Soc.* **1977**, *99*, 6179; P. C. Miller, P. P. Gaspar, *J. Org. Chem.* **1991**, *56*, 5101; P. B. Schreiner, W. L. Karney, P. v. R. Schleyer, W. T. Borden, T. P. Hamilton, H. F. Schaefer III, *J. Org. Chem.* **1996**, *61*, 7030.

[33] P. P. Gaspar, unpublished results

[34] J. D. Goddard, Y. Yoshioka, H. F. Schaefer III, *J. Am. Chem. Soc.* **1980**, *102*, 7644.

[35] R. S. Grev, G. E. Scuseria, A. C. Scheiner, H. F. Schaefer III, M. S. Gordon, *J. Am. Chem. Soc.* **1988**, *110*, 7337.

[36] R. T. Conlin, P. P. Gaspar, *J. Am. Chem. Soc.* **1976**, *98*, 868.

[37] R. West, M. J. Fink, J. Michl, *Science (Washington, D.C.)* **1981**, *214*, 1343; R. Okazaki, R. West, *Adv. Organomet. Chem.* **1996**, *39*, 231.

[38] N. Tokitoh, R. Okazaki, Multiply bonded germanium, tin and lead compounds, in *The Chemistry of Organic Germanium, Tin and Lead Compounds Vol. 2* (Ed. Z. Rappoport), Wiley, Chichester, **2002**, p. 843.

[39] P. P. Gaspar, G. S. Hammond, Spin states in carbene chemistry, in *Carbenes Vol. 2* (Eds. R. A. Moss, M. Jones, Jr.), Wiley, New York, **1975**, p. 207.

[40] W. Kutzelnigg, *Angew. Chem. Int. Ed. Engl.* **1984**, *23*, 272.

[41] P. P. Gaspar, M. Xiao, D. H. Pae, D. J. Berger, T. Haile, T. Chen, D. Lei, W. R. Winchester, P. Jiang, *J. Organomet. Chem.* **2002**, *646*, 68.

[42] Y. Apeloig, R. Pauncz, M. Karni, R. West, W. Steiner, D. Chapman, *Organometallics*, **2003**, *22*, 3250.

[43] M. S. Gordon, *Chem. Phys. Lett.* **1985**, *114*, 348.

[44] D. H. Pae, M. Xiao, M. Y. Chiang, P. P. Gaspar, *J. Am. Chem. Soc.* **1991**, *113*, 1281.
[45] R. S. Grev, H. F. Schaefer III, P. P. Gaspar, *J. Am. Chem. Soc.* **1991**, *113*, 5638.
[46] D. J. Berger, unpublished results.
[47] M. C. Holthausen, W. Koch, Y. Apeloig, *J. Am. Chem. Soc.* **1999**, *121*, 2623.
[48] P. P. Gaspar, A. M. Beatty, T. Chen, T. Haile, D. Lei, W. R. Winchester, J. Braddock-Wilking, N. P. Rath, W. T. Klooster, T. F. Koetzle, S. A, Mason, A. Albinati, *Organometallics* **1999**, *18*, 3921.
[49] M. Yoshida, N. Tamaoki, *Organometallics* **2002**, *21*, 2587.
[50] N. Wiberg, *Coord. Chem. Rev.* **1997**, *163*, 217; N. Wiberg, W. Niedermayer, *J. Organomet. Chem.* **2001**, *628*, 57.
[51] P. Jiang, P. P. Gaspar, *J. Am. Chem. Soc.* **2001**, *123*, 8622.
[52] Yoshida's calculations [49] predict that the σ_p triplet is 0.8 to 1.0 kcal/mol *higher* in energy than the σ^2 singlet, with a an uncertainty no greater than 1 kcal/mol.
[53] P. Jiang, D. Trieber II, P. P. Gaspar, *Organometallics* **2003**, *22*, 2233.
[54] Sekiguchi, T. Tanaka, M. Ichinohe, K. Akiyama, S. Tero-Kubota, *J. Am. Chem. Soc.* **2003**, *125*, 4962.
[55] P. P. Gaspar, D. Read, Ion-molecule reactions of $R\text{–}Si^+$, a new class of reactive intermediates with four valence electrons, XIII International Symposium on Organosilicon Chemistry, Guanajuato, Mexico, Aug. 25 – 30, **2002**.
[56] R. E. March, *J. Mass Spectrom.* **1997**, *32*, 351.
[57] M. S. Gordon, M. W. Schmidt, *Chem. Phys. Lett.* **1986**, *132*, 294.
[58] M. E. Colvin, J. Breulet, H. F. Schaefer III, *Tetrahedron* **1985**, *41*, 1429.
[59] M. Xiao, The chemistry of sterically congested organosilicon compounds and the quest for triplet ground state silylenes, doctoral dissertation, Washington University, St. Louis, December **1988**.
[60] J. L. Cichon, P. P. Gaspar, unpublished results.

New Molecular Systems with Silicon–Silicon Multiple Bonds

Mitsuo Kira,* Takeaki Iwamoto, Shintaro Ishida

Department of Chemistry, Graduate School of Science, Tohoku University
Aoba-ku, Sendai 980-8578, Japan
Tel.: +81 22 217 6585 — Fax: +81 22 217 6589
E-mail: mkira@si.chem.tohoku.ac.jp

Keywords: cyclic disilenes, spiropentasiladiene, trisilaallene, spiroconjugation

Summary: Synthesis, structure, and reactions of novel cyclic and conjugated disilenes are discussed. Photochemical and thermal interconversion among Si_4R_6 isomers including a cyclotetrasilene, a silylcyclotrisilene, and a bicyclo[1.1.0]tetrasilane occurred without apparent participation of the corresponding tetrasila-1,3-diene. The first spiropentasiladiene derivative, which is thermally very stable in contrast to spiropentadiene, which decomposes below –100 °C, shows modified but significant spiroconjugation between the two ring π systems. The first stable silicon compound with a formally *sp*-hybridized silicon atom, trisilaallene, has a remarkably bent and fluxional framework and exhibits a significant conjugation between the two trans-bent Si=Si bonds.

Introduction

Since the first isolation of tetramesityldisilene by West et al. in 1981, a number of stable disilenes have been synthesized, and their structures and properties have been extensively investigated up to date [1, 2]. However, known types of stable disilenes had been limited to simple acyclic derivatives with one Si=Si double bond until 1996, when the first stable cyclotetrasilene was synthesized [3]. Recently, the number of known skeletal types of disilenes has increased remarkably, and studies of their structures and reactions have made astonishing progress. The chemistry of disilenes is now approaching a stage to be compared with the extensive and profound chemistry of olefins. Here, we discuss the synthesis, structure, and reactions of novel cyclic and bicyclic disilenes and a trisilaallene, the first compound with a formally *sp*-hybridized silicon atom.

Persilyldisilenes, Cyclotetrasilene, Cyclotrisilene, and Spiropentasiladiene

Tetrakis(trialkylsilyl)disilenes formally made by the dimerization of the corresponding bis(trialkylsilyl)silylenes, whose singlet-triplet energy differences are much smaller than those of dialkyl- and diarylsilylenes [4], are expected to show distinctive structural and electronic properties and reactions, on the basis of the CGMT model [5]. Since the first synthesis of stable tetrakis(trialkylsilyl)disilenes **1a–c** by using a modified Kipping's coupling of the corresponding dibromosilanes with various reducing reagents [6a], we have continuously studied on various types of tetrasilyldisilenes, shown in Scheme 1. Disilenes **1a–e** showed interesting features in the structures around Si=Si double bonds [6], electronic spectra [6], anisotropic ^{29}Si NMR chemical shifts [7], and various reactions [8].

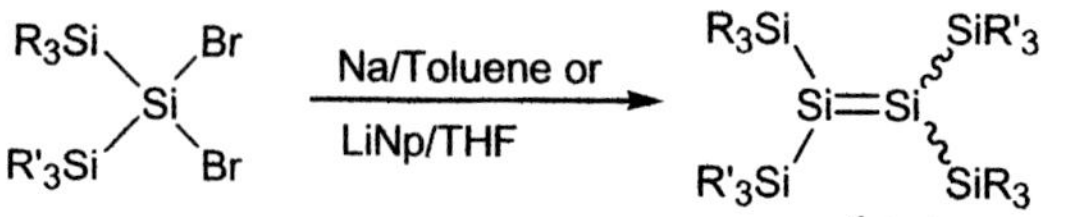

Na/Toluene or LiNp/THF → **1a–e**

a, $R_3Si = R'_3Si$ = *t*-BuMe$_2$Si; **b**, $R_3Si = R'_3Si$ = *i*-Pr$_2$MeSi;
c, $R_3Si = R'_3Si$ = *i*-Pr$_3$Si; **d**, R_3Si = *t*-BuMe$_2$Si, R'_3Si = *i*-Pr$_2$MeSi;
e, R_3Si = *t*-BuMe$_2$Si, R'_3Si = *i*-Pr$_3$Si.

Scheme 1. Synthesis of tetrakis(trialkylsilyl)disilenes.

As a reasonable extension of our work on trialkylsilyl-substituted disilenes, we have investigated a co-reductive coupling of dibromotrisilane **2** and 2,2,3,3-tetrabromotetrasilane **3** to synthesize a conjugated disilene **6**. However, the reaction gave the corresponding cyclotetrasilene **5** as the first stable cyclic disilene (14% yield) instead of the expected product **6** (Scheme 2) [3]. During the co-reduction, the initial product was found to be the corresponding bicyclo[1.1.0]tetrasilane **4**, which underwent facile thermal isomerization to **5** [9]. The first stable tetrasila-1,3-diene was recently synthesized by Weidenbruch et al. [10].

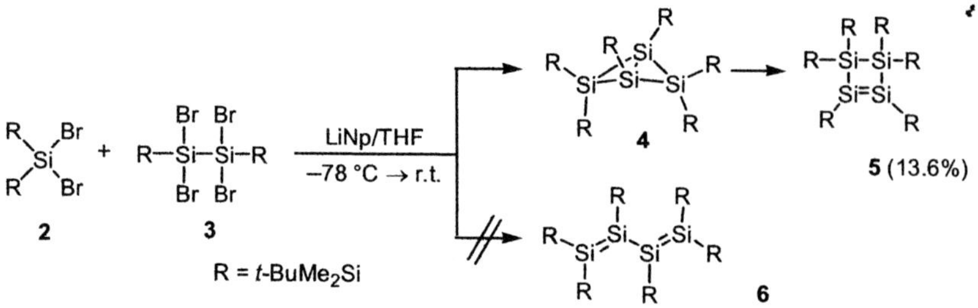

2 + **3** → LiNp/THF, −78 °C → r.t. → **4** → **5** (13.6%); ↛ **6**

R = *t*-BuMe$_2$Si

Scheme 2. Synthesis of cyclotetrasilene **5**.

Reduction of compound **7** resulted in an incredible variety of products, depending on the reducing conditions. While the reduction of **7** with Na in toluene gave cyclotetrasilene **5** in 64% yield, the reduction with potassium graphite in THF below −40 °C afforded the first cyclotrisilene **8** in 67% yield [11], in addition to a small amount of the fascinating spiropentasiladiene **9** (Scheme 3) [12].

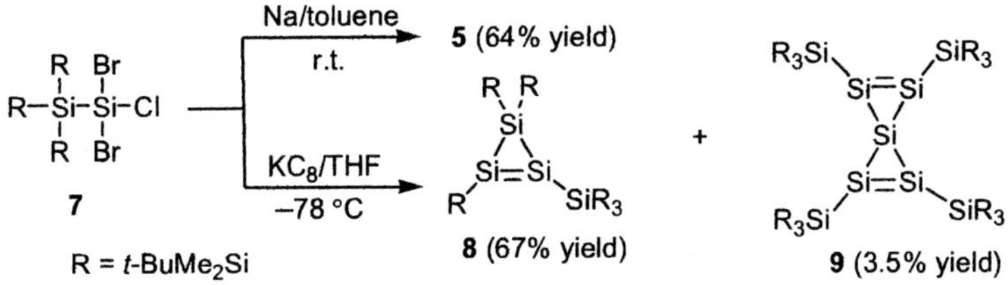

Scheme 3. Synthesis of cyclic disilenes using 1,1,1-trihalodisilane **7**.

The photochemical conversion of **5** to the corresponding bicyclo[1.1.0]tetrasilane **4** and its thermal reversion to **5** was observed [3, 9]. Upon irradiation of cyclotrisilene **8** in toluene with a low-pressure mercury arc lamp, the isomerization of **8** to bicyclo[1.1.0]tetrasilane **4** occurred in high yield [13]. In contrast to the well-studied electrocyclic interconversion of C_4H_6, no evidence was obtained for the intermediacy of tetrasila-1,3-butadiene **6** during the isomerizations of the Si_4R_6 isomers.

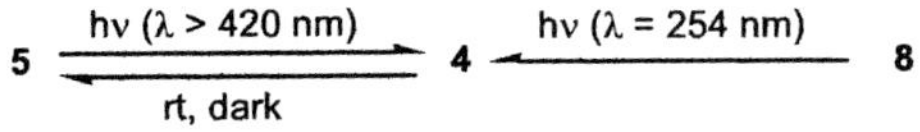

Scheme 4. Interconversion among Si_4R_6 isomers.

Spiropentasiladiene **9** is the first cyclic compound with two Si=Si double bonds, the first stable spiropentadiene comprising group-14 elements, and a nano-size cluster with twenty-one silicon atoms whose surface is covered by 36 alkyl groups. Although spiroconjugation is well investigated theoretically, no experimental inspection of the spiroconjugation in spiropentadiene, the smallest spiroconjugation system, has been performed, because of its very labile nature; spiropentadiene is known to decompose in several minutes even below –100 °C [14]. In contrast, **9** is thermally very stable and melts at 216 – 218 °C without decomposition, though it is sensitive to air.

The skeletal structure of **9** determined by X-ray single-crystal analysis holds D_2 symmetry (Fig. 1). The Si=Si double bond distance (2.186(3) Å) is slightly longer than that of known cyclotrisilenes (2.132 – 2.138 Å) [11, 15]. Two three-membered rings in **9** are not perpendicular to each other but slightly twisted, with a dihedral angle (θ) of 78.26°. The geometry around the Si=Si bonds in **9** is not planar but significantly twisted; the dihedral angle Si6-Si2-Si3-Si7 (δ) is 30.0(5)°, and the sum of the bond angles is 358.9° at Si2 and 358.0° at Si3, respectively. Because the geometry of model tetrasilylspiropentasiladiene **10** optimized at B3LYP/6-31G(d) level has local D_{2d} symmetry (θ = 90° and δ = 0°), the major reason for the deformation found for **9** is steric repulsion between four bulky tris(trialkylsilyl)silyl substituents.

Deformation in spiropentasiladiene **9** makes an interesting illustration of the effect of skeletal deformation on spirocojugation. As schematically shown in Fig. 2, the deformation from D_{2d} to D_2 causes splitting of the bonding π orbitals in addition to the splitting of the anti-bonding π* orbitals which appears in D_{2d} geometry. The qualitative MO feature coincides essentially with a conclusion obtained by more sophisticated calculations for the parent spiropentasiladiene **11** [12]. The UV-vis spectral feature of **9** in 3-methylpentane at room temperature was quite different from those of monocyclic cyclotrisilenes [11, 15]. Four major absorption bands in the visible region [λ_{max}/nm ($\varepsilon/10^4$) 560 0.253], 500 (0.364), 428 (1.17), 383 (1.81)] were observed for **9**, and the longest absorption maximum red-shifted by 78 nm relative to that for **8** (482 nm). The above-mentioned modified spiroconjugation would be responsible for the remarkable red shift and split π–π^* transitions, which may be allowed in 9 with D_2 symmetry.

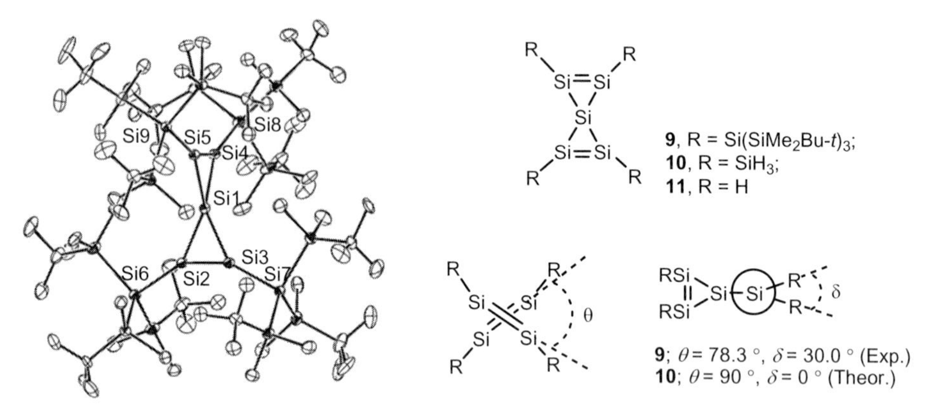

Fig. 1. Molecular structure of **9** and schematic representation of its structural deformation.

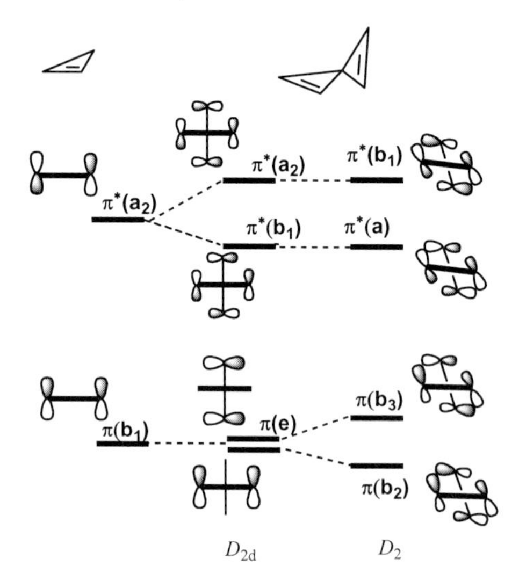

Fig. 2. Qualitative MO description for the spiroconjugation modified by skeletal deformation of spiropentasiladiene.

Spiropentasiladiene **9** is kinetically stabilized by steric protection by four bulky tris(*t*-butyldimethylsilyl)silyl groups, but the other main reason for the thermal stability is much smaller strain energy (SE) of the ring system than that of the carbon analog. The SE value of the parent spiropentasiladiene Si_5H_4 with D_{2d} symmetry was calculated to be almost a half of that of the corresponding spiropentadiene C_5H_4 (D_{2d}, 114.2 kcal/mol) [12].

Trisilaallene

Although the *sp*-hybridized carbon atoms are commonly found in acetylenes, nitriles, allenes, carbon dioxide, etc., heavy group-14 element analogs of these compounds are very rare [16, 17]. The first silicon compound with a formal *sp*-silicon atom, trisilaallene **12**, was synthesized as a dark green solid using two-step reactions from stable dialkylsilylene **13** [18] in overall 42% yield (Scheme 5) [19]. Trisilaallene **12** is sensitive to air but thermally rather stable, with a melting point of 198 – 200 °C.

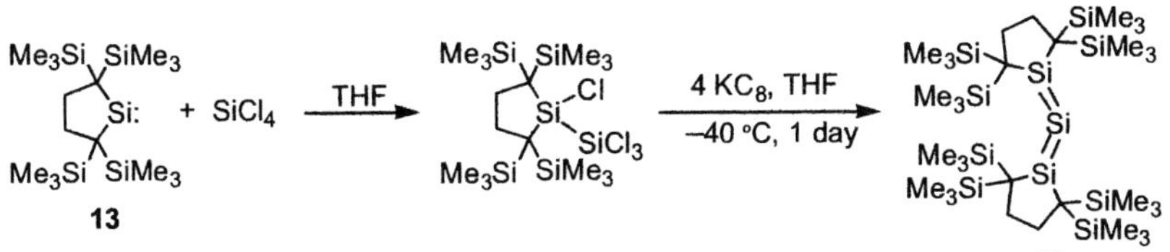

Scheme 5. Synthesis of trisilaallene **12**.

The allene skeleton of **12** determined by X-ray analysis (Fig. 3b) is not linear but bent. Curiously, the central silicon atom (Si^2) of **12** in the crystal was found at four positions labeled Si^{2A}-Si^{2D} at –50 °C, indicating that four structurally similar isomers exist, **A** to **D**; isomers **A** through **D** correspond to those whose Si^2 atom is located at Si^{2A} through Si^{2D}. The populations for **A** through **D** are independent of crystals but significantly temperature-dependent: 53, 22, 19, and 7% at –50 °C but 76, 18, 7, and 0% at –150 °C, respectively. The energy differences between isomers **A** – **D** are estimated to be within 1 kcal/mol, suggesting a dynamic disorder mediated by a facile swing and rotation of the Si^2 atom around the Si^1-Si^3 axis (Fig. 3c).

The detailed analysis of the molecular structure of isomer **A** of trisilaallene **12** revealed the bent-allenic nature with two trans-bent Si=Si double bonds that share a silicon atom; the Si^1-Si^2-Si^3 bond angle is 136.49 (6)°, the two Si=Si bond lengths are in the range of those for typical stable disilenes, the two terminal silicon atoms are pyramidalized, and the dihedral angles Si^3-Si^2-Si^1-C^2 and Si^1-Si^2-Si^3-C^4 are around 60°.

The observed skeletal structure for **12** was not reproduced by the *ab initio* MO calculations for a model trisilaallene $Me_2Si{=}Si{=}SiMe_2$ (**14**) (Fig. 4). The optimized structure of **14** (**14[opt]**) is characterized to be neither linear nor bent allenic but zwitterionic. The HOMO and LUMO for **14[opt]** are a symmetric and an antisymmetric π orbital delocalized to the trisilaallyl π system, respectively, and the HOMO-1 has the nature of a lone-pair orbital of silylene. The reason for the discrepancy between the structures of **12** and **14[opt]** may be ascribed to the severe steric hindrance between two bulky silacyclopentane rings caused when a zwitterionic structure similar to **14[opt]** is applied to **12**. MO calculations for **14[exp]**, where the coordinates of four carbon and three silicon atoms in **14** were fixed to those observed for **12**, revealed that bent pπ type orbitals at the terminal silicon atoms interact with both pπ- and in-plane-type orbitals at the central silicon atom, indicating the bent-allenic electronic structure in **14[exp]** and hence in **12**. Because **14[exp]** is only 8.8 kcal/mol higher in energy than **14[opt]** at the B3LYP/6-31+G(d,p) level, the distortion from a zwitterionic to a bent-allenic structure will be feasible by the steric effects. In other words, the bulky terminal substituents in **12** not only serve as sterically protecting groups for Si=Si bonds but also adjust the electronic structure.

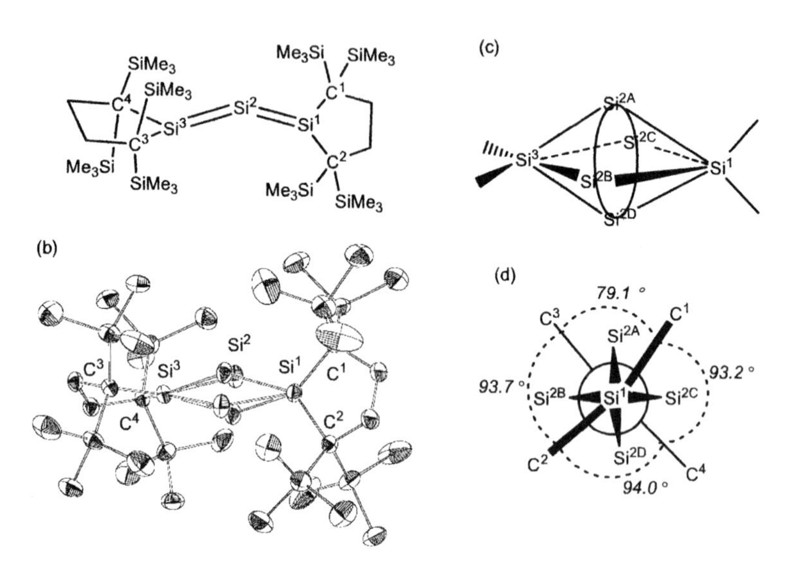

Fig. 3. (a) Schematic structure of trisilaallene **12**; (b) Molecular structure of **12** determined by X-ray crystallography at 0 °C; (c) Schematic representation of dynamic disorder of Si^2 atom in **12**; (d) Projection of the molecular frame to the plane perpendicular to Si^1–Si^3 axis.

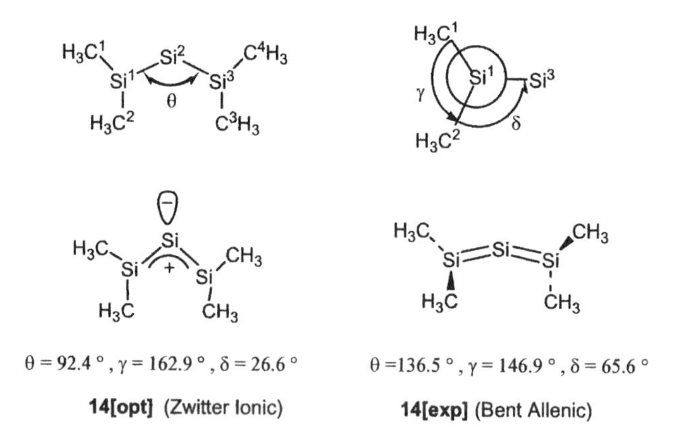

Fig. 4. Schematic structural descriptions of trisilaallene **14[opt]** and **14[exp]** calculated at the B3LYP/6-31+G(d,p) level.

The unusual bonding nature of **12** in the solid state is applicable also to that in solution. The ^{29}Si NMR resonances found at 157 and 197 ppm in benzene-d_6 are assigned to Si^2 and Si^1 (Si^3) nuclei, respectively, being in good accord with the theoretical values calculated at the GIAO/B3LYP/6-311+G(2df,p)//B3LYP/6-31+G(d,p) level for **14[exp]**. The fluxional nature of **12** in solution is evidenced by the highly symmetric ^{1}H, ^{13}C, and ^{29}Si NMR spectra; 72 ^{1}H, 24 ^{13}C, and 8 ^{29}Si nuclei in 8 trimethylsilyl groups and 8 ^{1}H and 4 ^{13}C in two methylene groups are all equivalent. The ^{1}H NMR spectrum for **12** was essentially independent of temperature and showed two singlets due to the methyl and methylene protons even at –80 °C, indicating low energy barriers for the interchange among **A** – **D** isomers. The UV-vis absorption maxima were found at 390 nm (ε 21 300) and 584 nm (ε 700) in hexane. The remarkable red shift of the longest absorption band compared with that for tetraalkyldisilenes as well as the two-splitting band feature for **12** are consistent with significant conjugation between the two Si=Si double bonds as suggested by the molecular orbital calculations; the extent of the conjugation is larger than that found for spiropentasiladiene **9**.

In conclusion, trisilaallene **12** has a fluxional bent framework and exhibits an effective conjugation between the two *trans*-bent Si=Si double bonds, in remarkable contrast to a carbon allene having rigid linear framework whose C=C π bonds are orthogonal to each other.

Acknowledgments: The authors are grateful to the Ministry of Education, Science, Sports, and Culture, Japan for financial support.

References

[1] For recent reviews, see: a) R. Okazaki, R. West, *Adv. Organomet. Chem.* **1996**, 39, 231; b) M. Weidenbruch, in *The Chemistry of Organic Silicon Compounds* (Eds.: Z. Rappoport, Y. Apeloig) Wiley, Chichester, **2001**, Vol. 3, p. 391.

[2] Accounts: a) M. Kira, T. Iwamoto, *J. Organomet. Chem.* **2000**, *611*, 236; b) M. Kira, *Pure Appl. Chem.* **2000**, *72*, 2333.

[3] M. Kira, T. Iwamoto, C. Kabuto, *J. Am. Chem. Soc.* **1996**, *118*, 10303.

[4] *Inter alia*: Y. Apeloig, in *The Chemistry of Organic Silicon Compounds* (Eds.: S. Patai, Z. Rappoport) Wiley, Chichester, **1989**, Vol. 1, Chapter 2.

[5] a) G. Trinquier, J.-P. Malrieu, *J. Am. Chem. Soc.* **1987**, *109*, 5303; b) J.-P. Malrieu, G. Trinquier, *J. Am. Chem. Soc.* **1989**, *111*, 5916; c) G. Trinquier, *J. Am. Chem. Soc.* **1990**, *112*, 2130.

[6] M. Kira, T, Maruyama, C. Kabuto, K. Ebata, H. Sakurai, *Angew. Chem. Int. Ed. Engl.* **1994**, *33*, 1489; b) M. Kira, S. Ohya, T. Iwamoto, M. Ichinohe, C. Kabuto, *Organometallics* **2000**, *19*, 1817; c) T. Iwamoto, J. Okita, C. Kabuto, M. Kira, *J. Organomet. Chem.* **2003**, *686*, 105.

[7] R. West, D. Cavalieri, J. J. Buffy, C. Fry, W. K. Zilm, C. J. Duchamp, M. Kira, T. Iwamoto, T. Müller, Y. Apeloig, *J. Am. Chem. Soc.* **1997**, *119*, 4972.

[8] a) T. Iwamoto, H. Sakurai, M. Kira, *Bull. Chem. Soc. Jpn.* **1998**, *71*, 2741; b) M. Kira, T. Ishima, T. Iwamoto, M. Ichinohe, *J. Am. Chem. Soc.* **2001**, *123*, 1676; c) M. Kira, T. Iwamoto, D. Yin, T. Maruyama, H. Sakurai, *Chem. Lett.* **2001**, 910.

[9] T. Iwamoto, M. Kira, *Chem. Lett.* **1998**, 277.

[10] M. Weidenbruch, S. Willms, W. Saak, G. Henkel, *Angew. Chem.* **1997**, *109*, 2612; *Angew. Chem. Int. Ed. Engl.* **1997**, *36*, 2503. See also: S. Willms, M. Weidenbruch, in *Organosilicon Chemistry IV — From Molecules to Materials* (Eds.: N. Auner, J. Weis), VCH, Weinheim, **2000**, p. 117.

[11] T. Iwamoto, C. Kabuto, M. Kira, *J. Am. Chem. Soc.* **1999**, *121*, 886 (Correction: **2000**, *122*, 12614).

[12] T. Iwamoto, M. Tamura, C. Kabuto, M. Kira, *Science* **2000**, *290*, 504.

[13] T. Iwamoto, M. Kira, *Chem. Lett.* **1998**, 277.

[14] T. Iwamoto, M. Tamura, C. Kabuto, M. Kira, *Organometallics* **2003**, *22*, 2342.

[15] a) W. E. Billups, M. M. Haley, *J. Am. Chem. Soc.* **1991**, *113*, 5084; b) R. K. Saini, V. A. Litosh, A. D. Daniels, W. E. Billups, *Tetrahedron Lett.* **1999**, *40*, 6157.

[16] M. Ichinohe, T. Matsuno, A. Sekiguchi, *Angew. Chem. Int. Ed.* **1999**, *37*, 2194.

[17] For a recent review for heavy allenes, see: J. Escudie, H. Ranaivonjatovo, L. Rigon, *Chem. Rev. (Washington D.C.)* **2000**, *100*, 3639.

[18] For a marginally stable tristannaallene, see: N. Wiberg, H.-W. Lerner, S.-K. Vasisht, S. Wagner, K. Karaghiosoff, H. Nöth, W. Ponikwar, *Eur. J. Inorg. Chem.* **1999**, 1211.

[19] M. Kira, S. Ishida, T. Iwamoto, C. Kabuto, *J. Am. Chem. Soc.* **1999**, *121*, 9722.

[20] S. Ishida, T. Iwamoto, C. Kabuto, M. Kira, *Nature* **2003**, *421*, 725.

1-[2,6-Bis(dimethylaminomethyl)phenyl]-2,2-bis(trimethylsilyl)silenes — Syntheses, Structures, and Reactions

*Ute Bäumer, Helmut Reinke, Hartmut Oehme**

Universität Rostock, Fachbereich Chemie, D-18051 Rostock, Germany
Tel.: +49 381 498 6400 — Fax: +49 381 498 6382
E-mail: hartmut.oehme@chemie.uni-rostock.de

Keywords: silaethenes, silanes, silenes, silicon, rearrangements

Summary: 1-[2,6-bis(dimethylaminomethyl)phenyl]silenes (**2a–d**) were prepared by treatment of the (dichloromethyl)oligosilanes $R^1(Me_3Si)_2Si–CHCl_2$ **1a–d** (**a**: R = Me; **b**: R = *tert*-Bu; **c**: R = Ph; **d**: R = Me_3Si) with 2,6-bis(dimethylaminomethyl)phenyllithium and were characterized by NMR studies and (in part) by X-ray structural analyses as well as by their reactions with water to give silanols. Treatment of **2a–d** with benzaldehyde produced 2,2-bis(trimethylsilyl)styrene (**4**) and a silanone polymer as the expected products. For the reaction of **2a** and **2c** with benzaldehyde, an interesting side reaction was observed leading to the 2-oxa-1-sila-1,2,3,4-tetrahydronaphthalenes **6a** and **6c**, respectively.

Introduction

Silenes (compounds with an Si=C double bond) are known to be extremely reactive compounds. Till now, only a few silenes, stabilized kinetically by bulky substituents, could be isolated under normal conditions and were characterized by X-ray structural analyses [1]. An effective stabilization of Si=C systems is also achieved by coordination of a base, such as amines, THF, or fluoride ions, to the electrophilic silene silicon atom [2], and recently we succeeded in synthesizing thermally unexpectedly stable intramolecularly donor-stabilized silenes [3]. In this paper, the synthesis and, in particular, the behavior of the 1-[2,6-bis(dimethylaminomethyl)phenyl]silenes **2a–d** are described.

Results and Discussion

The reaction of the (dichloromethyl)oligosilanes **1a–d** with organolithium compounds proved to be a versatile and straightforward method for the generation of silenes [3, 4]. The reaction path, which involves a complex sequence of deprotonation, elimination, isomerization, and addition steps, has been previously discussed in detail [4]. Following this general route, the silenes **2a–d** were prepared in a simple one-pot reaction by treatment of the (dichloromethyl)oligosilanes $R^1(Me_3Si)_2Si–CHCl_2$ **1a–d** (**a**: R^1 = Me; **b**: R^1 = *tert*-Bu; **c**: R^1 = Ph; **d**: R^1 = Me_3Si) with 2,6-bis(dimethylaminomethyl)-phenyllithium (molar ratio 1:2). Despite a really complex reaction path, which is outlined in an abbreviated form in Scheme 1, **2a–d** were isolated in yields between 49 and 78%.

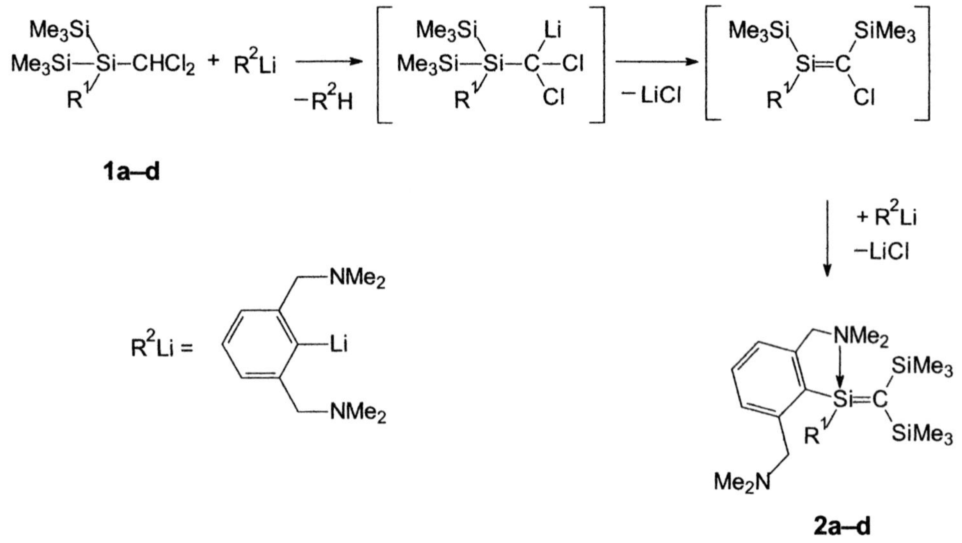

Scheme 1. The reaction of the (dichloromethyl)oligosilanes **1a–d** with 2,6-bis(dimethylaminomethyl)phenyl-lithium (molar ratio 1 : 2); **a**: R^1 = Me, **b**: R^1 = *tert*-Bu, **c**: R^1 = Ph, **d**: R = $SiMe_3$.

The intramolecularly donor-stabilized silenes **2a–d** are thermally stable compounds. **2a**, **2c** and **2d** are colorless solids; **2b** is an oil. X-ray structural analyses unambiguously revealed that in the solid state only one dimethylamino group of the ligand is coordinated to the silene silicon atom. Important structural parameters, e.g., the Si=C bond lengths and the Si–N distances, are in good agreement with data obtained for other intramolecularly amine-coordinated silenes [3].

In the NMR spectra of **2a–d** in solution at room temperature, only one signal was found for the two dimethylaminomethyl groups. At low temperatures, two different signals of the CH_2NMe_2 groups appeared. This indicates that in solution at low temperature the silene silicon atom is tetracoordinated through an N→Si interaction with one dimethylamino group as in the solid state structure, but at elevated temperatures rapid exchange processes lead to a magnetic equivalence of the two CH_2NMe_2 substituents.

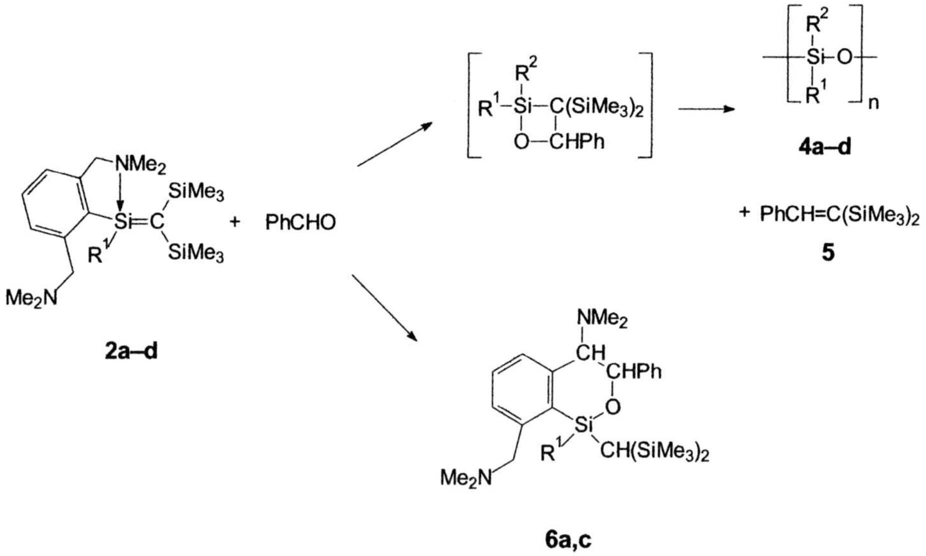

NMe_2 $SiMe_3$ $Si{=}C$ R^1 $SiMe_3$ Me_2N + PhCHO

2a–d

[R^2 R^1–Si–$C(SiMe_3)_2$ O–CHPh] → [R^2 –Si–O– R^1]$_n$

4a–d

+ $PhCH{=}C(SiMe_3)_2$

5

NMe_2 CH CHPh O Si R^1 $CH(SiMe_3)_2$ Me_2N

6a,c

Scheme 2. The reaction of the intramolecularly donor-stabilized silenes **1a–d** with benzaldehyde; **a**: R^1 = Me, **b**: R^1 = *tert*-Bu, **c**: R^1 = Ph, **d**: R^1 = $SiMe_3$; R^2 = 2,6-bis(dimethylaminomethyl)phenyl.

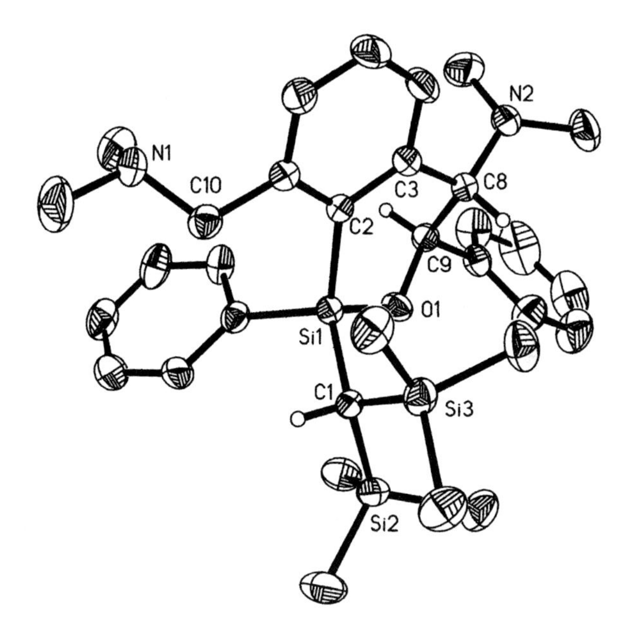

Fig. 1. Molecular structure of **6c** in the crystal (ORTEP, 30% probability level, H atoms omitted except C1H, C8H and C9H). Selected bond lengths [Å] and angles [°]: Si1–C1 1.874(4), Si1–O1 1.654(3), Si1–C2 1.881(4), C2–C3 1.418(5), C3–C8 1.556(5), C8–C9 1.541(5), C9–O1 1.440(4); O1-Si1-C2 102.65(15), C9-O1-Si1 123.4(2), C8-C9-O1 108.7(3), C3-C8-C9 112.3(2).

Compared with uncomplexed silenes, because of the strong N→Si interaction the chemical reactivity of **2a–d** is reduced dramatically. No dimers of **2a–c** could be obtained. Also, cycloaddition reactions with, e.g., 2,3-dimethylbutadiene, which were frequently used for the characterization of silenes [1], were unsuccessful. The components were always recovered unchanged. The reaction of **2a–d** with water expectedly led to the silanols $R^1[C_6H_3(CH_2NMe_2)_2]Si(OH)CH(SiMe_3)_2$ **3a–d**. As a characteristic structural feature of **3a–d**, strong hydrogen bonding between the silanol OH group and one amine nitrogen atom, leading to nonequivalence of the two dimethylaminomethyl groups in the NMR spectra of **3a–d**, may be mentioned.

Studies of the reaction of **2a–d** with benzaldehyde revealed a new and interesting side reaction. As expected, in all cases 2,2-bis(trimethylsilyl)styrene (**5**) was obtained as the main product of the reactions (60 – 70%) besides oily unidentified materials (**4a–d**), which are supposed to be the oligomers of the respective unstable silanones.

In case of the reaction of **2a** and **2c** with benzaldehyde, two crystalline compounds were isolated as by-products (12 and 24%, resp.), which were identified on the basis of NMR studies and X-ray analyses as the cyclosiloxanes **6a** and **6c** (Scheme 2).

The formation of **6a** and **6c** is supposed to proceed through an initial attack of the benzaldehyde carbonyl oxygen atom at the silene silicon atoms of **2a** or **2c** followed by a proton transfer from the benzyl CH_2 group to the negatively charged former silene carbon atom and subsequent ring closure of the resulting zwitterionic species. Studies concerning the factors influencing the formation of the 2-oxa-1-sila-1,2,3,4-tetrahydronaphthalene systems are in progress.

Acknowledgment: The support of these studies by the *Fonds der Chemischen Industrie* is gratefully acknowledged.

References

[1] Reviews: a) A. G. Brook, M. A. Brook, *Adv. Organomet. Chem.* **1996**, *39*, 71; b) T. Müller, W. Ziche, N. Auner in *The Chemistry of Organic Silicon Compounds, Vol. 2* (Eds.: Z. Rappoport, Y. Apeloig), Wiley, New York **1998**, p. 857.

[2] a) N. Wiberg, H. Köpf, *J. Organomet. Chem.* **1986**, *315*, 9; b) N. Wiberg, G. Wagner, G. Reber, J. Riede, G. Müller, *Organometallics* **1987**, *6*, 35.

[3] a) M. Mickoleit, K. Schmohl, R. Kempe, H. Oehme, *Angew. Chem.* **2000**, *112*, 1679; *Angew. Chem. Int. Ed.* **2000**, *39*, 1610; b) M. Mickoleit, R. Kempe, H. Oehme, *Chem. Eur. J.* **2001**, *7*, 987; c) M. Pötter, U. Bäumer, M. Mickoleit, R. Kempe, H. Oehme, *J. Organomet. Chem.* **2001**, *621*, 261; d) M. Mickoleit, M. Pötter, U. Bäumer, K. Schmohl, R. Kempe, H. Oehme in: *Organosilicon Chemistry V – From Molecules to Materials* (Eds.: N. Auner, J. Weis), VCH, Weinheim, **2003**, p. 82.

[4] K. Schmohl, H. Reinke, H. Oehme, *Eur. J. Inorg. Chem.* **2001**, 481.

Quantum Chemical DFT and Experimental Study of *ortho-*, *metha-*, and *para*-Tolyl-Substituted Methylsilene Rearrangements

S. L. Guselnikov,* V. V. Volkova, V. G. Avakyan, L. E. Gusel'nikov

Topchiev Institute of Petrochemical Synthesis, Russian Academy of Sciences
29 Leninsky Prospect 117912 GSP, Moscow, Russia
Tel.: +7 95 9525162 — Fax: +7 95 2302224
E-mail: guselnikovsl@ips.ac.ru

Keywords: silacyclobutanes, pyrolysis, silenes, rearrangement, 3,4-benzo-1-silacyclobutenes, 2-silaindanes, B3LYP/6-31G* calculations

Summary: Quantum chemical DFT and trapping studies have been carried out to ascertain pathways resulting in the formation of benzoannelated silacycles via silene rearrangements upon very low pressure pyrolysis of isomeric 1-tolyl-1-methyl-1-silacyclobutanes. Concerted rearrangement of **2a** into **3a** requires 52.3 kcal/mol, whereas a two-step mechanism involving an *o*-quinone intermediate formation is more favorable, with a highest activation energy of only 37.3 kcal/mol. Another two-step process involving silene-silylene isomerization of **2a** followed by intramolecular insertion of the silylene center of **7** into the C–H bond of the *ortho* methyl to form **4a'** requires 52.1 kcal/mol. The latter is a little lower-energy route than the concerted mechanism for the rearrangement of **2a** into **4a** with an activation energy of 54.1 kcal/mol. **2b** and **2c** rearrange into corresponding 3,4-benzo-1-methyl-1-silacyclobut-3-enes. The intermediacy of silenes and silylenes is proved by trapping with methanol-d_4 and 1,3-butadiene.

Introduction

The gas phase thermal 2 + 2 cycloreversion of monosilacyclobutanes is known to be a clean reaction for generation of transient silenes, $R_2Si{=}CH_2$ [1]. In the absence of trapping agents these silenes undergo a 2 + 2 cyclodimerization unless appropriate substituents at silicon facilitate their rearrangement. In particular, the investigation of silene rearrangements opens up a versatile tool for the synthesis of a variety of unsaturated and cyclic organosilicon compounds and also throws light on the nature and reactivity of the silicon carbon double bond. Thus, a few examples of silene rearrangements derived from the gas phase chemistry of arylsilacyclobutanes are known [2 – 4]. In particular, silacyclobutene mono- and diannelated benzenes were produced by the gas phase bis-silacycobutanes 2 + 2 cycloreversion - transient arylsilenes rearrangement sequence (Scheme 1) [2, 4].

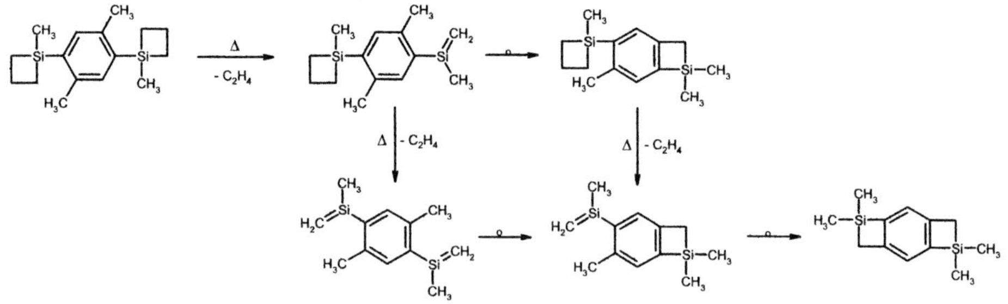

Scheme 1. Formation of silacyclobutene diannelated [a,d] benzene upon very low pressure pyrolysis of 2,5-bis (1-methyl-1-silacyclobut-1-yl)-*o*-xylene.

Here we report about 2+2 thermocycloreversions of 1-*o*-tolyl-1-methyl-1-silacyclobutane (**1a**), 1-*m*-tolyl-1-methyl-1-silacyclobutane (**1b**), and 1-*p*-tolyl-1-methyl-1-silacyclobutane (**1c**) resulting in ethylene and corresponding transient 1-*o*-tolyl-1-methylsilene (**2a**), 1-*m*-tolyl-1-methylsilene (**2b**), and 1-*p*-tolyl-1-methylsilene (**2c**) formations, respectively. Silenes **2a–c** rearrange thermally yielding appropriate 3,4-benzo-1-silacyclobut-3-enes **3a**, **3b**, **3b′**, and **3c** (Scheme 2).

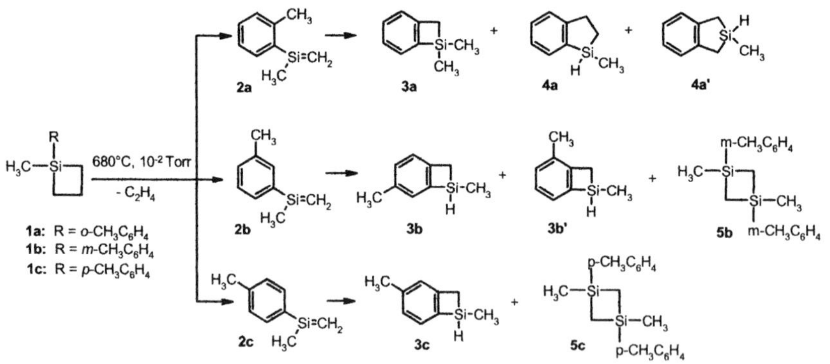

Scheme 2. Very low pressure pyrolysis products of 1-tolyl-1-methyl-1-silacyclobutanes.

Unlike **2b** and **2c**, a cyclodimerization for silene **2a** does not compete with the rearrangement. However, the rearrangement is accompanied by a minor formation of isomeric 1-methyl-1-silaindane (**4a**) and 2-methyl-2-silaindane (**4a′**). At first glance all intramolecular silene rearrangements look like involving 1,n- hydrogen atom migration from either methyl carbon of the *o*-tolyl group or *ortho* aryl carbon to sp^2 carbon or silicon atoms of the silicon–carbon double bond followed by ring closure. Therefore, DFT and trapping studies have been carried out to ascertain pathways resulting in the formation of benzoannelated silacycles via silene rearrangements upon very low pressure pyrolysis of **1a–c**.

Methods of Calculation

Density Functional Theory (DFT) with Becke's three-parameter hybrid functional using the LYP correlation functional (B3LYP) and the 6-31G(d) basis set (Gaussian 98 suite of programs) was applied to fully optimize all structures of interest. Subsequent frequency calculations were performed for the optimized structures of reactants and products to ensure the positivity of all vibrational frequences as a measure of a local minimum on the potential energy surface. A search of transition states (**TS**) was carried out using QST2 and QST3 procedures. The **TS** validity was evidenced by one imaginary frequency and the transition vector orientated from the reactant to the product. Also, IRC procedure was used to show that the transition states found are the maximal points on the paths of the reactions studied and connect the reactants and products.

Theoretical Study of Two Pathways of 1-*o*-Tolyl-1-methylsilene Rearrangement into 3,4-Benzo-1,1-dimethyl-1-silacyclobut-3-ene

In Scheme 3 and Fig. 1 two different pathways for the rearrangement of **2a** into **3a** are shown which were subjected to the DFT study. Being exothermic by 24.5 kcal/mol, the concerted 1,5-H shift (**TS$_3$**) requires an activation energy of 52.3 kcal/mol.

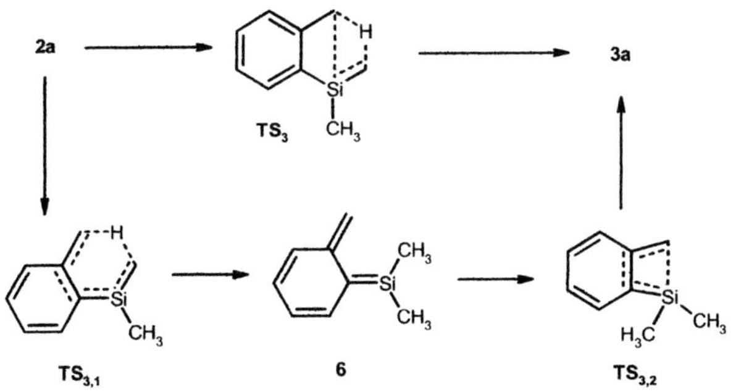

Scheme 3. Concerted and two-step mechanisms of isomerization of 1-*o*-tolyl-1-methylsilene into 3,4-benzo-1,1-dimethyl-1-silacyclobut-3-ene.

An alternative two-step pathway involves the intermediacy of *o*-quinone **6**, whose formation is exothermic by 18.8 kcal/mol and requires the activation energy **TS$_{3.1}$** to be equal to only 37.3 kcal/mol (cf. 52.3 kcal/mol for **TS$_3$**). Further cyclization of **6** to yield **3a** would require for **TS$_{3.2}$** a fairly negligible activation energy of 3.4 kcal/mol. Obviously, the two-step mechanism seems to be preferable to the concerted one.

Dearomatization can be followed by the change of bond distances in the benzene ring. In **2a** all C_{aryl}–C_{aryl} bond lengths are in the range 1.395 – 1.401 Å except for the bonds adjoining the silyl substituent, e.g., 1.421 Å (SiC_{aryl}–$C_{aryl}CH_3$ moiety). In **TS$_{3.1}$** the latter elongates by 0.025 Å, whereas C_{aryl}–CH_3 and C_{aryl}–Si bonds shorten by 0.074 and 0.054 Å, respectively. In **6** the corresponding changes become more pronounced by 0.062, 0.145, and 0.108 Å, respectively. The *o*-quinone structure of **6** is characterized by C=CH_2 and C=Si bond distances of 1.365 Å and 1.762 Å, respectively. In contrast, in $\mathbf{TS_{3.2}}$ the C_{aryl}–C_{aryl} bond in the SiC_{aryl}–$C_{aryl}CH_3$ moiety shortens by 0.032 Å, whereas either C_{aryl}–CH_3 and C_{aryl}–Si bonds elongate by 0.031 and 0.027 Å, respectively. Further changes result in intramolecular 2+2 cycloaddition yielding **3a**, whose formation is guided by the rearomatization of the C_6H_4 moiety.

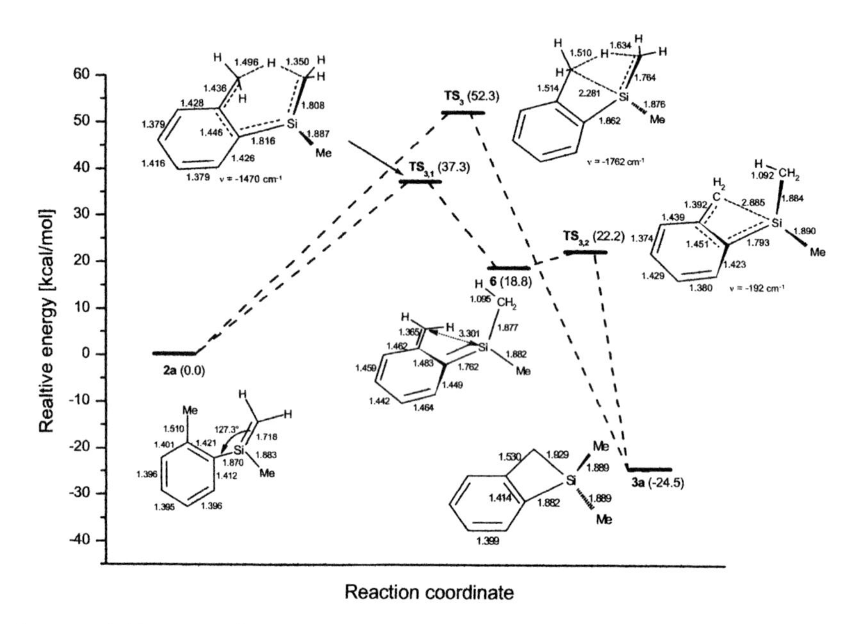

Fig. 1. Routes to 3,4-benzo-1,1-dimethyl-1-silacyclobut-3-ene. Relative energies and connections between species of the C_9SiH_{12} potential surface.

Theoretical Study of 1-*o*-Tolyl-1-methylsilene Rearrangement into Silaindanes

Reaction pathways leading to silaindanes **4a** and **4a′** are shown in Scheme 4 and Fig. 2. Thus, a concerted 1,4-H shift in **2a** to yield **4a** requires 54.1 kcal/mol ($\mathbf{TS_4}$), which is the largest activation energy for all the reactions studied. We are convinced that the isomerization of **2a** into **4a′** lacks any one-step mechanism. Therefore, a two-step mechanism involving silene-silylene isomerization and cyclization via intramolecular insertion of silylene **7** was calculated. Silene-silylene isomerization is endothermic by 8.4 kcal/mol and requires an activation energy of 42.6 kcal/mol ($\mathbf{TS_1}$). In turn, the intramolecular insertion of the silylene center into a C–H bond of **7** resulting in **4a′** is exothermic by 41.9 kcal/mol and requires the activation energy of 43.7 kcal/mol ($\mathbf{TS_2}$). Therefore, the highest energy point (52.1 kcal/mol) on the potential surface for the two-step process yielding **4a′** corresponds to $\mathbf{TS_2}$. It is by 2.0 kcal/mol lower than that of $\mathbf{TS_4}$. This is probably why **4a′** rather than **4a** dominates among the minor silaindanes formed upon very low pressure pyrolysis of **1a**. Obviously, neither of the isomerizations into silaindanes can efficiently compete with the two-step mechanism of 1-*o*-tolyl-1-methylsilene isomerization into 3,4-benzo-1,1-dimethyl-1-silacyclobut-3-ene, for which the highest energy point on the potential surface is only 37.3 kcal/mol.

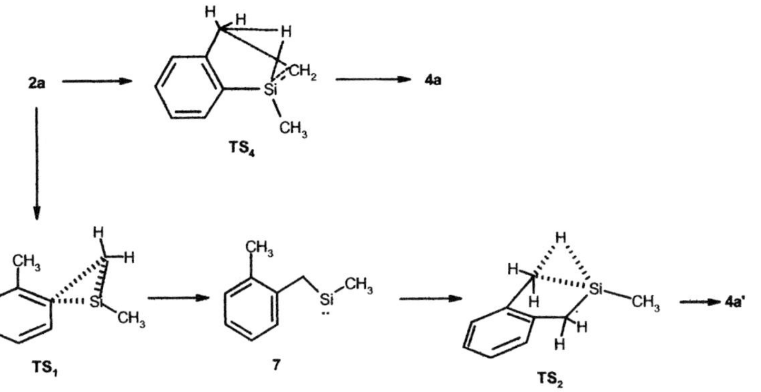

Scheme 4. Concerted and two-step mechanisms of isomerization of 1-*o*-tolyl-1-methylsilene into silaindanes.

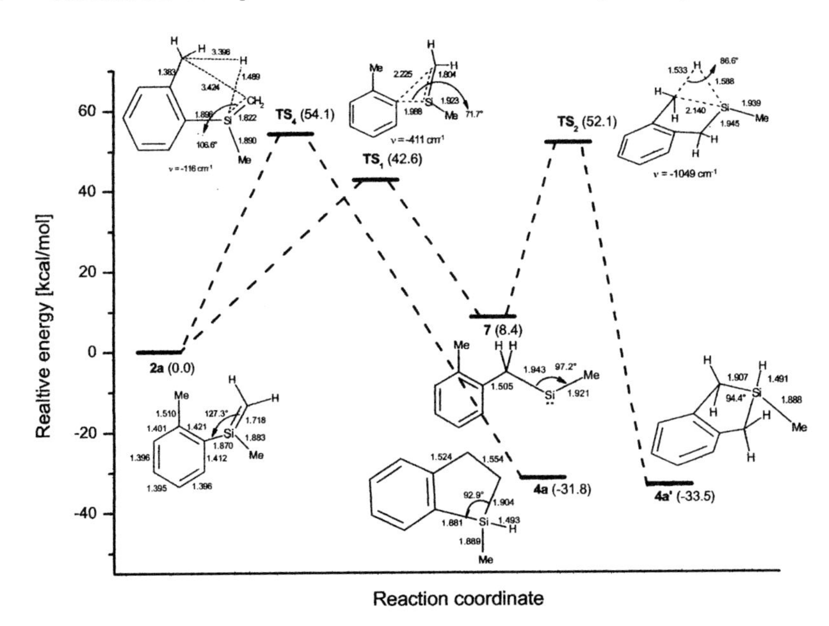

Fig. 2. Routes to silaindanes. Relative energies and connections between species of the C_9SiH_{12} potential surface.

Trapping the Intermediates

Pyrolysis of **1a** and **1b** in excess methanol-d_4 resulted in the expected addition products across the silicon-carbon double bond of silenes **2a** and **2b** (Scheme 5).

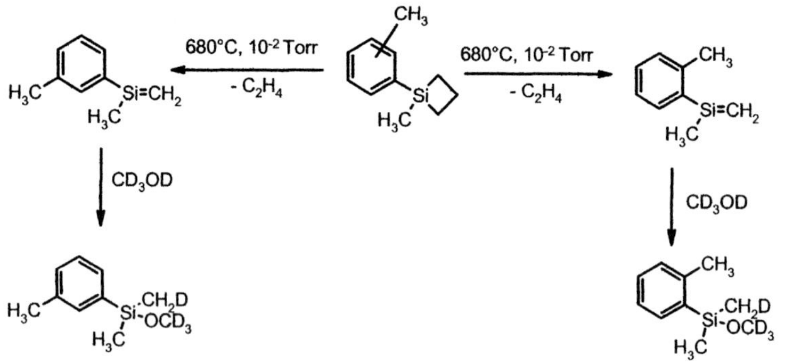

CH_3
680°C, 10^{-2} Torr
$- C_2H_4$
680°C, 10^{-2} Torr
$- C_2H_4$
H_3C Si=CH_2 H_3C
Si H_3C
CH_3 Si=CH_2 H_3C
CD_3OD
CD_3OD
H_3C Si CH_2D OCD_3 H_3C
CH_3 Si CH_2D OCD_3 H_3C

Scheme 5. Trapping silenes **2a** and **2b** with methanol-d_4.

Pyrolysis of **1c** in excess 1,3-butadiene yielded both silene and silylene cycloaddition products, as shown in Scheme 6.

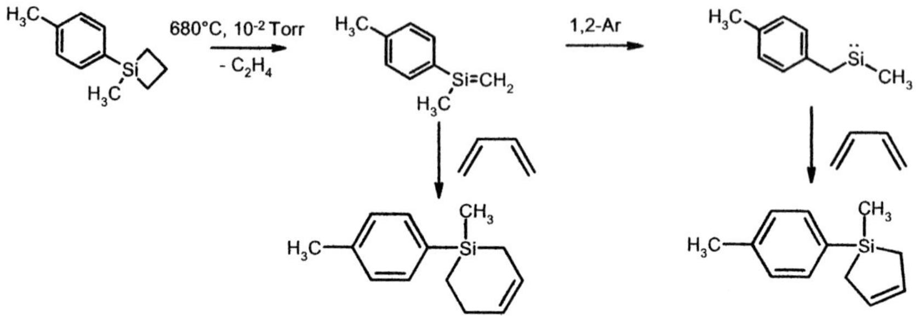

H_3C Si H_3C
680°C, 10^{-2} Torr
$- C_2H_4$
H_3C Si=CH_2 H_3C
1,2-Ar
H_3C Si CH_3
H_3C CH_3 Si
H_3C CH_3 Si

Scheme 6. Trapping silene **2c** and silylene **7** with 1,3-butadiene.

Acknowledgments: The financial support from the International Association for the Promotion of Cooperation with Scientists from the New Independent States of the Former Soviet Union (Grant INTAS 03-51-4164) and the Division of Chemistry and Material Sciences of RAS is gratefully acknowledged (Grant Nos.: 01-03-32833, 02-03-32329, 03-03-42886, and 04-03-32546).

References

[1] L. E. Gusel'nikov, *Coord. Chem. Rev.* **2003**, *244*, 149.

[2] V. V. Volkova, L. E. Gusel'nikov, E. A. Volnina, E. N. Buravtseva, *Organometallics* **1994**, *13*, 4661.

[3] L. E. Gusel'nikov, V. V. Volkova, B. D. Lavrukhin, *J. Organomet. Chem.* **1995**, *492*, C4.

[4] L. E. Gusel'nikov, V. V. Volkova, E. N. Buravtseva, A. S. Redchin, N. Auner, B. Herrschaft, B. Solouki, G. Tsantes, Yu. E. Ovchinnikov, S. A. Pogozhikh, F. M. Dolgushin, V. V. Negrebetsky, *Organometallics* **2002**, *21*, 1101 and references therein.

A 1,4-Disilapentalene and a Highly Stable Silicon Diradical

Irina Toulokhonova, Thomas C. Stringfellow, Robert West*

Organosilicon Research Center, University of Wisconsin,
1101 University Ave., Madison WI 53706 USA
Tel.: +1 608 2621873 — Fax: +1 608 2626143
E-mail: west@chem.wisc.edu

Keywords: siloles, EPR, free radicals

Summary: 1,1-dilithio-2,3,4,5-tetraphenylsilole (**1**) reacts with 1,1-dichloro-2,3-diphenylcyclopropene (**2**) yielding the novel 1,4-disila-1,4-dihydropentalene **3**, as well as an exceptionally stable diradical, **4**, for which the structure shown in Fig. 1 is suggested. The diradical is unreactive toward water, methanol and chloroform; upon heating it transforms into **3**. A reaction scheme describing the formation of **3** and **4** is presented.

Silole dianions, which are of current interest as aromatic species [1 – 3], behave as nucleophiles in simple displacement reactions, generally yielding simple trapping products [4 – 6]. However, the reaction of 1,1-dilithio-2,3,4,5-tetraphenylsilole (**1**) with 1,1-dichloro-2,3-diphenylcyclopropene (**2**) takes a most unexpected course, leading to ring-opening of the cyclopropene ring and ultimate formation of a tetracyclic disilapentalene and a stable bis-silole diradical [7].

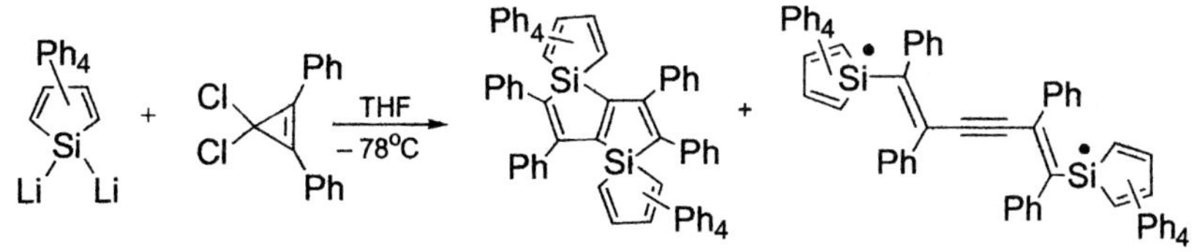

Eq. 1. Reaction of dilithiotetraphenylsilole with dichlorodiphenylcyclopropene.

The reaction of **1** with **2** in THF solution at –78 °C gave a bright red, strongly paramagnetic solution. The EPR signal of this solution persisted for many months at room temperature. The signal intensity corresponded to a 75% yield of a radical or diradical. Warming to room temperature and column chromatography to separate the products gave a 25% yield of diamagnetic, disilapentalene **3**, as bright red crystals. If the solution was heated to 80 °C for 24 h and then chromatographed, a 65% yield of **3** was obtained. Evidently the radical compound slowly transforms into the disilapentalene upon heating. The structure of **3** was established by X-ray crystallography, and is shown in Fig. 1. It is the first example of a 1,6-disilapentalene. The disilapentalene ring is nearly planar, and the two *spiro* tetraphenylsilole rings are perpendicular to the disilapentalene plane.

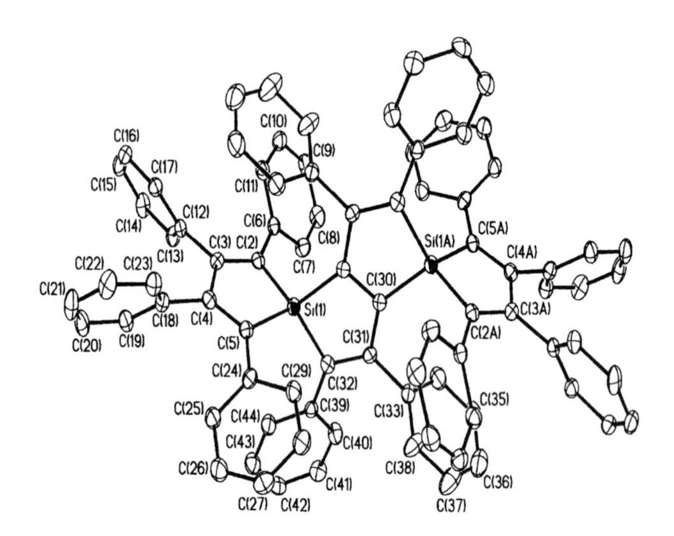

Fig. 1. Thermal ellipsoid diagram for disilapentalene **3**. 50% Probabilty ellipsoids, hydrogen atoms omitted.

The EPR evidence indicates that **4** is a triplet diradical. Especially important is the observation of a half-field (double-quantum) resonance for **4** at 1450 G. The room temperature EPR spectrum of the solid reaction mixture is also consistent with a triplet diradical. The central peak with 5.8 G width is accompanied by side peaks characteristic of a triplet state species [8], with D = 28 G, E = 2 G (Fig. 2).

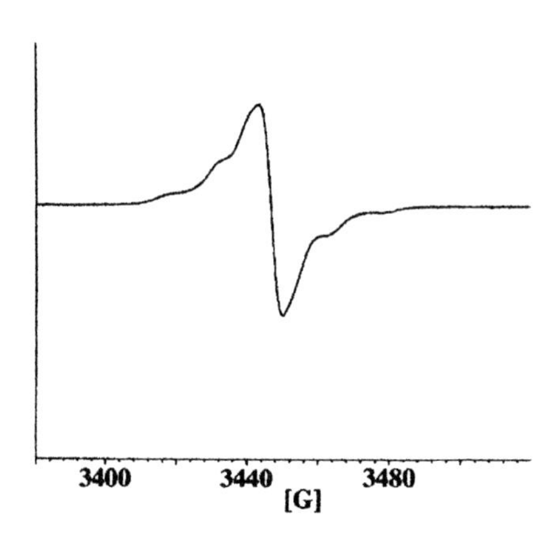

Fig. 2. EPR spectrum of solid reaction mixture at room temperature, showing triplet side bands.

From the *D* value the distance between the two unpaired electrons was determined as 998 pm. This is consistent with our formulation of the diradical as **4**, in which the electron-electron distance is approximately 1000 pm in the *trans-trans* stereoisomer [9].

The diradical **4** can exist as *trans-trans, trans-cis,* or *cis-cis* isomers. It seems likely that the *cis-cis* isomer may be the source of the disilapentalene, since it could undergo ring closure as shown in Scheme 1. The stable diradical is probably the *trans-trans* isomer, or perhaps a mixture of this with the *trans-cis* form. Upon heating, rotational isomerism may take place to give the *cis-cis* form which then undergoes ring closure as shown in the scheme.

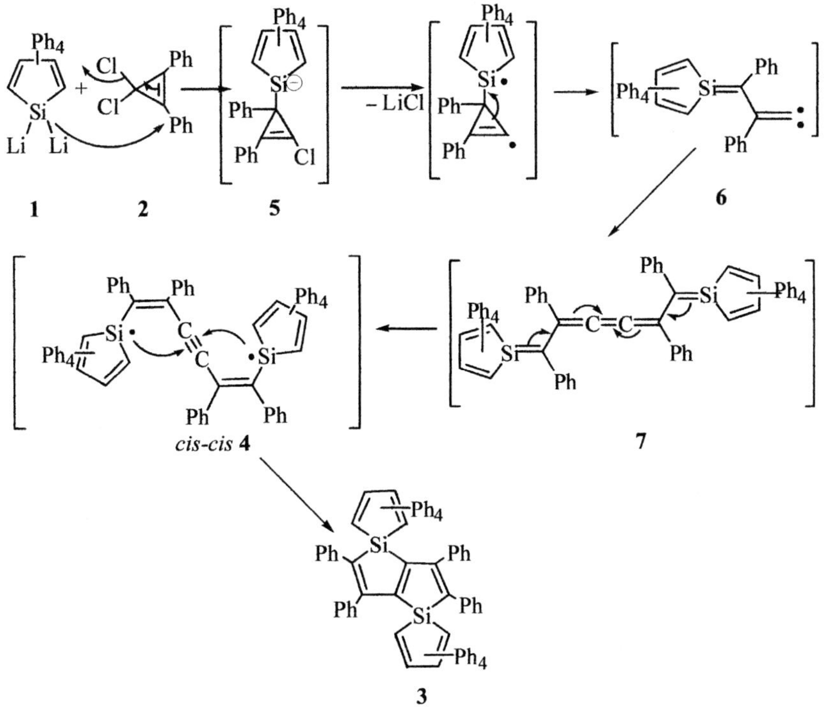

Scheme 1. Proposed mechanism for the formation of **3**.

Scheme 1 shows a possible mechanism for the formation of products **3** and **4**. In the first step the nucleophilic silicon attacks the double bond of the cyclopropene ring leading to intermediate **5**. Loss of chloride and ring opening would lead to carbene **6**, which could immediately dimerize to cumulene **7**. The cumulene could not exist in the planar form shown, because of strong steric interaction between the phenyl groups. Twisting of the silole rings would relieve the steric hindrance, and produce to the diradical **4**, shown here in the *cis-cis* form. This could close to the disilapentalene product **3**. The *trans-trans* and *trans-cis* forms would need to undergo rotation at one or both of the double bonds to generate the *cis-cis* isomer before ring closure to **3** could take place. Thus the 25% of **3** that is formed at low temperature probably arises from initially formed *cis-cis* **4**. Upon heating of the reaction mixture, the larger amounts of *trans-trans* and *trans-cis* **4** gradually isomerize to the *cis-cis* form, and produce more **3**.

The stability and inertness of diradical **4** are quite unprecedented for silicon radicals. **4** does not react at room temperature with water, alcohols, or chloroform, and only slowly with oxygen. The diradical is not indefinitely stable, but the half-time for its disappearance at room temperature is more than one year. The surprising, indeed amazing, inertness of **4** can be explained partly by steric effects, but probably also by specific stabilization due to the silole ring. The results of this study suggest that silole free radicals may form a previously unrecognized class of stable radical species.

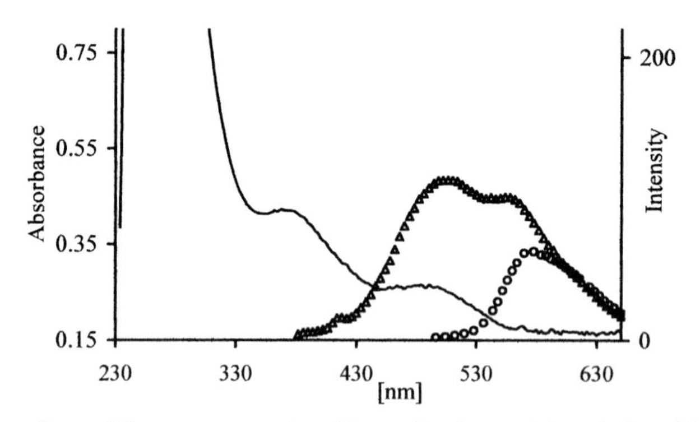

Fig. 3. Electronic absorption and fluorescence spectra of **3**. —, absorbance; ΔΔ, emission of **3** excited at 370 nm; oo, emission of **3** excited at 486 nm.

The electronic absorption and fluorescence spectra of **3** are shown in Fig. 3. Perphenylated siloles generally have two absorption bands in the UV near 250 nm (due to the phenyl rings) and 360 nm (due to the silole moiety) [10]. Compound **3** exhibits these two bands, and also a long-wavelength absorption at 484 nm, responsible for its red color. MO calculations allowed the assignment of the new long-wavelength band to a mixed transition, from both the HOMO and HOMO-2 to the LUMO. Both the HOMO and LUMO are π-type orbitals of the pentalene ring [7].

Experimental Section

Reaction of 1,1-dilithio-2,3,4,5-tetraphenylsilole with 1,1-dichloro-2,3-diphenylcyclopropene

1,1-Dichloro-2,3-diphenylcyclopropene (**2**) was prepared according to a literature procedure [11]. 1,1-Dilithio-2,3,4,5-tetraphenyl-1-silacyclopentadiene (**1**) was prepared from 1,1-dichloro-2,3,4,5-tetraphenyl-1-silole (0.5 g, 2.2 mmol) in 7.5 mL THF and Li (0.033 g, 4.8 mmol) in an argon atmosphere as described previously [2]. A solution of **2** (0.58 g, 2.2 mmol) in 10 mL THF was added to **1** (2.2 mmol in 10 mL THF) at –78 °C. The reaction mixture immediately turned dark red. After warming the reaction mixture to room temperature over 2 h, THF was evaporated *in vacuo* and the residue was dissolved in benzene and filtered to remove LiCl. Evaporation of the solvent gave a dark red powder. The EPR spectrum of the solid and its benzene solution showed strong signals assigned to diradical **4** [7].

Preparation of disilapentalene 3

A solution of **2** (0.58 g, 2.2 mmol) in 10 mL THF was added to **1** (2.2 mmol in 10 mL THF) at –78 °C. The reaction mixture immediately turned dark red. After heating for 24 h at 80 °C and evaporation of THF, the crude solid was dissolved in benzene and washed with a saturated aqueous solution of ammonium chloride. The organic layer was dried with $MgSO_4$ and the residue was purified by column chromatography (benzene-hexane (10:1) elution). 1.4 g (1.2 mmol, 60%) of dark red compound was obtained. Further purification by preparative GPC (toluene elution) and crystallization from benzene at room temperature afforded bright red crystals of **3**: mp 360 – 361 °C; ^{29}Si NMR ($CDCl_3$, 99.38 MHz): δ = –21.865; ^{1}H NMR ($CDCl_3$, 500.0 MHz): δ = 7.23 – 8.18 (br, m, Ph). ^{13}C NMR ($CDCl_3$, 125.38 MHz): δ = 130.33 – 135.06 (m, silole carbons); 136.36 – 147.30 (br, m, Ph). HRMS (MALDI) calcd for $C_{86}H_{60}Si_2$: 1149.57; found 1149.43. Anal. calcd. for $C_{86}H_{60}Si_2$: C, 89.85; H, 5.26; found: C, 89.67; H, 5.53.

Acknowledgement: We thank Drs. Sergey Ivanov and Artem Masunov for helpful theoretical calculations and the National Science Foundation for financial support.

References

[1] B. Goldfuss, P. von R. Schleyer, *Organometallics* **1994**, *13*, 3387.

[2] R. West, H. Sohn, U. Bankwitz, J. Calabrese, Y. Apeloig, T. Müller, *J. Am. Chem. Soc.* **1995**, *117*, 11608.

[3] Y. Liu, T. C. Stringfellow, D. Ballweg, I. A. Guzei, R. West, *J. Am. Chem. Soc.* **2002**, *49*, 1917.

[4] W. P. Freeman, T. D. Tilley, L. M. Liable-Sands, A. L Rheingold, *J. Am. Chem. Soc.* **1996**, *118*, 10457.

[5] S.-B. Choi, P. Boudjouk, P. Wei, *J. Am. Chem. Soc.* **1998**, *120*, 5814.

[6] S.-B. Choi, P. Boudjouk, J.-H. Hong, *Organometallics* **1999**, *18*, 2919.

[7] I. S. Touokhonova, T. C. Stringfellow, S. A. Ivanov, A. Masunov, R. West, *J. Am. Chem. Soc.* **2002**, *125*, 5767.

[8] L. Salem, C. Rowland, *Angew. Chem. Int. Ed. Engl.* **1972**, *11*, 92.

[9] For an extended discussion of the EPR evidence for **4**, and of the MO calculations on both **3** and **4**, see Ref. 7.

[10] H. J. Peng, Z. T. Liu, H. Y. Chen, Y. L. Ho, B. Z. Tang, M. Wong, H. C. Huang, H. Kwok, *J. Appl. Phys.* **2002**, *92*, 5735.

[11] S. Tobey, R. West, *J. Am. Chem. Soc.* **1964**, *86*, 4215.

Kinetic Stabilization of Polysilyl Radicals

Victoria Kravchenko, Dmitry Bravo-Zhivotovskii, Boris Tumanskii, Mark Botoshansky, Nadejda Segal, Gregory Molev, Monica Kosa, Yitzhak Apeloig**

Department of Chemistry and the Lise Meitner Minerva Center for Computational Quantum Chemistry, Technion-Israel Institute of Technology, Haifa 32000, Israel
E-mail: chrapel@tx.technion.ac.il , chrbrzh@tx.technion.ac.il

Keywords: silyl radicals, polysilanes, back-folded conformation, dissociation energy of Si–Si bonds

Summary: Octasilane $[(Me_3Si)_3Si]_2$ (**1a**), with a central Si–Si bond length of 2.405 Å, is thermally and photolytically stable. Substitution of the six Me_3Si groups by the larger Me_3SiMe_2Si groups has only a minor effect on the central Si–Si bond length, which in $[(Me_3SiMe_2Si)_3Si]_2$ (**2a**) is 2.409 Å. On the other hand, substitution of two Me_3Si groups in **1a** by the bulkier $(t\text{-}Bu)_2HSi$ groups extends the central Si–Si bond to 2.480 Å in $[(Me_3Si)_2(t\text{-}Bu_2HSi)Si]_2$ (**3a**). The X-ray structure of **2a** shows that the Me_3SiMe_2Si branches adopt a "discus-type" structure (Scheme 3 Structure A), with a ∠SiSiSiX torsion angle of 87.0°. According to DFT calculations the radicals $(Me_3SiMe_2Si)_3Si$• (**2b**), and $(t\text{-}BuMe_2Si)_3Si$• (**4b**) have "umbrella-type" back-folded conformations (Scheme 3, Structure B), with a ∠SiSiSiX or ∠CSiSiX (X denotes the radical center) dihedral angle in the range 17 – 19° (Scheme 1, Structure B). These back-folded conformations are less strained than the more open conformations such as those in **2b** (by 1 – 2 kcal/mol) or **4b** (by 7.8 kcal/mol). The bulky $t\text{-}Bu_2HSi$ substituents significantly stabilize kinetically the radical center by protecting the radical site from dimerization and other reactions. Radical **4b** is kinetically significantly more stable ($k_{dim} = 10^2\ s^{-1}M^{-1}$) than radicals **2b** and **3b** ($k_{dim} = 5 \times 10^3\ s^{-1}M^{-1}$ and $k_{dim} = 2 \times 10^3\ s^{-1}M^{-1}$ respectivelly). Radical **4b** exists at r.t. in equilibrium with its unstable dimer $[(t\text{-}BuMe_2Si)_3Si]_2$ (**4b**), ($\Delta H = -8.0 \pm 2$ kcal/mol), in contrast to dimers **2a** ($\Delta H = -56.0 \pm 2$ kcal/mol) and **3a** ($\Delta H = -58.0 \pm 2$ kcal/mol), where such equilibrium does not exist. The central Si–Si bond length in **4a** has been estimated by 3-layer ONIOM calculations (MP2/6-31G(d):B3LYP/6-31G(d):HF/3-21G) to be 2.630 Å.

Introduction

Polysilanes have been studied extensively in recent decades because of their interesting photophysical properties and potential applications [1]. Yet, much fundamental knowledge about these interesting molecules is still lacking. In particular, relatively little is known about the relationship between the strain around an Si–Si bond and its homolytic bond dissociation energy leading to the corresponding silyl radicals. Bock and coworkers found that in branched polysilanes a higher congestion around an Si–Si bond cause its lengthening, but the implications of this bond elongation on the dissociation energy of the congested Si–Si bond were not evaluated [2]. This can be contrasted with the detailed information available for the analogous dissociation of C–C bonds, where a linear relationship was found between the strain around a particular C–C bond and its homolytic bond dissociation energy [3]. In this paper we study the dissociation energy of the central Si–Si bond in three branched symmetrically substituted strained polysilanes of the type $R_3Si{-}SiR_3$ (**a**) and the kinetic stability of corresponding $R_3Si\bullet$ radicals (**b**).

The polysilanes of type **a** were prepared according to Eq. 1, as follows: reaction of the hydrosilanes R_3SiH with $t\text{-}Bu_2Hg$ yielded the corresponding silyl mercury compounds (**c**). Visible light irradiation of the silyl mercurials **c** yields the desired disilanes **a** via dimerization of the corresponding radicals **b** (Scheme 1, Eq. 1). Using this method we have synthesized polysilane dimers, $R_3Si{-}SiR_3$, with R_3Si = $(Me_3Si)_3Si$ (**1a**) $(Me_3SiMe_2Si)_3Si$ (**2a**) and $(t\text{-}Bu_2HSi(Me_3Si)_2Si$ (**3a**). A typical experimental procedure, e.g., to produce **1a**, is as follows: $(Me_3Si)_3SiH$ is stirred with an equimolar amount of $t\text{-}Bu_2Hg$ at 90 °C under a weak nitrogen flow until the evolution of isobutene gas stops (several hours); dry hexane is then added to the vacuumed reaction mixture and the produced silylmercurial **1c** is irradiated by visible light producing **1a** in an overall yield of 96%. Attempts to prepare by this method the polysilane dimer $R_3Si{-}SiR_3$ with R_3Si = $(t\text{-}BuMe_2Si)_3Si$ (**4a**) were not successful, but the corresponding radical **4b** was detected by EPR spectroscopy during irradiation of silyl mercurial **4c**, as shown in Scheme 1, Eq. 1.

Irradiation or thermolysis of polysilane dimers **a** yields the corresponding silyl radicals **b** (Scheme 1, Eq. 2).

Eq. 1.

$$\underset{}{R_3SiH} \xrightarrow{t\text{-}Bu_2Hg} \underset{\mathbf{c}}{(R_3Si)_2Hg} \xrightarrow[-Hg]{\text{daylight}} \underset{\mathbf{b}}{2R_3Si\bullet} \longrightarrow \underset{\mathbf{a}}{R_3Si{-}SiR_3}$$

Eq. 2.

$$R_3Si{-}SiR_3 \underset{}{\overset{h\nu \text{ or } \Delta}{\rightleftharpoons}} 2R_3Si\bullet$$

R_3Si = $(Me_3Si)_3Si$ (**1**), $(Me_3SiMe_2Si)_3Si$ (**2**), $(t\text{-}Bu_2HSi(Me_3Si)_2Si$ (**3**), $(t\text{-}BuMe_2Si)_3Si$ (**4**)

Scheme 1. Synthesis of polysilanes **a** and generation of the corresponding polysilyl radicals **b**.

In this study we aimed to evaluate the relationship between the central Si–Si bond length of the branched polysilanes **a** and the kinetic stability of corresponding silyl radicals **b**.

Results and Discussion

Structures of Hexasilyl Substituted Disilanes and Related Derivatives

Polysilane dimers **1a**, **2a**, and **3a** were synthesized according to Eq. 1 in Scheme 1. They were all successfully crystallized and their structures were determined by X-ray diffractometry. Selected geometrical parameters based on the X-ray molecular structures of **2a** and **3a** are reported below in Figs. 1 and 2, respectively. The structure of **1a** has been reported previously [2].

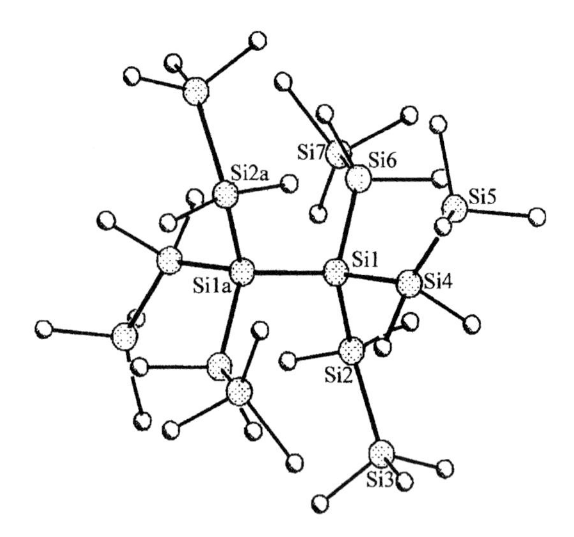

Fig. 1. ORTEP drawing and selected bond lengths [Å], bond angles [°] and torsion angles [°] of **2a** (hydrogen atoms are omitted for clarity): Si(1)–Si(1a) 2.409(2), Si(1)–Si(2) 2.387(2), Si(1)–Si(4) 2.385(2), Si(1)–Si(6) 2.385(6), Si(2)–Si(3) 2.387(2), Si(4)–Si(5) 2.385(2), Si(2)–C(1) 1.965(3), Si(3)–C(2) 1.877(4); Si(4)-Si(1)-Si(2) 104.0(1), Si(6)-Si(1)-Si(2) 104.9(1), Si(4)-Si(1)-Si(6) 104.0(1), Si(2)-Si(1)-Si(1a) 114.3(1), Si(1)-Si(2)-Si(3) 127.8(1), Si(1)-Si(4)-Si(5) 127.9(1), Si(2a)-Si(1a)-Si(1)-Si(2) 87.0.

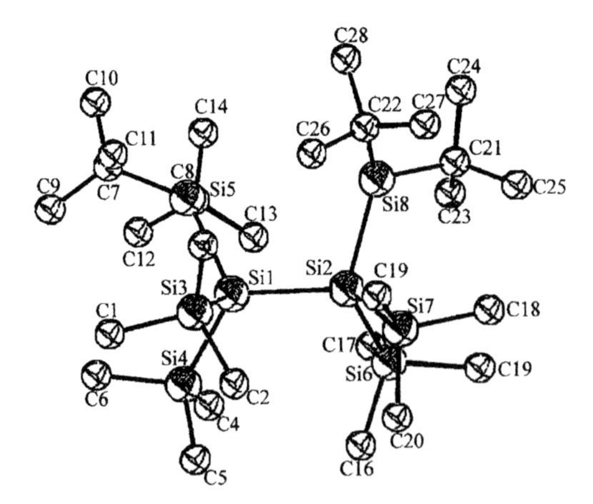

Fig. 2. ORTEP drawing and selected bond lengths [Å], bond angles [°] and torsion angles [°] of **3a** (hydrogen atoms are omitted for clarity): Si(1)–Si(2) 2.4896(8), Si(1)–Si(5) 2.4307(8), Si(2)–Si(8) 2.4342(8), Si(1)–Si(3) 2.4106(8), Si(1)–Si(4) 2.4083(9), Si(3)–C(1) 1.881(3), Si(5)–C(7) 1.926(3); Si(5)-Si(1)-Si(2) 109.09(3), Si(4)-Si(1)-Si(2) 116.32(3), C(7)-Si(5)-C(8) 113.69(11); Si(5)-Si(1)-Si(2)-Si(8) 17.88(4), Si(4)-Si(1)-Si(2)-Si(6) 14.78(4), Si(3)-Si(1)-Si(2)-Si(7) 15.11(5) [11].

The central Si–Si bond length in $[(Me_3Si)_3Si]_2$ (**1a**) is 2.405 Å. Substitution of the six Me_3Si substituents in **1a** by six Me_3SiMe_2Si groups has only a minor effect on the central Si–Si bond length, which in **2a** is 2.409 Å (Fig. 1), very similar to that in **1a**. However, when one of the Me_3Si groups on each of the disilane central silicon atoms is replaced by the bulkier *t*-Bu_2HSi substituent, the central Si–Si bond length is elongated considerably to 2.480 Å. As we have failed to synthesize **4a**, we had to rely on calculations to determine its central Si–Si bond length, which according to 3-layer ONIOM calculations (MP2/6-31G(d):B3LYP/6-31G(d):HF/3-21G) [4] is 2.630 Å, significantly longer than that in **1a** (see below). The central Si–Si bond lengths in **1a** – **4a** are shown in Scheme 2.

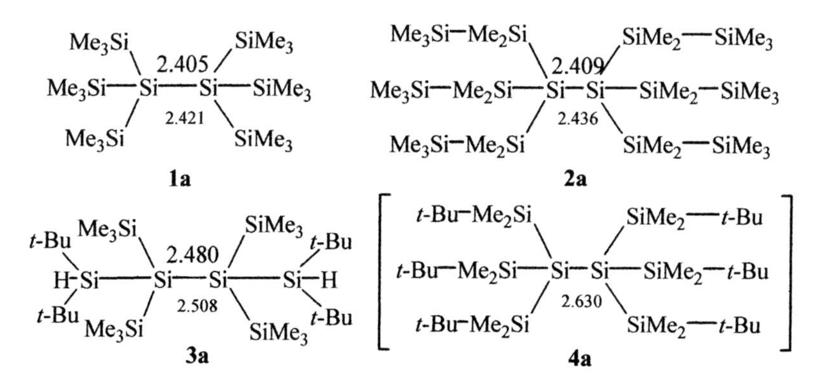

Scheme 2. Central Si–Si bond length of dimers **1a** – **4a** (experimental data for **1a**, **2a**, **3a** is given above the central Si–Si bond and calculated data for **1a**, **2a**, **3a** and **4a** is given below the central Si–Si bond).

The X-ray structure of silane **2a** shows that the six Me_3SiMe_2Si branches adopt a "discus-type" conformation as shown in structure A in Scheme 3. In **3a** the ∠Si3Si2Si1X torsion angle ϕ (for definition see Scheme 4) is 87.0°, where X is an imaginary point along the C_3 axis of the $(Me_3SiMe_2Si)_3Si$-fragment [in **2a** X coincides with Si(1a)]. We have reported previously that the disilylmercury compound $[(Me_3SiMe_2Si)_3Si]_2Hg$ (**3c**) [5] and the silyllithium compound $(Me_3SiMe_2Si)_3SiLi{\bullet}3THF$ (**5**) [6] having Me_3SiMe_2Si substituents adopt an "umbrella-type" conformation shown in structure B in Scheme 3 [5]. The "umbrella-type" conformation B differs from the "discus-type" conformation A in having a much smaller dihedral angle ϕ.

Interestingly, substitution of the six β-Me_3Si groups in **2c** by six *t*-Bu groups giving $R_3Si–Hg–SiR_3$, R=(*t*-$BuMe_2Si$) (**4c**), does not change the "umbrella-type" conformation of the silyl mercury compound. This conformation is also retained in (*t*-$BuMe_2Si)_3SiH$ (**6**). The ORTEP drawings of the silylmercury compound **4c** (Fig. 3) and of the corresponding monosilane, **6** (Fig. 4) show that the *t*-$BuMe_2Si$ branches adopt a back-folded "umbrella-type" conformation with a ∠CSiSiH torsion angle in **6** of 26° and a ∠CSiSiHg dihedral angle in **4c** of 33.8 – 34.5°. Thus, even the very large *t*-butyl groups prefer to cluster around the center of the molecule as structure B in Scheme 3, rather than to point outwards, as might be expected tentatively. The preference for a back-folded conformation in these molecules probably results from attractive Van der Waals interactions between hydrogen atoms of the *t*-butyl groups drawing them closer together [7]. Apparently, the back-folded "umbrella-type" conformation reduces strain in these molecules more effectively than in more open-type conformations.

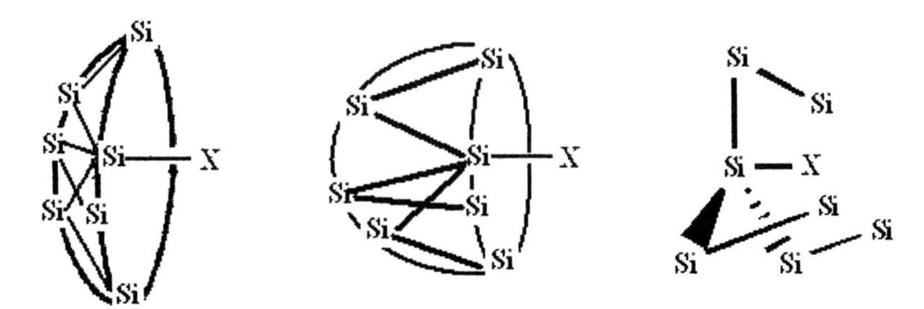

A –"discus-type" conformation B –"umbrella-type" conformation C – Radical conformation

Scheme 3. Schematic representation of types of arrangements of the silyl substituents around the central Si atom and the dihedral corresponding angles (see Scheme 4) in **2a–c** and **5**. A: X= $Si(SiMe_2SiMe_3)_3$ (**2a**), ϕ = 87.0° [5]; B: X = Li•3THF (**5**) (ϕ = 46.0°) [4]; X = $HgSi(SiMe_2SiMe_3)_3$ (**2c**) (ϕ= 47.6°) [5]; C: X = radical center (**2b**) (ϕ= 19.1°).

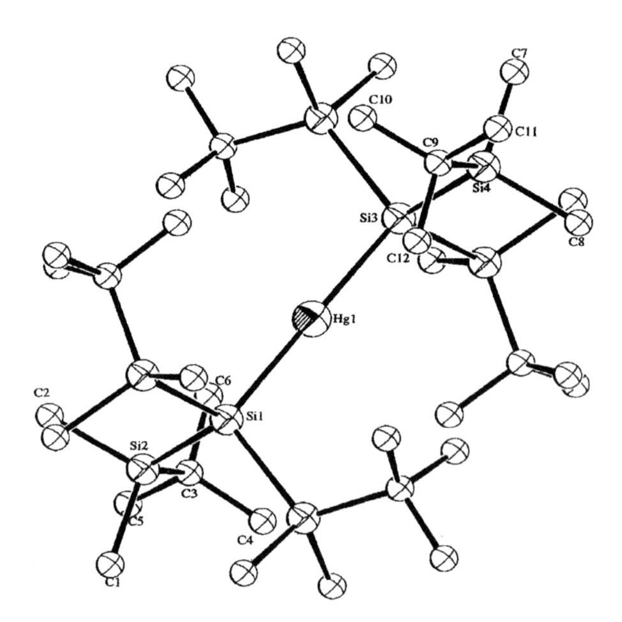

Fig. 3. ORTEP drawing and selected bond lengths [Å], bond angles [°] and torsion angles [°] of silyl mercurial **4c** (hydrogen atoms are omitted for clarity): Si(3)–Hg(1) 2.512(6), Si(1)–Hg(1) 2.517(6), Si(1)–Si(2) 2.397(5), Si(3)–Si(4) 2.407(5), Si(2)–C(3) 1.908(15), Si(4)–C(9) 1.920(15); Si(3)-Hg(1)-Si(1) 180.0, Si(2)-Si(1)-Hg(1) 111.27(17), Si(4)-Si(3)-Hg(1) 111.66(17), C(3)-Si(2)-Si(1) 114.7(4), C(9)-Si(4)-Si(3) 115.7(4); Hg(1)-Si(1)-Si(2)-C(3) 34.6(5), Hg(1)-Si(3)-Si(4)-C(9) 33.8(5) [11].

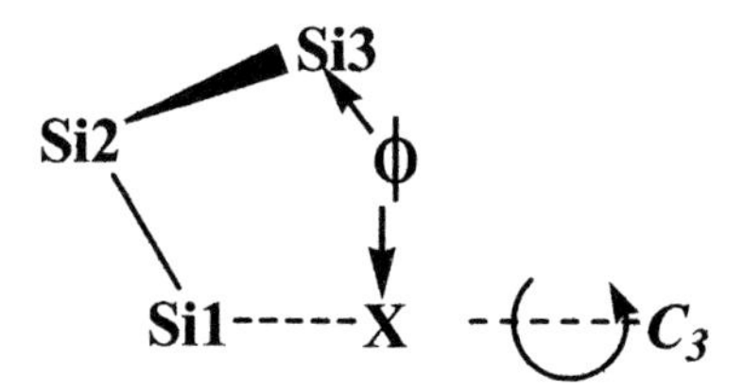

Scheme 4. Definition of the dihedral angle $\phi = \angle$XSi1Si2Si3. Imaginary point X is placed along the C_3 axis of the Si_3Si unit.

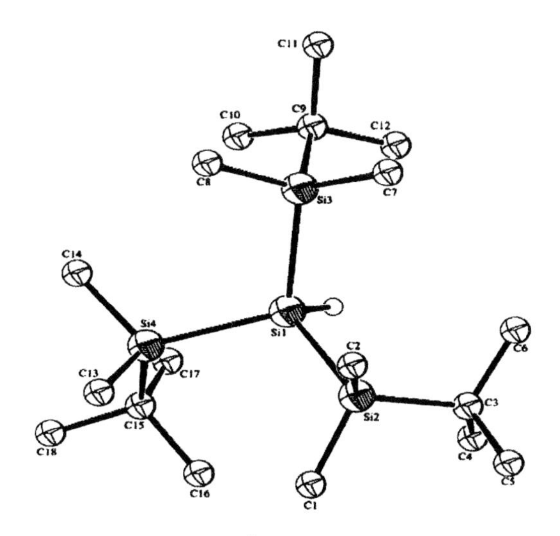

Fig. 4. ORTEP drawing and selected bond lengths [Å], bond angles [°], and torsion angles [°] of silane **6** (hydrogen atoms are omitted for clarity): Si(1)–Si(2) 2.389(4), Si(1)–Si(3) 2.388(5), Si(1)–Si(4) 2.378(3), Si(1)–H(1s1) 1.52(6), Si(2)–C(1) 1.835(15), Si(2)–C(2) 1.808(12), Si(2)–C(3) 1.922(10), C(1)–C(4) 1.488(7), C(1)–C(5) 1.497(7), C(1)–C(6) 1.511(8); Si(2)-Si(1)-Si(3) 113.63(7), Si(2)-Si(1)-H(1s1) 104.90(9), C(2)-Si(2)-Si(1) 112.1(2), C(3)-Si(2)-Si(1) 108.3(3), C(1)-Si(2)-Si(1) 111.4(2), C(1)-Si(2)-Si(1)-H 26.0 [11].

Dimerization Rates of Silyl Substituted Silyl Radicals

Several dimerization rates of alkyl-substituted silyl radicals were measured earlier [8]. However, the dimerization rates of silyl-substituted silicon-centered radicals have not previously been determined. In this study we have measured, using EPR spectroscopy, the rate constants for the recombination of four silyl radicals (**1b**, **2b**, **3b**, and **4b**), to produce the corresponding disilane dimers of type **a** (i.e. **1a**, **2a**, **3a**, and **4a** respectively). This dimerization reaction is shown as the backward reaction of Eq. 2 in Scheme 1. Radicals **1b**, **2b**, and **3b** were generated photochemically from the corresponding disilane dimers of type **a** (Scheme 1, Eq. 2), while radical **4b** was generated photochemically from the corresponding silylmercury compound **4c** (Scheme 1, Eq. 1).

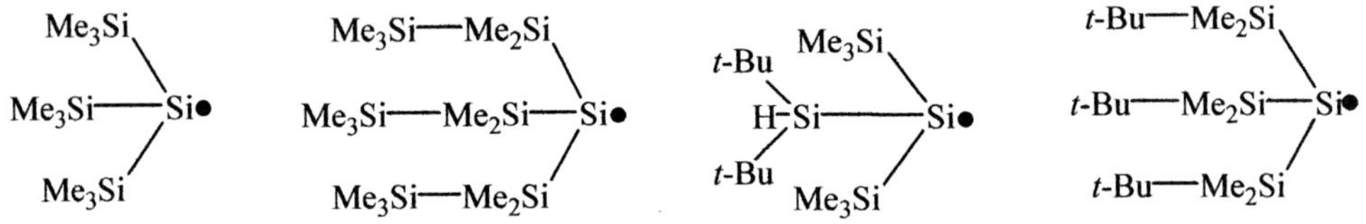

$(Me_3Si)_3Si\bullet$ **1b** $\quad$ $(Me_3Si{-}Me_2Si)_3Si\bullet$ **2b** $\quad$ $(t\text{-}Bu)_2HSi{-}Si(SiMe_3)_2\bullet$ **3b** $\quad$ $(t\text{-}Bu{-}Me_2Si)_3Si\bullet$ **4b**

In the kinetic measurements, the silyl radicals were generated by photolysis at 300 nm. Then the light was turned off (usually after a few seconds) and the decrease in the radical concentration (C) was measured as a function of time by following the decrease in the intensity of the radical's EPR signal, as shown in Fig. 5a for radical **3b**. Good linear curves of $1/C$ vs time were obtained in all cases, as shown for **3b** in Fig. 5b. The following second-order reaction rate constants (k_{dim}) at 290 K were measured: for **1b**, k_{dim} was approximately 1×10^8 $s^{-1}M^{-1}$; for **2b**, $k_{dim} = 5 \times 10^3$ $s^{-1}M^{-1}$; for **3b**, $k_{dim} = 2 \times 10^3$ $s^{-1}M^{-1}$; and for **4b** $k_{dim} = 2 \times 10^2$ $s^{-1}M^{-1}$. Thus, the relative dimerization rates are 1 (**4b**), 10 (**3b**), 25 (**2b**), and 5×10^5 (**1b**).

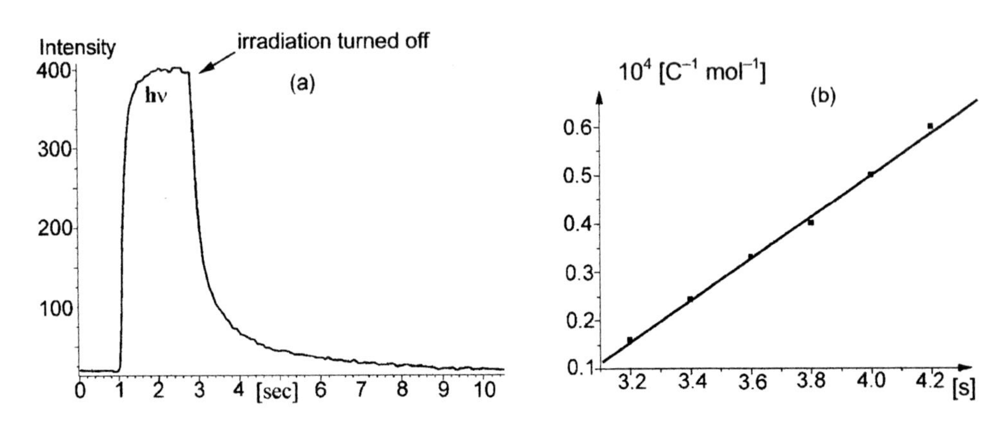

Fig. 5. Measurements of k_{dim} of silyl radical **3b**: (a) Decay curve of the EPR signal of radical **3b** at 290 K (the magnetic field is constant and corresponds to the signal maximum); (b) Plot of the reciprocal of the concentration $1/C$ vs time of radical **3b**.

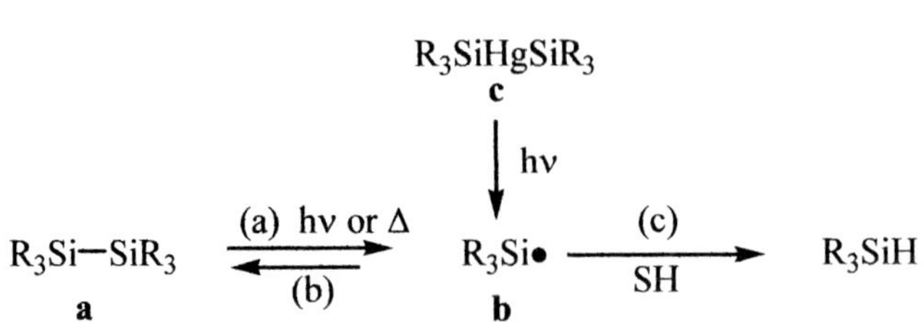

$R_3SiHgSiR_3$ **c** $\xrightarrow{h\nu}$ $R_3Si\bullet$ **b**

$R_3Si{-}SiR_3$ **a** $\underset{(b)}{\overset{(a)\ h\nu \text{ or } \Delta}{\rightleftharpoons}}$ $R_3Si\bullet$ **b** $\underset{SH}{\overset{(c)}{\longrightarrow}}$ R_3SiH

R = t-BuMe$_2$Si; SH= hydrogen atom donor

Scheme 5. Generation and reactions of silyl radicals of type **b**.

The fact that irradiation of disilanes **1a** – **3a** did not produce detectable products other than the corresponding silyl radicals indicates that the radicals produced photochemically undergo clean recombination to produce the starting silane, as shown in Scheme 5 path (b). An exception is the photochemical reaction of silyl mercurial compounds **4c**, which produce the hydrosilane **6** [Scheme 5, path (c)], while the corresponding dimer **4a** [Scheme 5, path (b)] is not detected. Apparently, for radical **4b**, hydrogen abstraction to produce silane **6** occurs significantly faster than its dimerization to **4a**.

The enthalpies, ΔH, of the homolytic cleavage of the central Si–Si bond in disilane dimers **a** can be calculated from temperature-dependent ESR experiments, using Eq. 3, in which C is the concentration of the radical, T is the absolute temperature and A is a constant. The change in the concentration of the radicals as a function of temperature could be followed by EPR as shown in Fig. 6a for radical **3b**. The concentrations of the thermally generated radicals (**2b** and **3b**) were determined by calibration of the height of the EPR signal of the silyl radicals in comparison with a 3×10^{-3} M toluene solution of TEMPO (2,2,6,6-tetramethyl-piperidinooxy).

$$\ln C^{2} = \frac{\Delta H}{RT} - A$$

Eq. 3.

The kinetic measurements yield ΔH values (see Fig. 6b) of 56 kcal/mol and 58 kcal/mol for the cleavage of the central Si–Si bond in **2a** and **3a,** respectively. These values are 9 – 11 kcal/mol lower than the Si–Si bond energy of 67 kcal/mol in $Me_3Si–SiMe_3$ (**7**) [9]. This weakening of the central Si–Si bond in **2a** and **3a** relative to that in **7** reflects the thermodynamic stabilization of the silyl radicals by α-silicon substituents relative to methyl groups [9]. The similar Si–Si bond cleavage energy in **1a** and **2a** reflects the fact that β-silyl substitution has practically no effect on the stability of the silyl radicals [5] and the fact that the β-silyl groups present in **2a** do not raise its ground state energy due to steric congestion.

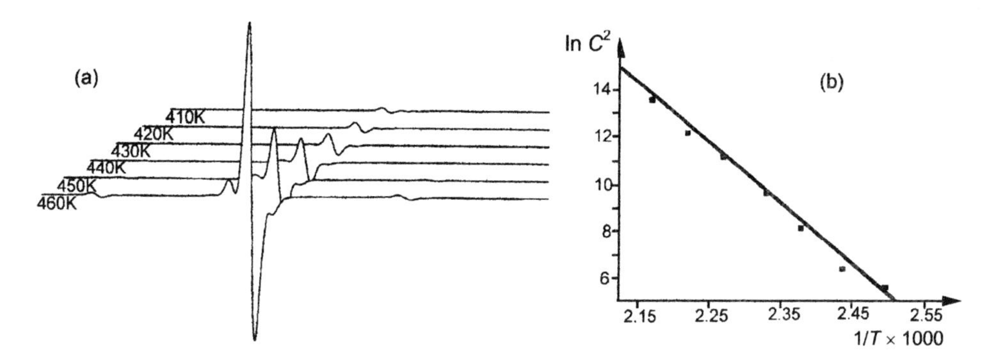

Fig. 6. (a) Temperature dependence of the intensity of the EPR signal of radical **3b**; (b) Plot of $\ln C^2$ vs $1/T$ for radical **3b**.

The EPR kinetic measurements of the decay of radical **4b** follow a second-order reaction kinetics, which strongly supports the conclusion that radicals **4b** dimerize to silane **4a**. However, the observed product of radical **4b** (produced from **4c**) is only silane **6** (90% yield). Dimer **4a** was not observed by NMR in the reaction mixture obtained upon photolysis of **4c**, indicating that if **4a** is present its concentration is less than 3%. The interpretation consistent with these facts is that **4a** undergoes a very fast reversible cleavage to produce radicals **4b**, which occasionally are captured by a hydrogen atom donor to yield silane **6**. The fact that the **4a** $\rightleftharpoons$ **4b** equilibrium does exist is clearly indicated by the fact that the decay kinetics of radical **4b** is second order in **4b**. Thus, the reaction mechanism for the production and decay of radical **4a** is described in Scheme 5, where the concentration of **4a** is lower than 3%. Using the experimental method presented in Fig. 6, we estimate that ΔH of the central Si–Si bond of **4a** is only about –8 kcal/mol.

Density functional calculations at the UB3LYP/6-31G(d) level of theory show that β-alkyl or β-silyl substitution has a negligible thermodynamic effect on the stability of silyl radicals [5]. Thus, the thermodynamic stabilities of **4b** and **1b** are expected to be nearly the same. *So, the reason for the fact that radical **4b** reacts 5×10^5 times slower than radical **1a** must be kinetic in origin.* The high kinetic stability of radical **4b** is most likely due to effective steric protection of the radical center by the large three *t*-$BuMe_2Si$ substituents. A similar, although smaller (**2b** reacts 2×10^4 slower than **1b**) kinetic stabilizing effect of the β-silyl substituents in radical **2b** was previously reported [5]. The efficient steric protection of the radical center in **4b** results, at least partially, from the "umbrella-type" conformation adopted by the *t*-butyl side-chain substituents, as can be clearly seen in its calculated structure shown in Fig. 7, where the ∠CSiSiX torsion angle is only 18.9°. According to 3-layer ONIOM (UMP2/6-31G(d):UB3LYP/6-31G(d):UHF/3-21G) calculations [4], the conformation shown in Fig. 7 is 7.8 kcal/mol more stable than the conformation in which the ∠CSiSiX torsion angle is 72.7°. This means that in radical **4b** the back-folded conformation involves less strain than other conformations, which tentatively appear to be less strained.

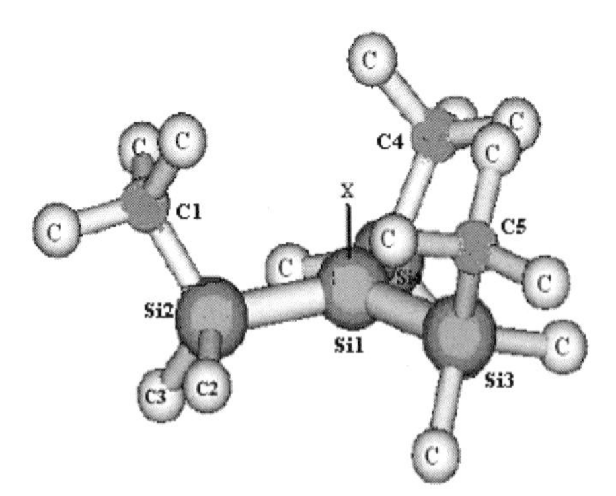

Fig. 7. Calculated structure (using 3-layer ONIOM calculations (UMP2/6-31G(d):UB3LYP/6-31G(d):UHF/3-21G) of radical **4b** (hydrogen atoms are omitted for clarity). Selected bond lengths [Å], bond angles [°], and torsion angles [°]: Si(1)–Si(2) 2.380, Si(2)–C(1) 1.925; Si(2)-Si(1)-Si(3) 119.0, Si(2)-Si(1)-C(1) 104.9; C(1)-Si(2)-Si(1)-X, X = radical, ϕ = 18.9°.

To estimate the central Si–Si bond length in silane **4a**, which, as stated above, could not be isolated, we calculated the geometries of dimers **1a** – **4a** using the ONIOM (MP2/6-31G(d):B3LYP/6-31G(d):HF/3-21G) method [4]. The ONIOM-calculated structures of silanes **1a** – **3a** are in a good agreement with their experimental geometries. The calculated central Si–Si bond length being (experimental values in parenthesis) 2.421 Å for **1a** (2.405 Å), 2.436 Å for **2a** (2.409 Å), and 2.508 Å for **3a** (2.480 Å), lending credibility to the ONIOM calculations for the unknown Si–Si bond length in **4a**. For the **4a**, the ONIOM calculations predict a central Si–Si bond length of 2.630 Å. This bond length is much longer than that in **2a** (2.409 Å) or **3a** (2.489 Å), consistent with the much lower estimated dissociation energy of **4a**. Based on our estimated central Si–Si bond dissociation energy in **4a** of only 8 kcal/mol (see above), the longest possible Si–Si bond in a branched polysilane is estimated to be around 2.65 Å. This bond length is significantly shorter than *r*(Si–Si) in *t*-Bu_3Si–Si(*t*-Bu)$_3$ (2.690 Å) [9]. At first, this may appear surprising, because *t*-Bu_3Si–Si(*t*-Bu)$_3$ is sterically significantly more congested than **4a**. However, on further consideration, we realize that this is yet another indication of the stabilizing effect of α-silyl substitution relative to α-alkyl substitution on the corresponding silyl radicals, thus making the Si–Si bond in **4a** longer and weaker than that in sterically more congested alkyl-substituted disilanes.

Conclusions

Table 1 collects the available Si–Si bond lengths in disilanes **1a** – **4a**, their dissociation energies, and structural and kinetic data of the corresponding silyl radicals. According to this data, the higher limit to the length of the central Si–Si bond in a polysilane of type **a** is around 2.65 Å, slightly longer than calculated for **4a**. For **4a**, with a central Si–Si bond length of 2.63 Å, the enthalpy for dissociation is only 8 kcal/mol. Silyl radicals **1b**, **2b**, and **3b** dimerize effectively to the corresponding dimers of type **a**, while radical **4b** decays by hydrogen abstraction. The bulky silyl substituents are responsible for the significant kinetic stabilization of silyl radicals **2b**, **3b**, and

Table 1. Structural, energetic, and kinetic data for polysilanes **1a** – **4a** and for silyl radicals **1b** – **4b**.

Polysilane	Central Si–Si bond length [Å][a]	Torsion angle $\phi^{a,b}$ [°]	ΔH [c]	Radical	Torsion angle ϕ^{d} [°]	k_{dim} [e]	Relative dimerization rate
1a	2.401	–	–	**1b**	–	1×10^8	5×10^5
2a	2.409	74.3	–58	**2b**	41.2	5×10^3	25
3a	2.480	87.7	–56	**3b**	19.1	2×10^3	10
4a	2.630[f]	–	–8	**4b**	18.9	2×10^2	1

a,b: Experimental data; b ϕ= SiSiSiX, X=SiR_3; c: Enthalpy (in kcal/mol) of cleavage of the central Si–Si bond; d: ϕ = SiSiSiX, X = radical (see Scheme 4), according to the B3LYP/6-31G*(d) calculations; e: Second order rate constant in $s^{-1}M^{-1}$ determined by EPR measurements; f: according to 3-layer ONIOM calculations (MP2/6-31G(d):B3LYP/6-31G(d):HF/3-21G) especially **4b**, relative to **1b**. In **4b**, the bulky β-*t*-Bu substituents provide more effective steric protection around the radical center than the β-Me_3Si substituents in **2b**.

Acknowledgments: This research was supported by the Israel Science Foundation administrated by the Israel Academy of Sciences and Humanities, the Fund for the Promotion of Research at the Technion, and the Minerva Foundation in Munich. B.T. and D.B.-Z. are grateful to the Ministry of Immigrant Absorption, State of Israel, for a Kamea scholarship.

References

[1] a) R. West, in *The Chemistry of Organic Silicon Compounds* (Eds.: Z. Rappoport, Y. Apeloig), **2001**, *3*, 541; b) M. Kira, in *The Chemistry of Organic Silicon Compounds* (Eds.: Z. Rappoport, Y. Apeloig), Vol. 2, Wiley, New York, **1998**, p. 1311; c) R. West, in *Comprehensive Organometallic Chemistry II* (Eds.: E. W. Abel, G. Stone, G. Wilkinson), Vol. 2, Oxford, Pergamon, **1995**, Chapter 3; d) R. D. Miller, J. Michl, *Chem. Rev.* **1989**, *89*, 1359.

[2] H. Bock, J. Meuret, K. Ruppert. *J. Organomet. Chem.* **1993**, *445*, 19.

[3] a) J. J. Brocks, H.-D. Beckhaus, A. L. J. Beckwith, C. Rüchardt, *J. Org. Chem.* **1998**, *63*, 1935; b) C. Rüchardt, H.-D. Beckhaus, *Top. Curr. Chem.* (*Synth. Org. Chem.*) **1986**, *130*, 1.

[4] T. Vreven, K. Morokuma, *J. Comput. Chem.* **2000**, *21*, 1419.

[5] Y. Apeloig, D. Bravo-Zhivotovskii, M. Yuzefovich, M. Bendikov, A. I. Shames, *Appl. Magn. Reson.* **2000**, *18*, 425.

[6] Y. Apeloig, D. Bravo-Zhivotovskii, M. Yuzefovich, M. Bendikov, D. Blaser, R. Boese, *Angew. Chem. Int. Ed.* **2001**, *40*, 3016.

[7] H. D. Thomas, K. Chen, N. L. Allinger, *J. Am. Chem. Soc.* **1994**, *116*, 5887.

[8] a) P. P. Gaspar, A. D. Haizlip, K. Y. Choo, *J. Am. Chem. Soc.* **1972**, *94*, 9032; b) J. C. J. Thynne, *J. Organomet. Chem.* **1969**, *17*, 155.

[9] I. M. T. Davidson, I. L. Stephenson, *J. Am. Chem. Soc.* **1968**, *2*, 282.

[10] N. Wiberg, H. Schuster, A. Simon, K. Peters, *Angew. Chem.* **1986**, *98*, 100.

[11] CCDC-257065, CCDC-257066, and CCDC-257067 contain the supplementary crystallographic data for, **3a**, **6**, and **4c**, respectively. These data can be obtained free of charge via www.ccdc.cam.ac.uk/conts/retrieving.html (or from the Cambridge Crystallographic Data Centre, 12 Union Road, Cambridge CB21EZ, UK; fax: (+44 1223 336 033; or deposit@ccdc.cam.ac.uk).

Out-of-Range δ^{29}Si Chemical Shifts of Diaminosilylenes — an *ab initio* Study

Michaela Flock,* Alk Dransfeld

Institute of Inorganic Chemistry, TU Graz, Stremayrg. 16, A-8010 Graz, Austria
Tel.: +43 316 8738704 — Fax: +43 316 8738701
E-mail: flock@anorg.tu-graz.ac.at

Keywords: silylene, ^{29}Si NMR, DFT

Summary: The effect of oxygen as internal electron donor on diaminosilylenes in a cyclic molecular framework with the amino substituents perpendicular to the N-Si-N plane is predicted to be remarkably large. A second oxygen-silicon coordination has a much smaller effect on stability and magnetic resonance. Geometries were optimized at the DFT B3LYP/6-31+G(d) level. The relative energies show a stabilization of 41 to 42 kJ/mol. The ^{29}Si NMR shifts calculated for the optimized geometries at IGLO SOS-DFPT PW91/IGLO-II and GIAO-MP2/TZVP levels indicate a considerable interaction, changing δ^{29}Si by 250 to 300 ppm upfield.

Introduction

Silylenes are very reactive molecules usually characterized by trapping reactions and ^{29}Si NMR. Examples are known for only half of the predicted δ^{29}Si range of silylenes [1]. Several diaminosilylenes were isolated [2 – 5]; others were predicted [6] to be good precursors for disilenes. Most known diaminosilylenes are characterized by NMR resonances in the range of 78 – 117 ppm [2 – 4]. Their stability can be attributed to the favorable allylic conjugation of the nitrogen lone pairs with the empty p orbital of the silylene silicon and steric protection by voluminous nitrogen substituents. For a diaminosilylene with the amino substituents twisted out of the N-Si-N plane by approximately 40°, an out-of-range δ^{29}Si of 223.9 ppm has been measured [5]. *Ab initio* calculations at the GIAO/MP2/6-311+G(2df,p)//B3LYP/6-311G(d,p) level for the parent compound $(H_2N)_2Si$ showed the ^{29}Si chemical shift changing from 107.0 ppm to 388.4 ppm, increasing the angle between the C-N-C and the N-Si-N to 90° [7]. However, it has not yet been possible to synthesize diaminosilylenes with nitrogen substituents completely perpendicular to the N-Si-N plane because of their increased reactivity. Preliminary calculations show that diaminosilylenes in a norbornane type molecular framework are kinetically stable (Chart 1).

In this work we extended our investigation to 1,4-diaza-7-silabicyclo[2.2.1]heptanes with inter- and intramolecular stabilization by an electron-donating group, such as oxygen, by density functional theory.

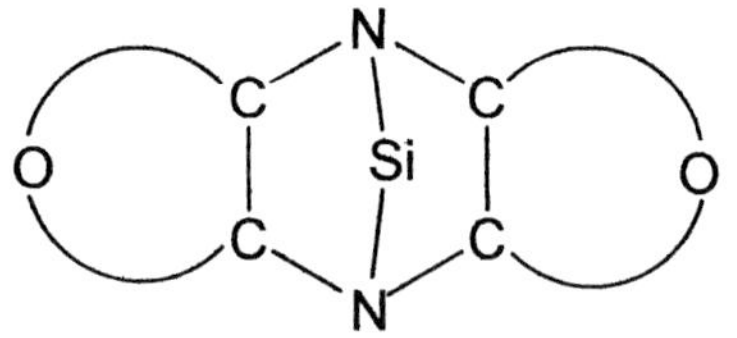

Chart 1.

Methods

Geometry optimizations and analytical frequency calculations have been carried out at the B3LYP/6-31+G(d) level using Gaussian98 [9]. The magnetic shieldings were computed at the GIAO-MP2/TZVP level with Turbomole [10] and at the IGLO SOS-DFPT PW91/IGLO-II level with the deMon program [11]. Tetramethylsilane (TMS), optimized at the same level of theory, has been used as reference molecule (σ_{MP2}= 391.5 ppm, σ_{DFT}= 366.1 ppm).

Geometries

In the parent compound, $(H_2N)_2Si$, the Si–N bond is 1.740 Å long and the N-Si-N angle is 99.8° in the planar conformer. With perpendicular amino substituents, the Si–N distances become longer (1.790Å) and the bond angle (N-Si-N 109.4°) wider [1]. In structure **1**, a bicyclic diaminosilylene, the amino substituents are forced to be perpendicular (Fig. 1).

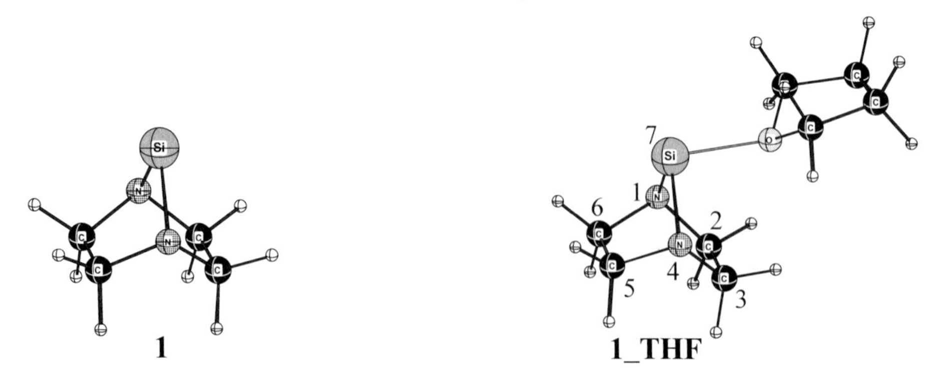

Fig. 1. The diaminosilylene, 1,4-diaza-7-silabicyclo[2.2.1]heptane, without and with THF.

The structure parameters given in Table 1 show an elongated Si-N bond (1.833Å) and a narrower N-Si-N bond angle (86°) enforced by the molecular framework. Nevertheless, the molecule does not spontaneously rearrange to relieve the strain. Intermolecular interaction with a free-electron donor, e.g., THF, "pushes" the Si atom. The angle α, defined by the middle of the C2–C3 bond,

X(C2,C3), the middle of the N atoms, X(N1,N4), and the Si atom, changes from 123° (in **1**) to 133° (**1_THF**). Angle α' is the angle between X(C5,C6), X(N2,N4), and Si.

Table 1. Selected bond lengths in Ångstroms, bond angles and ring folding α in degrees, and relative energies in kJ/mol for structures **1** to **3** obtained at the B3LYP/6-31+G(d) level.

	O–Si	N1–Si	C2–N1	N-Si-N	α/α'	ΔE
1	–	1.833	1.490	85.8	122.8/122.8	–
1_THF	2.070	1.861	1.481/1.491	83.9	133.2/113.9	–
2	2.334	1.879	1.470/1.482	82.7	121.4/122.9	0.0
2a	3.870	1.843	1.479/1.495	85.4	129.4/116.7	40.6
3	2.614	1.900	1.474	81.5	122.8/122.8	0.0
3a	2.308/3.858	1.896	1.472/1.475	81.9	119.4/126.2	18.4
3b	3.773	1.859	1.484	84.5	123.6/123.6	60.2

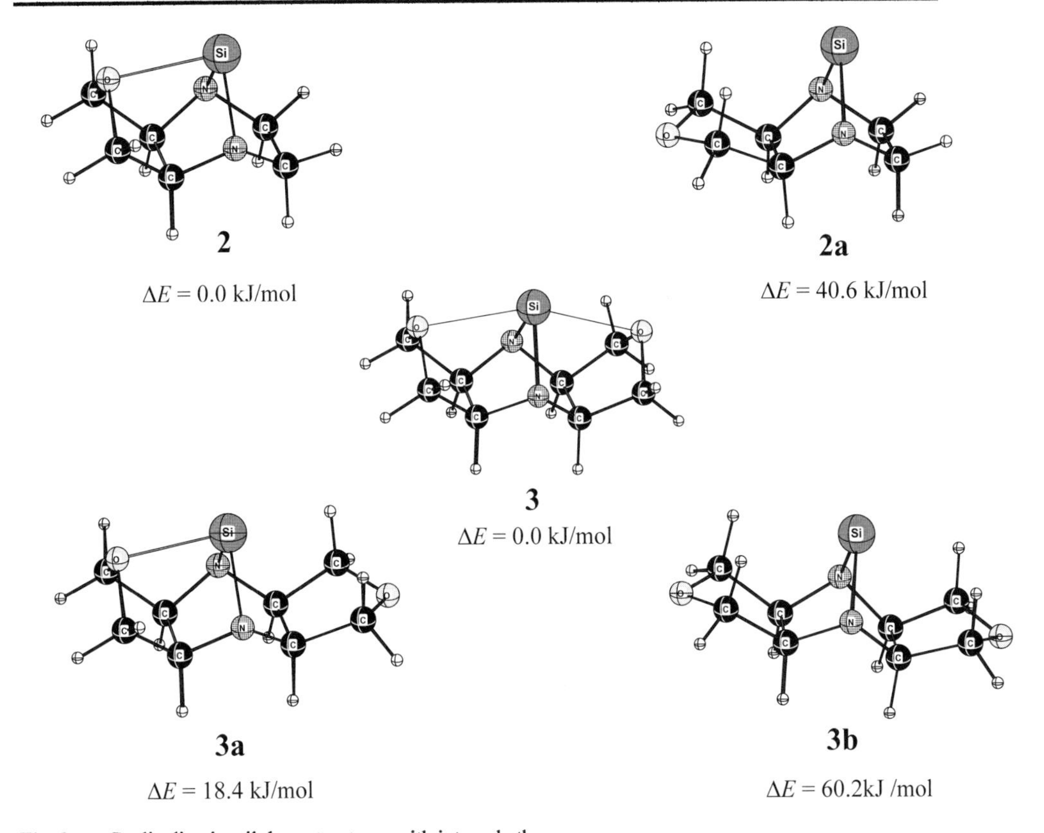

Fig. 2. Cyclic diaminosilylene structures with internal ether groups.

In molecule **2** a flexible ether group allows intramolecular Si-O interaction (see Fig. 2). The

Si–O distance is 0.26 Å longer than that in **1_THF**. In contrast to **1_THF**, the intramolecular Si–O interaction "pulls" at the Si atom (α decreases with respect to **1** and **2a**).

The relative energies in Table 1 show a distinct stabilization, ΔE_{stab}, of 40.6 kJ/mol caused by the formation of the dative oxygen-to-silicon bond. The N-Si-N angle becomes wider by 3° going from **2** to **2a**. In structures **3** the molecular framework is further extended with a second ether group. Formation of the first Si-O dative bond accounts for a stabilization (ΔE_{stab}) of 41.8 kJ/mol in **3a**, which is about as much as that in **2**. In **3b** the Si-O distances are both 0.3 Å longer than in **3a** or **2** with only one ether group interacting.

Si–O bond formation competes with dimerization: for **1**, the dimer is preferred by 106.9 kJ/mol, for molecule **2** the exothermicity of dimerization is already reduced to 55 kJ/mol. Noteworthy, for molecule **3** the dimerization becomes endothermic (E_{rel}(dimer) = 79.5 kJ/mol).

NMR Chemical Shifts

As mentioned above, the observed δ^{29}Si for the isolated and calculated diamino silylenes are 78 – 117 ppm for the planar species and 223.9 ppm for the acyclic diaminosilylene. δ^{29}Si of the parent molecule, $(H_2N)_2Si$, is 85.6 ppm for the planar conformer [1]. With the amino substituents perpendicular to the N-Si-N plane the Si chemical shift of $(H_2N)_2Si$ is 346.6 ppm as a consequence of the missing overlap between the nitrogen lone pairs and the pπ Si orbital.

Comparison of DFT- and MP2-computed δ^{29}Si values for **1** – **3** (Table 2) shows general agreement (the largest difference is 16 ppm for **2a**). For the molecules of this study we expect a better agreement between calculated and observed δ^{29}Si than in the previous study on a set of bromosilanes (rmse <32 ppm), since relativistic effects are less important. In the following discusssion we will refer to the DFT/NMR data.

Table 2. SOS-DFPT PW91/IGLO-II and GIAO MP2/TZVP calculated ^{29}Si chemical shifts for molecules **1** – **3**.

	Si–O bonds	Si pπ occ.	$\delta^{29}Si_{DFT}$	$\delta^{29}Si_{MP2}$
1	0	0.04	385.5	373.1
1_THF	1	0.15	114.4	109.3
2	1	0.14	145.8	158.9
2a	0	0.06	418.6	402.2
3	2	0.17	145.8	154.4
3a	1	0.16	170.7	177.6
3b	0	0.10	469.9	467.5

In the investigated set of heteronorbornanes the free silylenes have δ^{29}Si in the range 428 ± 42 ppm, while the molecules with one or two Si–O dative bonds show up in the region 143 ± 29 ppm. The difference of approximately 300 ppm corresponds to the reported effect of a donor group on bromoalkylsilylenes [1]. Within the free silylenes the Si chemical shift increases from **1** via **2a** to **3b**. Population analyses (NBO) show that the occupation of the Si pπ orbital and δ^{29}Si are related (the slope is 120 ppm/electron (population of pπ Si)). No such relation exists for the ether-stabilized silylenes **1_THF**, **2**, and **3a** (δ^{29}Si 114 < 146 < 171 ppm and Si pπ 0.15 > 0.14 <0.16). Remarkably, the δ^{29}Si difference between singly and doubly oxygen-coordinated silanorbornanes is small. The Si pπ population/δ^{29}Si trend in the set **3**, **3a**, **3b** is linear (slope approx. –250 ppm/electron). Although a relationship between the pπ Si occupation and the chemical shift exists in some of the molecule subsets it is not general.

Conclusion

Although intermolecular interaction is stronger, the use of an internal donor is reasonable as it is easier to keep in place and simultaneously provides steric protection. **3** should be sufficiently stable to prevent fast dimerization and should enable the isolation of a diaminosilylene with perpendicular arrangement of the amino substituents.

Acknowledgment: M. F. thanks the Austrian FWF for financial support (project T-101).

References

[1] M. Flock, A. Dransfeld, *Chem. Eur. J.* **2003**, *9*, 3320.

[2] M. Denk, R. Lennon, R. Hayashi, R. West, A. V. Belyakov, H. P. Verne, A. Haaland, M. Wagner, N. Metzler, *J. Am. Chem. Soc.* **1994**, *116*, 2691.

[3] a) R. West, M. Denk, *Pure Appl. Chem.* **1996**, *68*, 785; b) R. West, J. J. Buffy, M. Haaf, T. Müller, B. Gehrhus, M. F. Lappert, Y. Apeloig, *J. Am. Chem. Soc.* **1998**, *120*, 1639.

[4] B. Gehrhus, M. F. Lappert, J. Heinicke, R. Boese, D. Blaser, *J. Chem. Soc., Chem. Commun.* **1995**, 1931.

[5] G.-H. Lee, R. West, T. Müller, *J. Am. Chem. Soc.* **2003**, *121*, 9722.

[6] T. Müller, Y. Apeloig, *J. Am. Chem. Soc.* **2002**, *124*, 3457.

[7] T. Müller, *J. Organomet. Chem.* **2003**, *686*, 251.

[8] M. Kira, S. Ishida, T. Iwamoto, C. Kabuto, *J. Am. Chem. Soc.* **1999**, *121*, 9722.

[9] Gaussian 98 Rev A9, Gaussian Inc., Pittsburgh PA, 1999.

[10] R. Ahlrichs, M. Bär, M. Häser, H. Horn, C. Kölmel, *Chem. Phys. Lett.* **1989**, *162*, 165; Version 5.6, 2002.

[11] a) A. St-Amant, D. R. Salahub, *Chem. Phys. Lett.* **1990**, 169, 387; b) D. R Salahub, R. Fournier, P. Mlynarski, I. Papai, A. St-Amant, J. Ushio, *Density Functional Methods in Chemistry* (Eds.: J. Labanowski, J. Andzelm), Springer: New York, **1991**; c) V. G. Malkin, O. L. Malkina, L. A. Eriksson, D. R. Salahub, *Theoretical and Computational Chemistry Vol. 2* (Eds.: J. M. Seminario, P. Politzer), Elsevier Science B. V.: Amsterdam, **1995**.

Comparison of Heterocyclic Diaminosilylenes with Their Group 14 Homologs

J. Heinicke, O. Kühl*

Ernst-Moritz-Arndt-Universität Greifswald
Soldmannstr. 16, D-17487 Greifswald, Germany
Tel.: +49 3834 864337 — Fax: +49 3834 864319
E-mail: heinicke@uni-greifswald.de

Keywords: diaminosilylene, diaminocarbene, diaminogermylene

Summary: Anellation effects on *N*-heterocyclic silylenes, analogous carbenes, and germylenes are compared. Based on the trends observed, properties of new target molecules are estimated.

Suitably *N*-substituted heterocyclic diaminosilylenes such as a diazasilole-2-ylidene [1] and benzo- as well as pyrido[b]-anellated derivatives [2, 3] have been obtained by reduction of the respective $Si^{IV}Cl_2$ precursors with potassium in THF. Attempts to prepare analogously *N*-substituted diazasilole-2-ylidenes with more extended anellation, e.g., by naphthalene [3b], failed and raised the question whether this is due to preferred reduction of the anellated cycle by low-lying π^*-orbitals and subsequent easy formation of radical anions and alternative reaction pathways or to the influence of anellation on the properties and "stability" of diazasilole-2-ylidene type molecules. Quantum chemical and cyclovoltametric investigations on the reduction of related cyclic diaminocarbenes and their silicon and germanium homologs by Robert et al. (Table 1) [4] show that calculated or observed redox potentials increase in the order carbenes << silylenes < germylenes and that the influence of anellation depends on the two-valent element.

The strong increase in E^o from $\mathbf{1_C}$ to $\mathbf{1_{Si}}$ and $\mathbf{1_{Ge}}$ is due to a much lower loss of energy during the formation of the non-planar radical anions from the planar, more or less "aromatic" 6π-heterocycles **1** in the case of the larger E^{II} homologs as compared to $\mathbf{1_C}$, which gives evidence of a much stronger stabilization of imidazole-2-ylidenes as compared to its homologs and a considerable decrease of the π-stabilization with the size of the E^{II} atoms. The strong increase in E^o within the NHC series **1** – **4** indicates destabilization of NHC by anellation due to delocalization of the nitrogen electron lone pairs into the anellated rings on account of $N–C^{II}–N$ π-bonding interactions. In the case of NHSi and NHGe anellation, effects are markedly smaller, certainly because of the already lower strengths of $N–E^{II}–N$ π-interactions in the non-anellated systems **1**.

Table 1. Estimated standard redox potentials [a–c] E^{o} (in V vs SCE[d]) of NHE^{II} homologs. [4]

	1	2	3	4
		R = H		
E = C:	–4.41	–3.36	–3.10	–2.97
Si:	–2.97	–2.88	–2.69	–
Ge:	–2.88	–	–	–
		R = neopentyl [e]		
Ge:	–2.75	no redn.	no redn	–2 (–2.5)

[a] at the B3LYP/6-31 + G* level; solvent DMF; [b] at the B3LYP/6-311 +G** level; solvent DMF; [c] at the B3LYP/6-31 G* level; solvent THF; [d] the result obtained in Hartrees was first converted to Joules and then to V; [e] cyclovoltametric values in DMF, irreversible at slow scan rate.

The results of the electrochemical investigations [4] correlate well with some fundamental properties of the heterocycles: non-anellated unsaturated NHC (**1**$_C$) exist as monomers even with small *N*-alkyl substituents [5], while benzoanellated NHC (**2**$_C$) require bulkier groups like neopentyl to be persistent or even distillable [6], otherwise dimerization or a monomer-dimer equilibrium are observed [7]. Naphtho[b]- and in particular pyrido[b]-anellated NHC, **4**$_C$ and **3**$_C$, are somewhat less "stable" than **2**$_C$. The *N,N'*-dineopentyl derivatives, unambiguously characterized by ^{13}C NMR data (Table 2), are persistent at ordinary temperatures but decompose during attempts at high-vacuum distillation or recrystallization [8].

Table 2. ^{13}C NMR data of **3**$_C$ and **4**$_C$ (R = neopentyl, δ in C_6D_6).

	C-2	CH (anell.)	C_q (anell.)	NCH_2	CMe_3
3$_C$	234.7	116.7, 118.2, 142.1	128.2, 149.2	56.7, 59.6	28.5 (2s); 33.9, 34.1
4$_C$	239.8	107.3, 125.2, 129.0	130.3, 137.2	59.6	29.3; 34.7

4$_C$ (R = np) was obtained in high yield from the respective thione and two equivalents of potassium in THF without major side products (purity of crude product 85 – 90%). As the germylene **4**$_{Ge}$ (R = np) is even distillable (in vacuum), the silylene **4**$_{Si}$ (R = np) will also be persistent, at least under normal conditions, if suitable synthetic access is found.

3$_C$ (R = np) was formed in lower yield and purity (ca. 40%) because of partial reduction of **3**$_C$ by potassium yielding a mixture of unreacted initial thione, the carbene **3**$_C$ (R = np), and the respective NCH_2N heterocycle. The lower "stability" of **3**$_C$ (R = np) as compared to the benzo- and naphtho[b]-anellated systems **2**$_C$ and **4**$_C$ is attributed to the asymmetric π-charge distribution within the ring plane induced by the pyridine nitrogen atom and thus to kinetic effects. This was shown for the silylene $\mathbf{3}_{Si}$ [3a], which, in contrast to $\mathbf{2}_{Si}$ [2], suffers from extensive thermal decomposition during high-vacuum distillation. With increasing size of E^{II} this kinetic destabilization becomes smaller [3a].

Introduction of two nitrogen atoms in 1,4-position, i.e. anellation by pyrazine or quinoxaline, should avoid the kinetic destabilization, while the electron withdrawing effect is further increased. The competition of the anellated ring with E^{II} for the nitrogen electron lone pairs should increase the electrophilic and weaken the dominance of the nucleophilic properties of NHC and its homologs. The increased π-acidity should strengthen the back-bonding in complexes and alter the properties of respective transition metal catalysts but also destabilize the ligands themselves. These points are presently under investigation. This study was commenced with NHGe, as these are easily available and less destabilized by anellation than NHC and NHSi. The lower electronegativity of germanium vs carbon and the much less efficient Ge–N (p-p)π-bonding cause weaker interactions with the Ge-p_π orbital and thus favor π-interactions with anellated rings as well as π-back-bonding in transition metal complexes as compared to related NHC [9]. The π-interactions with the anellated ring in $\mathbf{2}_{Ge}$ and $\mathbf{3}_{Ge}$ is evident by shorter N–C and longer N–Ge bonds as compared to $\mathbf{1}_{Ge}$ and lengthening of the C–C bond within the five-membered ring (Table 3), which suggests a quinonediimine Ge^0 resonance structure. Nevertheless, the germanium remains sufficiently nucleophilic to avoid interactions with the basic pyridine nitrogen atom in $\mathbf{3}_{Ge}$. The crystal structure shows no interactions of these atoms (Fig. 1) [10].

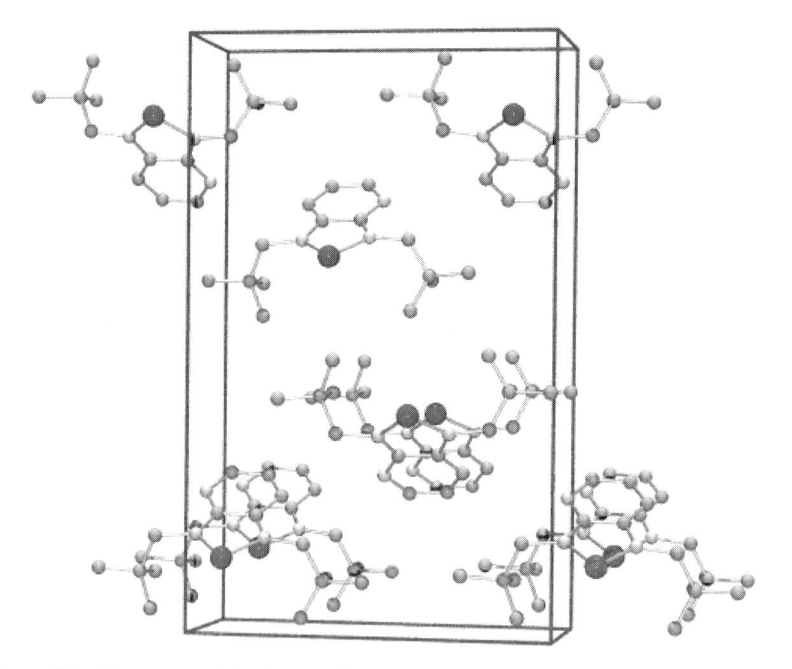

Fig. 1. Package of $\mathbf{3}_{Ge}$ (R = neopentyl) in the crystal.

Table 3. Characteristic bond lengths [pm] and angle [°] of $\mathbf{1}_{Ge}$, $\mathbf{2}_{Ge}$, $\mathbf{3}_{Ge}$ and *fac*-$(\mathbf{1}_{Ge})_3Mo(CO)_3$ (R = neopentyl).

	$\mathbf{1}_{Ge}$	$\mathbf{2}_{Ge}$	$\mathbf{3}_{Ge}$	*fac*-$(\mathbf{1}_{Ge})_3Mo(CO)_3$
N–Ge	180.8(1)	186.1(3) 186.6(3)	186.0(3) 186.6(3)	181.9(6) – 183.7(6)
N–C_{ring}	143.7(3)	139.3(5) 139.1(5)	137.6(4) 137.6(4)	137.7(10), 138.5(10), 136.2(8), 136.9(8), 137.4(10), 140.5(10)
C=C	138.6(7)	141.9(5) 138.3(6) – 139.9(5)	142.9(5) 133.9(4) – 138.8(5)	133.3 (11), 134.4 (9), 132.0 (12)
Ge–Mo	–	–	–	253.25(8), 253.39(10), 254.44(9)
N-Ge-N	87.75(6)	84.9(1)	84.6(1)	85.7(2) – 86.5(3)

The quinoxaline-anellated NHGe was obtained in solution but is highly sensitive towards moisture and air and as yet has not been isolated in pure form. A bis(quinoxaline)-anellated eight-membered NHGe, formed from 2-*tert*-butylamino-2-chloro-quinoxaline, *n*BuLi, and $GeCl_2$(dioxane), however, could be isolated as LiCl(dioxane) adduct. Li^+ is complexed by two quinoxaline nitrogen atoms and by two dioxane molecules forming a polymer backbone. Chloride is coordinated to Ge(II) (Fig. 2), indicating a switch from the dominance of the usually nucleophilic properties in NHGe to electrophilic properties [11]. The π-back-bonding in transition metal complexes becomes evident already in non-anellated NHGe, such as the *fac*-$[(\mathbf{1}_{Ge})_3Mo(CO)_3]$, by lengthened N–Ge bonds (Table 3, Fig. 2) and bathochromic shifts of the CO stretching vibrations (in nujol $\tilde{\upsilon}$ 1961, 1853 cm^{-1}) [12], but it should be stronger in electron-withdrawing anellated NHGe complexes. For *N*-heterocyclic silylenes we expect behavior similar to that of NHGe, and for NHC a rather dramatic increase of π-acceptor properties on anellation with quinoxalines.

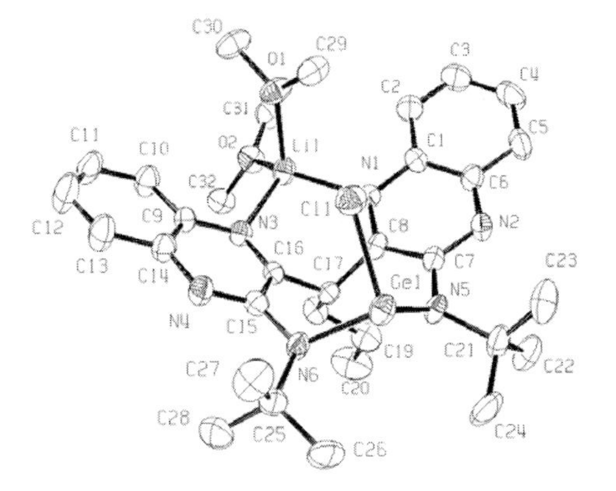

Fig. 2. Crystal structure of a bis(quinoxaline) anellated NHGe with LiCl(dioxane) [monomer unit with 2 half dioxane rings].

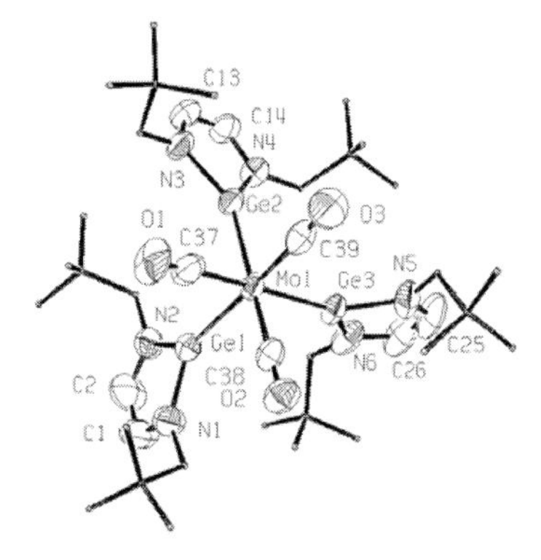

Fig. 3. Crystal structure of *fac*-$[(\mathbf{1}_{Ge})_3Mo(CO)_3]$.

Acknowledgment: We thank the Fonds der Chemischen Industrie for financial support.

References:

[1] M. Denk, R. Hayashi, R. West, *Chem. Commun.* **1994**, 33.

[2] B. Gehrhus, M. F. Lappert, J. Heinicke, R. Boese, D. Bläser, *Chem. Commun.* **1995**, 1931.

[3] a) J. Heinicke, A. Oprea, M. K. Kindermann, T. Karpati, L. Nyulászi, T. Veszprémi, *Chem. Eur. J.* **1998**, *4*, 537; b) J. Heinicke, A. Oprea, *Heteroatom Chem.* **1998**, *9*, 439.

[4] L. Pause, M. Robert, J. Heinicke, O. Kühl, *J. Chem. Soc., Perkin Trans.* **2001**, 1383.

[5] a) A. J. Arduengo, H. V. R. Dias, R. L. Harlow, M. Kline, *J. Am. Chem. Soc.* **1992**, *114*, 5530; b) Rev.: A. J. Arduengo, *Chem. unserer Zeit* **1998**, *32*, 6; W. A. Herrmann, C. Köcher, *Angew. Chem.* **1997**, *109*, 2256.

[6] a) W. M. Boesveld, B. Gehrhus, P. B. Hitchcock, M. F. Lappert, P. v. R. Schleyer, *Chem. Commun.* **1999**, 755; b) F. E. Hahn, L. Wittenbecher, R. Boese, D. Bläser, *Chem. Eur. J.* **1999**, *5*, 1931.

[7] a) F. E. Hahn, L. Wittenbecher, Duc Le Van, R. Fröhlich, *Angew. Chem.* **2000**, *39*, 541; b) Y. Liu, P. E. Lindner, D. M. Lemal, *J. Am. Chem. Soc.* **1999**, *121*, 10626.

[8] A. Oprea, PhD Thesis, Ernst-Moritz-Arndt-Universität Greifswald, **2000**.

[9] O. Kühl, *Coord. Chem. Rev.* (accepted waiting for publication).

[10] O. Kühl, P. Lönnecke, J. Heinicke, *Polyhedron 20*, **2001**, 2215.

[11] O. Kühl, P. Lönnecke, J. Heinicke, *New J. Chem.* **2002**, *26*, 1304.

[12] O. Kühl, P. Lönnecke, J. Heinicke, *Inorg. Chem.* **2003**, *42*, 2836.

Spectroscopic Evidence for the Formation of the Pentamethylcyclopentadienylsilicon Cation

P. Jutzi,* A. Mix, B. Rummel, B. Neumann
H.-G. Stammler, A. Rozhenko, W. W. Schoeller

Universität Bielefeld, Fakultät für Chemie
Universitätsstr. 25, D-33615 Bielefeld, Germany
Tel.: +49 521 1066181 — Fax: +49 521 1066026
E-mail: peter.jutzi@uni-bielefeld.de

Keywords: silyliumylidene, Si(II)-cation, divalent silicon, pentamethylcyclopentadienylsilicon cation

Summary: The formation of the highly reactive $Me_5C_5Si^+$ cation in the low-temperature reaction of decamethylsilicocene, $(Me_5C_5)_2Si$, with the protonation agent Et_2OH^+ $B(C_6F_5)_4^-$ is indicated by NMR spectroscopic investigations. Bonding parameters are discussed on the basis of theoretical calculations.

Silylene (SiH_2) and its derivatives with small organic substituents R represent highly reactive species, which are key intermediates in many thermal and photochemical reactions. They have been characterized by matrix and dilute gas phase studies [1]. The last two decades have seen great progress in the kinetic and/or thermodynamic stabilization of divalent silicon compounds and in the isolation of species persistent under ordinary conditions (Fig. 1) [2 – 6].

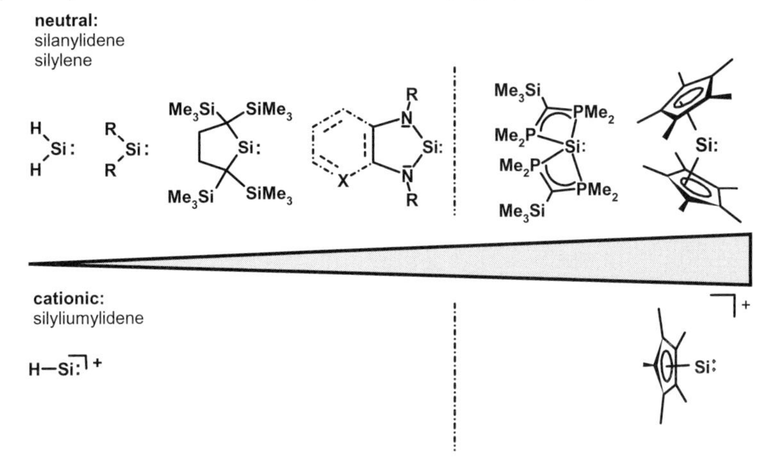

Fig. 1. Kinetic/thermodynamic stabilization of Si(II) species.

The even more reactive SiH^+ cation ("silyliumylidene") has been observed in the laboratory by discharge of SiH_4 [7] and also in the solar spectrum. If there is at all a chance to prepare a persistent derivative, then this would probably be with pentamethylcyclopentadienyl as π-bonded substituent (Fig. 1).

Our strategy to synthesize a compound containing the $Me_5C_5Si^+$ cation is based on observations from the chemistry of the heavier group 14 element metallocenes. Reaction of the decamethylmetallocenes $(Me_5C_5)_2El$ with several protic substrates led to the isolation of compounds containing the corresponding $Me_5C_5El^+$ cations (Scheme 1) [8].

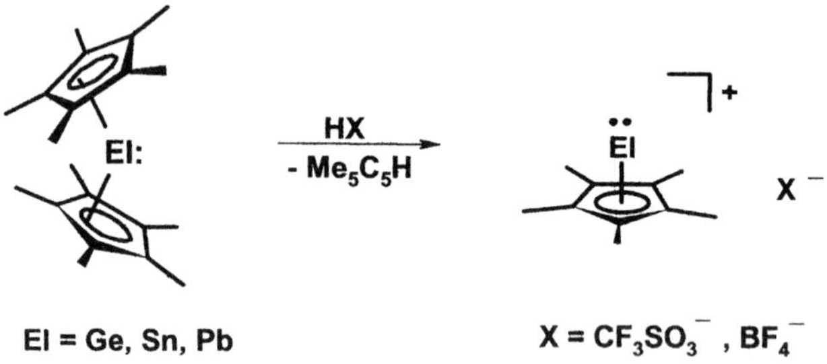

El:

HX

$- Me_5C_5H$

El

$]^+$

X^-

El = Ge, Sn, Pb

$X = CF_3SO_3^-$, BF_4^-

Scheme 1.

In contrast, the reaction of decamethylsilicocene with protic substrates follows three different pathways (Schemes 2 and 3) [6]. Oxidative addition is observed with the hydrogen halides HX, and the protonated silicocene is formed in the reaction with two equivalents of trifluorosulfonic acid. In the reaction with HBF_4, the wanted $Me_5C_5Si^+$ cation might be present together with the BF_4^- anion as a highly reactive intermediate which easily eliminates BF_3 under formation of the polymer $(Me_5C_5SiF)_n$ in a final step.

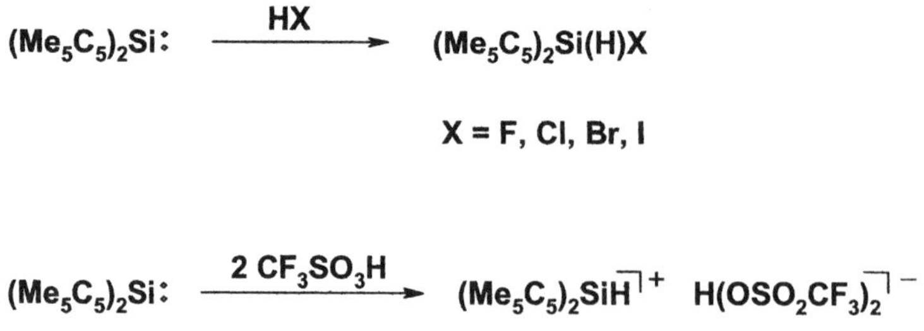

$$(Me_5C_5)_2Si: \xrightarrow{HX} (Me_5C_5)_2Si(H)X$$

X = F, Cl, Br, I

$$(Me_5C_5)_2Si: \xrightarrow{2\ CF_3SO_3H} [(Me_5C_5)_2SiH]^+ \ [H(OSO_2CF_3)_2]^-$$

Scheme 2.

$(Me_5C_5)_2Si:$ $\xrightarrow[-\,BF_3,\ -\,Et_2O,\ -\,Me_5C_5H]{HBF_4 \cdot OEt_2}$ $1/n\,[Me_5C_5SiF]_n$

$\downarrow$ $+\,HBF_4$, $-\,Me_5C_5H$

$[Me_5C_5Si:^+\ BF_4^-]$ $\xrightarrow{-\,BF_3}$ $1/2\,[Me_5C_5Si(\mu\text{-}F)_2SiC_5Me_5]$ $\uparrow$

Scheme 3.

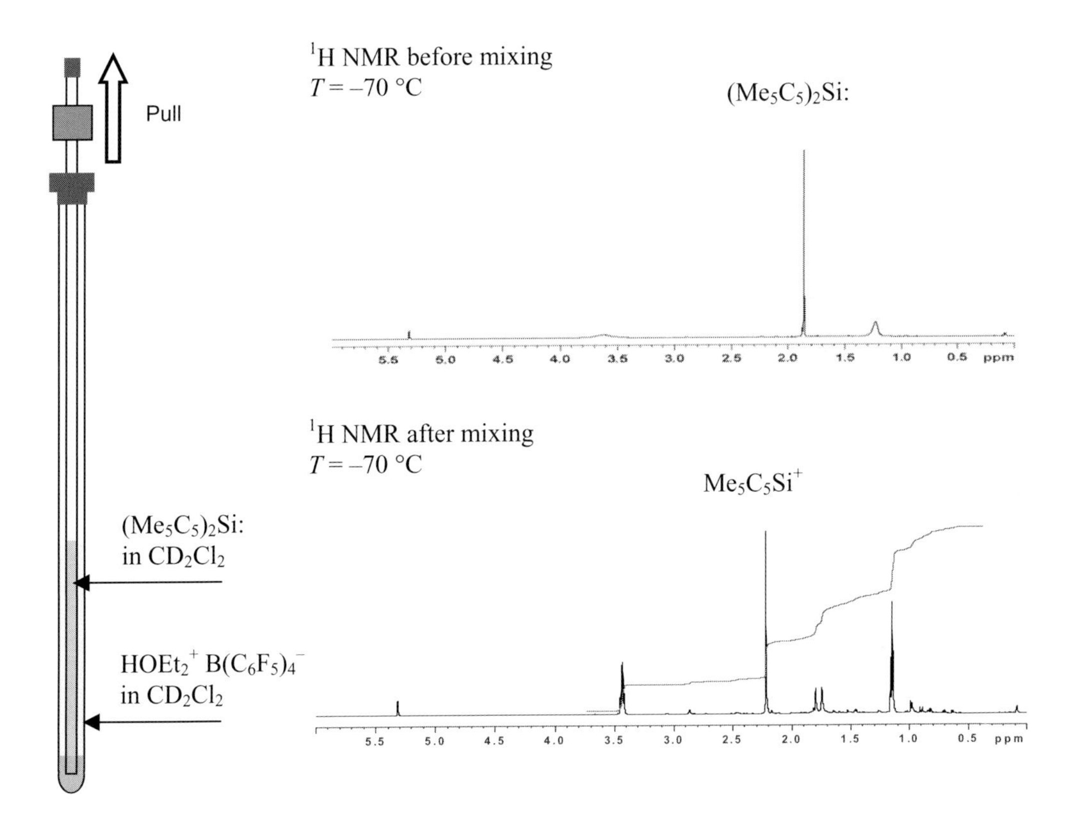

Fig. 2. Low temperature ^{1}H NMR experiment.

The first spectroscopic evidence for the existence of the $Me_5C_5Si^+$ cation in solution came from a low-temperature NMR experiment. In a special "two-chamber" NMR tube [9], solutions of $(Me_5C_5)_2Si$ in CD_2Cl_2 and of $HOEt_2^+$ $B(C_6F_5)_4^-$ [10] in CD_2Cl_2 were cooled to –70 °C. ^1H NMR spectra recorded before and after mixing of the two solutions (see spectra in Fig. 2) indicated a reaction according to Scheme 4. The presence of the $Me_5C_5Si^+$ cation in the reaction mixture is indicated by the appearance of a sharp singlet at $\delta = 2.23$ ppm.

$$(Me_5C_5)_2Si: + [HOEt_2]^+ [B(C_6F_5)_4]^- \xrightarrow[-70°C]{CD_2Cl_2} [Me_5C_5Si]^+ [B(C_6F_5)_4]^- + Me_5C_5H + OEt_2$$

Scheme 4.

Table 1 shows the ^1H NMR data of the Group 14 element sandwich [$(Me_5C_5)_2El$] and half-sandwich [$Me_5C_5El^+$] species (El = Si, Ge, Sn, Pb; Cp* = Me_5C_5) in CD_2Cl_2 solution. Comparable shift regions are observed within the sandwich and the half-sandwich series.

Table 1. ^1H NMR data of $(Me_5C_5)_2El$ and $Me_5C_5El^+$ species (El = Si, Ge, Sn, Pb).

	Cp*$_2$Si	Cp*Si$^+$	Cp*$_2$Ge	Cp*Ge$^+$	Cp*$_2$Sn	Cp*Sn$^+$	Cp*$_2$Pb	Cp*Pb$^+$
^1H NMR δ [ppm]	1.89	2.23	1.94	2.17	2.06	2.03*	2.17	2.40

*solvent CD_3CN

We investigated the $Me_5C_5Si^+$ cation by theoretical methods [b3lyb/6-311g(d,p)]. The calculations reveal a pentagonal pyramidal structure with the Me_5C_5 unit being η^5-bonded to the silicon atom (see Fig. 3). The ^1H NMR data are in agreement with the calculated structure.

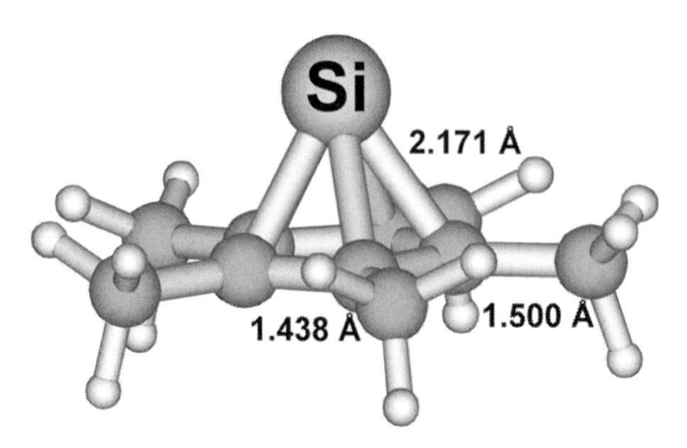

Fig. 3. Calculated structure of the $Me_5C_5Si^+$ cation.

Experiments to isolate the compound $Me_5C_5Si^+$ $B(C_6F_5)_4^-$ from the reaction of $(Me_5C_5)_2Si$ with

$HOEt_2^+$ $B(C_6F_5)_4^-$ have been unsuccessful so far, presumably because of decomposition processes initiated by unavoidable by-products in the synthetic procedure.

Acknowledgments: We gratefully acknowledge the financial support from the University of Bielefeld, the Deutsche Forschungsgemeinschaft, and the Fonds der Chemischen Industrie.

References

[1] P. P. Gaspar, R. West, *The Chemistry of Organosilicon Compounds*, Vol. 2 (Eds.: Z. Rappoport, Y. Apeloig).
[2] M. Kira, S. Ishida, T. Iwamoto, C. Kabuto, *J. Am. Chem. Soc.* **1999**, *121*, 9722.
[3] M. Haaf, A. Schwedake, R. West, *Acc. Chem. Res.* **2000**, *33*, 704.
[4] B. Gerhus, M. F. Lappert, *J. Organomet. Chem.* **2001**, *617 – 618*, 209.
[5] H. H. Karsch, U. Keller, S. Gamper, G. Müller, *Angew. Chem. Int. Ed. Engl.* **1990**, *29*, 295.
[6] T. Kühler, P. Jutzi, *Adv. Organomet. Chem.* **2003**, *49*, 1.
[7] A. E. Douglas, B. L. Lutz, *Can. J. Phys.* **1970**, *48*, 247.
[8] P. Jutzi, N. Burford, *Chem. Rev.* **1999**, *99*, 969.
[9] Construction: A. Mix.
[10] P. Jutzi, C. Müller, A. Stammler, H. G. Stammler, *Organometallics* **2000**, *19*, 1442.

Synthesis and Characterization of Bissilylated Onium Ions of Group 15 Elements

Robin Panisch, Thomas Müller*

Institut für Anorganische und Analytische Chemie der Goethe Universität Frankfurt
Marie Curie-Str. 11, D-60439 Frankfurt/Main, Germany
Tel.: +49 69 798 29166 — Fax: +49 69 798 29188
E-mail: dr.thomas.mueller@chemie.uni-frankfurt.de

Keywords: silyl cations, quantum mechanical calculations, NMR spectroscopy

Summary: Bissilylated onium ions of the elements N → Sb are formed by intramolecular addition of transient silylium ions to EPh_2 groups (E = N, P, As, Sb). Solutions of the onium salts in aromatic hydrocarbons are stable at room temperature for days, with the exception of the stibonium ion, which decomposes slowly. The cations were isolated as their $[B(C_6F_5)_4]^-$ salts and were characterized by NMR spectroscopy supported by quantum mechanical calculations.

Introduction

Recently, we suggested a model system, **1**, for the generation of silyl cationic species of different reactivity and stability [1]. In these cations **1**, the positive charge is divided between the two silicon atoms and the electron-donating group X. The nature of X determines the extent of charge transfer from the silicon atoms to the X group. We report on the synthesis and characterization of cyclic cations **2** with the Group 15 elements N, P, As, Sb as the electron-donating substituent. Additional quantum mechanical calculations describing the charge distribution and the bonding situation in onium cations **2** are presented.

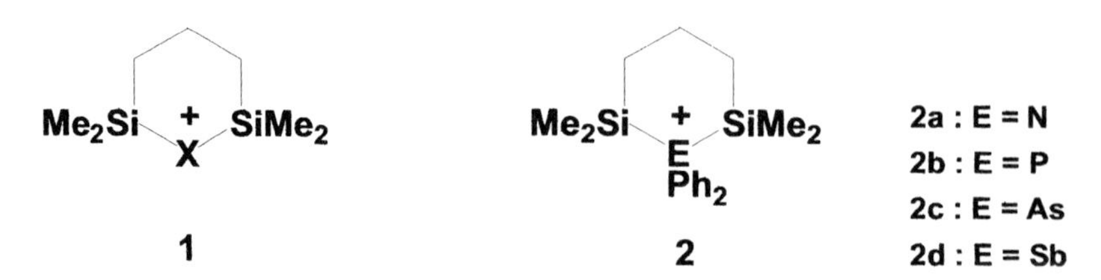

Fig. 1. The cyclic cations **1** and **2**.

Results and Discussion

Synthesis

The cations are prepared from 2,5-disilaheptanes, **3**, by hydride transfer reactions. The transient silylium ion **4** undergoes an intramolecular reaction to the more stable cyclic onium ion **2**. The precursor silanes, **3**, are synthesized by salt metathesis reaction from compounds of the type Ph_2ELi, where E is the relevant Group 15 element, and the silylchloride **5**. The salts **2** $[B(C_6F_5)_4]^-$ are isolated after washing with pentane as white to yellow microcrystalline powders in nearly quantitative yield. Solutions of the cations **2a–c** in aromatic solvents are stable at room temperature. The stibonium ion **2d** decomposes in solution slowly, yielding unidentified products.

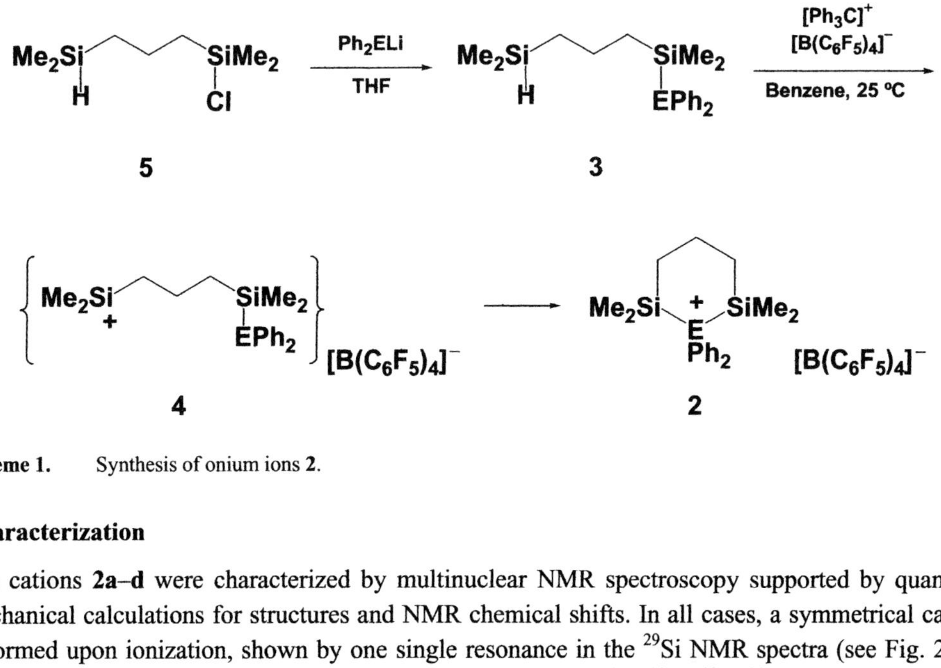

$Me_2Si(H)$–$SiMe_2(Cl)$ **5** —(Ph_2ELi, THF)→ $Me_2Si(H)$–$SiMe_2(EPh_2)$ **3** —($[Ph_3C]^+$ $[B(C_6F_5)_4]^-$, Benzene, 25 °C)→ { Me_2Si^+–$SiMe_2(EPh_2)$ } $[B(C_6F_5)_4]^-$ **4** ⟶ Me_2Si E^+Ph_2 $SiMe_2$ $[B(C_6F_5)_4]^-$ **2**

Scheme 1. Synthesis of onium ions **2**.

Characterization

The cations **2a–d** were characterized by multinuclear NMR spectroscopy supported by quantum mechanical calculations for structures and NMR chemical shifts. In all cases, a symmetrical cation is formed upon ionization, shown by one single resonance in the ^{29}Si NMR spectra (see Fig. 2 for an example). In toluene nearly the same NMR chemical shifts (^{1}H, ^{29}Si, ^{13}C, ^{31}P) are measured as in benzene solution; this indicates negligible interaction between the cations and the solvent under the applied experimental conditions. The significant down-field shift of the ^{29}Si NMR signal compared to the precursors **3**, $\Delta\delta^{29}$Si, indicates uptake of positive charge. $\Delta\delta^{29}$Si is largest for the ammonium ion **2a** ($\Delta\delta^{29}$Si = 37.2) and it decreases along the series **2d**, **2c**, **2b** with the smallest effect found for the phosphonium ion **2b** ($\Delta\delta^{29}$Si = 6.0, see also Table 1). The decrease of the ^{1}J(SiP) coupling constant in the phosphonium ion **2b** compared to the precursor silylphosphane **3b** (^{1}J(SiP) = 11.3 Hz (**2b**), ^{1}J(SiP) = 21.7 Hz (**3b**)) is in qualitative agreement with a weaker Si–P bond in the phosphonium ion (see also Fig. 2). Interestingly, quaternization of the phosphor in **2b** has only a small influence on its ^{31}P NMR chemical shift.

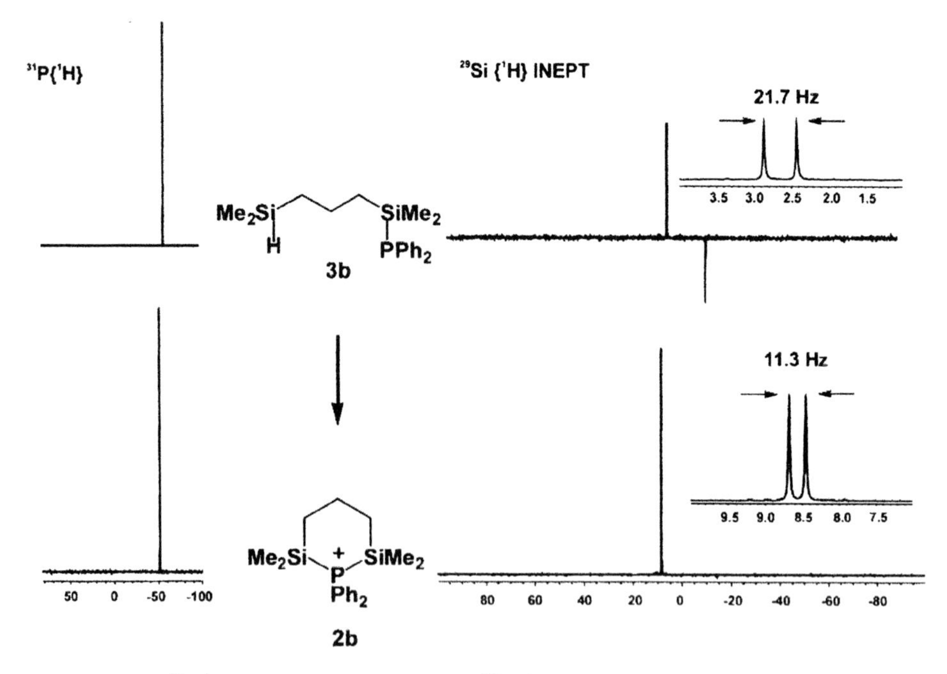

Fig. 2. 161.96 MHz $^{31}P\{^{1}H\}$ NMR (left) and 49.70 MHz $^{29}Si\{^{1}H\}$ INEPT NMR spectra (right) of silylphosphane **3b** in $CDCl_3$ (upper trace) and phosphonium ion **2b** in C_6D_6 (lower trace).

Table 1. Experimental and calculated ^{29}Si NMR chemical shifts for onium ions **2**, their chemical shift difference $\Delta\delta^{29}$Si to their precursors **3** and the calculated intramolecular stabilization energy ΔE^{intra} (in kcal mol^{-1}).

Cation	E	δ^{29}Si[a]	$\Delta\delta^{29}$Si[b]	δ^{29}Si^{calc}[c]	ΔE^{intra}[d]
2a	N	43.4	37.2	53.9	27.8
2b	P	8.6	6.0	13.8	50.9
2c	As	17.7	12.0	18.3	47.0
2d	Sb	17.5	15.3	22.9	40.3

[a] At room temperature in C_6D_6. [b] Calculated according to $\Delta\delta^{29}Si = \delta^{29}Si\ (\mathbf{2}) - \delta^{29}Si(\mathbf{3})$.
[c] Calculated at GIAO/MPW1PW91/6-311G(2d,p)//B3LYP/6-31G(d) vs. calculated TMS. σ(TMS, T_D) = 343.44. [d] Calculated at B3LYP/6-311+G(2d,p)//B3LYP/6-31G(d) + ΔZPVE.

Density functional calculations [2] at the B3LYP/6-31G(d)[3] predict for all onium ions **2a–d** nearly symmetrical structures (see Fig. 3 for an example). The 2,6-disilacyclohexane ring adopts in all cations a regular chair conformation and the silicon-element bonds in the tetrahedral-coordinated onium ions **2** are significantly elongated compared to the precursors **3**, where the element is tricoordinated, i.e. the Si–P bond in **2b** (r(Si–P) = 233.5 pm) is predicted to be longer by 4 pm than that in the silylphosphane **3b** (r(Si–P) = 229.5 pm, all values at B3LPY/6-31G(d)). GIAO NMR chemical shift calculations for the optimized structures of cations **2a–d** are in good agreement with the experimental data and confirm the validity of the computed "gas phase" structures also for the condensed phase. For example, the calculated ^{29}Si NMR chemical shift at the GIAO/MPW1PW91/6-311G(2d,p) level of theory for the phosphonium ion **2b** is 13.8, very close to the experimental value of $\delta^{29}Si$(**2b**) = 8.6 (see Table 1). Also the agreement in the phosphorus NMR data is good, i.e. $\delta^{31}P$(**2b**) = –47.7 calculated versus $\delta^{31}P$(**2b**) = –51.3 experimental.

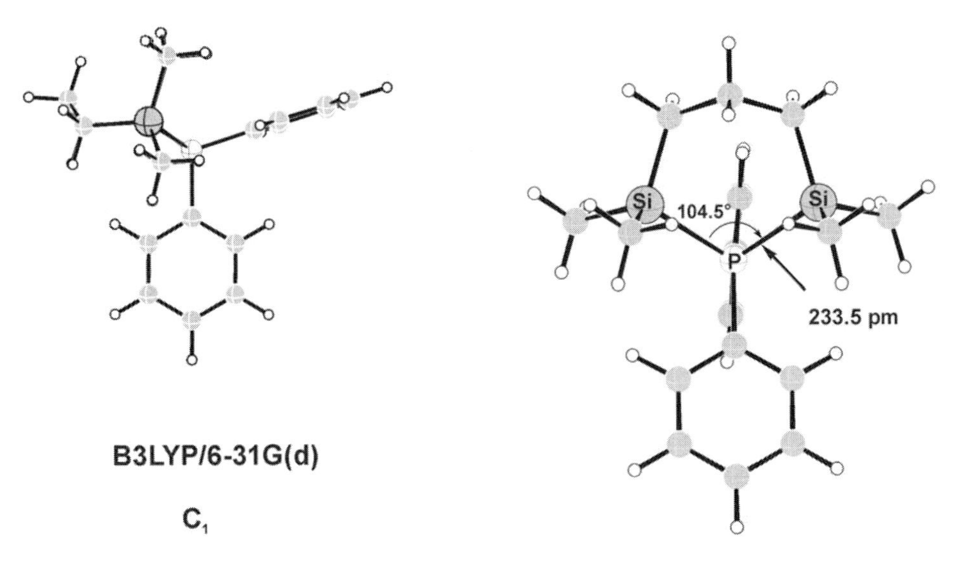

Fig. 3. Computed structure of phosphonium ion **2b** (B3LYP/6-31G(d): –1661.11273 Hartree, (0), ZPVE: 264.3 kcal mol^{-1}).

Bonding in Bissilylated Onium Ions

The intramolecular stabilization energy ΔE^{intra}, that is the energy difference between the open silylium ion **4a–d** and the cyclic onium ion, is predicted to be very large for the onium ions **2b–d**, with a maximum thermodynamic stabilization for the phosphonium ion **2b** (ΔE^{intra} = 50.9 kcal mol^{-1}, see Table 1 and Fig. 4). Interestingly, the ammonium ion, **2a**, is the least stabilized cation in this series, with ΔE^{intra} being less than 60% of the stabilization energy of the phosphonium ion **2b**. Two factors contribute to the comparatively small stabilization of **2a**. Steric crowding around the nitrogen atom destabilizes the cation **2a**, and on the other hand the silylamine-substituted silylium ion **4a** is stabilized by conjugative interactions between the nitrogen lone pair and the phenyl and silyl substituents, which are not present in the tetravalent onium ions **2a**. This conjugative interaction is much smaller in the silylphosphine, -arsine and -stibine-substituted silylium ions, **4b–d**; therefore the stabilization energy is decisively larger for the cations **2b–d**.

The charge distribution in the bissilylated onium ions will be discussed for close model compounds **6a–d**, where the phenyl groups of the cations **2** are replaced by methyl groups. According to a

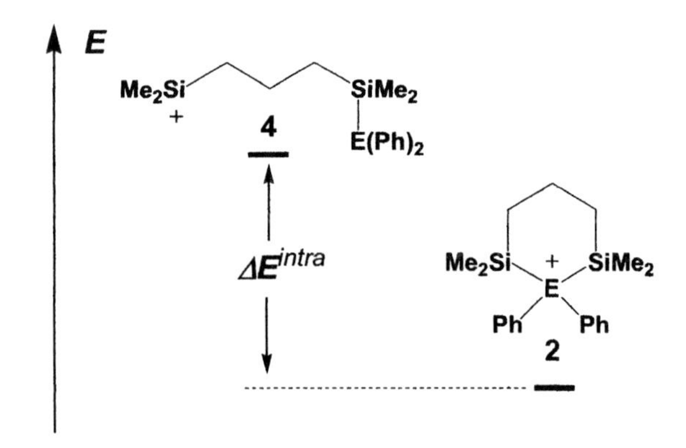

Fig. 4. Definition of the intramolecular stabilization energy ΔE^{intra}.

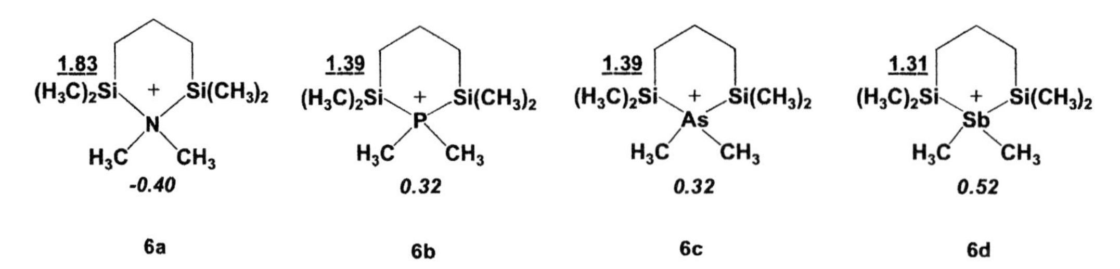

Fig. 5. Calculated charge distribution in cations **6a–d** according to an NBO analysis of the HF/6-311G(d) wave function. Group charges of the EMe_2 groups are given in italic; the atomic charges at silicon are underlined.

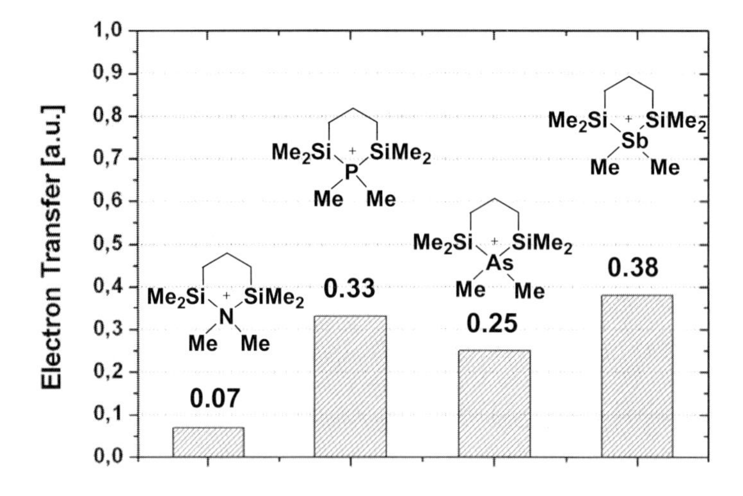

Fig. 6. Calculated electron transfer from the EMe_2 groups to the Me_2Si–R group in cations **6a–d** according to an NBO analysis of the HF/6-311G(d) wave function.

natural population analysis (NBO) [4], the atomic charge at silicon in the cations **6a–d** is largest for the ammonium ion **6a** and sharply decreases by ca. 0.5 a.u. for the phosphonium ion, while it remains nearly constant for the arsonium and stibonium ions (see Fig. 5). In qualitative agreement, the calculated electron transfer from the EMe_2 unit to the Me_2Si^+–R group is very small for the amino group, but is considerably larger for the phosphino group. It remains approximately constant for the arsonium and stibonium cations (see Fig. 6).

Conclusion

Cyclic bissilylated onium ions of the group 15 elements **2a–d** were synthesized and characterized by NMR spectroscopy supported by quantum mechanical calculations. All onium ions with the exception of the stibonium ion, **2d**, are stable in solution at room temperature. Experiments to test the potential of these cations in Lewis acid-catalyzed organic transformations are under way in our laboratories.

Acknowledgment: This work was supported by the Goethe Universität of Frankfurt and by the German Israeli Foundation for Scientific Research and Development (GIF).

References

[1] T. Müller, A model system for the generation of silyl cationic species of different reactivity and stability, in *Organosilicon Chemistry V — From Molecules to Materials* (Eds.: N. Auner, J. Weis), VCH, Weinheim, **2003**, p. 34.

[2] The GAUSSIAN 98 program was used. GAUSSIAN, Gaussian Inc., Pittsburgh, USA, **1998**.

[3] For Sb the Stuttgart-Dresden Pseudopotential as implemented in Gaussian 98 was used.

[4] NBO 4.0: E. D. Glendenning, J. K. Badenhoop, A. E. Reed, J. Carpenter, F. Weinhold, Theoretical Chemistry Institute, University of Wisconsin, Madison, WI, **1996**.

Structural and Spectroscopic Evidence for β-SiC Hyperconjugation in Vinyl Cations

Thomas Müller*

Institut für Anorganische und Analytische Chemie der Goethe Universität Frankfurt
Marie Curie-Str. 11, D-60439 Frankfurt/Main, Germany
Tel.: +49 69 798 29166 — Fax: +49 69 798 29188
E-mail: dr.thomas.mueller@chemie.uni-frankfurt.de

Mark Juhasz, Christopher A. Reed

Department of Chemistry, University of California, Riverside, CA 92521-0403, USA

Keywords: β-silyl effect, computational chemistry, hyperconjugation, NMR, X-ray structure

Summary: The synthesis and characterization of a stable β-bis-silyl-substituted vinyl cation is reported. Its spectral properties and its experimental molecular structure clearly reveal the consequences of Si–C hyperconjugation.

Introduction

Vinyl cations [1], the dicoordinated unsaturated analogs of trivalent carbenium ions, were first detected by Grob and coworkers in the early 1960s in solvolysis reactions of α-aryl vinyl halides [2]. The direct NMR detection of vinyl cations in superacidic media was achieved in 1992 at temperatures below −100 °C [3]. We recently reported a convenient synthesis of unusually stable vinyl cations at room temperature [4, 5]. One reason for the unusual high thermodynamic stabilization of these vinyl cations is the presence of two β-silyl substituents. [4]. We report here details of the X-ray structure of the vinyl cation **1** and discuss the structural and spectroscopic consequences of β-SiC hyperconjugation [6].

Results and Discussion

Synthesis and Characterization

Vinyl cation **1** was prepared by reaction of the alkinylsilanes **2** with trityl cation in benzene or toluene as described previously for related cations (see Scheme 1) [4, 5]. The counteranions were the tetrakispentafluorophenylborate, $[B(C_6F_5)_4]^-$, or the hexabromocarborane $[CB_{11}H_6Br_6]^-$, chosen for their low nucleophilicity and extreme inertness [7]. Suitable crystals for X-ray analysis were obtained by recrystallization of [**1**][$CB_{11}H_6Br_6$] from 1,2-dichlorobenzene. Cation **1** is under these conditions indefinitely stable at room temperature in solution and in the solid state.

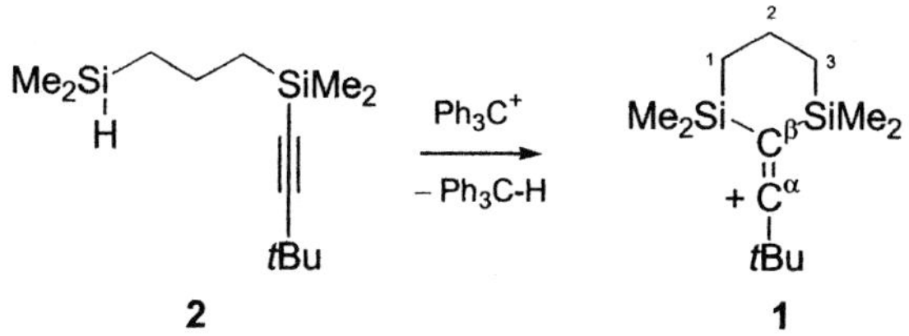

Scheme 1. Synthesis of vinyl cation **1**.

Vinyl cation **1** is characterized by NMR and IR spectroscopy and by X-ray diffraction data. One single ^{29}Si resonance at δ^{29}Si = 29.1 in the ^{29}Si NMR spectrum indicates the formation of a symmetric species on the NMR time scale. The C=C$^+$ group is easily identified in the ^{13}C NMR spectrum by the deshielded resonance of the positively charged C$^{\alpha}$ atom at δ^{13}C = 202.4 and by the high field signal for the sp^2-hybridized C$^{\beta}$ atom (δ^{13}C = 75.3, see Fig. 1).

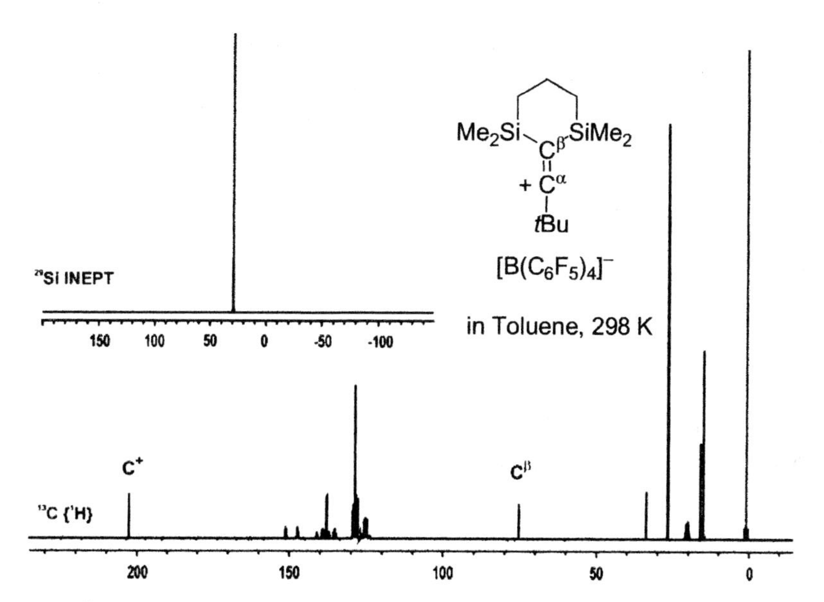

Fig. 1. 49.5 MHz ^{29}Si {^{1}H} INEPT NMR spectrum (upper trace) and 62.9 MHz ^{13}C {^{1}H} NMR spectrum of [**1**][$B(C_6F_5)_4$] in [D_8] toluene at 298K.

Si–C hyperconjugation in Vinyl Cation 1

The occurrence of Si–C hyperconjugation in vinyl cation **1**, as described by the resonance structures shown in Scheme 2, clearly influences important NMR parameters of **1**. That is, the ^{29}Si resonance at δ^{29}Si = 29.1 is markedly low-field shifted compared to the precursor alkinylsilane **2**. This is in agreement with significant delocalization of positive charge from the C$^{\alpha}$ atom to the β-positioned silyl groups. In addition the ^{1}J(SiC) coupling constant between the C$^{\beta}$ atom and the silicon atoms is extremely small (^{1}J(SiC) = 15.7 Hz, compared to regular ^{1}J(SiC) ≈ 60 Hz in trialkylvinylsilanes) (see Fig. 2). This suggests a small bond order between the silicon atoms and the C$^{\beta}$ atom.

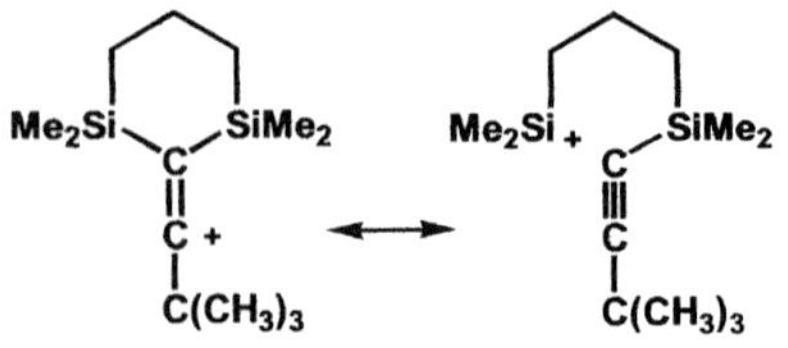

Scheme 2. Canonical resonance structures of cation **1**.

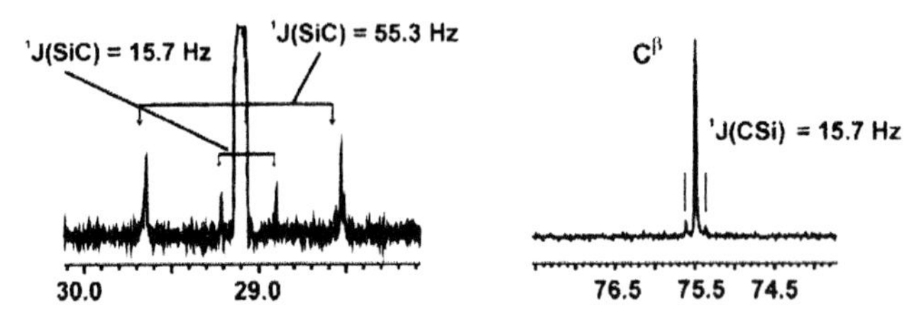

Fig. 2. Parts of the 49.5 MHz ^{29}Si NMR (left) and 62.9 MHz ^{13}C NMR (right) spectra of [**1**][B(C$_6$F$_5$)$_4$] in [D$_6$] benzene at 298K showing the ^{1}J(SiC) coupling.

The most characteristic absorption in the IR spectra of vinyl cation **1** is the band at ν = 1987 cm^{-1} (see Fig. 3), which is assigned to the C=C^{+} stretching vibration. This IR band is, because of the positive charge at C$^{\alpha}$, more intense than regular C=C stretch vibrations of C=C double bonds. Furthermore the position of the IR absorption is strongly shifted to higher energy, which suggest a bond order of the C=C^{+} bond in **1** which is significantly larger than 2.

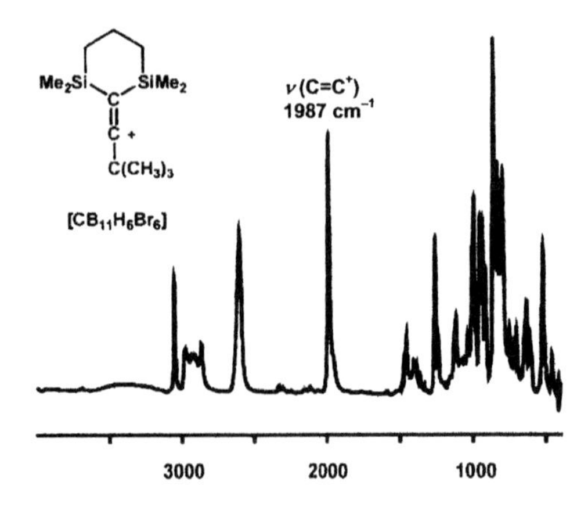

Fig. 3. FT-IR spectra of [**1**][CB$_{11}$H$_6$Br$_6$], KBr.

The molecular structure of **1** shown in Fig. 4 demonstrates that the vinyl cation is linear around the dicoordinated C^{α} atom, which indicates sp hybridization for the C^{α} atom. The C^{β}–C^{α} bond is unusually short (122.1 pm) and approaches the length of a regular C≡C triple bond. A quite remarkable feature of the molecular structure of **1** is the unusual length of the C^{β}–Si bonds (198.4 and 194.6 pm), around 10 pm longer than regular single bonds between sp^2-hybridized carbon and tetracoordinated silicon atoms.

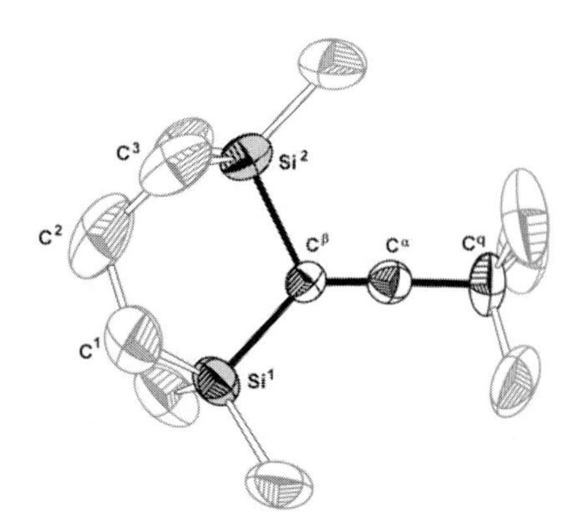

Fig. 4. Molecular structure of vinyl cation **1** (hydrogen atoms are omitted for clarity; thermal ellipsoids drawn at the 30% probability level). Selected bond lengths [pm] and bond angles [°]: C^{β}–C^{α} 122.0(9), C^{α}–C^{q} =145.3(9), Si^2–C^{β} = 198.4(6), Si^1–C^{β} = 194.6(6); C^{β}-C^{α}-C^{q} = 178.8(7), Si^2-C^{β}-C^{α} = 115.5(4), Si^1-C^{β}-C^{α} =133.0(5).

Although the NMR results indicate that **1** is symmetric in solution, in the solid state the two silicon atoms in **1** are clearly different. Quantum mechanical calculations [8] at the B3LYP/6-31G(d) and at the MP2/6-31G(d) level predict very similar structures, which closely match the experimental solid-state structure (see Fig. 5 for MP2 data) [6]. In particular, the unsymmetrical arrangement of the silicon atoms at the central $C{=}C^{+}$ unit is also found in the theoretical structures. This indicates that this arrangement is a result of the intrinsic bonding situation in cation **1** and not a consequence of crystal lattice or intermolecular interactions in the solid state. The C_s symmetrical structure of **1**, which has two identical silicon atoms, is the transition state for the transformation of two identical cations **1** and it is merely 0.7 kcal/mol higher in energy than **1** (at MP2/6-311G(d,p)//B3LYP/6-31G(d)). This suggests a time-averaged symmetry for cation **1** in solution and in the gas phase.

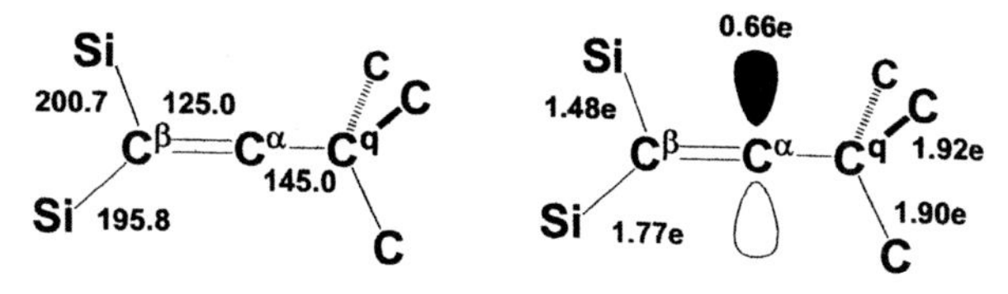

Fig. 5. Left: Part of the calculated structure of vinyl cation **1** (MP2/6-31G(d); bond lengths are given in pm. Right: Calculated orbital occupation according to an NBO analysis of the MP2/6-31G(d) wave function.

An NBO analysis [9] (see Fig. 5) gives additional support for the occurrence of hyperconjugation in vinyl cation **1**. Strongly depleted Si–C^{β} bonds (1.48e and 1.77e) and a formally empty C(2p) orbital at C^{α} with an occupation of 0.66e are the result of electron delocalization from σ-Si–C^{β} orbitals into the vacant C(2p).

Conclusion

Experimental (NMR, IR spectroscopic and structural) as well as theoretical data clearly indicate the occurrence of hyperconjugation between the occupied σ-Si–C^{β} orbitals and the empty 2p(C^{α}) orbital in vinyl cation **1**. This interaction can be described as resonance between the two Lewis structures depicted in Scheme 2.

Acknowledgment: This work was supported by the Goethe Universität Frankfurt and by the German Israeli Foundation for Scientific Research and Development (GIF).

References

[1] For a recent monograph on vinyl cations: *Dicoordinated Carbocations* (Eds.: Z. Rappoport, P. J. Stang), Wiley, **1997**.

[2] a) C. A. Grob, J. Csapilla, G. Cseh, *Helv. Chim. Acta* **1964**, *47*, 1590; b) C. A. Grob in *Dicoordinated Carbocations* (Eds.: Z. Rappoport, P. J. Stang), Wiley, **1997**, chapter 1, p. 1.

[3] H.-U. Siehl, in *Dicoordinated Carbocations* (Eds.: Z. Rappoport, P. J. Stang), Wiley, **1997**, chapter 5, p. 189.

[4] T. Müller, R. Meyer, D. Lennartz, H.-U. Siehl, *Angew. Chem.* **2000**, *112*, 3203; *Angew. Chem. Int. Ed.* **2000**, *39*, 3074.

[5] T. Müller, A model system for the generation of silyl cationic species of different reactivity and stability, in *Organosilicon Chemistry V — From Molecules to Materials* (Eds.: N. Auner, J. Weis), VCH, Weinheim, **2003**, p. 34.

[6] T. Müller, M. Juhasz, C. A. Reed, *Angew. Chem.* **2004**, *116*, 1569; *Angew. Chem. Int. Ed.* **2004**, *43*, 1543.

[7] C. A. Reed, *Acc. Chem. Res.* **1998**, *31*, 133.

[8] All calculations were performed with GAUSSIAN 98 (Revision A.7), Gaussian, Inc. Pittsburgh, PA, **1998** and GAUSSIAN 03 (Revision B.03) Gaussian, Inc. Pittsburgh, PA, **2003**.

[9] a) NBO 4.0. E. D. Glendening, J. K. Badenhoop, A. E. Reed, J. E. Carpenter, F. Weinhold, Theoretical Chemistry Institute, University of Wisconsin, Madison, WI, **1996**; b) A. E. Reed, L. A. Curtiss, F. Weinhold, *Chem. Rev.* **1988**, *88*, 899.

Tertiary Trisilyloxonium Ion and Silylenium Cation in Cationic Ring-Opening Polymerization of Cyclic Siloxanes

Marek Cypryk,* Julian Chojnowski, Jan Kurjata

Center of Molecular and Macromolecular Studies, Polish Academy of Sciences
Sienkiewicza 112, 90-363 Łódź, Poland
Tel.: +48 42 6803227 — Fax: +48 42 6847126
E-mail: mcypryk@bilbo.cbmm.lodz.pl

Keywords: cyclosiloxanes, ring-opening polymerization, silyloxonium ions

Summary: The mechanism of the cationic ring-opening polymerization (ROP) of cyclosiloxanes is discussed based on comparison with the results of polymerization of 1,4-dioxa-octamethyltetrasilacyclohexane ($^2D_2$), initiated with CF_3SO_3H with polymerization of cyclosiloxanes. The common mechanism involving tertiary silyloxonium ions as the active intermediates seems to be consistent with observations. The difference in polymerization kinetics between cyclosiloxanes and $^2D_2$ is interpreted in terms of a higher basicity of the latter, which results in more stable silyloxonium ions, as was supported by *ab initio* calculations. The existence of "free" silylenium ions in the polymerization systems in condensed phases is very unlikely. *Ab initio* calculations of the interactions of silylenium cations with model nucleophiles show that, in the presence of siloxanes, silylenium ions are practically completely converted into silyloxonium ions, unless they are substituted with extremely sterically demanding groups.

Introduction

The structure of the active propagation center in the cationic ring-opening polymerization (ROP) of cyclosiloxanes is still controversial. Trisilyloxonium ions generated from hexamethyl-cyclotrisiloxane, D_3, and octamethylcyclotetrasiloxane, D_4, were observed at low temperature by Olah *et al.* [1] and their participation as intermediates in cationic ROP of cyclosiloxanes was postulated. However, the main objection against this mechanistic concept is a relatively low rate of polymerization when an initiator able to form persistent tertiary oxonium ions is used [2].

Studies of the polymerization of a monomer able to form more stable tertiary trisilyloxonium ions could throw more light on this problem. Octamethyl-1,4-dioxatetrasilacyclohexane ($^2D_2$) may be such a model monomer. This monomer easily undergoes the cationic ROP initiated by a strong protic acid, such as CF_3SO_3H. Polymerization proceeds by exclusive cleavage and reformation of the siloxane bond, analogously to cyclosiloxanes [3], according to Scheme 1. Thus, it is reasonable to assume that the mechanism of polymerization of this monomer is analogous to that of cyclosiloxanes.

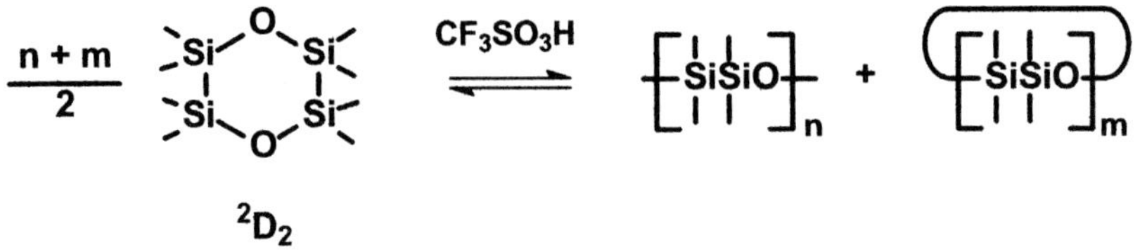

Scheme 1. Polymerization of 1,4-dioxa-octamethyltetrasilacyclohexane ($^2D_2$).

The kinetics and thermodynamics of this process show close similarities to the polymerization of D_4 initiated by CF_3SO_3H [3, 4]. Both processes involve simultaneous formation of cyclic oligomers and polymer and lead to equilibrium, with similar proportions of cyclic-to-linear fractions [3 – 5]. They also show similar thermodynamic parameters and a similar effect of water addition on the initial rate of polymerization. The specific feature of the polymerization of $^2D_2$ is that cyclic oligomers $^2D_3$ and $^2D_4$ are formed simultaneously with the polymer fraction, but they equilibrate with monomer much faster than the polymer fraction. This behavior is best understood assuming formation of the tertiary oxonium ion intermediate, which isomerizes by ring expansion-ring contraction [3]. These kinetic features of the polymerization of $^2D_2$ make this monomer an interesting model for deeper studies on the cyclic trisilyloxonium ion question.

Results and Discussion

Studies of the Basicity of Siloxanes

Despite close similarities of $^2D_2$ and D_4, the ROP of $^2D_2$ initiated by CF_3SO_3H is faster than that of D_4. Higher reactivity of $^2D_2$ may be explained by a higher basicity of this monomer and consequently easier formation of the tertiary silyloxonium ion intermediate. Quantum mechanical DFT calculations of proton affinity (negative enthalpy, PA) and basicity (negative Gibbs free energy, ΔG_{prot}) for the protonation reaction of hydrogen-substituted siloxanes in the gas phase predict that $^2D_2$ will be considerably more basic than the other model siloxanes [6]. The same is true for the corresponding thermodynamic parameters (SiA, ΔG_{sil}) of the reaction of addition of silylenium cation (silylation) (Eq. 1, Table 1). Moreover, the nucleophilicity toward silicon ("silicophilicity") of siloxanes is approximately linearly proportional to their basicity.

We performed calculations for model hydrogen-substituted analogs of the discussed monomers, because calculations for methyl siloxanes would require much more computational time. However, the results of basicity and ring strain calculations for hydrogen siloxanes reproduce reasonably well the trends observed for methyl analogs [7]. Another simplification is that the presented results are related to the isolated molecules in the gas phase. Because of close similarities in structures of the discussed compounds, the solvation effects are expected to be similar also, and should not much affect the order of basicities [7]. The hypothetical, very strained cyclodisiloxane, $D^H{}_2$, was calculated for reference purposes.

$$(\equiv Si)_2O + R^{\oplus} \rightleftharpoons (\equiv Si)_2O^{\oplus}{-}R \qquad R = H, SiH_3$$

Eq. 1.

Table 1. Calculated B3LYP/6-311+G(2d,p)//B3LYP/6-31G* affinities (PA and SiA), basicity (ΔG_{prot}) and silicophilicity (ΔG_{sil}) of model siloxanes in the gas phase.

Siloxane	PA [kcal mol^{-1}]	ΔG_{prot} [kcal mol^{-1}]	SiA [kcal mol^{-1}]	ΔG_{sil} [kcal mol^{-1}]	Ring strain [kcal mol^{-1}]
$H_3SiOSiH_3$	182.3[a]	174.2[a]	58.0	44.5	–
$(H_2SiO)_2$ ($D^H{}_2$)	189.0[a]	182.6[a]	–	–	29.4 – 30.3
$(H_2SiO)_3$ ($D^H{}_3$)	182.4[a]	178.3[a]	58.2	49.5	2.6 – 3.3
$(H_2SiO)_4$ ($D^H{}_4$)	185.7[a]	177.6[a]	60.2	46.7	–0.4 – 0.1
$(H_2SiH_2SiO)_2$ ($^2D^H{}_2$)	190.6[a]	183.6[a]	64.6	55.0	2.0 – 2.4
$(H_3SiH_2Si)_2O$	190.6	183.2	64.4	52.7	–

[a] Values from [6].

The results presented in Table 1 confirm that dioxatetrasilacyclohexane ($^2D^H{}_2$) is significantly more basic and more silicophilic than the other siloxanes, including very reactive cyclotrisiloxane. Its basicity is comparable to that of cyclodisiloxane, $D^H{}_2$, and the open-chain bis(disilanyl)oxane, $(H_3SiH_2Si)_2O$, which models the properties of oxygen atoms in the polyoxydisilylene chain. Thus, the higher basicity of $^2D_2$ is related to the SiSiOSiSi structure and not to the angular deformation of the ring, as is the case of $D^H{}_2$.

The reason for the higher basicity and nucleophilicity toward silicon ("silicophilicity") of $^2D_2$ compared to D_3 and D_4 is a different orbital interaction pattern involving the lone electron pair of oxygen, due to the difference in structures of their ring skeletons. Contribution from the hyperconjugation effect $n_O \rightarrow \sigma^*_{SiO}$ in polysiloxane significantly reduces the electron density on oxygen. This effect is less important in $^2D_2$ because the $n_O \rightarrow \sigma^*_{SiSi}$ interaction is much weaker [8].

The relatively high basicity of $^2D_2$ was confirmed by studies of its behavior as the proton acceptor in hydrogen bond formation with phenol in CCl_4. For example, the shifts of absorption maxima, $\Delta\nu$, from free to complexed $^2D_2$ and D_4 under these conditions are 211 and 159 cm^{-1}, respectively.

Ring Strain in Cyclosiloxanes

Ring strain is the most important factor determining both the thermodynamics and the kinetics of polymerization. Polymerization of strained D_3 is much faster than that of the other siloxanes and leads to a specific kinetic distribution of cyclics [9]. In contrast, polymerizations of unstrained D_4 and $^2D_2$ lead to equilibration. The ring strain in hydrogen-substituted analogs of these monomers was calculated by the DFT method as the negative enthalpy of the isodesmic reactions **A** – **D** (Scheme 2). Reaction **A** involves transformation of a cyclic substrate into open-chain fragments. If all types of bonds are preserved, the enthalpy change should predominantly come from the strain release. Reactions **B** – **D** involve conversions of smaller rings into larger ones and allow estimation of the strain in the substrates, assuming that the cyclic products are unstrained. The results are presented in Table 1. Since the strain was calculated from two different reactions, two values are given for each cyclic; however, the values are reasonably close to each other. The results for D^H_3 and D^H_4 correspond very well to the ring strains measured experimentally for permethylcyclosiloxanes, D_3 (2 – 4 kcal mol^{-1}) and D_4 (0.25 kcal mol^{-1}) [10, 11]. It seems reasonable to assume that the ring strain estimated for $^2D^H_2$ is a good approximation of the strain in $^2D_2$. However, polymerization of this monomer performed in the range 25 – 140 °C showed very little change in equilibrium concentration of $^2D_2$ with temperature, indicating that its ring strain is very small [4].

$$[(H_2Si)_mO]_n + n\ H_3SiOSiH_3 \rightarrow H_3SiO(H_2Si)_mOSiH_3\ (m = 2, n = 2; m = 1, n = 3, 4) \quad \textbf{(A)}$$

$$4\ (H_2SiO)_3 \rightarrow 3\ (H_2SiO)_4 \quad \textbf{(B)}$$

$$2\ (H_2SiO)_2 \rightarrow (H_2SiO)_4 \quad \textbf{(C)}$$

$$3\ [(H_2Si)_2O]_2 \rightarrow 2\ [(H_2Si)_2O]_3 \quad \textbf{(D)}$$

Scheme 2. Isodesmic reactions used for calculations of the ring strain in cyclosiloxanes.

As mentioned above, the higher reactivity of $^2D_2$ compared to D_4 is due to the higher nucleophilicity of the former, which results in higher concentration of the silyloxonium intermediate. However, some enhancement of the reactivity of $^2D_2$ because of a small ring strain cannot be excluded. On the other hand, relatively fast cyclics interconversion in the polymerization of $^2D_2$ results from the high basicity of oxygen atoms in the polyoxydisilylene chain (Table 1).

Synthesis of Tertiary Silyloxonium Ion Derived from $^2D_2$

The hydride transfer from a hydrosilane to the carbenium ion leads to the transient formation of a complex of tricoordinate silicon cation with solvent. Lambert *et al.* showed that if a stable counter-ion of very low nucleophilicity is used, the complex of silylenium ion with solvent is a persistent species [12]. Tetrakis(pentafluorophenyl)borate was successfully explored for this purpose, and this system was later used by Olah e*t al.* [1] for the generation of trisilyloxonium ions derived from D_4

and D_3, which were observed by NMR spectroscopy at low temperatures (–70 °C). Following this concept, we generated the cyclic trisilyloxonium ion derived from $^2D_2$ according to Eqs. 2 and 3.

The resulting trisilyloxonium ion and the ion being the product of addition of monomer (Eq. 4) were identified by ^{29}Si NMR at –45 °C and by quantum mechanical calculations of the chemical shifts [6]. Signals of silicon atoms at the oxonium center appear in the range 50 – 60 ppm, in accord with previous observations for trisilyloxonium ions derived from D_4 and D_3.

$$Ph_3CCl + KB(C_6F_5)_4 \longrightarrow Ph_3C^+ B(C_6F_5)_4^- \xrightarrow[-Ph_3CH]{HSiEt_3} [Et_3Si^+ B(C_6F_5)_4^-]$$

Eq. 2. Generation of the initiator.

$$[Et_3Si^+ B(C_6F_5)_4^-] + {}^2D_2 \longrightarrow Et_3SiO^{\oplus}\text{-ring } B(C_6F_5)_4^-$$

$Et_3Si^+\text{-}^2D_2$

Eq. 3. Formation of the silyloxonium ion.

$$Et_3SiO^{\oplus}\text{-ring } B(C_6F_5)_4^- + {}^2D_2 \longrightarrow Et_3SiOSiSiOSiSiO^{\oplus}\text{-ring } B(C_6F_5)_4^-$$

$\sim SiSi^{\pm}\text{-}^2D_2$

Eq. 4. Chain extension.

Mechanism of Cationic ROP of Cyclosiloxanes

A number of features of CROP of cyclotrisiloxanes point to the mechanism of chain extension involving fast activation-deactivation of the reactive silyloxonium centers [13]:

- The polymerization is a stepwise process; the molecular weight of the polymer and the concentration of the cyclic fraction increase linearly with monomer conversion
- Chain transfer to $Me_3SiOSiMe_3$ (M_2) proceeds with the formation of silyl ester, which initiates a new chain. This pathway competes with continuous initiation by acid
- The polymerization initiated by R_3SiH^+ $Ph_3C^+[B(C_6F_5)_4]^-$ occurs relatively slowly because $B(C_6F_5)_4^-$ is a very weak nucleophile and a poor deactivator of the trisilyloxonium ion
- A fast ester end group exchange occurs, according to ^{19}F NMR [3].

This mechanism is consistent with the observed features of polymerization of $^2D_2$. The kinetic differences between the polymerizations of D_3, D_4 and $^2D_2$ can be explained by differences in basicity and ring strain of these monomers, as discussed above.

May Silylenium Ion be the Active Propagation Center?

Tertiary silyloxonium cation may be considered as the complex of the silylenium ion with siloxane. Thus, there is a question about the position of the equilibrium, in which the "free" silylenium ion would exist in much lower concentration than its complex, but, being much more reactive, could significantly contribute to propagation.

Silylenium ions are more thermodynamically stable than carbenium ions, but are also much more reactive. Larger size of silicon atoms and longer bonds to substituents make them more accessible to nucleophilic attack than carbon. Thus, reactions at silicon are generally less sensitive to steric hindrance. Silicon is also more electropositive than carbon, and the positive charge in $\equiv Si^+$ is highly localized on silicon, while in $\equiv C^+$ it is largely dispersed over the substituents. This results in stronger electrostatic attraction forces between silylenium cation and electron-rich species [14].

Recent *ab initio* calculations show that silylenium ions such as Me_3Si^+ and Et_3Si^+ are strongly complexed even in solvents of such low nucleophilicity as CH_2Cl_2 [15]. In the presence of bases even as weak as siloxanes, they are practically completely converted into silyloxonium ions (Table 2).

Table 2. B3LYP/6-311+G(2d,p)//B3LYP/6-31G* enthalpies, ΔH^{298}_{compl}, (BSSE corrected) of complex formation between silylenium ions and selected nucleophiles.

	r(Si-Nu) [Å]	Wiberg bond order	ΔH^{298}_{compl} [kcal mol^{-1}]	ΔH^{298}_{exp} [a] [kcal mol^{-1}]
H_3Si^+-$O(SiH_3)_2$	1.868	0.351	–58.1	–
H_3Si^+-$(OSiH_2)_3$	1.791	–	–58.0	–
H_3Si^+-$(OSiH_2)_4$	1.794	–	–60.2	–
H_3Si^+-$(OSiH_2SiH_2)_2$	1.784	0.407	–64.1	–
Me_3Si^+-$(OSiMe_2SiMe_2)_2$ [b]	1.825	–	–43.0	–
Et_3Si^+-$(OSiMe_2SiMe_2)_2$ [b]	1.847	–	–35.2	–
Me_3Si^+-C_6H_6 [c]	2.229	0.335	–18.7	–23.9
Me_3Si^+-$NCCH_3$ [c]	1.893	0.426	–45.2	–
Me_3Si^+-OH_2 [c]	1.93	0.308	–30.0	–31
Me_3Si^+-NH_3 [c]	1.963	0.435	–46.3	–45 to –50
Me_3Si^+-CH_2Cl_2 [c]	2.407	0.422	–15.4	–
Et_3Si^+-OH_2 [c]	1.945	0.302	–24.7	–
Et_3Si^+-NH_3 [c]	1.964	0.427	–42.3	–

[a] Data taken from [16]; [b] B3LYP/6-31G* enthalpies, not corrected for BSSE; [c] Values from [15].

Conclusions

- $^2D_2$, because of its higher basicity and nucleophilicity, forms more stable oxonium ions than permethylcyclosiloxanes, such as D_4.
- In the system $^2D_2$ + Et_3SiH + Ph_3C^+ $B(C_6F_5)_4^-$ at low temperatures, not only is the generation of Et_3Si^+-$^2D_2$ observed but also its transformation to the corresponding tertiary silyloxonium ion at the chain end, i.e., $Et_3SiO(SiMe_2)_2(SiMe_2)_2^+$-$^2D_2$, may be followed. This species is relatively stable at 0 °C.
- Cationic polymerization of $^2D_2$ at room temperature initiated by Et_3SiH + Ph_3C^+ $B(C_6F_5)_4^-$ proceeds fast and chemoselectively with exclusive siloxane bond cleavage and reformation.
- Taking into account the similarity of cationic polymerization of $^2D_2$ to that of D_4, this reaction is a good model for study of the role of cyclic trisilyloxonium ions in the cationic polymerization of cyclosiloxanes.
- The mechanism of polymerization of cyclic siloxanes is consistent with the observed features of polymerization of $^2D_2$. The kinetic differences between the polymerizations of D_3, D_4 and $^2D_2$ can be explained by differences in basicity and ring strain of these monomers.
- Silylenium ions are unlikely to be the active centers of propagation, because, in the presence of bases even as weak as siloxanes, they are practically completely converted into silyloxonium ions.

Acknowledgments: The work was supported by the State Committee for Scientific Research (KBN), Grant No PBZ-KBN/01/CD/2000. Authors are grateful to Drs. D. Cardinaud and C. Priou from Rhodia Silicones for a kind gift of $K^+B(C_6F_5)_4^-$.

References

[1] G. A. Olah, X.-Y. Li, Q. Wang, G. Rasul, G. K. S. Prakash, *J. Am. Chem. Soc.* **1995**, *117*, 8962.
[2] G. Toskas, M. Moreau, M. Masure, P. Sigwalt, *Macromolecules* **2001**, *34*, 4730.
[3] J. Chojnowski, J. Kurjata, *Macromolecules* **1994**, *27*, 2302.
[4] J. Kurjata, J. Chojnowski, *Makromol. Chem.* **1993**, *194*, 3271.
[5] P. V. Wright, M. S. Beevers, in: *Cyclic Polymers*, (Ed.: J. A. Semlyen), Elsevier Applied Science Publ., London, **1986**, p. 85.
[6] M. Cypryk, J. Kurjata, J. Chojnowski, *J. Organomet. Chem.* **2003**, *686*, 373.
[7] M. Cypryk, Y. Apeloig, *Organometallics* **1997**, *16*, 5938.
[8] M. Cypryk, *Macromol. Theory Simul.* **2001**, *10*, 158.
[9] J. Chojnowski, M. Ścibiorek, J. Kowalski, *Makromol. Chem.* **1977**, *178*, 1351.
[10] W. A. Piccoli, G. G. Haberland, R. L. Merker, *J. Am. Chem. Soc.* **1960**, *82*, 1883.
[11] Y. A. Yuzhelevsky, V. V. Sokolov, L. V. Tagyeva, E. G. Kagan, *Vysokomol. Soedin. Ser. B*

1971, *2*, 95.
[12] J. B. Lambert, S. Zhang, C. L. Stern, J. C. Huffman, *Science* **1993**, *260*, 1917.
[13] J. Chojnowski, M. Cypryk, K. Kaźmierski, *Macromolecules* **2002**, *35*, 9904.
[14] C. Maerker, P.v.R. Schleyer, in: *The Chemistry of Organic Silicon Compounds*, (Eds.: Z. Rappoport, Y. Apeloig), Wiley, Chichester, **1998**, *Chapt. 10*, p. 510.
[15] M. Cypryk, *J. Organomet. Chem.* **2003**, *686*, 164.
[16] H. Basch, T. Hoz, S. Hoz, *J. Phys. Chem. A* **1999**, *103*, 6458.

Chapter II

Molecular Inorganic Silicon Chemistry

Dihalodimethylsilanes from Silicon Atoms and Methyl Halides

Günther Maier,* Jörg Glatthaar, Hans Peter Reisenauer

Institut für Organische Chemie der Justus-Liebig-Universität,
Heinrich-Buff-Ring 58, D-35392 Giessen, Germany
Tel.: +49 641 99 34301 — Fax: +49 641 99 34309
E-mail: Guenther.Maier@org.Chemie.uni-giessen.de

Keywords: matrix isolation, density functional calculations, reaction mechanisms

Summary: The reaction of silicon atoms with methyl halides **6a–d** has been studied in an argon matrix at 10 K. In the initial step, triplet n-adducts **T-5a–d** are formed. The next step can be induced photochemically. The primary photoproducts are the singlet halomethylsilylenes **S-1a–c**. In the case of methyl iodide (**6d**), the reaction with silicon atoms leads spontaneously to silylene **S-1d**. In a diluted argon matrix all silylenes **S-1a–d** can further be photoisomerized to the corresponding halosilenes **2a–d**. In the presence of an excess of a methyl halide, complexes of type **7a–d** are formed. Longer irradiation transforms these adducts into the dihalodimethylsilanes **8a–d**. The generation of the silanes **8a–d** starting from silicon and a methyl halide images the results of the Rochow-Müller synthesis. The relevance of our findings to this important technical process is discussed.

Introduction

During the past eight years, we have studied the reactions of thermally generated silicon atoms with a variety of low-molecular-weight reactants in an argon matrix. The reaction products were identified by means of IR and UV/Vis spectroscopy aided by comparison with calculated spectra. The method turned out to be very versatile and successful [1]. The selected substrates were mainly molecules with isolated, conjugated or aromatic π bonds, and compounds containing π bonds and at the same time free electron pairs.

Of special interest are molecules possessing n electrons only in combination with σ bonds. The reactions of silicon atoms with target molecules of this type are relevant for the understanding of the Rochow-Müller (R.-M.) synthesis [2]. In a recent essay Seyferth describes the enormous importance of this "direct synthesis" of dichlorodimethylsilane by reaction of a silicon/copper alloy with methyl chloride [3]. At the same time he points out that even today, more than 60 years after its discovery in 1940, the mechanism of this process is still not fully understood. It was our hope that investigation of the behavior of Si atoms in a matrix may perhaps throw some light upon the fundamental reactions occurring on the surface of the Si/Cu contact mass used in the R.-M. synthesis.

In this presentation, only the results of the reaction of silicon atoms with methyl chloride as the substrate molecule are discussed explicitly. In our full paper [1b] we show that matrix isolation techniques can be applied to uncover not only the *intramolecular* transformations of the triplet n-adducts **T-5a–d**, which are formed in the reaction of a Si atom with a single molecule of methyl halide, to the corresponding singlet halomethylsilylene **S-1a–d**. If the concentration of the target molecule is raised, also *intermolecular* processes, like the addition of a second methyl halide molecule to the silylene intermediate under formation of dihalodimethylsilanes **8a–d** via adducts **7a–d**, can be revealed.

Matrix Experiments

In the standard experiments, a gaseous mixture of a methyl halide **6a–d** and argon was deposited together with silicon atoms, generated by resistive heating of a silicon rod to a temperature of ca. 1380 °C, onto a CsI window at 10 K. FT-IR and UV/Vis spectra of the matrices were taken. Subsequent photochemical transformations were initiated by irradiating the matrices with light of the appropriate wavelength.

If silicon atoms are cocondensed at 10 K with a mixture **6b**:Ar = 1:250, the primary product is n-adduct **T-5b**. Already at 10 K a thermal transformation of **T-5b** into **S-1b** takes place. This thermal instability makes it difficult to identify **T-5b** spectroscopically. The C–Cl stretching vibration of **T-5b** is found at 679.1 cm^{-1} and corresponds to the calculated value at 667.3 cm^{-1}. A second band with less intensity (1/3; ^{37}Cl isotopomer) is found at 673.9 cm^{-1}. The UV absorption which can be attributed to **T-5b** lies at 260 nm. In addition, **T-5b** must possess an absorption in the long wavelength region at around 580 nm, since n-adduct **T-5b** disappears slowly upon irradiation with 578 nm light. But according to the IR spectrum the main component of the directly formed reaction product is chloromethylsilylene **S-1b**, which is characterized by strong IR bands at 484.4 and 1219.8 cm^{-1}. In the UV region **S-1b** absorbs at 405 nm. Therefore, with light of wavelengths >385 nm, **S-1b** can be transformed into chlorosilene **2b,** as shown by its IR spectrum and its UV absorption at 250 nm. The back reaction needs 254 nm light. The spectroscopic data for **S-1b** and **2b** confirm our results of an earlier study [4].

If the concentration of methyl chloride is raised (**6b:**Ar = 1:10), no IR bands of n-adduct **T-5b** or silylene **S-1b** can be detected. But **T-5b** has to be present in the matrix since in the UV spectrum the band at 260 (strong) is observable, together with a second absorption at 330 nm (weak), which we also attribute to n-adduct **T-5b**. Upon irradiation with 330 nm the band at 260 nm disappears. At the same time a new absorption, again at 330 nm, can be detected. If one compares with the IR results (see below), it has to be concluded that this band belongs to adduct **7b**, which is formed from the primarily generated uncomplexed silylene **S-1b**. Upon irradiation of the matrix with 330 nm light the IR bands of dichlorodimethylsilane **8b** appear. Obviously, during irradiation a sufficient amount of **T-5b** is isomerized to **S-1b**. In the concentrated matrix the silylene exists mainly as the complex **7b**, which in a subsequent photoreaction generates silane **8b**.

$H_2C{=}SiHX$ (**2a–d**) ⇌ (Ar, hν, 10 K) $H_3C{-}\dot{\ddot{Si}}{-}X$ (**S-1a–d**)

$\cdot\ddot{Si}\cdot$ + $H_3C{-}X$ (**6a–d**) → (Ar, 10 K) $H_3C{-}X{\rightarrow}Si$ (**T-5a–d**)

T-5a–d → **S-1a–d**: low conc. $H_3C{-}X$; **T-5a**: hν; **T-5b,c**: hν or Δ; **T-5d**: spontaneous

T-5a–d → **7a–d** (via **S-1a–d**): high conc. $H_3C{-}X$; **T-5a**: hν; **T-5b,c**: hν or Δ; **T-5d**: spontaneous

7a–d ($H_3C(X)\ddot{Si}{\leftarrow}X{-}CH_3$) → (Ar, hν, 10 K) $(H_3C)_2SiX_2$ (**8a–d**)

a: X = F; **b**: X = Cl; **c**: X = Br; **d**: X = I

Calculations

To get an overview of the relevant potential energy surfaces, several stationary points of each system were calculated with the 6-311+G** basis set and the B3LYP functional using the Gaussian package of programs [5]. For the structural identification of the expected species it was also necessary to obtain the calculated vibrational spectra.

The global minima on the CH_3SiX potential energy surfaces are the silylenes **S-1a–d**. Like all silylenes they have a singlet ground state. The excited triplet states **T-1a–d** lie considerably higher (35 – 45 kcal mol^{-1}) in energy. Between the silylenes of different multiplicities one finds the corresponding silenes **2a–d.** As the next series of isomers we expected the n-adducts **T-5a–d**, the primary products in the reactions of silicon atoms with the methyl halides **6a–d**. To our surprise the sum of the energies of the methyl radical **3** and a radical SiX **4a–d** is lower than those of the n-adducts **T-5a–d**. That means, a silicon atom can pull away a halogen atom from the methyl halides **6a–d** in a rather exothermic (5 – 25 kcal mol^{-1}) reaction. The next higher candidates are the n-adducts **T-5a–d**. The first three members **T-5a–c** are stabilized compared to the starting materials ^{3}Si atoms and methyl halides **6a–c**, which are the zero levels of our energy scale.

Another finding is that the n-adducts **T-5a–d** can use two different pathways for the insertion. For instance (Scheme 1), starting from **T-5b** the concerted mode leads in the first step under spin conservation to triplet silylene **T-1b**. Intersystem crossing then yields **S-1b**. The alternative reaction path proceeds via the breakage of the methyl-halogen bond and is (this statement is valid for all systems) preferred compared with the synchronous process. In case of n-adduct **T-5b** the splitting into a methyl radical and a radical SiCl is exothermic ($\Delta H_r = -12.8$ kcal mol^{-1}). The barrier amounts only to 4.3 kcal mol^{-1}. Recombination of the two fragments can lead to triplet silylene **T-1b** ($\Delta H_r = -28.4$ kcal mol^{-1}), or in a more exothermic reaction ($\Delta H_r = -65.9$ kcal mol^{-1}) directly to **S-1b**. The concerted pathway would afford an activation energy of 19.5 kcal mol^{-1}.

In order to get some information about the fate of silylenes **S-1a–d** in the presence of a second molecule of methyl halide **6a–d**, we checked whether silylenes **S-1a–d** can experience stabilization by donor/acceptor interaction with **6a–d**. Indeed, the two partners form weak donor/acceptor complexes **7a–d**. That means that if **S-1a–d** is created in a matrix at very low temperature in the presence of an excess of a methyl halide **6a–d**, complexes like **7a–d** should be present. Such a solvation might be the prerequisite for the direct synthesis of silanes **8a–d**. These compounds represent the global minima on the $C_2H_6SiX_2$ potential energy surfaces and are expected to be generated from the two educt molecules **S-1a–d** and methyl halide **6a–d** in rather exothermic ($\Delta H_r = 65 - 95$ kcal mol^{-1}) processes. In the chlorine series (Scheme 1) the value is 73.9 kcal mol^{-1}.

The existence of two alternative pathways for the first insertion (transformation of the n-adducts **T-5a–d** into the singlet silylenes **S-1a–d**) is in principle also possible for the second insertion into the C–X bond of the respective methyl halide (formation of the silanes **8a–d** from the complexes **7a–d**). Calculations showed that the diradical-type decay generating a methyl radical **3** and the radical CH_3SiX_2 has not to be considered. The activation energies for the concerted second insertion reaction **7a–d** → **8a–d** are rather high (**7a**: 25.8, **7b**: 33.0, **7c**: 31.7, **7d**: 33.1 kcal mol^{-1}).

Relevance to the Mechanism of the Rochow-Müller Synthesis

The first investigation of the mechanism of the R.-M. synthesis was already carried out as early as 1945 by Hurd and Rochow [6]. They discussed the possibility that a volatile and unstable copper methyl or even free methyl radicals may play a decisive role. As a key step the methyl radicals were believed to react with SiCl centers on the surface of the catalyst.

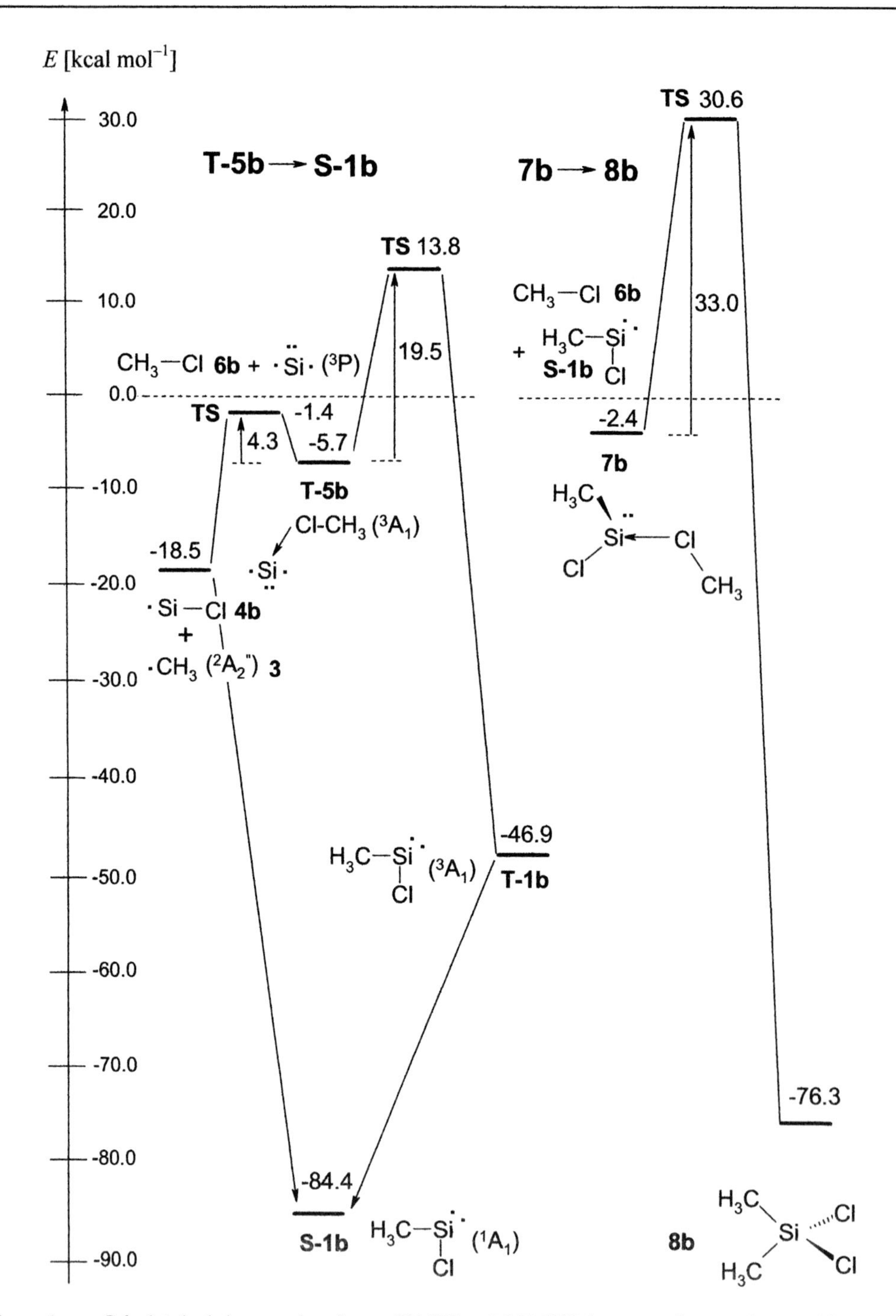

Scheme 1. Calculated relative energies of some CH_3SiCl and $C_2H_6SiCl_2$ isomers and connecting transition states.

Later on the "radical mechanism" was questioned [7]. A new hypothesis was developed according to which the direct synthesis of dihalodimethylsilane should be treated as a heterogeneous catalytic process in which the chemisorption of the methyl halide on the surface of the contact mass is of the utmost importance.

One conclusion can be drawn from the many studies [2c, 7] devoted to obtaining an understanding of the mechanism of the R.-M. synthesis: the process proceeds by way of surface-confined chloromethylsilylene. Being confronted with such a complex system one should keep in mind that the formation of dichlorodimethylsilane from silicon and methyl chloride does not necessarily need the presence of copper. For instance, it was reported as early as 1966 that *pyrophoric* silicon, when suspended in paraffin oil, reacts at 200 °C in the absence of any catalyst with methyl chloride with generation of chloromethylsilanes [8].

Is there anything we can learn in regard to the R.-M. synthesis from the results of our matrix studies? Surely, it is daring to compare the reactions of silicon atoms with those on the surface of the contact mass. There are obvious analogies. The physisorption state on the surface [9] corresponds to n-adduct **T-5b**, while the more stable chemisorption state [9] is comparable with radical pair **4b** + **3**. The dissociative chemisorption on the surface probably occurs by a nucleophilic attack of methyl chloride (**6b**) on a silicon atom positioned at the surface of the contact mass. It is also possible that the next step, the formation of the surface-bound silylene, follows a pathway similar to the sequence **T-5b** → methyl radical **3** + **4b** → **S-1b**. The rate-determining step should be the fragmentation of the surface-bound methyl chloride (**6b**) into a methyl radical (**3**) and surface-embedded SiCl. That means that the mechanism of the R.-M. synthesis has to be looked at as a combination of the "radical mechanism" in the sense of Rochow and the "chemisorption mechanism". Last but not least, the addition of the second methyl chloride may be comparable to the reaction path **S-1b** → **7b** → **8b**.

Conclusion

A direct synthesis of dihalodimethylsilanes **8a–d** can be achieved by the reaction of silicon atoms with solid methyl halides **6a–d** under concurrent irradiation with UV light (254 nm). Studies in argon matrices at 10 K uncovered all the details of these processes. The basic reactions are now well understood. The crucial step is the radical decay of the primarily formed n-adducts **T-5a–d** into a methyl radical and the corresponding Si–X radical. Hopefully, these studies can contribute to a better understanding of the decisive steps in the Rochow-Müller synthesis.

Acknowledgments: Support by Wacker-Chemie GmbH, Burghausen, the Deutsche Forschungsgemeinschaft, and the Fonds der Chemischen Industrie is gratefully acknowledged.

References

[1] a) G. Maier, H. P. Reisenauer, H. Egenolf, J. Glatthaar, L. Rösch, in *Silicon for the Chemical Industry VI* (Eds.: H. A. Øye, H. M. Rong, L. Nygaard, G. Schüssler, J. Kr. Tuset), Trondheim, Norway, **2002,** p. 285 – 297; b) G. Maier, J. Glatthaar, H. P. Reisenauer, *J. Organomet. Chem.* **2003,** *686*, 341.

[2] Summaries: a) R. J. H. Voorhoeve, *Organohalosilanes, Precursors to Silicones*, Elsevier, Amsterdam, **1967**; b) *Catalyzed Direct Reactions of Silicon* (Eds.: K. M. Lewis, D. G. Rethwisch), Elsevier, Amsterdam **1993**; c) L. N. Lewis, in *The Chemistry of Organic Silicon Compounds, Vol. 2* (Eds.: Z. Rappoport, Y. Apeloig), Wiley, New York, **1998**, Chap. 26, p. 1581 – 1597.

[3] D. Seyferth, *Organometallics* **2001**, *20*, 4978.

[4] G. Maier, G. Mihm, H. P. Reisenauer, D. Littmann, *Chem. Ber.* **1984**, *117*, 2369.

[5] Gaussian 98, Revision A.7; GAUSSIAN, Inc., Pittsburgh PA, **1998**.

[6] D. T. Hurd, E. G. Rochow, *J. Am. Chem. Soc.* **1945**, *67*, 1057.

[7] a) M. P. Clarke, *J. Organomet. Chem.* **1989**, *376*, 165; b) B. Pachaly, in *Handbook of Heterogeneous Catalysis, Vol. 4*, VCH, Weinheim, **1997**, p. 1786 – 1795.

[8] E. Bonitz, *Angew. Chem.* **1966**, *78*, 475; *Angew. Chem. Int. Ed. Engl.* **1966**, *5*, 462.

[9] J. Y. Lee, S. Kim, *Surf. Sci.* **2001**, *482 – 485*, 196.

Reactions of Silicon Atoms with Amines and Phosphine under Matrix Isolation Conditions

Günther Maier,* Jörg Glatthaar, Hans Peter Reisenauer

Justus-Liebig-University Giessen, Institute for Organic Chemistry
Heinrich-Buff-Ring 58, D-35392 Giessen, Germany
Tel.: +49 641 99 34301 — Fax: +49 641 99 34309
E-mail: Guenther.Maier@org.chemie.uni-giessen.de

Keywords: matrix isolation, silicon atom, photochemistry

Summary: Silicon atoms in solid argon react with amines (NR_3, R = H, CH_3) and phosphine (PH_3) to form donor-acceptor complexes, which are stable under matrix conditions. Upon subsequent photochemical excitation, rearrangements take place leading to the corresponding aminosilylenes or to phosphinosilylene (formal insertion products). On further irradiation, additional isomerizations and elimination reactions occur. In matrices containing high concentrations of amines the final products are diaminosilanes.

Introduction

Continuing our investigations [1] on the reactions of silicon atoms with industrially employed and prospective substrates for "direct synthesis", we explored the reaction of atomic silicon with amines and phosphine under matrix conditions.

Silicon Atoms and Ammonia

Co-condensation of silicon atoms, generated from a resistively heated, doped silicon rod (1350 – 1380 °C) with very diluted ammonia/argon gas mixtures (1:1000) at 10 K leads to the formation of a primary Si,NH_3 *n*-adduct **1**, which is stable under these conditions[2]. This compound can be identified by three prominent IR absorptions ($^{14}NH_3$: 3387.3, 1599.1, and 1185.6 cm^{-1}). The silicon–nitrogen stretching band is too low in intensity to be observed. These bands show the expected shifts when the co-deposition is repeated with isotopically labeled ammonia ($^{14}ND_3$: 2525.4, 1171.5, and 912.5 cm^{-1}; $^{15}NH_3$: 3378.8, 1596.1, and 1180.0 cm^{-1}).

On irradiation with 436 nm light, **1** is transformed into the isomeric aminosilylene **2** ($^{14}NH_3$: 3394.2, 1927.1, 1557.8, 938.6, 853.9, and 562.7 cm^{-1}; $^{15}NH_3$: 1926.8, 1552.7, 934.5, 838.3, and 559.3 cm^{-1}). The IR absorptions of **2** are strongly shifted to lower wavenumbers compared to the IR spectrum which we have obtained in photolysis studies of H_3SiN_3 (3408.7, 1975.8, 1582.6, 950.6, 866.4, and 570.4 cm^{-1}) [3]. These differences might be explained by the fact that on the one hand in the azidosilane photolysis the released nitrogen molecule is placed close to the matrix cage of the aminosilylene **2**; on the other hand, ammonia is a strong dipole, which also can cause band shifts. Irradiation of **2** with 313 nm leads to the formation of silaisonitrile HNSi **3** under loss of hydrogen. The IR absorptions of **3** measured in the present study again differ slightly ($^{14}NH_3$: 3570.7, 1196.7, and 514.8/510.0 cm^{-1}; $^{15}NH_3$:3561.3, 1173.1, and 512.3/507.4 cm^{-1}) from the results of the azidosilane photolysis ($^{14}NH_3$: 3585.6, 1202.9, 523.1 cm^{-1}) [3]. Since the latter values closely match the gas phase values [4] ($^{14}NH_3$: 3588.4 cm^{-1}) and since in the atomic silicon/ammonia experiment the NH stretching band of **3** is even split into two absorptions, the shift might be mainly due to some interaction with residual ammonia molecules. The co-condensation experiments with higher concentrations of ammonia show a very difficult IR pattern due to additional bands of higher aggregated ammonia molecules and additional, reactive silicon-containing species.

$$:\dot{\underset{\cdot}{Si}} \xrightarrow[\text{Ar, 10 K}]{+\,NH_3} :\dot{\underset{\cdot}{Si}}\leftarrow NH_3\ (\mathbf{1}) \xrightarrow{h\nu} HSi\text{—}NH_2\ (\mathbf{2}) \underset{h\nu}{\overset{h\nu}{\rightleftharpoons}} :Si{=}N\text{—}H + H_2\ (\mathbf{3})$$

Scheme 1. Reaction of atomic silicon with ammonia.

Silicon Atoms and Methylamine

Silicon atoms were co-deposited with a methylamine/argon gas mixture (1:1000) at 10 K. Again, in the first step a stable n-adduct **4** of a silicon atom and methylamine is formed (3358.9, 3305.1, 1590.7, 1461.3, 999.2, and 992.0 cm^{-1}). Upon irradiation with 490 nm light, the IR absorptions of **4** vanish completely, and, besides some unidentified bands (probably $H_2Si{=}N{-}CH_3$), new IR absorptions (3409.1, 3389.8, 1937.8, 1917.8, and 1912.3 cm^{-1}) are observed. They can be assigned to the *trans*-(*N*-methylamino)silylene **5-t**. There is no indication of the formation of the isomeric aminomethylsilylene **6** (well-known product of the $CH_3SiH_2N_3$ photolysis). Obviously, the migration of a hydrogen atom dominates the photochemistry of *n*-adduct **4**. Short-time irradiation with λ = 313 nm leads to the formation of the isomeric *cis*-(*N*-methylamino)silylene **5-c** (1967.4, 1960.3, 1951.8, and 1945.3 cm^{-1}). Upon longer irradiation the IR absorptions of **5-t** diminish faster than those of **5-c**, implying a photoequilibrium between these isomers. At the same time, a strong absorption at 2097.6 cm^{-1} appears. Together with the other, less prominent IR absorptions, this new compound can easily be identified as silylisocyanide H_3SiNC (**7**), which is a well-characterized compound. In more highly concentrated matrices, **7** is no longer the dominating final photoproduct. In our laboratory this species was obtained earlier when either H_3SiCN or $CH_3Si(N_3)_3$ was photolyzed with 254 nm light [5]. Isocyanide **7** was also observed by Sander during a 193 nm photolysis of $(CH_3)_2SiHN_3$ [6].

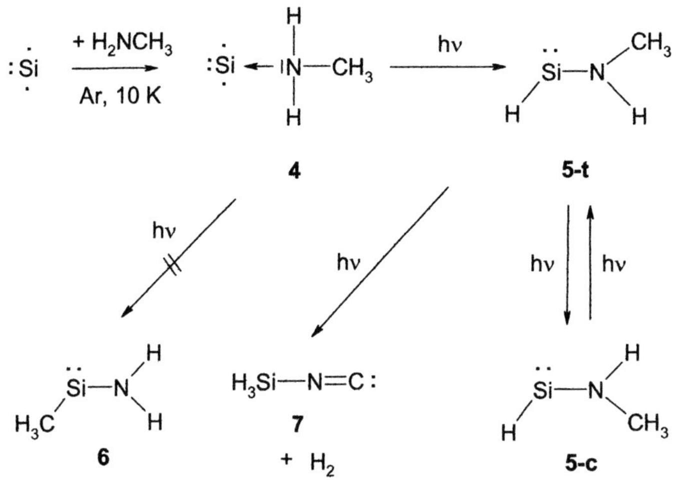

Scheme 2. Reaction of atomic silicon with methylamine.

Silicon Atoms and Dimethylamine

The co-condensation of silicon atoms and dimethylamine/argon gas mixtures (1 : 250) resulted in the formation of the stable *n*-adduct **8** (1117.8, 1014.0, 970.1, and 884.5 cm^{-1}). Upon irradiation with 436 nm light, all IR absorptions decreased and a set of new IR bands (1914.0, 1282.5, 1174.5, and 863 cm^{-1}) slowly increased. They can be assigned to the (*N*,*N*-dimethylamino)silylene **9**. Short wavelength photolysis (λ = 334 nm) leads to the disappearance of the IR absorptions of **9**. Besides methane, the IR bands of silylisocyanide **7** could be observed.

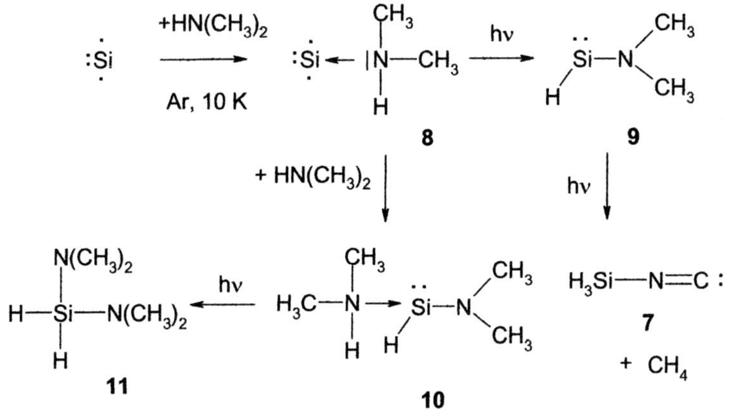

Scheme 3. Reaction of silicon atoms with dimethylamine.

The photochemical behavior is changed if one increases the ratio of dimethylamine in the argon gas mixtures. At ratios > 1:50 the IR absorptions of **7** were very weak; in pure dimethylamine matrices they could not be detected at all. A new reaction channel is opened up under these conditions.

Similarly to our findings on the reaction of silicon with dimethyl ether [1b], the (dimethylamino)silylene **9** reacts with additional dimethylamine under formation of a stable complex **10** (1928.0, 990.0, and 774.0 cm^{-1}). This complex **10** is not unspecifically destroyed under photolysis (254 or 313 nm), but bis(*N*,*N*-dimethylamino)silane **11** is exclusively formed.

Silicon Atoms and Trimethylamine

The reaction of silicon atoms with trimethylamine is similar to its behavior with the other amines. Co-condensation of silicon atoms and trimethylamine/argon gas mixtures (1 : 100) gives the stable *n*-adduct **12** (1401.1, 1248.7, 994.2, 807.6, and 487.5 cm^{-1}). In pure trimethylamine *n*-adduct **12** can be annealed up to 80 K before degradation begins.

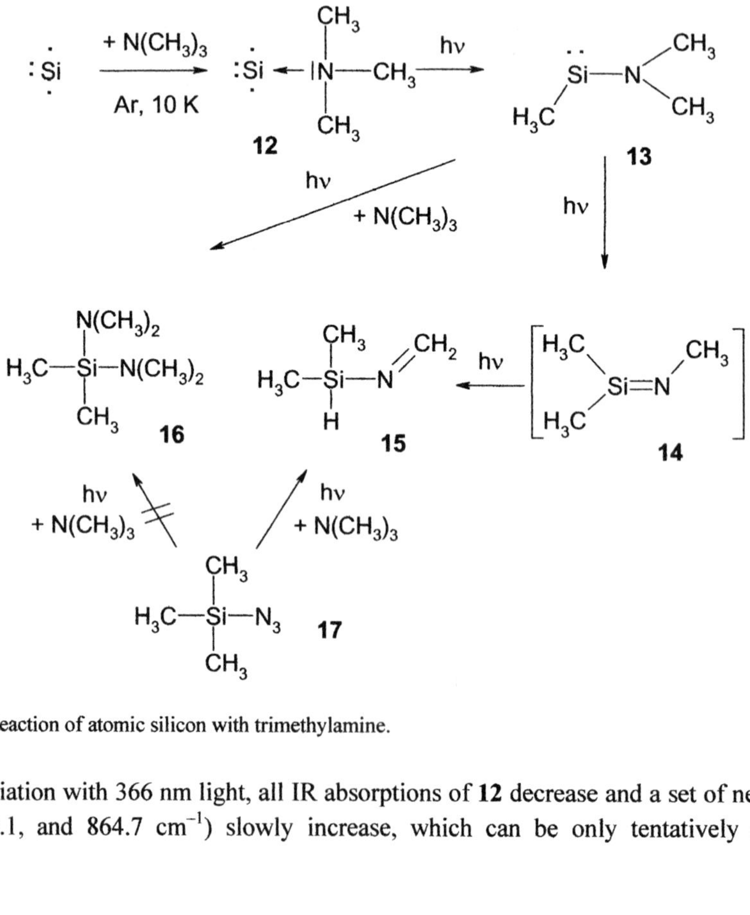

Scheme 4. Reaction of atomic silicon with trimethylamine.

Under irradiation with 366 nm light, all IR absorptions of **12** decrease and a set of new IR bands (1409.4, 1260.1, and 864.7 cm^{-1}) slowly increase, which can be only tentatively assigned to

(*N,N*-dimethylamino)methylsilylene (**13**). Photolysis with shorter-wavelength light (λ = 254, 313 nm) leads to the total disappearance of these IR absorptions. The new photoproduct (2155.4, 1169.5, and 916.9 cm^{-1}) is probably (*N*-dimethylsilyl)methanimine (**15**). The formation of **15** can be rationalized by the following two-step process: silylene **13** rearranges in the first step to silanimine **14** via a 1,2 migration of a methyl group. A subsequent, also photochemically induced, 1,3 migration of a hydrogen atom leads to the formation of **15**.

In pure trimethylamine, the strongly shifted IR absorptions of *n*-adduct **12** also disappeared upon irradiation with 254 nm light. By comparison with the IR spectra of an authentic sample it could be shown that under these conditions the only photoproduct was bis(*N,N*-dimethylamino)-dimethylsilane (**16**). The absence of the rearrangement product **15** could be demonstrated by comparison with the spectra of an authentic sample of **15** which was generated by photolysis of trimethylsilylazide **17** in pure NMe_3.

Silicon Atoms and Phosphine

In connection with our investigations on the reactivity of silicon atoms with ammonia (and amines), we were also interested in the system $Si+PH_3$, and this for several reasons. The parent H_3PSi-isomers are all unknown. From the results of density functional calculations (B3LYP/6-311G** [7]), one has to expect that silicon atoms will react with phosphine under matrix conditions under formation of a stable *n*-adduct **18** ($\Delta H_r = -19.0$ kcal mol^{-1}). $H_3P{\rightarrow}Si$ (**18**) has a triplet ground state, and of the four H_3PSi-isomers **18** – **21** it is the one highest in energy. The other three isomers, silylphosphinidene H_3SiP (**21**) (triplet ground state, –43.6 kcal mol^{-1}), phosphinosilylene $HSiPH_2$ (**19**) (–47.7 kcal mol^{-1}) and phosphasilene $H_2Si{=}PH$ (**20**) (global minimum; –60.3 kcal mol^{-1}) are much lower in energy.

After co-condensation of silicon atoms with a phosphine / argon gas mixture (1 : 200), a new prominent IR absorption at 1003.5 cm^{-1}, slightly shifted compared to the absorption of pure PH_3, is observed. This IR band, together with a couple of less intense IR absorptions, can easily be assigned to *n*-adduct **18**. Short-time irradiation of the matrix with light of wavelength 366 nm leads to the complete conversion of **18** into phosphinosilylene (**19**) (2379.8, 2360.6, and 2001.5 cm^{-1}). Upon longer irradiation, silylene **19** is isomerized into phosphasilene (**20**) (2293.4, 2207.2, and 2184.1 cm^{-1}). To our surprise the photochemical reaction does not stop here. With 366 nm light, **20** is isomerized into silylphosphinidene (**21**) (2177.9, 2172.1 and 884.9 cm^{-1}). The same kind of photochemical conversion cannot be reached in the Si + NH_3 system.

Substituted phosphinidenes RP are known as important transient species in phosphorus chemistry and can be stabilized through complexation with transition metal compounds [8]. Only a few examples of the direct detection of a free phosphinidene are known [9, 10]. The parent phosphinidene HP, which according to calculations has a triplet ground state (predicted S/T gap 22 kcal mol^{-1} [9]), is a well-known transient species. The triplet ground state of mesityl-phosphinidene was derived from ESR experiments [10]. So it is no surprise that the comparison between the calculated and experimental IR spectrum of silylphosphinidene (**21**) also indicates a triplet ground state.

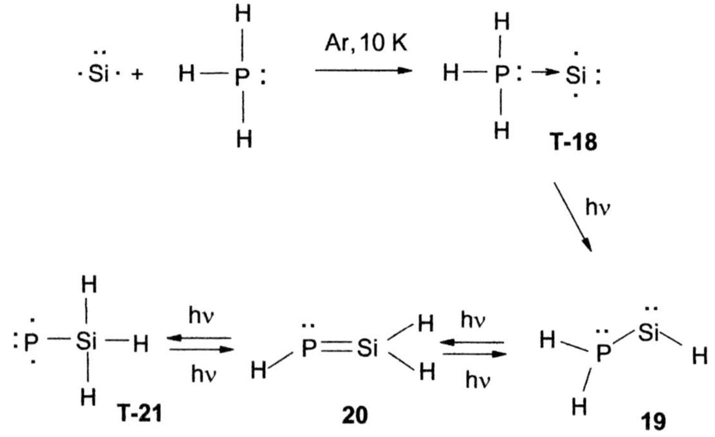

Scheme 5. Reaction of silicons atoms with PH_3.

Acknowledgements: Support by Wacker-Chemie GmbH, Burghausen, the Deutsche Forschungsgemeinschaft, and the Fonds der Chemischen Industrie is gratefully acknowledged.

References

[1] a) G. Maier, J. Glatthaar, H. P. Reisenauer, *J. Organomet. Chem.* **2003**, *686*, 341; b) G. Maier, J. Glatthaar, *Eur. J. Organomet. Chem.* **2003**, *17*, 3350.

[2] The reaction of silicon atoms and ammonia was the subject of an independent study: M. Chen, L. Aihua, H. Lu, M. Zhou, *J. Phys. Chem. A* **2002**, *106*, 3077.

[3] a) G. Maier, J. Glatthaar, H. P. Reisenauer, *Chem. Ber.* **1989**, *122*, 2403; b) G. Maier, J. Glatthaar, *Angew. Chem.* **1994**, *106*, 486; *Angew. Chem. Int. Ed. Engl.* **1994**, *33*, 473.

[4] M. Elhanine, B. Hanoune, G. Guelachvili, *J. Chem. Phys.* **1993**, *99*, 4970.

[5] G. Maier, J. Glatthaar in *Organosilicon Chemistry — From Molecules to Materials,* (Eds.: N. Auner, J. Weis), VCH Weinheim, **1994**, 131.

[6] W. Sander, A. Kuhn, *Organometallics* **1998**, *17*, 248.

[7] Other calculations on the H_3PSi–PES: a) M. R. Zachariah, C. F. Melius, *J. Phys. Chem. A* **1997**, *101*, 913; b) M. T. Nguyen, A. Van Keer, L. G. Vanquickenborne, *J. Org. Chem.* **1996**, *61*, 7077; c) H. B. Schlegel, A. G. Baboul, *J. Am. Chem. Soc.* **1996**, *118*, 8444.

[8] F. Mathey in *Multiple Bonds and Low Coordination on Phosphorus Chemistry* (Eds.: M. Regitz, J. Scherer), Georg Thieme Verlag, Stuttgart, **1990**.

[9] P. F. Cade, *Can. J. Phys.* **1968**, *46*, 1989.

[10] X. Li, S. I. Weissman, T-S. Lin, P. P. Gaspar, *J. Am. Chem. Soc.* **1994**, *116*, 7899.

New Insights into the Halophilic Reaction of a Stable Silylene with Halocarbons: A Matrix - Spectroscopic Study

Jörg Glatthaar,* Günther Maier

Justus-Liebig-University Giessen, Institute for Organic Chemistry
Heinrich-Buff-Ring 58, D-35392 Giessen, Germany
Tel.: +49 641 99 34301 — Fax: +49 641 99 34309
E-mail: Joerg.Glatthaar@org.chemie.uni-giessen.de

Robert West

Organosilicon Research Center of the University of Wisconsin, Madison
1101 University Avenue, Madison, Wisconsin 53706, USA

Keywords: matrix, silylene, photochemistry

Summary: The stable silylene **1** reacts with halocarbons like dichloro-, trichloro-, and tetrachloromethane at room temperature under formation of the disilane **3**. The mechanism of this reaction is still unknown. Results of recent *ab initio* calculations do not agree with the experimental observations. In this investigation we have studied the photochemical behavior of the stable silylene **1** in a diluted argon-tetrachloromethane matrix at 10 K as well as the thermal properties of **1** in pure solid CCl_4 in the temperature range of 10 – 200 K. The formation of a trichloromethyl radical in both types of experiments indicates that electron transfer may play an important role in these processes.

Introduction

Reports on the direct observation of reactions between silylenes and halocarbons [1] are rather scarce, although this reaction may play an important role in the direct synthesis of methylchlorosilanes.

The reaction of the stable silylene **1** with halocarbons (Scheme 1) was first observed by West and coworkers [2]. With *tert*-butyl chloride the expected insertion product **2** is formed. In the case of halocarbons like dichloro-, trichloro-, or tetrachloromethane, surprising results were obtained. Two silylene molecules and one halocarbon molecule combine very rapidly at room temperature yielding disilane **3**. With bromobenzene both reaction products **4** and **5** are generated in a slow process. The proportion of the products **4** and **5** depends on the ratio of the reactants; larger amounts of bromobenzene lead to more **4**. Once **4** is formed it cannot be transferred into **5** through reaction with an excess of silylene **1**.

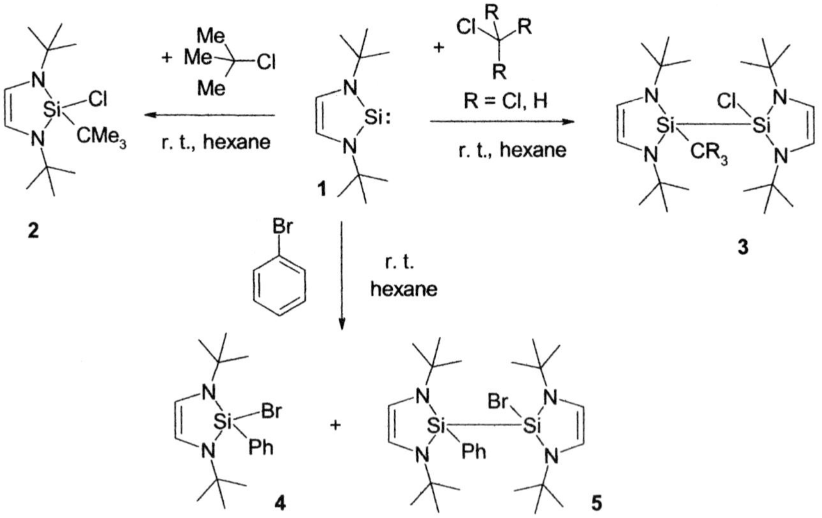

Scheme 1. Reaction of the stable silylene **1** with halocarbons.

Theoretical Approach

In the original paper [2], it was proposed that the results shown arise from a halophilic attack of the silylene nucleophile on the halogen atom of the reactants. On the other hand, Su has studied the "Lewis type" reaction of several silylenes with density functional theory [3, 4]. His basic idea is that the reaction may proceed in three steps. In the first, the halocarbon is expected to form a weak complex with the silylene. In the next, the complex is transferred into the insertion product. In the last step, the attack of a second silylene molecule leads to the formation of the final disilane product.

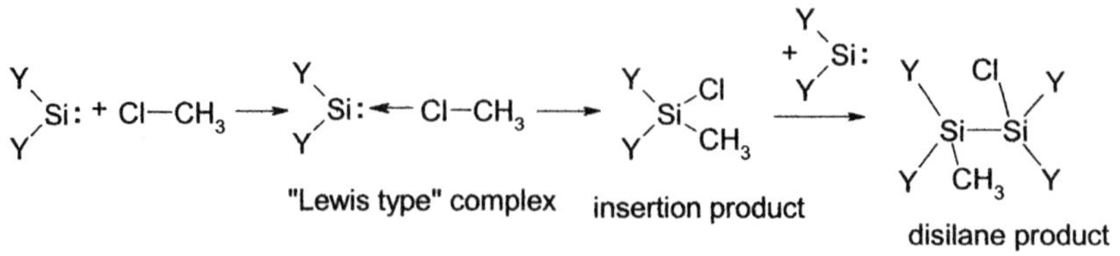

"Lewis type" complex insertion product disilane product

However, this mechanism should not occur in the case of silylene **1**. According to Su's calculations this stable silylene is not expected to form adducts with halocarbons. Moreover, the calculated activation energy for the insertion is 52.5 kcal mol^{-1}, inconsistent with the rapid reaction of **1** with chlorocarbons at room temperature.

Matrix Isolation Spectroscopy

In the Giessen group, the spectral properties and reactivity of "Lewis type" complexes between halomethylsilylenes and methyl halides, generated in co-condensation experiments between silicon atoms and methyl halides, have been studied [5]. These less stabilized silylenes add the methyl halides under transformation into the stable dihalodimethylsilanes. Encouraged by these results, we believed that matrix isolation spectroscopy may also be a very powerful tool to give new insights into the surprising behavior of the stable silylene **1** with halocarbons.

Several questions have to be addressed:

- Does tetrachloromethane form a complex with silylene **1** in the condensed phase?
- If yes, at which temperature does this reaction start?
- Are there any other possible reaction pathways besides the calculated concerted transformation of the "Lewis type complex" into the insertion product?

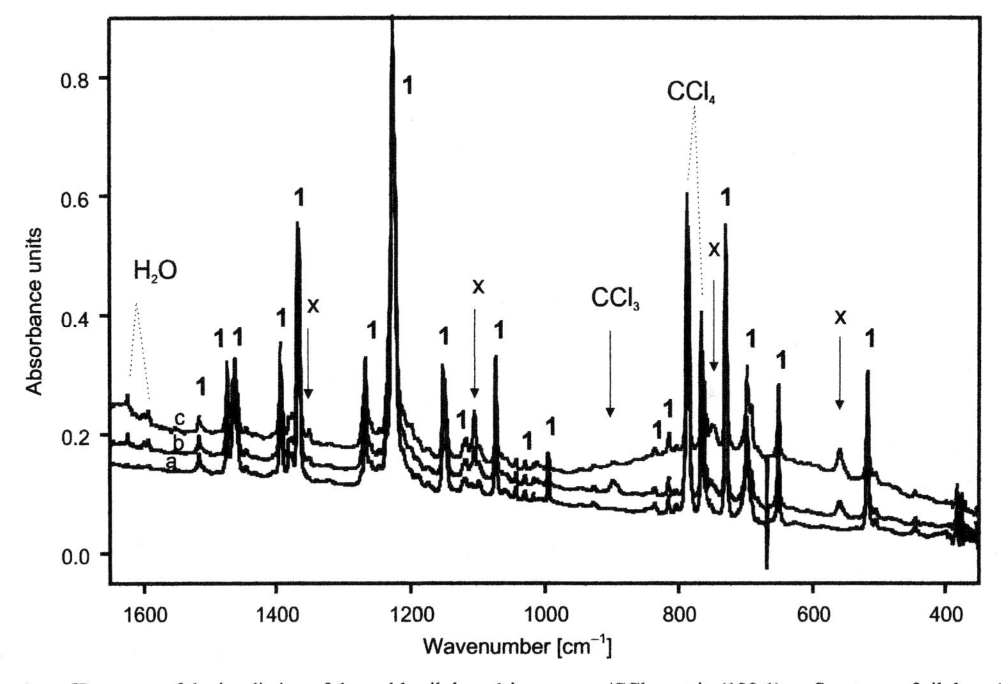

Fig. 1. IR spectra of the irradiation of the stable silylene **1** in an argon/CCl_4 matrix (100:1). a: Spectrum of silylene **1** after co-deposition and annealing to 35 K (0.5 h). b: After irradiation with 313 nm-light (3.5 h). c: After longer irradiation with 313 nm-light (31 h). IR bands of the new reactive species are marked with X.

In our first experiments we have recorded the IR and UV/Vis spectra of the matrix isolated silylene **1**. The measured IR bands of **1** fit the calculated IR spectrum very well.

The IR spectrum of **1** after co-deposition with an argon/CCl_4 mixture (100:1) did not show any band shifts even after annealing. In agreement with Su's calculations it is obvious that the stable silylene does not form any kind of complex with CCl_4. But upon irradiation of the matrix new IR absorptions were obtained (Fig. 1). One weak band at 900 cm^{-1} (curve b), which disappears on longer photochemical treatment (curve c), can be assigned to the CCl_3 radical – also independently generated from the reaction of silicon atoms and CCl_4 [6] – from comparison with the authentic spectrum of CCl_3 [7]. The assignment of additional new IR bands of X is still open. At least it is clear that they do not belong to disilane **3**.

The warm-up experiments with **1** in pure solid CCl_4 were even more interesting. At temperatures as low as 120 K large changes in the IR spectrum occur indicating that the activation energy of this reaction is rather low. CCl_3 radicals and the same reaction intermediates X as in the photolysis studies were observed.

From these findings a new mechanism can be derived: the reaction is initiated by a fast single electron transfer, followed by the migration of a chloride anion leading to the radical pair **6** + CCl_3. This radical pair can recombine under formation of the insertion reaction product **2** or can include the reaction with a second silylene molecule **1** yielding the final reaction product **3**.

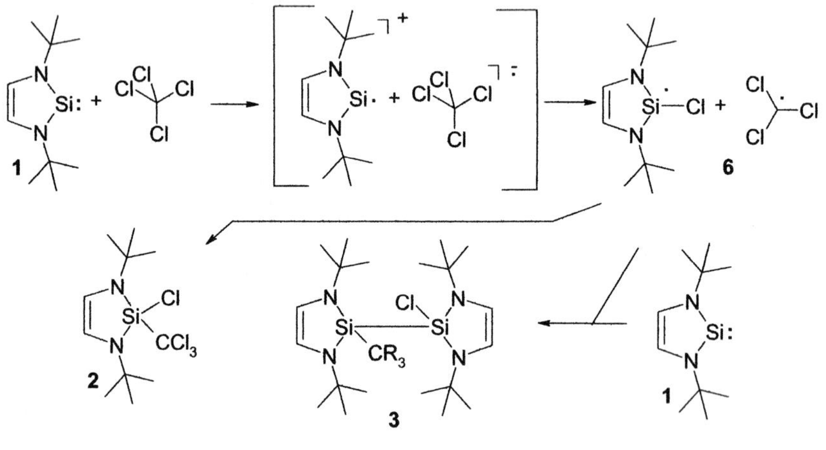

The intermediate appearance of a radical ion pair is a common feature in reactions of donor molecules. For instance, tri-*tert*-butylcyclobutadiene adds tetrachloromethane in a very similar fashion [8].

Conclusion

There are at least two different pathways for the reactions of silylenes with halocarbons.

Halomethylsilylenes form donor/acceptor complexes, which upon photoexcitation can be transformed into the corresponding silanes [5]. If an extremely good donor like silylene **1** is treated with a potent acceptor like carbon tetrachloride an electron transfer mechanism may be preferred.

References

[1] S. Ishida, T. Iwamoto, C. Kabuto, M. Kira, *Chem. Lett.* **2001**, *11*, 1102.
[2] D. F. Moser, T. Bosse, J. Olson, J. L. Moser, I. A. Guzei, R. West, *J. Am. Chem. Soc.* **2002**, *124*, 4186.
[3] M.-D. Su, *J. Am. Chem. Soc.* **2003**, *125*, 1714.
[4] M.-D. Su, *Chem. Phys. Lett.* **2003**, *46*, 1989.
[5] G. Maier, J. Glatthaar, H. P. Reisenauer, *J. Organomet. Chem.* **2003**, *686*, 341.
[6] G. Maier, J. Glatthaar, unpublished results.
[7] L. Andrews, *J. Chem. Phys.* **1968**, *48*, 972.
[8] G. Maier, W. Sauer, *Angew. Chem. Int. Ed. Engl.* **1977**, *16*, 51.

Compensation Effect in Direct Reactions of Silicon

J. Acker

Leibniz Institute for Solid State and Material Research Dresden
P.O. Box 27 01 16, D-01171 Dresden, Germany

H. Lieske

Institute for Applied Chemistry Berlin-Adlershof
Richard-Willstätter-Str. 12, D-12489 Berlin, Germany

K. Bohmhammel

Freiberg University of Mining and Technology
Department of Physical Chemistry, D-09596 Freiberg, Germany

Keywords: compensation effect, isokinetic effect, Rochow synthesis, transition metal silicide phases, selective energy transfer model

Summary: Both in the Rochow synthesis of methylchlorosilanes and in the reaction of transition metal silicides with HCl, catalytic reactions of silicon, bound as metal silicide, with gaseous reactants are involved. With both reactions, the kinetic parameters k_0 and E_A exhibit consequent compensation effects, with the isokinetic temperature positioned within the range of reaction temperatures investigated. In this paper, we apply the model of selective energy transfer from the catalyst to adsorbed species to the kinetic data. With Rochow synthesis Si–CH_3 rocking frequencies, and with hydrochlorination of silicides Si–H vibration frequencies could correspond to the isokinetic temperatures observed. An interpretation in terms of accessibility of the reactive silicon atom to reactant molecules is given.

Introduction

Analyzing kinetic data of closely related reactions, the so-called "compensation effect" or, better, the "isokinetic effect" has often been found in chemistry, especially also in gas-solid reactions and heterogeneous catalysis, e.g., [1, 2]. Provided validity of the Arrhenius equation

$$\ln k = \ln k_0 + E_A/RT$$

Eq. 1.

the effect can be described by

$$\ln k_0 = \ln k_{iso} + E_A / RT_{iso}$$

Eq. 2.

That means that the phenomenon "isokinetic effect" is indicated by the observation that the Arrhenius lines of the closely related reactions intersect in only one point, characterized by the isokinetic pre-exponential factor k_{iso} and by the isokinetic temperature T_{iso}, or, in other words, there is a linear relationship between ln k_0 and E_A with the slope $1/RT_{iso}$.

The compensation effect has frequently been found in heterogeneous catalysis, but has no satisfactory and generally accepted explanation up to now, e.g., [1, 2]. Among the models created from the point of view of heterogeneous catalysis, Larsson [3 – 5] developed the model of selective energy transfer. According to this model, a characteristic vibration ω of a "heath bath" (catalyst system) and a characteristic vibration ν of the absorbed reactant molecule form a damped oscillator system of classical mechanics. The closer the vibration frequency of the reactant molecules comes to the vibration of the catalyst system, the more effective is the energy transfer from the catalyst to the adsorbed molecules and the more probable is the reaction of the molecule. In case of resonance, $\omega = \nu$, there is a maximum efficiency of energy transfer, and it holds that

$$T_{iso} = Nhc/2R\nu$$

Eq. 3.

In this paper, we would like to present two examples of the compensation effect in direct reactions of silicon and to discuss both of them in terms of the concept of Larsson [3 – 5], which is one of the established theoretical concepts to explain the occurrence of the compensation effect.

We would like to emphasize that this attempt to apply the selective energy transfer model to our results does not mean that we reject other concepts for interpreting the compensation effect.

Results and Discussion

Compensation Effect with Rochow Synthesis of Methylchlorosilanes

Experimental details of the determination of kinetic data of the Rochow reaction are given in [6]. Briefly, the kinetic data of the synthesis reaction of silicon and methyl chloride (mc)

$$Si_{powder} + CH_3Cl_{gas} == \text{Cu catalyst} ==> (CH_3)_xH_ySiCl_z, \qquad x + y + z = 4$$

Eq. 4.

were determined in a vibrating glass microreactor with 10 g contact mass (c.m.) at atmospheric pressure, reaction temperatures between 280 and 360 °C and Si utilizations around 15%. The contact masses consisted of copper(II) chloride or copper(II) oxide as catalyst component, Zn, Sb, or Sn as promoters, and highly pure Si or technical-grade Si as silicon. Detailed contact mass compositions are given in Fig. 1. The contact masses were prepared by mechanically mixing the components. Initial reaction rates were taken as rate constants.

As seen from Fig. 1, the kinetic data ln k_0 and E_A strictly follow a compensation line. Two points are worth stressing: (a) Independently of the kind of contact mass variation (variation of promoters, copper component, copper content, or Si quality), all data follow the same compensation line. This line is characterized by an isokinetic rate constant of $k_{iso} = 7.1$ $mmol_{mc}/g_{c.m.}\cdot h$ and an isokinetic temperature of $T_{iso} = (603 \pm 20)$ K. (b) The isokinetic temperature is positioned within the range of reaction temperatures applied. That means that a ranking of contact mass reactivities is generally impossible and is possible only at a certain temperature.

In terms of the selective energy transfer model, i.e. according to Eq. 3, the value of $T_{iso} = (603 \pm 20)$ K corresponds to a characteristic frequency of $\nu = (850 \pm 30)$ cm^{-1}.

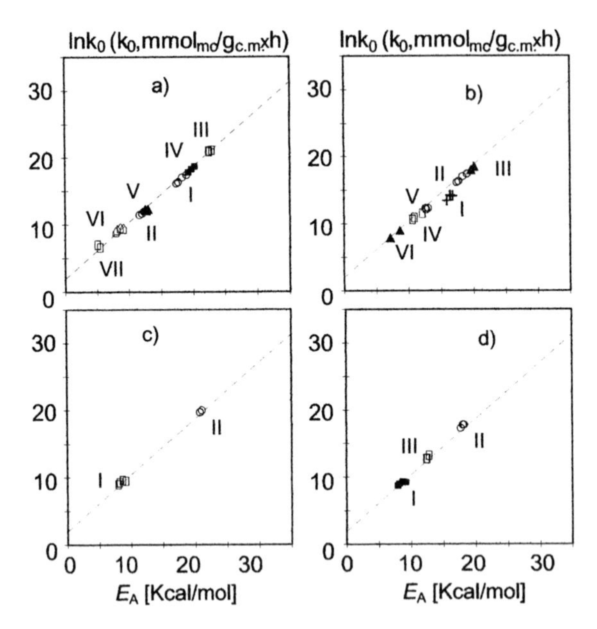

Fig. 1. Apparent activation parameters (individual, not averaged, measurements) on various contact masses as Constable-Cremer diagrams. (a) $CuCl_2/Si_{pure}$ 5 wt% Cu, with varied promoters; I: without promoters, II: 0.25 wt% Zn, III: 30 ppm Sb, IV: 30 ppm Sn, V: 0.25 wt% Zn + 30ppm Sb, VI: 0.25 wt% Zn + 30 ppm Sn, VII: 0.5 wt% Zn + 0.015 wt% Sn; (b) $CuCl_2/Si_{pure}$ with varied copper content; I: 2.5 wt% Cu, without promoter, II: 5.0 wt% Cu, without promoter, III: 7.5 wt% Cu, without promoter, IV: I + 0.25 wt% Zn + 30 ppm Sb, V: II + 0.25 wt% Zn + 30 ppm Sb, VI: III + 0.25 wt% Zn + 30 ppm Sb; (c) Si_{pure} + 0.25 wt% Zn + 30 ppm Sn, with varied copper component; I: 5 wt% Cu as $CuCl_2$, I: 5 wt% Cu as CuO; (d) 5 wt% Cu as $CuCl_2$ + 0.25 wt% Zn + 30ppm Sn; I: Si_{pure}, II: Si_{tech}, III: 50/50 mixture of Si_{pure}+Si_{tech}. (Reprinted from [6], with permission of Kluwer Academic Publishers.)

Compensation Effect in Reactions of Transition Metal Silicides and Hydrogen Chloride

The reaction between transition metal silicides and hydrogen chloride was studied by thermokinetic evaluation of isothermal calorimetric measurements performed by differential scanning calorimetry (DSC 111, Setaram, France) [7]. A constant volume of dry hydrogen chloride was passed by an argon flow into a through-flow calorimetric cell filled with a native silicide sample. A fresh silicide sample was used for each single measurement. The molar ratio between silicide phase and hydrogen chloride was chosen in such a way that silicon utilization ranged between 4% and 8% assuming complete conversion of hydrogen chloride.

The reaction rate constant was obtained by fitting the heat flow curves by the 3-dimensional Avrami-Erofeev equation, a thermokinetic model describing a spherical grain formation and growth of a new solid phase in a matrix phase. By this approach, the reaction is described in total only by a single parameter, the rate constant, which has no relation to the actual reaction mechanism. Because of single pulse dosing of hydrogen chloride to a fresh silicide sample, the measured calorimetric signal consists of the sum of the reaction enthalpies of all processes at the very early stage of hydrochlorination, like sorption processes, surface reactions and processes related to an initial bulk depletion. Hence, the kinetic data represent initial rate constants of the induction period and not of the steady state of reaction [7, 8].

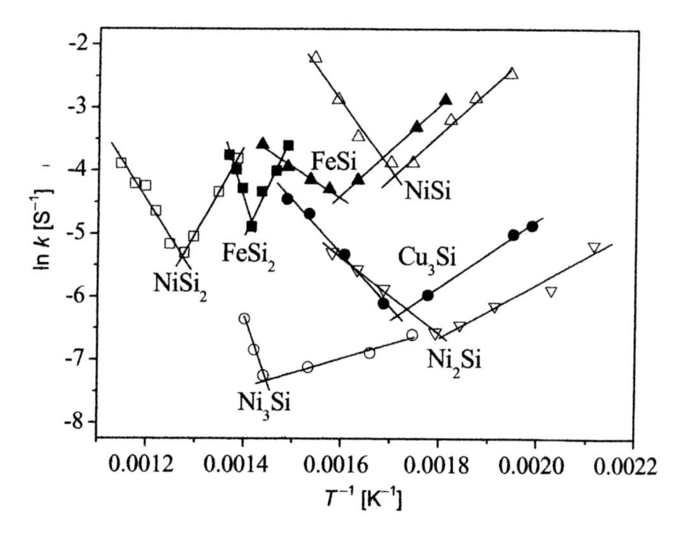

Fig. 2. Arrhenius plot of the reaction between transition metal silicide phases and hydrogen chloride. (Reprinted with permission from [7]. Copyright 2002 American Chemical Society.)

For each silicide phase investigated a v-shaped activation energy pattern was found (Fig. 2) [7]. The temperature at the minimum of the rate constants for a given silicide is designated as the reaction start temperature, T_{start}. The reaction start temperature, which depends on the nature of the metal, the metal-to-silicon ratio of the silicide phase, and the hydrogen chloride partial pressure, is considered as a parameter to describe the reactivity of silicide phases [7]. A positive apparent activation energy above T_{start} characterizes the chemical reaction that leads to the steady-state silane formation. A negative apparent activation energy below T_{start} is attributed to the chemisorption of hydrogen chloride and to the formation of a chlorine-containing reaction layer (precursor state formation) at the silicide surface without the formation of gaseous products. Silicon tetrachloride and trichlorosilane became detectable only above T_{start}, indicating the transition from the chemisorption-controlled to the reaction-controlled stage. This agrees with the observation during the induction period of the Rochow synthesis of a decomposition of methyl chloride and an enrichment of chlorine at the contact mass surface without formation of silanes [9 – 12].

By plotting activation energies vs pre-exponential factors for all silicide phases studied (Fig. 3), a linear dependence is obtained, which is characteristic of the compensation effect [8]. The linearity holds for positive and negative apparent activation energies; however, the order of the silicide phases changes from the positive to the negative apparent activation energy branch along the compensation line. The isokinetic temperature T_{iso} = (697 ± 22) K lies within the range of the reaction temperatures, as in the case of Rochow reaction (see above). The application of Larsson's model results in a vibration frequency of the reacting surface molecules of (969 ± 46) cm^{-1}.

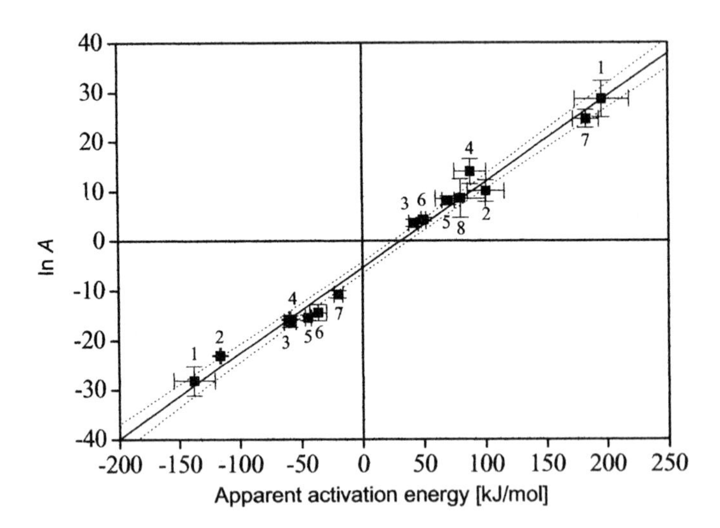

Fig. 3. Pre-exponential factor – apparent activation energy compensation plot of the reaction between transition metal silicides and hydrogen chloride under isothermal conditions (confidence interval 95%) [8]. Silicide phases are labeled as follows: 1: $FeSi_2$, 2: $NiSi_2$, 3: FeSi, 4: NiSi, 5: Cu_3Si, 6: Ni_2Si, 7: Ni_3Si, 8: Cu_3Si + 1.5 at% Zn.

Discussion

For two closely related reactions in silicon chemistry, the reaction between silicon and methyl chloride and the hydrochlorination of metal silicide phases, the compensation effect was found independently. The kinetic data from which the compensation effect is derived, were determined by entirely different methods. In Table 1 the parameters of compensation behavior are summarized.

Table 1. Parameters of the compensation effect in Rochow synthesis and in hydrochlorination of transition metal silicides.

	Rochow synthesis	**Hydrochlorination of transition metal silicides**
T_{iso} [K]	603 ± 20	696.9 ± 22.1
Heat bath frequency [cm^{-1}]	850 ± 30	969.3 ± 46.5
Range of reaction temperatues [K]	553 – 633	474 – 877
Range of apparent activation energies [kJ/mol]	22.6 – 95.1	41.6 – 196.3 –137.6 – –19.6
Kinetic evaluation	analysis of steady state gaseous products (integral reactor)	thermokinetic evaluation of isothermal calorimetric measurements
Ref.	6, 13	8

The application of Larsson's selective energy transfer model to the Rochow reaction results into a characteristic frequency $\nu = (850 \pm 30)\ cm^{-1}$, which is very close to the Si–CH_3 rocking vibration [13]. As a trial of interpretation, the Si–CH_3 rocking motion has been proposed to reduce the steric hindrance at the reacting surface silicon atom in order to facilitate the attack of a chlorine atom and to form dimethyldichlorosilane. The characteristic frequency in the hydrochlorination of silicides is positioned within the region of Si–H_x vibration modes between 1000 cm^{-1} and 850 cm^{-1} [7, 8]. The compensation effect in hydrochlorination might be related to a spill-over process of hydrogen chloride from an adsorbed state to a silylene precursor [7]. A more detailed mechanistic view can not be derived from the thermokinetic measurements. However, a model of steric hindrance analogous to the Rochow reaction is not ruled out. Another interpretation of compensation effect was suggested by Poco et al. [14]. He attributes the compensation effect to the anisotropy of the Rochow reaction [15, 16], i.e. to differences in the reactivity of the different crystallographic planes of the silicon lattice.

The basis for a common interpretation of the two compensation effects should be the control of the reactivity of silicon atoms by the nature of neighboring metal atoms as catalysts or promoters as well as by structural or morphological properties of the silicide phases involved. The reactivity of silicon atoms can vary in dependence on such influences; however, the essential step of the reactions is independent of them. The variation of the reactivity of silicon atoms with their environment always results in compensation behavior.

References

[1] A. K. Galwey, *Adv. Catal.* **1977**, *26*, 247.
[2] G. C. Bond, M. A. Keane, H. Kral, J. A. Lercher, *Catal. Rev.-Sci. Eng.* **2000**, *42(3)*, 323.
[3] R. Larsson, *J. Catal.* **1987**, *107*, 568.
[4] R. Larsson, *Chimica Scripta* **1987**, *27*, 371.
[5] R. Larsson, *J. Mol. Catal.* **1989**, *55*, 70.
[6] H. Lieske, R. Zimmermann, *Catal. Lett.* **1995**, 33, 413.
[7] J. Acker, K. Bohmhammel, *J. Phys. Chem. B* **2002**, *106*, 5105.
[8] J. Acker, K. Bohmhammel, *J. Organomet. Chem.* **2003**, *686*, 151.
[9] J. R. Anderson, B. H. McConkey, *J. Catal.* **1968**, *11*, 54.
[10] T. C. Frank, J. L. Falconer, *Langmuir* **1985**, *1*, 104.
[11] T. C. Frank, K. B. Kester, J. L. Falconer, *J. Catal.* **1985**, *95*, 396.
[12] T. C. Frank, K. B. Kester, J. L. Falconer, *J. Catal.* **1985**, *91*, 44.
[13] R. Larsson, H. Lieske, *Models in Chemistry* **2000**, *137*, 691.
[14] J. G. R. Poco, H. Furlan, R. Giudici, *J. Phys. Chem. B* **2002**, *106*, 4873.
[15] W. F. Banholzer, N. Lewis, W. L. Ward, *J. Catal.* **1986**, *101*, 405.
[16] W. F. Banholzer, M. C. Burrell, *J. Catal.* **1988**, *114*, 259.

On Reasons for Selectivity Losses in TCS Synthesis

U. Kürschner, J. Radnik, H. Lieske

Institute of Applied Chemistry Berlin-Adlershof
Richard-Willstätter-Str. 12, D-12489 Berlin, Germany

K. Hesse, U. Pätzold

Wacker-Chemie GmbH, PU-SV, PO Box 11 40, D-84479 Burghausen, Germany

Keywords: trichlorosilane synthesis, selectivity, XPS

Summary: The synthesis of trichlorosilane (TCS) from silicon and HCl produces considerable amounts of less desired chlorosilanes by side reactions, especially silicon tetrachloride (STC). The results of this paper support the view that the undesired STC is formed from TCS in a consecutive reaction, which is probably catalyzed by Si impurities and which is preferred at low space velocity and high temperatures. It seems that TCS selectivity losses are due to regions or spots in the industrial reactor with such conditions. XPS surface concentrations of Si impurities dramatically change with the proceeding synthesis reaction because of the mobility of the impurity species and do not correlate with results of Si bulk analysis.

Introduction

Trichlorosilane (TCS) is the basic material for the production of hyperpure silicon for the semiconductor industry according to the Siemens process. It is produced by hydrochlorination of technical-grade silicon: $Si + 3HCl \rightarrow SiHCl_3 + H_2$ [1 – 5]. The process aims at maximizing the selectivity toward the main product TCS and at minimizing the by-products like silicon tetrachloride (STC), dichlorosilane (DCS) and the so-called high boilers. Despite decades of industrial practice, the hydrochlorination of silicon is poorly understood.

Generally, it is known that metallurgical impurities in technical-grade silicon can strongly influence both the reactivity of the silicon and the selectivity of the reaction [1 – 5], but up to now no convincing relationships between selectivities and impurity contents have been established. As a continuation of our former work in this field [6], we tried to find out reasons for TCS selectivity losses in the industrial process by experiments with varying Si quality, by varying reaction conditions, and by XPS surface analysis of Si samples before and after the synthesis reaction.

Experimental

Materials

A sample of a highly pure silicon, Si_{pure}, and three samples (A, C, D) of technical-grade silicon, Si_{tech}, have been investigated with respect to their reactivity and to the selectivity of the TCS synthesis reaction. Si_{pure} was a monocrystalline semiconductor silicon. Impurities in the Si_{tech} samples A, C, and D, are given in Table 1. A grain size fraction between about 60 and 400 μm was used in all cases. HCl with high purity (UHP, 99.995 vol%) was taken from a cylinder.

Table 1. Impurities in the Si_{tech} samples A, C, D; bulk analysis (wt%).

Element	Sample A	Sample C	Sample D
Al	0.18	0.25	0.093
Ca	0.031	0.056	0.066
Fe	0.35	0.78	0.55
Mg	0.0024	0.0016	0.048

TCS Synthesis Experiments

The hydrochlorination of silicon was carried out in a test device consisting of feed supply, reactor, and product analysis. The reactor is made of glass and has an internal diameter of 35 mm. HCl feed is introduced to the silicon powder through a glass frit bottom of the reactor. The reaction temperature was controlled by a furnace with temperature controller and measured by a thermocouple inserted into an indent on the side of the reactor, reaching to the silicon bed. To adapt the conditions to the industrial fluidized bed reactor, the glass reactor was forced to vibrate by a commercial excentric drive.

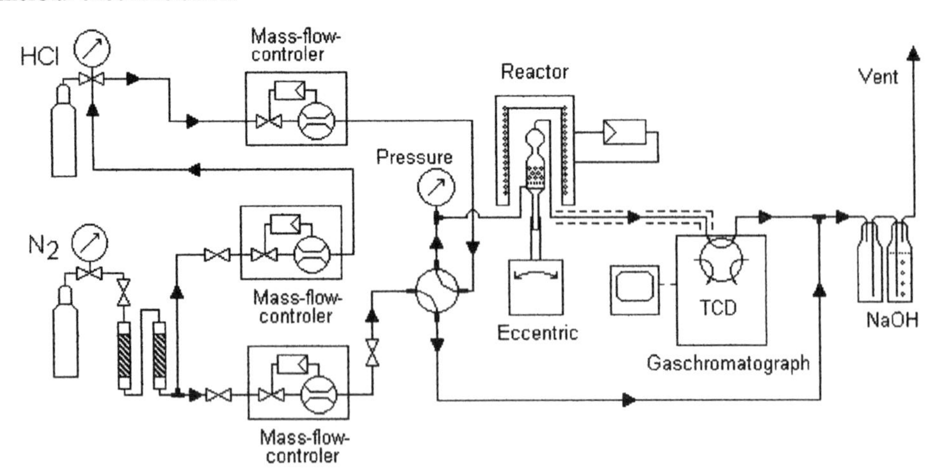

Fig. 1. Scheme of laboratory scale reactor system.

An electrical mass flow controller was used to control the flow rate of HCl. The gas flow through the reactor could be switched to nitrogen or HCl by a four-port valve.

Product analysis was carried out by a SIEMENS gas chromatograph PGC302-II, equipped with a crossbond trifluoropropylmethyl polysiloxane column 105 m long, 0.53 mm ID, 3.0 μm df (RTX-200, RESTEK GmbH) and TC detector.

A scheme of the reactor system is shown in Fig. 1.

XPS Surface Analysis

In order to get information on surface compositions of fresh and used silicon samples, XPS investigations were carried out. An ESCALAB 220 iXL device supplied by FiSONS was used. The radiation (1253.6 eV) was generated by a monochromatic X-ray tube.

Results and Discussion

Laboratory Scale Experiments with Samples of Si_{pure} and Si_{tech} under Industrially Relevant Conditions

For these experiments, reaction conditions near to those in the industrial reactor were chosen. Generally, the reaction temperature was 340 °C, except for the experiments with Si_{pure}. With this material, because of the higher starting temperature for the synthesis reaction (370 °C compared to 320 °C in the case of Si_{tech}), a reaction temperature as high as 380 °C was used.

In Figs. 2, 3, and 4, the HCl conversion and selectivity–time behavior of the Si_{tech} samples A, C, and D are shown. The HCl conversion always amounts to nearly 100%. The TCS selectivities are generally higher than is usual in industrial processes. The selectivities differ from sample to sample and, moreover, change with time on stream. Neither the differences between the samples nor their somewhat complicated selectivity–time behavior have so far been explained. Reliable correlations with the chemical bulk analysis values in Table.1 could not be established.

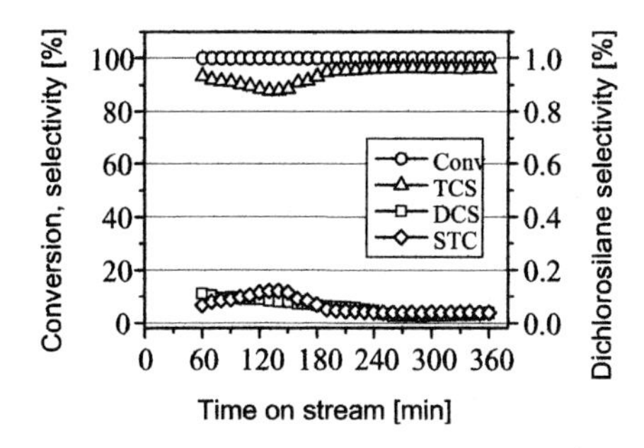

Fig. 2. Hydrochlorination of technical-grade Si; sample A; 340 °C; 1.26 L/h HCl.

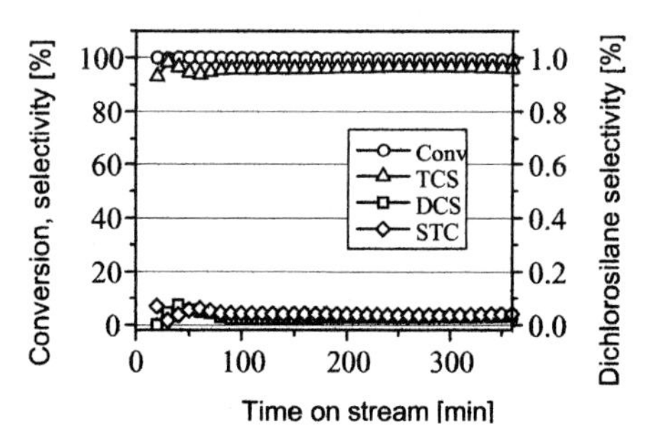

Fig. 3. Hydrochlorination of technical-grade Si; sample C; 340 °C; 1.26 L/h HCl.

In Fig. 5, analogous results with highly pure silicon are to be seen. Despite the higher reaction temperature, which generally enhances STC formation, the TCS selectivity is very high (> 95%) and constant over time on stream. The higher trichlorosilane selectivity of highly pure Si is in accordance with our model [6], which ascribes TCS losses by STC formation to a consecutive reaction of TCS, which is catalyzed by impurities in technical-grade Si.

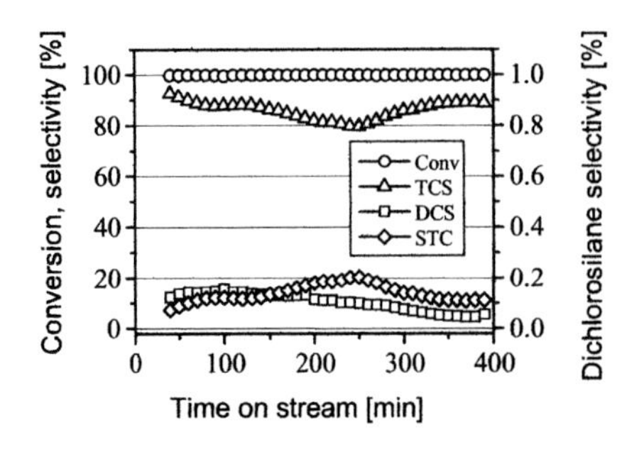

Fig. 4. Hydrochlorination of technical-grade Si; sample D; 340 °C; 1.26 L/h HCl.

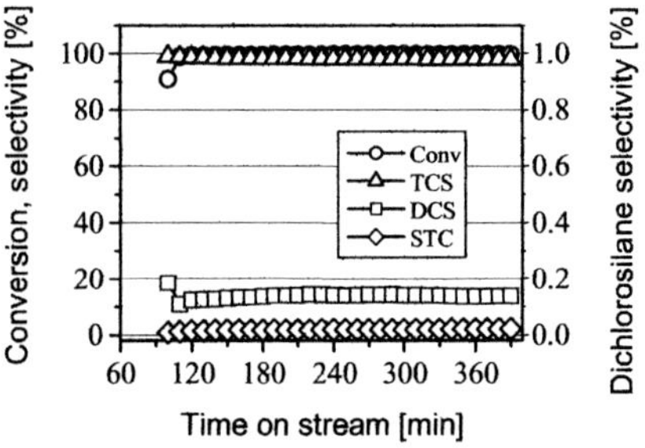

Fig. 5. Hydrochlorination of highly pure monocrystalline silicon; 380 °C; 1.26 L/h HCl.

Laboratory Scale Experiments under Conditions Deviating from the Nominal Industrial Ones

With these experiments, temperature and feed flow rate were varied to values beyond the industrially usual ones.

The influences of the reaction conditions are demonstrated by means of results of measurements with the technical-grade sample D. Comparison of the hydrochlorination results at 340 °C, (Fig. 6), and 420 °C, (Fig. 7), shows considerable differences in the by-product formation and, consequently, in the trichlorosilane selectivity. At 420 °C the trichlorosilane selectivity is markedly lower and decreases continuously with time on stream, reaching values less than 40%. Silicon tetrachloride has been obtained in nearly complementary amounts to trichlorosilane, it being the major by-product.

In further experiments, the HCl feed flow rate was varied in the wide range 0.3 – 5 L/h, to study the influence of HCl residence time on trichlorosilane selectivity and the formation of by-products. The HCl conversion remained near 100% and constant also under these conditions. At the low flow rate of 0.3 L/h HCl, a low trichlorosilane selectivity and a relatively high amount of silicon tetrachloride were observed (Fig. 8). With progressive time on stream the trichlorosilane selectivity decreases to 50%, while the conversion remains at nearly 100%. Applying high flow rates, the trichlorosilane selectivity is over 95% until nearly complete consumption of silicon (Fig. 9). The tetrachloride selectivity increases with increasing residence time, i.e. with decreasing space velocity. This is a typical pattern, which can always be observed if a secondary product is formed from a primary one in a consecutive reaction. This result, together with our former results concerning the role of impurities as catalysts, supports our thesis of STC formation as a catalytically enhanced consecutive reaction of the main product TCS [6].

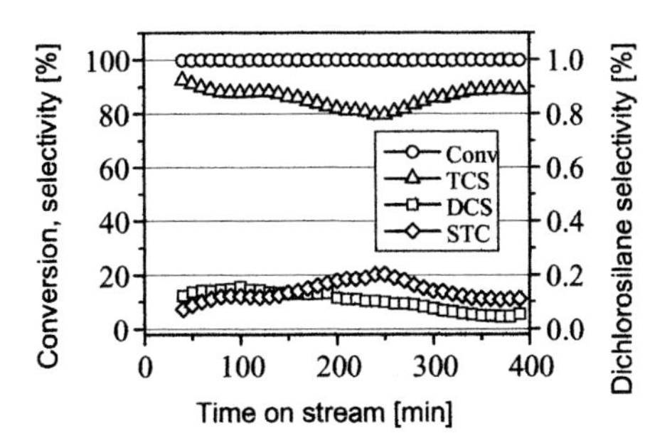

Fig. 6. Hydrochlorination at 340 °C; technical-grade Si; sample D; 1.26 L/h HCl.

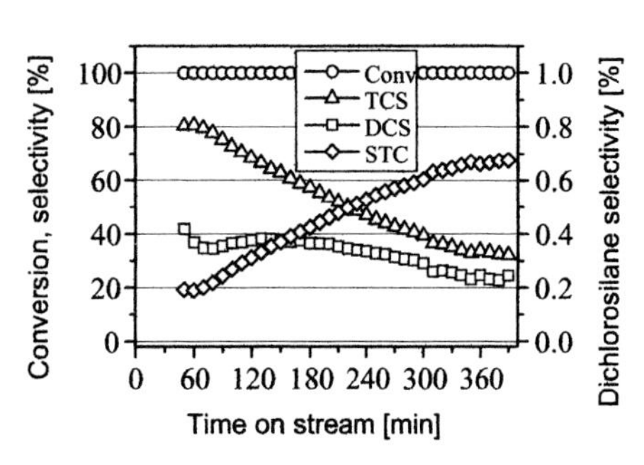

Fig. 7. Hydrochlorination at 420 °C; technical-grade Si; sample D; 1.26 L/h HCl.

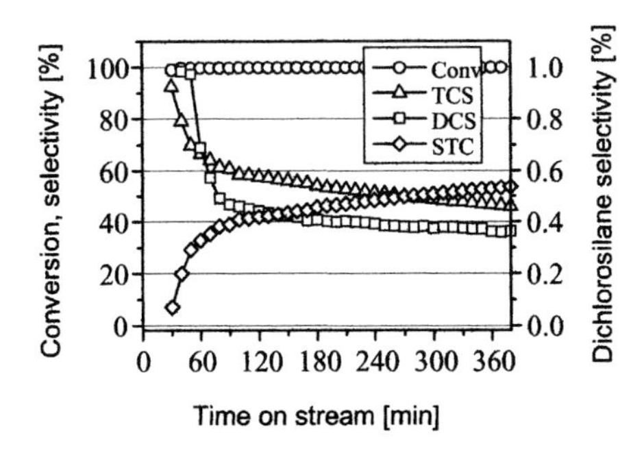

Fig. 8. Hydrochlorination at 0.32 L/h HCl; 340 °C; technical-grade Si; sample C.

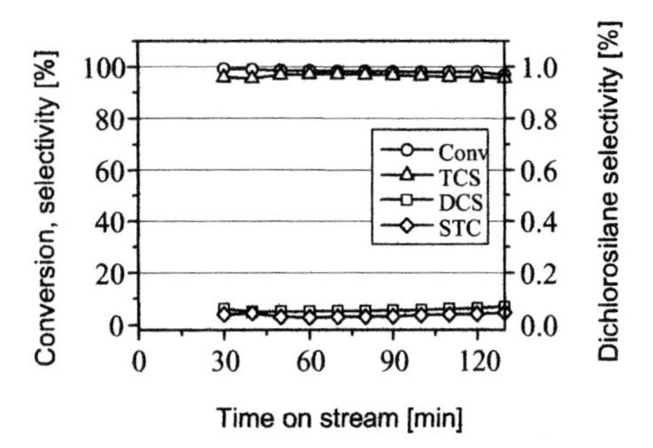

Fig. 9. Hydrochlorination at 5.04 L/h HCl; 340 °C; technical-grade Si; sample C.

Si Impurities and TCS Selectivity

Although a detrimental influence of impurities in Si_{tech} on TCS selectivity, i.e. their role as catalysts of the consecutive reaction TCS→STC, is in principle fairly certain, up to now in no case could a convincing correlation between bulk concentration of individual impurities and TCS selectivity be established.

In order to get a better insight into the reasons for TCS selectivity losses, we applied XPS surface analysis to Si samples with differing history, i.e. to samples of different purity, after reaction at varied temperature and varied time on stream in TCS synthesis. First we concentrated on Fe and Mg as impurities. This choice does not mean that we consider these two elements as more important than others.

Surface concentrations of the impurities Fe, Mg, Cl, and C in silicon samples are given in Fig. 10 as atomic ratios of the respective elements related to silicon. Unlike Fe, Cl, and C, the Mg surface concentrations up to now can be given only in arbitrary units.

At 340 °C and in the presence of an HCl atmosphere, the concentrations of Fe and Mg on the Si surface increase with time on stream. The chlorine concentration increases nearly proportionally to the contents of the metals. With highly pure silicon, only traces of Cl were detected, even after a longer time on stream. The proportional increase in Cl surface concentration with technical-grade silicon and increasing time as well as the absence of such increase in the case of highly pure silicon can be taken as a strong hint at the formation of metal chlorides on the silicon surface.

At 420 °C with the same HCl feed flow rate and consequently the same Si consumption rate (100% conversion in each case), the surface enrichment of the elements detected is about one order of magnitude higher than at 340 °C. This shows that the surface enrichments observed are not a consequence of silicon consumption, but must be caused by the migration of Fe and Mg species from special places with high impurity concentrations to free Si surface. The higher surface enrichment at the high temperature of 420 °C must be caused either by a higher formation rate of the migrating species and/or by their faster migration. In the absence of HCl, i.e. in an N_2 atmosphere, no significant increase of Mg and Fe content on the Si surface was observed (see Fig. 10). This finding strongly supports the view that the migrating species are chlorides.

On the whole, the investigations show that the actual surface concentrations of impurities are more strongly controlled by the synthesis reaction and its conditions than by the original Si bulk concentrations of the impurities. Hence, simple relationships between results of chemical bulk analysis of Si and results of the chlorosilane synthesis are not to be expected and indeed have not been found.

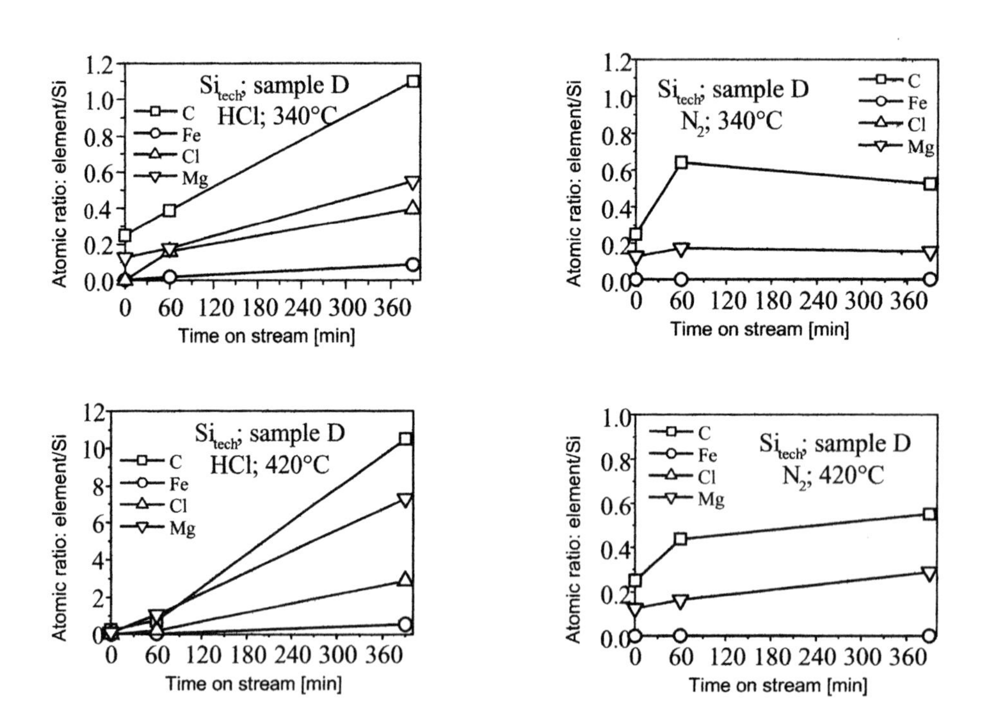

Fig. 10. XPS investigation of Si surface as a function of time, temperature, and ambient gas atmosphere.

Conclusions

- TCS synthesis laboratory scale experiments generally yield higher TCS selectivities than do industrial reactors, because of the more defined conditions over the whole of the reacting silicon bed in the laboratory scale reactor. The TCS selectivity losses in the industrial reactor result from regions of imperfect fluid-bed conditions, with significantly increased temperature and significantly lower HCl flow rate than desired.
- The increase in STC formation, linked with corresponding decrease of TCS selectivity, with decreasing HCl flow rate shows that STC is formed in a consecutive reaction of the main product TCS: $SiHCl_3 + HCl \rightarrow SiCl_4 + H_2$. This reaction is catalyzed by impurities in technical-grade silicon, and is preferred at high temperatures.
- Under the conditions of TCS synthesis, impurities in technical-grade silicon like Fe or Mg become mobile as chlorides and migrate from certain distinct places to free silicon surface, leading to an average surface enrichments of the impurities. These enrichments are significantly less controlled by the Si consumption of the synthesis reaction and by the original bulk contents of the impurities than by the reaction conditions, especially the presence of HCl and the reaction temperature. Hence, simple relationships between results of chemical bulk analysis of Si and selectivity of TCS synthesis are not to be expected and indeed have not been found.

References

[1] E. Enk, J. Nickl, H. Teich, DE 1105398, **1961**.

[2] A. I. Gorbunov, A. P. Belyi, S. A. Golubzov, N. S. Feldshtejn, *Proceedings of International Symposiumon Organosilicon Chemistry*, Prague **1965**, p. 395.

[3] R. Müller, *Chem. Tech.* **1950**, *2*, 7.

[4] I. Shiihada, J. Iyoda, *Bull. Chem. Soc. Japan* **1959**, *32*, 636.

[5] A. I. Gorbunov, A. V. Shchegolev, A. P. Belyi, S. A. Golubzov, *Zh. Fiz. Khim.* **1972**, *46*, 631.

[6] H. Ehrich, T. Lobreyer, K. Hesse, H. Lieske, *Stud. Surf. Sci. Catal.* **2000**, *130*, 2267.

Characterization of Trichlorosilane Direct Process Residue

Patrick J. Harder,* Arthur J. Tselepis

Dow Corning Corporation, 3901 S. Saginaw Rd., Midland MI 48640 USA
Tel.: +1 989 4961558 — Fax: +1 989 4966243
E-mail: pjharder@dowcorning.com a.j.tselepis@dowcorning.com

Keywords: trichlorosilane, tetrachlorosilane, direct process, direct process residue, characterization, shock-sensitive, hydrolysis, pyrophoric

Introduction

The direct process reaction of silicon metal with hydrogen chloride to form $SiH_nCl_{(4-n)}$ monomers produces a high-boiling residue (DPR). This material contains a mixture of chlorodisilanes and chlorosiloxanes. Historically, Dow Corning has experienced several safety incidents when handling this stream. This prompted an effort to better understand the hazards associated with direct process residue (DPR), and through better understanding we hope to improve the safe handling practices for this class of materials.

Trichlorosilane DPR from the Dow Corning Midland Plant was analyzed for composition, flash point, pyrophoricity, etc. New test methods were developed to quantify the shock-sensitive nature of the hydrolyzed gels which commonly form when handling this stream. This testing was performed on individual fractionated DPR species, and the hazards of each component were quantified. It was determined that SiH-containing chlorosiloxanes had played a key role in several past safety incidents.

Further work led to the discovery that these dangerous SiH-containing chlorosiloxanes can be made to rearrange to form $HSiCl_3$ and larger polysiloxanes. This was demonstrated both in the laboratory and in the manufacturing process. The effect of varying process conditions and the presence of catalytic species on this reaction are discussed. The role of this rearrangement chemistry and the conditions and properties that favor shock-sensitive gel formation are also presented. This work has resulted in an increased understanding of the hazards associated with handling this stream and the steps which can be taken to mitigate these hazards.

Trichlorosilane Process Overview

The process for the manufacture of trichlorosilane and silicon tetrachloride at Dow Corning's Midland site involves the reaction of metallic silicon with hydrogen chloride in a fluidized bed reactor. The chlorosilane monomers are removed from the system via distillation, and the DPR [1]

is concentrated as a bottoms stream. This residue consists of a mixture of disilanes and disiloxanes. There are several factors that influence the quantity of DPR produced; some of these include bed temperature, various catalysts in the bed, and bed isothermality, which may be influenced by means such as feed or cone configuration, mixing, superficial velocities, etc. Typical DPR levels in TCS FBR Crude lie between 0.1 and 1% by weight.

Over the past 40+ years of producing TCS by the Direct Process, there have been several incidents associated with the handling of DPR. These ranged from unexpected "pops" when sampling or otherwise handling DPR streams to small flash fires and larger scale explosions in quench areas. Because of these incidents and the unpredictable nature of the materials, a blanket designation of "pyrophoric" was given for DPR and disilanes in general. The best mitigation of these hazards has been achieved by handling materials only in dilute form, often resulting in the loss of the valuable monomer as waste. Recently, Dow Corning has performed work in the characterization of these materials to further understand and minimize these hazards.

Flammability and Composition Testing

The first analysis performed on the concentrated DPR streams was for pyrophoricity according to the UN Test Method for Division 4.2 (Spontaneously Combustible Materials). The results of this test came back Non-Pyrophoric. Flashpoint testing (Tag Open Cup ASTM D 1310) also gave unexpected results with a flash point of >100 °C. This places the material into the category of a Non-Combustible Liquid.

Using GC Mass Spectrometry, elutable components in DPR with a boiling point higher than STC were analyzed. Several species that were previously identified as Disilanes (e.g. $H_2Si_2Cl_4$), were determined to be disiloxanes (in this case Cl_5HSi_2O). This misidentification is prevalent in the literature, most likely because of the similar boiling points and cracking/cleavage products, two techniques used to identify components in a mixture prior to more modern analytical techniques.

Metallurgical Si is manufactured in industrial furnaces from quartzites and carbonaceous reductants containing various intermetallic impurities. These metallic impurities, which are concentrated at grain boundaries, induce precipitates in metallurgical Si during solidification [2]. Oxidized iron is a significant source of the oxygen which yields the various siloxanes in DPR.

$$FeO + 2\,HSiCl_3 \longrightarrow HCl_2Si{-}O{-}SiCl_2H + FeCl_2$$

Eq. 1. Disiloxane formation from iron oxide impurities in the incoming silicon feeds.

During the fractionation of the DPR, no SiH-containing disilanes were isolated. This is most probably because metal chlorides such as $AlCl_3$ will result in Si–Si bond cleavage. Gupper noted that ClH_2SiSiH_2Cl and $ClH_2SiSiHCl_2$ were easily separated and purified by fractional distillation.

Both disilanes were stable for weeks at room temperature, but underwent rapid H/Cl exchange reactions and SiSi cleavage upon the addition of traces of catalysts such as $AlCl_3$ [3].

Table 1. DPR compositions at feed to distillation and at distillation bottoms.

Elutable high boilers	DPR composition at distillation feeds [%]	DPR composition at distillation bottoms [%]
$Cl_4H_2Si_2O$	14	0
Cl_5HSi_2O	32	9
Cl_6Si_2O	18	19
Si_2Cl_6	37	72

As indicated in Table 1, the disiloxanes in the feed stream are reduced in concentration during the distillation process. H. F. Stewart of Dow Corning first discovered the redistribution of SiH and SiOSi bonds at 70 – 120 °C [4]. This reaction in the distillation system results in the conversion of SiH-containing siloxanes to usable monomer and higher siloxanes. It has the net effect of reducing the overall SiH content in the DPR and the recovery of valuable trichlorosilane monomer.

$$2\ Cl_3Si{-}O{-}SiHCl_2 \longrightarrow HSiCl_3 + HSiCl_2{-}O{-}SiCl_2{-}O{-}SiCl_3$$

Eq. 2. Cleavage of SiH siloxanes to form useful monomer and higher siloxanes.

In this reaction, dimer converts to trimer, tetramer, and so forth, eventually becoming a solid. The cleavage reaction is not limited to the distillation system, but can occur wherever the correct combination of temperature and residence time is experienced. Because of the generation of solids, this side reaction may be undesirable in some unit operations.

Formation of "Popping Gels"

Trichlorosilane DPR is known to produce shock-sensitive hydrolysis products. Because these "popping gels" contain all four necessary elements of the fire tetrahedron, fuel, oxidizer, sustainable reaction, and ignition source, they pose a significant hazard, particularly in classified areas or wherever flammable atmospheres may exist.

A test method was developed to evaluate various conditions and components in DPR with respect to their tendency to form shock-sensitive gels. The designed experiment looked at several fractionated components, contact phases, substrates, ambient conditions, film thickness, and age of gels. Ignition was attempted using strictly a mechanical means such as scraping. In most cases this was unsuccessful, and a semi-quantitative method using heat to determine auto-ignition was used.

Key variables that influenced shock sensitivity are a balance of the following three components:

- Age of gel.
- Si-H content of starting chlorosilane mix
- Si-Si content of starting chlorosilane mix

Mixtures that are high in Si–Si and Si–H can form shock-sensitive gels within 24 h, while pure HCDS was found to require over one month before becoming shock-sensitive. Gels that were shock-sensitive were also found to ignite under an inert atmosphere, indicating that atmospheric oxygen is not necessary to support the flame.

Our findings from laboratory scale generation of popping gels suggest that there are two interactive routes. One of the routes results in the generation of hydrogen or pyrophoric monomers trapped in the gel. These gases react with atmospheric oxygen when disturbed. The second route appears to be oxidation of Si-Si bond by bound oxygen such as uncondensed Si–OH.

Infrared analysis of hydrolyzed gels of hexachlorodisilane revealed the formation of silicon hydride upon mild heating.

Careful inspection of a freeze frame video of the ignition of gels produced from a chlorosilane mixture high in Si–Si and SiH-containing siloxanes reveals two flame fronts. The first is purple in color, indicating the burning of a hydrogen-rich gas. The second flame is yellow and is indicative of a rapid solid phase reaction generating significant amounts of heat (oxidation of Si–Si).

While the cause of the apparent pyrophoric nature of TCS DPR and the hazardous conditions to avoid have been determined, these materials are still highly unpredictable in generating shock-sensitive hydrolysis products. There is still a great deal to be learned before the mechanism is fully understood.

Conclusions

Dow Corning utilizes a cleavage reaction in its distillation system and other areas to minimize the concentration of Si–H containing siloxanes in its TCS DPR streams. This reaction allows the recovery of a portion of the DPR as usable TCS.

Based on material testing and other laboratory work, TCS DPR is not to be considered pyrophoric. This is primarily because of the absence of SiH-containing disilanes. However, these streams can form shock-sensitive hydrolysis products, which can be an ignition source for flammable atmospheres in the vicinity of the gel.

In order to minimize the formation of shock-sensitive gels, the following precautions should be considered when handling DPR-containing streams.

- Minimize contact with air and moisture.
- Dispose of all gels promptly and do not allow them to age undisturbed.
- Minimize the level of Si-H in the DPR stream.

- Inert conditions will NOT prevent the popping of gels if they are shock-sensitive, but it may reduce the risk of a secondary fire or explosion.

References

[1] M. G. Kroupa, *Silicon for the Chemical Industry VI* (Ed.: Harald A. Oye), Norwegian University of Science and Technology, Trondheim, Norway, June 17 – 21, **2002**, p. 201 – 207.

[2] T. Margaria, J. C. Anglezio, C. Servant, *Pechiney Electrometall.*, Paris, France. Symposium Series – South African Institute of Mining and Metallurgy, **1992**, S11(INFACON 6, Vol. 1), p. 209 – 214.

[3] A. Gupper, K. Hassler, *Eur. J. Inorg. Chem.* **2001**, *8*, 2007 – 2011.

[4] H. F. Stewart, *J. Organomet. Chem.* **1967**, *10*(*2*), 229.

Dichlorosilylene Transfer Reactions Using $Me_3GeSiCl_3$

Emma Seppälä,* Thorsten Gust, Cathleen Wismach, Wolf-Walther du Mont

Institut für Anorganische und Analytische Chemie der Technischen Universität
Postfach 3329, D-38023 Braunschweig, Germany
Tel.: +49 531 3915303 — Fax: +49 531 3915387
E-mail: e.seppaelae@tu-bs.de

Keywords: dichlorosilylene, trichlorosilyl germanes, phosphaalkenes, transfer reactions

Summary: Trichlorosilyl germanes $Me_3GeSiCl_3$ (**1**) and $Me_2Ge(SiCl_3)_2$ (**2**) can be considered as new sources for $SiCl_2/SiCl_3$ transfer reactions under mild, aprotic conditions. Transfer reactions with chloro- and diphosphanes (RR′PCl: R, R′ = ^{i}Pr **4a** and R = ^{t}Bu, R′ = ^{i}Pr **4b**; $(R_2P)_2$: R = ^{t}Bu **5a**, R = ^{i}Pr **5b**) and amine-catalyzed cross experiments with alkyl(chloro)germanes $R_{4-n}GeCl_n$ (R = Me, Et and n = 1, 2) lead by silylene/silyl transfer to trichlorosilyl phosphanes or germanes. Trapping experiments with *P*-phosphanylphosphaalkenes $(Me_3Si)_2C$=P-PRR′ (R = ^{t}Bu, R′ = ^{i}Pr **6a** and R, R′ = ^{i}Pr **6b**) lead in a very selective fashion to a complete P=C double-bond cleavage by unique double $SiCl_2$ addition with formation of stable *P*-phosphanylphosphadisiletanes **7a** and **7b**. In the case of compound **6b**, a new centrosymmetric diphosphene (**8**) containing two dichlorosilyl phosphane groups was also formed. The reaction of **1** with *P*-chlorophosphaalkene $(Me_3Si)_2C$=P–Cl (**9**) leads to formation of diphosphene **10**, transient *P*-silylphosphaalkene **11** and intermediate alkylidenediphosphane **12**, all known from the reaction of **9** with Si_2Cl_6. Besides these known products, a new species with a ^{31}P NMR pattern closely related to those of **7a** and **7b** was detected.

Introduction

Trichlorosilylmetal compounds, trichlorosilane, and hexachlorodisilane can be regarded as featuring hidden $SiCl_2$ functions (Scheme 1) [1, 2]. Trihalogenosilyl germanes are of interest to us as precursors for further transformation.

Chlorotrimethylgermane reacts with trichlorosilane and triethylamine to provide trichlorosilylgermane $Me_3GeSiCl_3$ (**1**) in a fair yield as distillable liquid. The formation of **1** is, however, followed by incomplete base-catalyzed decomposition reactions leading to the branched neopentane-like product $(Me_3Ge)_2Si(SiCl_3)_2$. Dichlorodimethylgermane reacts with trichlorosilane and triethylamine furnishing dimethylbis(trichlorosilyl)germane $Me_2Ge(SiCl_3)_2$ (**2**). Different from

1, compound **2** is stable in presence of triethylamine. This resistance allows its isolation from the reaction mixture in good yield. However, compound **2** dismutates in presence of $^{i}Pr_3P$ furnishing $Me_2(SiCl_3)Ge–Ge(SiCl_3)Me_2$ [3 – 5]. This observation led us to consider the stable compounds $Me_3GeSiCl_3$ (**1**) and $Me_2Ge(SiCl_3)_2$ (**2**) as potential nucleophile-induced sources of $SiCl_2$ moieties.

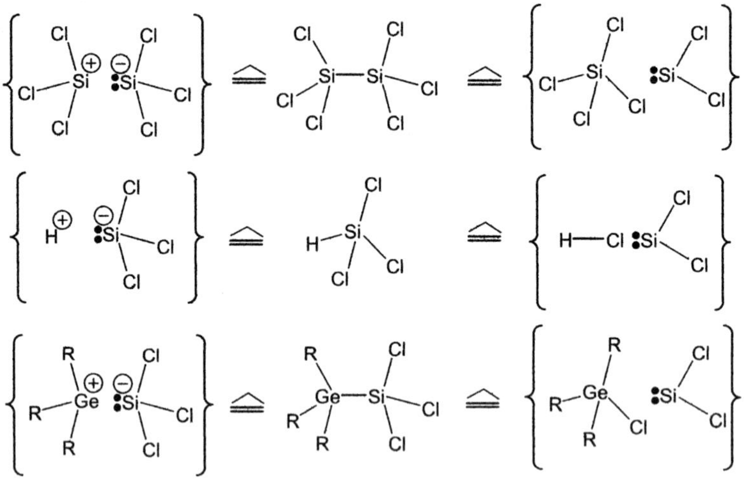

Scheme 1. $SiCl_2$ equivalents.

To test this hypothesis, several silylene/silyl transfer experiments were carried out with the compounds **1** and **2**. Here we report the cross experiments with alkyl(chloro)germanes, transfer experiments with chloro- and diphosphanes, and trapping reactions with three phosphaalkenes.

Cross Experiments

Cross experiments between $R_{4-n}Ge(SiCl_3)_n$ and $R_{4-n}GeCl_n$ (R = Me, Et and n = 1, 2) with catalytic amounts of triethylamine indicate a dynamic $SiCl_2$ (or $SiCl_3$) transfer equilibrium reaction (^{13}C and ^{29}Si NMR evidence). The equilibrium of these reactions is very dependent on the number and quality of the organic substituents bonded to germanium. In order to study this dependence, all 12 possible variations of these reactions between $R_{4-n}Ge(SiCl_3)_n$ and $R_{4-n}GeCl_n$ (R = Me, Et and n = 1, 2) were carried out and investigated with the help of ^{13}C and ^{29}Si NMR. One example of these reactions is presented in Scheme 2.

The reaction between $Me_3GeSiCl_3$ (**1**) and Et_3GeCl was observed to be slow, and equilibrium is reached when the reaction has proceeded approximately 50% (Scheme 2). Dismutation of $Me_3GeSiCl_3$ does not occur, but the presence of triethylamine induces $Cl/SiCl_3$ exchange reaction between the two kinds of trialkylgermyl groups. Adding Me_3GeCl to $Et_3GeSiCl_3$ (**3**) and a catalytic amount of triethylamine leads to the same equilibrium mixture.

$$\underset{\mathbf{1}}{Me_3GeSiCl_3} + Et_3GeCl \xrightleftharpoons{Et_3N\ (cat.)} Me_3GeCl + \underset{\mathbf{3}}{Et_3GeSiCl_3}$$

Scheme 2. Cross experiment between $Me_3GeSiCl_3$ (**1**) and Et_3GeCl.

Nucleophile-coordinated $SiCl_2$ can be regarded as one common intermediate of these reactions. The divalent silicon species is formed by nucleophile-induced α-elimination at the Si atoms of $Me_3GeSiCl_3$ or $Et_3GeSiCl_3$, and $SiCl_2$ can be intercepted by re-insertion into Ge–Cl bonds of Me_3GeCl or Et_3GeCl.

Reactions with Phosphanes

Hexachlorodisilane Si_2Cl_6 is known to react toward chloro- and diphosphanes as a trichlorosilylating agent [6 – 8]. Reactions of **1** and **2** with RR'PCl (R, R' = ^{i}Pr **4a** and R = ^{t}Bu, R' = ^{i}Pr **4b**) (Scheme 3) and with $(R_2P)_2$ (R = ^{t}Bu **5a**, R = ^{i}Pr **5b**) (Scheme 4) were investigated.

$$\underset{\mathbf{1}}{Me_3GeSiCl_3} + \underset{\mathbf{4a}}{{}^iPr_2PCl} \longrightarrow {}^iPr_2P\text{-}SiCl_3 + Me_3GeCl \quad (+\ SiCl_4 + {}^iPr_2PH + ({}^iPr_2P)_2)$$

Scheme 3. Silyltransfer reaction of **1** with chlorophosphane **4a**.

The silylation proceeds faster when using compound **1** instead of compound **2** (different dismutation behavior of the two trichlorosilyl germanes [3, 4]) and, of course, when the organic substituents bonded to phosphorus show lower steric strain (^{i}Pr$_2$ < ^{t}Bu^{i}Pr < ^{t}Bu$_2$). Besides the formation of desired trichlorosilyl phosphane in each experiment, formation of small amounts of diphosphane [6], and dialkylphosphane was observed. Small amounts of germylphosphane were additionally formed in the reactions using **2** as silylating agent. Concerning the diphosphanes however, Si_2Cl_6 was shown to be superior to **1** and **2** as silylating agent.

$$\underset{\mathbf{1}}{Me_3GeSiCl_3} + \underset{\mathbf{5b}}{({}^iPr_2P)_2} \longrightarrow {}^iPr_2P\text{-}SiCl_3 + Me_3GeCl \quad (+\ SiCl_4 + {}^iPr_2PH + ({}^iPr_2P)_2)$$

Scheme 4. Silyltransfer reaction of **1** with diphosphane **5b**.

Reactions with *P*-phosphanylphosphaalkenes

As a probe for the proposed $SiCl_2$ transfer, and as a new synthetic application of $SiCl_2$, we studied the reactions of compound **1** with two sterically different *P*-phosphanylphosphaalkenes $(Me_3Si)_2C$=P-PRR' (R = ^{t}Bu, R' = ^{i}Pr **6a** and R, R' = ^{i}Pr **6b**).

Reaction of **1** with **6a** leads in a very selective fashion to a complete P=C double-bond cleavage by unique double $SiCl_2$ addition with formation of a stable *P*-phosphanylphosphadisiletane (**7a**) with an exocyclic phosphanyl substituent (Scheme 5) [2, 9].

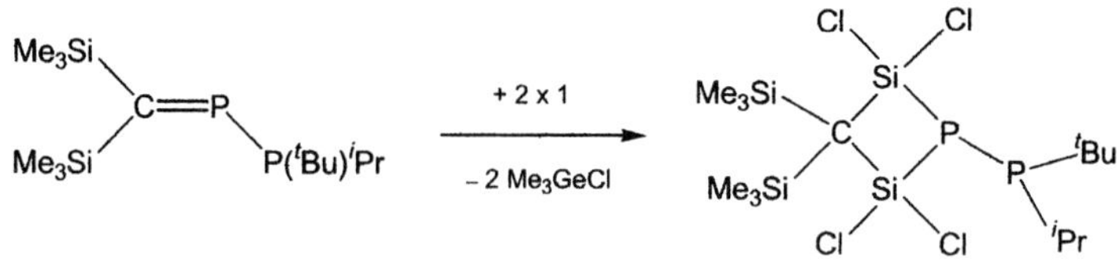

Me3Si
Me3Si
C=P
P(tBu)iPr
+ 2 x 1
- 2 Me3GeCl
Cl
Cl
Si
Me3Si
C
P
P
tBu
Me3Si
Si
Cl
Cl
iPr
6a
7a

Scheme 5. Formation of 2-phospha-1,3-disiletane **7a**.

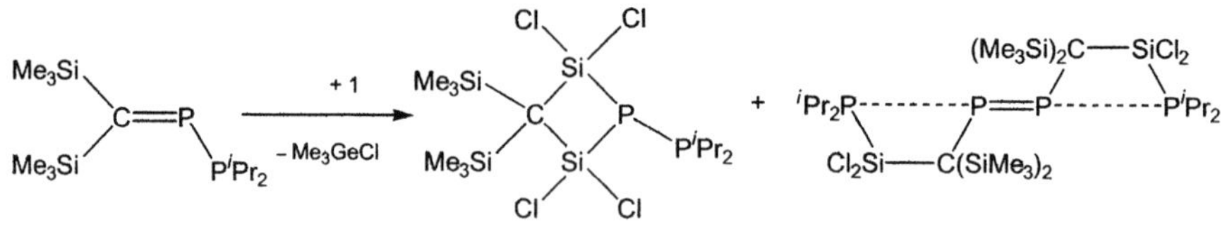

Me3Si
Me3Si
C=P
PiPr2
+ 1
- Me3GeCl
Cl
Cl
Me3Si
Si
C
P
PiPr2
Me3Si
Si
Cl
Cl
+
(Me3Si)2C—SiCl2
iPr2P--------P=P--------PiPr2
Cl2Si—C(SiMe3)2
6b
7b
8

Scheme 6. Formation of 2-phospha-1,3-disiletane **7b** and of diphosphene **8**.

In the reaction of compound **1** with the sterically less crowded **6b**, the formation of the heterocycle **7b** is accompanied by the formation of a new centrosymmetric diphosphene **8** (Scheme 6) [2, 9]. The ratio of the two formed compounds depends strongly on the reaction stoichiometry. In the reaction of **6b** with two equivalents of **1**, the heterocycle **7b** is predominantly obtained. The formation of the diphosphene **8**, however, cannot be suppressed completely.

Reaction with *P*-chlorophosphaalkene

The known reaction of $(Me_3Si)_2C{=}P{-}Cl$ **9** with Si_2Cl_6 leads to the formation of a diphosphene (**10**) as a final product. An alkylidenediphosphane (**12**) was also observed as an intermediate in the reaction (Scheme 7) [10]. Reinvestigation of the first step of the reaction by ^{31}P NMR monitoring allowed us to trace a transient +371 ppm signal, which suggests the formation of $(Me_3Si)_2C{=}P{-}SiCl_3$ (**11**).

With compound **1** as a source of $SiCl_2$ added to **9**, the transient +371 ppm ^{31}P NMR signal, which is apparently the precursor of the intermediate "unsymmetric dimer" **12** from Scheme 7, was again detected. The observed $^1J_{P,Si}$ coupling constant of 244 Hz for **11** is extraordinarily large. In the reaction mixture were also observed the diphosphene **10**, large amounts of alkylidenediphosphane **12** and another species that gives a ^{31}P NMR pattern closely related to those of the 2-phospha-1,3-disiletanes **7a** and **7b** (from Schemes 5 and 6) [2]. The suprising double $SiCl_2$ attack on the P=C double bond is achieved only by silylgermanes **1** and **2**, but not by Si_2Cl_6.

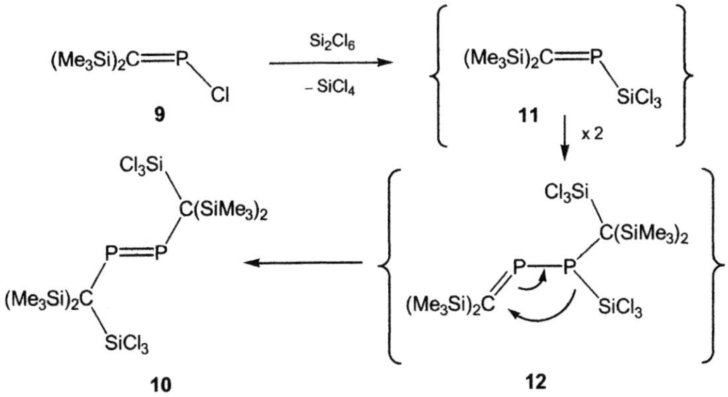

Scheme 7. Reaction of *P*-chlorophosphaalkene **9** with Si_2Cl_6.

Acknowledgments: We thank Professor H. C. Marsmann for ^{29}Si NMR measurements, Professor P. G. Jones for structure determinations and the Deutsche Forschungsgemeinschaft for financial support.

References

[1] G. Urry, *Acc. Chem. Res.* **1970**, *3*, 306.

[2] W.-W. du Mont, T. Gust, E. Seppälä, C. Wismach, *J. Organomet. Chem.* **2004**, *689*, 1331.

[3] W.-W. du Mont, E. Seppälä, T. Gust, J. Mahnke, L. Müller, *Main Group Metal Chemistry*, **2001**, *24*, 609.

[4] L. Müller, W.-W. du Mont, F. Ruthe, P. G. Jones, H. C. Marsmann, *J. Organomet. Chem.* **1999**, *579*, 156.

[5] E. Seppälä, W.-W. du Mont, T. Gust, J. Mahnke, L. Müller, in *Organosilicon Chemistry V* (Eds. N. Auner, J. Weis), Wiley-VCH, **2003**, 213.

[6] T. A. Banford, A. G. MacDiarmid, *Inorg. Nucl. Chem. Lett.* **1972**, *8*, 733.

[7] R. Martens, W.-W. du Mont, L. Lange, *Z. Naturforsch.* **1991**, *46b*, 1609.

[8] R. Martens, W.-W. du Mont, *Chem. Ber.* **1992**, *125*, 657.

[9] W.-W. du Mont, T. Gust, E. Seppälä, C. Wismach, P. G. Jones, L. Ernst, J. Grunenberg, H. C. Marsmann, *Angew. Chem.* **2002**, *114*, 3977-3979; *Angew. Chem. Int. Ed.* **2002**, *41*, 3829.

[10] A. Zanin, M. Karnop, J. Jeske, P. G. Jones, W.-W. du Mont, *J. Organomet. Chem.* **1994**, *475*, 95.

Synthesis of *B*-Alkylsilylborazines by Rhodium-Catalyzed Hydroboration of *N*-Alkylborazines

Christian Lehnert,* Gerhard Roewer

TU Bergakademie Freiberg, Department of Inorganic Chemistry
Leipziger Strasse 29, 09596, Freiberg, Germany
Tel.: +49 3731 393413 — Fax: +49 3731 394058
E-mail: Christian.Lehnert@student.tu-freiberg.de

Keywords: hydroboration, *B*-alkylsilylborazines, *N*-trialkylborazines, vinylsilanes

Summary: Hydroboration reactions of various vinylsilanes with *N*-trimethylborazine (N-TMB) and *N*-triethylborazine (N-TEB) had been used to prepare *B*-alkylsilylborazines. The *N*-trialkylborazines were easily accessible by addition of BH_3*SMe_2 to a solution of the primary amines RNH_2 (R= Me, Et) and subsequent thermolysis of the resulting borane-amine complexes at 200 °C. The $RhH(CO)(PPh_3)_3$-catalyzed hydroboration reactions of these borazine derivatives with various vinylsilanes gave mono-, di- and tri-*B*-alkylsilylborazines. The degree of substitution can be controlled by varying the borazine/silane ratios, and the regioselectivity of the reactions is influenced by the kind of silane used. ^{11}B, ^{13}C and ^{29}Si NMR spectroscopy confirmed the formation of the products.

Introduction

The synthesis of processable precursors for Si–B–N–C ceramics became a goal of intensive investigations as soon as the outstanding thermal and mechanical properties of this system were reported [1, 2]. The amorphous phase of Si–B–N–C ceramics can show excellent thermal stability up to 2000 °C without mass loss or crystallization. The role of boron is believed to be to increase the high-temperature stability and to prevent the crystallization and decomposition of silicon nitride above 1500 °C. Primarily, the atomic ratio and chemical environment of boron in Si–B–N–C precursors seem to affect the thermal behavior of resulting ceramic materials.

The examination of structure/property relations of molecular educts and resulting ceramics required the synthesis of stoichiometrically and structurally different precursors. A variety of synthesis routes for Si–B–N–C precursors from organosilanes, silazanes, and boron compounds have been reported in recent years [3 – 5]. As an example, Riedel obtained a polymeric precursor via hydroboration of methylvinyldichlorosilane and subsequent condensation of the hydroboration product with ammonia (Eq. 1). Pyrolysis led to silicoboron carbonitride ceramics exhibiting thermal stability up to 2000 °C [6].

3 $H_3C\text{–}Si(Cl)(CH{=}CH_2)\text{–}Cl$ + $H_3B\text{–}S(CH_3)_2$ —(Toluene / 0 °C; - $S(CH_3)_2$)→ $(R)_2B\text{–}CH_2\text{–}CH_2\text{–}Si(CH_3)Cl_2$

n $(R)_2B\text{–}CH_2\text{–}CH_2\text{–}Si(CH_3)Cl_2$ + 9n NH_3 —(THF / 0 °C; - 6n NH_4Cl)→ $[\text{–}Si(Cl)(CH_2CH_2B(R')_2)\text{–}NH\text{–}]_n$

R= $-CH(CH_3)\text{-}Si(CH_3)Cl_2$ R′= $-CH(CH_3)\text{-}Si(CH_3)\text{-}NH\text{-}$

Eq. 1. Hydroboration of methylvinyldichlorosilane and subsequent ammonolysis – a precursor for Si–B–N–C ceramic materials.

Borazine and its derivatives are also possible educts to synthesize precursors for Si–B–N–C ceramics. Sneddon and co-workers prepared Si–B–N–C preceramic precursors via the thermal dehydrocoupling of polysilazanes and borazines [7]. A further synthesis route is the hydroboration of borazines. The work group of Sneddon found that definite transition metal reagents catalyze hydroboration reactions with olefins and alkynes to give *B*-substituted borazines [8]. Recently, Jeon et al. reported the synthesis of polymer-derived Si–B–N–C ceramics even by uncatalyzed hydroboration reactions from borazines and dimethyldivinylsilane [9].

This paper reports the synthesis of *B*-alkylsilylborazines by rhodium catalyzed hydroboration with *N*-trialkylborazines. The characterization of these molecules by NMR spectroscopy gives valuable facts about the regioselectivity and completeness of the reactions and should enable a more facile assignment of NMR signals from precursors of Si–B–N–C ceramics based on borazines and vinylsilanes as educts.

Results and Discussion

Synthesis of *N*-Trialkylborazines

As shown by Framery and Vaultier, N-substituted borazines are easily accessible by thermolysis of primary amine-borane complexes $RNH_2{*}BH_3$ [10]. Based on this work, *N*-trimethylborazine (N-TMB) and *N*-triethylborazine (N-TEB) were synthesized pure and in excellent yields (Scheme 1).

First, the primary amine-borane complexes $MeNH_2{*}BH_3$ and $EtNH_2{*}BH_3$ were prepared by addition of the $H_3B{*}SMe_2$ complex to a solution of methylamine or ethylamine in thf at –78 °C.

The complexes were obtained after the removal of thf and SMe_2 under vacuum. Thermolysis of these complexes at 120 °C led to borazanes as intermediates. A further thermolysis at 200 °C gave the desired borazines, which were purified by distillation.

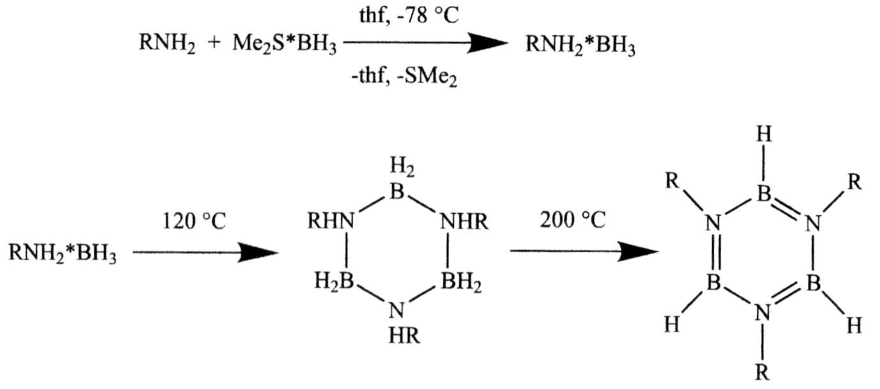

$$RNH_2 + Me_2S{*}BH_3 \xrightarrow[\text{-thf, -}SMe_2]{\text{thf, -78 °C}} RNH_2{*}BH_3$$

$$RNH_2{*}BH_3 \xrightarrow{120\ °C} (RHN\text{-}BH_2)_3 \xrightarrow{200\ °C} (RN\text{-}BH)_3$$

R= Me, Et

Scheme 1. Synthesis of *N*-alkyl-substituted borazines by thermolysis of primary amine-borane complexes

Hydroboration Reactions of *N*-Trialkylborazines with Vinylsilanes

The obtained borazines N-TMB and N-TEB were tested in hydroboration reactions with various vinylsilanes, including $ViSiCl_3$, $ViSiClMe_2$, $ViSiCl_2Ph$, $ViSiClPh_2$, $ViSiPh_3$ and $ViMeClSi\text{–}SiCl_2Me$. Whereas without using a catalyst no conversion of the educts occurs, the rhodium complex $RhH(CO)(PPh_3)_3$ catalyzed the desired reaction. Additionally, the reaction conditions are of importance. The reaction does not proceed in toluene or thf as solvent. However, the conversion of the educts was successful using CCl_4 as solvent. A typical reaction involved stirring a mixture of N-TMB or N-TEB and vinylsilane under reflux in CCl_4 as solvent in the presence of small amounts of the catalyst for several hours.

The reaction progress can be easily indicated by ^{13}C NMR spectroscopy. The signal intensity of the olefinic moiety (130 – 145 ppm) decreases during this process, its disappearance indicating the completeness of the hydroboration reactions. Simultaneously the signal intensity belonging to the alkyl bridge between borazine and silyl moiety increases.

In hydroboration reactions with vinylsilanes, the addition of the boron atom can take place at the end carbon atom of the vinyl moiety (anti-Markovnikov addition) or at the carbon atom bonded to the silicon atom (Markovnikov addition), as shown in Scheme 2. Under the employed conditions the anti-Markovnikov product was formed predominantly.

Reactions with $ViSiPh_3$, $ViSiPh_2Cl$, and $ViSiPhCl_2$ gave this product exclusively. But reactions with $ViSiCl_3$, $ViSiClMe_2$, and $ViMeClSi\text{–}SiClMe_2$ also yield the Markovnikov product as indicated by using the ^{13}C APT pulse sequence. These results suggest that the regioselectivity depends on the steric properties of the vinylsilane. The ^{29}Si NMR data of hydroboration reactions in a borazine /silane ratio 1:1 are presented in Table 1.

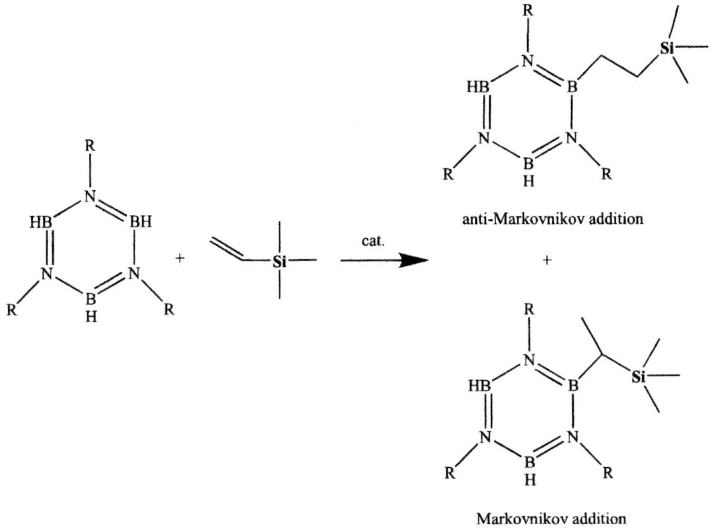

cat.

anti-Markovnikov addition

Markovnikov addition

R= Me, Et

Scheme 2. Regioselectivity of hydroboration reactions between *N*-alkylsubstituted borazines and vinylsilanes.

Table 1. ^{29}Si NMR data of products obtained by hydroboration reactions between vinylsilanes and *N*-trialkylborazines; borazines/silane ratio 1:1.

	Vinylsilane	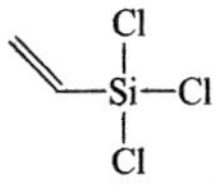 $CH_2=CH-SiCl_3$	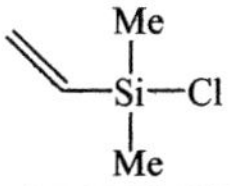 $CH_2=CH-SiMe_2Cl$	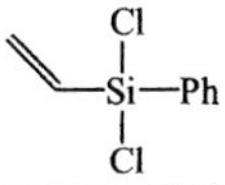 $CH_2=CH-SiCl_2Ph$	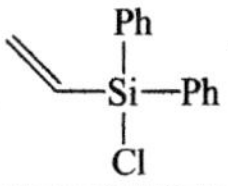 $CH_2=CH-SiClPh_2$	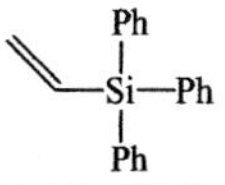 $CH_2=CH-SiPh_3$	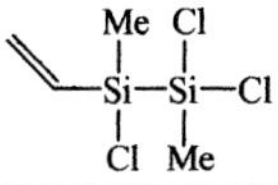 $CH_2=CH-SiMeCl-SiMeCl_2$
N-trialkyl-borazine	Educt	3.0	17.9	3.6	−1.6	−16.9	1.8; 23.5
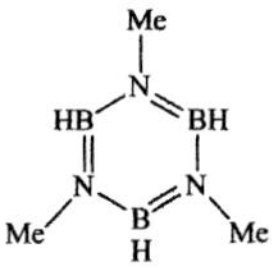 $(MeNBH)_3$	Anti-Markovnikov product	12.1	22.4	16.8	10.8	−11.9	20.7; 31.0
	Markovnikov product	14.8	22.8	–	–	–	21.7; 32.2
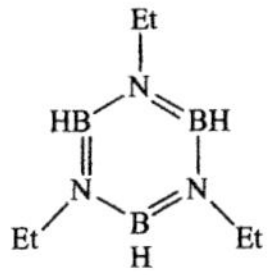 $(EtNBH)_3$	Anti-Markovnikov product	11.9	22.0	18.0	11.2	−11.5	20.9; 31.4
	Markovnikov product	14.6	22.6	–	–	–	–

The conversion of the vinyl moieties into alkyl substituents gives rise to a signal downfield shift of 5 – 20 ppm in the ^{29}Si NMR spectra compared to the educt silanes. Varying the silane/borazine ratio to 2:1 or 3:1 led in several cases to complex ^{29}Si and ^{13}C NMR spectra due to the formation of mixed anti-Markovnikov and Markovnikov products.

The degree of substitution can be controlled by varying the reactant ratios (Eq. 2).

Eq. 1. Hydroboration reaction between N-TMB and vinyltrichlorosilane.

Mono-, di- and tri-*B*-alkylsilylborazines as well as the unsubstituted *N*-trialkylborazine show characteristic ^{11}B NMR spectra (Fig. 1).

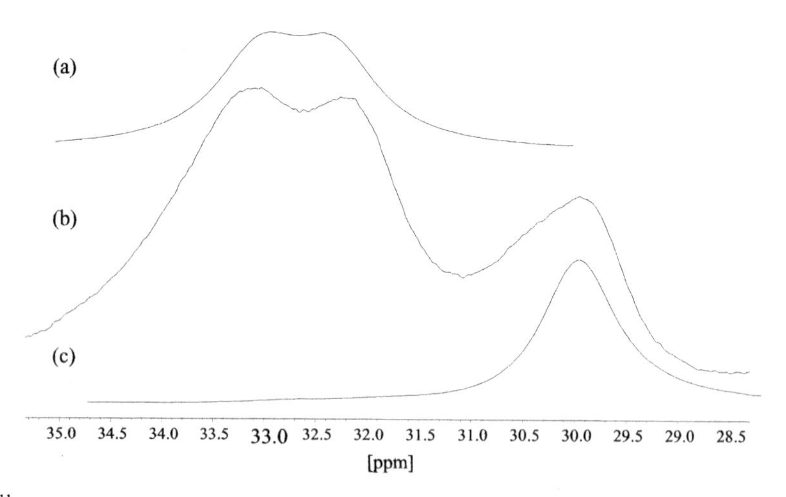

Fig. 1. ^{11}B NMR spectrum of (a) N-TMB, (b) TMB + $ViSiCl_3$, ratio 1:1, (c) TMB + $ViSiCl_3$, ratio 1:3.

Whereas the ^{11}B NMR spectrum of the pure *N*-trimethylborazine consists of a broad doublet with its center near 33 ppm, the spectra of mono-*B*-alkylsilylborazines show a singlet between 29 and 30.5 ppm beside the doublet. In the spectra of two fold-substituted borazines, the signal intensity of the singlet increases and the doublet signal shows less intensity. In the ^{11}B NMR spectra of tri-*B*-alkylsilylborazines, only a singlet appears near 30 ppm. A borazine/silane ratio 1:3 gave exclusively completely substituted borazines according to ^{11}B NMR spectroscopy.

Experimental

Synthesis of N-TMB and N-TEB: both borazine derivatives were synthesized as described in Ref. [10]. The reactions of 0.2 mol $MeNH_2$ or 0.2 mol $EtNH_2$ with 0.2 mol H_3B*SMe_2 gave 6.5 g N-TMB (yield: 80%; ^{11}B NMR: $\delta = 32.8$; ^{13}C NMR: $\delta = 38.1$; b.p. (760 Torr) 132 °C) and 8.5 g N-TEB (yield: 77%; ^{11}B NMR: $\delta = 32.2$; ^{13}C NMR: $\delta = 20.9, 46.1$; b.p. (15 Torr) 63 °C).

Hydroboration reactions: all preparations were carried out under an inert atmosphere (Ar). Carefully dried glassware and dry solvents were used. In a typical reaction, 50 mg of $RhH(CO)(PPh_3)_3$ was dissolved in 10 mL CCl_4. 1.0 g N-TMB (8.16 mmol) or 1.0 g N-TEB (6.03 mmol) were added to the yellow solution. The respective vinylsilane was added dropwise together with 10 mL of CCl_4 to these mixtures in borazine/silane ratios of 1:1, 1:2 or 1:3. The resulting reaction mixtures were kept under reflux for several hours. After this, the solvent was removed under vacuum and the catalyst was filtered off. The remaining oils were characterized by NMR spectroscopy.

Conclusions

The $RhH(CO)(PPh_3)_3$-catalyzed hydroboration reactions of N-TMB and N-TEB with various vinylsilanes in CCl_4 as solvent led to mono-, di- and tri-*B*-alkylsilylborazines. The degree of substitution can be controlled by varying the borazines/silane ratios. The regioselectivity of the reactions is influenced by the steric properties of the vinylsilanes. However, the formation of anti-Markovnikov products is preferred under the employed reaction conditions. In several cases the product mixtures were complicated because of the formation of mixed anti-Markovnikov and Markovnikov products and different degrees of substitution. The aim of further examinations must be the separation of the formed products.

References

[1] D. Seyferth, W. S. Rees, *Chem. Mater.* **1991**, *3*, 1106.
[2] J. Bill, F. Aldinger, *Adv. Mater.* **1995**, *7*, 775.
[3] D. Seyferth, H. Plenio, *J. Am. Ceram. Soc.* **1990**, *73*, 2131.
[4] M. Jansen, B. Jäschke, T. Jäschke, *Structure and Bonding* **2002**, *101*, 137.
[5] M. Weinmann, T. W. Kamphowe, J. Schuhmacher, K. Müller, F. Aldinger, *Chem. Mater.* **2000**, *12*, 2116.
[6] R. Riedel, A. Kienzle, V. Szabo, J. Mayer, *J. Mater. Sci.* **1993**, *28*, 3931.
[7] T. Wideman, K. Su, E. E. Remsen, G. A. Zank, L. G. Sneddon, *Chem. Mater.* **1995**, *7*, 2203.
[8] A. T. Lynch, L. G. Sneddon, *J. Am. Chem. Soc.* **1989**, *111*, 6201.
[9] J. K. Jeon, Q. D. Nghiem, D. P. Kim, J. Lee, *J. Organomet. Chem.* **2004**, *689*, 2311.
[10] E. Framery, M. Vaultier, *Heteroat. Chem.* **2000**, *3*, 218.

Hydrosilylation of Ethynylborazines and Their Use for the Formation of a Highly Functionalized Silica Gel

*J. Haberecht, H. Rüegger, R. Nesper, H. Grützmacher**

ETH Zurich, ETH Hönggerberg-HCI H131
W.-Pauli-Str. 10, CH-8093 Zurich, Switzerland
Tel.: +41 1 6322855 — Fax: +41 1 6331032
E-mail: gruetzmacher@inorg.chem.ethz.ch

Keywords: hydrosilylation, silica gel, NMR

Summary: We report that triethynylborazine (**1**) can be easily functionalized by a heterogeneously catalyzed hydrosilylation reaction to give mixtures of α- and β-tri(silylvinyl) borazines (**2** – **5**). The reaction with trichlorsilane is particularly useful, as one isomer, the β-tris-(trichlorosilylvinyl)borazine (R = Cl), is formed preferentially (~80% yield). The resulting tri(silylvinyl)borazines are suitable precursors for the formation of silica gels which are highly functionalized. Especially, the trichlorosilyl derivative, where the Si–Cl bonds are more rapidly hydrolyzed than the B_3N_3 ring, allows the preparation of a silica gel with intact borazine and vinyl moieties. BET measurements show that this material has a specific surface area of approximately 315 m^2/g.

Introduction

The design of porous solids has become of great interest in both chemical research and industry because of their high potential for catalysis and material science. For example, inorganic-organic precursors for porous solid structures have been published; borazines of the type $[B(CH_2SiCl_2R)_3(NH)_3]$ were also recognized as suitable single-source precursors for the formation of multinary non-oxide inorganic materials [1, 2]. The aim of our work was the formation of a silica gel with well-defined structural units based on borazine precursor molecules.

Hydrosilylation Reactions of Triethynylborazine

Starting from triethynylborazine (**1**), we studied the possibility of performing platinum-catalyzed hydrosilylations of the C≡C triple bonds [3]. The regioselectivity of the addition of chlorosilanes to terminal alkynes with platinum on charcoal as heterogeneous catalyst was studied recently, and a high selectivity for the formation of β-*trans* vinylsilanes was observed [4]. In our experiments, we used trichlorosilane, $HSiCl_3$, and trialkoxysilanes, $HSi(OR^1)_3$ (R^1 = Me, Et, *i*Pr), as reagents and the triethynylborazine **1** as substrate (Scheme 1).

After 36 – 48 h at T = 120 °C in toluene, the hydrosilylation of **1** was almost complete; in experiments with the alkoxysilanes, less than 5% of non-hydrosilylated C≡C triple bond units were detected by ^{13}C NMR spectroscopy. In all cases, the addition of the silane to the C≡C triple bond led to *cis*-stereoselective reaction products; these were tri-β-*trans* isomers β-**2** – β-**5**. With $HSiCl_3$, more than 80% of this isomer, β-**5**, was formed. With the alkoxysilanes, $HSi(OR^1)_3$, the reactions were less regioselective, and complex product mixtures with α- and β-silylated vinyl groups were obtained. From an integration of the ^{13}C NMR resonances, we estimated that approximately 60% of the C≡C triple bond units were hydrosilylated in the β-position, independently of the nature of the substituents (R^1 = Me, Et, *i*Pr). ^{29}Si-NMR investigations (see Table 1) reveal typical shifts in the range of –53 ppm to –63 ppm [5].

H–C≡C– ... HSiX$_3$, Pt/C 1% → ... + ...

1

β-**2** X = Cl

β-**3** X = OMe

β-**4** X = OEt

β-**5** X = O^{i}Pr

α/β-**2** – **5**

Scheme 1. Catalytic hydrosilylation of triethynylborazine.

Table 1. ^{29}Si (in $CDCl_3$) and ^{13}C (in C_6D_6) NMR shifts of the *B,B′,B″*-tri(silylvinyl)borazines. Shifts are mean values of isomer mixtures.

X=	$\delta(^{29}Si)$ Si–C$_\alpha$C	$\delta(^{29}Si)$ Si–C$_\beta$C	$\delta(^{13}C)$ B–C$_\alpha$C	$\delta(^{13}C)$ B–CC$_\beta$
Cl	0.2	–3.3	153.6	137.4
OMe	–53.3	–56.2	152.4	135.5
OEt	–55.4	–59.2	152.1	138.3
O*i*Pr	–59.5	–63.0	151.7	140.4

Using Pt on charcoal (1% weight) as catalyst, the hydrosilylation reaction is selective for the C≡C triple bond; no further hydrosilylation or any reaction with the borazine can be observed.

Silica Gel Formation

Evidently, the *B*-tri(silylvinyl)borazines **2** – **5** are attractive reagents for sol-gel processes to give highly functionalized silica gels, which contain Lewis-acidic (the boron centers) and Lewis-basic (the nitrogen and olefinic) sites. This property would make these borazines interesting supports. To our knowledge, silica gels comparable to the ones we targeted have not been described in the literature. Only related composite materials formed from cyclo /polyphosphazenes and metal oxides (e.g., SiO_2, TiO_2, ZrO_2, Al_2O_3) via a sol-gel process have been described in the literature [6]. As mentioned above, borazines are comparatively stable; however, they are susceptible to hydrolysis upon prolonged exposure to moisture. Therefore we have chosen the borazine β-**2**, containing highly reactive trichlorosilyl groups, for our experiments. In this compound, the chloro substituents should be hydrolyzed much more rapidly than the B_3N_3 ring. Indeed, when a solution of β-**2** in tetrahydrofurane (THF) containing tri(*n*-butyl)amine as HCl binding reagent has reacted with an equimolar amount of water under classical sol-gel conditions, a colorless clear gel was formed in daytime (see equation in Scheme 2). After extraction with THF to remove the ammonium chloride [nBu$_3$HN]Cl and drying in vacuum, a yellow amorphous silica gel was obtained (see Fig. 1).

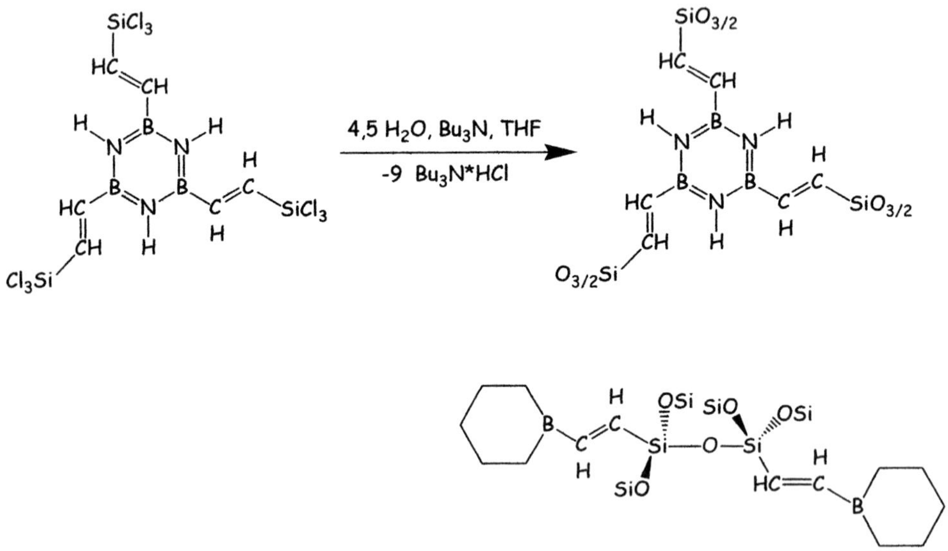

Scheme 2. Formation of the silica gel from **2** and the building block of it characterized by NMR and IR spectroscopy.

In Fig. 2, the IR spectrum of the crude silica gel after purification compared to that of the precursor molecule **2** are shown. Obviously, in both spectra, E' N–B–N vibration modes at about 1470 cm^{-1} and 1320 cm^{-1}, typical for borazines, can be observed. The C=C stretching vibration is present at about 1590 cm^{-1}. The H–C=C bending mode is observed together with the intense Si–O–Si stretching modes at about 1000 cm^{-1}. These data indicate that all functional groups, especially the borazine rings, have been retained, and the silicon atoms are connected by Si–O–Si bridges during silica gel formation (see Scheme 2). In the ^{29}Si MAS NMR spectrum of the silica gel (Fig. 3), only one signal at –72 ppm is observed, which is very typical of an SiO_3C environment around the silicon nucleus. It demonstrates that all chloro substituents in β-**2** have been replaced by oxygen substituents [5].

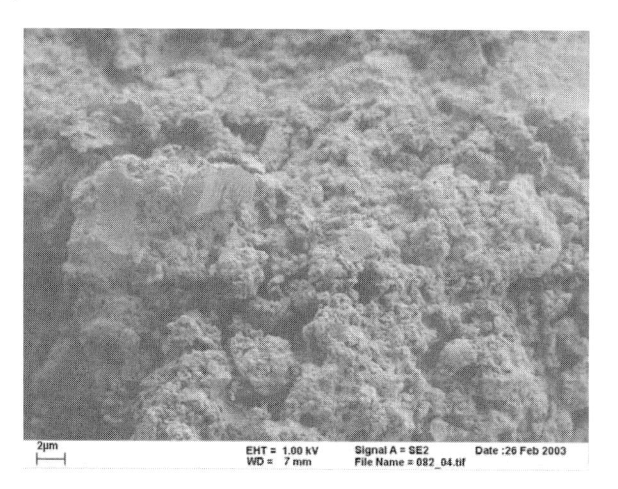

Fig. 1. SEM image of the silica gel.

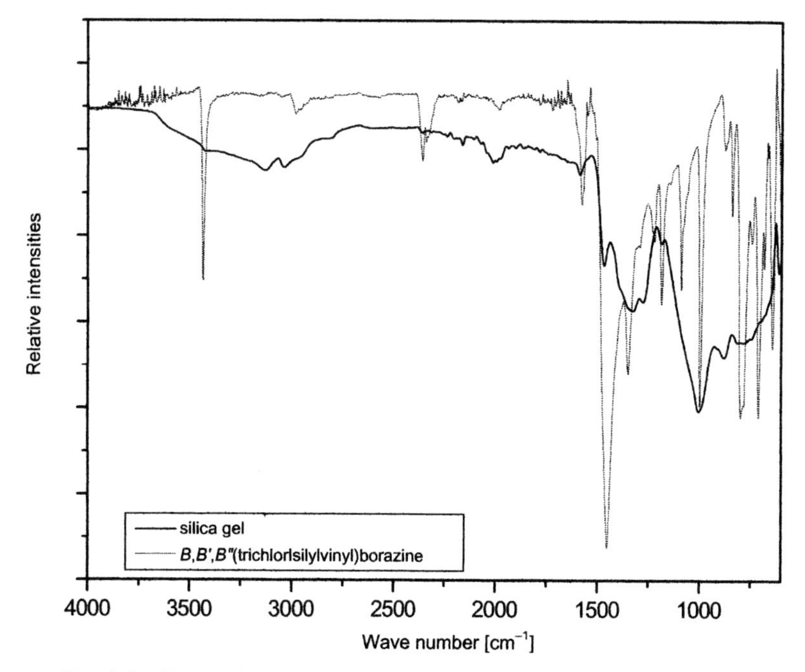

Fig. 2. IR spectra of **2** and the silica gel for comparison.

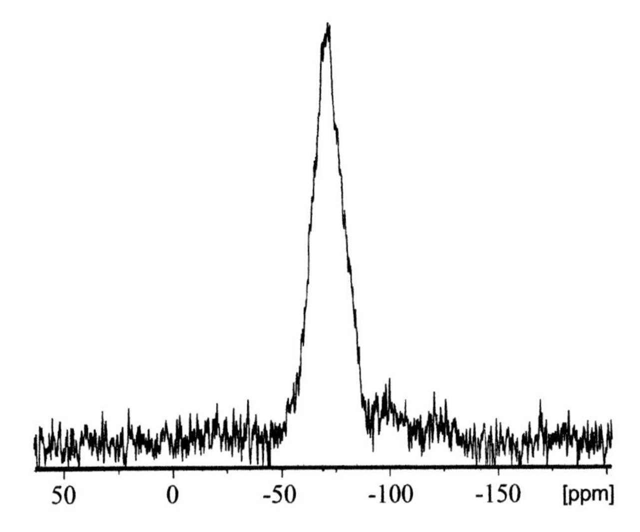

Fig. 3. The ^{29}Si MAS NMR spectrum of the silica gel reveals a [SiO_3C] coordination [5].

The thermal stability of the silica gel was investigated by a TG measurement. A first weight loss occurs at about 200 °C and amounts to 5% at about 400 °C. In a second step, another 5.4% of the weight is lost between 400 °C and 800 °C. Finally, at a temperature of 1500 °C, a total of about 13.5% of the starting weight has been lost. Repeated BET measurements of the purified and dried silica gel give a specific surface area of about 315 m^2/g.

We also tested silica gel formation using the trialkoxy-substituted tri(silylvinyl)borazines **3** – **5**. The gel formation was much slower (4 – 5 d) and not complete, i.e. residual alkoxy groups were detected by IR spectroscopy. Also, these materials had a significantly lower specific surface area (~230 m^2/g).

Conclusion

With the data presented in this work, we showed that trialkynylborazines can easily be functionalized by a heterogeneously catalyzed hydrosilylation reaction. The resulting tri(β-silylvinyl)borazines are suitable precursors for the formation of silica gels which are highly functionalized. In particular, trichlorosilyl derivatives allow the preparation of a silica gel with well-defined structural units, because the Si–Cl bonds are more rapidly hydrolyzed than the B_3N_3 rings. Remarkably, stirring a suspension of the silica gel in water for several hours did not lead to a color change and notable decomposition, indicating that the borazine rings are quite stable against hydrolysis when immobilized in the gel. The vinyl groups are additional functional groups to which transition metals may be bonded, and this aspect is currently being investigated in our laboratories.

Acknowledgments: We thank Dr. C. Stinner for BET measurements and Dr. F. Krumeich for SEM investigations.

References

[1] S. Nitsche, E. Weber, K. Trommer, G. Roewer, Synthesis and characterization of new precursors based on organosilicon for construction of porous solid structures in *Organosilicon Chemistry IV — From Molecules to Materials* (Eds.: N. Auner, A. Weiss), VCH, Weinheim, **2000**, p. 312.

[2] M. Jansen, T. Jaeschke, B. Jaeschke, *Amorphous multinary ceramics in the Si–B–N–C system* in: *Structure and Bonding, Vol. 101* (Ed.: M. Jansen), Springer-Verlag, Berlin, Heidelberg, **2002**, p. 137.

[3] J. Haberecht, A. Krummland, F. Breher, B. Gebhardt, H. Rüegger, R. Nesper, H. Grützmacher, *Dalton Trans.* **2003**, 2126.

[4] M. Chauhan, B. J. Hauck, L. P. Keller, P. Boudjouk, *J. Organomet. Chem.* **2002**, *645*, 1.

[5] *The Chemistry of Organic Silicon Compounds, Vol. 3*, (Eds.: Z. Rappoport, Y. Apeloig), John Wiley, Chichester, **2001**.

[6] M. Guglielmi, G. Brusatin, G. Facchin, M. Gleria, *Appl. Organomet. Chem.* **1999**, *13*, 339.

Influence of the Tri(*tert*-butyl)silyl Substituents on the Molecular Structures of Phosphanylalanes, -gallanes and -indanes

Sabine Weinrich, Mathias Krofta, Axel Schulz

Department Chemie, Ludwig-Maximilians-Universität München
Butenandtstr. 9, D-81377 Munich, Germany

Matthias Westerhausen*

Institut für Anorganische und Analytische Chemie der Friedrich-Schiller-Universität
August-Bebel-Straße 2, D-07743 Jena, Germany
Tel.: +49 3641 948129 — Fax: +49 3641 948102
E-mail: m.we@uni-jena.de

Keywords: gallium, heteroallyl cations, indium, phosphanide, phosphorus heterocycles

Summary: The metathesis reaction of $GaCl_3$ and $KP(H)SitBu_3$ as well as the metalation of $H_2PSitBu_3$ with $GaEt_3$ yield $[R_2GaP(H)SitBu_3]_2$, with R = Cl $(\mathbf{1})_2$ and R = Et $(\mathbf{2})_2$. Reaction of $(\mathbf{1})_2$ with *t*BuLi or thermolysis of $(\mathbf{2})_2$ give the tetramers $(R'GaPSitBu_3)_4$ with R′ = H $(\mathbf{3})_4$ and R′ = Et $(\mathbf{4})_4$, respectively. The metathesis reaction of $(\mathbf{1})_2$ with $KP(H)SitBu_3$ yields $[tBu_3SiP(H)–Ga(\mu\text{-}PSitBu_3)]_2$ $(\mathbf{5})_2$ with triply coordinated Ga and P atoms, whereas the metathesis with $AlCl_3$ gives $[(toluene)K]^+$ $[(Cl_2Al)_3(\mu_3\text{-}PSitBu_3)_2]^-$ (**6**) with a central trigonal Al_3P_2 bipyramid. The metalation of $H_2PSitBu_3$ with $AlEt_3$ yields $[Et_2AlP(H)SitBu_3]_2$ $(\mathbf{7})_2$. The metalation of $H_2PSitBu_2$ with $In[N(SiMe_3)_2]_3$ gives the dimer $[tBu_3SiP(H)–In(\mu\text{-}PSitBu_3)]_2$ $(\mathbf{8})_2$. At elevated temperatures $InCl_3$ reacts with $KP(H)SitBu_3$ to the KCl adduct of $(ClInPSitBu_3)_4$ $(\mathbf{9})_4$ with a central In_4P_4 heterocubane moiety.

Introduction: Brief Overview of Compounds of the Type $[R_2E^{III}–P(SiMe_3)_2]_n$

The intramolecular steric strain defines the size of the $(E^{III}P)_n$ cycles in compounds such as $(X_2E^{III}–PR_2)_n$ and of cages in $(XE^{III}\text{-}PR)_n$, with E^{III} being an "earth metal" (element of the boron group) and X an alkyl or halide group. The elimination of $ClSiMe_3$ from $Cl_3E^{III}–P(SiMe_3)_3$ yields the dimers $[Cl_2E^{III}–P(SiMe_3)_2]_2$ of aluminum [1], gallium [2] and indium [3]. The interest in these phosphanylalanes, gallanes and indanes resulted from the expectation that these compounds were synthons for the preparation of AlP [4], GaP [5] and InP [3], respectively. Whereas $E^{III}Cl_3$ and $P(SiMe_3)_3$ formed a 1:1 adduct [2, 6], phosphanylalanes were also obtained from the metathesis reaction of dialkylaluminum halide and lithium bis(trimethylsilyl)phosphanide [7, 8]. Metalation reactions of trialkylalane [9] or dialkylalane [10] with $HP(SiMe_3)_2$ yielded cyclic $[R_2AlP(SiMe_3)_2]_2$, which were monomerized with strong Lewis bases [11]. The metalation of H_2PSiPh_3 with diisobutylalane at 80 °C in toluene gave $(i\text{BuAlPSiPh}_3)_4$, with a central Al_4P_4 heterocubane cage [12]. Smaller substituents at the Al atom led to the formation of hexamers with hexagonal Al_6P_6 prisms [13, 14].

Similar investigations led to the synthesis of $[R_2GaP(SiMe_3)_2]_2$ [15]. The partial substitution of $SiMe_3$ groups by smaller substituents led to an enlargement of the ring size, and a Ga_3P_3 cycle was formed [16]. An arylgallium triphenylsilylphosphanediide with a Ga_2P_2 moiety showed pyramidally coordinated phosphorus atoms with Ga–P distances of 233.8 pm and angle sums at the P atoms of 325.5° [17].

The influence of intramolecular steric strain can easily be seen from $[R_2InP(SiMe_3)_2]_n$ with n = 3 for R = Me [18] and n = 2 for an ethyl group at the earth metal [19]. The intramolecular metalation of $[Et_2InP(H)SiR_3]_2$ led to cages with In_4P_4 heterocubane (SiR_3 = Si*i*Pr_3) or hexagonal prismatic In_6P_6 (SiR_3 = $Si(Me_2)CMe_2$*i*Pr) moieties [20].

Synthesis of Tri(*tert*-butyl)silylphosphanylalanes, -gallanes and -indanes

The metathesis reaction of potassium tri(*tert*-butyl)silylphosphanide [21] with $GaCl_3$ in toluene gives $[Cl_2GaP(H)Si\textit{t}Bu_3]_2$ $(\mathbf{1})_2$ with the *trans* isomer as the major product. The metalation of $H_2PSi\textit{t}Bu_3$ with $GaEt_3$ yields $[Et_2GaP(H)Si\textit{t}Bu_3]_2$ $(\mathbf{2})_2$ according to Eq. 1.

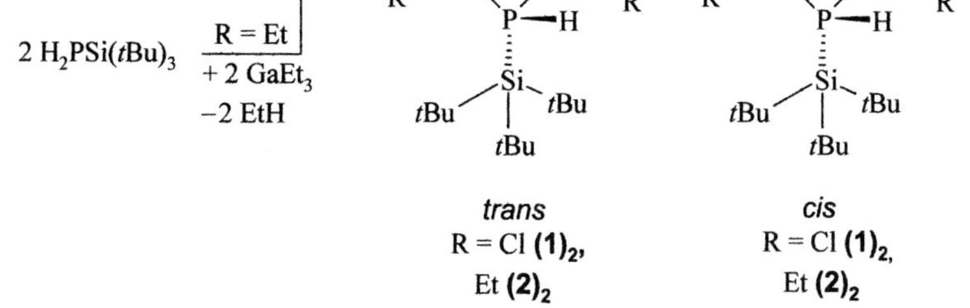

Eq. 1. Synthesis of $[R_2GaP(H)Si\textit{t}Bu_3]_2$ via metathesis [R = Cl, $(\mathbf{1})_2$] and metalation reaction [R = Et, $(\mathbf{2})_2$].

The reaction of $(\mathbf{1})_2$ with *t*BuLi leads to a metalation of the phosphanide substituent and the elimination of LiCl. Another equivalent of the lithiation reagent yields LiCl, butane, and the tetramer $[HGaPSi\mathit{t}Bu_3]_4$ $(\mathbf{3})_4$ according to Eq. 2. Heating of $(\mathbf{2})_2$ leads to an intramolecular metalation and dimerization which finally yields $[EtGaPSi\mathit{t}Bu_3]_4$ $(\mathbf{4})_4$. The central moieties of these compounds are Ga_4P_4 heterocubane fragments.

In order to enhance the intramolecular steric strain on the Ga_2P_2 cycle, compound $(\mathbf{1})_2$ is reacted with excess of $KP(H)Si\mathit{t}Bu_3$ according to Eq. 3 to give $[\mathit{t}Bu_3SiP(H)\text{–}Ga(\mu\text{-}PSi\mathit{t}Bu_3)]_2$ $(\mathbf{5})_2$. The dimeric nature of this compound can clearly be deduced from the ^{31}P NMR spectrum.

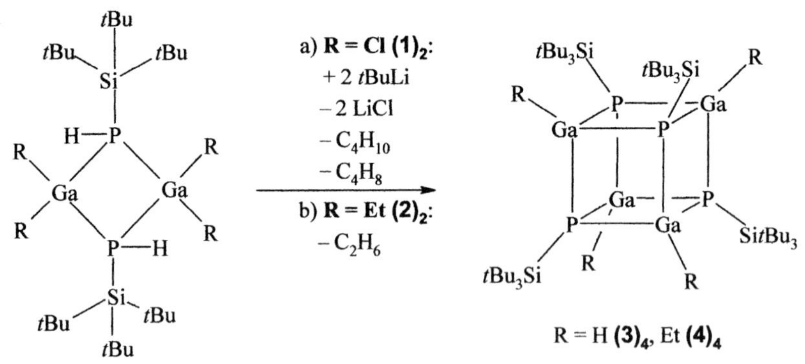

Eq. 2. Synthesis of tri(*tert*-butyl)silylphosphanediides of gallium.

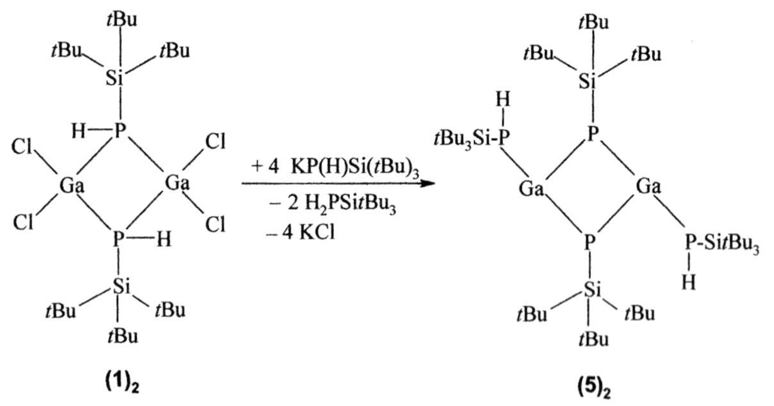

Eq. 3. Synthesis of dimeric tri(*tert*-butyl)silylphosphanylgallium tri(*tert*-butyl)silylphosphanediide $(\mathbf{5})_2$.

The metathesis reaction of $KP(H)Si\mathit{t}Bu_3$ with $AlCl_3$ yields the solvent-separated ion pair $[(toluene)K]^+$ $[(Cl_2Al)_3(\mu_3\text{-}PSi\mathit{t}Bu_3)_2]^-$ $(\mathbf{6})$ with a central trigonal Al_3P_2 bipyramid according to Eq. 4. The metalation of tri(*tert*-butyl)silylphosphane with $AlEt_3$ gives $[Et_2AlP(H)Si\mathit{t}Bu_3]_2$ $(\mathbf{7})_2$ as a mixture of the *cis* and *trans* isomers similar to the gallium derivative $(\mathbf{2})_2$.

The metalation of $H_2PSi\mathit{t}Bu_3$ with $In[N(SiMe_3)_2]_3$ in a molar ratio of 2:1 leads to the formation of the dimer $[tBu_3SiP(H)–In(\mu\text{-}PSitBu_3)]_2$ **(8)**$_2$. A metathesis reaction of KP(H)Si*t*Bu$_3$ with $InCl_3$ in a molar ratio of 2 : 1 yields, under elimination of $H_2PSitBu_3$, the KCl adduct of the tetramer $(ClInPSitBu_3)_4$ **(9)**$_4$ with a central In_4P_4 heterocubane moiety according to Eq. 5. The metathesis reaction of the sterically less crowded KP(H)Si*i*Pr$_3$ with $InCl_3$ in a molar ratio of 4 : 1 gives polymeric potassium tetrakis[tri*iso*propylsilylphosphanyl]indate $[KIn(\mu\text{-}P\{H\}SiiPr_3)_4]_\infty$ [22].

$$4\ KP(H)SitBu_3 + 3\ AlCl_3 \xrightarrow[-3\ KCl]{-2\ H_2PSiR_3} (\text{toluene})K^+\ [(Cl_2Al)_3(\mu_3\text{-}PSiR_3)_2]^-$$

Eq. 4. Synthesis of the solvent-separated ion pair $[(\text{toluene})K]^+\ [(Cl_2Al)_3(\mu_3\text{-}PSitBu_3)_2]^-$ **(6)**.

$$8\ KP(H)SitBu_3 + 4\ InCl_3 \xrightarrow[-7\ KCl]{-4\ H_2PSiR_3} (ClInPSiR_3)_4 \cdot KCl$$

Eq. 5. Synthesis of the KCl adduct of $(ClInPSitBu_3)_4$ **(9)**$_4$.

Characteristic NMR Spectroscopic Data

The compounds with the general formula $[R_2E^{III}–P(H)SitBu_3]_2$ show remarkably large $^2J(PP)$ and $^1J(PH)$ coupling constants (Table 1) which are accessible from the AA'XX' pattern of the ^{31}P NMR spectra. Usually, the tri(*tert*-butyl)silylphosphanides show $^1J(PH)$ coupling constants of less than 200 Hz [21]; for $H_2PSitBu_3$ a value of 185.6 Hz was observed [23].

The $^2J(PP)$ coupling constants show values in the region of $^1J(PP)$ values (P_2H_4: –108 Hz). In four-membered M_2P_2 cycles of the alkaline earth metals $^2J(PP)$ coupling constants up to 135 Hz were detected [24].

Table 1. Characteristic ^{31}P NMR parameters of the compounds with the general formula $[R_2E^{III}\text{-}P(H)Si\textit{t}Bu_3]_2$. Values of *trans* isomers; values of the *cis* isomers are given in parentheses.

	(1)₂	**(2)₂**	**(7)₂**
E^{III}/R	Ga/Cl	Ga/Et	Al/Et
$\delta(^{31}P)$	–159.1 (–139.1)	–245.1 (–229.7)	–257.9 (–249.6)
^1J(PH)	276.6 (274.5)	248.4 (243.2)	243.7 (239.7)
^2J(PP)	236.9 (267.9)	236.8 (273.7)	206.8 (241.1)
^3J(PH)	–12.0 (–3.1)	–3.1 (–2.0)	–3.5 (–2.9)

Molecular Structures

The molecular structures of $(HGaPSi\textit{t}Bu_3)_4$ **(3)**$_4$, $(EtGaPSi\textit{t}Bu_3)_4$ **(4)**$_4$ and $(ClInPSi\textit{t}Bu_3)_4$ **(9)**$_4$ are represented in Fig. 1. For the heterocubane moieties in **(3)**$_4$ and **(4)**$_4$ Ga–P distances of 241.5 to 242.2 pm and 243.2 to 244.8 pm, respectively, are observed. For **(9)**$_4$ In–P bond lengths of 256.8 pm are found, which are shorter than in $[Et_2InP(SiMe_3)_2]_2$ (265 pm [19]) because of the more electronegative substituents at the earth metal as well as the larger electrostatic attraction between the dianions and dications.

The molecular structures of *trans*-$[Cl_2GaP(H)Si\textit{t}Bu_3]_2$ **(1)**$_2$, *cis*-$[Et_2AlP(H)Si\textit{t}Bu_3]_2$ **(7)**$_2$ and $[\textit{t}Bu_3SiP(H){-}Ga(\mu\text{-}PSi\textit{t}Bu_3)]_2$ **(5)**$_2$ as well as of the anion $[(Cl_2Al)_3(\mu_3\text{-}PSi\textit{t}Bu_3)_2]^-$ of **(6)** are represented in Fig. 2. The Ga–P bond lengths of **(1)**$_2$ (238.8 pm) compare well to those published for $[Cl_2GaP(SiMe_3)_2]_2$ (237.9 pm [2]); however, the intramolecular strain leads to a flattening of the Ga_2PSi fragment (GaPSi 130.1 and 130.4°, GaPGa' 90.0°) and to an elongation of the P–Si bond (231.4 pm). Less electronegative substituents at the earth metal E^{III} lead to longer Ga–P bonds, as for example in $[Et_2GaP(SiMe_3)_2]_2$ (245.6 pm [15]). For *cis*-**(7)**$_2$ Al–P distances of 242.7 to 246.5 pm are observed which are even larger than the Ga–P values of *trans*-**(1)**$_2$.

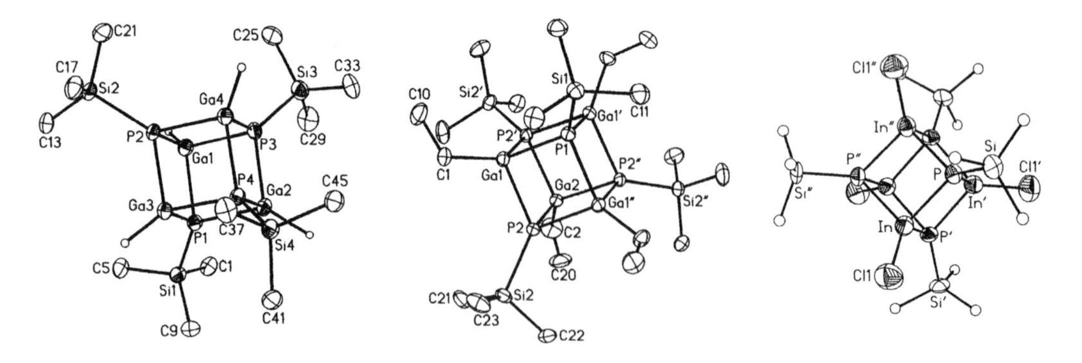

Fig. 1. Molecular structures of $(HGaPSi\textit{t}Bu_3)_4$ **(3)**$_4$ and $(EtGaPSi\textit{t}Bu_3)_4$ **(4)**$_4$ and $(ClInPSi\textit{t}Bu_3)_4$ **(9)**$_4$, with central $E^{III}{}_4P_4$ heterocubane moieties. The methyl groups are not drawn for clarity reasons.

Special attention has to be drawn to $(\mathbf{5})_2$ with triply coordinated gallium and phosphorus atoms. The Ga atoms are in a distorted trigonal planar environment. The terminal P atoms show angle sums of 299° for P1 and 304° for P3. Whereas the ring atom P4 displays an angle sum of 298° the other endocyclic phosphorus atom P2 is in a planar environment. Furthermore, the four-membered cycle contains the Ga1–P2–Ga2 fragment with Ga–P bond lengths of 226 pm, whereas the other endocyclic Ga–P4 bonds are, with values of 233 pm, significantly longer. The short bonds are comparable to those of a digallanylphosphane published by Petrie and Power already ten years ago [25]. In addition, the P2–Si2 bond length of 224.6 pm is smaller than the other P–Si values (228.7 to 229.9 pm) in this molecule. Therefore, this molecule can be regarded as a R'Ga=P(R)–GaR' heteroallyl cation bonding to a phosphanidyl substituent.

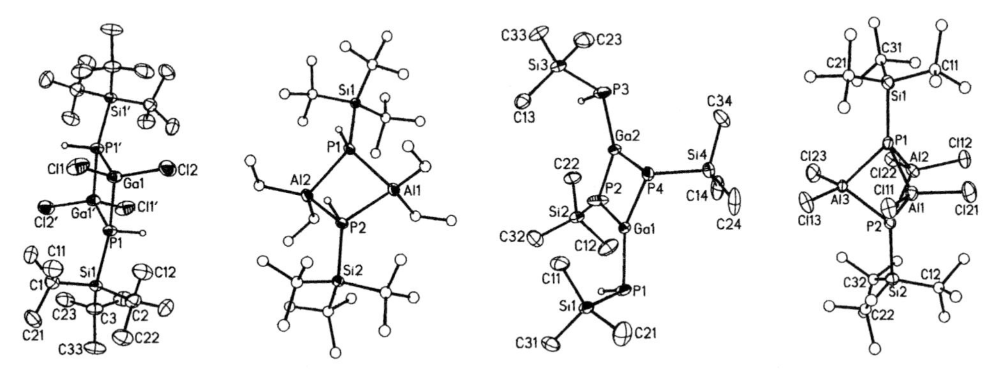

Fig. 2. Structures of *trans*-$[Cl_2GaP(H)Si\textit{t}Bu_3]_2$ $(\mathbf{1})_2$, *cis*-$[Et_2AlP(H)Si\textit{t}Bu_3]_2$ $(\mathbf{7})_2$ and $[\textit{t}Bu_3SiP(H){-}Ga(\mu\text{-}PSi\textit{t}Bu_3)]_2$ $(\mathbf{5})_2$ (methyl groups are neglected for clarity reasons) with central four-membered $E^{III}{}_2P_2$ rings as well as of the anion $[(Cl_2Al)_3(\mu_3\text{-}PSi\textit{t}Bu_3)_2]^-$ of (**6**).

Scheme 1. Valence bond representation of $[R'P(H){-}Ga(\mu\text{-}PR')]_2$ $(\mathbf{5})_2$ with R' = SiH_3, $SiMe_3$ and $Si\textit{t}Bu_3$ (bond lengths in **A** are given in pm).

In order to verify this interpretation and in order to estimate the importance of steric shielding, *ab initio* calculations on the B3LYP/6-31G(d,p) niveau were performed for $[RP(H){-}Ga(\mu{-}PR)]_2$ with R = SiH_3, $SiMe_3$ and $SitBu_3$. These calculations show that different isomers resulting from inversion of the terminally bonded phosphanide ligands are similar in energy. However, with increasing size of the silyl groups the differences Δd between the endocyclic bond lengths increase as well (SiH_3: Δd = 1.7 pm, $SiMe_3$: Δd = 3.2 pm, $SitBu_3$: Δd = 4.7 pm; found for $(\mathbf{5})_2$: Δd = 7.8 pm), whereas the endocyclic distances of the Ga–P multiple bonds decrease. These findings show that the large $SitBu_3$ substituent strongly supports the formation of a Ga–P–Ga heteroallylic bond system. Therefore, the mesomeric forms as shown in Scheme 1 are an excellent description of the bonding situation. The mesomeric forms **A**, **B** and **C** represent the main contributions; however, the small weight of **D** and **E** explains why the exocyclic Ga–P bonds are shorter than the endocyclic Ga1/2–P4 bonds.

The NBO analysis shows the following partial charges: P2 with –0.94, P4 with –0.80, Ga1 and Ga2 with +0.86*e*. These charges contradict the formulas shown in Scheme 1; however, a strong polarization of the Ga–P σ-bond as well as π-bond systems leads to a negative charge at the P atoms and a positive charge at the earth metals. This electronic situation is quite similar to the boron trihalides, where intramolecular donor-acceptor interactions lead to partial B–X double bonds [26].

Acknowledgments: We thank the Deutsche Forschungsgemeinschaft (DFG) and the Fonds der Chemischen Industrie for financial support.

References

[1] a) G. Fritz, R. Emuel, *Z. Anorg. Allg. Chem.* **1975**, *416*, 19; b) L. Rösch, W. Schmidt-Fritsche, *Z. Anorg. Allg. Chem.* **1976**, *426*, 99.

[2] a) R. L. Wells, M. F. Self, A. T. McPhail, S. R. Aubuchon, R. C. Woudenberg, J. P. Jasinski, *Organometallics* **1993**, *12*, 2832; b) J. F. Janik, R. A. Baldwin, R. L. Wells, W. T. Pennington, G. L. Schimek, A. L. Rheingold, L. M. Liable-Sands, *Organometallics* **1996**, *15*, 5385.

[3] M. D. Healy, P. E. Laibinis, P. D. Stupik, A. R. Barron, *J. Chem. Soc., Chem. Commun.* **1989**, 359.

[4] See for example: J. F. Janik, R. L. Wells, P. S. White, *Inorg. Chem.* **1998**, *37*, 3561.

[5] a) A. H. Cowley, R. A. Jones, *Angew. Chem.* **1989**, *101*, 1235; *Angew. Chem. Int. Ed. Engl.* **1989**, *28*, 1208; b) O. I. Micic, J. R. Sprague, C. J. Curtis, K. M. Jones, J. L. Machol, A. J. Nozik, H. Giessen, B. Flügel, G. Mohs, N. Peyghambarian, *J. Phys. Chem.* **1995**, *99*, 7754.

[6] R. L. Wells, A. T. McPhail, J. A. Laske, P. S. White, *Polyhedron* **1994**, *13*, 2737.

[7] R. L. Wells, A. T. McPhail, M. F. Self, J. A. Laske, *Organometallics* **1993**, *12*, 3333.

[8] R. L. Wells, E. E. Foos, A. L. Rheingold, G. P. A. Yap, L. M. Liable-Sands, P. S. White, *Organometallics* **1998**, *17*, 2869.

[9] L. K. Krannich, C. L. Watkins, S. J. Schauer, C. H. Lake, *Organometallics* **1996**, *15*, 3980.

[10] L. K. Krannich, C. L. Watkins, S. J. Schauer, *Organometallics* **1995**, *14*, 3094.

[11] F. Thomas, S. Schulz, M. Nieger, *Eur. J. Inorg. Chem.* **2001**, 161.

[12] H. Cowley, R. A. Jones, M. A. Mardones, J. L. Atwood, S. G. Bott, *Angew. Chem.* **1990**, *102*, 1504; *Angew. Chem. Int. Ed. Engl.* **1990**, *29*, 1409.

[13] a) M. Driess, S. Kuntz, K. Merz, H. Pritzkow, *Chem. Eur. J.* **1998**, *4*, 1628; b) M. Driess, S. Kuntz, C. Monsé, K. Merz, *Chem. Eur. J.* **2000**, *6*, 4343.

[14] C. v. Hänisch, F. Weigend, *Z. Anorg. Allg. Chem.* **2002**, *628*, 389.

[15] a) D. Wiedmann, H.-D. Hausen, J. Weidlein, *Z. Anorg. Allg. Chem.* **1995**, *621*, 1351; b) R. L. Wells, R. A. Baldwin, P. S. White, W. T. Pennington, A. L. Rheingold, G. P. A. Yap, *Organometallics* **1996**, *15*, 91; c) R. L. Wells, E. E. Foos, R. A. Baldwin, A. L. Rheingold, G. P. A. Yap, *Heteroat. Chem.* **1998**, *9*, 147; d) R. J. Jouet, R. L. Wells, A. L. Rheingold, C. D. Incarvito, *J. Organomet. Chem.* **2000**, *601*, 191.

[16] A. Schaller, H.-D. Hausen, J. Weidlein, P. Fischer, *Z. Anorg. Allg. Chem.* **2000**, *626*, 616.

[17] A. H. Cowley, R. A. Jones, M. A. Mardones, J. Ruiz, J. L. Atwood, S. G. Bott, *Angew. Chem.* **1990**, *102*, 1169; *Angew. Chem. Int. Ed. Engl.* **1990**, *29*, 1150.

[18] A. Schaller, H.-D. Hausen, W. Schwarz, G. Heckmann, J. Weidlein, *Z. Anorg. Allg. Chem.* **2000**, *626*, 1047 and literature cited therein.

[19] C. v. Hänisch, *Z. Anorg. Allg. Chem.* **2001**, *627*, 68.

[20] C. v. Hänsch, B. Rolli, *Z. Anorg. Allg. Chem.* **2002**, *628*, 2255.

[21] M. Westerhausen, S. Weinrich, B. Schmid, S. Schneiderbauer, M. Suter, H. Nöth, H. Piotrowski, *Z. Anorg. Allg. Chem.* **2003**, *629*, 575.

[22] M. Westerhausen, S. Weinrich, H. Piotrowski, *Z. Naturforsch.* **2001**, *56*, 576.

[23] a) N. Wiberg, H. Schuster, *Chem. Ber.* **1991**, *124*, 93; b) N. Wiberg, A. Wörner, H.-W. Lerner, K. Karaghiosoff, D. Fenske, G. Baum, A. Dransfeld, P. v. R. Schleyer, *Eur. J. Inorg. Chem.* **1998**, 833.

[24] M. Westerhausen, M. Krofta, N. Wiberg, J. Knizek, H. Nöth, A. Pfitzner, *Z. Naturforsch.* **1998**, *53b*, 1489.

[25] M. A. Petrie, P. P. Power, *Inorg. Chem.* **1993**, *32*, 1309.

[26] a) G. Frenking, S. Fau, C. M. Marchand, H. Grützmacher, *J. Am. Chem. Soc.* **1997**, *119*, 6648; b) C. Aubauer, T. M. Klapötke, A. Schulz, *J. Mol. Struct. (Theochem)* **2001**, *543*, 285.

Trifluoromethyl Silicon Compounds with Geminal Nitrogen Donor Centers

Krunoslav Vojinović, Norbert W. Mitzel*

Institut für Anorganische und Analytische Chemie, Westfälische Wilhelms-Universität Münster, Corrensstrasse 30, 48149 Münster, Germany
Tel.: +49 251 83 36006 — Fax: +49 251 83 36007
E-mail: Mitzel@uni-muenster.de

Martin Korth, Roland Fröhlich, Stefan Grimme

Organisch-chemisches Institut, Westfälische Wilhelms-Universität Münster, Corrensstrasse 40, 48149 Münster, Germany

Keywords: hypercoordination, donor acceptor interaction, three-membered ring, hydroxylamine, phase-dependent structure

Summary: The synthesis and structural characterization of $(F_3C)F_2SiONMe_2$ by X-ray diffraction and quantum chemical calculations, revealed a new three-membered ring compound with a strong donor-acceptor interaction between silicon and nitrogen atoms. The replacement of one F atom in $F_3SiONMe_2$ by an F_3C group leads to further increase in the already distinct Si···N interaction. The crystal structure of $(F_3C)F_2SiONMe_2$ shows that the SiN distance [1.904(2) Å] is even shorter than the covalent SiC bond [1.912(3) Å]. The compound is unstable at ambient temperature and decomposes, probably with formation of difluorcarbene.

We have demonstrated that silicon compounds with nitrogen centers in the β-position form three-membered ring systems with one bond being a donor-acceptor interaction. The strongest interactions so far were observed in $F_3SiONMe_2$ [1] and $ClH_2SiONMe_2$ [2] [<SiON = 77.1(1)° and 79.7(1)° in the crystals]. In our search for silicon compounds with strong intramolecular donor-acceptor interactions we have synthesized and structurally characterized the first trifluoromethylsilylhydroxylamine $(F_3C)F_2SiONMe_2$. The compound was prepared under similar conditions to those for $F_3SiONMe_2$, using Me_2NOLi and an excess of F_3CSiF_3 (Eq. 1) [3].

$$Me_2NOLi + F_3CSiF_3 \longrightarrow LiF + (F_3C)F_2SiONMe_2$$

Eq. 1.

The compound was isolated and purified by fractional condensation through a series of cold traps and appears as a colorless liquid with a melting point of –43 °C. $(F_3C)F_2SiONMe_2$ decomposes slowly even at –30 °C, probably with formation of difluorocarbene. The compound was identified by multinuclear NMR spectroscopy (^1H, ^{13}C, ^{15}N, ^{19}F, and ^{29}Si) and gas-phase IR spectroscopy. Particularly informative was the triplet of quartets in the ^{29}Si NMR spectrum (Fig. 1) resulting from the $^1J(SiF)$ (252.2 Hz) and $^2J(SiF)$ (64.0 Hz) couplings. The silicon atom is strongly deshielded by its electronegative substituents, which results in a low field shift (–116.8 ppm) comparable to that of $F_3SiONMe_2$ (–117.2 ppm).

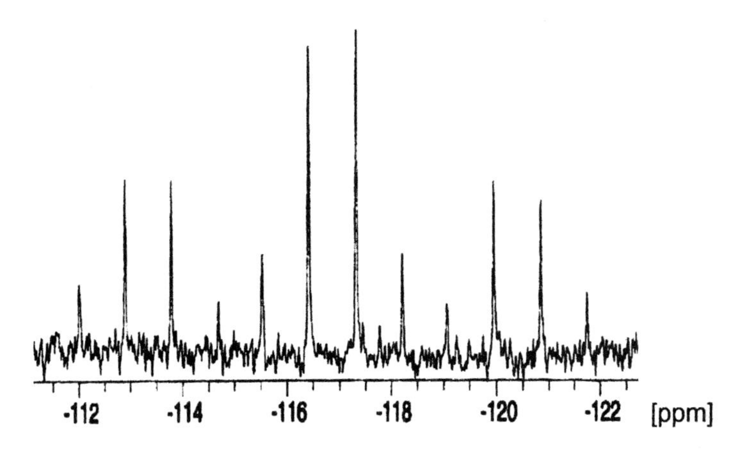

Fig. 1. Section from the ^{29}Si NMR spectrum of $(F_3C)F_2SiONMe_2$ showing a triplet of quartets.

A single crystal could be grown *in situ* on the diffractometer in a sealed Duran® glass capillary, which made it possible to determine the structure in the solid state (Fig. 2).

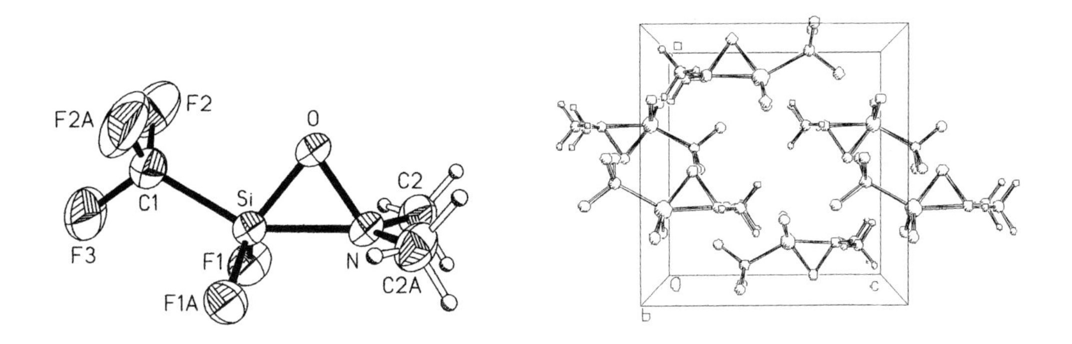

Fig. 2. Molecular structure and packing diagram of $(F_3C)F_2SiONMe_2$ obtained by low-temperature X-ray crystallography.

The structural data listed in Table 1 show that the substitution of only one F atom in $F_3SiONMe_2$ by an F_3C group leads to a further strengthening of the intramolecular donor-acceptor interaction between the Si and N atoms. The F_3C group has an electronegativity similar to that of an F atom but no possibility of back bonding to the silicon atom. It can thus exhibit a stronger electron-withdrawing effect than a fluorine substituent. The Si-O-N angle in solid $(F_3C)F_2SiONMe_2$ is 74.1(1)°, which is smaller than the Si-O-N angle found in $F_3SiONMe_2$ [77.1(1)°]. It is interesting that the Si···N distance [1.904(2) Å] resulting from a strong donor-acceptor interaction is even shorter than the covalent SiC bond [1.912(3) Å]. The coordination geometry of the silicon atom is distorted trigonal bipyramidal. The formation of the three-membered SiON ring leads to a substantial lengthening of the O–N bond [1.512(2) Å].

Table 1. Bond lengths and angles of $(F_3C)F_2SiONMe_2$ determined by X-ray diffraction.

Parameter	Value
Si···N	1.904(2)
Si–O	1.642(2)
O–N	1.515(2)
Si–C(F)	1.912(3)
Si–F	1.574(1)
N–C	1.465(2)
Si-O-N	74.1(1)
C(F)-Si-O	99.6(1)
F_3-C(F)-Si	114.2(2)
F-C(F)-Si	112.5(2)

Quantum chemical calculations at different levels of theory were carried out for the structures of $F_3SiONMe_2$ and $(F_3C)F_2SiONMe_2$ and are shown in Table 2. The results show that correlated QM methods are necessary to describe the Si···N interaction quantitatively. Hartree-Fock as well as DFT methods including a fraction of HF exchange (e.g. B3LYP) underestimate the interaction, because it has a significant dispersive (van der Waals) component. The variations of the predicted Si-O-N angle are relatively large because of the flatness of the corresponding potential. For $F_3SiONMe_2$ with a properly polarized TZV basis set (TZVPP), the values at the MP2 level are close to what can be estimated for the basis set limit (about 88°, cc-pVQZ basis). This well-known overestimation of dispersive interactions by MP2 is almost completely corrected by the new SCS-MP2 method, which yields 92.6° in very good agreement with the GED value of 94.3°. The pure density functionals PBE and BP give values of 87.3° and 93.1°, which can be considered as quite accurate, suggesting them as the methods of choice for the study of larger systems. Except for HF and B3LYP, all methods predict a smaller bond angle for the compound where one fluorine atom has been replaced by a CF_3 group. Qualitatively, this is in line with the results of the XRD data, which yield a difference in the bond angles of 3 degree between $F_3SiONMe_2$ and $(F_3C)F_2SiONMe_2$ (theory provides a range of 3.0 – 7.9°).

Table 2. Quantum chemical calculations of the Si-O-N bond angle for $(F_3C)F_2SiONMe_2$ and $F_3SiONMe_2$ at different levels on theory.

Method	$(F_3C)F_2SiONMe_2$	$F_3SiONMe_2$
XRD	74.1(1)°	77.1(1)°
GED	–	94.3(9)°
HF/TZVPP	111.5°	106.6°
B3LYP/TZVP	102.3°	101.6°
B3LYP/TZVPP	103.9°	102.5°
BP/TZVP	84.7°	86.2°
BP/TZVPP	88.7°	93.1°
PBE/TZVPP	84.3°	87.3°
MP2/TZVP	79.4°	83.6°
MP2/TZVPP	79.2°	86.5°
MP2/cc-pVQZ	79.2°	88.1°
SCS-MP2/cc-pVTZ	–	92.6°

The results underline on one hand the importance of a proper high-level theoretical treatment of such systems, which are difficult to describe because of the shallow-angle bending potential, and on the other hand they also underline the importance of experimental structure determinations in the gas phase (Table 2).

Acknowledgments: This work has been supported by Deutsche Forschungsgemeinschaft and Fonds der Chemischen Industrie. We are grateful to PD Dr. H. Beckers (Wuppertal) for the generous gift of a sample of F_3CSiF_3.

References

[1] N. W. Mitzel, U. Losehand, *J. Am. Chem. Soc.* **1998**, *120*, 7320.

[2] N. W. Mitzel, U. Losehand, A. Wu, D. Cremer, D. W. H. Rankin, *J. Am. Chem. Soc.* **2000**, *122*, 4471.

[3] H. Bürger, R. Eujen, P. Moritz, *J. Organomet. Chem.* **1991**, *401*, 249.

Alkali Metal Cyantrimethylsilylamides — $M[NCNSi(CH_3)_3]$

Edwin Kroke

University of Konstanz, Chemistry Department, Chemical Material Science
Universitätsstrasse 10, 78464 Konstanz, Germany
Tel.: +49 7531 884415 — Fax: +49 7531 884406
E-mail: edwin.kroke@uni-konstanz.de

Keywords: carbodiimide, reduction, alkali metal

Summary: The aim of the present work was to test a concept which states that bis(trimethylsilyl)carbodiimide ($(CH_3)_3SiNCNSi(CH_3)_3$ = BTSC) behaves similarly to water. In analogy to the formation of alkali hydroxides from alkali metals and water, alkali metal-cyantrimethylsilylamides $M[NCNSi(CH_3)_3]$ have been prepared by direct reactions of Li, Na, K, and Cs metal or their naphthalenides with BTSC. The products were characterized by FTIR, Raman, and ^1H and ^{29}Si NMR spectroscopy. In addition, thermal gravimetric analysis (TGA) and bulk calcination experiments show that dialkali cyanamides $M_2(NCN)$ are formed by quantitative evolution of BTSC. The latter reaction is analogous to oxide formation upon calcination of hydroxides. Therefore, the present results support a concept which describes alkali metal-cyantrimethylsilylamides as pendants to alkali hydroxides and BTSC as a water analogue.

The carbodiimide-group can be introduced into silicon compounds by reactions of bis(trimethylsilyl)carbodiimide (BTSC) with chlorosilanes (Eq. 1) [1]:

$$n\,R_xSiCl_{4-x} + n(4-x)/2\ (H_3C)_3Si{-}NCN{-}Si(CH_3)_3 \longrightarrow \left[R_xSi(NCN)_{(4-x)/2}\right]_n + n(4-x)\,(H_3C)_3SiCl$$

Eq. 1.

This reaction is analogous to the hydrolysis of chlorosilanes. In both cases, cyclic and linear oligomers are formed if dichlorosilanes are used [1]. Further analogies between hydrolyses and "carbodiimidolyses" by BTSC were reported for reactions with trichlorosilanes $RSiCl_3$ (R = H, Cl, Alkyl, Aryl) [2, 3] and other element halides [3 – 6]. A characteristic sol-gel transition was observed:

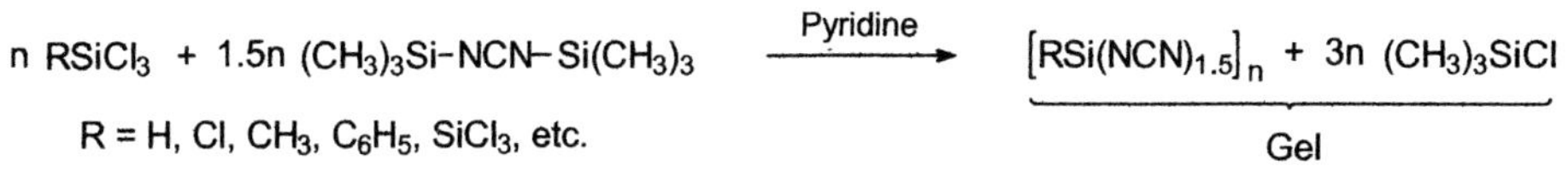

$$n\ RSiCl_3 + 1.5n\ (CH_3)_3Si\text{-}NCN\text{-}Si(CH_3)_3 \xrightarrow{\text{Pyridine}} \underbrace{[RSi(NCN)_{1.5}]_n}_{\text{Gel}} + 3n\ (CH_3)_3SiCl$$

$R = H, Cl, CH_3, C_6H_5, SiCl_3$, etc.

Eq. 2.

Additionally, it was shown that several parallels to oxide sol-gel processes exist [1, 5 – 8]. Reactions of BTSC with $SiCl_4$ furnished the first crystalline Si/C/N-phases (Scheme 1) [9], where SiC_2N_4 (= $[Si(N{=}C{=}N)_2]_n$) shows a close structural relationship to the SiO_2 phase cristobalite, and Si_2CN_4 (= $[Si_2N_2(NCN)]_n$) to Si_2N_2O [9].

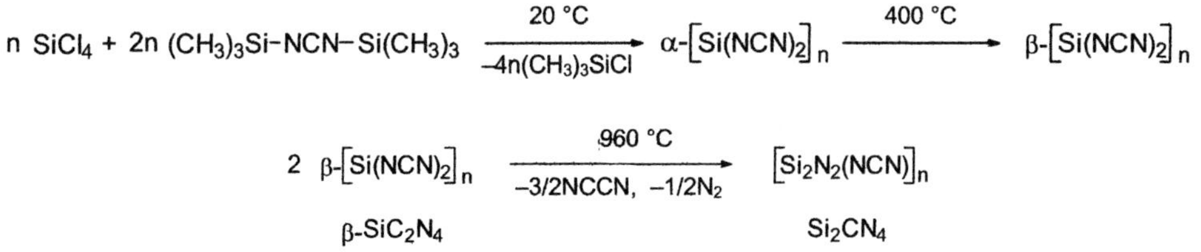

$$n\ SiCl_4 + 2n\ (CH_3)_3Si\text{-}NCN\text{-}Si(CH_3)_3 \xrightarrow[-4n(CH_3)_3SiCl]{20\ °C} \alpha\text{-}[Si(NCN)_2]_n \xrightarrow{400\ °C} \beta\text{-}[Si(NCN)_2]_n$$

$$\underset{\beta\text{-}SiC_2N_4}{2\ \beta\text{-}[Si(NCN)_2]_n} \xrightarrow[-3/2NCCN,\ -1/2N_2]{960\ °C} \underset{Si_2CN_4}{[Si_2N_2(NCN)]_n}$$

Scheme 1.

In order to understand this resemblance between S–O- and Si–(NCN)-compounds, it is useful to recall that carbodiimides are known as typical pseudochalcogenides [10]. Furthermore, the group electronegativity of the trimethylsilyl group [11] as well as the carbodiimide unit [12] are very close to the electronegativity of oxygen [13] and hydrogen [13], respectively. This heuristic approach led us to formulate a "BTSC-water analogy concept" (Fig. 1), which will be discussed and compared to related concepts elsewhere [14].

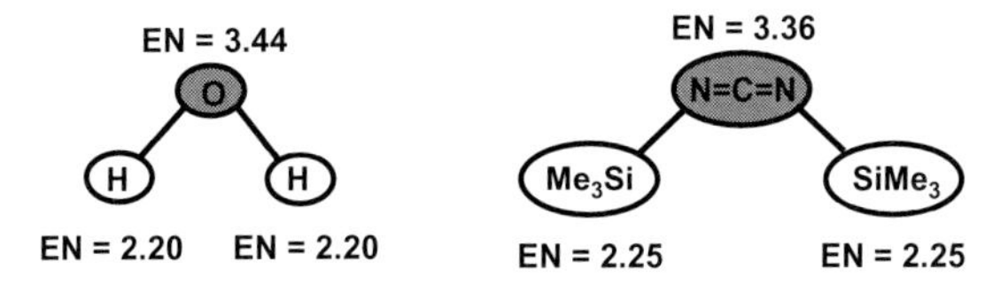

Fig. 1. Schematic representation of the "BTSC-water analogy concept". The group electronegativities for the trimethylsilyl as well as the carbodiimide group were taken from the literature [11, 12]. The values are very close to the respective electronegativity values of hydrogen and oxygen [13]. In addition to this analogy, there are a number of further parallels between bis(trimethylsilyl)carbodiimide (BTSC) and water (see text).

The objective of the present study was to check this concept with respect to the chemical properties of BTSC. Alkali metals were reacted with BTSC (see experimental part, Scheme 2 and

Table 1). Cyantrimethylsilylamides are obtained which correspond to the formation of hydroxides.

The volatile side product hexamethyldisilane was detected by a GC comparison with an authentic sample and unambiguously identified by GC-MS. This molecule corresponds to the H_2 formed upon reactions of alkali metals with water. Alkali metal naphthalenides showed the same reaction behavior.

$$M + (CH_3)_3Si\text{-}NCN\text{-}Si(CH_3)_3 \longrightarrow M[NCN\text{-}Si(CH_3)_3] + 0.5(CH_3)_3Si\text{-}Si(CH_3)_3$$

$$M + H\text{-}O\text{-}H \longrightarrow M[O\text{-}H] + 0.5\ H\text{-}H$$

M = alkali metal

$$MX + (CH_3)_3Si\text{-}NCN\text{-}Si(CH_3)_3 \longrightarrow M[NCN\text{-}Si(CH_3)_3] + X\text{-}Si(CH_3)_3$$

$$MX + H\text{-}O\text{-}H \longrightarrow M[O\text{-}H] + X\text{-}H$$

M = alkali metal; X = H, CH_3, nC_4H_9, OCH_3, PH_2 (see Table 1)

Scheme 2.

Table 1. Results of reactions of bis(trimethylsilyl)carbodiimide (BTSC) with alkali metals as well as naphtalenides, alkyls, phosphides and amides.

Reactant	Main product	Volatile side product	Yield	Literature
Li	$Li[NCNSi(CH_3)_3]$	$(H_3C)_3SiSi(CH_3)_3$ [a]	~90% [c]	t.s. [b]
Na	$Na[NCNSi(CH_3)_3]$	$(H_3C)_3SiSi(CH_3)_3$ [a]	~90% [c]	t.s. [b]
K	$K[NCNSi(CH_3)_3]$	$(H_3C)_3SiSi(CH_3)_3$ [a]	~60% [c]	t.s. [b]
Cs	$Cs[NCNSi(CH_3)_3]$	$(H_3C)_3SiSi(CH_3)_3$ [a]	~92% [c]	t.s. [b]
$Li/C_{10}H_8$	$Li[NCNSi(CH_3)_3]$	$(H_3C)_3SiSi(CH_3)_3$ [a]	~88% [c]	t.s. [b]
$Na/C_{10}H_8$	$Na[NCNSi(CH_3)_3]$	$(H_3C)_3SiSi(CH_3)_3$ [a]	~67% [c]	t.s. [b]
Li(nBu)	$Li[NCNSi(CH_3)_3]$	$(H_3C)_3Si(nBu)$ [a]	~92%	t.s. [b]
$LiCH_3$	$Li[NCNSi(CH_3)_3]$	$Si(CH_3)_4$ [a]	~95%	[15] and t.s. [b]
$Li(PH_2)$	$Li[NCNSi(CH_3)_3]$	$(H_3C)_3SiPH_2$	~82%	[15]
$LiN[Si(CH_3)_3]_2$	$Li[NCNSi(CH_3)_3]$	$[(H_3C)_3Si]_3N$	~78%	[16]

[a] Detected by GC and/or GC-MS; [b] t.s. = this study; [c] estimated from NMR spectra.

A CAS literature search showed that the Li compound has been obtained before by reactions of methyl lithium [15] as well as lithium-bis(trimethylsilyl)amide [16] with BTSC. Spectroscopic investigations and a single-crystal X-ray structure of a THF adduct proved the formation of an alkali metal cyantrimethylsilylamide [15]. Reactions of phosphanides furnished the same Li compound [15]. These transformations proceed with water in an analogous manner (see Table 1 and Scheme 2).

The alkali metal cyantrimethylsilylamides may be prepared in an inert organic solvent like *n*-hexane or toluene. Alternatively, an excess of the liquid BTSC can be used. In all cases the alkali metal compound is obtained as a white solid. Purification was performed by washing with hexane, drying in vacuum, and/or re-crystallization from THF or pyridine.

The alkali metal cyantrimethylsilylamides are insoluble in non-polar organic solvents, but are moderately soluble in DMSO and DMF. As expected, the ^{1}H NMR Spectra in d_6-DMSO show a singlet at –0.1 ppm for the $Si(CH_3)_3$ group. ^{13}C NMR spectra show signals at 2.8 and 127.6 ppm for the $Si(CH_3)_3$– and the cyanamide group, respectively. Singlet at –10.6 ppm are found in the ^{29}Si NMR spectra. These chemical shift data are very close to the values reported for the Li compound dissolved in d_8-THF [15].

The vibrational spectra clearly indicate the presence of cyanamide groups, with characteristic absorptions at 2095 – 2120 cm^{-1} [17]. Also, the bands for C–H and Si–CH_3 groups at 2890 – 2960 cm^{-1} and 1250 cm^{-1} are detected.

The melting and decomposition points of the alkali metal cyantrimethylsilylamides are close to 300 °C. Thermogravimetric examinations indicated that no undefined decomposition to a large number of product molecules takes place. Quantitative formation of BTSC and the corresponding cyanamide was observed instead. The mass loss is equal to the theoretical values, and the products were identified by FTIR and NMR (BTSC) as well as FTIR and powder XRD (cyanamide). This reaction formally corresponds to the calcination of hydroxides to oxides (see Fig. 2).

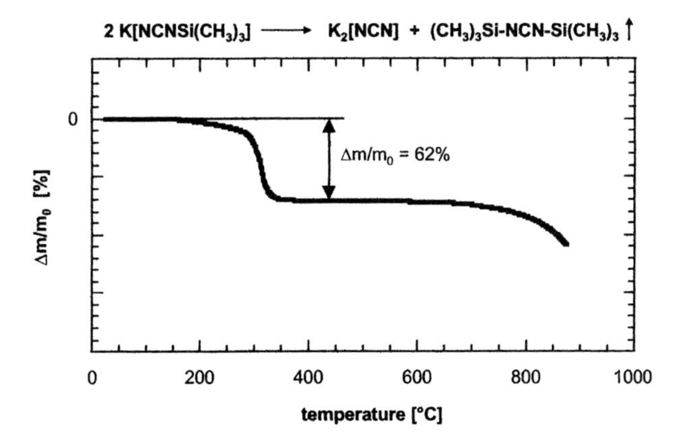

Fig. 2. Thermogravimetric examination (TGA) of potassium cyantrimethylsilylamide. The mass loss at $T = 300$ °C of 62% is identical with the theoretically expected value for the depicted reaction. The volatile bis(trimethylsilyl)carbodiimide (BTSC) was detected by mass spectrometry. Bulk calcination experiments proved the quantitative formation of pure BTSC (NMR) and potassium cyanamide (powder XRD). This reaction is analogous to calcination reactions of hydroxides forming oxides.

Elemental analyses indicated that the cyantrimethylsilylamides contain less than the theoretical amount of silicon, and the content of alkali metals is too high. This may be explained by the above described thermal decomposition to form BTSC and the alkali cyanamide.

The presented reactions can be described as reductions of BTSC. They show that BTSC indeed behaves analogously to water in reactions with nucleophiles and/or reductive agents. The question whether these reactions are also mechanistically similar is difficult to answer based on the presented results. The LUMO of BTSC, which dictates the attack of a nucleophile according to the frontier orbital theory, is located at the central C atom as a π^*-orbital. In contrast, the LUMO of the water molecule is the σ^*-orbital. In a very simple approach, such reactions may be described as proton transfers. Similarly, it is possible that in BTSC the trimethylsilyl group is attacked nucleophilically.

Further examinations will show whether the BTSC-water analogy concept is also applicable to reactions, e.g., with halogens. Furthermore, it should be determined whether the silylcyanamides may be used as "pseudo hydroxides" in the synthesis of novel interesting classes of compounds such as silicate or metallate analogous cyanamides. The alkali cyanotrimethylsilyl amides may be useful reagents for the introduction of the $(CH_3)_3Si$–NCN group, i.e. the preparation of asymmetric organic and inorganic carbodiimides or cyanamides. Also, the presented reactions may be used in a more general way to prepare alkali metal derivatives of silylated compounds.

Experimental Part

All manipulations were performed under inert gas using a glove box and/or Schlenk technique. Bis(trimethylsilyl)carbodiimide [18] (BTSC) was prepared according to the literature. All other chemicals were purchased and used without further purification. Water-free THF, *n*-hexane, DMF, pyridine, and d_6-DMSO were used.

Synthesis of cyantrimethylsilylamides from elemental alkali metals. In a typical experiment 0.5 g alkali metal was cut into small pieces and mixed with an excess, e.g. 10 mL, BTSC. The reaction mixture became cloudy after a few minutes to several hours, depending on the alkali metal and the temperature (20 – 135 °C). The reaction was stopped after 5 h to 3 d. Products consisted of fine white powders which were washed several times with *n*-hexane and dried at 50 – 150°C / 10^{-2} mbar. The lithium and sodium compound were recrystallized from THF and hot pyridine, respectively.

Synthesis of cyantrimethylsilylamides from elemental alkali metal naphthalenides. Small pieces of lithium or sodium were added to a stirred solution of 5 g naphthalene in 120 mL THF, generating a dark green suspension. This alkali metal naphthalenide reagent was slowly added to a stirred mixture of 10 mL BTSC and 10 mL *n*-hexane at –40 °C. Isolation and purification were performed as described above after complete separation of the naphthalene.

Synthesis of cyantrimethylsilylamides from alkali metal alkyls. A methyl lithium solution (25 mL, 1.6 mol, FLUKA) or *n*-buthyl lithium (8 mL, 2.5 mol, CHEMETALL) in diethylether or toluene were slowly added to a mixture of 10 mL BTSC and 15 mL *n*-hexane at 0 °C. A white precipitate formed which was purified as described above.

Calcination of lithium and potassium cyantrimethylsilylamide. In addition to the TG-MS studies, alkali cyantrimethylsilylamides were calcined in Schlenk flasks at 320 – 350 °C. Pure BTSC was quantitatively formed and condensed in a cooling trap. The compound was identified by ^{1}H- and ^{13}C-NMR spectra. The residue, a fine gray powder, was examined with FTIR and XRD. The data are identical to the literature, indicating that $Li_2(NCN)$ and $K_2(NCN)$ were obtained.

The following instruments and analytical techniques were used: FT-IR: Perkin-Elmer FTIR 1750, KBr pellets, Raman: Bruker IFS 55; ^{1}H, ^{13}C and ^{29}Si NMR spectroscopy: Bruker WM 300, solvent: d_6-DMSO, internal standard: $Si(CH_3)_4$); TGA-MS: Netzsch STA 429, coupled with a Balzers QMG 420 mass spectrometer.

Acknowledgments: The author thanks R. Riedel (TU Darmstadt) and F. F. Lange (University of California, Santa Barbara, USA) for their support. A. Gabriel and A. Greiner are acknowledged for many useful discussions on the BTSC-water analogy concept. The Alexander von Humboldt Foundation is acknowledged for granting a Feodor Lynen scholarship. The research was financially supported by Deutsche Forschungsgemeinschaft (DFG, Bonn, Germany) and the Fonds der Chemischen Industrie (FCI, Frankfurt, Germany).

References

[1] Overview: R. Riedel, E. Kroke, A. Greiner, A. O. Gabriel, L. Ruwisch, J. Nicolich, P. Kroll, *Chem. Mater.* **1998**, *10*, 2964.

[2] a) A. O. Gabriel, R. Riedel, *Angew. Chem.* **1997**, *109*, 371; *Angew. Chem. Int. Ed. Engl.* **1997**, *36*, 384; b) A. O. Gabriel, R. Riedel, S. Storck, W. F. Maier, *Appl. Organomet. Chem.* **1997**, *11*, 833.

[3] E. Kroke, A. O. Gabriel, D. S. Kim, R. Riedel, *Organosilicon Chemistry IV* (Eds.: N. Auner, J. Weis) Wiley-VCH, Weinheim, **2000**, p. 812.

[4] D. S. Kim, E. Kroke, R. Riedel, A. O. Gabriel, S. C. Shim, *Appl. Organomet. Chem.* **1999**, *13*, 495.

[5] E. Kroke, W. Voelger, A. Klonczynski, R. Riedel, *Angew. Chem.* **2001**, *113*, 1751; *Angew. Chem. Int. Ed. Engl.* **2001**, *40*, 1698.

[6] C. Balan, W. Voelger, E. Kroke, R. Riedel, *Macromolecules* **2000**, *33*, 3404.

[7] Y.-L. Li, E. Kroke, A. Klonczynski, R. Riedel, *Adv. Mater.* **2000**, *12*, 956.

[8] E. Kroke, Novel sol-gel routes to non-oxide ceramics, 9th CIMTEC – World Ceramics Congress, Ceramics: Getting into the 2000s – Part C (Ed.: P. Vincenzini), Techna Srl., **1999**, p. 123.

[9] R. Riedel, A. Greiner, G. Miehe, W. Dreßler, H. Fueß, J. Bill, F. Aldinger, *Angew. Chem.* **1997**, *109*, 657; *Angew. Chem. Int. Ed. Engl.* **1997**, *36*, 603.

[10] Overview: L. Jäger, H. Köhler, *Sulfur Reports* **1992**, *12*, 159.

[11] Y. Vignollet, J. C. Maire, *J. Organomet. Chem.* **1969**, *17*, 43.

[12] H.-D. Schädler, L. Jäger, I. Senf, *Z. Anorg. Allg. Chem.* **1993**, *619*, 1115.

[13] *CRC Handbook of Chemistry and Physics*, 79th. Edition (Ed.: D. R. Lide) CRC Press, Boca Raton, **1999**, p. 9 – 74.

[14] a) E. Kroke, R. Riedel, *to be published*; b) E. Kroke, *Precursortechnik – Molekülchemische Konzepte zur Darstellung nicht-oxidischer, keramischer Materialien durch Sol-Gel-Verfahren und Hochdrucksynthesen*, Habilitationsschrift, Tenea Verlag, Berlin, **2004**.

[15] G. Becker, K. Hübler, J. Weidlein, *Z. Anorg. Allg. Chem.* **1994,** *620*, 16.

[16] E. H. Amonoo-Neizer, R. C. Golesworthy, R. A. Shaw, B. C. Smith, *J. Chem. Soc.* **1965**, 5452.

[17] J. Weidlein, U. Müller, K. Dehnicke, *Schwingungsspektroskopie*, Thieme Stuttgart, 2nd. edition **1988**.

[18] I. A. Vostokov, Y. I. Dergunov, *Zh. Obshch. Khim.* (*J. Gen. Chem. USSR*) **1977**, *47*, 1769.

Amination in Supercritical Ammonia — Continuous Production of Aminoalkylsilanes

Andreas Bauer, Johann Weis*

Consortium für elektrochemische Industrie GmbH
Zielstattstraße 20, 81379, Munich, Germany
Tel.: +49 89 74844 0 — Fax: +49 89 74844 242

*Jochen Rauch**

Wacker-Chemie GmbH, Johannes-Hess-Strasse 24, 84489, Burghausen, Germany
Tel.: +49 8677 83 7637 — Fax: +49 8677 83 5706
E-mail: andreas.bauer@wacker.com

Keywords: aminoalkylsilanes, supercritical ammonia, continuous production process

Summary: The newly developed continuous process [1] realized in the pilot plant shown is superior to the state-of-the-art batch process currently in use. The main advantages are a higher space time yield and a simplified work-up resulting in a more cost efficient production of primary aminoalkyl silanes.

Introduction

Primary aminoalkylsilanes are important compounds in application fields like adhesion promotion, surface modifying, and crosslinking, but are also used for endcapping of various substrates. Currently in focus is the recently developed endcapper (Fig. 1) which is the key molecule for the preparation of thermoplastic silicone elastomers (TPSE). Modification of silicone resins and fumed silica with **1** leads also to products with interesting economical potential.

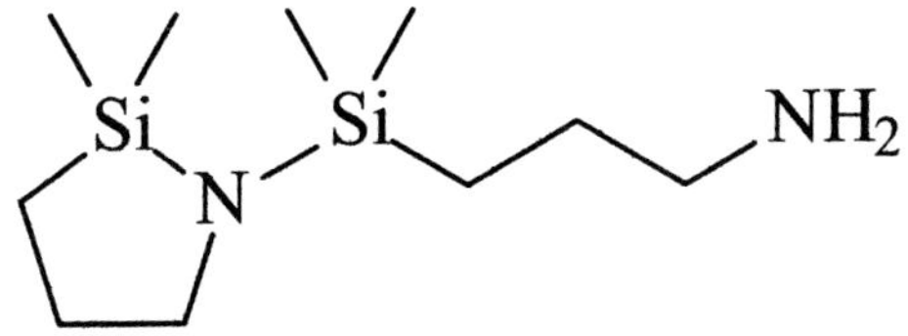

Si Si N NH_2

1

Fig. 1. Structure of endcapper **1**.

Chemistry

The desired primary aminoalkylsilanes can be synthesized starting from the corresponding chloroalkyl precursors. The educts are reacted with excess ammonia (molar ratios 1 : 20 to 1 : 150) at pressures ranging from 30 to 200 bar and temperatures of 40 to 200 °C. The large ammonia excess is necessary to supress side reactions yielding higher alkylated amines (Eq. 1).

$$RCl + 2\,NH_3 \longrightarrow RNH_2 + NH_4Cl$$

$$\xrightarrow{RCl} R_2NH + R_3N + NH_4Cl$$

Eq. 1. Reaction of a chloroalkylsilane with ammonia.

Reaction Kinetics

The kinetics of the amination reaction was determined for a series of chloroalkylsilanes in the laboratory. The conversion of 3-chloropropyltriethoxysilane to 3-aminopropyltriethoxysilane was found to be the slowest among the investigated reactions (Fig. 2). In this case, the Arrhenius parameter A for the pseudo-first order reaction (50-fold molar excess of ammonia) is $e^{4.41}$, and the activation energy E_a amounts to 25.5 kJ mol^{-1}. In supercritical ammonia at a temperature of 180 °C the calculated reaction time to complete conversion is less than one hour, which is sufficient for the realization of a continuous production process.

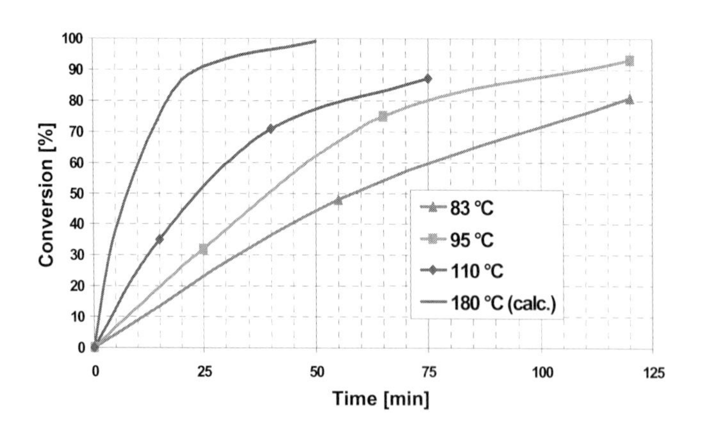

Fig. 2. Amination of 3-chloropropyltriethoxysilane at various temperatures.

"Liquid Salt"

Under certain temperature and pressure conditions, i.e. in the supercritical state of ammonia, the appearence of an additional fluid phase in the final reaction mixture is observed. This phase consists mainly of ammonium chloride, ammonia, and traces of silane. A separation of this phase from the excess ammonia is easily achieved because of the higher density of the so-called "liquid salt". The vast majority of the product aminoalkylsilane is contained in the remaining ammonia phase and can be separated via pressure reduction.

Pilot Plant Pressure Amination

Based on the results obtained in the laboratory, a continuous pilot plant was designed (Fig. 3). The process concept includes reaction, gravity separation of "liquid salt" utilizing density differences and silane separation in a similar way. The surplus of ammonia is recycled and subsequently reused in the reaction. Because of the physical phenomena in supercritical ammonia, no solid handling is necessary in the high pressure areas of the pilot plant.

The pilot plant is designed for temperatures and pressures up to 300 °C and 300 bar. Because of the highly corrosive environment inside the plant, only Ni alloy steels are used. Special fittings and instruments had to be developed to ensure high reliability of the pilot plant.

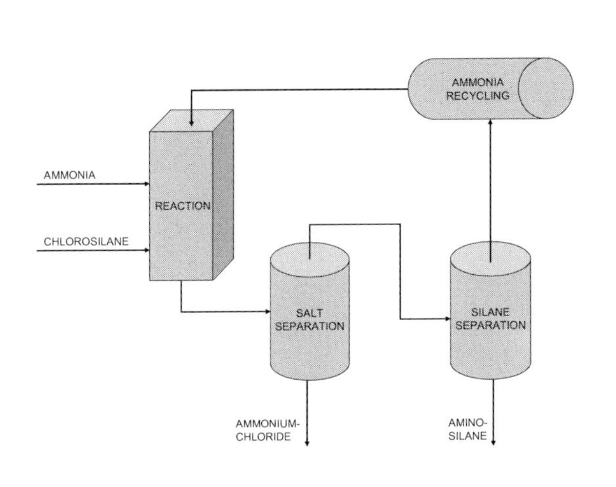

Fig. 3. Block flow scheme of the pilot plant.

Fig. 4. Picture of the pilot plant.

References

[1] A. Bauer, H. Jekat, J. Rauch, P. John, W. Kohlmann, V. Frey, B. Pachaly, EP1 327 634A1, **2002**.

Mono- and Bis(hydroxylamino)silanes — Reactions, Rearrangements, and Structures

Christina Ebker,* Friedhelm Diedrich, Uwe Klingebiel

Institute of Inorganic Chemistry, Georg-August-University of Goettingen
Tammannstr. 4, D-37077 Goettingen, Germany
Tel.: +49 551 39 3052 — Fax: +49 551 39 3373
E-mail: uklinge@gwdg.de

Keywords: silylhydroxylamines, bis(hydroxylamino)silanes, X-ray structures

Summary: The first crystal structure of a monosilylhydroxylamine is presented. In the reaction of dichlorosilanes and hydroxylamine hydrochloride, bis(hydroxylamino)-silanes can be prepared. Starting from lithiated bishydroxylamines, *N,O*-bis(silyl)-*N*-stannylhydroxylamines can be obtained. A hydrolysis product is presented.

Introduction

Pioneering work in the synthesis of tri(organyl)silylhydroxylamines and in studies of rearrangements was done by the groups of Wannagat [1] and West [2].

In contrast to organylhydroxylamines, silylhydroxylamines are able to undergo anionic, neutral, radical, and thermal intramolecular rearrangements [1 – 9].

Monosilylhydroxylamines

We are now able to present the first crystal structure of a monosilylhydroxylamine, the *tert*-butyldiphenylsilylhydroxylamine [10]:

$$R_3Si{-}Cl \quad + \quad HONH_3Cl \quad \xrightarrow[-(H_2CNH_3Cl)_2]{+(H_2CNH_2)_2} \quad R_3Si{-}ONH_2$$

Eq. 1.

In the crystal there are no hydrogen bridges connecting the monomers. With the N(1)-hydrogen atoms H(1) and H(2) the sum of the angles amounts to 304.7°, indicating a pyramidal environment at N(1).

At 110.1°, the Si-O-N-angle in *tert*-butyldiphenyl-silylhydroxylamine is the smallest so far known in molecules containing only hydrogen atoms at the nitrogen atom, thereby determining a short nonbinding Si---N-distance of 258 pm.

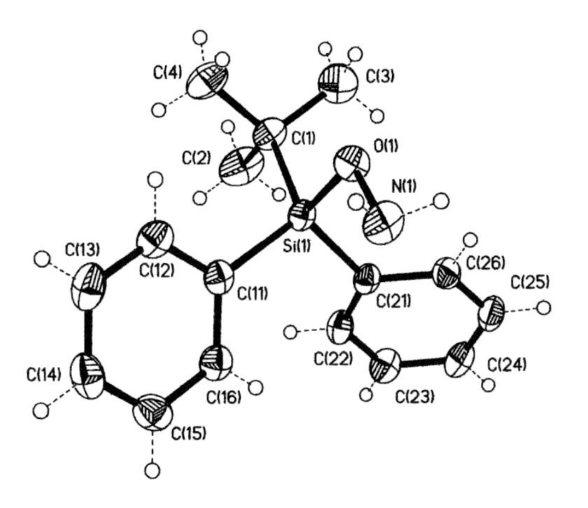

Selected bond lengths [pm] and angles [°].

Si(1)–O(1)	166.3(5)
O(1)–N(1)	147.8(7)
Si(1)–C(1)	189.3(9)
Si(1)–C(11)	187.8(9)
N(1)-O(1)-Si(1)	110.1(3)
O(1)-Si(1)-C(11)	109.2(3)
O(1)-Si(1)-C(1)	101.8(3)

Fig. 1. Crystal structure of a monosilylhydroxylamine.

O,O′-Bis(hydroxylamino)silanes

In analogy to Bottaro's [11] method for the synthesis of monosilylhydroxylamines, two equivalents of ethylenediamine are added to a suspension of two equivalents of hydroxylamine hydrochloride in CH_2Cl_2 (Eq. 2).

$$RR'SiCl_2 \xrightarrow[-2\,(H_2CNH_3Cl)_2]{+2\,(H_2CNH_2)_2,\ +2\,HONH_3Cl} RR'Si(ONH_2)_2 \qquad R, R' = \text{alkyl, aryl}$$

Eq. 2.

The basis of the reaction is the *in situ* release of HO–NH_2, which together with the HCl acceptor base ethylenediamine forms a second phase above the heavier dichloromethane. The dialkyldichlorosilane is added dropwise and reacts with the hydroxylamine.

The chlorine atoms are substituted by ONH_2 groups. The bis(hydroxylamino)silane moves through the phase boundary and is protected against further reactions in the upper, basic phase.

Di-*tert*-butyl-*O,O′*-bis(hydroxylamino)silane is the first stable *O,O′*-bis(hydroxylamino)silane [12]:

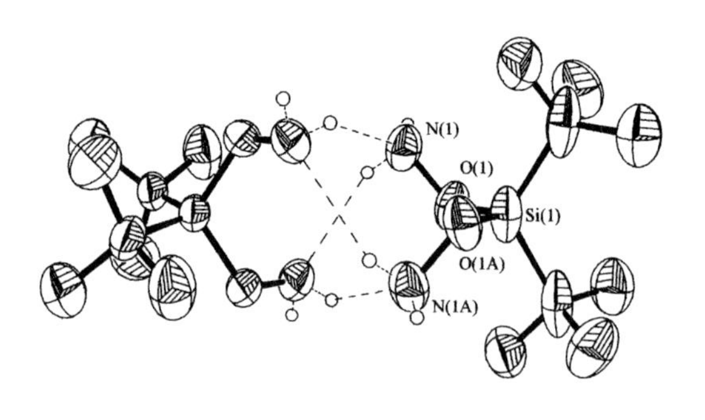

Selected bond lengths [pm] and angles [°].

Si(1)–O(1)	165.7(3)
Si(1)–O(1A)	165.7(3)
O(1)–N(1)	147.0(6)
Si(1)–C(4)	199.4(8)
N(1)-O(1)-Si(1)	112.3(3)
O(1A)-Si(1)-O(1)	112.3(2)
O(1)-Si(1)-C(4)	112.4(2)
O(1)-Si(1)-C(4A)	100.7(2)

Fig. 2. Crystal structure of a dimeric *O,O′*-bis(hydroxylamino)silane.

The *O,O′*-bis(hydroxylamino)silane forms a dimer via NH---N hydrogen bridges in the crystal. The nitrogen atoms show a remarkably small sum of angles (296.9°). They exhibit a pyramidal environment. The Si-O-N angles amount to 112.3°. The non-binding N---Si distances are found to be 260 pm, which is noticeably longer than those found and calculated in silylhydroxylamines methylated at the nitrogen atom [13, 14].

Isomeric *N,N′*- and *O,O′*-Bis(silylhydroxylamino)silanes

In a substitution reaction it is possible to prepare *N,N′*-bis(-*O*-silylhydroxylamino)silanes, e.g., in the reaction of $Me_3C(Me)SiCl_2$ and Me_3SiONH_2 [12]:

$$Me_3C(Me)SiCl_2 \xrightarrow[-2\,HCl\cdot NEt_3]{+2\,NEt_3,\ +2\,Me_3SiONH_2} Me_3C(Me)Si(NH{-}OSiMe_3)_2$$

Eq. 3.

The synthesis of *O,O′*-bis(hydroxylamino)silanes now allows the preparation of isomeric *O,O′*-bis(-*N*-silyl-hydroxylamino)silanes:

$$Me_3C(Me)Si(ONH_2)_2 \xrightarrow[-2\,HCl\cdot NEt_3]{+2\,NEt_3,\ +2\,Me_3SiCl} Me_3C(Me)Si(O{-}NHSiMe_3)_2$$

Eq. 4.

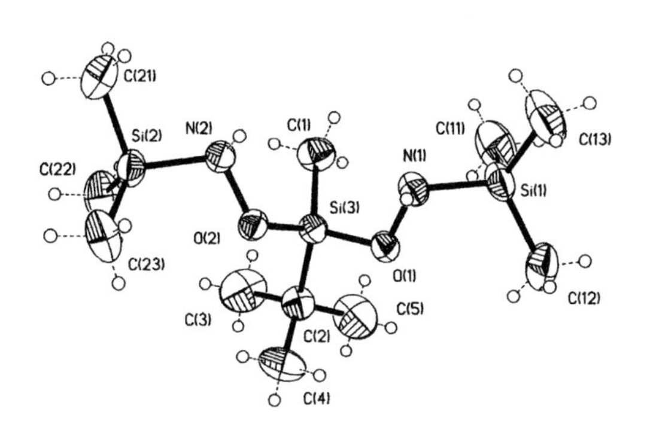

Selected bond lengths [pm] and angles [°].

Si(1)–N(1)	175.7(2)
Si(3)–O(1)	166.0(2)
O(1)–N(1)	147.9(2)
Si(1)–O(1)	259.4(3)
Si(3)–N(1)	262.3(2)
N(1)–H(1)	81.6(5)
N(1)-O(1)-Si(3)	113.29(12)
O(1)-N(1)-Si(1)	106.24(13)
H(1)-N(1)-Si(1)	115.90(3)
H(1)-N(1)-O(1)	105.80(2)

Fig. 3. Crystal structure of an *O,O'*-bis(-*N*-silyl-hydroxylamino)silane.

In the solid state it forms a dimer via NH---O bridges. The orientation of the monomeric units to each other is found to be about 90°. The nitrogen atoms show a tetrahedral environment. The sum of angles at N(1) is 327.9° and at N(2) 329.4°. The hydrogen atoms of the NH groups are in the *trans* position to each other. The non-binding distances Si(3)---N(1) and Si(1)---O(1) are found to be 262.3 pm and 259.4 pm.

N,O-Bis(silyl)-*N*-stannylhydroxylamines

For the preparation of stannylhydroxylamines *N,O*-bis(silyl)hydroxylamines were used, which form after treatment with *n*-BuLi dimeric, trimeric, or tetrameric oligomers depending on the bulkiness of the substituents and the solvent. In any case, $(R_3Si)_2N–O^-$ anions are formed, which means that a silyl group migration from oxygen to nitrogen occurred. The lithium is bonded to the oxygen atom, and, in the subsequent reaction under LiCl-formation, the new stannyl moiety is expected to be attached to the oxygen atom. However, in reactions of chlorotrimethylstannane and lithium derivatives only the *N,O*-bis(silyl)-*N*-trimethylstannyl-hydroxylamines were isolated [15]:

$$(R_3Si)_2N\text{–}OLi \xrightarrow[-\,LiCl]{Me_3SnCl} (R_3Si)_2N\text{–}O\text{–}SnMe_3 \longrightarrow \left[R_3Si\text{–}N\text{–}O \text{ (bridged by } SnMe_3 \text{ and } SiR_3) \right]^{\ddagger} \longrightarrow (R_3Sn)(R_3Si)N\text{–}O\text{–}SiMe_3$$

Eq. 5.

Under the reaction conditions the primarily formed *O*-stannylhydroxylamine could not be detected. The formation of *N,N*-$Me_3Sn(R_3Si)N–O–SiR_3$ (**1**) can be explained by the Lewis concept. The soft Lewis base tin tries to be bonded to the soft Lewis base nitrogen. This result implies a 1,2-anionic stannyl group migration from oxygen to nitrogen and a silyl group migration from nitrogen to the oxygen via a dyotropic transition state.

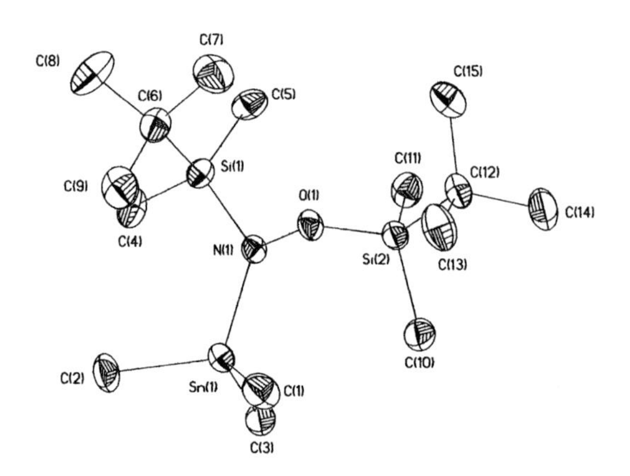

Selected bond lengths [pm] and angles [°].

Sn(1)–N(1)	208.8(3)
N(1)–O(1)	149.7(3)
N(1)–Si(1)	174.9(3)
O(1)–Si(2)	166.6(2)
Sn(1)–C(3)	214.3(4)
O(1)-N(1)-Si(1)	105.36(17)
Si(1)-N(1)-Sn(1)	126.57(14)
O(1)-N(1)-Sn(1)	108.59(16)
N(1)-O(1)-Si(2)	115.70(18)
N(1)-Sn(1)-C(1)	105.76(12)

Fig. 4. Crystal structure of an *N,O*-bis(silyl)-*N*-stannylhydroxylamine.

The N atom has an almost pyramidal environment, with the widest angle 126.5° between the silyl and stannyl groups (sum of angles N(1) = 340.5°). The distance to the plane of the neighboring atoms O(1), Si(1) and Sn(1) is measured at 45.26 pm.

A Tricyclic Tetraoxatetrastannane: a Hydrolysis Product of a Bis(Silyl)-stannylhydroxylamine

As a consequence of hydrolysis, 2,2,4,4,6,6,8,8-octamethyl-3,7-bis[(3-*tert*-butyl-1,1,3,3-tetramethyl)-1,3-disiloxano]-4,8-bis(*tert*-butyldimethylsiloxy)-1,3,5,7-tetraoxa-2,4,6,8-tetrastanna-$[4.2.0.0^{2,5}]$-tricyclooctane is obtained [15] (Fig. 5).

The compound has an inversion center and is a dimer in the crystal.

In the tricyclic ring system, every tin atom is pentacoordinated and oxygen adopted a trigonal planar coordination sphere ($\Sigma° = 360.0°$). As shown by the figure above, the ladder-like ring system is planar, with the sum of the angles in the center ring (as well as that of the outer ones) $\Sigma° =$ 360.0°.

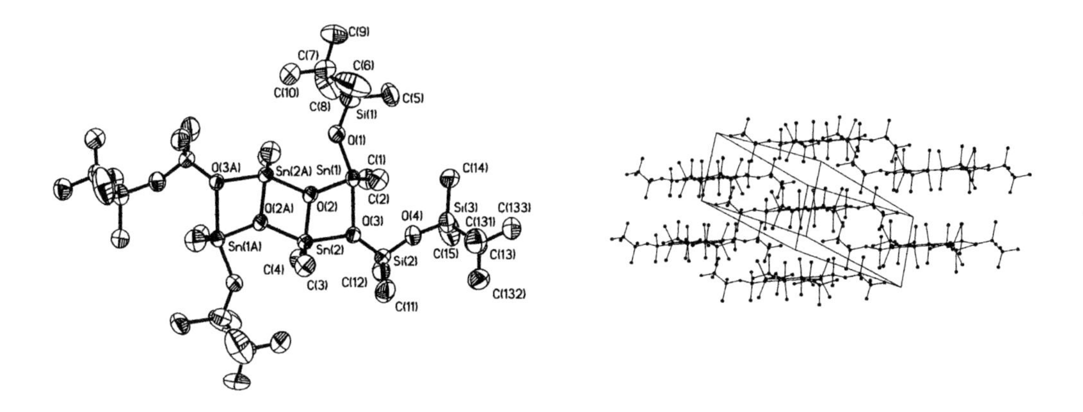

Selected bond lengths [pm] and angles [°].

Sn(1)–O(2)	198.9(7)	Sn(1)–O(1)	201.4(7)
Sn(1)–O(3)	240.1(7)	Si(1)–O(1)	161.0(8)
O(2)–Sn(2A)	214.6(7)	O(2)–Sn(2)	202.0(7)
Sn(2)–O(2A)	214.6(7)	Sn(2)–O(3)	214.5(6)
Si(2)–O(3)	161.7(7)	Si(2)–O(4)	162.4(8)
Si(3)–O(4)	164.2(8)		
O(2)-Sn(1)-O(1)	89.1(3)	O(2)-Sn(1)-C(2)	111.5(4)
O(1)-Sn(1)-C(1)	100.4(4)	O(1)-Sn(1)-O(3)	159.8(3)
O(2)-Sn(1)-O(3)	70.7(3)	C(2)-Sn(1)-O(3)	87.5(4)
Si(1)-O(1)-Sn(1)	137.6(5)	Sn(1)-O(2)-Sn(2)	116.4(3)
Sn(1)-O(2)-Sn(2A)	136.8(4)	Sn(2)-O(2)-Sn(2A)	106.8(3)
O(2)-Sn(2)-O(2A)	73.2(3)	O(2)-Sn(2)-O(3)	75.9(3)
O(4)-Si(2)-O(3)	107.1(4)	O(3)-Sn(2)-O(2A)	149.1(3)
Si(2)-O(3)-Sn(2)	135.9(4)	Si(2)-O(3)-Sn(1)	127.1(4)
Sn(2)-O(3)-Sn(1)	97.0(3)	Si(2)-O(4)-Si(3)	153.9(6)

Fig. 5. Crystal structure of the tricyclic tetraoxatetrastannane.

Acknowledgment: This work was supported by the Deutsche Forschungsgemeinschaft and the Fonds der Chemischen Industrie.

References

[1] U. Wannagat, J. Pump, *Monatsh. Chem.* **1963**, *94*, 141 – 150.

[2] P. Boudjouk, R. West, *Intra-Sci. Chem. Rep.* **1973**, *7*, 65 – 82.

[3] R. Wolfgramm, T. Müller, U. Klingebiel, *Organometallics* **1998**, *17*, 3222 – 3226.
[4] S. Schmatz, F. Diedrich, C. Ebker, U. Klingebiel, *Eur. J. Inorg. Chem.* **2002**, *4*, 876 – 885.
[5] F. Diedrich, U. Klingebiel, F. Dall'Antonia, C. Lehmann, M. Noltemeyer, T. R. Schneider, *Organometallics* **2000**, *19*, 5376 – 538.
[6] F. Diedrich, U. Klingebiel, M. Schäfer, *J. Organomet. Chem.* **1999**, *588*, 242 – 246.
[7] F. Diedrich, C. Ebker, U. Klingebiel, *Phosphorus, Sulfur and Silicon* **2001**, *169*, 253 – 256.
[8] R. Wolfgramm, U. Klingebiel, *Z. Anorg. Allg. Chem.* **1998**, *624*, 1031 – 1034.
[9] R. Wolfgramm, U. Klingebiel, *Z. Anorg. Allg. Chem.* **1998**, *624*, 1035 – 1040.
[10] A. O. Stewart, J. G. Martin, *J. Org. Chem.* **1989**, *54(5)*, 1221 – 1223.
[11] C. Ebker, F. Diedrich, U. Klingebiel, M. Noltemeyer, S. Schmatz, *Organometallics* **2003**, *22(13)*, 2594 – 2598.
[12] J. C. Bottaro, C. D. Bedford, A. Dodge, *Synth. Comm.* **1985**, *15(14)*, 1333 – 1335.
[13] N. W. Mitzel, M. Hofmann, E. Waterstradt, P. v. Ragué Schleyer, H. J. Schmidbaur, *J. Chem. Soc., Dalton Trans.* **1994**, 2503.
[14] N. W. Mitzel, U. Losehand, *J. Am. Chem. Soc.* **1998**, *120*, 7320.
[15] S. Schmatz, C. Ebker, T. Labahn, H. Stoll, U. Klingebiel, *Organometallics* **2003**, *22(3)*, 490 – 498.

From Silyldiamines to Mono-NH-SiF-functional Cyclodisilazanes — Synthesis, Reactions, and Crystal Structures

Clemens Reiche, Uwe Klingebiel*

Institute of Inorganic Chemistry
Georg-August-University Goettingen,
Tammannstr. 4, D-37077 Goettingen, Germany
Tel.: +49 551 393052 — Fax: +49 551 393373
E-mail: uklinge@gwdg.de

Keywords: cyclodisilazanes, spirocycle, aldehyde insertion

Summary: Treatment of geminal silyldiamines with monohalosilanes leads to the formation of 1-amino-1,3-disilazanes. Mono-NH-SiF-functional cyclodisilazanes can be isolated by ring closure of their lithium derivatives in the reaction with trifluorosilanes. Lithium salts of these cyclodisilazanes react with H-acidic compounds like alcohols as amides and with unsaturated compounds like aldehydes as silaimines.

1-Amino-1,3-disilazanes

The treatment of alkali metalated diamines with halosilanes leads to the formation of 1-amino-1,3-disilazanes.

$$Me_3C\text{—}Si(R)(NH_2)\text{—}NH_2 \xrightarrow[-C_4H_{10},\ -LiHal]{+C_4H_9Li,\ +HalSiR_1R_2R_3} Me_3C\text{—}Si(R)(NH_2)\text{—}NH\text{—}Si(R_1)(R_3)\text{—}R_2$$

R, R_1, R_2 = alkyl, aryl
R_3 = alkyl, H, Hal

Eq. 1.

Synthesis of NH-SiF-functional Disilazanes

For the synthesis of mono-NH-SiF-functional cyclodisilazanes, two pathways are imaginable. First, silyldiamines can be dilithiated and treated with trihalosilanes and once again substituted with a silyl group. Second, a cyclization of dilithiated 1-amino-1,3-disilazanes may be achieved by using trihalosilanes. The treatment of dilithiated diamines with trifluorosilanes leads to the formation of four-membered cyclodisilazanes, e.g. [1]:

$$(Me_3C)_2Si(NHLi)_2 + F_3SiN(SiMe_2CMe_3)_2 \xrightarrow{-2\,LiF} (Me_3C)_2Si(\mu\text{-}NH)_2Si(F)N(SiMe_2CMe_3)_2$$

Eq. 2.

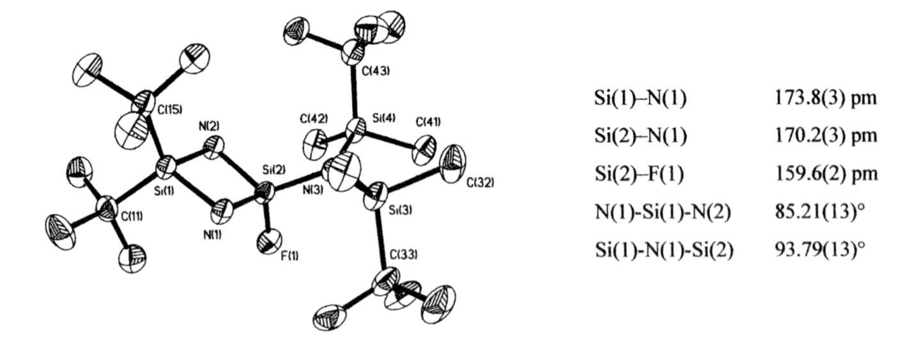

Fig. 1.

Following the second pathway, in the reaction of dilithiated 1-amino-1,1-di-*tert*-butyl-3,3,3-trimethyl-1,3-disilazane with a trifluorosilane a spirocyclic compound formed by elimination of a lithiumamide and LiF is isolated [2]:

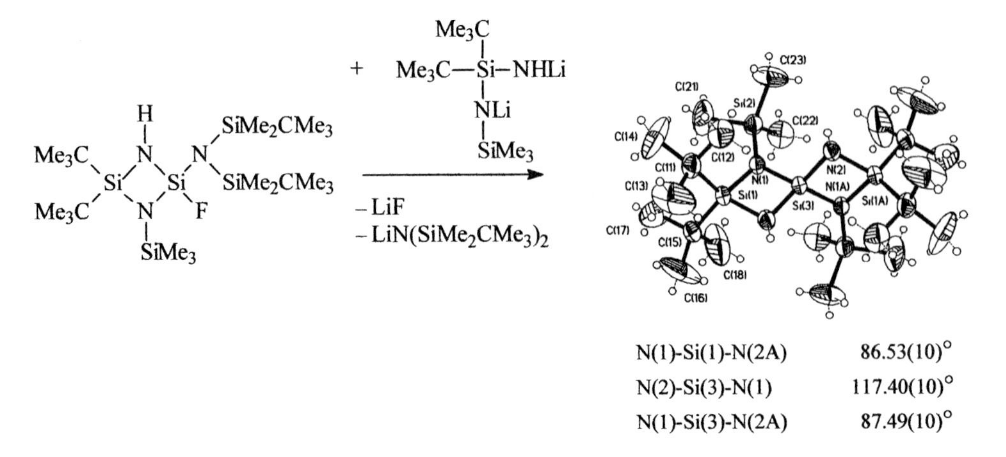

Scheme 1.

The crystal structure shows two planar four-membered ring elements which are 83.4° twisted with respect to each other. The synthesis of spirocyclic compounds can be prevented by starting from more bulky substituted 1-amino-1,3-disilazanes. Thus, the first mono-NH-SiF-functional cyclodisilazanes were isolated, e.g.:

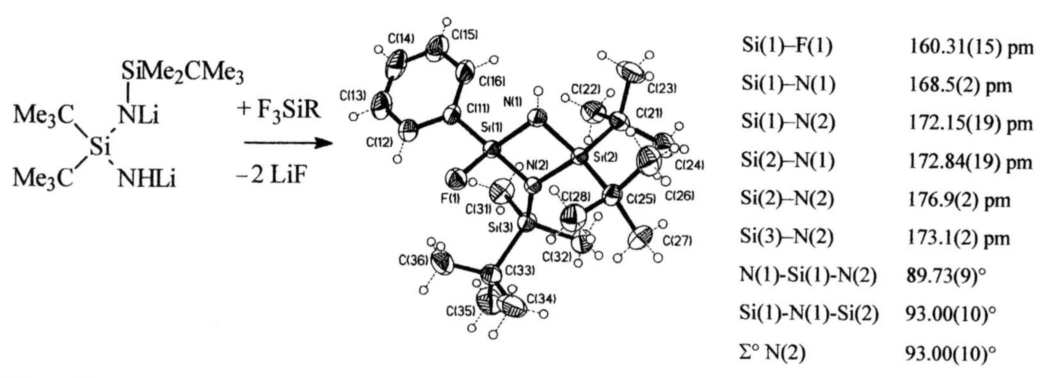

Si(1)–F(1)	160.31(15) pm
Si(1)–N(1)	168.5(2) pm
Si(1)–N(2)	172.15(19) pm
Si(2)–N(1)	172.84(19) pm
Si(2)–N(2)	176.9(2) pm
Si(3)–N(2)	173.1(2) pm
N(1)-Si(1)-N(2)	89.73(9)°
Si(1)-N(1)-Si(2)	93.00(10)°
Σ° N(2)	93.00(10)°

Scheme 2.

Reactivity of Mono-NH-SiF-functional Cyclodisilazanes

The treatment of cyclic fluorosilylamines with *n*-butyllithium leads to the formation of lithium amides, e.g.:

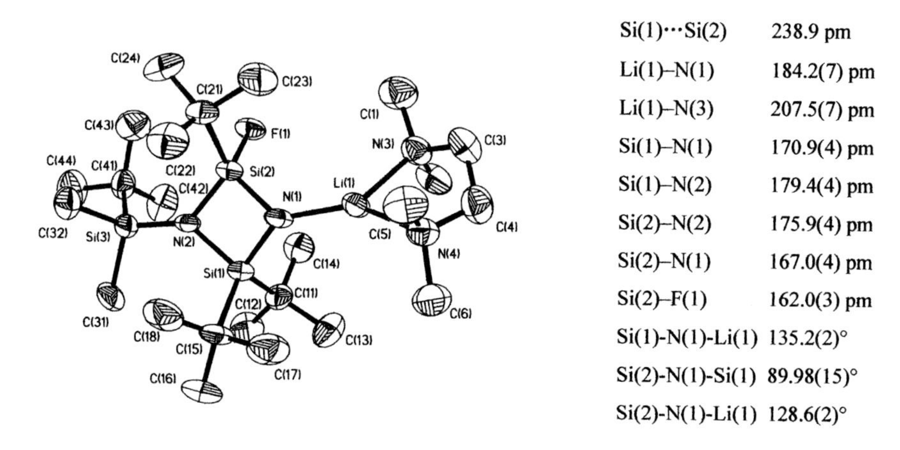

Si(1)···Si(2)	238.9 pm
Li(1)–N(1)	184.2(7) pm
Li(1)–N(3)	207.5(7) pm
Si(1)–N(1)	170.9(4) pm
Si(1)–N(2)	179.4(4) pm
Si(2)–N(2)	175.9(4) pm
Si(2)–N(1)	167.0(4) pm
Si(2)–F(1)	162.0(3) pm
Si(1)-N(1)-Li(1)	135.2(2)°
Si(2)-N(1)-Si(1)	89.98(15)°
Si(2)-N(1)-Li(1)	128.6(2)°

Fig. 2.

In the presence of a coordinating solvent, no oligomerization of the lithium derivative is observed. The cyclodisilazane unit is 50° twisted with respect to the Li-TMEDA ring unit. Compared to a Si–Si bond length of 235 pm, the transannular Si···Si distance is very short

(238.9 pm).

Based on these lithium derivatives, a substitution of the N atom can be achieved in the reaction with Me_3SiCl as well as in the reaction with Me_3SiOTf:

$$(Me_3C)_2Si[\mu\text{-}N(SiMe_2CMe_3)][\mu\text{-}N(H)]Si(F)(CMe_3) \xrightarrow[-C_4H_{10}]{+C_4H_9Li} \xrightarrow[-LiCl \;/\; -LiOTf]{+Me_3SiCl \;/\; +Me_3SiOTf} (Me_3C)_2Si[\mu\text{-}N(SiMe_2CMe_3)][\mu\text{-}N(SiMe_3)]Si(F)(CMe_3)$$

Eq. 3.

With alcohols, lithium salts of cyclic flourosilylamines react as amides, which means that the F atom is substituted by an alkoxy group, e.g.:

$$(Me_3C)_2Si[\mu\text{-}N(SiMe_2CMe_3)][\mu\text{-}N(Li(TMEDA))]Si(CMe_3)(F) \xrightarrow[-LiF,\ -TMEDA]{+MeOH} (Me_3C)_2Si[\mu\text{-}N(SiMe_2CMe_3)][\mu\text{-}N(H)]Si(CMe_3)(OMe)$$

Eq. 4.

If metalated again and treated with difluorosilanes, a rearrangement is prevented and the attacking group is not bonded to the O atom but to the N atom because of the bulkiness of the substituents.

$$(Me_3C)_2Si[\mu\text{-}N(SiMe_2CMe_3)][\mu\text{-}N(H)]Si(CMe_3)(OMe) \xrightarrow[-C_4H_{10},\ -LiF]{+C_4H_9Li,\ +F_2SiMe_2} (Me_3C)_2Si[\mu\text{-}N(SiMe_2CMe_3)][\mu\text{-}N(SiMe_2F)]Si(CMe_3)(OMe)$$

Eq. 5.

Their reactivity toward unsaturated compounds like aldehydes is comparable to silaimines, e.g.:

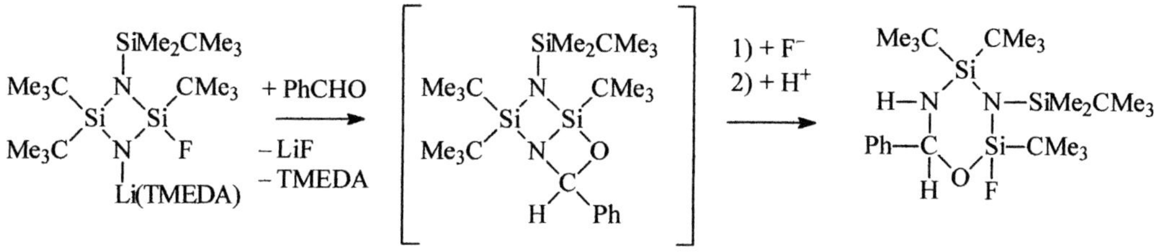

$SiMe_2CMe_3$ / Me_3C / N / CMe_3 / Si / Si / Me_3C / N / F / Li(TMEDA) + PhCHO, − LiF, − TMEDA → [$SiMe_2CMe_3$ / Me_3C / N / CMe_3 / Si / Si / Me_3C / N / O / C / H / Ph] 1) + F^- 2) + H^+ → Me_3C / CMe_3 / Si / H—N / N—$SiMe_2CMe_3$ / Ph—C / Si—CMe_3 / H / O / F

Scheme 3.

An (SiNCO)-unit arises as intermediate product as result of a [2 + 2]-cycloaddition. Thermal reaction with LiF leads to a cyclohexane derivative. An alternative pathway might be the formation of an alcoxide starting from the aldehyde which attacks the silicon atom.

Acknowledgment: This work has been supported by the Fonds der Chemischen Industrie.

References

[1] H. J. Rakebrandt, U. Klingebiel, M. Noltemeyer, U. Wenzel, D. Mootz, *J. Organomet. Chem.* **1996**, *524*, 237.

[2] C. Reiche, U. Klingebiel, *Z. Naturforsch.* **2003**, *58b*, 939.

From Lithium Halogenosilylamides to Four- and/or Fourteen-Membered Rings and New 14π-Electron Systems

Annette Wand, Sibylle Kucharski,* Uwe Klingebiel*

Institute of Inorganic Chemistry, Georg-August-University Goettingen
Tammannstr. 4, 37077 Goettingen, Germany

Keywords: 1,3-bis(2-pyridyl)-cyclodisilazanes, 1,8-dioxa-3,10-diaza-2,9-disila-cyclotetradecane, 4,4′-bis(fluorosilylimino)-biphenyl

Summary: In contrast to organyl-substituted cyclodisilazanes, only a few halogeno substituted Si–N rings are known. Normally LiF elimination from lithium fluorosilylamides leads to the formation of four-membered Si–N rings. Using THF as solvent, a THF cleavage occurs and by-products are observed. Lithiated aminotrichlorosilanes react with THF under formation of new fourteen-membered heterocycles. An unknown 14π-electron system was isolated after thermal treatment of lithiated 2,6-diisopropylphenylfluorosilylamine.

Cyclizations of Fluoro- and Chlorosilylamines

Lithium salts of aromatic amines, e.g., 2-aminopyridine or 2,6-diisopropylaniline, react with halosilanes to form halosilylamines. In particular, in the case of di- or trifluorosilanes, a migration of the lithium from the lithium amide to the fluorosilylamine occurs to yield bis(fluorosilyl)amines (e.g., Eq. 1).

Eq. 1.

Halosilylorganylamides are suitable reagents to synthesize bis(amino)silanes and Si–N ring systems. Here, fluorosilylamines and the analogous chlorosilylamines react differently.

Fluorosilylamines

Me_3C–Si(R')(F)–N(R)(Li) $\xrightarrow{-\,LiF}$

Δ, hexane, R' = CMe_3 → 1/2 [$(Me_3C)_2Si$–N(R)]$_2$ (cyclodisilazane: R–N, N–R; Me_3C, CMe_3 on each Si)

hexane, THF, R' = F → R–N(H)–Si(CMe_3)(F)–O–$(CH_2)_3$–CH_3

R = 2-pyridyl

Eq. 2.

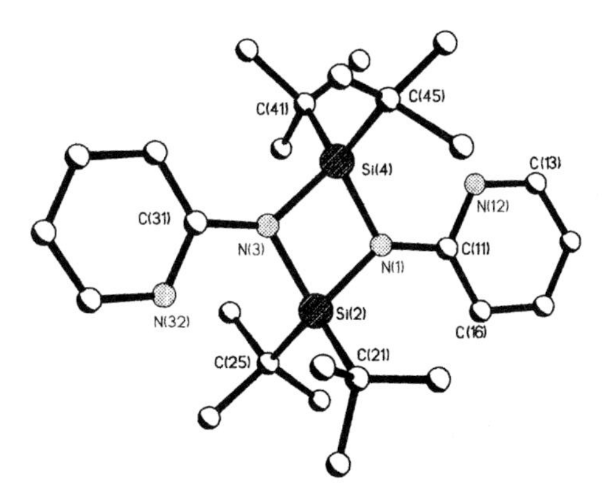

Fig. 1. Crystal structure of 1,3-bis(2-pyridyl)-2,2,4,4-tetra-*tert*-butyl-cyclodisilazane.

The exocyclic N(1)–C(11) single bond is found to be 8.2 pm shorter than calculated (147 pm). The Si–N ring system is planar (Σ(N1) = 359.6°). The nitrogen atoms of the pyridyl rings are in *trans* position. They are orientated to one silicon atom of the cyclodisilazane. The result is a smaller Si(4)-N(1)-C(11) angle than the Si(2)-N(1)-C(11) angle and a smaller N(1)-C(11)-N(12) angle than the N(12)-C(11)-C(16) angle. The short Si(4)–N(12) distance leads to an enlargement of the endocyclic Si(4)-N(1)-Si(2) angle (97.3°) and a reduction of the endocyclic N(3)-Si(4)-N(1) angle (82.5°). This is the smallest N-Si-N angle in cyclodisilazanes with a four-coordinated silicon atom.

Cyclization in THF leads to the formation of a by-product under THF cleavage.

Table 1. Selected bond lengths and angles.

N(3)–Si(4)	178.2 pm	C(11)–N(12)	133.0 pm
Si(4)–N(1)	176.4 pm	N(12)–C(13)	133.8 pm
N(1)–C(11)	138.8 pm		
N(3)-Si(4)-N(1)	82.5°	N(1)-C(11)-N(12)	114.2°
Si(4)-N(1)-Si(2)	97.3°	N(1)-C(11)-C(16)	124.7°
Si(4)-N(1)-C(11)	124.8°	C(11)-N(12)-C(13)	118.6°
Si(2)-N(1)-C(11)	137.1°		

Another pathway to the formation of cyclodisilazanes is the reaction of lithiated bis(amino)silanes with halosilanes (e.g., Eq. 3).

$$\mathrm{Me\text{–}Si(CMe_3)_2\text{–}N(Li)\text{–}SiMe_2\text{–}N(Li)\text{–}Si(CMe_3)_2\text{–}Me} \xrightarrow[-\,2\mathrm{LiF}]{+\,\mathrm{SiF_4}} \mathrm{Me\text{–}Si(CMe_3)_2\text{–}N\langle SiMe_2, SiF_2\rangle N\text{–}Si(CMe_3)_2\text{–}Me}$$

Eq. 3.

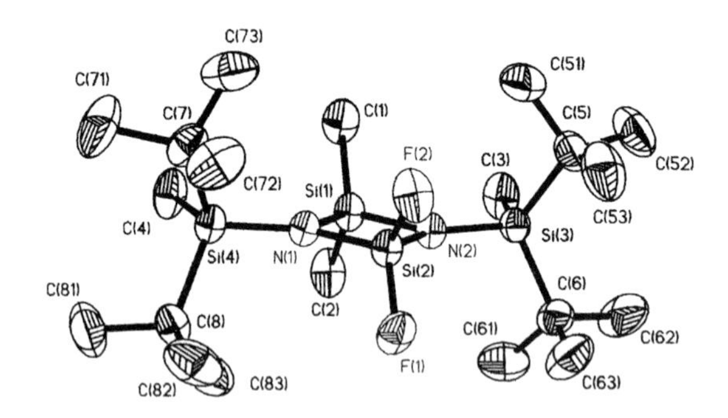

Fig. 2. Crystal structure of 1,3-bis(di-*tert*-butylmethylsilyl)-2,2-difluoro-4,4-dimethyl-cyclodisilazane.

Table 2. Selected bond lengths and angles.

Si(1)–N(1)	176.0 pm	Si(2)–N(2)	168.6 pm
Si(1)–N(2)	176.1 pm	N(1)–Si(4)	173.8 pm
Si(2)–N(1)	168.7 pm	N(2)–Si(3)	173.9 pm
N(1)-Si(1)-N(2)	88.2°	Si(2)-N(1)-Si(1)	89.3°
N(2)-Si(2)-N(1)	93.2°	Si(2)-N(2)-Si(1)	89.3°

The planar four-membered Si–N ring has several irregularities. Only one endocyclic angle is larger than 90° (N(2)-Si(2)-N(1): 93.2°). Because of the electron-withdrawing effect of the fluorine atoms, two short and two long Si–N bonds are found in the ring system. It is unusual that the bulky *tert*-butyl groups are not staggered but eclipsed to each other.

Chlorosilylamines

In hexane, LiCl elimination from lithium trichlorosilylamides leads to cyclodisilazanes. In THF, a new type of THF cleavage occurs.

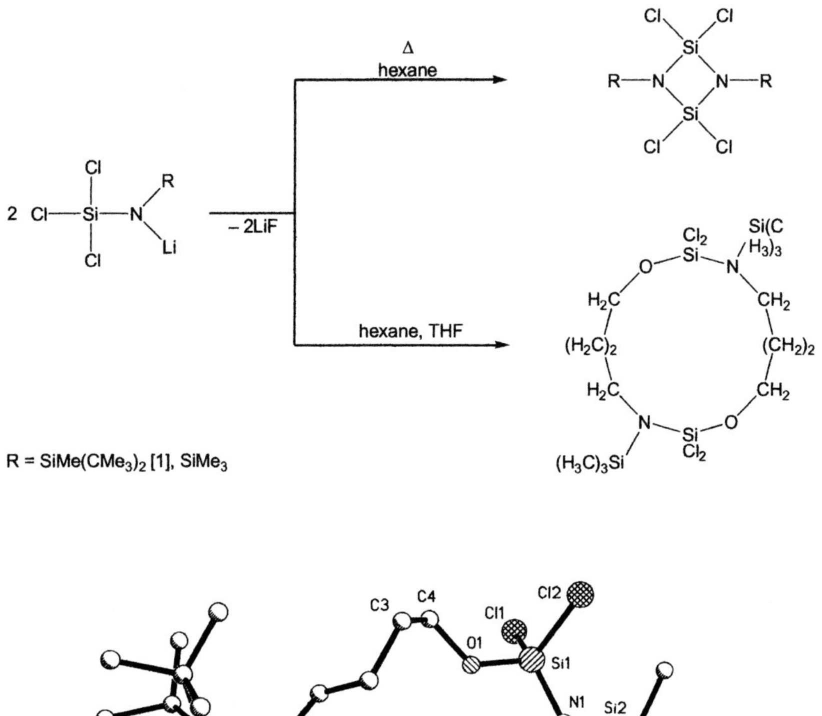

R = SiMe(CMe$_3$)$_2$ [1], SiMe$_3$

Eq. 4.

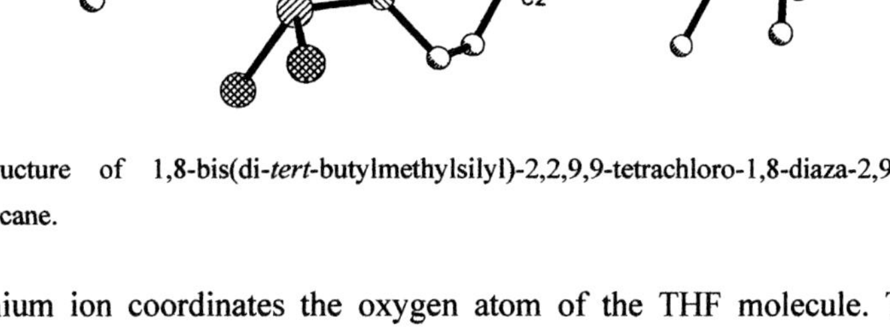

Fig. 3. Crystal structure of 1,8-bis(di-*tert*-butylmethylsilyl)-2,2,9,9-tetrachloro-1,8-diaza-2,9-disila-3,10-dioxa-cyclotetradecane.

At first the lithium ion coordinates the oxygen atom of the THF molecule. This leads to a cleavage of the H_2C–O bond and the formation of a H_2C–N bond. The formed lithium salt reacts intermolecularly to a fourteen-membered macrocycle.

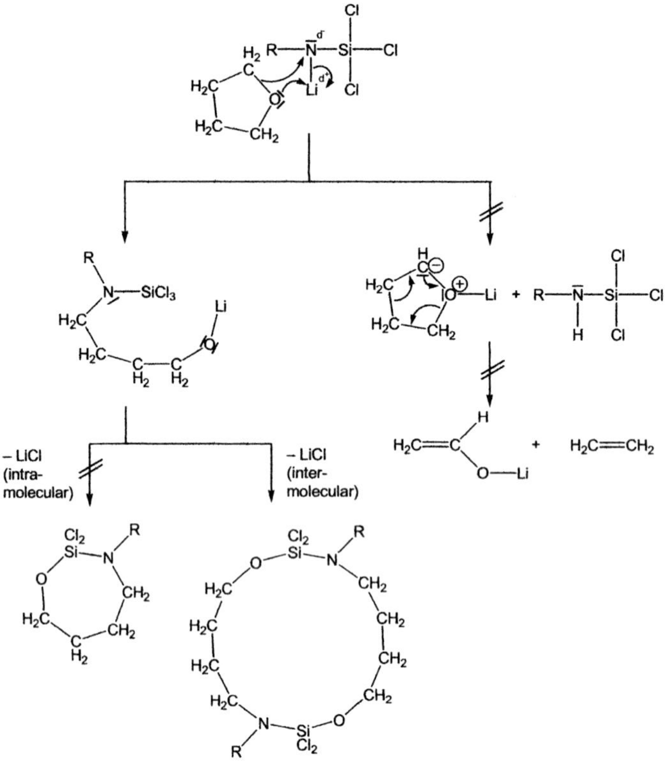

Eq. 5.

Synthesis of a New 14π-Electron System

A new 14π-electron system is obtained in the reaction of (2,6-di*iso*propylphenyl)-amino-di-*tert*-butylfluorosilane with butyllithium.

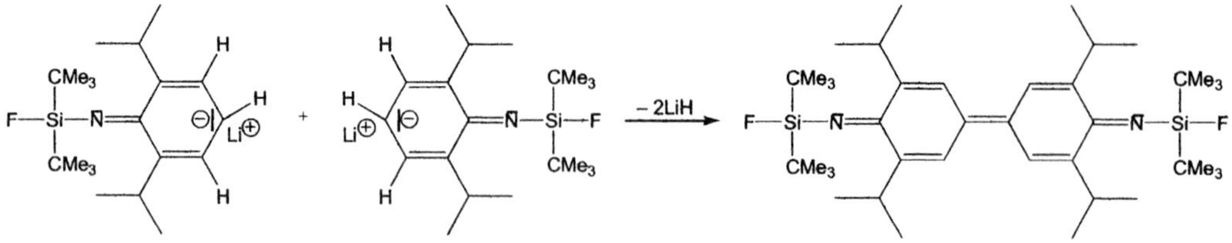

Eq. 6.

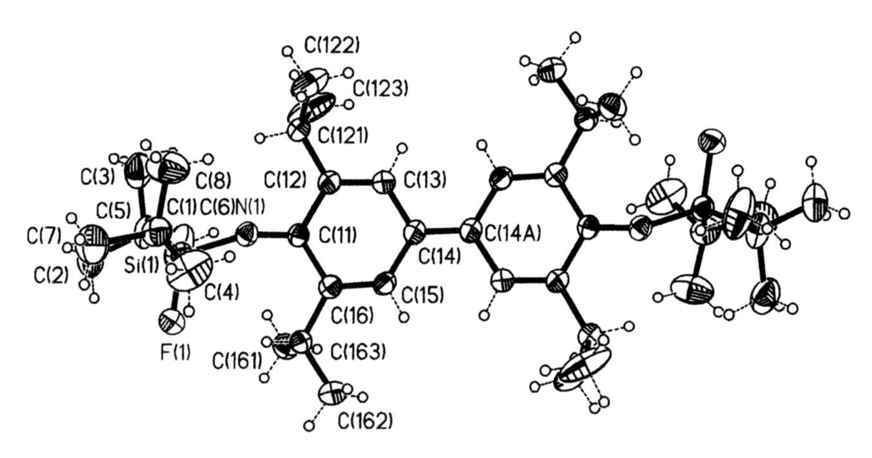

Fig. 4. Crystal structure of 4,4′-bis[(di-*tert*-butylfluorosilyl)imino]-3,3′,5,5′-tetra(*iso*propyl)-biphenyl.

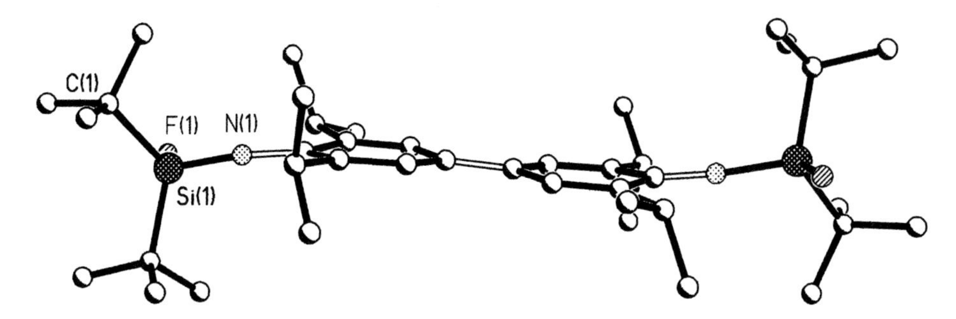

Fig. 5. Alternative view of Fig. 4.

Table 3. Selected bond lengths and angles.

Si(1)–N(1)	167.9 pm	C(12)–C(13)	135.2 pm
N(1)–C(11)	127.9 pm	C(13)–C(14)	144.7 pm
C(11)–C(12)	148.7 pm	C(14)–C(14A)	140.7 pm
C(11)-N(1)-Si(1)	159.7°		

This molecule with four coordinated silicon has the shortest Si–N imino bond as far as we know.

Acknowledgment: This work has been supported by the Fonds der Chemischen Industrie.

Reference

[1] B. Jaschke, R. Herbst-Irmer, U. Klingebiel, P. Neugebauer, T. Pape, *Z. Anorg. Allg. Chem.* **1999**, 625, 1683.

Silyl-Enolethers and -Ethers — New Results of Keto-Enol Tautomerism

Thomas Büschen,* Uwe Klingebiel

Institute of Inorganic Chemistry, Georg-August-University Goettingen
Tammannstr. 4, D-37077 Goettingen, Germany
Tel.: +49 551 39 3052 — Fax: +49 551 39 3373
E-mail: uklinge@gwdg.de

Keywords: enolfluorosilanes, fluorosiloxane-enolethers, 4,6-hexadienephenone

Summary: Substitution of acid hydrogens by silyl groups in ketones led to the formation of silylenolethers. We found, that depending on lithium-organyls and fluorosilanes, silylenolethers or silylethers are formed. By changing the molar ratio, fluorosilyldienolethers can be obtained. In an aldol addition reaction, a new type of fluorosilyl-ketone-ether was isolated, and as a side product a new six-membered ring system was isolated by aldol condensation reactions.

Introduction

When two constitutional isomers are present in a rapidly established equilibrium, the two molecular types are designated tautomers and the phenomenon tautomerism. Tautomers differ in the position of a double bond and a flexible group. For the keto-enol tautomerism, the tautomeric forms are the keto and the enol form. Mostly, the equilibrium is on the left hand side.

$$R_2R_3CH{-}C(=O)R_1 \rightleftharpoons R_2R_3C{=}C(OH)R_1$$

keto form — enol form

Eq. 1. Equilibrium of the keto and the enol form.

The keto form is often more stable than the enol form, because the C=O double bond, the C–C single bond and the C–H single bond are more stable than the C=C double bond, the C–O single bond and the O–H single bond [1].

Substitution of acid hydrogens by silyl groups in ketones leads to the formation of silyl-enolethers. In competition with the synthesis of the silyl-enolethers is an addition reaction in which the alkyl group of the lithiated reagent adds at the carbonyl carbon. From this reaction, silyl-ethers can be isolated.

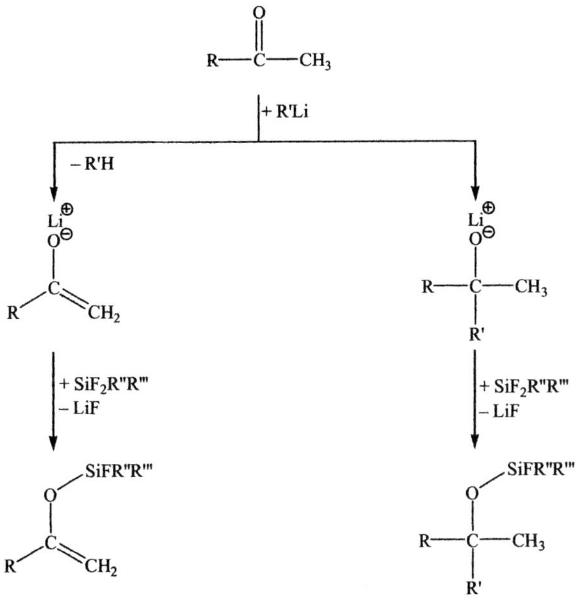

Scheme 1. Pathways of the reaction of ketones with a lithiated reagent and a difluorosilane.

Fluorosilyl-Enolethers

The first fluorofunctional silyl-enolethers were isolated in the reaction of the ketones 3,3-dimethyl-2-butanone / *n*-butyllithium [2] and 2,4-dimethyl-3-pentanone / *tert*-butyllithium with di- and trifluorosilanes, e.g.,

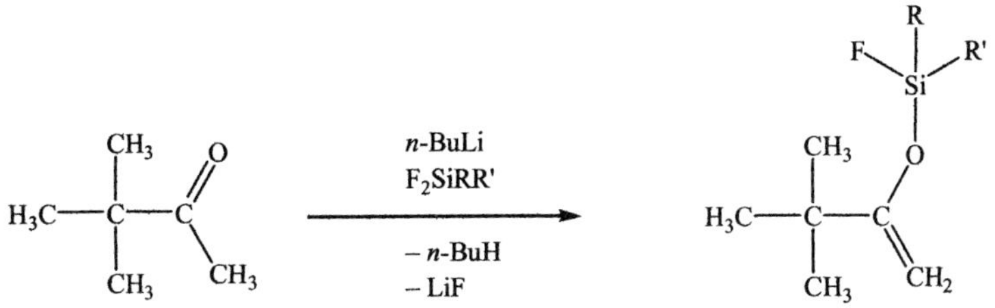

3,3-dimethyl-2-butanone

Eq. 2. Synthesis of fluorosilyl-enolethers with 3,3-dimethyl-2-butanone.

The silyl group stabilizes the enol form and the fluorosilyl-enolethers are isolable. The formation of the silyl-enolethers depends on the lithiated reagent.

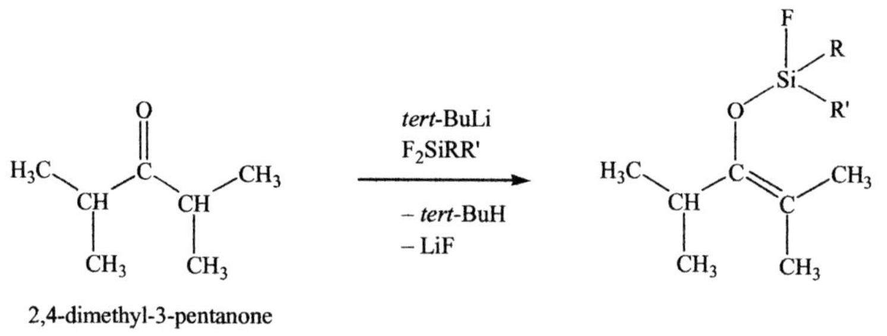

Eq. 3. Synthesis of fluorosilyl-enolethers with 2,4-dimethyl-3-pentanone.

Fluorosilyl-Ethers

For the synthesis of the silyl-ethers, the ketones were lithiated by organyllithium compounds. The lithium salt is converted with di- or trifluorosilanes.

The addition of the alkyl group from the lithiated reagent to the carbonyl carbon is affected by the dimension of the alkyl group. Sterically undemanding groups favor the addition reaction and prevent the formation of the silyl-enolethers.

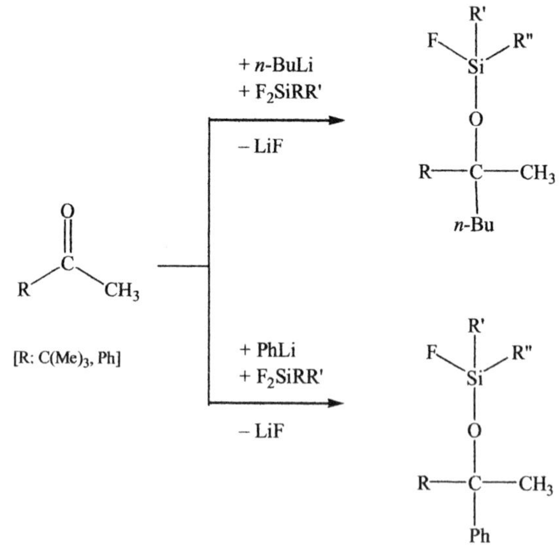

Eq. 4. Synthesis of different fluorosilyl-ethers by changing the lithiated reagent.

The silyl-ethers were synthesized with the ketones 3,3-dimethyl-2-butanone and phenylmethylketone, the lithiated reagents *n*-butyllithium and phenyllithium and with different fluorosilanes. Further fluorosilyl-ethers were isolated in the reaction of the ketone 2,4-dimethyl-3-pentanone and *tert*-butyllithium as lithiated reagent.

$$(H_3C)HC(CH_3)\text{-}C(=O)\text{-}CH(CH_3)CH_3 \xrightarrow[-\,LiF]{tert\text{-}BuLi,\ F_2SiRR'} (H_3C)_2HC\text{-}C(OSiFRR')(C(CH_3)_3)\text{-}CH(CH_3)_2$$

Eq. 5. Synthesis of fluorosilyl-ethers of 2,4-dimethyl-3-pentanone.

Bis(enol)fluorosilanes

By changing the molar ratio in further reaction of the fluorosilyl-enolether with the lithiumenolate, it was possible to synthesize the first bis(enol)fluorosilanes.

$$R'RSiF\text{-}O\text{-}C(=CH_2)C(CH_3)_3 + LiO\text{-}C(=CH_2)C(CH_3)_3 \xrightarrow{-\,LiF} R'RSi[O\text{-}C(=CH_2)C(CH_3)_3]_2$$

[**R**: Me, Ph ; **R'**: Me, F]

Eq. 6. Synthesis of bis(enol)fluorosilanes.

Synthesis of Siloxanes from Fluorosilyl-Enolethers and Fluorosilyl-Ethers

Fluorosilyl-enolethers and -ethers are useful educts for the synthesis of unknown bis(silylenol)-siloxanes and bis(silylether)siloxanes, e.g.,

$$2\ R'RSiF\text{-}O\text{-}C(=CH_2)C(CH_3)_3 + 2\,OH^- \xrightarrow[-\,H_2O,\ -\,2KF]{+\,2KOH} (H_3C)_3C\text{-}SiF(OC(=CH_2)C(CH_3)_3)\text{-}O\text{-}SiF(OC(=CH_2)C(CH_3)_3)\text{-}C(CH_3)_3$$

Eq. 7. Synthesis of siloxanes from fluorosilyl-enolethers.

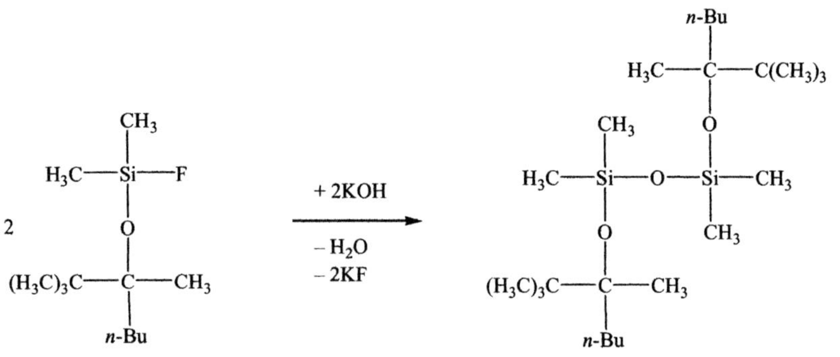

Eq. 8. Synthesis of siloxanes from fluorosilyl-ethers.

Syntheses of Fluorosilyl-Ketone-Ether

In an aldol addition reaction, a new type of fluorosilyl-ketone-ether was isolated.

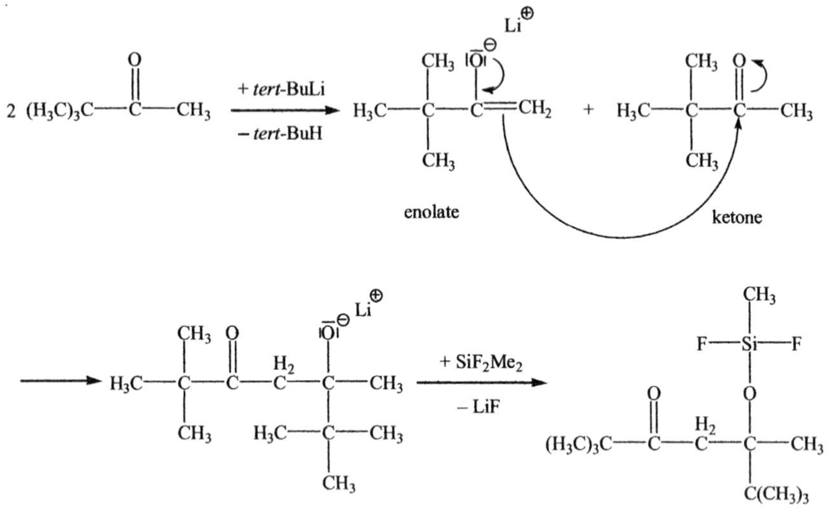

Scheme 2. Pathway of the aldol addition reaction of 3,3-dimethyl-2-butanone with lithium-3,3-dimethyl-2-butanolate.

Syntheses of Six-membered Ring System by Aldol Condensation Reactions

As a side product, a new six-membered ring system was obtained. Three aldol condensation reactions of phenylmethylketone occurred. The last one led to an intramolecular ring closure.

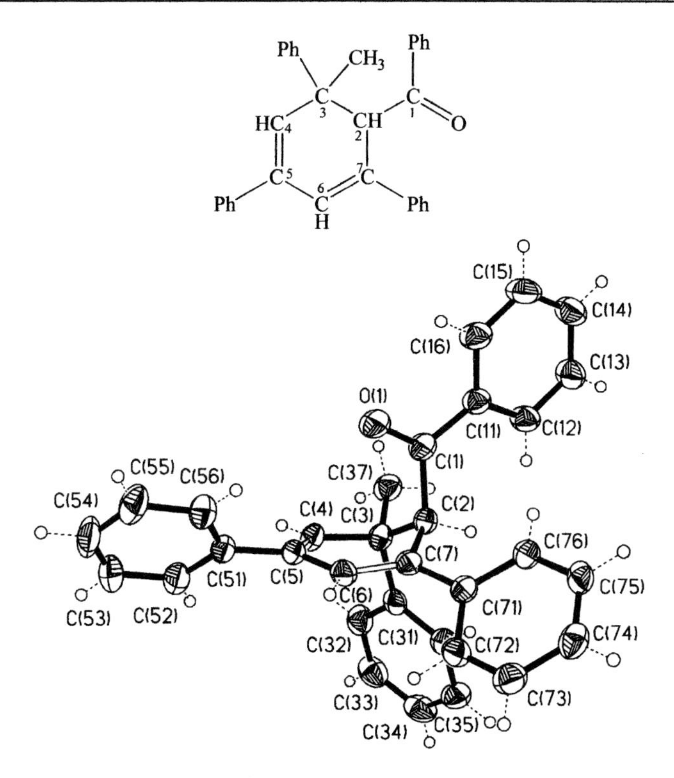

Fig. 1. Six-membered ring system as a product of aldol condensation reactions of phenylmethylketone.

Acknowledgment: We thank the Fonds der Chemischen Industrie for their financial support.

References

[1] J. Clayden, N. Greeves, S. Warren, P. Wothers, *Organic Chemistry*, Oxford University Press **2001**.

[2] U. Klingebiel, S. Schütte, D. Schmidt-Bäse, *Studies in Inorganic Chemistry Vol. 14*, **1992**.

“Hypervalent” Molecules — Low-Valency Candidates for Materials?

Hans H. Karsch,* Thomas Segmüller

Anorganisch-chemisches Institut, Technische Universität München
Lichtenbergstraße 4, D-85747 Garching, Germany
Tel.: +49 89 289 13132 — Fax: +49 89 14421
E-mail: Hans.H.Karsch@lrz.tum.de

Keywords: silicon amidinates, germanium amidinates, hypervalency, low-valency

Summary: A series of silicon and germanium compounds with amidinate ligands have been synthesized and investigated spectroscopically and structurally. In contrast to silicon, germanium also forms stable compounds with amidinate ligands in the +2 valence state with remarkable stuctural features. Bis-anionic amidinate ligands at silicon centers provide bridging bonding modes which differ from those described in earlier reports.

Introduction

Amidinate ligands play a prominent role not only in transition metal and lanthanoid coordination, but also in main group chemistry. Aside from the continuing studies in our laboratories, examples of silicon and germanium compounds with amidinate ligands are rather scarce, however [1 – 3]. This perhaps is due to the fact that amidinate complexes have been synthesized mainly for catalytic purposes, but silicon and germanium compounds are not generally regarded as relevant for this application. In earlier studies, we have introduced amidinate ligands into silicon and germanium compounds in the context of a new and different concept: the formation of strong element-to-nitrogen bonds together with the introduction of *two* nitrogen atoms with *one* anionic ligand enables the generation of silicon and germanium centers with high coordination numbers, where the central elements are embedded in a nitrogen-rich environment. In this way, amidinate complexes of silicon and germanium may provide valuable precursors for CVD purposes (Si/(C)/N). In the form of *di*anionic ligands, i.e. $[N–C(R)–NR']^{2-}$, they even may provide easily tuneable structural units in polymeric arrangements with pseudo-chalcogenide bridges, resembling silicate structures with O^{2-} bridges, which may further be converted to Si/C/N/X high-temperature-resistant materials. Aside from any application purposes, silicon compounds with high coordination numbers (“hypervalent” compounds) are of general interest, and examples with four-membered rings, as formed by chelating amidinate ligands, are almost unknown. Finally, with the aid of these ligands, low-valent complexes of these elements (Si(II), Ge(II)) seem accessible.

Results

Aside from our own contributions in silicon amidinate chemistry [4, 5], there are only two earlier reports on silicon amidinates, which were not characterized structurally, however: the formation of an eight-membered ring with two silicon centers and two di-anionic amidinate ligands $[PhC(NMe)N]^{2-}$ in the system $Me_2SiCl_2/Li[PhC(NMe)NH]$ [1], and the isolation of $[PhC(NSiMe_3)_2]SiCl_3$ (**1**) with a presumably penta-coordinate silicon center [2], both characterized by NMR spectroscopy. We re-investigated both systems, and we were successful in the isolation of **1** as colorless crystals suitable for X-ray investigation. The result is depicted in Fig. 1 and Table 1, confirming the previously proposed structure. Compound **1** was successfully converted to the triazido derivative **2**, which is remarkably stable toward impact and thermal treatment: **2** melts without decomposition at 141 °C. The structure of **2** may be inferred from Fig. 2 and Table 1.

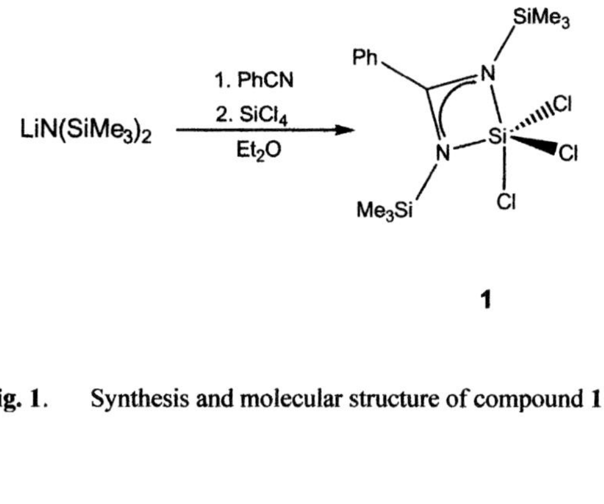

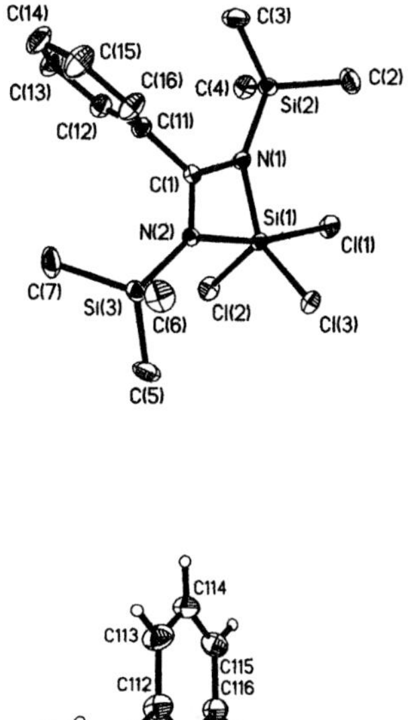

Fig. 1. Synthesis and molecular structure of compound **1**.

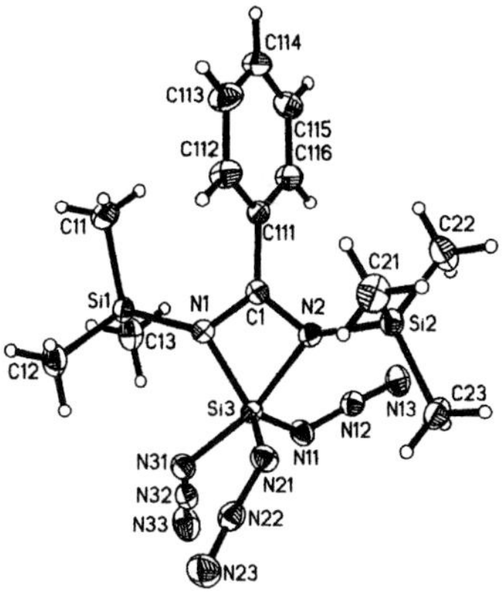

Fig. 2. Synthesis and molecular structure of compound **2**.

Table 1. Selected distances [Å] and angles [°] for **1** and **2**.

	1		**2**
Si(1)–N(2)	1.919(2)	Si(3)–N(2)	1.930(1)
Si(1)–N(1)	1.797(2)	Si(3)–N(1)	1.805(1)
Si(1)–Cl(1)	2.133(1)	Si(3)–N(11)	1.761(1)
Si(1)–Cl(3)	2.0843(9)	Si(3)–N(21)	1.758(1)
Si(1)–Cl(2)	2.0744(9)	Si(3)–N(31)	1.816(1)
N(1)–Si(2)	1.797(2)	N(2)–C(1)	1.317(2)
N(2)–Si(3)	1.783(2)	N(1)–C(1)	1.354(2)
N(2)–C(1)	1.311(3)	N(11)–N(12)	1.229(2)
N(1)–C(1)	1.366(3)	N(12)–N(13)	1.131(2)
Cl(2)-Si-Cl(3)	113.49(4)	N(11)-N(12)-N(13)	174.5(2)
Cl(1)-Si-Cl(3)	94.10(4)	N(1)-Si(3)-N(2)	71.4(2)
Cl(1)-Si-Cl(2)	94.03(4)	N(2)-Si(3)-N(11)	95.35(5)
N(1)-Si(1)-N(2)	71.6(1)	N(2)-Si(3)-N(21)	89.37(6)
N(2)-Si(1)-Cl(2)	92.18(7)	N(2)-Si(3)-N(31)	166.86(6)
N(2)-Si(1)-Cl(3)	92.80(7)	N(1)-C(1)-N(2)	109.4(1)
N(2)-Si(1)-Cl(1)	168.02(8)	N(1)-Si(3)-N(31)	95.58(6)
N(1)-C(1)-N(2)	108.6(2)	C(1)-N(1)-Si(3)	91.69(8)

In addition, an analogous species (**3**) containing a silicon–hydrogen bond was obtained by treating $HSiCl_3$ with 1 equiv. of a lithium amidinate ligand (Fig. 3, Table 2). As may be expected, the hydrogen atom adopts an equatorial position at the pentacoordinate silicon atom.

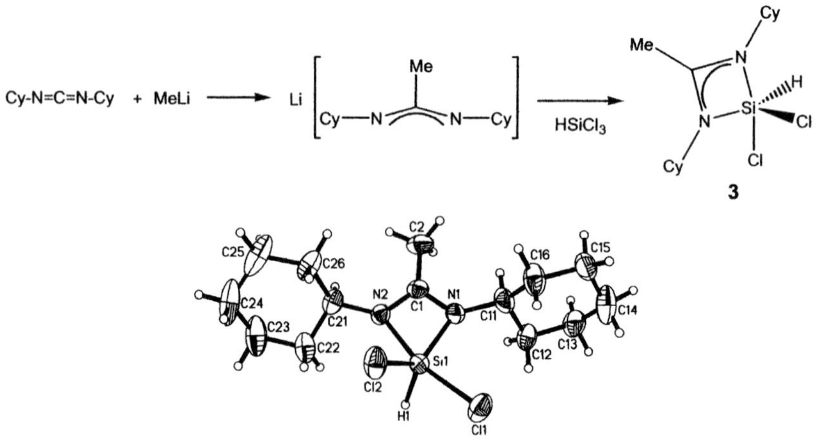

Fig. 3. Synthesis and molecular structure of compound **3**.

Table 2. Selected distances [Å] and angles [°] for **3**.

		3	
Si(1)–N(2)	1.937(2)	N(1)–C(1)	1.352(3)
Si(1)–N(1)	1.792(2)	N(1)-Si(1)-N(2)	69.1(1)
Si(1)–Cl(1)	2.1662(9)	N(2)-Si(1)-Cl(1)	165.52(7)
Si(1)–Cl(2)	2.086(1)	N(2)-Si(1)-Cl(2)	124.06(8)
Si(1)–H(1)	1.54(3)	N(2)-Si(1)-H(1)	95(1)
N(2)–C(1)	1.302(3)	N(1)-C(1)-N(2)	105.9(2)

The monoanionic, bidentate four-electron-donor amidinates are also suitable for stabilizing low oxidation states and/or high coordination numbers at germanium centers. The monosubstituted and mononuclear compound **4**, as well as compound **5**, have been isolated and were characterized spectroscopically and structurally.

$Li[RC(NR')_2]$ $\xrightarrow[GeCl_2 \cdot Dioxan]{Et_2O}$

R= Ph, R'=$SiMe_3$: **4**
R= ^{t}Bu, R'=Cy : **5**

4/5 (dimer)

Fig. 4. Synthesis of compounds **4** and **5**.

Both compounds **4** and **5** are, at least at first glance, monomeric in the solid state (Fig. 4, Table 3). They are packed strictly parallel, the Ge–Cl vector pointing to the neighboring Ge atom at a distance equivalent to the sum of the van der Waals radii. Thus, infinite, almost linear, chains are formed. This may account for the unexpected observation that dimers and dimer fragments of **4** and as well of **5** are observed by mass spectroscopy.

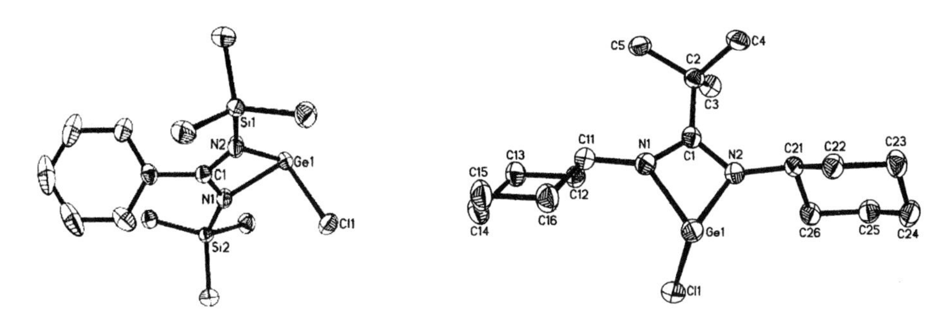

Fig. 5. Molecular structure of compounds **4** and **5**.

Table 3. Selected distances [Å] and angles [°] for **4** and **5**.

4		**5**	
Ge(1)–N(1)	2.028(2)	Ge(1)–Cl(1)	2.3033(5)
Ge(1)–N(2)	2.049(2)	Ge(1)–N(1)	1.982(2)
Ge(1)–Cl(1)	2.2651(6)	Ge(1)–N(2)	2.000(2)
C(1)–N(1)	1.333(2)	N(1)–C(1)	1.349(2)
C(1)–N(2)	1.331(2)	N(2)–C(1)	1.451(2)
N(1)-Ge(1)-N(2)	65.78(6)	Cl(1)-Ge(1)-N(1)	98.15(5)
Cl(1)-Ge(1)-N(1)	95.93(5)	Cl(1)-Ge(1)-N(2)	96.69(5)
Cl(1)-Ge(1)-N(2)	97.68(5)	N(1)-Ge(1)-N(2)	65.64(7)
Ge(1)-N(1)-C(1)	90.6(1)	N(1)-C(1)-N(2)	107.2(2)
Ge(1)-N(2)-C(1)	89.7(1)	N(1)-C(1)-C(2)	127.6(2)
Ge(1)-N(2)-Si(1)	134.28(9)	N(2)-C(1)-C(2)	125.0(2)
Ge(1)-N(1)-Si(2)	131.99(9)	Ge(1)-N(1)-C(1)	93.2(1)
N(1)-C(1)-N(2)	112.6(2)	Ge(1)-N(1)-C(11)	131.2(1)

Compound **6** has been obtained by reaction of **4** with $LiNPh_2$. The molecular framework corresponds to that of **4** and **5**.

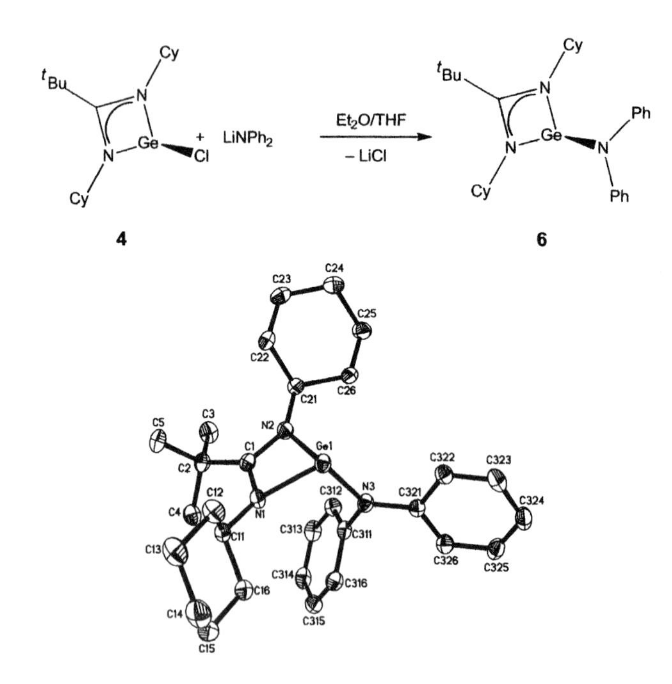

Fig. 6. Synthesis and molecular structure of compound **6**.

Table 4. Selected distances [Å] and angles [°] for **6**.

	6		
Ge(1)–N(1)	2.011(1)	Ge(1)-N(1)-C(1)	93.3(1)
Ge(1)–N(2)	2.004(1)	Ge(1)-N(1)-C(11)	128.8(1)
Ge(1)–N(3)	1.941(1)	C(1)-N(1)-C(11)	129.8(1)
N(1)–C(1)	1.350(2)	Ge(1)-N(2)-C(1)	94.2(1)
N(2)–C(1)	1.463(2)	Ge(1)-N(2)-C(21)	130.4(1)
N(1)-Ge(1)-N(2)	65.01(6)	C(1)-N(2)-C(21)	134.9(1)
N(1)-Ge(1)-N(3)	97.67(6)	C(1)-Ge(1)-N(3)	99.60(6)
N(2)-Ge(1)-N(3)	95.21(6)	Ge(1)-N(3)-C(311)	120.5(1)
N(1)-C(1)-N(2)	107.2(1)	Ge(1)-N(3)-C(321)	120.9(1)
N(1)-C(1)-C(2)	123.8(2)	C(311)-N(3)-C(321)	118.3(1)

In addition, we reinvestigated the system Me_2SiCl_2/Li[PhC(NMe)NH] (Fig. 7), where the formation of an eight-membered ring silicon species with bridging *di*anionic amidinate ligands $[N–C(R)–NR']^{2-}$ was reported [1]. The X-ray investigation of compound **7** (Fig. 8, Table 5) is at variance to this previous finding.

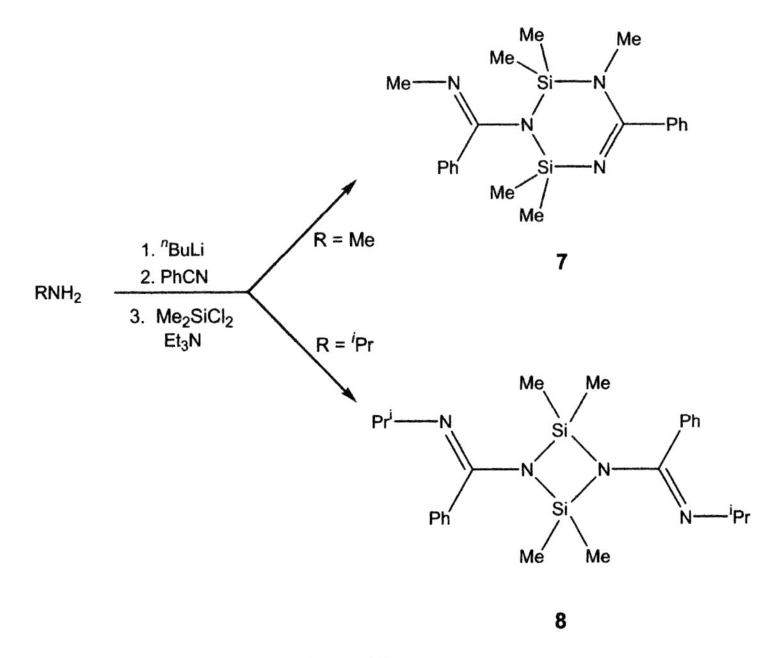

Fig. 7. Synthesis and constitution of compounds **7** and **8**.

Surprisingly, a six-membered ring (**7**) was formed. Moreover, with the more sterically demanding $[PhC(N^iPr)N]^{2-}$ ligand, a four-membered ring compound **8** (Figs. 7, 8, and Table 5) was obtained.

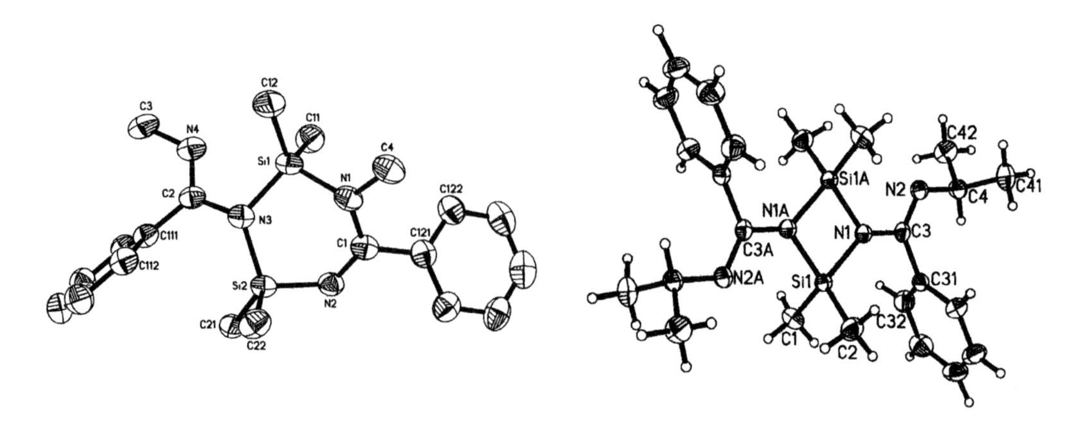

Fig. 8. Molecular structure of compounds **7** and **8**.

Table 5. Selected distances [Å] and angles [°] for **7** and **8**.

		7		**8**	
Si(1)–N(1)	1.778(2)	N(1)-Si(1)-N(3)	102.89(7)	Si(1)–N(1)	1.761(1)
Si(1)–N(3)	1.759(2)	C(1)-N(1)-Si(1)	118.8(1)	Si(1)–N(1A)	1.758(1)
Si(2)–N(2)	1.732(2)	N(1)-C(1)-N(2)	123.0(2)	N(1)–C(3)	1.383(2)
Si(2)–N(3)	1.758(2)	C(1)-N(2)-Si(2)	123.5(1)	N(2)–C(3)	1.283(2)
N(1)–C(1)	1.389(2)	N(2)-Si(2)-N(3)	106.27(7)	Si(1)···Si(1A)	2.6181(7)
N(2)–C(1)	1.289(2)	Si(1)-N(3)-Si(2)	118.16(8)	N(1)-Si(1)-N(1A)	83.88(5)
N(3)–C(2)	1.404(2)	N(3)-C(2)-N(4)	116.6(2)	Si(1)-N(1)-Si(1A)	96.12(5)
C(2)–N(4)	1.285(2)	Si(1)-N(1)-C(4)	120.4(1)	N(1)-C(3)-N(2)	119.2(1)
		N(1)-Si(1)-N(3)	102.89(7)	Si(1)-N(1)-C(3)	135.8(1)
		C(1)-N(1)-Si(1)	118.8(1)	Si(1A)-N(1)-C(3)	127.53(9)

Conclusion

This study demonstrates the utility of amidinate ligands for the synthesis of hypervalent silicon and germanium compounds, and also, for the synthesis of low-valent germanium compounds. In addition, surprising results with *di*anionic amidinate ligands $[N–C(R)–NR']^{2-}$ at silicon centers are documented.

Acknowledgments: We thank Prof. Dr. N. Mitzel, Dr. A. Schier, and S. Nogai for performing the X-ray determinations and the *Fonds der Chemischen Industrie* for financial support.

References

[1] J. Scherer, O. Hornig, *Chem. Ber.* **1968**, *101*, 2533.

[2] H. W. Roesky, B. Meller, M. Noltemeyer, H. G. Schmidt, U. Scholz, G. M. Sheldrick, *Chem. Ber.* **1988**, *121*, 1403.

[3] S. R. Foley, C. Bensimon, D. S. Richeson, *J. Am. Chem. Soc.* **1997**, *119*, 10359.

[4] H. H. Karsch, P. A. Schlüter, M. Reisky, *Eur. J. Inorg. Chem.* **1998**, 433.

[5] H. H. Karsch, P. A. Schlüter in *Organosilicon Compounds IV — From Molecules to Materials* (Eds.: N. Auner, J. Weis), Wiley-VCH, Weinheim **2000**, 28.

Strong Evidence for an Unconventional 1,2-(C→P)-Silyl Migration: Formation and Reactions of a *P*-Silyl Phosphaalkene Complex

Emanuel Ionescu, Hendrik Wilkens, Rainer Streubel*

Institut für Anorganische Chemie, Rheinische Friedrich-Wilhelms-Universität Bonn
Gerhard-Domagk-Strasse 1, 53121 Bonn, Germany
Tel:. +49 228 73 5345 — Fax: +49 228 73 9616
E-mail: r.streubel@uni-bonn.de

Keywords: silatropy, phosphinidene complexes, phosphaalkene complexes

Summary: Thermal rearrangement of the short-lived phosphinidene complex $[(OC)_5W\{PCH(SiMe_3)_2\}]$ yielded stereoselectively an η^1-*E*-phosphaalkene tungsten complex via 1,2-silyl shift; this product was isolated and unambiguously characterized by multinuclear NMR spectroscopy. We also report preliminary studies on the reactivity of this η^1-*E*-phosphaalkene complex toward $(Me_2N)_3P$, cyclopentadiene and DMAD.

Silyl shifts are among the most important synthetic methods in modern organoelement chemistry [1, 2], as illustrated by the syntheses of the first stable compounds with a P=C [3] or Si=C [4] double bond via 1,3-silyl shift and/or the generation of short-lived silaalkenes from reactive singlet carbenes [5, 6] via 1,2-silyl shift. There are only a very few examples of 1,2-silyl shifts to nucleophilic singlet carbene centers in organoelement chemistry, e.g., the shifts to yield **1** [7] and **2** [8, 9] (Scheme 1), but even more scarce is the knowledge about silyl shifts to electrophilic singlet carbenes and/or carbene-like centers such as silylenes, e.g., the rearrangement to **3** [10]. To the best of our knowledge, there are no examples in organophosphorus chemistry that mimic rearrangements of the latter silylene.

Although the chemistry of short-lived electrophilic terminal phosphinidene tungsten complexes [11, 12] has received increased attention during the recent years, partially because of the versatility of 2*H*-azaphosphirene complexes [13, 14], only a single example of a rearrangement yielding a *P*-Cl–substituted η^1-phosphaalkene complex – via a 1,2-chlorine shift – has been reported so far [15].

As we discovered that bulky electrophilic terminal phosphinidene tungsten complexes with C_5Me_5 or $CH(SiMe_3)_2$ at the phosphorus do not dimerize in the absence of trapping reagents [16, 17], we became interested in their fate.

Heating *diluted ortho*-xylene solutions of 2*H*-azaphosphirene complex **4** [18] afforded almost quantitatively (ca. 90 – 95% by ^{31}P NMR spectroscopy) and stereoselectively the *P*-trimethylsilyl-substituted η^1-*E*-phosphaalkene complex **6**. The rearrangement of the thermally generated short-lived phosphinidene complex $[(OC)_5W\{PCH(SiMe_3)_2\}]$ to complex **6** via 1,2-(C→P)-trimethylsilyl shift (Scheme 2) is very reasonable, but was completely unexpected!

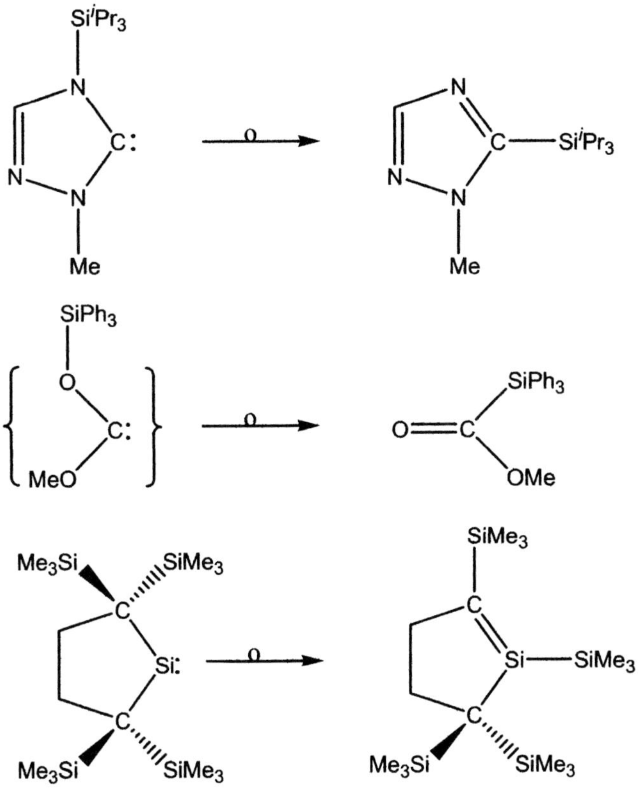

Scheme 1. Examples of 1,2-silyl shifts to nucleophilic carbene centers (**1**, **2**) and an electrophilic silylene center (**3**).

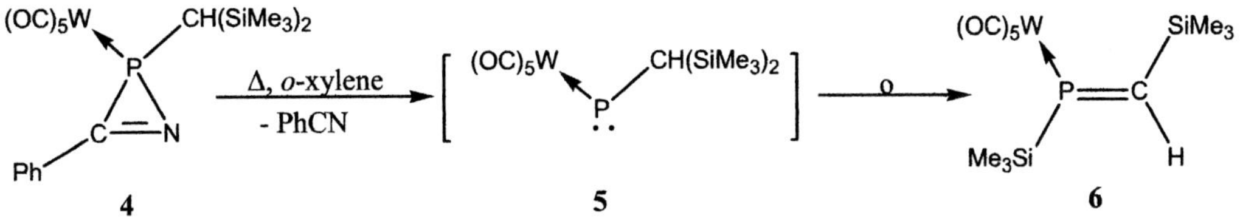

Scheme 2. Proposed reaction course for the formation of **6**.

It was remarkable that we neither obtained evidence for products pointing to radical reactions, e.g., diphosphabutadiene complexes, nor did we observe formation of an isomeric *P*-H-substituted

η^1-phosphaalkene complex.

^{31}P NMR monitoring of the thermolysis of the 2*H*-azaphosphirene tungsten complex **4** in *o*-xylene (various concentrations, temperatures and thermolysis times) revealed, in all cases, the formation of the complexes **6**, **7** [19], and **8** [20] (Scheme 3) in varying amounts.

As already mentioned, we obtained **6** as the major product in diluted solution, at high(er) temperatures (105 – 110 °C), and after short reaction times (ca. 10 min.), but in concentrated solutions, at lower temperatures, and after long(er) reaction times (ca. 20 – 40 min.) **7** or complex **8** were the main products (Table 1).

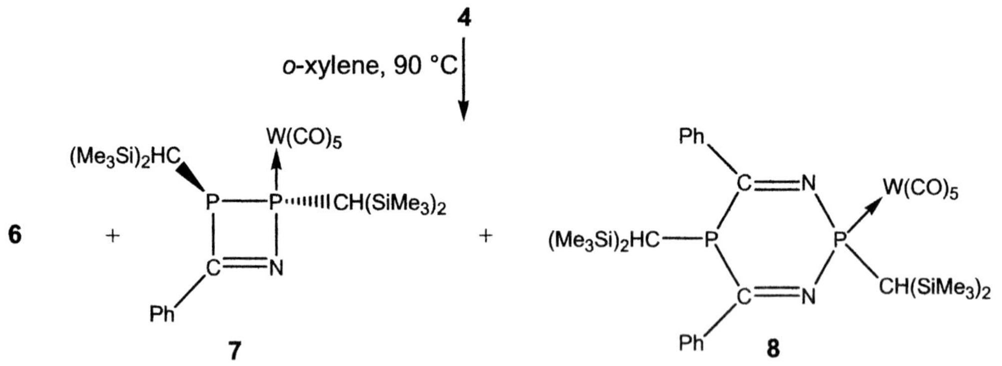

4
o-xylene, 90 °C
6 + 7 + 8
W(CO)5
(Me3Si)2HC
P—P····CH(SiMe3)2
C=N
Ph
Ph
C=N
(Me3Si)2HC—P
P
W(CO)5
CH(SiMe3)2
C=N
Ph

Scheme 3. Products of the thermolysis of complex **4**.

Table 1. Product contents of the thermolysis of **4** in *o*-xylene (after 10 min at 110 °C, various concentrations).[a]

Concentration [mmol/mL]	**6** [%]	**7** [%]	**8** [%]
0.08	95	0	0
0.14	85	2	5
0.21	65	1	15
0.28	45	35	10
0.34	45	35	15
0.4	30	40	25
0.46	30	45	20
0.53	15	55	20

[a] C_6D_6; determined by ^{31}P NMR integration.

We have also observed that, if the reaction solution of **6** was kept at room temperature, the amount of **6** decreased after ca. 18 h from ca. 85% to 22% and the amount of **7** and **8** increased from ca. 1% to 16% (**7**) and from ca. 14% to 35% (**8**)! This provides the first evidence that **6** could be regarded as the precursor of **7** and **8**; at one stage or another a benzonitrile-induced 1,2-(P→C)-trimethylsilyl shift seems to occur.

The constitution of complex **6** was unambiguously established by ^{13}C NMR spectroscopy and MS spectrometric investigations. Using DEPT-135 experiments the resonance at δ = 193.3 ppm (^{1}J(P,C) = 12.7 Hz) was assigned to the carbon atom of the P=C(H) moiety. The P=C double bond has *E* configuration, according to the observed NOE effects in the ^{1}H NMR spectrum. Noteworthy are also the ^{1}H resonances of the P=C(H) moiety at δ = 9.62 ppm (^{2}J(P,H) = 31.8 Hz) and that of the *P*-trimethylsilyl group, the latter showing a significant solvent-induced shift dependence at ambient temperature ([D_6]benzene: –0.01, [D_2]dichloromethane: 0.43 ppm). The heteronuclear NMR experiments showed a ^{31}P resonance at δ = 295.5 ppm (^{1}J(W,P) = 217.0 Hz), which is well outside the usual range of *P*-trimethylsilyl-substituted η^{1}-phosphaalkene tungsten complexes (ca. 70 – 90 ppm) [21, 22] and ^{29}Si resonances at δ = 0.3 (J(P,Si) = 9.1 Hz) and 3.0 ppm (J(P,Si) = 8.6 Hz) that displayed phosphorus-silicon coupling constants of similar magnitudes, one of which must represent a ^{1}J(P,Si) coupling, which then would be quite small [23].

Because we became curious about the mechanism of a donor-induced rearrangement, we studied the reaction of **6** with several donors such as carbonitriles and tertiary phosphanes. Here, we present preliminary results on the reaction of **6** with tris(dimethylamino)phosphane (Scheme 4).

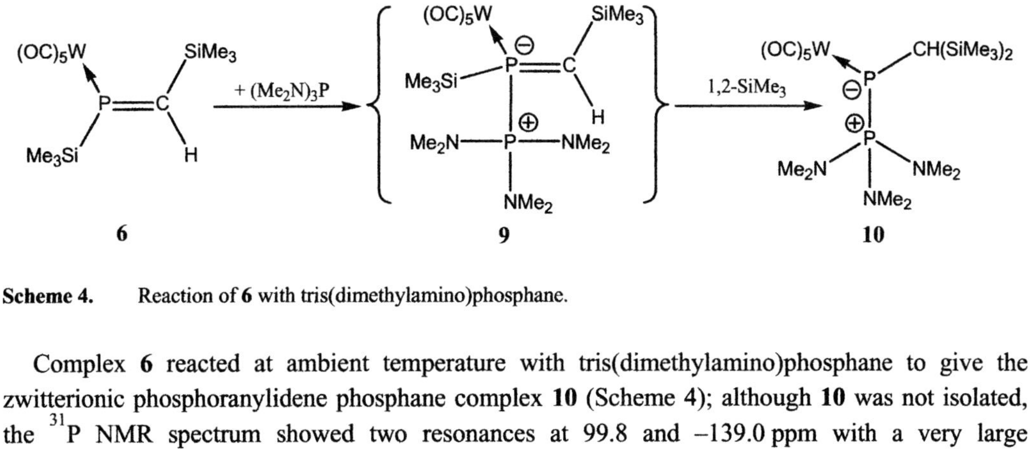

Scheme 4. Reaction of **6** with tris(dimethylamino)phosphane.

Complex **6** reacted at ambient temperature with tris(dimethylamino)phosphane to give the zwitterionic phosphoranylidene phosphane complex **10** (Scheme 4); although **10** was not isolated, the ^{31}P NMR spectrum showed two resonances at 99.8 and –139.0 ppm with a very large phosphorus-phosphorus coupling constant magnitude of 508.5 Hz. Such data are typical for such a class of compound, i.e. $(OC)_5W\{(Ph)P{=}P(^{n}Bu)_3\}$ displays two resonances at 30.9 and –100.0 ppm with ^{1}J(P,P) = 444.3 Hz [24]. Our interpretation of the reaction course, as shown in Scheme 4, is that tris(dimethylamino)phosphane undergoes nucleophilic attack at the phosphorus atom of **6**, thus forming the donor adduct **9**, which then rearranges to complex **10** via a 1,2-(C→P) silyl shift.

Complex **6** reacted with cyclopentadiene at *ambient* temperature (!) in toluene in a [4+2]-cyclo-addition reaction to give phospha-norbornene complex **11** (^{31}P NMR: –69.1 ppm, ^{1}J(P,W) = 204.5 Hz), which reacted during the column chromatography to furnish the *P*-H-substituted phospha-norbornene complex **12**. Although complex **12** was fully characterized by NMR and MS spectroscopy, the confirmation was only tentatively assigned (Scheme 5).

Surprisingly selective was the reaction of **6** with DMAD in toluene at 60 °C that yielded the 2-phospha-butadiene complex **13** (Scheme 5), which was purified by column chromatography and characterized by NMR spectroscopy.

It is remarkable that the ^{1}H resonance of the P=C(H) moiety was found at 7.89 ppm, thus showing a significant up-field shift as compared to **6**, with a ^{2}J(P,H)coupling constant of only 5.2 Hz, which is quite small compared to the respective coupling constant of **6** (31.9 Hz). The ^{31}P NMR measurement showed a resonance at 250.2 ppm, with a ^{1}J(W,P) coupling constant of 270.5 Hz.

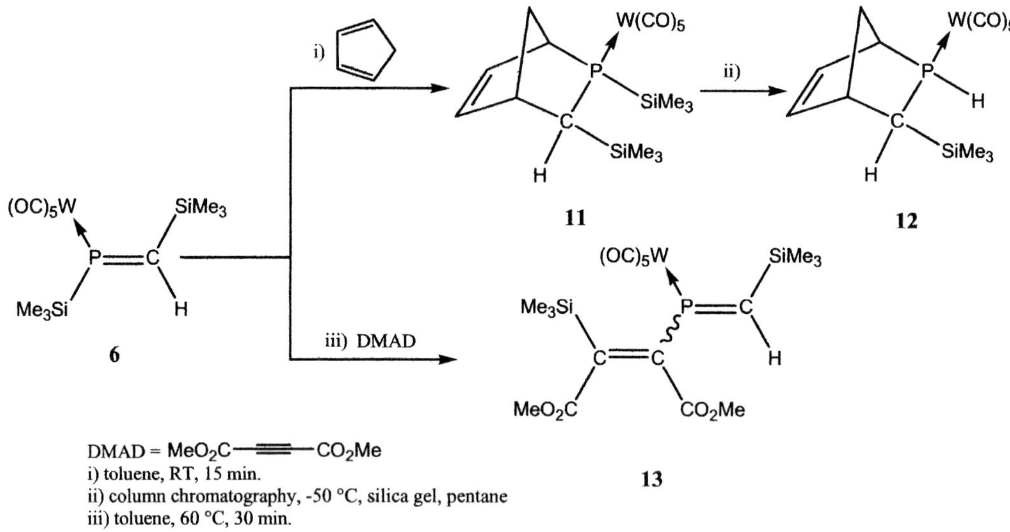
i)
ii)
iii) DMAD
W(CO)5
SiMe3
H
(OC)5W
Me3Si
MeO2C
CO2Me
P
C
11
12
13
6
DMAD = MeO2C—≡—CO2Me
i) toluene, RT, 15 min.
ii) column chromatography, -50 °C, silica gel, pentane
iii) toluene, 60 °C, 30 min.

Scheme 5. Addition reactions of **6**.

Experimental part

All reactions and manipulations were carried out under an atmosphere of deoxygenated dry argon, using standard Schlenk techniques with conventional glassware. Solvents were dried according to standard procedures and reagents were used as purchased. NMR spectra were recorded on a Bruker AX 300 spectrometer (121.5 MHz for ^{31}P, 75.0 MHz for ^{13}C and 300.1 MHz for ^{1}H) using TMS (^{1}H and ^{13}C) and 85% H_3PO_4 (^{31}P) as external standard.

{[*E*-1,2-bis(trimethylsilyl)methylenephosphane-κP]pentacarbonyltungsten(0)} (**6**): 307 mg (0.5 mmol) of complex **4** in 4.5 mL of *o*-xylene was stirred for 10 min at 110 °C. After removing the solvent *in vacuo*, the oily dark brown residue was purified by column chromatography (silanized SiO_2, –50 °C, *n*-pentane). Evaporating the solvent of the first fraction gave **6** as a dark-yellow oil; yield: 165 mg (66%). Selected NMR data of **6**: ^{1}H (C_6D_6): δ –0.01 (d, ^{3}J(P,H) = 5.7 Hz, 9H, P-$SiMe_3$), 0.15 (s, 9H, $SiMe_3$), 9.62 (d, ^{2}J(P,H) = 31.8 Hz, ^{3}J(W,H) = 9.5 Hz, 1H, PC*H̲*); ^{13}C{^{1}H} (C_6D_6): δ –0.6 (d, ^{2}J(P,C) = 8.3 Hz, $SiMe_3$), 0.4 (d, ^{2}J(P,C) = 5.5 Hz, $SiMe_3$), 193.3 (d, ^{1}J(P,C) = 12.7 Hz, P=C̲H), 196.5 (d, ^{2}J(P,C) = 8.2 Hz, *cis*-CO), 199.7 (d, ^{2}J(P,C) = 22.6 Hz, *trans*-

CO); $^{31}P\{^1H\}$ (C_6D_6): δ 295.5 (^1J(P,W) = 217.0 Hz).

[(2-Trimethylsilyl-1-phospha-norbornene-κP)pentacarbonyltungsten(0)] (11): A solution of 257 mg (0.5 mmol) of **6** and 33 mg (0.5 mmol) of cyclopentadiene was stirred for 15 min in 2 mL toluene at 25 °C. The end of the reaction was controlled by ^{31}P NMR (^{31}P NMR (C_6D_6): δ –69.1 ppm, ^1J(P,W) = 204.5 Hz). After removing the solvent *in vacuo*, the residue was purified by column chromatography (silica gel, –50 °C, *n*-pentane:diethylether 20:80). The product which was obtained was not **11**, but **12**. Selected NMR data of **12**: $^{13}C\{^1H\}$ ($CDCl_3$): δ –0.6 (d, ^3J(P,C) = 2.4 Hz, $SiMe_3$), 20.7 (d, ^1J(P,C) = 11.6 Hz, <u>C</u>HSiMe$_3$), 43.8 (d, ^{2+3}J(P,C) = 18.0 Hz, <u>C</u>H$_2$), 45.2 (d, ^2J(P,C) = 3.2 Hz, <u>C</u>HCHSiMe$_3$), 46.7 (d, ^1J(P,C) = 18.8 Hz, <u>C</u>HPH), 133.8 (d, ^3J(P,C) = 11.9 Hz, CH<u>C</u>HCHP), 135.6 (d, ^3J(P,C) = 7.6 Hz, <u>C</u>HCHCHP), 197.0 (d, ^2J(P,C) = 6.6 Hz, *cis*-CO), 198.7 (d, ^2J(P,C) = 22.0 Hz, *trans*-CO); ^{31}P ($CDCl_3$): δ –42.8 (^1J(W,P) = 217.1 Hz, ^1J(P,H) = 328.4 Hz).

[(3,4-Bis(methoxycarbonyl)-1,4-bis(trimethylsilyl)-2-phosphabuta-1,3-diene-κP)penta-carbonyltungsten(0)] (13): To a solution of 170 mg (0.33 mmol) of **6** in 1.5 mL toluene, 46.8 mg (0.33 mmol) of DMAD was added and the solution stirred for 30 min at 60 °C. After removing the solvent *in vacuo*, the oily dark brown residue was purified by column chromatography (silica gel, –50 °C, *n*-pentane:diethylether 1:1). Evaporating the solvent from the second fraction yielded **14** as a brown oil; yield: 150 mg (70%). ^1H ($CDCl_3$): δ 0.12 (s, 9H, $SiMe_3$), 0.34 (s, 9H, $SiMe_3$), 3.78 (s, 3H, CO_2Me), 3.80 (s, 3H, CO_2Me), 7.89 (d, ^2J(P,H) = 5.2 Hz, PC<u>H</u>); $^{13}C\{^1H\}$ ($CDCl_3$): δ 0.4 (s, $SiMe_3$), 0.5 (d, ^2J(P,C) = 5.6 Hz, $SiMe_3$), 52.2 (s, CO$_2$<u>C</u>H$_3$), 53.1 (s, CO$_2$<u>C</u>H$_3$), 143.9 (d, ^2J(P,C) = 6.7 Hz, <u>C</u>CPC), 161.1 (d, ^xJ(P,C) = 9.3 Hz, <u>C</u>O$_2$CH$_3$), 163.7 (d, ^xJ(P,C) = 7.7 Hz, <u>C</u>O$_2$CH$_3$), 170.2 (d, ^1J(P,C) = 16.3 Hz, C<u>C</u>PC), 171.1 (d, ^1J(P,C) = 25.6 Hz, CCP<u>C</u>), 195.1 (d, ^2J(P,C) = 9.5 Hz, *cis*-CO), 198.8 (d, ^2J(P,C) = 31.5 Hz, *trans*-CO); $^{31}P\{^1H\}$ (C_6D_6): δ 250.2 (^1J(P,W) = 270.5 Hz).

Acknowledgments: We are grateful to the Deutsche Forschungsgemeinschaft and the Fonds der Chemischen Industrie for financial support.

References

[1] *The Silicon-Heteroatom Bond* (Eds.: S. Patai, Z. Rappoport), John Wiley and Sons, Chichester, **1991**.

[2] *Organosilicon Chemistry III — From Molecules to Materials* (Eds.: N. Auner, J. Weis), Wiley-VCH, Weinheim, **1998**.

[3] G. Becker, *Z. Anorg. Allg. Chem.* **1976**, *423*, 242.

[4] A. G. Brook, F. Abdesaken, B. Gutekunst, G. Gutekunst, R. K. M. R. Kallury, *J. Chem. Soc., Chem. Commun.* **1981**, 191.

[5] A. Sekiguchi, W. Ando, *Organometallics* **1987**, *6*, 1857; and references there.

[6] W. Ando, T. Hagiwara, T. Migita, *J. Am. Chem. Soc.* **1973**, *95*, 7518.

[7] S. Solé, H. Gornitzka, O. Guerret, G. Bertrand, *J. Am. Chem. Soc.* **1998**, *120*, 9100.
[8] J. P. Pezacki, P. G. Loncke, J. P. Ross, J. Warkentin, T. A. Gadosy, *Org. Lett.* **2000**, *2*, 2733.
[9] P. G. Loncke, T. A. Gadosy, G. H. Peslherbe, *Can. J. Chem.* **2002**, *80*, 302.
[10] M. Kira, *Pure Appl. Chem.* **2000**, *72*, 2333.
[11] Recent review: F. Mathey, N. G. Tran Huy, M. Marinetti, *Helv. Chim. Acta* **2001**, *84*, 2938.
[12] Recent review: K. Lammertsma, M. J. M. Vlaar, *Eur. J. Org. Chem.* **2002**, 1127.
[13] R. Streubel, *Coord. Chem. Rev.* **2002**, *227*, 172.
[14] R. Streubel, *Top. Curr. Chem.* **2002**, *223*, 91.
[15] See the CuCl-catalyzed 1,2-chlorine shift in: B. Deschamps, F. Mathey, *J. Organomet. Chem.* **1988**, *354*, 83.
[16] H. Wilkens, A. Ostrowski, J. Jeske, F. Ruthe, P. G. Jones, R. Streubel, *Organometallics* **1999**, *18*, 5627.
[17] R. Streubel, U. Schiemann, N. Hoffmann, Y. Schiemann, P. G. Jones, D. Gudat, *Organometallics* **2000**, *19*, 475.
[18] R. Streubel, A. Ostrowski, S. Priemer, U. Rohde, J. Jeske, P. G. Jones, *Eur. J. Inorg. Chem.* **1998**, 257.
[19] E. Ionescu, P. G. Jones, R. Streubel, *J. Chem. Soc., Chem. Commun.* **2002**, 2204.
[20] R. Streubel, H. Wilkens, F. Ruthe, P. G. Jones, *J. Chem. Soc., Chem. Commun.* **2000**, 2453.
[21] A. Marinetti, L. Ricard, F. Mathey, M. Slany, M. Regitz. *Tetrahedron* **1993**, *49*, 10279.
[22] L. Weber, M. Meyer, H.-G. Stammler, B. Neumann, *Chem. Eur. J.* **2001**, *7*, 5401.
[23] L. Weber, S. Kleinebekel, A. Rühlicke, H.-G. Stammler, B. Neumann, *Eur. J. Inorg. Chem.* **2000**, *6*, 1185.
[24] P. Le Floch, A. Marinetti, L. Ricard, F. Mathey, *J. Am. Chem. Soc.* **1990**, *112*, 2407.

Strong Evidence for an Unconventional 1,2-(C→P)-Silyl Migration: DFT Structures and Bond Strengths (Compliance Constants)

Gerd von Frantzius, Jörg Grunenberg, Rainer Streubel*

Institut für Anorganische Chemie, Rheinische Friedrich-Wilhelms-Universität Bonn,
Gerhard-Domagk-Strasse 1, 53121 Bonn, Germany,
Tel.: +49 228 735345 — Fax: +49 228 739616
E-mail: r.streubel@uni-bonn.de

Keywords: 1,2-silyl migration, double bond formation, phosphinidene metal complexes

Summary: We report on DFT calculations (geometries, activation energies, reaction enthalpies) and bond strength description of an isomerization of bis(trimethylsilyl)-methyl-substituted phosphinidene complexes to corresponding phosphaalkene complexes under formation of a C–P double bond via neutral 1,2-(C→P)-silyl migration.

Introduction: Silyl Migrations

Experimentally, 1,2-silyl migrations with formation of double bonds between the migration origin and the cationic or neutral target have been found to take place intramolecularly, e.g., σ^4,λ^4-C→σ^3,λ^3-C [1], σ^4,λ^4-C→σ^2,λ^2-C [2], σ^4,λ^4-C→σ^2,λ^2-Si [3], σ^4,λ^4-Si→σ^2,λ^2-Si [4], and intermolecularly, e.g., σ^3,λ^3-N→σ^2,λ^2-C [5]. While in the latter three cases the silyl group shifted toward a silylene or a carbene center, the ability of silicon to migrate over π-frameworks encouraged Barton et al. to investigate the opposite direction, that is the thermal rearrangement of an olefin to a carbene on the basis of a supposed unique stabilization of the energy surface of olefin isomerization by a migrating silyl group [6]. Creary and Butchko assumed a stabilizing interaction of silicon with the lone pair of the carbene as well as with the vacant carbene orbital [2].

Theoretically transient three-membered rings containing hypervalent silicon as a structural motif (Fig. 1) have been studied at the SCF and DFT level and the relevant bond lengths are shown in Table 1 (substituents at C and E are not included in Table 1). Characterization of the transition state as early or late can be assumed from the difference between C–E and E–Si bond lengths. From Table 2 the method dependence of the barrier to selected 1,2-silyl-shifts can be taken. Generally, inclusion of electron correlation substantially diminishes the barrier while zero-point correction has only a minor effect. Barriers to silyl migration are in general much smaller than those to related

1,2-hydrogen and 1,2-methyl migrations [3].

Table 1.

E	R_1	R_2	C–E [Å]	C–Si [Å]	E–Si [Å]
C [3]	Me	Me	1.383	2.068	2.250
	Me	^{t}Bu	1.380	2.086	2.240
Si [7]	H	H	2.577	1.779	2.423

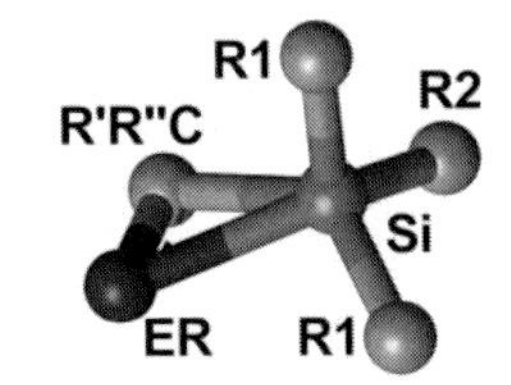

Fig. 1.

Table 2. Activation energies of selected silyl migrations ("+ZPE" = zero-point-corrected).

Method	Type	Educt	E_a →	[kJ/mol] ←	Product
HF/6-31G(d)	C→C [9]	HC–$CH_2(SiH_3)$	286.8	310.2	$H_3Si(H)C{=}CH_2$
MP2/6-31G(d)			297.7	286.8	
MP4/6-311G(d,p)			285.1	265.4	
HF/3-21	C→Si [7]	HSi–$CH_2(SiH_3)$	180.5	121.4	$H_3Si(H)Si{=}CH_2$
HF/6-31			164.5	134.0	
HF/6-31*			145.7	126.0	
MP2/6-31*			98.4	110.5	
MP3/6-31*			103.8	109.7	
Experiment	C→Si [3]	Me_3Si $SiMe_3$ / Si: / Me_3Si $SiMe_3$	88.3 (293 K)	–	Me_3Si / Si–$SiMe_3$ / Me_3Si SiM_3
MP3/6-31*//HF/6-31	Si→Si [10]	HSi-$SiH_2(SiH_3)$	35.6	76.2	$H_3Si(H)Si{=}SiH_2$
" " +ZPE			35.2	72.0	
B3LYP/6-311**	C→P [11]	$(CO)_5W$–P–$CH(SiMe_3)_2$ *syn*-periplanar **1b**	75.72	128.58	$(CO)_5W$, H, P=, Me_3Si, $SiMe_3$ (Z) **2b**
B3LYP/6-311**+ZPE			74.23	127.14	
B3LYP/6-311**	C→P [11]	$(CO)_5W$–P–$CH(SiMe_3)_2$ *anti*-periplanar **1a**	56.07	146.49	$(CO)_5W$, $SiMe_3$, P=, Me_3Si, H (E) **2a**
B3LYP/6-311**+ZPE			53.57	144.13	

Accompanying experimental work [8], we report on DFT calculations of the 1,2-(σ^4,λ^4-C→σ^1,λ^2-P)-silyl migration of pentacarbonyltungsten phosphinidene complexes **1a** (*anti*-periplanar) and **1b** (*syn*-periplanar) to the corresponding phosphaalkene complexes **2a** and **2b**. The formation of the C–P double bonds proceeds via transition state complexes **TS a** and **TS b** (Fig. 2; hydrogens except P=CH are omitted). Relative bond strengths are described by compliance constants (Table 3).

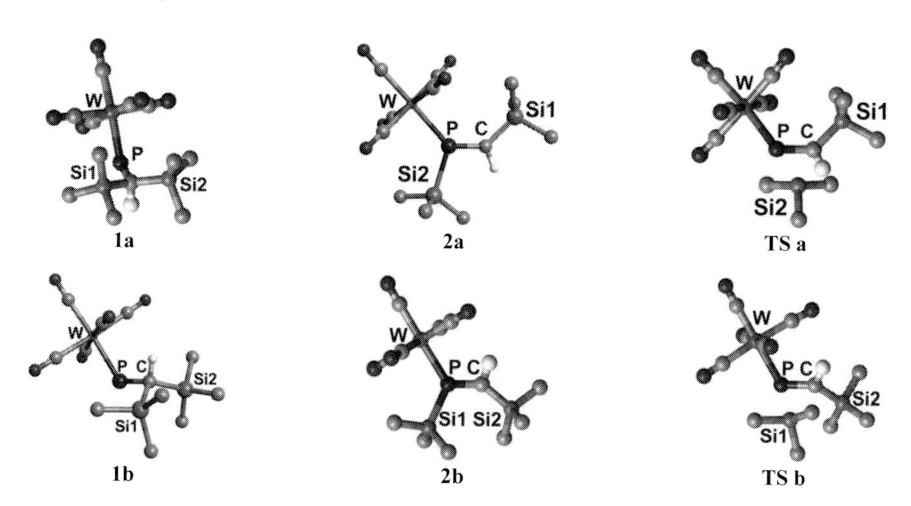

Fig. 2. Phosphinidene (**1a**, **1b**) and phosphaalkene tungsten complexes (**2a**, **2b**) and transition states (**TS a**, **TS b**).

Table 3. Lengths and compliance constants (COCO [Å²/aJ]) of selected bonds. Dark grey: C-Si bonds involved in the transition state (cf. text).

	P–C		P–Si		C–Si1		C–Si2		W–P	
	[Å]	COCO	[Å]	COCO	[Å]	COCO	[Å]	COCO	[Å]	COCO
1a	1.837	0.422	–	–	1.933	0.512	1.948	0.540	2.460	0.646
1b	1.825	0.403	–	–	1.949	0.532	1.938	0.498	2.457	0.622
TS a	1.741	0.234	2.508	0.870	1.914	0.429	2.303	0.363	2.528	0.839
TS b	1.735	0.237	2.439	1.041	2.487	0.124	1.910	0.430	2.543	0.869
2a	1.678	0.189	2.314	0.646	1.894	0.391	–	–	2.546	0.900
2b	1.678	0.190	2.322	0.681	–	–	1.891	0.387	2.546	0.883

Results and Discussion

While free phosphinidenes P–R exist only as triplet intermediates — singlet-triplet gaps (ΔSTs) of P–H: 92.1 kJ/mol and P–Me: 109 kJ/mol [12] — the transition metal-complexed species [M]P–R prefer singlet ground states [13]. Pentacarbonylchromium phosphinidene complexes $(OC)_5CrP–R$ (e.g., R = H, CH_3) have been studied by Nguyen on the CASSCF(12,12)/CASPT2 (12,12)//B3LYP-/6-31** niveau with singlet-triplet gaps of the order 24.6 kJ/mol (B3LYP) and 36.6 kJ/mol (CASPT2) respectively for $(OC)_5CrP–H$ and 24.4 and 35.8 kJ/mol respectively for $(OC)_5CrP–Me$.

While the B3LYP functional has been found reliable concerning the geometries, ΔSTs were about 10 – 20 kJ/mol higher for the free PR species and generally underestimated for the complexed phosphinidenes compared with multireference methods [14]. A recent DFT study (LDA Vosko-Wilk-Nusair/Becke88/Perdew86) of pentacarbonyltungsten phosphinidene complexes $(OC)_5WP–R$ yielded a ΔST (R = H, Ph) of ca. 42 kJ/mol in favor of the singlet state; splittings for the non-coordinated species agreed excellently with high-level ab initio data showing the reliability of the DFT method [15]. In our case, ΔST of **1b** is about 40 kJ/mol in favor of the singlet state (without zero-point correction). Furthermore, the wavefunctions of the singlet ground states of **1a,b** as well as those of **TS a** and **TS b** are stable with respect to an RHF/UHF instability [16]. Computations of the singlet excited state (restricted and un-restricted) of non-complexed **1** and the transition state had to cope with this instability and are thus not yet available.

Studies of the principal bonding features of transition metal coordination to methylenephosphanes (η^1- and η^2-mode, Fig. 3) showed a comparability of results obtained from B3LYP/SBK(d) DFT calculations with those from the CCSD(t) niveau. Exothermic coordination of a d^6-$M(CO)_5$ fragment (M = Cr, Mo, W) to methylenephosphanes generally preserves the structure of the uncomplexed compound [17].

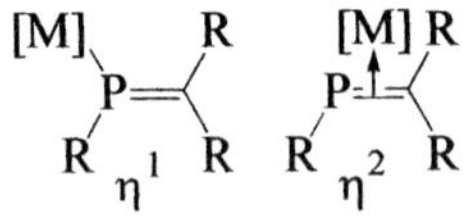

Fig. 3. Complexation modes of phosphaalkenes.

Relative energies of the 1,2-silyl migration of **1a,b** to corresponding **2a,b** can be taken from Fig. 4 and Table 4. *Anti*-periplanar phosphinidene complex **1a** rearranges via transition state **TS a** to the *E*-phosphaalkene complex **2a**. C_1 symmetric phosphinidenes **1a,b** exhibit a long and a short C–Si bond (Table 3); the former is the one involved in forming the three-membered ring of the transition state which contains a distorted square-planar hypervalent silicon (Fig. 1, ER = $PW(CO)_5$).

Table 4. HF energies and zero-point-corrected energies (ZPE) (B3LYP/6-311g(d,p)/LanL2DZ at W) of **1**, **2**, and **TS**.

		HF energy [Hartree]	ZP energy [Hartree]	Δ-HF [kJ/mol]	Δ-ZPE [kJ/mol]
Phosphinidenes					
anti-periplanar	**1a**	−1833.492190	−1833.209508	90.42	90.59
syn-periplanar	**1b**	−1833.501372	−1833.218831	66.32	66.11
Phosphaalkenes					
E-isomer	**2a**	−1833.526631	−1833.244012	0.00	0.00
Z-Isomer	**2b**	−1833.521504	−1833.238986	13.46	13.20
Transition states					
anti → *E*	**TS a**	−1833.470835	−1833.189114	146.49	144.13
syn → *Z*	**TS b**	−1833.472530	−1833.190559	142.04	140.34

From an estimated ΔST of about 40 kJ/mol under thermal reaction conditions phosphinidene complex **1a** can choose from two concurrent pathways: either undergo a fast reaction from the triplet excited state or — in the absence of suitable reaction partners — rearrange to *E*-phosphaalkene complex **2a**; the reverse 1,2-silyl shift is hindered by a substantial barrier of about 145 kJ/mol. These findings agree well with experimental results where reaction conditions could be optimized to synthesize pentacarbonyltungsten phosphaalkene complex **2a** from a precursor of phosphinidene complex **1**.

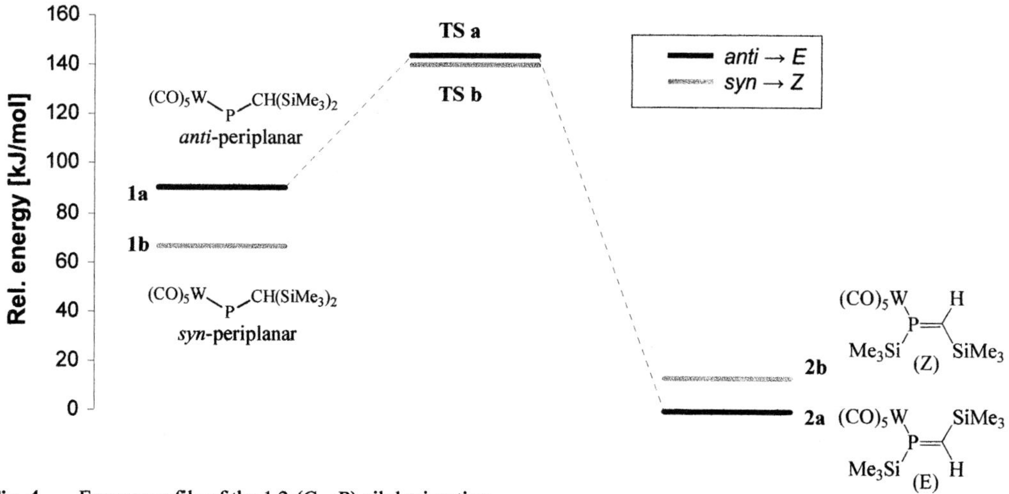

Fig. 4. Energy profile of the 1,2-(C→P)-silyl migration.

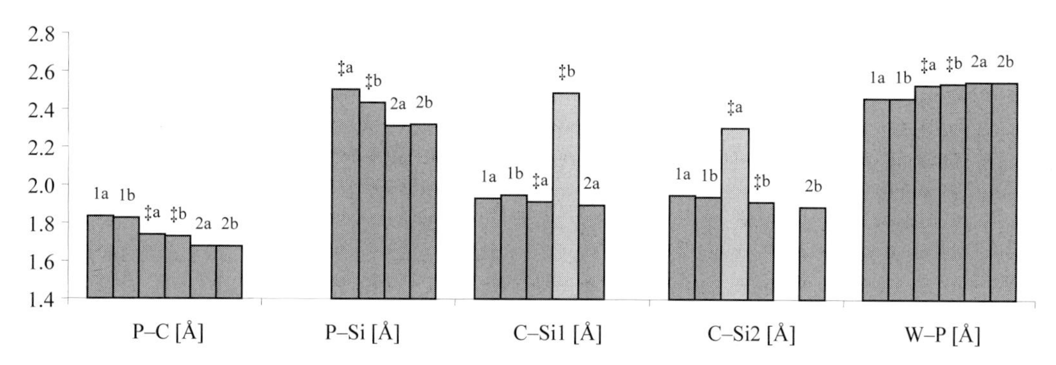

Fig. 5. Selected bond lengths in the course of the 1,2-silyl migration; **‡a**, **‡b**, are **TS a**, **TS b**. Light grey: C–Si bonds involved in the transition state.

From Figs. 5 and 6 the variations of selected bond lengths and strengths can be followed through the silyl migration. While the P–C bond shortens by about 10% and becomes stronger, the W–P bond elongates by about 4% and becomes weaker. The C–Si bond not involved in the transition state shortens by about 3% but becomes considerably stronger. Apart from a smaller barrier

(54 kJ/mol **1a**→**2a** vs 74 kJ/mol for **1b**→**2b**, Table 2) the major difference between **TS a** and **TS b** can be seen by the difference of the P–Si and C–Si bond lengths involved in the transition state. While **TS a** is early (9% difference; a pronounced interaction of the silyl group with a carbon p orbital has been found in the HOMO-4, cf. "Methods" section), in **TS b** the silicon is almost in between the originating carbon and the targeted phosphorus, although the compliance constants indicate a C–Si1 interaction which is much stronger than the P–Si. Since compliance constants have not yet been used for a discussion of bond strengths in transition states, this finding needs further investigation to exclude the possibility of computational artifacts.

What remains to be done is a comparative study of the rearrangement of the non-complexed species using multireference methods.

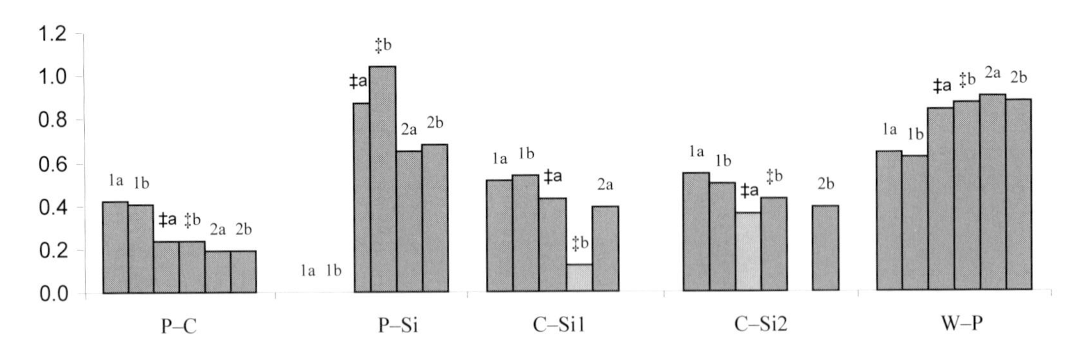

Fig. 6. Relative strengths of selected bonds in the course of the 1,2-silyl migration as measured by compliance constants in [Å²/aJ]. Compliance constants are inversely proportional to bond strengths.

Methods

All ground-state geometries (singlet, closed-shell) were obtained from full optimizations using the B3LYP functional [18] with the 6-311G(d,p) [19, 20] basis set for all atoms except tungsten, which has been described by the Los Alamos LanL2DZ ECP [21]. Minima have been characterized by a number of zero imaginary frequencies, transition states by one imaginary frequency. Rearrangement **1a** → **2a** has been checked by a reaction path following (IRC) with Si1=SiH_3.

All calculations were performed using GAUSSIAN 03 RevB.03 on the IBM Regatta p690 cluster ("JUMP") of the John von Neumann-Institut for Computing (NIC) at the Forschungszentrum Jülich [22].

Bond strengths have been described using complete matrices of compliance constants [23, 24] for each ground-state structure using the INTC/FCTINT-algorithms of Fogarasi and Pulay [25]. Energy second derivatives (Hessians) in cartesian coordinates coming out of a GAUSSIAN 03 calculation have been transformed to Hessians in non-redundant internal coordinates; the non-redundant Hessians were inverted by standard methods to yield full compliance matrices, the diagonal elements of which are the compliance constants (unit: [Å²/aJ] in case of a bond length), each belonging to a particular internal coordinate (bond length, angle or dihedral). Compliance constants are inversely proportional to bond strengths.

Acknowledgments: We are grateful to the Deutsche Forschungsgemeinschaft and the Fonds der Chemischen Industrie for financial support; we thank the Rechenzentrum of the Technische Universität Braunschweig and the John von Neumann-Institut for Computing (NIC) at the Forschungszentrum Jülich.

References

[1] K.-T. Kang, J. C. Lee, J. Sun U, *Tetrahedon Lett.* **1992**, 4953.
[2] X. Creary, M. A. Butchko, *J. Org. Chem.* **2002**, *67*, 112.
[3] M. Kira, *Pure Appl. Chem.* **2000**, *72*, 2333.
[4] H. Sakurai, Y. Nakadaira, H. Sakaba, *Organomet.* **1983**, *2*, 1484.
[5] S. Solé, H. Gornitzka, O. Guerret, G. Bertrand, *J. Am. Chem. Soc.* **1998**, *120*, 9100.
[6] T. J. Barton, J. Lin, S. Ijadi-Maghsoodi, M. D. Power, X. Zhang, Z. Ma, H. Shimizu, M. S. Gordon, *Organosilicon Chemistry* **1997**, *3*, 17.
[7] S. Nagase, T. Kudo, *J. Chem. Soc., Chem. Commun.* **1984**, 1392.
[8] E. Ionescu, H. Wilkens, R. Streubel, 2nd Silicon Days **2003**, poster contribution 098 in *Organosilicon Chemistry VI — From Molecules to Materials.*
[9] T. J. Barton, J. Lin, S. Ijadi-Maghsoodi, M. D. Power, X. Zhang, Z. Ma, H. Shimizu, M. S. Gordon, *J. Am. Chem. Soc.* **1995**, *117*, 11695.
[10] S. Nagase, T. Kudo, *J. Chem. Soc., Chem. Commun.* **1984**, 141.
[11] This work.
[12] M. T. Nguyen, A. Van Keer, L. G. Vanquickenborne, *J. Org. Chem.* **1996**, *61*, 7077.
[13] K. Lammertsma, M. J. M. Vlaar, *Eur. J. Org. Chem.* **2002**, 1127.
[14] S. Creve, K. Pieloot, M. T. Nguyen, L. G. Vanquickenborne, *Eur. J. Inorg. Chem.* **1999**, 107.
[15] A. W. Ehlers, K. Lammertsma, E. J. Baerends, *Organomet.* **1998**, *17*, 2738.
[16] R. Bauernschmitt, R. Ahlrichs, *J. Chem. Phys.* **1996**, *104*, 9047.
[17] W. W. Schoeller, A. B. Rozhenko, S. Grigoleit, *Eur. J. Inorg. Chem.* **2001**, 2891.
[18] A. D. Becke, *J. Chem. Phys.* **1993**, *98*, 5648.
[19] R. Krishnan, J. S. Blinkley, R. Seeger, J. A. Pople, *J. Chem. Phys.* **1980**, *72*, 650.
[20] A. D. McLean, G. S. Chandler, *J. Chem. Phys.* **1988**, *53*, 1995.
[21] P. J. Hay, W. R. Wadt, *J. Chem. Phys.* **1988**, *82*, 270.
[22] http://jumpdoc.fz-juelich.de/.
[23] J. C. Decius, *J. Chem. Phys.* **1962**, *38*, 241.
[24] J. Grunenberg, *Angew. Chem.* **2001**, *113*, 4150; J. Grunenberg, R. Streubel, G. von Frantzius, W. Marten, *J. Chem. Phys.* **2003**, *119*, 165.
[25] G. Fogarasi, X. Zhou, P. W. Taylor, P. Pulay, *J. Am. Chem. Soc.* **1992**, *114*, 8191.

Silyl Group Migrations between Oxygen and Nitrogen in Aminosiloxanes

Susanne Kliem,* Uwe Klingebiel

Institute of Inorganic Chemistry of the Georg-August-University,
Tammannstr. 4, D-37077 Goettingen, Germany
Tel.: +49 551 393052 — Fax: +49 551 393373
E-mail: uklinge@gwdg.de

Keywords: silyl group migration, aminosiloxanes, lithium salts

Summary: In contrast to carbon chemistry, compounds with two or three H-acidic electronegative groups like NH_2 or OH at one silicon atom can be stabilized kinetically [1 – 5]. In the 1980s the first aminosilanol [6], $(Me_3C)_2Si(NH_2)OH$, was prepared. Its alkaline metal derivatives form an aminosilanolate anion that is 15.9 kcal/mol more stable than the isoelectronic amidosilanol anion [7].

$$(Me_3C)_2Si(NH_2)O^{\ominus}; \qquad (Me_3C)_2Si(OH)NH^{\ominus}$$

1-Aminosiloxanes

The stability of the aminosilanolate salts explains the position of substituents. For example, in reactions with halosilanes, numerous mono-, bis-, tris- and tetrakis-1-amino-siloxanes are formed [7 – 10], e. g.,

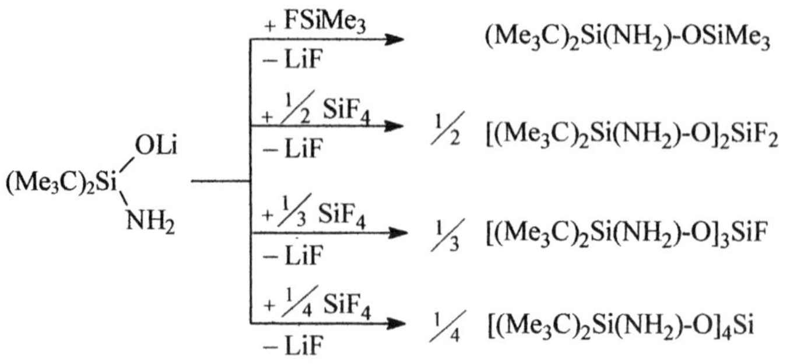

$$(Me_3C)_2Si(OLi)(NH_2) \xrightarrow[-\,LiF]{+\,FSiMe_3} (Me_3C)_2Si(NH_2)\text{-}OSiMe_3$$

$$(Me_3C)_2Si(OLi)(NH_2) \xrightarrow[-\,LiF]{+\,\frac{1}{2}\,SiF_4} \tfrac{1}{2}\ [(Me_3C)_2Si(NH_2)\text{-}O]_2SiF_2$$

$$(Me_3C)_2Si(OLi)(NH_2) \xrightarrow[-\,LiF]{+\,\frac{1}{3}\,SiF_4} \tfrac{1}{3}\ [(Me_3C)_2Si(NH_2)\text{-}O]_3SiF$$

$$(Me_3C)_2Si(OLi)(NH_2) \xrightarrow[-\,LiF]{+\,\frac{1}{4}\,SiF_4} \tfrac{1}{4}\ [(Me_3C)_2Si(NH_2)\text{-}O]_4Si$$

Scheme 1.

The tetrakis-(1-amino-siloxane) condenses thermally, yielding a spirocyclic six-membered ring and NH_3 [11].

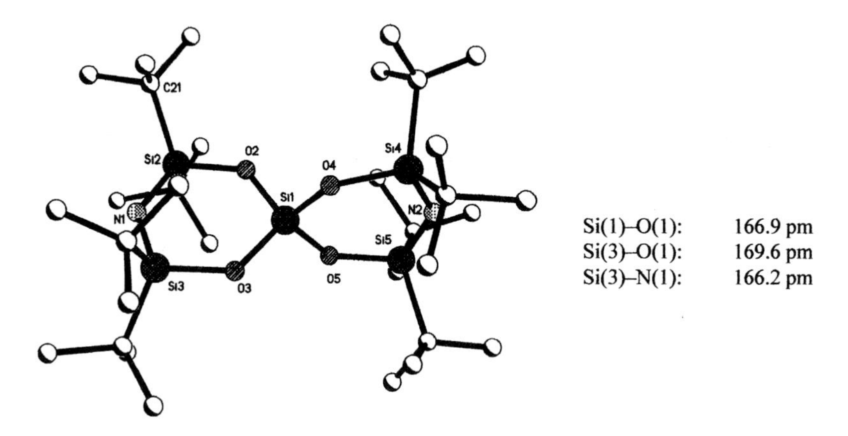

Fig. 1.

Starting with the 1-amino-siloxanes, further substitutions turn out to be more complicated. Lithium salts of these compounds form, depending on the properties and bulkiness of the silyl groups, 1,3-disilazane-1-olates or the less stable 1-amido-1,3-disiloxanes. The first case includes a 1,3-silyl group migration from the oxygen to the nitrogen atom.

In order to understand the driving force of this isomerization, we calculated the energies of the following isomeric lithium salts.

Me_2Si / O / Li / $N(H)-SiMe_3$	Me_3Si-O / Me_2Si / Li / NH	and	FMeSi / F / Li / O / $SiMe_2$ / N(H)	FMeSi / F / Li / NH / $SiMe_2$ / O
Silanolate	Silylamide		Silanolate	Silylamide
0 kcal/mol	**0.7 kcal/mol**		**0 kcal/mol**	**6.4 kcal/mol**

The silanolates are more stable than the silylamides. Using fluorosilyl groups, lithium is coordinated to the fluorine atom.

1-Silylamino-1,3-disiloxanes

Lithium salts of amino-siloxanes react with fluorosilanes to give in the absence of strong steric or electronic restraints, the isomeric silylamino-1,3-disiloxane. A 1,3-(O→N)-silyl group migration has occurred. The analogous raction with difluorosilanes leads to both isomers, and with trifluorosilanes no isomerization or silyl group migration is observed.

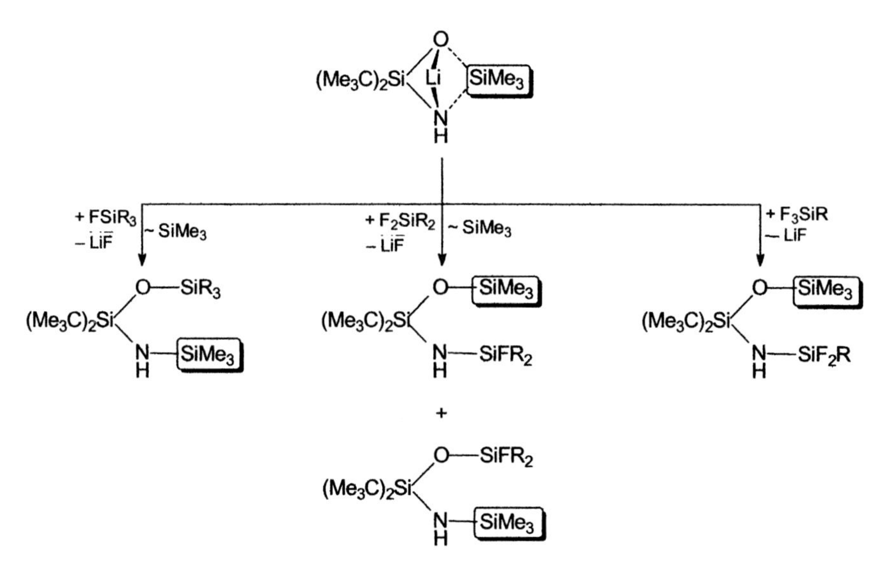

Scheme 2.

Bulky groups prevent kinetically the silyl group migration from the oxygen to the nitrogen atom and for that reason the isomerization in the lithiation reaction [11].

$(Me_3C)_2Si(OSiMe_3)NH_2$ $\xrightarrow[-\,n\text{-BuH},\ -\,LiF]{+\,n\text{-BuLi},\ +\,F_3SiR}$ $(Me_3C)_2Si(O{-}SiMe_3)(NH{-}SiF_2R)$ (not: $(Me_3C)_2Si(O{-}SiF_2R)(NH{-}SiMe_3)$)

Eq. 1.

Experiments, quantum chemical calculations of model compounds, and crystal structures prove that the N–SiF_2R-substituted compounds are more stable than the isomeric N–$SiMe_3$-substituted compounds [11].

1,1-Bis(silylamino)-1,3-disiloxanes

Chemical calculations and the experiments show that in further reactions lithium salts of 1-silylamino-1,3-disiloxanes react with fluorosilanes under retention of the configuration to give 1,1-bis(silylamino)-1,3-disiloxanes.

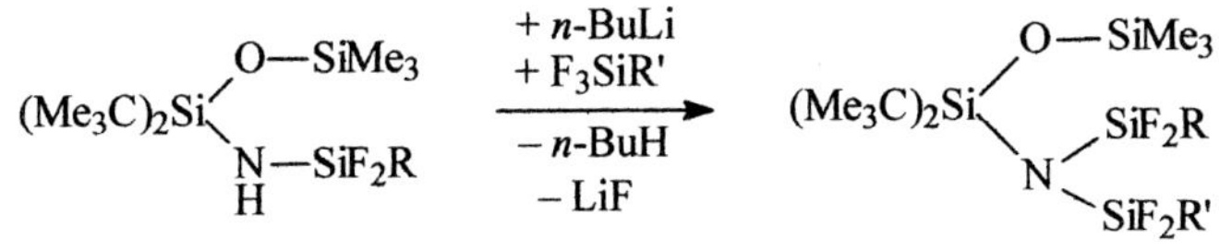

$(Me_3C)_2Si(O{-}SiMe_3)(NH{-}SiF_2R) \xrightarrow[-\,n\text{-}BuH,\ -\,LiF]{+\,n\text{-}BuLi,\ +\,F_3SiR'} (Me_3C)_2Si(O{-}SiMe_3)N(SiF_2R)(SiF_2R')$

Eq. 2.

Lithium-1-silylamido-1,3-disiloxanes

Quantum chemical calculations of model compounds show that amides are more stable isomers than the olates, e.g.,

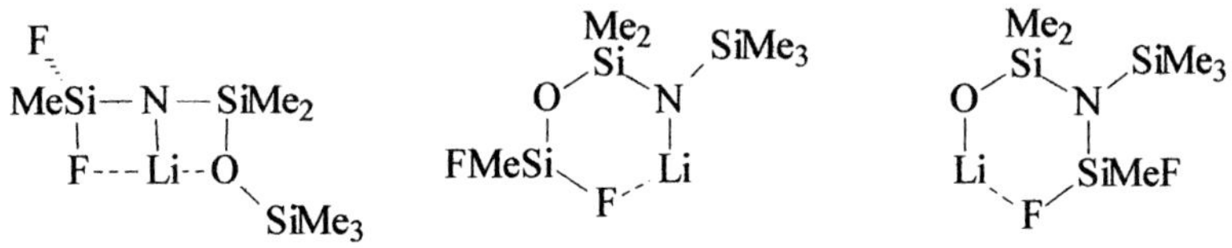

Silylamide	Silylamide	Silanolate
0 kcal/mol	**6.0 kcal/mol**	**12.1 kcal/mol**

We succeeded in the preparation of some lithium salts of 1-silylamino-1,3-disiloxanes. The next two lithium salts could be isolated as dimers. The first one forms a tricyclic compound existing as two (SiOSiN)-four-membered rings connected by an $(LiFSiN)_2$-eight-membered ring, and the second salt forms a spiro-pentacyclic compound existing as two (SiOSiN), two (SiF_2Li)-four-membered rings which are connected by an $(LiNSiF_2)_2$-eight-membered ring system.

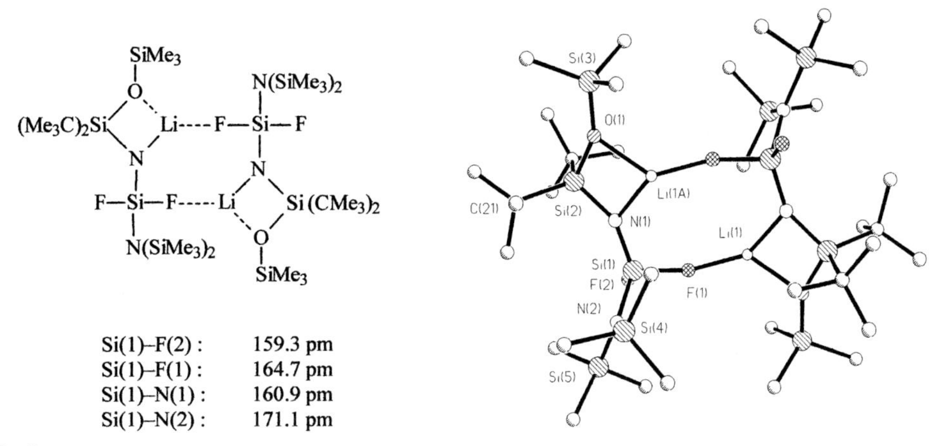

Fig. 2.

The crystal structure shows some irregularities: One Si–F bond length is, because of the Li---F-contact, 5.3 pm longer than the other. The Si(1)–N(1) bond length is in the range of a double bond length. The Li---O contact lengthens the Si(2)–O(1) bond. The Li–F contact is unusually short — even shorter than the Si–N bond. Lithium is only three coordinated (Σ°Li = 349.7°).

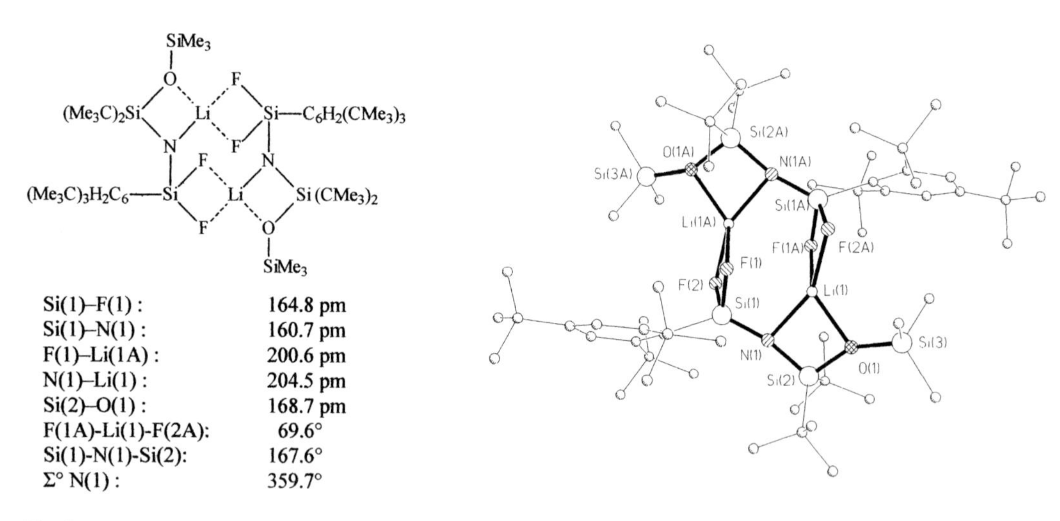

Si(1)–F(1) : 164.8 pm
Si(1)–N(1) : 160.7 pm
F(1)–Li(1A) : 200.6 pm
N(1)–Li(1) : 204.5 pm
Si(2)–O(1) : 168.7 pm
F(1A)-Li(1)-F(2A): 69.6°
Si(1)-N(1)-Si(2): 167.6°
Σ° N(1) : 359.7°

Fig. 3.

In this example lithium has a tetrahedral coordination: two fluorine, one oxygen, and one amido contact. This coordination of the lithium ion was hitherto unknown. The Si–F bonds are equal (164.7 pm).

The structure of the lithium salt depends on the solvents used, in *n*-hexane the salt crystallizes as dimer and in thf as monomer.

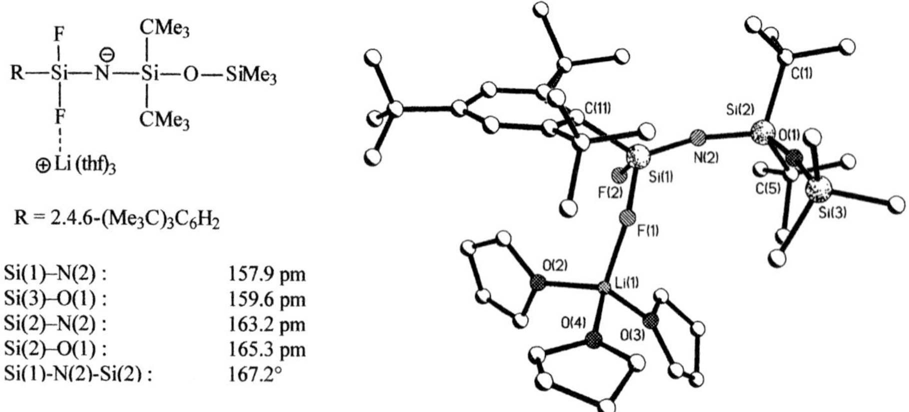

Fig. 4.

Crystallization from thf leads to a complete rupture of the Li---N contact. Lithium now has migrated to the stronger Lewis base fluorine. The result is a short Li---F bond (186.6 pm) and a long Si–F bond (165.9 pm). The Si–N bond is, at 157.9 pm, a double bond length, as far as we know, the shortest Si–N bond with four coordinate silicons. The Si-N-Si angle of 167.2° is typical of an imine. Therefore, the molecule must be considered as a Li–F adduct of an iminosilane, demonstrated in the next figure.

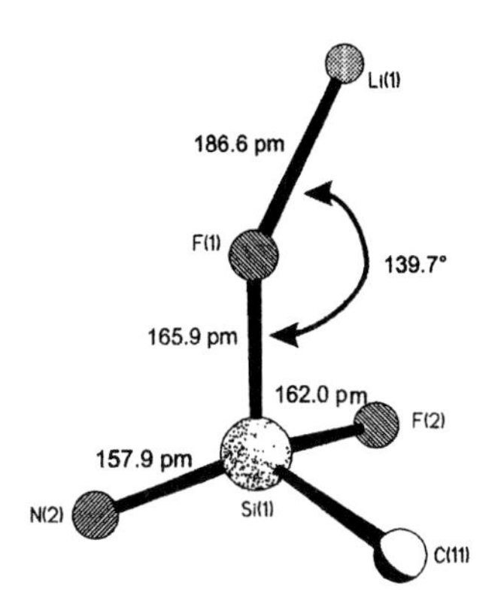

Fig. 5.

Acknowledgments: We are grateful to the Fonds der Chemischen Industrie for the financial support of this work.

References

[1] L. H. Sommer, J. Tyler, *J. Am. Chem. Soc.* **1954**, *76*, 1030.

[2] P. D. Lickiss, *Adv. Inorg. Chem.* **1995**, *42*, 147; *The Chemistry of Organic Silicon Compounds* (Eds.: Z. Rappoport and Y. Apeloig), John Wiley & Sons, New York, **2001**, *Vol. 3*, 695.

[3] K. Ruhlandt-Senge, R. A. Bartlett, M. M. Olmstead, P. P. Power, *Angew. Chem.* **1993**, *105*, 459; *Angew. Chem. Int. Ed. Engl.* **1993**, *32*, 425.

[4] R. Murugavel, M. Bhattadarjee, H. W. Roesky, *Appl. Organomet. Chem.* **1999**, *13*, 227.

[5] O. Graalmann, U. Klingebiel, *J. Organomet. Chem.* **1984**, *275*, C1.

[6] O. Graalmann, U. Klingebiel, W. Clegg, M. Haase, G. M. Sheldrick, *Angew. Chem.* **1984**, *96*, 904; *Angew. Chem. Int. Ed. Engl.* **1984**, *23*, 891.

[7] C. Reiche, S. Kliem, U. Klingebiel, M. Noltemeyer, C. Voit, R. Herbst-Irmer, S. Schmatz, *J. Organomet. Chem.* **2003**, *667*, 24.

[8] D. Schmidt-Bäse, U. Klingebiel, *J. Organomet. Chem.* **1989**, *364*, 313.

[9] K. Dippel, U. Klingebiel, G. M. Sheldrick, D. Stalke, *Chem. Ber.* **1987**, *120*, 611.

[10] U. Klingebiel, M. Noltemeyer, *Eur. J. Inorg. Chem.* **2001**, 1889.

[11] S. Kliem, U. Klingebiel, *Silicon for the Chemical Industry VI* (Ed.: H. A. Oye), **2002**, 139, Trondheim, Norway.

Terphenyl Phosphanosilanes

Rudolf Pietschnig*

Institut für Chemie, Karl-Franzens-Universität Graz
Schubertstraße 1, A-8010, Graz, Austria
Tel.: +43 316 380 5285 — Fax: +43 316 380 9835
E-mail: rudolf.pietschnig@uni-graz.at

Jürgen J. Tirrée

Institut für Anorganische Chemie, Rheinische Friedrich-Wilhelms-Universität Bonn
Gerhard-Domagk-Str. 1, D-53121 Bonn, Germany

Keywords: terphenylsilane, phosphanosilane, phosphasilyne

Summary: Our investigations show that the presence of a terphenyl ligand increases the thermal stability of trisphosphanosilanes in solution. Nevertheless, standard work-up procedures induce condensation to Si–P rings and cages under elimination of PH_3. The P–P and P–H coupling patterns in the ^{31}P NMR spectra allow facile assignment of the major products to characteristic ring and cage structures in solution. Moreover, moderate heating leads to a rearrangement in which formally a silicon–carbon bond is broken and replaced by a phosphorus–carbon bond. This rearrangement might involve intermediates such as a terphenyl phosphasilyne. Generally, the increased acidity of phosphanyl groups adjacent to silane units leads to partial lithiation of the initially formed phosphanosilanes by the phosphanylating reagent $LiPH_2$. However, alternative reagents to introduce phosphano groups lead to incomplete displacement of the fluorine atoms in the starting terphenyl trifluoro silane.

Introduction

Despite their limited stability, phosphanosilanes have been fascinating synthetic targets in the past. A landmark in this respect has been the isolation of the otherwise unsubstituted tetraphosphino silane $Si(PH_2)_4$ [1]. Bulky substituents do not necessarily provide sufficient steric protection for an adjacent trisphosphanosilyl unit. For instance $Cp^*Si(PH_2)_3$ condenses above –30 °C with elimination of PH_3 to form diphosphino–diphosphadisilacyclobutanes [2, 3]. Because of our continuing interest in the chemistry of terphenylsilanes [4 – 6], we investigated the formation of the analogous terphenyl-substituted phosphanosilanes. These should be suitable starting materials for synthesizing Si–P clusters, in which bulky ligands could provide control and protection. Moreover, the lability of the Si–P bond in a sterically crowded coordination environment might also be a useful feature to generate unusual multiply bonded Si–P π-systems.

Results and Discussion

To synthesize terphenyl phosphanosilanes, we investigated the reactivity of 2,6-dimesityl phenyl trifluorosilane (**1**) with several phosphanylating reagents. The best results were obtained using $LiPH_2$ (**2**) at low temperature. While stoichiometric amounts of **2** generally resulted in the formation of only partially defluorinated phosphanosilanes, a complete displacement of all fluorine atoms can be achieved using an excess of $LiPH_2$. Because of this excess and the increased acidity of phosphanyl groups adjacent to silane units, the primary product which can be identified is not the expected 2,6-dimesityl phenyl trisphosphanosilane (**3**), but its singly deprotonated derivative **4**.

$$\mathbf{(1)}\ \xrightarrow[-3\ LiF]{+3\ LiPH_2}\ [\mathbf{(3)}]\ \xrightarrow[-\ PH_3]{+\ LiPH_2}\ \mathbf{(4)}$$

Scheme 1. Synthesis of **4**.

In the proton-decoupled ^{31}P NMR spectra, these show two signals, a doublet at –172.6 ppm for the PH_2 units and a triplet at –215.4 ppm for the PHLi unit, both connected by a $^2J(PP)$ coupling constant of 12 Hz. The signal of the phosphanyl groups is shifted by roughly 40 – 60 ppm to a lower field compared to the related trisphosphanosilanes [3, 7]. When proton coupling is admitted, the ^{31}P NMR spectra of **4** show a typical splitting pattern which is characteristic of an $(AX_2)_2BY$-system (A, B = ^{31}P; X, Y = ^1H) (Fig. 1).

By fitting the simulated NMR parameters to the measured spectra, the parameters depicted in Scheme 2 can be elucidated [8]. These values compare to coupling constants of $^1J(PH)$: 175 – 194 Hz, $^2J(PP)$: 3.8 – 16.8 Hz, $^3J(PH)$: 4.4 – 26.1 Hz, $^4J(HH)$: 0.4 – 9.5 Hz in related compounds [1–3, 7]. It should be pointed out that our terphenyl-substituted derivatives show well-resolved ^{31}P NMR spectra at room temperature, while the corresponding Cp* phosphanosilanes show broad ^{31}P resonances in solution, because of dynamic behavior under these conditions. Figure 2 shows the proton-decoupled and the coupled ^{31}P NMR spectrum of the crude reaction mixture.

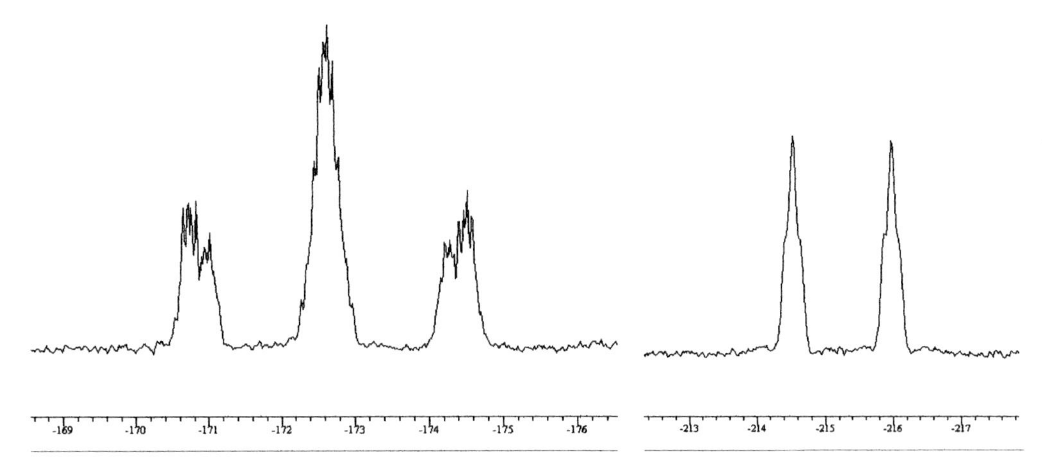

Fig. 1. A and B part of the ^{31}P NMR spectrum of **4** without proton decoupling.

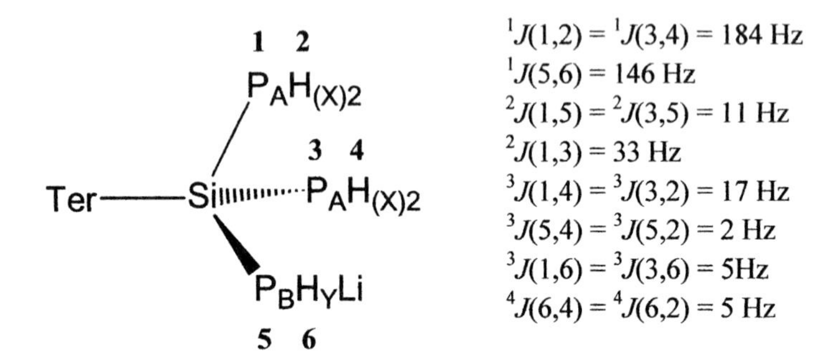

Scheme 2. Coupling constants in **4** obtained by simulation and iteration.

In the reaction mixture of **4**, minor amounts of the corresponding dilithiated trisphosphanosilane **6** ($Mes_2C_6H_3$–Si$(PHLi)_2PH_2$) can also be detected. The proton decoupled ^{31}P NMR spectra of the latter show two signals, a doublet at –221.5 ppm for the PHLi units and a triplet at –178.8 ppm for the PH_2 unit, both connected by a $^{2}J(PP)$ coupling constant of 70 Hz. The (compared to **4**) significantly higher $^{2}J(PP)$ coupling can be attributed to a *syn* orientation of the lone pairs at the phosphorus atoms [9].

The singly deprotonated 2,6-dimesitylphenyl trisphosphanosilane (**4**) is stable at room temperature in the reaction mixture, which means that the presence of a terphenyl ligand increases the thermal stability of the adjacent phosphano silyl unit. On the other hand, removal of the solvent under reduced pressure results in the evolution of PH_3 and formation of Si–P rings. For the major products, an assignment to the respective ring and cage structures in solution was possible because of the characteristic shift values and P–P and P–H coupling patterns.

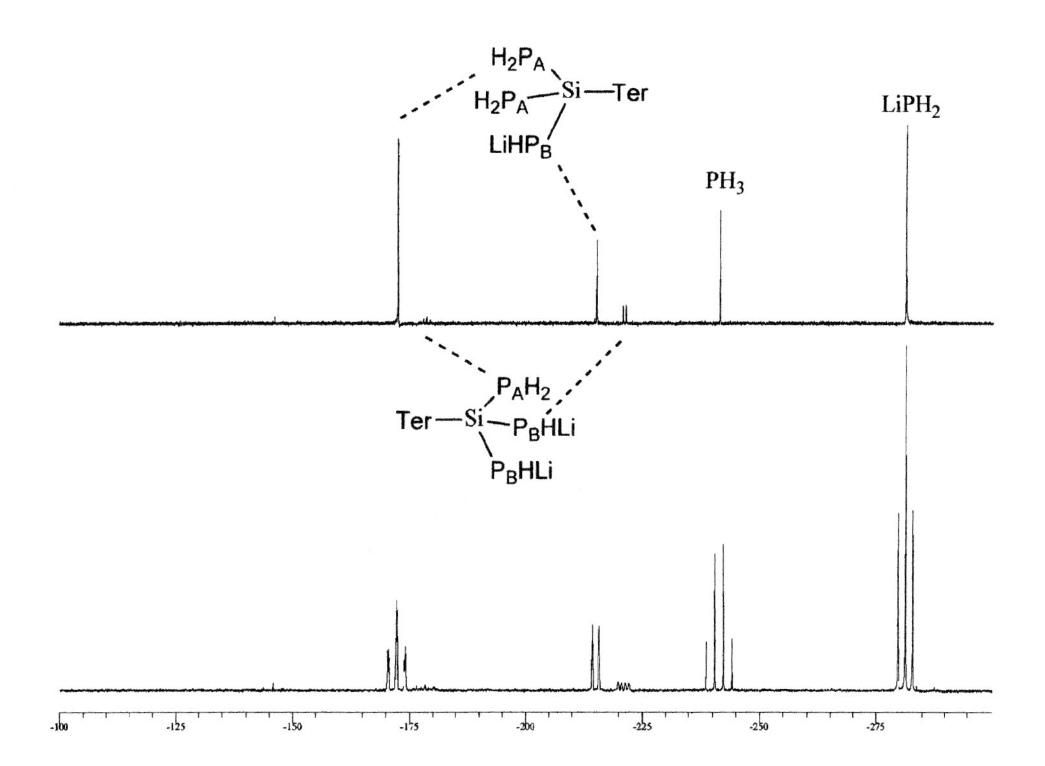

Fig. 2. Comparison of the proton decoupled and the coupled ^{31}P NMR spectrum of the crude reaction mixture.

The major product again is not the expected monocyclic diphosphadisilacyclobutane $[TerSi(PH)PH_2]_2$ (**6**), but its dilithiated derivative **7**, which can be identified on the basis of the (AX)(A′X′)(BY)(B′Y′) spin system in the ^{31}P NMR spectra (A, B = ^{31}P; X, Y = ^{1}H). The alternative isomer of this product, in which the bridging rather than the terminal phosphano groups are lithiated and which thermodynamically should be more stable, should give rise to an $AA'(BX_2)(B'X'_2)$ spin system (A, B = ^{31}P; X = ^{1}H). This, however, is not observed, which means that the kinetic product is formed under these conditions because of the sterically hindered situation.

In the proton-decoupled ^{31}P NMR spectra, compound **7** shows two signals, a triplet at −160.4 ppm for the endocyclic PH units and a triplet at −187.9 ppm for the exocyclic PHLi units, both connected by a ^{2}J(PP) coupling constant of 9.3 Hz.

Very surprisingly, the completely silicon-free 2,6-dimesityl phenyl phosphane (**9**) could also be identified as further by-product (3% yield) [10]. Moreover, if a solution of the initially formed **4** is heated to approximately 60 °C, **9** is one of the major products (33% yield). The formation of **9** necessarily involves a fission of the terphenyl carbon–silicon bond. However, under the reaction conditions (room temperature), a homolytic Si–C bond cleavage is highly unlikely. A plausible though speculative explanation for the formation of **9** could involve the terphenyl phosphasilyne (**8a**), which, in analogy to its nitrogen congener [11], should rearrange to the corresponding terphenyl isophosphasilyne (**8b**). This could then be cleaved to form terphenyl phosphane **9**, for instance by traces of moisture or solvent.

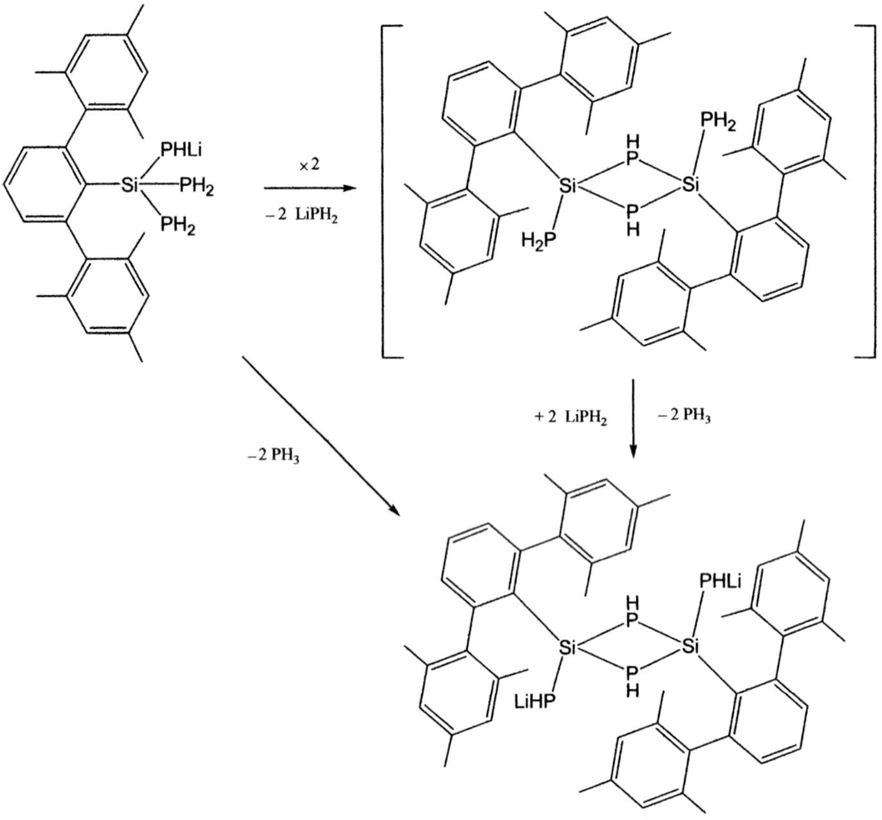

Scheme 3. Condensation of **4** to **7** with elimination of PH_3.

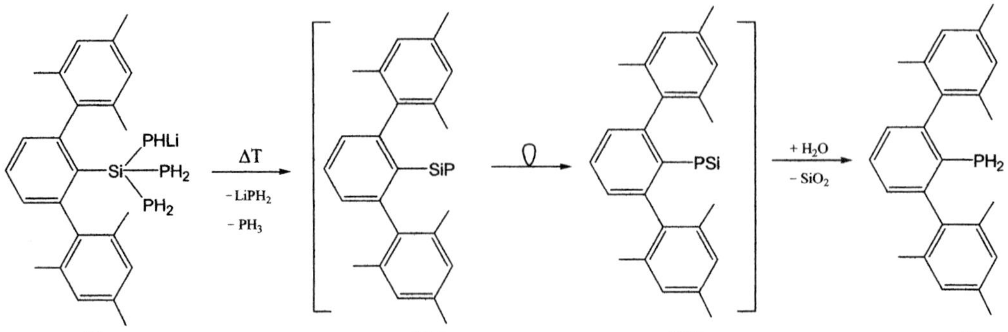

Scheme 4. Tentative mechanism of the formation of **9**.

Generally, we found that in order to achieve complete displacement of all fluorine atoms in the starting silane, an excess of $LiPH_2$ is necessary. To avoid this issue, we also investigated the reactivity of some alternative phosphanylating reagents toward terphenyl trifluorosilane (**1**). However, reagents such as Na_3P, $NaPH_2$, and $LiPTMS_2$ also resulted in no or only partial displacement of the fluorine atoms in **1** even if an excess of the phosphanylating reagents was supplied. Moreover, no evidence for the successful attachment of phosphano groups to the terphenylsilane unit could be found with these reagents, except for $LiPTMS_2$.

Acknowledgment: The authors gratefully acknowledge financial support by the Fonds der Chemischen Industrie and the Karl-Franzens-University, Graz.

References

[1] M. Driess, C. Monsé, R. Boese, D. Bläser, *Angew. Chem.* **1998**, *110*, 2389.
[2] M. Waltz, M. Nieger, D. Gudat, E. Niecke, *Z. Anorg. Allg. Chem.* **1995**, *621*, 1951.
[3] M. Baudler, W. Oehlert, B. Tillmanns, *Z. Naturforsch.* **1992**, *47*, 379.
[4] R. Pietschnig, D. R. Powell, R. West, *Organometallics* **2000**, *19*, 2724.
[5] R. Pietschnig, K. Merz, *J. Chem. Soc., Chem. Commun.* **2001**, 1210.
[6] R. Pietschnig, S. Schäfer, *Silicon* **2003**, *2(3 – 4)* 131.
[7] M. Baudler, G. Scholz, W. Oehlert, *Z. Naturforsch.* **1989**, *44b*, 627.
[8] Simulation was done with the program Calm (Version 2.0).
[9] J. Hahn, Higher order ^{31}P NMR spectra of polyphosphorus compounds, in *Phosphorus-31 NMR Spectroscopy in Stereochemical Analysis* (Eds.: J. G. Verkade, L. D. Quin), VCH, Weinheim, **1987**, p. 331.
[10] E. Urnezius, J. D. Protasiewicz, *Main Group Chem.* **1996**, *1*, 369.
[11] H. Bock, R. Dammel, *Angew. Chem.* **1985**, *97*, 128.

Preparations and X-Ray Structures of some Silicon-Phosphorus and Silicon-Arsenic Cages

Karl Hassler, Günter Tekautz, Judith Baumgartner*

Institute of Inorganic Chemistry, University of Technology Graz
Stremayrgasse 16/IV, A-8010 Graz, Austria
E-mail: hassler@anorg.tu-graz.ac.at

Keywords: silicon-phosphorus cage, silicon-arsenic cage, synthesis

Summary: The preparation of two novel silicon-arsenic cages with a bicycloheptane structure (**1**) and a tricyclononane structure (**2**) is described. Furthermore we report the synthesis of a new anionic silicon-phosphorus cage with a bicyclooctane skeleton (**3**). Its reaction with chlorotrimethylsilane gives **4**. X-ray structures and NMR data of all newly synthesized compounds are presented.

Introduction

The reactions of chlorooligosilanes with sodium potassium phosphide or arsenide are known to yield cage-like structures composed of SiSi and SiE bonds (E=P, As, Sb, Bi), as shown in Fig. 1 [1].

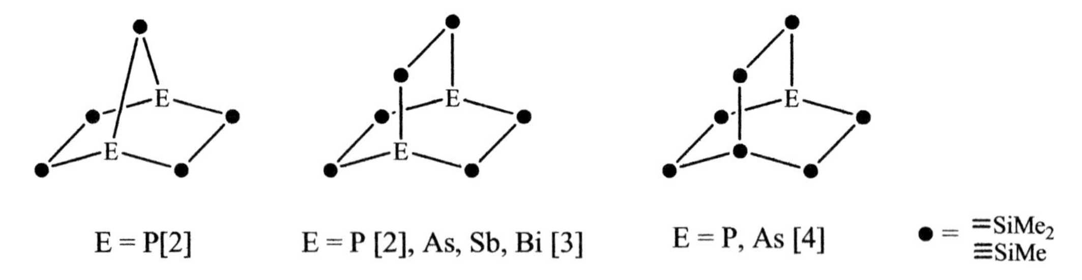

Fig. 1. Cages described in the literature.

Here we report on the synthesis of some new cage compounds from various chlorooligosilanes and $(Na/K)_3P$ or $(Na/K)_3As$ under carefully chosen reaction conditions. They were characterized by ^{29}Si and ^{31}P NMR spectroscopy, elemental analysis, and IR and Raman spectroscopy. The crystal structures were elucidated by X-ray crystallography.

Preparation

The compounds synthesized in this work are highly sensitive to oxygen and moisture. Therefore all operations were carried out under a nitrogen atmosphere. Solvents were distilled from sodium potassium alloy and saturated with dry nitrogen prior to use.

Decamethyl-1,4-diarsa-2,3,5,6,7-pentasila-bicyclo[2.2.1]heptane (1)

Analogous to the reaction of 2-chlorodimethylsilyl-1,3-dichloropentamethyltrisilane (**5**) with Na/K phosphide described in the literature [2], the reaction with the Na/K arsenide in monoglyme yields **1** as the main product. **1** can be crystallized from *n*-heptane.

$$ClMe_2Si{-}SiMe(SiMe_2Cl)_2 \; (\mathbf{5}) + (Na/K)_3As \longrightarrow \mathbf{1} \qquad \bullet = {=}SiMe_2,\ {\equiv}SiMe$$

Eq. 1. Reaction scheme for the formation of **1**.

Dodecamethyl-1,5-diarsa-2,3,4,6,7,8,9-heptasilatricyclo-[3.3.1.0^{3,7}]nonane (2)

To a suspension of sodium potassium arsenide in DME, a solution of 2,3-bis(chlorodimethylsilyl)-1,4-dichlorohexamethyltetrasilane (**6**) in DME is added. The reaction mixture is refluxed for 12 h. After filtration of the salts the solvent is removed *in vacuo*, and **2** is recrystallized from *n*-heptane.

$$(ClMe_2Si)_2MeSi{-}SiMe(SiMe_2Cl)_2 \; (\mathbf{6}) + (Na/K)_3As \longrightarrow \mathbf{2} \qquad \bullet = {=}SiMe_2,\ {\equiv}SiMe$$

Eq. 2. Reaction scheme for the formation of **2**.

Sodium decamethyl-1,3,5-triphospha-2,4,6,7,8-pentasilabicyclo[3.2.1]octane-3-ide (3)

A suspension of sodium potassium phosphide in DME (50% excess) is added to a refluxing solution of **5** in DME. After refluxing for 24 h no reaction occurs and the solid is removed by filtration. The yellow filtrate is again refluxed and a freshly prepared phosphide suspension is added in a molar ratio of 1.5 once more. As can be easily followed by ^{29}Si NMR spectroscopy the reaction immediately proceeds. From the resulting yellowish grey suspension the salts were removed by filtration. After removal of the solvent *in vacuo* **3** was crystallized from heptane/DME.

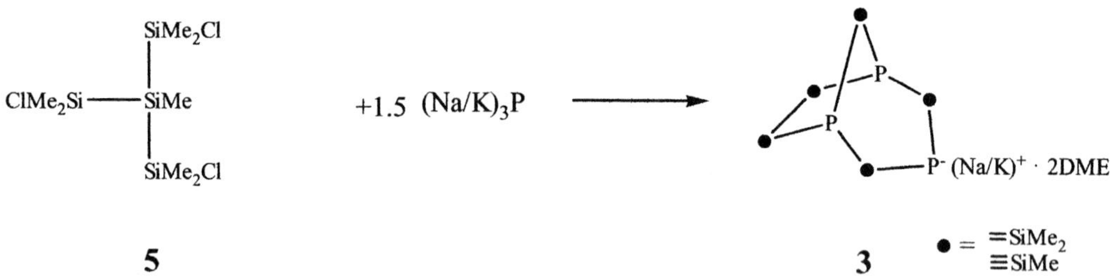

$SiMe_2Cl$
$ClMe_2Si$—SiMe + 1.5 $(Na/K)_3P$ → P^- $(Na/K)^+ \cdot$ 2DME
$SiMe_2Cl$

5 **3** ● = $=SiMe_2$ / $\equiv SiMe$

Eq. 3. Reaction scheme for the formation of **3**.

Decamethyl-3-trimethylsilyl-1,3,5-triphospha–2,4,6,7,8-pentasilabicyclo[3.2.1]octane (4)

A pentane solution of the DME adduct **3** is cooled to –80 °C. Chlorotrimethylsilane is added dropwise. After completion of the reaction the solvent is removed *in vacuo*. The residue is suspended in *n*-heptane and the salts are filtrated. **4** is crystallized from this solution.

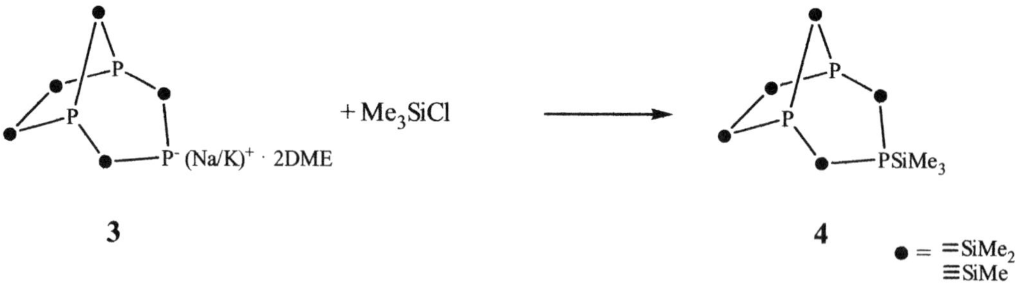

P^- $(Na/K)^+ \cdot$ 2DME + Me_3SiCl → $PSiMe_3$

3 **4** ● = $=SiMe_2$ / $\equiv SiMe$

Eq. 4. Reaction scheme for the formation of **4**.

X-Ray Structures

The crystal structures of compounds **1** – **4** have been elucidated and are presented in Fig. 2.

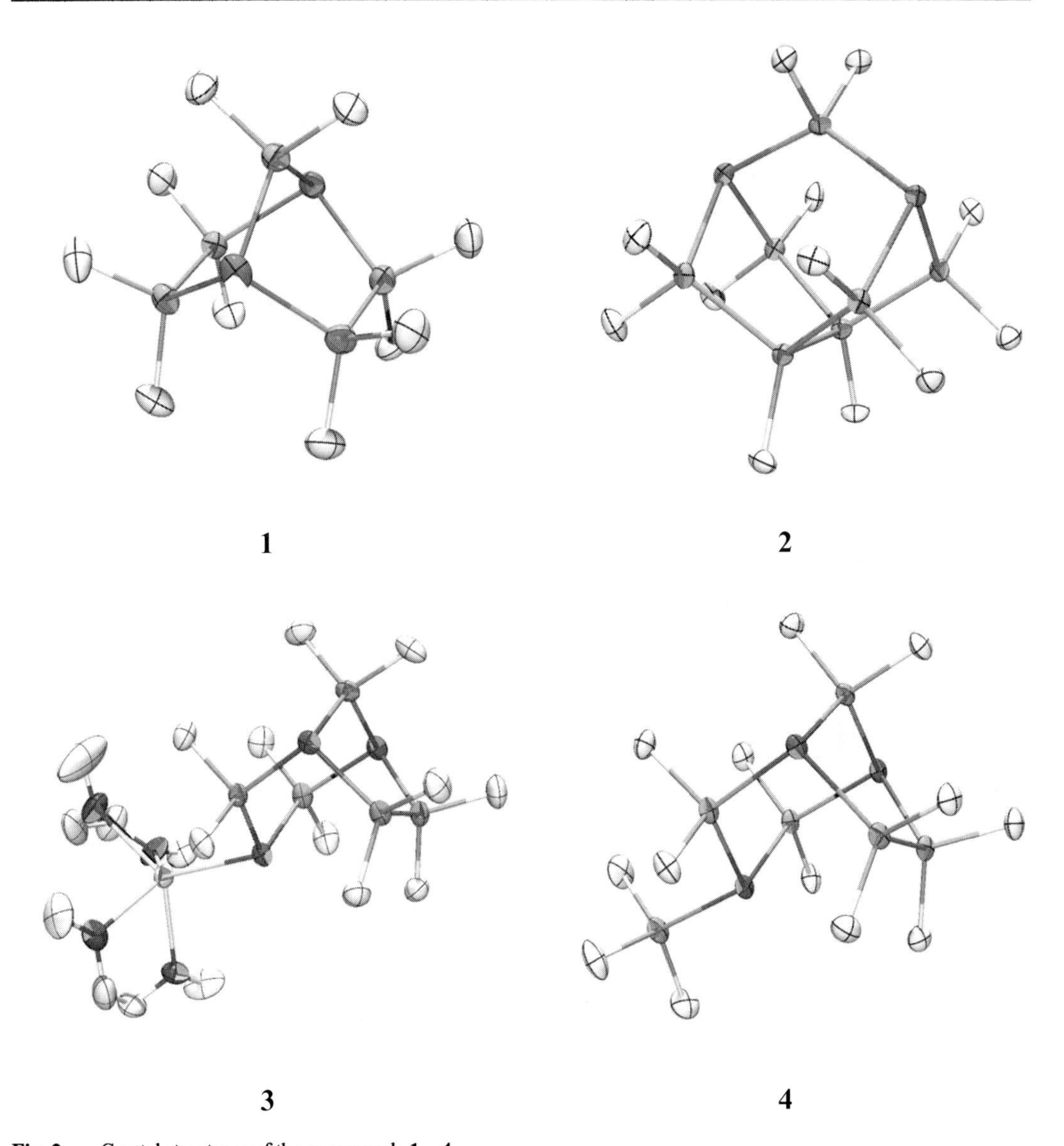

Fig. 2. Crystal structures of the compounds **1** – **4**.

NMR Spectroscopy

^{29}Si INEPT and ^{31}P NMR spectroscopy have been carried out for further characterization. The NMR data are collected in Table 1.

Table 1. ^{29}Si and ^{31}P NMR Data for the compounds **1** – **4**.[a]

	δ(Si)	δ(Si*)	δ(Si**)	δ(Si***)	δ(P)	δ(P*)
	–3.2	–1.7	–73.8			
	1.2	26.6				
	–4.9(dd) (85.9Hz; 39.9Hz)	16.9(t) (50Hz)	4.2(dd) (39.6Hz; 5.9Hz)		–247.0(d) (1.8Hz)	–262.0(t) (2.2Hz)
	–4.7(ddd) (40.4Hz; 4.9Hz; 3.2Hz)	13.9(dt) (48.0Hz; 2.3Hz)	2.0(t) (43.4Hz)	0.8(d) (45.9Hz)	–231.1(d) (8.1Hz)	–238.9(t) (8.3Hz)

[a] Values in parentheses are the values for ^1J and ^2J(SiP)

References

[1] K. Hassler, Silicon-phosphorus, -arsenic, -antimony, and –bismuth cages: syntheses and structures, in *Organosilicon Chemistry — From Molecules to Materials* (Eds: N. Auner, J. Weis), VCH Weinheim, **1996**, p. 203.

[2] G. Kollegger, U. Katzenbeisser, K. Hassler, C. Krüger, D. Brauer, R. Gielen, *J. Organomet. Chem.* **1997**, *543(1 – 2)*, 103.

[3] K. Hassler, S. Seidl, *J. Organomet. Chem.* **1988**, *347(1 – 2)*, 27.

[4] K. Hassler, G. Kollegger, H. Siegl and G. Klintschar, *J. Organomet. Chem.* **1997**, *533(1 – 2)*, 51.

Homo- and Heterometallic Bismuth Silanolates

Michael Mehring,* Dirk Mansfeld, Christof Nolde, Markus Schürmann

Universität Dortmund, Anorganische Chemie II, D-44221, Dortmund, Germany
Tel.: +49 231 755 3835 — Fax: +49 231 755 5049
E-mail: michael.mehring@uni-dortmund.de

Keywords: silanolates, bismuth, sodium, metal oxo clusters

Summary: Attempts to prepare pure $Bi(OSiMe_3)_3$ starting from $BiCl_3$ and $NaOSiMe_3$ failed but gave three novel heterometallic bismuth sodium oxo clusters instead. The molecular structures of the heterometallic oxo clusters $[Na_4Bi_2O(OSiMe_3)_8]$ (**1**), $[Na_5Bi_{10}O_7(OSiMe_3)_{15}(OH)_6]$ (**2**) and $[Na_4Bi_{18}O_{20}(OSiMe_3)_{18}]$ (**3**) were determined by single crystal X-ray diffraction analysis. In addition, the molecular structures of the novel sodium trimethylsilanolate clusters $[Na_8(OSiMe_3)_4(OH)_4(HOSiMe_3)_4(thf)_4]$ (**4**) and $[Na_{11}(CO_3)(OSiMe_3)_3(OH)_6(HOSiMe_3)_9]$ (**5**) are discussed. The μ_{12}-carbonate anion in **5** serves as a template, and hydrogen bonding stabilizes the structure. The bismuth silanolates $Bi(OSiR_2R')_3$ (**6**, R = Ph, R′ = *t*Bu; **7**, R = Me, R′ = *t*Bu; **8**, R = R′ = *i*Pr) were prepared starting from $Bi(OtBu)_3$ and the corresponding silanols. Compound **6** was also prepared by the reaction of $BiPh_3$ with $Ph_2tBuSiOH$.

Introduction

Bismuth-containing materials are associated with a variety of applications such as high-T_c superconductivity, oxide ion conduction, and catalysis. They are found in special glasses and pigments, and they play a key role in some thermoelectric, ferroelectric and dielectric devices [1 – 4]. Homo- and heterometallic alkoxides are well-known precursors for heterometallic metal oxide materials [5 – 8], but, in contrast to the alkoxides, the corresponding bismuth silanolates and their potential as precursors for metal oxide materials have been studied only briefly [9 – 13]. The first report on the synthesis of bismuth tris(trimethysilanolate), $Bi(OSiMe_3)_3$, dates back to 1968 [9]. The bismuth silanolate was prepared by the reaction of $NaOSiMe_3$ with $BiCl_3$ and isolated as an air-sensitive compound. In this communication we report our attempts to prepare $Bi(OSiMe_3)_3$ and present the structures of the heterometallic sodium bismuth oxo clusters $[Na_4Bi_2O(OSiMe_3)_8]$ (**1**), $[Na_5Bi_{10}O_7(OSiMe_3)_{15}(OH)_6]$ (**2**) and $[Na_4Bi_{18}O_{20}(OSiMe_3)_{18}]$ (**3**) which were isolated instead. Furthermore, the molecular structures of the sodium silanolate clusters $[Na_8(OSiMe_3)_4(OH)_4$-$(HOSiMe_3)_4(thf)_4]$ (**4**) [14] and $[Na_{11}(CO_3)(OSiMe_3)_3(OH)_6(HOSiMe_3)_9]$ (**5**), and the syntheses of the bismuth silanolates $Bi(OSiR_2R')_3$ (**6**, R = Ph, R′ = *t*Bu [12]; **7**, R = Me, R′ = *t*Bu [13]; **8**, R = R′ = *i*Pr) are described.

Results and Discussion

Heterometallic Bismuth Oxo Clusters

So far our attempts to prepare $Bi(OSiMe_3)_3$ by the reaction of $BiCl_3$ with $MOSiMe_3$ (M = Li, Na, K) have not been successful, but have produced three novel heterometallic bismuth oxo clusters in low yield. All of these compounds are highly air sensitive and only sparingly soluble in common organic solvents. The use of excess $NaOSiMe_3$ gave the oxo cluster $[Na_4Bi_2O(OSiMe_3)_8]$ (**1**) (Eq. 1, Fig. 1), in which the sodium and bismuth atoms are disordered and each position is occupied by 2/3 Na and 1/3 Bi. Currently, the origin of the oxygen atom in **1** is under investigation. The formation of $Me_3SiOSiMe_3$ was observed, which either results from hydrolysis and subsequent condensation of Me_3SiOH or is formed in course of the reaction.

$$10\ NaOSiMe_3 + 2\ BiCl_3 \xrightarrow[-\ Me_3SiOSiMe_3]{-\ 6NaCl} \underset{\mathbf{1}}{Na_4Bi_2(O)(OSiMe_3)_8}$$

Eq. 1.

Several examples of μ_6-oxo-centered structures such as $[Na_4M_2O(OtBu)_8]$ (M = Bi, Sb), $[K_4Sb_2O(OtBu)_8(thf)_4]$, $[Na_4Bi_2(O)(OC_6F_5)_8(thf)_4]$ and $[Na_4Sb_2O(OSiMe_3)_8]$ have been reported previously [15 – 17]. Notably, the antimony cluster $[Na_4Sb_2O(OSiMe_3)_8]$ was prepared from $Sb(OSiMe_3)_3$ and $NaOSiMe_3$ [17]. Thus, it is most likely that the bismuth oxo cluster **1** forms analogously from *in situ* prepared $Bi(OSiMe_3)_3$.

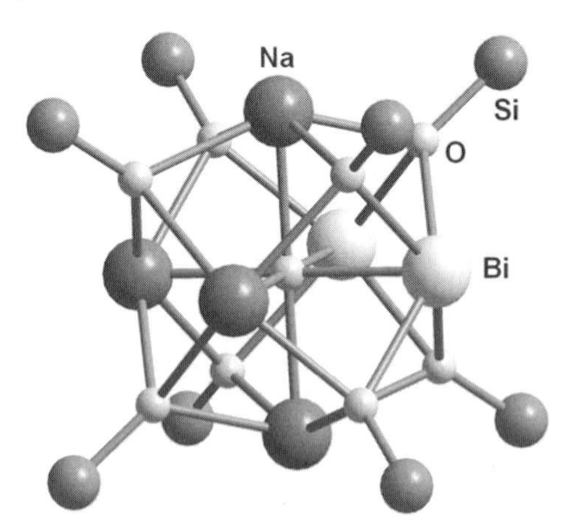

Fig. 1. Core structure of $[Na_4Bi_2O(OSiMe_3)_8]$ (**1**). Methyl groups are omitted.

The use of the appropriate stoichiometry of $BiCl_3$ and $NaOSiMe_3$ gave yellowish suspensions, which were filtered several times to give clear solutions. After these solutions had been kept for several weeks at –20 °C, colorless crystals deposited to give $[Na_5Bi_{10}O_7(OSiMe_3)_{15}(OH)_6]$ (**2**) from

THF/toluene and $[Na_4Bi_{18}O_{20}(OSiMe_3)_{18}]$ (**3**) from toluene. The structures of the sodium bismuth oxo clusters are shown in Figs. 2 and 3. Both bismuth and sodium show a flexible coordination geometry and widespread M–O distances. The Na–O distances range from 232 to 267 pm and 215 to 301 pm and the Bi–O distances range from 201 to 301 pm and 205 to 329 pm for clusters **2** and **3**, respectively.

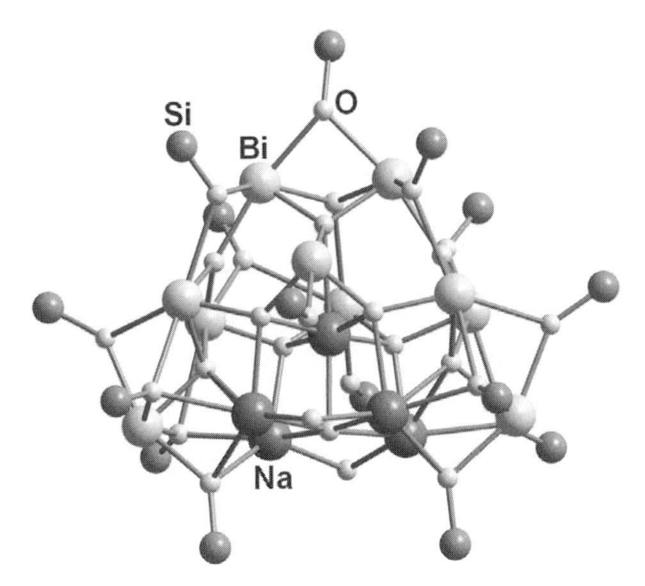

Fig. 2. Core structure of $[Na_5Bi_{10}O_7(OSiMe_3)_{15}(OH)_6]$ (**2**). Methyl groups are omitted.

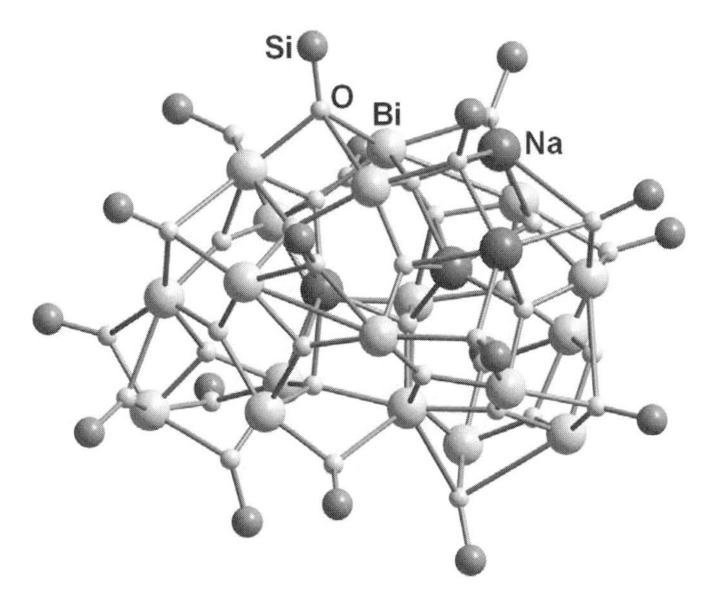

Fig. 3. Core structure of $[Na_4Bi_{18}O_{20}(OSiMe_3)_{18}]$ (**3**). Methyl groups are omitted.

Sodium Trimethylsilanolate Clusters

The formation of the bismuth oxo clusters might result from partial hydrolysis, and we therefore started to investigate the partial hydrolysis of $NaOSiMe_3$. The preparation of anhydrous $NaOSiMe_3$[18, 19] was reported 50 years ago, and a single-crystal X-ray structural analysis of $[NaOSiMe_3]_4{\cdot}4HMPA$ was reported recently [20]. However, the affinity of sodium silanolates to water is documented by the single-crystal X-ray structural analysis of $NaOSiMe_3{\cdot}3H_2O$ [21], $Na_{11}(OSiMe_3)_{10}(OH)$ [22, 23], and $[NaOSiPh_3]_4{\cdot}3H_2O$ [24]. In the course of our studies we isolated two additional examples of sodium silanolate clusters, namely $[Na_8(OSiMe_3)_4(OH)_4(HOSiMe_3)_4(thf)_4]$ (**4**) [14] and $[Na_{11}(CO_3)(OSiMe_3)_3(OH)_6(HOSiMe_3)_9]$ (**5**). The latter was obtained upon addition of water and a small quantity of Na_2CO_3. The structures of the sodium silanolate clusters are shown in Figs. 4 and 5.

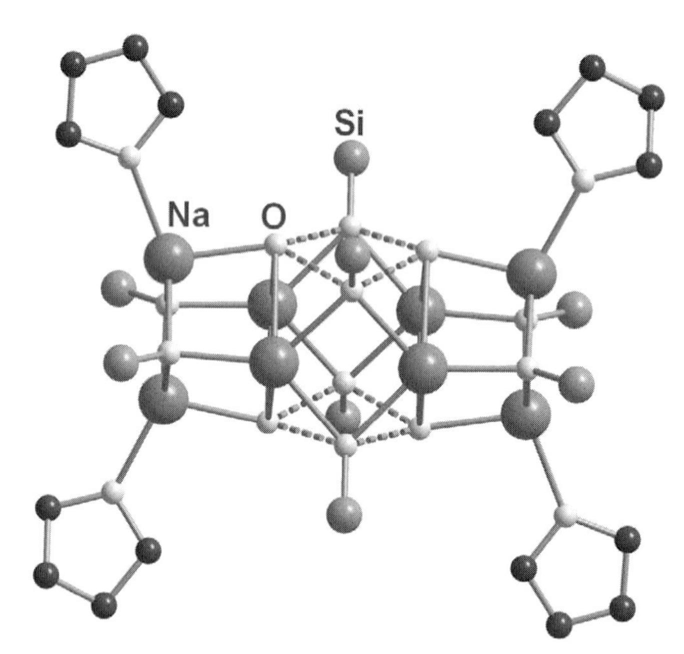

Fig. 4. Core structure of $[Na_8(OSiMe_3)_4(OH)_4(HOSiMe_3)_4(thf)_4]$ (**4**). Methyl groups are omitted.

The structure of $[Na_8(OSiMe_3)_4(OH)_4(HOSiMe_3)_4(thf)_4]$ (**4**) is best described as being composed of two Na_4O_4 cubes, similar to those obtained in tetrameric $[NaOSiMe_3]_4{\cdot}4HMPA$. In **4**, two such cubes are connected via bridging $HOSiMe_3$ and via hydrogen bonding between $HOSiMe_3$ and OH. The O(H)···O distances range from 265 to 270 pm. Notably, compound **4** might also be formulated as $[Na_8(OSiMe_3)_8(H_2O)_4(thf)_4]$ since the positions of the hydrogen atoms were not determined. The Na–O distances are found to be in the range from 234 to 276 pm, which is comparable with the distances observed in the heterometallic oxo clusters **2** and **3**.

The key structural feature of $[Na_{11}(CO_3)(OSiMe_3)_3(OH)_6(HOSiMe_3)_9]$ (**5**) is the μ_{12}-CO_3 group.

Nine sodium atoms are connected via the carbonate, with Na–O bond distances in the range from 234 to 249 pm. The outer sphere of the cluster is formed by three $OSiMe_3$, nine $HOSiMe_3$, and six OH groups which are connected via hydrogen bonding, with O(H)···O distances in the range from 243 to 253 pm. The positions of the hydrogen atoms were not determined, and compound **5** could be formulated alternatively as $[Na_{11}(CO_3)(OSiMe_3)_9(HOSiMe_3)_3(H_2O)_6]$.

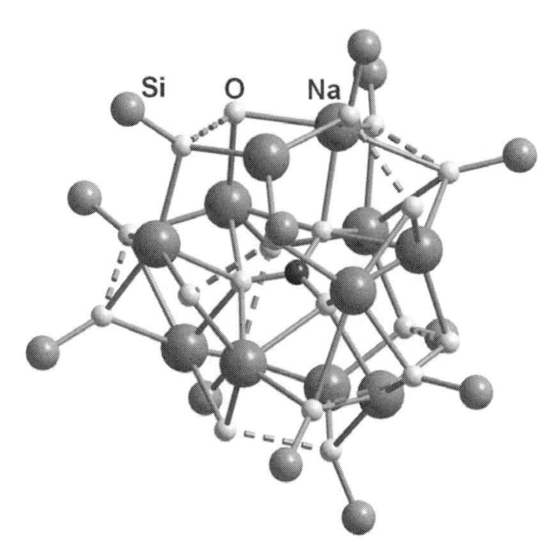

Fig. 5. Core structure of $[Na_{11}(OSiMe_3)_3(HOSiMe_3)_9(OH)_6(CO_3)]$ (**5**). Methyl groups are omitted.

Bismuth Silanolates

A useful route to the novel bismuth silanolates $Bi(OSiR_2R')_3$ (**6**, R = Ph, R′ = *t*Bu [12]; **7**, R = Me, R′ = *t*Bu [13]; **8**, R = R′ = *i*Pr) is based on the reaction of $Bi(OtBu)_3$ with the corresponding silanols (Eq. 2). Alternatively, compound **6** was also prepared by the reaction of $BiPh_3$ with $Ph_2tBuSiOH$ in moderate yield using a prolonged reaction time at temperatures above 150 °C (Eq. 3). Attempts to prepare the bismuth trialkylsilanolates **7** and **8** analogously to **6** failed.

$$3\ R_2R'SiOH + Bi(OtBu)_3 \xrightarrow[-\ 3tBuOH]{} Bi(OSiR_2R')_3$$

6 R = Ph, R' = *t*Bu
7 R = Me, R' = *t*Bu
8 R = R' = *i*Pr

Eq. 2.

$$3\ R_2R'SiOH + BiPh_3 \xrightarrow[-\ 3PhH]{} Bi(OSiR_2R')_3$$

6 R = Ph, R' = *t*Bu

Eq. 3.

The bismuth silanolate $Bi(OSiPh_2tBu)_3$ (**6**) was characterized by single-crystal X-ray diffraction analysis (Fig. 6). Coordination of the bismuth atom to a phenyl ring of a neighboring molecule results in a dimer. The Bi–C distances range from 357 to 366 pm, which is in the upper range usually observed for the coordination of aryl groups to bismuth.

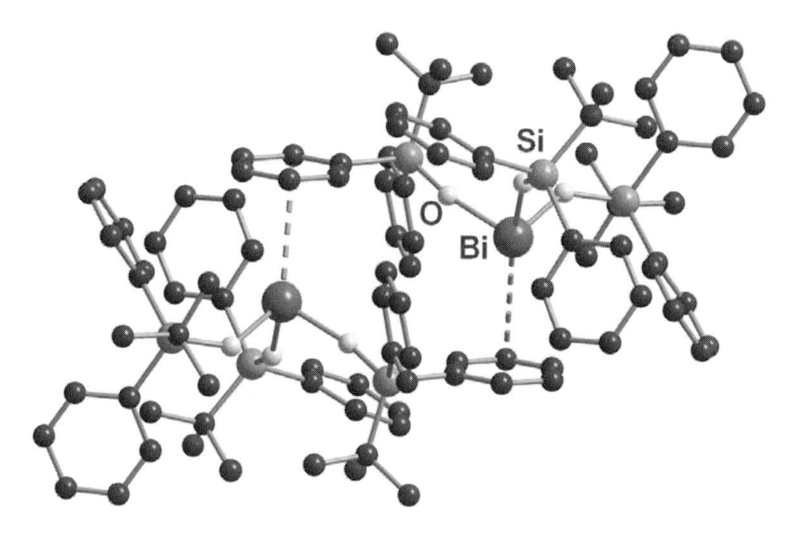

Fig. 6. Molecular structure of $Bi(OSiPh_2tBu)_3$ (**6**).

Conclusion

The metathesis route to bismuth silanolates proved not to be successful for the compounds studied in this work. Instead, $Bi(OSiR_2R')_3$ (**6**, R = Ph, R′ = *t*Bu [12]; **7**, R = Me, R′ = *t*Bu [13]; **8**, R = R′ = *i*Pr) was prepared starting from $Bi(OtBu)_3$ and the corresponding silanols. In the case of silanols which (i) are thermally stable up to temperatures of 160 °C and (ii) have a high boiling point, reaction of $BiPh_3$ with the corresponding silanol might be an alternative. The novel bismuth silanolates **7** and **8** are interesting since they are potential precursors for MOCVD applications. So far we have not been able to prepare pure $Bi(OSiMe_3)_3$, but have isolated the novel heterometallic oxo clusters $[Na_4Bi_2O(OSiMe_3)_8]$ (**1**), $[Na_5Bi_{10}O_7(OSiMe_3)_{15}(OH)_6]$ (**2**), and $[Na_4Bi_{18}O_{20}(OSiMe_3)_{18}]$ (**3**) instead. The origin of the oxygen atoms within the clusters is not yet completely understood. Most likely, $Bi(OSiMe_3)_3$ is formed *in situ* and reacts with $NaOSiMe_3$ to give the metal oxo clusters. However, the easy uptake of water by $NaOSiMe_3$ is also demonstrated by $[Na_8(OSiMe_3)_4(OH)_4(HOSiMe_3)_4(thf)_4]$ (**4**) [14] and $Na_{11}(OSiMe_3)_3(HOSiMe_3)_9(OH)_6(CO_3)$ (**5**).

Acknowledgments: The Deutsche Forschungsgemeinschaft, the Fonds der Chemischen Industrie, and Prof. Dr. K. Jurkschat are gratefully acknowledged for support of this work.

References

[1] C. A. Paz de Araujo, J. D. Cuchiaro, L. D. McMillan, M. C. Scott, J. F. Scott, *Nature* **1995**, *374*, 627.

[2] B. H. Park, B. S. Kang, S. D. Bu, T. W. Noh, J. Lee, W. Jo, *Nature* **1999**, *401*, 682.

[3] K. H. Whitmire, *Chemtracts-Inorg. Chem.* **1995**, *7*, 167.

[4] *Organobismuth Chemistry* (Eds.: H. Suzuki, Y. Matano), Elsevier, Amsterdam, **2001**.

[5] P. A. Williams, A. C. Jones, M. J. Crosbie, P. J. Wright, J. F. Bickley, A. Steiner, H. O. Davies, T. J. Leedham, G. W. Crichtlow, *Chem. Vapor Deposition* **2001**, *7*, 205.

[6] S. Parola, R. Papiernik, L. G. Hubert-Pfalzgraf, C. Bois, *J. Chem. Soc., Dalton Trans.* **1998**, 737.

[7] C. Limberg, M. Hunger, W. Habicht, E. Kaifer, *Inorg. Chem.* **2002**, *41*, 3359.

[8] J. H. Thurston, K. H. Whitmire, *Inorg. Chem.* **2002**, *41*, 4194.

[9] H. Schmidbaur, M. Bergfeld, *Z. Anorg. Allg. Chem.* **1968**, 85.

[10] K. W. Terry, K. Su, T. D. Tilley, A. L. Rheingold, *Polyhedron* **1998**, *17*, 891.

[11] M.-C. Massiani, R. Papiernik, L. G. Hubert-Pfalzgraf, J.-C. Daran, *Polyhedron* **1991**, *10*, 437.

[12] D. Mansfeld, M. Mehring, M. Schürmann, *Z. Anorg. Allg.* **2004**, *630*, 1795.

[13] D. Mansfeld, M. Mehring, M. Schürmann, *Angew. Chem.* **2005**, *117*, 250; *Angew. Chem. Int. Ed.* **2005**, *44*, 245.

[14] M. Mehring, C. Nolde, M. Schürmann, *Appl. Organometal. Chem.* **2004**, *18*, 487.

[15] M. Veith, E.-C. Yu, V. Huch, *Chem. Eur. J.* **1995**, *1*, 26.

[16] J. L. Jolas, S. Hoppe, K. H. Whitmire, *Inorg. Chem.* **1997**, *36*, 3335.

[17] M. Baier, P. Bissinger, J. Blümel, H. Schmidbaur, *Chem. Ber.* **1993**, *126*, 947.

[18] W. S. Tatlock, E. G. Rochow, *J. Org. Chem.* **1952**, *17*, 1555.

[19] J. F. Hyde, O. K. Johannson, W. H. Daudt, R. F. Fleming, H. B. Laudenslager, M. P. Roche, *J. Am. Chem. Soc.* **1953**, *75*, 5615.

[20] O. V. Kononov, V. D. Lobkov, V. A. Igonin, S. V. Lindeman, V. E. Shklover, Yu. T. Struchkov, *Organomet. Chem. USSR* **1991**, *4*, 784.

[21] I. L. Dubchak, V. E. Shklover, Yu. T. Struchkov, V. M. Kopylov, P. L. Prikhod'ko, *Zh. Strukt. Khim* **1983**, *24*, 59.

[22] I. L. Dubchak, V. E. Shklover, M. Yu. Antipin, Yu. T. Struchkov, V. M. Kopylov, A. M. Muzafarov, P. L. Prikhod'ko, A. A. Zhdanov, *Zh. Strukt. Khim* **1982**, *23*, 63.

[23] M. Mehring, C. Nolde, M. Schürmann, *Appl. Organometal. Chem.* **2004**, *18*, 489.

[24] A. Mommertz, K. Dehnicke, J. Magull, *Z. Naturforsch. B* **1996**, 1584.

The Effect of Silyl Anion Substituents on the Stability and NMR Characteristics of Cyclic Polyphosphines – an *ab initio*/NMR Study

Alk Dransfeld,* Karl Hassler

Institut für Anorganische Chemie der Technischen Universität Graz
Stremayrgasse 16, A-8010 Graz, Austria
Tel.: +43 316 873 8704 — Fax: +43 316 873 8701
E-mail: dransfeld@anorg.tu-graz.ac.at

Keywords: polyphosphines, *ab initio*, NMR, ring strain

Summary: *Ab initio* energy-optimized geometries of monocyclic silicon comprising polyphosphines show that the exocyclic silyl anion group mainly activates four-membered polyphosphine rings. For all exo- and endocyclic $H_nP_{(n-1)}Si$ anion isomers, negative ring strain (stabilization) is obtained for ring sizes of 5 to 7. The ^{29}Si NMR chemical shift in the exodocyclic isomers is predicted to occur at –22 to –104 ppm, with upfield trend related to increasing ring size.

Introduction

The transformation of silyl substituents to a silyl anion group is a standard procedure. While the effect of F, Cl, and Br substituents on phosphirane ring opening [1, 2] was reported earlier, the effect of anionic groups on the ring isomerization of cyclopolyphosphines has not been reported. The presented *ab initio* computed geometric and magnetic parameters of some models show trends which are helpful in modifying and monitoring these compounds.

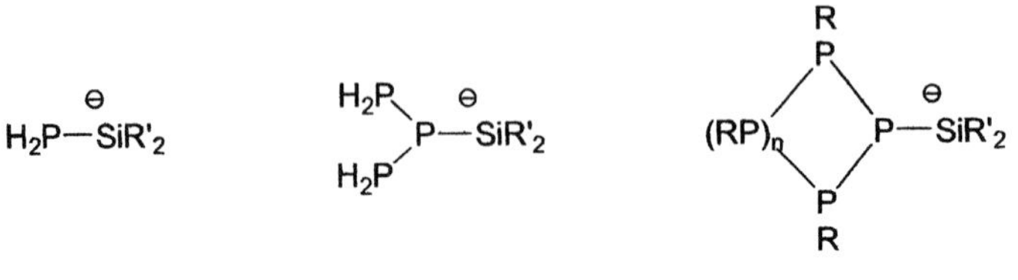

$H_2P{-}SiR'_2{}^{\ominus}$ (**1**) $(H_2P)_2P{-}SiR'_2{}^{\ominus}$ (**2**) $(RP)_n(PR)_2P{-}SiR'_2{}^{\ominus}$ (**3** – **6** (n = 0 – 3))

Fig. 1. Acyclic molecules **1** and **2** and cyclic molecules **3** – **6**.

Cyclic molecules comprising tricoordinate phosphorus, σ^3-P, have been studied previously with the aim of understanding polymorphism of polyphosphines, $H_{(n-x)}P_n$. These studies for 3- to 6-membered $(PH)_n$ rings indicate a preference for n = 3 over 4 and for n ≥ 5. The SiR_2 building block has been studied in the context of the technically interesting tailoring of polysilanes [3]. From the facility to turn a silyl-substituted polyphosphine into a molecule with an exocylclic, anionic SiR_2 group and its observed isomerization arises the question whether the ring size preferences in neutral cyclopolyphosphines is transferable to the corresponding anions and their heteroanalogs.

Methods

The presented conclusions are based on results for B3LYP/6-31+G* optimized geometries and their relative energies and ring strain energies for isomers of $(PH)_n$, $(PH)_{(n-1)}(P^-)$, $(PH)_{(n-1)}(SiH^-)$, $(PH)_{(n-2)}(P^-)(SiH_2)$, and $(PH)_{(n-2)}(P\text{–}SiH_2^-)$. Ring strain energies, E_{RiS}, were determined with Eq. 1, which is a generalization of equations used earlier in this context [4].

$$n\text{–}1\ H_2P\text{–}PH\text{–}PH_2 \ + \ H_2P\text{–}X\text{–}PH_2 \longrightarrow c\text{-}(PH)_{(n-1)}(X) \ + \ n\ H_2P\text{–}PH_2$$

Eq. 1. Homodesotic equation for evaluation of ring strain in A_nX rings.

Results

Ring Strain in $(PH)_n$, $(PH)_{(n-1)}(P^-)$, and $(PH)_{(n-2)}(P\text{–}SiH_2^-)$

Previous empirical rules and computations [4, 5] showed that in the $(PH)_n$ homologs the 3- and

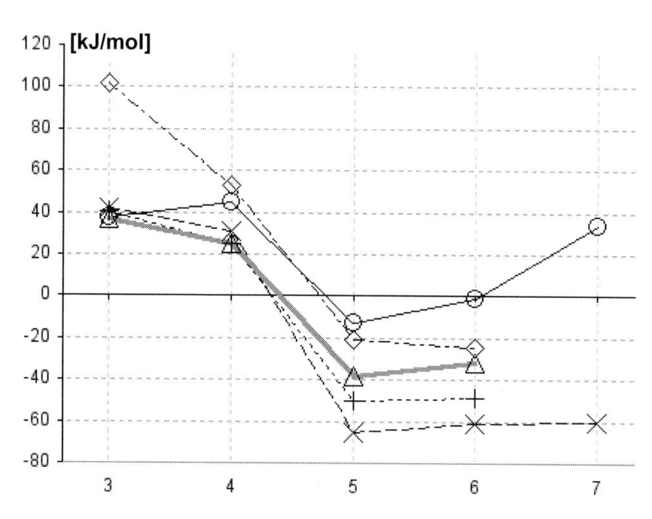

Fig. 2. Strain energies in kJ/mol for cyclic molecules with ring size *r* ranging from 3 to 7 at B3LYP/6-31+G* (○ for $(PH)_n$ with r = n, × for $(PH)_{(n-1)}(P^-)$, + for $(PH)_{(n-1)}(SiH^-)$, ◇ for $(PH)_{(n-2)}(P^-)(SiH_2)$ with r = n–1, and △ for $(PH)_{(n-2)}$ $(P\text{–}SiH_2^-)$ with r = n–2).

5-membered rings are preferred. As is to be expected from electrostatic considerations, the ring strain, E_{RiS}, of the deprotonated cyclophosphines decreases with the ring size, r, until $r = 5$ (Fig. 2). In contrast to the neutral $(PH)_n$, the strain in the $(PH)_{(n-1)}(SiH^-)$ does not have a maximum at $r = 4$.

With $r \geq 5$, ring strain does remain strongly stabilizing for the anions but increases in the $(PH)_n$ rings back to about +34 kJ/mol ($E_{RiS}(r = 3) \approx E_{RiS}(r = 7)$, Fig. 2). Astonishingly, E_{RiS} in c-$(PH)_2(P^-)$ is only slightly smaller than that for the neutral c-$(PH)_3$. The exocyclic, anionic SiH_2 group affects the ring strain in a similar way to deprotonation. With n = 2 and 3 the strain of c-$(PH)_n(P^-)$ is almost equal to that of c-$(PH)_n(P–SiH_2^-)$. Comparison of the five- and six-membered rings shows that stability (negative ring strain) in $(PH)_{(n-2)}(P–SiH_2^-)$ lies above E_{RiS}(c-$(PH)_n$) and below E_{RiS}(c-$(PH)_{(n-1)}(P^-)$).

Ring Strain in $H_nP_{(n-1)}Si^-$ isomers

The anionic $P_{(n-1)}H_nSi^-$ molecules are isoelectronic to the neutral $(PH)_n$ with the same n. Within the tautomers of $(PH)_2(SiH^-)$ the structure with an intact SiH_2 group and a phosphanide moiety has an outstandingly high ring strain without extraordinary deformation of the bond angles. The difference between the ring strains of the $–PH–SiH^-–$ compared to the $–P^-–SiH_2–$ comprising compounds is probably due to the choice of references (X = $–SiH^-–$ and X = $–P^-–SiH_2–$ in Eq. 1) in determining the ring strain. Nevertheless, it is noteworthy that stabilization remains, even with increasing ring size from 5 to 6 and 7 for all anions in this set.

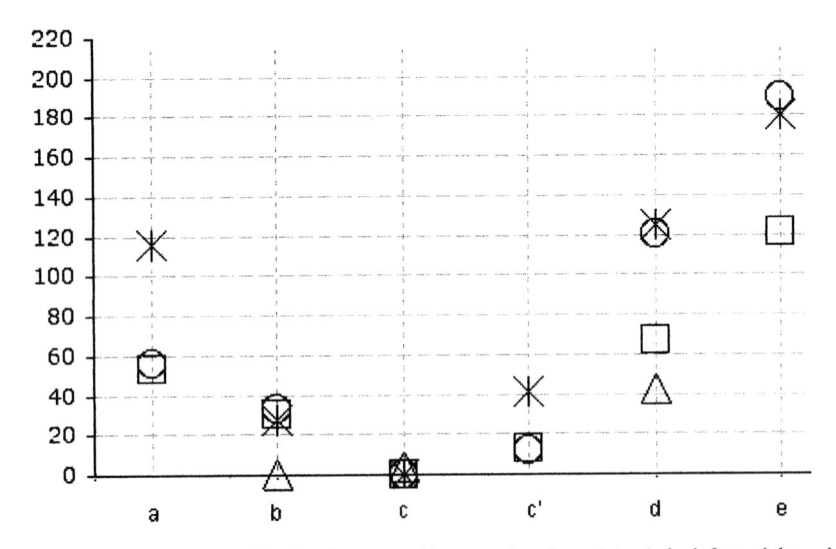

Fig. 3. Relative energies of $H_nP_{(n-1)}Si$ anion isomers with n ranging from 3 to 6; in left to right order, **3a – 5a**: c-$(PH)_{(n-1)}(P–SiH_2^-)$, **3b – 5b,7**: c-$(PH)_n(SiH^-)$, **3c – 5c,8**: c-$(PH)_{(n-1)}(P^-)(SiH_2)$, **3c′–5c′**: c-$(PH)_{(n-2)}(P^-)(PH)$-(SiH_2), **3d–5d,9**: n-$H_2Si(^-)–P=P–(PH)_{(n-3)}–PH_2$, **3e–5e**: n-$H_2Si=P–P–(PH)_{(n-3)}–(PH^-)$; **7** = c-$(PH)_2(SiH^-)$, **8** = c-$(PH)(P^-)(SiH_2)$, and **9** = $(H_2Si^-)–P=PH$.

Relative Energies of Anion Isomers

Figure 3 shows that in all sets of $H_nP_{(n-1)}Si^-$ isomers the structures with endocyclic –SiH⁻– or $-SiH_2-$ groups are preferred over those with an exocyclic $-SiH_2^-$. Furthermore, direct proximity of the phosphanide and the silylene groups are preferred by up to 42 kJ/mol. Although acyclic isomers should allow a better distribution of the negative charge, these isomers generally have high relative energies (between 42 and 189 kJ/mol). Within the acyclic isomers, the isomers with a silyl anion moiety and a –P=P– double bond are preferred over those with $H_2Si=P-$ and a terminal phosphanide group.

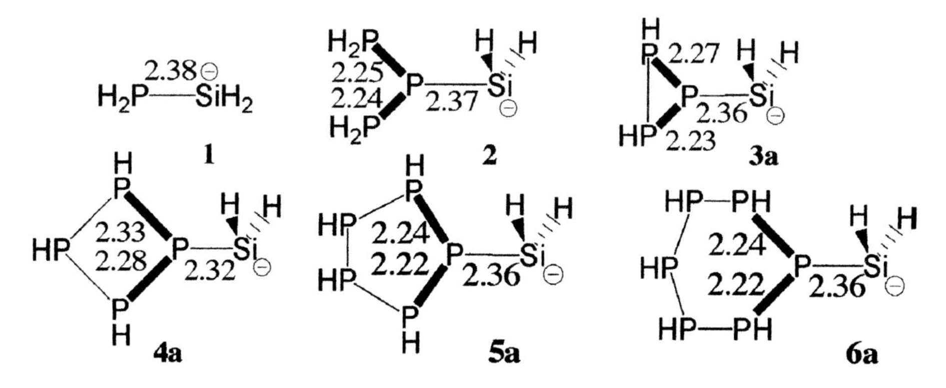

Fig. 4. Geometries (B3LYP/6-31+G*) of acyclic and cyclic R_2P-SiH_2 anions.

The exocyclic silyl anion group means not only a less stable structure but also an activation of the P_n ring opening. Compared to the *trans* P–P bond length in c-$(PH)_5$ (2.23 Å), some endocyclic bonds of **3a** – **6a** are widened. While negative hyperconjugation classically elongates the *cis*-coplanar bond, the effect in **3a** – **6a** is widening of the P–P *trans*-coplanar to the Lp(Si). The effect is most pronounced in **3a** and its strength correlates with the P–Si bond length. The Si–P bond in acyclic isomers with β-phosphanido group is similarly long (2.368 Å (**3d′**), 2.360 (**4d′**), and 2.362 (**5d′**) in H_2Si—PH⋯P⁻⋯PR) as in adducts of silylene with triphosphanide-allyl [6] moieties.

²⁹Si NMR Chemical Shifts of Exocyclic $H_nP_{(n-1)}Si$ Anion Isomers

In the homologous set of molecules **3a** – **6a**, the ^{29}Si NMR chemical shifts change toward higher field with increasing ring size (–22 (**3a**), –64 (**4a**), –93 (**5a**), –104 (**6a**)). The α-P NMR resonance in the cyclophosphanides ($(PH)_{(n-1)}(P^-)$ with n = 2 – 5) and the ($(PH)_{(n-1)}(P^-)$ with same range of n, **3a** – **6a**, show a similar change of chemical shift with increased ring size (c-$P_nH_{(n-1)}^-$: –190 ppm, 164, 7, and –193; **3a** – **6a**: –170, 36, 59, and –211 ppm).

Acknowledgment: A. D. and K. H. thank the FWF for financial support (Project S7 905 and P11 878).

References

[1] J. Matrai, A. Dransfeld, M. T. Nguyen, in *Quantum Chemistry in Belgium 4*, **1999**.

[2] M. T. Nguyen, A. Dransfeld, L. Landuyt, L. G. Vanquickenborne, P. v. R. Schleyer, *Europ. J. Inorg. Chem.* **2000**, 103.

[3] R. Fischer, D. Frank, W. Gaderbauer, C. Kayser, C. Mechtler, J. Baumgartner, C. Marschner, *Organometallics* **2003**, *22*, 3723.

[4] W. W. Schoeller, *Theochem (J. Mol. Struct.)* **1993**, *284*, 61.

[5] S. Böcker, M. Häser, *Z. Anorg. Allg. Chem.* **1995**, *621*, 258.

[6] N. Wiberg, A. Wörner, H.-W. Lerner, K. Karaghiosoff, D. Fenske, G. Baum, A. Dransfeld, P. v. R. Schleyer, *Europ. J. Inorg. Chem.* **1998**, 833.

Silanols as Precursors to Cyclo- and Polysiloxanes

M. Veith,[†] A. Rammo, F. O. Schütt, P. P. Spaniol, V. Huch*

Institut für Anorganische Chemie der Universität des Saarlandes
Im Stadtwald, D-66041 Saarbrücken, Germany

[†] Institut für Neue Materialien, Im Stadtwald, D-66041 Saarbrücken, Germany
Tel.: +49 681 3023415 — Fax: +49 681 3023995
E-mail: veith@mx.uni-saarland.de

Keywords: functionalized siloxanes, cyclosiloxanes, silanols, silylamines, aminosiloxanes, metalamidosiloxanes

Summary: Silanols of the general formula $X_2Si(OH)_2$ or $XSi(OH)_3$, with X = organic or inorganic group, are versatile starting molecules for the synthesis of cyclo-, poly- and alumosiloxanes. Most of these silanols (especially those with small organic ligands X) tend to condense to cyclo- and polysiloxanes. This is not the case for $(Me_3Si)_2NSi(OH)_3$ which condenses only to $(Me_3Si)_2NSi(OH)_2OSi(OH)_2N(SiMe_3)_2$ or for ${}^tBuSi(OSiMe_2$ $-OH)_3$, which can be crystallized. In ${}^tBuSi(OSiMe_2OH)_3$ the hydroxo groups may be formally exchanged by amino groups, and again the primary amine ${}^tBuSi(OSiMe_2NH_2)_3$ is stable up to –25 °C with respect to auto-condensation. The cage compound ${}^tBuSi(OSiMe_2O)_3Si{}^tBu$ is obtained from ${}^tBuSi(OH)_3$ and ${}^tBuSi(OSiMe_2Cl)_3$. Further examples of aminosiloxanes with the general formula ${}^tBuSi(OSiMe_2N(R)H)_3$ as well as their transformation into metal amides are shortly reviewed.

Introduction

Whereas stable silanetriols of the general formula $RSi(OH)_3$ with bulky organic or amino ligands at the silicon atoms are well established (for example: R = Ph [1], ${}^tBu$ [2], $Si(Me_3Si)_3$, $C(Me_3Si)_3$ [3], Cp* [4], $C_6H_2({}^tBu)_3$ [5], $N(SiMe_3)(Ph)$ [6]), stable low molecular branched species of the type $RSi(OR'_2SiOH)_3$ (which may be derived from the triols by formal triple insertion of $-R'_2SiO-$ units into the Si–OH bonds) have so far been only rarely addressed [7]. On the same line it has been found that 1,3-siloxanediols $HOSiR_2OSiR_2OH$ (obtained by formal insertion of $-OSiR_2-$ units in silanediols $R_2Si(OH)_2$) are well established even with small organic ligands R [8 – 12], while 1,1,3,3-siloxanetetraols $(HO)_2Si(R)O(R)Si(OH)_2$ (obtained by formal insertion of a $-OSi(R)(OH)-$ unit in a silanetriol) are hardly known [13]. We report here about the syntheses and structural characterizations of two new compounds, $[(Me_3Si)_2N]Si(OH)_2OSi(OH)_2[N(SiMe_3)_2]$ and ${}^tBuSi(OSiMe_2OH)_3$, and their transformation to other derivatives. In a second chapter we review shortly the use of the $[{}^tBuSi(OSiMe_2N(R))_3]^{3-}$ anion as a chelating ligand toward different metallic elements.

New Branched Silanols of the Type R′Si(OH)₂OSi(OH)₂SiR′ and R′Si(OSiR₂OH)₃

When bis(trimethylsilyl)aminosilicontrichloride, $(Me_3Si)_2NSiCl_3$ [14], is allowed to react with a stoichiometric amount of water in dioxane in the presence of the hydrogen chloride scavenger triethylamine, the trisilanol $(Me_3Si)_2NSi(OH)_3$ is not formed, but instead the condensation product $(Me_3Si)_2NSi(OH)_2OSi(OH)_2N(SiMe_3)_2$ is isolated in 55% yield (Eq. 1).

$$(Me_3Si)_2NSiCl_3 + 2.5\,H_2O + 3\,NEt_3 \longrightarrow 0.5\,(Me_3Si)_2NSi(OH)_2OSi(OH)_2N(SiMe_3)_2 + 3\,(HNEt_3)Cl$$

Eq. 1.

The reaction presumably has $(Me_3Si)_2NSi(OH)_3$ as an intermediate, which we were neither able to isolate nor to identify spectroscopically. The molecule $(Me_3Si)_2NSi(OH)_2OSi(OH)_2N(SiMe_3)_2$ can be crystallized as its THF/dioxane-adduct and is stable under ordinary conditions when water and oxygen are excluded. Both bases THF and dioxane are coordinated through hydrogen bridges to the OH-groups of the molecule forming a three dimensional network in the crystal. In Fig. 1 as part

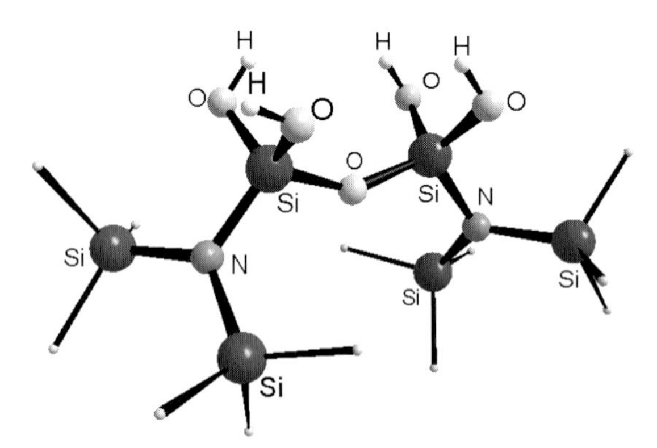

Fig. 1. Ball and stick representation of $(Me_3Si)_2NSi(OH)_2OSi(OH)_2N(SiMe_3)_2 \cdot THF \cdot 0.5$ dioxane derived from an X-ray single crystal structure analysis. The methyl substituents on the silicon atoms are drawn as sticks. The coordinating bases have been omitted for clarity. The Si–O distances to oxygen atoms of the hydroxides ($Si\text{–}O_{mean}$ = 1.617(6) Å) are somewhat larger than those in the Si–O–Si bridge ($Si\text{–}O_{mean}$ = 1.610(2) Å); the Si–N distances are normal ($Si\text{–}N_{mean}$ = 1.728(9) Å) [17].

of the structure a drawing of the silanol is shown omitting the coordinating bases for clarity.

The tetrahydroxydisiloxane $(Me_3Si)_2NSi(OH)_2OSi(OH)_2N(SiMe_3)_2$ seems to be the first of such compounds without bulky organic groups at the nitrogen or at the silicon atoms, as may be seen from comparison with other siloxanes of the type $RSi(OH)_2OSi(OH)_2R$ structurally characterized so far [13, 15, 16].

The branched siloxane $^tBuSi(OSiMe_2OH)_3$ is obtained from $^tBuSi(OH)_3$ in a two-step reaction, as may be derived from Eq. 2.

$$^tBuSi(OH)_3 + 3\,Me_2SiCl_2 + 3\,Et_3N \longrightarrow {}^tBuSi(OSiMe_2Cl)_3 + 3\,Et_3NHCl$$
$$^tBuSi(OSiMe_2Cl)_3 + 3\,H_2O + 3\,Et_3N \longrightarrow {}^tBuSi(OSiMe_2OH)_3 + 3\,Et_3NHCl$$

Eq. 2.

The intermediate $^tBuSi(OSiMe_2Cl)_3$ can be isolated in over 70% yield and has been completely characterized by spectroscopic and analytical means [18, 19]. Its subsequent reaction with water is run in diethylether as a solvent, and the branched siloxanetriol $^tBuSi(OSiMe_2OH)_3$ is obtained in 90% yield by crystallization. From toluene, single crystals can be obtained which are stable at ordinary conditions in nitrogen atmosphere and which can easily be stored for longer periods. As a result of the X-ray structure analysis, a graphic representation of $^tBuSi(OSiMe_2OH)_3$ is shown in Fig. 2.

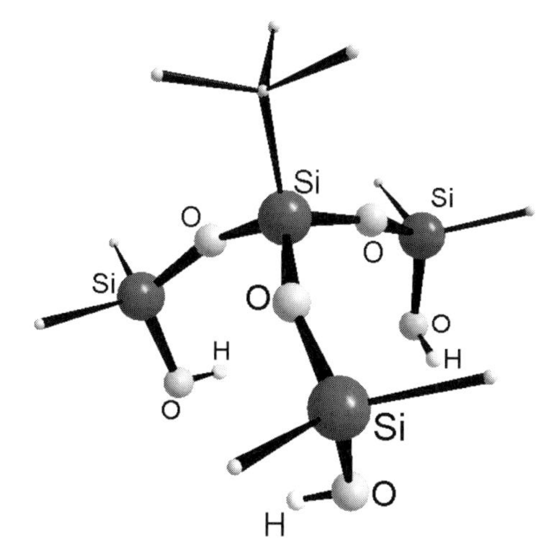

Fig. 2. Ball and stick representation of $^tBuSi(OSiMe_2OH)_3$·toluene as result of an X-ray single crystal structure analysis. The methyl substituents on the carbon and silicon atoms are drawn as sticks. The Si–O distances in the Si–OH parts of the molecule ($Si\text{–}O_{mean}$ = 1.622(6) Å) are distinctly larger than those in the Si–O–Si bridges ($Si\text{–}O_{mean}$ = 1.570(5) Å). In the crystal the molecules form dimers through hydrogen bridging with a crystallographic inversion center in the middle of the aggregate. The toluene molecule serves as space filler in the crystal lattice [18].

The monomeric molecule ^{t}BuSi(OSiMe$_2$OH)$_3$ has almost C_3 point symmetry in the crystal. The monomers are connected in pairs around an inversion center of the lattice through six hydrogen bridges (O···O range: 2.675(9) to 2.693(9) Å), the whole dimeric entity approaching S_6 point symmetry. All oxygen atoms of the hydroxyl groups are therefore involved in the hydrogen bonds as acceptors and donors at the same time and are threefold coordinated (each oxygen atom has one silicon and two hydrogen atoms as neighbors).

When ^{t}BuSi(OSiMe$_2$OH)$_3$ is heated, poly-condensation is immediately started, forming many different products. In order to synthesize low-molecular condensates, we have reacted ^{t}BuSi(OSiMe$_2$Cl)$_3$ with one equivalent of the triol ^{t}BuSi(OH)$_3$ using triethylamine as HCl captor (compare also Eq. 2): the cage like ^{t}BuSi(OSiMe$_2$O)$_3$Si^{t}Bu is formed in 60% yield. In Fig. 3 the result of an X-ray structure analysis is depicted. Bicyclic molecules with the same skeleton of atoms but with different substituents are known [20, 21].

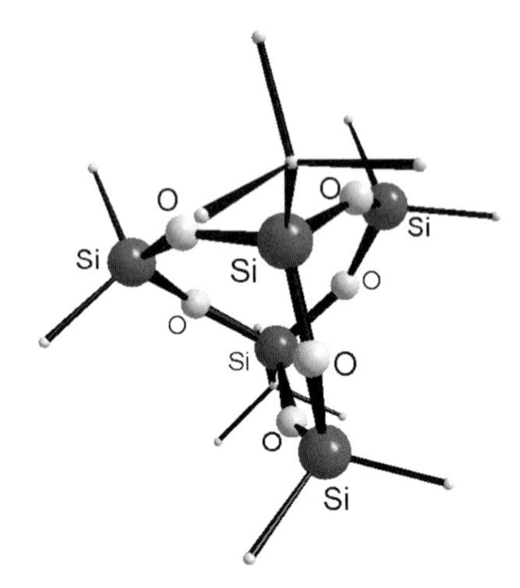

Fig. 3. Ball and stick representation of ^{t}BuSi(OSiMe$_2$O)$_3$Si^{t}Bu as the result of an X-ray single crystal structure analysis. The methyl substituents on the carbon and silicon atoms are drawn as sticks. The molecule has crystallographic C_3 point symmetry. The Si–O distances starting from ^{t}Bu–Si are slightly shorter (Si–O$_{mean}$ = 1.612(2) Å) than those of the dimethylsilyl groups (Si–O$_{mean}$ = 1.630(3) Å) [18].

New Branched Aminosiloxanes of the Type R'Si(OSiR$_2$N(R)H)$_3$

Using ammonia with the chloro-derivative ^{t}BuSi(OSiMe$_2$Cl)$_3$ (see Eq. 2) in the place of water gives the amino-derivative ^{t}BuSi(OSiMe$_2$NH$_2$)$_3$, which can be stored under exclusion of air for longer periods at –25 °C without decomposition (Eq. 3) [19].

$$^t\text{BuSi(OSiMe}_2\text{Cl)}_3 + 6\,\text{H}_3\text{N} \longrightarrow {}^t\text{BuSi(OSiMe}_2\text{NH}_2\text{)}_3 + 3\,\text{NH}_4\text{Cl}$$

Eq. 3.

The ^{15}N NMR spectrum of $^tBuSi(OSiMe_2NH_2)_3$ displays one signal at –359.1 ppm and in the infrared spectrum two bands at 3400 and 3476 cm^{-1} can be found, consistent with NH_2-groups. Unfortunately we did not succeed in growing crystals for a single X-ray diffraction analysis [19].

The siloxaneamine $^tBuSi(OSiMe_2NH_2)_3$ is considerably stabilized with respect to auto-condensation when one of the two hydrogen atoms of the amino-groups is replaced by organic or silyl substituents. The siloxazanes $^tBuSi(OSiMe_2N(R)H)_3$ with R = CH_3, $C(CH_3)_3$, $Si(CH_3)_3$, and C_6H_5 have thus been obtained using a route similar to Eq. 3 or by reaction of $^tBuSi(OSiMe_2Cl)_3$ with the corresponding lithium amides RN(H)Li under elimination of lithium chloride (see [19]). In these amines the amino-hydrogen atoms may be easily withdrawn by strong organic bases like n-$C_4H_9^-$ forming n-butane on one side and the tripodal anion $[^tBuSi(OSiMe_2N(R))_3]^{3-}$ on the other. This anion may bind either to monovalent metallic elements like Li, Na or Cu [19] or to a divalent germanium atom [22]. In the first case molecules of the general formula $[^tBuSi(OSiMe_2N(R))_3]M_3$ are obtained and in the second case the $[^tBuSi(OSiMe_2N(Ph))_3]Ge^-$ anion is formed, which in the crystal and in solution forms an ion pair with $[Li(THF)_3]^+$ with a direct Li–Ge link [22]. Both compounds have polycyclic structures; the lithium compound $[^tBuSi(OSiMe_2N(^tBu))_3]Li_3$ is sketched in Fig. 4.

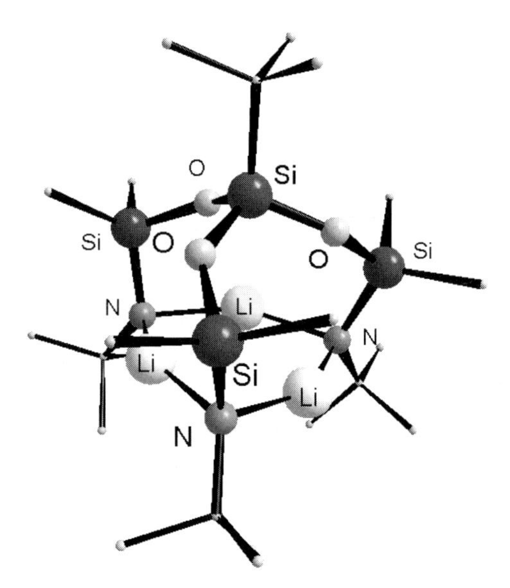

Fig. 4. Ball and stick representation of $^tBuSi(OSiMe_2(^tBu)N)_3Li_3$ as the result of an X-ray single crystal structure analysis. The methyl substituents on the carbon and silicon atoms are drawn as sticks. The Si–O distances can be divided into short ones (silicon at the top, Si–O_{mean} = 1.597(6) Å) and larger ones (silicon atoms with methyl ligands; Si–O_{mean} = 1.655(5) Å). The larger Si–O bond lengths are compensated by short Si–N bonds (1.691(3) Å). The cage-like polycycle has an Li_3N_3 basal six-membered ring to which three similar eight-membered $Si_3O_2N_2Li$ rings are fused [19].

Conclusions

The stabilities and facile syntheses of silanols and aminosiloxanes presented in this review are either due to bulky substituents at silicon (like *tert*-butyl), special electronic properties and bulk (like hexamethyldisilazyl), or branched Si–O structures. The new compounds may be used in condensation reactions similar to other silanols such as $Ph_2Si(OH)_2$ or $(HO)Ph_2SiOSiPh_2(OH)$, which we have widely used in the synthesis of molecular siloxanes incorporating aluminum and gallium (molecular alumosilicates) [23 – 26]. They also may be transformed to stannates $R_2Si(OSnR'_3)_2$, which are versatile starting compounds to cyclic siloxanes [26 – 29].

Acknowledgments: We thank the Fonds der Chemischen Industrie as well as the Deutsche Forschungsgemeinschaft (Schwerpunktprogramm: Spezifische Phänomene in der Silicium-Chemie and SFB 277) for financial support.

References

[1] D. S. Korkin, M. I. Buzin, E. V. Matukhina, L. N. Zherlitsyna, N. Auner, O. I. Shchegolikhina, *J. Organomet. Chem.* **2003**, *686*, 313.

[2] N. Winkhofer, H. W. Roesky, M. Noltemeyer, W. T. Robinson, *Angew. Chem.* **1992**, *104*, 670; *Angew. Chem. Int. Ed. Engl.* **1992**, *31*, 599.

[3] S. S. Al-Juaid, N. H. Buttrus, R. I. Damja, Y. Derouich, C. Eaborn, P. B. Hitchcock, P. D. Lickiss, *J. Organomet. Chem.* **1989**, *371*, 287.

[4] P. Jutzi, G. Strassburger, M. Schneider, H. Stammler, B. Neumann, *Organomet.* **1996**, *15*, 2842.

[5] N. Winkhofer, A. Voigt, H. Dorn, H. W. Roesky, A. Steiner, D. Stalke, A. Reller, *Angew. Chem.* **1994**, *106*, 1414; *Angew. Chem. Int. Ed. Engl.* **1994**, *33*, 1352.

[6] R. Murugavel, V. Chandrasekhar, A. Voigt, H. W. Roesky, H. G. Schmidt, M. Noltem, *Organometallics* **1995**, *14*, 5298.

[7] H. Uchida, Y. Kabe, K. Yoshino, A. Kawamata, T. Tsumuraya, S. Masamune, *J. Am. Chem. Soc.* **1990**, *112*, 7077.

[8] W. Clegg, *Acta Crystallogr. Sect. C* **1983**, *39*, 901.

[9] V. E. Shklover, Yu. T. Struchkov, I. V. Karpova, V. A. Odinets, A. A. Zhdanov, *Zh. Strukt. Khim.* **1985**, *26*, 125.

[10] M. A. Hossain, M. B. Hursthouse, *J. Cryst. Spectr. Res.* **1988**, *18*, 227.

[11] A. P. Polishuk, T. V. Timofeeva, M. Yu. Antipin, N. N. Makarova, N. A. Golovina, Yu. T. Struchkov, O. D. Lavrentovich, *Metalloorg. Khim. (Organometallic Chem. in USSR)* **1991**, *4*, 147.

[12] P. D. Lickiss, A. D. Redhouse, R. J. Thompson, W. A. Stanczyk, K. Rozga, *J. Organomet. Chem.* **1993**, *463*, 41.

[13] P. D. Lickiss, S. A. Lister, A. D. Redhouse, C. J. Wisener, *J. Chem. Soc.,Chem. Comm.* **1991**, 173.
[14] U. Wanngat, K. Behmel, H. Bürger, *Chem. Ber.* **1964**, *97*, 2029.
[15] C. E. F. Rickard, W. R. Roper, D. M. Salter, L. J. Wright, *J. Am. Chem. Soc.* **1992**, *114*, 9682.
[16] R. Murugavel, P. Böttcher, A. Voigt, M. G. Walawalkar, H. W. Roesky, E. Parisini, M. Teichert, M. Noltemeyer, *J. Chem. Soc., Chem. Comm.* **1996**, 2417.
[17] M. Veith, P. P. Spaniol, V. Huch, in preparation.
[18] M. Veith, F. O. Schütt, V. Huch, in preparation.
[19] M. Veith, O. Schütt, V. Huch, *Z. Anorg. Allg. Chemie* **1999**, *625*, 1155.
[20] G. Menczel, J. Kiss, *Acta Cryst.* **1975**, *B31*, 1214.
[21] V. E. Shklover, N. G. Bokii, Y. T. Struchkov, K. A. Andrianov, M. N. Ermakova, N. A. Dmitricheva, *Dokl. Akad. Nauk SSSR* **1975**, *220(6)*, 1321.
[22] M. Veith, O. Schütt, V. Huch, *Angew. Chem.* **2000**, *112*, 608; *Angew. Chem. Int. Ed.* **2000**, *39*, 601.
[23] M. Veith, M. Jarczyk, V. Huch, *Angew. Chem.* **1997**, *109*, 140; *Angew. Chem. Int. Ed. Engl.* **1997**, *36*, 117.
[24] M. Veith, M. Jarczyk, V. Huch, *Angew. Chem.* **1998**, *110*, 109; *Angew. Chem. Int. Ed.* **1998**, *37*, 105.
[24] M. Veith, H. Vogelgesang, V. Huch, *Organometallics* **2002**, *21*, 380.
[25] M. Veith, H. Hreleva, J. Biegler, V. Huch, A. Rammo, *Phosphorus, Sulfur, and Silicon*, in press.
[26] M. Veith, A. Rammo, in *Silicon Chemistry* (Eds.: P. Jutzi, U. Schubert) Wiley-VCH, Weinheim **2003**.
[27] M. Veith, A. Rammo, M. Gießelmann, *Z. Anorg. Allg. Chem.* **1998**, *624*, 419.
[28] M. Veith, A. Rammo, *Phosphorus, Sulfur, and Silicon* **1997**, *123*, 75.
[29] M. Veith, A. Rammo, J. Huppert, J. David, *C. R. Chimie* **2003**, 117.

The Origin of Ring Strain and Conformational Flexibility in Tri- and Tetrasiloxane Rings and Their Heavier Group 14 Congeners

Jens Beckmann

Institut für Anorganische und Analytische Chemie
Freie Universität Berlin, Fabeckstr. 34-36, 14195 Berlin, Germany
Fax: +49 30 83852440
E-mail: beckmann@chemie.fu-berlin.de

Dainis Dakternieks, Allan E. K. Lim,* Kieran F. Lim

Centre for Chiral and Molecular Technologies
Deakin University, Geelong 3217, Australia

Klaus Jurkschat*

Lehrstuhl für Anorganische Chemie II, Universität Dortmund
D-44221 Dortmund, Germany
Fax: +49 231 755 5048
E-mail: kjur@platon.chemie.uni-dortmund.de

Keywords: siloxane rings, group 14 congeners, DFT calculations, ring strain, ring puckering

Summary: A simple theoretical model was developed to examine the influence of the Group 14 metals on the ring strain and ring flexibility of various cyclic Group 14 organoelement oxides. This model exclusively uses bending potential energy functions, which were derived from relaxed potential energy surface (PES) scans of simple model compounds, namely $H_3MOM'H_3$ and $H_2M(OH)_2$ (M, M′ = Si, Ge, Sn), at the B3LYP/(v)TZ level of theory. Trends in ring properties were related to the angular strain components of the various M–O–M′ and O–M–O linkages present within these rings, and subsequently linked to the different electronegativities of Si, Ge, and Sn. The results obtained are compared with available experimental data, and their implications for the design of new ROP precursors are discussed.

Introduction

The kinetically controlled ring-opening polymerization (ROP) of strained trisiloxane rings, *cyclo*-$(R_2SiO)_3$ (R = alkyl, aryl), is the most powerful process for the preparation of high-molecular-weight silicone polymers. In contrast, the thermodynamically controlled ROP of *cyclo*-$(R_2SiO)_3$, or the ROP of (virtually) strain-free tetrasiloxane rings, *cyclo*-$(R_2SiO)_4$, provides equilibrium mixtures consisting of low-molecular-weight polymers and substantial amounts of unwanted cyclic oligomers. The ring strain energies of the *cyclo*-siloxanes have been estimated using calorimetric measurements to be 12.6 kJ mol^{-1} for *cyclo*-$(Me_2SiO)_3$ and ca. 1.0 kJ mol^{-1} for *cyclo*-$(Me_2SiO)_4$ [1]. This difference in strain energy is sufficient to cause the differences between the ROP behavior of six- and eight-membered *cyclo*-siloxanes mentioned above. However, the origin of this difference is not clearly understood, and even less information concerning the ring strain energies is available for the related *cyclo*-metallasiloxanes, *cyclo*-germoxanes and *cyclo*-stannoxanes [2]. We have previously observed that successive replacement of the silicon atoms in *cyclo*-tetrasiloxanes, *cyclo*-$(R_3SiO)_4$, with tin atoms leads to more puckered and less flexible rings, where flexibility was measured as the magnitude of the planarization energy of the rings [3]. Furthermore, the Si–O–Sn and Sn–O–Sn linkages favored narrower angles compared to the Si–O–Si linkages in the eight-membered rings [3].

M–O–M′ and O–M–O Bending Potential Energy Functions

All calculations were performed with the Gaussian 98 suite of programs [4] using the hybrid DFT/HF B3LYP method. The all-electron 6-311+G(2df) basis set was used for H, O, Si, and Ge, whereas the valence triple zeta pseudopotential basis set of Stoll et al. [5] was used for Sn. The Sn basis set was augmented with diffuse function exponents one-third in size of the outermost valence exponents, the two-membered d-polarization set of Huzinaga, [6] and f(ζ)Sn = 0.286. This basis set combination will be called "(v)TZ" throughout this paper.

The calculated energies from the relaxed PES scans of the H_3MOMH_3 (M = Si, Ge, Sn) model compounds are plotted vs the M–O–M angle in Fig. 1.

All three curves are anharmonic potential wells with energy minima corresponding to the equilibrium bond angles. A quadratic potential with a Lorentzian hump, $V_{M\text{-}O\text{-}M'}(\rho)$, was fitted to the calculated data (Eq. 1).

$$V_{M\text{-}O\text{-}M'}(\rho) = \frac{Hf(\rho^2-\rho_e^2)^2}{f\rho_e^4+(8H-f\rho_e^2)\rho^2}$$

Eq. 1. The M–O–M′ bending potential function. $\rho = 180° - \theta_{M\text{-}O\text{-}M'}$; $\rho_e = 180° - \theta_{M\text{-}O\text{-}M'}$(equilibrium); f is the harmonic force constant and H is the linearization barrier.

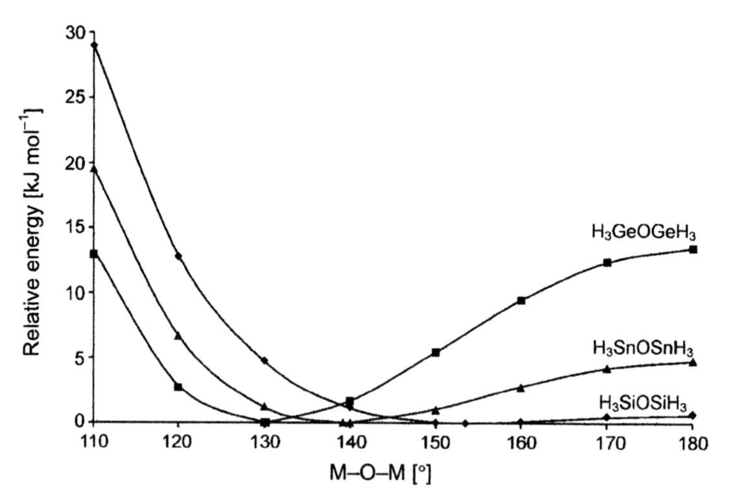

Fig. 1. Calculated M–O–M bending potential energy curves for H_3MOMH_3 (M = Si, Ge, Sn).

The equilibrium M–O–M angles decrease, and the linearization barriers increase as the electronegativity of M increases ($\chi_{Si} = 1.90 < \chi_{Sn} = 1.93 < \chi_{Ge} = 2.00$). This agrees with previous computational studies by Gillespie et al. [7, 8], demonstrating that trends in equilibrium angles at O and the E–O–E linearization barriers are intimately related to the E–O electronegativity difference. They attributed the deviations of these angles from the VSEPR tetrahedral angle to the ionic character of the E–O bonds [7, 8].

The calculated O–M–O bending PES curves (Fig. 2) are significantly more harmonic than the M–O–M bending PES curves. Furthermore, the calculated equilibrium O–M–O angles are close to the VSEPR tetrahedral angle (109.47°), and all lie within a narrow range of less than 4°.

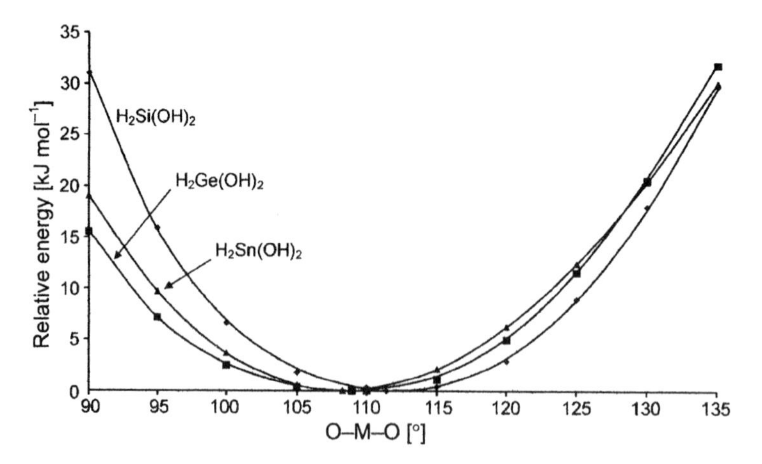

Fig. 2. Calculated O–M–O bending potential energy curves for $H_2M(OH)_2$ (M = Si, Ge, Sn).

A quintic polynomial, $V_{O\text{-}M\text{-}O}(\theta)$, was fitted to the calculated data (Eq. 2).

$$V_{O\text{-}M\text{-}O}(\theta) = \tfrac{1}{2}[k + k_3(\theta-\theta_0) + k_4(\theta-\theta_0)^2 + k_5(\theta-\theta_0)^3](\theta-\theta_0)^2$$

Eq. 2. The O–M–O bending potential function. $\theta = \theta_{O\text{-}M\text{-}O}$; $\theta_0 = \theta_{O\text{-}M\text{-}O}$(equilibrium); k is the harmonic force constant and k_3, k_4 and k_5 are anharmonic force constants.

A Simple Ring Strain Model

The ring strain energy, E_{RS}, is approximated as the total sum of endocyclic angular strain contributions (Eq. 3).

$$E_{RS} = \sum_{i=1}^{x} V_i(\theta_i)$$

Eq. 3. Ring strain energy, E_{RS}, for an x-membered ring with endocyclic angles θ_1 ... θ_x. $V_i(\theta_i)$ is the bending potential function corresponding to the triatomic linkage describing angle θ_i.

Preliminary studies have indicated that torsional contributions are negligible for the Group 14 oxides, and substituent effects may be neglected if the substituents are not bulky [9].

The constraint in Eq. 4 for an x-membered ring is a consequence of three-dimensional geometry. The "puckering factor" a has a maximum value of 1 for a planar ring and decreases with increasing ring puckering.

$$\sum_{i=1}^{x} \theta_i = a(x-2) \times 180°$$

Eq. 4. A geometrical constraint on the sum of endocyclic angles of an x-membered ring. For example, a six-membered ring will have a maximum angular sum of 720° when it is planar ($a = 1$), and ring puckering will result in a lower sum.

Equation 3 can be solved for minimum ring strain energy (E_{RSmin}) subject to the geometrical constraint associated with Eq. 4. The flexibility (i.e. ease of deformation) of a ring can thus be qualitatively assessed by examining the relationship between E_{RSmin} and a.

Strain Energies and Conformational Preferences of Six-Membered Rings

The relationships between E_{RSmin} and the puckering factor a for the homometallic six-membered rings, *cyclo*-$(H_2MO)_3$ (M = Si, Ge, Sn), are depicted in Fig. 3.

It is apparent from Fig. 3 that all three rings favor planar configurations, since this is where the total angular strain contributions are minimized. The curves associated with the heterometallic six-membered rings are intermediate between the three curves in Fig. 3 [9]. The endocyclic angles associated with the planar strain-free solutions fall in the following ranges: $\theta_{M\text{-}O\text{-}M'}$: 128° – 139°;

$\theta_{O\text{–}M\text{–}O}$: 107° – 111° [9].

Therefore, the ring strain values in the six-membered rings are attributable to compression of the M–O–M′ linkages to angles narrower than their equilibrium values. The minimum strain conformations are planar, as ring puckering would cause further angular compression.

The global minimum ring strain energies for the homometallic rings are 13.0 (*cyclo*-$(H_2SiO)_3$), 1.7 (*cyclo*-$(H_2SnO)_3$) and 0 kJ mol^{-1} (*cyclo*-$(H_2GeO)_3$), respectively. More electropositive elements, such as Si, favor wider M–O–M′ linkages, and are therefore associated with greater angular compression in six-membered rings and higher strain energies. The global minimum ring strain energies for the heterometallic rings are 7.0 (*cyclo*-$H_2Sn(OSiH_2)_2O$), 4.9 (*cyclo*-$H_2Ge(OSiH_2)_2O$), 3.4 (*cyclo*-$H_2Si(OSnH_2)_2O$), 0.8 (*cyclo*-$H_2Si(OGeH_2)_2O$), 0.7 (*cyclo*-$H_2Ge(OSnH_2)_2O$) and 0.1 kJ mol^{-1} (*cyclo*-$H_2Sn(OGeH_2)_2O$), respectively.

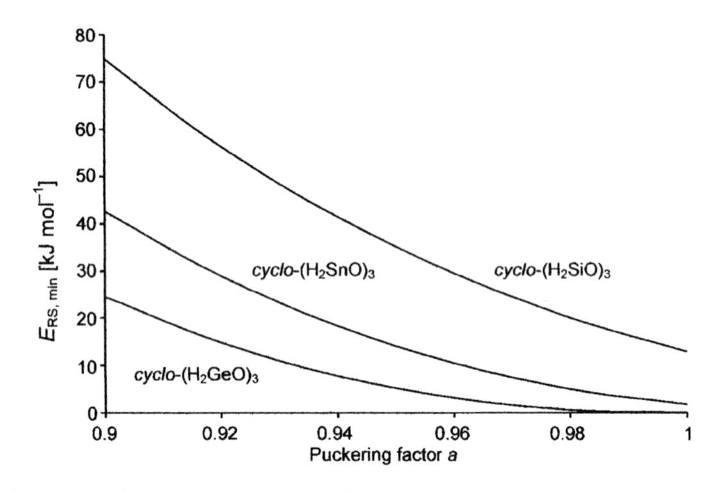

Fig. 3. Plots of E_{RSmin} vs *a* for *cyclo*-$(H_2MO)_3$ (M = Si, Ge, Sn).

This observation is consistent with experimental findings that silicon-rich six-membered cyclic Group 14 organoelement oxides are susceptible to ROP and ring-expansion reactions (Eq. 5) [2, 3].

M = Si; R = Ph
M = Sn; R = *t*-Bu

x n

M = Si; R = Ph
M = Sn; R = *t*-Bu
M= Sn; R = Cp_2Fe
M = Ge; R = Me

x 2

M = Sn; R = Cp_2Fe
M = Ge; R = Me

Eq. 5. Experimentally observed ROP and ring-expansion reactions of silicon-rich six-membered cyclic Group 14 organoelement oxides.

Six-membered *cyclo*-distannasiloxane rings have been observed to undergo a redistribution to form less strained rings upon crystallization from solution (Eq. 6). In contrast, rings possessing more electronegative Group 14 elements, such as *cyclo*-(*t*-$Bu_2SnO)_3$, *cyclo*-$Ph_2Si(OGePh_2)_2O$, *cyclo*-*t*-$Bu_2Sn(OGePh_2)_2O$, and *cyclo*-$(Ph_2GeO)_3$ do not undergo ROP or ring-expansion reactions.

Eq. 6. Experimentally observed redistribution reaction involving a six-membered *cyclo*-distannasiloxane ring.

Strain Energies and Conformational Preferences of Eight-Membered Rings

The relationships between $E_{RS,\,min}$ and the puckering factor a for the homometallic eight-membered rings, *cyclo*-$(H_2MO)_4$ (M = Si, Ge, Sn), are depicted in Fig. 4. In each case, puckering to achieve a strain-free ring conformation is possible. The more electronegative Group 14 elements favor narrower M–O–M angles and thus lead to more puckered eight-membered rings.

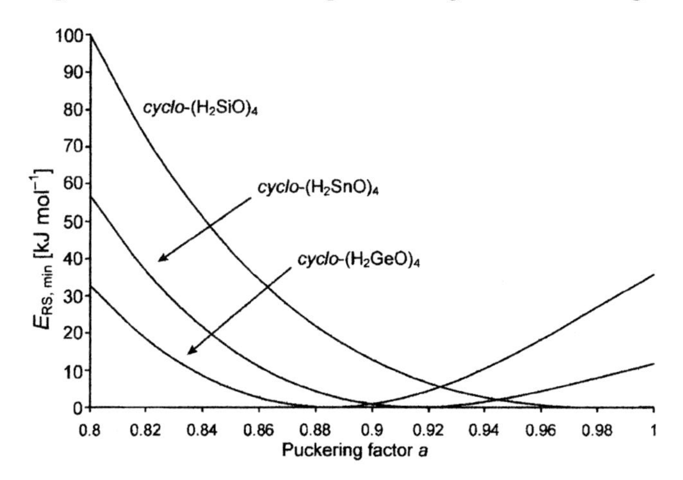

Fig. 4. Plots of $E_{RS,\,min}$ vs. a for *cyclo*-$(H_2MO)_4$ (M = Si, Ge, Sn).

The strain-free puckering factors decrease in the order: *cyclo*-$(H_2SiO)_4$ (0.983) > *cyclo*-$H_2Sn(OSiH_2O)_2SiH_2$ (0.959) > *cyclo*-$(H_2SiO)_2(H_2SnO)_2$ (0.942) > *cyclo*-$H_2Si(OSnH_2O)_2SiH_2$ (0.935) > *cyclo*-$(H_2SnO)_4$ (0.916) > *cyclo*-$(H_2GeO)_4$ (0.885). Eight-membered *cyclo*-tetrasiloxanes favor (nearly) planar conformations, and, as silicon is successively replaced by more electronegative elements, the ring becomes increasingly puckered and the energy barrier to planarity increases.

This difference is clearly seen in the ring geometries of *cyclo*-$(Ph_2SiO)_4$ (a = 0.995) and *cyclo*-$(Ph_2GeO)_4$ (a = 0.903) (Fig 5) [10, 11]. The experimentally known solid-state geometries of the eight-membered stannasiloxane rings also agree with this trend [3].

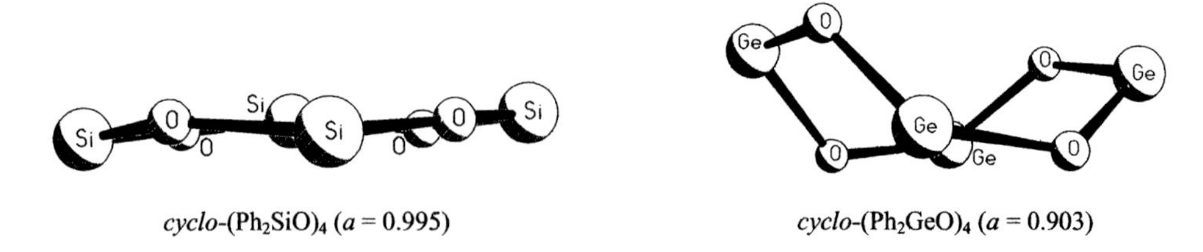

cyclo-$(Ph_2SiO)_4$ (a = 0.995) *cyclo*-$(Ph_2GeO)_4$ (a = 0.903)

Fig. 5. Comparison of the solid-state ring geometries of *cyclo*-$(Ph_2SiO)_4$ and *cyclo*-$(Ph_2GeO)_4$.

Summary and Concluding Remarks

Replacing silicon with a more electronegative element leads to reduced ring strain in six-membered rings, and increased ring puckering in eight-membered rings. These results indicate that only six-membered rings are suitable precursors for kinetically controlled ROP processes. For the design of heterometallic rings of type $M(OSiR_2)_2O$ that are suitable for kinetically controlled ROP processes, M is supposed to be a sufficiently electropositive element.

Acknowledgments: The Australian Research Council and the Deutsche Forschungsgemeinschaft are thanked for financial support.

References

[1] W. A. Piccoli, G. G. Haberland, R. L. Merker, *J. Am. Chem. Soc.* **1960**, *82*, 1883.
[2] J. Beckmann, K. Jurkschat, *Coord. Chem. Rev.* **2001**, *215*, 267.
[3] J. Beckmann, K. Jurkschat, M. Schürmann, D. Dakternieks, A. E. K. Lim, K. F. Lim, *Organometallics* **2001**, *20*, 5125.
[4] *Gaussian 98*, Revision A.9, Gaussian Inc., Pittsburgh PA, **1998**.
[5] A. Bergner, M. Dolg, W. Kuechle, H. Stoll, H. Preuss, *Mol. Phys.* **1993**, *80*, 1431.
[6] S. Huzinaga, *Gaussian Basis Sets for Molecular Calculations*, Elsevier, New York, **1984**.
[7] R. J. Gillespie, E. A. Robinson, *Angew. Chem. Int. Ed. Engl.* **1996**, *35*, 495.
[8] R. J. Gillespie, S. A. Johnson, *Inorg. Chem.* **1997**, *36*, 3031.
[9] J. Beckmann, D. Dakternieks, A. E. K. Lim, K. F. Lim, K. Jurkschat *Organometallics* **2005**, submitted.
[10] Y. E. Ovchinnikov, Y. T. Struchkov, M. I. Buzin, V. S. Papkov, *Vysokomol. Soedin., Ser. A Ser. B* **1997**, *39*, 430.
[11] L. Ross, M. Dräger, *Z. Naturforsch.* **1984**, *39B*, 868.

^{29}Si NMR Chemical Shift Tensors in Organosilicon Chalcogenides

Uwe Herzog, Uwe Böhme*

Institut für Anorganische Chemie, TU Bergakademie Freiberg
Leipziger Straße 29, D-09596 Freiberg, Germany
Tel.: +49 3731 394343 — Fax: +49 3731 394058
E-mail: Uwe.Boehme@chemie.tu-freiberg.de

Erica Brendler

Institut für Analytische Chemie TU Bergakademie Freiberg
Leipziger Straße 29, D-09596 Freiberg, Germany
E-mail: Erica.Brendler@chemie.tu-freiberg.de

Keywords: silthiane, ^{29}Si NMR, chemical shift tensor

Summary: ^{29}Si CP MAS NMR spectra of the organosilicon sulfur compounds spiro-$[C_6H_4(S)_2]_2Si$ (**2**), $S(SiMe_2SiMe_2)_2S$ (**3**), $Me_2Si(S)_2Si_2Me_2(S)_2SiMe_2$ (**4**), and $Me_4Si_2(S)_2SiMe{-}SiMe(S)_2Si_2Me_4$ (**5**) have been recorded at different spinning frequencies. Analysis of the spinning side bands yielded the principal values of the ^{29}Si NMR chemical shift tensors. The data are compared with the results of GIAO calculations based on the geometries observed in the crystal structures. GIAO calculations on spiro-$[C_2H_4(S)_2]_2Si$ (**1**) revealed that the calculated large anisotropy of 88 ppm is caused by the partial planarization of the SiS_4 tetrahedron (spiro angle of 74.4° in the crystal structure). If this spiro angle is fixed at 90° the calculated anisotropy decreases to 23 ppm.

In the course of our investigations on organosilicon chalcogenides, we observed a significant influence of the ring size (e.g., five or six) on the ^{29}Si NMR chemical shifts. Usually the incorporation of silicon into five-membered rings results in a low-field shift, while six-membered rings show a less pronounced high-field shift. For instance, δ_{Si} of the S–SiMe$_2$–SiMe$_2$–S unit in acyclic BuS–SiMe$_2$–SiMe$_2$–SBu is −1.6 ppm [1], in the five-membered ring compound $Me_4Si_2(S)_2SiMe{-}SiMe(S)_2Si_2Me_4$ (**5**) it is 12.5 ppm [2], and in the six-membered ring compound $S(SiMe_2{-}SiMe_2)_2S$ (**3**) –4.8 ppm [2] is observed.

In order to get a better understanding of these effects, ^{29}Si CP MAS NMR measurements at different spinning frequencies of several organosilicon sulfur compounds (**1–5**) with known crystal structures (see Figs. 1 – 3) were carried out (Table 1); one example is shown in Fig. 4.

From these experiments it was possible to obtain the principal values of the chemical shifts as well as the anisotropy (span) and the skew (Herzfeld-Berger convention [3]).

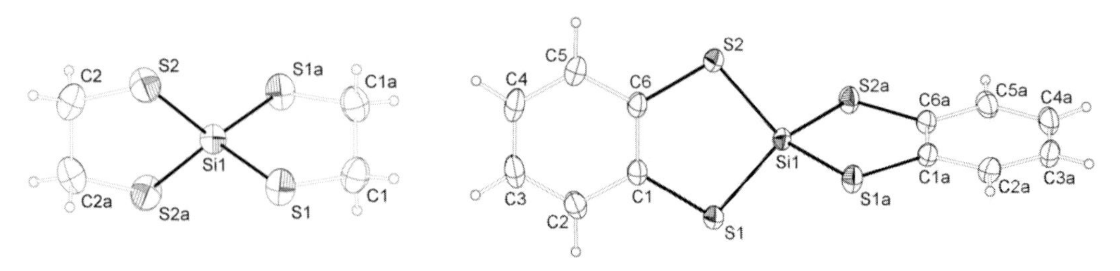

Fig. 1. Molecular structures of spiro-$[C_2H_4(S)_2]_2Si$ (**1**) [4] (left) and spiro-$[C_6H_4(S)_2]_2Si$ (**2**) [5] (right).

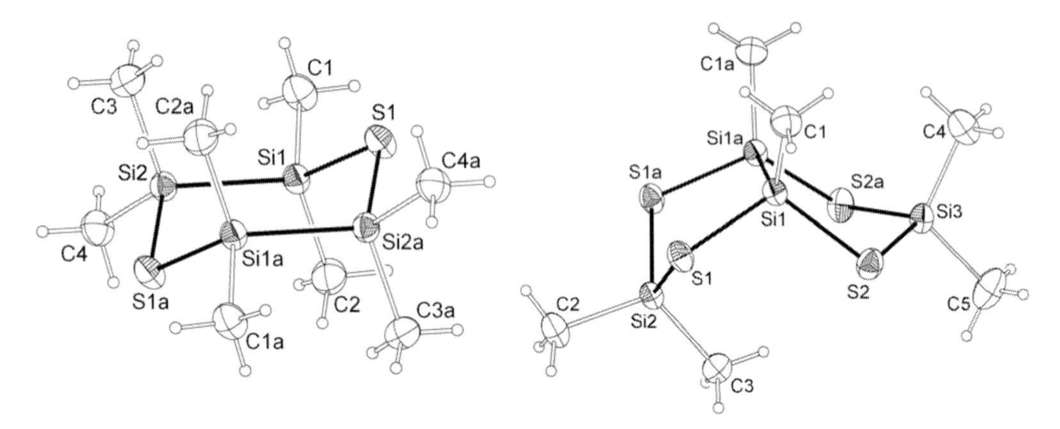

Fig. 2. Molecular structures of $S(Me_2Si\text{–}SiMe_2)_2S$ (**3**) [2] (left) and $Me_2Si(S)_2Si_2Me_2(S)_2SiMe_2$ (**4**) [6, 7] (right).

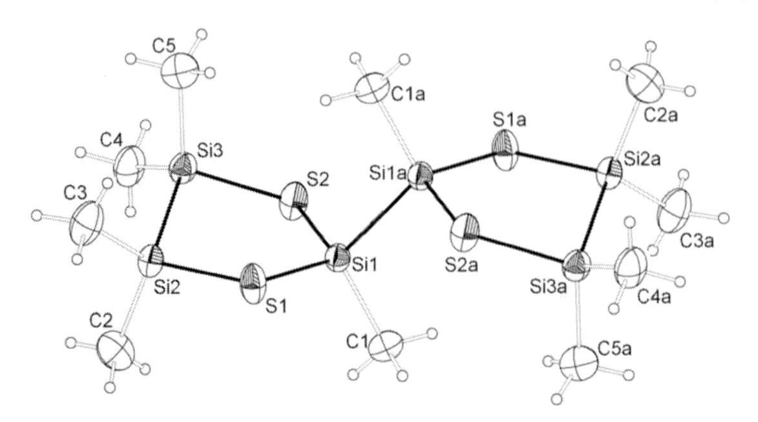

Fig. 3. Molecular structure of $Me_4Si_2(S)_2SiMe\text{–}SiMe(S)_2Si_2Me_4$ (**5**) [2].

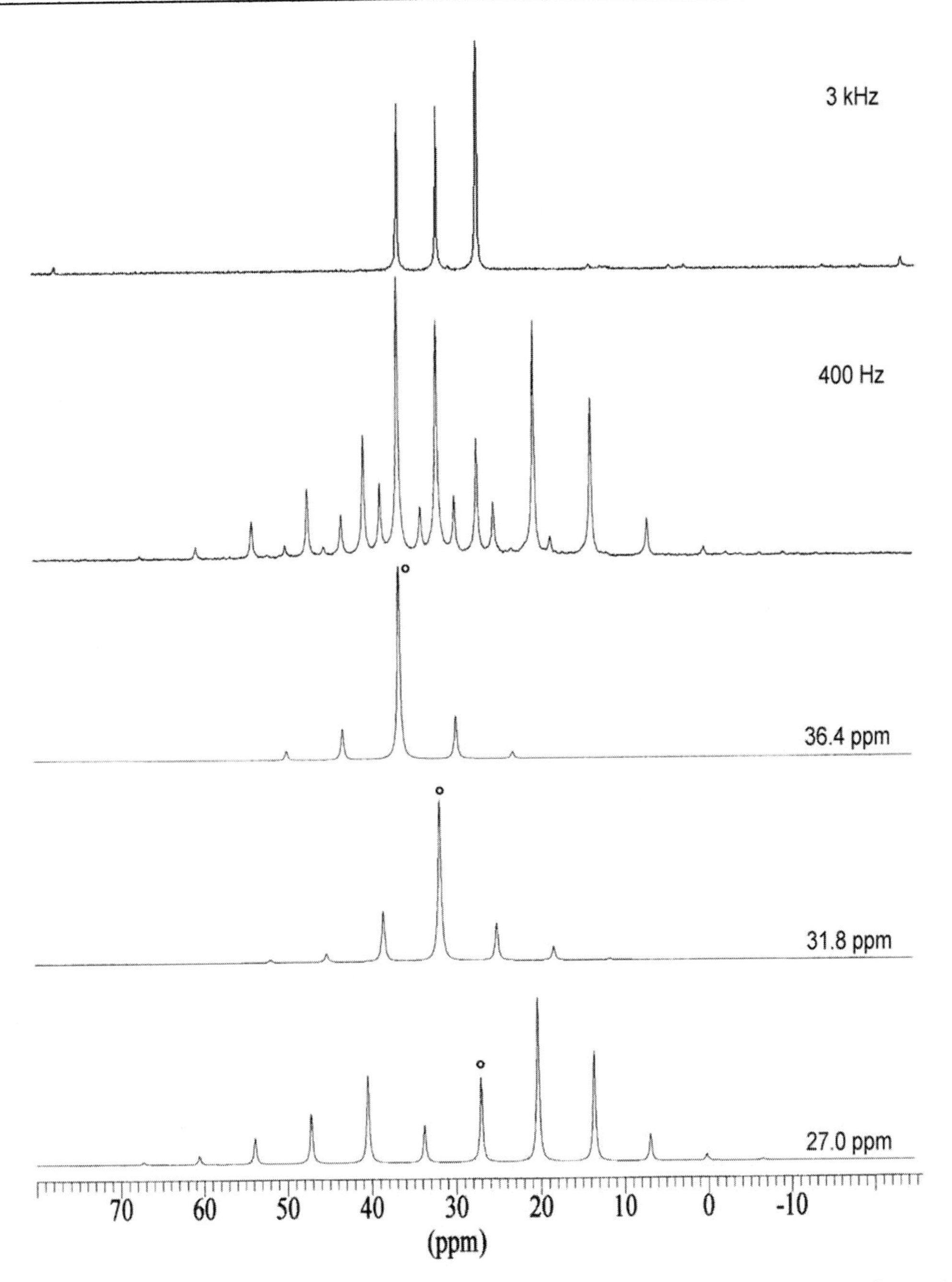

Fig. 4. ^{29}Si CP MAS NMR spectra (59.6 MHz) of $Me_2Si(S)_2Si_2Me_2(S)_2SiMe_2$ (**4**) at different spinning frequencies (3 kHz, 400 Hz). The spectrum at 400 Hz shows three overlapping sets of spinning sidebands. Below are calculated subspectra [8] of the three signal sets centered at 36.4, 31.8, and 27.0 ppm.

Table 1. ^{29}Si MAS NMR data of organosilicon sulfur compounds.

Compound	Unit	δ_{sol}[a]	δ_{iso}	δ_{11}	δ_{22}	δ_{33}	Ω[b]	κ[b]
1	SiS_4	57.3	53.8	[c]	[c]	[c]		
2	SiS_4	45.4	44.9	85.7	45.2	3.6	82.1	0.01
3	$SiMe_2$	–4.8	–4.8	27.9	–15.4	–27.0	54.9	–0.58
3	$SiMe_2$	–4.8	–4.5	28.1	–14.6	–27.1	55.2	–0.55
4	SiMe	29.8	27.0	57.0	16.8	7.2	49.8	–0.61
4	$SiMe_2$	35.3	36.4	46.6	33.9	28.6	18.0	–0.42
4	$SiMe_2$	35.3	31.8	41.3	34.0	20.0	21.3	0.10
5	SiMe	20.2	22.6	62.4	6.9	–1.7	64.1	–0.73
5	$SiMe_2$	12.5	15.6	36.1	27.6	–16.8	52.9	0.68
5	$SiMe_2$	12.5	14.4	36.2	20.0	–13.1	49.3	0.34

[a] Chemical shift in solution. [b] Herzfeld-Berger convention [3], $\Omega = \delta_{11} - \delta_{33}$, $\kappa = 3(\delta_{22} - \delta_{iso}) / \Omega$. [c] Analysis of spinning side bands not possible because of low melting point.

Besides MAS NMR experiments, we calculated the ^{29}Si NMR parameters by the GIAO method based on the geometries of the crystal structures (see Table 2). In general, there is a good correlation between the experimental and the calculated data.

Table 2. GIAO calculation (HF/6-311+G(2d,p)) of ^{29}Si NMR chemical shifts of **1** – **5** (geometries of the crystal structures).

Compound	Unit	δ_{iso}	δ_{11}	δ_{22}	δ_{33}	Ω	κ
1 [a]	SiS_4	63.7	106.7	79.4	5.0	101.7	0.46
1 [b]	SiS_4	66.7	106.4	81.5	12.3	94.1	0.47
1 [c]	SiS_4	69.8	82.3	72.6	54.5	27.8	0.30
2	SiS_4	54.8	90.9	60.0	13.6	77.3	0.20
3	$SiMe_2$	–3.9	20.1	–11.0	–20.9	41.0	–0.67
3	$SiMe_2$	–2.4	23.1	–13.3	–17.1	40.2	–0.81
4	SiMe	27.4	53.3	15.9	13.1	40.2	–0.86
4	$SiMe_2$	39.0	40.5	39.2	37.3	3.2	0.19
4	$SiMe_2$	35.4	42.6	39.5	24.0	18.6	0.66
5	SiMe	23.7	38.7	25.6	6.8	31.9	0.18
5	$SiMe_2$	14.2	33.5	21.7	–12.5	46.0	0.49
5	$SiMe_2$	11.5	30.7	16.4	–12.7	43.4	0.34

[a] Geometry of the crystal structure (spiro angle at Si: 74.4°). [b] Fully optimized structure (B3LYP/6-31G*, spiro angle at Si: 77.2°). [c] Calculated structure (B3LYP/6-31G*, spiro angle at Si fixed at 90°).

A comparison of the NMR parameter of the $SiMe_2$ units in **3** and **5** (same first coordination sphere at silicon, $SiSC_2$, but in a six- and a five-membered ring, respectively) reveals that as well as a difference in δ_{iso} of some 20 ppm the anisotropy Ω remains almost unchanged, but the relative position of δ_{22} varies, resulting in a significant change of the skew κ.

On the other hand, if the SiMe units in **4** and **5** (first coordination sphere at silicon: SiS_2C) are considered, the experimental δ_{iso} values differ by 4.4 ppm but Ω and κ are rather similar. In both compounds the SiMe units are incorporated into five-membered rings but in different ways.

Despite a first coordination sphere of four sulfur atoms, the calculated ^{29}Si NMR chemical shift tensors of **1** and **2** show relatively large anisotropies of 101.7 and 77.3 ppm, respectively. These large anisotropies are mainly caused by the partial planarization of the SiS_4 tetrahedron resulting in spiro angles (angle between the two SiS_2 planes) of 74.4° in **1** and 83.4° in **2**.

If the spiro angle in **1** is fixed at 90°, the calculated anisotropy Ω is reduced to 27.8 ppm (see Table 2). Figures 5 – 7 give an idea of the spatial orientation of the ^{29}Si NMR chemical shift tensors in **1** and **2**. As can be seen from Figs. 5 and 6, a change of the spiro angle in **1** mainly influences the chemical shift in the *z* direction.

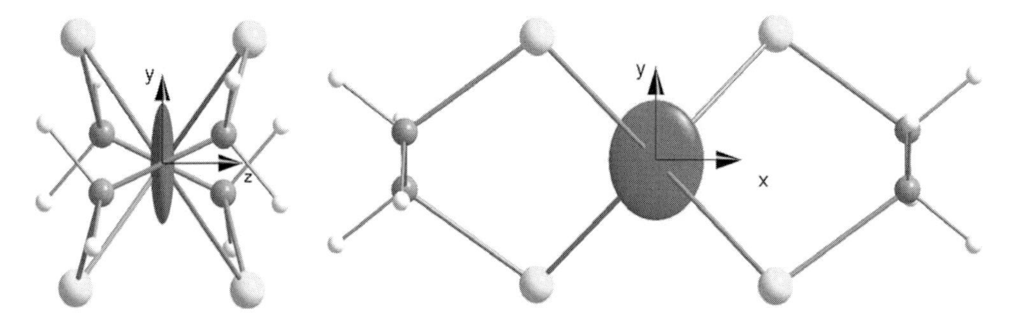

Fig. 5. Calculated ^{29}Si NMR chemical shift tensor (dark gray) in spiro-$[C_2H_4(S)_2]_2Si$ (**1**); geometry of the crystal structure with spiro angle at Si of 74.4°. For better visualization the chemical shifts with respect to TMS = 0 ppm were used as principal values in all tensor drawings.

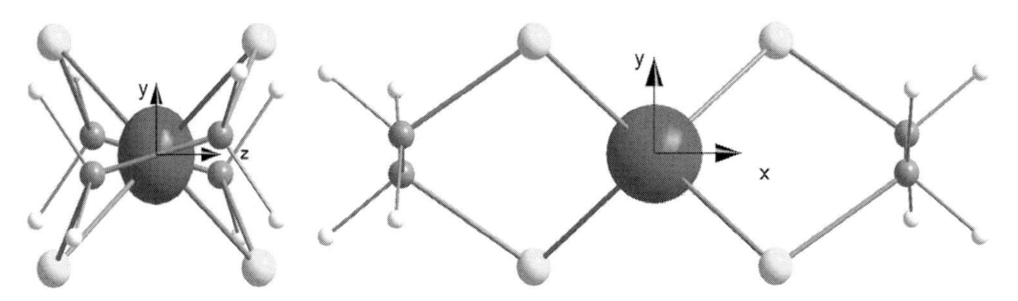

Fig. 6. Calculated ^{29}Si NMR chemical shift tensor (dark gray) in spiro-$[C_2H_4(S)_2]_2Si$ (**1**); calculated geometry with a spiro angle at Si of 90°.

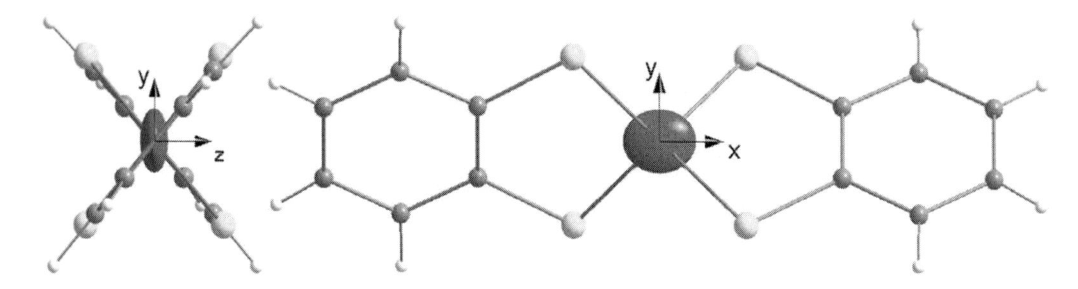

Fig. 7. Calculated ^{29}Si NMR chemical shift tensor (dark gray) in spiro-$[C_6H_4(S)_2]_2Si$ (**2**); calculated geometry of the crystal structure with spiro angle at Si of 83.4°.

Acknowledgments: The authors thank the Deutsche Forschungsgemeinschaft for financial support. Special thanks are given to the Computing Centre of the TU Bergakademie Freiberg for supplying disk space and computing time.

References

[1] U. Herzog, G. Roewer, *Main Group Metal Chem.* **1999**, *22*, 579.
[2] U. Herzog, G. Rheinwald, *J. Organomet. Chem.* **2001**, *627*, 23.
[3] J. Herzfeld, A. E. Berger, *J. Chem. Phys.* **1980**, *73*, 6021.
[4] W. Wojnowski, K. Peters, M. C. Böhm, H. G. v. Schnering, *Z. anorg. allg. Chem.* **1985**, *523*, 169.
[5] U. Herzog, U. Böhme, G. Rheinwald, *J. Organomet. Chem.* **2000**, *612*, 133.
[6] U. Herzog, U. Böhme, G. Roewer, G. Rheinwald, H. Lang, *J. Organomet. Chem.* **2000**, *602*, 193.
[7] U. Herzog, U. Böhme, E. Brendler, G. Rheinwald, *J. Organomet. Chem.* **2001**, *630*, 139.
[8] M. Braun, *HBMAS*, Friedrich-Schiller-Universität Jena.

Hypersilyltelluro-substituted Silanes and $(Ph_2SiTe)_3$

H. Lange, U. Herzog, G. Roewer

Institut für Anorganische Chemie, TU Bergakademie Freiberg
Leipziger Str. 29, D-09596 Freiberg, Germany
Tel.: +49 3731 393583 — Fax: +49 3731 394058
E-mail: Gerhard.Roewer@chemie.tu-freiberg.de

H. Borrmann

Max-Planck-Institut für Chemische Physik fester Stoffe / Dresden
Nöthnitzer Str. 40, D-01187 Dresden, Germany

Keywords: hypersilyl, silanes, tellurides

Summary: Treatment of hypersilylpotassium, prepared by reaction of $Si(SiMe_3)_4$ with KO^tBu in THF, with tellurium, yields potassium hypersilyltellurolate. Reactions with organochlorosilanes $R_{4-x}SiCl_x$ ($x = 1 - 4$, R = Me, Ph) and $ClSiMe_2SiMe_2Cl$ yielded hypersilyl-substituted silanes $[(Me_3Si)_3SiTe]_xSiR_{4-x}$ and $(Me_3Si)_3SiTeSiMe_2SiMe_2Te$-$Si(SiMe_3)_3$. All compounds have been characterized by ^1H, ^{13}C, ^{29}Si and ^{125}Te NMR spectroscopy. In the preparation of $[(Me_3Si)_3SiTe]_2SiPh_2$ (**1**) we observed the formation of $(Ph_2SiTe)_3$ (**2**) as a by-product. The molecular structures of **1** and **2** are reported. **2** is the first example of a crystal structure of a compound with an Si_3Te_3 six-membered ring. In the solid state **2** forms a dimer with an intermolecular Te···Te contact of 3.585 Å.

Recently, we have shown that hypersilylthio- and hypersilylseleno-substituted silanes and stannanes can be prepared by reactions of potassium hypersilylthiolate and -selenolate with organochlorosilanes and -stannanes [1]. In order to extend these investigations to the related tellurium compounds we have prepared potassium hypersilyltellurolate by insertion of tellurium into the Si–K bond of hypersilylpotassium prepared from $(Me_3Si)_4Si$ and KO^tBu [2]:

$$(Me_3Si)_2Si(SiMe_3)_2 \xrightarrow[-Me_3SiO^tBu]{KO^tBu} (Me_3Si)_2Si^-(SiMe_3)\ K^+ \xrightarrow{Te} (Me_3Si)_2Si(Te^-)(SiMe_3)\ K^+$$

Eq. 1. Formation of potassium hypersilyltelluride.

Subsequent reactions with organochlorosilanes and 1,2-dichlorotetramethyldisilane yielded hypersilyltelluro-substituted silanes and disilanes (see Scheme 1).

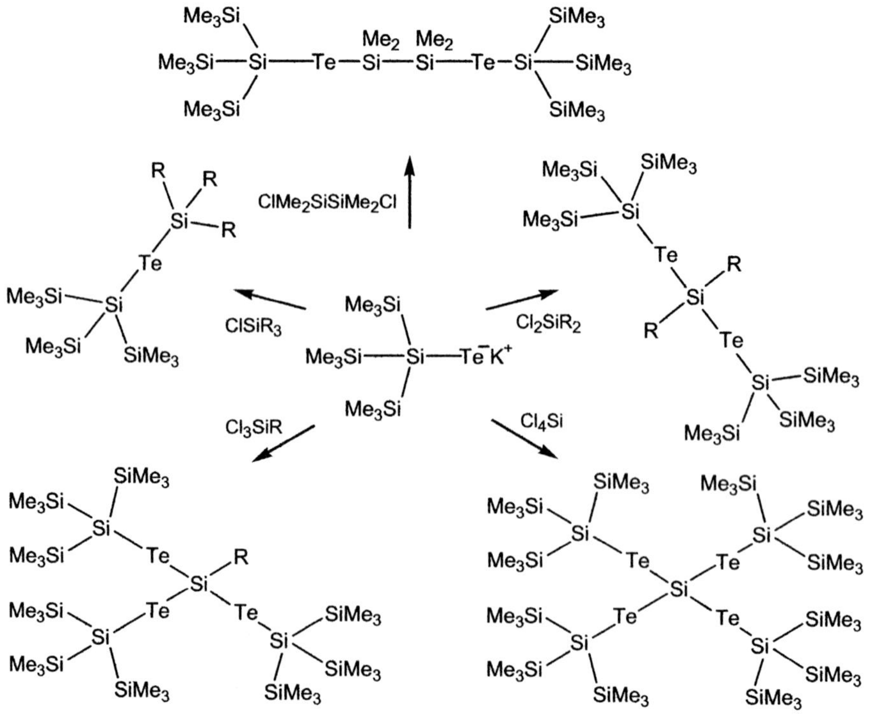

Scheme 1. Formation of $[(Me_3Si)_3SiTe]_xSiR_{4-x}$ (R = Me, Ph) and $(Me_3Si)_3SiTeSiMe_2SiMe_2TeSi(SiMe_3)_3$.

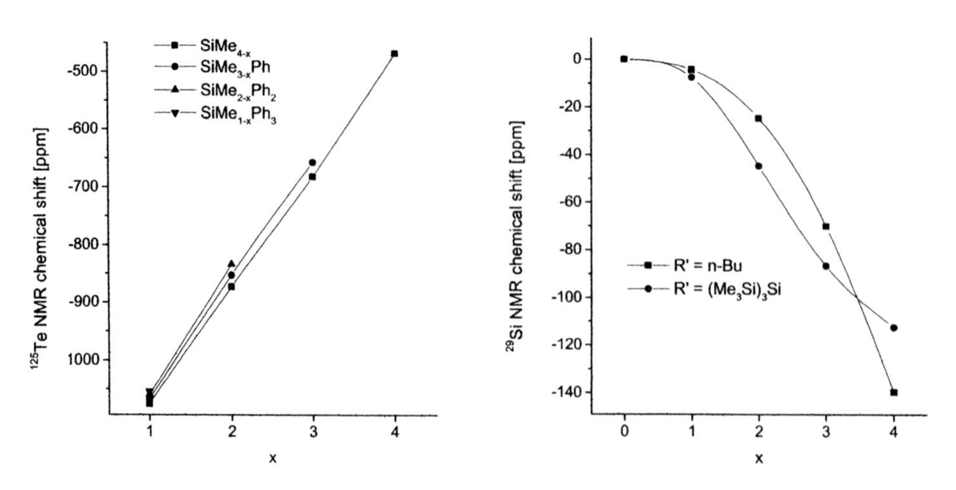

Fig. 1. δ_{Te} in $[(Me_3Si)_3Si\underline{Te}]_xSiR_{4-x}$; R = Me, Ph.

Fig. 2. δ_{Si} in $(R'Te)_x\underline{Si}Me_{4-x}$; R′ = n-Bu, $(Me_3Si)_3Si$.

Table 1. ^{125}Te and ^{29}Si NMR chemical shifts and coupling constants of $[(Me_3Si)_3SiTe]_xSiR_{4-x}$ (R = Me, Ph) and $(Me_3Si)_3SiTeSiMe_2SiMe_2TeSi(SiMe_3)_3$.

Compound	δTe	δSi^A	$^1J(Si^ATe)$	δSi^B	$^1J(Si^BTe)$	δSi^C	$^1J(Si^ASi^C)$
$(Me_3Si^C)_3Si^ATeSi^BMe_3$	–1076	–100.5	288.7	–7.5	318.8	–11.9	56.4
$(Me_3Si^C)_3Si^ATeSi^BPhMe_2$	–1067	–99.1	286.7	–12.6	338.2	–11.6	56.4
$(Me_3Si^C)_3Si^ATeSi^BPh_2Me$	–1060	–98.3	285.7	–14.0	356.6	–11.3	56.8
$(Me_3Si^C)_3Si^ATeSi^BPh_3$	–1055	–96.1	292.5	–12.8	–	–11.0	54.0
$[(Me_3Si^C)_3Si^ATe]_2Si^BMe_2$	–873	–93.6	294.5	–44.6	404.3	–11.4	54.4
$[(Me_3Si^C)_3Si^ATe]_2Si^BPhMe$	–853	–91.8	295.9	–47.4	419.9	–11.1	54.4
$[(Me_3Si^C)_3Si^ATe]_2Si^BPh_2$	–834	–90.6	–	–41.6	–	–10.8	53.9
$[(Me_3Si^C)_3Si^ATe]_3Si^BMe$	–682	–86.4	280.9	–86.6	–	–10.6	53.0
$[(Me_3Si^C)_3Si^ATe]_3Si^BPh$	–657	–84.9	283.3	–62.1	–	–10.4	53.7
$[(Me_3Si^C)_3Si^ATe]_4Si^B$	–468	–79.6	269.7	–112.5	–	–9.9	–
$[(Me_3Si^C)_3Si^ATeSi^BMe_2]_2$	–1107	–98.1	298.8	–28.8	367.9[a]	–11.8	54.1

[a] $^2J(Si^BTe)$: 70.4 Hz

As can be seen from Table 1 and Fig. 1 the ^{125}Te NMR resonances as well as those of the central silicon atoms of the hypersilyl units (Si^A) are shifted downfield with increasing x (number of hypersilyltelluro units at Si^B).

A comparison of $\delta_{Si}{}^B$ of the hypersilyltelluro-substituted silanes with δ_{Si} in $(BuTe)_xSiMe_{4-x}$ [3] (Fig. 2) reveals deviations which may be attributed to the steric influence of the bulky hypersilyl units.

In the case of $[(Me_3Si)_3SiTe]_2SiPh_2$ (**1**), a crystal structure analysis has been carried out; the result is shown in Fig. 3. Important bond lengths and angles are given in Table 2.

Table 2. Selected bond lenghts and angles of $Ph_2Si[TeSi(SiMe_3)_3]_2$ (**1**).

Atoms	[Å]	Atoms	[°]
Si(1)–Te(1)	2.501(2)	Te(1)-Si(1)-Te(2)	104.71(7)
Si(1)–Te(2)	2.501(2)	C(1)-Si(1)-C(7)	110.40(3)
Si(2)–Te(1)	2.555(2)	Si(1)-Te(1)-Si(2)	111.91(7)
Si(3)–Te(2)	2.538(2)	Si(1)-Te(2)-Si(3)	113.40(6)

Because of the bulky hypersilyl units, the bond angles at the tellurium atoms are relatively large. Furthermore, it can be seen that the bonds between the tellurium atoms and the middle silicon atoms of the hypersilyl units (Si2, Si3) are slightly longer than the Si–Te bonds at the central silicon atom (Si1).

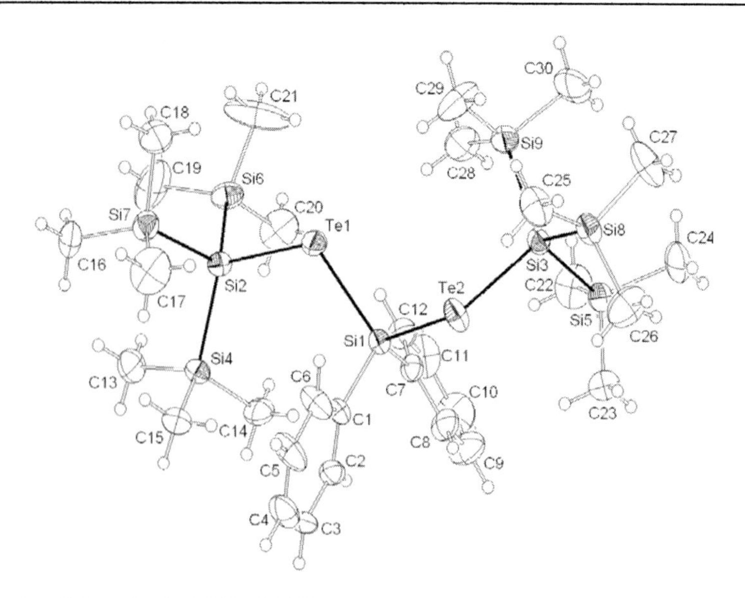

Fig. 3. ORTEP plot of $[(Me_3Si)_3SiTe]_2SiPh_2$ (**1**) [4].

In reactions of potassium hypersilyltellurolate with chlorosilanes as shown in Scheme 1 sometimes hypersilyl chloride and organosilicon tellurides have been observed as by-products, e. g., the preparation of **1** yielded the six membered ring compound $(Ph_2SiTe)_3$ (**2**) as a minor by-product. **2** can also be prepared from Ph_2SiCl_2 by treatment with Li_2Te (from $LiBEt_3H$ and Te):

$$Ph_2SiCl_2 \xrightarrow[\text{B) } + Li_2Te,\ -2\ LiCl]{\text{A) } + (Me_3Si)_3SiTeK,\ -(Me_3Si)_3SiCl,\ -KCl} \tfrac{1}{3}\ (Ph_2SiTe)_3$$

δ_{Si}: −19.56 ppm; δ_{Te}: − 673 ppm; ^1J(SiTe): 373.2 Hz

Eq. 3. Formation of $(Ph_2SiTe)_3$ (**2**).

The crystal structure analysis of **2** (Fig. 4) revealed a central Si_3Te_3 six-membered ring in a slightly twisted boat conformation. This is the first crystal structure of a compound with an Si_3Te_3 ring. Important bond lengths and angles are given in Table 3.

The Si–Te bond lengths of $(Ph_2SiTe)_3$ are similar to the central Si–Te bonds in **1**, while the bond angles Si-Te-Si are approximately 10° smaller than in **1**.

In contrast to the related tin compounds $(Me_2SnTe)_3$ [5], $[(PhCH_2)_2SnTe]_3$ [6], $[(Me_3SiCH_2)_2SnTe]_3$ [7], and $(Ph_2SnTe)_3$ [8], a weak intermolecular Te···Te contact of 3.585 Å can be observed in **2**, which leads to a dimerization in the crystal structure (see Fig. 5). This Te···Te contact may also be the reason for the fact that in **2** the Si_3Te_3 ring adopts almost a boat conformation to allow this contact, while twisted boat conformations have been observed in $(PhMeSiS)_3$ [9] and all related tin and lead compounds $(R_2ME)_3$ (E = S, Se, Te).

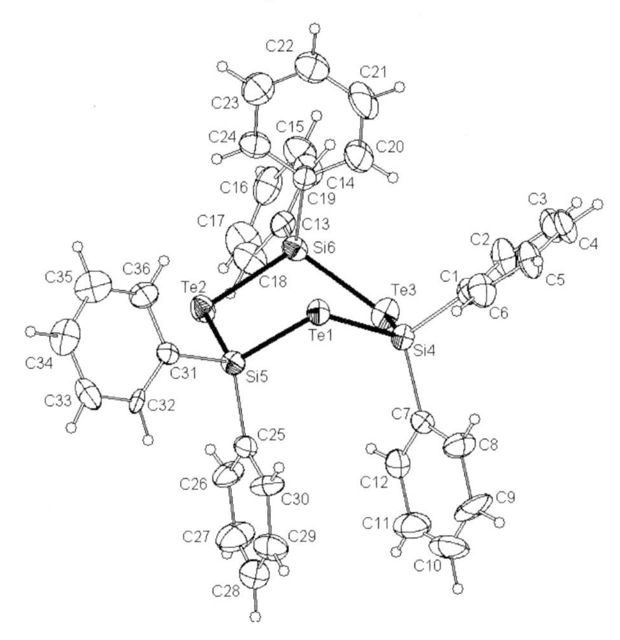

Fig. 4. ORTEP plot of (Ph$_2$SiTe)$_3$ (**2**) [10].

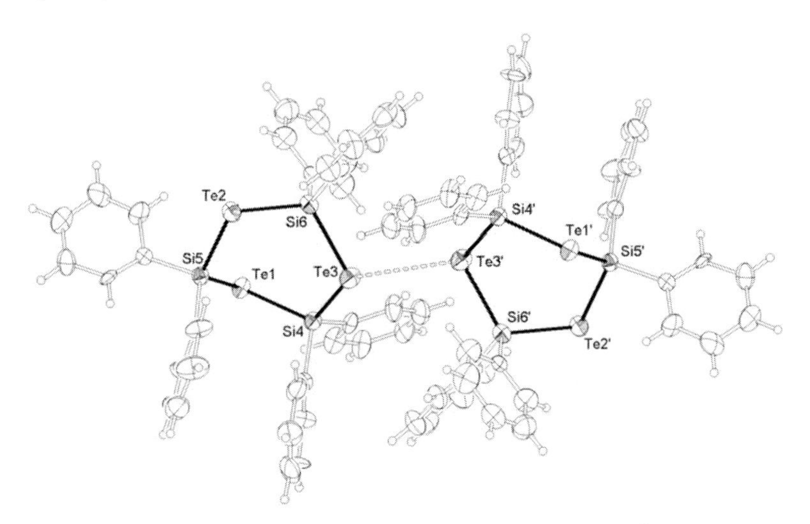

Fig. 5. ORTEP plot of **2** showing the intermolecular Te···Te contact, Te3–Te3': 3.585 Å.

Table 3. Selected bonds lengths, angles, and torsions angles of $(Ph_2SiTe)_3$ (**2**).

Atoms	[Å]	Atoms	[°]	Atoms	[°]
Si(4)–Te(1)	2.508(5)	Te(1)-Si(5)-Te(2)	113.28(2)	Si(4)-Te(1)-Si(5)-Te(2)	–71.34
Si(5)–Te(2)	2.506(5)	Te(2)-Si(6)-Te(3)	112.29(2)	Si(5)-Te(2)-Si(6)-Te(3)	53.80
Si(6)–Te(2)	2.496(5)	Te(3)-Si(4)-Te(1)	115.96(2)	Si(6)-Te(3)-Si(4)-Te(1)	13.08
Si(4)–Te(3)	2.484(5)	Si(4)-Te(1)-Si(5)	100.61(2)	Te(1)-Si(5)-Te(2)-Si(6)	21.13
Si(5)–Te(1)	2.490(5)	Si(5)-Te(2)-Si(6)	102.39(2)	Te(2)-Si(6)-Te(3)-Si(4)	–72.30
Si(6)–Te(3)	2.481(5)	Si(6)-Te(3)-Si(4)	101.68(2)	Te(3)-Si(4)-Te(1)-Si(5)	50.40

Acknowledgments: The authors thank the Deutsche Forschungsgemeinschaft for financal support.

References

[1] H. Lange, U. Herzog, *J. Organomet. Chem.* **2002**, *660*, 36.

[2] C. Marschner, *Eur. J. Inorg. Chem.* **1998**, 221.

[3] U. Herzog, *Main Group Metal Chemistry* **2001**, *24*, 31.

[4] **1** crystallizes triclinic, space group: $P\bar{1}$, *a*: 10.0818(5) Å, *b*: 15.4719(8) Å, *c*: 17.5220(12) Å, α: 75.22(1)°, β: 72.00(1)°, γ: 72.82(1)°, $Z = 2$.

[5] A. Blecher, M. Dräger, *Angew. Chem.* **1979**, *91*, 740; R. J. Batchelor, F. W. B. Einstein, C. H. W. Jones, *Acta Cryst.* **1989**, *C45*, 1813.

[6] P. Boudjouk, M. P. Remington Jr., D. G. Grier, W. Triebold, B. R. Jarabek, *Organometallics* **1999**, *18*, 4534.

[7] F. W. B. Einstein, I. D. Gay, C. H. W. Jones, A. Riesen, R. D. Sharma, *Acta Cryst.* **1993**, *C 49*, 470.

[8] H. Lange, U. Herzog, U. Böhme, G. Rheinwald, *J. Organomet. Chem.* **2002**, *660*, 43.

[9] L. Pazdernik, F. Brisse, R. Rivest, *Acta Cryst.* **1977**, *B33*, 1780.

[10] **2** crystallizes monoclinic, space group: $P2_1/n$, *a*: 9.684(1) Å, *b*: 20.828(2) Å, *c*: 18.031(2) Å, β: 101.704(5)°, $Z = 4$.

Novel Dimeric Pentacoordinate Silicon Complexes: Unusual Reactivity of Electron-Rich Aminosilane Intermediates

Nicoleta Dona,* Klaus Merz, Matthias Driess

Chair of Inorganic Chemistry I: Cluster- and Coordination Chemistry
Ruhr-University Bochum, Universitätsstrasse 150, D-44780 Bochum, Germany
Tel.: +49 234 3224178 — Fax: +49 234 3214378
E-mail: nicoleta.dona@ruhr-uni-bochum.de

Keywords: silicon complexes, reactive intermediates, hypercoordination

Summary: The synthesis of the first silicon complexes with the tridentate diketoamine $HN[CH_2C(t\text{Bu})=O]_2$ (**1**) is described. The reaction of $YSiX_3$ (X = Cl, Br; Y = H, Cl, Br, Ph, Vinyl) with 5-aza-2,2,8,8-tetramethylnonane-3,7-dione (**1**) in the presence of NEt_3 as an auxiliary base furnished the unusual dinuclear silicon complexes (**6** – **11**) having pentacoordinate silicon centers. Theoretical calculations (B3LYP/6-31G**) carried out for the monomeric chloro-substituted silane **7A** are consistent with a folded structure and a relatively short and highly polarized Si–N bond (1.742 Å; Charge: Si 0.96; N –0.53). The calculated value for the dimerization energy of **7A** is –29 kcal/mol, reflecting the presence of a partial Si–N π-bond. Thus, the novel dimeric silicon compound **7** is the result of a head-to-tail dimerization of formally two Si=N bonds in **7A**.

Introduction

Many studies in the coordination chemistry of the main group elements have been centered on creating unusual electronic features. We have extensively investigated the ability of different diketoamines $RN[CH_2C(t\text{Bu})=O]_2$ (R = H, alkyl, aryl) as suitable mono-, di-, and trianionic chelate ligands for the stabilization of main group elements in unusual low oxidation states [1]. Thus, the ligand $HN[CH_2C(t\text{Bu})=O]_2$ (**1**) is able to stabilize Groups 14 and 15 elements in the oxidation states +1 and +2. Not only that, the corresponding trianionic amido-dienolate form of **1** is able to reduce main group metal atoms M after initial coordination and intramolecular L→M two-electron-transfer (π-donation). However, only a few examples are so far known which show intramolecular electron transfer in main group element complexes. Intriguing examples are the 3,7-di(*tert*-butyl)-5-aza-2,8-dioxa-1-pnictabicyclo[3.3.0]octa-2,4,6-trienes (ADPnO) [2], representing complexes of the heavier Group 15 elements (pnictogens) with the tridentate diketoamine ligand **1**, which can exist in the form of two different valence isomers (Scheme 1).

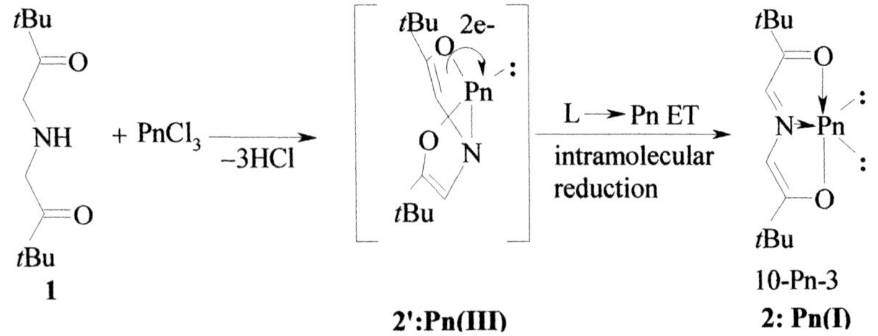

Scheme 1. Valence isomers of ADPnO complexes (Pn = Pnictogen atom).

The most important structural feature of these compounds is their preference for the planar structure **2** rather than the bent structure **2′**, due to an intramolecular two-electron transfer from the ligand backbone to the pnictogen atom. At the same time, the pnictogen atom (P, As, Sb) undergoes reduction from the formal oxidation state +3 to +1, leading to a so-called hypervalent 10-Pn-3 system [2]. The preferred non-classical planar structure **2** can also be simply described as donor-stabilized carbene-analogs of Group 15 elements (phosphinidenes, arsinidenes, and stibanidenes). The latter results prompted us to investigate the coordination properties of the ligand **1** toward the heavier Group 14 elements (Si, Ge, Sn, and Pb).

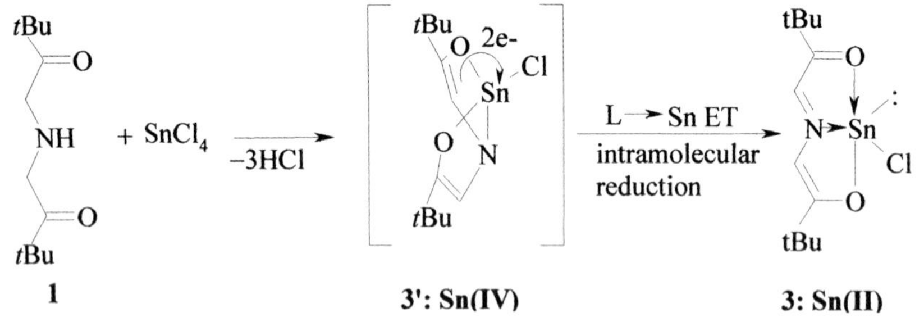

Scheme 2. The non-classical vs classical diketoamido-tin complexes **3** and **3′**.

The system should be in principle suitable for the synthesis of novel hypervalent carbene homologs (10-E-4 systems) of the Group 14 elements. Indeed, previous investigations have shown that the reaction of **1** with $SnCl_4$ in the presence of NEt_3 as an auxiliary base leads exclusively to the hypervalent 10-Sn-4 stannylene **3** instead of **3′** (Scheme 2) [3]. Surprisingly, the conversion of **1** with $GeCl_4$ readily consumes two molar equivalents $GeCl_4$ and results in the formation of the neutral $GeCl_3$ complex **5** and $GeCl_2$. (Scheme 3) [4].

The Si homologs have been hitherto unknown. However, related silicon compounds with higher coordination numbers than four have been of considerable interest in the last two decades [5, 6].

Here we describe the results of the reaction of the diketoamine ligand **1** with several halosilanes, which furnish the novel silicon dimers **6 – 11**.

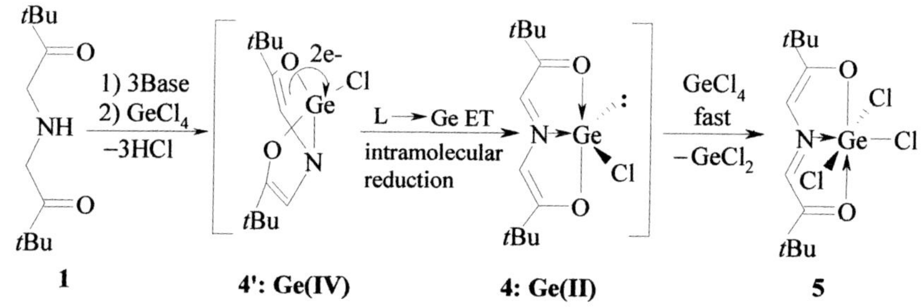

Scheme 3. Synthesis of hypervalent Ge(IV) complex **5**.

Results and Discussion

Following the same reaction route as that for stannylene **3** and the germanium complex **5**, the reaction of the diketoamine ligand **1** with different tetra- and trihalosilanes in the presence of NEt_3 as an auxiliary base furnished the hypervalent dimeric silanes **6 – 10** in high yields (> 85%) (Scheme 4). The Cl/F- ligand exchange reaction of **7** with ZnF_2 leads to the respective fluorosilane **11** in quantitative yield. The silanes **6 – 11** can be isolated in the form of air-stable solids with melting points in the range 110 – 130 °C. Their hydrolysis leads to the "free" ligand **1** and silica or sesquisiloxanes.

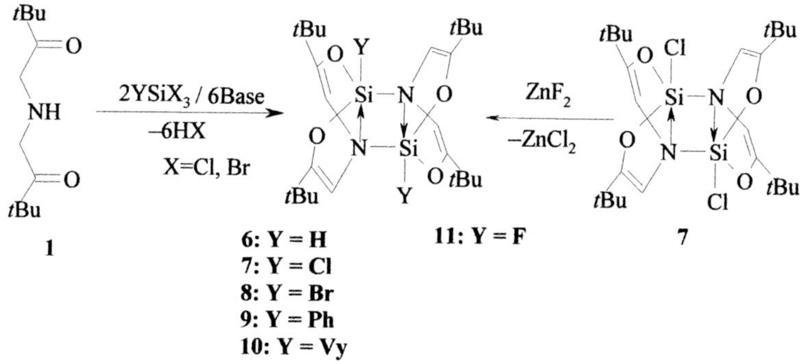

Scheme 4. The synthesis of the novel pentacoordinate silicon complexes **6 – 11**.

The constitution and compositions of **6 – 11** have been proven by NMR spectroscopy, mass spectrometry, and correct elemental analyses. Previous studies show that the ^{29}Si NMR chemical shift is susceptible to the coordination number around the silicon center (between +24 and –76 ppm for four-coordinate silicon and –47 to –110 ppm for five-coordinate silicon) [7], and this can be used to determine the coordination number of silicon in **6 – 11**.

In addition, the pentacoordination of silicon is reflected by the relatively large $^{1}J_{Si-H}$ coupling constant of 276.6 Hz in **6**, which is diagnostic for Si–H compounds with a pentacoordinate silicon atom. The compounds **6 – 11** exhibit ^{29}Si NMR resonances at –81.6 ($^{1}J_{Si-H}$ = 276.6 Hz), –85.9, –106.7, –38.3, –34.3, and –100.1 ($^{1}J_{Si-F}$ = 197.4 Hz) ppm. The latter reflect the common feature that electronegative substituents increase the shielding of the ^{29}Si nuclei. There were many precedents reported which show a higher shielding for the hypercoordinate silicon atom bearing electronegative substituents [8] and deshielding for those silicon nuclei with electropositive substituents [9]. The ^{1}H and ^{13}C NMR chemical shifts of the derivatives **6 – 11** are listed in Table 1. A comparison of the solution and CP-MAS solid state ^{29}Si and ^{13}C NMR data reveals that the dimeric structure is retained in solution.

Table 1. ^{1}H and ^{13}C NMR chemical shifts and coupling constants for **6 – 11**.

	^{1}H			^{13}C			
Complex	**CH_3**	**N*CH***	**Si–*H***	***CH_3***	***$C(CH_3)_3$***	***CN***	***CO***
6	1.18	5.67 (d, $^{4}J_{H-H}$ = 1.48 Hz)	4.22 (br. s)	27.56	34.87	106.01	164.17
7	1.12	5.55 (Si sat, J_{Si-H} = 12.9 Hz)	–	27.19	33.84	107.63	163.83
8	1.07	5.48	–	27.38	34.47	107.63	163.81
9	1.17	5.55	–	28.73	34.97	112.39	149.67
10	1.20	5.48	–	28.84	34.52	111.87	147.85
11	1.11	5.47	–	27.45	33.90	107.07	162.52

Another interesting feature of the dimeric compounds **6 – 11** is their low reactivity toward nucleophiles; for example, the dimer **7** is hydrolyzed in wet THF at room temperature within two days and does not react with MeLi, in contrast to the high reactivity of other penta- and hexacoordinate silicon compounds [10].

Recrystallization of **7** from dichloromethane affords single crystals which are suitable for X-ray diffraction analysis. Unfortunately, Because of a twinning problem, the results of the crystal measurements were not satisfactory for the determination of precise geometric parameters. However, the molecular array of **7** has been elucidated and the results are shown in Fig. 1. The analysis revealed that **7** is a new type of silicate dimeric species.

The compound **7** crystallizes in the monoclinic space group $P2_1/n$. Its structure consists of a planar four-membered Si_2N_2 ring as a central structural motif, which results from the head-to-tail Si–N dimerization of two monomeric units. The nitrogen atoms are tetrahedral and the silicon atoms trigonal-bipyramidal coordinated. The chlorine and nitrogen atoms of the chelating ligand prefer the axial positions.

What is the driving force for the dimerization? Although many dialkoxyamino-substituted silane compounds have been reported [11], they possess exclusively monomeric structures. In order to understand the high tendecy for dimerization, we carried out theoretical calculations of the monomeric structure of the chloro-substituted silane **7A**. A restricted hybrid HF-DFT calculation was performed using the Spartan .02 Quantum Mechanics Program [12]. Initial studies were carried out with the Hartree-Fock (HF) method with 3-21G and 6-31G basis sets, followed by density-functional theory (DFT) [13] calculations. The gas-phase structure was optimized at the B3LYP/6-311G** level of theory, leading to a folded geometry for the monomeric **7A** (Fig. 2) and a strongly distorted tetrahedral coordination around the silicon atom.

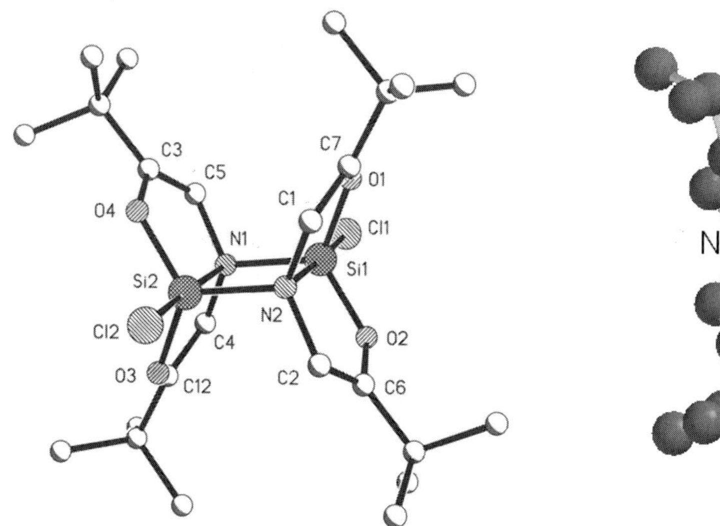

Fig. 1. Molecular structure of 7.

Fig. 2. DFT calculated structure **7A**.

Table 2. Selected calculated distances [Å] and angles [°] for **7A**.

Bond	Length [Å]	Angle	Angle [°]
Si–N	1.742	Cl-Si-N	125.37
Si–O	1.669	O-Si-O	124.61
Si–C	2.049	N-Si-O	96.61
N–C	1.432	Cl-Si-N	107.46
C–C	1.346	C-N-C	122.42
C–O	1.414		

The theoretical calculations revealed an Si–N distance of 1.742 Å. This bond length indicates some multiple bond character (Si=N distance in iminosilanes 1.611 Å [14], Si–N distance in aminosilanes 1.75 Å [15]). This is also supported by the calculated charge-density distribution in the highest occupied MOs. The HOMO of **7A** (Fig. 3), being a little extended, exhibits a symmetrical π charge distribution around the C=C bonds and an unsymmetrical π charge distribution around the plane between the nitrogen and silicon atoms. The LUMO in **7A** is mainly localized on the silicon atoms, represented by a large contribution of the *p*-type orbital of Si and minor contributions of N and O atoms. The calculated charge distributions in **7A** shows an increase in the positive charge on the silicon atom and an increased negative charge at the nitrogen atom (Si: 0.96; N: –0.53).

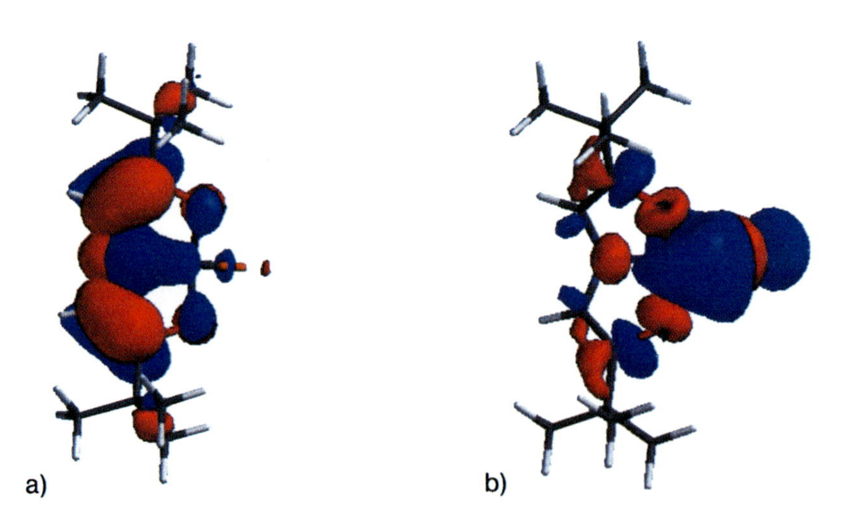

Fig. 3. Qualitative HOMO (a) and LUMO (b) representations for 7**A** calculated by means of the B3LYP level of theory, basis set 6-311G**.

The calculated value for the energy of dimerization of **7A** is, at –29 kcal/mol, remarkably exothermic. We assume that, because of the increased negative charge of the nitrogen atom, the latter undergoes nucleophilic attack by the silicon atom of another molecule, affording, in a head-to-tail dimerization, the new silicon complex **7** (Scheme 5), with a central planar four-membered Si_2N_2 ring.

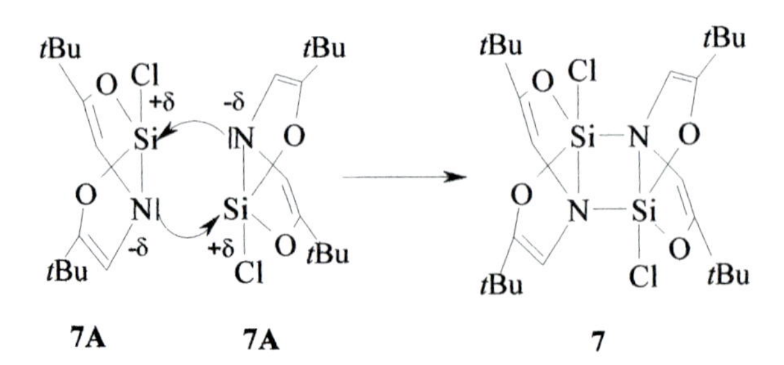

Scheme 5. Dimerization of 7**A** to 7.

The pathway is similar to that proposed by Verkade et al. for the formation of the dimeric silicon compound $[Si(HNCH_2CH_2)_2(NCH_2CH_2)N]_2$ [16].

Conclusions and Outlook

The simple diketoamine ligand **1** is predestined for stabilization of the main group elements in unusual low oxidation states. Thus, the conversion of the diketoamine ligand **1** with different silicon halides afforded the novel dimeric pentacoordinate silicon complexes **6** – **11**. In this case, monomeric silicon complexes having a highly polarized Si–N π-bond have been proposed as reactive intermediates. The latter undergo head-to-tail Si–N dimerization to afford the silicate complexes **6** – **11**. Synthesis of a kinetically stabilized monomeric silicon compound is the main target for the future. Further efforts are being directed toward determining whether the introduction of more bulky ligands attached to silicon will give monomomeric silicon compounds with silicon atoms in the formal oxidation state +2 (silylene-like).

References

[1] M. Driess, N. Dona, K. Merz, *J. Chem. Soc. Dalton Trans*. **2004**, 3176.

[2] a) S. A. Culley, A. J. Arduengo, *J. Am. Chem. Soc*. **1984**, *106*, 1164; b) A. J. Arduengo, A. C. Steward, *J. Am. Chem. Soc*. **1987**, *109*, 627; c) A. J. Arduengo, A. C. Steward *Chem. Rew*. **1994**, *94*, 1215; d) The N-X-L nomenclature system has been described: C. W. Perkins, J. C. Martin, A. J. Arduengo III, W. Lau, A. Algeria, J. K. Kochi, *J. Am. Chem. Soc*. **1980**, *102*, 7753. N valence electrons about X, bonded to L ligands.

[3] G. Bettermann, A. J. Arduengo, *J. Am. Chem. Soc*. **1988**, *110*, 877.

[4] M. Driess, N. Dona, K. Merz, *Chem. Eur. J.* accepted.

[5] a) C. Chuit, R. J. P. Corriu, C. Reye, J. C. Young, *Chem. Rev*. **1993**, *93*, 1371; b) R. J. P. Corriu, J. C. Young, in *The Chemistry of Organic Silicon Compounds* (Eds.: S. Patai, Z. Rappoport), Wiley, NY, **1989**, 1241 – 1289; c) R. R. Holmes, *Chem. Rev*. **1990**, *90*, 17; d) I. Kalikham, S. Krivonos, A. Ellern, D. Kost, *Organometallics* **1996**, *15*, 5073; e) S. P. Narula, R. Shankar, M. Kumar, R. K. Chadha, C. Janiak, *Inorg. Chem*. **1997**, *36*, 3800; f) W. Ziche, B. Ziemer, John, B. N. Auner, *J. Organomet. Chem*. **1997**, *531*, 245; g) M. Tasaka, M. Hirotsu, M. Kojima, S. Utsuno, Y. Yoshikawa, *Inorg. Chem*. **1996**, *35*, 6981.

[6] W. S. Sheldrick, W. Z. Wolfsberger, *Naturforschung* **1977**, *32b*, 22; b) D. G. Anderson, A. J. Blake, *Angew. Chem., Int. Ed. Engl*. **1986**, *25*, 107; c) Y. Wan, J. G. Verkade, *Organometallics* **1996**, *15*, 5769; d) L. M. Englehardt, C. P. Junk, W. C. Patalinghug, S. E. Rodney, C. L. Raston, *J. Chem. Soc., Chem. Commun*. **1991**, 930; d) A. G. Blake, E. A. V. Ebsworth, A. J .Welch, *Acta Crysallogr., Sect. C* **1984**, *40,* 895.

[7] Y. Takeuchi, T. Takayama, in *The Chemistry of Organosilicon Compounds*, (Eds.: Z. Rappoport, Y. Apeloig), John Wiley & Sons, Chichester **1998**, Vol. 2, Chapter 6, p. 267 – 354.

[8] K. Tamao, M. Asahara, A. Kawachi, A. Toshimitsu, *J. Organomet. Chem*. **2002**, *643*, 479; b) B. J. Helmer, R. West, R. J. P. Corriu, M. Poitier, G. Royo, A. de Saxcé, *J. Organomet. Chem*. **1983**, *251*, 285; c) E. A. Williams, in *The Chemistry of Organosilicon Compounds*

(Eds.: Z. Rappoport, Y. Apeloig), John Wiley & Sons, Chichester **1989**, Vol. 1, Chapter 8, p. 511.

[9] F. Carré, G. Cerveau, C. Chuit, R. J. P. Corriu, N. K. Nayyar, C. Reyé, *Organometallics* **1990**, *9*, 1989; b) K. Tamao, M. Asahara, T. Saeki, A. Toshimitsu, *Chem. Lett.* **1999**, 35.

[10] J. Boyer, C. Brelière, R. J. P. Corriu, A. Kpoton, M. Poirier, G. Royo, *J. Organomet. Chem.* **1976**, *311*, C39; b) P. Arya, R. J. P. Corriu, K. Gupta, G. F. Lanneau, Z. Yu, *J. Organomet. Chem.* **1990**, *399*, 11.

[11] a) U. Wannagat, H. Bürger, P. Geymayer, G. Torper, *Monatsh. Chem.* **1964**, *95(1)*, 39; b) U. Wannagat, W. Veigl, H. Bürger, *Monatsh. Chem.* **1965**, *96(2)*, 593; c) J. Grobe, H. Schröder, N. Auner, *Z. Naturforsch. B* **1990**, *45(6)*, 785; d) W. Uhlig, C. Tretner, *J. Organomet. Chem.* **1994**, *467*, 31; f) U. Wannagat, H. Bürger, *Z. Anorg. Allg. Chem.* **1964**, *629*, 309.

[12] The RHF (Restricted Hartree-Fock) and B3LYP-DFT (Density Functional Theory) calculations were performed with 6-31G** basis sets using the Spartan .02 Quantum Mechanics Program package.

[13] R. G. Parr, W. Yang, *Density-Functional Theory of Atoms and Molecules*, Oxford University Press: New York, **1989**.

[14] J. Niesmann, U. Klingebiel, M. Shafer, R. Boese, *Organometallics* **1998**, *17*, 947.

[15] B. Jaschke, R. Herbst-Irmer, U. Klingebiel, T. Pape, *J. Chem. Soc., Dalton Trans.* **2000**, 1827.

[16] Y. Wan, J. G. Verkade, *Organometallics* **1996**, *15*, 5769.

Unique Switching of Coordination Number with Imine and Enamine Complexes of Group 14 Elements

Jörg Wagler,* Uwe Böhme, Gerhard Roewer

Institut für Anorganische Chemie, Technische Universität Bergakademie Freiberg
Leipziger Str. 29, D-09596 Freiberg, Germany
E-mail: joerg.wagler@chemie.tu-freiberg.de

Keywords: chelates, enamines, hypervalent compounds, schiff bases, silicon

Summary: Novel pentacoordinate Si complexes with enamine-functionalized salen type ligands react with various acids to give complexes with a hexacoordinate Si atom in high yield. These investigations were extended to analogous Ge- and Sn-containing precursors ($PhMCl_3$, M = Si, Ge, Sn). Interesting differences in the coordination behavior of the observed group 14 elements were shown by the X-ray structures of their complexes.

Results

Formation of Pentacoordinate Enamine Complexes

Recently we reported an easy route obtaining enamine-functionalized silicon compounds with a pentacoordinate silicon atom from organotrichlorosilanes and ethylene-*N,N'*-bis(2-hydroxyaceto-phenoneimine) (salen*H_2) [1].

Phenyltrichlorosilane and phenyltrichlorogermane show a similar reactivity pattern toward the salen type ligand **1** in presence of an excess of triethylamine to yield **2a** and **2b** (Scheme 1). Both complexes crystallize in the non-centrosymmetric monoclinic space group $P2_1$ with only one enantiomer in the asymmetric unit. Both the silicon atom and the germanium atom have a distorted trigonal bipyramidal coordination sphere (Fig. 1). The angle O2-Ge-N1 is significantly smaller than the analogous one O2-Si-N1 because of the bigger size of the germanium atom (see Table 1).

Phenyltrichlorostannane reacts with **1** surplus amine to yield the hexacoordinate tin compound **2c** only. An elimination of HCl from **2c** seems to be impossible. Even reaction with a proton sponge, *N,N,N',N'*-tetramethylnaphthalene-1,8-diamine, failed. **2c** crystallizes in a non-centrosymmetric monoclinic space group (Cc). The molecule involves the all-*cis* configuration of the salen type ligand. The angle (93.9°) of C19-Sn-Cl indicates a real *cis* arrangement of the chlorine atom and the phenyl group (Fig. 2) [2].

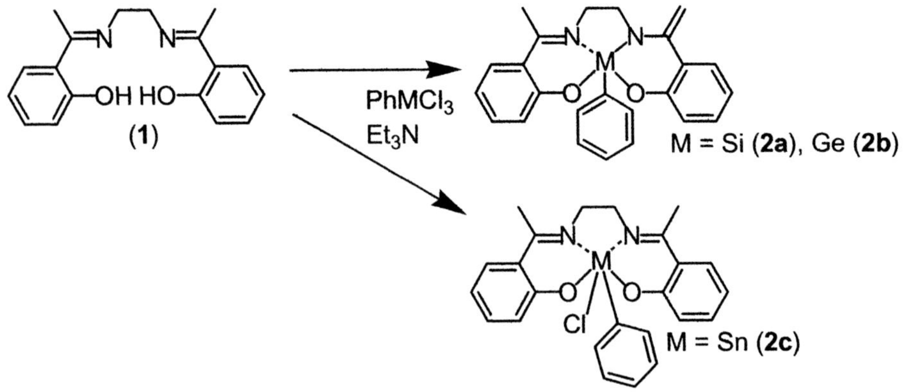

Scheme 1. Base-supported reactions between ligand **1** and phenyltrichlorosilane, -germane and -stannane.

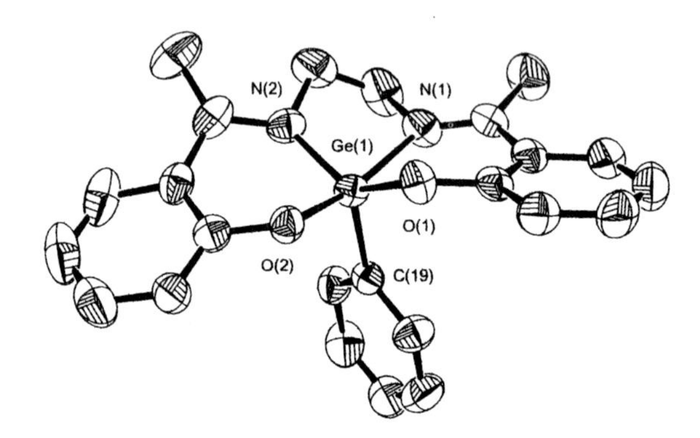

Fig. 1. Molecular structure of **2b** (ORTEP plot with 50% probability ellipsoids, hydrogen atoms omitted).

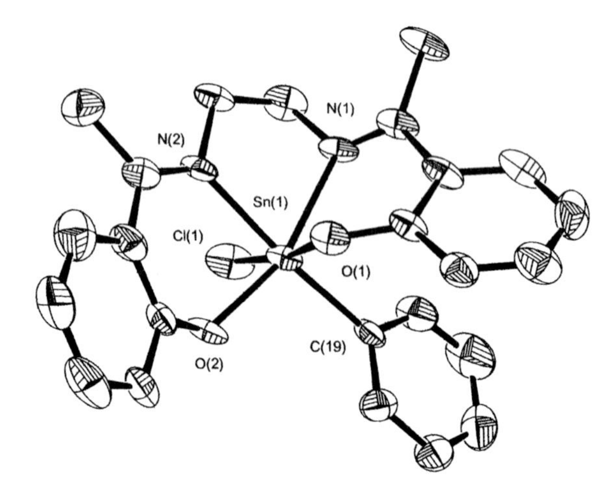

Fig. 2. Molecular structure of **2c** (ORTEP plot with 50% probability ellipsoids, hydrogen atoms omitted).

Table 1. Selected bond distances [Å] and angle [deg] of **2a**, **2b**, and **2c**.

	M–C(19)	M–N(1)	M–N(2)	M–O(1)	M–O(2)	N(1)-M-O(2)
2a	1.87	2.01	1.76	1.68	1.72	167.9
2b	1.93	2.09	1.86	1.82	1.85	162.1
2c	2.14	2.22	2.22	2.08	2.02	156.6

Reactions of Hypercoordinate Enamine Complexes with Acids

2a reacts with a variety of acids to give the adducts **3a** – **3f** (Scheme 2). Hexacoordination of the Si atom was proven by ^{29}Si CP/MAS NMR in all cases (Table 2). The formation of the enamine-functionalized complexes **2a** and **2b** proceeds quite similarly, but an amazing difference between Si and Ge atoms is found in view of the coordination behavior toward the picrate anion (see **3b** and **3g** in Figs. 3 and 4).

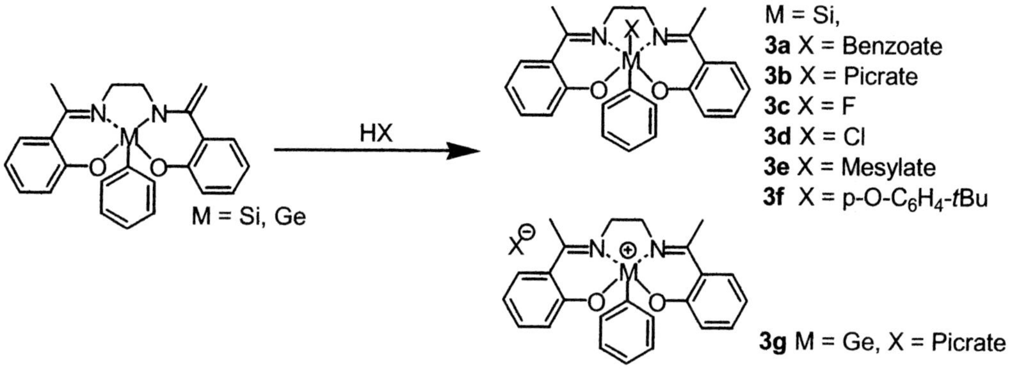

M = Si, Ge

HX

M = Si,
3a X = Benzoate
3b X = Picrate
3c X = F
3d X = Cl
3e X = Mesylate
3f X = p-O-C_6H_4-*t*Bu

3g M = Ge, X = Picrate

Scheme 2. Reactions of the enamine complexes **2a**, **2b** with Brønsted acids.

Table 2. ^{29}Si CP/MAS NMR data of compounds **3a** – **3f**.

Compound	3a	3b	3c	3d	3e	3f
$\delta_{iso}(^{29}Si)$ [ppm]	–186.9	–173.5	(–179.9, –182.8)*	–168.4	–174.3	–183.1

* 2 peaks because of *J*(SiF) coupling, 300 MHz CP/MAS NMR spectrum, ν_{rot} = 4 kHz.

The molecular structure analyses of **3a** and **3b** (Fig. 3) confirm the hexacoordination of the Si atom. The coordination sphere around the Si atom is distorted octahedral. In both cases the monodentate ligands (phenyl group as well as benzoate and picrate) are situated in the *trans* position. This is probably typical for Salen-Si complexes [3, 4], but their configuration is quite different from that of the hexacoordinate tin compound **2c**.

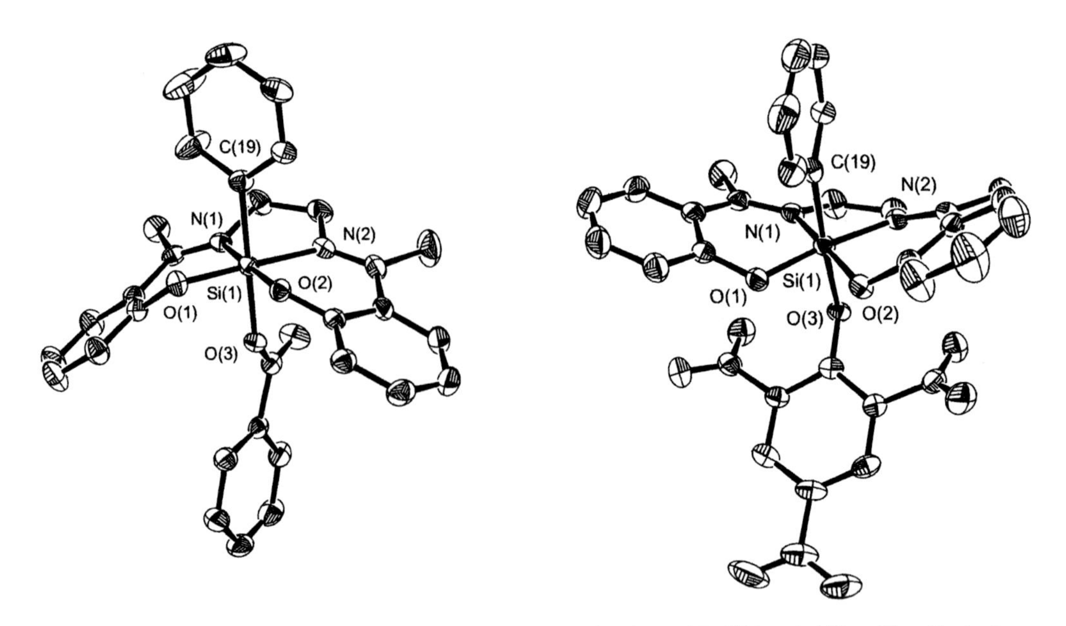

Fig. 3. Molecular structures of **3a** (left) and **3b** (right) (ORTEP plots with 50% probability ellipsoids, hydrogen atoms and solvent molecules omitted).

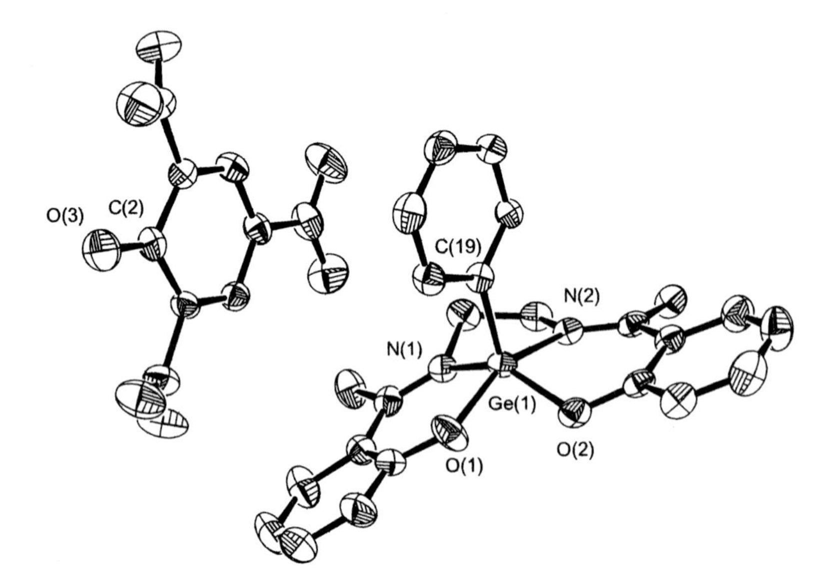

Fig. 4. Molecular structure of **3g** (ORTEP plot with 50% probability ellipsoids, hydrogen atoms omitted).

There are no significant differences between **3a** and **3b** regarding the coordination geometry of the salen type ligand or the phenyl group (Si–N ≈ 1.93 Å, Si–O(1),O(2) ≈ 1.73 Å, Si–C(19) ≈ 1.95

Å **3a**, 1.91 Å **3b**). More interesting is the different coordination behavior of the added ligands benzoate and picrate respectively, which is indicated by the different Si–O bond lengths (Si–O(3) ≈ 1.82 Å **3a**, 1.90 Å **3b**). This effect is expected to be the main influence on the shielding of the ^{29}Si nucleus (see Table 2: ^{29}Si NMR data).

There is also an interesting difference in the coordination behavior of the picrate ligand in compounds **3b** and **3g**. The Si atom attains coordination with the picrate oxygen atom O(3), but the Ge atom in **3g** prefers to be pentacoordinate although it is bigger, thus providing more space for ligands than the Si-atom (Fig. 4). The coordination of picrate in **3b** gives rise to a slight stretching of the O(3)–C(2) bond (1.29 Å in **3b**, 1.24 Å in **3g**).

In contrast to the neutral pentacoordinate germanium compound **2b**, the coordination sphere of the Ge atom in the cation of **3g** is approximately square-pyramidal, with the phenyl group on the top. This must be due to the similar strength of both Ge–N bonds (1.98 Å, 2.02 Å).

Conclusions

Si and Ge yield enamine complexes with special salen type ligands, but Sn does not! Salen ligands are able to adapt to the sterical requirements of the other substituents in their environment, but the coordinated acid anions hardly disturb the chelating effect of the salen ligand. Regarding the variety of anions, their coordination strength toward the (salen*)PhSi-unit differs in each case

Cationic salen germanium complexes seem to be more stable than analogous ones of silicon. Therefore, even picrate coordinates with the silicon atom of **3b**, although an unusually long Si–O bond results.

Acknowledgments: This work was financially supported by Deutsche Forschungsgemeinschaft and Fonds der Chemischen Industrie. We gratefully acknowledge the X-ray structure analysis of **3a**, **3b** and **3g** by Sigrid Goutal, Institut für Organische Chemie, Technische Universität Dresden, Bergstr. 66, D-01069 Dresden (Germany).

References

[1] J. Wagler, U. Böhme, G. Roewer, *Angew. Chem. Int. Ed.* **2002**, *41*, 1732.

[2] To the best of our knowledge, this is the first X-ray structure of a truly *cis* configurated hexacoordinate tin(VI) compound with tetradentate salen type ligand and two monodentate ligands (September 2003). Several hexacoordinate tin compounds with tetradentate salen type ligands and two monodentate substituents at the Sn atom have been studied by X-ray structure analysis. The angle A-Sn-B between the monodentate substituents is significantly larger than 100° in each case, e.g., Me-Sn-Me ca. 160° in (salen)$SnMe_2$: M. Calligaris, G. Nardin, L. Randaccio, *J. Chem. Soc., Dalton Trans.* **1972**, 2003. For further examples see: S.-G. Teoh,

G.-Y. Yeap, C.-C. Loh, L.-W. Foong, S.-B. Teo, H.-K. Fun, *Polyhedron* **1997**, *16*, 2213; D. K. Dey, M. K. Das, H. Nöth, *Z. Naturforsch.* **1999**, 145; D. K. Dey, M. K. Saha, M. K. Das, N. Bhartiya, R. K. Bansal, G. Rosair, S. Mitra, *Polyhedron* **1999**, *18*, 2687.

[3] F. Mucha, J. Haberecht, U. Böhme, G. Roewer, *Monatsh. Chem.* **1999**, *130*, 117; F. Mucha, U. Böhme, G. Roewer, *Chem. Commun.* **1998**, 1289.

[4] J. Wagler, Th. Doert, G. Roewer, *Angew. Chem. Int. Ed.* **2004**, *43*, 2441.

[5] M. S. Singh, P. K. Singh, *Main Group Met. Chem.* **2000**, *23(3)*, 183.

Structures of Novel Diorgano-Substituted Silicon Complexes with Hexacoordinate Silicon Atom

Gerhard Roewer,* Jörg Wagler

Institut für Anorganische Chemie, TU Bergakademie Freiberg
Leipziger Straße 29, D-09596 Freiberg, Germany
Tel.: +49 3731 39 3174 — Fax: +49 3731 39 4058
E-mail: gerhard.roewer@chemie.tu-freiberg.de

Keywords: chelates, hypervalent compounds, Schiff bases, silicon

Summary: Our investigations on the coordination behavior of ethylene-*N,N'*-bis(2-oxy-4-methoxybenzophenoneiminate) **1**, a tetradentate chelating ligand of the salen type, toward diorgano-substituted silicon atoms led to the syntheses and X-ray structure analyses of novel hexacoordinate silicon complexes. The first X-ray structures of a metal-free hexacoordinate dimethylsilane and of a hexacoordinate *cis*-configurated silicon complex with a salen type ligand are presented.

Introduction

Hypercoordinate silicon complexes with tetradentate (O, N, N, O)-chelating ligands of the salen type are expected to exhibit unusual chemical and physical properties because of the higher coordination number of the silicon atom [1, 2]. Therefore, several attempts were made to synthesize such compounds [2, 3]. Starting from easily available silicon compounds such as $SiCl_4$ or other chlorosilanes, conversion with salen type ligands mostly yielded complexes with a hexacoordinate [2, 3] and, in some cases, pentacoordinate silicon atom [4]. Unfortunately, there are only a few examples where the coordination geometry has been confirmed by X-ray structure analysis [2, 4].

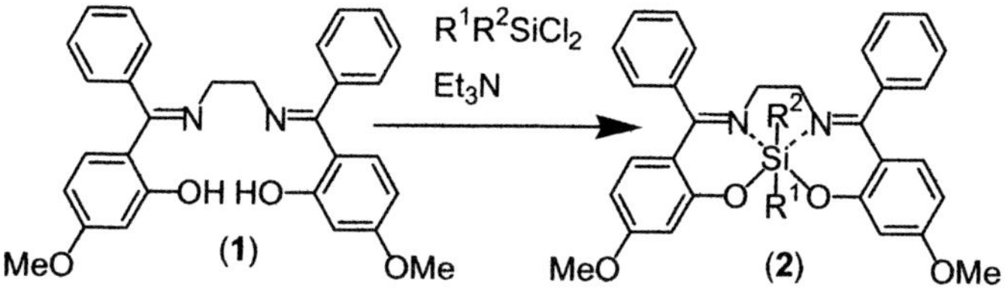

$R^1R^2SiCl_2$, Et_3N; (1) → (2)

$R^1 = R^2 = Ph$ (**2a**)
$R^1 = Ph, R^2 = Me$ (**2b**)
$R^1 = R^2 = Me$ (**2c**)
$R^1 = Ph, R^2 = Et$ (**2d**)
$R^1 = Ph, R^2 = Cy$ (**2e**)
$R^1 = Vi, R^2 = Me$ (**2f**)
$R^1 = Vi, R^2 = Ph$ (**2g**)
$R^1 = Ph, R^2 = CH_2$ (dimeric complex) (**2h**)

Scheme 1. Formation of hexacoordinate silicon complexes with ligand **1**.

Generally, the hypercoordination is promoted by electron-withdrawing substituents (e. g. –F,

–Cl, –CF_3), which create a higher Lewis-acidity at the silicon atom. For this reason, there are so far only a few structurally confirmed examples of hexacoordinate dialkylsilanes [5].

Now we have succeeded in preparing and crystallizing novel hexacoordinate diorgano-substituted Si-complexes with a special salen type ligand (Scheme 1).

Results

The Ligand: This salen type ligand is easily obtained by condensation between ethylenediamine and 2-hydroxy-4-methoxybenzophenone. The methoxy groups offer various possibilities for an arrangement of the molecule in the crystal lattice. In contrast with this flexible part, the rotation of the phenyl groups is retarded in solution. This is easily proven by ^{13}C NMR spectroscopy of compounds **2b, 2d, 2e, 2f**, and **2g**. There are always two more aromatic ^{13}C NMR signals due to diastereotopic *o*- and *m*- positions at the phenyl groups.

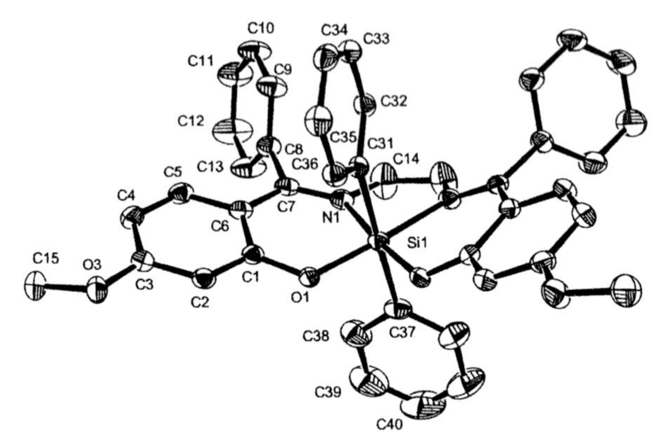

Fig. 1. Molecular structure of **2a** (ORTEP plot with 50% probability ellipsoids, hydrogen atoms and chloroform molecule omitted).

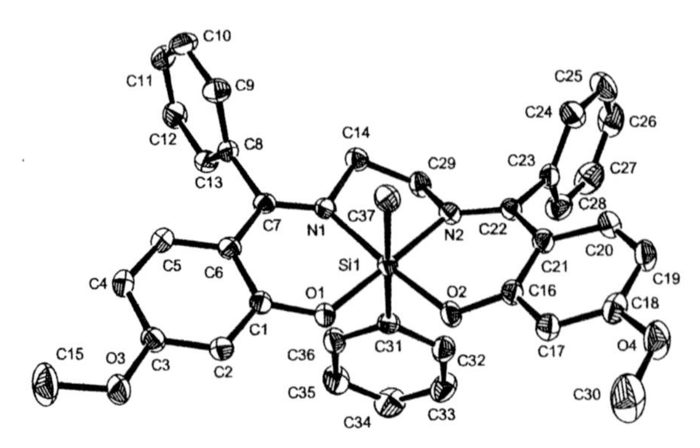

Fig. 2. Molecular structure of **2b** (ORTEP plot with 50% probability ellipsoids, hydrogen atoms and chloroform molecules omitted).

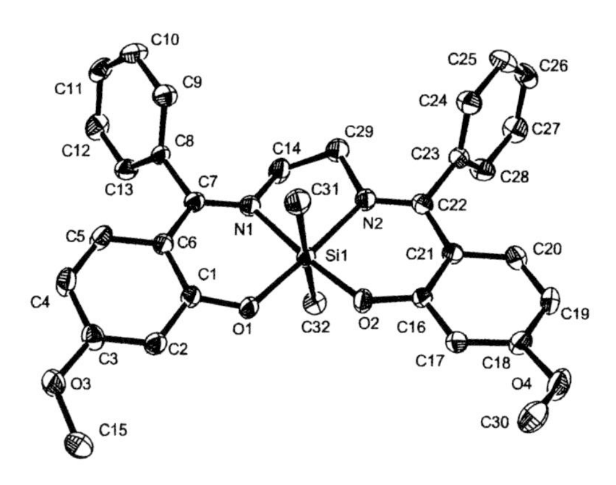

Fig. 3. Molecular structure of **2c** (ORTEP plot with 50% probability ellipsoids, hydrogen atoms omitted).

Coordination sphere of the Si-atom: Compounds **2a**, **2b**, and **2c** were characterized by X-ray single-crystal structure analysis. They represent typical hexacoordinate silicon complexes (Figs. 1 – 3). Selected bond distances and angles are given in Tables 1 and 2. The Si–N and Si–O bond distances differ only slightly because of a restricted coordination sphere of the Si atom within the chelate. Owing to this chelating tetradentate ligand system, the dative Si–N bonds are forced to be quite short (1.96···2.00 Å). In these three structures the coordination sphere around the silicon atom is nearly an ideal octahedron, with the donor atoms of the salen type ligand in equatorial positions, but the carbon substituents (Ph, Me) in axial positions.

Table 1. Selected bond distances [Å] of **2a**, **2b** and **2c**.

	2a*	2b	2c
Si–C31	1.97	1.96	1.94
Si–C32	–	–	1.93
Si–C37	1.94	1.94	–
Si–N1	1.96	1.99	1.97
Si–N2	N1 = N2	1.99	2.00
Si–O1	1.76	1.76	1.77
Si–O2	O1 = O2	1.78	1.78
N1–C7	1.30	1.30	1.30
N2–C22	≡ N1–C7	1.29	1.30

* **2a** is situated on a crystallographic bisecting plane

In spite of the similar bond distances and angles, the ^{29}Si NMR properties of compounds **2a** – **2c** differ significantly (see Table 3). The hybridization of the Si-linked carbon atoms is expected to create the main influence on the ^{29}Si chemical shift. This suggestion is also supported by the NMR data of compounds **2d** – **2h**. The presence of Si-linked sp^2-hybridized carbon atoms gives rise to an upfield shift of the ^{29}Si NMR signal. The resonances of the complexes with one Si-linked sp^3-hybridized C atom plus one sp^2-hybridized C atom can be found in a narrow range (–167.6 ppm to –172.4 ppm), while the chemical shifts of **2a**, **2g** (2 sp^2-hybridized Si-linked C atoms) are significantly different from that of **2c** (2 sp^3-hybridized Si-linked C atoms).

Table 2. Selected bond angles [deg] of **2a**, **2b**, and **2c**.

	2a*	2b	2c
C31–Si–C32	–	–	175.4
C31–Si–C37	178.2	175.1	–
N1–Si–O2	174.8	174.5	175.5
N2–Si–O1	174.8	174.0	175.8
N1–Si–C31	89.1	90.1	88.1
N2–Si–C31	89.1	85.7	89.8
N1–Si–C32	–	–	88.2
N2–Si–C32	–	–	87.0
N1–Si–C37	89.5	86.7	–
N2–Si–C37	89.5	90.2	–
N1–Si–N2	82.3	82.9	83.4

* Molecule **2a** is situated on a crystallographic bisecting plane.

Table 3. ^{29}Si NMR chemical shifts of compounds **2a**, **2b**, **2c**, **2d**, **2e**, **2f**, **2g**, and **2h** (in $CDCl_3$).

Compound	2a	2b	2c	2d	2e	2f	2g	2h
δ(^{29}Si $CDCl_3$) / ppm	–177.6	–171.1	–165.3	–169.6	–170.4	–172.4	–178.2	–167.6

In a similar synthesis starting with ligand **1** and 1,1-dichlorosilacyclobutane, complex **2i** (Fig. 4) is formed in high yield. **2i**, recrystallized from mesitylene, crystallizes in the orthorhombic space group Pbcn with 8 molecules in the unit cell. The cyclobutane ring system remains intact in the molecule. Therefore, a significant distortion of the tetradentate ligand occurs. The coordination sphere of the Si atom represents a distorted octahedron with an all-*cis*-configurated salen type ligand. In spite of the ligand's distortion, the Si–O, Si–C, and even the Si–N bond distances are similar to those of the *trans* configurated compounds **2a** – **2c** (see Table 4).

The ^{29}Si chemical shift of **2i** (–153.7 ppm, solution in $CDCl_3$) differs from the corresponding values for compounds **2a** – **2h**. The strong downfield shift of +11.6 ppm compared to the dimethyl compound **2c** is expected to originate in ring strain effects. The ethylene bridge between the nitrogen atoms of ligand **1** represents the flexible part of the resulting complexes. It enables an extreme distortion of the tetradentate ligand to even form compounds with all-*cis*-positioned donor

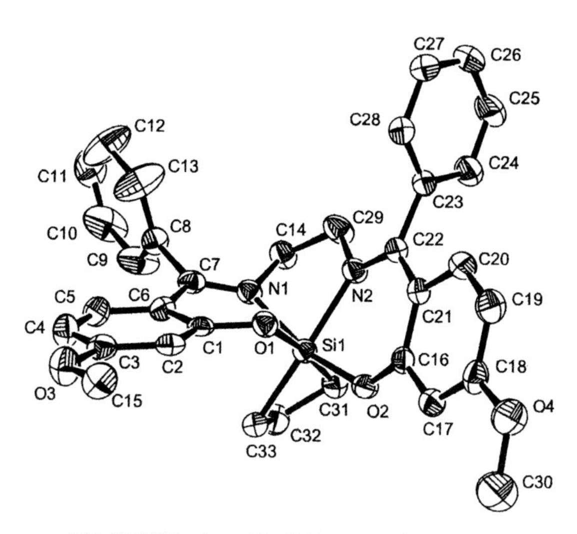

Fig. 4. Molecular structure of **2i** (ORTEP plot with 50% probability ellipsoids, hydrogen atoms and mesitylene molecule omitted).

Table 4. Selected bond distances [Å] and angles [deg] of **2i**.

Bond	d [Å]	Angle	[deg]
Si1–N1	1.98	**N1–Si1–O2**	164.8
Si1–N2	1.98	**N2–Si1–C33**	167.9
Si1–O1	1.81	**O1–Si1–C31**	172.2
Si1–O2	1.77	**C31–Si1–C33**	73.6
Si1–C33	1.92	**Si–C33–C32**	90.8
Si1–C31	1.94	**C33–C32–C31**	97.7
C7=N1	1.30	**C32–C31–Si1**	90.0
C22=N2	1.30		

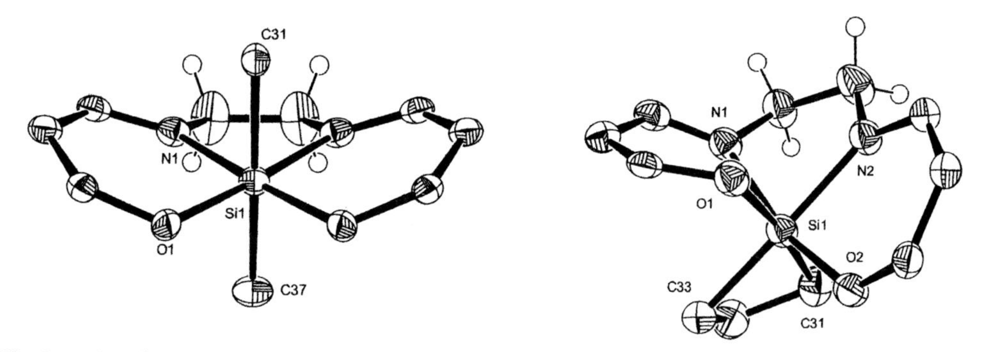

Fig. 5. Coordination sphere of the Si atom in **2a** (left) and **2i** (right).

atoms. Unlike this, the backbone of the six-membered (O,N)-chelates remains nearly unchanged because of conjugated π electron systems. Therefore, the silicon atom in **2i** is not situated in the same plane as the atoms O1, C1, C6, C7, N1 (see Fig. 5).

Conclusions

The chelating properties of ligand **1** provide access to many novel hypercoordinate silicon complexes, even with two alkyl groups at the Si atom. The first X-ray structure of a truly hexacoordinate dimethylsilane was obtained.

The ligand's flexible methoxy groups support solvent- and temperature-dependent dissolution/crystallization, unlike the ligand's backbone, which is rather rigid.

The large variety of accessible diorgano-substituted Si complexes enables the systematic investigation of their spectroscopic properties, depending on the organic substituents.

Usually in the case of salen Si complexes, the *trans* configuration is formed predominantly [2]. Now the first X-ray structure of a *cis*-configurated Si complex with salen type ligand has been presented.

Acknowledgments: This work was financially supported by Deutsche Forschungsgemeinschaft and Fonds der Chemischen Industrie. We gratefully acknowledge the X-ray structure analysis by Sigrid Goutal, Institut für Organische Chemie, Technische Universität Dresden, Bergstr. 66, D-01069 Dresden (Germany).

References

[1] C. Chuit, R. J. P. Corriu, C. Reye, J. C. Young, *Chem. Rev.* **1993**, *93*, 1371 – 1448.

[2] F. Mucha, J. Haberecht, U. Böhme, G. Roewer, *Monatsh. Chem.* **1999**, *130*, 117 – 132; F. Mucha, U. Böhme, G. Roewer, *Chem. Commun.* **1998**, 1289 – 1290.

[3] M. S. Singh, P. K. Singh, *Main Group Met. Chem.* **2000**, *23(3)*, 183 – 188.

[4] J. Wagler, U. Böhme, G. Roewer, *Angew. Chem. Int. Ed.* **2002**, *41*, 1732 – 1734.

[5] Up to September 2003, the only published X-ray structures of hexacoordinate dialkylsilanes without Si metal coordination are described in: H. H. Karsch, R. Richter, E. Witt, *J. Organomet. Chem.* **1996**, *521* (*1* – *2*), 185 – 190 (a hexacoordinate silacyclobutane); J.-Y. Zheng, K. Konishi, T. Aida, *Inorg. Chem.* **1998**, *37* (*10*), 2591 – 2594 (a bis-(trimethylsilyl-methyl)silane with tetraphenylporphyrin ligand); K. Hensen, F. Gebhardt, M. Bolte, *Z. Anorg. Allg. Chem.* **1997**, *623* (*4*), 633 – 636 (a hexacoordinate silacyclopentane cation).

Novel Hypercoordinate Silicon Complexes from Silicon Tetrahalides and Bidentate <O,N> Donor Ligands

Michael Schley,* Uwe Böhme, Gerhard Roewer

Institut für Anorganische Chemie, TU Bergakademie Freiberg
Leipziger Straße 29, D-09596 Freiberg, Germany
Tel.: +49 3731 393194 — Fax: +49 3731 394058
E-mail: Gerhard.Roewer@chemie.tu-freiberg.de

Erica Brendler

Institut für Analytische Chemie, TU Bergakademie Freiberg
Leipziger Straße 29, D-09596 Freiberg, Germany
Tel.: +49 3731 393194 — Fax: +49 3731 394058

Keywords: hypervalent silicon, Schiff bases, 8-hydroxyquinoline

Summary: Treatment of SiX_4 (X = Cl, Br) with bidentate ligands 2-[(*N*-benzylimino)-phenylmethyl]-5-methoxyphenol L^1H and 8-hydroxyquinoline L^2H leads to hyper-coordinate bis- or tris-chelates of silicon. These complexes were characterized by means of multinuclear NMR spectroscopy. The X-ray crystal structure of $(L^1)_2SiCl_2$ was determined. Furthermore the molecular structure of dichlorobis(8-oxoquinolinato)-silicon $(L^2)_2SiCl_2$ is presented, which is concluded from the comparison of calculated with experimental ^{29}Si CP/MAS NMR data.

Introduction

Interest in hypercoordinate silicon compounds has grown considerably in the last thirty years. 8-hydroxyquinoline L^2H and 2-[(*N*-benzylimino)phenylmethyl]-5-methoxyphenol L^1H are potential ligands to react with SiX_4 (X = Cl, Br) in solution. Various products are formed depending on solvent properties, X (reactivity of silicon source), and stoichiometric ratio of starting materials. The synthetic pathways yielding the investigated novel silicon complexes are given in Scheme 1.

The characterization of these compounds was carried out with ^{29}Si NMR spectroscopy. Some of the products, mainly those with the 8-oxoquinolinato ligand, are nearly insoluble in most organic solvents. Thus, they had to be characterized by solid-state ^{29}Si CP/MAS NMR spectroscopy. The spectroscopic data obtained are presented in Table 1.

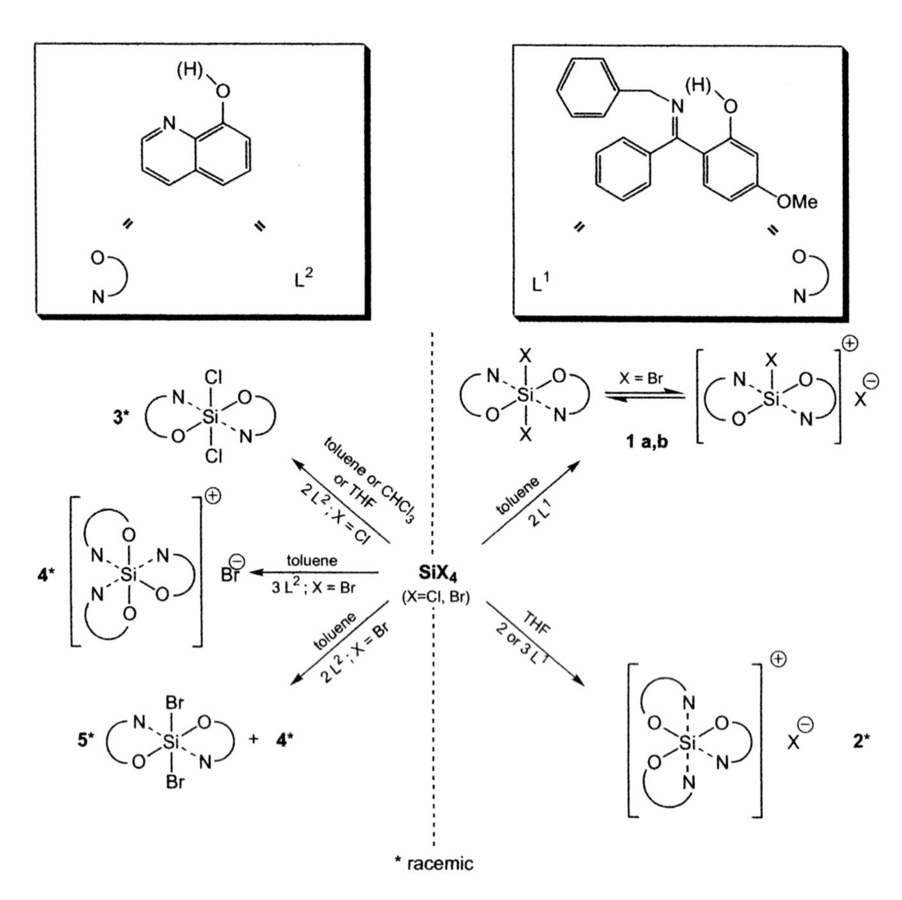

Scheme 1. Synthesis of hypercoordinate silicon compounds with 2-[(benzylimino)phenylmethyl]-5-methoxyphenolato- (right, L^1) and 8-oxoquinolinato- (left, L^2) ligand.

Table 1. ^{29}Si NMR data of hypercoordinate silicon compounds.

Compound	X	^{29}Si chemical shift [ppm]	
		Solution ($CDCl_3$)	Solid state
1a	Cl	–179.2	–
1b	Br	–134.2	–180[a]
2	Cl, Br	–186.5	–186.3
3	Cl	–	–158.3
4	Cl, Br	–	–150.5, –151.9
5	Br	–	–170[a]

[a] broad signal due to residual dipolar coupling to ^{79,81}Br.

Several configurations (*cis*/*trans*/*fac*/*mer*) of the novel hexacoordinate Si complexes could be expected. Some of these configurations were identified by ^{1}H NMR spectroscopy and, in the case of **1a**, by X-ray structure analysis.

Results and Discussion

Hypercoordinate Silicon Compounds with 2-[(*N*-benzylimino)phenylmethyl]-5-methoxyphenolato-ligand (L^1)

Dihalido silicon complexes **1a** and **1b** can be obtained in toluene. In contrast, synthesis in THF with equal stoichiometric ratio of educts, 2/1 (SiX_4/chelating ligand), leads to the threefold ligand-substituted hypercoordinate silicon compound **2**. The spectrum of products is probably influenced by the properties of the solvent. Because of a higher dielectric constant and donor number (compared with toluene), THF favors the ionic dissociation of an Si–X bond. Thus, a third ligand L^1 is able to coordinate with the silicon atom. The ionic structure of complex **2** in solution is confirmed by identical ^{29}Si-chemical shifts of the products from the reaction of $SiCl_4$ and $SiBr_4$ with 2-[(*N*-benzylimino)phenylmethyl]-5-methoxyphenol (L^1H), respectively. The ^{1}H NMR spectrum of compound **2** shows three signals for O–CH_3 protons and six doublets of diastereotopic methylene protons caused by the meridional configuration. The hypercoordinate bromosilicon compound **1b** dissociates in solution, indicated by an unusual low field shift of the ^{29}Si resonance compared with its value in the solid state. Analogous examples with other ligands have been published by D. Kost et al. [1].

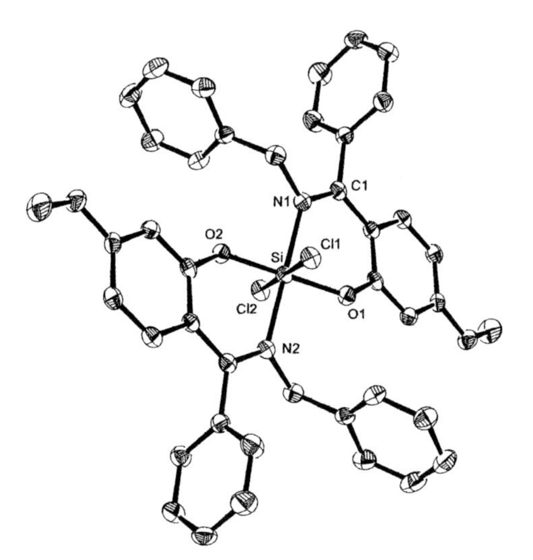

Fig. 1. Molecular structure of **1a** (ORTEP plot with 50% probability ellipsoids, hydrogen atoms omitted).

Table 2. Crystallographic data of **1a**.

	Bond length [Å]	Bond angle [°]
Si–Cl1	2.26	–
Si–N1	1.70	–
Si–O1	1.94	–
N1–C1	1.31	–
X1-Si-X2 (X=Cl, N, O)	–	180
N1-Si-O1	–	91.66
N1-Si-O2	–	88.34

1a was characterized by X-ray structure analysis (Fig. 1). It crystallizes in the monoclinic space group $P2_1/n$ with 2 molecules in the unit cell. The molecule has an inversion center at the Si atom, which is almost ideally octahedrally coordinated.

Hypercoordinate Silicon Compounds with the 8-Oxoquinolinato Ligand (L^2)

Compound **3** was obtained in toluene, THF, and $CHCl_3$ from reaction of $SiCl_4$ and 8-hydroxyquinoline in the stoichiometric ratio of 1 : 2. In contrast, the analogous reaction with $SiBr_4$ (in toluene) yields the tris-chelate complex **4**. Consequently, the spectrum of the products is predominantly influenced by the reactivity of the silicon source, but a significant dependence on the used solvents was not observed. However, this conclusion is contrary to the results for ligand L^1H (reactions with 2-[(*N*-benzylimino)phenylmethyl]-5-methoxyphenol) mentioned above. Compounds **3** and **4** were isolated, but we do not know exactly the configuration of the silicon-ligand sphere. We were not able to isolate the dibromo complex **5** from the mixture with **4**.

At first sight, the dihalido-bis-chelates **3** and **5** exhibit good solubility in DMSO, but the characterization of these compounds indicates their decomposition in this solvent. One product of this decomposition is a silicon complex with an ^{29}Si chemical shift at −151.8 ppm, which corresponds to the chemical shift of the analogous tris-chelate **4**. The solid-state ^{29}Si NMR spectrum of **4** shows two tightly situated signals probably caused by the meridional and facial product. This supposition is confirmed by the fact that 36 peaks of the aromatic carbon atoms result in the ^{13}C NMR (solution) spectrum. Probably, a transformation between facial and meridional configuration takes place in solution within the NMR time scale of the ^{29}Si and ^{13}C NMR spectroscopy.

Dichlorobis(8-oxoquinolinato)silicon (3) with Halido Ligands in *cis* or in *trans* Position ?

In the known molecular structure of a hexacoordinate silicon compound with the 8-oxoquinolinato ligand published by Klebe et al. [2] (Fig. 2 (left)), the methyl group and the chloro ligand are located in *cis* position to each other. In the case of compound **3** we expected to obtain also the *cis* isomer.

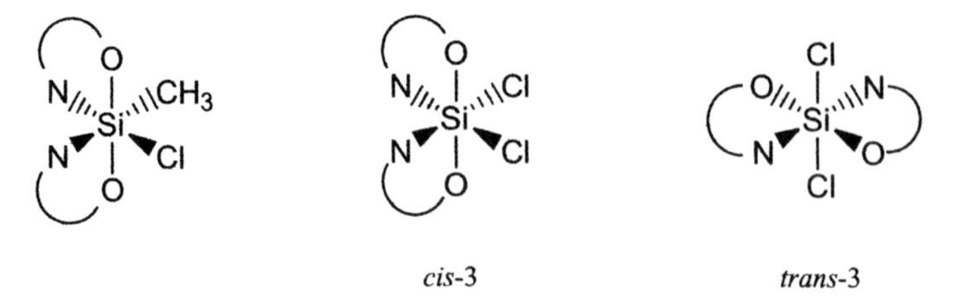

Fig. 2. Configuration of $(L^2)_2Si(Me)Cl$ published by Klebe [2] (left) and suggested configurations of the complex **3** (middle and right).

We were not able to conclude the configuration of **3** from NMR data, but suitable crystals for X-ray structure analysis have not been available. For this reason, both possible configurations of **3** were fully optimized by a DFT method (B3LYP/6-31G*), and the presence of global minima was verified by calculating the Hessian matrices. The principal values of the ^{29}Si NMR chemical shift tensor were calculated from the optimized molecules by GIAO calculation (RHF/6-311+G(2d,p)). Additionally, the solid-state ^{29}Si CP/MAS NMR spectrum of **3** was recorded at a spinning frequency of 1.5 kHz (Fig. 3). The principal values of the ^{29}Si NMR chemical shift tensor were obtained from the spectrum using the program HB-MAS [4]. The experimental and the calculated NMR data are listed in Table 3. The value of the isotropic shift (δ_{iso}) as well as the principal values of the chemical shift tensor (δ_{11}, δ_{22}, δ_{33}) agree better for the *trans* isomer. Surprisingly, we find a very good match between the experimental and the calculated spectrum of the *trans* isomer. Most likely, electronic effects lead to the preferred formation of the *trans* isomer.

The comparison of the calculated IR frequencies with the experimental IR spectrum gives further support for the hypothesis that the *trans* isomer of **3** is formed [3].

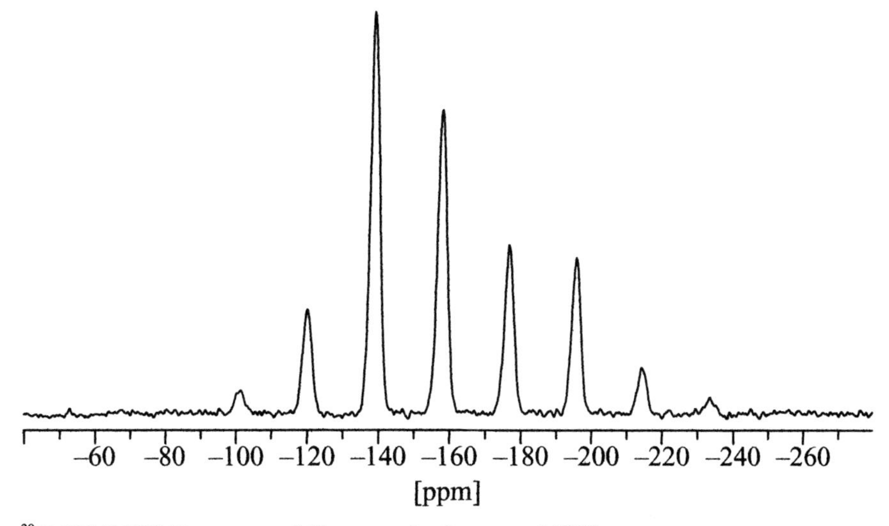

Fig. 3. ^{29}Si CP/MAS NMR spectrum of silicon complex **3** at υ_{spin} = 1.5 kHz.

Table 3. Principal and isotropic values of the ^{29}Si NMR shift tensor of calculated *trans*-**3**, *cis*-**3** and measured **3**.

	δ_{11} [ppm]	δ_{22} [ppm]	δ_{33} [ppm]	δ_{iso} [ppm]
cis-**3** (calculated)	–121.2	–132.7	–170.6	–141.6
trans-**3** (calculated)	–125.2	–137.8	–201.0	–154.7
3 (experimental)	–116.9	–140.0	–216.7	–157.9

Acknowledgments: This work was financially supported by Deutsche Forschungsgemeinschaft and Fonds der Chemischen Industrie. We gratefully acknowledge the X-ray structure analysis of **1a** by Sigrid Goutal, Institut für Organische Chemie, Technische Universität Dresden, Bergstr. 66, D-01069 (Germany).

References

[1] I. Kalikhman, B. Gostevskii, O. Girshberg, S. Krivonos, D. Kost, *Organometallics* **2002**, *21*, 2551 – 2554.

[2] G. Klebe, D. T. Qui, *Acta Cryst.* **1984**, *C40*, 476 – 479.

[3] Diploma thesis M. Schley, TU Bergakademie Freiberg, Fakultät für Chemie und Physik **2003**.

[4] M. Braun, FSU Jena, Germany.

Steric Effect on the Formation, Structure, and Reactions of Pentacoordinate Siliconium Ion Salts

Inna Kalikhman, Boris Gostevskii, Akella Sivaramakrishna, Daniel Kost*

Department of Chemistry, Ben-Gurion University of the Negev
Beer-Sheva 84105 Israel
Tel.: +972 8 6461192 — Fax: +972 8 6472943
E-mail: Kostd@bgumail.bgu.ac.il

Nikolaus Kocher, Dietmar Stalke

Institut für Anorganische Chemie, Universität Würzburg, Würzburg 97074 Germany
Tel: +49 931 888 4783 — Fax: 49 931 888 4619
E-mail: Dstalke@chemie.uni-wuerzburg.de

Keywords: hexacoordinate, hypervalent compounds, model Berry pseudorotation, dative-bond dissociation

Summary: Bulky monodentate ligands affect the structure and reactivity of siliconium ion complexes in several ways: they enhance ionization, they promote a methyl halide elimination reaction, and they severely distort complex geometries. The latter effect enabled the assembly of a reaction coordinate model for the Berry pseudorotation, composed of crystal structures with varying NSiN and OSiO bond angles. In a competition between opposing effects of electron withdrawal by CF_3 and steric bulk of a cyclohexyl ligand in the same molecule, a *nonionic* dissociation of the dative N$\rightarrow$Si bond was observed.

Introduction

Recent studies on hydrazide-based hypercoordinate silicon complexes demonstrated the unusual flexibility of these compounds: their tendency to reversibly transform between penta- and hexacoordinate compounds [1, 2] on the one hand and to irreversibly rearrange to more stable complexes [3] on the other. Thus, neutral hexacoordinate bis-chelate complexes (**1**) undergo reversible ionization in solution (Eq. 1), which is strongly dependent on a variety of factors: temperature (ionization is enhanced at *low* temperatures), solvent (ionization takes place in hydrogen-bond donor solvents such as $CHCl_3$, CH_2Cl_2, and $CHFCl_2$), the nature of the anion, the monodentate ligand X, and the remote substituent R.

The present article summarizes three profound steric effects of the ligand X on the equilibrium between **1** and **2** and the structure and reactivity of **2**.

Steric Effect on Ionization

When X is a bulky group (X = cyclohexyl, *t*-butyl), the equilibrium reaction (Eq. 1) is shifted completely toward the ionic side even at room temperature. This is evident from the ^{29}Si NMR chemical shifts of the mixtures of **1** and **2** (Table 1), which are characteristic of pentacoordination and prove that the equilibrium leans heavily toward **2**. Apparently, the proximity of the bulky *t*-butyl and dimethylamino groups in **1** to the halogeno-ligand forces the latter to dissociate preferentially even at room temperature, in contrast to complexes with small X ligands, which ionize at lower temperature.

X, R = Me, Ph, $PhCH_2$, *t*-Bu, *i*-Bu

Eq. 1.

Table 1. ^{29}Si NMR chemical shifts (ppm) for complexes **2** and **3** ($CDCl_3$, 300 K).

Compound	X = *t*-Bu			X = C_6H_{11}		
R	Me	Ph	*t*-Bu	Me	Ph	*t*-Bu
2	–61.6	–61.1	–61.1	–66.0	–66.0	–65.0
3	–71.8	–71.7	–73.1	–74.1	–73.4	–73.4

Methyl Halide Elimination

In the presence of bulky X ligands, a facile methyl halide elimination reaction is observed (Eq. 2) [3]. In this elimination the siliconium ion complex **2**, with its two N→Si dative bonds, is converted into a neutral pentacoordinate complex **3**, with only one remaining dative bond (Fig. 1, Table 1). The reaction is probably driven by partial release of steric interaction, caused by the removal of one of the *N*-methyl groups. This is indicated by a decrease in elimination rate in the presence of less bulky ligands, cyclohexyl and isobutyl, and the failure to observe elimination when X = methyl. The reactivity order of the halide ions follows their nucleophilicities: $I^- > Br^- > Cl^-$, while the less nucleophilic triflate ion does not react at all.

$$[\text{2}]^{+}\,\text{Hal}^{-} \xrightarrow{-\text{MeHal}} \text{3}$$

X = *t*-Bu, C_6H_{11}
R = Me, Ph, *t*-Bu

Eq. 2.

Interestingly, the stereochemistry of the elimination product **3** is very different from that of **2**; while in **2** the nitrogen ligands are in axial positions, in **3** the dative and the covalent N-Si bonds occupy equatorial positions. This is in contrast to all previously studied pentacoordinate silicon complexes sharing the same ligand-atom framework, $SiCO_2N_2$, in which the nitrogen ligands are essentially axial [1]. This is very likely also the result of a steric effect, in which the bulky *t*-butyl group prefers the smaller O-ligands in its proximity over the bulkier NMe_2 and NMe groups. This leads directly to the third profound effect of steric bulk, namely the effect on the geometry of the complexes.

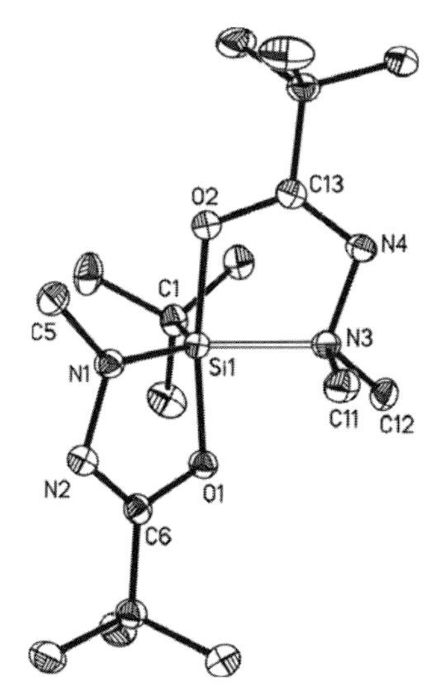

Fig. 1. Crystal structure of 3 (R = X = *t*-Bu) featuring *equatorial* nitrogen ligands [3].

Steric Effect on Geometry: a Model Pseudorotation

Bulky X-ligands, and in particular the *t*-butyl group, cause severe geometrical distortions of pentacoordinate silicon complexes. The most dramatic example is that mentioned above, in which the nitrogen ligands in **3** have moved from axial to equatorial positions upon elimination of methyl halide from **2**. The crystallographic evidence for this geometry is depicted in Fig. 1. The extent of structural distortion varies from one complex to another: from an essentially pure trigonal bipyramid (TBP1, Fig. 2) with axial nitrogens, through intermediate structures with bond angles N-Si-N < 180° and O-Si-O > 120°, to a square-pyramidal mid-point geometry where N-Si-N = O-Si-O, and finally to the completely inverted TBP2, observed in **3**. The geometrical progress demonstrated in Fig. 2 follows the changes associated with the Berry pseudorotation mechanism [6], and the sequence constitutes a crystallographic reaction-coordinate model for the Berry process in the Bürgi-Dunitz sense [7].

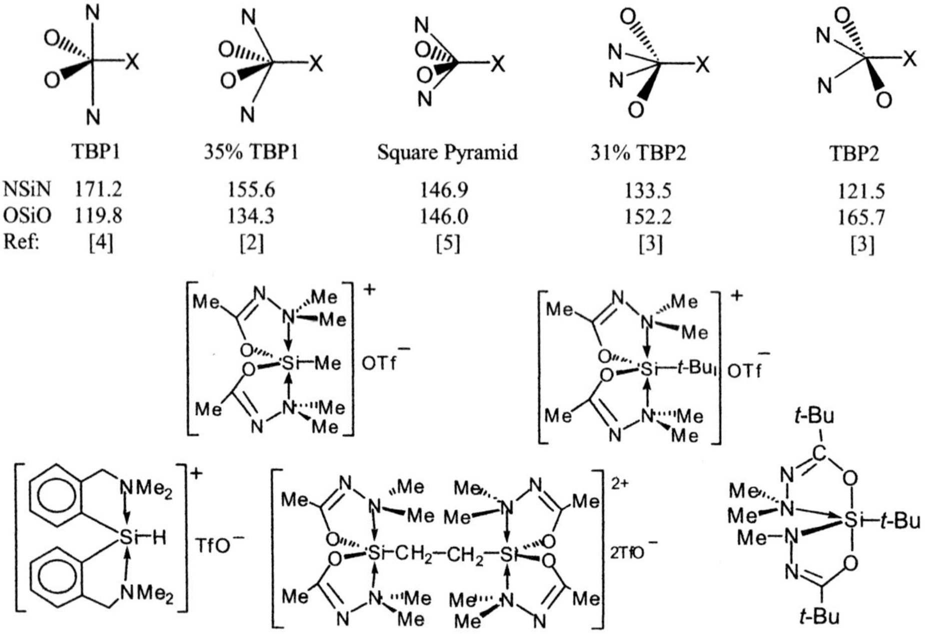

	TBP1	35% TBP1	Square Pyramid	31% TBP2	TBP2
NSiN	171.2	155.6	146.9	133.5	121.5
OSiO	119.8	134.3	146.0	152.2	165.7
Ref:	[4]	[2]	[5]	[3]	[3]

Fig. 2. A model Berry pseudorotation reaction coordinate made up of crystal structures of pentacoordinate silicon complexes with varying N-Si-N and O-Si-O angles.

Competing Steric and Electronic Effects: Neutral N–Si Dissociation

In contrast to the ionization enhancement caused by bulky ligands, prevention of ionization was observed in complexes **1** possessing strongly electron-withdrawing substituents or ligands: when X = halogen or R = CF_3, equilibrium ionization was essentially prevented [2]. This was attributed to the electron withdrawal by the halogen or CF_3 group, causing increased partial positive charge on silicon, which tends to suppress formation of a silicon cation.

We have combined the two opposing structural effects, CF_3 substituent and a bulky cyclohexyl ligand (**4**), in an attempt to study the outcome of the *simultaneous* effect: resistance to ionization by the CF_3 group and enhancement of ionization by the bulky cyclohexyl ligand (Eq. 3). It was expected that one of the two factors might prevail, resulting in either ionization (as in Eq. 1) or resistance to ionization in the form of sustained **4**.

high temp.

low temp.

4 **5**

Eq. 3.

However, measurement of the temperature dependence of the ^{29}Si NMR chemical shift of **4** in solution provided evidence that *neither* of the two options prevailed (Fig. 3): in fact, the evidence suggested that a third, unexpected response took place, namely the neutral dissociation of the dative N→Si bond (Eq. 3). The ^{29}Si NMR spectra of **4** in toluene-d$_8$ solution as a function of temperature are shown in Fig. 3. There is a distinct temperature dependence of the chemical shift, such that at higher temperatures the ^{29}Si resonance moves to lower field, in accord with a decrease in coordination number. This temperature dependence of **4** is in sharp contrast with that observed for **1** (where R = *t*-Bu, X = *i*-Bu) because of ionization (Eq. 1). This is shown in Fig. 3.

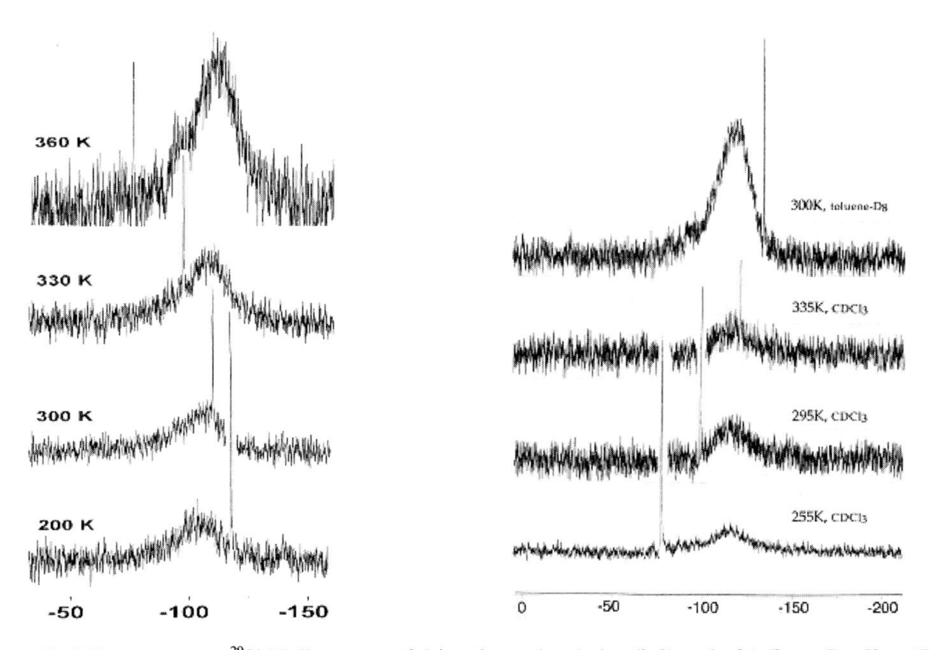

Fig. 3. Variable temperature ^{29}Si NMR spectra of **4** in toluene-d$_8$ solution (left) and of **1** (R = *t*-Bu, X = *i*-Bu) in $CDCl_3$ or toluene-*d*$_8$ solutions (right).

The temperature dependence depicted on the right-hand side of Fig. 3 corresponds to H-bond donor solvent-driven ionization (Eq. 1) of **1** (R = *t*-Bu, X = *i*-Bu) as the temperature is decreased, as reported previously [2]. The opposite trend found for **4** in Fig. 3 (left) must be the result of a different process, which at higher temperature corresponds to increased population of the pentacoordinate species in a rapidly exchanging system. This is likely due to the *non*-ionic dissociation process shown in Eq. 3.

In addition to the ^{29}Si NMR evidence presented in Fig. 3, supporting the nonionic dissociation, the ^{13}C NMR also supports this reaction: the ^{13}C chemical shift of the chelate-ring carbon has an average value between those of the open and closed ring forms.

The fact that only one spectral set is observed for *both* of the chelate rings in **4** – **5** in the ^{1}H and ^{13}C NMR spectra is evidence that (a) the equilibrium between **4** and **5** is rapid relative to the NMR time scale and, (b), that the open and closed chelate moieties in **5** are equivalent, i.e. they also undergo simultaneous rapid equilibration. The equilibration of the chelate rings must follow a rapid "flip-flop" type mechanism, by which one NMe_2 group displaces the other, as described previously for other systems [8].

The observed N→Si dissociation satisfies the tendencies of *both* opposing effects: the bulk of the cyclohexyl forces cleavage of one of the bonds, but the electron-withdrawal prevents ionic cleavage.

Acknowledgments: This work was supported by the German Israeli Foundation for Scientific R&D (GIF), grant I-628-58.5/1999, by INTAS grant number 03-51-4164, by the Deutsche Forschungsgemeinschaft and the Fonds der Chemischen Industrie.

References

[1] D. Kost, I. Kalikhman, Hydrazide-based hypercoordinate silicon compounds in advances, in *Organometallic Chemistry* (Eds.: R. West, A. F. Hill) Elsevier, Amsterdam **2004**, pp. 1 – 106.

[2] D. Kost, V. Kingston, B. Gostevskii, A. Ellern, D. Stalke, B. Walfort, I. Kalikhman, *Organometallics* **2002**, *21*, 2293.

[3] D. Kost, B. Gostevskii, N. Kocher, D. Stalke, I. Kalikhman, *Angew. Chem. Int. Ed.* **2003**, *40*, 1023.

[4] J. Belzner, D. Schär, B. O. Kneisel, R. Herbst-Irmer, *Organometallics* **1995**, *14*, 1840.

[5] I. Kalikhman, V. Kingston, B. Gostevskii, V. Pestunovich, D. Stalke, B. Walfort, D. Kost, *Organometallics* **2002**, *21*, 4468.

[6] a) R. S. Berry, *J. Chem. Phys.* **1960**, *32*, 933; b) R. R. Holmes, J. A. Deiters, *J. Am. Chem. Soc.* **1977**, *99*, 3318.

[7] H. B. Bürgi, J. Dunitz, *Acc. Chem. Res.* **1983**, *16*, 153.

[8] a) R. Probst, C. Leis, S. Gamper, E. Herdtweck, C. Zybill, N. Auner, *Angew. Chem., Int. Ed. Engl.* **1991**, *30*, 1132; b) H. Handwerker, C. Leis, R. Probst, P. Bissinger, A. Grohmann, P. Kiprof, E. Herdtweck, J. Blumel, N. Auner, C. Zybill, *Organometallics* **1993**, *12*, 2162.

Synthesis and Structural Characterization of Novel Neutral Hexacoordinate Silicon(IV) Complexes with $\mathit{Si}O_2N_4$ Skeletons

Oliver Seiler, Markus Fischer, Martin Penka, Reinhold Tacke*

Institut für Anorganische Chemie, Universität Würzburg,
Am Hubland, D-97074 Würzburg, Germany
Tel.: +49 931 888 5250 — Fax: +49 931 888 4609
E-mail: r.tacke@mail.uni-wuerzburg.de

Keywords: coordination chemistry, hexacoordination, silicon, cyanato-*N* ligands, thiocyanato-*N* ligands

Summary: Treatment of $Si(NCO)_4$ or $Si(NCS)_4$ with 4-aminopent-3-en-2-ones yielded novel neutral hexacoordinate silicon(IV) complexes with an $\mathit{Si}O_2N_4$ framework, compounds **3** – **6**. These silicon(IV) complexes were characterized in the solid state by single-crystal X-ray diffraction and ^{29}Si VACP/MAS NMR spectroscopy. Compounds **3** – **5** crystallized as the (*OC*-6-12)-isomer, and **6** was isolated as the *trans*-isomer.

Introduction

In two recent publications, we have reported on the synthesis and structural characterization of the neutral hexacoordinate silicon(IV) complexes **1** [1] and **2** [2]. These compounds contain an $\mathit{Si}O_4N_2$ skeleton, with two bidentate acetylacetonato(1–) ligands and two monodentate cyanato-*N* or thiocyanato-*N* ligands bound to the silicon coordination center, the monodentate ligands occupying *cis* positions. We have now succeeded in synthesizing a series of neutral hexacoordinate silicon(IV) complexes with $\mathit{Si}O_2N_4$ skeletons, compounds **3** – **6**. From a formal point of view, the chelate complexes **3** – **5** can be regarded as derivatives of **1** or **2** (O/NR exchange; R = Bn, Ph); however, in contrast to **1** and **2** (each *cis* configuration), the cyanato-*N* and thiocyanato-*N* ligands in **3** – **5** occupy *trans* positions. We report here on the synthesis of compounds **3** – **5** [(*OC*-6-12)-configuration each] and **6** (*trans* configuration) and their structural characterization in the solid state by single-crystal X-ray diffraction and ^{29}Si VACP/MAS NMR spectroscopy. The studies presented here were performed as part of our systematic investigations on higher-coordinate silicon compounds (for recent publications, see Refs. [1 – 13]).

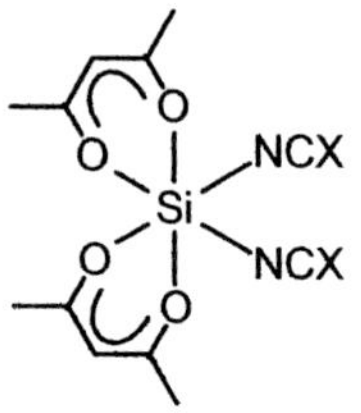

1: X = O
2: X = S

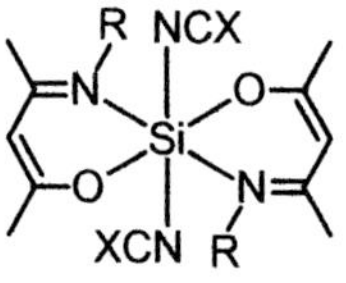

3–5

	R	X
3	Bn	O
4	Ph	O
5	Ph	S

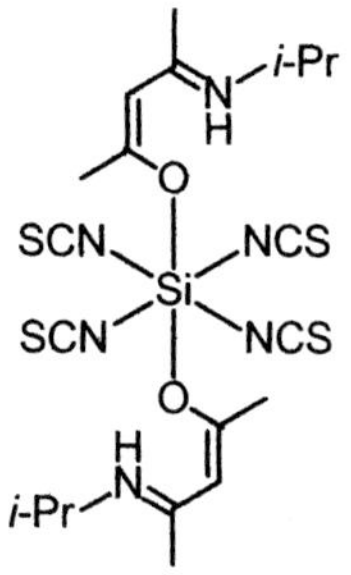

6

Results and Discussion

Compounds **3** – **6** were synthesized according to Scheme 1 by treatment of tetra(cyanato-*N*)silane (→ **3**, **4**) or tetra(thiocyanato-*N*)silane (→ **5**, **6**) with two molar equivalents of the respective 4-aminopent-3-en-2-ones Me–C(NRH)=CH–C(O)–Me (R = Bn, Ph, *i*-Pr). The products were isolated as colorless crystalline solids (yields 65 – 87%). Surprisingly, the reaction of $Si(NCS)_4$ with 4-(isopropylamino)pent-3-en-2-one does not lead to a chelate complex (such as **3** – **5**) under the reaction conditions used; instead, we isolated compound **6**.

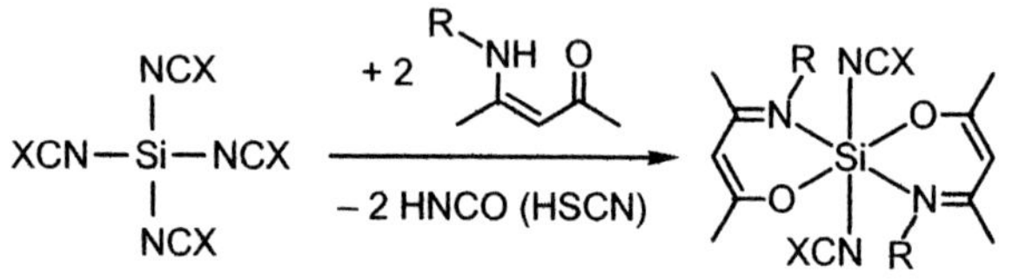

	R	X
3	Bn	O
4	Ph	O
5	Ph	S

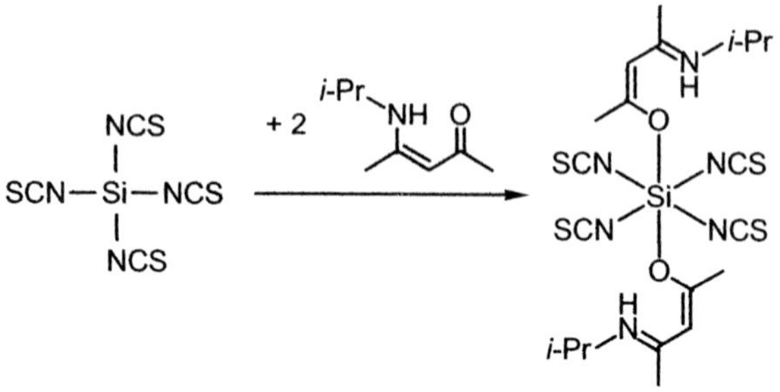

Scheme 1. Syntheses of the neutral hexacoordinate silicon(IV) complexes **3** – **6**.

Compounds **3** – **6** were structurally characterized by single-crystal X-ray diffraction. The molecular structures in the crystal are shown in Figs. 1 – 4; selected bond distances and angles are given in the respective figure captions.

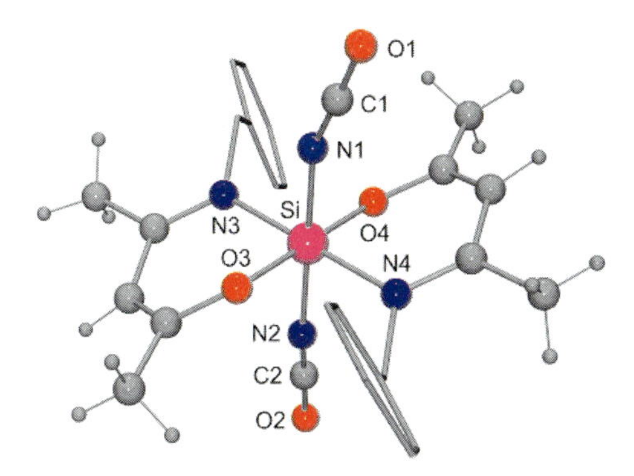

Fig. 1. Molecular structure of **3** in the crystal. The benzyl groups are represented as stick models for clarity. Selected bond lengths (Å) and angles (deg): Si–O3 1.7385(10), Si–O4 1.7406(10), Si–N1 1.8299(13), Si–N2 1.8356(13), Si–N3 1.9131(13), Si–N4 1.9171(13); O3-Si-O4 179.21(5), O3-Si-N1 89.11(5), O3-Si-N2 90.68(5), O3-Si-N3 90.95(5), O3-Si-N4 88.90(5), O4-Si-N1 91.59(5), O4-Si-N2 88.62(5), O4-Si-N3 88.70(5), O4-Si-N4 91.45(5), N1-Si-N2 179.78(6), N1-Si-N3 89.12(5), N1-Si-N4 90.39(5), N2-Si-N3 90.94(5), N2-Si-N4 89.55(5), N3-Si-N4 179.49(5), Si-N1-C1 157.33(12), Si-N2-C2 164.13(12), N1-C1-O1 177.47(16), N2-C2-O2 178.10(18).

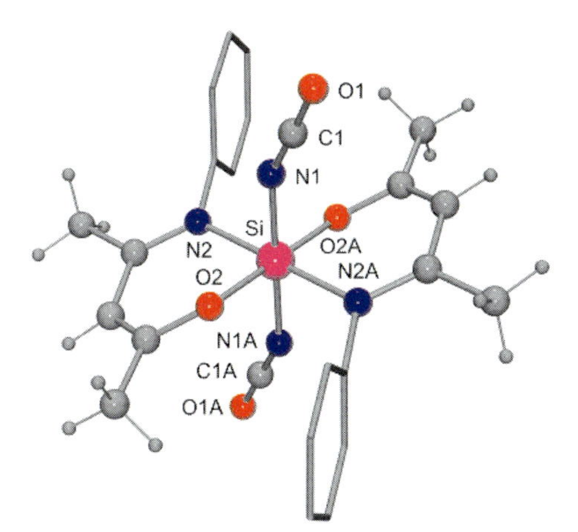

Fig. 2. Molecular structure of **4** in the crystal. The phenyl groups are represented as stick models for clarity. Selected bond lengths (Å) and angles (deg): Si–O2 1.7388(10), Si–N1 1.8199(11), Si–N2 1.9328(11); O2-Si-O2A 180.0, O2-Si-N1 91.21(5), O2-Si-N1A 88.79(5), O2-Si-N2 91.46(5), O2-Si-N2A 88.54(5), N1-Si-N1A 180.0, N1-Si-N2 90.11(5), N1-Si-N2A 89.89(5), N2-Si-N2A 180.0, Si-N1-C1 151.62(12), N1-C1-O1 177.96(18).

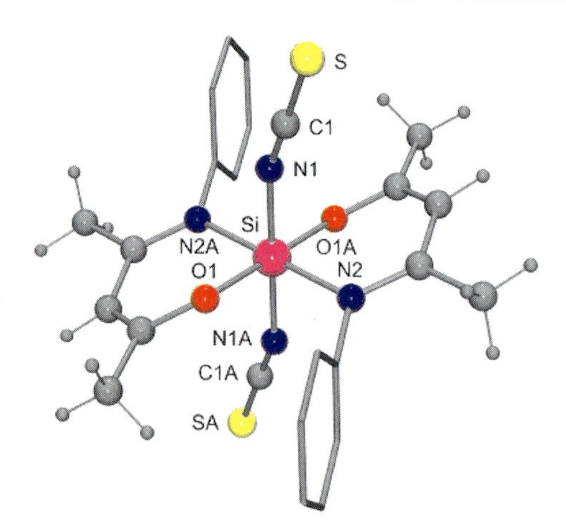

Fig. 3. Molecular structure of **5** in the crystal. The phenyl groups are represented as stick models for clarity. Selected bond lengths (Å) and angles (deg): Si–O1 1.7318(10), Si–N1 1.8401(12), Si–N2 1.9138(12); O1-Si-O1A 180.0, O1-Si-N1 88.76(5), O1-Si-N1A 91.24(5), O1-Si-N2 88.74(5), O1-Si-N2A 91.26(5), N1-Si-N1A 180.0, N1-Si-N2 89.21(5), N1-Si-N2A 90.79(5), N2-Si-N2A 180.0, Si-N1-C1 156.68(12), N1-C1-S 179.22(13).

Because of the presence of two identical unsymmetric bidentate ligands and two identical monodentate ligands in the chelate complexes **3** – **5**, five stereoisomers have to be considered for these compounds (Fig. 5). As shown by the crystal structure analyses, all compounds crystallized as the (*OC*-6-12)-isomer (for the nomenclature system, see Ref. [14]), with slightly distorted octahedral Si-coordination polyhedra.

The Si–O [Si–N] bond distances in the chelate rings of **3** – **5** amount to 1.7318(10) – 1.7406(10) Å [1.9131(13) – 1.9328(11) Å], and the Si–O bond lengths of **6** [1.7732(12) Å] are only slightly longer than those observed for **3** – **5**. The Si–NCO bond distances of **3** and **4** [1.8199(11) – 1.8356(13) Å] are slightly longer than those of **1** [1.8024(12) Å], and the Si–NCS distances of **5** [1.8401(12) Å] are also slightly longer than those of **2** [1.8093(17) – 1.8145(16) Å]. The Si–NCS bond lengths of **6** [1.8135(16) – 1.8153(16) Å], however, are very similar to those observed for **2**. Interestingly, the Si–NCX (X = O, S) moieties of the chelate complexes **3** – **5** are significantly more bent [Si-N-C 151.62(12) – 164.13(12)°] than the Si–NCS groups of **6** [Si-N-C 172.66(15) – 176.99(16)°].

The isotropic ^{29}Si chemical shifts obtained in solid-state VACP/MAS NMR studies of **3** – **6** clearly indicate the presence of hexacoordinate silicon atoms in all compounds (**3**, δ = –206.2; **4**, δ = –207.6; **5**, δ = –209.2; **6**, δ = –233.0). The ^{29}Si chemical shifts of the chelate complexes **3** – **5** (SiO_2N_4 skeleton) are very similar to the isotropic ^{29}Si chemical shifts observed for **1** (δ = –206.6) and **2** (δ = –210.2), although the latter two compounds contain an SiO_4N_2 framework. The isotropic ^{29}Si chemical shift of **6** (δ = –233.0) differs significantly from those observed for **3** – **5**, although all compounds contain an SiO_2N_4 skeleton. This high-field shift can be explained by the presence of four thiocyanato-*N* ligands and correlates well with the isotropic ^{29}Si chemical shift observed for the $[Si(NCS)_6]^{2-}$ dianion in $[K(18\text{-crown-}6)]_2[Si(NCS)_6]$ (δ = –253.0) [2], which contains six thiocyanato-*N* ligands.

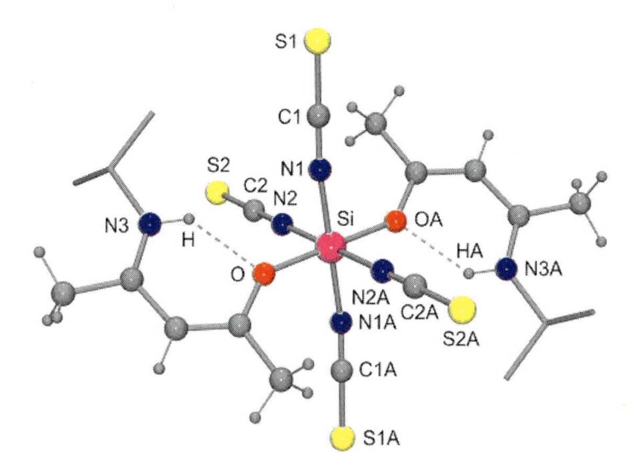

Fig. 4. Molecular structure of **6** in the crystal. The isopropyl groups are represented as stick models for clarity. Selected bond lengths (Å) and angles (deg): Si–O 1.7732(12), Si–N1 1.8153(16), Si–N2 1.8135(16); O-Si-OA 180.0, O-Si-N1 87.38(6), O-Si-N1A 92.62(6), O-Si-N2 87.45(7), O-Si-N2A 92.55(6), N1-Si-N1A 180.0, N1-Si-N2 90.27(7), N1-Si-N2A 89.73(7), N2-Si-N2A 180.0, Si-N1-C1 172.66(15), Si-N2-C2 176.99(16), N1-C1-S1 178.78(16), N2-C2-S2 179.06(17). The dashed lines indicate N–H···O hydrogen bonds [N3–H 0.86(2), H···O 1.98(2), N3···O 2.667(2), N3–H···O 135.9(19)].

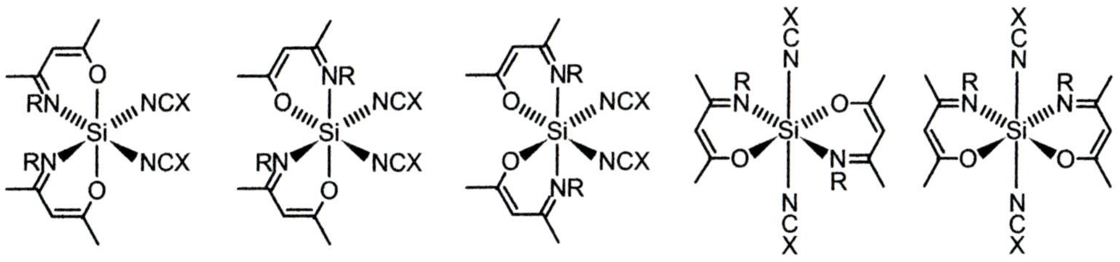

(*OC*-6-13) (*OC*-6-32) (*OC*-6-33) (*OC*-6-12) (*OC*-6-22)

Fig. 5. Possible stereoisomers for the chelate complexes **3** – **5**.

Acknowledgment: This work was supported by the Fonds der Chemischen Industrie.

References

[1] R. Tacke, R. Bertermann, M. Penka, O. Seiler, *Z. Anorg. Allg. Chem.* **2003**, *629*, 2415.

[2] O. Seiler, R. Bertermann, N. Buggisch, C. Burschka, M. Penka, D. Tebbe, R. Tacke, *Z. Anorg. Allg. Chem.* **2003**, *629*, 1403.

[3] R. Tacke, M. Penka, F. Popp, I. Richter, *Eur. J. Inorg. Chem.* **2002**, 1025.

[4] A. Biller, C. Burschka, M. Penka, R. Tacke, *Inorg. Chem.* **2002**, *41*, 3901.

[5] I. Richter, M. Penka, R. Tacke, *Inorg. Chem.* **2002**, *41*, 3950.

[6] I. Richter, M. Penka, R. Tacke, *Organometallics* **2002**, *21*, 3050.

[7] O. Seiler, C. Burschka, M. Penka, R. Tacke, *Z. Anorg. Allg. Chem.* **2002**, *628*, 2427.

[8] S. Dragota, R. Bertermann, C. Burschka, J. Heermann, M. Penka, I. Richter, B. Wagner, R. Tacke, *Silicon Chem.* **2002**, *1*, 291.

[9] O. Seiler, C. Burschka, M. Penka, R. Tacke, *Silicon Chem.* **2002**, *1*, 355.

[10] R. Tacke, R. Bertermann, A. Biller, C. Burschka, M. Penka, *Can. J. Chem.* **2003**, *81*, 1315.

[11] R. Bertermann, A. Biller, M. Kaupp, M. Penka, O. Seiler, R. Tacke, *Organometallics* **2003**, *22*, 4104.

[12] R. Tacke, O. Seiler, Higher-coordinate silicon compounds with SiO_5 and SiO_6 skeletons, in *Silicon Chemistry: From the Atom to Extended Systems* (Eds.: P. Jutzi, U. Schubert), Wiley-VCH, Weinheim, **2003**, p. 324 – 337.

[13] O. Seiler, M. Penka, R. Tacke, *Inorg. Chim. Acta* **2004**, *357*, 1955.

[14] W. Liebscher, E. Fluck, *Die systematische Nomenklatur der Anorganischen Chemie*, Springer-Verlag, Berlin/Heidelberg, **1999**, p. 292 – 297.

Vinyloligosilyl Anions — a New Class of Compounds

Jelena Markov, Judith Baumgartner, Christoph Marschner*

Institut für Anorganische Chemie, Technische Universität Graz
Stremayrgasse 16, A-8010 Graz, Austria
Tel.: +43 316 873 8209 — Fax: +43 316 873 8701
E-mail: christoph.marschner@tugraz.at

Hartmut Oehme, Thoralf Gross

Fachbereich Chemie der Universität Rostock
Albert-Einstein-Str. 3a, D-18059 Rostock, Germany

Keywords: oligosilanes, vinylsilanes, silyl anions

Summary: Bulky tris(trimethylsilyl)vinylsilanes (**3a**, **3b**, **5**) react with potassium *tert*-butoxide in a fast manner to quantitatively yield vinyloligosilyl anions, a class of compounds first described here. The obtained anions can be further utilized for the synthesis of other types of functionalized vinylsilanes (**7**).

The chemistry of silyl anions has recently become a rapidly developing field [1]. Over the last five years, beginning with the first reported selective access to larger oligosilyl anions [2], we have developed a new approach which allows high-yielding syntheses of larger homo- and hetero-substituted oligosilanes. Our recent efforts have demonstrated easy access to α,ω-oligosilyl dianions [3] and subsequent synthesis of various cyclic silanes, as well as the synthesis of alkynylsilyl anions [4] and their application in the synthesis of rigid spacer-bridged oligosilanes. In this study we wish to report on vinyloligosilyl anions, obtained in a high-yielding, selective manner, their spectroscopic and structural properties, and first examples of further functionalization.

Triethylborane-induced stereoselective radical addition of R_3SiH to carbon-carbon triple bonds was investigated by Oshima and Utimoto [5]. Even though the use of triethylborane frequently proved to cause reproducibility problems, the reaction represents a simple, convenient route to alkenylsilanes (Eq. 1), including bulky tris(trimethylsilyl)vinylsilanes (**3a**, **3b**).

The selective cleavage of one terminal trimethylsilyl group [2] with potassium *tert*-butoxide is a superior protocol for the synthesis of silyl anions. Also in the case of vinylsilanes, the method proved to be equally selective and efficient (Eq. 2), providing quantitative yield of the first vinylsilyl anions (**4a**, **4b**).

R–≡ + H–Si($SiMe_3$)$_3$ —(Et_3B)→ (R)CH=CH–Si($SiMe_3$)$_3$

1a, b **2** R = $SiMe_3$ (**1a, 3a**), ^{t}Bu (**1b, 3b**) **3a, b**

Eq. 1. Hydrosilylation of alkynes to vinylsilanes.

(R)CH=CH–Si($SiMe_3$)$_3$ —(KO^tBu; D = 2 THF, 18-Cr-6)→ (R)CH=CH–Si($SiMe_3$)$_2$–K·D

3a,b R = $SiMe_3$ (**3a, 4a**), ^{t}Bu (**3b, 4b**) **4a,b**

Eq. 2. Selective cleavage of a terminal trimethylsilyl group.

The reaction can be conducted either in THF or in an aromatic solvent, where it requires the presence of one equivalent of crown ether. The crown ether adduct of the product **4a** was subjected to X-ray diffraction analysis and represents the first fully characterized vinyloligosilyl anion [6]. The crystal structure and selected bond lengths and bond angles are given in Fig. 1 and Table 1, respectively.

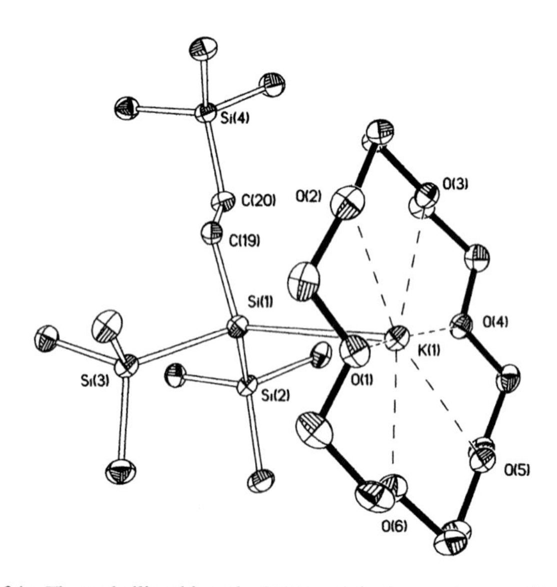

Fig. 1. Crystal structure of **4a**. Thermal ellipsoids at the 30% level (hydrogen atoms omitted for clarity).

The crystal structure shows the expected E geometry of the double bond. The length of the silicon-potassium bond (3.45 Å) lies within the range observed for similar anionic compounds [3]. The length of the carbon-carbon olefinic bond is not significantly affected by the adjacent anionic silicon atom, in contrast to the spectroscopic behavior of this bond (see below).

Table 1. Selected bond lengths [Å] and bond angles [°] of **4a**.

Bond lengths	K(1)–Si(1)	3.4485
	Si(4)–C(20)	1.854
	C(19)–C(20)	1.346
	Si(1)–C(19)	1.909
	Si(1)–Si(3)	2.342
	Si(1)–Si(2)	2.349
Bond angles	C(19)-Si(1)-Si(3)	99.69
	C(19)-Si(1)-K(1)	115.64
	Si(2)-Si(1)-K(1)	113.46
	Si(1)-C(19)-C(20)-Si(4)	–172.36

Trans-1,2-bis[tris(trimethylsilyl)silyl]ethylene **5** was obtained from the reaction of tris(trimethylsilyl)silyllithium with formic acid methyl ester [7]. The corresponding dianion **6** was obtained as the only reaction product in a fast reaction of **5** with two equivalents of potassium *tert*-butoxide in the presence of two equivalents of crown ether as the first example of a vinylidene-bridged oligosilyl α,ω-dianion (Eq. 3), which is in good analogy with our previously synthesized alkynylidene- and alkylidene-bridged compounds [3, 4, 8].

$SiMe_3$ Si $SiMe_3$ $SiMe_3$ Me_3Si Me_3Si Si $SiMe_3$ — 2 eq KO^tBu → $SiMe_3$ Si $K{\cdot}D$ $SiMe_3$ Me_3Si $D{\cdot}K$ Si $SiMe_3$

5 D = 18-Cr-6 **6**

Eq. 3. Synthesis of the vinylidene-bridged dianion.

The reactivity of the newly synthesized anions was studied. As expected, they can easily be protonated and alkylated, in close analogy to other silyl anions we have investigated earlier [3, 4]. Furthermore, oligosilyl anions were successfully used for the formation of interesting silicon-heteroatom bonds [9]. In a similar manner, vinylsilyl potassium compound **4b** can be transmetalated into the Mg analog [10], and the respective anion can then be used as a nucleophile in the reaction with diethylaminodichlorophosphine (Eq. 4). The monosilylated derivative **7** can be selectively obtained as the only reaction product.

$$\mathbf{4b} \xrightarrow[\text{2) } Et_2NPCl_2]{\text{1) } MgBr_2} \mathbf{7}$$

Eq. 4. Functionalization of the anion provides easy access to new silicon-heteroatom bonds.

Compounds like **7** can be further derivatized employing the chloride or the amino group at the phosphorus, by reaction with the olefinic side chain, or by the generation of another silyl anion by a subsequent trimethylsilyl group cleavage. It is therefore evident that the synthetic methodology presented in this account provides a convenient tool for the synthesis of organosilicon compounds with a fairly high degree of functionalization.

A comparison of selected NMR spectroscopic properties of the reported compounds is given in Table 2, where ^{29}Si as well as the ^{13}C olefinic chemical shifts are provided.

Table 2. Selected ^{29}Si and ^{13}C NMR data (α and β refer to the position of the oligosilyl group).

Compound	δ ^{29}Si [ppm]	^{13}C olefinic chemical shifts [ppm]
3a	–8.8; –13.2; –82.1	143.0 (β); 154.8 (α)
4a	–8.7; –8.9; –104.7	134.0 (β); 171.3 (α)
3b	–13.1; –85.1	113.5 (α); 160.4 (β)
4b	–8.1; –118.8	133.5 (α); 147.2 (β)
5	–12.9; –80.9	145.5
6	–8.7; –113.6	150.7
7	–13.2; –48.9, $^{1}J(SiP)$ = 85.9 Hz	117.5 (α); 160.5 (β)

The ^{29}Si NMR chemical shifts, as well as the ^{13}C olefinic shifts, nicely illustrate the chemical transformation of the vinylsilanes into the vinylsilyl anions. ^{29}Si shifts of the central silicon atom of the anions show the expected, characteristic upfield shift of approximately 20 – 30 ppm in comparison to the starting vinylsilanes. Also, ^{13}C olefinic shifts display characteristic NMR behavior for the double bond adjacent to an anionic silicon species — typical is the larger downfield shift of the α carbon as well as an upfield shift of the β carbon atom.

Acknowledgments: This study was supported by the Fonds zur Förderung der wissenschaftlichen

Forschung via the START project Y120. We thank Wacker Chemie GmbH, Burghausen, for the generous donation of various chlorosilanes as starting materials.

References

[1] a) P. D. Lickiss, C. M. Smith, *Coord. Chem. Rev.* **1995**, *145*, 75; b) K. Tamao, A. Kawachi, *Adv. Organomet. Chem.* **1995**, *38*, 1; c) J. Belzner, U. Dehnert, in *The Chemistry of Organic Silicon Compounds* (Eds.: Z. Rappoport, Y. Apeloig), J. Wiley & Sons, New York, **1998**, *Vol. 2*, p. 779; d) A. Sekiguchi, V. Y. Lee, M. Nanjo, *Coord. Chem. Rev.* **2000**, *210*, 11.

[2] Ch. Marschner, *Eur. J. Inorg. Chem.* **1998**, 221.

[3] R. Fischer, D. Frank, W. Gaderbauer, Ch. Kayser, Ch. Mechtler, J. Baumgartner, Ch. Marschner, *Organometallics* **2003**, *22*, 3723.

[4] Ch. Mechtler, M. Zirngast, J. Baumgartner, Ch. Marschner, *Eur. J. Inorg. Chem.* **2004**, 3254.

[5] K. Utimoto, K. Oshima, K. Miura, *Bull. Chem. Soc. Jpn.*, **1993**, *66*, 2356.

[6] Crystallographic data for **4a**: $C_{23}H_{53}KO_6Si_4$, monoclinic space group $P2_1/n$, a = 11.681(3) Å, b = 15.999(3) Å, c = 18.838(4) Å, β= 97.15(3)°, $R1$ = 0.0462, $wR2$ = 0.1307, GoF = 1.053. Data have been deposited at the Cambridge Crystallographic Data Centre. CCDC 238 897.

[7] T. Gross, R. Kempe, H. Oehme, *J. Organomet. Chem.* **1997**, *534,* 229.

[8] Ch. Mechtler, Ch. Marschner, *Tetrahedron Lett.* **1999**, *40*, 7777.

[9] J. Markov, R. Fischer, H. Wagner, N. Noormofidi, J. Baumgartner, Ch. Marschner, *J. Chem. Soc., Dalton Trans.* **2004**, 2166.

[10] J. D. Farwell, M. F. Lappert, Ch. Marschner, Ch. Strissel, T. D. Tilley, *J. Organomet. Chem.* **2000**, *603,* 185.

Oligosilyl-1,2-dipotassium Compounds: a Comparative Study

Roland Fischer, Tina Konopa, Judith Baumgartner, Christoph Marschner*

Institut für Anorganische Chemie, Technische Universität Graz
Stremayrgasse 16, A-8010 Graz, Austria
Tel.: +43 316 873 8209 — Fax: +43 316 873 8701
E-mail: christoph.marschner@tugraz.at

Keywords: oligosilyl dianion, NMR spectroscopy, crystal structure

Summary: A series of mono- and dimetalated oligosilyl compounds with trimethylsilyl- and phenyl substituents has been prepared. Multinuclear NMR spectroscopic data and crystal structures are compared. Trends concerning chemical shift π-polarization effects are pointed out.

Introduction

Recently we have developed a convenient and high-yielding access to monometalated (**1a** – **3a**) and 1,2-dipotassium (**1b** – **3b**) oligosilyl compounds starting from trimethylsilyl- and phenyl-substituted disilanes (**1** – **3**) [1]. Cleavage of trimethylsilyl–silicon bonds by a combination of potassium *tert*-butoxide/ 18-crown-6 in aromatic solvents was found to be completely regioselective and to occur stepwise [2], allowing the directed synthesis of mono- and dimetallated oligosilanes (Eq. 1). This is in sharp contrast to previous investigations which report exclusive fission of the central silicon-silicon bond employing methyllithiums as metalating agents [3].

R^3–Si(1)(R^1)(Me_3Si(3))–Si(2)(R^2)(Si(4)Me_3)–R^4 ←(1 eq KO*t*Bu/ 1 eq 18-Cr-6)— R^3–Si(1)(R^1)(Me_3Si(3))–Si(2)(R^2)(Si(4)Me_3)–R^4 —(2 eq KO*t*Bu/ 2 eq 18-Cr-6)→ R^3–Si(1)(R^1)(Me_3Si(3))–Si(2)(R^2)(Si(4)Me_3)–R^4

1a R^1, R^2, R^4 = $SiMe_3$; R^3 = K·18-Cr-6
2a R^1, R^4 = $SiMe_3$; R^2 = Ph; R^3 = K·18-Cr-6
3a R^1, R^2 = Ph; R^4 = $SiMe_3$; R^3 = K·18-Cr-6

1 R^1, R^2, R^3, R^4 = $SiMe_3$
2 R^1, R^3, R^4 = $SiMe_3$; R^2 = Ph
3 R^1, R^2 = Ph; R^3, R^4 = $SiMe_3$

1b R^1, R^2 = $SiMe_3$ R^3, R^4 = K·18-Cr-6
2b R^1 = $SiMe_3$; R^2 = Ph; R^3, R^4 = K·18-Cr-6
3b R^1, R^2 = Ph; R^3, R^4 = K·18-Cr-6

Eq. 1. Formation of mono- and 1,2-dimetalated oligosilanes **1a** – **3a** and **1b** – **3b**, respectively.

Interestingly, **2a** is the sole reaction product obtained from **2** upon cleavage of one trimethylsilyl group. Metalation was found to occur selectively to give the higher silylated oligosilyl anion. This may be understood in terms of anion stabilities rather than kinetic control.

NMR Spectroscopic Investigations

Multinuclear NMR spectroscopy was found to be a tool well suited to the investigation of oligosilyl anions. Comparison of ^{29}Si NMR resonances observed for compounds **1** – **3**, **1a** – **3a** and **1b** – **3b** reveals a pronounced upfield shift for the metalated silicon center. This is in contrast to the trend for Si(2) for compounds **1a** – **3a**, where a slight downfield shift of 0.6 to 6.3 ppm is found. The anionic silicon center of **1a** resonates at –192.6 ppm, which is almost identical with the shift for the central silicon atom in $(Me_3Si)_3SiK$. Upon the second metalation step the resonance in **1b** is found to be shifted to a lower field by 9.2 ppm. Similar behavior is found for compounds **2b**, **3b** and other oligosilyl dianions [4] studied so far. The resonances for the trimethylsilyl groups remaining at the metalated silicon center are also found to be shifted to a lower field for all mono- and dianions by 0.3 to 4.4 ppm. In compounds **1a** – **3a**, trimethylsilyl groups in β-position to metalated Si(1) were found to resonate further upfield by 1.2 to 2.0 ppm (Table 1).

Table 1. ^{29}Si NMR resonances (shifts in ppm relative to TMS) of compounds **1** – **3**, **1a** – **3a** and **1b** – **3b**.

	R^1	R^2	R^3	R^4	Si(1)	Si(2)	Si(3)	Si(4)
1	$SiMe_3$	$SiMe_3$	$SiMe_3$	$SiMe_3$	–129.5	–129.5	–9.0	–9.0
1a	$SiMe_3$	$SiMe_3$	$SiMe_3$	K·18-Cr-6	–192.6	–128.9	–5.6	–10.7
1b	$SiMe_3$	$SiMe_3$	K·18-Cr-6	K·18-Cr-6	–183.4	–183.4	–5.4	–5.4
2	$SiMe_3$	Ph	$SiMe_3$	$SiMe_3$	–128.1	–69.0	–9.4	–12.5
2a	$SiMe_3$	Ph	K·18-Cr-6	$SiMe_3$	–186.5	–62.3	–5.6	–14.5
2b	$SiMe_3$	Ph	K·18-Cr-6	K·18-Cr-6	–176.9	–81.8	–5.0	–10.5
3	Ph	Ph	$SiMe_3$	$SiMe_3$	–70.8	–70.8	–12.1	–12.1
3a	Ph	Ph	K·18-Cr-6	$SiMe_3$	–93.6	–64.6	–10.0	–13.3
3b	Ph	Ph	K·18-Cr-6	K·18-Cr-6	–77.2	–77.2	–11.8	–11.8

^{13}C NMR spectroscopy was found to be a useful tool for investigating charge delocalization of lone pairs of carbanions and silyl anions to an attached aromatic ring system [5]. Among the oligosilyl anions given in Table 2, *ipso*-carbon atoms of the phenyl groups exhibit large downfield shifts compared to the respective neutral species. *Para*-carbon atoms in contrary show a smaller but still evident upfield shift. Shift differences of *ortho*- and *meta*-carbon atoms are less dramatic but still show a clear trend as the *ortho* position is always found to resonate at a lower field than the *meta*-carbon atoms. This is in sharp contrast to carbanions where the trend for *meta* and *ortho* positions is reversed.

Table 2. ^{13}C NMR resonances of aromatic carbon atoms (shifts in ppm relative to TMS).

	δ_{ipso}	δ_{ortho}	δ_{meta}	δ_{para}	$\Delta\delta_{ipso}$ [d]	$\Delta\delta_{ortho}$ [d]	$\Delta\delta_{meta}$ [d]	$\Delta\delta_{para}$ [d]
Ph TMS_2 Si Si TMS_3 **2**	137.7	137.0	127.9	128.1	–	–	–	–
Ph TMS_2 Si Si TMS_2 K [a] **2a**	147.9	137.3	127.5	126.2	10.2	0.3	–0.4	–1.9
Ph TMS_2 Si Si TMS_2 K [b] **2a**	146.2	138.0	126.6	125.6	8.5	1.0	–1.3	–2.5
K Ph TMS Si Si TMS_2 K [b] **2b**	169.4	139.2	124.3	119.2	31.7	2.2	–3.6	–8.9
Ph TMS_2 Si Si TMS_2 Ph **3**	136.6	137.6	128.0	128.2	–	–	–	–
Ph TMS_2 Si Si TMS Ph* K [b] **3a**	157.7	138.5	125.9	122.3	21.1	0.9	–2.1	–5.9
Ph* TMS_2 Si Si TMS Ph K [b] **3a**	145.7	138.0	126.7	125.7	9.1	0.4	–1.3	–2.5
K Ph TMS Si Si TMS Ph K [b] **3c**	169.4	137.4	125.1	117.9	32.8	–0.2	–2.9	–10.3
Ph TMS_3 Si [c]	135.4	136.7	128.2	127.8	–	–	–	–
Ph TMS_2 Si K [a, c]	157.2	136.3	127.8	123.2	21.8	–0.4	–0.4	–4.6

[a] THF adduct; [b] 18-crown-6 complex. All values are reported for benzene solutions; [c] See [2] and ref. cited therein; [d] $\Delta\delta = \delta_{\text{silyl anion}} - \delta_{\text{neutral species}}$

Most dramatic effects are of course found for phenyl-substituted 1,2-dianions with shift differences in the *ipso* position as high as 31.7 (**2b**) and 32.8 (**3b**) ppm and –8.9 (**2b**) and –10.3 (**3b**) ppm for *para* positions.

The pattern of chemical shift differences found for phenyl-substituted silyl anions is typical for π-polarization effects, whereas resonance effects are responsible for electronic effects in carbanions. UV spectroscopic results support this explanation, as carbanions show bathochromic shifts compared to silyl anions [5].

X-Ray Structure Analyses

The central silicon-silicon bond lengths of compound **1b** [1, 6] were found to be 2.448(2) and 2.450(2) Å and thus this compound is markedly elongated compared to its congener **1** for which bond lengths ranging from 2.403(2) to 2.405(5) have been reported [7]. Central silicon–silicon bond distances for compounds **2b** and **3b** were determined to be 2.4485(17) and 2.432(5), respectively. Monoanion **2a** exhibits a central silicon–silicon bond length of 2.3677(17), being much shorter [2]. Also the other silicon–silicon bonds in **2a** evolving from Si(1) are shortened compared to the Si–Si bonds at the phenylated Si(2). Silicon–potassium bond lengths from 3.477(2) to 3.5994(19) were found without any obvious trend to be seen.

Substituents around the metalated silicon centers enclose bond angles which add up to a value of typically around 295° with the only exception being again **2a**, where angles add up to 316°. Hence, the lone pair in silyl anions is strongly localized, and metalated silicon centers are highly pyramidalized, with steeper cone angles for phenylated anionic silicon centers. This is in good agreement with results from ^{13}C NMR spectroscopy as discussed earlier. $Ph_3CLi{\cdot}Et_2O$ was found to be planar around the central carbon atom, again clearly indicating the influence of charge delocalization in carbanions [8].

Table 3. Selected bond lengths and angles for **1b**, **2a**, **2b** and **3b**.

	1b	**2a**	**2b**	**3b**		**1b**	**2a**	**2b**	**3b**
		Bond lengths [Å]					**Bond angles [°]**		
Si(1)–Si(2)	$2.448^{(2)}$	$2.3677^{(17)}$	$2.4485^{(17)}$	$2.432^{(5)}$	R^1-Si(1)-Si(2)	$99.43^{(8)}$	$106.48^{(7)}$	$101.99^{(6)}$	$104.2^{(3)}$
Si(1)–R^3	$3.477^{(2)}$	$3.5994^{(19)}$	$3.4790^{(16)}$	$3.557^{(3)}$	Si(2)-Si(1)-Si(3)	$103.41^{(8)}$	$108.64^{(7)}$	$96.21^{(6)}$	$97.04^{(15)}$
Si(1)–Si(3)	$2.358^{(2)}$	$2.3428^{(19)}$	$2.3345^{(17)}$	$2.337^{(4)}$	R^1-Si(1)-Si(3)	$94.51^{(8)}$	$100.92^{(8)}$	$99.31^{(6)}$	$93.5^{(3)}$
Si(1)–R^1	$2.362^{(2)}$	$2.352^{(2)}$	$2.3411^{(17)}$	$1.960^{(9)}$	Σ	297.4	316.04	297.5	294.7
Si(2)–R^4	$3.489^{(2)}$	$2.3686^{(18)}$	$3.5895^{(16)}$	$3.557^{(3)}$	R^2-Si(2)-Si(1)	$93.43^{(8)}$	$108.55^{(16)}$	$102.70^{(15)}$	$104.2^{(3)}$
Si(2)–Si(4)	$2.353^{(2)}$	$2.3612^{(18)}$	$2.3550^{(19)}$	$2.337^{(4)}$	R^2-Si(2)-Si(4)	$100.95^{(8)}$	$103.24^{(15)}$	$91.51^{(14)}$	$97.04^{(15)}$
Si(2)–R^2	$2.361^{(2)}$	$1.932^{(4)}$	$1.951^{(5)}$	$1.960^{(9)}$	Si(1)-Si(2)-Si(4)	$103.39^{(8)}$	$108.80^{(7)}$	$99.46^{(6)}$	$93.5^{(3)}$
					Σ	297.8	320.6	293.7	294.7
		Dihedral angles [°]			R^3-Si(1)-Si(2)	$140.64^{(7)}$	$130.11^{(6)}$	$135.43^{(6)}$	$134.68^{(15)}$
R^3-Si(1)-Si(2)-R^4	$172.18^{(7)}$		$-177.31^{(6)}$	180	R^3-Si(1)-Si(3)	$102.83^{(7)}$	$102.50^{(6)}$	$108.47^{(5)}$	$97.60^{(10)}$
R^3-Si(1)-Si(2)-R^2	$163.32^{(8)}$		$-178.15^{(5)}$	$63.9^{(3)}$	R^3-Si(1)-R^1	$107.18^{(7)}$	$104.63^{(6)}$	$109.58^{(5)}$	$117.4^{(3)}$
					R^4-Si(2)-Si(1)	$136.52^{(7)}$	$123.48^{(7)}$	$133.63^{(6)}$	$134.68^{(15)}$
					R^4-Si(2)-Si(4)	$92.19^{(7)}$	$107.21^{(7)}$	$106.32^{(5)}$	$97.60^{(10)}$
					R^4-Si(2)-R^2	$118.59^{(7)}$	$103.72^{(15)}$	$114.23^{(14)}$	$117.4^{(3)}$

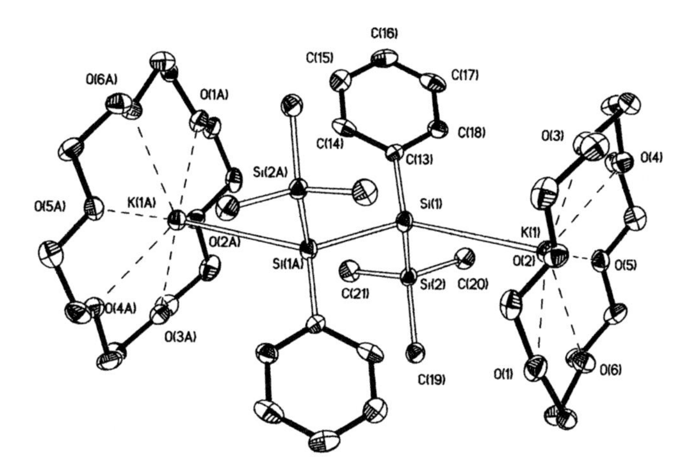

Fig. 1. Molecular structure of **3b** with 30% probability ellipsoids. All hydrogen atoms have been omitted for clarity.

All 1,2-oligosilyl dipotassium compounds studied were found to acquire an approximate *trans* conformation of the potassium-crown ether moieties, but only **3b** was found to adopt an exact *trans* conformation [9]. Although **3b** can acquire both a *meso-* and a *rac*-type configuration, only the *meso* form was found to exist in the crystal.

Acknowledgments: This study was supported by the Fonds zur Förderung der wissenschaftlichen Forschung via the START project Y120. We thank Wacker Chemie GmbH, Burghausen, for the generous donation of various chlorosilanes as starting materials.

References

[1] R. Fischer, T. Konopa, J. Baumgartner, C. Marschner, *Organometallics* **2004**, *23*, 1899.

[2] C. Kayser, R. Fischer, J. Baumgartner, C. Marschner, *Organometallics* **2002**, *21*, 1023.

[3] a) H. Gilman, R. L. Harrell Jr., *J. Organomet. Chem.* **1967**, *9*, 67; b) Y. Apeloig, M. Yuzefovich, M. Bendikov, D. Bravo-Zhivotovskii, *Organometallics* **1997**, *16*, 1265.

[4] a) C. Kayser, G. Kickelbick, C. Marschner, *Angew. Chem.* **2002**, *114*, 1031, *Angew. Chem. Int. Ed.* **2002**, *41*, 989; b) R. Fischer, D. Frank, W. Gaderbauer, C. Kayser, C. Mechtler, J. Baumgartner, C. Marschner, *Organometallics* **2003**, *23*, 3723; c) R. Fischer, T. Konopa, S. Ully, J. Baumgartner, C. Marschner, *J. Organomet. Chem.* **2003**, *685*, 79.

[5] a) G. Olah, R. J. Hunadi, *J. Am. Chem. Soc.* **1980**, *102*, 6989; b) E. Buncel, T. K. Venkachalam, B. Eliasson, U. Edlund, *J. Am. Chem. Soc.* **1985**, *107*, 303.

[6] Crown ether complex of **1b** was found to crystallize with two independent molecules of dianion and four molecules of benzene in the unit cell. Structural data for only one of the 1,2-dianion moieties are given in Table 3.

[7] a) S. P. Mallela, I. Bernal, R. A. Geanangel, *Inorg. Chem.* **1992**, *31*, 1626; b) F. P. Fronczek, P. D. Lickiss, *Acta Cryst.* **1993**, *C49*, 331; c) H. Bock, J. Meuret, K. Ruppert, *J. Organomet. Chem.* **1993**, *445*, 19.

[8] R. A. Bartlett, H. V. R. Dias, P. P. Power, *J. Organomet. Chem.* **1988**, *341*, 1.

[9] Crystallographic data for **3b**: $C_{42}H_{76}K_2O_{12}Si_4$, orthorhombic space group *Pbca*, a = 17.760(4) Å, b = 14.984(3) Å, c = 20.070(4) Å, V = 5340.9(19) Å^3, Z = 4, ρ_{calcd} = 1.198 Mg/m^3, $R1$ = 0.0984, $wR2$ = 0.1826, GOF = 1.168. Data have been deposited at the Cambridge Crystallographic Data Centre. CCDC 238892.

Heteroatom-Substituted Silyl Anions

Roland Fischer, Pravin R. Likhar, Judith Baumgartner, Christoph Marschner*

Institut für Anorganische Chemie, Technische Universität Graz
Stremayrgasse 16, A-8010 Graz, Austria
Tel.: +43 316 873 8209 — Fax: +43 316 873 8701
E-mail: marschner@anorg.tu-graz.ac.

Keywords: silyl anion, disilene, silylenoid

Summary: Reaction of α-heteroatom-substituted oligosilanes with potassium alkoxides is of interest to obtain insight into the mechanism of the Wurtz type polymerization of halosilanes. It also provides access to building blocks with unique reactivity. Reactions with fluoro- and methoxy-substituted silanes exhibited initial formation of silylenoid species which can undergo self-condensation. This property is less pronounced with the alkoxysilanes, which allowed for the isolation and structural characterization of an α-methoxy-silyl potassium compound (**3**).

Introduction

The chemistry of functionalized silyl anions has experienced an exciting development within the last decade. In particular, Tamao et al. have investigated the class of silylenoid compounds [1]. These are discussed to be involved in the Wurtz-type polymerization of diorganodihalosilanes. In addition they also can be useful as building blocks in the main group and transition metal chemistry of silanes. Our investigations concerning the formation and properties of silyl anions [2] have brought about an interest in the possibility of obtaining α-heteroatom-substituted oligosilyl potassium compounds.

Results and Discussions

Recently we found that the reaction of oligosilanes with potassium alkoxides is a valuable method for the generation of oligosilyl potassium compounds [2]. Besides the generation of alkylated, arylated, and other oligosilanes, this method also permits a simple access to α-heteroatom-substituted oligosilanes. These compounds could be subjected again to the metalation reaction, and the structure and reactivity of the obtained anions were explored.

We found that the properties and behavior of hydrogen- [3] and nitrogen-bearing [4] silyl anions very much resemble the chemistry of "ordinary" silyl anions. This means that while the heteroatom can be considered as a useful handle for further functionalization, it is not introducing new reactivity patterns which had not been observed before for silyl-, alkyl- and aryl-substituted anions. The situation, however, was different when we investigated the reaction of fluorotris-(trimethylsilyl)silane with potassium *tert*-butoxide (Scheme 1) [5]. In this case a clean conversion occurred, and it was found that a condensation product (**1**) bearing a fluoride in the β-position with respect to the anionic silicon atom, was formed.

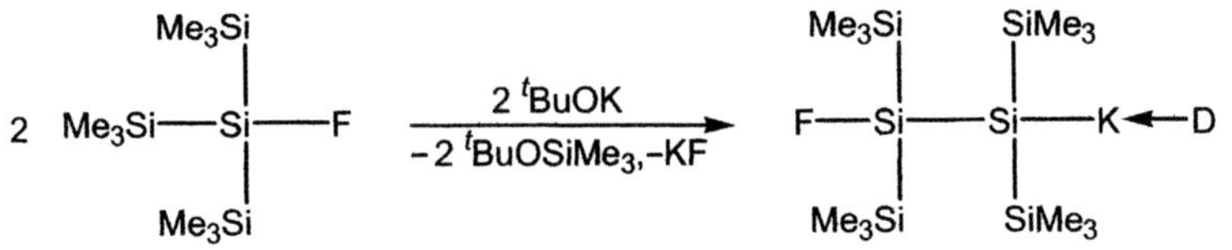

$$2\ (Me_3Si)_3Si{-}F \xrightarrow[-2\ {}^tBuOSiMe_3,\ -KF]{2\ {}^tBuOK} F{-}Si(SiMe_3)_2{-}Si(SiMe_3)_2{-}K \leftarrow D \quad (\mathbf{1})$$

D = THF, DME, 18-Cr-6

Scheme 1. Formation of 2-fluoro-1,1,2,2-tetrakis(trimethylsilyl)disilanylpotassium (**1**).

It was assumed that an α-fluorosilyl potassium species was formed initially, and that this subsequently underwent a self-condensation reaction. The eventual product (**1**) displayed both nucleophilic as well as electrophilic character, which was demonstrated in various derivatization reactions. Although it contains fluorine and potassium atoms in close proximity, the compound displayed a remarkable thermal stability. Even at 80 °C, potassium fluoride elimination occurred only sluggishly. Attempted transmetalation reactions with various metal halides, though, caused an immediate elimination of metal fluoride and the formation of tetrakis(trimethylsilyl)disilene. The latter can be trapped in cycloaddition reactions [5] or, in the absence of trapping reagents, it dimerizes to a cyclotetrasilane (Scheme 2) [6].

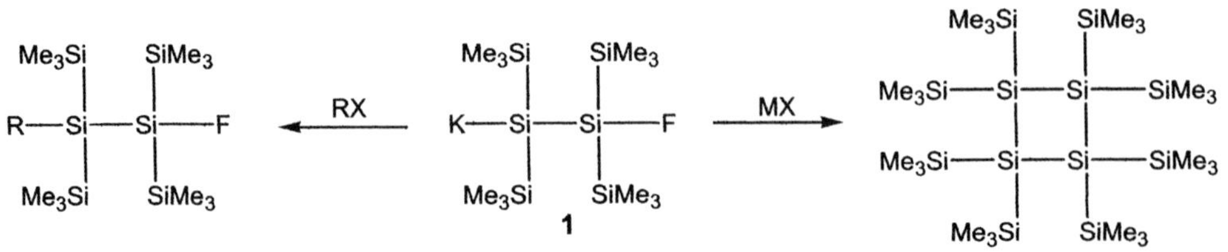

$$R{-}Si(SiMe_3)_2{-}Si(SiMe_3)_2{-}F \xleftarrow{RX} K{-}Si(SiMe_3)_2{-}Si(SiMe_3)_2{-}F\ (\mathbf{1}) \xrightarrow{MX} [(Me_3Si)_2Si]_4$$

RX = Me_3SiCl, Me_2SO_4, EtBr MX = $MgBr_2$ or $CpZrCl_2$

Scheme 2. Derivatization reactions of **1**.

Reactions of the respective chloro- and bromotris(trimethylsilyl)silanes with potassium *tert*-butoxide did not give other β-halosilyl potassium compounds but a metalated cyclotetrasilane (Scheme 3) [7]. Most likely this can be explained by the assumption of chemistry similar to that discussed above, which, however, does not involve stable disilene precursors but exhibits fast salt elimination and dimerization steps so that part of the potassium alkoxide can react with the formed octakis(trimethylsilyl)cyclotetrasilane.

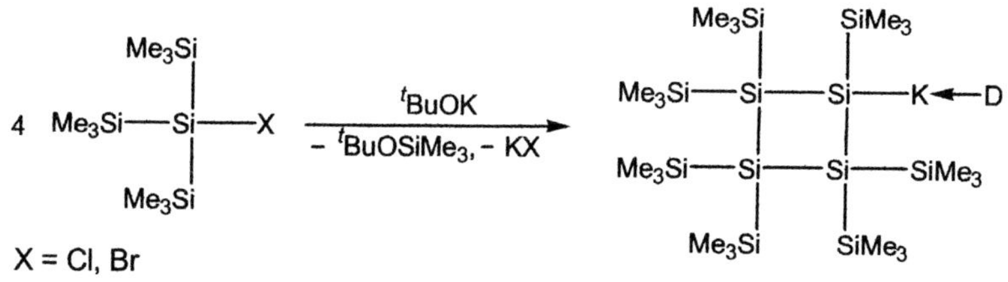

Scheme 3. Reaction of chloro- or bromotris(trimethylsilyl)silanes with potassium *tert*-butoxide.

Changing the heteroatom to oxygen (the element between nitrogen and fluorine), it was intriguing to observe that it exhibited reactivity patterns of both its neighbors [4, 8]. Conducting the reaction of methoxytris(trimethylsilyl)silane with potassium *tert*-butoxide in benzene in the presence of 18-crown-6 led to the formation of the α-alkoxysilyl anion **2**. The compound was meta-stable at room temperature and displayed slow self-condensation. But it could be derivatized with electrophiles like ethyl bromide and trimethylchlorosilane to give the respective methoxysilanes.

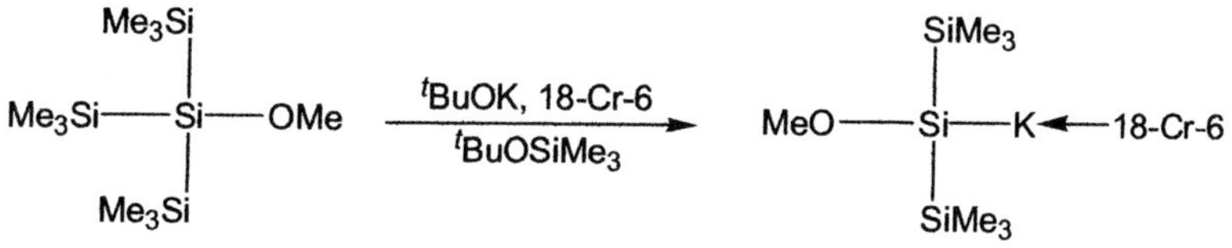

Scheme 4. Formation of the crown ether adduct of methoxybis(trimethylsilyl)silyl potassium (**2**).

If the reaction was carried out in THF without any crown ether, the self-condensation reaction was observed at a much faster pace leading to the formation of a β-methoxysilyl anion (**3**).

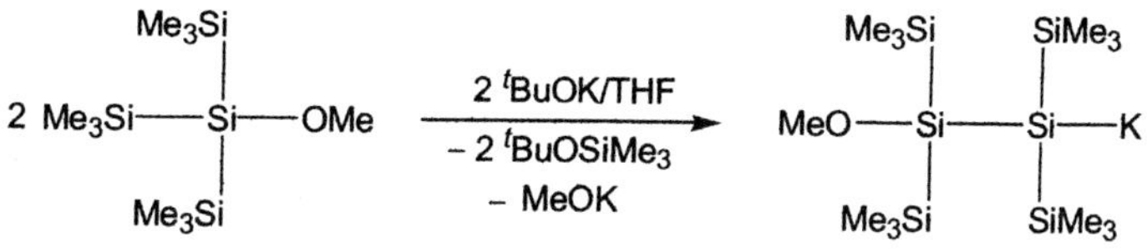

Scheme 5. Self-condensation of the initially formed α-alkoxysilyl anion.

Both anions could be characterized by multinuclear NMR spectroscopy and single-crystal X-ray diffraction analysis [9]. ^{29}Si NMR spectroscopy (Table 1) reveals the silylenoid character of the first reaction product (**2**) by a rather large downfield shift of the resonance of the anionic silicon atom of some 30 ppm compared to the shift of methoxytris(trimethylsilyl)silane. This behavior is in full accordance with the observations of Tamao and Kawachi [8] on related compounds. Once the self-condensation has taken place, the expected shift values for the anionic silicon atom (**3**) are found.

Table 1. NMR Data of α- and β-substituted silyl potassium compounds.

	1	**2**	**3**
^{1}H	3.37 (s, 24H), 0.71 (s, 18H), 0.56 (s, 18H)	3.71 (s, 3H), 3.23 (s, 24H), 0.73 (s, 18H)	3.15 (s, 3H), 0.50 (s, 18H), 0.44 (s, 18H)
^{13}C	70.5, 8.1, 1.4	70.1, 62.6, 4.8	54.9, 7.6, 2.1
^{29}Si	58.6, –6.5, –17.5, –169.3	32.3, –14.7	33.2, –14.4, –17.9, –170.4

Acknowledgments: This study was supported by the Austrian *Fonds zur Förderung der wissenschaftlichen Forschung* via the START project Y120. We also thank Wacker AG, Burghausen, for the generous donation of various silanes.

References

[1] K. Tamao, A. Kawachi, M. Asahara, A. Toshimitsu, *Pure Appl. Chem.* **1999**, *71*, 393.

[2] a) Ch. Marschner, *Eur. J. Inorg. Chem.* **1998**, 221; b) Ch. Kayser, R. Fischer, J. Baumgartner, Ch. Marschner, *Organometallics* **2002**, *21*, 1023.

[3] T. Iwamoto, J. Okita, C. Kabuto, M. Kira, *J. Am. Chem. Soc.* **2002**, *124*, 11604 and references therein.

[4] a) K. Tamao, A. Kawachi, Y. Ito, *J. Am. Chem. Soc.* **1992**, *114*, 3989; b) A. Kawachi, H. Maeda, K. Tamao, *Organometallics* **2002**, *21*, 1319; c) A. Kawachi, K. Tamao, *J. Am. Chem. Soc.* **2000**, *122*, 1919.

[5] R. Fischer, J. Baumgartner, G. Kickelbick, Ch. Marschner, *J. Am. Chem. Soc.* **2003**, *125*, 3414.

[6] Y.-S. Shen, P. Gaspar, *Organometallics* **1982**, *1*, 1410.

[7] R. Fischer, T. Konopa, J. Baumgartner, Ch. Marschner, *Organometallics* **2004**, *23*, 1899.

[8] a) K. Tamao, A. Kawachi, *Angew. Chem. Int. Ed. Engl.* **1995**, *34*, 818; b) A. Kawachi, K. Tamao, *Organometallics* **1996**, *15*, 4653.

[9] P. R. Likhar, M. Zirngast, J. Baumgartner, Ch. Marschner, *Chem. Commum.* **2004**, 1764.

Reactions of Hypersilyl Potassium with Rare Earth Bis(trimethylsilylamides): Addition Versus Peripheric Deprotonation

M. Niemeyer*

Institut für Anorganische Chemie, Universität Stuttgart, Pfaffenwaldring 55
D-70569 Stuttgart, Germany
Tel.: +49 711 6854217 — Fax: +49 711 6854241
E-mail: niemeyer@iac.uni-stuttgart.de

Keywords: bis(trimethylsilylamides), hypersilyl, rare earth metal amides, rare earth metal silyls

Summary: The scope of hypersilyl potassium, KHyp (Hyp = $Si(SiMe_3)_3$) as a silylation agent for some rare earth bis(trimethylsilyl)amides has been explored. Thus, the reaction with $Yb\{N(SiMe_3)_2\}_2$ affords the addition product $KYbHyp\{N(SiMe_3)_2\}_2$, which contains a three-coordinate ytterbium atom, therefore representing the first example of a lanthanide silyl with a coordination number lower than six. In contrast, peripheric deprotonation is observed with the trisamides $Ln\{N(SiMe_3)_2\}_3$, which results in the formation of compounds of the type $[K][CH_2Si(Me)_2N(SiMe_3)Ln\{N(SiMe_3)_2\}_2]$. Depending on the reaction conditions used σ-donor-free complexes may be obtained.

Introduction

The use of charged π-donor ligands and the principle of steric saturation by associated neutral σ-donors or the addition of anionic ligands have dominated organolanthanoid [1] chemistry for some decades. More recently, however, there has been a growing interest in low-coordinate σ-bonded lanthanoid complexes with bulky substituted organyls [2] or bonding to the heavier group 14 elements [3]. Moreover, the avoidance of coordinating solvents often leads to novel structures and unusual lanthanoid–ligand interactions. Examples of the latter include agostic-type interactions with alkyl-substituted ligands [4 – 10] or metal–π-arene interactions [11 – 16] in aryl-substituted ligand systems. In contrast to the growing number of well-characterized σ-bonded rare earth organyls, there is only a very limited number of structurally authenticated lanthanide silyls [17 – 23]. Part of the interest in these compounds is due to their proposed role in the dehydropolymerization of primary silanes [24] or in the transfer of organyl groups to hydrosilanes [25]. In this paper the scope of hypersilyl potassium, KHyp (Hyp = $Si(SiMe_3)_3$) (**1**), as a silylation agent for some rare earth bis(trimethylsilyl)amides is explored.

Synthesis

The preparation of compounds **2** – **4** is summarized in Scheme 1. The required starting material, donor solvent-free hypersilyl potassium (**1**) [26], was prepared by thermally induced desolvation of KHyp(thf)$_x$ which itself is easily accessible from tetrakis(trimethylsilyl)silane and potassium *tert*-butoxide [27].

$$Yb\{N(SiMe_3)_2\}_2 + KSi(SiMe_3)_3 \ (\mathbf{1}) \longrightarrow (Me_3Si)_3Si{-}Yb(\mu\text{-}N(SiMe_3)_2)_2K \ (\mathbf{2})$$

$$Ln\{N(SiMe_3)_2\}_3 + KSi(SiMe_3)_3 \xrightarrow{-\,HSi(SiMe_3)_3} [K][CH_2Si(Me)_2N(SiMe_3)Ln\{N(SiMe_3)_2\}_2] \quad \mathbf{3}\ (Ln = Y),\ \mathbf{4}\ (Ln = Yb)$$

$$\mathbf{3} \xrightarrow{\text{heptane/benzene}} [(C_6H_6)_2K][CH_2Si(Me)_2N(SiMe_3)Y\{N(SiMe_3)_2\}_2] \quad \mathbf{3a}$$

Scheme 1. Synthesis of compounds **2** – **4**.

The reaction of **1** with Yb{N(SiMe$_3$)$_2$}$_2$ in a benzene/*n*-heptane mixture afforded deep orange crystals of the adduct KYbHyp{N(SiMe$_3$)$_2$}$_2$ (**2**) in good yield (82%). A quite different result is found for the reaction of **1** with bis(trimethylsilyl)amides of the late trivalent rare earth metals (Ln{N(SiMe$_3$)$_2$}$_3$ with Ln = Y, Yb). Herein no addition or substitution reaction occurs, which is obviously because of the steric interference with the third amido substituent. As a consequence, peripheric deprotonation and formation of σ-donor-free ate complexes with the composition [K][CH$_2$Si(Me)$_2$N(SiMe$_3$)Ln{N(SiMe$_3$)$_2$}$_2$] (Ln = Y (**3**), Yb (**4**)) is observed. Crystallization of **3** from an *n*-heptane/benzene (20:1) mixture at –60 °C gave the benzene solvate [(C$_6$H$_6$)$_2$K][CH$_2$Si(Me)$_2$N(SiMe$_3$)Y{N(SiMe$_3$)$_2$}$_2$] (**3a**). Products similar to **3**, **3a**, and **4** have been obtained earlier by C–H σ-bond metathesis of transition metal [28 – 31] or rare earth metal bis(trimethylsilyl)amides [32] with strong bases such as LiN(SiMe$_3$)$_2$, NaN(SiMe$_3$)$_2$, or Li*n*Bu. Donor-stabilized compounds of the type [(donor)M^I][CH$_2$Si(Me)$_2$N(SiMe$_3$)M^{III}{N(SiMe$_3$)$_2$}$_2$] (with (donor)M^I/M^{III} = (12-crown-4)$_2$Na/Ti [29]; (12-crown-4)Li/Ti [30]; (thf)$_3$Na/Ln where Ln = Sc, Yb [32]), which are closely related to **3a**, have been structurally characterized by the Dehnicke group.

Compounds **2** – **4** are highly air sensitive, especially **2**, which instantly burns on contact with air. With the exception of **3a**, which rapidly loses coordinated solvent at ambient temperature, they show considerable thermal stability, with decomposition points in the range 145 – 195 °C. Solutions of **2** and **3a** in C_6D_6 are stable for weeks.

Solid-State Structures

In the solid-state structure of the adduct **2** (Fig. 1), a polymeric chain structure is observed in which the potassium cations connect adjacent ytterbate anions via interactions with two amido nitrogen atoms and five trimethylsilyl carbon atoms. The ytterbium atom shows a trigonal planar coordination by the silyl and amido ligands, with a Yb–Si1 bond length of 303.87(10) pm. In addition, there are two short intramolecular contacts to trimethylsilyl carbon atoms.

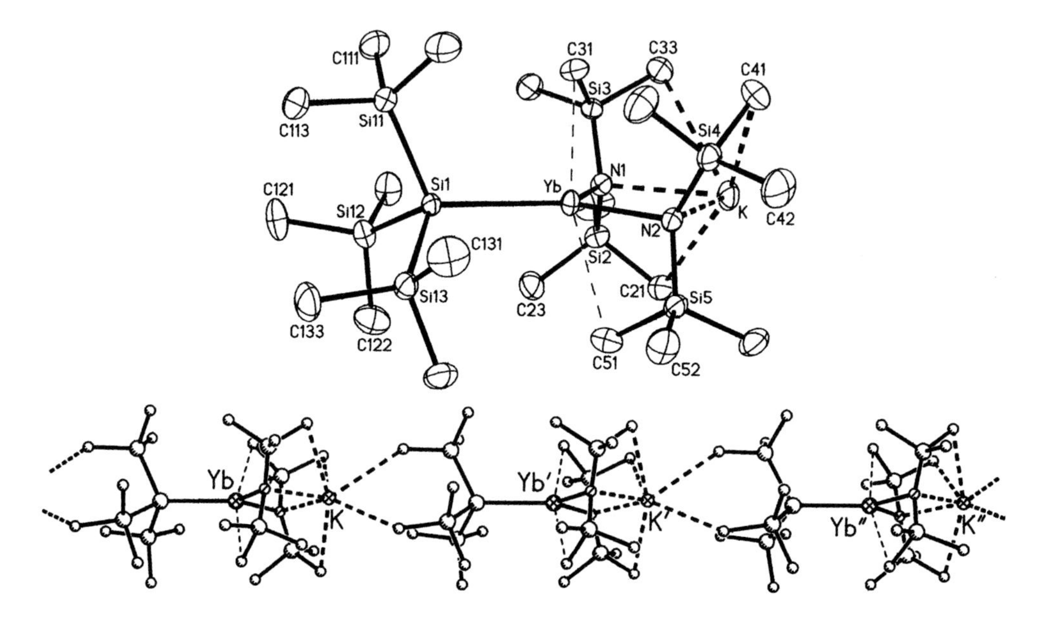

Fig. 1. Molecular (top) and solid-state structure (bottom) of **2**. Important bond distances [pm]: K–N1 290.7(3), K–N2 291.0(3), Yb–N1 236.9(3), Yb–N2 237.6(3), Yb–Si1 303.87(10), Yb···Si3 319.10(11), Yb···Si5 320.49(11), Yb···C31 285.4(4), Yb···C51 288.5(4), K···C21 326.3(4), K···C41 327.8(4), K···C33 349.7(4), K···C113' 349.8(4), K···C133' 344.9(4).

The structure of the deprotonation product **3a**, in which the yttrium atom is part of a four-membered Y–C11–Si1–N1 ring, is shown in Fig. 2. The carbanionic C11 atom shows a distorted trigonal bipyramidal coordination, the axial and equatorial positions being occupied by Y, K {with K-C11-Y: 176.36(18)°}, and the two H atoms and Si1, respectively. The potassium cation exhibits a pseudo-tetrahedral coordination by the C11 atom, the centroids of two benzene rings, and one intermolecular agostic contact to the carbon atom of an adjacent trimethylsilyl group.

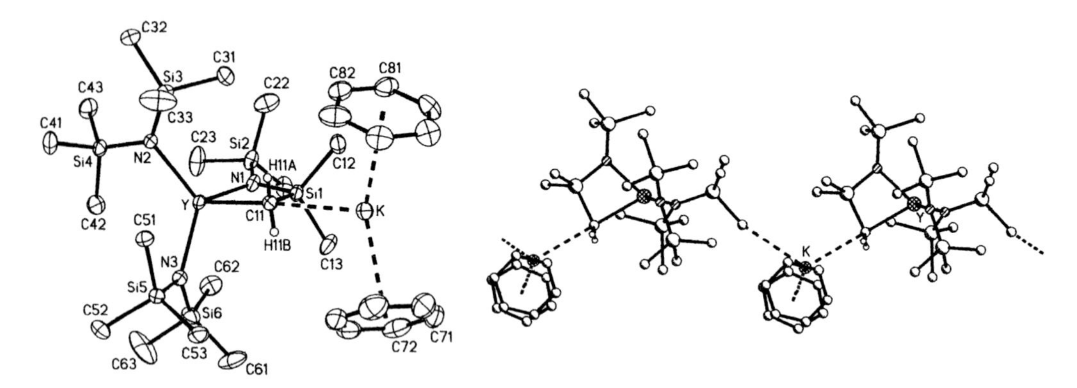

Fig. 2. Molecular (left) and solid-state structure (right) of **3a**. Important bond parameters [pm, °]: K–C11 300.0(4), av. K···C(benzene) 329, av. K···centroid 299, Y–C11 244.9(4), Y–N1 222.2(3), Y–N2 229.0(3), Y–N3 228.7(4), Y···Si1 295.80(17), Si1–C11 183.5(4), N1–Si1 172.1(3), K···C41' 322.7(4) , K-C11-Y 176.36(18).

A rather complicated solid-state structure is observed for the donor-free deprotonation product **4**. Two independent molecular units (Fig. 3, left) are connected via polyagostic contacts between the potassium cations and trimethylsilyl carbon atoms. Two free coordination sides (indicated by dotted arrows in Fig. 3, right) are used for a further linkage, thus resulting in a polymeric layer structure. In contrast to the structure of **3a**, the two independent molecular units in **4** show an average Ln-C-K angle of 90°.

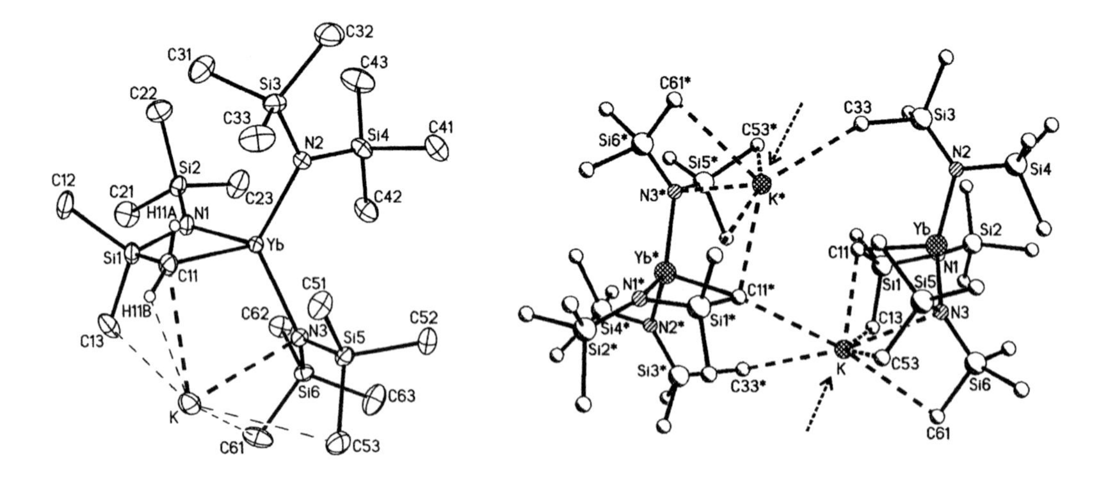

Fig. 3. One of the two independent molecular units (left) and solid-state structure (right) of **4**. Important bond parameters [pm, °]: K–C11 302.2(5), K–N3 305.0(4), Yb–C11 236.0(5), Yb–N1 219.4(4), Yb–N2 223.1(4), Yb–N3 227.6(4), Yb···Si1 292.84(14), Si1–C11 186.1(5), N1–Si1 173.4(4), K-C11-Yb 92.04(16).

Spectroscopic Characterization

The diamagnetic compounds **2** and **3** were characterized by ^{1}H, ^{13}C, ^{29}Si, and ^{171}Yb (**2** only) NMR spectroscopy in C_6D_6 solution. As expected, two sets of signals are observed for the $N(SiMe_3)_2$ and $Si(SiMe_3)_3$ groups in the ^{1}H and ^{13}C NMR spectra of **2** with corresponding $^{1}J(^{13}C{-}^{29}Si)$ coupling constants of 52.3 and 41.0 Hz in the latter. The ^{29}Si NMR spectrum comprises three resonances at δ −148.6, −13.9, and −4.7, the first assigned to $\mathit{Si}(SiMe_3)_3$ showing satellites with $^{1}J(^{29}Si{-}^{171}Yb)$ = 716 and $^{1}J(^{29}Si{-}^{29}Si)$ = 27.9 Hz, the second $\{N(\mathit{Si}Me_3)_2\}$ displaying $^{1}J(^{29}Si{-}^{13}C)$ = 52.2 and $^{2}J(^{29}Si{-}^{171}Yb)$ = 5.0 Hz, and the last $\{Si(\mathit{Si}Me_3)_3\}$ revealing $^{2}J(^{29}Si{-}^{171}Yb)$ =8.9 Hz and confirming the previously measured $^{1}J(^{29}Si{-}^{13}C)$ and $^{1}J(^{29}Si{-}^{29}Si)$ coupling. The $^{1}J(^{29}Si{-}^{171}Yb)$ coupling of 716 Hz may be compared with the corresponding values in the silyls $Yb(Cp^*)(Hyp)(thf)_2$ [33] (**A**) and $Yb(Si(SiMe_3)_2SiMe_2Pph)_2(thf)_2$ (Pph = 2',3',4',5',6'-pentamethylbiphenyl-C^2) [34] (**B**) which are 829 and 728 Hz, respectively. The ^{29}Si NMR chemical shift for the central metal-bound silicon atom in **2** (−148.6 ppm) lies somewhere in the middle of the values reported for **A** (−158.3 ppm) and **B** (−140.5 ppm) and is considerably smaller than the −185.7 ppm found in dimeric donor-free potassium hypersilyl [26]. In the ^{171}Yb NMR spectrum of **2**, a single resonance at δ 1057 is observed.

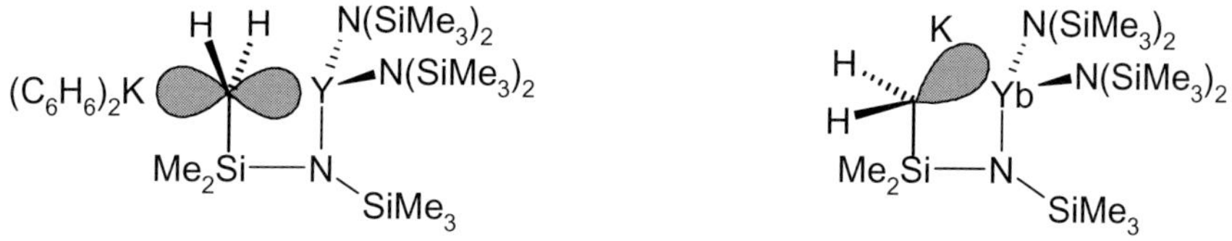

	3a	4
Si–CH$_2$	183.5(4) pm	186.1(5) pm
K-C-Ln	176°	90°

Scheme 2. Different bonding modes in compounds **3a** and **4**.

The ^{1}H, ^{13}C, and ^{29}Si NMR spectra of **3a** in C_6D_6 solution are in accordance with the solid-state structure. Therefore, they show three signals for the $N(SiMe_3)_2$, $NSiMe_3$ and $NSiMe_2$ fragments, respectively. In addition, duplets for the metal-bound CH_2 group are observed in the ^{1}H NMR at δ −1.27 with $^{2}J(^{1}H{-}^{89}Y)$ = 2.6 Hz and in the $^{13}C\{^{1}H\}$ NMR spectrum at δ 23.6 with $^{1}J(^{89}Y{-}^{13}C)$ = 22.9 Hz. A comparison of these values with the corresponding parameters in other yttrium alkyls, which contain either trimethylsilylmethyl or bis(trimethylsilyl)methyl substituents shows that the $^{1}J_{CY}$ coupling is unusually small [35]. Additional information about the bonding situation in the central metalla cyclus is provided by the evaluation of further coupling constants. Besides the small value for $^{1}J_{YC}$, several other findings are remarkable: a $^{3}J(^{89}Y{-}^{13}C)$ coupling to the $SiMe_2$ group (1.8 Hz), which seems to be the first of this kind, a relatively large $^{2}J(^{89}Y{-}^{29}Si)$ coupling of 3.9 Hz, [36] and finally a $^{1}J(^{13}C{-}^{1}H)$ coupling of 120.2 Hz. The latter is ca. 20–30 Hz larger than the corresponding coupling in other $Y{-}CH_2SiMe_3$ or $Y{-}CH(SiMe_3)_2$ compounds. These observations are in accordance with a high s character for the bonding in the CH_2Si fragment (i.e. sp^2 hybridization) and a high p contribution for the bonding from C(11) to the yttrium and potassium cations (Scheme 2), which corresponds well with the structural data and computational results [37] on suitable model compounds.

References

[1] M. N. Bochkarev, L. N. Zakharov, G. S. Kalinina, *Organoderivatives of Rare Earth Elements*, Kluwer Academic, Dordrecht, **1995**.

[2] a) S. A. Cotton, *Coord. Chem. Rev.* **1997**, *160*, 93; b) F. T. Edelmann, D. M. M. Freckmann, H. Schumann, *Chem. Rev.* **2002**, *102*, 1851.

[3] F. Nief, *Coord. Chem. Rev.* **1998**, *178-180*, 13.

[4] J. M. Boncella, R. A. Andersen, *Organometallics* **1985**, *4*, 205.

[5] P. B. Hitchcock, M. F. Lappert, R. G. Smith, R. A. Bartlett, P. P. Power, *J. Chem. Soc., Chem. Commun.* **1988**, 1007.

[6] P. B. Hitchcock, J. A. K. Howard, M. F. Lappert, S. Prashar, *J. Organomet. Chem.* **1992**, *437*, 177.

[7] W. J. Evans, R. Anwander, J. W. Ziller, *Inorg. Chem.* **1995**, *34*, 5927.

[8] M. Westerhausen, M. Hartmann, W. Schwarz, *Inorg. Chim. Acta* **1998**, *269*, 91.

[9] W. T. Klooster, R. S. Lu, R. Anwander, W. J. Evans, T. F. Koetzle, R. Bau, *Angew. Chem.* **1998**, *110*, 1326.

[10] W. T. Klooster, L. Brammer, C. J. Schaverien, P. H. M. Budzelaar, *J. Am. Chem. Soc.* **1999**, *121*, 1381.

[11] a) G. B. Deacon, Q. Shen, *J. Organomet. Chem.* **1996**, *506*, 1; b) M. N. Bochkarev, *Chem. Rev.* **2002**, *102*, 2089.

[12] M. Niemeyer, S.-O. Hauber, *Z. Anorg. Allg. Chem.* **1999**, *625*, 137.

[13] G. B. Deacon, C. M. Forsyth, P. C. Junk, B. W. Skelton, A. H. White, *Chem. Eur. J.* **1999**, *5*, 1452.

[14] G. Heckmann, M. Niemeyer, *J. Am. Chem. Soc.* **2000**, *122*, 4227.

[15] M. Niemeyer, *Eur. J. Inorg. Chem.* **2001**, 1969.

[16] M. Niemeyer, *Acta Cryst.* **2001**, *E57*, m578.

[17] $[Li(dme)_3][Lu(Cp)_2(SiMe_3)_2]$: H. Schumann, J. A. Meese-Marktscheffel, F. E. Hahn, *J. Organomet. Chem.* **1990**, *390*, 301 and H. Schumann, S. Nickel, E. Hahn, M. J. Heeg, *Organometallics* **1985**, *4*, 800.

[18] $\{Sm(Cp^*)_2SiH(SiMe_3)_2\}_2$: N. S. Radu, T. D. Tilley, A. L. Rheingold, *J. Am. Chem. Soc.* **1992**, *114*, 8293.

[19] $Yb(SiPh_3)_2(thf)_4$: L. N. Bochkarev, V. M. Makarov, Y. N. Hrzhanovskaya, L. N. Zakharov, G. K. Fukin, A. I. Yanovsky, Y. T. Struchkov, *J. Organomet. Chem.* **1994**, *467*, C3.

[20] $Yb(Cp^*)\{Si(SiMe_3)_3\}(thf)_2$: M. M. Corradi, A. D. Frankland, P. B. Hitchcock, M. F. Lappert, G. A. Lawless, *Chem. Commun.* **1996**, 2323.

[21] $Sm_3(Cp^*)_6(\mu\text{-}SiH_3)(\mu^3\text{-}\eta^1,\eta^1,\eta^2\text{-}SiH_2SiH_2)$: N. S. Radu, F. J. Hollander, T. D. Tilley, A. L. Rheingold, *Chem. Commun.* **1996**, 2459.

[22] Lu(Cp*)$_2$SiH$_2$(2-MeC$_6$H$_4$): I. Castillo, T. D. Tilley, *Organometallics* **2001**, *20*, 5598.

[23] For the structures of two related silylene complexes Ln(Cp)$_3${Si(NN)} with Si(NN) = Si[{N(CH$_2$*t*Bu)$_2$}$_2$C$_6$H$_4$-1,2] and Ln = Y, Yb: X. Cai, B. Gehrhus, P. B.Hitchcock, M. F. Lappert, *Can. J. Chem.* **2000**, *78*, 1484.

[24] N. S. Radu, T. D. Tilley, A. L. Rheingold, *J. Organomet. Chem.* **1996**, *516*, 41.

[25] W.-S. Jin, Y. Makioka, T. Kitamura, Y. Fujiwara, *Chem. Commun.* **1999**, 955.

[26] K. W. Klinkhammer, *Chem. Eur. J.* **1997**, *3*, 1418.

[27] C. Marschner, *Eur. J. Inorg. Chem.* **1998**, 221.

[28] P. Berno, R. Minhas, S. Hao, S. Gambarotta, *Organometallics* **1994**, *13*, 1052.

[29] M. A. Putzer, J. Magull, H. Goesmann, B. Neumüller, K. Dehnicke, *Chem. Ber.* **1996**, *129*, 1401.

[30] M. A. Putzer, B. Neumüller, K. Dehnicke, *Z. Anorg. Allg. Chem.* **1998**, *624*, 1087.

[31] R. Messere, M.-R. Spirlet, D. Jan, A. Demonceau, A. F. Noels, *Eur. J. Inorg. Chem.* **2000**, 1151.

[32] M. Karl, K. Harms, G. Seybert, W. Massa, S. Fau, G. Frenking, K. Dehnicke, *Z. Anorg. Allg. Chem.* **1999**, *625*, 2055.

[33] M. M. Corradi, A. D. Frankland, P. B. Hitchcock, M. F. Lappert, G. A. Lawless, *Chem. Commun.* **1996**, 2323.

[34] M. Niemeyer, S.-O. Hauber, unpublished results.

[35] Comparable small $^1J_{YC}$ coupling constants around 20 Hz are typical for binuclear yttrium compounds with bridging alkyl groups. See for example: C. J. Schaverien, *Organometallics* **1994**, *13*, 69.

[36] Typical values for other known $^2J_{Y(C)Si}$ coupling constants are found below 2.3 Hz. There seem to be no reports of an observable $^2J_{Y(N)Si}$ coupling. It is notable, that a relatively high $^2J_{Y(O)Si}$ coupling of 7.7 Hz was measured in the alkoxide [Y(μ-OSiPh$_3$)(OSiPh$_3$)$_2$]$_2$ only for the terminal OSiPh$_3$ groups and not for the bridging ones, which has been explained by the larger Y-O-Si angle (higher s character!) in the former. See: P. S. Coan, L. G. Hubert-Pfalzgraf, K. G. Caulton, *Inorg. Chem.* **1992**, *31*, 1262.

[37] The DFT calculations were performed on [K][CH$_2$Si(Me)$_2$N(SiMe$_3$)Y{N(SiMe$_3$)$_2$}$_2$] (**3M**) model systems with two different K-C-Y angles of 90° (**3M$_I$**) and 176° (**3M$_{II}$**) using the B3PW91 functional, 6-31G* basis sets, and Stuttgart type pseudopotentials for Y, K and Si. According to the results of a natural bond orbital (NBO) analysis the contributions of the natural atomic orbitals for the carbanionic lone pair are as follows: s (30%), p (70%) {**3M$_I$**}; s (18%), p (82%) {**3M$_{II}$**}.

Synthesis of Organosilicon Polymers from Silyl Triflates and (Aminosilyl)lithium Compounds

Wolfram Uhlig

ETH Zürich, Laboratory of Inorganic Chemistry
ETH-Hönggerberg, CH-8093 Zürich, Switzerland
Tel.: +41 16334505 — Fax: +41 16321149
E-mail: uhlig@inorg.chem.ethz.ch

Keywords: aminosilanes, oligosilanes, silyl triflates

Summary: Diethylamino-substituted silyllithium compounds have been prepared *in situ* from (dialkylamino)phenylchlorosilanes and lithium. These reagents undergo coupling reactions with triflate derivatives of silanes and oligosilanes. Exchange processes analogous to metal halogen exchange and Si–Si bond cleavage were not observed. Based on this coupling reaction and the amino-to-triflate transformation, functionalized tri-, tetra-, penta-, and hexasilanes have been synthesized. α,ω-Triflate-substituted oligosilanes containing π electron systems have also been obtained. These compounds are useful building blocks for new organosilicon polymers. The formation of the silicon polymers at low temperatures, in short reaction times, and with high yields is reported. The ^{29}Si NMR spectra indicate a regular, alternating arrangement of the building blocks in the polymer backbone.

Introduction

Polysilanes have received much attention because of their interesting chemical and physical properties. To investigate these properties, synthetic methodologies for structurally well-defined oligosilanes have been developed. The standard method for the Si–Si chain elongation is based on a combination of two steps: (1) coupling of (organosilyl)lithium compounds with silanes bearing good leaving groups (halogen, triflate) and (2) conversion of the organic groups (phenyl, *p*-tolyl, *p*-anisyl, allyl) on the silicon atoms in good leaving groups [1 – 2]. Based on this methodology, several groups were successful in the syntheses of dendritic polysilanes with regular three-dimensional structures [3].

Synthetically useful silyl anions have long been limited to only a few simple triorganosilyl anions such as Ph_3Si^-, Ph_2MeSi^-, $PhMe_2Si^-$ [4], Me_3Si^- [5] and $(Me_3Si)_3Si^-$ [6], although three functional silyl anions, Cl_3Si^- [7], $(RO)_nMe_{3-n}Si^-$, and HPh_2Si^- have been reported. Stable, functional substituted silyl anions were first described by the groups of Tamao [8] and Roewer [9].

(Dialkylamino)phenyl-substituted chlorosilanes are able to form silyllithium compounds. Because the generated lithium species have an unusually high stability compared with the other known silyllithium compounds, they are useful synthons for the synthesis of oligosilane structures. (Aminosilyl)lithium compounds readily react with chlorosilanes to yield a variety of amino-substituted oligosilanes. By treatment with acyl chloride [8] or hydrogen chloride, these amino-oligosilanes can be converted into chloro-oligosilanes, which work as precursors for further Si–Si chain elongation. In the reaction of (aminosilyl)lithium compounds with chloro-oligosilanes, however, cleavage of the Si–Si bond often occurs to lower the yields of the coupling products. Moreover, metal halogen exchange processes cannot be suppressed. Therefore, Tamao proposed the use of the mixed reagent $(R_2N)R_2SiLi/R^1MgBr$ [10]. The authors assume the *in situ* formation of (aminosily)alkylmagnesium species. The selectivity of the silylation increases as the alkyl group R^1 became bulkier. Side reactions were completely suppressed in the case of R^1 = *i*-propyl. This method is highly effective if the chloro-oligosilane does not contain any other functional groups. However, *i*-PrMgBr is regenerated during the silylation process and can attack functional groups, for instance Si–H bonds. Our investigations in the field of silyl triflates [11] have shown that organolithium and organomagnesium compounds react rapidly and completely with silyl triflates at low temperatures. Interestingly, it was found that no exchange processes analogous to metal–halogen exchange occur. Therefore we investigated systematic, stepwise syntheses of oligosilanes and organosilicon polymers from silyl triflates and (aminosilyl)lithium anions.

Results and Discussion

The highly reactive silyl triflates and silyl bis(triflates) are obtained by reaction of the corresponding aminosilanes with triflic acid in diethylether as shown in Schemes 1 and 2 [12]. The substitution patterns at the silicon atoms are variable. Functional substituted silanes, e.g., vinyl, allyl, or hydrogen derivatives can also be obtained. (Diethylamino)diphenylsilyllithium is formed from (diethylamino)diphenylchlorosilane with lithium in THF. Yields of both reactions are high (90 – 95%). The products can be used for following reaction without further purification. Silyl triflates and (aminosilyl)lithium compounds react to give the amino-substituted trisilanes **1a** – **1c** (Scheme 1).

It must be considered that silyl triflates are not stable in THF for an extended time, because they can initiate the cationic ring-opening polymerization (ROP) of cyclic ethers. Therefore, the silyl triflate (in diethyl ether) is added in drops to the (aminosilyl)lithium compound dissolved in THF. The reaction proceeds very quickly, and ROP of THF is not observed under these conditions. The trisilanes **1a** – **1c** can be converted by reaction with triflic acid into the silyl bis (triflates) **2a** – **2c**. These compounds are precursors for further chain elongation. The reaction with (diethylamino)-diphenylsilyllithium leads to the pentasilanes **3a** – **3c**. The reaction of bis(diethylamino)phenylsilyl-lithium [9] with silyl triflates proceeds analogously. The formation of the disilane **4** is shown in Scheme 2 as an example. Conversion with triflic acid and chain elongation with $LiSiPh_2(NEt_2)$ leads to the branched tetrasilane **6**.

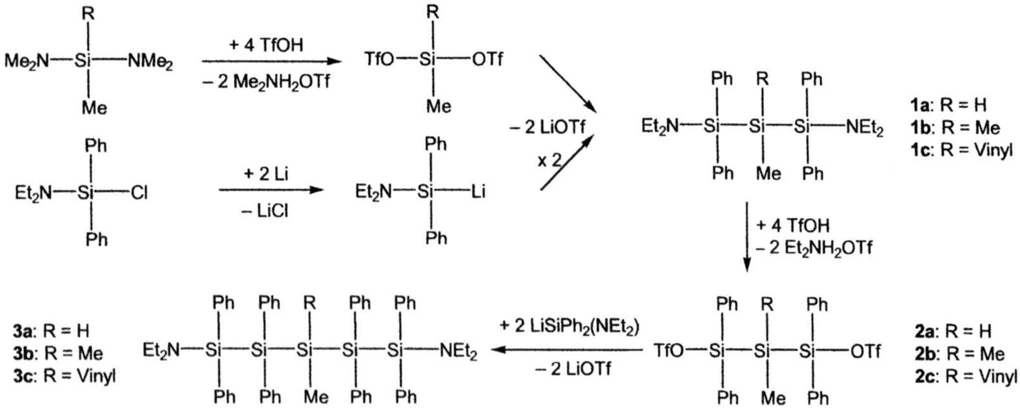

Scheme 1. Synthesis of α,ω-amino-substituted pentasilanes **3a** – **3c**.

The described reaction pathways are useful for the synthesis of new structurally well-defined organosilicon polymers. Recently, we reported on the synthesis of hydrogen-rich organosilicon polymers from α,ω-bis(trifluoromethylsulfonyloxy)-substituted organosilicon compounds and organometallic dinucleophiles [13]. The polymers were obtained in excellent yields. Exchange processes analogous to metal halogen exchange or other side reactions were not observed.

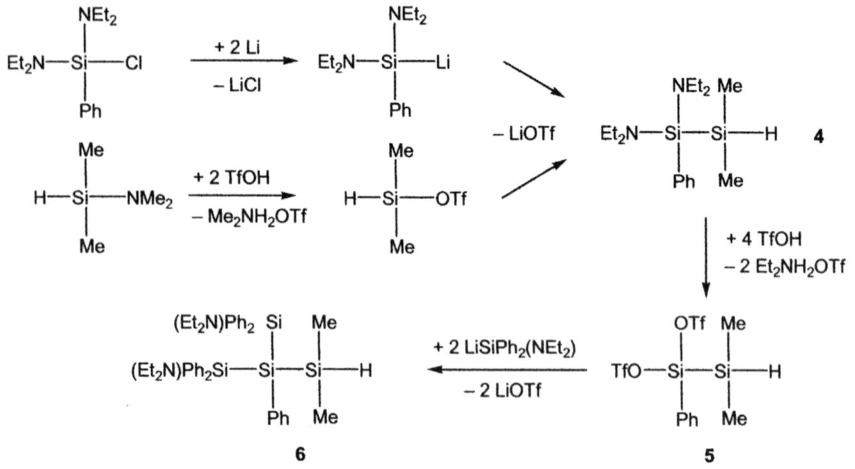

Scheme 2. Synthesis of the branched amino-substituted tetrasilane **6**.

Consequently, a regular alternating arrangement of organosilicon and organic groups was found in the polymeric backbone, indicated by narrow signals in the ^{29}Si NMR spectra. The terminal triflate-substituted disilanes, disilylalkynes, and 1,4-disilylbenzenes can be modified by reaction with (diethylamino)diphenylsilyllithium. We obtained α,ω-amino-substituted derivatives of 1,2-bis-(disilyl)ethyne, tetrasilane, and 1,4-bis(disilyl)benzene (Scheme 3). The terminal amino groups of the compounds can be exchanged by triflic acid. It should be emphasized that these compounds contain several different bonds, which can be easily cleaved by triflic acid, e.g., Si–Ph and Si–H. But the amino substituent is the best leaving group [11]. We obtained selectively the triflate-substituted derivatives **7 – 9** (Scheme 3). These compounds are new useful precursors for organosilicon polymers.

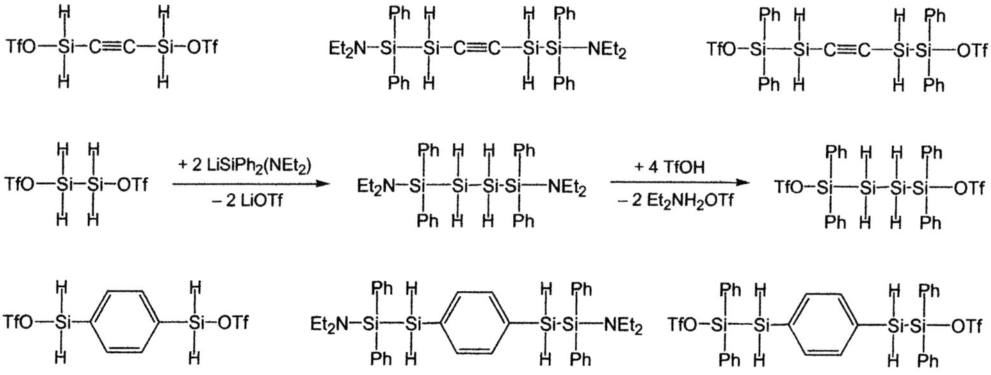

TfO–SiH₂–C≡C–SiH₂–OTf → Et₂N–SiPh₂–SiH₂–C≡C–SiH₂–SiPh₂–NEt₂ → TfO–SiPh₂–SiH₂–C≡C–SiH₂–SiPh₂–OTf **7**

TfO–SiH₂–SiH₂–OTf —(+ 2 LiSiPh₂(NEt₂), – 2 LiOTf)→ Et₂N–SiPh₂–SiH₂–SiH₂–SiPh₂–NEt₂ —(+ 4 TfOH, – 2 Et₂NH₂OTf)→ TfO–SiPh₂–SiH₂–SiH₂–SiPh₂–OTf **8**

TfO–SiH₂–C₆H₄–SiH₂–OTf → Et₂N–SiPh₂–SiH₂–C₆H₄–SiH₂–SiPh₂–NEt₂ → TfO–SiPh₂–SiH₂–C₆H₄–SiH₂–SiPh₂–OTf **9**

Scheme 3. Synthesis of α,ω-bis(trifluoromethylsulfonyloxy)-substituted organosilicon compounds **7 – 9**.

Further synthetic modifications of **7** and **8** are summarized in Schemes 4 and 5 [14]. Conversions of **9** are described in [15]. The reaction with $LiSiPh_2(NEt_2)$ leads to the elongation of the silicon chain (**10**, **14**). The reduction with the graphite intercalation compound C_8K gave the copoly-

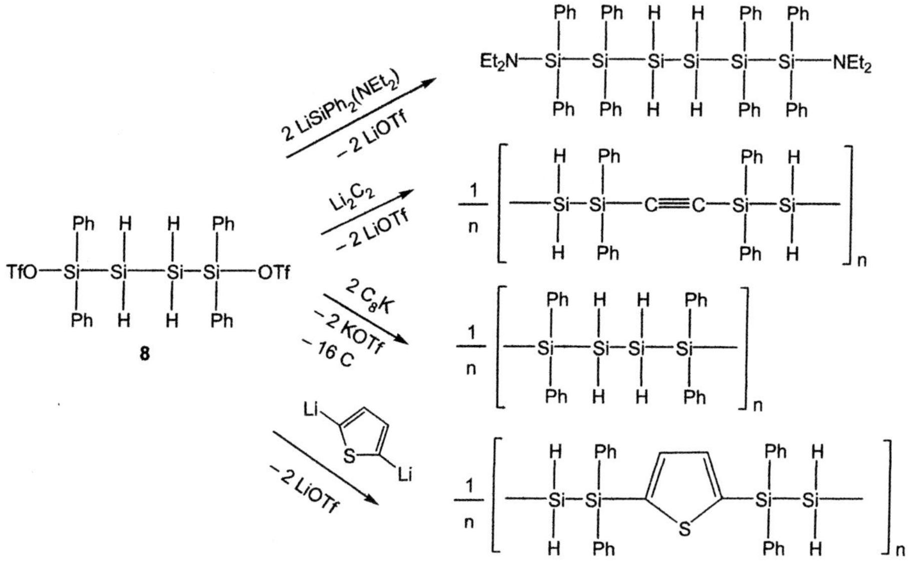

TfO–SiPh₂–SiH₂–SiH₂–SiPh₂–OTf **8**

—(2 LiSiPh₂(NEt₂), –2 LiOTf)→ Et₂N–SiPh₂–SiPh₂–SiH₂–SiH₂–SiPh₂–SiPh₂–NEt₂ **10**

—(Li₂C₂, –2 LiOTf)→ 1/n [–SiH₂–SiPh₂–C≡C–SiPh₂–SiH₂–]ₙ **11**

—(2 C₈K, –2 KOTf, –16 C)→ 1/n [–SiPh₂–SiH₂–SiH₂–SiPh₂–]ₙ **12**

—(2,5-dilithiothiophene, –2 LiOTf)→ 1/n [–SiH₂–SiPh₂–(2,5-thienylene)–SiPh₂–SiH₂–]ₙ **13**

Scheme 4. Synthesis of organosilicon oligomers and polymers **10 – 13**.

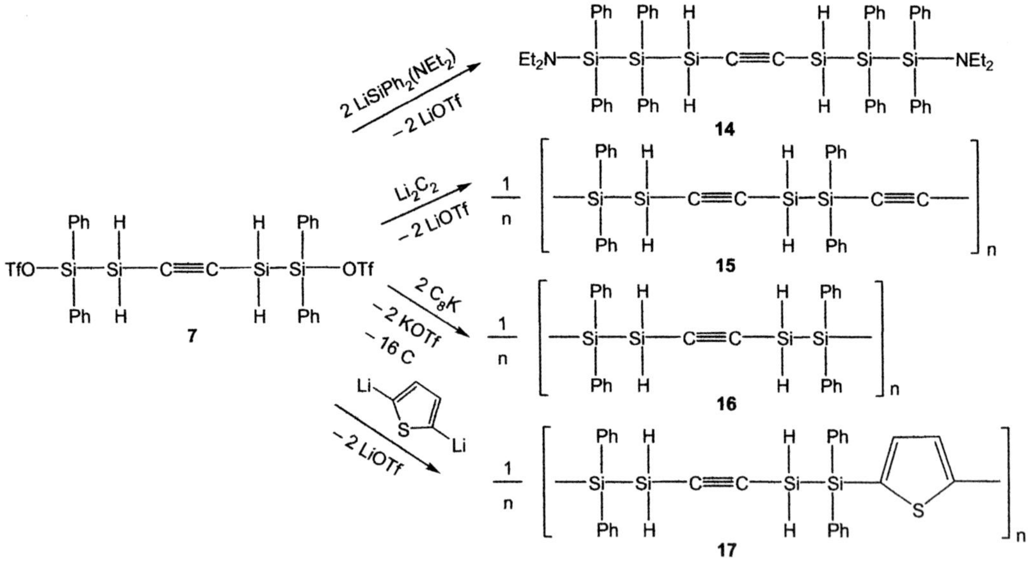

Scheme 5. Synthesis of organosilicon oligomers and polymers **14** – **17**.

silane **12** and the copolymer **16**, which contains an alternating arrangement of tetrasilane units and ethyne groups in the backbone. The best results were obtained using the "inverse reducing method" (addition of C_8K in portions to the silyl triflate solution). Otherwise, an excess of the strong reducing agent C_8K leads to Si–Si bond cleavage and reduction of triflate ions.

The organometallic dinucleophiles 1,2-dilithioethyne LiC≡CLi and 2,5-dilithiothienylene $Li(C_4H_2S)Li$ react with **7** and **8** to give the copolymers **11**, **13**, **15**, and **17**. These compounds are characterized by a regular alternating arrangement of oligosilane units and unsaturated organic groups. The syntheses are carried out using a small excess of the organolithium compound. Under these conditions we obtained polymers containing terminal ethyne or thienylene groups after hydrolysis of the reaction mixture. In contrast, a small excess of the bis(triflate) derivative leads to copolymers with terminal triflate groups. After hydrolysis, these polymers always involve small proportions of undesired siloxy units in the backbone.

We were able to confirm the formation of **10** – **17** at low temperatures, in short reaction times, and with high yields. The structural characterization was mainly based on NMR spectroscopy. ^{29}Si NMR chemical shifts are particularly useful (see Table 1). The ^{1}H, ^{13}C, and ^{29}Si NMR spectra of all compounds are consistent with the proposed structures of the polymer chain. As expected, one observes relatively broad signals, which are typical for organosilicon polymers. However, the half-band widths of the ^{29}Si NMR signals, 1.5 – 3.0 ppm, are much narrower than those in the case of organosilicon polymers prepared from chlorosilanes. The narrower signals of **11** – **13** and **15** – **17** indicate the regular alternating arrangement of the building blocks in the polymer backbone resulting from the fact that the condensation reactions are not accompanied by exchange processes analogous to metal halogen exchange. The ^{13}C NMR data also support this view. Thus, only isolated ethynyl groups are detectable in the spectra of **11** and **15**. Diethynyl units –C≡C–C≡C–, which are produced in variable yields (5 – 15%) by the reaction of Li_2C_2 with dihalogenosilanes, are not observed.

Table 1. ^{29}Si NMR data and molecular weights of the polymers **10 – 17**.

No.	Compound	δ ^{29}Si (ppm)	M_w	M_w:M_n
10	$[(Et_2N)Ph_2Si–Si^*Ph_2–Si^\circ H_2–]_2$	–16.5; –37.1*; –82.7°	–	–
11	$[–SiH_2–Si^*Ph_2–C≡C–Si^*Ph_2–SiH_2–]_n$	–38.2*; –82.0	8300	2.5
12	$[–SiH_2–Si^*Ph_2–Si^*Ph_2–SiH_2–]_n$	–36.0*; –84.7	11100	3.1
13	$[–SiH_2–Si^*Ph_2–C_4H_2S–Si^*Ph_2–SiH_2–]_n$	–21.3*; –82.4	12400	2.8
14	$[(Et_2N)Ph_2Si–Si^*Ph_2–Si^\circ H_2–C≡]_2$	–15.8; –38.6*; –79.8°	–	–
15	$[–SiH_2–Si^*Ph_2–C≡C–Si^*Ph_2–SiH_2–C≡C–]_n$	–36.5*; –79.0	15600	2.7
16	$[–SiH_2–Si^*Ph_2–Si^*Ph_2–SiH_2–C≡C–]_n$	–37.2*; –81.5	10400	2.9
17	$[–SiH_2–Si^*Ph_2–C_4H_2S–Si^*Ph_2–SiH_2–C≡C–]_n$	–20.8*; –78.9	13900	2.6

Weight-average molecular weights in the range of M_w = 8000 – 16000, relative to polystyrene standards, were found by GPC. They correspond to polymerization degrees of n = 15 – 30. The polydispersities (M_w/M_n) were found in the range 2.5 – 3.1. It must be emphasized, that the molecular weights are strongly determined by the reaction conditions. Higher values of M_w were obtained using more concentrated solutions of the reactants. The exact compliance with the stoichiometric ratio of 1:1 is another important requirement. It is therefore necessary to determine the content of the organometallic compounds quantitatively before use. However, molecular weights can also fall below 5000 when diluted solutions are used. Other changes such as the use of different solvents and reaction temperatures are currently being investigated. Thus, molecular weights reported in the experimental section are those found under the conditions specified there.

The described polymers are solids. They are soluble in the usual organic solvents such as toluene, chloroform, and THF. Definite melting points are not observed. The solids become highly viscous fluids, with concentration of volume, at temperatures between 170 and 220 °C. These conversions occur within temperature intervals of about 20 °C. Moreover, decomposition (hydrogen elimination) is observed during the melting process. Further work will be directed to investigations of the thermal behavior of the compounds.

In general, polymers composed of alternating silylene units and π electron systems are insulators. However, when the polymers are treated with an oxidizing agent, they became conducting. Preliminary doping studies showed, when cast films of selected polymers were exposed to iodine vapor under reduced pressure, that the conductivity of the films increased and reached almost constant values after 10 h. The conductivity at this point was found to be 3.6×10^{-5} S cm^{-1} (**13**) and 8.9×10^{-5} S cm^{-1} (**17**).

Obviously, the synthetic methods described here may currently appear to be too expensive for technical application. However, the essential advantage of the silyl triflate chemistry consists in the possibility of preparing, with relatively little effort, small amounts of numerous differently structured organosilicon polymers for investigations in the field of materials and surface science.

Acknowledgments: This work was supported by the ETH Zürich, by Schweizer Nationalfonds zur Förderung der Wissenschaften, by Wacker-Chemie GmbH, Burghausen, and by Siemens (Schweiz) AG. Furthermore the author thanks Prof. R. Nesper (ETH Zürich) and Prof. R. Schwarzenbach (ETH Zürich) for support of this investigation.

References

[1] K. Tamao, A. Kawachi, *Adv. Organomet. Chem.* **1995**, *38*, 1.
[2] P. D. Lickiss, C. M. Smith, *Coord. Chem. Rev.* **1995**, *145*, 75.
[3] L. B. Lambert, J. L. Pflug, C. L. Stern, *Angew. Chem.* **1995**, *107*, 106; *Angew. Chem. Int. Ed. Engl.* **1995**, *34*, 98.
[4] H. Gilman, G. D. Lichtenwalter, *J. Am. Chem. Soc.* **1958**, *80*, 608.
[5] H. Sakurai, A. Okada, M. Kira, K. Yonezawa, *Tetrahedron Lett.* **1971**, 1511.
[6] J. Dickhaut, B. Giese, *Org. Synth.* **1991**, *70*, 164.
[7] R. A. Benkeser, *Acc. Chem. Res.* **1971**, *4*, 94.
[8] K. Tamao, A. Kawachi, Y. Ito, *J. Am. Chem. Soc.* **1992**, *114*, 3989.
[9] K. Trommer, U. Herzog, U. Georgi, G. Roewer, *J. Prakt. Chem.* **1998**, *340*, 557.
[10] A. Kawachi, K. Tamao, *J. Organomet. Chem.* **2000**, *601*, 259.
[11] W. Uhlig, *Chem. Ber.* **1996**, *129*, 733.
[12] W. Uhlig, C. Tretner, *J. Organomet. Chem.* **1994**, *467*, 31.
[13] W. Uhlig, *Silicon Chem.* **2002**, *1*, 129.
[14] W. Uhlig, *J. Organomet. Chem.* **2003**, *685*, 70.
[15] W. Uhlig, *Z. Naturforsch.* **2003**, *58b*, 183.

Polyhydroxyoligosilanes — Synthesis, Structure, and Coordination Chemistry

*C. Krempner**

The Ohio State University, Department of Chemistry
100 W. 18th Ave, 43210, Columbus, Ohio, USA
Tel.: +1 614 6883625 — Fax: +1 614 2920368
E-mail: ckrempne@chemistry.ohio-state.edu

Keywords: silane, silanol, oligosilane, siloxide

Summary: A variety of polyhydroxyoligosilanes with different chain lengths, substitution patterns and degrees of branching have been prepared by selective coupling of oligosilylpotassium reagents with simple chlorosilanes followed either by hydrolysis or protodesilylation and subsequent hydrolysis. The UV-Vis spectra of a series of substituted oligosilanes with longest chains of 6 or 7 silicon atoms and the molecular structures of a tetrasilanol and two trisilanols are reported. Novel tridentate Ti and Zr siloxides have been synthesized from the reaction of triol **8a** with $Zr(NEt_2)_4$ and $Ti(OEt)_4$, respectively.

Introduction

Oligosilanes, molecules with long backbones of silicon atoms, have been shown to display unusual electronic and optical properties because of the delocalization of σ-electrons along the oligosilane backbone (σ-conjugation) [1]. Their absorption maxima in the UV-Vis were found to be red-shifted upon increasing the number of silicon atoms in the chain and strongly depend on the conformation of the silicon backbone. In this regard, we have initiated a study of how the long-wavelength absorption of oligosilanes is perturbed by the introduction of various substituents, especially those which are commonly believed to interact with Si σ-orbitals. To investigate the influence of such substituents, straightforward synthetic procedures for the selective introduction of functional groups (X = OH, OR, OAc) on the oligosilane backbone have been developed. Herein we wish to report on the synthesis, structure, electronic properties and coordination behavior of hydroxy-substituted oligosilanes.

Results and Discussion

Synthesis

One possible method for the Si–Si backbone elongation and functionalization is based on a combination of two steps: (a) a coupling of organosilylmetal compounds with silanes bearing leaving groups (X = Cl, Br, OAc), and (b) conversion of the organic groups (Ph, OH) on the silicon atoms into good leaving groups. Therefore, we used as starting materials oligosilylpotassium derivatives, which can be prepared very easily by selective Si–Si cleavage of oligosilanes of general formula $RSi(SiMe_3)_3$ (R = Ph, $SiMe_3$, Me) with *t*BuOK [2].

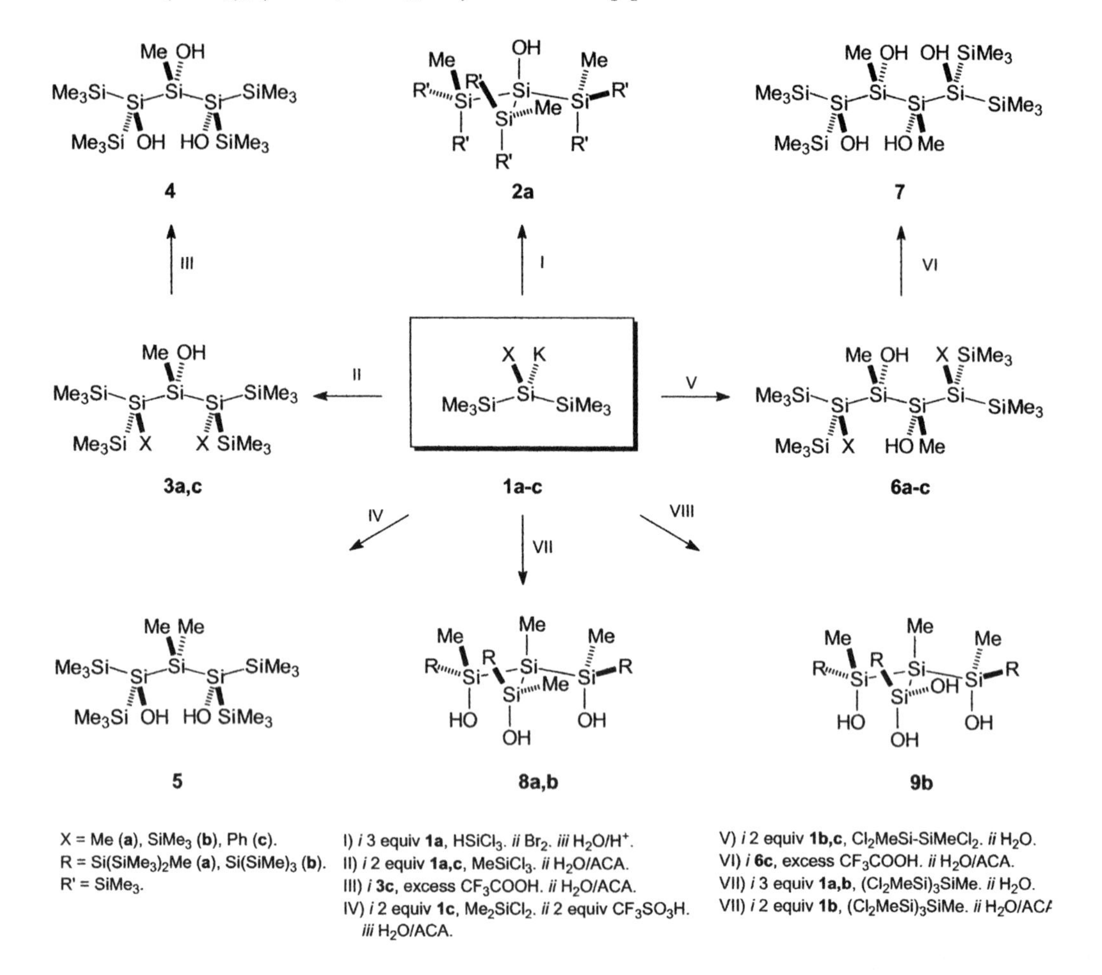

Scheme 1. Synthesis of polyhydroxyoligosilanes **2** – **9**.

Treatment of two or three equiv of **1** with simple chlorosilanes such as $MeSiCl_3$, Me_2SiCl_2, $Cl_2MeSi–SiMeCl_2$, $HSiCl_3$ and $(Cl_2MeSi)_3SiMe$ leads to the formation of a variety of branched oligosilanes with longest chains of 5 – 7 silicon atoms. The remaining functional groups in these molecules can be exchanged either directly by hydrolysis (X = Cl, Br) [3] to yield the corresponding hydroxyoligosilanes or by protodesilylation (X = Ph) using CF_3COOH as an acid and subsequent hydrolysis of the formed trifluoroacetoxy-functionalized oligosilanes. Using this synthetic methodology a series of hydroxyl-functionalized oligosilanes with different chain lengths and degrees of branching has been produced as illustrated in Scheme 1.

The obtained polyhydroxyoligosilanes are relatively thermally stable, colorless solids that can be purified by crystallization or precipitation from pentane or acetone, except for the liquid silanol **3a**, which was purified by distillation under vacuum. The NMR spectra of the compounds are rather straightforward and are in full agreement with the structures proposed.

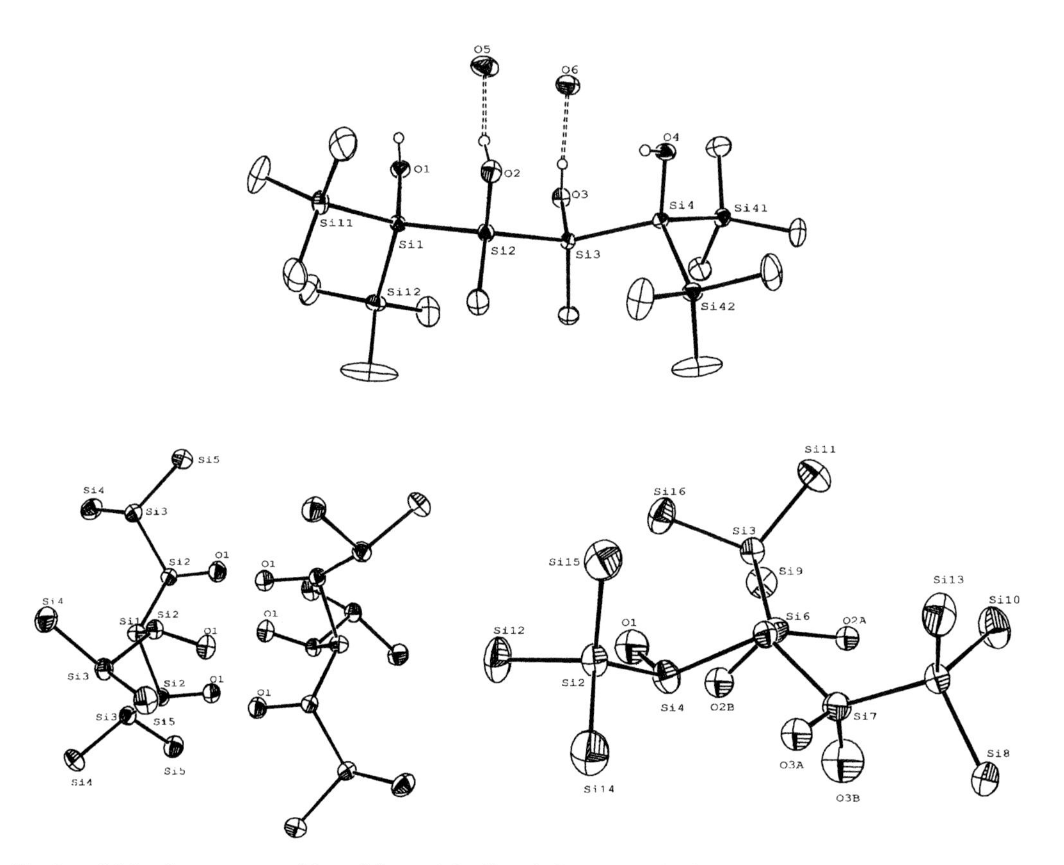

Fig. 1. Molecular structures of **8a** and (lower left, all methyl groups omitted), **8b** (lower right, all H atoms omitted) and of **7** (top, all H atoms omitted, except H atoms attached to the oxygen).

X-ray Structures

For the tetraol **7** and the triols **8a,b** the molecular structures were determined by X-ray crystallography. Because of their relatively high acidities and basicities, silanols tend to undergo hydrogen bond interaction in the solid state, which may lead to the formation of novel structural motifs. For example, the tetraol **7** (Fig. 1, top) forms an adduct with two molecules of water, connected to each other by intermolecular hydrogen bonds. Interestingly, the hexasilane chain Si11–Si1–Si2–Si3–Si4–Si41 in the molecule adopts approximately an all-*anti* conformation, which is believed to be optimal for σ-conjugation.

A structural comparison of the related dendrimeric triols **8a,b** (Fig. 2) clearly shows that the capability to form intra- and intermolecular interactions of OH groups strongly depends on the steric requirements in the molecule. Whereas **8a** forms a dimer in the solid state in which the silanol functionalities interact with each other via hydrogen bonds, the OH groups in **8b** are fully enclosed by the bulky $(Me_3Si)_3Si$ substituents. Thus, neither intra- nor intermolecular hydrogen bonds were found in the solid state.

UV-Vis Spectra

The UV-Vis spectra of a series of hexasilanes and heptasilanes have been measured at room temperature in heptane. As can be seen in Fig. 2, the permethylated hexasilanes **9** [5] and **10** exhibit intense and relatively sharp absorption maxima which are strongly shifted to shorter wavelengths

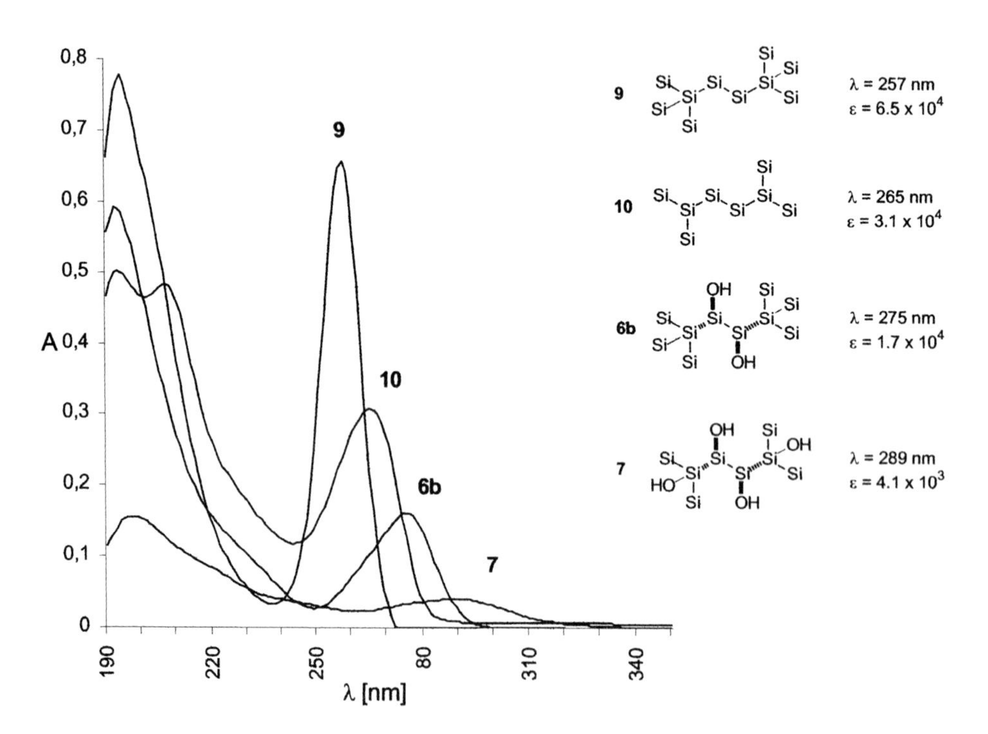

Fig. 2. UV-Vis spectra of the hexasilane series measured in heptane (10^{-5} M). For the structural formulae all methyl groups attached to silicon are omitted for clarity.

than those of the hydroxy-substituted hexasilanes **6b** and **7**. However, the intensity of the absorption observed for **6b** and **7** is significantly reduced, an effect which was also found for a series of substituted trisilanes [4].

The same effect was observed for the heptasilane series (Fig. 3). The absorption maxima for the hydroxy-substituted oligosilanes **8a,b** are significantly red-shifted in comparison to those of the permethylated derivatives [6], coupled with some reduction of intensity. It seems very likely that the interaction of the lone pairs of the hydroxy groups attached to the silicon leads to an electronic perturbation of the conjugation along the oligosilane backbone. That may increase the energy of the HOMO of the molecule, resulting in a lower σ-σ^* transition for oligosilane chains. This leads to a red shift of the absorption maximum compared to the permethylated compounds.

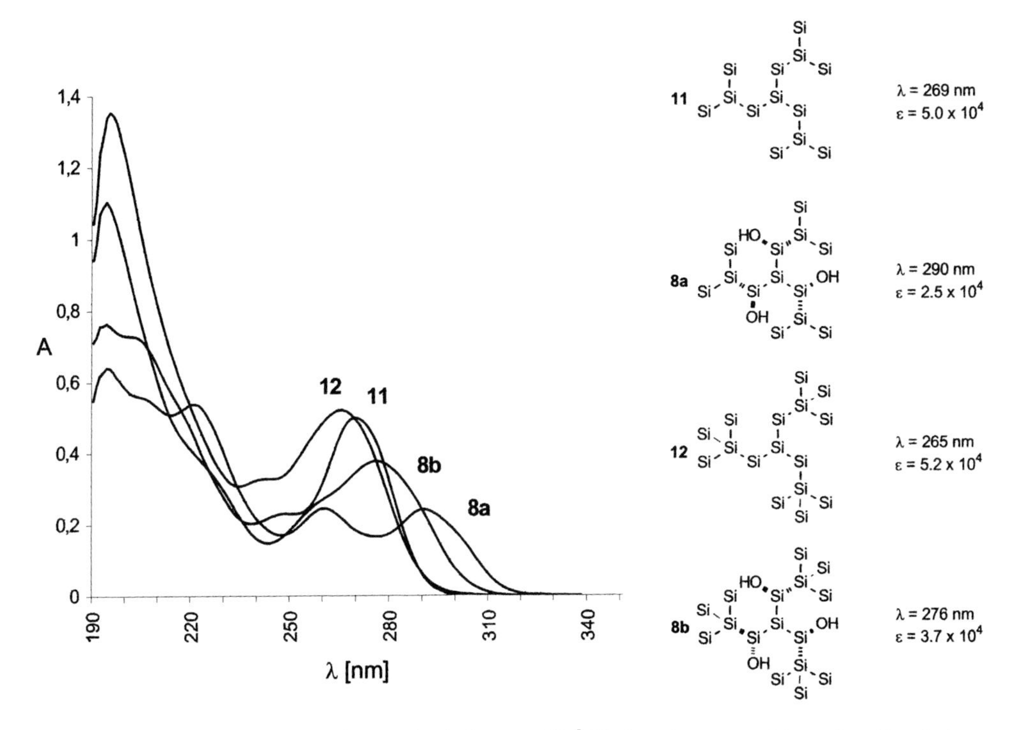

Fig. 3. UV-Vis spectra of the hexasilane series in heptane (10^{-5} M). For the structural formulae all methyl groups attached to silicon are omitted for clarity.

Coordination Chemistry

With the intent to investigate the ability of polyhydroxysilanes to act as supporting anionic ligands for metals, we focused our studies on the synthesis of novel Ti and Zr siloxide complexes from triol **8a** (Scheme 2). When $Ti(OEt)_4$ was reacted with one equivalent of **8a** in heptane for 3 h, a yellow microcrystalline material could be isolated in a yield of 35%. An X-ray structure analysis revealed the formation of the bimetallic Ti-siloxide **14**, in which both Ti atoms are connected to each other by an oxygen bridge.

In contrast, the reaction of **8a** with $Zr(NEt_2)_4$ in heptane led to the formation of the spirocyclic Zr-siloxide **13** as a colorless solid in a yield of 50%. Although single crystals suitable for X-ray analysis were not obtained, the proposed structure is in full agreement with the IR and NMR data and the elemental analysis.

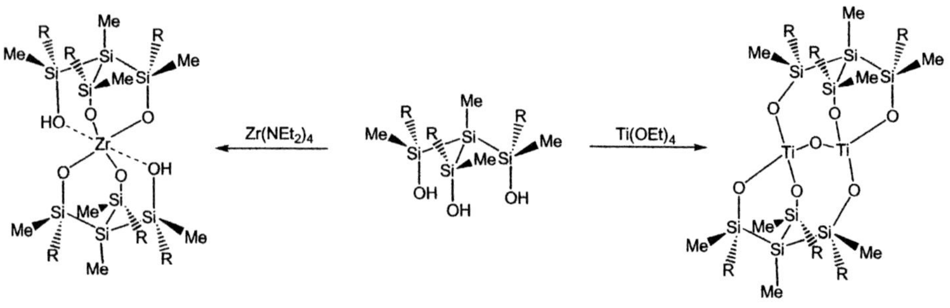

Scheme 2. Synthesis of Zr- and Ti-siloxides **13** and **14**.

Conclusion

A variety of polyhydroxyoligosilanes with different chain lengths, substitution patterns, and degrees of branching were prepared. The results of an X-ray structure analysis of **7** and **8a,b** revealed a strong influence of intra- and intermolecular hydrogen bond interactions on the structure of silanols depending on the number of hydroxy groups and the steric environment in the molecule. The hydroxy functionalities in the molecules significantly perturb the conjugation of σ-electrons along the oligosilane backbone. This results in a remarkable red shift of the absorption maximum in the UV-Vis and a reduction in intensity compared to the permethylated oligosilanes. Finally, by simple reaction of the triol **8a** with $Ti(OEt)_4$ and $Zr(NEt_2)_4$, tridentate metal siloxides of Zr and Ti could be prepared very easily.

Acknowledgment: I gratefully acknowledge the support of our work by the Fonds der Chemischen Industrie and I thank Prof. H. Oehme and Prof. M. H. Chisholm for their generous support. I thank Prof. J. Kopf, Prof. H. Reinke, and A. Spannenberg for carrying out the X-ray structure analysis.

References

[1] a) J. Michl, R. D. Miller, *Chem. Rev.* **1989**, *89*, 1359; b) R. West in *The Chemistry of Organic Silicon Compounds* (Eds.: S. Patai, Z. Rappoport), Wiley, Chichester **1989**, p. 1207.

[2] a) C. Marschner, *Eur. J. Inorg. Chem.* **1998**, 221; b) C. Kayser, R. Fischer, J. Baumgartner, C.

Marschner, *Organometallics* **2002**, *21*, 1023.

[3] C. Krempner, H. Reinke, *J. Organomet. Chem.* **2003**, *685*, 134.

[4] C. G. Pitt, *J. Am. Chem. Soc.* **1969**, *91*, 6613.

[5] S. M. Whittaker, M.-C. Brun, F. Cervantes-Lee, K. H. Pannell, *J. Organomet. Chem.* **1995**, *499*, 247.

[6] a) J. B. Lambert, J. L. Pflug, C. L. Stern, *Angew. Chem., Int. Ed. Engl.* **1995**, *34*, 98; b) A. Sekiguchi, M. Nanjo, C. Kabuto, H. Sakurai, *J. Am. Chem. Soc.* **1995**, *117*, 4195; c) J. B. Lambert, J. L. Pflug, J. M. Denari, *Organometallics* **1996**, *15*, 615.

Synthesis, Structure, and Reactivity of Novel Bidentate Metal Disiloxides

C. Krempner*

The Ohio State University, Department of Chemistry
100 W. 18th Ave, 43210, Columbus, Ohio, USA
Tel.: +1 614 6883625 — Fax: +1 614 2920368
E-mail: ckrempne@chemistry.ohio-state.edu

H. Reinke, K. Weichert

Fachbereich Chemie, Abteilung Anorganische Chemie, Universität Rostock
A.-Einstein-Str. 3a, D-18059 Rostock, Germany

Keywords: silicon, silanol, siloxide, zirconium, titanium

Summary: The bidentate metal disiloxides of general formula $LMCl_2$ [M = Ti, Zr; L = (**1**) – 2H] (**3**) were readily prepared from the reaction of $(Me_3Si)_2HOSi–SiMe_2–SiOH(SiMe_3)_2$ [LH_2] (**1**) with $M(NEt_2)_4$ and subsequent addition of MCl_4 (M = Ti, Zr), respectively. The solid-state structures of **3** were determined by X-ray crystallography.

Introduction

The chemistry of transition metal siloxide complexes has continued to attract considerable attention in the field of material science [1] and catalysis [2], particularly over the last 20 years. The synthesis, reactivity, and bonding of such complexes in a wide variety of supporting ligand environments continues to be explored. In this regard, different types of silanols, e.g., silanediols [$R_2Si(OH)_2$], disilanols [$R_2Si(OH)–O–(OH)SiR_2$], and silanetriols [$RSi(OH)_3$] have been investigated [3]. Recently, we have developed disilane-1,2-diols [$R_2(OH)Si–Si(OH)R_2$] as new supporting ligands for d^0 metal centres [4]. Herein we report on the synthesis of bidentate titanium and zirconium disiloxides using the novel ligand $(Me_3Si)_2HOSi–SiMe_2–SiOH(SiMe_3)_2$.

Results and Discussions

A suitable entry into the chemistry of Ti and Zr complexes is the use of highly reactive metal amides. In fact, by treatment of $M(NEt_2)_4$ (M = Ti, Zr) with two equiv of the $(Me_3Si)_2HOSi-SiMe_2-SiOH(SiMe_3)_2$ (**1**), spirocyclic metal disiloxides of type **2** (M = Ti, Zr) can be obtained as moisture sensitive, colorless solids in nearly quantitative yields. Without isolation, these metal disiloxides can be transformed into the corresponding Ti and Zr dichlorides **3** in good yields by subsequent addition of $TiCl_4$ or $ZrCl_4$ x 2 THF at low temperatures (Scheme 1). The NMR spectra of the new complexes **2** and **3** are rather straightforward and are in full agreement with the structures proposed.

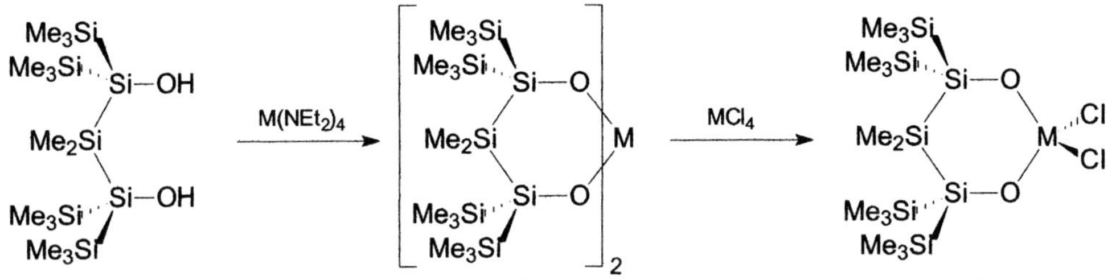

Scheme 1. Synthesis of the metal disiloxides **2** and **3** (M = Ti, Zr).

In addition, the molecular structure of the zirconium dichloride **3**, which forms a THF adduct, was determined by X-ray crystallography (Fig. 1). As expected, the central zirconium atom is

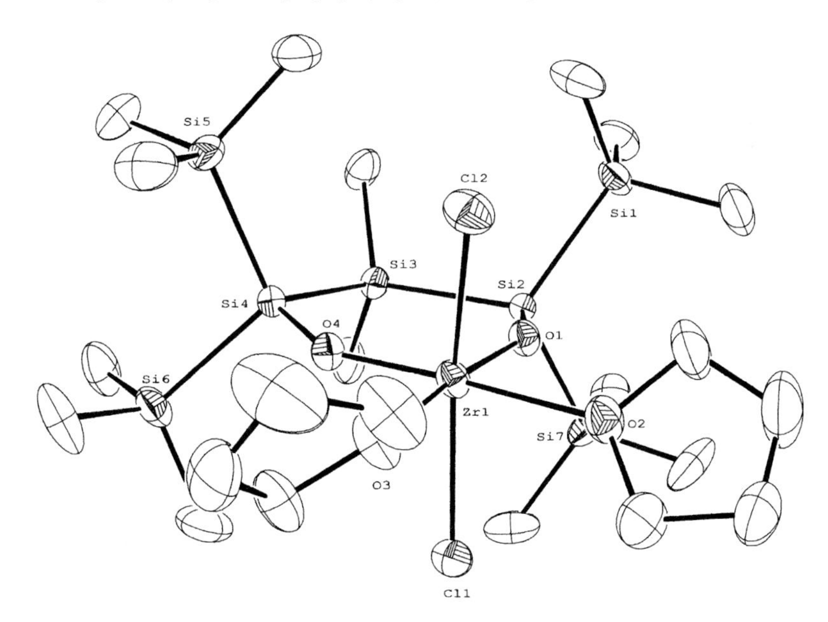

Fig. 1. Molecular structure of **3** (M = Zr) in the crystal. The thermal ellipsoids correspond to 50% probability. Hydrogen atoms are omitted for clarity. Selected bond lengths [Å] and angles [°]: Cl1–Zr1 2.4742(14), Cl5–Zr1 2.4805(13), O1–Si4 1.665(3), O1–Zr1 1.924(3), O2–Si2 1.668(3), O2–Zr1 1.917(3), O3–Zr1 2.315(3), O4–Zr1 2.307(3); Si4-O1-Zr1 147.55(18), Si2-O2-Zr1 148.57(18), O2-Si2-Si3 109.22(11), Si4-Si3-Si2 106.84(6), O1-Si4-Si3 109.35(12), O2-Zr1-O1 96.18(12), O2-Zr1-O4 173.06(13), O1-Zr1-O3 174.38(12), O4-Zr1-O3 83.73(13), Cl1-Zr1-Cl5 162.66(5).

bonded in a chelate fashion [O1-Zr1-O2 96.18(12)°] to both siloxide groups of the ligand, giving a nearly planar six-membered ring. The geometry around the zirconium atom is described best as distorted octahedral with the Cl1 and Cl5 atoms occupying axial positions, respectively. The Zr–O distances [O1–Zr1 1.924(3), O2–Zr1 1.917(3) Å] within the ring are comparable to those found in related compounds containing Si–O–Zr units and are significantly shorter than the Zr–O_{THF} distances [O3–Zr1 2.315(3), O4–Zr1 2.307(3) Å].

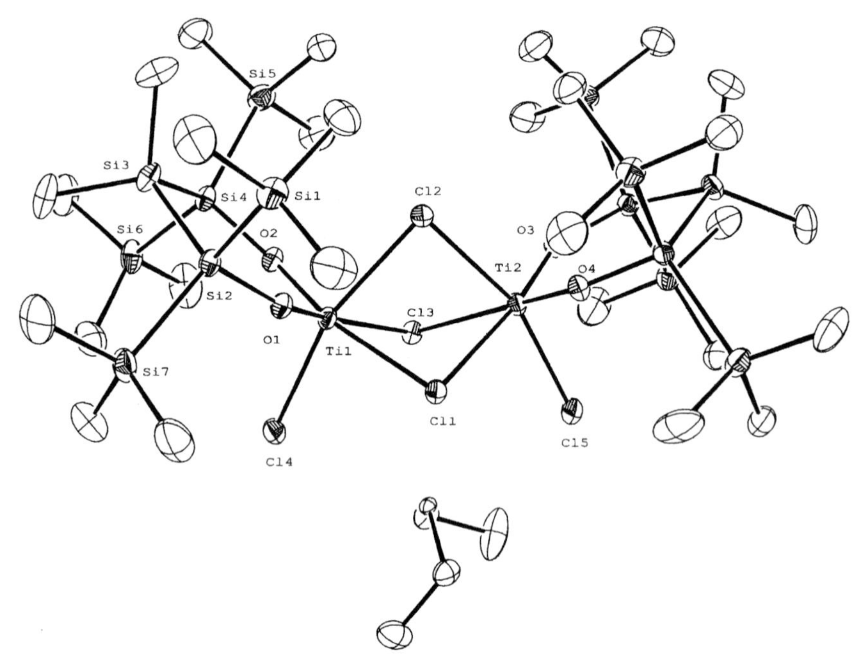

Fig. 2. Molecular structures of **3** (M = Ti) in the crystal. The thermal ellipsoids correspond to 50% probability. Hydrogen atoms are omitted for clarity. Selected bond lengths [Å] and angles [°]: Ti1–O2 1.7606(16), Ti1–O1 1.7687(16), Ti1–Cl4 2.3323(8), Ti1–Cl2 2.5206(7), Ti1–Cl1 2.5439(7), Ti1–Cl3 2.5955(7), Si2–O1 1.6834(16), Si4–O2 1.6827(17); O2-Ti1-O1 98.11(8), Cl4-Ti1-Cl3 88.30(2), Cl2-Ti1-Cl3 77.06(2), Cl1-Ti1-Cl3 78.27(2), Ti1-Cl1-Ti2 86.22(2), Ti2-Cl2-Ti1 89.15(2), Ti2-Cl3-Ti1 85.47(2), Si2-O1-Ti1 149.26(11), Si4-O2-Ti1 151.26(11).

Single crystals of the titanium derivative **3** suitable for an X-ray structure determination were grown from pentane at –40 °C. The results reveal the molecule to be dimeric in the solid state as a consequence of chlorine bridging bonds between both titanium atoms (Fig. 2), with Ti–$Cl_{bridging}$ bonds [Ti1–Cl2 2.5206(7), Ti1–Cl1 2.5439(7), Ti1–Cl3 2.5955(7) Å] significantly longer than the non-bridging Ti–Cl bonds with 2.3323(8) and 2.3403(8) Å. The geometry around both titanium atoms can be regarded as distorted octahedral with the Cl4, Cl5, and bridging Cl2 atom occupying axial positions. Unexpectedly, five chlorine atoms, two non-bridging and three bridging, were found in the molecule, suggesting an anionic structure of the complex. However, a cation could not be found in the difference map. Therefore, we propose compound **3** to be a dimeric HCl adduct, in which the chlorine is coordinated to both titanium atoms and the proton is weakly bonded by the oxygen atoms. The unusual formation of the compound can be explained by the fact that the $TiCl_4$ solution used for the synthesis of **3** (see Scheme 1) is contaminated by small amounts of HCl. Careful NMR-spectroscopic investigations have shown that the titanium dichloride complex **3** is slightly contaminated by the HCl adduct of **3**, for which the solid state structure is available.

Acknowledgment: We gratefully acknowledge the support of our work by the Fonds der Chemischen Industrie and we thank Prof. H. Oehme for his generous support.

References

[1] G. Perego, G. Ballussi, C. Corno, M. Tamarasso, F. Buonomo, A. Esposito, New developments in zeolite science and technology, in *Studies in Surface Science and Catalysis* (Eds.: Y. Murakami, A. Iijima, J. W. Word), Elsevier, Amsterdam, **1986**; b) A. Bhaumik, R. J. Kumar, J. *Chem. Soc., Chem. Commun.* **1995**, 869; c) B. Notari, *Stud. Surf. Sci. Catal.* **1988**, *37*, 413.

[2] H. C. L. Abbenhuis, S. Krijnen, R. A. van Santen, *J. Chem. Soc., Chem. Commun.* **1997**, 331; b) I. E. Buys, T. W. Hambley, D. J. Houlten, T. Maschmeyer, F. A. Masters, A. K. Smith, *J. Mol. Catal.* **1994**, *86*, 309; c) M. Crocker, R. H. M. Herold, A. G. Orpen, *Chem. Commun.* **1997**, 2411; d) T. Maschmeyer, M. C. Klunduk, C. M. Martin, D. S. Shepard, J. M. Thomas, B. F. G. Johnson, *Chem. Commun.* **1997**, 1847.

[3] R. Murugavel, A. Voigt, M. G. Walawalkar, H. W. Roesky, *Chem. Rev.* **1996**, *96*, 2205; b) L. King, A. C. Sullivan, *Coord. Chem. Rev.* **1999**, 189, 19; c) V. Lorenz, A. Fischer, S. Gießmann, J. W. Gilje, Y. Gun'ko, K. Jacob, F. T. Edelmann, *Coord. Chem. Rev.* **2000**, *206 – 207*, 321; d) B. Marciniec, H. Maciejewski, *Coord. Chem. Rev.* **2001**, *223*, 301; e) R. Duchateau, *Chem. Rev.* **2002**, *102*, 3525; f) P. D. Lickiss, *Adv. Inorg. Chem.* **1995**, *42*, 147.

[4] D. Hoffmann, H. Reinke, C. Krempner, *J. Organomet. Chem.* **2002**, *662*, 1 – 8; b) D. Hoffmann, H. Reinke, C. Krempner in *Organosilicon Chemistry V — From Molecules to Materials* (Eds.: N. Auner, J. Weis), VCH, Weinheim, **2003**, p 420 – 424.

Electronic Excitation in Decamethyl-*n*-tetrasilane

M. C. Piqueras, R. Crespo*

Departament de Química Física, Universitat de València,
Doctor Moliner 50, E-46100-Burjassot, Spain
Tel.: +34 963543344 — Fax: +34 963544564
E-mail: Raul.Crespo@uv.es

Keywords: oligosilanes, vertical excitation energies, TD-DFT, MS-CASPT2, UV spectra

Summary: Time-Dependent Density Functional Theory (TD-DFT) calculations of the low-energy valence excited states of transoid, ortho, and gauche conformers of decamethyl-*n*-tetrasilane using the PBE0 functional and the cc-pVTZ/6-311G*/6-311G basis set are reported. The computed values are compared with our previous calculations performed using the high level *ab initio* Multistate Complete Active Space Second-Order Perturbation Theory (MS-CASPT2) method. PBE0 functional gives vertical excitation energies that are in qualitative good agreement with those obtained at the MS-CASPT2 level. TD-DFT energy values are larger than those computed by the MS-CASPT2 method. Oscillator strengths calculated at the TD-DFT level present values smaller than their corresponding MS-CASPT2 data. PBE0 provides a good description of the low-energy region of the individual UV absorption spectrum of the transoid, ortho, and gauche conformers of *n*-Si_4Me_{10} as well as the UV absorption spectra of *n*-Si_4Me_{10} conformer equilibrium mixtures at room temperature and 77K.

Introduction

Conformational effects on the electronic spectra of oligosilanes are especially interesting for the understanding of the fundamental photophysics of their parent polymers. Recently, we reported calculations on the low-energy valence excited states of transoid, ortho, and gauche conformers of *n*-Si_4Me_{10} [1] performed using the high level *ab initio* Multistate Complete Active Space Second-Order Perturbation Theory (MS-CASPT2) method [2]. We found that although the results confirm the main features of the previously developed description of the dependence of the UV absorption spectrum of the tetrasilane chromophore on the silicon backbone, at small dihedral angles the situation seems to be more complicated because of additional avoided crossings. Hence, computations on a larger number of dihedral angles are required for a complete description of the angular dependence of the UV absorption spectrum of the tetrasilane chromophore. The calculation of the electronic excited states is a demanding task for theoretical methods, especially for medium and large size molecules of chemical interest as n-Si_4Me_{10}. Thus, an inexpensive yet accurate method for calculating vertical excitation energies of silicon-containing compounds would be very useful.

Density Functional Theory (DFT) has been remarkably successful at computing a variety of ground state properties at the same level of accuracy as post-Hartree-Fock (HF) methods such as second order Moller-Plesset Perturbation Theory (MP2) [3]. Since DFT methods rectify many problems of the HF approach at a comparable computational cost, Time-Dependent generalization of the DFT (TDDFT) [4] offers an excellent possibility for the calculation of vertical electronic excitations in a rigorous way which is less expensive than other high-level *ab initio* methods such as MS-CASPT2 [2]. A number of papers [5] have shown that TD-DFT provides results for low excitation energies usually superior to those obtained by Time-Dependent HF methods (TDHF) [6] or the Configuration Interaction with Singles excitations (CIS) approach [7]. Hybrid HF/DFT methods have also been proposed for the calculation of excitation energies [8]. Tests have shown that B3LYP functional represents a further improvement over conventional DFT methods [9]. However, it gives vertical excitation energies to Rydberg states of poor quality compared to valence states [10]. Also, a number of functionals constructed to satisfy physical constraints have been developed lately. Of particular interest is the PBE0 functional [11], which provides vertical excitation energies that are in fairly good agreement with experimental data and of similar quality to those obtained by more time-consuming post-Hartree-Fock methods [12].

The present study analyzes the performance of the PBE0 functional on the calculation of the vertical excitations of decamethyl-*n*-tetrasilane. We focus on the transoid, ortho, and gauche conformers of n-Si_4Me_{10} and compare the values calculated at the TD-DFT level with our previous results obtained using the high-level *ab initio* MS-CASPT2 method [1]. We will interpret and assign the two observed absorption bands and discuss the nature of the low-lying valence excited states and the previously proposed assignments within the TD-DFT/PBE0 method. The general validity of the computational level necessary to obtain accurate results is also analyzed. The selection of MS-CASPT2 values as reference data is made on the basis of the excellent agreement found in our previous MS-CASPT2 studies on the electronic spectra of trisilane [13] and *n*-tetrasilane [14].

Computational Details

TD-DFT calculations have been performed using the Perdew-Burke-Ernzerhof generalized gradient functional [15] with a predefined amount of exact exchange, PBE0 [11], as implemented in the Gaussian 98 program [16]. The calculations were carried out at the equilibrium geometries of the ground state of the transoid, ortho, and gauche conformers of n-Si_4Me_{10}, obtained previously [1] at the MP2(fc) level using the cc-pVTZ basis set for the silicon atoms and the 6-31G* and 6-31G basis set for the carbon and the hydrogen atoms, respectively. The calculations were done under the C_2 symmetry point group constraints with z as the twofold symmetry axis.

A number of different basis sets have been used to investigate the performance of the functional:

3-21G*, 6-31G*, 6-31G**, 6-311G*/6-31G*/6-31G, 6-311G*/6-311G*/6-31G, 6-311G(2d)/6-311G*/6-31G, 6-311G*, 6-311G(2d)/6-311G*/6-311G, cc-pVTZ/6-311G*/6-311G, and 6-311+G(2d)/6-311G*/6-311G. These have allowed us to investigate the effect of using a double zeta or a triple zeta basis set on the values computed for the energies and oscillator strengths, the changes produced in the calculated values by the addition of polarization functions on the Si, C, and H atoms, and also the influence of the inclusion of diffuse functions on the calculated properties. It will also help us to select the basis set that gives accurate results using minimum computational resources for its future use in the study of more extended systems.

Results and Discussion

The vertical excitation energies and oscillator strengths of the singlet valence states of the transoid conformer of n-Si_4Me_{10} calculated at the TD-DFT level using the PBE0 functional are listed in Table 1. We only show the values calculated using the cc-pVTZ/6-311G*/6-311G basis set, as this represents a good compromise between accuracy of results and cost of computational resources [17]. It is also to be remarked that double zeta basis sets such as 3-21G*, 6-31G*, and 6-31G** do not give a correct characterization of the states. This mixing is very pronounced for the values calculate for the 3 ^1A state of the transoid conformer. To avoid this strong mixing, triple-zeta basis sets need to be used. The addition of a diffuse function to the basis set produces a strong decrease of the energy of the excited states due to a strong Rydberg-valence mixing.

Table 1. TD-DFT/PBE0 vertical excitation energies (E, eV) and oscillator strengths (f) for the transoid conformer of n-Si_4Me_{10}, calculated using the cc-pVTZ/6-311G*/6-311G basis set.[a]

MS-CASPT2			PBE0		
State	E	f	State	E	f
1 ^1B ($\sigma_1\sigma_1$*)	5.28	0.585	2 ^1A ($\sigma_1\pi_1$*)	5.49	0.003
2 ^1A ($\sigma_1\pi_1$*)	5.38	0.001	1 ^1B ($\sigma_1\sigma_1$*)	5.58	0.347
3 ^1A ($\sigma_1\sigma_2$*)	5.87	0.0001	3 ^1A ($\sigma_1\sigma_2$*)	5.98	0.0000
2 ^1B ($\sigma_1\pi_2$*)	5.99	0.027	2 ^1B ($\sigma_1\pi_2$*)	6.37	0.030
4 ^1A	6.46	0.036	3 ^1B ($\sigma_3\sigma_1$*)	6.58	0.019
3 ^1B	6.46	0.007	4 ^1A ($\sigma_2\sigma_1$*)	6.62	0.019
4 ^1B	6.51	0.221	4 ^1B	6.67	0.003
5 ^1A	6.54	0.247	5 ^1A ($\sigma_3\pi_1$*)	6.76	0.068

[a] Values for MS-CASPT2 method taken from [1].

The values for the first ionization potential of the three conformers of n-Si_4Me_{10} lie near 7.5 eV, and we therefore list only states with excitation energies lower than 7.0 eV. All of these are valence states. The electronic transitions are all dipole-allowed. At the MS-CASPT2 level they are located in three well-defined energy intervals near to 5.3, 6.0, and 6.5 eV. At the TD-DFT level there is a small shift of about 0.2 eV, and the states lie in three intervals located around 5.5, 6.2, and 6.7 eV. TD-DFT values of the energy of the excited states are always located about 0.1 – 0.4 eV higher in energy than those calculated at the MS-CASPT2 level. The values calculated for the oscillator strengths are always smaller and less reliable than the values calculated for the energies of the electronic transitions. TD-DFT method fails to account for the oscillator strength of the 4 ^{1}B and 5 ^{1}A excited states. The analysis of the wave functions shows that valence excited states are well characterized by PBE0 functional. There is a clear correlation between both methods in the nature of the orbitals that determine the electronic transitions. Seven TD-DFT excitations are classified as electronic promotions from a σ occupied orbital to a σ* or π* virtual orbital, depending on the predominant composition of the molecular orbital. The 4 ^{1}B state is mixed so strongly that it contains comparable contributions of σ or π types, and hence it has not been labeled. Although the symbols σ or π are not strictly applicable to a molecule without a plane of symmetry, they can be applied qualitatively, and we have labeled the molecular orbitals of all conformers accordingly.

Table 2. TD-DFT/PBE0 vertical excitation energies (*E*, eV) and oscillator strengths (*f*) for the ortho conformer of *n*-Si_4Me_{10}, calculated using the cc-pVTZ/6-311G*/6-311G basis set.[a]

MS-CASPT2			PBE0		
State	*E*	*f*	State	*E*	*f*
2 ^{1}A ($\sigma_1\sigma_2$*)	5.34	0.053	2 ^{1}A ($\sigma_1\sigma_2$*)	5.62	0.004
1 ^{1}B ($\sigma_1\sigma_1$*)	5.38	0.326	1 ^{1}B ($\sigma_1\sigma_1$*)	5.62	0.116
3 ^{1}A ($\sigma_1\pi_1$*)	5.93	0.004	3 ^{1}A ($\sigma_1\pi_1$*)	6.11	0.004
2 ^{1}B ($\sigma_2\sigma_2$*)	5.94	0.215	4 ^{1}A ($\sigma_2\sigma_1$*)	6.31	0.016
4 ^{1}A	6.11	0.043	2 ^{1}B ($\sigma_2\sigma_2$*)	6.34	0.122
3 ^{1}B	6.29	0.015	3 ^{1}B ($\sigma_1\pi_2$*)	6.60	0.073
4 ^{1}B	6.58	0.119	5 ^{1}A ($\sigma_3\sigma_2$*)	6.76	0.021
5 ^{1}A	6.68	0.015	4 ^{1}B	6.80	0.030

[a] Values for MS-CASPT2 method taken from [1].

Table 2 displays the vertical excitation energies and oscillator strengths of the singlet valence states of the ortho conformer of *n*-Si_4Me_{10} calculated at the PBE0 level employing the cc-pVTZ/6-311G*/6-311G basis set. Similar features to those observed for the transoid conformer are observed here. This conformer also shows a strong absorption peak in each of the three spectral regions calculated for the transoid conformer. The first two are well developed at 5.4 and 6.0 eV, respectively, and there is an indication of the third one near 6.5 eV at the MS-CASPT2 level and at around 5.6, 6.3, and 6.7 at the TD-DFT level. All the electronic transitions are dipole-allowed and present valence character. As observed previously for the transoid conformer, the energies of the electronic transitions are computed to lie at higher values than those calculated at the MS-CASPT2 level. The deviations between both methods lie in the 0.1 – 0.4 eV interval. The TD-DFT method fails to account for the oscillator strength of the 4 ^{1}B excited state. Except for the 4 ^{1}B state, which shows a strong mixing, all the excitations have been classified as electronic promotions from a σ orbital to a σ^* or π^* orbital, depending on the predominant composition of the molecular orbital. The analysis of the nature of the excitations shows that there is a clear correlation between TD-DFT and MS-CASPT2 methods in the assignment of the character of the orbitals that determine the electronic transition.

The analysis of the values calculated for the gauche conformer of *n*-Si_4Me_{10} shows three regions for the electronic transitions which are located around 5.4, 5.9, and 6.5 eV, at the MS-CASPT2 level, as can be observed in Table 3. However the TD-DFT level shows only two regions at around 6.2 and 6.8 eV, as it does not describe correctly the oscillator strength of the quite strong 2 ^{1}A absorption. This electronic transition is calculated at 5.42 eV at the MS-CASPT2 level with a value of the oscillator strength of 0.115, and is observed experimentally at 5.3 eV. As observed for the other conformers, all the electronic transitions are dipole-allowed and show valence character. There are only six TD-DFT excitations where the orbitals that determine the electronic transition have been characterized, as 3 ^{1}B and 4 ^{1}B states show a quite strong mixing of σ and π character. The deviations between the energies computed by both methods are about 0.1 – 0.4 eV.

Table 3. TD-DFT/PBE0 vertical excitation energies (*E*, eV) and oscillator strengths (*f*) for the gauche conformer of *n*-Si_4Me_{10}, calculated using the cc-pVTZ/6-311G*/6-311G basis set.[a]

MS-CASPT2			PBE0		
State	E	f	State	E	f
2 ^{1}A ($\sigma_1\sigma_2$*)	5.42	0.115	1 ^{1}B ($\sigma_1\sigma_1$*)	5.67	0.033
1 ^{1}B ($\sigma_1\sigma_1$*)	5.53	0.069	2 ^{1}A ($\sigma_1\sigma_2$*)	5.78	0.013
2 ^{1}B ($\sigma_2\sigma_2$*)	5.91	0.316	3 ^{1}A ($\sigma_2\sigma_1$*)	6.20	0.013
3 ^{1}A ($\sigma_2\sigma_1$*)	5.96	0.015	2 ^{1}B ($\sigma_2\sigma_2$*)	6.21	0.124
4 ^{1}A	6.10	0.001	4 ^{1}A ($\sigma_1\pi_1$*)	6.27	0.002
3 ^{1}B	6.34	0.001	3 ^{1}B	6.60	0.050
4 ^{1}B	6.55	0.120	5 ^{1}A ($\sigma_1\pi_2$*)	6.79	0.001
5 ^{1}A	6.64	0.039	4 ^{1}B	6.80	0.125

[a] Values for MS-CASPT2 method taken from [1].

The TD-DFT simulation of the low-energy region of the UV absorption spectra of *n*-Si_4Me_{10} conformer equilibrium mixtures at room temperature and 77K are shown in Fig. 1. The simulation has been performed using the computed conformer populations [1], assuming that the spectral line shapes are Gaussian functions with a full width at half-maximum of 0.3 eV, and neglecting solvent effects. The correlation between experimental [18] and PBE0-computed spectra is quite good with respect to the position of the first absorption band, which only shows a shift toward higher energies of about 0.3 eV, at both temperatures. As expected, PBE0 relative intensities of the first absorption band are smaller than the observed ones. The second absorption band is not well reproduced by PBE0 calculations, as the method fails to account for the oscillator strength of the 4 ^{1}B and 5 ^{1}A states of the transoid conformer and of the 4 ^{1}B state of the ortho conformer, which lie in this region.

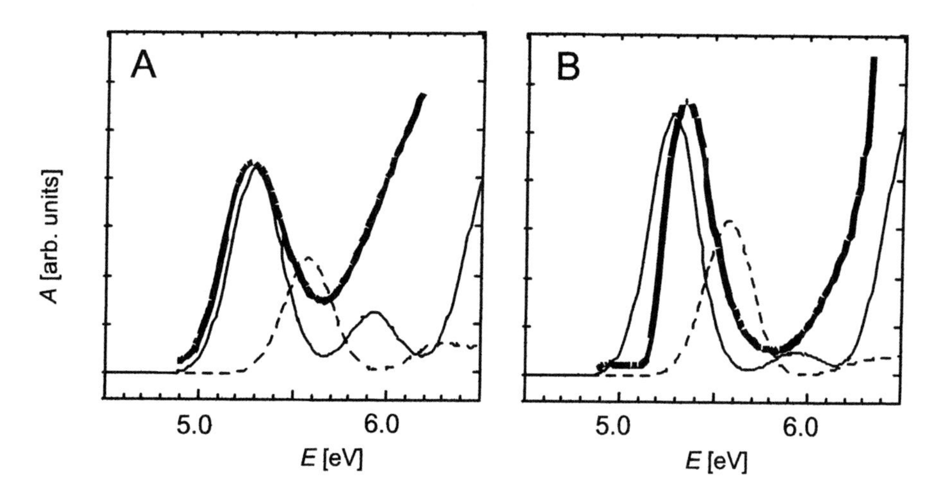

Fig. 1. UV absorption spectra of the equilibrium conformer mixture of *n*-Si_4Me_{10} at room temperature (A) and 77K (B): TD-DFT (dashed line) and MS-CASPT2 (solid line) simulations of the low-energy region assuming a 0.3 eV full width at half-maximum, and observed spectra (thick solid line) [18].

Conclusions

PBE0 functional provides a quite satisfactory description of the first valence excited states of the transoid, ortho, and gauche conformers of *n*-Si_4Me_{10}, and also of the first band of the UV absorption spectra of their equilibrium conformer mixtures at room temperature and at 77 K. The analysis of the dependence of the nature of the electronic excitation on the dihedral angle shows that there is an interchange of the $\sigma\sigma^*$ and $\sigma\pi^*$ character of the states upon going between the three conformers. These features suggest that at small dihedral angles additional avoided crossings are produced. A complete description of the angular dependence will require computations on a larger number of dihedral angles. Also, other functionals should be checked to evaluate and discriminate between the different approximations to the exchange correlation energy.

Acknowledgments: This work was supported by Generalitat Valenciana, MEC and European Commission (IST-2001-32243).

References

[1] M. C. Piqueras, R. Crespo, J. Michl, *J. Phys. Chem.* **2003**, *107*, 4661.

[2] J. Finley, P. A. Malmqvist, B. O. Roos, L. Serrano-Andres, *Chem. Phys. Lett.* **1998**, *288*, 299.

[3] B. G. Johnson, P. M. W. Gill, J.A. Pople, *J. Chem. Phys.* **1993**, *98*, 5612.

[4] M. E. Casida, in *Recent Advances in Density Functional Methods, Part I* (Ed.: D. P. Chong), World Scientific, Singapore, **1995**.

[5] M. K. Casida, C. Jamorski, K. C. Casida, D. R. Salahub, *J. Chem. Phys.* **1998**, *108*, 4439; D. J. Tozer, N. C. Handy, *J. Chem. Phys.* **1998**, *109*, 1018.

[6] T. H. Dunning, V. McKoy, *J. Chem. Phys.* **1967**, *47*, 1735.

[7] J. B. Foresmen, H. Head-Gordon, J. A. Pople, M. J. Frisch, *J. Phys. Chem.* **1992**, *96*, 135.

[8] R. Bauernschmitt, R. Alhrichs, *Chem. Phys. Lett.* **1996**, *256*, 454; R. E. Stratmann, G. E. Scuseria, M. J. Frisch, *J. Chem. Phys.* **1998**, *109*, 8218.

[9] K. B. Wiberg, R. E. Stratmann, M. J. Frisch, *Chem. Phys. Lett.* **1998**, *297*, 60.

[10] M. K. Casida, C. Jamorski, K. C. Casida, D. R. Salahub, *J. Chem. Phys.* **1998**, *108*, 4439.

[11] M. Ernzerhof, G. E. Scuseria, *J. Chem. Phys.* **1999**, *110*, 5029; C. Adamo, V. Barone, *J. Chem. Phys.* **1999**, *110*, 6158.

[12] C. Adamo, G. E. Scuseria, V. Barone, *J. Chem. Phys.* **1999**, *111*, 2889.

[13] M. C. Piqueras, R. Crespo, J. Michl, *Mol. Phys.* **2002**, *100*, 747.

[14] M. C. Piqueras, M. Merchan, R. Crespo, J. Michl, *J. Chem. Phys. A* **2002**, *106*, 9868 ; R. Crespo, M. Merchan, J. Michl, *J. Chem Phys. A* **2000**, *104*, 8593.

[15] J. P. Perdew, K. Burke, M. Ernzerhof, *Phys. Rev. Lett.* **1996**, *77*, 3865.

[16] M. J. Frisch, et al. *Gaussian 98*, Revision A.11, Gaussian Inc., Pittsburgh, PA, **1998**.

[17] M. C. Piqueras, R. Crespo, to be published.

[18] D. W. Rooklin, T. Schepers, M. K. Raymond-Johansson, J. Michl, *Photochem. Photobiol. Sci.* **2003**, *2*, 1.

Preparation and Structural Studies on Cyclohexasilane Compounds

Roland Fischer, Tina Konopa, Stephan Ully, Andreas Wallner Judith Baumgartner, Christoph Marschner

Institut für Anorganische Chemie Technische Universität Graz
Stremayrgasse 16, A-8010 Graz, Austria
Tel.: +43 316 873 8209 — Fax: +43 316 873 8701
E-mail: christoph.marschner@tugraz.at

Keywords: cyclohexasilanes, silyl anions, crystal structures

Summary: The use of oligosilyl dianions greatly facilitates the synthesis of cyclohexasilanes with a specific substituent pattern. A number of such compounds differently substituted in 1,4-positions are compared according to their structural data. While almost all cyclohexasilanes studied so far prefer a chair conformation, we find the unusual example of a twist conformation for 1,1,4,4-tetrakis(trimethylsilyl)-octamethyl-cyclohexasilane (**2**).

Introduction

Cyclosilanes [1] are among the oldest known compounds with silicon-silicon bonds. Their origins date back to the work of F. S. Kipping [2]. In our laboratories these compounds have been studied over the last decades. Especially, Hengge et al. have contributed tremendously to the advancement of cyclosilane chemistry [1b, 1d]. As shown recently, the cleavage of trimethylsilyl groups from the silicon backbone proves to be a highly selective and generally applicable method for the synthesis of oligosilanyl mono- and dianions [3]. Starting from 1,4-dianion **1** we were able to provide a selective access to differently substituted cyclohexasilanes [4].

Results and Discussion

Synthesis

The reaction of 1,1,1,4,4,4-hexakis(trimethylsilyl)tetramethyltetrasilane with two equivalents of potassium *tert*-butoxide and 18-crown-6 affords the 1,4-dianion **1** in almost quantitative yield [3]. When **1** is reacted with 1,2-dibromotetramethyldisilane, 1,1,4,4-tetrakis(trimethylsilyl)octamethyl-cyclohexasilane (**2**) is obtained (Scheme 1). Reaction of **2** with one or two equivalents of potassium *tert*-butoxide generates the respective cyclohexasilanyl mono- and dipotassium compounds (**3**, **4**).

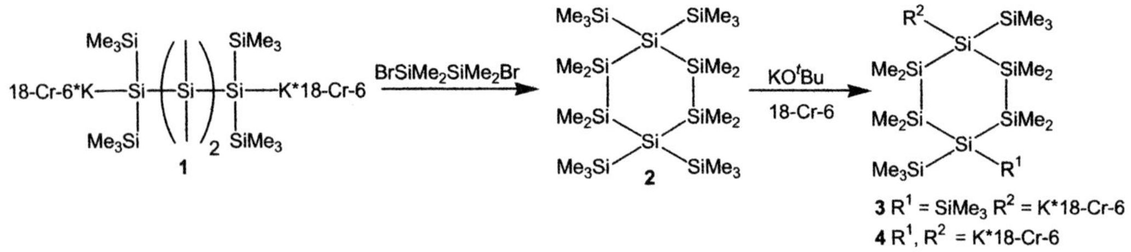

BrSiMe$_2$SiMe$_2$Br
KO^{t}Bu
18-Cr-6
3 R^1 = SiMe$_3$ R^2 = K*18-Cr-6
4 R^1, R^2 = K*18-Cr-6

Scheme 1. Synthesis of 1,1,4,4-tetrakis(trimethylsilyl)octamethylcyclohexasilane (**2**) and the 1-potassio (**3**) and 1,4-dipotassio (**4**) derivatives.

3 can easily be derivatized with sulfuric acid, dimethyl sulfate or ethyl bromide to give the corresponding hydro- (**5**), methyl- (**6**) or ethyl- (**7**) compounds (Scheme 2).

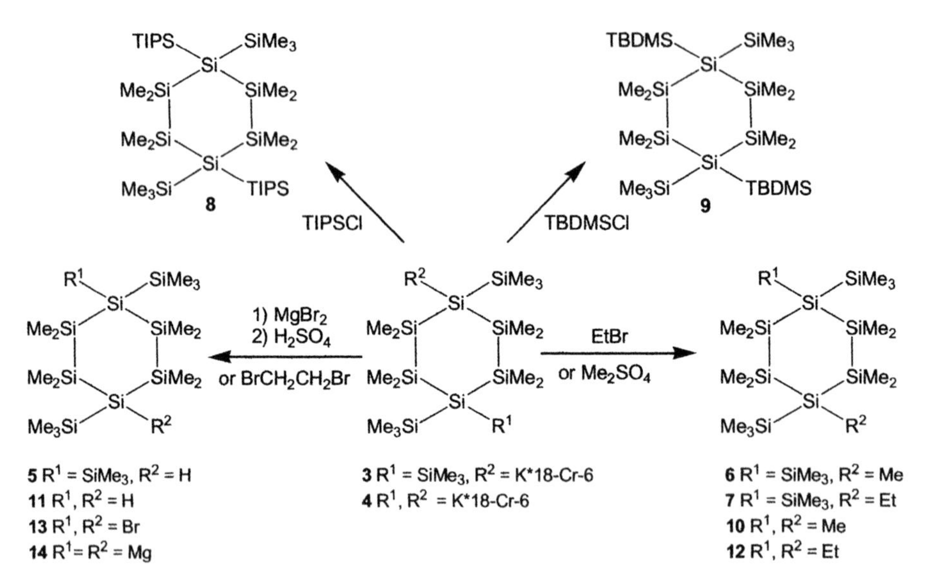

Scheme 2. Derivatization reactions of **3** and **4**.

4 was reacted with tri*iso*propylchlorosilane, *tert*-butylchlorodimethylsilane, dimethyl sulfate or ethyl bromide to give the corresponding 1,4-bis(tri*iso*propylsilyl)- (**8**), 1,4-bis(*tert*-butyldimethylsilyl)- (**9**), 1,4-dimethyl- (**10**) or 1,4-diethyl- (**12**) compounds (Scheme 2). The reaction of **4** with an excess of 1,2-dibromoethane provides the dibromo-compound **13**. The reaction of **4** with one equivalent of magnesium bromide yields a bicyclic compound where the 1- and 4-positions are bridged by a magnesium atom (**14**). Hydrolysis of **14** with sulfuric acid gives the *cis*-dihydrosilane (*cis*-**11**) whereas the trans isomer (*trans*-**11**) is obtained by hydrolysis of **4** (Scheme 2) [4].

Structural Studies

The preparation and structural characterization of several cyclohexasilanes with different substituents in the 1,4-position [4] (Tables 1 and 2) provided us with the unique opportunity to study the influence of the substituents on the conformation of the cyclohexasilane ring. Almost all known cyclohexasilanes adopt a chair conformation. The crystal structure of the dibromo compound **13** (Fig. 2 and Table 2) [5] provides a rather typical example. We decided to compare the flattening of the chair as a function of the substituents. This property can be described by two angles (α, β) between the actual seat plane and the planes spanned by the atoms above and below and the seat edges (Fig. 1 and Table 1).

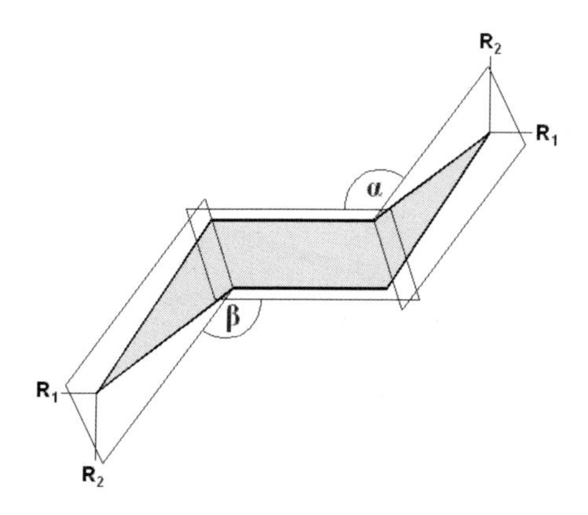

Fig. 1. The chair conformation of cyclohexasilane with selected angles (α, β).

Table 1. The angles [°] between the planes of cyclohexasilanes with different substituents (Fig. 1).

R_1	R_2	α	β	Note
$SiMe_3$	K*18cr6	125.4	123.4	K*18-cr-6 in axial positions Relatively long Si–K bond
$SiMe_3$	Br	131.6	131.6	Bromine atoms in axial positions
$SiMe_2{}^tBu$	$SiMe_3$	145.8	145.4	
$SiMe_3$	$SiMe_3$	–	–	Twist conformation
$SiMe_3$	H	132.5	128.2	
Me	Me	131.9	131.9	
cyclohexane			128.5	Reference

As expected, we find that large substituents tend to occupy the equatorial positions. This behavior is found for the *trans*-1,4-dihydro (*trans*-**11**), -bis(*tert*-butyldimethylsilyl) (**9**), and also dibromo-1,4-bis(trimethylsilyl)octamethylcyclohexasilane (**13**). Also the respective 1,4-dipotassium compound exhibits this conformation (**4**). We assume that the large potassium-crown ether substituent with weak silicon potassium interaction cannot be considered to be sterically very demanding because of its rather remote position. It is striking that the 1,4-dihydro- (*trans*-**11**) and the respective dibromocyclohexasilane (**13**) display almost the same conformational angles. A comparison with cyclohexane as a reference compound reveals a very similar conformational situation. This seems to indicate that, provided that the substituents are not too sterically overcrowded, the degree of planarity of the ring does not change much. Also dodecamethylcyclohexasilane, the first structurally characterized cyclohexasilane, fits this scheme nicely [6].

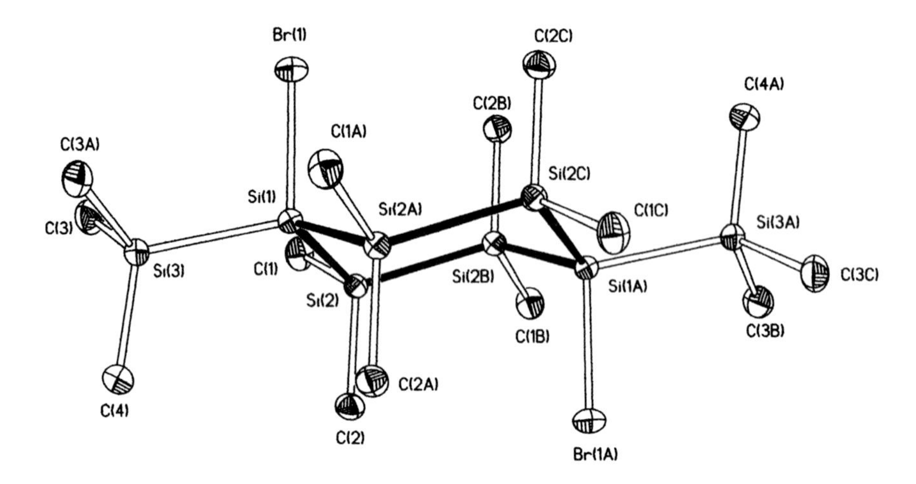

Fig. 2. Crystal structure of **13**. Thermal ellipsoids at the 30% level (hydrogen atoms omitted for clarity).

Table 2. Selected bond lengths [Å] and bond angles [°] of **13**.

Bond lengths	Br(1)–Si(1)	2.288
	Si(1)–Si(2A)	2.350
	Si(1)–Si(3)	2.351
	Si(2)–Si(2B)	2.346
Bond angles	Br(1)-Si(1)-Si(2A)	106.44
	Si(2A)-Si(1)-Si(2)	112.15
	Br(1)-Si(1)-Si(3)	102.69
	Si(2)-Si(1)-Si(3)	114.03
	Si(2B)-Si(2)-Si(1)	111.73

As mentioned above, usually a chair conformation with the large substituents in equatorial position is typical for the 1,4-substituted cyclohexasilanes. 1,1,4,4-tetrakis(trimethylsilyl)-

octamethylcyclohexasilane (**2**), however, is an exception of this frequently found pattern [7]. As all four trimethylsilyls have the same steric demand the system seems to avoid the chair but adopts a rather unusual twist conformation (Fig. 3).

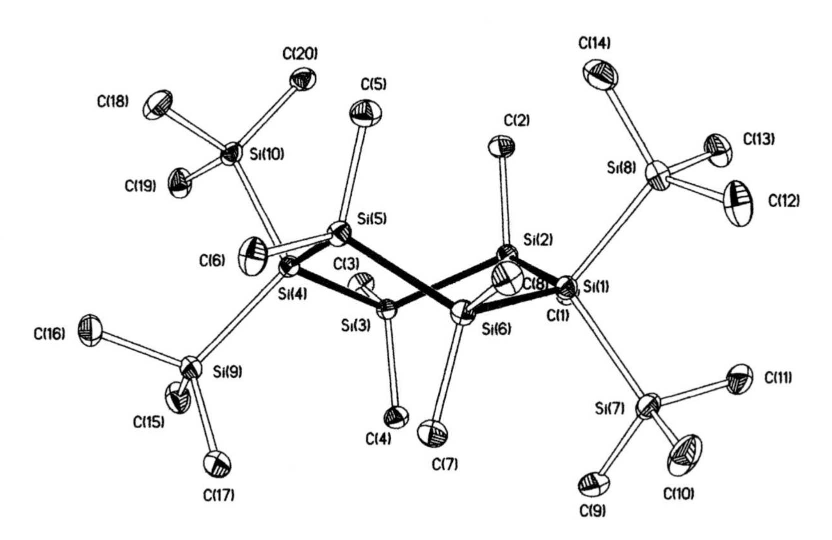

Fig. 3. Crystal structure of **2**. Thermal ellipsoids at the 30% level (hydrogen atoms omitted for clarity).

Table 3. Selected bond lengths [Å] and bond angles [°] of **2**.

	Si(1)–Si(2)	2.357
	Si(1)–Si(7)	2.357
Bond lenghts	Si(1)–Si(8)	2.359
	Si(1)–Si(6)	2.362
	Si(2)–Si(3)	2.346
	Si(2)-Si(1)-Si(7)	110.81
	Si(2)-Si(1)-Si(8)	108.39
	Si(7)-Si(1)-Si(8)	105.95
Bond angles	Si(2)-Si(1)-Si(6)	111.78
	Si(3)-Si(2)-Si(1)	112.45
	Si(2)-Si(3)-Si(4)	113.31

Acknowledgments: This research was supported by the Fonds zur Förderung der wissenschaftlichen Forschung via the START project Y120. We thank Wacker Chemie GmbH, Burghausen, for the generous donation of various chlorosilanes as starting materials.

References

[1] For reviews on cyclosilanes see: a) R. West, in *Comprehensive Organometallic Chemistry* (Eds.: G. Wilkinson, F. G. A. Stone, E. W. Abel), Pergamon, Oxford, **1982**, *Vol. 2*, p. 365; b) E. Hengge, R. Janoschek, *Chem. Rev.* **1995**, *95*, 1495; c) R. West, in *Comprehensive Organometallic Chemistry II* (Eds.: E. W. Abel, F. G. A. Stone, G. Wilkinson), Pergamon, Oxford, **1995**, *Vol. 2*, p. 77; d) E. Hengge, H. Stüger, in *The Chemistry of Organic Silicon Compounds* (Eds.: Z. Rappoport, Y. Apeloig), Wiley, New York, **1998**, *Vol. 2*, Chapter 37, p. 2177.

[2] F. S. Kipping, J. E. Sands, *J. Chem. Soc.* **1921**, *119*, 830.

[3] a) Ch. Marschner, *Eur. J. Inorg. Chem.* **1998**, 221; b) R. Fischer, D. Frank, W. Gaderbauer, Ch. Kayser, Ch. Mechtler, J. Baumgartner, Ch. Marschner, *Organometallics* **2003**, *22*, 3723.

[4] R. West, A. Indrikson, *J. Am. Chem. Soc.* **1992**, *94*, 6110.

[5] R. Fischer, T. Konopa, S. Ully, J. Baumgartner, Ch. Marschner, *J. Organomet. Chem.* **2003**, *685*, 79.

[6] Crystallographic data for **13**: $C_{14}H_{42}Br_2Si_8$, monoclinic space group *C2/m*, a = 13.309(3) Å, b = 11.429(2) Å, c = 10.646(2) Å, β= 110.70(3)°, $R1$ = 0.0241, $wR2$ = 0.0584, GoF = 1.093. Data have been deposited at the Cambridge Crystallographic Data Centre. CCDC 238 998.

[7] H. L. Carrell, J. Donohue, *Acta Cryst.* **1972**, *B28*, 1566.

[8] Crystallographic data for **2**: $C_{20}H_{60}Si_{10}$, monoclinic space group *C2/c*, a = 16.236(3) Å, b = 10.616(2) Å, c = 44.116(9) Å, β= 95.81(3)°, $R1$ = 0.0445, $wR2$ = 0.0982, GoF = 1.220. Data have been deposited at the Cambridge Crystallographic Data Centre. CCDC 238 999.

Synthesis and Photoluminescence of Cyclohexasilanes Bearing Siloxy- and Amino Side Groups

Harald Stüger, Gottfried Fürpass, Karin Renger*

Institut für Anorganische Chemie, Technische Universität Graz
Stremayrgasse 16, A-8010 Graz, Austria
Tel.: +43 316 873 8708 — Fax: +43 316 8701
E-mail: Stueger@anorg.tu-graz.ac.at

Keywords: cyclopolysilanes, photoluminescence, siloxane

Summary: The reaction of chloropermethylcyclohexasilanes $Si_6Me_{12-n}Cl_n$ (n = 1 – 3) with $LiOSiMe_2{}^tBu$ affords the siloxy derivatives $Si_6Me_{12-n}(OSiMe_2{}^tBu)_n$. All compounds exhibit bathochromically shifted first UV absorption maxima and room temperature photoluminescence in solution with emission maxima near 340 nm and remarkably enhanced luminescence intensities as compared to Si_6Me_{12} or to the corresponding phenyl derivatives. The solution photoemission spectra of the corresponding aminocyclohexasilanes $Si_6Me_{12-n}(NR_2)_n$ (n = 1, 2; R = H, $-SiMe_3$), which are accessible from $Si_6Me_{12-n}Cl_n$ with NH_3, Na/NH_3 or $LiN(SiMe_3)_2$, respectively, do not show any luminescence above 300 nm. The stability of the amino derivatives $Si_6Me_{12-n}(NH_2)_n$ toward self condensation is found to depend on the method of preparation and on the number of NH_2 groups attached to the cyclohexasilane ring. ^{29}Si NMR data of formerly unknown compounds are given and the crystal structures of selected compounds are discussed.

Introduction

Crystalline silicon is the most widely used semiconductor material today, with a market share of above 90%. Because of its indirect electronic band structure, however, the material is not able to emit light effectively and therefore cannot be used for key applications like light-emitting diodes or lasers. Selected one- or two-dimensional silicon compounds like linear or branched polysilylenes [1] or layered structures like siloxene [2], however, possess a direct band gap and therefore exhibit intense visible photoluminescence. Siloxene, a solid-state polymer with a sheet-like layered structure and an empirical formula $Si_6H_{6-n}(OH)_n$, in particular, is considered as an alternative material for Si-based luminescent devices. Detailed studies of structural and photophysical properties of the material, however, are strongly impeded by its insolubility in organic solvents.

In a preceding study, we were able to show that polysiloxane polymers with two-dimensional siloxene-like structures containing cyclosilanyl subunits are strongly photoluminescent [3]. Novel, luminescent materials, therefore, might be accessible on the basis of cyclohexasilanyl rings and oxygen-containing side groups. Thus, we prepared a series of siloxy-substituted cyclohexasilanes of the general formula $Si_6Me_n(OSiR_3)_{6-n}$ and examined their photoluminescence properties. In order to investigate the influence of further heterosubstituents on the luminescence behavior of cyclohexasilanes, the study additionally has been extended to the corresponding amino derivatives $Si_6Me_n(NR_2)_{6-n}$.

Synthesis

The reaction pathway used for the synthesis of methylsiloxycyclohexasilanes is shown in Scheme 1.

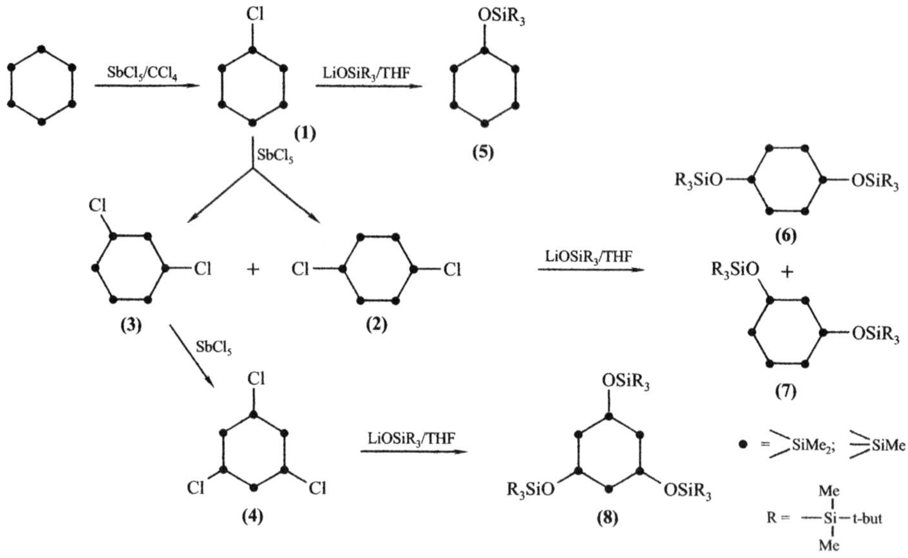

Scheme 1. Synthesis of siloxy-substituted permethylcyclohexasilanes.

With $SbCl_5$, either one, two, or three of the methyl groups present in Si_6Me_{12} can be replaced by chlorine depending on the stoichiometric ratio of the reactants [4]. The reaction of the partially chlorinated cyclohexasilanes **1** – **4** thus obtained with lithium silanolates easily affords the corresponding siloxy derivatives **5** – **8**.

The di- and trisubstituted compounds **6** – **8** are obtained as statistical mixtures of *cis/trans* isomers. Isolation and purification of the least soluble isomer can be achieved by crystallization from 2-propanol. Single-crystal X-ray diffraction studies of **6** and **7** exhibit the cyclohexasilane ring in a slightly distorted chair conformation and the bulky $-OSiMe_2{}^tBu$ groups in equatorial sites in order to minimize non-bonding interactions (Fig. 1).

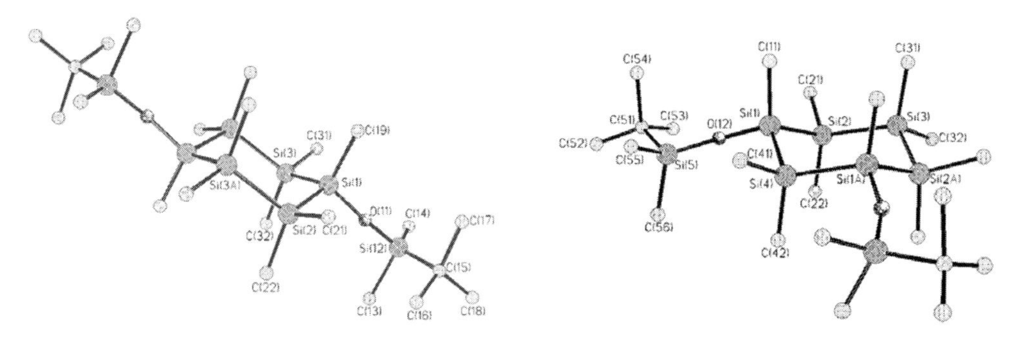

Fig. 1. Crystal structure of compounds **6** and **7**.

Two general methods are suitable for the synthesis of permethylaminocyclohexasilanes [5]. As shown in Scheme 2, the reaction of **1** with NH_3 or Na/NH_3, respectively, cleanly affords aminoundecamethylcylohexasilane (**9**), which can be further silylated yielding the silazanes **10** and **11** after lithiation with *n*-BuLi. The air-stable compound **12** is obtained in good yield from **1** and $LiN(SiMe_3)_2$.

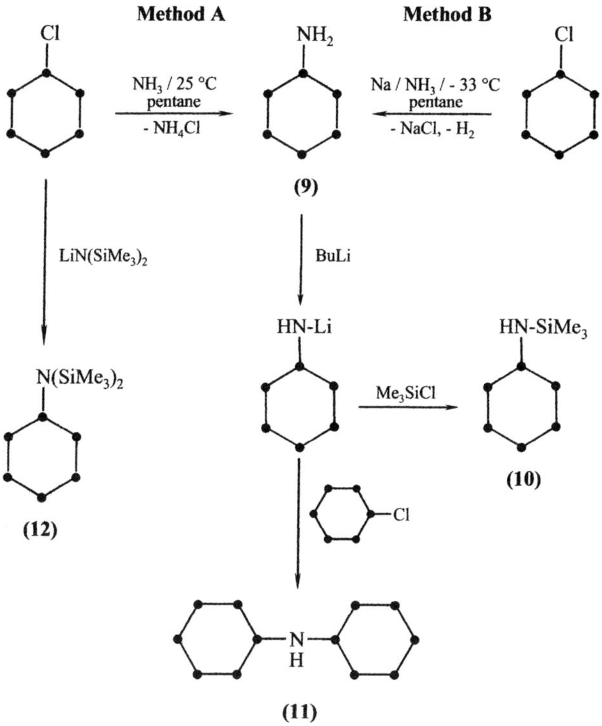

Method A **Method B**

Cl — NH_3 / 25 °C, pentane, - NH_4Cl → NH_2 ← Na / NH_3 / - 33 °C, pentane, - NaCl, - H_2 — Cl

(9)

$LiN(SiMe_3)_2$ ↓ BuLi ↓

$N(SiMe_3)_2$ HN-Li — Me_3SiCl → HN-SiMe$_3$

(12) **(10)**

—Cl ↓

N H

(11)

Scheme 2. Synthesis of undecamethylcyclohexasilanylamino derivatives **9** – **12**.

However, the results obtained depend on whether method A or method B is used, starting from the appropriate dichlorocyclohexasilanes **2** and **3** (Scheme 3). Both methods turned out to be equally suitable for the preparation of the 1,3-diamino derivative **14**. When **14** prepared according to method A, however, is stored dissolved in pentane at room temperature for several weeks, it slowly condenses to cyclohexasilanylsilazane oligomers, while **14** made via method B turned out to be completely stable. The attempted synthesis of the 1,4-analog **13** by method A actually failed because of competing inter- and intramolecular condensation reactions yielding a complex mixture of products from which **13** cannot be isolated. Pure **13**, however, is easily accessible by method B, although only the *trans* isomer is obtained in this case according to ^{13}C and ^{1}H NMR spectroscopy. The synthesis of **14** by method B also affords only one of the two possible isomers, while method A results in the formation of the expected statistical *cis/trans* mixture. Different reaction mechanisms, therefore, may be assumed to be working in the course of methods A and B.

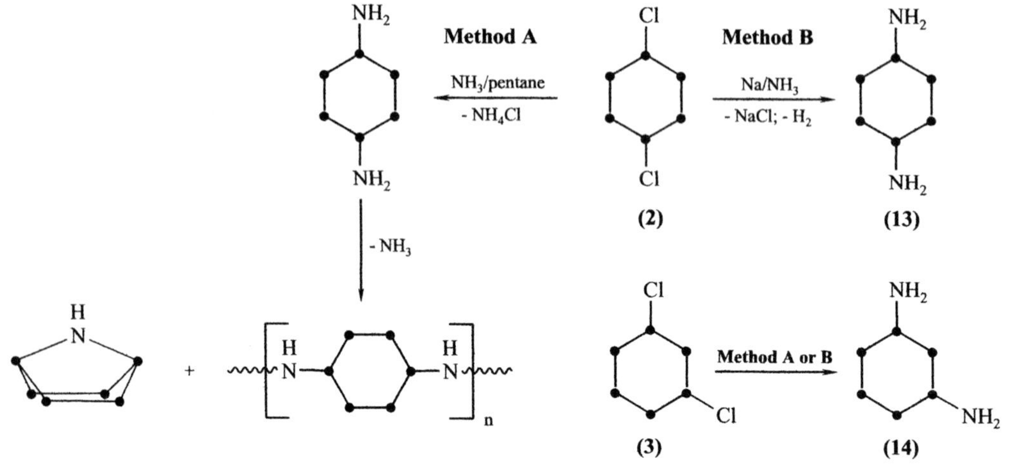

Scheme 3. Synthesis and condensation of diaminodecamethylcyclohexasilanes **12** and **13**.

The limited stability of **13** and **14** prepared according to method A is very likely caused by the presence of residual traces of NH_4Cl still present after recrystallization and capable of catalyzing condensation of the primarily formed aminosilane. This assumption is substantiated by the condensation behavior of the less reactive monoamino derivative **9** (Scheme 4). Heating a hydrocarbon solution of **9** prepared by method A in a sealed tube up to 150 °C results in the quantitative formation of the condensation product **11**.

However, performing the same procedure for **9** prepared by method B leaves the starting material unaffected even up to 250 °C, while subsequent addition of catalytic amounts of NH_4Cl again leads to the clean formation of **11**, proving the presence of NH_4Cl to be crucial for condensation.

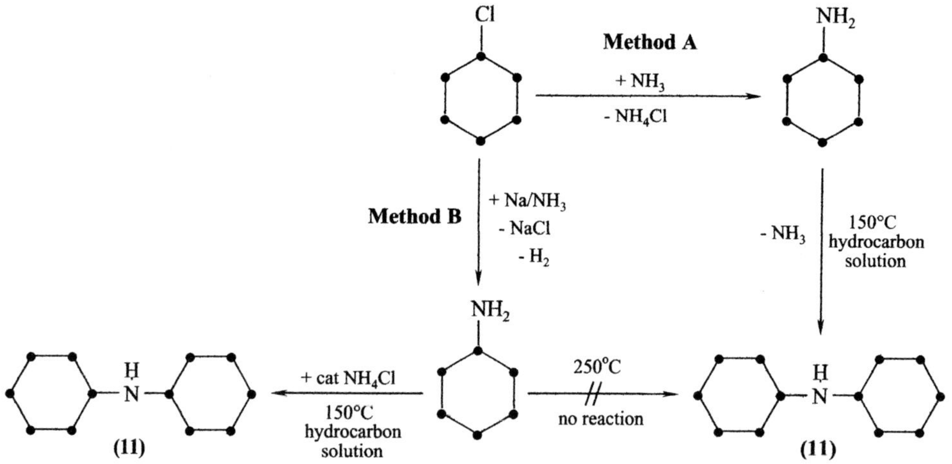

Scheme 4. Thermal condensation of aminoundecamethylcyclohexasilane **9**.

The formerly unknown compounds **5** – **13** were fully characterized by standard techniques. ^{29}Si NMR data including assignment of the resonance lines are given in Table 1.

Table 1. ^{29}Si NMR data of compounds **1** – **13** (assignment of the resonance lines are given in parentheses).

Compound	$\delta^{29}Si$ [C_6D_6, ext. TMS, ppm]
5	11.3 ($SiMe_2{}^tBu$); 6.1 ($Si_{Ring}MeOR$); –41.9, –42.2, –43.1 ($Si_{Ring}Me_2$)
cis/trans-**6**	11.6/11.4 ($SiMe_2{}^tBu$); 6.6/6.1 ($Si_{Ring}MeOR$); –44.0/–44.4 ($Si_{Ring}Me_2$)
cis/trans-**7**	11.6/11.0 ($SiMe_2{}^tBu$); 6.0/5.9 ($Si_{Ring}MeOR$);–43.0/–43.5, –44.0/–44.9, –45.7/–48.6 ($Si_{Ring}Me_2$)
cis,cis-**8**	11.7($SiMe_2{}^tBu$); 4.9 ($Si_{Ring}MeOR$); –44.3 ($Si_{Ring}Me_2$)
9	–15.3 ($Si_{Ring}MeNH_2$); –42.3, –42.6, –44.1 ($Si_{Ring}Me_2$)
10	3.9 ($SiMe_3$); –14.3 ($Si_{Ring}MeN$); –42.5, –42.6, –43.6 ($Si_{Ring}Me_2$)
11	–10.6 ($Si_{Ring}MeN$); –42.5, –42.5, –43.3 ($Si_{Ring}Me_2$)
12	2.3 ($SiMe_3$); –15.7 –38.6, –42.0, –43.7 ($Si_{Ring}Me_2$)
trans-**13**	–15.2 ($Si_{Ring}MeNH_2$); –44.6 ($Si_{Ring}Me_2$)
cis/trans-**14**	–14.1/ –15.1 ($Si_{Ring}MeNH_2$); –42.7 (x2); –43.8/–44.0; –45.5/–46.2 ($Si_{Ring}Me_2$)

Absorption and Photoluminescence Spectra

UV absorption and photoluminescence emission maxima of compounds **5** – **8** are listed in Table 2 together with the corresponding data for dodecamethylcyclohexasilane. The first maxima appearing at the low-energy side of the absorption spectra are shifted bathochromically with increasing number of siloxy groups because of enhanced σ(Si–Si)→n(O) conjugation.

Table 2. Absorption and emission data for Si_6Me_{12} [6] and **5** – **8** in C_6H_{12} (c= 5 × 10^{-4} mol/L, λ_{ex} = 288 nm).

Compound	$\lambda_{max, abs}$ [nm] (ε [L/mol·cm])	$\lambda_{max, em}$ [nm]
Si_6Me_{12}	258 (1200); 231 (5500)	–
5	269 (870); 233 (5200)	330 sh; 342; 360 sh
trans-**6**	275 (890); 250 (2950); 228 (7800)	340
cis-**7**	280 (700); 253 (2250); 236 (5940)	339
cis,cis-**8**	295 (390); 255 (2250); 234 (7900)	333; 345

When isomerically pure samples of **5** – **8** are excited near the wavelength of the first UV maximum around 290 nm, distinct photoluminescence is observed. Around 340 nm several overlapping emission bands appear, the position of which is not influenced markedly by the number of siloxy groups attached to the cyclohexasilane ring. Although the trends apparent in the luminescence spectra of **5** – **8** cannot be interpreted straightforwardly so far, the emission intensity of siloxy-substituted cyclohexasilanes turns out to be considerably higher than that of the corresponding phenyl derivatives, while the fluorescence of Si_6Me_{12} is rather weak (see Fig. 2). The fact that also the aminocyclohexasilanes **10** – **12** do not show any photoluminescence above 300 nm additionally demonstrates the importance of the oxygen-containing subsituents for the exciting photoluminescence behavior of compounds **5** – **8**. Although further studies including additional model substances will be necessary in order to generalize the trends observed in this study, our results clearly demonstrate that novel, luminescent materials might be accessible on the basis of cyclohexasilanyl rings bearing –OR side groups.

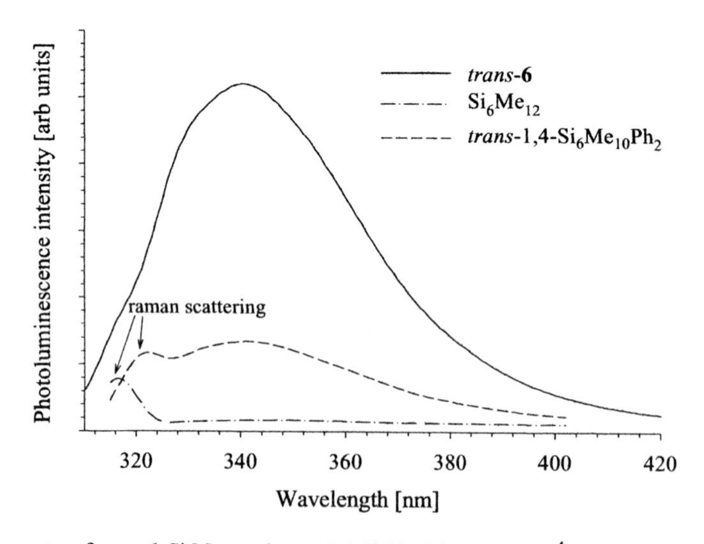

Fig. 2. Emission spectra of *trans*-**6**, Si_6Me_{12} and *trans*-1,4-$Si_6Me_{10}Ph_2$ (c= 5 × 10^{-4}mol/L, λ_{ex} = 288 nm) in C_6H_{12}.

Acknowledgments: We thank the FWF (Wien, Austria) for financial support within the Forschungsschwerpunkt "Novel Approaches to the Formation and Reactivity of Compounds containing Silicon–Silicon Bonds" and the Wacker Chemie GmbH (Burghausen, Germany) for the donation of silane precursors.

References

[1] a) R. D. Miller, J. Michl, *Chem. Rev.* **1989**, *89,* 1359; b) N. Matsumoto, K. Takeda, H. Teramae, M. Fujino, in *Silicon-Based Polymer Science, Adv. Chem. Ser. Vol. 224*, (Eds.: J. M. Zeigler, F. W. G. Fearon), ACS, **1990**, p. 515; c) Y. Kanemitsu, K. Suzuki, S. Kyushin, H. Matsumoto, *Synth. Met.* **1995**, *69*, 585.

[2] M. S. Brandt, T. Puchert, M. Stutzmann, in *Tailor-Made Silicon-Oxygen Compounds* (Eds.: R. Corriu, P. Jutzi), Vieweg, Wiesbaden, **1994**, p. 117.

[3] A. Kleewein, H. Stüger, *Monatsh. Chem.* **1999**, *130*, 69.

[4] a) E. Hengge, M. Eibl, *J. Organomet. Chem.* **1992**, *428*, 335; b) M. Eibl, U. Katzenbeisser, E. Hengge, *J. Organomet. Chem.* **1993**, *444*, 29; c) A. Spielberger, P. Gspaltl, H. Siegl, E. Hengge, K. Gruber, *J. Organomet. Chem.* **1995**, *499*, 241.

[5] A. Ackerhans, P. Böttcher, P. Müller, H. W.Roesky, I. Uson, H. G. Schmidt, M. Noltemeyer, *Inorg. Chem.* **2001**, *40*, 3766.

[6] E. A. Carberry, R. West, G. E. Glass, *J. Am. Chem. Soc.* **1969**, *91*, 446.

Conformational Properties of 1,1,2,2-Tetrakis(trimethylsilyl)disilanes and of Tetrakis(trimethylsilyl)diphosphine: a Comparative Vibrational Spectroscopic and *ab initio* Study

Günter Tekautz,* Karl Hassler

Institute of Inorganic Chemistry, University of Technology Graz
Stremayrgasse 16/IV, A-8010 Graz, Austria
E-mail: tekautz@anorg.tu-graz.ac.at

Keywords: disilane, diphosphine, conformation

Summary: An investigation of the conformational properties of the title compounds is presented. Energy profiles for rotation around the central bonds were computed at HF level. Frequency calculations for the optimized geometries were carried out. Temperature-dependent Raman spectroscopy was utilized to revise the computed results. While the results of the *ab initio* calculations predict up to four minima on the potential energy surfaces (PES), Raman spectroscopy gives unambiguous evidence for more than one rotamer for tetrakis(trimethylsilyl)disilane only.

Introduction

During the last decade, the investigation of the conformational properties of the oligosilanes became a significant research area in organosilicon chemistry. Deviations from the usual 180° and ±60° dihedral angles were found to be frequent in silicon chains with substituents bigger than hydrogen. Moreover, additional minima exist. Therefore a new systematic nomenclature for conformations of linear chains, cisoid (~40°), ortho (~90°), deviant (~150°), and transoid (~160°) was suggested [1].

Theoretical Calculations

The energy profiles for rotation around the central M–M bond (M = Si, P) in steps of 10° for the title structures, 1,1,2,2-tetrakis(trimethylsilyl)dimethyldisilane (**1**), 1,1,2,2-tetrakis(trimethylsilyl)-disilane (**2**) and tetrakis(trimethylsilyl)diphosphine (**3**) were investigated by *ab initio* calculations at the RHF level of theory using 6-31G* basis sets. For all compounds C_2-symmetry was adopted for the calculations.

In the NBO picture, as advocated by Weinhold [2], the rotational barrier in ethane-like molecules arises from vicinal interactions between bonding and antibonding orbitals. The energy-lowering $\delta E\sigma$ from a $\sigma\sigma^*$ interaction can be calculated approximately using second-order perturbation theory. The sum of all vicinal interactions at the central bond reflects the electronic part of the barrier. Van der Waals and steric interactions superimpose on the electronic part, shifting minima or creating new ones depending on the size of the substituents.

Results

The resulting energy profiles are shown in Fig. 1. For **1**, cisoid (46°), ortho (82°), deviant (145°) and transoid (160°) conformations show up as minima. Cisoid (59°), ortho (92°) and transoid (156°) are the calculated minima for **2**. For **3**, only an ortho (105°) minimum appears. Further calculations at the MP2 level are in progress.

The calculated barriers between the minima in **2** are so low that a rapid interconversion of the rotamers is expected. In **1**, the barrier between gauche/ortho and transoid is significantly above 2.5 kJ/mol (= RT at room temperature), which should make two distinct rotamers detectable with vibrational spectroscopy if the spectra of the rotamers differ sufficiently.

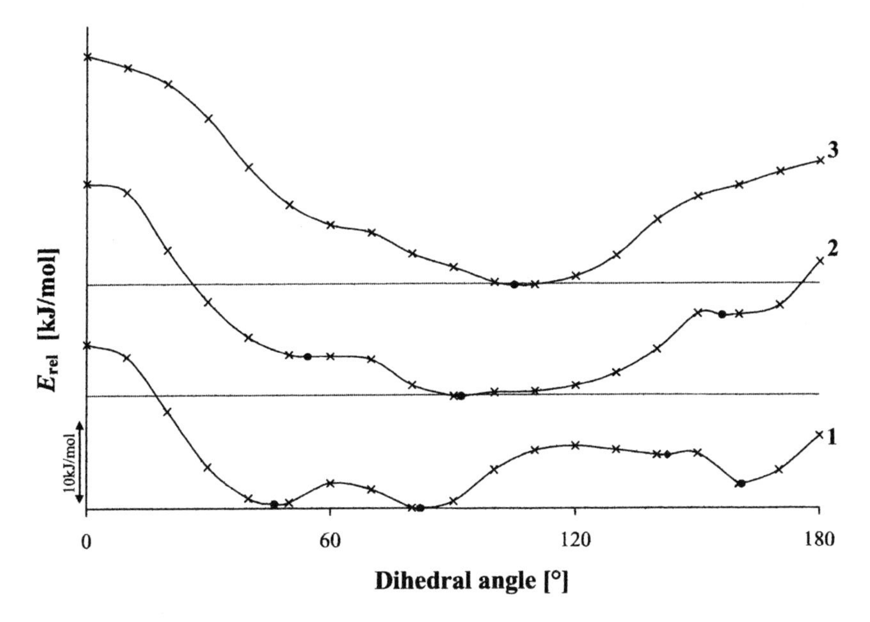

Fig. 1. Energy profiles for the rotation around the central bond for **1**, **2**, and **3**. Minima are labeled with a dot.

The curves in Fig. 2 (left) show the energy-lowering δE_σ from $\sigma\sigma^*$ interactions. *A* represents the σ(SiC) – σ(SiSi)* interaction for **1**, *B* displays σ(SiH) – σ(SiSi)* for **2**. *C* gives the lone pair – σ(SiP)* interaction for **3**. All interactions vary with the dihedral angle between the bonds. The diagram shows that σ(SiH) – σ(SiSi)* (**1**) is slightly more stabilizing than σ(SiC) – σ(SiSi)* in **2** and that the strongest interaction is between the lone pair and the antibonding SiP orbital in **3**.

The sum of all vicinal σσ* interactions, which represents the electronic part of the barrier, is also displayed in Fig. 2 (right). The curves for the disilanes show the expected characteristic with stabilization of gauche and anti. The profile for the diphosphine displays just one maximum for gauche.

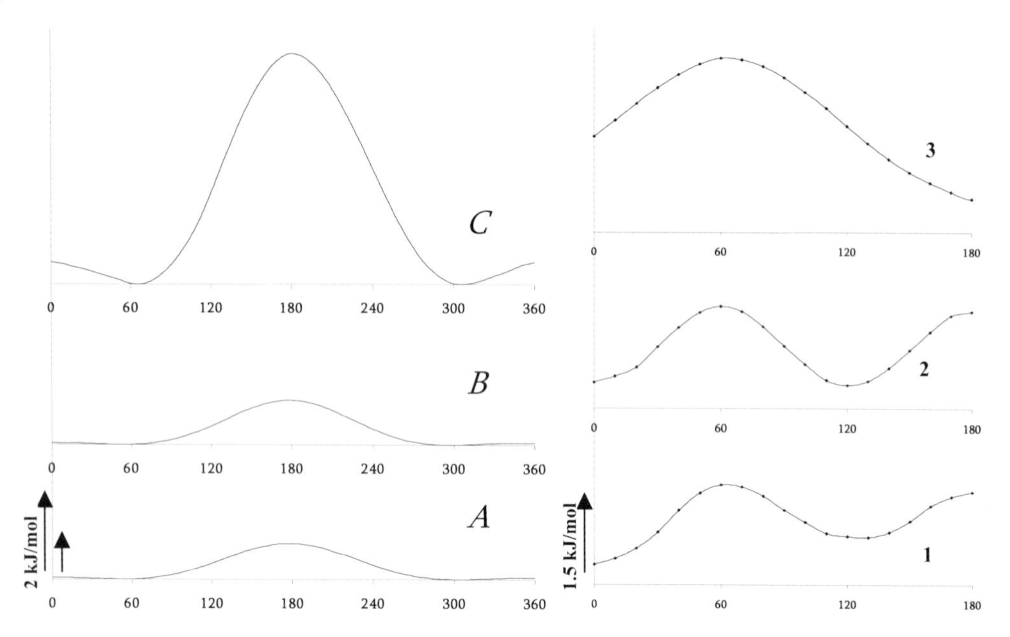

Fig. 2. δE_σ from σσ* interactions (left) and the sum of all vicinal σσ* interactions (right) for **1**, **2**, and **3**.

Spectroscopy

Temperature-dependent Raman spectra for **1**, **2**, and **3** were recorded. The spectra of **1** and **2** for various temperatures are shown in Fig. 3.

As expected for one rotamer, the spectra of **3** do not change with temperature. In the Raman spectra of **1** one would expect from the calculations to see some changes with temperature. However the temperature-dependent spectra of **1** give no unambiguous evidence for the existence of more than one rotamer.

In the spectra of **2** some band intensities change with temperature, which is not expected from the calculated energy profile at HF level because of the low barriers. As it is known from the literature, MP2 calculations usually give more distinct minima and higher barriers [3].

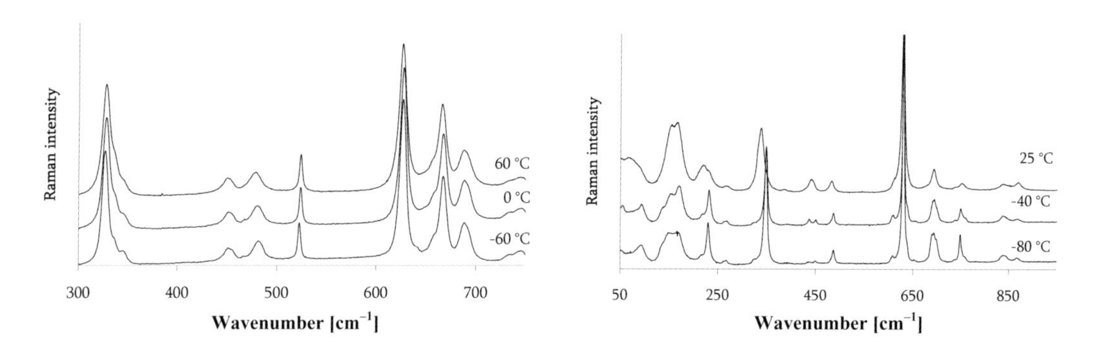

Fig. 3. Raman spectra of a solution of **1** in toluene (left) and of **2** at different temperatures.

Figure 4 shows the region of the SiC deformations, displaying some significant intensity changes. The right hand side of Fig. 4 shows a van't Hoff plot of the band pair 231/218. A good linear correlation can be found. The value of ΔH, which can be calculated from the slope, is 18.4 kJ/mol, with a margin of at least ±10 kJ/mol inherent in the method [4]. There are two rotamers detectable with vibrational spectroscopy. Obviously, the barrier separating the cisoid from the ortho conformer is so small that either a rapid interconversion of the rotamers occurs or alternatively the lowest torsional level lies above the barrier. In both cases a single averaged rotamer is observed in the Raman spectrum.

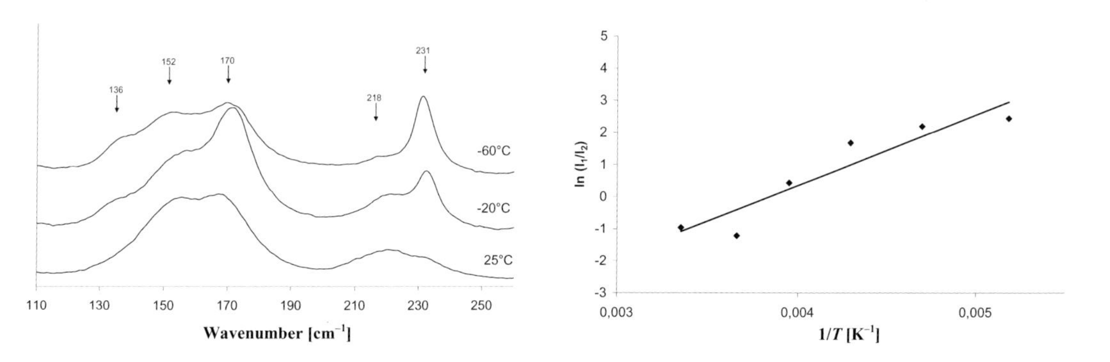

Fig. 4. Section of the Raman spectra of **2** (left) and the van't Hoff plot for the band pair 231/218 cm^{-1}(right).

Besides SiC deformation vibrations, SiSi stretching and SiSiH deformation vibrations are expected to be most sensitive to conformations. The latter, however, lie in the region of strong $\nu_{as}SiC_3$ and ρCH_3 vibrations and cannot be assigned.

In the region of the $\nu_{(s,as)}SiSi_2$ vibrations more than four bands are expected for a single conformer. The wavenumbers for these vibrations and assignments are shown in Table 1 assuming C_2 symmetry.

Table 1. Calculated and observed wavenumbers [cm^{-1}] for the νSiSi and the δSiSiH vibrations for **2**.

Vibration	Calculated (scaled by 0.95)			Observed		Point group
	cisoid	ortho	transoid	Ra (liquid)	IR (liquid)	
δSiSiH	662	712	755	–		A
δSiSiH	780	705	625	–		B
νSiSi	484	456	448	483(m)		A
$\nu_{as}SiSi_2$	437	422	427	449(w)/441(s)/436(w)	439(m)/449sh	B
$\nu_{as}SiSi_2$	448	416	403	–	400vvw, sh	A
$\nu_{s}SiSi_2$	387	363	360	386 (vw)	357(m)/385(w)	B
$\nu_{s}SiSi_2$	327	322	335	332(w)/338(s)/348(w)	336(m)	A

Acknowledgment: The authors thank the Austrian "Fonds zur Förderung der wissenschaftlichen Forschung" for financial support (project P15 366).

References

[1] J. Michl, R. West, *Acc. Chem. Res.* **2000**, *33*, 82.
[2] A. E. Reed, F. Weinhold, *Isr. J. Chem.* **1991**, *13*, 277.
[3] R. Zink, G. Tekautz, A. Kleewein, K. Hassler, *Chemphyschem* **2001**, *6*, 377.
[4] P. Klaboe, *Vib. Spectrosc.* **1995**, *9*, 3.

Reactions of Octasilacubane

Masafumi Unno,* Hideyuki Matsumoto

Department of Nano-Material Systems, Graduate School of Engineering
Gunma University, 1-5-1 Tenjin-cho, 376-8515, Japan
Tel.: +81 277 30 1293 — Fax: +81 277 30 1291
E-mail: unno@chem.gunma-u.ac.jp

Keywords: cage polysilane, silicon polyhedranes, reactions

Summary: Octasilacubane is an intriguing polyhedral polysilane because of its unique electronic properties arising from its highly strained Si–Si σ-bonded framework. The first octasilacubane, octakis(t-butyldimethylsilyl)octasilacubane was reported by our group in 1988, as the first example of Platonic solid compounds comprising higher Group 14 elements. We then prepared the alkyl-substituted $(ThexSi)_8$ (Thex = 1,1,2-trimethylpropyl) in 1992. This octasilacubane is air-stable and soluble in organic media, and possesses low oxidation potential. These striking properties make this compound a good choice for the study of the reactions with electrophiles. In this paper, the reactions of octasilacubanes: oxidation, photooxidation, halogenation, regeneration of octasilacubane, and photolysis are summarized.

Introduction

Cubane was synthesized as the first polyhedral hydrocarbon in 1964 by Eaton et al. [1], and more than 24 years were to elapse before its higher Group 14 analogs appeared [2]. Since then, a great deal of progress has been recorded in constructing polyhedral silicon compounds [3]. However, regarding their reactions, only limited reports [4] have ever been published, even though their strained frameworks were expected to show unique reactivity. This is simply explained by three facts: (1) all these compounds possess very bulky substituents for stabilization, which also diminish reactivity, (2) because of high symmetry, solubility in organic media is sometimes poor, and (3) the reaction products are often unstable; thus separation may be difficult. Fortunately, the octasilacubanes we have reported are free from these problems and exhibited various reactions. In this paper, we describe the synthesis and reactions of octasilacubanes.

Synthesis of Octasilacubane

In the mid-80s, our laboratory observed that silyl substituents were very effective for the facile synthesis of strained cyclic oligosilanes [5]. Therefore, we naturally examined silyl substituents for cage compounds. Synthesis of silyl-substituted octasilacubane (**1**) is illustrated in Scheme 1. The starting tribromosilane or tetrabromodisilane was prepared from triphenylsilane or tetraphenyldisilane, respectively, in high yields. The reductive coupling was effected with Na in toluene at 120 °C. After filtration of NaBr, recrystallization of the crude solid from methylcyclohexane gave pure octasilacubane, $(t\text{-BuMe}_2\text{SiSi})_8$ (**1**) in fair yields (18% from tribromosilane and 24% from tetrabromodisilane) [3a]. The obtained octasilacubane can be handled in the air for a short time, but is slowly oxidized. The crystal structure of **1** is shown in Fig. 1 [6]. The strained framework is protected effectively with eight bulky silyl substituents.

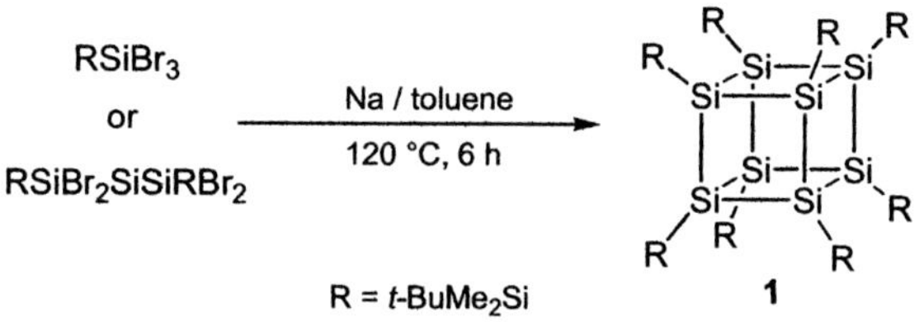

$RSiBr_3$ or $RSiBr_2SiSiRBr_2$ —(Na / toluene, 120 °C, 6 h)→ **1**

R = *t*-BuMe$_2$Si

Scheme 1. Preparation of silyl-substituted octasilacubane **1**.

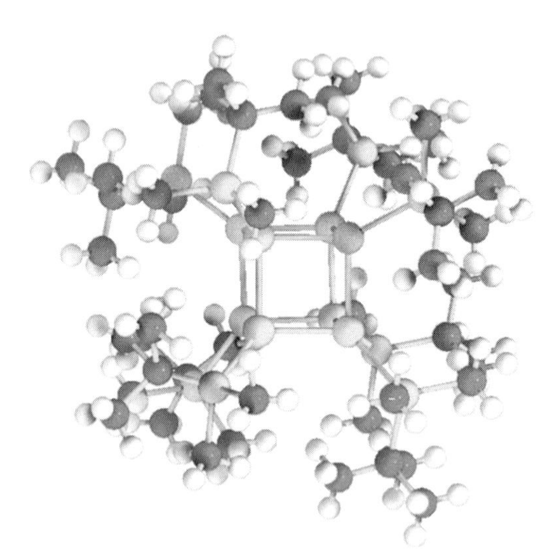

Fig. 1. Molecular structure of $(t\text{-BuMe}_2\text{SiSi})_8$ (light gray: Si, dark gray: C, white: H).

In the hope of preparing a more stable octasilacubane, we then synthesized thexyl-(1,1,2-trimethylpropyl)-substituted octasilacubane, **2** [3b]. Because of the shorter length of Si–C than that of Si–Si, the thexyl group was expected to shield the cubane framework more effectively than silyl substituents without much sacrificing its reactivity. The procedure is shown in Scheme 2. Starting thexyltrichlorosilane was obtained by the reaction of trichlorosilanes with 2,3-dimethyl-2-butene, and the reductive coupling proceeded similarly to that in the case of **1**. Yield was less satisfactory (3%); however, we were pleased that the obtained red crystals of **2** are stable in the air and handled without any precautions. Surprisingly, octasilacubane (**2**) survived in reverse-phase HPLC (ODS column, eluent: MeOH-THF) even after several times of recycling. This feature is promising as we have to separate the products after the reaction in most cases.

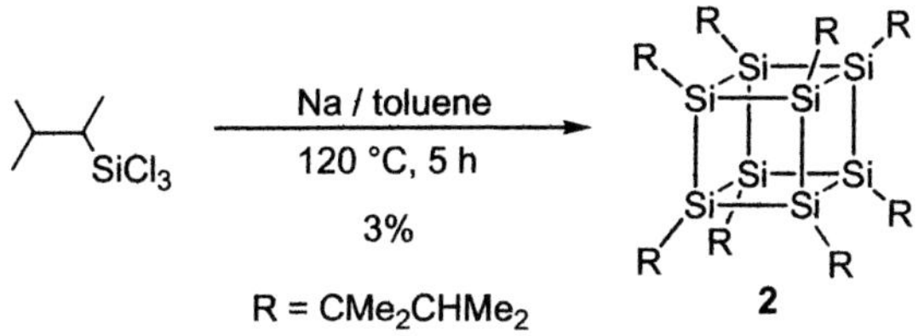

Na / toluene
120 °C, 5 h
3%
R = CMe_2CHMe_2

Scheme 2. Preparation of thexyl-substituted octasilacubane (**2**).

Reaction of Octasilacubane

Complete Oxidation

As expected, octasilacubane possesses high reactivity toward electrophiles. As shown in Scheme 3, *m*CPBA oxidation of silyl-substituted **1** resulted in a high-yield formation of octasilsesquioxane (**3**) (T_8). Silsesquioxanes are well studied, but were previously only prepared by the hydrolysis-dehydration of halo- or alkoxysilanes. When the substituents are bulky, the reaction proceeds no further than the silanol stage, and silsesquioxanes are not obtained. For example, hydrolytic condensation of *t*-butyldimethylsilyltrichlorosilane gave only partly hydrolyzed silanols. Thus, **3** is the octasilsesquioxane with the bulkiest substituents, and the only silyl-substituted one ever reported. The structure was determined by X-ray crystallography [7], and the bond lengths and angles are similar to those of known octasilsesquioxanes.

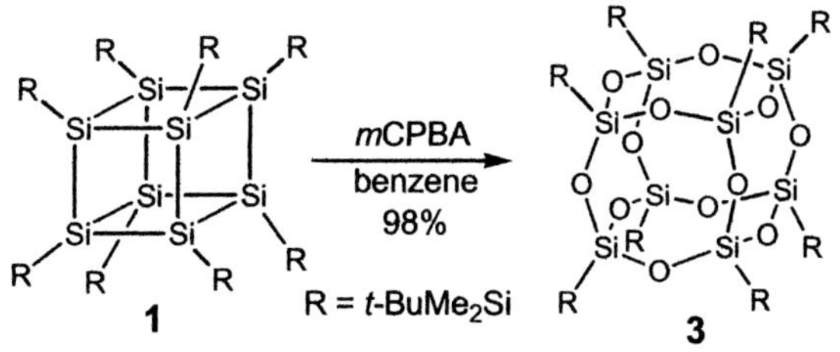

*m*CPBA
benzene
98%
R = *t*-$BuMe_2Si$

Scheme 3. *m*CPBA oxidation of octasilacubane (**1**).

Interestingly, only cage Si–Si bonds were reacted, and exocyclic Si–Si bonds remained intact even with prolonged reaction with excess *m*CPBA from **3**. It is also noteworthy that the UV spectrum of **3** indicates the absorption maxima at 250 and 285 nm. In Fig. 2, UV spectra of **3** with those of $(Me_3SiSi)_8O_{12}$ and $(Me_3SiC_2H_4Si)_8O_{12}$ are shown. Usually, silsesquioxanes with alkyl substituents show no particular absorption maxima, as in the case of $(Me_3SiC_2H_4Si)_8O_{12}$, and only silyl-substituted T_8 shows distinct absorption maxima. Thus we reasoned that this feature is due to the spatial allocation of eight Si–Si bonds with oxygen as a spacer.

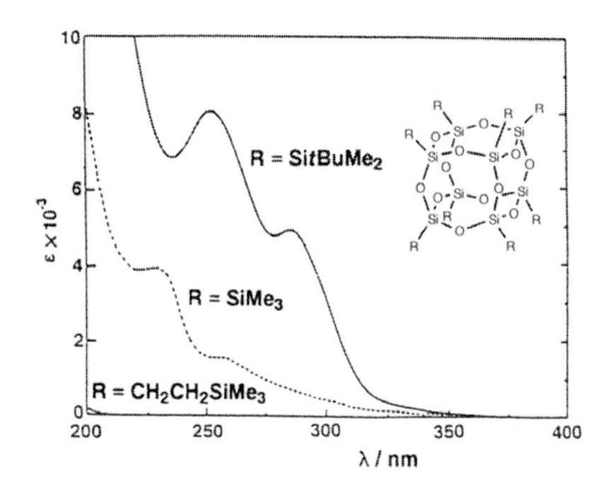

Fig. 2. UV spectrum of octasilsesquioxane (3) (and model compounds) in cyclohexane.

Photooxidation

On the other hand, thexyl-substituted octasilacubane (**2**) resists oxidation and is stable in the air. Indeed, *m*CPBA oxidation gave only a mixture of products even under severe conditions (50 eq. *m*CPBA in toluene 115 °C, 50 h). Then we tried photooxidation (Scheme 4). Irradiation of **2** with a high-pressure mercury lamp (> 300 nm) in the presence of DMSO afforded octasilaoxahomocubane (**4**) and octasiladioxabishomocubane (**5**) in good yields [8].

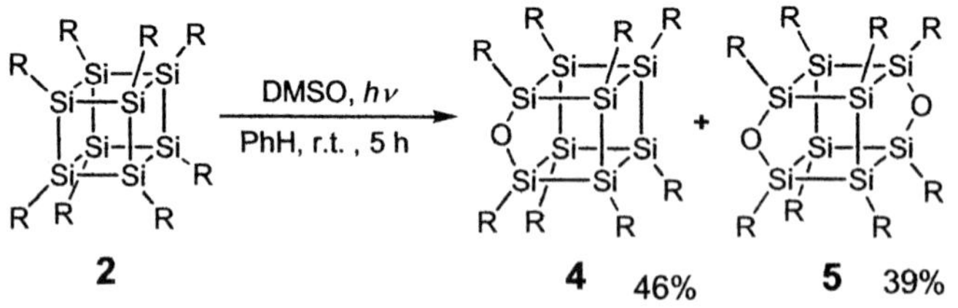

Scheme 4. Photooxidation of octasilacubane (**2**).

The structures of both compounds were determined by X-ray crystallography. Selected bond lengths are illustrated in Scheme 5. Most of the bonds are similar to those of octasilacubane (average 2.421 Å), but one of the bonds in **4** is extremely expanded (diagonal to Si–O–Si, 2.56 Å). This is explained by the steric hindrance of thexyl groups as a result of the insertion of oxygen and a second oxygen inserted in the longest (reactive) Si–Si bond. As expected, an independent experiment showed that the oxidation of **4** with *m*CPBA gave **5** exclusively.

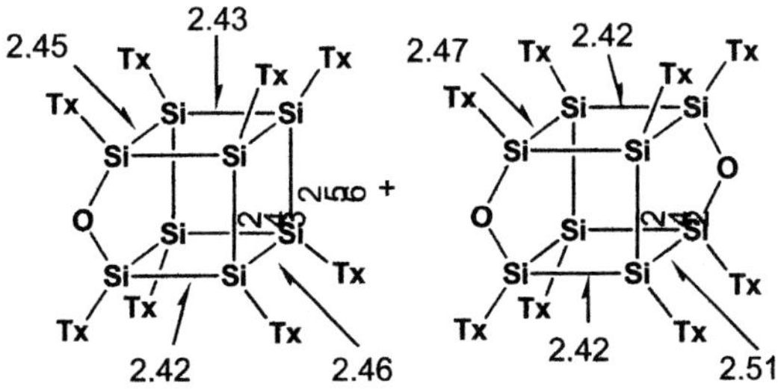

Scheme 5. Selected bond lengths (Å) for **4** and **5**.

Halogenation and Regeneration

With its high electron-donating ability, octasilacubane smoothly reacts with halogens. The treatment of octasilacubane (**2**) with Cl_2, PCl_5, Br_2, or I_2 furnished dihalides **6** in good yields [8]. All three isomers (*endo,endo-*, *endo,exo-*, and *exo,exo-*forms) were generated, and these are separated with recycle-type reverse-phase HPLC (ODS, MeOH-THF). Because of the effective steric protection of thexyl groups, even diiodides could survive in this separation procedure. Surprisingly, obtained dihalides were not the expected one-bond fission compounds. Instead, skeletal rearrangement was observed and isomeric dihalides were formed (Scheme 6). The reaction plausibly proceeded through the intermediacy of silyl cation and a simple one-bond rearrangement to form more stable five-membered rings. All the structures were unequivocally determined by X-ray crystallography and NMR spectra. Notably, the ratio of isomers varied with halogens. Smaller chlorine favors *exo*-isomer, whereas larger iodine prefers *endo*-isomer.

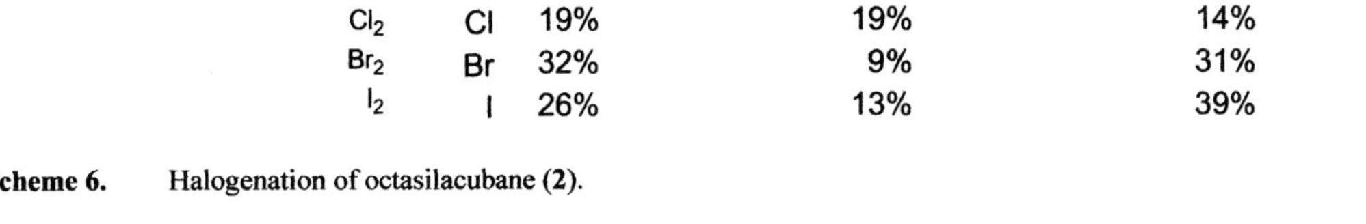

R=Me_2CCHMe_2

[X] =	X =	*exo,exo*-**6**	*endo,exo*-**6**	*endo,endo*-**6**
PCl_5	Cl	37%	33%	15%
Cl_2	Cl	19%	19%	14%
Br_2	Br	32%	9%	31%
I_2	I	26%	13%	39%

Scheme 6. Halogenation of octasilacubane (**2**).

Similar reaction occurs under radical conditions [9]. Thus, heating **2** in CCl_4 led to the formation of *endo,exo*- and *exo,exo*-**6**, while reaction of **2** with CBr_4 or $Br_2C_2H_4Br$ with irradiation by a high-pressure Hg lamp furnished *exo,exo*-**6** exclusively. Interestingly, the reaction proceeded in non-degassed benzene in the dark, and oxasecocubane **7** in addition to *exo,exo*-**6** were obtained. The generation of **7** is explained by that oxygen insertion occurred faster than the rearrangement. This compound is the first example of an octasilasecocubane structure determined by X-ray crystallography.

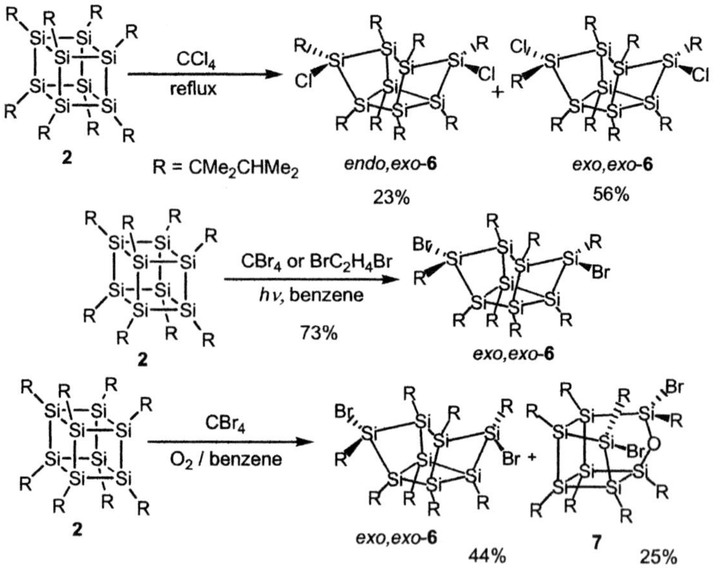

Scheme 7. Radical halogenation of octasilacubane (**2**).

Quite interestingly, when the all isomers of **6** were subjected to the condition similar to the synthesis of octasilacubane (Na/toluene, 120 °C), the octasilacubane **2** was regenerated. In addition, the reduced product **8** was obtained as a minor product. The choice of the reaction conditions is crucial for this reaction: (1) other metals such as lithium gave no reaction, (2) prolonged reaction time caused decomposition of **2**, and the yield was decreased. To overcome the energy difference, drastic conditions were necessary. Similar reaction is observed for thexyl-substituted octagermacubane [10]; the regeneration occurred only with Na/toluene at 120 °C, and there was no reaction with Mg/MgBr, which was utilized for the synthesis of this octagermacubane.

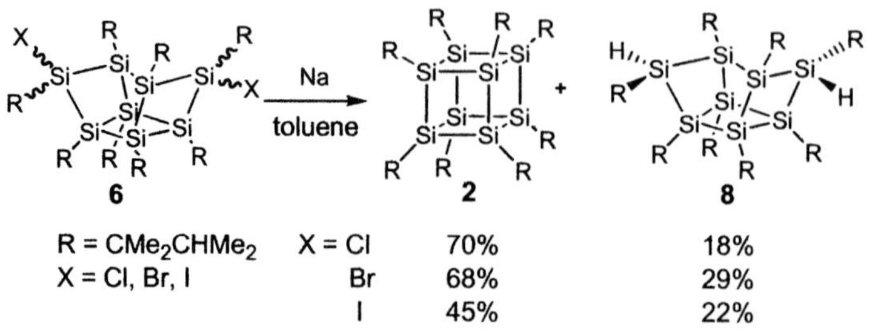

Scheme 8. Regeneration of octasilacubane (**2**).

Photolysis

Because of the strain and fused-ring system, octasilacubane shows absorbances in a visible region. Irradiation of octasilacubane with a 400 W high-pressure Hg lamp resulted in the appearance of new absorbances at 345, 471, and 714 nm (Fig. 3). Warming up the glass matrix to room temperature gave rise to the loss of these peaks, but they did not decay for several hours at 77 K. Additionally, the absorption spectrum after warming was exactly same as that of **2**, and HPLC analysis after steady-state photolysis at 77 K shows only the peak of **2**. These results clearly indicate that the intermediate restores the starting molecule and could be ascribable to the photoisomer of **2**. However, in spite of the multiple trials of trapping experiments with alcohol, water, or diene, no particular compounds were obtained.

We also measured time-resolved transient spectra and the decay profile, and the results indicated no formation of silylene or silyl radicals; instead, closed-shell species were suggested. With all these results and INDO/S-CI calculation, we currently assign the intermediate responsible for the transient absorption spectrum to **9**, one of the isomers of **2** [11]. In fact, after the measurement of transient absorption spectra in O_2-saturated cyclohexane, the formation of bishomocubane **5** was confirmed. When oxygen atoms were inserted in the elongated bridge bonds, **5** can be generated. Additionally, **9** is also supported by the trapping experiments using alcohol or diene, both of which can trap the intermediate species containing unsaturated bonds.

Similar features were observed for octasilacubane **1**, which showed 370, 485, and 685 nm absorptions after photolysis. We are currently investigating more direct proof of the formation of **9**.

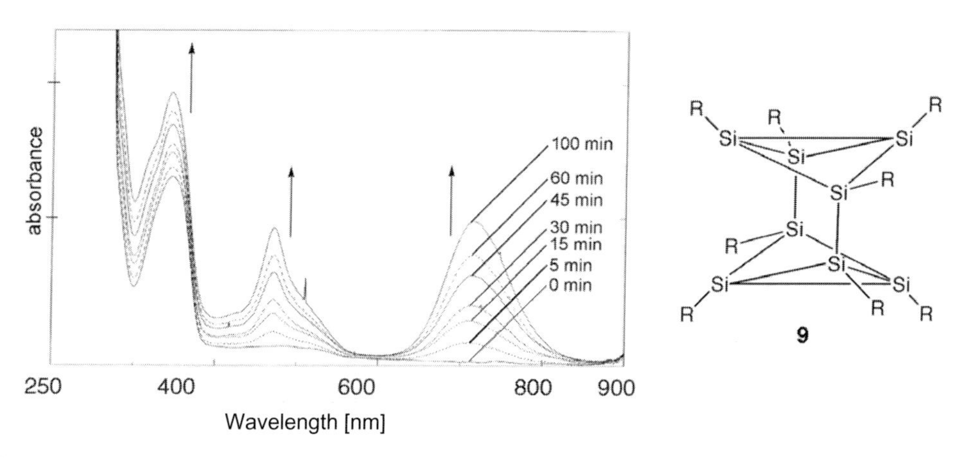

Fig. 3. Photolysis of octasilacubane **2** at 77 K in 3-methylpentane, and plausible intermediate **9**.

In summary, in addition to its aesthetic quality, octasilacubane is also unique for its unusual reactions ascribable to the strain and polysilane cage structure. We believe that these results are only a part of its diversity, and further investigation is now in progress.

Acknowledgments: The study of octasilacubane started at the laboratory of late Prof. Yoichiro Nagai, and we are much indebted to him. None of the research could have been realized without the efforts of the co-authors shown in the references. This work was supported by the grants from the Ministry of Education, Culture, Sports, Science and Technology of Japan and CREST-JST.

References

[1] P. E. Eason, T. W. Cole, Jr., *J. Am. Chem. Soc.* **1964**, *86*, 962 and 3157; Review: P. E. Eaton, *Tetrahedron* **1975**, *35*, 2189.

[2] Reviews: A. Sekiguchi, H. Sakurai, *Adv. Organomet. Chem.* **1995**, *37*, 1; A. Sekiguchi, S. Nagase, Polyhedral silicon compounds, in *The Chemistry of Organic Silicon Compounds* (Eds.: S. Patai, Z. Rappoport), Vol. 2, John Wiley & Sons, Inc., Chichester, **1998**, p. 119.

[3] a) H. Matsumoto, K. Higuchi, Y. Hoshino, H. Koike, Y. Naoi, Y. Nagai, *J. Chem. Soc., Chem. Commun.* **1988**, 1083; b) H. Matsumoto, K. Higuchi, S. Kyushin, M. Goto, *Angew. Chem. Int. Ed. Engl.* **1992**, *31*, 1354; c) K. Furukawa, M. Fujino, and N. Matsumoto, *Appl. Phys. Lett.* **1992**, *60*, 2744; d) A. Sekiguchi, T. Yatabe, H. Kamatani, C. Kabuto, H. Sakurai, *J. Am. Chem. Soc.* **1992**, *114*, 6260; e) A. Sekiguchi, T. Yatabe, C. Kabuto, H. Sakurai, *J. Am. Chem. Soc.* **1993**, *115*, 5853; f) N. Wiberg, C. M. M. Finger, K. Polborn, *Angew. Chem. Int. Ed. Engl.* **1993**, *32*, 1054.

[4] Ref. 3e and N. Wiberg, H. Auer, H. Nöth, J. Knizek, K. Polborn, *Angew. Chem. Ed. Engl.* **1998**, *37*, 2869; M. Ichinohe, N. Takahashi, A. Sekiguchi, *Chem. Lett.* **1999**, 553.

[5] H. Matsumoto, A. Sakamoto, Y. Nagai, *J. Chem. Soc., Chem. Commun.* **1986**, 1768; H. Matsumoto, N. Yokoyama, A. Sakamoto, Y. Aramaki, R. Endo, Y. Nagai, *Chem. Lett.* **1986**, 1643.

[6] M. Unno, T. Matsumoto, K. Mochizuki, K. Higuchi, M. Goto, H. Matsumoto, *J. Organomet. Chem.* **2003**, *685*(*1–2*), 156.

[7] M. Unno, T. Yokota, H. Matsumoto, *J. Organomet. Chem.* **1996**, *521*, 409.

[8] a) M. Unno, K. Higuchi, M. Ida, H. Shioyama, S. Kyushin, M. Goto, H. Matsumoto, *Organomet.* **1994**, *13*, 4633; b) M. Unno, H. Shioyama, M. Ida, H. Matsumoto, *Organomet.* **1995**, *14*, 4004.

[9] M. Unno, H. Masuda, H. Matsumoto, *Silicon Chem.* **2004**, *1*(*5*/*6*), 377.

[10] M. Unno, K. Higuchi, K. Furuya, H. Shioyama, S. Kyushin, M. Goto, H. Matsumoto, *Bull. Chem. Soc. Jpn.* **2000**, *73*, 2093.

[11] H. Horiuchi, Y. Nakano, T. Matsumoto, M. Unno, H. Hiratsuka, H. Matsumoto, *Chem. Phys. Lett.* **2000**, *322*, 33.

Chapter III

Transition Metals in Organosilicon-Based Chemistry

Transition Metals in Organosilicon Chemistry

T. Don Tilley

University of California, Berkeley
Department of Chemistry, Berkeley, California 94720-1460, USA
Tel.: +1 510 642 8939 — Fax: +1 510 642 8940
E-mail: tdtilley@socrates.berkeley.edu

Keywords: metal silyl, metal silylene, hydrosilation

Summary: This contribution describes two general approaches to the development of new transition metal-based silicon chemistry. In one approach, the focus is on early transition metals and the chemistry of d^0 metal-silicon bonds. A second approach targets metal-silicon multiple bonds, which are expected to exhibit rich reaction chemistry. Various new compounds and chemical processes are discussed.

Introduction

Our research in transition metal chemistry began with the expectation that transition metal reagents and catalysts should play a prominent role in the development of new silicon chemistry. This perception was largely based on analogies between silicon and carbon, and the great impact that transition metals have had in the evolution of carbon-based chemistry. In attempting to develop new chemical processes involving transition metal silicon chemistry, we have pursued two different strategies directed toward the exploitation of reactive metal-silicon bonds. One bond of this type involves the early transition metals, since d^0 metal-silicon single bonds are unsupported by metal-to-silicon π-backbonding and are therefore weaker than other transition metal-silicon bonds. A second type of reactive metal-silicon bond involves multiple bonding, as might exist in a silylene complex of the type $L_nM{=}SiR_2$.

Transition metals have already established a prominent role in synthetic silicon chemistry [1 – 5]. This is well illustrated by the Direct Process, which is a copper-mediated combination of elemental silicon and methyl chloride to produce methylchlorosilanes, and primarily dimethyldichlorosilane. This process is practiced on a large, worldwide scale, and is the basis for the silicones industry [6]. Other transition metal-catalyzed reactions that have proven to be synthetically useful include hydrosilation [7], silane alcoholysis [8], and additions of Si–Si bonds to alkenes [9]. However, transition metal catalysis still holds considerable promise for enabling the production of new silicon-based compounds and materials. For example, transition metal-based catalysts may promote the direct conversion of elemental silicon to organosilanes via reactions with organic compounds such as ethers. In addition, they may play a strong role in the future development of new silicon-based polymers. A more thorough understanding of the mechanisms by which metal complexes interact with and activate silicon compounds will advance progress toward the discovery of catalytic processes.

Early Metal (d^0) Silyl Complexes

Complexes that possess a d^0 metal-silicon bond were targeted as potentially reactive species that could activate small molecules and lead to new reaction pathways involving silicon. Such bonds were thought to represent a fundamentally new type of transition metal-silicon interaction, since the absence of d electrons at the metal corresponds to a lack of any metal-silicon π-bonding that would serve to strengthen the bond and make it correspondingly less reactive. In addition, given the more electropositive nature of the early transition metals, it was expected that d^0 metal-silicon bonds would exhibit a "reversed polarity", M(δ+)—Si(δ-), and react via nucleophilic transfer of the silyl group.

The first early metal silyl complex obtained in our laboratories, $Cp_2Zr(SiMe_3)Cl$ [10], was prepared by the reaction of Cp_2ZrCl_2 with Rösch's aluminum reagent $Al(SiMe_3)_3{\bullet}OEt_2$ [11]. Investigations with this complex, and related derivatives, indicated that in fact the metal-silicon bonds are relatively long and quite reactive. Thus, this compound was observed to undergo the first insertion of carbon monoxide to produce the silaacyl complex $Cp_2Zr(\eta^2\text{-}COSiMe_3)Cl$ [12]. This type of CO insertion reaction is quite general, and was used in the first synthesis of a formylsilane, $(Me_3Si)_3SiCHO$ [13]. For the tantalum silyl complex $Cp^*Ta(SiMe_3)Cl_3$ ($Cp^* = \eta^5\text{-}C_5Me_5$), insertions of ketones and aldehydes occur via nucleophilic transfer of silicon to carbon to give alkoxy species of the type $Cp^*Ta(OCRR'SiMe_3)Cl_3$ [14].

Early in our investigations on the reactivity of d^0 metal silyl compounds, it was observed that such species generally react rapidly with hydrogen to form the corresponding hydrides of the metal and silicon. For example, $Cp_2Zr(SiMe_3)Si(SiMe_3)_3$ undergoes hydrogenolysis (1 atm of H_2) within a minute at room temperature to give $[Cp_2ZrH_2]_n$, $HSiMe_3$, and $HSi(SiMe_3)_3$ [15]. For comparison, the analogous dimethyl zirconocene complex reacts extremely slowly (over days) with hydrogen under similar conditions. These results provided the first indication that d^0 M–Si bonds are highly reactive toward other σ bonds. By analogy to related and well-studied hydrogenolysis reactions involving d^0 M–C bonds, it is presumed that these processes occur via concerted, 4-center transition states in which the M–Si and H–H bonds are cleaved as the new M–H and Si–H bonds form (a process known as σ-bond metathesis). Subsequently, it became clear that this rich reactivity extends to Si–H bonds, and this theme achieved greater significance with the publication of a landmark paper by Harrod and coworkers on the Cp_2TiMe_2-catalyzed, dehydrogenative oligomerization of $PhSiH_3$ to $(SiHPh)_n$ polysilanes [16].

In studies on stoichiometric reactions of hydrosilanes with zirconocene and hafnocene derivatives containing M–Si or M–H bonds, it was found that σ-bond metathesis processes readily occur. Thus, 4-center transition states containing a d^0 metal center, hydrogen, and silicon easily form and mediate facile bond-making and bond-breaking processes involving these components.

For example, the hydride $CpCp^*HfHCl$ reacts with $PhSiH_3$ to form H_2 and $CpCp^*Hf(SiH_2Ph)Cl$, and the latter silyl derivative was observed to react with a second equivalent of $PhSiH_3$ to regenerate the hafnium hydride with formation of an Si–Si bond. Based on observations such as these, we proposed a mechanism for the dehydropolymerization of silanes to polysilanes, as catalyzed by early transition metal complexes, involving two σ-bond metathesis steps [17]. More recent investigations into the kinetic behavior of the polymerization of $PhSiH_3$ by $CpCp^*HfHCl$ revealed a rate law of the form: $-(d[PhSiH_3]/dt) = (K_{eq}k_2/[H_2])[Hf][PhSiH_3]^2$. This implies a mechanism involving reversible activation of silane by the hydride, and irreversible Si–Si bond formation (Scheme 1) [18].

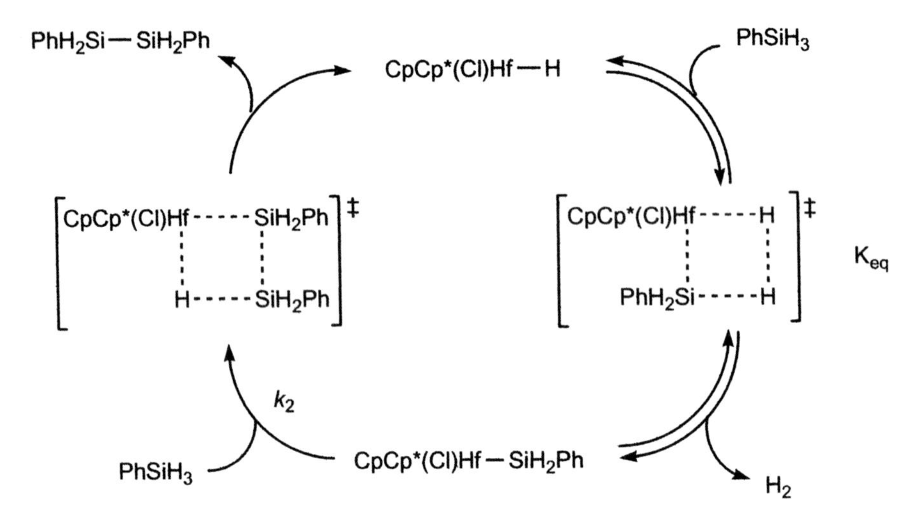

Scheme 1. Mechanism for the dehydropolymerization of $PhSiH_3$, as catalyzed by $CpCp^*HfHCl$.

This catalytic process illustrates the potential for σ-bond metathesis as the basis for new catalytic processes involving element-element bond formations. To explore such possibilities in the context of hydrocarbon activation and Si–C bond formation, we have examined the reactivity of cation-like metallocene species toward σ-bonds. This direction was suggested by the fact that cationic zirconocene derivatives of the type Cp_2ZrR^+ are the active species in olefin polymerization catalysis. As the latter process involves an electrocyclic, 4-center transition state similar to that leading to σ-bond metathesis, it seemed that related cationic species might be highly active in σ-bond activation processes. This was confirmed, for example, in studies on the reaction of $PhSiH_3$ with the cationic complex $[CpCp^*HfMe][MeB(C_6F_5)_3]$, which gives the corresponding hafnium hydride and $PhMe_2SiH$. This process, which appears to proceed via a 4-center transition state for transfer of a methyl group from hafnium to silicon, occurs over 3 h at room temperature. For comparison, the analogous neutral dimethyl complex $CpCp^*HfMe_2$ does not react with $PhSiH_3$ over the course of 5 days under analogous conditions. This higher reactivity for σ bonds in cation-like d^0 complexes is manifested in activation of the Si–C bond of $PhSiH_3$ upon reaction with

$CpCp^*(H)Hf(\mu\text{-}H)B(C_6F_5)_3$. This rather complex reaction involves the dehydrocoupling of $PhSiH_3$, as observed for analogous neutral catalysts, but also redistribution at silicon, which converts $PhSiH_3$ to SiH_4 and Ph_2SiH_2 via σ-bond metathesis [18].

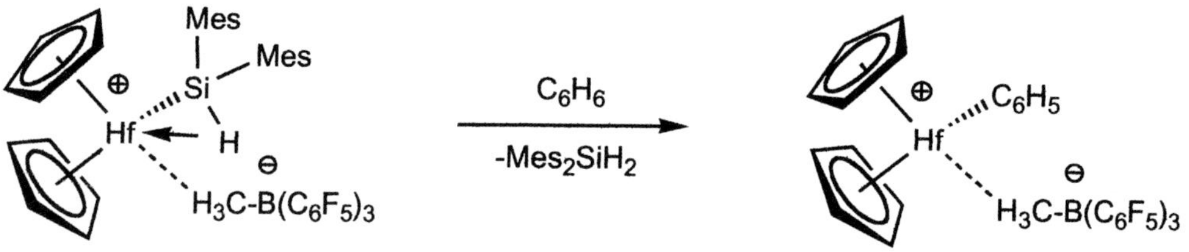

Eq. 1.

Additional investigations on the reactivity of cationic complexes have produced a silyl derivative which mediates a promising new type of C–H activation reaction. This new silyl complex, $Cp_2Hf(SiHMes_2)(\mu\text{-}Me)B(C_6F_5)_3$ (Mes = 2,4,6-$Me_3C_6H_2$), was obtained by reaction of $B(C_6F_5)_3$ with $Cp_2Hf(SiHMes_2)Me$. Interestingly, it appears to possess a strong α-H agostic interaction, which results in a reduced J_{SiH} coupling constant (57 Hz) and a low ν_{SiH} stretching frequency (1414 cm^{-1}). This silyl complex reacts smoothly with benzene at room temperature to produce the corresponding phenyl derivative, with elimination of Mes_2SiH_2 (Eq. 1). Kinetic and mechanistic experiments point to a 4-center transition state, which maintains the agostic interaction, for the C–H bond cleavage. Indeed, this agostic interaction appears to be crucial for the C–H activation, as related cationic species without this agostic interaction do not react with benzene [19].

A search for other highly reactive metal-silicon bonds focused on scandium derivatives that are isoelectronic with the cationic hafnium complex of Eq. 1. These efforts have produced the complexes $Cp^*_2Sc\text{-}SiH_2R$ (R = Mes, $SiPh_3$, $Si(SiMe_3)_3$), obtained by the reaction of Cp^*_2ScMe [20] with $RSiH_3$. These silyl complexes activate benzene, but more dramatically react with methane to produce $Mes(Me)SiH_2$ and Cp^*_2ScH. The latter reaction, which requires several days, appears to proceed through the intermediates Cp^*_2ScMe and $MesSiH_3$. However, current evidence indicates that the key C–H activation step occurs via the interaction of methane with the scandium hydride. In fact, as shown in Eq. 2, this hydride serves as a slow catalyst for the methylation of Ph_2SiH_2 with methane. It is believed that this catalysis proceeds via the activation of methane by the scandium hydride to produce Cp^*_2ScMe, which then transfers its methyl group to silicon [21].

$$Ph_2SiH_2 + CH_4\ (7\ atm) \xrightarrow[C_6D_{12}]{Cp^*_2ScH} Ph_2MeSiH + H_2$$

Eq. 2.

Transition Metal Silylene Complexes

Silylene ligands have been the focus of considerable interest and speculation in transition metal–silicon chemistry. For example, it is thought that silylene species occur as surface-bound intermediates in the Direct Process [6]. Silylene complexes have also been proposed as intermediates in metal-mediated redistribution reactions [22 – 24] and in metal-catalyzed oligomerizations and polymerizations of silanes [25, 26]. Our first success in obtaining silylene complexes produced a base-stabilized example via abstraction of a group from silicon in an electron-rich ruthenium silyl complex. The resulting product, $[Cp^*(Me_3P)_2RuSiPh_2(NCMe)][BPh_4]$, was characterized by X-ray crystallography [27]. Around the same time, base-stabilized silylene complexes of iron were reported by Zybill [28] and Ogino [29]. Whereas such species may reversibly dissociate the base to provide kinetic access to a base-free silylene complex, the ground state structures appear to involve considerable metal-silicon single-bond character, with a positive charge localized primarily on the silicon-bound base.

Cationic, base-free silylene complexes may also be obtained by the abstraction route, and early examples of such species include $[(Cy_3P)_2(H)Pt{=}Si(SEt)_2][BPh_4]$ [30], $[Cp^*(Me_3P)_2Ru{=}SiMe_2]$-$[BPh_4]$, and $[Cp^*(Me_3P)_2Ru{=}SiPh_2][BPh_4]$ [31]. A characteristic spectroscopic feature of such complexes is a highly downfield-shifted ^{29}Si NMR resonance; for example, the latter three derivatives exhibit ^{29}Si NMR shifts at 309, 299, and 311 ppm, respectively. More recently, we developed a route to neutral silylene complexes of the type $L_2Pt{=}SiMes_2$ (L = PCy_3, P^iPr_3), by trapping the photochemically-generated silylene :$SiMes_2$ with PtL_2 [32]. Now that stable, isolable silylenes are available, a variety of silylene complexes may be prepared by an analogous procedure involving the coordination of such species to a transition metal fragment [33, 34].

Reactivity studies on silylene complexes have revealed a number of unexpected and interesting transformations. For example, it was found that the reaction system generated by a combination of related silyl and silylene complexes, $Cp^*(Me_3P)_2RuSi(STol)_3$ + $[(\eta^5\text{-}C_5Me_4Et)(Me_3P)_2$-$Ru{=}Si(STol)_2]^+$, involves a bimolecular exchange of the thiolate groups at silicon. This process suggests a new silylene-based mechanism for the commonly observed, metal-catalyzed redistribution at silicon [35]. An investigation of the reactions of alkyl halides with the silylene complex $[Cp^*(Me_3P)_2Os{=}SiMe_2]^+$ revealed that silicon-halogen bond formation occurs via abstraction of a halogen atom by the metal-bound silylene ligand to produce an oxidized $OsSiMe_2X$ complex. Interestingly, this process involves the synergetic action of a transition metal redox center with a Lewis acidic silylene ligand in cleaving a carbon-halogen bond. Thus, a molecular species was shown to model a potentially key step in the Direct Process, which has previously been proposed to involve reaction of a metal-bound silylene fragment with the alkyl chloride [36].

Key questions regarding the reactivity of silylene complexes concern their potential role in metal-catalyzed transformations. For the participation of intermediate silylene complexes in a catalytic cycle, low-energy chemical pathways must exist for the conversion of simple silanes to silylene ligands via activation processes at the metal center. Most probably, a key step in such silylene-forming processes would be the α-migration of a group from silicon to the metal. In search of such a reaction, we prepared the methyl silyl complex shown in Eq. 3. This complex is quite stable, but it reacts readily with the Lewis acid $B(C_6F_5)_3$ to produce a cationic, silylene hydride. The abstraction of methide appears to generate a reactive, three-coordinate platinum complex that undergoes rapid migration to form the square-planar silylene complex (Eq. 3) [37]. Thus, the α-H migration is much more rapid for the complex with a less favored coordination geometry. This theme was also demonstrated for α-migration in an iridium system, which requires abstraction to generate a reactive, 5-coordinate intermediate. In this conversion, *fac*-Ir(H)(Me)($SiHMes_2$)(PMe_3)$_3$ reacts with $B(C_6F_5)_3$ to give [$(Me_3P)_3$(H)Ir=$SiMes_2$][MeBB$(C_6F_5)_3$] [38].

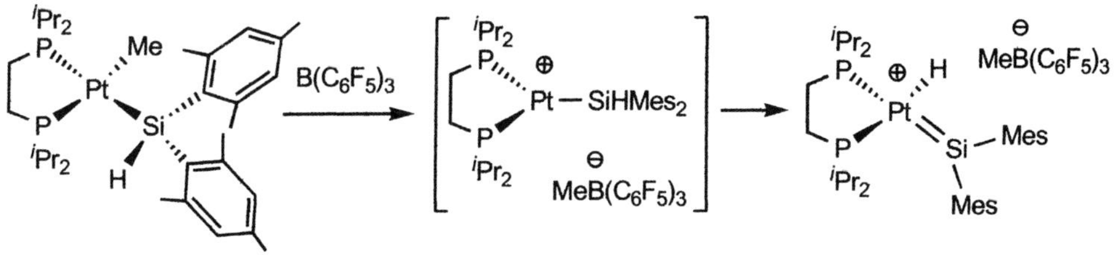

Eq. 3.

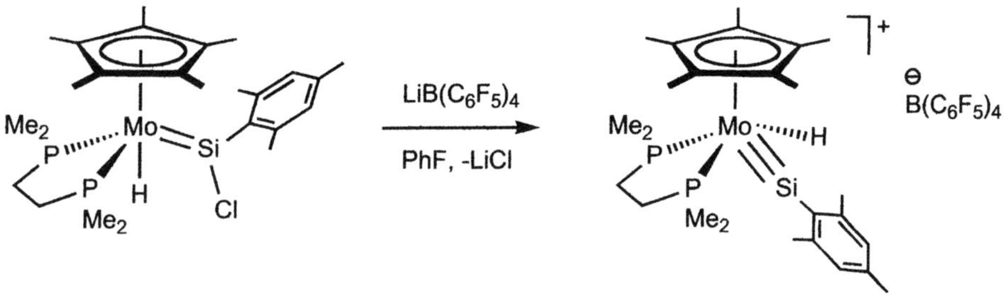

Eq. 4.

Chemical reactions related to those mentioned above involve the extrusion of a silylene ligand from a silane, and likely consist of two steps – an oxidative addition of an Si–H bond followed by α-H migration. An example of this process is seen in the reaction of a complex with a doubly-metalated Cp* ligand, [(η^7-$C_5Me_3(CH_2)_2$)(dmpe)W(H)$_2$][B$(C_6F_5)_4$], with hydrosilanes of the type R_2SiH_2 to produce the silylene complexes [Cp*(dmpe)H_2W=SiR_2][B$(C_6F_5)_4$] [39]. Other reactions of this type involve a metal complex with an organic group that promotes the net extrusion by functioning as a leaving group and removing hydrogen from the system. For example, the η^3-allyl complex [$PhBP_3$]Ir(H)(η^3-C_8H_{13}) (where [$PhBP_3$] = PhB$(CH_2PPh_2)_3^-$) reacts with secondary silanes to give silylene complexes of the type [$PhBP_3$](H)$_2$Ir=SiR_2 with loss of cyclooctene [38]. In addition, the molybdenum benzyl complex Cp*(dmpe)Mo(η^3-CH_2Ph) reacts with primary and secondary silanes via loss of toluene and formation of silylene complexes of the type Cp*(dmpe)Mo(H)(=SiRR'). Structural characterizations of these species, including a neutron diffraction study, indicate that the hydride ligand interacts weakly with the silicon center [40]. A novel silylene complex obtained by this method, Cp*(dmpe)Mo(H)(=SiClMes), possesses a chloride substituent at silicon. This chloride may be abstracted, and the resulting product appears to have considerable silylyne character, as indicated by a nearly linear Mo–Si–C arrangement (171°) and a low J_{SiH} coupling constant of 15 Hz (Eq. 4). However, this complex, [Cp*(dmpe)Mo(H)(≡SiMes)][B(C_6F_5)$_4$], may possess a weak, interligand H···Si interaction [41].

An initial attempt to develop a new type of catalytic reaction featuring a metal-mediated silylene extrusion focused on hydrosilations. This possibility was suggested by the potential for silylene complexes to participate in cycloaddition reactions, as indicated by the observation of metal-catalyzed transfers of silylene fragments to unsaturated carbon-carbon bonds [42 – 44]. A screening of several isolated silylene complexes for reactivity toward olefins revealed that the osmium silylene complex [Cp*(^{i}Pr$_3$P)(H)$_2$Os=Si(H)(TRIP)][B(C_6F_5)$_4$] (TRIP = 2,4,6-^{i}Pr$_3$C$_6$H$_2$) reacted cleanly with 1-hexene to give the hexyl-substituted silylene [Cp*(^{i}Pr$_3$P)(H)$_2$Os=Si(Hex)-(TRIP)][B(C_6F_5)$_4$]. This Si–C bond-forming reaction seems to be consistent with hydrosilation activity; however, numerous attempts to observe catalytic hydrosilations with osmium silylene catalysts proved futile.

The search for a silylene complex that functions as a hydrosilation catalyst was eventually successful with the synthesis of the base-stabilized silylene complex [Cp*(^{i}Pr$_3$P)(H)$_2$-Ru=Si(H)Ph•Et$_2$O][B(C_6F_5)$_4$]. This compound, which was obtained via the reaction of Cp*(^{i}Pr$_3$P)Ru(H)(Cl)(SiH$_2$Ph) with Li[B(C_6F_5)$_4$], possesses a loosely bound ether. The silylene complex reacts with 1-hexene to give [Cp*(^{i}Pr$_3$P)(H)$_2$Ru=SiPhHex] [B(C_6F_5)$_4$], and serves as a catalyst for the hydrosilation of various olefins. Interestingly, this catalysis is highly selective toward the silane substrate (only primary silanes undergo reaction), but is rather insensitive toward the steric properties of the olefin reactant. For this catalysis, the clean hydrosilation of difficult substrates such as 1-methyl-cyclohexene, the observed selectivity for monosubstituted silanes, the absence of unsaturated products, and the exclusive anti-Markovnikov regiochemistry of addition are unusual [45].

The mechanism of this hydrosilation seems to require a silylene ligand, since closely related complexes with no silylene ligand, such as Cp*(^{i}Pr$_3$P)(H)RuSi(H)Ph(OTf) and Cp*(^{i}Pr$_3$P)Ru(H)(Cl)(SiH$_2$Ph), do not catalyze this reaction. A mechanism based on Lewis acid catalysis [46], with the silylene center acting as the Lewis acid, appears to be ruled out by the exclusive *cis*-hydrosilation of 1-methylcyclohexene and the lack of reactivity for a range of secondary and tertiary silanes. A mechanism involving the [$2\pi + 2\pi$] cycloaddition of the metal–silicon bond with the alkene may be considered, but this seems unlikely, as the resulting ruthenium(VI) metallasilacyclobutane intermediate would possess a highly crowded conformation to provide the observed anti-Markovnikov regiochemistry. In addition, a Chalk-Harrod type mechanism, in which the silylene complex represents a resting state of the catalyst, does not explain the observed selectivity toward primary silanes and the insensitivity of the catalyst to steric properties of the alkene.

Given the considerations mentioned above, the mechanism of Scheme 2 is proposed. This mechanism features a concerted addition of the Si–H bond to the C=C bond of the substrate, in a manner analogous to the B–C bond-forming step in the hydroboration of alkenes. Indeed, three-coordinate, cationic silicon centers are isoelectronic with monomeric boranes (refer to the second resonance structure given in Scheme 2). The proposed mechanism involves the activation of two Si–H bonds of the silane substrate, direct addition of an (sp^2)Si–H bond to the alkene, and finally 1,2-H migration and reductive elimination steps [45].

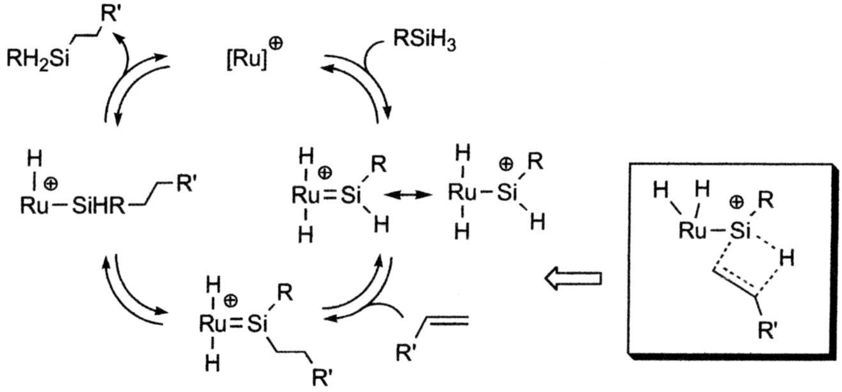

Scheme 2. Possible mechanism for a ruthenium-catalyzed hydrosilation.

Concluding Remarks

Transition metal-silicon chemistry will continue to play an important role in the development of new silicon-based technologies. A number of potential advances are possible, including the discovery of new catalysts that will allow efficient and selective Si–C bond-forming reactions using simple and readily available starting materials (such as methane and benzene). In general, one can envision numerous transformations at silicon for which selectivity (e.g., chemo-, regio-, and enantioselectivities) may be controlled with a transition metal reagent or catalyst. Along these lines, it may be possible to develop new, metal-based routes to silicon compounds from inexpensive starting materials such as silica or elemental silicon. New polymerizations to form silicon-containing materials will undoubtedly be based on transition metal catalysis. An emerging field of silicon chemistry involves the creation and use of nanostructures, and this activity is likely to expand considerably with the motivation to produce smaller and more advanced electronic devices. Certainly transition metals can play an important role in the development of new silicon-based nanoscience, and this potential is well illustrated by recent reports that gold nanoparticles may be used to produce nanowires of silicon [47].

Acknowledgment: The author wishes to thank the many talented and devoted coworkers who have contributed to the work described above. The National Science foundation is thanked for their financial support of this research.

References

[1] J. Y. Corey, J. Braddock-Wilking, *Chem. Rev.* **1999**, *99*, 175.
[2] H. K. Sharma, K. H. Pannell, *Chem. Rev.* **1995**, *95*, 1351.
[3] J. A. Reichl, D. H. Berry, *Adv. Organomet. Chem.* **1999**, *43*, 197.
[4] T. D. Tilley, Transition-metal silyl derivatives, in *The Silicon-Heteroatom Bond* (Eds.: Z. Rappoport, Y. Apeloig), J. Wiley and Sons, New York, **1991**, p. 245.
[5] M. S. Eisen, Transition-metal silyl derivatives, in *The Chemistry of Organic Silicon Compounds* (Eds.: S. Patai, Z. Rappoport), J. Wiley and Sons, New York, **1998**, p. 2037.
[6] *Catalyzed Direct Reactions of Silicon* (Eds.: K. M. Lewis, D. G. Rethwisch), Elsevier, Amsterdam, **1993**.
[7] I. Ojima, Z. Li, J. Zhu, Recent advances in the hydrosilylation and related reactions, in *The Chemistry of Organic Silicon Compounds* (Eds.: S. Patai, Z. Rappoport), J. Wiley and Sons, New York, **1998**, p. 1687.
[8] A. J. Chalk, *J. Chem. Soc., Chem. Commun.* **1970**, 847.
[9] M. Suginome, Y. Ito, *Chem. Rev.* **2000**, *100*, 3221.
[10] T. D. Tilley, *Organometallics* **1985**, *4*, 1452.
[11] L. Rösch, G. Altnau, *J. Organomet. Chem.* **1980**, *195*, 47.
[12] T. D. Tilley, *J. Am. Chem. Soc.* **1985**, *107*, 4084.
[13] F. H. Elsner, H.-G. Woo, T. D. Tilley, *J. Am. Chem. Soc.* **1988**, *110*, 5872.
[14] J. Arnold, T. D. Tilley, *J. Am. Chem. Soc.* **1987**, *109*, 3318.
[15] B. K. Campion, J. Falk, T. D. Tilley, *J. Am. Chem. Soc.* **1987**, *109*, 2049.
[16] C. Aitken, J. F. Harrod, E. Samuel, *J. Organomet. Chem.* **1985**, *279*, C11.
[17] a) H.-G. Woo, J. F. Walzer, T. D. Tilley, *J. Am. Chem. Soc.* **1992**, *114*, 7047; b) T. D. Tilley, *Acc. Chem. Soc.* **1993**, *26*, 22.
[18] A. Sadow, T. D. Tilley, *Organometallics* **2001**, *20*, 4457.
[19] A. Sadow, T. D. Tilley, *J. Am. Chem. Soc.* **2002**, *124*, 6814.
[20] M. E. Thompson, S. M. Baxter, A. R. Bulls, B. J. Burger, M. C. Nolan, B. D. Santarsiero, W. P. Schaefer, J. E. Bercaw, *J. Am. Chem. Soc.* **1987**, *109*, 203.
[21] A. D. Sadow, T. D. Tilley, *Angew. Chem., Int. Ed.* **2003**, *42*, 803.
[22] M. D. Curtis, P. S. Epstein, *Adv. Organomet. Chem.* **1981**, *19*, 213.
[23] K. H. Pannell, J. Cervantes, C. Hernandez, J. Cassias, S. Vincenti, *Organometallics* **1986**, *5*, 1056.
[24] T. Kobayashi, T. Hayashi, H. Yamashita, M. Tanaka, *Chem. Lett.* **1988**, 1411.
[25] T. D. Tilley, *Comments Inorg. Chem.* **1990**, *10*, 37.
[26] S. M. Katz, J. A. Reichl, D. H. Berry, *J. Am. Chem. Soc.* **1998**, *120*, 9844.
[27] D. A. Straus, T. D. Tilley, A. L. Rheingold, S. J. Geib, *J. Am. Chem. Soc.* **1987**, *109*, 5872.
[28] C. Zybill, G. Müller, *Angew. Chem., Int. Ed. Engl.* **1987**, *26*, 669.
[29] K. Ueno, H. Tobita, M. Shimoi, H. Ogino, *J. Am. Chem. Soc.* **1988**, *110*, 4092.
[30] S. D. Grumbine, T. D. Tilley, F. Arnold, A. L. Rheingold, *J. Am. Chem. Soc.* **1993**, *115*, 7884.
[31] S. D. Grumbine, T. D. Tilley, F. Arnold, A. L. Rheingold, *J. Am. Chem. Soc.* **1994**, *116*, 5495.

[32] a) J. Feldman, G. P. Mitchell, J.-O. Nolte, T. D. Tilley, *J. Am. Chem. Soc.* **1998**, *120,* 11184; b) J. D. Feldman, G. P. Mitchell, J.-O. Nolte, T. D. Tilley, *Can. J. Chem.* **2003**, *81*, 1127.
[33] M. Haaf, T. A. Schmedake, R. West, *Acc. Chem. Res.* **2000**, *33*, 704.
[34] B. Gehrhus, M. F. Lappert, *J. Organomet. Chem.* **2001**, *617*, 209.
[35] S. K. Grumbine, T. D. Tilley, *J. Am. Chem. Soc.* **1994**, *116,* 6951.
[36] P. W. Wanandi, P. B. Glaser, T. D. Tilley, *J. Am. Chem. Soc.* **2000**, *122*, 972.
[37] G. P. Mitchell, T. D. Tilley, *Angew. Chem. Int. Ed. Engl.* **1998**, *37*, 2524.
[38] a) J. C. Peters, J. D. Feldman, T. D. Tilley, *J. Am. Chem. Soc.* **1999**, *121,* 9871; b) J. D. Feldman, J. C. Peters, T. D. Tilley, *Organometallics* **2002**, *21*, 4065.
[39] a) B. V. Mork, T. D. Tilley, *J. Am. Chem. Soc.* **2001**, *123,* 9702; b) B. V. Mork, T. D. Tilley, *J. Am. Chem. Soc.* **2004**, *126*, 4375.
[40] B. V. Mork, T. D. Tilley, A. J. Schultz, J. A. Cowan, *J. Am. Chem. Soc.* **2004**, *126*, 10428.
[41] B. V. Mork, T. D. Tilley, *Angew. Chem. Int. Ed.* **2003**, *42,* 357.
[42] D. Seyferth, M. L. Shannon, S. C. Vick, T. F. O. Lim, *Organometallics* **1985**, *4*, 57.
[43] W. S. Palmer, K. A.Woerpel, *Organometallics* **1997**, *16*, 4824.
[44] K. Yamamoto, H. Okinoshi, M. Kumada, *J. Organomet. Chem* **1971**, *27*, C31.
[45] P. B. Glaser, T. D. Tilley, *J. Am. Chem. Soc.* **2003**, *125*, 13640.
[46] M. Rubin, T. Schwier, N. Gevorgyan, *J. Org. Chem.* **2002**, *67,* 1936.
[47] Y. Xia, P. Yang, Y. Sun, Y. Wu, B. Mayers, B. Gates, Y. Yin, F. Kim, H. Yan, *Adv. Mater.* **2003**, *15,* 353.

Hydrosilation (or is it Hydrosilylation?): A Personal Perspective on a Scientifically and Technologically Fascinating Chemical Methodology

John F. Harrod

Chemistry Department, McGill University,
801, Sherbrooke St. W., Montreal, QC Canada H3A 2K6
E-mail: john.harrod@McGill.ca

Keywords: hydrosilation, chiral catalyst, metal-catalyzed processes

The history of hydrosilation chemistry is more or less contemporaneous with the independent research career of the author. I will use the term "hydrosilation" throughout this piece since it was the term coined by Alan Chalk and me, in analogy to the already widely used "hydroboration". In the intervening years the term "hydrosilylation" has also been widely used. In this brief essay, the history of hydrosilation will be highlighted by references to the works of both the author and his collaborators, and to a number of works by others that are seen as milestones in the development of the field. The limitations of time and space compel the omission of reference to many significant contributions that have been made by others.

My road into silicon chemistry was one that was traveled without foreknowledge of the destination. It began with doctorate research on the recently announced topic of "Ziegler Catalysis". After three years of stumbling around in the labyrinth of archaeo-organometallic chemistry, I left Birmingham with a PhD degree and an enduring obsession to understand how transition metal compounds catalyze chemical reactions. Where better to start than the catalytic activation of hydrogen, the simplest of molecules? A newly arrived Lecturer at Birmingham, Dr. Brian Gowenlock, advised me to seek a post-doctoral position with a young professor at the University of British Columbia, Jack Halpern, who was building a reputation for himself on precisely that subject [1]. The period of my doctoral and post-doctoral studies coincided with an explosion of discovery and understanding in the areas of homogeneous transition metal catalysis and of transition organometallic chemistry in general. The discoveries of Ziegler catalysis and of ferrocene are often cited as the starting gates of this historical process. However, another important factor was the vast storehouse of raw material, accumulated in Germany before and during the Second World War, on useful conversions of coal derived raw materials such as acetylenes, CO, H_2, and olefins, which began to appear in the open literature by the early 1950s. This material was sequestered by the allied powers at the end of the war and stimulated intensive research on the catalytic chemistry of metal carbonyls.

My post-doctoral work was concerned mainly with establishing the mechanisms of the catalytic activation of H_2 by simple chloro-complexes of Pd, Rh, and Ru in aqueous medium. The latter reactions were studied using the reduction of some simple inorganic substrates, but we never lost sight of the fact that the big prize was the discovery of a simple homogeneous catalyst for the hydrogenation of organic substrates. We did in fact succeed in this goal, but only with some unsaturated carboxylic acids, such as maleic/fumaric acid [2]. The most significant aspect of this work for me was that it introduced me to the elementary catalytic processes of activation of a simple covalent bond, coordination of an unsaturated substrate to a metal hydride center, insertion of an unsaturated substrate into an M–H bond, displacement of a saturated product from a metal alkyl by reaction with H_2, and regeneration of the catalytic M–H species. All of these concepts, so excitingly novel at the time, now form the basis of much of the maturing field of coordination catalysis.

Shortly after I joined the Polymer and Interface Studies Section at the GE Research Laboratory in 1960, I attended a seminar given by Harry Lamoureux, of the Silicone Products Division of GE, on his efforts to find an alternative to the recently announced Dow Corning hydrosilation catalyst technology. My attendance at this seminar had two immediate results. In the first instance it fired up my enthusiasm for Si–H chemistry and in the second instance I met my colleague Alan Chalk for the first time. I was aware that Alan had completed a PhD under the supervision of Jack Halpern at U B C, but unaware that he had joined the Organic Chemistry Section at G E R L shortly before my arrival in Schenectady. After some excited discussions regarding the possible analogies between H_2 and Si–H chemistry, we asked for and were granted managerial approval to start a basic research project on reactions of Si–H compounds with transition metal complexes.

The cornerstone of the new Dow Corning technology was the discovery by John Speier, and his colleagues, of the extraordinary activity of chloroplatinic acid for the catalytic addition of Si–H compounds to olefins [3]. By 1960, the Speier group had completed, and published, a very thorough survey of many features of this family of reactions. However, in common with the overwhelming majority of organic chemists of the day, their unfamiliarity with the current developments in coordination chemistry constrained their ability to interpret the underlying catalytic chemistry. Our initial approach was to investigate the reactions of some simple silanes with a variety of late transition metal (TM) organometallic complexes and to test these complexes as catalysts. This quickly resulted in the discovery that olefin complexes of Pt^{II} and Rh^{I} are active catalysts for hydrosilations [4]. We also investigated the reactions of silanes with $Co_2(CO)_8$ in the expectation that this carbonyl dimer would cleave in the presence of silanes, by analogy with the reaction of H_2. This indeed appeared to be the case, and $Co_2(CO)_8$ also proved to be an effective catalyst for hydrosilylation of alkenes [5]. Our initial interpretation of the stoichiometry of the reaction of $RSiH_3$ with $Co_2(CO)_8$ (Eq. 1) was later shown by Marko et al. to be incorrect [6], but the production of $R_3SiCo(CO)_4$ still stands as the prototype of a reaction that has been widely used to generate Si–M complexes

Our thinking regarding the mechanism of Si–H activation was dramatically redirected by the appearance of the seminal communication by Vaska and DiLuzio in which they demonstrated the facile, reversible addition of H_2 to a square planar Ir^{I} complex to give an octahedral Ir^{III} dihydride

[7]. We immediately repeated this experiment with a variety of tertiary silanes instead of H_2 and observed a similar oxidative addition reaction. This provided the key to the proposition of a catalytic cycle, which explained the large body of qualitative observations made by the Dow Corning group, and has subsequently become known as the Chalk-Harrod (C-H) mechanism. At the same time as we were developing this mechanistic view of hydrosilation, Nesmeyanov and coworkers published a number of communications describing the hydrosilation of ethylene in the presence of an $Fe(CO)_5$ catalyst [8]. A very significant finding was the production of large amounts of vinylsilane product in addition to the "normal" ethylsilane product. The possibility of insertion of alkene into the M–Si bond, as well as into the M–H bond, was later incorporated into the C–H mechanism in order to account for the vinylsilane product in the Nesmeyanov reactions [9]. Eventually, a number of specific examples of such insertions in model reactions were reported and the M–Si insertion loop became known as the "Alternative Chalk-Harrod Mechanism" [10].

At the time we began work on hydrosilation, only a single compound containing a TM–Si bond had been reported. Today, there are over 500 characterized complexes described in the literature [11]. Our work with Co–Si and Ir–Si complexes not only provided an underpinning to proposed mechanisms of hydrosilation, but it also pointed out the potential of reactions between Si–H and TM complexes as a general synthetic route to TM–Si complexes. In addition, following up this chemistry eventually led the author to the discovery of a number of other interesting metal-catalyzed processes involving the TM–Si bond [12].

The hydrosilation of simple olefinic bonds is mature technology, and platinum-based catalysts (such as **1** [13] and **2** [14]) are now available of such a level of efficiency that the catalyst cost is

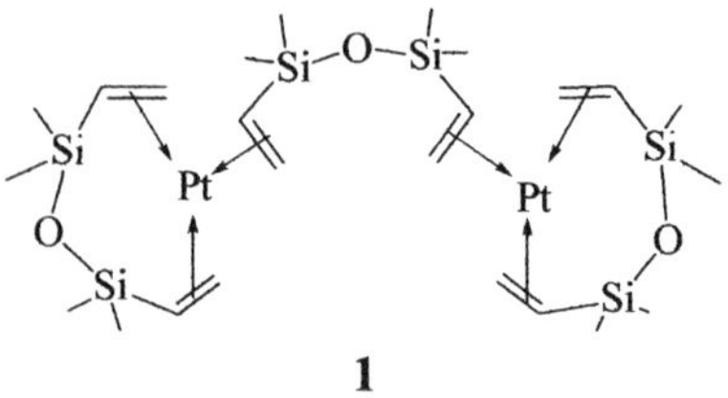

1 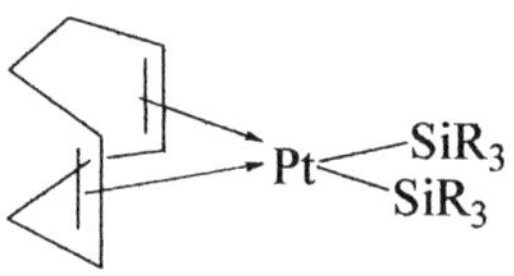 **2** 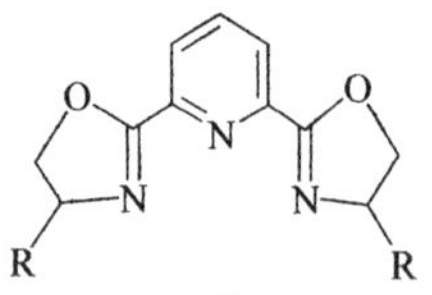 **3**

usually negligible compared to the value of the product. These platinum complexes are the catalysts of choice for the elaboration of polysiloxanes and for functionalization of other polymers. However, there is still a pressing need for a deeper understanding of how to control undesirable side reactions, such as olefin isomerization, and how to increase activity for olefinic derivatives with reactive functional groups.

In the 1970s, the potential value of asymmetric hydrosilation as a tool for organic synthesis began to interest a large number of researchers around the world, and the subject is still a lively one to the present day [15]. Although the first report of an attempted asymmetric hydrosilation of ketones used a chiral platinum catalyst, complexes of rhodium and ruthenium quickly became the focus of interest because of the easy synthesis of large families of catalyst precursors. Initially, chiral phosphine complexes were the main target, and many interesting results were reported, but by the 1980s it was evident that other classes of ligand, particularly those with coordinating N, could give higher rates and enantioselectivities. By the end of the decade, enantioselectivities of well over 90% had been achieved with the "pybox" ligand, **3**, and its analogs [16]. The utility of hydrosilations of ketones, ketimines, and olefins results from the simple and efficient protocols for converting the initial hydrosilation products to alcohols, or amines, with excellent retention of chirality.

A significant advance in the enantioselective hydrosilation of prochiral ketones and imines resulted from the development of C2 chiral titanium catalysts. Although several groups, including the author's, were active in the field, the most spectacular results were achieved by Buchwald and his coworkers [17]. They initially used BuLi to reduce $LTiCl_2$ (L = 1,2-Bis(tetrahydroindenyl)-ethane) as the catalyst. It was later shown that the cheap, air- and water-stable $LTiF_2$, which is activated directly by Si–H compounds, obviated the use of pyrophoric RLi reagents. These catalysts, in conjunction with optimization of reaction conditions give *ee*s well in excess of 90% with a broad range of aromatic ketones and imines. A general protocol for similarly enantioselective hydrosilation of fully aliphatic prochiral ketones and ketimines is yet to be realized.

The hydrosilation of prochiral olefins provides synthetic routes to both chiral organosilicon compounds and chiral alcohols. The latter are produced by a stereoselective oxidation of the former. Numerous attempts during the 1980s and early 1990s with Pt and and Rh complexes to hydrosilate simple olefins were generally disappointing. The major breakthrough came in 1994 when Hayashi et al. reported the very highly enantioselective Pd/MOP catalyst for these reactions [18].

$[Pd(\eta^3\text{-}C_3H_5)Cl_2$ (0.1%), (*S*)-(–)-MOP (0.02%); 2) EtOH, Et_3N; 3) H_2O_2

R = C_6H_{13} + $HSiCl_3$ → R–C(H)(OH)–Me

71% 95% *ee* (R)

OMe, PPh_2

S-(–)-MOP

For most of the 1980s I was preoccupied with the catalysis of various reactions of hydrosilanes by titanocene and zirconocene derivatives. Part of this work involved the synthesis and characterization of complexes of the general type $Cp'_2Ti(SiR_3)L$, where Cp′ is a ligand containing a cyclopentadienyl group and L is a classical donor ligand. This led to the synthesis of a number of titanocene silyl pyridine complexes. The preferred synthesis involved the co-reaction of Cp'_2TiMe_2, a pyridine, and a hydrosilane. A number of graduate students had carried out this reaction, but it took the sharp eye of my post-doc, Dr. (now Professor) Hee-Gweon Woo, to notice that unusual organic products were always present in the mother reaction liquor after isolation of the crystalline titanium complex. It did not take long to demonstrate that these new products were a mixture of *N*-silyl-di- and -tetrahydropyridines [19]. This discovery led us to a lot of interesting chemistry, but, for me, the most significant aspect was the fact that we had discovered the first example of a catalytic hydrosilation of a highly aromatic substrate. Even though we had the advantage of a coordinating nitrogen atom, it should be noted that the resonance stabilization energy of pyridine is almost equal to that of benzene. An interesting extension of this concept would be a hydrosilation of benzene itself. In this case, re-aromatization of the aryl group would probably occur by H transfer to yield an arylsilane, thus:

L_3SiH + (benzene) —Catalyst→ (L_3Si, H-substituted cyclohexadiene) —$-H_2$→ (L_3Si-benzene)

For the time being, our attention is focused in another direction. Results obtained for the hydrosilation of pyridines led us to investigate the reduction of nicotinamide. This study was also motivated by an interesting ring vs substituent specificity we observed for the hydrosilation of nicotinic and isonicotinic esters. Because the catalyst and solvent systems were incompatible with the unprotected NH_2 group, we turned to studying the hydrosilation of *N,N*-dialkylnicotinamides. It was clear from the outset that a facile reaction was occurring, but we were unable to identify what appeared to be a complex mixture of products. To simplify matters, we backed off and removed the pyridine nitrogen from the equation, i.e. we carried out the reaction with *N,N*-dialkylbenzamides. Again, a facile reaction was observed, and we were able to easily identify the unexpected product shown in the following equation [20]:

(benzamide, O, N, R^1, R^2, R^3) + $PhMeSiH_2$ —$[Cp_2TiF_2]$ or $[Cp_2TiMe_2]$, Toluene, 30 min; 80 °C→ (1,2-diamine: R^2–N(R^1), N(R^1)–R^2, R^3, R^3) + $H(PhMeHSiO)_nH$

85–96% *rac/meso* ~50/50

We were initially excited by the discovery of this completely new method for the synthesis of 1,2-diamines. Unfortunately, the stereoselectivity remained more or less unchanged by the many variations we tried on the substituents, the reactants, and the catalyst. We were forced to conclude that the reaction is most likely going by a radical mechanism and there was little hope of achieving significant stereocontrol.

A natural application for a new reaction for the formation of a C–C bond is the synthesis of a polymer, and we are presently exploring the prospect of using this chemistry to prepare unusual families of polymers. In the case of the *N,N*-dialkylbenzamides, the isolated yields of product are essentially quantitative, a prerequisite for obtaining a high molecular weight. However, our latest results with *N,N,N′N′*-tetramethylterephthaldiamide show that the reaction rate is much slower than expected, possibly because of a deactivating influence of the *p*-amide and *p*-aminomethyl substituents. In addition, there is a detectable side reaction of reduction of the carbonyl group to methylene, which is not observed under the same conditions with the dialkylbenzamides.

Nevertheless, we have observed oligomers of the type shown below:

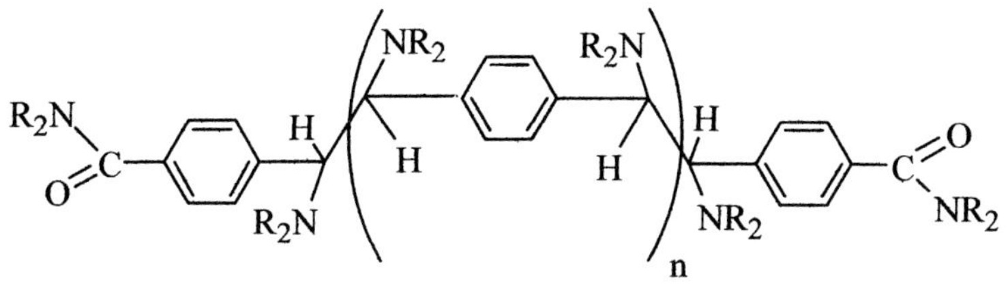

Samples of such oligomers with values of n up to 10 have been characterized by MS and NMR spectroscopies. This polymer, which would be very difficult to make by any other route, is expected to exhibit interesting chemical reactivity. For example, autoxidation could remove the H atoms from the bridging ethylene units to give a polyconjugated species. A similar result could be achieved by deamination of the backbone. Our further research into this chemistry will involve a search for more active and more selective catalysts and a study of the chemical properties of these polymeric 1,2-diamines.

Conclusions

Although it is now almost fifty years since Speier and his colleagues first announced the chloroplatinic acid-catalyzed hydrosilation of olefins, we are still far from complete control of the chemistry. A particular problem is the suppression of double bond migration. A solution of this problem will require a more detailed understanding of the factors affecting the relative rates of β-hydride elimination from an alkyl group and of the reductive elimination of Si–H from a platinum silyl hydride complex. Another factor which is poorly understood is suppression of the irreversible reduction of the platinum catalyst to Pt^0 metal. Both of these problems can greatly increase costs of production of certain products.

Hydrosilation chemistry is presently having, and will continue to have, an important impact on certain areas of materials chemistry. The rapidity and cleanliness of the reaction, and the stability of the resulting Si–C bonds, have been used to produce dendritic polymers and block copolymers with well-defined structures [21]. The reaction can also be employed to attach molecules to surface Si–H groups, notably the Si–H present on the surfaces of aqueous etched single crystal, or nanoparticular Si [22], or other siliceous substrates, with substantial modification to their physical and chemical properties [23].

Acknowledgments: The author thanks the organizers of the 2nd European Si Days for the invitation to present this brief history. Thanks also to my students, post-docs, and collaborators, and to all those in the industrial and academic world of Si chemistry who have been so supportive over the years.

References

[1] J. Halpern, *Ann. Rev. Phys. Chem.* **1965**, *16I*, 103.

[2] J. Halpern, J. F. Harrod, B. R. James, *J. Am. Chem. Soc.* **1961**, *83*, 753.

[3] J. Speier, *Adv. Organomet. Chem.* **1979**, *17*, 407.

[4] A. J. Chalk, J. F. Harrod, US. Patent 3 271 362 (**1966**).

[5] J. F. Harrod, A. J. Chalk, *J. Am. Chem. Soc.* **1965**, *87*, 1133.

[6] A. Sisak, F. Ungvăry, L. Markŏ, *Organometallics* **1986**, *5*, 1019.

[7] L. Vaska, DiLuzio, *J. Am. Chem. Soc.* **1962**, *84*, 679.

[8] R. K. Friedlina, E. C. Chukovskaya, J. Tsao, A. N. Nesmeyanov, *Doklady. Akad. Nauk. SSSR* **1960**, *132*, 374; A. N. Nesmeyanov, R. K. Friedlina, E. C. Chukovskaya, R. G. Petrova, A. B. Belavsky, *Tetrahedron* **1961**, *17*, 61.

[9] J. F. Harrod, A. J. Chalk, in: *Organic Syntheses via Metal Carbonyls, Vol II* (Ed.: Pino, Wender), Wiley, **1977**, p. 692.

[10] F. Seitz, M. S. Wrighton, *Angew. Chem. Int. Ed. Engl.* **1988**, *27*, 289; R. S. Tanke, R. H. Crabtree, *Organometallics* **1991**, *10*, 415; S. H. Bergens, P. N. Whelan, *J. Am. Chem. Soc.* **1992**, *114*, 2128; S. B. Duckett, R. N. Perutz, *Organometallics* **1992**, *11*, 90; T. R. Yasue, *Organometallics* **1996**, *15*, 2098; A. M. LaPointe, F. C. Rix, M. J. Brookhart, *J. Am. Chem. Soc.* **1997**, *119*, 906.

[11] J. Y. Corey, K. Braddock-Wilkin, *Chem. Rev.* **1999**, *99*, 175.

[12] J. F. Harrod, *Coord. Chem. Rev.* **2000**, *206*, 541.

[13] B. D. Karstedt, US. Patent 3 775 452 (**1973**).

[14] A. K. Roy, R. B. Taylor, *J. Am. Chem. Soc.* **2002**, *124*, 9510.

[15] H. Nishiyama, K. Itoh, in *Catalytic Asymmetric Synthesis, Second Edition* (Ed.: I. Ojima), Wiley-VCH, **2000**, Ch. 2.

[16] H. Brunner, A. Kürzinger, *J. Organomet. Chem.* **1988**, *346*, 413; H. Nishayama, H. Sakaguchi, T. Nakamura, M. Horihata, M. Kond, K. Itoh, *Organometallics* **1989**, *8*, 846.

[17] J. Yun, S. L. Buchwald, *J. Am. Chem. Soc.* **1999**, *121*, 5640; M. Hanson, S. L. Buchwald, *Org. Lett.* **2000**, *2*, 713.

[18] T. Hayashi, Y. Uozumi, *J. Am. Chem. Soc.* **1991**, *113*, 9887.

[19] R. Shu, J. F. Harrod, *Can. J. Chem.* **2001**, *79*, 1075.

[20] K. Selvakumar, J. F. Harrod, *Angew. Chem. Int. Ed.* **2001**, *40*(*11*), 2129.

[21] H. Lang, B. Luhrmann, *Adv. Mater.* **2001**, *13*(*20*), 1523.

[22] L. H. Lie, M. Duersin, E. M. Tuite, A. Houlton, B. R. Horrocks, *J. Electroanal. Chem.* **2002**, *538 – 539*, 183.

[23] J. J. Pesek, M. T. Matyska, *J. Chromatog.* **1996**, *736*, 255.

DFT Calculations on the Activation of Silanes by Platinum Complexes with Hemilabile P,N Ligands

Dietmar Sturmayr, Ulrich Schubert*

Institute of Materials Chemistry, Vienna University of Technology
Getreidemarkt 9/165, A-1060 Wien, Austria
Tel.: +43 1 58801 15320 — Fax: +43 1 58801 15399
E-mail: uschuber@mail.zserv.tuwien.ac.at

Keywords: metal silyl complexes, oxidative addition, reductive elimination

Summary: The primary steps in the reactions between hydrogeno- and chlorosilanes and platinum complexes $(P\cap N)Pt(X)CH_3$ ($X = CH_3$, H, Cl, SiH_3) with a hemilabile $P\cap N$ ligand were investigated by DFT calculations, especially the stereochemical outcome of the initial oxidative addition reaction of the silane and the relaxation of the Pt(IV) intermediate by a reductive elimination.

Introduction

We have shown in a series of articles that the reactivity of Pt(II) complexes towards silanes is greatly enhanced by hemilabile chelating ligands of the type κ^2-Ph_2P–Z–NR_2 (denoted as $P\cap N$; Z = $(CH_2)_n$, $C_6H_4CH_2$, etc.). Several interesting stoichiometric and catalytic reactions of organosilanes were observed which are not undergone by the corresponding complexes with chelating bis(phosphine) ligands κ^2-R_2P–Z–PR_2 [1 – 4].

We have analyzed the mechanisms of some of the observed reactions of the complexes $(P\cap N)PtXY$ by DFT calculations to shed light on the complex reactions occurring [4 – 6]. The calculations were done with simplified silanes (the organic substituents replaced by H) and $P\cap N$ ligands simplified by either $H_2PCH_2CH_2NH_2$, or by $PH_3 + NH_3$ in place of the chelating ligand (this simplification does not affect the outcome of the calculations very much). Each of the structures shown in the following schemes was optimized.

The most stable isomer of the complexes $(P\cap N)PtXY$ is the one in which the ligand with the smaller trans influence is trans to the PR_3 ligand [7]. Thus, the thermodynamically preferred isomers of the methyl-substituted complexes $(P\cap N)Pt(X)CH_3$ are

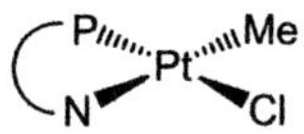

for X = Cl, and 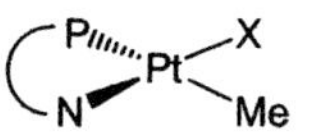 for X = H, CH_3 or SiH_3.

Results and Discussion

In this article we concentrate on the first step of the reactions, *viz.* the initial reaction of silanes with the square-planar Pt(II) complexes (P⌒N)Pt(X)CH_3 (X = CH_3, H, Cl, SiH_3). Only stereochemical aspects of the initial oxidative addition/reductive elimination reactions (Scheme 1) are considered in this contribution. Energetic aspects and details of the reaction mechanisms were discussed elsewhere [4 – 6]. Some of the conversions of the Pt(II) complexes shown in Scheme 1 are endothermic, and therefore only occur if succeeded by another energy-releasing reaction, for example in a catalytic cycle.

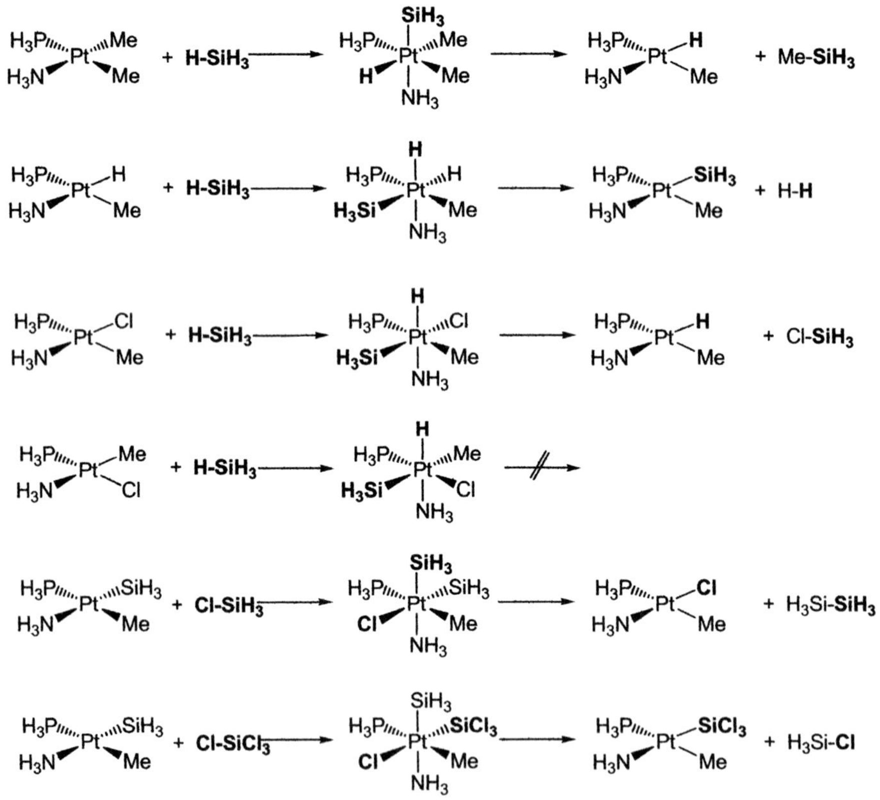

H_3P, H_3N, Pt(Me)(Me) + **H-SiH$_3$** → H_3P, **H**, Pt(**SiH$_3$**)(Me)(Me), NH_3 → H_3P, H_3N, Pt(**H**)(Me) + **Me-SiH$_3$**

H_3P, H_3N, Pt(H)(Me) + **H-SiH$_3$** → H_3P, **H$_3$Si**, Pt(**H**)(H)(Me), NH_3 → H_3P, H_3N, Pt(**SiH$_3$**)(Me) + H-**H**

H_3P, H_3N, Pt(Cl)(Me) + **H-SiH$_3$** → H_3P, **H$_3$Si**, Pt(**H**)(Cl)(Me), NH_3 → H_3P, H_3N, Pt(**H**)(Me) + Cl-**SiH$_3$**

H_3P, H_3N, Pt(Me)(Cl) + **H-SiH$_3$** → H_3P, **H$_3$Si**, Pt(**H**)(Me)(Cl), NH_3 ↛

H_3P, H_3N, Pt(SiH$_3$)(Me) + **Cl-SiH$_3$** → H_3P, **Cl**, Pt(**SiH$_3$**)(SiH$_3$)(Me), NH_3 → H_3P, H_3N, Pt(**Cl**)(Me) + H$_3$Si-**SiH$_3$**

H_3P, H_3N, Pt(SiH$_3$)(Me) + **Cl-SiCl$_3$** → H_3P, **Cl**, Pt(SiH$_3$)(**SiCl$_3$**)(Me), NH_3 → H_3P, H_3N, Pt(**SiCl$_3$**)(Me) + H$_3$Si-**Cl**

Scheme 1. The investigated reactions. The given structures represent the optimized geometries. The groups introduced by the precursor silanes are drawn in bold letters.

The experimental findings are as follows:

- The reaction of (P⌒N)PtMe_2 with $HSiR_3$ ($HSiEt_3$ or $HSiMePh_2$) results in the initial formation of (P⌒N)Pt(Me)H and $MeSiR_3$ (1st reaction in Scheme 1). The hydrido complex (P∩N)Pt(Me)H then reacts with a second equivalent of the silane to give (P∩N)Pt(Me)SiR_3 and H_2 (2nd reaction in Scheme 1). DFT calculations (with the simplified ligand and silane) show that the first reaction is exothermic by 9.7 kcal/mol and the second by 6.9 kcal/mol [4].

- No reaction is observed between *trans*-(P∩N)Pt(Cl)Me (Cl trans to P) and $HSiR_3$ (4th reaction in Scheme 1). DFT calculations show that oxidative addition of the silane would be endothermic by 16.4 kcal/mol, and – more important – no feasible reductive elimination pathway was found. However, it was shown that the energetically less favored isomer *cis*-(P∩N)Pt(Cl)Me (Cl *cis* to P) is the active species in catalytic C-Cl/Si-H exchange reactions. The energy difference between the cis and trans isomer is 9.0 kcal/mol. The initial step of the catalytic cycle is the formation of (P∩N)Pt(Me)H from SiH_4 and *cis*-(P∩N)Pt(Cl)Me (3rd reaction in Scheme 1) [3, 6].
- The reaction of (P∩N)Pt(Me)SiH_3 and H_3SiCl (5th reaction in Scheme 1) was postulated as the disilane-forming step in the complex reaction of (P∩N)$PtMe_2$ with $Me_2PhSiCl$, eventually leading to (P∩N)Pt$(Me)_3$Cl, (P∩N)Pt(Me)Cl and $Me_4Ph_2Si_2$ as the final products [2, 5].
- The reaction of $(H_3P)(H_3N)$Pt(Me)SiH_3 and $SiCl_4$ (6th reaction in Scheme 1) results in an exchange of the silyl ligand and is slightly exothermic (–1.5 kcal/mol). Although this particular reaction was not investigated in practice, similar exchange reactions were previously observed for $(R_3P)_2Pt(SiPhMe_2)_2$ + $HSiR_3$ [8].

The theoretical calculations show that in the first five reactions the most stable Pt(IV) intermediate results from a cis oxidative addition of the silane. In the reaction of $(H_3P)(H_3N)$Pt(Me)SiH_3 with $SiCl_4$, the initially formed transition state with cis Cl and $SiCl_3$ relaxes to the trans compound. The energy profile (with the nitrogen ligand de-coordinated) is shown in Fig. 1.

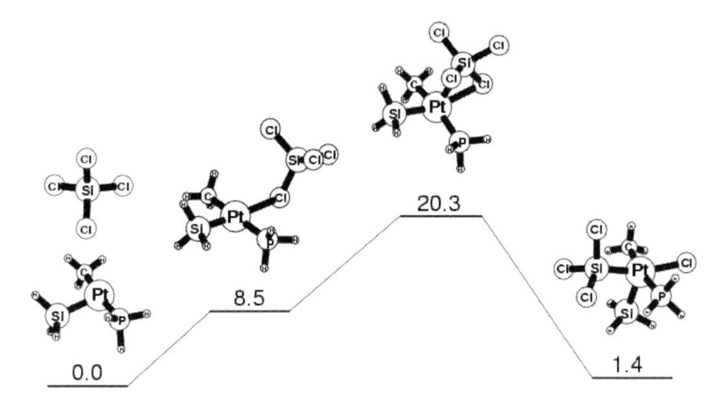

Fig. 1. Reaction profile for the oxidative addition of $SiCl_4$ to (H_3P)Pt(Me)SiH_3 (energies in kcal/mol). Reprinted with permission of Wiley-VCH from Ref. [5].

The stereochemical outcome of the *oxidative addition reactions* can be rationalized, in a simplified manner, by the different trans influence of the involved ligands (PR_3~SiR_3~H>CH_3>>Cl >NR_3):

- In each case the geometry of the T-shaped (H_3P)Pt(Me)X fragment of the starting complex is retained. In the cis oxidative additions, the nitrogen donor atom moves to an axial position in the formed octahedral Pt(IV) complex. As a consequence, the group being trans to P in the starting complex (Cl or CH_3) stays in that position, even if another ligand with a weaker trans influence (in the case of the Si-Cl additions) becomes available by the oxidative addition.
- In the resulting octahedral Pt(IV) complex, the ligand with the second largest trans influence (H or SiR_3) is located trans to the nitrogen donor (P with the largest trans influence cannot be trans to N, because both atoms are tied together by virtue of the P⌒N ligand).

The Pt(IV) complexes formed by oxidative elimination of the silane then undergo a *reductive elimination reaction* by which a new Pt(II) complex is formed. Note again that the resulting Pt(II) complex need not necessarily be stable but instead may be another intermediate in a more complex reaction. It is remarkable that in the overall reaction, i.e. the substitution of X in (P⌒N)Pt(Me)X by Y, the T-shaped (P⌒N)PtMe fragment is always retained. Starting with T-shaped (P⌒N)PtCl (4th reaction in Scheme 1) does not result in an overall substitution. Most eliminations are cis, and only the elimination of H_3SiCl in the 3rd reaction is trans. While all oxidative additions are possible without de-coordination of the nitrogen donor atom, the reductive eliminations proceed via 5-coordinate intermediates formed by de-coordination of the nitrogen donor atom. A typical example is shown in Fig. 2.

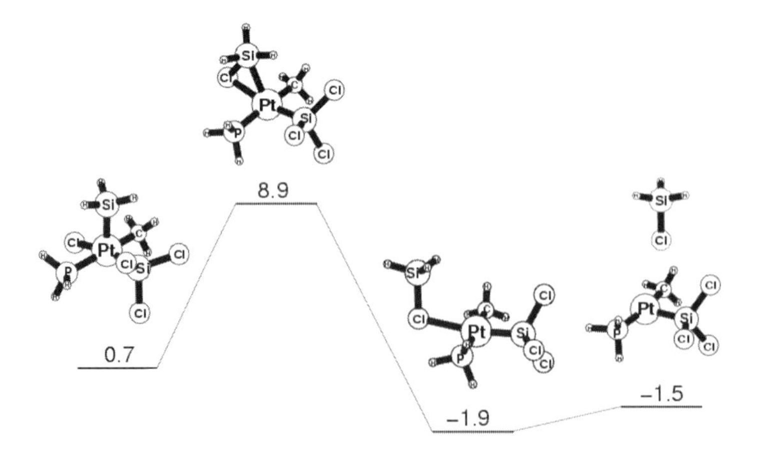

Fig. 2. Reaction profile for the reductive elimination of H_3SiCl from the 5-coordinate intermediate formed by elimination of the N donor (energies in kcal/mol). Reprinted with permission of Wiley-VCH from Ref. [5].

Conclusion

The main influence of the P∩N ligand in the oxidative addition reactions of the silanes to complexes of the type (P∩N)Pt(X)CH_3 is to direct the incoming ligands in certain positions, i.e. the stereochemistry of the resulting Pt(IV) complexes is mainly determined by the different trans influence of the P and N donor atoms of the P∩N ligand. The subsequent reductive elimination reactions only proceed after de-coordination of the nitrogen atom, i.e., the hemilabile nature of the P∩N ligand is more important for this step of the overall reaction. The combination of the reversible de-coordination of the amino group and the different trans effect of the phosphorus and nitrogen donor group is responsible for the unique reactivity of the P∩N-substituted Pt(II) complexes toward silanes.

Acknowledgment: This work was supported by the Fonds zur Förderung der wissenschaftlichen Forschung.

References

[1] U. Schubert, J. Pfeiffer, F. Stöhr, D. Sturmayr, S. Thompson, *J. Organomet. Chem.* **2002**, *646*, 53, and references cited therein.
[2] F. Stöhr, D. Sturmayr, G. Kickelbick, U. Schubert, *Eur. J. Inorg. Chem.* **2002**, 2305.
[3] F. Stöhr, D. Sturmayr, U. Schubert, *Chem. Commun.* **2002**, 2222.
[4] S. M. Thompson, F. Stöhr, D. Sturmayr, G. Kickelbick, U. Schubert, *J. Organomet. Chem.* **2003**, *686*, 183.
[5] D. Sturmayr, U. Schubert, *Eur. J. Inorg. Chem.* **2004**, 776.
[6] D. Sturmayr, U. Schubert, *Eur. J. Inorg. Chem.* **2002**, 2658.
[7] D. Sturmayr, U. Schubert, *Monatsh. Chem.* **2003**, *134*, 791.
[8] U. Schubert, D. Kalt, H. Gilges, *Monatsh. Chem.* **1999**, *130*, 207; D. Kalt, U. Schubert, *Inorg. Chim. Acta* **2000**, *306*, 211.

Hydrosilylation of Ethylene, Cyclohexene, and Allyl Chloride in the Presence of Pt(0) and Pt(2) Catalysts

E. A. Chernyshev, Z. V. Belyakova, S. P. Knyazev, P. A. Storozhenko*

State Research Institute for Chemistry and Technology of Organoelement Compounds,
38, Shosse Entuziastov, 111123, Moscow, Russia
Tel.: +7 95 2734953 — Fax: +7 95 9132538
E-mail: eos@eos.incotrade.ru

Keywords: hydrosilylation, catalysts, ethylene, cyclohexene, allyl chloride

Summary: Hydrosilylation of ethylene, cyclohexene, and allyl chloride was studied. A Pt(0) catalyst was shown to be more effective than one based on Pt(2). Modeling of Pt(2) and Pt(0) complexes with olefin ligands in a GAUSSIAN 98 program modeling calculation using Lan L2DZ/B3LYP basis was carried out. As a methylsilane molecule approaches the catalytic complexes, the variation in general energy dependence on Pt–H bond length was found. The potential barrier is lower with the Pt(0) complex than with the Pt(2) complex. A potential energy minimum on the surface was found that indicates a higher activity of the Pt(0) complex than that of the Pt(2) complex in the hydrosilylation reaction.

It is well known that hydrosilylation of ethylene by trichloro- and methyldichlorosilanes easily proceeds in the presence of Speier catalyst (SC), wherein platinum is in the divalent state, at room temperature with practically quantitative adduct yield. Under similar conditions, triethoxysilane does not bond to ethylene.

Ethylene hydrosilylation with triethoxysilane has not yet been described in the literature. There are data concerning ethylene and propylene hydrosilylation in the presence of rhodium catalyst $[RhCl(CO)_2]_2$. However, it is known that a side reaction, dehydrogenating silylation, is typical for rhodium catalysts (Scheme 1).

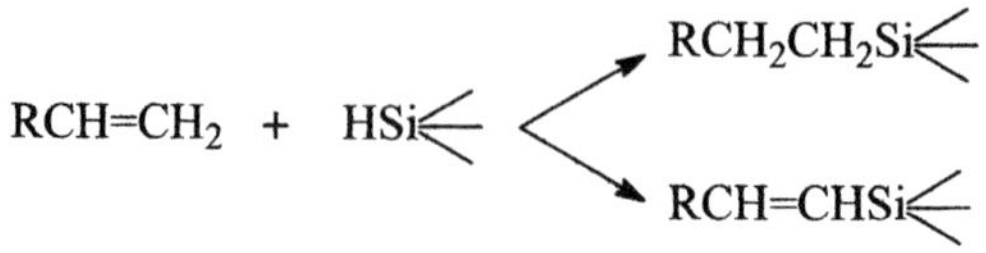

$$RCH{=}CH_2 + HSi\lessdot \;\rightarrow\; \begin{cases} RCH_2CH_2Si\lessdot \\ RCH{=}CHSi\lessdot \end{cases}$$

Scheme 1.

Thus, for example, when the reaction proceeds in xylene at 148 – 225 °C, the $EtSi(OEt)_3$ yield amounts to only 28%, but of $VinSi(OEt)_3$ it is 62% [1]. At 20 °C the $EtSi(OEt)_3$ yield is 95%, but the synthesis time is 65 h.

We have studied as catalysts: (1) Speier catalyst (SC), a 0.1 M solution $H_2PtCl_6{\cdot}H_2O$ in isopropyl alcohol; (2) carbenium salts (CS), $[(Me_2N)_2CCl]_2^{+}PtCl_6^{-}$; (3) Karstedt catalyst (KC), being a Pt(0) complex with tetramethyldivinyldisiloxane (**1**); (4) catalyst produced from hexavinyldisiloxane and $H_2PtCl_6{\cdot}H_2O$, being a Pt(0) complex with hexavinyldisiloxane of general formula possibly **2** (GC); (5) catalyst on a support (C/S) produced by CS interaction with porous glass treated with diallylaminopropyltriethoxysilane.

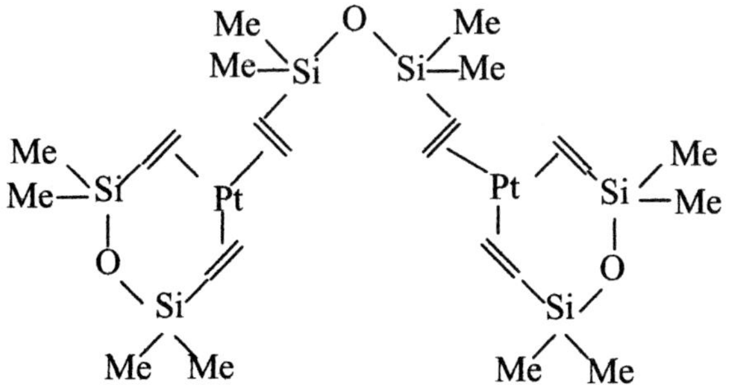

1 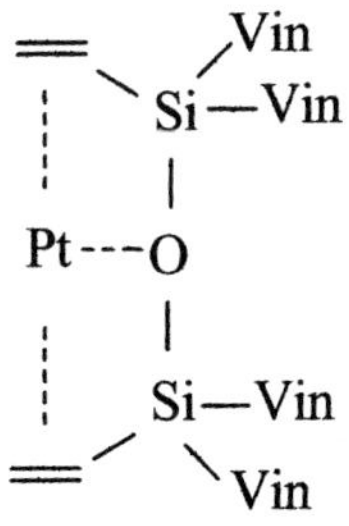 **2**

In the presence of SC the synthesis did not proceed at temperatures below 80 °C.

In the presence of CS the reaction did not begin at 25, 40, 50, 60, or 70 °C. At 80 – 85 °C instant catalyst conversion from yellow insoluble salt to brown solution, evidently of a Pt(0) complex, takes place. On this catalyst, ethylene was adsorbed to full triethoxysilane conversion. The catalyst proved to be stable in the presence of ethylene, thus testifying to the combined character of the compound. The next day, when a new portion of triethoxysilane was added to the previous synthesis, the reaction proceeded quantitatively, but a little more slowly (lower platinum concentration). In the absence of ethylene on the next day a black residue of platinum metal precipitated where synthesis did not proceed.

In the presence of catalyst on the support, evidently being a platinum complex with diallylamine fixed on the porous glass, the reaction did not proceed at 70 – 80 °C, but at 90 – 100 °C ethylene was actively absorbed and the ethyltriethoxysilane content in the mixture amounted to 94.3%. When the synthesis was repeated on the same catalyst, the reaction proceeded more slowly, but the triethoxysilane was completely consumed.

In the presence of KC the reaction began at ambient temperature 2 – 5 min after provision of the ethylene (or propylene) supply and finished in 2.5 – 3 h. The temperature increased to 70 °C. In the presence of GC the reaction proceeded in a similar way.

Starting with absence of triethoxysilane in the reaction mixture in all syntheses the final ethyltriethoxysilane content amounted to 94.1 – 95.8%. In neither synthesis was vinyltriethoxysilane found, a dehydrogenating silylation product, either by GLC or NMR methods, which testifies to the absence of a dehydrosilylation side reaction.

On the basis of the obtained results we can conclude that ethylene hydrosilylation reaction by triethoxysilane proceeds on Pt(0) compounds.

Ethylene hydrosilylation with trimethoxysilane readily proceeds on KC and GC catalysts, synthesis being complete in 1.5 – 2 h, the ethyltrimethoxysilane content in the mixture being 66 – 76%. Vinyltrimethoxysilane was not identified in either experiment.

Cyclohexene hydrosilylation is not as easy as that of linear olefins. For comparison of cyclohexene and hexene activity we carried out competitive hydrosilylation of these unsaturated compounds by trichlorosilane. After boiling the mixture for 16 h in the presence of GC, the hexyltrichlorosilane yield was 57.6% and the cyclohexyltrichlorosilane yield was 0.6%.

We studied cyclohexene hydrosilylation by trichloro- and methyldichlorosilane in the presence of SC, KC and GC. In SC, after boiling for 5 h, no addition actually proceeded. When the mixture is boiled in KC and GC for 1.5 h, the yield amounts to 70 – 80%. In the case of prolonged standing (1 – 1.5 months) at room temperature, adduct yields amount to 93 – 95%.

It is known that allyl chloride hydrosilylation (1) is followed by side reduction reaction (2) (Scheme 2).

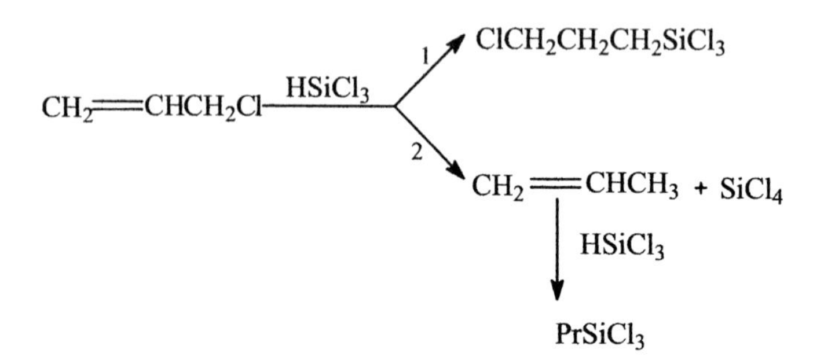

Scheme 2.

For trichlorosilane, the ratio of the products of addition and reduction (A/R) in the presence of SC is equal to approximately 5.0. However in the presence of platinum complexes with tetramethyldivinyldisiloxane or hexavinyldisiloxane, the A/R ratio is increased to about 12 – 16, and in contrast to SC the liberated propylene is quantitatively converted to propyltrichlorosilane, and the γ-chloropropyltrichlorosilane yield increases from 62 to 67 – 75%

In order to explain the obtained results, Pt(2) and Pt(0) activities were computed in GAUSSIAN 98 software. Methylsilane was chosen as the hydrosilane for simplicity. Pt(2) – Zeise salt complex approach coordinates with methylsilane were calculated in B3LYP approximation with the employment of LanL2MB basis, which is intended for the computation of compounds containing heavy element atoms.

With the purpose of evaluating the correctness of the chosen method, we compared Zeise salt calculation results with literature data of X-ray diffraction analysis. A simplified Karstedt catalyst variant-Pt(0) complex with three ethylene molecules was taken as a model for the calculation of the Pt(0) complex when approached by methylsilane. We then compared the calculation results for the said complex with literature data for Karstedt catalyst X-ray diffraction analysis. Differences from calculated results do not exceed 5 – 7%, i.e. they are within the experiment uncertainty (10%). Data obtained in the results of Pt(2) and Pt(0) approached by methylsilane are presented in the Fig. 1.

The plot demonstrates that when the Pt(2) complex approach coordinates with methylsilane are calculated, the Pt–H bond is stronger and the complex energy is lower. Therefore we can conclude that Si–H bond activation along this coordinate does not result in intermediate compound formation.

The potential minimum of the full potential energy surface was determined during the calculation of the Pt(0) complex approach coordinates with methylsilane.

On the basis of the obtained data we can say that the Pt(0) complex requires less activation energy for the Si–H bond. Therefore the Pt(0) complex is more active in hydrosilylation reactions.

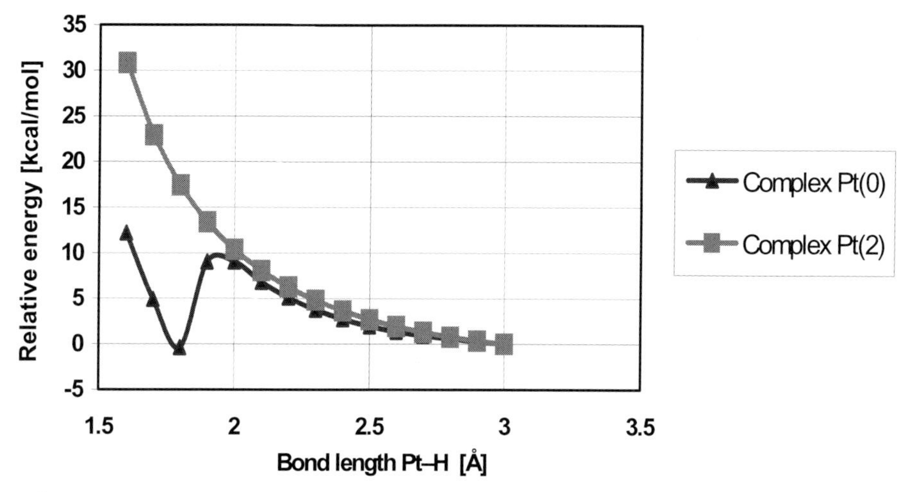

Fig. 1. Energy of Pt(2) and Pt(0) complex with methylsilane as a function of the Pt–H bond length.

Acknowledgment: This research was funded by the President of RF for young scientists and a leading scientific school (grant SS-1811.2003.3).

References

[1] US Patent 4 668 812 (**1987**).

[2] M. Yu. Gol`tsev, *Complex Organosilicon Platinum-Containing Catalyst for Fast-Curing Compositions*, Author's abstract of his Thesis, Moscow (**2001**).

Synthesis of Glycidoxypropyl-Silanes and -Siloxanes via Rhodium Siloxide-Catalyzed Hydrosilylation

Hieronim Maciejewski, Bogdan Marciniec, Paulina Błażejewska-Chadyniak*

Adam Mickiewicz University, Faculty of Chemistry
Grunwaldzka 6, 60-780 Poznań, Poland
Tel.: +48 61 8291366 — Fax: +48 61 8291508
E-mail: metalorg@amu.edu.pl

Izabela Dąbek

Poznań Science and Technology Park, AMU Foundation
Rubież 46, 61-612 Poznań, Poland
Tel.: +48 61 8227369 — Fax: +48 61 8227369
E-mail: centech@amu.edu.pl

Keywords: hydrosilylation, rhodium-siloxide complexes, glycidoxypropyltrialkoxysilane, epoxy-functional (poly)siloxane

Summary: Rhodium siloxide complexes such as dimeric $[\{Rh(cod)(\mu\text{-}OSiMe_3)\}_2]$ (**1**), monomeric phosphine-containing $[Rh(cod)(PCy_3)(OSiMe_3)]$ (**2**), $[Rh(cod)(PPh_3)(O\text{-}SiMe_3)]$ (**3**) and non-phosphine-containing $[Rh(cod)(OSiMe_2CH{=}CH_2)]$ (**4**) appeared to be active catalysts (even at room temperature) of hydrosilylation of allyl glycidyl ether by triethoxysilane and hydro(poly)siloxanes to yield glycidoxypropyltrialkoxy-silanes, which are important silane coupling agents or epoxy-functional polysiloxanes valuable for the synthesis of organic block copolymers and ionic surfactants.

Glycidoxypropyl-Silanes and -Siloxanes

Epoxy-functional carbosilanes are some of the most important silane coupling agents. The main representative of this group is 3-glycidoxypropyltrialkoxysilane [1]. The epoxy ring reacts with many organic functionalities, and therefore these compounds are suitable as adhesion promoters in polysulfide, urethane, epoxy, and acrylic caulks. On the other hand, the epoxy functionality is stable in waterborne environments, which permits the use of glycidoxypropylsilane for the production of waterborne acrylic and vinylic sealants [2]. Because of trialkoxy functionality, the polymer bonds via silane to inorganic substrates to provide excellent wet or dry adhesion [1, 2].

Many commercial polymer matrices can be processed into single-component, moisture-curable systems by means of silane cross-linking (Scheme 1) [3].

$$\text{polymer}\!-\!NH_2 + \overset{O}{CH_2\!-\!CH}\!-\!CH_2O(CH_2)_3\!-\!Si(OR)_3 \quad (RO)_3Si\!-\!(CH_2)_3OCH_2\overset{O}{CH\!-\!CH_2} + H_2N\!-\!\text{polymer}$$

$$\downarrow$$

$$\text{polymer}\!-\!NHCH_2\underset{|}{\overset{OH}{C}}HCH_2O(CH_2)_3Si(OR)_3 \quad (RO)_3Si\!-\!(CH_2)_3OCH_2\overset{OH}{C}HCH_2NH\!-\!\text{polymer}$$

$$\downarrow H_2O$$

$$\text{polymer}\!-\!NHCH_2\overset{OH}{C}HCH_2O(CH_2)_3\!-\!\underset{O}{\overset{O}{Si}}\!-\!O\!-\!\underset{O}{\overset{O}{Si}}\!-\!(CH_2)_3OCH_2\overset{OH}{C}HCH_2NH\!-\!\text{polymer}$$

Scheme 1.

Significant improvements in polymer properties such as tensile strength and tear resistance can be achieved by silylating matrix polymers. Glycidoxypropylsilanes are also modifying agents for inorganic surfaces, and can be used, e.g., as components of corrosion-preventing metal coatings or fire- and abrasion-resistant finishes [3, 4].

Many unique properties of silicones such as low surface tension, gas permeability, excellent dielectric properties, physiological inertness, and moisture resistance make them valuable in a myriad of end products. On the other hand, modifications of polysiloxane side groups are extensively explored in order to obtain polymers with special properties or to make them chemically active, thus giving access to new industrial applications. Depending on the properties expected of modified polysiloxanes, a great variety of functional groups may be introduced into the side chains of polysiloxanes [4, 5]. An exemplary group of modified polysiloxanes are the epoxy group-containing silicones. Because of the high reactivity of the epoxy ring, the silicones are valuable for synthesis of various organic block co-polymers. Epoxy-functional siloxanes are capable of UV curing, which is achieved via a cationic photopolymerization. The main application of these polysiloxanes is in release papers, but they are also used for the protection of optical fibers and as paper coatings [5]. The epoxy side groups can also be transformed into ionic substituents, as illustrated in Scheme 2. The resulting ionic silicone surfactants have wide applications in cosmetics and textiles [6].

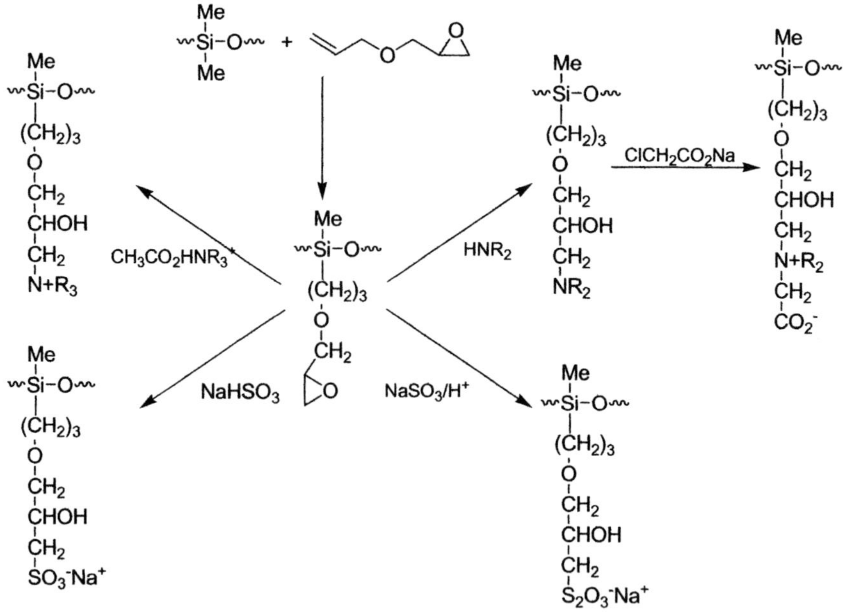

Scheme 2.

The aim of this work was to examine the catalytic activity of rhodium siloxide complexes in the hydrosilylation of allyl glycidyl ether by triethoxysilane and hydro(poly)siloxanes, leading to optimization of procedures for the synthesis of epoxy-functional silanes and siloxanes.

Catalysis of Hydrosilylation

Platinum compounds and complexes are the most important and commonly used catalysts for hydrosilylation processes [7 – 9]. Platinum catalysts tolerate a variety of functional groups, but some impurities may interact with them leading to catalyst poisoning [10]. This has stimulated much research aimed at employing other transition metal compounds as potential catalysts. For example, Rh(I) complexes are selective and active hydrosilylation catalysts [11] and more resistant to poisoning than the platinum ones [12].

Our contribution to this field was the synthesis, isolation, and full characterization a few new rhodium siloxide complexes, both dimeric [13] and monomeric (see Fig. 1) [14].

Siloxides, like alkoxides, have been employed as ancillary ligands of transition metal complexes, markedly influencing the reactivity of a metal center by electronic and steric effects of the substituents at the silicon [15, 16].

Catalytic activity of synthesised Rh(I) siloxide complexes has been demonstrated in some reactions, i.e. in the hydrosilylation of alkenes [17] and allyl alkyl ethers [14, 18, 19] and in the silylative coupling of vinylsilanes with alkenes [20].

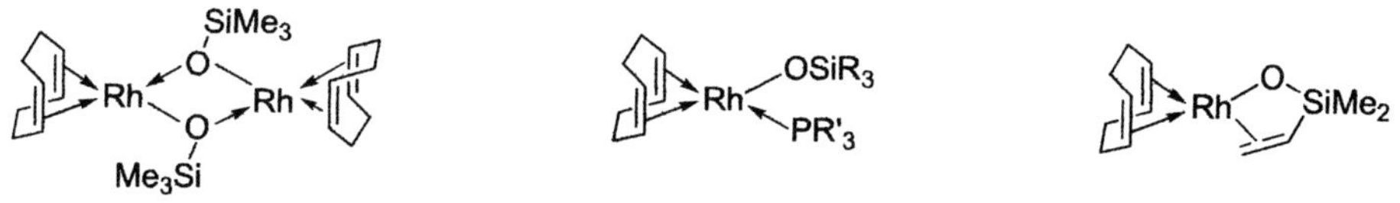

1

2 R=Me, R'=Cy
3 R=Me, R'=Ph

4

Fig. 1.

Epoxy-functional silanes and siloxanes may be prepared by hydrosilylation of unsaturated epoxides or epoxidizing unsaturated silicon compounds. Hydrosilylation is the most convenient and widely used method for the synthesis of these kinds of compounds. The reaction in the presence of the most popular platinum catalysts, i.e. Speier's (H_2PtCl_6 in isopropanol) or Karstedt's ($[Pt_2\{(CH_2{=}CHSiMe_2)_2O\}_3]$) occurs at high temperatures (80 – 150 °C), giving the desired product with 50 – 80% yield [21]. All the rhodium-siloxide complexes **1** – **4** were tested in the hydrosilylation of allyl glycidyl ether by triethoxysilane. The reaction gives the hydrosilylation product (I) with very high yield accompanied by products of the dehydrogenative silylation (II + III) according to Eq. 1.

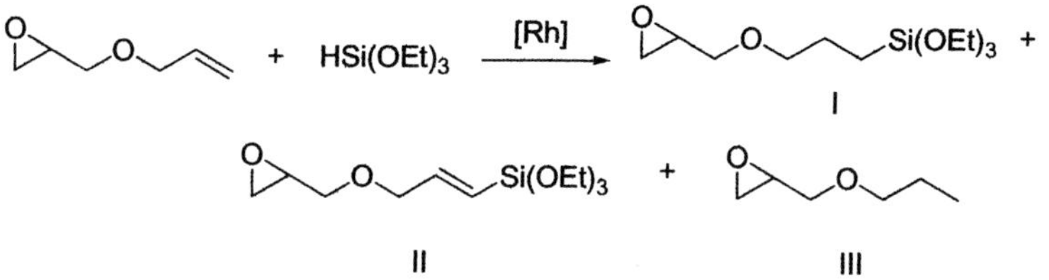

Eq. 1.

High catalytic activity of dimeric complex **1** in the hydrosilylation of allyl esters and allyl ethers (as well allyl glycidyl ether) has been shown in Ref. [18]. The catalytic activity of rhodium-siloxide-phosphine complexes depends on the steric effects of the siloxy group and stereoelectronic effects of the trisubstituted phosphine [14]. Comparison of catalytic activity of dimeric (**1**) and monomeric, phosphine (**2, 3**) and non-phosphine (**4**) rhodium siloxide complexes in the examined reaction is presented in Table 1.

All complexes have shown high catalytic activity, even at room temperature (in contrast to platinum catalysts). Hydrosilylation in the presence of phosphine-rhodium complexes occurred in air, because "real catalyst" (active intermediate) was formed after oxygenation and/or dissociation of phosphine, as reported previously [14]. The non-phosphine complexes **1** and **4** are also very efficient catalysts for the hydrosilylation of allyl glycidyl ether. Irrespective of the starting precursor, a tetracoordinated Rh–H species, responsible for catalysis, is generated under reaction conditions, as illustrated in Scheme 3.

Table 1. The hydrosilylation of allyl glycidyl ether by triethoxysilane catalyzed by rhodium siloxide complexes.

Catalyst	[Rh]:[HSi(OEt)$_3$]	Conversion of [HSi(OEt)$_3$] [%]	Yield [%] I	Yield [%] II + III
		25 °C, 15 min		
1	5×10^{-3}:1	100	90	10
2*		100	96	4
3*		72	70	2
4		100	98	2
		25 °C, 4 h		
1	10^{-3}:1	100	98	2
2*		100	99	1
3*		100	100	0
4		100	98	2
		40 °C, 24 h		
1	10^{-4}:1	100	99	1
2*		100	98	2
3*		100	99	1
4		100	99	1
		60 °C, 1 h		
1	5×10^{-5}:1	98	98	0
2* (2)		99 (50)	98 (50)	1 (0)
3*(3)		41(4)	41 (4)	traces
4		100	99	1

[HSi(OEt)$_3$]:[ether] = 1:1.5, in argon; *in air.

Table 2. The conversion of triethoxysilane and the yield of products obtained in the three processes with the same proportion of the catalyst.

No	Time [h]	Conversion of HSi(OEt)$_3$ [%]	Yield [%] I	Yield [%] I + II
1	3	57	57	0
	24	90	81	9
2	3	54	48	6
	24	90	81	9
3	3	36	30	6
	6	53	47	5
	24	86	76	8

[HSi(OEt)$_3$]:[ether]:[cat] = 1:1:10^{-5}; T = 80 °C.

The activity of rhodium siloxide complexes is not suppressed in the hydrosilylation process, which means that the same catalyst can be used a few times. Table 2 shows typical activities of [Rh(cod)(PCy_3)($OSiMe_3$)] (**2**) three times reused in the reaction examined.

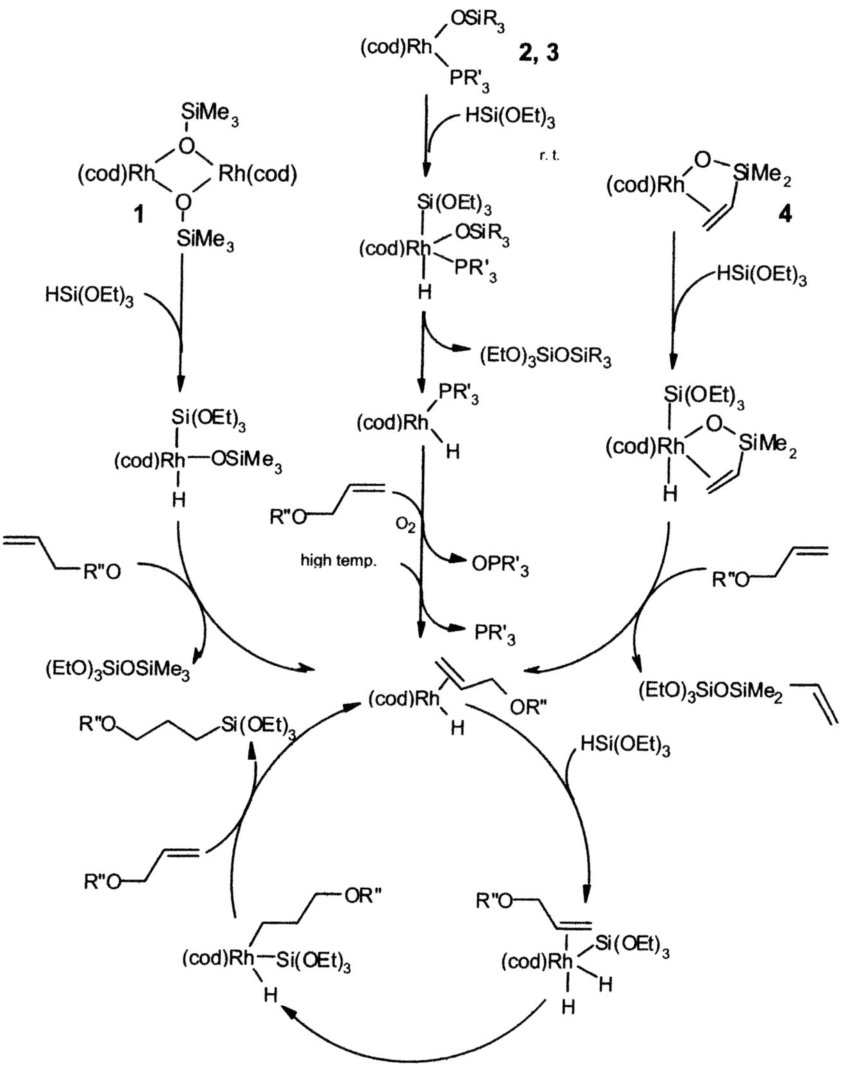

Scheme 3. Mechanism of catalysis of hydrosilylation by rhodium siloxide complexes.

Hydrosilylation is also one of the basic methods for synthesis of modifying silicones. The rhodium siloxide complexes have been used in the reaction between heptamethyltrisiloxane and allyl glycidyl ether, which is a modeling reaction of the polymeric system (Eq. 2).

Similarly to the results obtained in the reaction of hydrosilylation by triethoxysilane, all complexes are highly catalytically active, especially complexes **1**, **2**, and **4**, and allow us to obtain a desired product with high yield (over 90%) at room temperature in 2 h. The same result for the reaction catalyzed by Pt (Karstedt) has been obtained after 24 h and for the process run at 90 °C.

$$H_3C{-}Si(CH_3)_2{-}O{-}SiH(CH_3){-}O{-}Si(CH_3)_2{-}CH_3 + CH_2{=}CHCH_2OCH_2\text{-}(epoxy) \longrightarrow H_3C{-}Si(CH_3)_2{-}O{-}Si(CH_3)[(CH_2)_3OCH_2\text{-}(epoxy)]{-}O{-}Si(CH_3)_2{-}CH_3$$

Eq. 2.

The use of rhodium siloxide complexes in a polymeric system (Eq. 3) has not affected the activity of the catalysts, as checked in the reaction with trisiloxane. The conversion of poly(hydro)siloxane was almost quantitative under very mild reaction conditions (25 – 40 °C, 2 h), and a valuable epoxy-functional polysiloxane was formed selectively.

$$(CH_3)_3Si{-}[O{-}Si(CH_3)_2]_{50}{-}[O{-}SiH(CH_3)]_{25}{-}O\,Si(CH_3)_3 + CH_2{=}CHCH_2OCH_2\text{-}(epoxy) \longrightarrow (CH_3)_3Si{-}[O{-}Si(CH_3)_2]_{50}{-}[O{-}Si(CH_3)(CH_2(CH_2)_2OCH_2\text{-}(epoxy))]_{25}{-}O{\cdot}Si(CH_3)_3$$

Eq. 3.

In respect of the activity, selectivity, and economic issues (price, reuse of catalyst, installation design, and cost) the above-mentioned homogeneous catalytic systems may compete with the platinum system commonly used in this type of reaction.

On the basis of the obtained results, the technologies of synthesis of epoxy-functional silane and (poly)siloxanes have been worked out and will be implemented in the very near future.

Acknowledgments: The work was supported by the State Committee for Scientific Research, Project No. 3 T09B 121 26.

References

[1] E. P. Pluedemann, *Silane Coupling Agents*, Plenum Press, New York, **1991**.

[2] E. P. Pluedemann, in *Silanes and other Coupling Agents* (Ed.: K. L. Mittal), VSP, Utrecht, **1992**.

[3] *Kirk-Othmer Encyclopedia of Chemical Technology* (Eds.: I. J. Kroschwitz, M. Howe-Grant),

4th Edn. Wiley, New York, **1997**.
[4] J. G. Tesoro, Y. Wu, in *Silanes and other Coupling Agents* (Ed.: K. L. Mittal), VSP, Utrecht, **1992**.
[5] *Silicon-Containing Polymers* (Eds.: R. G. Jones, W. Ando, J. Chojnowski,), Kluwer Academic Publishers, Dordrecht, **2000**.
[6] *Novel Surfactants* (Ed.: K. Holmberg), Marcel Dekker Inc., New York, **1998**.
[7] *Comprehensive Handbook on Hydrosilylation* (Ed.: B. Marciniec), Pergamon Press, Oxford, **1992**.
[8] I. Ojima, Z. Li, J. Zhu, in *The Chemistry of Organic Silicon Compounds* (Eds.: Z. Rappoport, Y. Apeloig), John Wiley & Sons, Chichester, **1998**, Chapter 29.
[9] B. Marciniec, J. Guliński, H. Maciejewski, *Hydrosilylation in the Encyclopedia of Catalysis* (Ed.: I. Horvath,), J. Wiley & Sons, New York, **2003**.
[10] M. A. Brook, *Silicon in Organic, Organometallic and Polymer Chemistry*, Wiley, New York, **2000**.
[11] F. de Charentenay, J. A. Osborn, G. Wilkinson, *J. Chem. Soc. A.* **1968**, 787.
[12] Pat. USA 4 804 768, **1989**.
[13] B. Marciniec, P. Krzyzanowski, *J. Organomet. Chem.* **1995**, *493*, 261; P. Krzyzanowski, M. Kubicki, B. Marciniec, *Polyhedron* **1996**, *15*, 1; B. Marciniec, P. Krzyzanowski, M. Kubicki, *Polyhedron* **1996**, *15*, 4233.
[14] B. Marciniec, P. Blazejewska-Chadyniak, M. Kubicki, *Can. J. Chem.* **2003**, *81*, 1292.
[15] P. T. Wolczanski, *Polyhedron* **1995**, *14*, 3335; F. J. Feher, T. A. Budzichowski, *Polyhedron* **1995**, *14*, 3239.
[16] B. Marciniec, H. Maciejewski, *Coord. Chem. Rev.* **2001**, *223*, 301.
[17] B. Marciniec, P. Krzyzanowski, E. Walczuk-Gusciora, W. Duczmal, *J. Mol. Catal.* **1999**, *144*, 263.
[18] B. Marciniec, E. Walczuk, P. Blazejewska-Chadyniak, D. Chadyniak, M. Kujawa-Welten, S. Krompiec, in *Organosilicon Chemistry V — From Molecules to Materials* (Eds.: N. Auner, J. Weiss), Wiley VCH, Weinheim, **2003**.
[19] Polish Pat. PL-351 449 (**2001**), PL-351 450 (**2001**); PL-351 451 (**2001**).
[20] B. Marciniec, E. Walczuk-Gușciora, P. Blazejewska-Chadyniak, *J. Mol. Catal.* **2000**, *160*, 165; B. Marciniec, E. Walczuk-Guściora, C. Pietraszuk, *Organometallics* **2001**, *20*, 3423.
[21] Europ. Pat. 288 286 (**1988**); German Pat. 1 937 404 (**1970**).

New Functionalization of Vinyl-Substituted Organosilicon Compounds via Ru-Catalyzed Reactions

Bogdan Marciniec,* Yujiro Itami, Dariusz Chadyniak, Magdalena Jankowska

Department of Organometallic Chemistry, Faculty of Chemistry
Adam Mickiewicz University, Grunwaldzka 6, 60-780 Poznań, Poland
Tel.: +4861 8291366 — Fax: +4861 8291508
E-mail: marcinb@main.amu.edu.pl

Keywords: cross-metathesis, silylative coupling, *trans*-silylation, ruthenium catalysts, vinylsilicon compound

Summary: Two catalytic reactions, i.e. silylative coupling (*trans*-silylation) (SC) catalyzed by complexes containing or generating Ru–H and/or Ru–Si bonds (**I, II, V, VI**) and cross-metathesis (CM) catalyzed by ruthenium–carbene (i.e. 1st and 2nd generation ruthenium Grubbs catalyst (**III, IV**)) of vinyl and allyl-substituted hetero(N,S,B)organic compounds with commercially available vinyltrisubstituted silanes, siloxanes, and silsesquioxane have been overviewed. They provide a universal route toward the synthesis of well-defined molecular compounds with vinylsilicon functionality.

Introduction

Substituted vinylsilanes such as *E,Z*-$RCH{=}CHSiR'_3$ and $R(SiR'_3)C{=}CH_2$ are a class of organosilicon compounds commonly used in organic synthesis. Many efficient stereo- and regio-selective methodologies for the synthesis of substituted vinylsilanes involving classical stoichiometric routes from organometallic reagents and, more recently, transition metal-catalyzed transformations of alkynes, silylalkynes, alkenes, and simple vinylsilanes have been reported [1 – 3].

In the last 15 years we have developed two new catalytic reactions between the same parent substances, i.e. silylative coupling (SC) (also called *trans*-silylation or silyl group transfer) and cross-metathesis (CM) of alkenes, which have provided an universal route for the synthesis of well-defined molecular compounds with vinylsilicon functionality. While the cross-metathesis is catalyzed by well-defined Ru and Mo carbenes, the silylative coupling is catalyzed by complexes initiating or generating M–H or M–Si bonds (where M = Ru, Rh, Ir). For recent reviews see Refs. [4 – 6].

The mechanism of catalysis of SC proved by Wakatsuki et al., [7] and corrected by us [8], proceeds via insertion of vinylsilanes into the M–H bond and β-silyl transfer to the metal with elimination of ethylene to generate M–Si species, followed by insertion of alkene and β-H transfer to the metal with elimination of substituted vinylsilane. A notable peculiarity of this reaction is the silylation of vinyl alkyl ether with vinylsilanes [9] to give β-alkoxysubstituted vinylsilanes, which are difficult to synthesize via other TM-catalyzed reactions. A functionalization of vinyl-substituted cyclosiloxane and cyclosilazane [10] as well as novel starburst compounds based on highly stereo- and regio-selective reactions of tris(dimethylvinylsilyl)benzene with substituted styrenes to give a new core for dendritic compounds [11] are recent examples of this new synthetic route.

Cross-metathesis

$$CH_2{=}CH{-}SiR_3' + CH_2{=}CH{-}R \xrightarrow{[Ru]=C} CH_2{=}CH_2 + R{-}CH{=}CH{-}SiR_3' \; (E+Z)$$

Silylative Coupling (*trans*-silylation)

$$CH_2{=}CH{-}SiR_3' + H{-}CH{=}C(H){-}R \xrightarrow{[Ru]\text{-}H/[Ru]\text{-}Si} CH_2{=}CH_2 + R{-}CH{=}CH{-}SiR_3' \; (E+Z) + CH_2{=}C(R)SiR_3'$$

Scheme 1.

On the other hand, our recent study on the highly efficient cross-metathesis of vinyltrialkoxy- and vinyltrisiloxy-silanes with various olefins, for example, with styrene [12] allyl ethers [13] and esters [14] as well as octavinylsilsesquioxane [15] with several olefins have opened a new opportunity for the use of alkene-cross-metathesis in the synthesis of unsaturated organosilicon compounds (see also Refs. [5] and [6]). In this paper new examples of the two reactions involving hetero(N,S,B)organic olefins have been overviewed.

Reactions of Vinyl-Substituted Silanes with Vinyl-Substituted Hetero(N,S,B)organic Derivatives

Previous reports on the reaction of vinyl-substituted silanes [9] and silsesquioxane [15] with vinyl alkyl ethers catalyzed by ruthenium complexes containing or generating Ru–H and/or Ru–Si bonds show that the process proceeds according to the nonmetallacarbene mechanism as in SC and yields (usually in 5-fold excess of alkene) 1-silyl-2-alkoxy-ethenes with a high preference for the *E*-isomer (Eq. 1).

The first generation Grubbs complex $(PCy_3)_2Cl_2Ru(=CHPh)$ also appeared to be a very active catalyst in these processes, but additional experiments provided evidence for the initiation of Ru–H bond formation on the Grubbs catalyst heating at 60 °C [16].

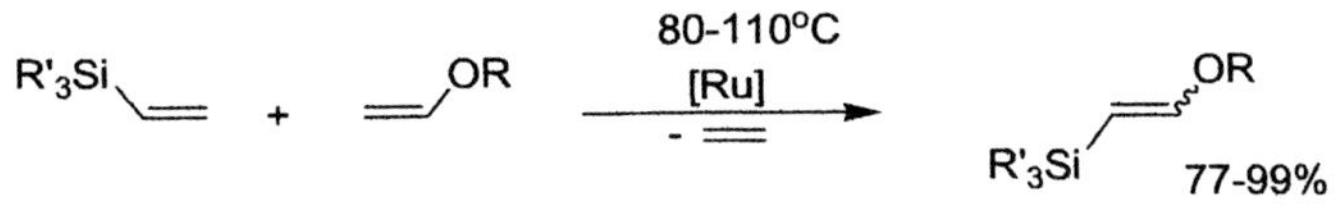

R'_3Si + OR → (80-110°C, [Ru], − $CH_2{=}CH_2$) R'_3Si–CH=CH–OR 77-99%

where [Ru] is [RuHCl(CO)(PPh$_3$)$_3$], [RuCl(SiMe$_3$)(CO)(PPh$_3$)$_2$], [RuHCl(CO)(PCy$_3$)$_2$] [9, 15]
R is Et, Pr, Bu, ^{t}Bu, Hex, SiMe$_3$
R'$_3$ is Me, (OEt)$_3$, Me$_2$Ph, silsesquioxane

Eq. 1.

The reactions of vinyl-substituted silanes and silsesquioxane with vinyl-substituted hetero(N,S,B)organic compounds proceed according to the equations presented in Scheme 2.

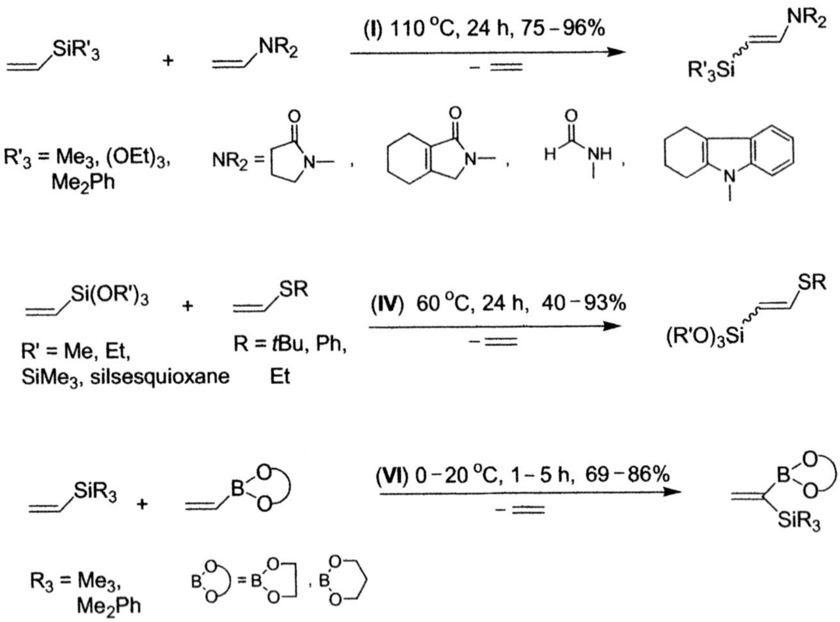

SiR'$_3$ + NR$_2$ → (I) 110 °C, 24 h, 75 - 96% (− $CH_2{=}CH_2$) R'$_3$Si–CH=CH–NR$_2$

R'$_3$ = Me$_3$, (OEt)$_3$, Me$_2$Ph
NR$_2$ = N-pyrrolidonyl, N-phthalimidyl, HC(O)NH–, carbazolyl

Si(OR')$_3$ + SR → (IV) 60 °C, 24 h, 40 - 93% (− $CH_2{=}CH_2$) (R'O)$_3$Si–CH=CH–SR

R' = Me, Et, SiMe$_3$, silsesquioxane
R = *t*Bu, Ph, Et

SiR$_3$ + B(OR)$_2$ → (VI) 0 - 20 °C, 1 - 5 h, 69 - 86% (− $CH_2{=}CH_2$) $CH_2{=}C(B(OR)_2)SiR_3$

R$_3$ = Me$_3$, Me$_2$Ph

Scheme 2.

The following ruthenium catalysts are most effective for the synthetic procedures: RuHCl(CO)(PPh$_3$)$_3$ (**I**), RuHCl(CO)(PCy$_3$)$_2$ (**II**), Ru(=CHPh)Cl$_2$(PCy$_3$)$_2$ (**III**), Ru(=CHPh)Cl$_2$-(PCy$_3$)H$_2$IMes) (**IV**), Ru(SiMe$_3$)Cl(CO)(PPh$_3$)$_3$ (**V**) and RuHCl(CO)(PCy$_3$)$_2$ + CuCl (**VI**).

Vinylamides such as *N*-vinylpyrrolidone, *N*-vinylphtalimide and *N*-vinylformamide undergo, in the presence of (**I**), stereoselective silylative coupling with various vinylsilanes up to E/Z = 99/1 [17] as well as with octavinylsilsesquioxane [15] (E/Z = 95/5). Extended heating of the reaction mixture allowed exclusive isolation of the E-isomer because of $Z \rightarrow E$ isomerization of the products. In contrast to the cross-metathesis with vinyl ethers, no reaction of *N*-vinylpyrrolidinone with vinylsilane was observed in the presence of ruthenium-carbene complexes (**III** and **IV**) even at higher temperatures. This can be explained by intramolecular coordination of the carbonyl group blocking the coordination site of the Ru-complex according to Scheme 3 [18].

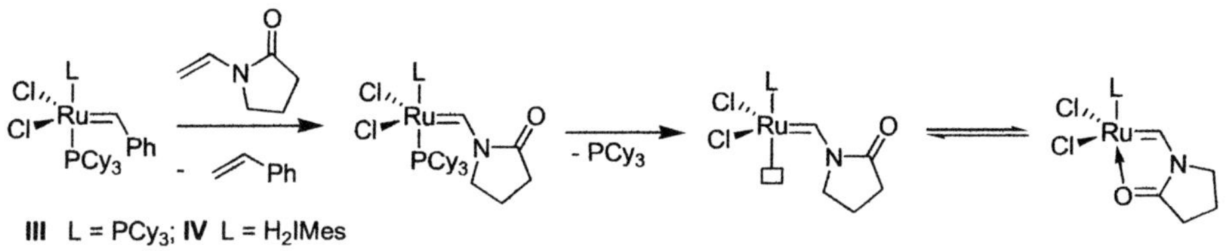

III L = PCy_3; **IV** L = H_2IMes

Scheme 3.

1-vinylcarbazole and 1-vinylimidazole undergo both SC and CM reactions in the presence of **I** and **III**, **IV** catalysts, respectively. No catalytic effect is observed in vinylimidazole because of strong coordination of amines to the ruthenium atom in all complexes (**I**, **III**, **IV**) used [18]. Such an effect of inactivity of **III** and **IV** in cross-metathesis has been observed and explained earlier [20]. However, a strong steric effect at the nitrogen in 9-vinylcarbazole reduces its coordination power, favoring the coordination of the vinyl group to the **I** and/or **IV** complex, initiating both SC and CM processes with vinylsilanes to yield effectively and stereoselectivily *E*-(silyl)vinyl-9H-carbazoles [19]. Yet, the silylative coupling process can be more useful for synthetic procedures.

It has been assumed that the reaction of vinylsilanes with vinyl alkyl sulfides would also occur in the same manner as the reaction with vinyl ether [9] or vinyl pyrrolidinone [17], exhibiting active silylative coupling and inactive cross-metathesis. Nevertheless, we have observed the reverse.

Cross-metathesis of trialkoxy- and trisiloxy-substituted vinylsilanes [21] as well as octavinylsilsesquioxane [15] with vinyl sulfides proceeds efficiently but only in the presence of the 2nd generation Grubbs catalysts (**IV**) to offer a new and very attractive route for syntheses of [alkyl(aryl)]sulfide-substituted vinylsilanes and vinyl-silsesquioxane with high preference for the *E*-isomer, as illustrated by exclusive isolation of such isomers. The Fischer-type ruthenium carbene complex $Ru(=CHSPh)Cl_2(PCy_3)_2$ has recently been reported as an effective catalyst in the ring opening/cross-metathesis of norbornene derivatives with vinyl sulfide [22], suggesting that these carbenes can be reactive in the cross-metathesis.

Contrary to expectations based on the efficient silylative coupling of the heteroatom (O, N, Si and B), functionalized alkenes with vinylsilanes catalyzed by Ru–H and/or Ru–Si-containing complexes do not undergo this transformation, which has been explained by formation of an Ru–S complex into which no insertion of vinylsilanes (a step necessary in the catalytic cycle of SC) was observed.

Organic compounds containing silicon [23] or boron [24] atoms in the molecule play an important role in organic synthesis. Olefins bearing silyl and boryl group are potentially useful organic intermediates, as they combine the synthetic potentials of both functionalities. While cross-metathesis of vinylboronates with vinylsilanes in the presence **III** or **IV** does not occur, the silylative coupling is catalyzed efficiently by $RuHCl(CO)(PCy_3)_2/CuCl$ catalyst **V** to yield exclusively 1,1-boryl silyl ethenes.

Stoichiometric study of the $Ru(SiMe_2Ph)Cl(CO)(PCy_3)_2$ with vinylboronate confirms the insertion of olefin into the Ru–Si bond and supports the non-metallacarbene mechanism of this reaction [25].

Cross-Metathesis vs Silylative Coupling of Vinylsilane with Allyl-substituted Hetero(N,S,B)organic Derivatives

The reactions of vinyltrisubstituted silanes with allyl substituted heteroorganic compounds proceed in the presence of ruthenium alkylidene complexes (catalyst **IV**) as well as catalysts including an Ru–H bond (**I**) according to the following equation, giving two isomeric products (*E* + *Z*) and ethylene (Scheme 4).

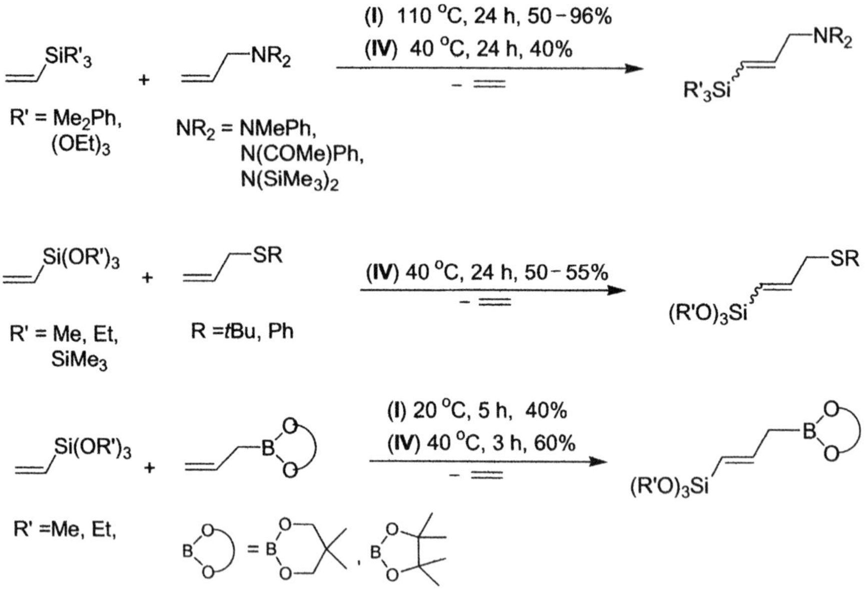

(I) 110 °C, 24 h, 50–96%
(IV) 40 °C, 24 h, 40%

SiR'_3 + NR_2 → R'_3Si ... NR_2 − ethylene

R' = Me_2Ph, $(OEt)_3$

NR_2 = NMePh, N(COMe)Ph, $N(SiMe_3)_2$

$Si(OR')_3$ + SR → (IV) 40 °C, 24 h, 50–55% → $(R'O)_3Si$ ... SR − ethylene

R' = Me, Et, $SiMe_3$

R = *t*Bu, Ph

$Si(OR')_3$ + B(O)(O) → (I) 20 °C, 5 h, 40%; (IV) 40 °C, 3 h, 60% → $(R'O)_3Si$ ... B(O)(O) − ethylene

R' = Me, Et,

Scheme 4.

Mechanistic implications of a general cross-metathesis of vinylsilicon with allyl-substituted heteroorganic compounds have been studied in detail for the reaction with allyl alkyl ethers [13]. The detailed NMR study of the stoichiometric reaction of Grubbs catalyst with allyl-*n*-butyl ether has provided information on individual steps of the catalytic cycle. A general mechanism of the cross-metathesis of vinyltri(alkoxy, siloxy)silanes (as well as octavinylsilsesquioxane) with 3-heteroatom-containing 1-alkenes in the presence of ruthenium carbene is shown in Scheme 5.

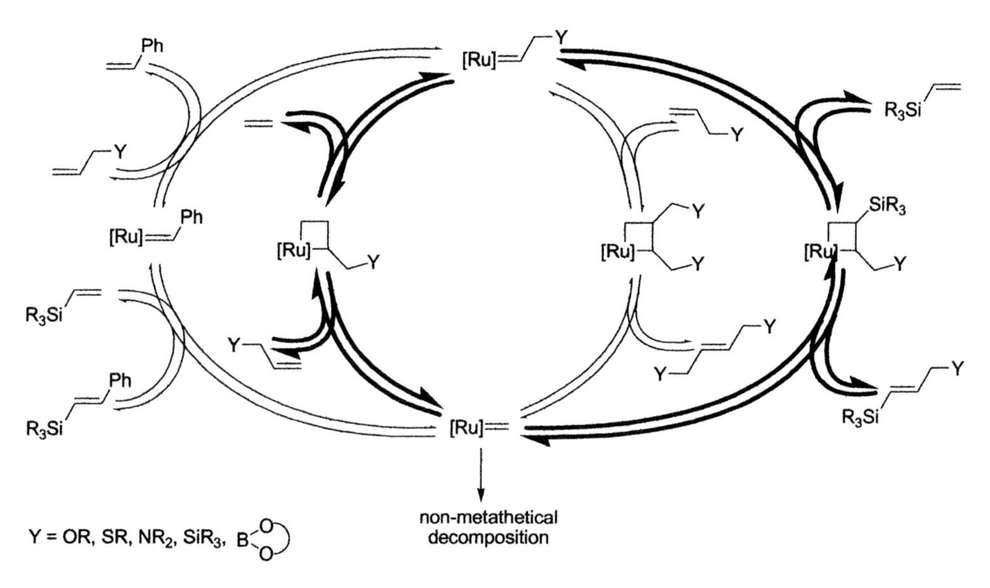

Scheme 5.

According to this Scheme, this process leads to effective formation of the $[Ru]=CH_2$ complex, which nevertheless readily undergoes non-metathetical decomposition. Thus, an addition of some amount of the allyl compound during the reaction would enhance the final yield of the cross-metathesis product, which has been confirmed experimentally for allyl ether [13].

It is well known that silylation of allyl derivatives with vinylsilane catalyzed by a ruthenium hydride complex is accompanied by isomerization of propen-1-yl to propen-2-yl derivatives as well as homo-coupling of vinylsilane when equimolar amounts of the initial substances are used. If catalyst **I** was used in the SC of allyl amide and allyl amine with vinylsilanes, a 5-fold excess of olefin to vinylsilane was used to stop homocoupling of vinylsilane, but simultaneously no more than 5% of isomerization of allyl compound was observed [19, 26]. When allyl boronate is used instead of allylamine under mild conditions (20 – 40 °C), the two reactions catalyzed by **I** and **IV** yield stereoselectively *E*-product (see Scheme 4) [26].

The inactivity of Ru–H (and Ru–Si) complexes in the SC of allyl sulfides with vinylsilanes is due to analogous formation of ruthenium sulfide, reported previously [27], and elimination of propene according to the following equation.

[Ru]–H ⟶ (S*t*Bu) [[Ru] … S*t*Bu ; [Ru] … *t*BuS] ⟶ (– propene) [Ru]–S*t*Bu

Eq. 2.

Conclusions

- The two reactions catalyzed by ruthenium complexes, i.e. silylative coupling (SC) (*trans*-silylation) catalyzed by **I**, **II**, **V**, and **VI** and cross-metathesis (CM) (catalysts **III** and **IV**) of vinyl- and allyl-substituted hetero(N,S,B)organic compounds with commercially available vinyltrisubstituted silanes, siloxanes, and silsesquioxanes provide a universal route toward the synthesis of well-defined molecular compounds with vinylsilicon functionality.
- Vinylsilanes undergo productive cross-metathesis (CM) and silylative coupling (SC) with allyl-substituted (N, B)functionalized alkenes to yield 1-silyl-3,*N*,*B*-substituted propenes with preference (for *N*-derivatives) and exclusive formation (for boronates) of the *E*-isomer.
- The reaction of vinyl-substituted silanes and octavinylsilsesquioxane with vinyl-substituted amides, amines (carbazole) as well as boronates catalyzed by **I** proceeds effectively to yield under optimum conditions stereo- and/or regio-selectively 1-silyl-2-*N*- and 1,1-silylboryl-substituted ethenes. 1-silylvinyl carbazole can also be obtained via cross-metathesis of vinylsilane with vinylcarbazole, but only in the presence of the 2nd generation Grubbs catalyst (**IV**).
- Vinyl and allyl sulfides undergo an unusual cross-metathesis transformation with vinylsilane and vinylsilsesquioxane in the presence of the 2nd generation Grubbs catalyst (**IV**) to yield 1,2-silylvinyl sulfide or 1,3-silylpropenyl sulfide, respectively. The SC reactions of these substrates performed in the presence of Ru–H (**I**) and Ru–Si (**II**) give no products, which is explained by formation of the ruthenium sulfide species – inactive in the SC reactions since there is no insertion of olefin into the Ru–S bond.

Acknowledgment: The work was supported by The State Committee for Scientific Research, Project No. 3 T03A 145 26.

References

[1] K. Oshima, *Science of Synthesis* **2002**, *34*, 713.
[2] C. Laza, E. Dunach, *C. R. Chimie* **2003**, *6*, 185.
[3] B. Marciniec, C. Pietraszuk, I. Kownacki, M. Zaidlewicz, in *Comprehensive Organic Functional Group Transformations II (COFGT-II)*, (Eds.: A. Katritzky, R. J. K. Taylor), Elsevier, Oxford, **2005** Chapter 2.18.
[4] B. Marciniec, in *Applied Homogeneous Catalysis with Organometallic Compounds*, (Eds.: B. Cornils, W. A. Hermann), **1996**, *Chapter 2.6.*
[5] B. Marciniec, C. Pietraszuk, *Curr. Org. Chem.* **2003**, *7*, 691.
[6] B. Marciniec, C. Pietraszuk, in *Handbook of Metathesis*, Vol. 1–3, (Ed.: R. H. Grubbs), Wiley-VCH, New York, **2003**.
[7] Y. Wakatsuki, H. Yamazaki, M. Nakano, Y. Yamamoto, *J. Chem. Soc., Chem. Commun.*

1991, 703.
[8] B. Marciniec, C. Pietraszuk, *Chem. Commun.* **1995**, 2003.
[9] B. Marciniec, M. Kujawa, C. Pietraszuk, *Organometallics* **2000**, *19*, 1677.
[10] Y. Itami, B. Marciniec, M. Majchrzak, M. Kubicki, *Organometallics* **2003**, *22*, 1835.
[11] Y. Itami, B. Marciniec, M. Kubicki, *Organometallics* **2003**, *22*, 3717.
[12] C. Pietraszuk, B. Marciniec, H. Fischer, *Organometallics* **2000**, *19*, 913.
[13] M. Kujawa-Welten, C. Pietraszuk, B. Marciniec, *Organometallics* **2002**, *21*, 840.
[14] M. Kujawa-Welten, B. Marciniec, *J. Mol. Catal.* **2002**, *190*, 27.
[15] Y. Itami, B. Marciniec, M. Kubicki, *Chem. Eur. J.* **2004**, *10*, 1239.
[16] B. Marciniec, M.Kujawa, C. Pietraszuk, *New. J. Chem.* **2000**, *24*, 671.
[17] B. Marciniec, D. Chadyniak, S. Krompiec, *Tetrahedron Lett.* **2004**, *45*, 4065.
[18] J. Louis, R. H. Grubbs, *Organometallics* **2002**, *21*, 2153.
[19] D. Chadyniak, B. Marciniec, S. Krompiec, W. Prukala, *Proc. of XXIII Poland-German Colloquy on Organometallic Chemistry*, Wierzba (Poland), April **2003**, O-11.
[20] M. Sanford, J. A. Love, R. H. Grubbs, *Organometallics* **2001**, *21*, 5314.
[21] B. Marciniec, D. Chadyniak, S. Krompiec, *J. Mol. Catal.* **2004**, *224*, 111.
[22] K. Katayama, N. Nagao, F. Ozawa, *Organometallics* **2003**, *22*, 586.
[23] E. W. Colvin, *Silicon Reagents in Organic Synthesis*, Academic Press London, **1988.**
[24] A. Pelter, K. Smith, H. C. Brook, *Borano Reagents*, Academic Press London, **1988**.
[25] M. Jankowska, B. Marciniec, C. Pietraszuk, J. Cytarska, M. Zaidlewicz, *Tetrahedron Lett.* **2004**, *45*, 6615.
[26] M. Jankowska, B. Marciniec, J. Cytarska, M. Zaidlewicz (unpublished results).
[27] N. Kuznik, S. Krompiec, T. Bieg, S. Baj, K. Skutil, A. Chrobok, *J. Organomet. Chem.* **2003**, *665*, 167.

Hydrosilylation Using Ionic Liquids

B. Weyershausen,* K. Hell, U. Hesse

R&D Department, Oligomers/Silicones, Degussa AG
Goldschmidtstr. 100, 45127 Essen, Germany
Tel.: +49 201 1731655 — Fax: +49 201 1731839
E-mail: bernd.weyershausen@degussa.com

Keywords: ionic liquids, hydrosilylation, polyethersiloxanes

Summary: The use of an ionic liquid in hydrosilylation reactions enables standard hydrosilylation catalysts to be easily recovered and subsequently reused after separation from the product at the end of the reaction. Remarkably, the recovered catalyst/ionic liquid solution does not need to be purified or treated before its reuse. Employing this method, a variety of organomodified polydimethylsiloxanes were synthesized.

Introduction

Hydrosilylation

One of the most fundamental and elegant methods of laboratory and industrial synthesis of organomodified polydimethylsiloxanes is the transition metal-catalyzed hydrosilylation of CC-double bond-containing compounds with SiH-functional polydimethylsiloxanes [1]. Regardless of the broad applicability of the hydrosilylation reaction, its technical use still suffers from substantial difficulties. Usually, the reaction is homogeneously catalyzed, which means that after completion of the reaction the catalyst either remains within the product or has to be removed at high cost. From an economic and ecological point of view, in homogenous catalysis, the separation, immobilization, and the reuse of the expensive precious metal catalyst in a subsequent reaction represent serious problems [2]. Therefore, there has been no lack of attempts to reduce the amount of catalyst, which in most cases leads to relatively long reaction times. A technical process based on homogenous catalysis is economically efficient only if, in combination with acceptable reaction times, the catalyst losses can be kept as small as possible. Hence, there is a demand for processes which allow for the recycling of the catalyst without catalyst losses and the lowest possible stress for the products. In the past, intensive work has been done on immobilization, heterogenization, and anchoring of homogenous catalysts for an easy catalyst separation from the products and recovery of the catalyst. Multiphasic reactions, e.g., the extraction of the catalyst or its adsorption at ion exchangers, represent another possibility for the separation of product and catalyst phase. In recent years, biphasic reactions employing ionic liquids have gained increasing importance [3]. The ionic liquid generally forms the phase in which the catalyst is dissolved and immobilized.

Ionic Liquids

Ionic liquids (IL) are salts melting at low temperatures, and represent a novel class of solvents with non-molecular ionic character. In contrast to a classical molten salt, which is a high-melting, highly viscous, and very corrosive medium, an ionic liquid is already liquid at temperatures below 100 °C and is of relatively low viscosity [4]. In most cases, ionic liquids consist of combinations of cations such as ammonium, phosphonium, imidazolium, or pyridinium with anions such as halides, phosphates, borates, sulfonates, or sulfates. The combination of cation and anion has a great influence on the physical properties of the resulting ionic liquid. By careful choice of cation and anion it is possible to fine tune the properties of the ionic liquid and provide a tailor-made solution for each task (Fig. 1), and this is why ionic liquids are often referred to as designer solvents or materials.

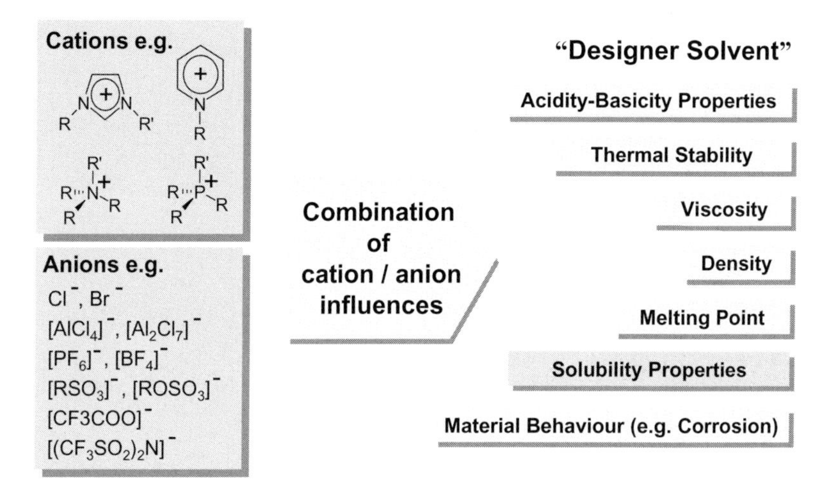

Fig. 1. Properties of ionic liquids.

With many organic product mixtures, ionic liquids form two phases, from which arises the possibility of carrying out biphasic reactions for the separation of the homogenous catalyst [5]. Ionic liquids are not any longer a class of esoteric compounds, but are already used in a multitude of different applications (Fig. 2) and since 1999 have been commercially available [6].

In 1972, Parshall used an ionic liquid for the first time for the immobilization of a transition metal catalyst in a biphasic reaction set-up [7]. He described the hydrogenation of CC-double bonds with $PtCl_2$ dissolved in tetraethylammonium chloride associated with tin dichloride ($[Et_4N][SnCl_3]$, m.p. 78 °C) at temperatures between 60 and 100 °C. "A substantial advantage of the molten salt medium [over conventional organic solvents] ... is that the product may be separated by decantation or simple distillation". The use of ionic liquids as novel media for transition metal catalysis started to receive increasing attention when in 1992 Wilkes reported on the synthesis of novel non-chloroaluminate, room temperature liquid salts with significantly enhanced stability toward hydrolysis, such as tetrafluoroborate salts [8]. Some of the first successful examples of catalytic reactions in non-chloroaluminate ionic liquids include the rhodium-catalyzed hydrogenation [9] and hydroformylation [10] of olefins. However, very often catalyst leaching into the organic and/or the product phase is observed, as the transition metal catalyst is not completely retained in the polar ionic liquid phase. Current approaches to circumvent this problem include the modification of the catalyst, the modification of the ligands [10], and the development of task-specific ionic liquids [11].

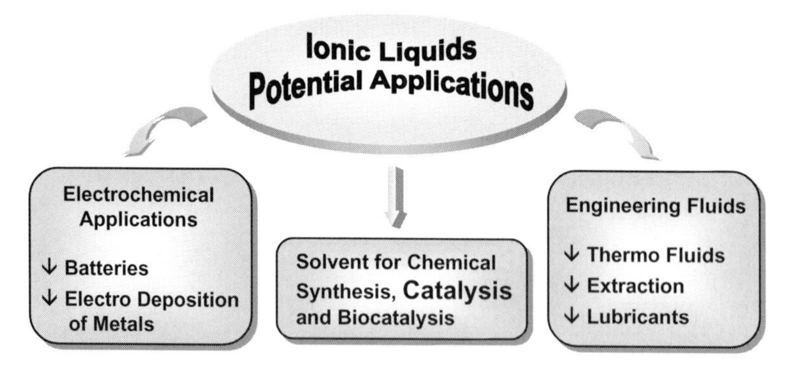

Fig. 2. Potential applications of ionic liquids.

Organically Modified Polydimethylsiloxanes

The hydrosilylation of CC-double bond-containing compounds with SiH-functional polydimethylsiloxanes is a widely applied reaction in industrial synthesis for the production of organosilicon compounds (Fig. 3) on an industrial scale [1].

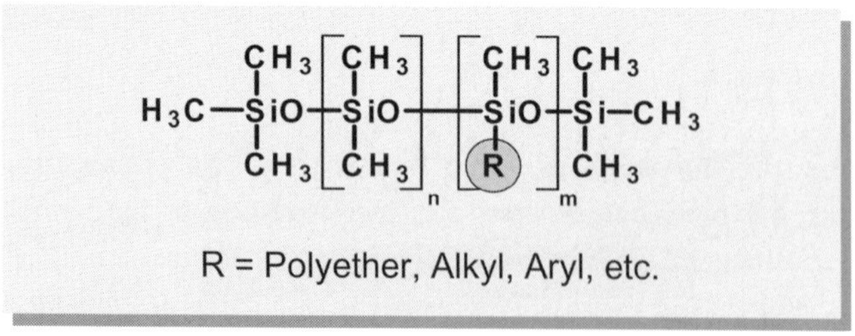

$H_3C-Si(CH_3)_2O-[Si(CH_3)_2O]_n-[Si(CH_3)(R)O]_m-Si(CH_3)_2-CH_3$

R = Polyether, Alkyl, Aryl, etc.

Fig. 3. Organically modified polydimethylsiloxanes (OMS) offer high synthetical flexibility.

Apart from our general investigations of the hydrosilylation reaction, we became interested in ionic liquids and their potential to be used in hydrosilylation reactions as a means for catalyst heterogenization. In particular, we aimed at the synthesis of polyethersiloxanes. Polyethersiloxanes constitute an important class of surface-active compounds, which find use in a broad range of industrial applications (Fig. 4) [12].

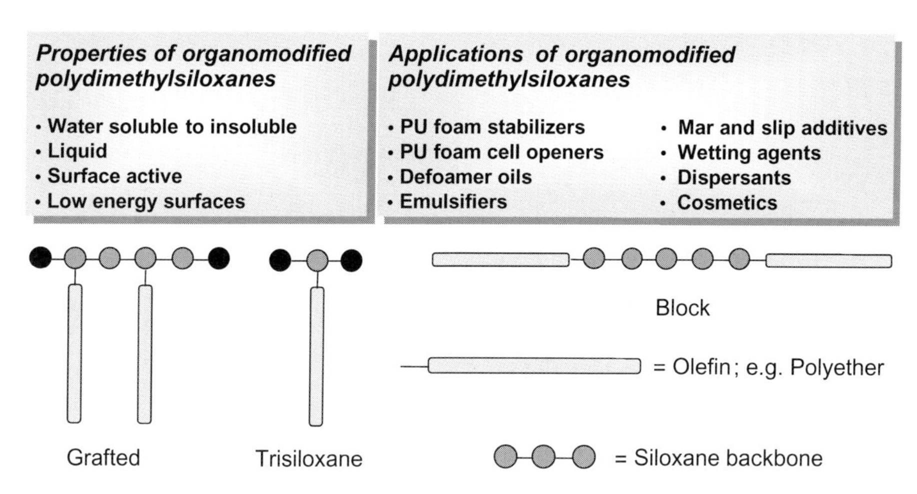

Fig. 4. Properties of organomodified polydimethylsiloxanes (OMS).

Results and Discussion

Herein, we report on a novel process for the synthesis of organomodified polydimethylsiloxanes employing ionic liquids for the heterogenization and/or immobilization of the precious metal catalyst [13]. The advantage of this novel hydrosilylation process is that standard hydrosilylation catalysts can be used without the need for prior modification to prevent catalyst leaching. To the best of our knowledge, this is the first example of a hydrosilylation of olefinic compounds using ionic liquids (Scheme 1). However, a method for the transition metal-catalyzed hydroboration and hydrosilylation of alkynes in ionic liquids has recently been described [14].

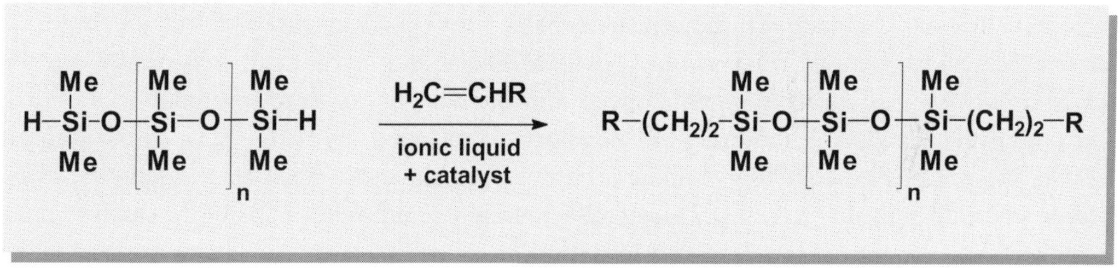

$$H{-}SiMe_2{-}O{-}[SiMe_2{-}O]_n{-}SiMe_2{-}H \xrightarrow[\text{ionic liquid + catalyst}]{H_2C{=}CHR} R{-}(CH_2)_2{-}SiMe_2{-}O{-}[SiMe_2{-}O]_n{-}SiMe_2{-}(CH_2)_2{-}R$$

Scheme 1. Hydrosilylation of olefins (e.g., polyethers).

Employing the novel hydrosilylation process using ionic liquids, a broad range of different organomodified polydimethylsiloxanes were synthesized. The procedure is fairly simple and can be described as a one-pot synthesis (Fig. 5).

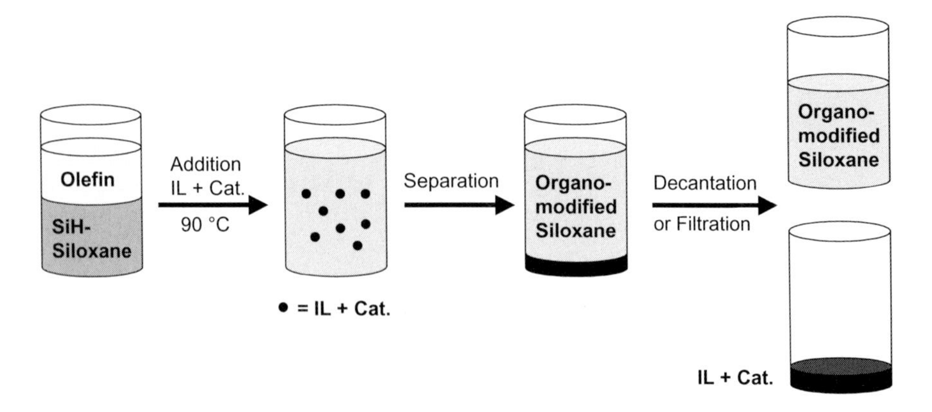

Fig. 5. Schematic representation of the hydrosilylation process using ionic liquids.

Typically, the reaction is performed in a liquid-liquid biphasic system where the substrates and products (upper phase) are not miscible with the catalyst/ionic liquid solution (lower phase). The SiH-functional polydimethylsiloxane and the olefin are placed in the reaction vessel and heated up to 90 °C. Then the precious metal catalyst (20 ppm) and the ionic liquid (1 %) are added. After complete SiH conversion, the reaction mixture is cooled to room temperature and the products are removed from the reaction mixture by either simple decantation or filtration (in case of non-room-temperature ionic liquids). The recovered catalyst/ionic liquid solution can be reused several times without any significant change in catalytic activity. A treatment or workup of the ionic liquid-catalyst solution after each reaction cycle is not necessary. The metal content of the products was analyzed by ICP-OES (*Inductively coupled plasma optical emission spectroscopy*) and the chemical identity of the organomodified polydimethylsiloxane was verified by NMR spectroscopy.

We chose SiH-functional polydimethylsiloxanes with different chain lengths and functionality patterns and polyethers with different ethylene (EO) and propylene (PO) oxide contents, as well as one small non-polyether olefin (AGE) in order to evaluate the influence of the hydrophilicity/hydrophobicity of the substrates and corresponding products on the catalytic performance of the various catalyst/ionic liquid solutions and, even more importantly, on the separation behavior of the ionic liquid at the end of the reaction. A clean separation of the ionic liquid from the products is the necessary condition to examine the partitioning of the catalyst between the two phases. Naturally, without a clean separation and a much better solubility of the catalyst in the ionic liquid phase than in the product phase a complete retention of the catalyst in the ionic liquid and the desired recovery of the catalyst cannot be achieved. The hydrophobicity of the organosilicon products increases with increasing chain length and decreasing SiH functionalization of the SiH-functional polydimethylsiloxane. The polyethers are more hydrophobic the higher is the content of propylene oxide. Now, one would expect that the ionic liquid because of its ionic nature, would separate more easily from the organosilicon products the more hydrophobic they are. During our studies, this anticipation was proven to be right. Figure 6 depicts some of the various polydimethylsiloxanes, olefins, ionic liquids, and catalysts used in our investigations; only the ionic liquids and catalysts are shown which gave satisfactory results.

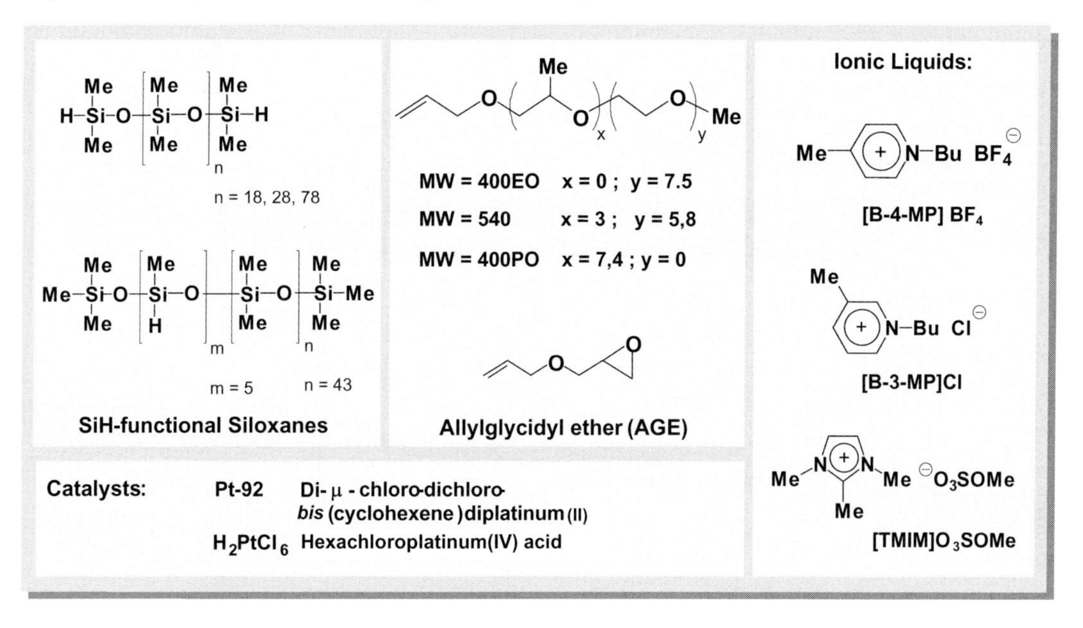

Fig. 6. Raw materials.

It turned out that for successful recovery of the catalyst and its reusability it is crucial to find an appropriate combination of a catalyst and an ionic liquid, which has to be harmonized with the hydrophilicity/hydrophobicity of the product. First of all, not every catalyst is soluble in every ionic liquid, and secondly, not every ionic liquid separates as readily as desired from the product phase. However, we were able to identify at least one suitable catalyst/ionic liquid combination for the synthesis of each polyethersiloxane (Table 1).

In some cases, more than one catalyst/ionic liquid combination gave good results (see entries Nos. 3 and 4, Table 1). It is noteworthy that all of the polyethersiloxanes synthesized in this way exhibit very different polarities.

The hydrosilylation reaction can also be conventionally conducted by reaction of an olefin and an SiH-functional polydimethylsiloxane in the presence of a standard transition metal catalyst, and after the reaction the catalyst can be extracted with an ionic liquid. In some cases, the use of an ionic liquid in the hydrosilylation process even improved the quality of the polyethersiloxanes with respect to color compared to the standard process. An explanation might be the avoidance of catalyst reduction leading to the formation of colloidal metal particles, which tend to color the product slightly brownish. In other words, the ionic liquid seems to have a stabilizing effect on the catalyst.

Hydrosilylation reactions with ionic liquids derived from 1-alkylimidazole, i.e. in 2-position unsubstituted 1,3-dialkylimidazolium salts, did not give the desired polyethersiloxanes. We assume that the proton in 2-position of the 1,3-dialkylimidazolium salts react with the SiH group of the

SiH-functional polydimethylsiloxanes with formation of the corresponding 1,3-dialkylimidazol-ylidene, which deactivates the hydrosilylation catalyst by coordination to the metal center.

Table 1. Results of hydrosilylation reactions of olefins with SiH-functional polydimethyl-siloxanes using ionic liquids.

Entry No.	Catalyst [20 ppm]	IL	SiH-siloxane	Olefin	SiH-conv. [%/h]	Detect. metal cont. [ppm]
1	Pt-92	TMIMS	n = 18	400EO	> 99/5	< 1
2	H_2PtCl_6	B-3-MP Cl	n = 18	540	> 99/3	< 1
3	H_2PtCl_6	B-4-MP TFB	n = 18	400PO	> 99/1	< 1
4	H_2PtCl_6	TMIMS	n = 18	400PO	93/5	< 1
5	H_2PtCl_6	TMIMS	n = 78	AGE	> 99/3	< 1
6	Pt-92	TMIMS	n = 28	AGE	> 99/1	< 1
7	H_2PtCl_6	TMIMS	m = 5	400PO	> 99/1	< 1
8	H_2PtCl_6	TMIMS	m = 5	AGE	> 99/3	2

Conclusions

A novel transition metal-catalyzed hydrosilylation process is described. The use of an ionic liquid in this process allows for the immobilization, heterogenization, and recovery of the expensive precious metal catalyst as well as its direct reuse in a subsequent hydrosilylation reaction. From an economic and ecological point of view, this process perfectly fits in the concept of "Sustainable Chemistry". Future research activities will aim at the prolongation of the catalyst life-time. For this, it is necessary to gain a deeper understanding of the catalytically active species in the catalyst/ionic liquid solution.

References

[1] B. Marcniec in *Applied Homogeneous Catalysis with Organometallic Compounds* (Eds.: B. Cornils, W. A. Herrmann), VCH, Weinheim, **1996**, *Vol. 1*, p. 487.

[2] B. Cornils, W. A. Herrmann in *Applied Homogeneous Catalysis with Organometallic Compounds* (Eds.: B. Cornils, W. A. Herrmann), VCH, Weinheim, **1996**, *Vol. 2*, p. 573.

[3] P. Wasserscheid, Transition metal catalysis in ionic liquids, in *Ionic Liquids in Synthesis* (Eds.: P. Wasserscheid, T. Welton), VCH, Weinheim, **2002**, p. 213 – 257.

[4] K. R. Seddon *J. Chem. Technol. Biotechnol.* **1997**, *68*, 351 – 356; P. Wasserscheid, T. Welton in: *Ionic Liquids in Synthesis*, VCH, Weinheim, **2002**.

[5] J. D. Holbrey, K. R. Seddon, *Clean Products and Processes* **1999**, *1*, 223; T. Welton, *Chem. Rev.* **1999**, *99*, 2071; P. Wasserscheid, W. Keim, *Angew. Chem. Int. Ed.* **2000**, *39*, 3772.

[6] http://www.solvent-innovation.de
[7] G. W. Parshall, *J. Am. Chem. Soc.* **1972**, *94*, 8716 – 8719.
[8] J. S. Wilkes, M. J. Zaworotko, *J. Chem. Soc., Chem. Commun.* **1992**, 965 – 967.
[9] P. A. Z. Suarez, J. E. L. Dullius, S. Einloft, R. F. de Souza, *Polyhedron* **1996**, *15*, 1217 – 1219.
[10] Y. Chauvin, L. Mußmann, H. Olivier, *Angew. Chem.* **1995**, *107*, 2941 – 2943.
[11] J. H. Davis, Jr., Synthesis of task-specific ionic liquids, in *Ionic Liquids in Synthesis* (Eds.: Peter Wasserscheid, T. Welton), VCH, Weinheim, **2002**, p. 33 – 40.
[12] http://www.degussa.com/en/structure/performance_materials/oligomers_silicones.html
[13] B. Weyershausen, K. Hell, U. Hesse, Production of organopolysiloxanes using ionic liquids, *Eur. Pat. Appl.*, EP1 382 630, 21.1.**2004**, Goldschmidt AG.
[14] S. Aubin, F. Le Floch, D. Carrie, J. P. Guegan, M. Vaultier, *ACS Symposium Series* **2002**, *818* (Ionic Liquids), 334 – 346.

On-Line FT-Raman Spectroscopy for Process Control of the Hydrosilylation Reaction

Frank Baumann,* Thomas List

Wacker-Chemie GmbH, Johannes-Hess-Strasse 24, D-844892 Burghausen, Germany
Tel.: +49 8677 83 1662 — Fax: +49 8677 83 6265
E-mail: frank.baumann@wacker.com

Keywords: FT-raman spectroscopy, hydrosilylation, on-line process control

Summary: The use of the on-line FT-Raman spectroscopy for monitoring a multi-step hydrosilylation reaction combines all the advantages of an on-line analytical tool (like "real time" measuring results, a direct view into the reaction, and no off-line sample collection) with the requirements for the application of technology in production plants, e. g., low calibration effort within a wide temperature range, stable calibration, "simple system handling" for the operator, small sized equipment at the reaction vessel, and no contact with the reaction media.

Introduction

There are two main success factors for a company in the market from the point of view of the supply chain. One is to bring the product with the right quality at the right time to the customer. The other is to have the best and most cost-effective processes. Therefore, there is always pressure to optimize the processes with respect to both product quality and cost effectiveness

One excellent tool to fulfill these requirements is an on-line analytical method which allows a direct view into the running process and give the plant manager a chance to optimize, for example, the general process design, the running time of the processes, the raw material usage factor, the catalyst concentration, the overall yield of the process, and the product quality.

Despite this potential benefits, there are some special requirements for a production plant which have to be fulfilled by the on-line analytical method to control a chemical reaction, like low calibration effort with a wide temperature range, "simple system handling" for the operator, small and robust equipment at the reaction vessel, and preferable no contact of the measurement equipment with the reaction media.

Therefore in this presentation we will focus on an on-line FT-Raman Spectroscopy method for process control on the pilot plant scale.

Phenomenological Description of the Raman Effect

The Raman effect can be described using the energy level model of quantum mechanics. By using, for example a "green" photon, this photon raises the energy of the molecule from the ground state to a "virtual" state. This virtual state is not a stationary energy state of the molecule in the quantum mechanical sense. Rather, it is just a distortion of the electron distribution of a covalent bond. The molecule immediately relaxes back to the original electronic state by emitting a photon. There are now three options:

- The molecule returns to the same vibrational energy level from where it started; therefore the wavelength of the emitted photon is the same as the wavelength of the initial photon.
- The molecule returns to a higher vibrational energy level than that from where it started; therefore the emitted photon has a longer wavelength than the wavelength of the initial photon. This event is called Stokes Raman scattering.
- The molecule returns to a lower vibrational energy level than that from where it started; therefore the emitted photon has a shorter wavelength than the wavelength of the initial photon. This event is called the anti-Stokes Raman scattering.

In general, Stokes Raman scattering is used for spectroscopy. The intensity of the Raman scattering light is direct proportional to the number of molecules or molecule parts which are illuminated, i. e. to the concentration of molecules or molecule parts of one kind [1, 2].

Example of the Application of the FT-Raman Spectroscopy on the Pilot Plant Scale

Characteristics of the FT-Raman Spectrometer

The FT-Raman spectrometer used HP-532/1-50/1-4CH is a commercially available system from Kaiser Optical Systems. The system contains a green LASER with a wavelength of 532 nm and an output power of 35 mW. The power output at the probe head is about 10 mW. The focused probe head has a sapphire window and can be used up to 280 °C and 60 bar pressure. The probe head is connected with a 100 µm optical fiber with a CCD detector for the Raman scattered light and an FT-spectrometer. The FT-spectrometer has a spectral coverage from 100 to 4400 cm^{-1} and a spectral resolution of 5 cm^{-1} [3].

The Reaction Monitoring for the Hydrosilylation Reaction

In this paper we focus only on the hydrosilylation reaction of siloxanes. During this reaction the Si–H bond is converted to a Si–CH_2 bond. This process is widely used for the preparation of organofunctional silanes and siloxanes in the chemical industry.

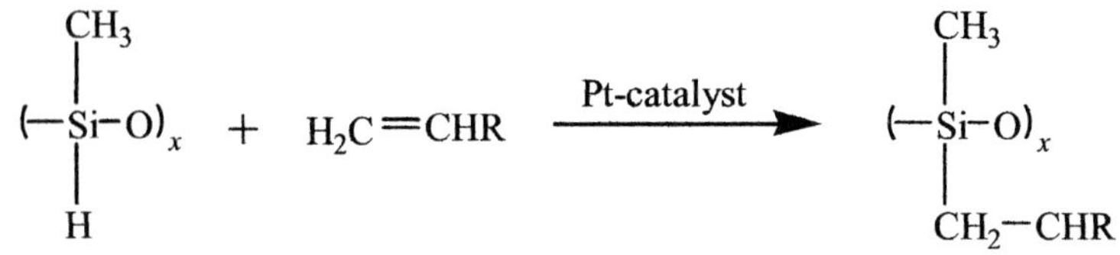

$$(-\underset{\mathrm{H}}{\overset{\mathrm{CH_3}}{\mathrm{Si}}}-\mathrm{O})_x + \mathrm{H_2C{=}CHR} \xrightarrow{\text{Pt-catalyst}} (-\underset{\mathrm{CH_2-CHR}}{\overset{\mathrm{CH_3}}{\mathrm{Si}}}-\mathrm{O})_x$$

R: Acyl group, Aryl group, Chloracyl group

Fig. 1. Reaction model.

The decrease in the Si–H concentration in the reaction mixture could easily be observed via Raman spectroscopy because there is a single, isolated peak at about 2160 cm^{-1}, Si–H stretch bending (see Fig. 2). Also the vinyl peak can clearly be detected at about 1600 cm^{-1} [3].

Furthermore, at wave numbers between 2850 and 3100 cm^{-1}, the Si–CH_3 stretch bending could be observed, which will be used for the internal calibration of the Si–H peak of the measured system. This peak remains constant in intensity during the hydrosilylation reaction (see Fig. 2).

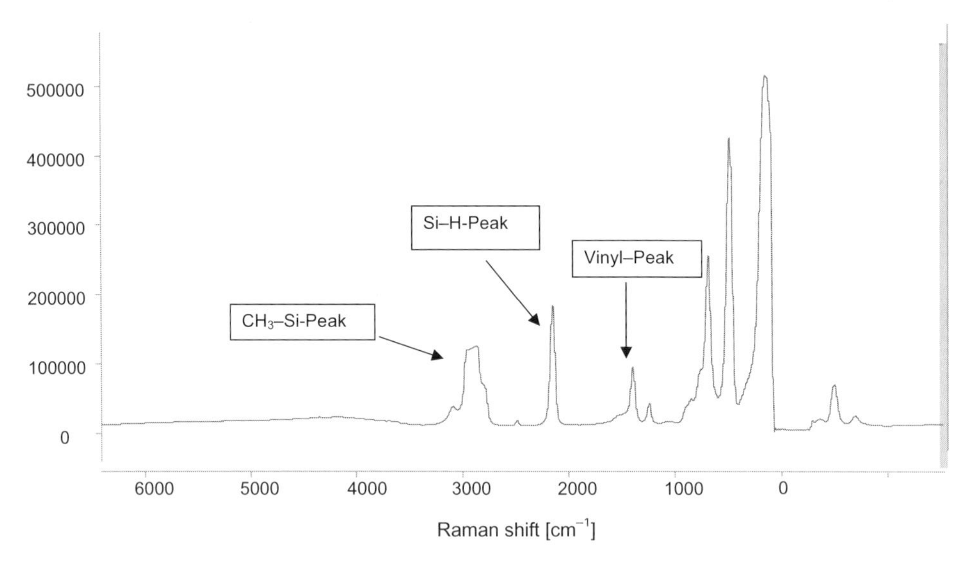

Fig. 2. Example of a Raman spectrum of a reaction mixture at the start of the reaction.

Description of the Hydrosilylation Process in the Pilot Plant

The process which is on-line controlled is a two-step continuous hydrosilylation of Si–H groups containing polydimethylsiloxane with two different α-olefins. In the first step of the hydrosilylation, styrene is bound to the siloxane backbone. During the second step, the olefin 1-dodecene is connected to the siloxane backbone.

The process is under control if the residual content of Si–H groups is lower than 3 mol% relative to the new Si–CH_2 bonds.

For the experiments, the molar ratio of Si–H to the total amount of used vinyl groups, the product throughput through the reactor, and the quality of the educts and the catalyst were constant within natural variation.

Calibration of the Raman Spectrometer

The end product of the hydrosilylation reaction is mixed with an exact amount of the Si–H-containing educt (6 samples). These samples are analyzed in parallel by ^{1}H NMR and an FT-Raman spectrometry probe head which is connected to the spectrometer via a fiber optical cable. For the FT-Raman measurement, the same parameters are used which will be used later for the real measurement, e.g., measuring time 60 s and background measuring before every measurement.

Finally, both results are correlated and the calibration is checked with another independent sample.

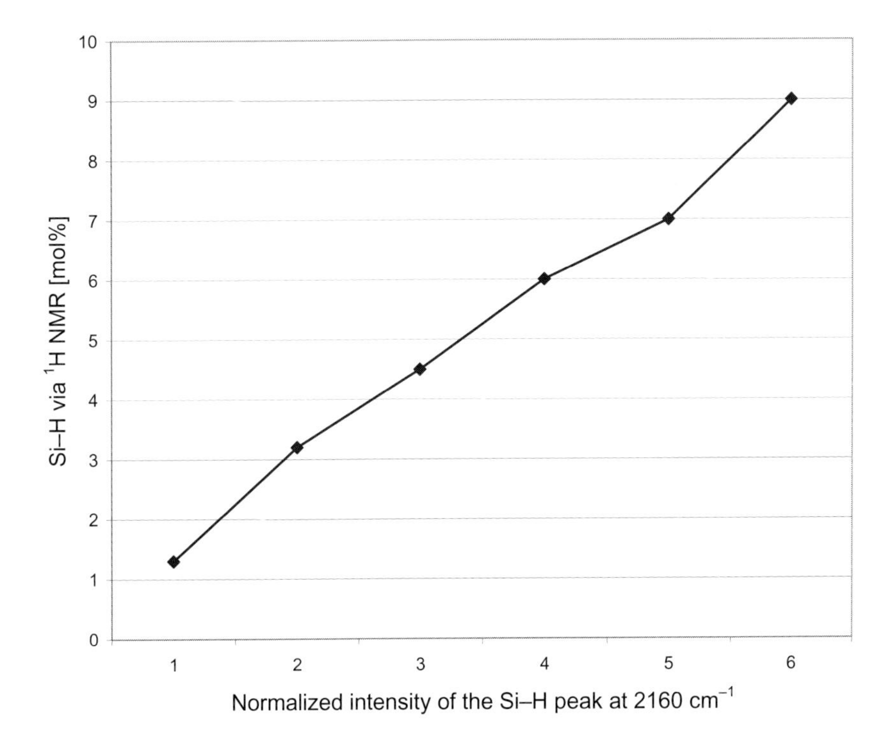

Fig. 3. Example of a calibration.

On-line FT-Raman Spectrometry Measurement Results for the Continuous Hydrosilylation Process

The first diagram (Fig. 4) shows an optimal process for the continuous two-step hydrosilylation as described above. The process runs for about 12 h well below the upper specification limit of 3 mol% Si–H (gray line in the diagram).

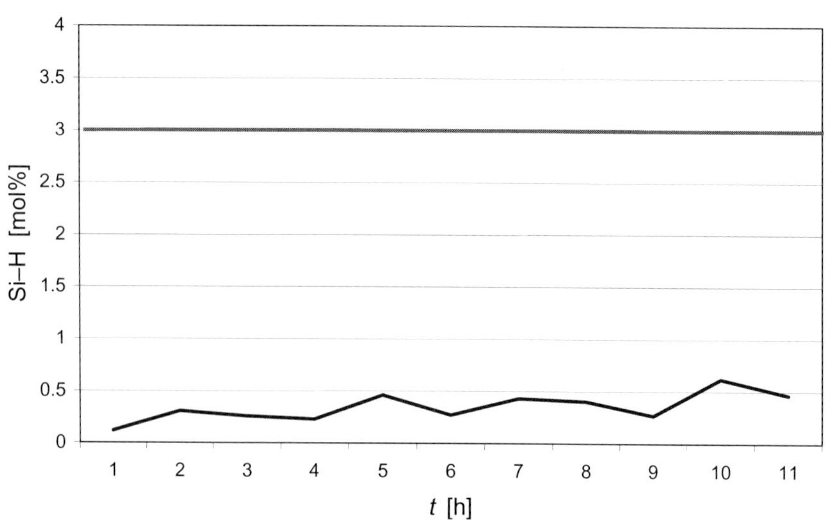

Fig. 4. Optimum of the reaction parameters.

Deviation of the reactor temperature from the optimum reaction temperature is directly detected by the on-line Raman FT-spectrometer as shown in Fig. 5.

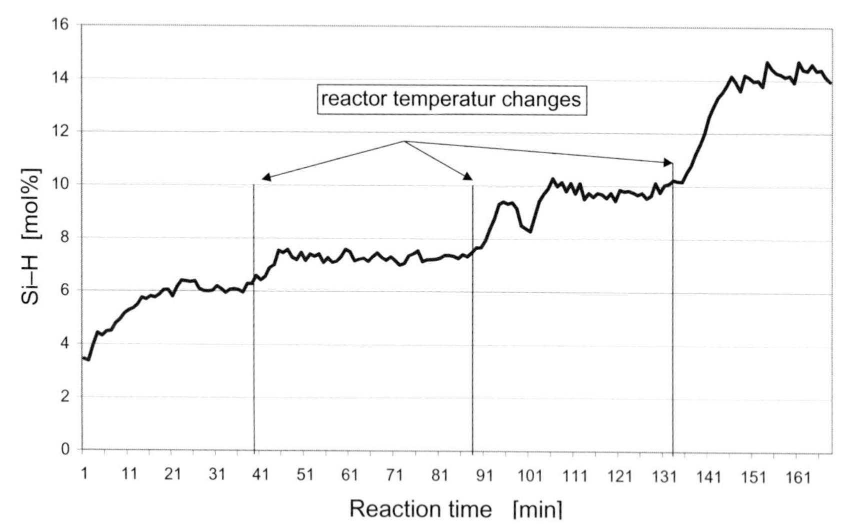

Fig. 5. Temperature effect on the reaction.

Figure 6 shows a typical application of the on-line FT-Raman spectrometer in a pilot plant for a process optimization. During the running production, the amount of used catalyst is reduced step-wise. The operator gets a direct feedback of the reaction system. In this case the amount of catalyst can be reduced by about 20%.

This last example shows that the on-line FT-Raman spectrometer is a powerful tool to optimize processes in both ways: you can create a cost-effective process, which produces a high quality product.

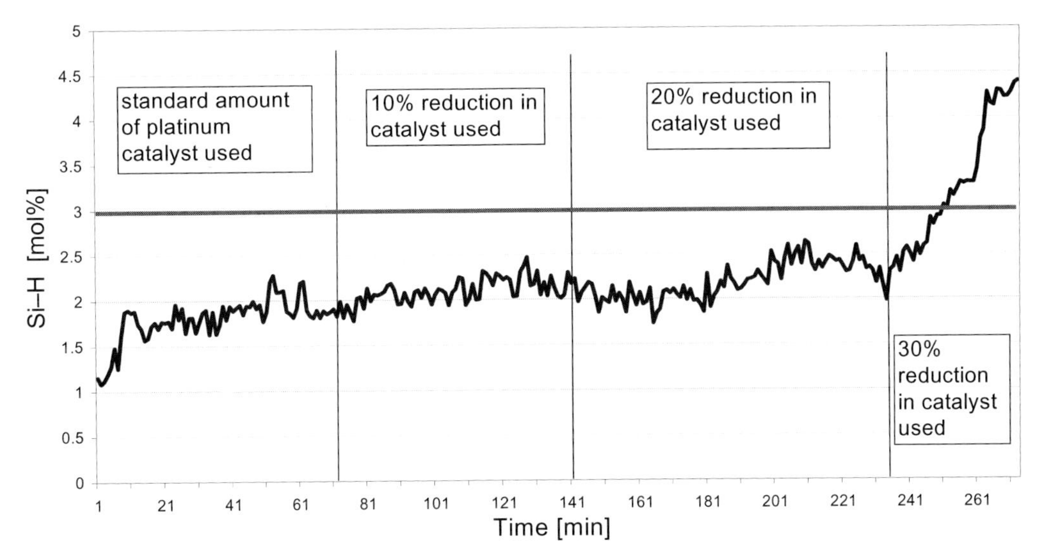

Fig. 6. Reduction of the amount of catalyst.

References

[1] R.W. Wood, *Nature* **1928**, *122*, 349.

[2] M. J. Pelletier, in *Analytical Application of Raman Spectroscopy*, Blackwell Science, **1999**, p. 3 ff.

[3] E. L. Lipp, M. A. Leugers, in *Analytical Application of Raman Spectroscopy*, Blackwell Science, **1999**, p. 108 ff.

[2+2]-Cycloadditions of $(OC)_4Fe{=}SiMe_2$ — Theoretical Study

*Uwe Böhme**

Institut für Anorganische Chemie, TU Bergakademie Freiberg,
Leipziger Str. 29, 09596 Freiberg, Germany
E-mail: uwe.boehme@chemie.tu-freiberg.de

Keywords: silylene complexes, iron silylene, density functional calculations

Summary: The mechanisms of [2+2]-cycloadditions of (OC)4Fe=$SiMe_2$ with the reagents ethylene, acetylene, HCN, and OCH_2 have been investigated by the B3LYP method. An effective core potential and valence double zeta basis set for iron and 6-31G* basis set for all main group elements were used. The *in situ* generation of the donor-free iron silylene complex $(OC)_4Fe{=}SiMe_2$ is possible with an activation energy of 86.9 kJ/mol starting from $(OC)_4Fe{=}SiMe_2(HMPA)$. The dimerization of $(OC)_4Fe{=}SiMe_2$ is the dominant reaction which always must be taken into account. Ethylene and acetylene are mainly coordinated at the silicon atom in the transition states. The addition of the polar reagents HCN and formaldehyde occurs via the facile formation of adducts. Only the reactions of $(OC)_4Fe{=}SiMe_2$ with acetylene and with formaldehyde are able to compete with the dimerization because of the low activation barriers and the strong exothermic reactions.

Introduction

Transition metal silylene complexes have been considered as important intermediates in the dehydrogenative coupling of silanes by late transition metal catalysts in the hydrosilylation process, in the chemical vapor deposition of silicides from volatile precursors, in the "Direct Process" for the production of methylchlorosilanes, and in silylene transfer reactions to unsaturated organic compounds [1, 2]. Two general approaches to the synthesis of silylene complexes of tetracarbonyl iron are known. The first is the reaction of $Na_2Fe(CO)_4$ with dichlorodiorganosilanes in the presence of HMPA (HMPA = hexamethylphosphoramide, $(Me_2N)_3P{=}O$). The second uses the *in situ* generation of the carbonylate anion by deprotonation of $H_2Fe(CO)_4$ with NEt_3 [2].

Silylene complexes show high reactivities toward nucleophiles such as water, alcohols, ketones, and isocyanates [3, 4]. Some typical reactions of $(OC)_4Fe{=}SiR_2(HMPA)$ are summarized in Scheme 1 [2]. The dimerization forming the diferradisilacyclobutane is the only known [2+2]-cycloaddition starting from **1**. The photochemical reaction of **1** with 2,3-dimetylbutadiene gives the product of a [4+1]-cycloaddition. This reaction proceeds under cleavage of the Si–Fe bond.

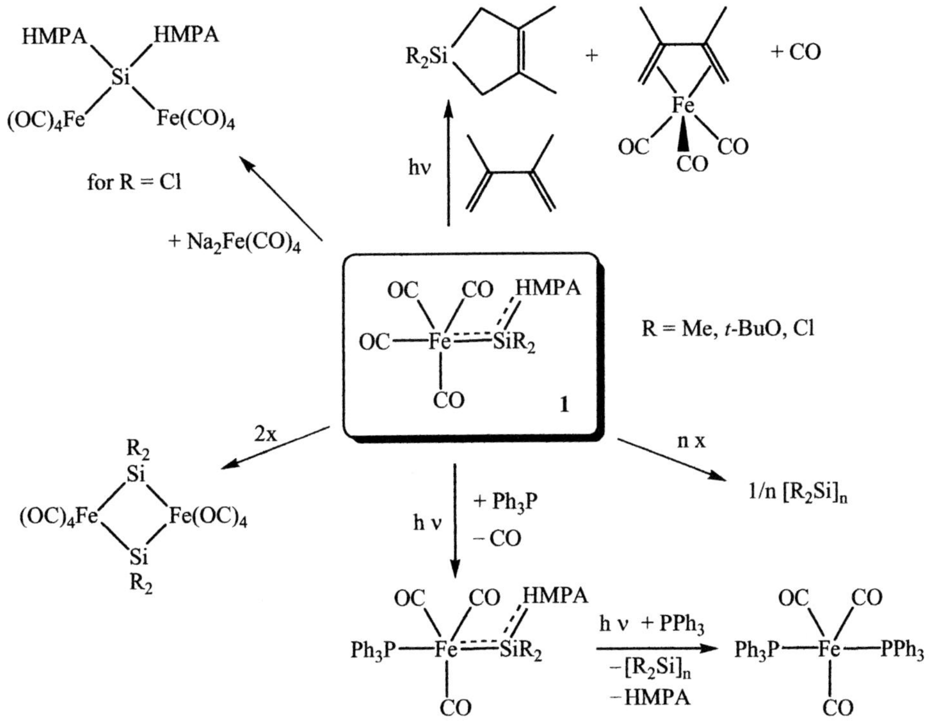

Scheme 1. Reactions of $(OC)_4FeSiR_2(HMPA)$.

Results and Discussion

Very little is known about [2+2]-cycloadditions with transition metal silylene complexes. The mechanisms of [2+2]-cycloadditions of $(OC)_4Fe{=}SiMe_2$ with the reagents ethylene, acetylene, HCN, and OCH_2 have been investigated by the B3LYP method [5, 6]. All geometries have been fully optimized. Geometry optimization and frequency calculations have been performed with an effective core potential and valence double zeta basis set for iron [7] plus an additional polarization function [8] and the 6-31G* basis set for all main group elements [9]. The structures of all molecules have been identified as minima with 0 imaginary frequencies or as transition states with one imaginary frequency by calculating the Hessian-Matrices. The energies are in kJ/mol and have been calculated from the total energies of the optimized molecules including zero point corrections.

$(OC)_4Fe{=}SiMe_2$ is only stable as an adduct with suitable donor ligands. HMPA is often used as the donor, but other ligands like THF or 2-*N,N*-dimethylaminomethylphenyl derivatives have also been reported [10]. HMPA stabilizes the silylene complex by 86.9 kJ/mol. Trimethylamine stabilizes the silylene complex by 72.1 kJ/mol and THF by 63.3 kJ/mol (see Scheme 2).

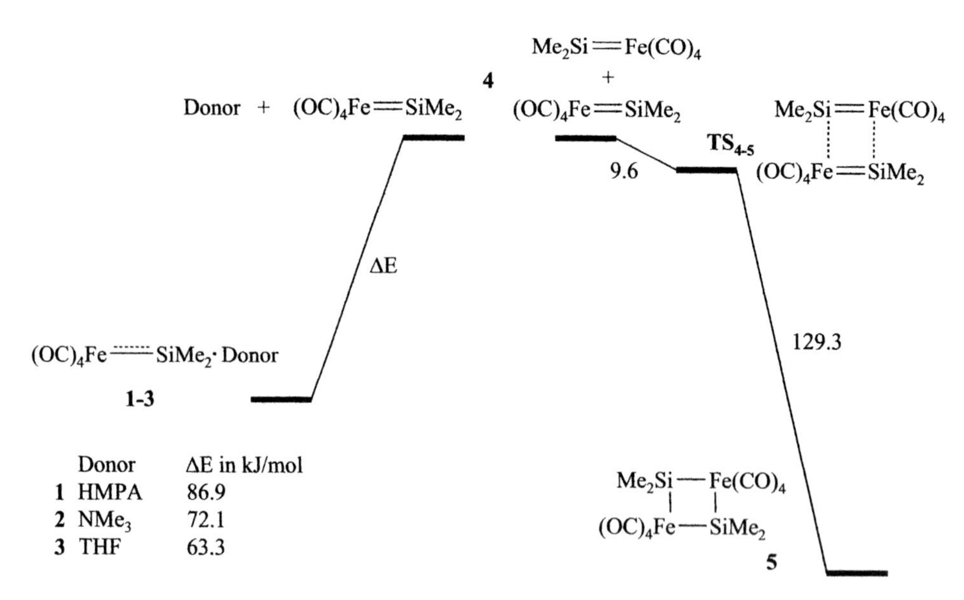

Scheme 2. Relative energies in kJ/mol for the formation (left) and the dimerization of $(OC)_4Fe{=}SiMe_2$ (right).

$(OC)_4Fe{=}SiMe_2$ (**4**) has C_{2v} symmetry as shown in Fig. 1. The double bond between Fe and Si is weakened in the donor adduct **1**. The pyramidal distortion of the substituents around the silicon atom allows better interaction between the empty p-orbital and the donor atom (see Fig. 1). The removal of the stabilizing ligand is accompanied by a rearrangement of the CO ligands from C_s symmetry to the arrangement shown in Fig. 2.

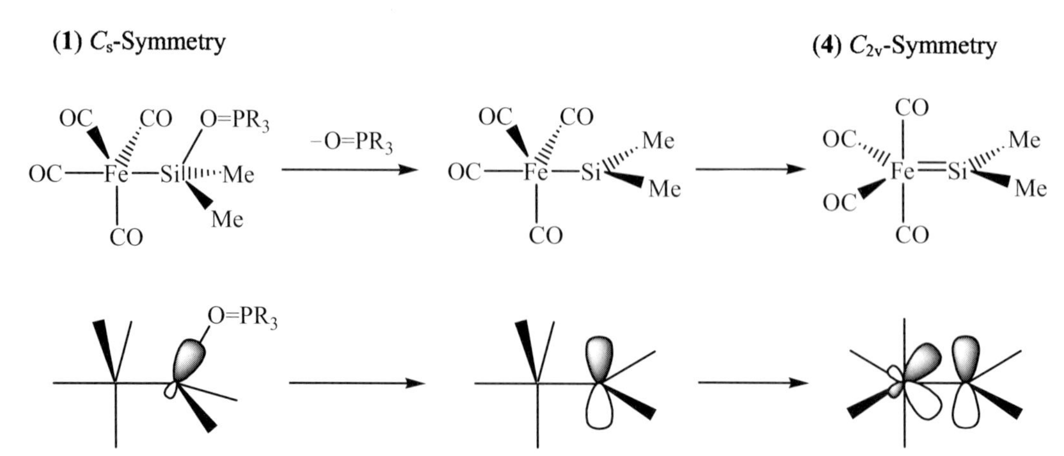

Fig. 1. Schematic representation of the orbital interactions between iron, silicon, and the stabilizing ligand during the transformation from $(OC)_4FeSiMe_2(HMPA)$ (left) to $(OC)_4FeSiMe_2$ (right).

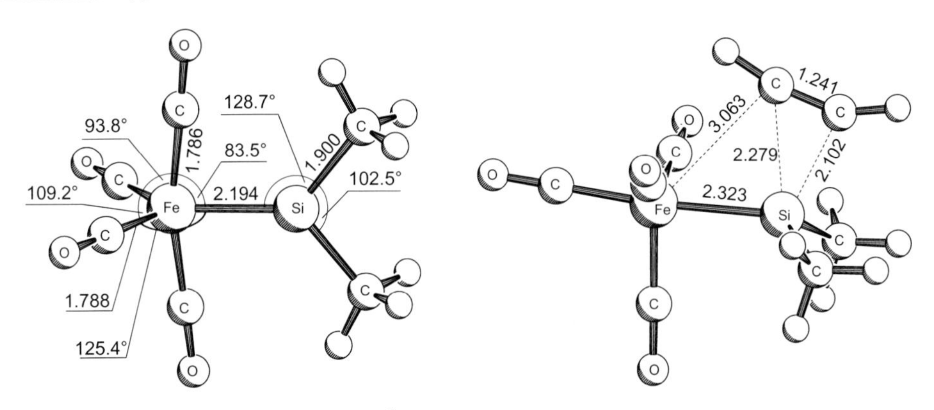

Fig. 2. Calculated geometries (bond lengths in Å, angles in degrees) of **4** (left) and $\mathbf{TS_{4\text{–}6}}$ (right).

Dimerization of $(OC)_4Fe{=}SiMe_2$

The free silylene complex **4** is a highly reactive intermediate. Therefore the dimerization reaction occurs without activation energy, leading to the transition state $\mathbf{TS_{4\text{–}5}}$. The dimerization is exothermic by 138.9 kJ/mol (Scheme 2).

Cycloaddition with Nonpolar Reagents

Moderate activation energies of 20.7 and 33.7 kJ/mol are necessary for the cycloaddition of ethylene and acetylene, respectively (Scheme 3). The reagent is mainly coordinated at the

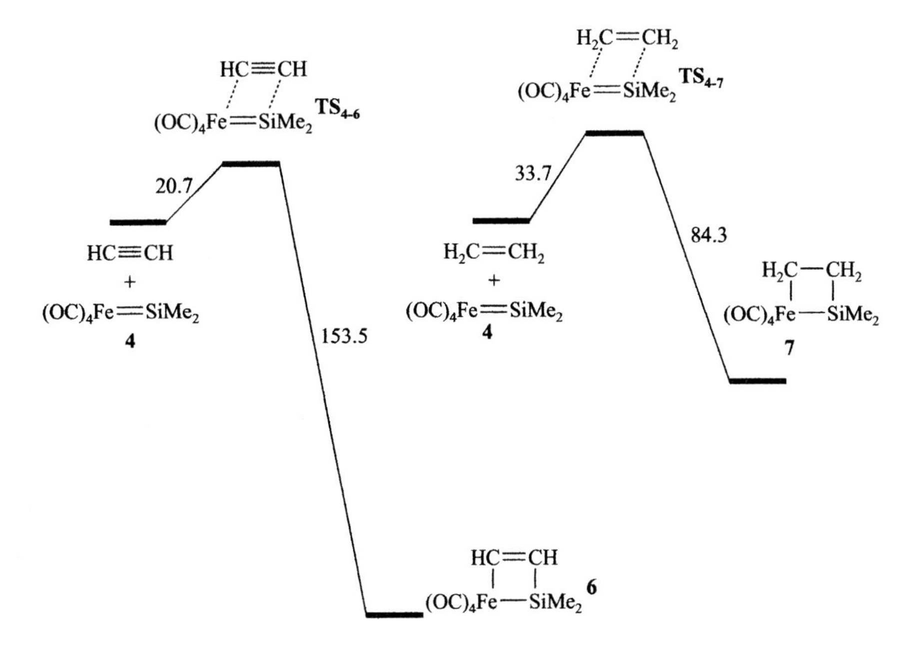

Scheme 3. Energy diagram (in kJ/mol) for the cycloaddition of nonpolar reagents to **4**.

unsaturated silicon atom in the transition state as shown in Fig. 2. Both reactions are exothermic, the addition of acetylene by 132.8 and the addition of ethylene by 50.6 kJ/mol. We have to consider the formation of two molecules of product if we compare the energy differences with the energy to form the dimer **5**. The addition of ethylene will probably not occur (101.2 kJ for two mol **7**), since the dimerization of **4** is more exothermic. Only the addition of acetylene might compete with the dimerization (265.6 kJ for two mol **6**).

Cycloaddition with Polar Reagents

The addition reactions of the polar reagents HCN and formaldehyde occur via the facile formation of the adducts $\mathbf{AD_{4\text{-}8}}$ and $\mathbf{AD_{4\text{-}9}}$. Moderate activation energies of 17.7 and 49.3 kJ/mol respectively lead to the transition states of the cycloaddition (Scheme 4). The optimized structures of $\mathbf{AD_{4\text{-}9}}$ and

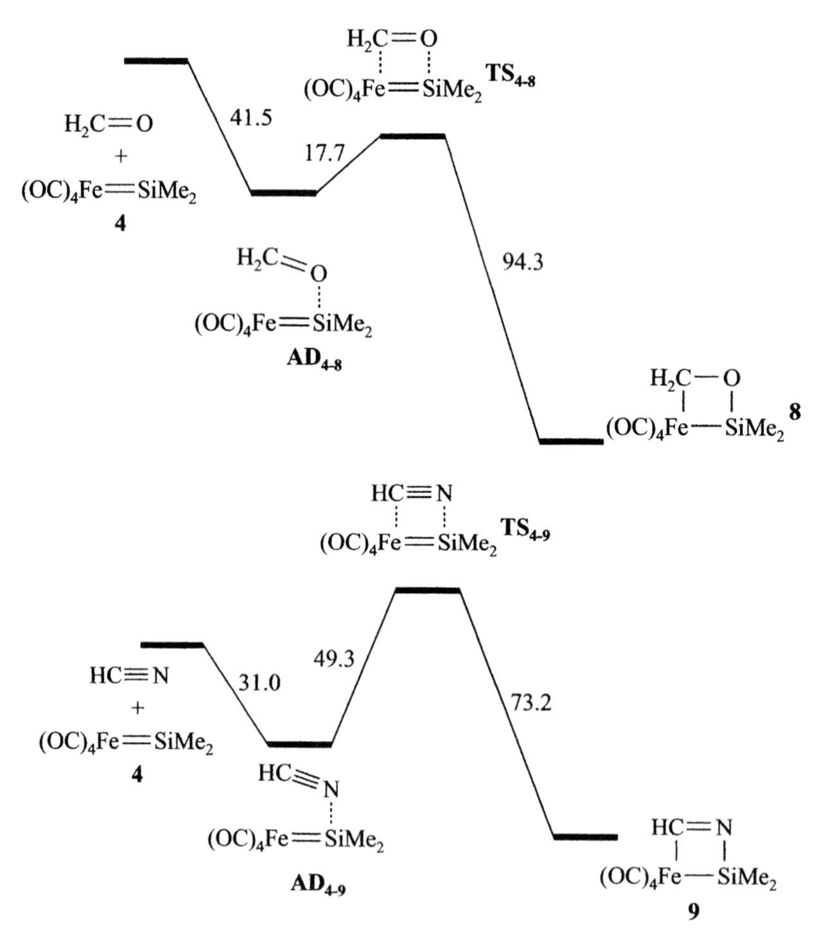

Scheme 4. Energy diagram (in kJ/mol) for the cycloaddition of polar reagents with **4** (top: addition of OCH_2; bottom: addition of HCN).

$TS_{4\text{–}9}$ are shown as examples in Fig. 3. The nitrogen atom of HCN is coordinated at the silicon atom in **$AD_{4\text{–}9}$**. This leads to a pyramidalization of the silicon atom. The angle C3-N-Si is 169.7°. This angle is reduced to 111.9° in the transition state **$TS_{4\text{–}9}$**. The distance N–Si remains almost unchanged.

The transformation from **4** to **8** is exothermic by 118.1 kJ/mol, the transformation from **4** to **9** by 54.9 kJ/mol. Only the formation of **8** is more exothermic than the dimerization (236.2 kJ for two mol **8**).

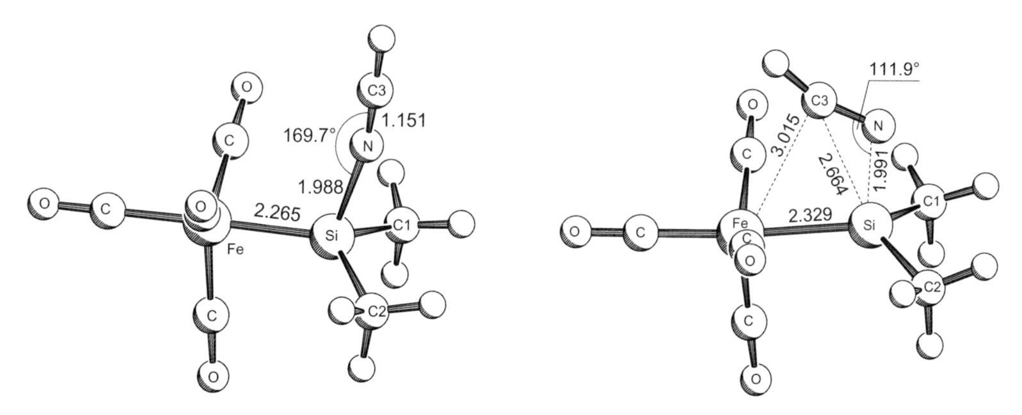

Fig. 3. Calculated geometries (bond lengths in Å, angles in degrees) of **$AD_{4\text{–}9}$** (left) and **$TS_{4\text{–}9}$** (right).

Conclusions

The *in situ* generation of the donor-free iron silylene complex **4** is possible with a moderate activation energy. **4** is a highly reactive intermediate which undergoes a number of subsequent reactions. The dimerization to **5** is the dominant reaction which always must be taken into account. Cycloaddition reactions with **4** might be possible if two conditions are fulfilled: (a) the activation energy reaching the transition state is not too high (kinetic reason); (b) the overall reaction is at least as exothermic as the dimerization of **4** (thermodynamic reason).

Both conditions are fulfilled for the addition of acetylene leading to **6** and for the addition of formaldehyde leading to **8**. Further experimental work is necessary to prove the statements of this theoretical investigation.

Acknowledgments: I wish to thank the computing centers of the TU Bergakademie Freiberg and the TU Chemnitz for supplying disk space and computing time.

References

[1] T. D. Tilley, Transition-metal silyl derivatives, in *The Chemistry of Organic Silicon*

Compounds, (Eds.: S. Patai, Z. Rappoport), J. Wiley & Sons, Chichester **1989**, 1415.
[2] C. Zybill, *Top. Curr. Chem.* **1991**, *160*, 1.
[3] M. Okazaki, H. Tobita, H. Ogino, *Dalton Trans.* **2003**, 493.
[4] G. Raabe, J. Michel, Multiple bonds to silicon, in *The Chemistry of Organic Silicon Compounds* (Eds.: S. Patai, Z. Rappoport), J. Wiley & Sons, Chichester **1989**, 1015.
[5] A. D. Becke, *J. Chem. Phys.* **1993**, *98*, 5648.
[6] P. J. Stevens, F. J. Devlin, C. F. Chablowski, M. J. Frisch, *J. Phys. Chem.* **1994**, *98*, 11623.
[7] P. J. Hay, W. R. Wadt, *J. Chem. Phys.* **1985**, *82*, 299.
[8] A. W. Ehlers, M. Böhme, S. Dapprich, A. Gobbi, A. Höllwarth, V. Jonas, K. F. Köhler, R. Stegmann, A. Veldkamp, G. Frenking, *Chem. Phys. Lett.* **1993**, *208*, 111.
[9] P. C. Hariharan, J. A. Pople, *Theoret. Chimica Acta* **1973**, *28*, 213.
[10] M. Eisen, Transition-metal silyl complexes, in *The Chemistry of Organic Silicon Compounds*, Vol. 2 (Eds.: Z. Rappoport, Y. Apeloig), J. Wiley & Sons, Chichester **1998**, 2037.

Reactions of Undecamethylcyclohexasilyl-potassium with Transition Metal Compounds

Florian Hoffmann,* Uwe Böhme, Gerhard Roewer

Institut für Anorganische Chemie, Technische Universität Bergakademie Freiberg
Leipziger Str. 29, 09596 Freiberg, Germany
Tel.: +49 3731 39 3603 — Fax: +49 3731 39 4058
E-mail: Florian.Hoffmann@chemie.tu-freiberg.de

Keywords: undecamethylcyclohexasilylpotassium, transition metal compounds, silyl complexes

Summary: The reactions of undecamethylcyclohexasilylpotassium (**1**) with various transition metal compounds were examined. Depending on the metal compound, **1** can act in four ways: (a) as a nucleophile, (b) as a donor ligand, (c) as a strong base, and (d) as a reducing agent. Only (a) and (b) lead to transition metal silyl complexes, while (c) and (d) bring on unwanted side reactions.

Introduction

Some years ago Hengge et al. published a feasible synthesis of KSi_6Me_{11} (**1**) and did basic research on the reactivity of this silyl anion including the preparation of transition metal complexes [1 – 3]. Our group contributed two crystal structures of such compounds [4, 5]. In an effort to get further insight into the reactivity of **1** and the structures of its transition metal complexes, we reacted it with various transition metal compounds.

$CpFe(CO)_2I$ (2), $EtMe_4C_5Fe(CO)_2I$ (3), and $CpNiPPh_3Cl$ (4)

2 and **4** are easily accessible organo transition metal halides which should react with **1** under formation of silyl complexes. Unfortunately they do not.

1 is protonated in the reaction with **4** giving $Me_{11}Si_6H$ (**5**) as the major silicon-containing product (Scheme 1). The proton source is probably **4**, although participation of the ether solvent cannot be excluded. The ^{29}Si NMR spectrum also featured several small signals showing silicon–phosphorus coupling. This indicates that to a minor extent silicon–phosphorus or even silicon–phosphorus–nickel compounds are formed. But the amounts were too small to determine their structure. Thus one can conclude that this reaction is not effective in the synthesis of nickel silyl complexes.

2 is deprotonated by **1** as well as reduced to $[CpFe(CO)_2]_2$ (**6**) (Scheme 1). Thus, the silicon-containing products are **5** and $(Me_{11}Si_6)_2$ (**7**). The ^{13}C NMR spectrum showed the signals of a ring-substituted CpFe-complex besides that of **6**, which proves that **2** and/or **6** are the proton source.

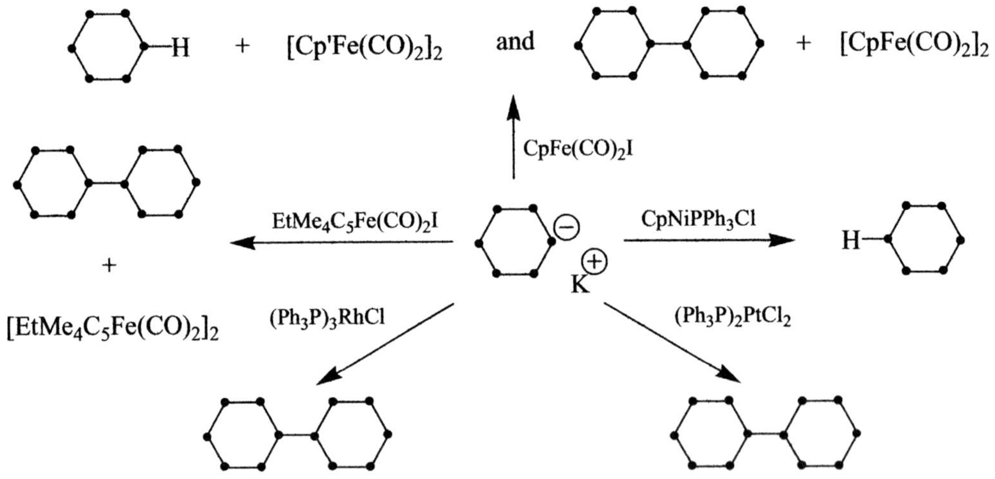

Scheme 1. Reactions of **1** with various transition metal halides.

If deprotonation of the educt compound is impossible, like in **3**, only reduction occurs, **7** and $[EtMe_4C_5Fe(CO)_2]_2$ (**8**) being the reaction products (Scheme 1). Variation of solvent and temperature did not change this result.

Obviously the used cyclopentadienyl donor ligand late transition metal halides are unsuitable reagents to form silyl complexes with **1**. In order to determine the molecular structure of $CpFe(CO)_2Si_6Me_{11}$ (**9**), it was finally prepared from $KCpFe(CO)_2$ and $Me_{11}Si_6I$ [6]. The results of the X-ray structure analysis are shown in Table 1 and Fig. 1 [7].

Table 1. Details of the X-ray structure analysis of **9**.

Space group	$P2_1/n$	R (I > 2σ(I))	0.023
Z	4	R (all data)	0.039
a	1150.9 pm	α	90°
b	1467.1 pm	β	105.6°
c	1704.8 pm	γ	90°

The $CpFe(CO)_2$ moiety occupies an equatorial position on the ring. The Cp and CO ligands lie in the voids between the methyl groups of the Si_6Me_{11} ring, thus minimizing the steric repulsion. The Fe–Si bond (236.4 pm) is markedly longer than the average value of the currently known Fe–Si bond lengths (230.8 pm [8]). The Si–Si bonds of the cyclohexasilane ring (234.1 – 235.9 pm) are somewhat elongated in comparison with Si_6Me_{12} (233.7 ± 0.5 pm) [9]. This elongation is most distinct next to the iron-bearing silicon atom and decreases with increasing distance.

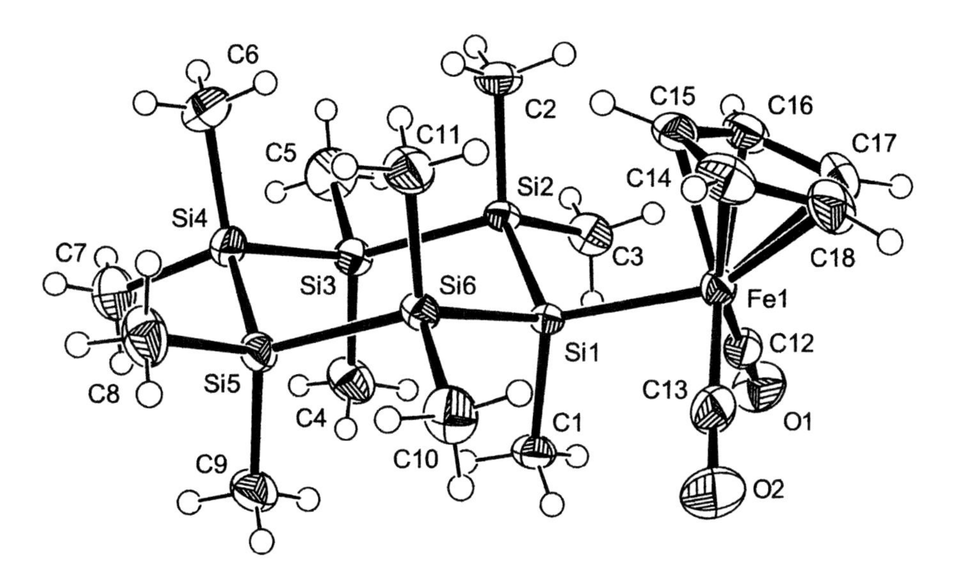

Fig. 1. Molecular structure of **9**.

$(Ph_3P)_3RhCl$ (10) and $(Ph_3P)_2PtCl_2$ (11)

In order to synthesize cyclohexasilyl complexes of noble metals, we reacted **10** and **11** with **1**. But unfortunately, redox reactions and decomposition took place, (**7**) and PPh_3 (**12**) being the major products (Scheme 1).

$Mn(CO)_5Br$ (13) and $Re(CO)_5Br$ (14)

The reaction of **13** with **1** giving $(OC)_5MnSi_6Me_{11}$ (**15**) was first reported by Hengge et al [2]. **14** reacts analogously with formation of $(OC)_5ReSi_6Me_{11}$ (**16**), obtained as colorless crystals (Eq. 1).

⊖ K⊕ + OC CO Br−Re−CO OC CO —(-KBr)→ OC CO −Re−CO OC CO

Eq. 1. Synthesis of **16**.

In both syntheses reduction of the metal center occurs as a side reaction. In the case of manganese this is negligible, the reduction products being $Mn_2(CO)_{10}$ (**17**), $Mn(CO)_5^-$ (**18**), and $Mn_3(CO)_{14}^-$ (**19**), identified by IR spectroscopy. In contrast, $Re(CO)_5Br$ gives rise to considerable amounts of $Re_2(CO)_{10}$ (**20**) and additionally to small amounts of higher rhenium carbonyl clusters which could not be identified. $Re(CO)_5^-$ (**21**) was not observed.

The IR and ^{29}Si NMR data of **15** and **16** as well as of $(OC)_5MnSi_5Me_9$ (**22**) and $(OC)_5ReSi_5Me_9$ (**23**), which were observed as by-products because the used **1** contained minor amounts of KSi_5Me_9 (**24**), are shown in Tables 2 and 3.

Table 2. IR data of **15**, **16**, **30**, **33**, and **34** (solution).

Compound	νCO [cm^{-1}]
15	2089 (m), 1993 (s), 1960 (w)
16	2109 (w), 2003 (s), 1995 (m,sh)
30	2005 (m), 1916 (m), 1884 (s), 1867 (m, sh)
33	1980 (s), 1895 (m, sh), 1874 (s), 1861 (m, sh)
34	1988 (s), 1918 (s)

Table 3. ^{29}Si NMR data of **15**, **16**, **22**, **23**, **30**, and **33** – **37**.

Compound	Si$_\alpha$ [ppm]	Si$_\beta$ [ppm]	Si$_\gamma$ [ppm]	Si$_\delta$ [ppm]
15	–43.2[a]	–28.7	–38.6	–42.5
16	–80.1	–30.4	–39.2	–42.2
22	–45.8	–28.5	–42.4	–
23	–86.5	–28.6	–41.9	–
30	–48.7	–31.1	–41.0	–42.7
33	–22.4	–32.7	–41.0	–42.6
34	–56.5	–33.9	–41.1	–41.7
35	–53.6	–28.6	–42.7	–
36	–23.6	–33.0	–40.9	–
37	–63.8	–30.1	–41.0	–

[a] The value reported in [2] is probably in error.

$M(CO)_6$ (M=Cr (25), Mo (26), W (27)), $Fe(CO)_5$ (28), and $Ni(CO)_4$ (29)

Because silyl anions are much more reactive in substitution reactions than the isosteric phosphanes [10], we examined the reaction of **1** with transition metal carbonyls.

Indeed, one CO ligand is substituted by $Me_{11}Si_6^-$, giving anionic silyl complexes (Eq. 2). The Mo and W compounds **31** and **32** are already known and characterized [4, 5]. The IR and ^{29}Si NMR data of the hitherto unknown complexes **30**, **33**, and **34**, as well as of the by-products $(OC)_5CrSi_5Me_9^-$ (**35**), $(OC)_4FeSi_5Me_9^-$ (**36**), and $(OC)_3NiSi_5Me_9^-$ (**37**) are shown in Tables 2 and 3.

The ^{29}Si NMR spectra of **30**, **33**, and **34** feature a considerable downfield shift of the Si_α atoms compared to **1** (^{29}Si NMR: Si_α –109.1 ppm, Si_β –34.2 ppm, Si_γ –40.9 ppm, Si_δ –41.1 ppm [1]), especially in **33**. The shifts of the other Si atoms are almost unchanged. A similar effect is observed for the Si_5Me_9 complexes **35** – **37**.

$$K^{+}\,[Si_6Me_{11}]^{-} + M(CO)_n \xrightarrow[-CO]{DME} K(DME)_x^{+}\,[Si_6Me_{11}-M(CO)_{n-1}]^{-}$$

M = Cr n = 6 (**30**)
M = Mo n = 6 (**31**)
M = W n = 6 (**32**)

M = Fe n = 5 (**33**)
M = Ni n = 4 (**34**)

Eq. 2. Reaction of **1** with transition metal carbonyls.

In order to determine the crystal structure of the iron derivative, **33** was converted into $NEt_4^+(OC)_4FeSi_6Me_{11}^-$ (**38**) (colorless crystals) because the $K(DME)_x^+$ salts are unstable, easily decomposing with loss of DME. The results of the X-ray crystal structure analysis are shown in Table 4 and Fig. 2 [7].

Also in **38**, the transition metal moiety occupies an equatorial position on the ring. The coordination polyhedron around the iron atom is a trigonal bipyramid, the silyl ligand occupying an axial position. The three carbonyl ligands in the equatorial position lie in the voids between the methyl groups to minimize steric repulsion. Nevertheless, they are somewhat bent toward the silyl ring. *Ab initio* calculations indicate that this is not a packing effect but inherent to the molecule [11]. The Fe–Si bond is considerably longer than the average (230.8 pm) [8]. The Si–Si bonds of the cyclohexasilane ring (233.6 – 236.2 pm) are again somewhat elongated in comparison with Si_6Me_{12} (233.7 ± 0.5 pm) [9]. This elongation is most distinct next to the iron-bearing silicon atom and decreases with increasing distance. The tetraethyl-ammonium cation is disordered.

Table 4. Details of the X-ray structure analysis of **38**.

Space group	$Pca2_1$	R (I>2σ(I))	0.031
Z	4	R (all data)	0.055
a	1592.9 pm	α	90°
b	1720.2 pm	β	90°
c	1316.8 pm	γ	90°
Fe–Si	240.5 pm	C1-Fe1-Si1	83.8°
C4-Fe1-Si1	86.7°	C3-Fe1-Si1	82.4°

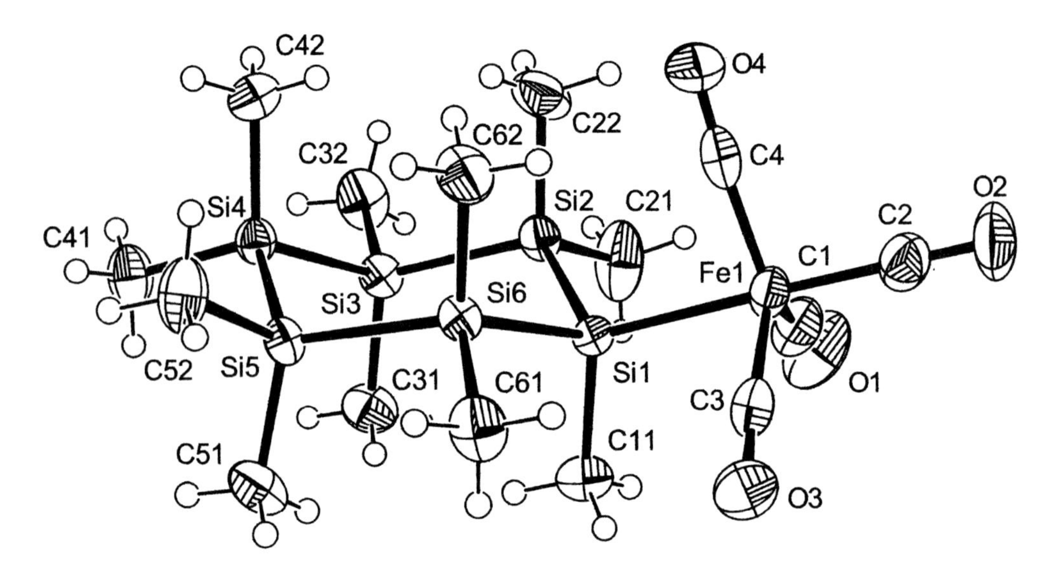

Fig. 2. Molecular structure of the anion of **38**.

Conclusions

1 is a valuable reagent for the synthesis of transition metal cyclohexasilyl compounds by nucleophilic or ligand substitution (Scheme 2).

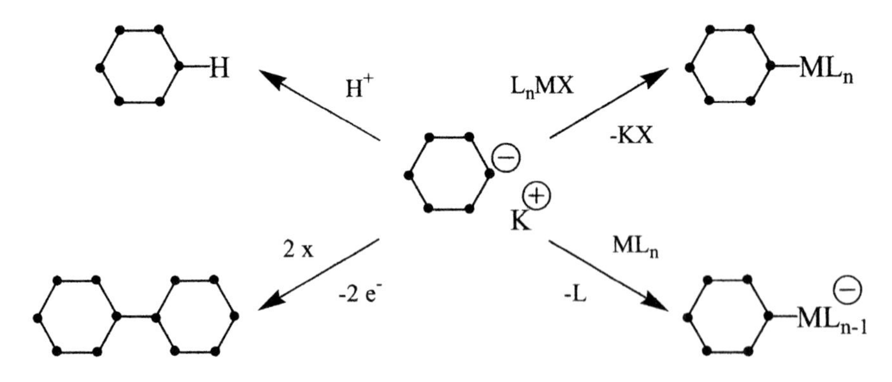

Scheme 2. Reaction pathways of **1**.

However, its applicability has some limitations. Depending on the transition metal compound, deprotonation and/or redox reactions can interfere severely or even prevent silyl complex formation. Obviously the structure of the metal complexes plays a decisive role.

Acknowledgments: This work was supported by the Deutsche Forschungsgemeinschaft and the Fonds der Chemischen Industrie. The X-ray structure analyses were carried out by S. Goutal at the Institut für Organische Chemie, Technische Universität Dresden.

References

[1] F. Uhlig, P. Gspaltl, M. Trabi, E. Hengge, *J. Organomet. Chem.* **1995**, *493*, 33.

[2] E. Hengge, E. Pinter, M. Eibl, F. Uhlig, *Bull. Soc. Chim. Fr.* **1995**, *132*, 509.

[3] E. Hengge, P. Gspaltl, E. Pinter, *J. Organomet. Chem.* **1996**, *521*, 145.

[4] W. Palitzsch, U. Böhme, C. Beyer, G. Roewer, *Organometallics* **1998**, *17*, 2965.

[5] W. Palitzsch, C. Beyer, U. Böhme, B. Rittmeister, G. Roewer, *Eur. J. Inorg. Chem.* **1999**, 1813.

[6] E. Hengge, M. Eibl, F. Schrank, *J. Organomet. Chem.* **1989**, *369*, C23.

[7] CCDC 217338 and 217339 contain the supplementary crystallographic data for this paper. These data can be obtained free of charge via www.ccdc.cam.ac.uk/conts/retrieving.html (or from the CCDC, 12 Union Road, Cambridge CB2 1EZ, UK; fax: +44 1223 336033; e-mail: deposit@ccdc.cam.ac.uk).

[8] Mean value of 152 compounds with 235 observed Fe–Si bonds; source: Cambridge Structural Database, August 2003.

[9] H. L. Carrell, J. Donohue, *Acta Cryst., Sect. B* **1972**, *28*, 1566.

[10] Th. Kruck, E. Job, U. Klose, *Angew. Chem.* **1968**, *80*, 360.

[11] U. Böhme, unpublished results.

Cp-Free Hafnium Silyl Substituted Compounds

Dieter Frank,* Judith Baumgartner, Christoph Marschner

Institut für Anorganische Chemie, Technische Universität Graz
Stremayrgasse 16, A-8010 Graz, Austria
Tel.: +43 316 873 8209 — Fax: +43 316 873
E-mail: christoph.marschner@tugraz.at

Keywords: oligosilyl anions, oligosilyl dianions, Group 4 silyl compounds

Summary: Oligosilanes with trimethylsilyl groups react with one or two molar equivalents of potassium *tert*-butoxide in the presence of tetramethylethylenediamine (tmen) to yield quantitatively oligosilyl mono- and dianions (**1**, **2**). Conversion of these compounds with Group 4 metal tetrachlorides leads to a novel type of Cp-free oligosilyl Group 4 metal chloride tmen complexes.

The chemistry of Group 4 silyl compounds is an area receiving increased recent attention [1]. However, most of this chemistry studied so far involves compounds with cyclopentadienyl (Cp) or related ligands. Although there are a few examples of Cp-free silyl Zr and Hf complexes [2], the seemingly most straightforward way to access this class of compounds, namely the reaction of a silyl anion with a Group 4 tetrahalide, has not been reported until recently [3]. A likely reason for this is that, in this particular reaction, ethereal solvents, present from the generation of the respective silyl reagents, are activated by the Lewis acidic metal halide toward a nucleophilic attack by the silyl anion [3].

In the course of our studies of oligosilyl anions we have investigated alternative conditions for the generation of oligosilyl potassium compounds [4]. For example, we have employed tetramethylethylenediamine (tmen) / toluene solvent mixtures to obtain the tris(trimethylsilyl)silyl potassium (**1**) and 1,1,4,4-tetrakis(trimethylsilyl)-2,2,3,3-tetramethyltetrasilanyl 1,2-dipotassium (**2**) tmen adducts (Eqs. 1, 2).

$$(Me_3Si)_3Si{-}SiMe_3 \xrightarrow[-\,^{t}BuOSiMe_3]{KO^{t}Bu,\ tmen,\ toluene} (Me_3Si)_3Si{-}K(tmen)_2 \quad \mathbf{1}$$

Eq. 1. Cleavage of one trimethylsilyl group to yield oligosilyl anion **1**.

KO^{t}Bu, tmen, toluene

- ^{t}BuOSiMe$_3$

2

Eq. 2. Selective cleavage of two trimethylsilyl groups to give oligosilyl dianion **2**.

1 was studied by single-crystal X-ray diffraction analysis and represents the first fully characterized oligosilyl potassium tmen adduct (Fig. 1 and Table 1) [5].

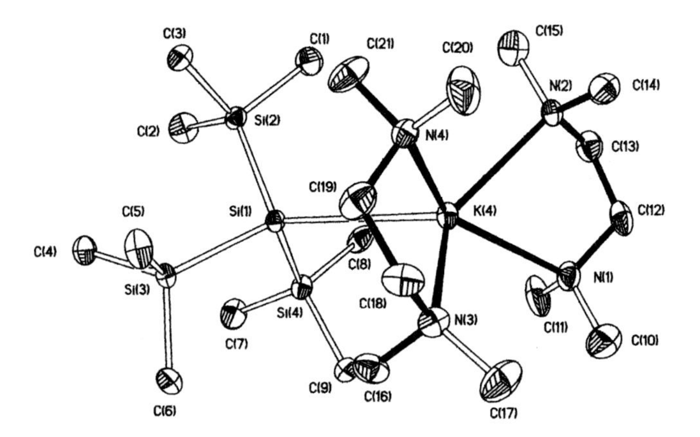

Fig. 1. Crystal structure of **1**. Thermal ellipsoids at the 30% level (hydrogen atoms omitted for clarity).

Table 1. Selected bond lengths [Å] and bond angles [°] of **1**.

Bond lengths	K–Si(1)	3.394
	K–N(1)	2.929
	K–N(2)	2.801
	K–N(3)	2.818
	K–N(4)	2.886
	Si(1)–Si(2)	2.319
Bond angles	Si(4)-Si(1)-Si(3)	102.47
	N(2)-K-Si(1)	123.33
	N(3)-K-Si(1)	100.15

The structure of **1** reveals a trigonal bipyramidal coordination around the potassium atom. N(1) and N(4), occupying the axial positions, exhibit elongated nitrogen–potassium bond distances. The length of the silicon–potassium bond (3.394 Å) is slightly shorter than that of the related crown ether adducts (about 3.50 Å) [4c].

The reaction of **1** with hafnium- or zirconium tetrachloride in toluene gave octahedrally coordinated hafnium- (**3**) and zirconium- (**4**) trichlorotris(trimethylsilyl)silyl tmen complexes (Eq. 3) [3].

$$(Me_3Si)_3Si{-}K(tmen)_2 + MCl_4 \xrightarrow[-\,KCl]{toluene,\ -36\ °C\ to\ r.t.} (tmen)MCl_3{-}Si(SiMe_3)_3$$

1 M = Zr, Hf **3, 4**

Eq. 3. Synthesis of hafnium and zirconium trichlorotris(trimethylsilyl)silyl complexes (**3**, **4**).

In a similar reaction, conversion of **2** with hafnium tetrachloride in toluene gave the hafnium cyclopentasilanyl tmen complex **5** (Eq. 4).

$$\mathbf{2} + HfCl_4 \xrightarrow[-\,2\ KCl]{toluene,\ -80\ °C\ to\ r.t.} \mathbf{5}$$

2 **5**

Eq. 4. Preparation of a hafnium dichlorocyclopentasilanyl tmen complex (**5**).

Product **5** was subjected to X-ray diffraction analysis (Fig. 2 and Table 2) [6]. The structure was found to be an octahedrally coordinated metal dichloro complex with the two bidentate ligands occupying the equatorial positions. The observed stereochemistry can be considered to be caused by the "trans-effect" of the tmen group [3], which activates the two chloro atoms of the hafnium tetrachloride in equatorial position and makes them susceptible to the nucleophilic attack by the silyl dianion.

Compared to a related Cp-containing hafnium structure [4b], the lengths of the silicon–hafnium bonds and the silicon–silicon bonds of **5** are in the same range. The absence of Cp groups minimizes sterical interactions with the trimethylsilyl groups, resulting in a less strained ring. This is illustrated by a widened Si(1)-(Hf)-Si(4) angle of 103.5°, compared to 96.4° in the Cp-containing product [4b].

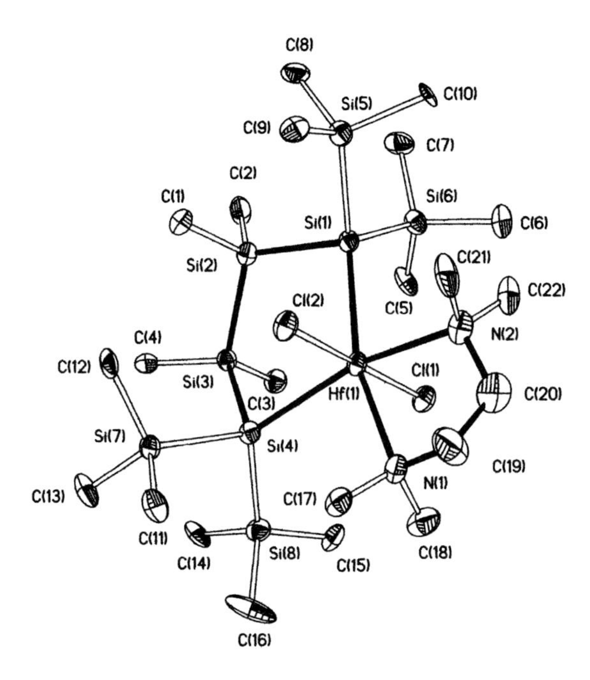

Fig. 2. Crystal structure of **5**. Thermal ellipsoids at the 30% level (hydrogen atoms omitted for clarity).

Table 2. Selected bond lengths [Å] and bond angles [°] of **5**.

	Hf–Cl(1)	2.402
	Hf–N(1)	2.491
	Hf–Si(1)	2.794
Bond lengths	Hf–Si(4)	2.793
	Si(1)–Si(5)	2.367
	Si(1)–Si(2)	2.378
	Si(2)–Si(3)	2.330
	Cl(2)-Hf-Cl(1)	178.3
	Si(4)-Hf-Si(1)	103.5
	Si(2)-Si(1)-Hf	99.4
Bond angles	N(1)-Hf-Si(1)	164.6
	N(1)-Hf(1)-N(2)	72.4
	Cl(1)-Hf(1)-Si(4)	93.02

Acknowledgments: This research was supported by the Fonds zur Förderung der wissenschaftlichen Forschung via the START project Y120. We thank Wacker Chemie GmbH, Burghausen, for the generous donation of various chlorosilanes as starting materials.

References

[1] a) M. S. Eisen, in *The Chemistry of Organic Silicon Compounds* (Eds.: Z. Rappoport, Y. Apeloig), John Wiley & Sons Ltd, **1998**, *Vol. 2*, Ch. 35, p. 2037; b) T. D. Tilley, in *The Silicon Heteroatom Bond*, (Eds.: S. Patai, Z. Rappoport), John Wiley & Sons Ltd, Chichester, **1991**, Ch. 10, p. 309; c) T. D. Tilley, in *The Chemistry of Organic Silicon Compounds*, (Eds.: S. Patai, Z. Rappoport), John Wiley & Sons Ltd, Chichester, **1989**, Ch. 24, p. 1415.

[2] X. Yu, L. A. Morton, Z.-L. Xue, *Organometallics*, **2004**, *23*, 2210 and references therein.

[3] D. Frank, J. Baumgartner, Ch. Marschner, *Chem. Comm.* **2002**, 1190.

[4] a) Ch. Marschner, *Eur. J. Inorg. Chem.* **1998**, 221; b) Ch. Kayser, G. Kickelbick, Ch. Marschner, *Angew. Chem. Int. Ed.* **2002**, *41*, 989; c) R. Fischer, D. Frank, W. Gaderbauer, Ch. Kayser, Ch. Mechtler, J. Baumgartner, Ch. Marschner, *Organometallics* **2003**, *22*, 3723.

[5] Crystallographic data for **1**: $C_{21}H_{59}KN_4Si_4$, monoclinic space group $P2_1/c$, a = 12.254(3) Å, b = 14.689(3) Å, c = 22.270(7) Å, β = 119.77(2)°, $R1$ = 0.0847, $wR2$ = 0.1737, GoF = 1.044. Data have been deposited at the Cambridge Crystallographic Data Centre. CCDC 238 890.

[6] Crystallographic data for **5**: $C_{29}H_{72}Cl_2HfN_2Si_8$, triclinic space group $P\bar{1}$, a = 12.609(3) Å, b = 14.112(3) Å, c = 15.357(3) Å, α = 66.51(3)°, β= 82.91(3)°, γ= 71.32(3)°, $R1$ = 0.0987, $wR2$ = 0.2458, GoF = 1.001. Data have been deposited at the Cambridge Crystallographic Data Centre. CCDC 238 891.

Metal- and Cyclopentadienyl-Bound Silanol Groups in Tungsten Complexes†

*Holger Bera, Siegfried Schmitzer, Dirk Schumacher, Wolfgang Malisch**

Institut für Anorganische Chemie der Universität Würzburg
Am Hubland, 97074 Würzburg
Tel.: +49 931 888 5277 — Fax: +49 931 888 4618
E-mail: Wolfgang.Malisch@mail.uni-wuerzburg.de

Keywords: oxofunctionalization, half-sandwich complexes, metallo-silanols

Summary: The tungsten dimethylsilanol $Cp(OC)_2(Me_3P)WSiMe_2OH$ (**3**) is synthesized by oxofunctionalization of the tungsten silane $Cp(OC)_2(Me_3P)WSiMe_2H$ (**2**) with dimethyldioxirane (DMD). **3** is stable with respect to self-condensation, but controlled co-condensation with $Me_2Si(R)Cl$ is realized to give the tungsten-substituted disiloxanes $Cp(OC)_2(Me_3P)WSiMe_2OSiMe_2R$ [R = H (**4a**), Me (**4b**)]. $[HMe_2Si$-$[\eta^5$-$(HMe_2Si$-$C_5H_4)](OC)_2(Me_3P)WSiMe_2H$ (**6**) is treated with DMD to give primarily the tungsten silanol $[\eta^5$-$(HMe_2Si$-$C_5H_4)](OC)_2(Me_3P)WSiMe_2OH$ (**7**) and the bis-silanol $[\eta^5$-$(HOMe_2Si$-$C_5H_4)](OC)_2(Me_3P)WSiMe_2OH$ (**A**) in the following step. **A** undergoes rapid intramolecular condensation to yield the novel disiloxane $[\eta^1$-$SiMe_2OMe_2Si$-$(\eta^5$-$C_5H_4)](Me_3P)(OC)_2W$ (**8**). The X-ray analysis of **8** is presented.

Introduction

Metallo-silanols represent a special class of silanols which is characterized by a remarkably high stability with respect to self-condensation. This property, which is even valid for silanetriol derivatives [2], and the generally stable metal-silicon bond makes these compounds useful precursors for controlled condensation with chlorosilanes to build up unusual arrangements of functionalized siloxanes at metal centers. The metal fragment is responsible for highly hydridic Si–H units, creating the conditions for electrophilic oxygenation with DMD [3].

Recent investigations have shown that molybdenum and tungsten silanols with the η^5-cyclopentadienyl ligand as spacer between the metal and the silanol unit are stable with respect to self-condensation [1, 4, 5]. Extension of this work is now focused on compounds of this type containing, in addition, a metal-bound silanol function.

---

† Part 35 of the series "*Metallo-Silanols and Metallo-Siloxanes*". In addition, Part 64 of the series "*Synthesis and Reactivity of Silicon Transition Metal Complexes*". Part 63/34, see Ref. [1].

Results and Discussion

The Si–H-functionalized tungsten silane **2**, which serves as starting material for the synthesis of the tungsten silanol **3**, is prepared by the nucleophilic metalation of chlorodimethylsilane with the tungsten metalate **1** under heterogeneous conditions in cyclohexane (Scheme 1). **2** is isolated after a reaction time of 2 h as a beige microcrystalline solid in moderate yield (48%).

The tungsten dimethylsilanol **3** is obtained by treatment of **2** with DMD in acetone in 43% yield. Storage of **3**, isolated as a pale beige solid, at room temperature for several weeks neither leads to self-condensation nor to significant decomposition. The ^{29}Si NMR resonance at 50.09 ppm reveals a strong influence of the transition metal fragment on the silicon atom [6].

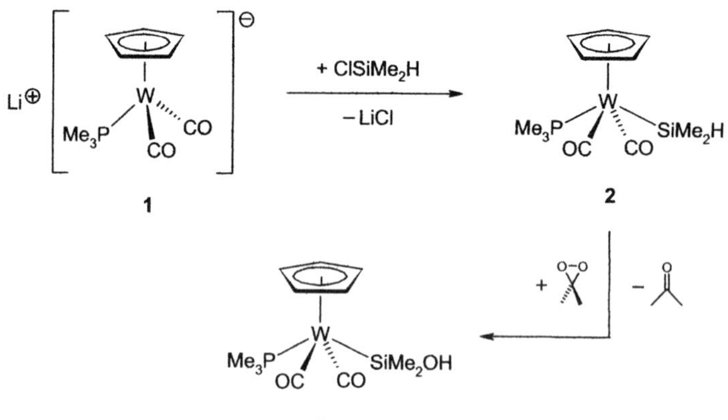

Li⊕ [W(Me$_3$P)(CO)$_2$Cp]⊖ **1** —(+ ClSiMe$_2$H, − LiCl)→ Cp(Me$_3$P)(OC)(CO)W–SiMe$_2$H **2** —(+ DMD, − acetone)→ Cp(Me$_3$P)(OC)(CO)W–SiMe$_2$OH **3**

Scheme 1. Synthesis of the tungsten silanol **3**.

The silanol **3** shows the expected reactivity toward di- and trimethylchlorosilane, respectively. In the presence of NEt_3 as an auxiliary base, the corresponding metallo-siloxanes **4a,b** are obtained in good yields [72% (**4a**), 63% (**4b**)] as pale yellow powders (Eq. 1).

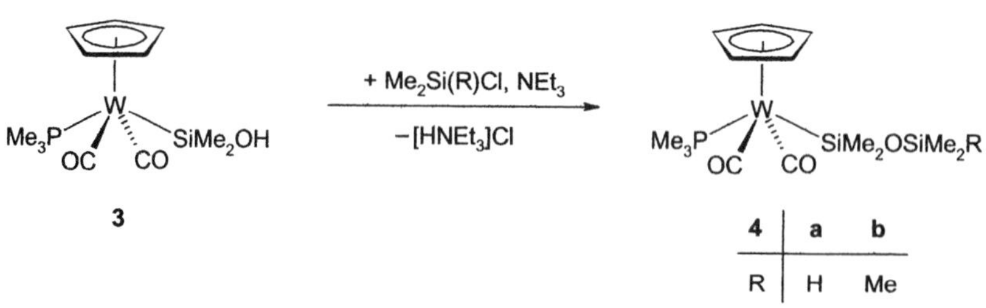

Cp(Me$_3$P)(OC)(CO)W–SiMe$_2$OH **3** —(+ Me$_2$Si(R)Cl, NEt$_3$; − [HNEt$_3$]Cl)→ Cp(Me$_3$P)(OC)(CO)W–SiMe$_2$OSiMe$_2$R **4**

4	a	b
R	H	Me

Eq. 1. Synthesis of the tungsten-substituted siloxanes **4a,b**.

Compared to the silanol **3**, the ^{29}Si NMR values of the α-Si atoms are shifted about 6 and 9 ppm to a higher field [43.87 (**4a**), 41.03 (**4b**) ppm]. The resonances for the γ-Si atoms appear at −10.33

(**4a**) and 3.04 (**4b**) ppm.

The synthetic approach to a tungsten complex bearing both a metal- and a cyclopentadienyl-fixed silanol function involves treatment of the double Si–H-functionalized complex **6** [6] with DMD. A crucial charateristic of **6** is the different electronic situation of the silicon atoms, which allows successive transformation of the Si–H functions into silanol groups. Reaction of **6** with one equivalent of DMD at –78 °C in acetone occurs regioselectively at the more hydridic tungsten-bonded Si–H group to give the tungsten silanol **7**, which is isolated in 71% yield as a pale yellow solid (Scheme 2).

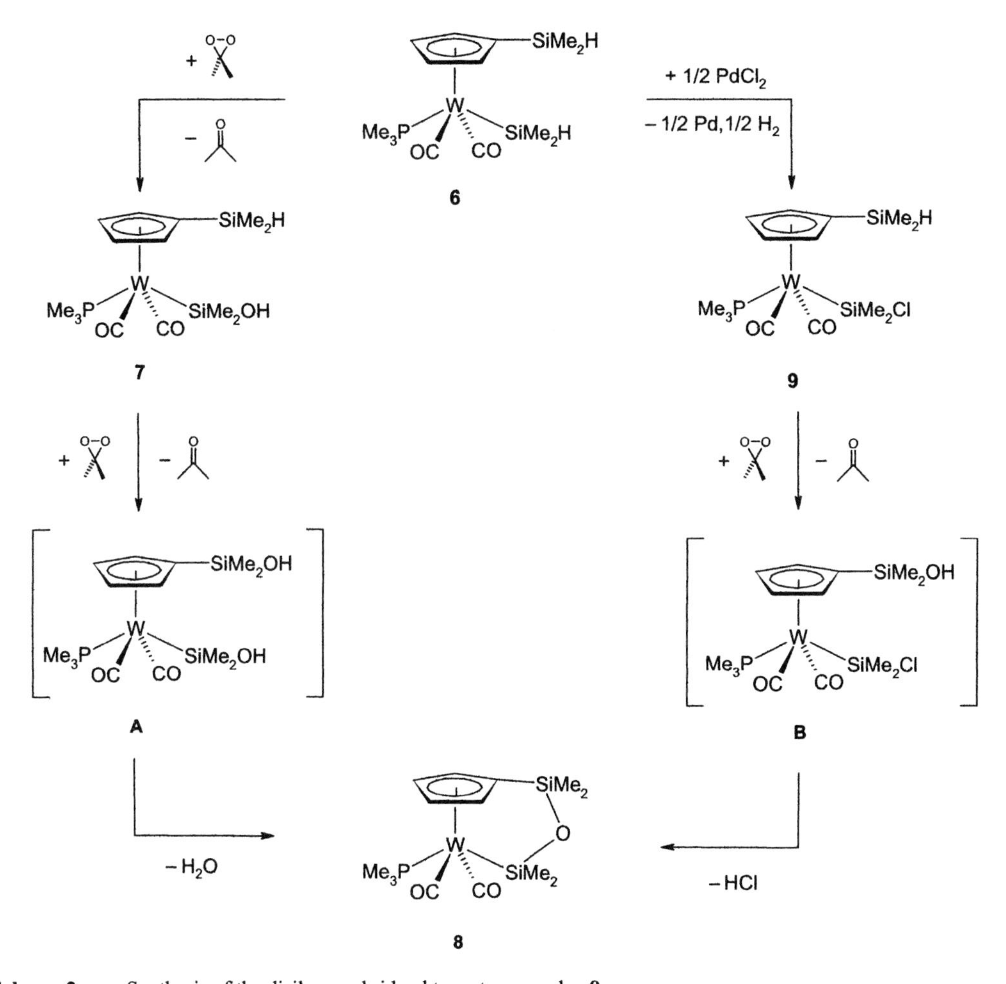

Scheme 2. Synthesis of the disiloxane-bridged tungsten complex **8**.

Treatment of **7** with DMD leads to the short-lived bis-silanol **A**. However, this intermediate undergoes intramolecular condensation, resulting immediately in the formation of the tungsten complex **8** bearing a disiloxane unit bridging the cyclopentadienyl ligand and the metal center. **8** is obtained in 58% yield as an ochre-colored solid. The ^{29}Si NMR resonance at 44.38 ppm indicates a strong metal influence on the tungsten-bonded silicon atom in comparison to that attached at the cyclopentadienyl ligand, detected at 4.93 ppm.

An alternative synthesis of the tungsten bis-silanol **A** starts with the regioselective chlorination of the metal-bound Si–H function of **6** with $PdCl_2$. The corresponding tungsten chlorosilane **9** is isolated in 85% yield as a yellow solid (Scheme 2).

The treatment of **9** with DMD in acetone at –78 °C leads to the tungsten chlorosilane with a cyclopentadienyl-fixed silanol unit (**B**), which as the bis-silanol **A** shows fast intramolecular condensation to **8**.

X-ray Analysis

The molecular structure of **8** (Fig. 1) reveals a pseudo-tetragonal pyramidal arrangement of the ligands at the tungsten atom with the cyclopentadienyl ligand in the apical position. The two carbonyl ligands are found in a *trans* arrangement as well as the phosphine and the silyl unit. **8** shows an envelope conformation of the five-membered ring with the atoms W1, Si2, O2 and Si1 lying in one plane. The carbon atom C1 lies above this plane, illustrated by the torsion angle of 21.7(3)° (C1-Si1-O1-Si2). The disiloxane angle amounts to 125.21(16)°, which is significantly smaller than that in open-chained organodisiloxanes. The distance W1–Si2 [2.5896(11) Å] lies in the expected range for tungsten-substituted silanes [7].

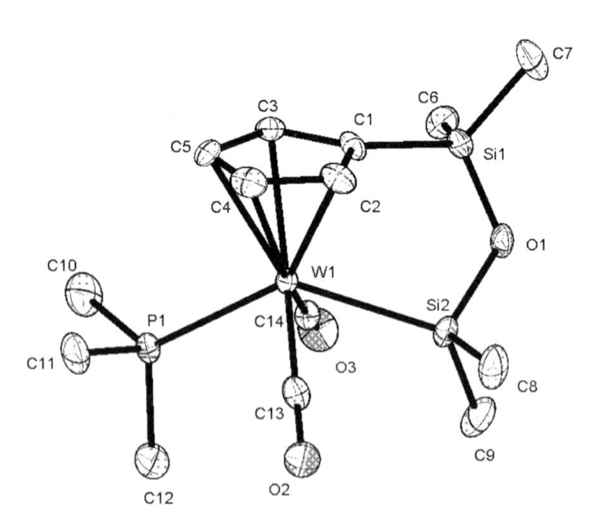

Fig. 1. Molecular structure of [η^1-$SiMe_2OMe_2Si(\eta^5$-$C_5H_4)](Me_3P)(OC)_2W$ (**8**). Selected bond lengths [Å], bond and torsion angles [°]: W1–C13 1.945(4), W1–C14 1.935(4), W1–P1 2.4297(10), W1–Si2 2.5896(11), C1–Si1 1.869(4), Si1–O1 1.631(3), Si2–O1 1.672(3); Si(1)-O(1)-Si(2) 125.21(16), O1-Si2-W1 104.94(10), O1-Si1-C1 105.23(16); Si1-O1-Si2-W1 0.3(2), C1-Si1-O1-Si2 21.7(3), C4-C2-C1-Si1 -171.32(29).

Acknowledgment: We gratefully acknowledge financial support from the Deutsche Forschungsgemeinschaft (Schwerpunktprogramm "Spezielle Phänomene in der Siliciumchemie").

References

[1] H. Bera, A. Sohns, W. Malisch, in *Organosilicon Chemistry VI — From Molecules to Materials* (Eds.: N. Auner, J. Weis), VCH, Weinheim **2005**, submitted.

[2] W. Malisch, R. Lankat, S. Schmitzer, J. Reising, *Inorg. Chem.* **1995**, *34*, 5702.

[3] a) W. Adam, U. Azzena, F. Prechtl, K. Hindahl, W. Malisch, *Chem. Ber.* **1992**, *125*, 1409; b) W. Malisch, K. Hindahl, H. Käb, J. Reising, W. Adam, F. Prechtl, *Chem. Ber.* **1995**, *128*, 963; c) S. Möller, H. Jehle, W. Malisch, W. Seelbach, in *Organosilicon Chemistry III — From Molecules to Materials* (Eds.: N. Auner, J. Weis), VCH, Weinheim, **1998**, 267.

[4] F. Javier de la Mata, P. Giner, P. Royo, *J. Organomet. Chem.* **1999**, *572*, 155.

[5] a) A. Sohns, H. Bera, D. Schumacher, W. Malisch, in *Organosilicon Chemistry V — From Molecules to Materials* (Eds.: N. Auner, J. Weis), VCH, Weinheim, **2003**, 348; b) A. Sohns, D. Schumacher, W. Malisch, in: *Organosilicon Chemistry VI — From Molecules to Materials* (Eds.: N. Auner, J. Weis), VCH, Weinheim, **2005**, submitted.

[6] W. Malisch, H. Jehle, C. Mitchel, W. Adam, *J. Organomet. Chem.* **1998**, *566*, 259.

[7] S. Schmitzer, *Doctoral Thesis*, University of Würzburg **1992**.

Half-Sandwich Iron Complexes with a Silanol-Functionalized Cyclopentadienyl Ligand†

*Andreas Sohns, Dirk Schumacher, Wolfgang Malisch**

Institut für Anorganische Chemie der Universität Würzburg
Am Hubland, D-97074 Würzburg, Germany
Tel.: +49 931 888 5277 — Fax : +49 931 888 4818
E-mail: Wolfgang.Malisch@mail.uni-wuerzburg.de

Martin Nieger

Institut für Anorganische Chemie der Universität Bonn
Gerhard-Domagk-Strasse 1, D-53121 Bonn, Germany

Keywords: iron, silanols, condensation, siloxanes

Summary: The complexes $[(X)R_2Si–C_5H_4](OC)_2Fe–R'$ (X = OMe, H; R = Me, *i*Pr, Ph; R′ = Me, $SiMe_3$) (**4a–e**) are synthesized from the corresponding silyl-functionalized iron anions $Li[Fe(CO)_2(C_5H_4–SiR_2X)]$ (X = H, OMe; R = Me, *i*Pr, Ph) (**2b–e**) by treatment with methyl iodide or chlorotrimethylsilane, respectively. Hydrolysis of $(XMe_2Si–C_5H_4)(OC)_2Fe–R$ (X = OMe, Cl; R = Me, $SiMe_3$) (**4a**, **5a,b**) or reaction of $(HMe_2Si–C_5H_4)(OC)_2Fe–R$ [R = Me (**4b**), $SiMe_3$ (**4c**)] with $Co_2(CO)_8$ followed by treatment with H_2O yields the silanols $[(HO)Me_2Si–C_5H_4](OC)_2Fe–R$ [R = Me (**6a**), $SiMe_3$ (**6b**)]. **6a,b** undergo condensation reactions with chlorodimethylsilane or titanocene dichloride, respectively, to yield the iron fragment-substituted disiloxanes $(HMe_2SiOMe_2Si–C_5H_4)(OC)_2Fe–R$ [R = Me (**7a**), $SiMe_3$ (**7b**)] or the heterosiloxanes $[(Cl)Cp_2TiOMe_2Si–C_5H_4](OC)_2Fe–R$ [R = Me (**8a**), $SiMe_3$ (**8b**)].

Introduction

The preparation of catalysts by immobilizing active transition metal complexes on solid supports such as inorganic surfaces or organic polymers has attracted widespread attention in recent years [2]. Most studies are focused on phosphines as ligands for anchoring the catalytically active metal [3]. Although a great number of organometallic compounds contain cyclopentadienyl as a ligand, anchoring via cyclopentadienyl has only marginally been studied. Our interest is mainly focused on the generation of model compounds that represent promising precursors for the immobilization of

† Part 33 of the series "*Metallo-Silanols and Metallo-Siloxanes*". In addition, Part 62 of the series "*Synthesis and Reactivity of Silicon Transition Metal Complexes*". Part 61/32, see Ref. [1].

half-sandwich complexes on silica surfaces. In contrast to metallo-silanols with a metal-to-silicon bond [4], which are stable toward self-condensation, spacer group-separated metallo-silanols [5] can be envisaged for fixation on silica because of their enhanced SiOH reactivity.

Results

The generation of the precursors for cyclopentadienyl-silanol-functionalized iron complexes involves the formation of the corresponding iron anion in a first step [6]. **2a** is obtained by reductive cleavage of the methoxysilyl-cyclopentadienyl-substituted iron dimer **1** with sodium amalgam in THF (Scheme 1). This reaction is restricted to alkoxysilyl-cyclopentadienyl-functionalized iron anions because of the limited access to the corresponding Si–H-functionalized iron dimers.

2	a	b	c	d	e
R	Me	Me	Me	*i*Pr	Ph
X	OMe	OMe	H	H	H
M	Na	Li	Li	Li	Li

3	a	b	c	d
R	Me	Me	*i*Pr	Ph
X	OMe	H	H	H

Scheme 1. Synthesis of silyl-cyclopentadienyl-functionalized iron anions.

An alternative access to cyclopentadienyl-silyl-functionalized iron anions is the base-induced rearrangement of the ferrio-silanes **3a–d** induced by a strong base like LDA (Scheme 1). The initial deprotonation of the cyclopentadienyl ligand is followed by silyl migration from the metal center to the cyclopentadienyl ligand to give **2b–e**. The advantage of this method in comparison to the reductive cleavage of the dimeric silyl-functionalized complexes is the possibility of generating Si–H-functionalized iron anions. **2b–e** are isolated as yellow pyrophoric powders in quantitative yields.

The molecular structure of **1** (Fig. 1) shows considerable agreement in bond lengths and angles with the well-known dimeric iron complex $[Cp(CO)(\mu\text{-}CO)Fe]_2$ [7]. The Fe1–Fe1A distance of 2.5463(7) Å lies as well in the expected range as the bond lengths Fe1–C1B [1.757(3) Å] for the terminal and Fe1–C1A [1.941(3) Å] for the bridging carbonyl ligand. The carbon atoms of the bridging CO ligands and the iron atoms lie in a plane (C1AA-Fe1-C1A-Fe1A of 0.0°). The methoxy group stands almost vertical to the plane that is spanned by the cyclopentadienyl ligand (C2-C1-Si1-O1 of 92.7°).

Reaction of the lithium metalates **2b–e** with methyl iodide or chlorosilanes in cyclohexane leads in an established manner to the neutral half-sandwich methyl- or silyl-iron complexes **4a–e** which are isolated as brown or red oils, respectively, in yields between 72 and 89% (Eq. 1).

The generation of the silanols **6a,b** can be realized by acid-assisted hydrolysis of the methoxy-silyl-functional complexes. An alternative access to **6a–d** is offered by chlorination of the corresponding Si–H-functionalized complexes **4a–d** followed by NEt_3-assisted hydrolysis (Scheme 2).

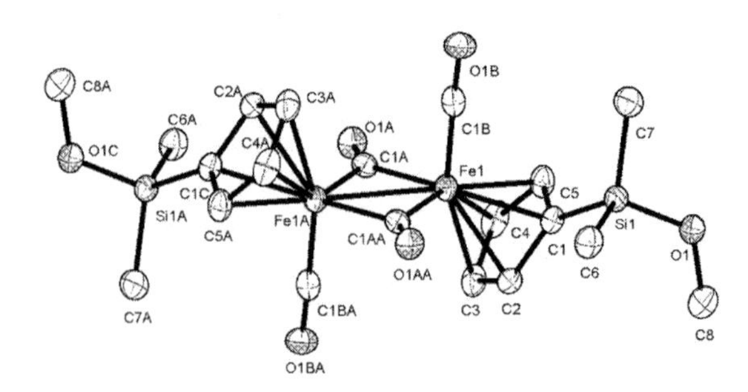

Fig. 1. Molecular structure of *trans*-{[(MeO)Me_2Si–C_5H_4](CO)(μ-CO)Fe}$_2$ (Fe–Fe) (**1**). Selected bond lengths [Å], bond and torsion angles [°]: Fe1–C1B 1.757(3), Fe1–C1A 1.941(3), Fe1–Fe1A 2.5463(7), C1A–O1A 1.179(3), C1B–O1B 1.154(4), C1–Si1 1.873(3); O1A-C1A-Fe1 138.0(2), O1B-C1B-Fe1 177.9(3); C2-C1-Si1-O1 92.7, C1AA-Fe1-C1A-O1A-179.0, C1AA-Fe1-C1A-Fe1A 0.0.

$Li^{\oplus}$ [C_5H_4–SiR_2H)Fe(CO)$_2$]$^{\ominus}$ $\xrightarrow[-\ LiX'\ (X' = I,\ Cl)]{+\ CH_3I\ \text{or}\ Me_3SiCl}$ (C_5H_4–SiR_2X)Fe(CO)$_2$R'

2	b	c	d	e
R	Me	Me	*i*Pr	Ph
X	OMe	H	H	H

4	a	b	c	d	e
R	Me	Me	Me	*i*Pr	Ph
R'	Me	Me	$SiMe_3$	Me	Me
X	OMe	H	H	H	H

Eq. 1. Reaction of the lithium metalates with electrophiles.

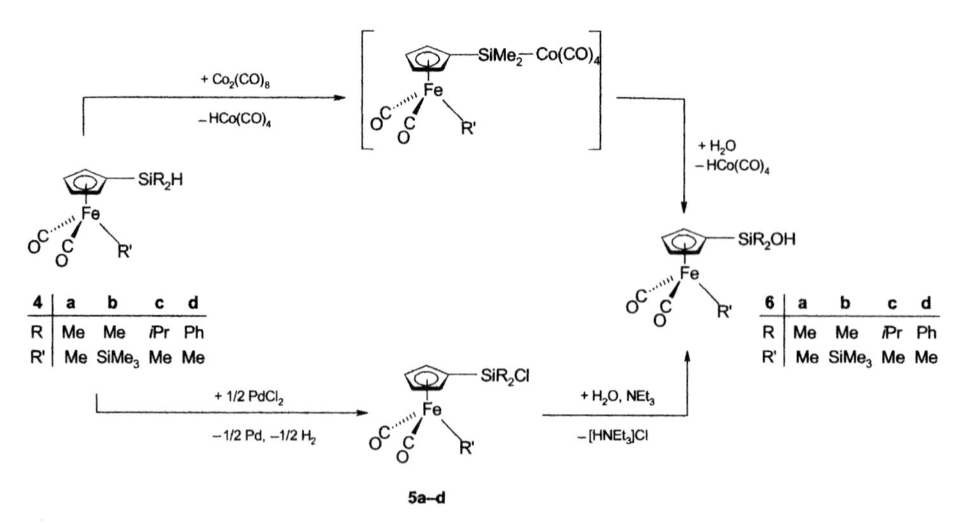

Scheme 2. Preparation of silanol-cyclopentadienyl-functionalized iron complexes.

The silanols **6a,b** can in addition be obtained by reaction of the corresponding Si–H-functionalized complexes **4a,b** with one equivalent of $Co_2(CO)_8$ followed by treatment with H_2O. Catalytic amounts of $Co_2(CO)_8$ as described for oxygenation of metal-bound Si–H functions in the literature [8] are not efficient in the case of Si–H functions separated by a cyclopentadienyl spacer from the metal fragment.

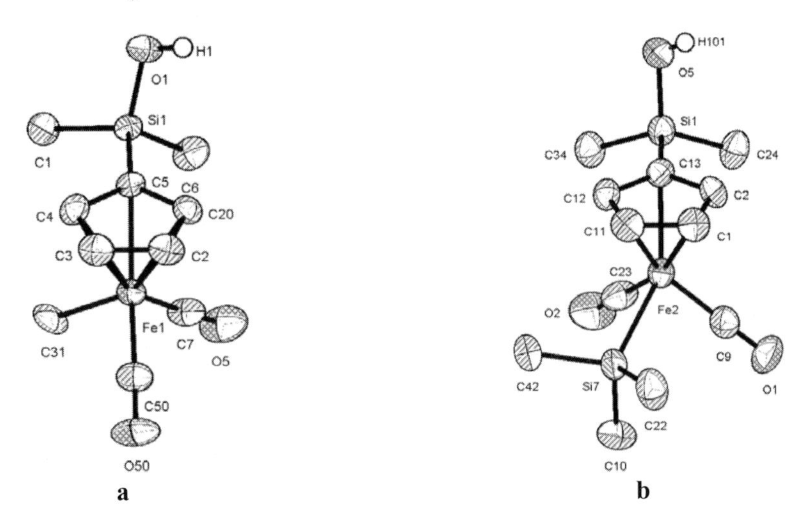

Fig. 2. **a**: Dicarbonyl[η^5-(dimethylhydroxysilyl)cyclopentadienyl]methyl iron (**6a**). Selected bond lengths [Å], bond and torsion angles [°]: Fe1–C7 1.758(2), Fe1–C50 1.778(4), Fe1–C31 2.022(11), Si1–O1 1.6557(16), Si1–C1 1.848(2), Si1–C6 1.860(2), Si1–C5 1.869(2); O1-Si1-C1 108.30(10), O1-Si1-C6 108.16(10), C1-Si1-C6 111.68(13), O1-Si1-C5 108.46(9), C1-Si1-C5 108.94(11), C6-Si1-C5 111.21(10); O1-Si1-C5-C20 75.8(2). **b**: Dicarbonyl[η^5-(dimethylhydroxysilyl)cyclopentadienyl]trimethylsilyl iron (**6b**). Selected bond lengths [Å], bond and torsion angles [°]: Fe2–C9 1.737(4), Fe2–C23 1.738(4), Fe2–Si7 2.3336(12), C13–Si1 1.867(4), Si1–O5 1.655(3); C13-Si1-O5 108.17(17), C13-Si1-C24 109.39(17), Si7-Fe2-C9 148.45(10), Si7-Fe2-C23 82.45(14).

The iron atoms of **6a,b** show pseudo-octahedral coordination with the cyclopentadienyl substituent considered as a tridentate ligand. The asymmetric unit of **6b** contains 3 molecules which only differ insignificantly in their structural parameters. For this reason only one molecule is discussed, shown in Fig. 2. The bond lengths Si1–C5 of 1.869(2) Å (**6a**) and Si1–C13 of 1.867(4) Å (**6b**) are in the range reported in the literature [9] for comparable cyclopentadienyl-silyl compounds. The silicon atoms are coordinated tetrahedrally in both molecules **6a** and **6b** indicated by bond angles between 108.16(10)° (O1-Si1-C6) and 111.68(13)° (C1-Si1-C6) in the case of **6a** and between 108.17(17)° and 109.39(17)° in the case of **6b**. The torsion angle O1-Si1-C5-C20 amounts to 75.8(2)° (**6a**), which means that the hydroxyl group stands almost vertically above the plane that is spanned by the carbon atoms of the cyclopentadienyl ligand. The same situation is valid for **6b** with the Si–OH group above the cyclopentadienyl ligand indicated by the torsion angle C2-C13-Si1-O5 of –92.7(3)°. The bond length Fe2–Si7 of 2.3336(12) Å (**6b**) lies in the expected range for Fe–Si bonds.

The main difference between **6a** and **6b** is the arrangement in the solid state with respect to hydrogen bonding. While the molecules of **6a** show no intermolecular interaction, the structure of **6b** is determined by hydrogen bonds between six molecules forming a chair-like ring arrangement (Fig. 3).

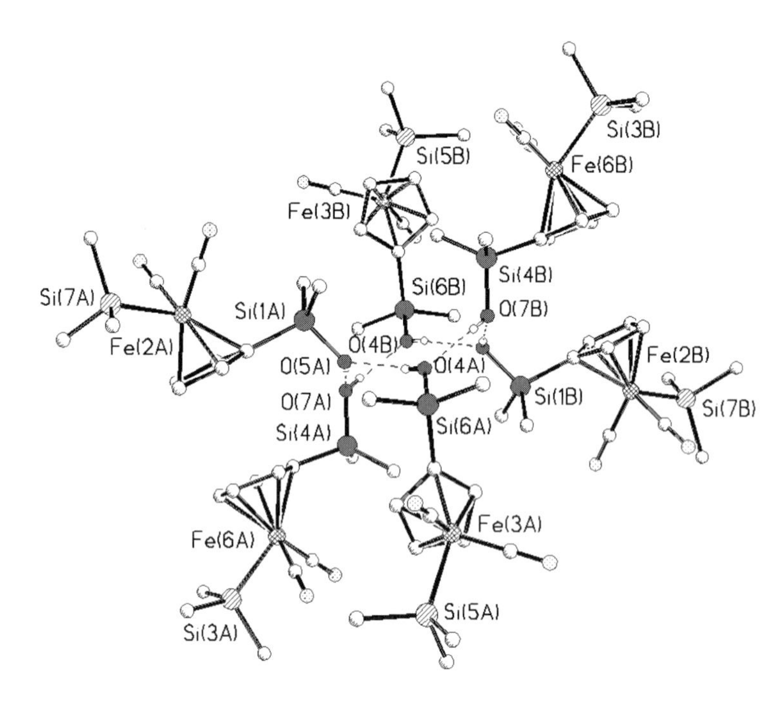

Fig. 3. Hydrogen bonding of **6b** in the solid state.

Treatment of **6a,b** with chlorodimethylsilane in the presence of NEt_3 yields the disiloxanes **7a,b** in almost quantitative yields. Analogous reaction of **6a,b** with titanocene dichloride results in the formation of the titana-siloxanes **8a,b**, isolated as yellow air-sensitive solids in yields of 72 – 79% (Scheme 3).

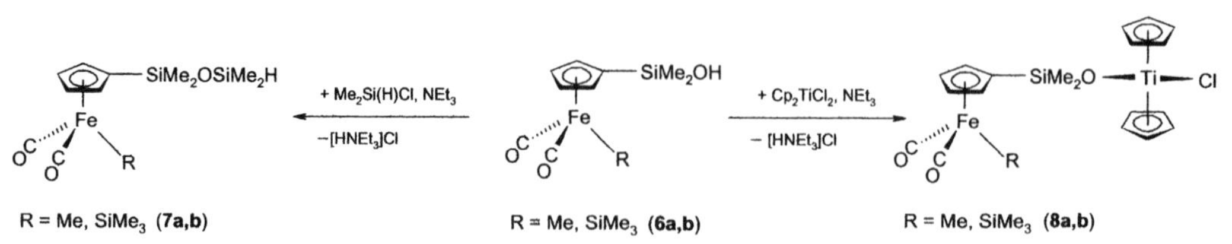

$SiMe_2OSiMe_2H$ ← $+ Me_2Si(H)Cl, NEt_3$ / $-[HNEt_3]Cl$ — $SiMe_2OH$ — $+ Cp_2TiCl_2, NEt_3$ / $-[HNEt_3]Cl$ → $SiMe_2O$—Ti—Cl

R = Me, $SiMe_3$ (**7a,b**) R = Me, $SiMe_3$ (**6a,b**) R = Me, $SiMe_3$ (**8a,b**)

Scheme 3. Reaction of the silanols **6a,b** forming disiloxanes **7a,b** and heteronuclear siloxanes **8a,b**.

Acknowledgment: We gratefully acknowledge financial support from the Deutsche Forschungsgemeinschaft (Schwerpunktprogramm "Spezielle Phänomene in der Siliciumchemie").

Reference

[1] D. Schumacher, W. Malisch, N. Söger, M. Binnewies, in *Organosilicon Chemistry VI — From Molecules to Materials* (Eds.: N. Auner, J. Weis) Wiley-VCH, Weinheim **2005**, submitted.

[2] M. Capka, *Collect. Czech. Chem. Commun.* **1990**, *55*, 2803.

[3] L. L. Murrell, in *Advanced Materials in Catalysis* (Eds.: J. J. Burton, R. L. Garten), **1977**, Academic Press, New York, p. 236.

[4] a) W. Malisch, M. Neumayer, O. Fey, W. Adam, R. Schuhmann, *Chem. Ber.* **1995**, *128*, 1257; b) W. Malisch, M. Vögler, D. Schumacher, M. Nieger, *Organometallics* **2002**, *21*, 2891.

[5] a) W. Malisch, M. Hofmann, M. Nieger, W. W. Schöller, A. Sundermann, *Eur. J. Inorg. Chem.* **2002**, 3242; b) M. Hofmann, W. Malisch, H. Hupfer, M. Nieger, *Z. Naturforsch. B* **2003**, *58b*, 36.

[6] a) W. Malisch, A. Sohns, D. Schumacher, G. Thum, M. Nieger, *Organometallics*, in preparation; b) A. Sohns, H. Bera, D. Schumacher, W. Malisch, in *Organosilicon Chemistry V — From Molecules to Materials* (Eds.: N. Auner, J. Weis) Wiley-VCH, Weinheim **2003**, p. 348; c) W. Malisch, M. Hofmann, M. Vögler, D. Schumacher, A. Sohns, H. Bera, H. Jehle, in *Silicon Chemistry — From Small Molecules to Extended Systems* (Eds.: P. Jutzi, U. Schubert) Wiley-VCH, Weinheim **2003**, p. 348.

[7] P. Greene, R. F. Bryan, *J. Chem. Soc. A*: *Inorganic, Physical, Theoretical* **1970**, *18*, 3064.

[8] W. Malisch, H. Jehle, C. Mitchel, W. Adam, *J. Organomet. Chem.* **1998**, *566*, 259.

[9] a) S. Ciruelos, T. Cuenca, P. Gomez-Sal, A. Manzanero, P. Royo, *Organometallics* **1995**, *14*, 177; b) H. Sun, X. Zhou, X. Yao, H. Wang, *Polyhedron* **1996**, *15*, 4489; c) Y. Zhang, F. Cervantes-Lee, K. H. Pannell, *J. Organomet. Chem.* **2001**, *634*, 102; d) G. Jimenez, E. Rodriguez, P. Gomez-Sal, P. Royo, T. Cuenca, *Organometallics* **2001**, *20*, 2459; e) G. W. Stowell, R. R. Whittle, C. M. Whaley, D. P. White, *Organometallics* **2001**, *20*, 1050.

Synthesis and Reactivity of Polychlorinated Metallo-Siloxanes[†]

Dirk Schumacher, Wolfgang Malisch*

Institut für Anorganische Chemie der Universität Würzburg
Am Hubland, D-97074 Würzburg, Germany
Tel.: +49 931 888 5277 — Fax: +49 931 888 4618
E-mail: Wolfgang.Malisch@mail.uni-wuerzburg.de

Nicola Söger, Michael Binnewies

Institut für Anorganische Chemie der Universität Hannover
Callinstr. 9, D-30167 Hannover, Germany
Fax: +49 511 762 19032
E-mail: binn@aca.uni-hannover.de

Keywords: iron, silanols, condensation, siloxanes

Summary: Reaction of the metalates $Na[Fe(CO)_2C_5R_5]$ [R = H (**1a**), Me (**1b**)] and $Li[M(CO)_2(PMe_3)Cp]$ [M = Mo (**1c**), W (**1d**)] with the chlorosiloxanes $Cl_3SiOSiCl_3$ (**2a**) and $Cl_3SiOSiCl_2OSiCl_3$ (**2b**), respectively, leads to the corresponding metallo-siloxanes $L_nM–SiCl_2OSiCl_3$ [L_nM = $Cp(OC)_2Fe$ (**3a**), $Cp(OC)_2(Me_3P)Mo$ (**4a**), [$Cp(OC)_2(Me_3P)W$ (**4b**)], and $C_5R_5(OC)_2Fe–SiCl_2OSiCl_2OSiCl_3$ [R = H (**3b**), Me (**3c**)]. $Cp(OC)_2Fe–SiCl_2OSiCl_3$ (**3a**) can be converted into the bis-metalated siloxane $Cp(OC)_2Fe–SiCl_2OSiCl_2Fe(CO)_2Cp$ (**8a**) by reaction with a second equivalent of $Na[Fe(CO)_2Cp]$ (**1a**). **8a** is succesfully transformed into the bis(ferrio)-tetrahydroxydisiloxane (**10**) by reaction with water. In addition, the polychlorinated ferrio-trisiloxane $Cp(OC)_2Fe–SiMe_2–OSiCl_2OSiCl_3$ (**13**) is obtained by reaction of the ferrio-silanol $Cp(OC)_2Fe–SiMe_2OH$ (**12**) with hexachlorodisiloxane (**2a**). **13** can be converted into the (bis)ferrio-tetrasiloxane (**15**) by addition of a second equivalent of **12**.

Introduction

Silanols bearing a Si-bonded CpL_nM-fragment are characterized by a high stability toward self-

[†] Part 32 of the series "*Metallo-Silanols and Metallo-Siloxanes*". In addition, Part 61 of the series "*Synthesis and Reactivity of Silicon Transition Metal Complexes*". Part 60/31, see [1].

condensation due to their strongly reduced H-acidity compared to organosilanols [1 – 9]. This property, which is mainly a consequence of the electron donating capacity of the transition metal, offers the possibility to perform controlled reactions involving the SiOH unit. It has been demonstrated that transition metal substituted silanols are easily accessible via hydrolysis of metallo-chlorosilanes and by oxygenation of metallo-silanes with dimethyldioxirane or the catalytic system MTO/UHP [2]. Especially SiH-functionalized metallo-siloxanes can be successfully used for the generation of dinuclear, siloxane-bridged complexes characterized by a metal-silicon bond [10]. We have now extended our studies to transition metal substituted polychlorinated siloxanes, which are accessible via metalation of perchloro-siloxanes [11].

Synthesis of Polychlorinated Metallo-Siloxanes via Metalation of Polychlorosiloxanes

An easy approach to metallo-siloxanes is given by the reaction of the chlorosiloxanes $Si_nO_{n-1}Cl_{2n+2}$ [n = 2 (**2a**); 3 (**2b**)] or the cyclic octachloro-tetrasiloxane (**2c**) with the metalates $Na[Fe(CO)_2C_5R_5]$ [R = H (**1a**), Me (**1b**)] or $Li[M(CO)_2(PMe_3)(C_5H_5)]$ [M = Mo (**1c**), W (**1d**)], respectively. In accordance with the established heterogeneous metalation of chlorosilanes, a suspension of metalates in cyclohexane is reacted with an equimolar amount of the chlorosiloxanes for 16 to 72 h. Under these conditions the Si_4O_4-network of the octachlorotetrasiloxane **2c** is not affected by ring-opening reaction.

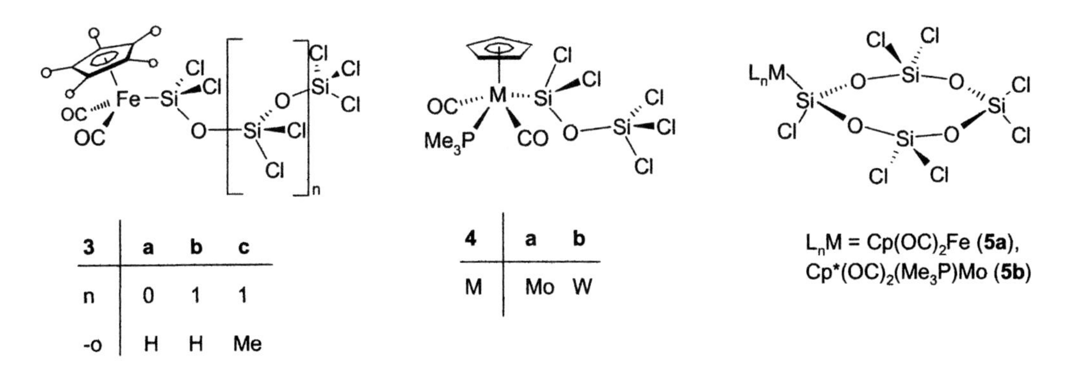

Fig. 1. Polychlorinated metallo-siloxanes **3** – **5**.

3 – **5** are isolated as beige, moisture and air sensitive solids which can be stored for several weeks under an atmospöhere of nitrogen at –20 °C. The substitution of one silicon atom by a transition metal fragment is evident by low-field resonance of the ^{29}Si NMR spectrum [38.5 (**3a**), 51.7 (**3b**), 51.7 (**4a**), 26.4 ppm (**4b**), compared to –46.0 ppm for $(Cl_3Si)_2O$], which is the characteristic of the silicon atom bound to metal atom. The γ-Si-atoms in the cyclic siloxane **5a** are not influenced by the iron fragment, indicated by the resonance of ^{29}Si NMR at –69.5 ppm which is close to that of the ε-Si-atom (–72.3 ppm).

Methanolysis Reaction of the Metallo-Polychlorosiloxanes

The metallo-siloxane **3a** bears the possibility of a regiospecific exchange reaction as the electrophilicity of the α-Si-atoms is decreased due to the electron donating influence of the transition metal fragment. For this reason, the methanolysis reaction of **5a** with three equivalents of MeOH involves the γ-silicon-atom to yield preferentially the metallo-trimethoxydisiloxane **6**, which can be converted into the pentamethoxydisiloxane **7** by using methanol as solvent.

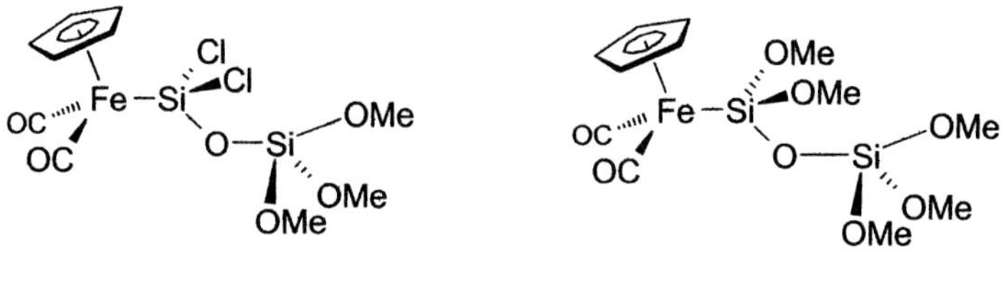

Fig. 2. Methanolysis products **6**, **7**.

Synthesis of Bis-metalated Metallo-Siloxanes

The ferrio-siloxanes **3a,b** can be easily transformed into the corresponding bis-metalated species **8a,b** by treatment with a second equivalent of sodium ferrate **2a** under heterogeneous reaction conditions.

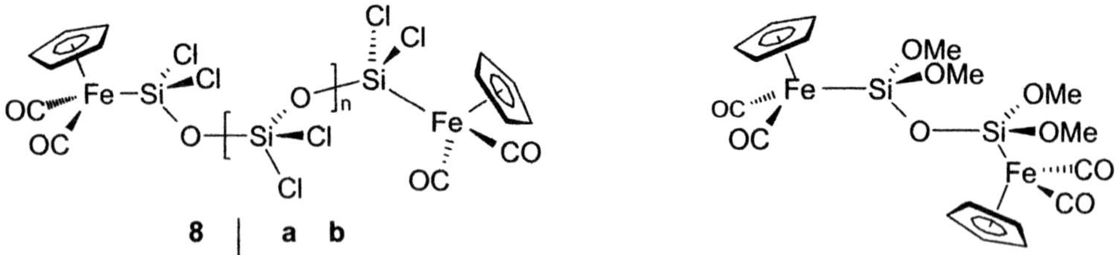

8	a	b
n	0	1

Fig. 3. Synthesis of the bis(ferrio)silanols **8** and **9**.

The bis(ferrio)tetrachlorodisiloxane **8a** was the target for further investigations concerning Cl-exchange. In a nucleophilic substitution reaction with MeOH the tetramethoxy-derivative **9** can be isolated as a beige, microcrystalline solid, showing good solubility in non-polar solvents like *n*-pentane or diethylether.

8a is a good precursor for the generation of a stable tetrahydroxy-disiloxane since the metal fragments guarantee stability with respect to self-condensation. The reaction of **8a** with an excess of water in the presence of NEt_3 offers the corresponding bis(ferrio)tetrahydroxysiloxane **10** in a yield of 84%. **10** is isolated as a beige, microcrystalline powder, which shows a limited solubility in apolar solvents (*n*-pentane or diethylether), but a good solubility in polar solvents, such as tetrahydrofurane or acetonitrile.

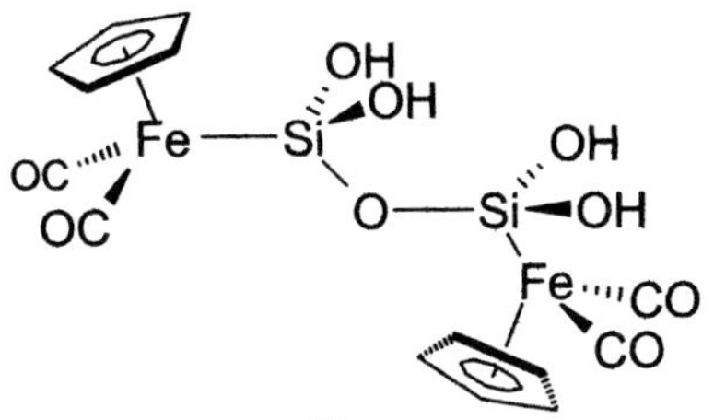

Fig. 4. Bis(ferrio)tetrahydroxydisiloxane **10** and bis(ferrio)hexasiloxane **11**.

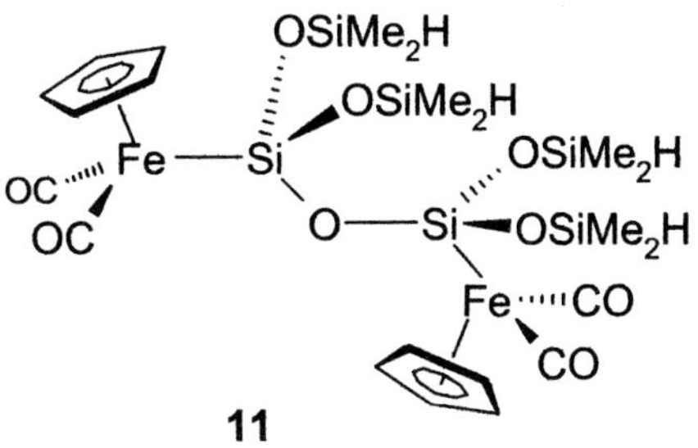

10 shows the expected reactivity concerning condensation with organochlorosilanes, e.g. $Me_2Si(H)Cl$, which results in the formation of the branched hexasiloxane **11**.

Synthesis of Polychlorinated Metallo-Siloxanes via Reaction of the Ferrio-Silanol $Cp(OC)_2SiMe_2OH$ (12) with Hexachlorodisiloxane (2a)

Metallo-siloxanes with three or more Si-atoms in the siloxane framework are in addition accessible by reaction of the ferrio-dimethylsilanol **12** with hexachlorodisiloxane (**2a**) in diethylether. **13** is obtained in a yield of 75%.

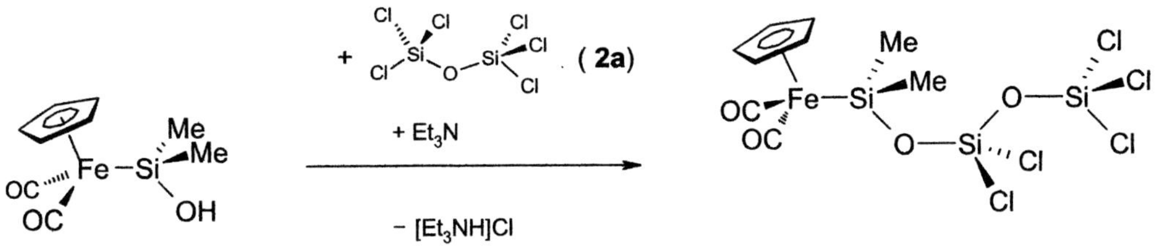

Eq. 1. Ferrio-dimethylpentachloro-trisiloxane **13**.

13 can be isolated as a brown oil and shows a high moisture and air sensitivity. Nevertheless it can be stored for several weeks under an atmosphere of nitrogen at –20 °C without any indication of decomposition. The ^{29}Si NMR signal of the α-$SiMe_2$ group (73.5 ppm) is found in the range of the ferrio-dimethylsilanols **12** (66.6 ppm).

The methanolysis reaction of the ferrio-siloxane **13** in the presence of NEt_3 shows clearly that the remaining SiCl functions are still reactive toward a nucleophilic attack. After a reaction time of 2 h the fully methoxy-substituted compound **14** can be isolated in a yield of 79%.

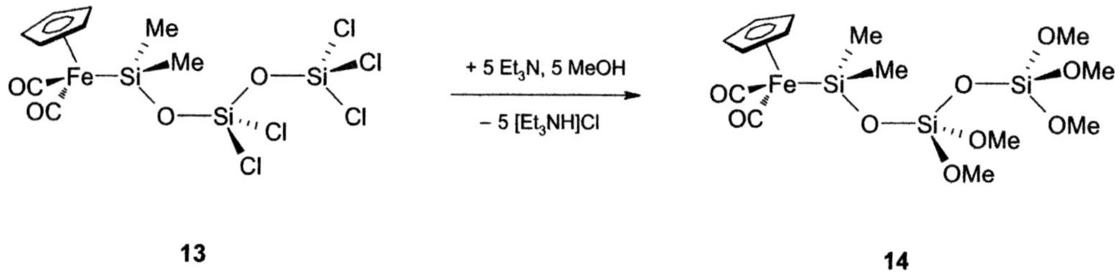

$+ 5\ Et_3N,\ 5\ MeOH$ / $- 5\ [Et_3NH]Cl$

13 **14**

Eq. 2. Methanolysis of the ferrio-siloxane **13**.

The transformation of the pentachloro-substituted ferrio-dimethyltrisiloxane **13** to the penta-methoxy-substituted ferrio-dimethyltrisiloxane **14** results in a significant high-fieldshift of the γ-Si and the ε-Si-signal in the ^{29}Si NMR spectrum.

The ferrio-siloxane **13** can also be transformed into the corresponding bis-metalated species **15** by reaction with another equivalent of the ferrio-dimethylsilanol (**12**).

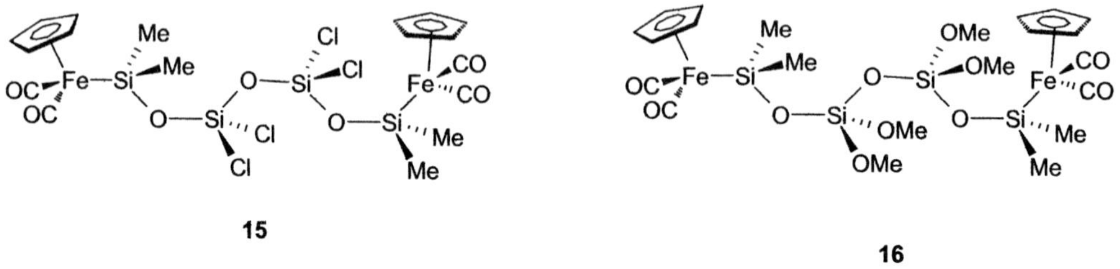

15 **16**

Fig. 5. Tetrachloro-bis(ferrio)tetramethyltetrasiloxane **15** and bis(ferrio)-tetramethoxytetramethyltetrasiloxane **16**.

The methanolysis reaction of **15** in MeOH yields the tetramethoxy-substituted (bis)metallo-tetrasiloxane **16** as a brown oil, showing good solubility in *n*-pentane, diethylether and benzene.

Acknowledgment: This work has been generously supported by the Deutsche Forschungsgemeinschaft (Schwerpunktprogramm: Spezifische Phänomene in der Siliciumchemie).

References

[1] W. Malisch, D. Schumacher, B. Schmiedeskamp, H. Jehle, D. Eisner, W. W. Schöller, M. Nieger, *Eur. J. Inorg. Chem.* **2003**, 2133.

[2] W. Malisch, H. Jehle, C. Mitchel, W. Adam, *J. Organomet. Chem.* **1998**, *566*, 259.

[3] a) C. S. Cundy, M. F. Lappert, C. K. Yuen, *J. Chem. Soc., Dalton Trans.* **1978**, 427; b) L. S. Chang, M. P. Johnson, M. J. Fink, *Organometallics* **1991**, *10*, 1219; c) W. Adam, U. Azzena, F. Prechtl, K. Hindahl, W. Malisch, *Chem. Ber.* **1992**, *125*, 1409; d) H. Handwerker, C. Leis, R. Probst, P. Bissinger, A. Grohmann, P. Kiprof, E. Herdtweck, J. Bluemel, N. Auner, C. Zybill, *Organometallics* **1993**, *12*, 2162; e) W. Malisch, S. Schmitzer, G. Kaupp, K. Hindahl, H. Käb, U. Wachtler, in *Organosilicon Chemistry I — From Molecules to Materials* (Eds.: N. Auner, J. Weis), VCH, Weinheim, **1994**, p. 185; f) W. Malisch, S. Schmitzer, R. Lankat, M. Neumayer, F. Prechtl, W. Adam, *Chem. Ber.* **1995**, *128*, 1251; g) W. Malisch, K. Hindahl, H. Käb, J. Reising, W. Adam, F. Prechtl, *Chem. Ber.* **1995**, *128*, 963; h) R. Goikhman, M. Aizenberg, H.-B. Kraatz, D. Milstein, *J. Am. Chem. Soc.* **1995**, *117*, 5865; i) S. Möller, O. Fey, W. Malisch, W. Seelbach, *J. Organomet. Chem.* **1996**, *507*, 239; j) W. Malisch, S. Möller, R. Lankat, J. Reising, S. Schmitzer, O. Fey, in *Organosilicon Chemistry II — From Molecules to Materials* (Eds.: N. Auner, J. Weis), VCH, Weinheim, **1996**, p. 575; k) S. H. A. Petri, D. Eikenberg, B. Neumann, H.-G. Stammler, P. Jutzi, *Organometallics* **1999**, *18*, 2615.

[4] W. Malisch, M. Neumayer, O. Fey, W. Adam, R. Schuhmann, *Chem. Ber.* **1995**, *128*, 1257.

[5] W. Malisch, R. Lankat, O. Fey, J. Reising, S. Schmitzer, *J. Chem. Soc., Chem. Commun.* **1995**, 1917.

[6] a) C. E. F. Rickard, W. R. Roper, D. M. Salter, L. J. Wright, *J. Am. Chem. Soc.* **1992**, *114*, 9682; b) W. Malisch, R. Lankat, S. Schmitzer, J. Reising, *Inorg. Chem.* **1995**, *34*, 5701.

[7] a) W. Adam, U. Azzena, F. Prechtl, K. Hindahl, W. Malisch, *Chem. Ber.* **1992**, *125*, 1409; b) W. Malisch, M. Hofmann, G. Kaupp, H. Käb, J. Reising, *Eur. J. Inorg. Chem.* **2002**, 3235.

[8] a) W. Malisch, M. Vögler, in *Organosilicon Chemistry IV — From Molecules to Materials* (Eds.: N. Auner, J. Weis), VCH, Weinheim, **2000**, p. 442; b) W. Malisch, H.-U. Wekel, I. Grob, F. H. Köhler, *Z. Naturforsch.* **1982**, *37b*, 601; c) H.-U. Wekel, W. Malisch, *J. Organomet. Chem.* **1984**, *264*, C10.

[9] W. Malisch, M. Vögler, D. Schumacher, M. Nieger, *Organometallics* **2002**, *21*, 2891.

[10] W. Malisch, M. Hofmann, G. Kaupp, H. Käb, J. Reising, *Eur. J. Inorg. Chem.* **2002**, 3235.

[11] W. Malisch, H. Jehle, D. Schumacher, M. Binnewies, N. Söger, *J. Organomet. Chem.* **2003**, *667*, 35.

Half-Sandwich Tungsten Complexes with Silyl-Functionalized η^5-Cyclopentadienyl Ligand†

Holger Bera, Andreas Sohns, Wolfgang Malisch*

Institut für Anorganische Chemie der Universität Würzburg
Am Hubland, 97074 Würzburg
Tel.: +49 931 888 5277 — Fax: +49 931 888 4618
E-mail: Wolfgang.Malisch@mail.uni-wuerzburg.de

Keywords: tungsten, half-sandwich complexes, metallo-silanols

Summary: Treatment of tungsten-silanes $Cp(L)(OC)_2WSiMe_2H$ [L = CO (**1a**), PMe_3 (**1b**)] with *n*BuLi or LiN(*i*Pr)$_2$, respectively, induces a silyl shift to the cyclopentadienyl ring, resulting in the formation of the tungsten-centered anions $\{[HMe_2Si(C_5H_4)](L)(OC)_2W\}Li$ [L = CO (**2a**), PMe_3 (**2b**)]. **2a,b** are transformed to the neutral complexes $[HMe_2Si(C_5H_4)](L)(OC)_2WR$ [L = CO, PMe_3; R = H, Me, $SiMe_2H$, $SiMe_3$, (**3**, **4a–c**)] by interaction with HOAc, MeI, $Me_2Si(H)Cl$ and Me_3SiCl. The tetramethyl derivative $[ClMe_2Si(C_5Me_4)](OC)_3WH$ (**6**) is synthesized by reaction of $ClMe_2SiC_5Me_4H$ (**5**) with $(MeCN)_3W(CO)_3$. Hydrolysis of the chlorosilyl tungsten complexes $[ClMe_2Si(C_5R_4)](OC)_3WR'$[R = H, Me; R′ = H, Cl (**6** – **8**)] leads to the novel tungsten complexes $[HOMe_2Si(C_5R_4)](OC)_3WR'$[R = H, Me; R′ = H, Cl (**9a–d**)], characterized by a η^5-cyclopentadienyl-bonded silanol group. The molecular structure of the silanol **9d** is presented.

Introduction

The condensation of organosilanols is an important step in the synthesis of siloxanes and silicones. This process can be inhibited by bulky organic groups at the silicon, which in some cases offers access even to stable silanediols, $R_2Si(OH)_2$, and silanetriols, $RSi(OH)_3$. Apart from the sterical aspect, another kind of stabilization of silanol is achieved by use of a σ-bonded metal fragment as substituent. In this context silanols of the general type $L_nM\text{–}SiR_{3-n}(OH)_n$ (n = 1 – 3) [2 – 7] have been realized, in which the metal fragment acts as strong electron releasing group inhibiting self-condensation. Our interest has now been focused on silanols, with the SiOH unit separated from the metal fragment by a η^5-cyclopentadienyl ligand, for which we expect a higher tendency toward

† Part 34 of the series "*Metallo-Silanols and Metallo-Siloxanes*". In addition, Part 63 of the series "*Synthesis and Reactivity of Silicon Transition Metal Complexes*". Part 62/33, see [1].

self- and co-condensation. Furthermore, these silanols offer vast variation of the ligand sphere at the metal center. However, only a few examples of such organometallic silanols have been described [8 – 11]. This report presents a general synthetic approach to these species containing tungsten as the central metal, as well as the first structure characterized by single-crystal X-ray diffraction.

Results and Discussion

Silylation of the Cyclopentadienyl Ligand

The base-induced migration of metal-bound ligands to the Cp ring [12 – 15] is a well known reaction of silicon-metal half-sandwich complexes. Treatment of the tungsten-silanes **1a,b** with $LiN(iPr)_2$ or *n*BuLi, respectively, leads via the lithiocyclopentadienyl complex $(LiC_5H_4)(L)(OC)_2WSiMe_2H$ (L = CO, PMe_3) (**A**) to a silatropic rearrangement, resulting in the formation of the tungsten-centered anions $\{[HMe_2Si(C_5H_4)](L)(OC)_2W\}Li$ [L = CO (**2a**), PMe_3 (**2b**)] (Scheme 1). The reactions were accompanied by desilylation of the complexes (about 6 – 35 %). In all cases the use of $LiN(iPr)_2$ proved to be more successful than *n*BuLi, which is in accordance with the observations for base-induced germyl-, stannyl- and plumbyl-migrations in analogous tungsten complexes [15]. For the tris-carbonyl-substituted tungsten-silane **1a** the silyl shift occurs more controlled than for the phosphine derivative **1b**. Possible reasons for this finding are a weaker W–Si bond, more acidic Cp-hydrogens, and a more stabilized tungsten anion of **1a** in relation to **1b**.

1	a	b
L	CO	PMe_3

+ LiR

– RH

R = *n*Bu, *t*Bu, $(iPr)_2N$

2	a	b
L	CO	PMe_3

A

Scheme 1. Metal-to-ligand silatropic shift.

The tungsten-centered anions of **2a,b** offer reaction with various electrophilic reagents as HOAc, MeI, $ClSiMe_2H$, or $ClSiMe_3$, which lead to the neutral complexes **3** and **4a–c** (Fig. 1).

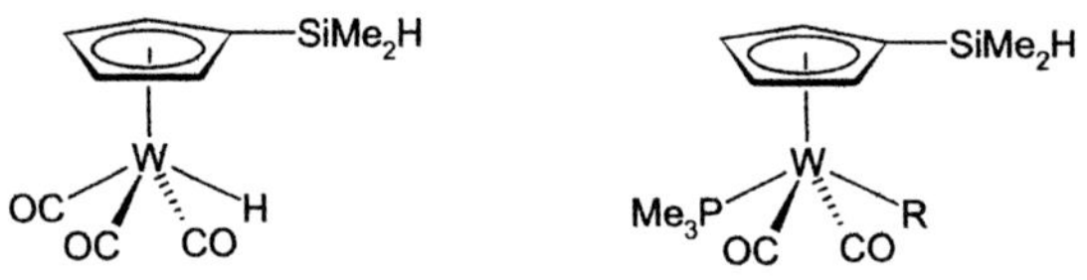

4	a	b	c
R	Me	$SiMe_2H$	$SiMe_3$

Fig. 1. Neutral silylcyclopentadienyl complexes derived from **2a,b**.

For the synthesis of complexes with a methyl-substituted cyclopentadienyl ligand, the silyl cyclopentadiene $ClMe_2SiC_5Me_4H$ (**5**) was reacted with $(MeCN)_3W(CO)_3$ to give the tungsten hydrido complex $[ClMe_2Si(C_5Me_4)](OC)_3WH$ (**6**) (Eq. 1). **6** was obtained quantitativly and shows a higher stability compared to the hydrogen-substituted analogue $[ClMe_2Si(C_5H_4)](OC)_3WH$ [10].

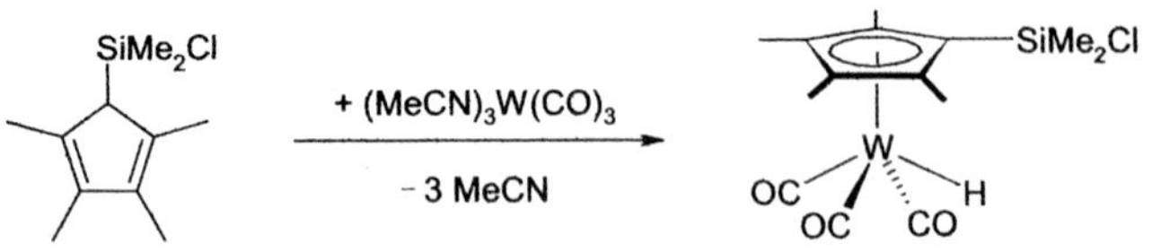

Eq. 1. Synthesis of $[ClMe_2Si(C_5Me_4)](OC)_3WH$ (**6**).

Chlorination of Silylcyclopentadienyl Tungsten Complexes

Treatment of **3** with $PdCl_2$ resulted in an H/Cl-exchange at both the silicon and the tungsten, leading to the known complex **7a** [10] (Scheme 2). **7b** was obtained by reaction of **6** with a high excess of CCl_4 in 87% yield.

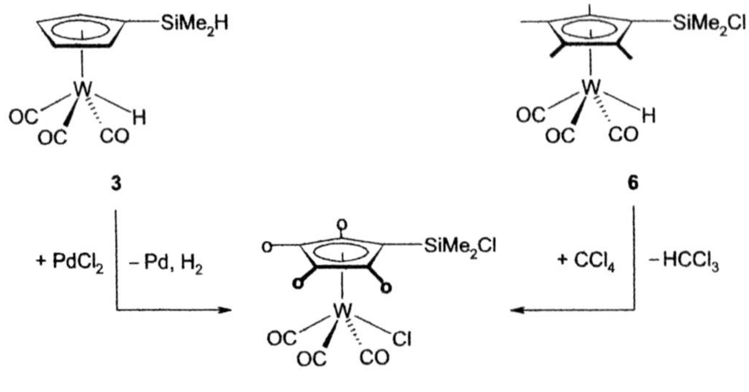

7	a	b
o-	H	Me

Scheme 2. Chlorination of tungsten-silanes **3** and **6**.

Generation of Tungsten-Silanols

The Si–Cl-functionalized complexes **6** – **8** can be easily transformed into the corresponding silanols **9a–d** by hydrolysis in the presence of NEt_3 as auxiliary base (Fig. 2). **9a–d** were obtained as yellow and orange solids in yields of 78 – 90%.

The hydrolysis at 0 °C during a period of 40 min led, in the case of **7a**, to a mixture of the condensation product $O\{[Me_2Si(C_5H_4)](OC)_3WCl\}_2$ (**10**) and small amounts of the desilylated complex $Cp(OC)_3WCl$. The targeted tungsten-silanol **9b** was obtained, when hydrolysis was performed at lower temperature and for shorter time. Apparently the silanol group shows increased reactivity by exchanging the metal-bound hydrogen (**8**) against chlorine.

Tetramethyl substitution of the cyclopentadienyl ligand in **6**, **7b** decreases the condensation tendency, leading to the stable hydrolysis products $[HOMe_2Si(C_5Me_4)](OC)_3WR$ [R = H (**9c**), Cl (**9d**)]. The ^{29}Si NMR resonances appear between 4.31 (**9b**) and 6.34 (**9d**) ppm. For the tungsten-disiloxane **10**, a significant shift to high-field (–0.50 ppm) in comparison with **9b** is observed, typical for SiOSi-units. All ^{29}Si NMR values reveal a decreased influence of the transition metal fragment on the silicon atom attached at the Cp-ring. For example, the signal of **9a** is shifted approximately 19 ppm to higher field, compared to its counterpart $Cp(OC)_2(Ph_3P)WCH_2SiMe_2OH$ with a methylene spacer (23.43 ppm) [15] and nearly 45 ppm compared to $Cp(OC)_2(Ph_3P)WSiMe_2OH$ (49.6 ppm) [6], containing a directly metal-bonded silanol group.

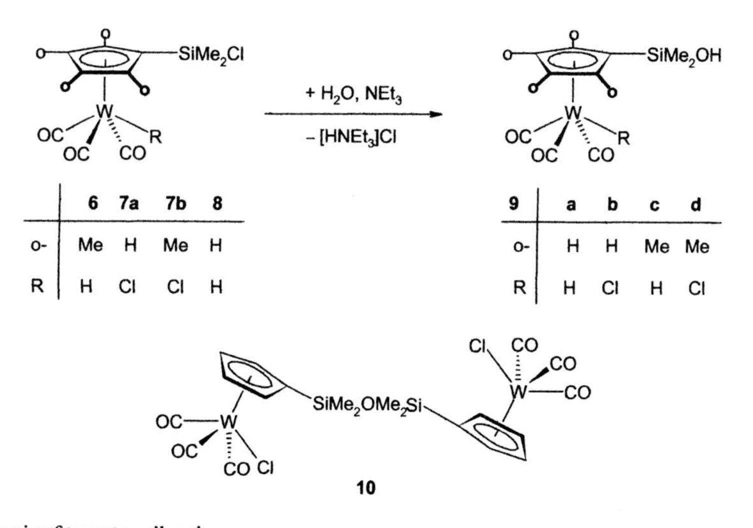

Fig. 2. Synthesis of tungsten-silanols.

X-ray Analysis

The molecular structure of **9d** (Fig. 3) reveals a pseudo-tetragonal pyramidal piano stool arrangement of the ligands at the central tungsten atom with the cyclopentadienyl ligand in the apical position, typical for (η^5-C_5H_5)$(OC)_3MX$ (M = Cr, Mo, W; X = Cl, Br, I) complexes. The Si1–C2 bond length with 1.908(10) Å is very similar to those of the silicon methyl groups [Si1–C16 1.907(11), Si1–C17 1.807(15) Å]. The silicon atom shows a tetrahedral arrangement of its substituents with angles between 107.6(5)° (C17-Si1-C2) and 112.5(6)° (C16-Si1-C2). The bond distance Si1–O3 of 1.622(9) Å is in accordance with those for literature-known metallo-silanols [17].

The silyl- and the Cp-bonded methyl groups lie above the cyclopentadienyl plane [C25-C1-C2-Si1 170.4(8)°, C15-C1-C2-C3 –170.7(11)°, C26-C25-C1-C15 6.9(18)°] while the silanol unit is found beneath. This arrangement might be responsible for the fact that no intermolecular hydrogen bond interaction between the silanol groups is observed, in contrast to other metallo-diorganosilanols, e.g., $C_5Me_5(OC)_2FeSiMe_2OH$ [18], $C_5Me_5(OC)_2RuSi(o\text{-}Tol)_2OH$ [19], and $C_5H_5(OC)_2FeCH_2SiMe_2OH$ [20], which prefer discrete dimeric or tetrameric arrangements.

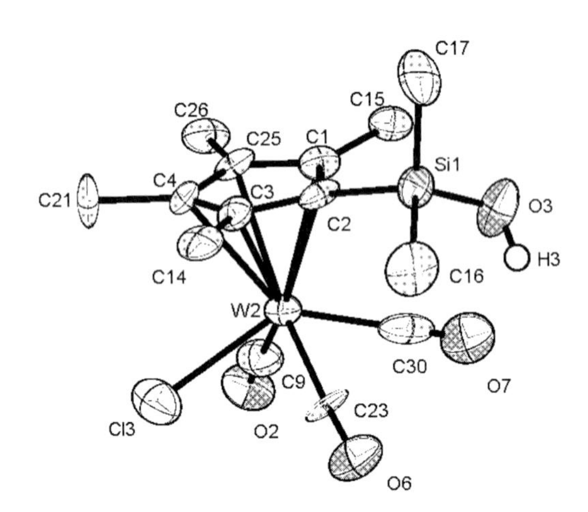

Fig. 3. Molecular structure of **9d**. Selected bond lengths [Å] and bond and torsion angles [°]: Si1–C2 1.908(10), Si1–C17 1.807(15), Si1–C16 1.907(11), Si1–O3 1.622(9); O3-Si1-C17 111.1(7), O3-Si1-C16 108.7(6), C17-Si1-C16 109.1(7), O3-Si1-C2 107.6(5), C17-Si1-C2 107.7(6), C16-Si1-C2 112.5(6), Si1-C2-W2 126.3(5); C25-C1-C2-Si1 –170.4(8).

Acknowledgment: We gratefully acknowledge financial support from the Deutsche Forschungsgemeinschaft (Schwerpunktprogramm "Spezielle Phänomene in der Siliciumchemie").

References

[1] A. Sohns, D. Schumacher, W. Malisch, in *Organosilicon Chemistry VI — From Molecules to Materials* (Eds.: N. Auner, J. Weis), VCH, Weinheim, **2005**, in press.

[2] a) C. S. Cundy, M. F. Lappert, C. K. Yuen, *J. Chem. Soc., Dalton Trans.* **1978**, 427; b) L. S. Chang, M. P. Johnson, M. J. Fink, *Organometallics* **1991**, *10*, 1219; c) C. E. F. Rickard, W. R. Roper, D. M. Salter, L. J. Wright, *J. Am. Chem. Soc.* **1992**, *114*, 9682; d) H. Handwerker, C. Leis, R. Probst, P. Bissinger, A. Grohmann, P. Kiprof, E. Herdtweck, J. Bluemel, N. Auner, C. Zybill, *Organometallics* **1993**, *12*, 2162; e) R. Goikhman, M. Aizenberg, H.-B. Kraatz, D. Milstein, *J. Am. Chem. Soc.* **1995**, *117*, 5865.

[3] a) W. Adam, U. Azzena, F. Prechtl, K. Hindahl, W. Malisch, *Chem. Ber.* **1992**, *125,* 1409; b) W. Malisch, K. Hindahl, H. Käb, J. Reising, W. Adam, F. Prechtl, *Chem. Ber.* **1995**, *128,* 963; c) S. Möller, O. Fey, W. Malisch, W. Seelbach, *J. Organomet. Chem.* **1996**, *507*, 239; d) W. Malisch, H. Jehle, S. Möller, C.-S. Möller, W. Adam, *Eur. J. Inorg. Chem.* **1999**, 1585.

[4] W. Malisch, R. Lankat, O. Fey, J. Reising, S. Schmitzer, *J. Chem. Soc., Chem. Commun.* **1995**, 1917.

[5] W. Malisch, R. Lankat, S. Schmitzer, J. Reising, *Inorg. Chem.* **1995**, *34*, 5701.

[6] W. Malisch, H. Jehle, C. Mitchel, W. Adam, *J. Organomet. Chem.* **1998**, *259*, 566.

[7] W. Malisch, S. Schmitzer, R. Lankat, M. Neumayer, F. Prechtl, W. Adam, *Chem. Ber.* **1995**, *128(12)*, 1251.

[8] F. Amor, E. Jesús, I. Ana, P. Royo, A. V. Miguel, *Organometallics* **1996**, *15*, 365.

[9] S. Ciruelos, T. Cuenca, R. Gómez, P. Gómez-Sal, A. Manzanero, P. Royo, *Polyhedron* **1998**, *17(7)*, 1055.

[10] F. J. Mata, P. Giner, P. Royo, *J. Organomet. Chem.* **1999**, *572*, 155.

[11] A. Sohns, H. Bera, D. Schumacher, W. Malisch, in *Organosilicon Chemistry V — From Molecules to Materials* (Eds.: N. Auner, J. Weis), VCH, Weinheim, **2003**, 348.

[12] W. K. Dean, W. A. G. Graham, *Inorg. Chem.* **1977**, *16*, 1061.

[13] a) P. Pasman, J. J. M. Snel, *J. Organomet. Chem.* **1986**, *301*, 329; b) G. L. Crocco, C. S. Young, K. E. Lee, J. A. Gladysz, *Organometallics* **1988**, *7*, 2158.

[14] a) S. R. Berryhill, B. Sharenow, *J. Organomet. Chem.* **1981**, *221*, 143; b) G. Thum, W. Ries, D. Greissinger, W. Malisch, *J. Organomet. Chem.* **1983**, *252*, C67; c) S. R. Berryhill, G. L. Clevenger, F. Y. Burdurlu, *Organometallics* **1985**, *4*, 1509; d) K. H. Pannell, S. P. Vicenti, R. C. Scott, *Organometallics* **1987**, *6*, 1593; e) S. Sharma, J. Cervantes, J. L. Mata-Mata, M. C. Brun, F. Cervantes-Lee, K. H. Pannell, *Organometallics* **1995**, *14*, 4269.

[15] J. Cervantes, S. P. Vicenti, R. N. Kappor, K. H. Pannell, *Organometallics* **1989**, *8*, 744.

[16] M. Hofmann, W. Malisch, H. Hupfer, M. Nieger, *Z. Naturforsch.* **2003**, *58b*, 36.

[17] a) K. Jörg, W. Reich, A. Meyer, U. Schubert, W. Malisch, *Angew. Chem.* **1986**, *98*(*1*), 103; b) B. P. Bir`yukov, Y. T. Struchkov, *J. Chem. Soc., Chem. Commun.* **1968**, 667.

[18] W. Malisch, M. Hofmann, G. Kaupp, H. Käb, J. Reising, *Eur. J. Inorg. Chem.* **2002**, 3235.

[19] S. Möller, O. Fey, W. Malisch, W. Seelbach, *J. Organomet. Chem.* **1996**, *507*, 239.

[20] W. Malisch, M. Hofmann, M. Nieger, W. Schöller, A. Sundermann, *Eur. J. Inorg. Chem.* **2002**, 3242.

Chapter IV

Silicon in Organic and Bioorganic Chemistry

Norbornylsilanes: New Organosilicon Protecting Groups

Dieter K. Heldmann,* Jürgen Stohrer, Rafael Zauner

Consortium für elektrochemische Industrie GmbH, Central Research Company of Wacker-Chemie GmbH, Zielstattstr. 20, 81379 Munich, Germany

Tel.: +49 89 748 44221 — Fax: +49 89 748 44242

E-mail: dieter.heldmann@wacker.com

Keywords: protecting groups, hydrosilylation

Summary: The use of readily available norbornylsilanes as protecting groups for alcohols and diols is reported. Stabilities of exemplary norbornylsilyl compounds toward various reagents and deprotection conditions are compared with *tert*-butyldimethylsilyl-, *iso*-propyldimethyl, and trimethylsilyl groups. The norbornyl-substituted compounds have been found to be more stable than the corresponding *iso*-propyl derivatives.

Introduction

The ideal protecting group for an active-hydrogen moiety such as an alcohol or a carboxylic acid should attach in high yield, be stable toward a large number of reaction conditions, and at the same time be selectively removable in the presence of other functional groups containing different protecting groups. While no single silyl group can fulfill all of these conditions in all cases, the availability of different silyl groups offers an appropiate answer to nearly every protection-deprotection challenge [1]. The range of organic groups available on the silicon atom changes both steric and electronic characteristics of the protecting group and thereby causes different stabilities of the silyl compound to a wide variety of reaction and deprotection conditions. It is this versatility which allows the synthetic chemist to select a silyl group that can, for example, be selectively removed in the presence of another silyl or other protecting group. The ease and high-yield introduction and deprotection contribute significantly to the popularity and utility of silyl protecting groups [2]. Besides trimethylsilyl (M3), especially *tert*-butyldimethylsilyl (TBM2) is often used as a monovalent protecting group.

The possibility to protect both hydroxy functionalities of a diol with one molecule of a divalent organosilicon compound is a further advantage. A five, six, or even seven-membered ring is formed. Very popular examples are di-*iso*-propyl- and di-*tert*-butylsilylene groups.

Silyl compounds bearing a sterically demanding residue like *iso*-propyl or *tert*-butyl are usually prepared by metal-organic reactions (lithium or Grignard-type) [3]. Our objective focused on exploring the possibilities to prepare bulky silyl compounds by cost-effective, transition metal-catalyzed hydrosilylation techniques [4], thereby omitting the use of expensive reaction steps.

Synthesis

The hydrogen atom of a silane H–SiR_3 is transferred to the more highly substituted carbon atom of the double bond ("anti-Markownikow") by using transition metal-catalyzed hydrosilylation. Therefore, the reaction of dimethylchlorosilane and 2-methylpropene does not yield *tert*-butyldimethylchlorosilane.

The use of disubstituted alkenes is the only way to achieve hydrosilylation in the desired manner. Norbornene is such an alkene, and it is cheap and commercially available. Additionally, because of the angle strain, norbornene is much more reactive in hydrosilylation reactions than cyclopentene or open-chain alkenes. It is known that platinum-catalyzed hydrosilylation of norbornene (**1**) with dimethylchlorosilane (**2**) provides 2-norbornyldimethylsilane (**3**, "Silane NM2") in high yields [5]. We routinely used [COD]$PtCl_2$ as catalyst. We wondered, if this silane could be used advantageously as an organosilicon protecting group [6].

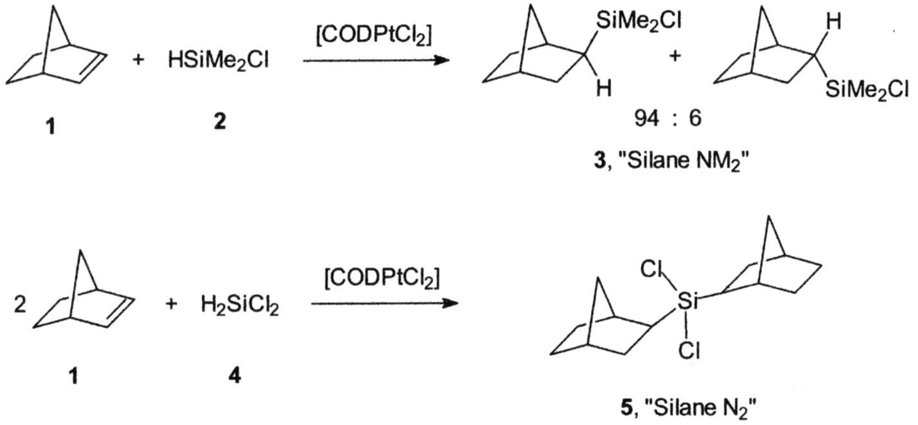

Scheme 1. Syntheses of the norbornylsilanes **3** and **5**.

Further, we prepared dinorbornyldichlorosilane (**5**, "Silane N2") as an alternative substitute for commonly used silylene protecting groups like di-*iso-propyl*- and di-*tert*-butylsilylene. This new silane was prepared by a one-pot reaction starting from dichlorosilane (**4**) and norbornene (two equivalents). The first hydrosilylation step takes places during warming the reaction mixture in toluene (caution: dichlorosilane is a volatile, highly flammable compound) from –20 °C to 80 °C (norbornyldichlorosilane has been identified as the first major reaction product by GC/MS). The second reaction step proceeds to completion upon heating the reaction mixture for 2 – 3 h at 80 °C.

Concentration and distillation provide the pure silane **5**, which was characterized by NMR- and mass-spectrometric investigations (GC/MS). Because of the exo/endo-composition of **3**, it is assumed that dinorbornyldichlorosilane is a mixture of isomers also. However, we did not succeed in achieving a suitable gas-chromatographic separation. The general properties of both norbornylsilanes are summarized in Table 1.

Table 1. General properties of norbornylsilanes.

	Silane NM2	**Silane N2**
molecular weight	188.77	289.32
Molecular formular	$C_9H_{17}ClSi$	$C_{14}H_{22}Cl_2Si$
bp [°C/mbar]	120 °C/10 mbar	118 °C/1.0 mbar
optical appearance	colorless liquid	colorless liquid
Exo/endo ratio	94/6	mixture, n.d.
chemical purity (GC)	>99%	97%
status	pilot plant	lab scale

Comparison of the Properties as a Protecting Group

The NM2 Protecting Group

Silane NM2 is capable of masking primary, secondary, and tertiary alcohols by using standard protocols (imidazole or NEt_3 as base, ambient temperature for primary/secondary alcohols). However, in line with the TBM2 group, the introduction of the more sterically demanding NM2 group requires harsher reaction conditions (elevated reaction temperatures, prolonged reaction times) for the formation of NM2 silyl ethers of tertiary alcohols.

Alternatively, the NM2 silyl triflate **7** can be used as a very powerful silylating reagent capable of silylating even tertiary alcohols under mild reaction conditions (2,6-lutidine as base, ambient temperature, 2 h, aqueous work-up). The NM2 silyl triflate can be simply prepared by reacting neat silane NM2 **3** with triflic acid **6** at 60 °C for 10 h. The triflate is a colorless liquid that can be purified by direct distillation (bp 100 – 105 °C/1 mbar) from the reaction flask in 83% yield. The transformation from chloride to triflate does not alter the exo/endo ratio according to GC/NMR.

$$\text{SiMe}_2\text{Cl (3)} + F_3CSO_3H \text{ (6)} \xrightarrow{60\ °C} \text{SiMe}_2\text{OSO}_2\text{CF}_3 \text{ (7)} + HCl$$

Scheme 2. Synthesis of the NM2 silyl triflate.

In order to determine suitable deprotection conditions, silyl ethers of *n*-butanol, 2-butanol, cyclohexanol, *tert*-butanol and phenol were synthesized. Table 2 shows various reaction and cleavage conditions that are commonly used in desilylation reactions for secondary alcohols. The progress of the reaction was analyzed using GC techniques and compared to a blank sample. For direct comparison, the corresponding silylethers of *iso*-propyldimethylsilyl (**8**, IPM2) and TBM2 **10** were treated in the same way as the NM2 silyl ether **9**.

Table 2. $t_{1/2}$ for cleavage of the Si–O–bond in silyl ethers.

8 **9** **10**

Reagent	*iso*-Propyl [IPM2]	2-Norbornyl [NM2]	*tert*-Butyl [TBM2]
2M BuMgCl in THF	stable	stable	stable
*n*BuLi in hexane/THF	stable	stable	stable
$LiAlH_4$ in THF	stable	stable	stable
PCC in CH_2Cl_2	1 h	2 h	stable
0.05M NaOH in MeOH	stable	stable	stable
KF in MeOH (25 °C)	stable	stable	stable
KF in MeOH (65 °C)	2 h	7 h	stable
nBu_4NF in THF	<< 1 [min]	< 1 [min]	> 30 [min]

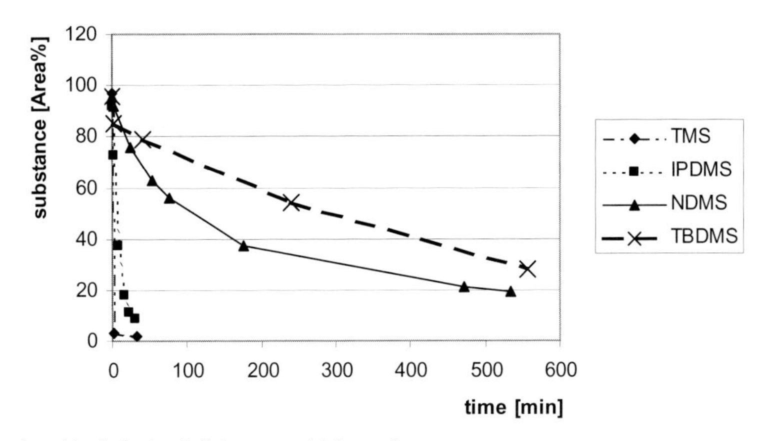

Fig. 1. Acid-catalyzed hydrolysis of silyl-protected 1-butanol.

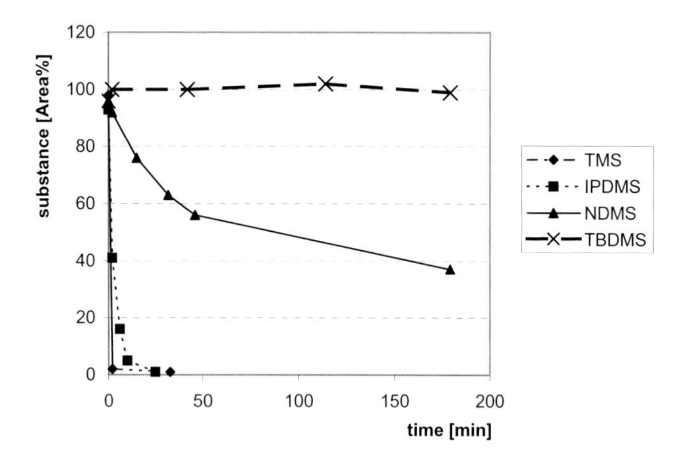

Fig. 2. Acid-catalyzed hydrolysis of silyl-protected 2-butanol.

The progress of the acid-catalyzed hydrolysis was followed by GC over an appropriate period of time. As an example, Figs. 1 and 2 show the acid hydrolysis of 1- and 2-butanol protected by M3, IPM2, NM2, TBM2. The NM2-ethers are cleaved 2 to 3 times more slowly than the IPM2 silyl ether under otherwise identical conditions. Despite its steric bulk, the secondary alkyl residue does not reach the stability of a tertiary alkyl group (TBM2). Therefore, the NM2 silyl group is always the choice if the protecting group has to withstand an aqueous work-up, which is often difficult with the trimethylsilyl group, and strong resistance to dilute acids is not needed or is even prohibitive.

The N2 Protecting Group

Silane N2 can be used as a silylene-type protecting group for diol compounds. The introduction can be easily achieved by using 1-hydroxybenzotriazole (HOBt) as catalyst and triethylamine as base in acetonitrile at ambient temperature [7]. Protection can be achieved under milder conditions compared to the di-*tert*-butyl silylene group (heating is necessary). The additional preparation of a highly reactive (and sensitive) ditriflate, which is commonly used for the di-*tert*-butylsilyl derivative, is not necessary. This method is suitable for aliphatic and aromatic 1,2-diols as well as for 1,3-diols.

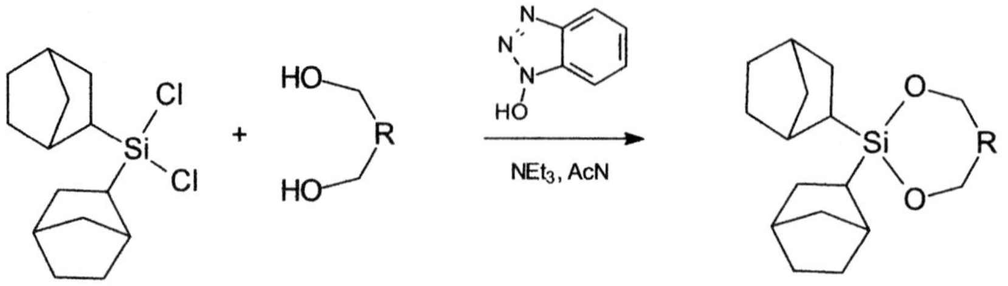

Scheme 3. Use of Silane N2 for the preparation of protected diols.

The identity of the protected diols (Fig. 3) was established by NMR and GC-MS investigations. 1,3-diols are more stable than 1,2-diols. Aliphatic 1,2-diols give cleaner reactions than aromatic diols.

The stability of the dinorbornylsilyl-protected diols is superior to that of the commonly used di*iso-propyl*silyl derivative, but does not reach that of the di-*tert*-butyl derivative. Major advantages are therefore (i) easy introduction without the need for highly reactive triflates and (ii) the good stability toward aqueous work-up conditions without significant hydrolysis for many diols (sufficient for many applications).

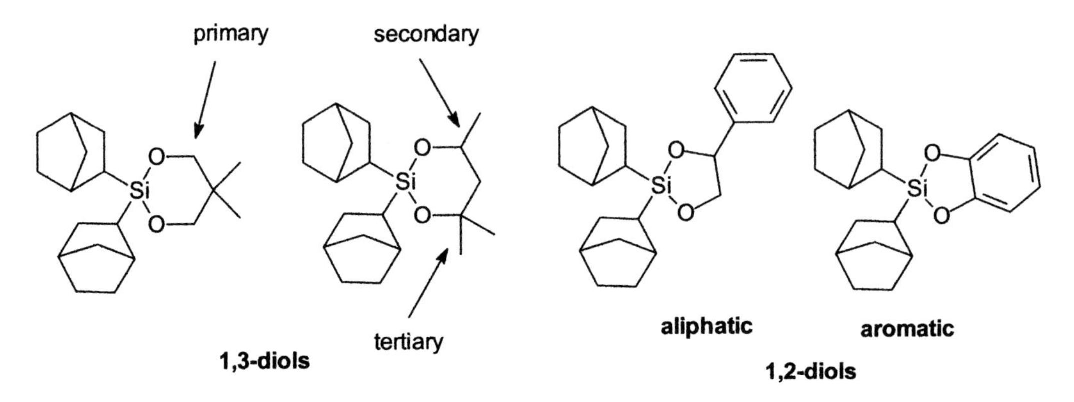

Fig. 3. Exemplary structures of dinorbornylsilyl-protected diols.

References

[1] (a) Greene, T. W.; Wuts, P. G. M. *Protective Groups in Organic Synthesis*, 3rd edn., J. Wiley & Sons, New York, **1999**. (b) Kocieński, P. J. *Protecting Groups*; Thieme: Stuttgart, **1994**.

[2] Lalonde, M.; Chan, T. H. *Synthesis* **1985**, 817.

[3] (a) Corey, E. J.; Venkateswarlu, A. *J. Am. Chem. Soc.* **1972**, *94*, 6190. (b) Morrison, R. J.; Hall, R. W.; Dover, B. T.; Kamienski, C. W.; Engel, J. F. Eur. Pat. EP 525 880, **1998**; *Chem. Abstr.* **1993**, *118*, 192001. (c) Shirahata, A. Eur. Pat. 405 560, **1998**; *Chem. Abstr.* **1991**, *114*, 164495. (d) Winterfeld, J.; Abele, B. C.; Stenzel, O. PCT-WO 2000/64909, **2000**; *Chem. Abstr.* **2000**, *132*, 279350.

[4] The hexyldimethylsilyl protecting group introduced by Oertle and Wetter is prepared by $AlCl_3$-mediated hydrosilylation of 2,3-dimethyl-2-butene with dimethylchlorosilane. Oertle, K.; Wetter, H. *Tetrahedron Let.* **1985**, *26*, 5511.

[5] Eddy, V. J.; Hallgren, J. E. *J. Org. Chem.* **1987**, *52*, 1903. (b) Green, M.; Spencer, J. L.; Stone, F. G. A.; Tsipsis, C. A. *J. Chem. Soc., Dalton Trans.* **1987**, *52*, 1519.

[6] Heldmann, D. K.; Stohrer, J.; R. Zauner *Synlett* **2002**, 1919.

[7] Trost, B. M.; Caldwell, C. G.; Murayama, E.; Heissler, D. *J. Org. Chem.* **1983**, *48*, 3252.

Synthesis and Reactivity of an Enantiomerically Pure *N*-Methyl-2-Silyl-Substituted Pyrrolidine

K. Strohfeldt, T. Seibel, P. Wich, C. Strohmann*

Institut für Anorganische Chemie, Universität Würzburg
Am Hubland, 97074 Würzburg, Germany
Tel.: +49 931 8884613 — Fax: +49 931 8884605
E-mail: mail@carsten-strohmann.de

Keywords: lithiumorganyle, (aminomethyl)silane, (lithiomethyl)silane, stereogenic center

Summary: Our research on α-metalated organosilanes currently focuses on (aminomethyl)silanes containing a defined stereogenic center next to the silicon center. A synthetic route based on the preparation of 2-silyl-substituted pyrrolidines was developed. The racemic product could be synthesized by metalation of *N*-Boc-pyrrolidine in the presence of TMEDA and conversion with the corresponding chlorosilane, whereas the enantioenriched form was achieved by metalation in the presence of the chiral amine (–)-sparteine. Subsequent metalation and transformation reactions yielded the formation of the corresponding (aminomethyl)(lithiomethyl)silane.

As part of our studies on α-metalated organosilanes, we use (aminomethyl)(lithiomethyl)silanes of the general type **A** [1]. Systems of types **B** and **C** containing defined stereochemical information could be synthesized starting from either chiral amines or from *Si*-chiral silanes. A new structural motif (**D**) arises if the stereogenic center is adjacent to silicon (preferably as part of a rigid system). The goal of these studies was the synthesis of the enantiomerically pure, lithiated 2-silyl-substituted pyrrolidine **E**.

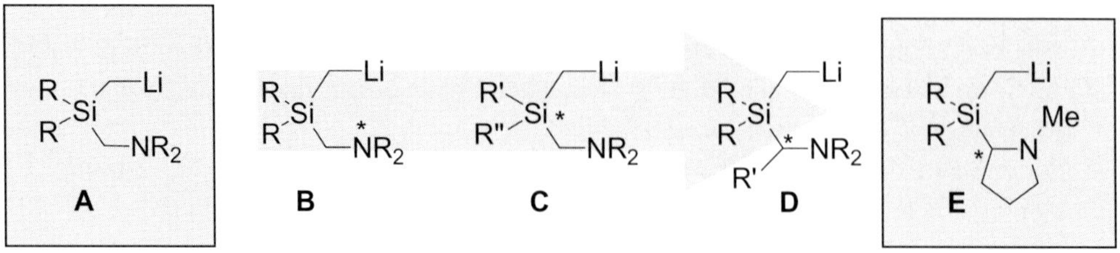

Scheme 1. Structural motifs of (aminomethyl)(lithiomethyl)silanes.

According to the the work of P. Beak and coworkers on 2-substituted pyrrolidines [2], the following synthetic route, containing *asymmetric deprotonation* and *transformation* of the protecting group, was chosen for the preparation of the (aminomethyl)silanes **3** and **5**. The (aminomethyl)(lithiomethyl)silane **4**, a molecule of type **E** (Scheme 1), was achieved by subsequent *metalation* of **3** (Scheme 2).

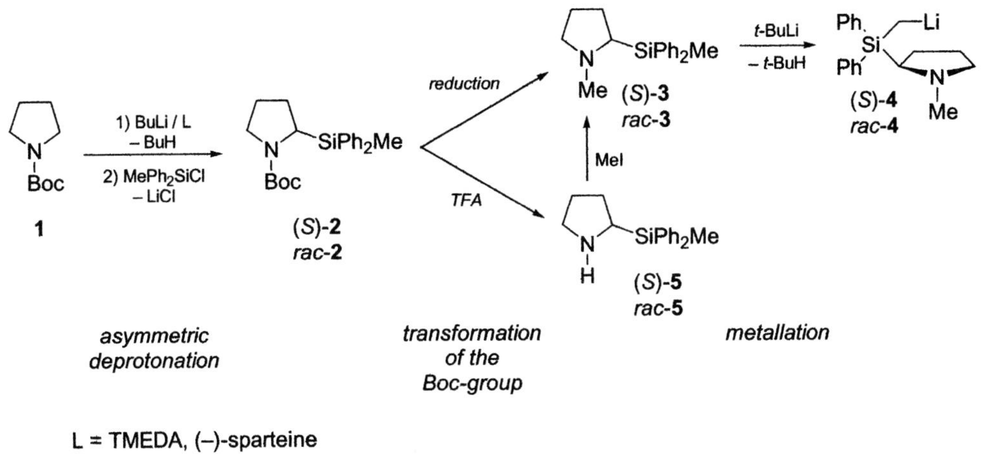

asymmetric deprotonation — *transformation of the Boc-group* — *metallation*

L = TMEDA, (–)-sparteine

Scheme 2. Preparation of the (aminomethyl)silanes **3** and **5** and of (aminomethyl)(lithiomethyl)silane **4**.

The *asymmetric deprotonation* of *N*-Boc-pyrrolidine (**1**) was performed by using *s*-BuLi/(–)-sparteine, resulting in the formation of the corresponding enantioenriched lithiumorganyle. After reaction of this lithium compound with chloromethyldiphenylsilane, the enantioenriched compound (*S*)-2-(methyldiphenylsilyl)-*N*-Boc-pyrrolidine [(*S*)-**2** (e. r. = 94:6), Scheme 2] was isolated in 88% yield. The use of TMEDA instead of (–)-sparteine afforded *rac*-**2** (70% yield).

The crucial reagents for the synthesis of the enantioenriched lithium compound (*S*)-**2** are (–)-sparteine complexes of simple alkyllithium bases. P. Beak and cowokers showed that *s*-BuLi and *i*-PrLi complexed by (–)-sparteine could be efficiently used for an asymmetric deprotonation reaction, whereas complexes of the chiral amine and *t*-BuLi or *n*-BuLi showed hardly any stereoselectivity or reactivity (Scheme 3) [2].

An explanation for the lack of reactivity or stereoselectivity might be received from an analysis of the molecular structure in the solid state.

- *t*-BuLi·(–)-sparteine crystallizes as a monomer [3]. In spite of high reactivity low stereoselectivities can be observed. This is understood by quantum-chemical studies on a *t*-BuLi/*N*-Boc-pyrrolidine complex, indicating little difference in the energies of the corresponding transition states during the deprotonation reaction [4].
- *n*-BuLi complexed with (–)-sparteine forms a symmetric dimer in the solid state [5] as well

as in diethyl ether solution [6]. Together with the lower basicity, the steric shielding of the lithium centers by the (–)-sparteine ligands is responsible for the low reactivity.

- In solution [7] and solid state (–)-sparteine-coordinated *i*-PrLi exists as an unsymmetric aggregate containing two lithium alkyls [5]; this dissociates during the reaction under formation of a precoordinated complex. Apart from *s*-BuLi, *i*-PrLi is the only known simple alkyllithium compound which can, in combination with (–)-sparteine, efficiently be used for asymmetric deprotonation of *N*-Boc-pyrrolidine [2].

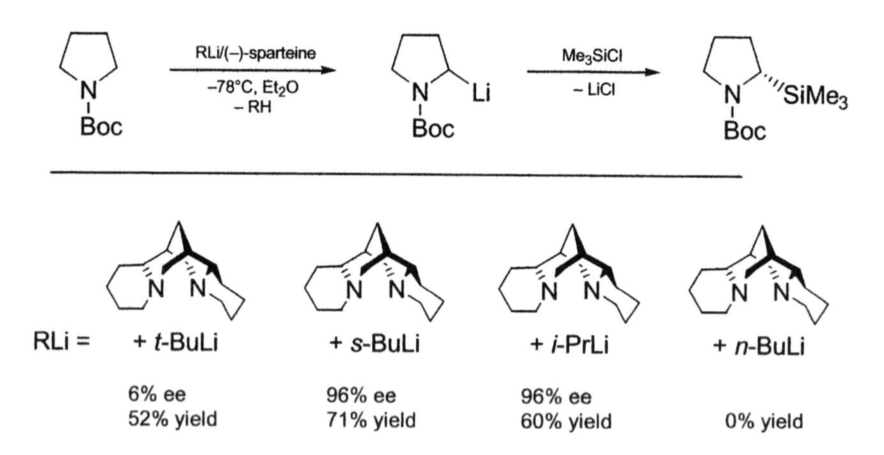

Scheme 3. Deprotonation of *N*-Boc-pyrrolidine with different alkyllithium bases in the presence of (–)-sparteine [2].

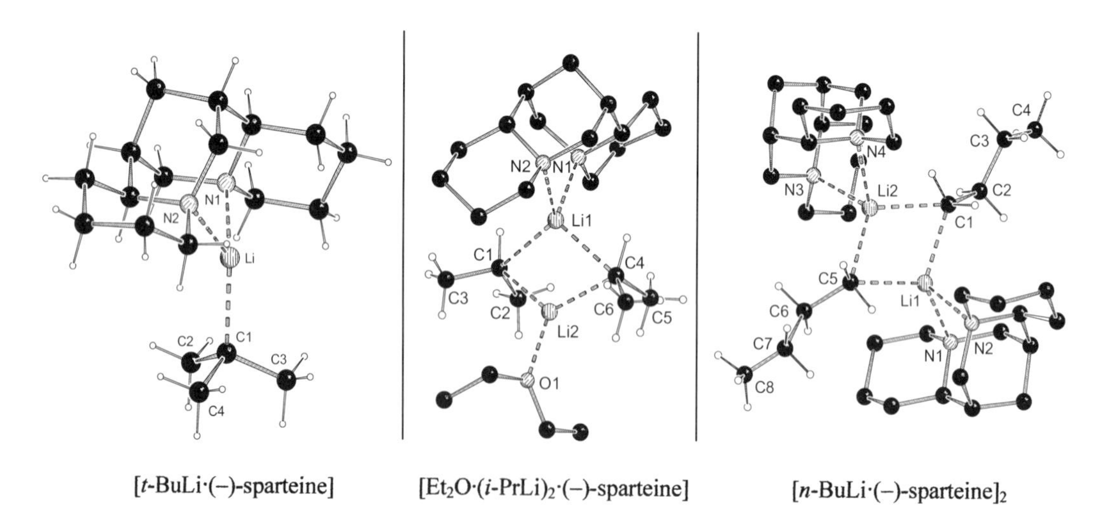

Fig. 1. Molecular structures of [*t*-BuLi·(–)sparteine] (left), [Et_2O·(*i*-PrLi)$_2$·(–)-sparteine] (middle) and [*n*-BuLi·(–)-sparteine]$_2$ (right) in the crystal [3, 5].

The *reduction* of the protecting group of compound **2** with *i*-Bu_2AlH (Scheme 4) resulted in the formation of the enantioenriched compound *N*-methyl-(*S*)-2-(methyldiphenylsilyl)pyrrolidine [(*S*)-**3** (e. r. = 94:6)] and the racemic product *rac*-*N*-methyl-2-(methyldiphenylsilyl)pyrrolidine (*rac*-**3**) for (*S*)-**2** and *rac*-**2**, respectively [8].

The isolation of the *enantiomerically pure* compound (*S*)-**3** (e. r. ≥ 99:1) was achieved by optical resolution with (*R*)-(–)-mandelic acid, and the absolute configuration determined by subsequent X-ray diffraction and ^1H NMR analysis. In this manner the enantiomerically pure compound (*S*)-**3** (e. r. ≥ 99:1), to be used as a starting-point for further metalation studies, was synthesized.

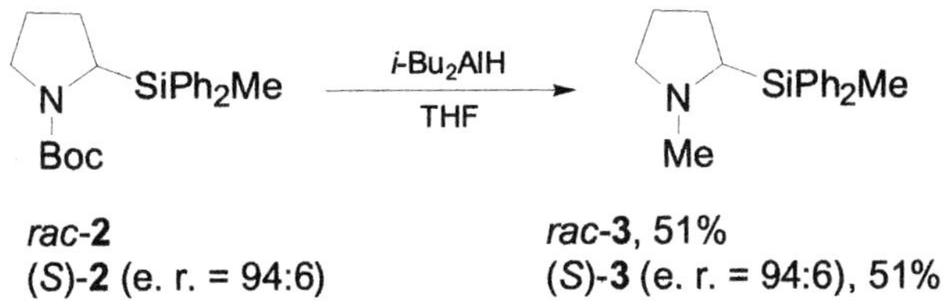

N(Boc) SiPh$_2$Me —[*i*-Bu$_2$AlH, THF]→ N(Me) SiPh$_2$Me

rac-**2** → *rac*-**3**, 51%

(*S*)-**2** (e. r. = 94:6) → (*S*)-**3** (e. r. = 94:6), 51%

Scheme 4. Reduction of the protecting group of compound **2**.

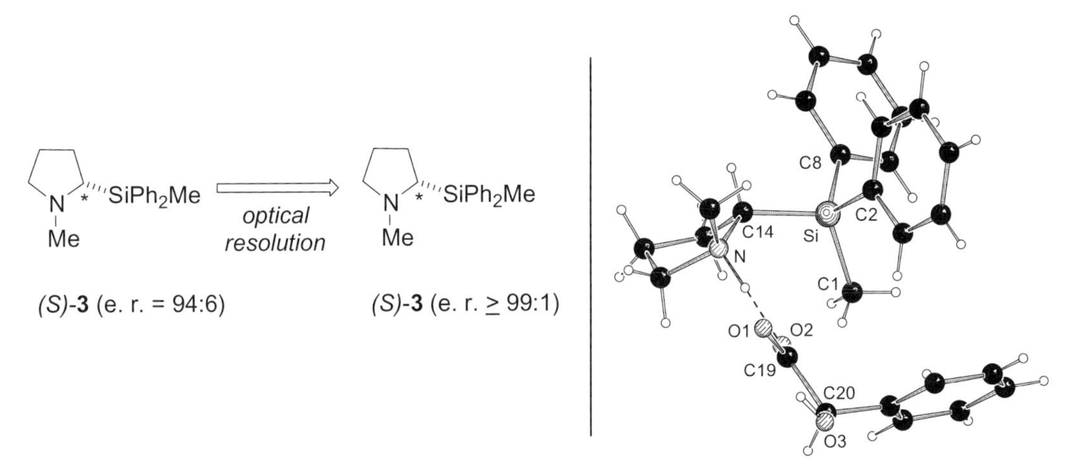

Fig. 2. Optical resolution of (*S*)-**3** (e. r. > 94:6) (left) and molecular structure of compound (*S*)-**3**·(*R*)-(–)-mandelic acid in the crystal; SCHAKAL plot (right) [9]. Selected bond lengths [Å] and angles [°]:C(1)–Si 1.863(2), C(2)–Si 1.883(2), C(8)–Si 1.876(2), C(14)–N 1.504(3), C(14)–Si 1.910(2), C(18)–N 1.473(4); N-C(14)-Si 117.68(15), C(1)-Si-C(8) 111.56(10), C(1)-Si-C(2) 109.37(10), C(8)-Si-C(2) 108.10 (9), C(1)-Si-C(14) 110.97(10), C(8)-Si-C(14) 104.20(10), C(2)-Si-C(14) 112.54(10).

The *metalation* of the 2-silyl-substituted pyrrolidine **3** with *t*-BuLi resulted in the formation of the corresponding (aminomethyl)(lithiomethyl)silane **4**, which represents a new structural motif for (aminomethyl)(lithiomethyl)silanes.

In the presence of THF it was possible to isolate crystals of $[rac\text{-}\mathbf{4}\cdot THF]_2$ and to determine the solid-state structure by X-ray diffraction analysis. [*rac*-**4**·THF]$_2$ crystallizes as a C_2-symmetric dimer with a central four-membered Li–C–Li–C-ring. One lithium center is coordinated by two molecules of THF and the other by two nitrogen centers [1h].

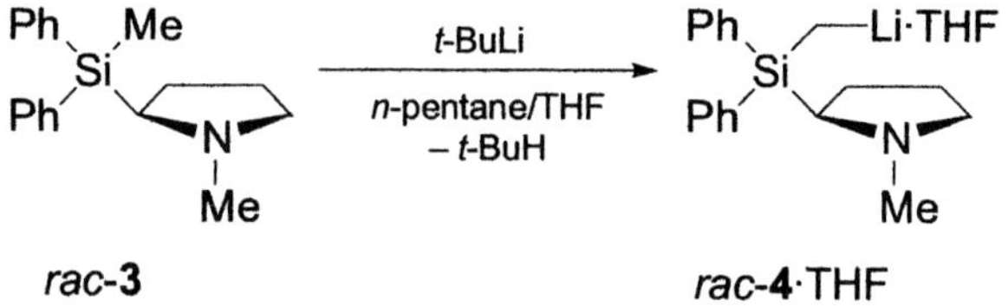

Scheme 5. Metalation of the (aminomethyl)silane *rac*-**3**.

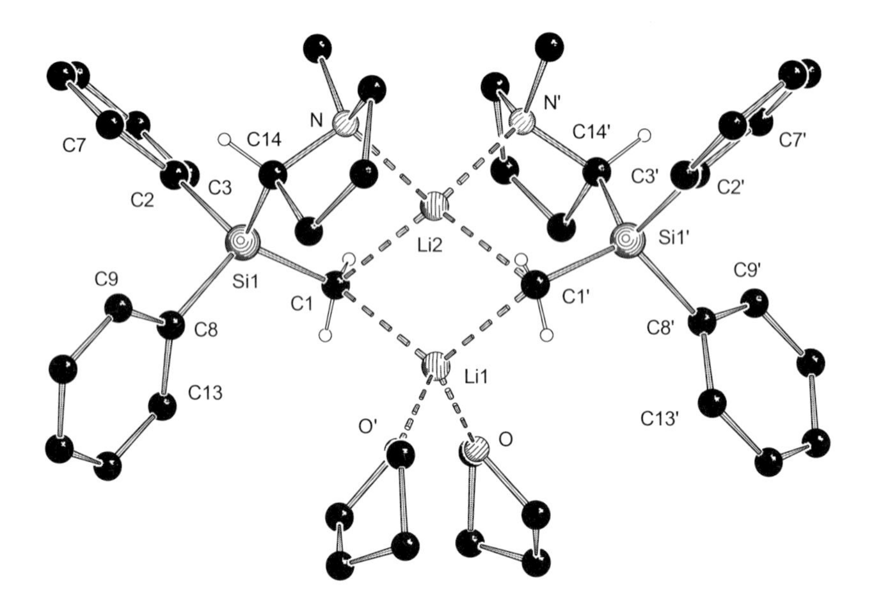

Fig. 3. Molecular structure of compound [*rac*-**4**·THF]$_2$ in the crystal; SCHAKAL plot [9]. Selected bond lengths [Å] and angles [°]: Si(1)–C(1) 1.805(3), Si(1)–C(2) 1.894(3), Si(1)–C(14) 1.898(4), Si(1)–C(8) 1.903(4), Li(1)–O 1.979(6), Li(1)–C(1) 2.252(6), Li(2)–N 2.211(6), Li(2)–C(1) 2.244(6); N'-Li(2)-N 107.9(4), N'-Li(2)-C(1) 124.57(15), N-Li(2)-C(1) 94.98(12), C(1)-Li(2)-C(1') 112.5(4).

Our ongoing research on (aminomethyl)(lithiomethyl)silanes focuses on modification of the substituents at the nitrogen center. The "unprotected" 2-silyl-substituted pyrrolidine **5** is a very useful starting point for the synthesis of different *N*-substituted pyrrolidines like *rac*-**6**. The racemic compound *rac*-**5** and the enantioenriched compound (*S*)-**5** could be synthesized by reaction of the *N*-Boc-protected compound *rac*-**2** or (*S*)-**2** trifluoroacetic acid (TFA) [10]. In subsequent reactions it was possible to introduce a methyl group at the nitrogen center by conversion of *rac*-**5** with methyl iodide, affording *rac*-**3** (Scheme 6).

N-BOC 2-SiMePh$_2$ pyrrolidine —(TFA, CH_2Cl_2, $-CO_2$)→ N-H 2-SiMePh$_2$ pyrrolidine —(MeI, −HI)→ N-Me 2-SiMePh$_2$ pyrrolidine

rac-**2**
(*S*)-**2** (e. r. = 94:6)

rac-**5**, 40%
(*S*)-**5** (e. r . = 94:6), 35%

rac-**3**, 50%

Scheme 6.

As a result of the numerous possibilities for modification at both the nitrogen and the silicon centers, this class of (aminomethyl)silanes has considerable potential for use in synthesis of pharmaceutically interesting compounds [11]. The metalated species can be used as reagents for the transfer of stereogenic centers [1a] as well as building blocks for novel ligands of organometallic systems [12].

Acknowledgments: We are grateful to the Institut für Anorganische Chemie der Universität Würzburg, the Deutsche Forschungsgemeinschaft (DFG) the Graduiertenkolleg 690 and the Fonds der Chemischen Industrie (FCI) for financial support. We acknowledge Wacker-Chemie GmbH for providing us with special chemicals.

References

[1] a) D. Schildbach, M. Arroyo, K. Lehmen, S. Martín-Barrios, L. Sierra, F. Villafañe, C. Strohmann, *Organometallics* **2004**, *23*, 3228; b) C. Strohmann, D. H. M. Buchold, K. Wild, D. Schildbach in *Organosilicon Chemistry V — From Molecules to Materials* (Eds.: N. Auner, J. Weis), VCH, Weinheim, **2003**, p. 155; c) C. Strohmann, D. H. M. Buchold, T. Seibel, K. Wild, D. Schildbach, *Eur. J. Inorg. Chem.* **2003**, 3453; d) C. Strohmann, D. Schildbach, *Acta Cryst. C* **2002**, *C58*, m447; e) C. Strohmann, B. C. Abele, K. Lehmen, F. Villafañe, L. Sierra, S. Martín-Barrios, D. Schildbach, *J. Organomet. Chem.* **2002**, *661*, 149; f) C. Strohmann, K. Lehmen, K. Wild, D. Schildbach, *Organometallics* **2002**, *21*, 3079; g) C. Strohmann, K. Lehmen, A. Ludwig, D. Schildbach, *Organometallics* **2001**, *20*, 4138; h) C. Strohmann, B. C. Abele, D. Schildbach, K. Strohfeldt, *Chem. Commun.* **2000**, 865; i) C. Strohmann, B. C. Abele, *Organometallics* **2000**, *19*, 4173; k) W. Uhl, L. Cuypers, K. Schüler, T. Spies, C. Strohmann, K. Lehmen, *Z. Anorg. Allg. Chem.* **2000**, *626*, 1526; l) B. C. Abele, C. Strohmann, in: *Organosilicon Chemistry III — From Molecules to Materials* (Eds.: N. Auner, J. Weis), VCH, Weinheim, **1997**, p. 206; m) C. Strohmann, B. C. Abele, *Angew. Chem. Int. Ed. Engl.* **1996**, *35*, 2378.

[2] P. Beak, S. T. Kerrick, S. Wu, J. Chu, *J. Am. Chem. Soc.* **1994**, *116*, 3231.

[3] C. Strohmann, T. Seibel, K. Strohfeldt, *Angew. Chem. Int. Ed.* **2003**, *42*, 4531.

[4] a) K. B. Wiberg, W. F. Bailey, *J. Mol. Struct.* **2000**, *556*, 239; b) K. B. Wiberg, W. F. Bailey, *J. Am. Chem. Soc.* **2001**, *123*, 8231.

[5] C. Strohmann, K. Strohfeldt, D. Schildbach, *J. Am. Chem. Soc.* **2003**, 125, 13672.

[6] J. L. Rutherford, D. Hoffmann, D. B. Collum, *J. Am. Chem. Soc.* **2002**, *124*, 264.

[7] a) D. J. Gallagher, S. T. Kerrick, P. Beak, *J. Am. Chem. Soc.* **1992**, *114*, 5872; b) D. J. Gallagher, P. Beak, *J. Org. Chem.* **1995**, *60*, 7092.

[8] R. E. Gawley, Q. Zhang, *J. Am. Chem. Soc.* **1993**, *115*, 7515.

[9] E. Keller, *SCHAKAL99, A Computer Program for the Graphic Representation of Molecular and Crystallographic Models*, Universität Freiburg **1999**.

[10] a) P. Beak, W. K. Lee, *J. Org. Chem.* **1993**, *58*, 1109.; b) G. Pandey, T. D. Bagul, A. K. Sahoo, *J. Org. Chem.* **1998**, *63*, 760.

[11] S. McN. Sieburth, H. K. O'Hare, J. Xu, Y. Cheng, G. Liu, *Org. Lett.* **2003**, *5*, 1859.

[12] a) M. Knorr, S. Kneifel, C. Strohmann, I. Jourdain, F. Guyon, *Inorg. Chim. Acta* **2003**, *350*, 455; b) M. Knorr, S. Kneifel, C. Strohmann, in *Organosilicon Chemistry III — From Molecules to Materials* (Eds.: N. Auner, J. Weis), VCH, Weinheim, **1997**, p. 211.

Diastereomerically Enriched α-Lithiated Benzylsilanes

*D. Schildbach, M. Bindl, J. Hörnig, C. Strohmann**

Institut für Anorganische Chemie, Universität Würzburg,
Am Hubland, 97074 Würzburg, Germany
Tel.: +49 931 8884613 — Fax: +49 931 8884605
E-mail: mail@carsten-strohmann.de

Keywords: organosilane, benzylsilane, alkyllithium compound, stereochemistry, density functional calculation

Summary: When non-chiral benzyldimethyl(piperidinomethyl)silane is lithiated, the resulting benzyllithium forms a tetramer in the solid state, where the "carbanionic" unit is fully planar. Lithiation of chiral benzylmethylphenyl(piperidinomethyl)silane results in **3**, which could be crystallized as $(l\text{-}\mathbf{3})_2$ and *u*-**3**·TMEDA. The molecular structures of the dimer $(l\text{-}\mathbf{3})_2$ and the monomer *u*-**3**·TMEDA in the solid state show that the absolute configurations at the metalated carbon center change according to the absolute configuration at the silicon center. Only the *unlike* diastereomer *u*-**3**·TMEDA can be detected in the crystal, while the dimer $(l\text{-}\mathbf{3})_2$ exists as the *like* diastereomer as a consequence of steric interactions in the dimer. The monomeric TMEDA adduct *u*-**3**·TMEDA shows a slightly increased sum of the bond angles of 348°, but is still far from planar. $(l\text{-}\mathbf{3})_2$ forms a dimer with a strongly pyramidalized "carbanionic" unit, proven by a sum of the bond angles of 341°. These structural parameters – planar or pyramidalized lithiated carbon centers – have a significant impact on the stereochemical course of reactions with these diastereomerically enriched α-lithiated benzylsilanes.

Highly diastereomerically enriched α-lithiated benzylsilanes are ideally suitable for transferring a stereogenic metalated carbon center stereospecifically and stereoselectively to metal fragments, organic electrophiles, or heteroelement systems. In our work group, non-chiral as well as chiral lithiated (aminomethyl)silanes have been synthesized [1]. Especially by the systematic structural investigations in the solid state by "conventional" single-crystal X-ray diffraction methods, we realized that significant structural variations arise from the differences in the ligand sphere around the lithium and from varying substituents at the organosilane unit.

When non-chiral benzyldimethyl(piperidinomethyl)silane is lithiated, the resulting alkyllithium **1** forms a tetramer in the solid state, not being coordinated by a further external solvent molecule. The "carbanionic" unit is fully planar, the sum of the bond angles amounting to 360° (Fig. 1) [1g].

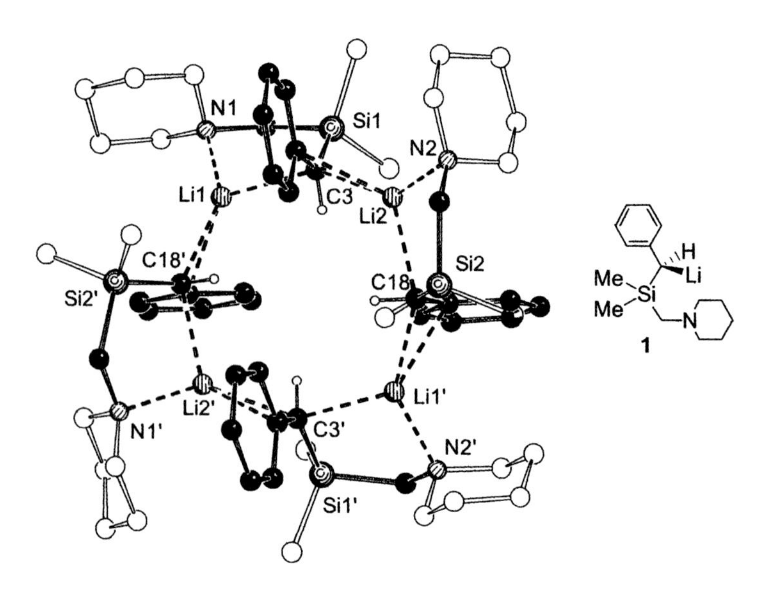

Fig. 1. Molecular structure of tetrameric benzyllithium compound **1** [1g] in the crystal; SCHAKAL plot [2].

If one methyl group is exchanged by a phenyl group, the starting organosilane becomes chiral. Interestingly, when the racemic compound *rac*-**2** is lithiated, the product **3** forms a dimer in the solid state with a strongly pyramidalized "carbanionic" unit, proven by a sum of the bond angles of 341°. Thus, profound structural differences can be observed by this substitution at the silicon center.

After treating *rac*-**2** with *tert*-butyllithium, both benzyllithium compounds (*l*-**3**)$_2$ and *u*-**3**·TMEDA could be crystallized at –30 °C and subjected to X-ray structure analysis (Scheme 1).

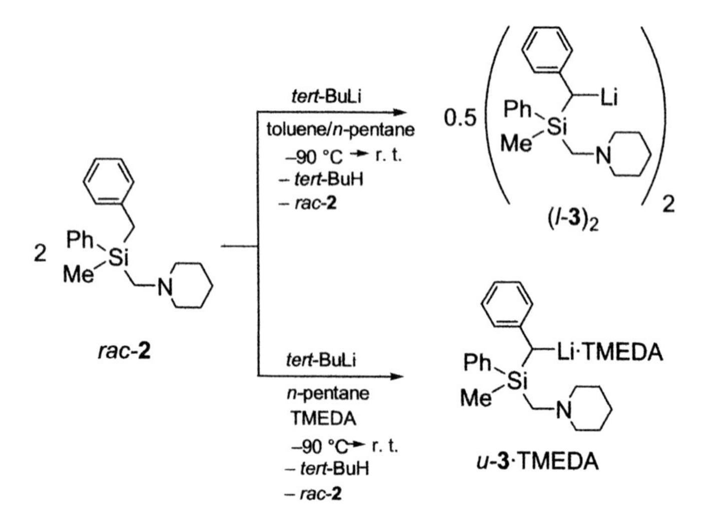

Scheme 1. Synthesis of benzyllithium compounds (*l*-**3**)$_2$ and *u*-**3**·TMEDA.

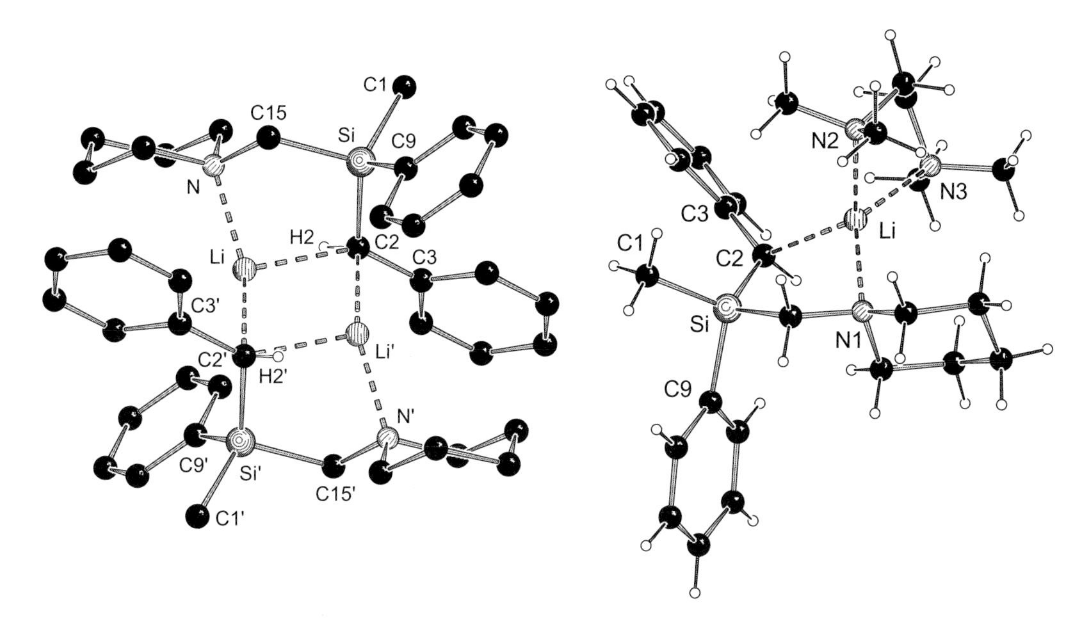

Fig. 2. Molecular structures of the dimer $(l\text{-}\mathbf{3})_2$ and the monomer *u*-**3**·TMEDA in the crystal; SCHAKAL plots [2].

The molecular structures of the dimer $(l\text{-}\mathbf{3})_2$ and the monomer *u*-**3**·TMEDA show that the absolute configurations at the metalated carbon center C(2) change according to the absolute configuration at the silicon center Si. Whereas $(l\text{-}\mathbf{3})_2$ consists of two lithium alkyls, (*R*,*R*)-**3** and (*S*,*S*)-**3**, both with *like* (*l*) configuration, *u*-**3**·TMEDA has the opposite – *unlike* (*u*) – configuration. The *like* configuration of $(l\text{-}\mathbf{3})_2$ minimizes steric interactions between the four phenyl groups. The metalated carbon center C(2) is – for benzyllithium compounds – highly angled. The sum of angles at the "carbanionic" unit, consisting of Si, C(2), C(3) and H(2), is 341° in the case of $(l\text{-}\mathbf{3})_2$ and 348° in the case of *u*-**3**·TMEDA. A literature search for known (lithiomethylphenyl)silanes found no other instance of such strong pyramidalization in similar benzyllithium compounds [3].

Metathesis reactions of the lithiated *rac*-**2** with chlorotributylstannane and chlorotrimethylsilane were carried out to determine regioselectivity and stereoselectivity, and the solvent- and temperature-dependence thereof. Scheme 2 shows the metathesis reactions observed with the trapping reagent chlorotributylstannane [1g].

The relative product ratios of the regioisomers **6** and *rac*-**5** varied according to the reaction conditions. The highest (relative) amount of compound **6** (11:1 product ratio) was found using the polar solvent THF, and the product ratio most in favor of *rac*-**5** (2.4:1) was found in the non-polar solvent *n*-pentane. We made a similar observation during the lithiation of benzyldimethyl-(piperidinomethyl)silane [1g]. When benzyldimethyl(piperidinomethyl)silane was treated with 1 equiv of *tert*-butyllithium in *n*-pentane or toluene at –90 °C, one of the methyl groups was lithiated selectively. Selective metalation of the benzyl position under formation of **1** occurs only when the same reaction is carried out in a polar donor solvent, such as THF.

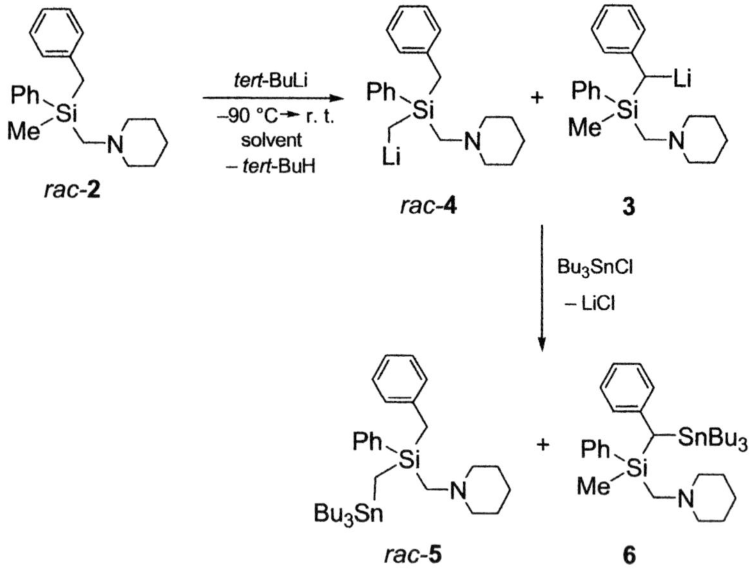

Scheme 2. Reaction products of the lithiation of *rac*-**2** and metathesis reaction with chlorotributylstannane.

Furthermore, it was found that the diastereomeric ratios of **6** depend on the solvent and temperature of the trapping reaction. At low temperatures, in THF, diastereomeric ratios up to 73:27 were observed, decreasing (with increasing temperature) to 53:47 at room temperature. Similar diastereomeric ratios were also obtained in diethyl ether by adding coordinating solvents such as TMEDA or triethylamine. By treating *rac*-**2** in non-polar solvents (e.g. *n*-pentane) with *tert*-butyllithium, the dimer $(l\text{-}\mathbf{3})_2$ was formed instead. In this case an enrichment of the other diastereomer was observed in trapping reactions (d. r. = 42:58).

By means of quantum chemical studies of the monomeric benzyllithium compounds *u*-**3**·TMEDA and *l*-**3**·TMEDA it was shown that **MIN-1** (the energy minimum of *u*-**3**·TMEDA) is about 4 kJ/mol more stable than **MIN-2** (*l*-**3**·TMEDA). This energy difference is not large, but sufficient for the crystallization of diastereomer *u*-**3**·TMEDA at –30 °C as the thermodynamically most stable product. The experimentally observed diastereomeric ratio of trapping products at low temperatures (–30 °C) indicates that no thermodynamically controlled equilibrium exists under these conditions, hence the relative amounts of *u*-**3**·TMEDA and *l*-**3**·TMEDA are formed by kinetic control. Figure 3 shows the B3LYP/6-31+G(d)-optimized structures of **MIN-1** (the energy minimum of *u*-**3**·TMEDA) and **MIN-2** (the energy minimum of *l*-**3**·TMEDA). The TMEDA ligand can adopt a second conformation resulting in the corresponding structures of **MIN-3** (other conformer of *u*-**3**·TMEDA) and **MIN-4** (other conformer of *l*-**3**·TMEDA) (Fig. 4). The almost similar energy of **MIN-1** and **MIN-3** – energy minimized *u*-**3**·TMEDA – is also observed by crystallization of *u*-**3**·TMEDA at room temperature, where we can find crystals of *u*-**3**·TMEDA with a disordered TMEDA ligand.

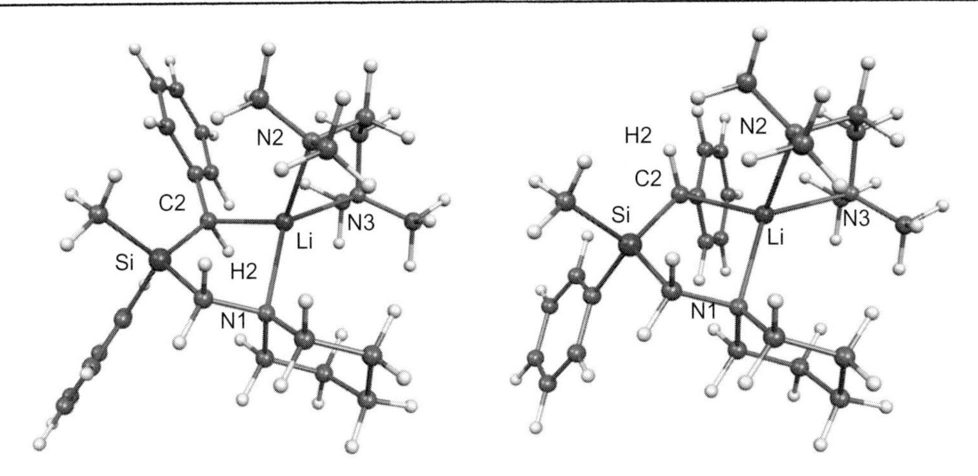

Fig. 3. MOLEKEL plots [4] of the B3LYP/6-31+G(d)-optimized structures of **MIN-1** (the energy minimum of *u*-**3**·TMEDA) and **MIN-2** (the energy minimum of *l*-**3**·TMEDA).

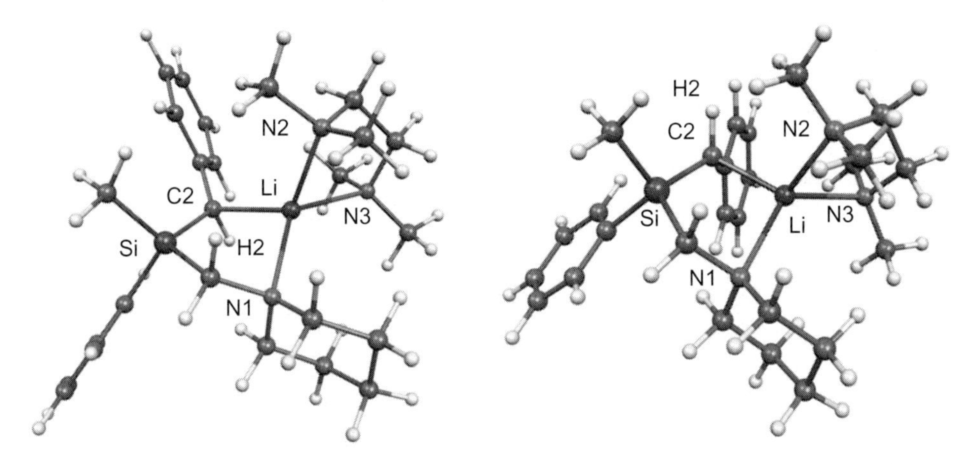

Fig. 4. MOLEKEL plots [4] of the B3LYP/6-31+G(d)-optimized structures of **MIN-3** (the energy minimum of *u*-**3**·TMEDA) and **MIN-4** (the energy minimum of *l*-**3**·TMEDA). The TMEDA ligand adopts a different conformation compared to **MIN-1** and **MIN-2**.

Table 1. Relative energies E_{rel} of the B3LYP/6-31+G(d)-optimized stationary points [kJ/mol] of **MIN-1** to **MIN-4**.

	Unlike		**Like**	
	MIN-1	**MIN-3**	**MIN-2**	**MIN-4**
E_{rel}	0	1	6	6
E_{rel} + zero point correction (ZPVE)	0	1	4	5

The highly enantiomerically enriched (aminomethyl)benzylsilane (*S*)-**7**, chiral by the (*S*)-2-(methoxymethyl)pyrrolidinomethyl (SMP) substituent, can be stereoselectively deprotonated to give the highly diastereomerically enriched alkyllithium (*R*,*S*)-**8** [1f]. The absolute configuration at the metalated stereogenic carbon center was determined by single-crystal X-ray diffraction. In the crystal, the compound forms a coordination polymer with infinite chains. In reactions with various electrophiles (among them methyl iodide, trimethyltin chloride, and palladium(II) chloride), high diastereoselectivities have been observed, mainly with inversion of configuration at the stereogenic metalated carbon center (Scheme 3) [1c, 1e, 1f]. Moreover, (*R*,*S*)-**8** has been crystallized in the presence of coordinating donor molecules like TMEDA, DABCO, and quinuclidine, showing (*R*) configuration in all these cases [1c].

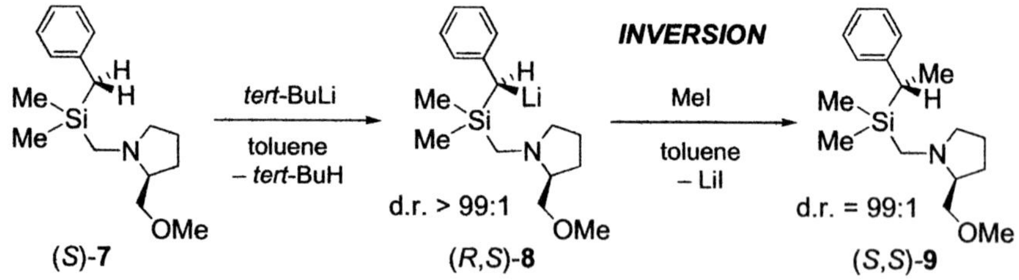

(*S*)-**7** —[*tert*-BuLi; toluene; – *tert*-BuH]→ (*R*,*S*)-**8** (d.r. > 99:1) —[INVERSION; MeI; toluene; – LiI]→ (*S*,*S*)-**9** (d.r. = 99:1)

Scheme 3. Observed stereochemical reaction pathway of (*R*,*S*)-**8** with MeI [1f].

The HOMO of *u*-**3**·TMEDA, showing a pyramidalized "carbanionic" unit, was visualized (Fig. 5) and compared with the HOMO of the benzyllithium compound (*R*,*S*)-**8**, which displays a planar "carbanionic" unit. Whereas the planar system (*R*,*S*)-**8** reacts under inversion, in the case of the pyramidal system *u*-**3**·TMEDA retention of configuration should be expected.

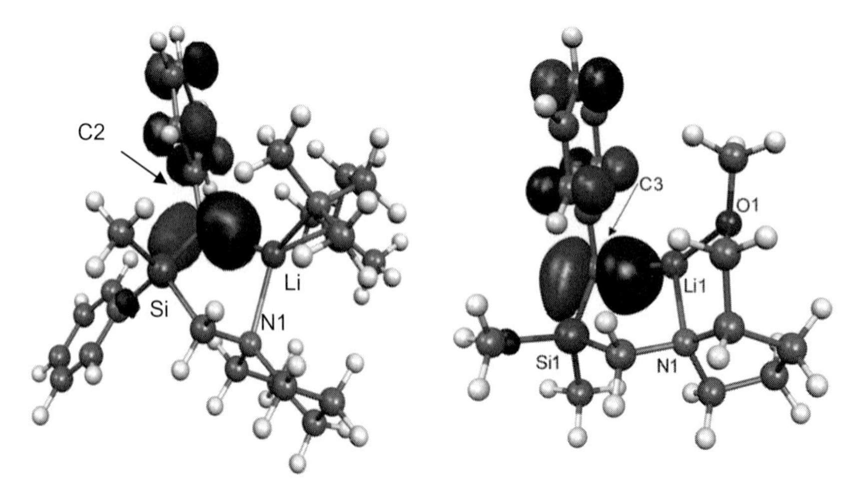

Fig. 5. MOLEKEL plots [4] of the B3LYP/6-31+G(d)-optimized structures, with HOMOs, of *u*-**3**·TMEDA and (*R*,*S*)-**8**.

Overall, the geometry of lithiated carbon center has a significant impact on the stereochemical course of reactions with these diastereomerically enriched α-lithiated benzylsilanes. The factors determining whether a monomer or a dimer is formed and whether a planar or pyramidal geometry is adopted by the benzyl unit, and the manipulation of them, are surely worthy of further study.

Acknowledgments: We are grateful to the Institut für Anorganische Chemie der Universität Würzburg, the Deutsche Forschungsgemeinschaft (DFG), the priority programme 1178, and the Fonds der Chemischen Industrie (FCI) for financial support. We acknowledge Wacker-Chemie GmbH for providing us with special chemicals.

References

[1] a) D. Schildbach, M. Arroyo, K. Lehmen, S. Martín-Barrios, L. Sierra, F. Villafañe, C. Strohmann, *Organometallics* **2004**, *23*, 3228; b) C. Strohmann, D. H. M. Buchold, K. Wild, D. Schildbach in *Organosilicon Chemistry V — From Molecules to Materials* (Eds.: N. Auner, J. Weis), VCH, Weinheim, **2003**, p. 155; c) C. Strohmann, D. H. M. Buchold, T. Seibel, K. Wild, D. Schildbach, *Eur. J. Inorg. Chem.* **2003**, 3453; d) C. Strohmann, D. Schildbach, *Acta Cryst. C* **2002**, *C58*, m447; e) C. Strohmann, B. C. Abele, K. Lehmen, F. Villafañe, L. Sierra, S. Martín-Barrios, D. Schildbach, *J. Organomet. Chem.* **2002**, *661*, 149; f) C. Strohmann, K. Lehmen, K. Wild, D. Schildbach, *Organometallics* **2002**, *21*, 3079; g) C. Strohmann, K. Lehmen, A. Ludwig, D. Schildbach, *Organometallics* **2001**, *20*, 4138; h) C. Strohmann, B. C. Abele, D. Schildbach, K. Strohfeldt, *Chem. Commun.* **2000**, 865; i) C. Strohmann, B. C. Abele, *Organometallics* **2000**, *19*, 4173; k) W. Uhl, L. Cuypers, K. Schüler, T. Spies, C. Strohmann, K. Lehmen, *Z. Anorg. Allg. Chem.* **2000**, *626*, 1526; l) B. C. Abele, C. Strohmann, in *Organosilicon Chemistry III — From Molecules to Materials* (Eds.: N. Auner, J. Weis), VCH, Weinheim, **1997**, p. 206; m) C. Strohmann, B. C. Abele, *Angew. Chem. Int. Ed. Engl.* **1996**, *35*, 2378.

[2] E. Keller, *SCHAKAL99, A Computer Program for the Graphic Representation of Molecular and Crystallographic Models*, Universität Freiburg **1999**.

[3] See also: T. Stey, D. Stalke in *The Chemistry of Organolithium Compounds* (Eds.: Z. Rappoport, I. Marek), John Wiley & Sons Ltd, Chichester (UK) **2004**, p. 47.

[4] P. Flükiger, H. P. Lüthi, S. Portmann, J. Weber, *MOLEKEL 4.0*, Swiss Center for Scientific Computing, Manno (Schweiz), **2000**.

An Enantiomerically Enriched Silyllithium Compound and the Stereochemical Course of its Transformations

*J. Hörnig, D. Auer, M. Bindl, V. C. Fraaß, C. Strohmann**

Institut für Anorganische Chemie, Universität Würzburg
Am Hubland, 97074 Würzburg, Germany
Tel.: +49 931 8884613 — Fax: +49 931 8884605
E-mail: mail@carsten-strohmann.de

Keywords: benzyl halide, density functional calculation, organosilane, silyllithium compound, stereochemistry

Summary: Starting from enantiomerically pure disilane (*R*)-**12**, the enantiomerically enriched silyllithium compound **4** was synthesized by Si–Si cleavage with lithium metal. Reactions of **4** with chlorosilanes resulted in the formation of disilanes with high e. r. values under retention, but also substitution reactions with organohalides were performed. By the reaction of **4** with benzyl chloride and benzyl bromide, the expected benzylsilane **5** was obtained in high yields and selectivities. Surprisingly, the two stereoisomers (*R*)-**5** and (*S*)-**5** with opposite configurations at the stereogenic silicon center were observed. In the case of benzyl chloride, retention of the configuration was found [(*R*)-**5**], while in the case of benzyl bromide, inversion of the configuration [(*S*)-**5**] occured. The absolute configuration of (*S*)-**5** was clarified by single-crystal X-ray structural analysis of the methiodide (*S*)-**6**. For cyclopropylmethyl halides (Cl, Br, I), both a closed-cyclopropyl ring and an opened-ring product were observed, indicating the additional presence of a radical mechanism. Overall, competing reaction mechanisms are the cause of the variable product distribution. The lowest ratio of radical-pathway to non-radical-pathway products was observed using the chloride.

Chiral silyllithium compounds were synthesized and their behavior in reactions was studied for the first time by Sommer et al. [1], Corriu et al. [2], and later Kawakami et al. [3]. Subsequent transformations with H_2O or H_2O/HCl succeeded in retaining the configuration at the silicon center. Apart form our working group [4], only Kawakami et al. [3] studied transformations with chlorosilanes or chlorostannes, observing retention of the configuration at the stereogenic silicon center. Corresponding reactions with organic electrophiles are not known so far. Here, we describe reactions of silyllithium compound **4** [accessible by Si–Si cleavage of disilane (*S*)-**12** or (*R*)-**12**] with benzyl [5] and cyclopropylmethyl halides resulting in unexpected stereochemical courses.

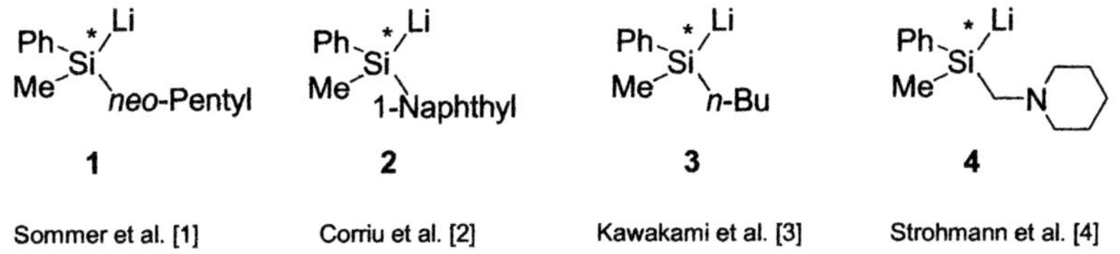

Sommer et al. [1] Corriu et al. [2] Kawakami et al. [3] Strohmann et al. [4]

Scheme 1. Known enantiomerically enriched silyllithium compounds [1 – 4].

The focus of this work was the reaction stereochemistry of the highly enantiomerically enriched silyllithium compound **4**, which is accessible through Si–Si bond cleavage of disilane (*S*)-**12**. Assuming that the Si–Si cleavage of (*S*)-**12** occurs with retention of configuration to give (*R*)-**4**, the subsequent reaction with benzyl chloride, yielding the benzylsilane (*S*)-**5**, proceeds with retention, but the analogous reaction with benzyl bromide affords (*R*)-**5**, the product of an inversion (Scheme 2). The resulting enantiomeric ratios, of 6:94 in the case of benzyl chloride and 95:5 with benzyl bromide, were determined by ^{1}H NMR spectroscopy after treatment of the NMR samples with (*S*)-mandelic acid.

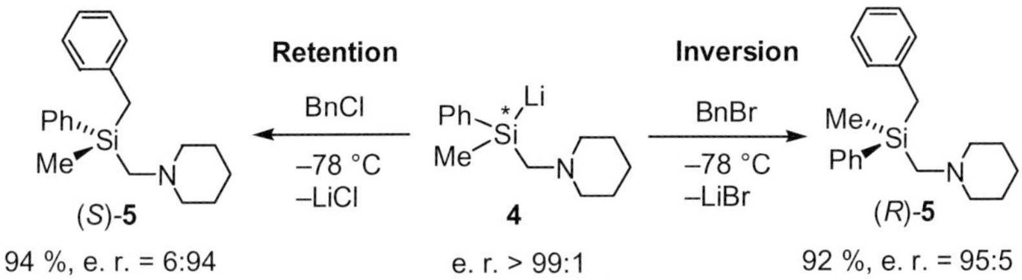

Scheme 2. Stereochemical course of highly enantiomerically enriched **4** reacted with benzyl halides.

The assignment of the absolute configurations of (*S*)-**5** and (*R*)-**5** was carried out by single-crystal X-ray structure analysis of the methiodide (*S*)-**6** (Scheme 3). Compound (*S*)-**6** was synthesized from an enantiomerically pure sample of (*S*)-**5** prepared either by resolution of *rac*-**5** or by repeated recrystallization as hydrochloride from the 94:6 enantiomerically enriched (*S*)-**5**.

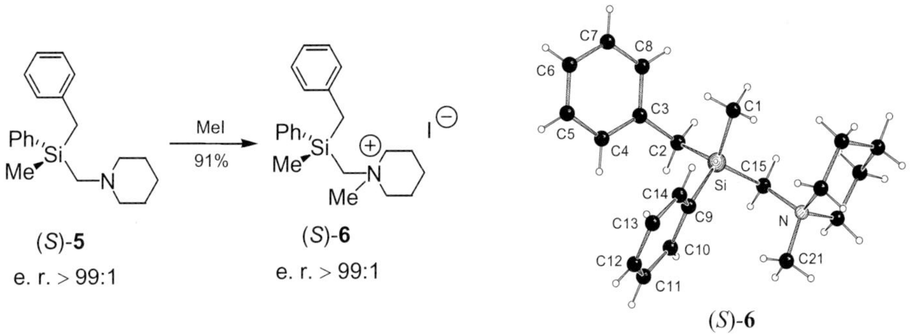

Scheme 3. Determination of the absolute configuration of (*S*)-**5**; SCHAKAL plot of (*S*)-**6** [6].

Assuming that the Si–Si cleavage of the disilane (*S*)-**12** occurs with retention of configuration to give (*R*)-**4**, two different reaction mechanisms can be proposed to explain the unexpected stereochemistry. The individual steps must be very selective to give products with such high yields and enantiomeric ratios. For the configuration-retaining reaction with benzyl chloride, the nucleophilic silicon center of the silyllithium compound (*R*)-**4** attacks the electrophilic carbon center in benzyl chloride (S_N2 mechanism at carbon center). The pentacoordinate transition state (*S*)-**7** ejects chloride to give (*S*)-**5** with no change of configuration (Scheme 4).

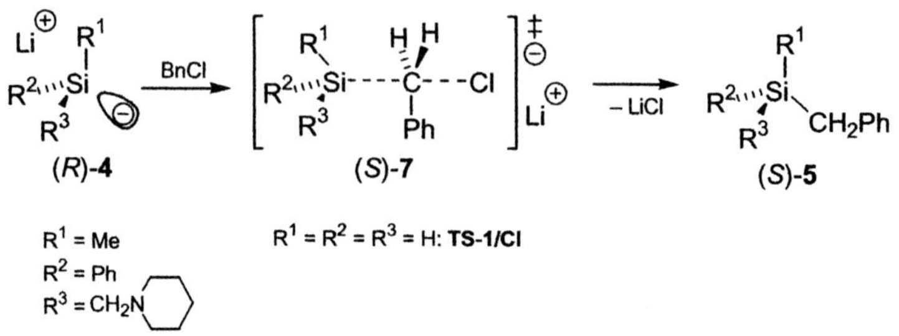

Scheme 4. Proposed mechanism for the observed stereochemical reaction pathway of (*R*)-**4** with benzyl chloride: S_N2-reaction with retention at the silicon center.

The reaction of silyllithium compound (*R*)-**4** with benzyl bromide, however, runs with inversion of configuration at the stereogenic silicon center. A bromine-lithium exchange, by way of the ate-complex like transition state (*S*)-**8**, gives the bromosilane (*S*)-**9** (with retention of configuration) and the lithium alkyl **10**. Then an addition-elimination reaction proceeds, involving the pentacoordinate intermediate **11** which inverts the configuration at the silicon (Scheme 5).

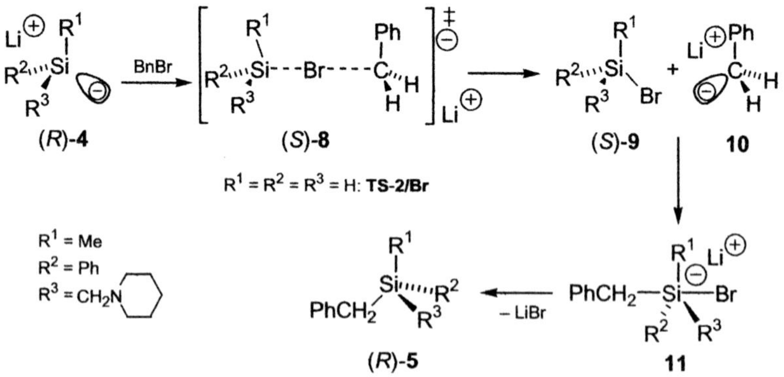

Scheme 5. Proposed mechanism for the observed sterochemical reaction pathway of (*R*)-4 with benzyl bromide: bromine-lithium exchange and substitution via the pentacoordinate intermediate stage **11** (*S*)-**8** with inversion at the silicon center.

In order to support the proposed mechanisms, DFT calculations were carried out on the model systems $H_3Si^-/ClCH_2Ph$ and $H_3Si^-/BrCH_2Ph$. Stationary points in THF solution were calculated at B3LYP/6-31+G(d), using the self-consistent reaction field (SCRF) method, based on the Onsager model, and the polarizable conductor calculation model (CPCM) (Fig. 1).

In the case of benzyl chloride, the transition state **TS-2/Cl** for a chlorine-lithium exchange is situated about 12.9 kJ/mol higher in energy than the alternative path through direct substitution of the halogen (**TS-1/Cl**). However, for benzyl bromide these energy relationships are reversed, and the bromine-lithium exchange (**TS-2/Br**) is favored by 9.5 kJ/mol over the substitution (**TS-1/Br**). Likewise, the ate-complex **MIN-1/Br** can be characterized as a minimum.

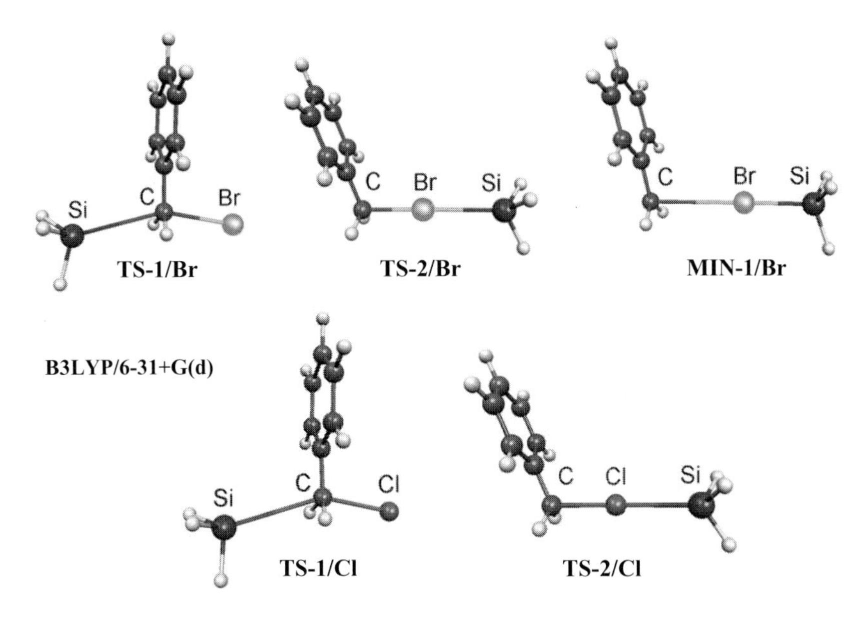

Fig. 1. MOLEKEL plots [7] of the calculated stationary points for the studied model systems $H_3Si^-/ClCH_2Ph$ and $H_3Si^-/BrCH_2Ph$ [calculated at B3LYP/6-31+G(d) with SCRF, CPCM].

Table 1. Relative energies E_{rel} of the calculated stationary points from Fig. 1.

	Relative energies E_{rel} [+ zero point correction (ZPVE)] [kJ/mol]				
Method	**TS-1/Br**	**TS-2/Br**	**MIN-1/Br**	**TS-1/Cl**	**TS-2/Cl**
Onsager	19.3	0	–27.2	0	1.0
CPCM[a]	9.5	0	–38.4	0	12.9

[a] values for the zero-point energies (ZPVE) derived from frequency calculations with the Onsager model.

Cyclopropylmethyl halides were employed to determine whether, in addition to the aforementioned mechanisms, a reaction pathway via radical intermediates was possible (Scheme 6). It is well known that the cyclocarbinyl radical reacts very fast and practically irreversibly by ring opening to the 3-butenyl radical [8]. Hence, if radicals are involved in the process, ring-opening products should be observed.

For all cyclopropylmethyl halides, both the closed-ring (**14** and **16**) and opened-ring (**15** and **17**) products are observed, indicating the presence of a radical mechanism. The lowest ratio of radical-pathway to non-radical-pathway products was observed using the chloride (Scheme 6).

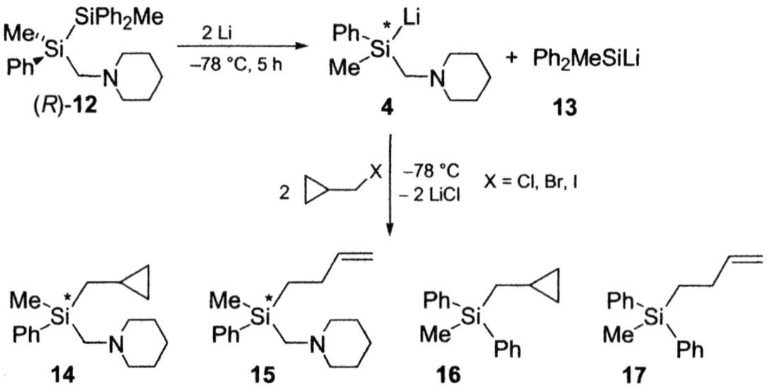

relative product ratios	X = Cl	X = Br	X = I
14:**15**	82:18	57:43	67:33
16:**17**	92:8	52:48	74:26

Scheme 6. Reactions of the enantiomerically enriched silyllithium compound **4** with cyclopropylmethyl halides.

The reaction of the enantiomerically enriched silyllithium compound **4** [prepared from disilane (*R*)-**12**] with cyclopropylmethyl chloride occurs with retention of the configuration [(*R*)-**14**], while, for cyclopropylmethyl bromide and iodide, mainly inversion of the configuration [(*S*)-**14**] was observed. The products of the radical reaction (**15**) indicate a racemization at the silicon center.

Table 2. Yields and enantiomeric ratios of **14** and **15** from the trapping reactions of **4** with cyclopropylmethyl halides.

Trapping reagent	Yield of 14	Yield of 15	e. r. of 14	e. r. of 15
Cyclopropylmethyl chloride	61%	13%	6:94	49:51
Cyclopropylmethyl bromide	46%	34%	87:13	56:44
Cyclopropylmethyl iodide	43%	21%	63:37	56:44

Absolute configuration was clarified by single-crystal X-ray structural analysis of (*R*)-**18**.

Only the stereochemical probe at the silicon center permitted us to obtain an insight into the described processes, since the products of both main reaction pathways – S_N2-reaction and halogen-lithium exchange – differ merely in their absolute configuration at the silicon center.

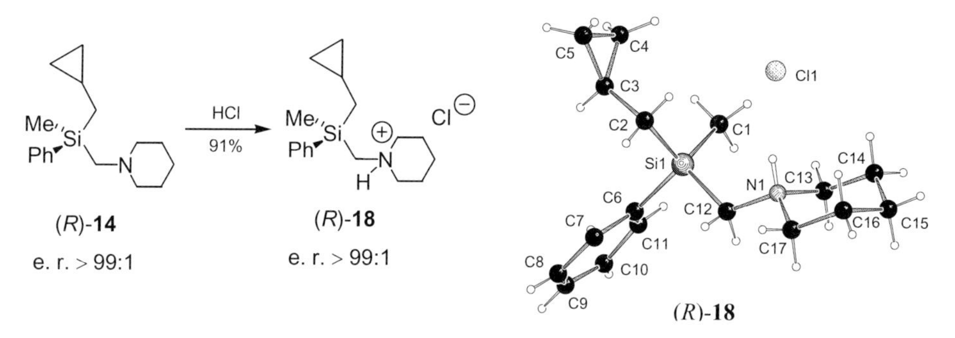

Scheme 7. Determination of the absolute configuration of (*R*)-**14**; SCHAKAL plot of (*R*)-**18** [6].

Acknowledgment: We are grateful to the Institut für Anorganische Chemie der Universität Würzburg, the Deutsche Forschungsgemeinschaft (DFG) the Graduiertenkolleg 690 and the Fonds der Chemischen Industrie (FCI) for financial support. We acknowledge Wacker-Chemie GmbH for providing us with special chemicals.

References

[1] L. H. Sommer, R. Mason, *J. Am. Chem. Soc.* **1965**, *87*, 1619.

[2] a) E. Colomer, R. J. P. Corriu, *J. Organomet. Chem.* **1977**, *133*, 159; b) E. Colomer, R. Corriu, *J. Chem. Soc., Chem. Commun.* **1976**, 176.

[3] a) M. Omote, T. Tokita, Y. Shimizu, I. Imae, E. Shirakawa, Y. Kawakami, *J. Organomet. Chem.* **2000**, *611*, 20; b) H. Oh, M. Omote, K. Suzuki, I. Imae, Y. Kawakami, *Polymer Preprints* **2001**, *42*, 194.

[4] a) C. Strohmann, J. Hörnig, D. Auer, *J. Chem. Soc., Chem. Commun.* **2002**, 766; b) D. Auer, J. Hörnig, C. Strohmann in *Organosilicon Chemistry V — From Molecules to Materials* (Eds.: N. Auner, J. Weis), VCH, Weinheim, **2003**, p. 167.

[5] C. Strohmann, M. Bindl, V. C. Fraaß, J. Hörnig, *Angew. Chem.* **2004**, 116, 1029; *Angew. Chem. Int. Ed.* **2004**, *43*, 1011.

[6] E. Keller, *SCHAKAL99, A Computer Program for the Graphic Representation of Molecular and Crystallographic Models*, Universität Freiburg **1999**.

[7] P. Flükiger, H. P. Lüthi, S. Portmann, J. Weber, *Swiss Center for Scientific Computing, MOLEKEL 4.0*, Manno (Schweiz), **2000**.

[8] a) T. Linker, M. Schmittel, *Radikale und Radikalionen in der Organischen Synthese*, Wiley-VCH, Weinheim, **1998**, p. 129 – 133; b) P. Dowd, W. Zhang, *Chem. Rev.* **1993**, *93*, 2091; c) A. Effio, D. Griller, K. U. Ingold, A. L. J. Beckwith, A. K. Serelis, *J. Am. Chem. Soc.* **1980**, *102*, 1734; d) P. J. Krusic, P. J. Fagan, J. S. Filippo Jr., *J. Am. Chem. Soc.* **1977**, *99*, 250; e) J. S. Filippo Jr., J. Silbermann, P. J. Fagan, *J. Am. Chem. Soc.* **1978**, *100*, 4834.

Reactions of $CF_3Si(CH_3)_3$ and $C_6F_5Si(CH_3)_3$ with Perfluoroolefins and Perfluoroimines

Masakazu Nishida, Yoshio Hayakawa, Taizo Ono*

National Institute of Advanced Industrial Science and Technology (AIST)
2266-98 Shimoshidami, Moriyama-ku, Nagoya 463-8560, Japan
Tel.: +81 52 7367329 — Fax: +81 52 7367304
E-mail: m-nishida@aist.go.jp

Keywords: fluorinated silane, transfer reagent, nucleophilic substitution, steric hindrance

Summary: A silicon-based technique of using $R_FSi(CH_3)_3$ reagents is quite useful for introducing fluorinated substituents into perfluoroolefins and perfluoroimines, especially for constructing the highly branched derivatives of perfluoroolefins and perfluoroimines, which are otherwise difficult to synthesize. The scope and limitations of these reactions are described, and comparison of the reactivities of the perfluoroolefin and the perfluoroimine with $R_FSi(CH_3)_3$ are made.

Introduction

Fluorinated organosilicon compounds have been used not only as precursors for various functional materials but also as transfer reagents of fluorinated substituents into organic molecules. The synthesis of an organic-inorganic composite is a current topic of the functional material, and recently the fluorinated organosilicon compound has been applied for the synthesis of highly ordered mesoporous hybrid silica [1]. A transfer reagent of fluorinated substituent, $CF_3Si(CH_3)_3$ (**1a**), is well known as a useful trifluoromethylating reagent for various organic molecules [2]: a recent report has described the trifluoromethylation of less reactive carboxylic esters, thiocyanates, and selenocyanates [3]. An aromatic analog, $C_6F_5Si(CH_3)_3$ (**1b**), which is prepared by reaction of C_6F_5Br, $ClSi(CH_3)_3$, and $P[N(CH_3)_2]_3$ [4], can also react with various substrates, and the pentafluorophenylation proceeded similarly to the corresponding reaction of **1a** [5]. In order to study more details of and limitations affecting such transfer reagents of fluorinated substituents, reactions of highly electronegative and sterically hindered substrates were investigated. Here, we target a highly branched perfluoroolefin, *F*-(4-methyl-3-isopropyl-2-pentene) (**T-2**) [6], and cyclic perfluoroimines, *F*-(5,6-dihydro-2*H*-oxazine) (**C-6**) and *F*-(3,4-dihydro-2*H*-pyrrole) (**C-5**) [7]. The results of the comparative study on reactivities of the perfluoroolefin and the perfluoroimine with $CF_3Si(CH_3)_3$ (**1a**) and $C_6F_5Si(CH_3)_3$ (**1b**) are presented.

Perfluoroolefins

Perfluoroolefins have an activated C=C double bond in the presence of nucleophilic reagents owing to the electronegativity of fluorinated substituents. Although a hexafluoropropene trimer, *F*-(4-methyl-3-*iso*propyl-2-pentene) (**T-2**), has several bulky fluorinated substituents in addition, $CF_3Si(CH_3)_3$ (**1a**) easily reacted with **T-2** to give CF_3-substituted fluoroolefins in high yields (Scheme 1) [8].

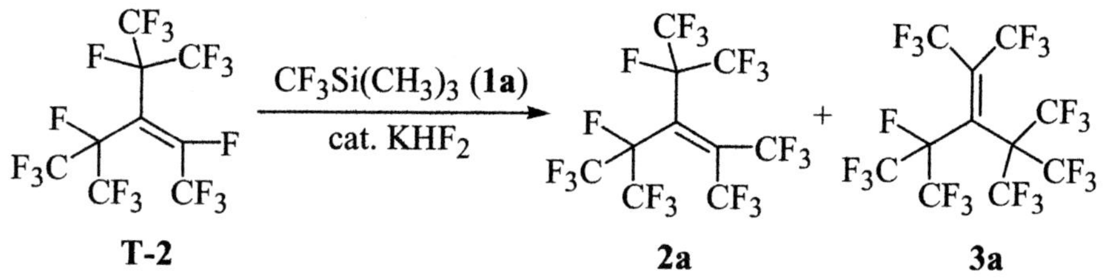

$CF_3Si(CH_3)_3$ (**1a**), cat. KHF_2

T-2 → **2a** + **3a**

Scheme 1. Reactions of *F*-(4-methyl-3-isopropyl-2-pentene) (**T-2**) with $CF_3Si(CH_3)_3$ (**1a**).

This reaction was initiated by a catalytic fluoride anion. All fluoride anion sources such as CsF, KF, and KHF_2 are usable, but KHF_2 was most beneficial because KHF_2 is not hygroscopic unlike other fluoride anion sources. The trifluoromethylation proceeded in an aprotic polar solvent, such as DMF, 1,3-dimethylimidazolidinone (DMI), tetraglyme (TG), etc. The fluoroolefin **T-2** was immiscible with these solvents; thus, the trifluoromethylation required the two layers consisting of the hydrocarbon and the fluorocarbon to be vigorously stirred. Since the mixture remained as a two-layer liquid after the reaction, the trifluoromethylated products were easily isolated from the lower layer. A mono-CF_3-substituted fluoroolefin **2a** was selectively obtained in 79% yield in DMI. However, in DMF and TG, using 1.1 mol-equiv. of **1a** against **T-2**, a bis-CF_3-substituted olefin **3a** was also obtained (**2a**: 62% yield and **3a**: 13% yield in DMF; **2a**: 80% yield and **3a**: 6.9% yield in TG). Since **3a** is severely hindered, no further reactions occurred in the reaction with a large excess of **1a**.

Pentafluorophenylation of the fluoroolefin **T-2** with $C_6F_5Si(CH_3)_3$ (**1b**) proceeded in an aprotic solvent in the presence of catalytic fluoride anion, similarly to the reaction of **T-2** with **1a**; however, successive nucleophilic attacks of **1b** on the C_6F_5 ring were also observed (Scheme 2) [9].

Under the conditions of using 1.1 mol-equiv. of **1b** against **T-2** in both DMF and DMI, a nonafluorobiphenylyl olefin **4** was obtained in addition to a mono-C_6F_5-substituted fluoroolefin **2b** (**2b**: 56% yield and **4**: 18% yield in DMF; **2b**: 59% yield and **4**: 7.6% yield in DMI). Furthermore, a large excess of **1b** (2.2 mol-equiv. against **T-2**) provided a tridecafluoroterphenylyl olefin **5**, while a bis-C_6F_5-substituted fluoroolefin **3b** could not be detected by ^{19}F NMR and GC-MS analyses. Treatment of **T-2** with 0.5 mol-equiv. of **1b** in DMF gave the mono-C_6F_5-substituted olefin **2b** selectively without any **3b**-like bis-adducts, which is contrast with the fact that **3a** was always produced in the reaction of **T-2** with **1a** except the case using DMI as a solvent.

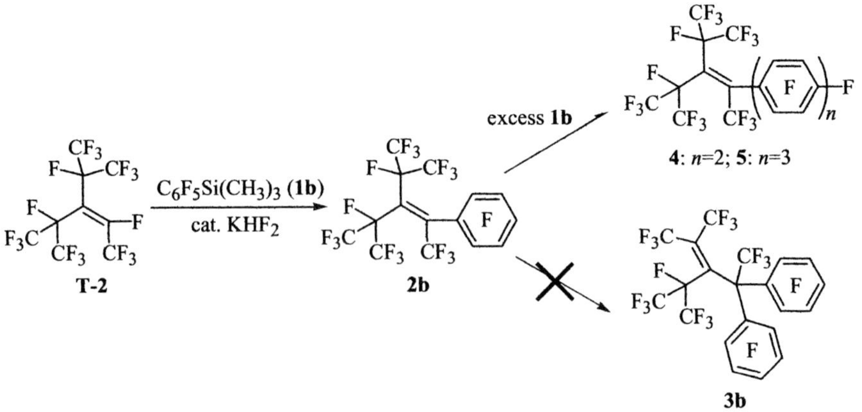

Scheme 2. Reactions of *F*-(4-methyl-3-isopropyl-2-pentene) (**T-2**) with $C_6F_5Si(CH_3)_3$ (**1b**).

Perfluoroimines

Perfluoroimines are more reactive than perfluoroolefins because of a fluorine atom connected with the C=N bond. In order to investigate the effects of the electron density and the ring size, a six-membered cyclic perfluoroimine, *F*-(5,6-dihydro-2*H*-oxazine) (**C-6**) and a five-membered cyclic perfluoroimine *F*-(3,4-dihydro-2*H*-pyrrole) (**C-5**) were subjected to reaction with the fluorinated organosilanes $R_FSi(CH_3)_3$ (**1**) [10, 11]. The cyclic perfluoroimines **C-6** and **C-5** are more sensitive to moisture, and the reactions of **C-6** and **C-5** with **1** were performed in the presence of spray-dried KF. Similarly to the reaction of the perfluoroolefin **T-2** with **1**, aprotic solvents were suitable for the reactions of **C-6** and **C-5** with **1**. The products were handled in anhydrous conditions under an inert atmosphere and/or with a vacuum system. Since the boiling points of trifluoromethylated products and a resulting by-product $FSi(CH_3)_3$ were very close to each other, the yields were determined by ^{19}F NMR or GC analyses of a mixture of a few components, without complete purification.

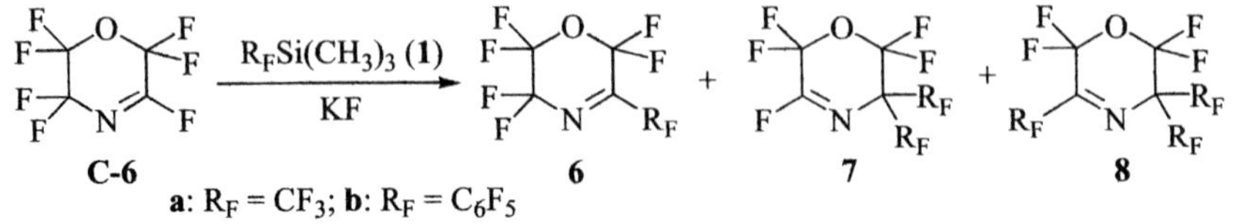

Scheme 3. Reactions of *F*-(5,6-dihydro-2*H*-oxazine) (**C-6**) with $R_FSi(CH_3)_3$ (**1**).

The results of the reactions of **C-6** with $R_FSi(CH_3)_3$ (**1**) are given in Scheme 3 and Table 1. Even using 1.0 mol-equiv. of $CF_3Si(CH_3)_3$ (**1a**) for **C-6**, not only a mono-CF_3 imine **6a** but also a bis-CF_3 imine **7a** and a tris-CF_3 imine **8a** were obtained; a selective formation of **6a** was not observed. Using 3.0 mol-equiv. of **1a** for **C-6** also gave a mixture of **6a-8a** along with a large amount of unreacted **1a**. In a case of using DMF, only **7a** and **8a** were obtained as volatile CF_3-substituted imines, whereas unknown high boiling products were also obtained; therefore, total yields of the CF_3-substituted imines were rather low (Table 1, entry 1). The reaction of **C-6** with $C_6F_5Si(CH_3)_3$ (**1b**) proceeded well in acetonitrile and tetraglyme (TG), but the C_6F_5-substituted products happened to have boiling points similar to TG, so that acetonitrile was the choice of solvent for this reaction. In the case of using an equal amount of **1b**, similarly to the reaction of **C-6** with **1a**, three C_6F_5-substituted products **6b-8b** were obtained (Table 1, entry 6). Interestingly, three molar equivalents of **1b** for **C-6** gave only the tris-C_6F_5 product **8b** (Table 1, entry 8). In the reactions of **C-6** with **1**, neither peaks of positional isomers of **6** and **7** nor peaks of morpholine dimers were detected by GC-MS and ^{19}F NMR analyses.

Table 1. Reaction products of *F*-(5,6-dihydro-2*H*-oxazine) (**C-6**) with $CF_3Si(CH_3)_3$ (**1a**) and $C_6F_5Si(CH_3)_3$ (**1b**).

Entry	$R_FSi(CH_3)_3$ [a]	Solvent	Yield (%)
1	**1a** (1.0)	DMF	**7a** (6.6), **8a** (6.8)
2	**1a** (1.0)	C_6H_5CN	**6a** (29), **7a** (15), **8a** (4.4)
3	**1a** (3.0)	C_6H_5CN	**7a** (26), **8a** (20)
4	**1a** (1.0)	TG [c]	**6a** (37), **7a** (18), **8a** (16)
5	**1a** (3.0)	TG [c]	**6a** (10), **7a** (14), **8a** (17)
6	**1b** (1.0)	CH_3CN	**6b** (37), **7b** (5.9), **8b** (8.4)
7	**1b** (2.0)	CH_3CN	**6b** (29), **7b** (3.3), **8b** (52)
8	**1b** (3.0)	CH_3CN	**8b** (40)

[a] Figures in parentheses show molar ratios of $R_FSi(CH_3)_3$ to the cycloimine **C-6**.
[b] tetraglyme.

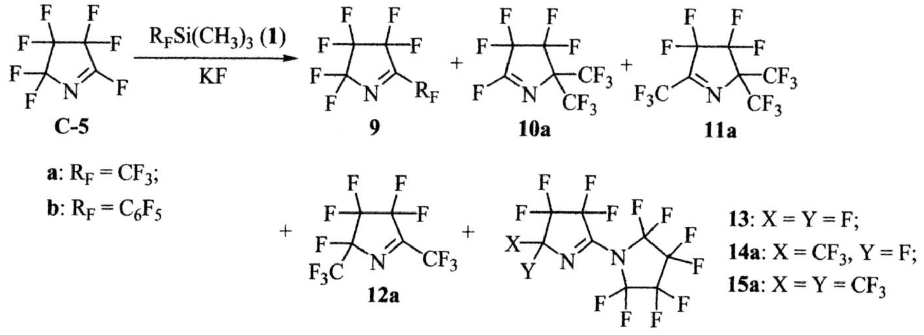
F F
F F F F
R_FSi(CH3)3 (1)
KF
C-5
9
10a
11a
12a
CF3
F3C
R_F
N
X
Y
a: R_F = CF3;
b: R_F = C6F5
13: X = Y = F;
14a: X = CF3, Y = F;
15a: X = Y = CF3

Scheme 4. Reactions of *F*-(3,4-dihydro-2*H*-pyrrole) (**C-5**) with $R_FSi(CH_3)_3$ (**1**).

In the reactions of **C-5** with $R_FSi(CH_3)_3$ (**1**), regardless of the molar ratio between **1a** and **C-5**, a mixture of several CF_3-substituted imines **9a** – **12a** was obtained. Unlike the reaction of **C-6**, dimerization of **C-5** was a significant competing process to the trifluoromethylation, and thus several pyrrolidine dimers **13**, **14a**, and **15a** with/without the CF_3 groups accompanied the trifluoromethylated products (Scheme 4).

Table 2. Reaction products of *F*-(3,4-dihydro-2*H*-pyrrole) (**C-5**) with $CF_3Si(CH_3)_3$ (**1a**) and $C_6F_5Si(CH_3)_3$ (**1b**).

Entry	$R_FSi(CH_3)_3$ [a]	Solvent	Yield [%]
1	**1a** (1.0)	DMF	**9a** (1.9), **11a** (8.5), **12a** (1.4), **13** (8.7), **14a** (9.2), **15a** (3.5)
2	**1a** (1.0)	C_6H_5CN	**9a** (5.2), **10a** (11), **11a** (16), **12a** (2.2), **13** (9.2), **14a** (6.3)
3	**1a** (3.0)	C_6H_5CN	**9a** (5.6), **10a** (7.0), **11a** (18), **12a** (3.0), **13** (4.5), **14a** (2.7)
4	**1a** (1.0)	TG [c]	**9a** (7.8), **11a** (20), **12a** (9.0), **13** (8.7), **14a** (9.2), **15a** (3.5)
5	**1a** (3.0)	TG [c]	**9a** (2.5), **10a** (1.2), **11a** (15), **12a** (3.6), **13** (4.8), **14a** (7.1), **15a** (1.5)
6	**1b** (3.0)	CH_3CN	**9b** (2.1), **13** (40)

[a] Molar ratio of $R_FSi(CH_3)_3$ to the cycloimine **C-5**. [b] tetraglyme.

The results of the reactions of **C-5** with $R_FSi(CH_3)_3$ (**1**) are given in Table 2. In the reactions of **C-5** with **1a**, the yields of the CF_3-substituted products **9a** – **12a** were rather lower owing to the dimerization process. Although compound **12a**, a positional isomer of **10a**, was obtained, no positional isomers of **9a** were detected even by GC-MS and ^{19}F NMR analyses. Solvent DMF dramatically lowered the product yield again as in the reaction of **C-6** (Table 2, entry 1). On the other hand, in the reaction of **C-5** with **1b**, a trace amount of a mono-C_6F_5 derivative **9a** was formed, and a main product was the pyrrolidine dimer **13** (Table 2, entry 6).

Discussion

The trifluoromethylation and pentafluorophenylation described above are considered to proceed via the addition-elimination mechanism (Ad_N-E) shown in Scheme 5. The attack of the R_F anion upon the C=C or C=N double bond gives the anionic intermediate **16**, which converts the R_F-substituted products by releasing the fluoride anion on the carbon connected with the R_F group.

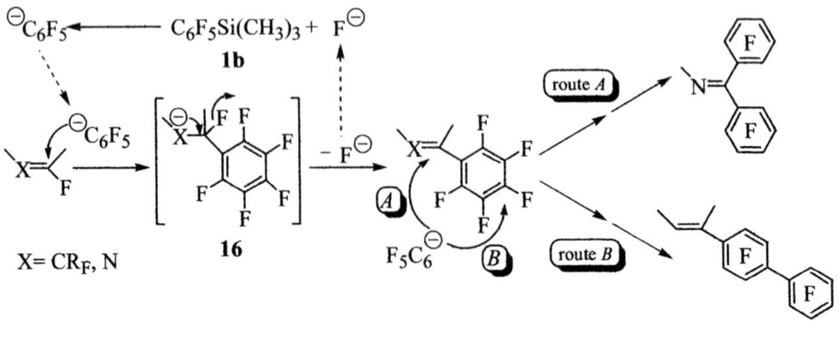

Scheme 5. Addition-elimination mechanism of formation of perfluoroaryl derivatives.

With addition of a large excess of $C_6F_5Si(CH_3)_3$ (**1b**), the mono-C_6F_5-substituted compound can react with **1b** by two possible routes, A and B. Because the mono-C_6F_5-substituted olefin **2b** has two bulky substituents (C_6F_5 and CF_3) on the active site of the C=C double bond, the excess **1b** reacts on the *para*-position of C_6F_5 group to provide the nonafluorobiphenyl olefin **4** (route B). On the other hand, the double bond of the mono-C_6F_5-substituted imine **6b** is not so crowded that the excess **1b** can react on the C=N double bond to provide the bis-C_6F_5-substituted imine **7b** (route A).

As described above, the reactions of **C-6** and **C-5** with **1a**, similar to the reactions with **1b**, gave a mixture of mono, bis, and tris compounds. The fact that even an equimolar use of the reagents could not suppress the formation of bis and tris compounds is rather surprising for us because a CF_3 group is known to be as bulky as an isopropyl group [12]. This result is in contrast with the reaction of **C-5** with (polyfluoroalkyoxy)trimethylsilane in which a mono-substituted polyfluoroalkyoxylated product was obtained exclusively without formation of bis- and tris-substituted products [13]. Many factors are included in the observed phenomena, so that no conclusion could yet be drawn. However, we can point out that the introduction of CF_3 group(s) into the heterocyclic perfluoroimine systems activates the resulting double bond enough for the subsequent trifluoromethylation, overcoming the effect of the steric hindrance of the CF_3 group(s).

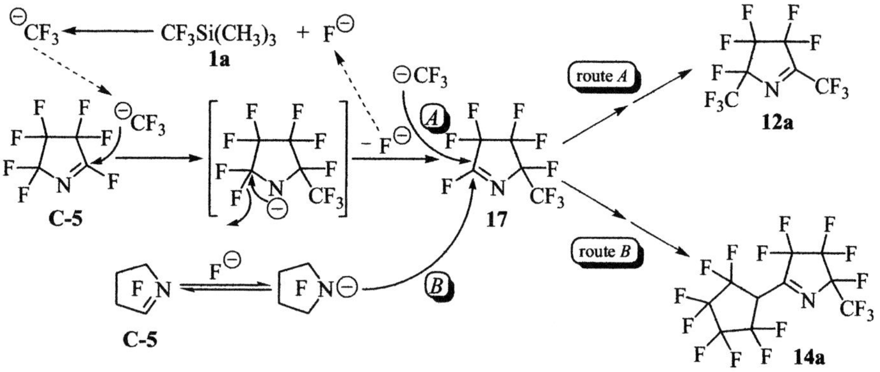

Scheme 6. Mechanism of formation of **12a** and **14a**.

The significant difference between the reaction of $R_FSi(CH_3)_3$ (**1**) with **C-6** and that of **1** with **C-5** was the formation of positional isomer **12a** and pyrrolidine dimers **13**, **14a**, and **15a**. These results could be explained by the reaction mechanism shown in Scheme 6. An anionic intermediate formed by the attack of the CF_3 anion releases a fluorine anion opposite to the CF_3 group about the ring nitrogen atom to give the positional isomer **17**, which immediately reacts with excess CF_3 anions to give **12a** (route A). A pyrrolidide anion, which can be formed by the nucleophilic attack of the fluoride anion upon **C-5**, reacts with **17** to provide the CF_3-substituted pyrrolidine dimer **14a** (route B). A previous report showed that treatment of a mixture of **C-6** and **C-5** with fluoride anions gave only **13**, and most of the **C-6** was recovered [11]. This result indicates that **C-5** easily reacts with fluoride anion to give pyrrolidide anion, which attacks only another **C-5** molecule, and that **C-6** was less reactive against both fluoride and pyrrolidide anions. This difference in reactivity between **C-6** and **C-5** for the nucleophiles is considered to be one of the reasons for these different product profiles emphasized by the fluoride anion release routes and dimer formations.

References

[1] B. Lebeau, C. Marichal, A. Mirjiol, G. J. de A. A. Soler-Illia, R. Buestrich, M. Popall, L. Mazerolles, C. Sanchez, *New J. Chem.* **2003**, *27*, 166.
[2] G. K. S. Prakash, A. K. Yudin, *Chem. Rev.* **1997**, *97*, 757.
[3] R. P. Singh, J. M. Shreeve, *Tetrahedron* **2000**, *56*, 7613.
[4] N. R. Patel, R. L. Kirchmeier, J. M. Shreeve, *Inorg. Chem.* **1993**, *32*, 4802.
[5] M. Nishida, A. Vij, R. L. Kirchmeier, J. M. Shreeve, *Inorg. Chem.* **1995**, *34*, 6085.
[6] W. Dmowski, W. T. Flowers, R. N. Hazeldine, *J. Fluorine Chem.* **1977**, *9*, 94.
[7] M. Nishida, H. Fukaya, T. Abe, *J. Fluorine Chem.* **1996**, *76*, 3.
[8] T. Ono, to be submitted for publication.
[9] M. Nishida, in preparation.
[10] M. Nishida, T. Ono, T. Abe, *Nippon Kagaku Kaishi*, **2001**, 281 (in Japanese).
[11] M. Nishida, T. Ono, T. Abe, *Nippon Kagaku Kaishi*, **2000**, 817 (in Japanese).
[12] M. Schlosser, D. Michel, *Tetrahedron*, **1996**, *53*, 99.
[13] M. Nishida, T. Ono, T. Abe, *J. Fluorine Chem.* **2001**, *110*, 63.

Bis(trimethylsilyl)mercury: a Powerful Reagent for the Synthesis of Amino Carbenes

Michael Otto, Valentyn Rudzevich, Vadim D. Romanenko, Guy Bertrand**

UCR-CNRS Joint Research Chemistry Laboratory, UMR 2282, Department of Chemistry, University of California, Riverside, California 92521-0403, USA
E-mail: MichaelOtto@carbene.com, guy.bertrand@ucr.edu

Keywords: amino carbene, bis(trimethylsilyl)mercury, chloroiminium salt

Summary: Bis(trimethylsilyl)mercury cleanly reacts at low temperature with chloroiminium chlorides to form stable metal-free cyclic and acyclic diaminocarbenes as well as aryl-, oxy-, chloro-, hydrogeno- and alkyl-amino carbenes. The aryl-, chloro- and hydrogeno-amino carbenes were formed as transient intermediates that undergo dimerization into the corresponding alkenes, which were isolated in good yields. Alkyl- and oxy-amino carbenes were observed at low temperature by ^{13}C NMR spectroscopy.

Introduction

In the last 15 years, considerable progress has been made in the understanding of carbene chemistry by the direct observation of short-lived species or through the synthesis of stable versions. Our group published the synthesis of a phosphino silyl carbene, the first stable singlet carbene, in 1988 [1]. Three years later, Arduengo et al. reported the preparation and X-ray crystal structure of the first stable N-heterocyclic carbene [2].

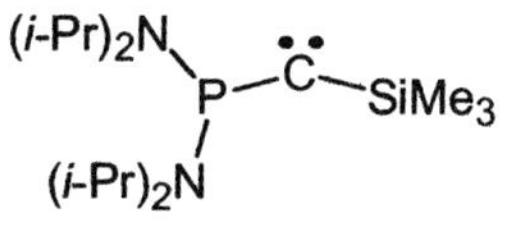

$(i\text{-}Pr)_2N$, $(i\text{-}Pr)_2N$, P, C, $SiMe_3$

Bertrand, 1988

Ad, N, N, Ad

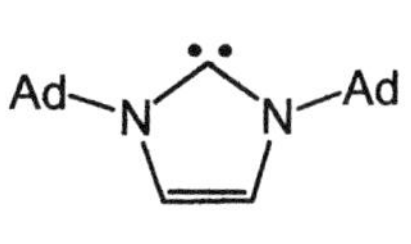

Arduengo, 1991

Several excellent reviews covering stable carbene chemistry have been published since the first one by Herrmann and Köcher in 1997 [3]. They include the influence of the substituents on the stability of carbenes, the synthetic methods available, structural data, reactivity, coordination behavior, and the catalytic properties of the corresponding complexes.

Deprotonation of formamidinium salts is by far the most popular way to generate diaminocarbenes [3]. A major advantage of this method is that deprotonation is a rapid reaction, even at low temperatures. However, hindered strong anionic bases are required, especially for open-chain diaminocarbenes, which have higher pK_a values [4] and are more susceptible to addition of nucleophilic species than imidazolium ions; yet side reactions are sometimes observed [5]. Moreover, for non-hindered diaminocarbenes, the use of strong anionic bases carrying alkali metal counterions produces complexed carbenes [6]. All attempts so far to remove the metal from bis(dimethylamino)carbene have resulted in the destruction of the carbene [6a]; it has also been reported that even using crown ethers, tertiary complexes involving the metal are formed [6a]. Alkali metal coordination to carbenes is particularly important with regard to the rate and mechanism of the dimerization. Alder suggested that metal ions might act as Lewis acid catalysts for dimerization, as is observed for protons [7], but strong complexation might also eventually suppress dimerization [7a, 7b]. All the other methods known to generate diaminocarbenes, which includes desulfurization of thioureas [8], involved drastic experimental conditions and are not easily applicable to unhindered carbenes.

In a search for new methods to generate carbenes, it appeared that readily prepared (from commercially available urea and amides) chloro-amidinium and iminium salts could be interesting precursors [9].

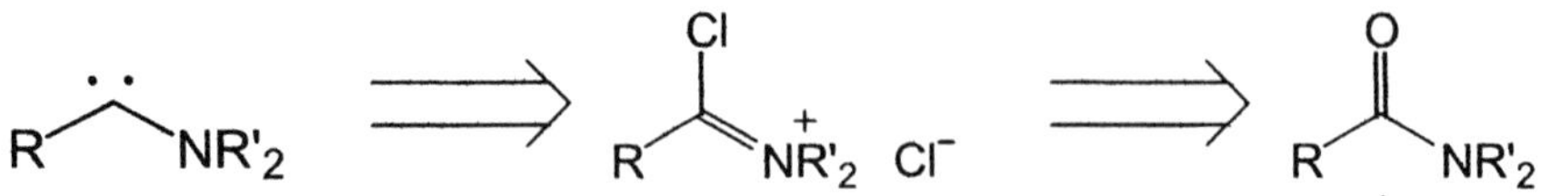

R–C(:)–NR'$_2$ ⟹ R–C(Cl)=$\overset{+}{N}$R'$_2$ Cl$^-$ ⟹ R–C(=O)–NR'$_2$

Scheme 1. Retrosynthetic approach to aminocarbenes from chloroiminium chlorides and carboxamides.

Bis(trimethylsilyl)mercury [10] (also: disilylmercury), one of the strongest silylating agents known in literature, was chosen as the dechlorinating agent. This reagent was already successfully used for the dehalogenation of dihalo organoboranes by Eisch and Becker [11]. Similar reactions with diorganotin halides and related compounds were reported by Mitchell et al. [12]. Note that Seyferth et al. established an alternative route toward dichloro carbenes starting from (phenyl)(trihalomethyl)mercury [13].

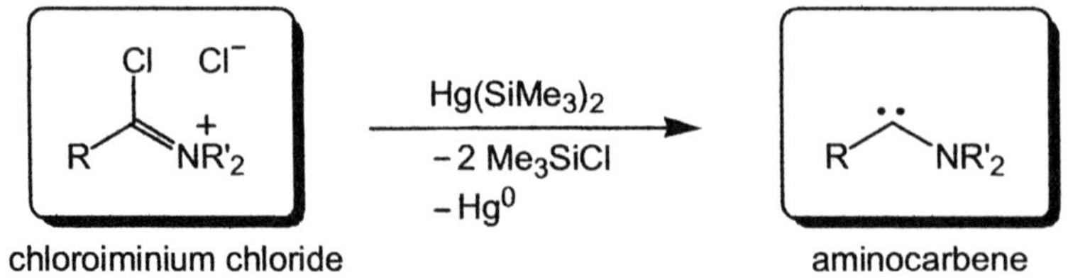

R–C(Cl)=$\overset{+}{N}$R'$_2$ Cl$^-$ (chloroiminium chloride) $\xrightarrow[-2\ Me_3SiCl,\ -Hg^0]{Hg(SiMe_3)_2}$ R–C(:)–NR'$_2$ (aminocarbene)

Scheme 2. General reaction of a chloroiminium chloride with bis(trimethylsilyl)mercury.

An anticipated advantage of our new approach resided in the fact that besides the carbene, only elemental mercury and chlorotrimethylsilane were expected to be formed, and both of these products seemed easily removed from the reaction mixture.

Diaminocarbenes (Acyclic and Cyclic)

To probe our synthetic strategy, we first synthesized Alder's stable bis(diisopropylamino) carbene [14]. Starting from bis(diisopropylamino) chloroiminium chloride, we isolate the desired carbene in quantitative yield.

For this reaction, we added a solution of bis(trimethylsilyl)mercury (in THF) to the suspension of the chloroiminium salt in THF at low temperature and allowed the reaction mixture to warm to room temperature. After about one hour, we obtained a clear yellow solution and a drop of mercury. After separation of the metal and evaporation of the solvent the carbene was isolated.

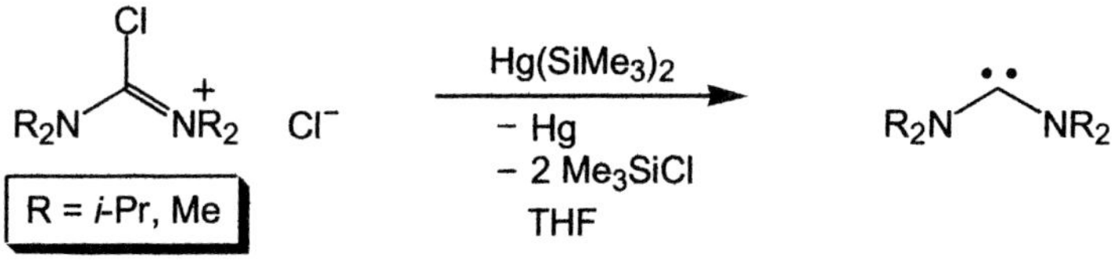

Cl
R_2N $\overset{+}{N}R_2$ Cl^-
R = *i*-Pr, Me
$Hg(SiMe_3)_2$
– Hg
– 2 Me_3SiCl
THF
R_2N NR_2

Scheme 3. Formation of acyclic diaminocarbenes.

A similar reaction with bis(dimethylamino) chloroiminium chloride was performed at temperatures below –20 °C and led to the formation of the bis(dimethylamino) carbene with a conversion of about 60%. For the first time, the bis(dimethylamino) carbene was formed without complexation with metal cations, which usually occurs in the deprotonation method. The ^{13}C chemical shift of the carbene carbon atom was significantly shifted toward low field compared to the complexed version. To our surprise, the carbene was not found to dimerize at higher temperatures, but instead a complex mixture was obtained with no traces of dimer being detectable. This raises the question whether the cations are necessary for the dimerization process, as already discussed in literature [7].

Similarly, the 1,3-dimethyl-hexahydropyrimid-2-ylidene, a 6-membered cyclic diamino carbene, was obtained with a conversion of about 50%.

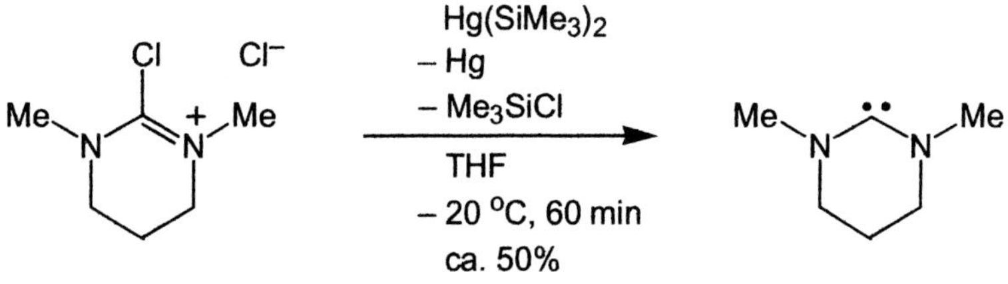

Cl Cl^-
Me N $\overset{+}{N}$ Me
$Hg(SiMe_3)_2$
– Hg
– Me_3SiCl
THF
– 20 °C, 60 min
ca. 50%
Me N N Me

Scheme 4. Formation of 1,3-dimethyl-hexahydropyrimid-2-ylidene by the "mercury method".

If the reaction mixture was not filtered after 50% conversion, the formation of the corresponding dication was observed, reducing the yield of formation of the carbene. Surprisingly, this carbene is not only stable at room temperature but it can be sublimated at 40 °C under vacuum.

Monoamino Carbenes

After the successful synthesis of several diamino carbenes, we tried to broaden the scope of application of the "mercury method" to monoamino carbenes. Note that our group using the deprotonation method reported the first examples of stable aryl amino carbene [15].

Aryl Amino Carbenes

In our quest to prepare new aryl amino carbenes, we reacted *N*-diisopropyl-*C*-phenyl chloroiminium chloride with bis(trimethylsilyl)mercury in THF. The reaction was monitored by ^{13}C NMR spectroscopy at –78 °C, but the carbene was not observed. Instead, the dimer was formed as a 90/10 mixture of *E* and *Z* isomers.

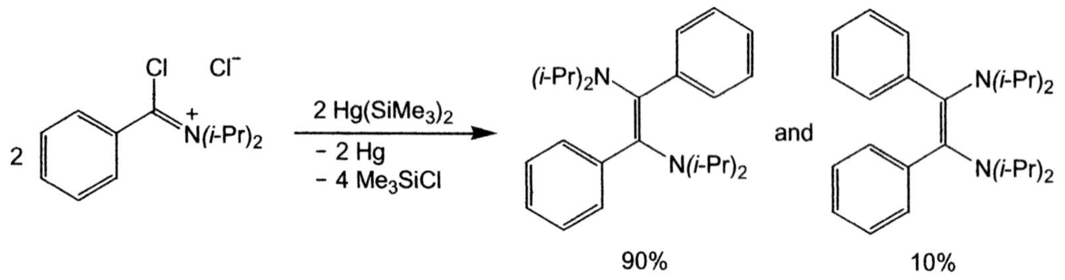

Scheme 5. Reaction of *N*-dialkyl-*C*-aryl chloroiminium chloride with bis(trimethylsilyl)mercury.

We performed similar reactions with various methoxy substituted aryl precursors in order to obtain a more stable carbene, but all our experiments led exclusively to the formation of the corresponding dimers in good yields.

Hydrogeno and Chloro Aminocarbenes

In a more reckless attempt to prepare an amino hydrogeno carbene, we performed a reaction between bis(trimethylsilyl)mercury and *N*-diisopropyl-*C*-chloro aldiminium chloride. Not surprisingly, even at low temperature we observed the direct formation of the dimer .

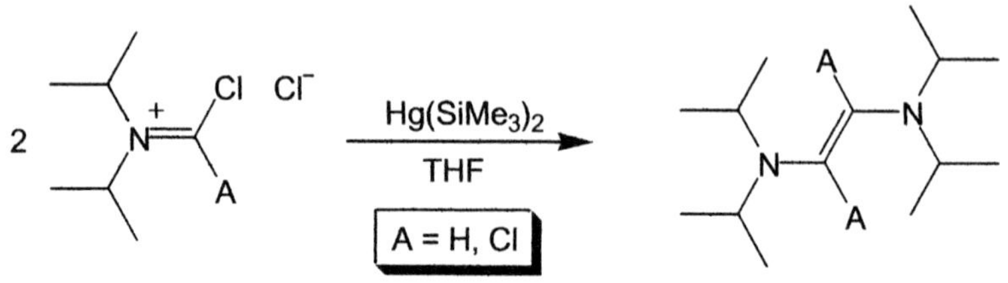

Scheme 6. Formation of the dimers of hydrogeno and chloro aminocarbenes.

Similarly, reaction with *N*-diisopropyl-*C*-dichloro iminium chloride leads to the formation of the corresponding dimer without the observation of the desired chlorocarbene.

Amino Oxy Carbenes

In the experiment with the naphthoxy substituted iminium salt, the corresponding carbene was formed at –80 °C. The ^{13}C NMR spectrum clearly showed a signal at 258.1 ppm, which was assigned to the carbene carbon. Alder reported similar chemical shifts for other amino oxy carbenes [16].

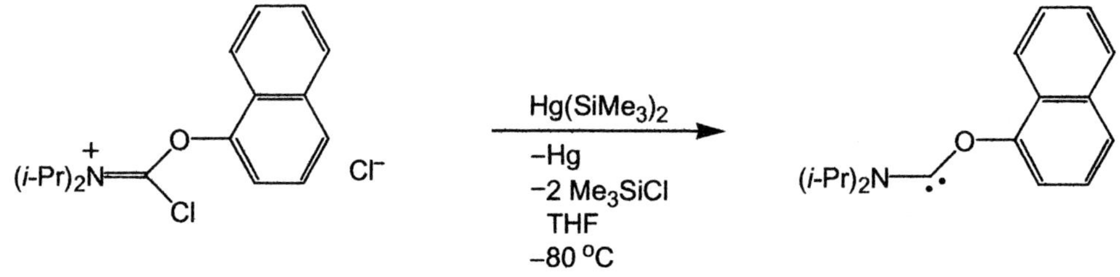

$(i\text{-}Pr)_2N^+$ O Cl Cl^- — $Hg(SiMe_3)_2$, –Hg, $-2\ Me_3SiCl$, THF, –80 °C → $(i\text{-}Pr)_2N$ O

Scheme 7. Formation of the naphthoxy amino carbene as an example of amino oxy carbene.

After the reaction mixture was allowed to warm to room temperature, the ^{13}C carbene signal disappeared, and a complex mixture was obtained.

In the reaction with the 3,5-dimethylphenoxy iminium salt, the carbene was observed at –80 °C with a chemical shift of 266.0 ppm (^{13}C NMR), but again was not isolated.

Alkyl Amino Carbenes

Stable alkyl amino carbenes are still unknown. In contrast to the difficulties met in the synthesis of the precursor for the deprotonation method, the corresponding chloro iminium salt was readily available by chlorination of *tert*-butyl diisopropylamide with oxallyl chloride.

Starting from the chloroiminium salt we performed the reaction with bis(trimethylsilyl)mercury at low temperature in THF.

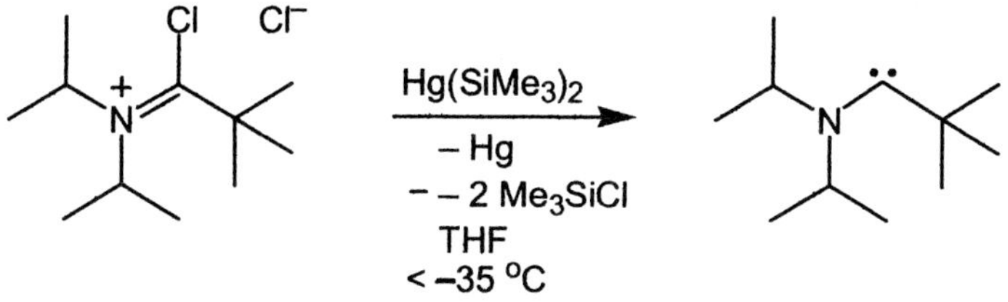

Cl Cl^- N^+ — $Hg(SiMe_3)_2$, – Hg, – – 2 Me_3SiCl, THF, < –35 °C → N

Scheme 8. Formation of the first observed alkyl amino carbene.

We were able to characterize at –25 °C the corresponding alkyl amino carbene by ^{13}C NMR spectroscopy with a signal at 325.3 ppm. At a conversion of about 40%, the spectra already showed the formation of several new compounds, which could not be identified.

Conclusion

In conclusion, we have shown that the "mercury way" is another route to generate amino carbenes. In the case of diaminocarbenes, this synthetic approach provides us with "metal-free" carbenes, which are usually not accessible by the deprotonation procedure. The fact that we never observed the formation of dimers in our experiments with diaminocarbenes showed that metal cations are also of importance for the dimerization.

With the "mercury way", we were able to characterize for the first time an alkyl amino carbene, although we were not able to isolate it.

With the synthesis of amino oxy carbenes, we could also verify the applicability of the new method, and show that the mercury method might be an alternative to traditional ways for preparing stable amino carbenes. Aryl-, chloro- and hydrogeno-amino carbenes were formed as transient intermediates in our experiments and underwent dimerization. The dimers were generally obtained in good yields.

Acknowledgment: We are grateful to the ACS/PRF (38192- AC4) and RHODIA for financial support of our work.

References

[1] Igau, A.; Grutzmacher, H.; Baceiredo, A.; Bertrand, G. *J. Am. Chem. Soc.* **1988**, *110*, 6463.

[2] Arduengo, A. J., III; Harlow, R. L.; Kline, M. *J. Am. Chem. Soc.* **1991**, *113*, 361.

[3] a) Herrmann, W. A.; Köcher, C. *Angew. Chem. Int. Ed. Engl.* **1997**, *36*, 2162. b) Kirmse, W. *Angew. Chem. Int. Ed.* **2004**, *43*, 1767. c) Perry, M. C.; Burgess, K. *Tetrahedron: Asymmetry* **2003**, *14*, 951. d) Hillier, A. C.; Grasa, G. A.; Viciu, M. S.; Lee, H. M.; Yang, C.; Nolan, S. P. *J. Organomet. Chem.* **2002**, *653*, 69. e) Herrmann, W. A. *Angew. Chem. Int. Ed.* **2002**, *41*, 1290. f) Alder, R. W. In *Carbene Chemistry* (Ed.: Bertrand, G.), Marcel dekker, New York **2002**, p. 153 – 171. g) Jafarpour, L.; Nolan, S. P. *Adv. Organomet. Chem.* **2001**, *46*, 181. h) Enders, D.; Gielen, H. *J. Organomet. Chem.* **2001**, *617 – 618*, 70. i) Bourissou, D.; Guerret, O.; Gabbaï, F. P.; Bertrand, G. *Chem. Rev.* **2000**, *100*, 39.

[4] a) Alder, R. W.; Blake, M. E.; Oliva, J. M. *J. Phys. Chem.* **1999**, *103*, 11200. b) Alder, R. W.; Allen, P. R.; Williams, S. J. *J. Chem. Soc., Chem. Commun.* **1995**, 1267. c) Dixon, D. A.; Arduengo, A. J. *J. Phys. Chem.* **1991**, *95*, 4180. d) Kim, Y. J.; Streitwieser, K. *J. Am. Chem. Soc.* **2002**, *124*, 5757.

[5] Alder, R. W. in *Carbene Chemistry: From Fleeting Intermediates to Powerful Reagents* (Ed.:

Bertrand, G.), Marcel Dekker, New York, **2002**, p. 153 – 176.

[6] a) Alder, R. W.; Blake, M. E.; Bortolotti, C.; Bufali, S.; Butts, C. P.; Linehan, E.; Oliva, J. M.; Orpen, A. G.; Quayle, M. J. *Chem. Commun.* **1999**, 241. b) Boche, G.; Hilf, C.; Harms, K.; Marsch, M.; Lohrenz, J. C. W. *Angew. Chem. Int. Ed. Engl.* **1995**, *34*, 487.

[7] a) Alder, R. W.; Blake, M. E. *Chem. Commun.* **1997**, 1513. b) Alder, R. W.; Butts, C. P; Orpen, A. G. *J. Am. Chem. Soc.* **1998**, *120*, 11526. c) Arduengo, A. J.; Goerlich, J. R.; Marshall, W. J. *Liebigs. Ann.* **1997**, 365. d) Chen, Y. T.; Jordan, F. *J. Org. Chem.* **1991**, *56*, 5029. e) Liu, Y. F.; Lemal, D. M. *Tetrahedron Lett.* **2000**, *41*, 599.

[8] a) Kuhn, N.; Kratz, T. *Synthesis* **1993**, 561. b) Denk, M. K.; Thadani, A.; Hatano, K.; Lough, A. J. *Angew. Chem. Int. Ed. Engl.* **1997**, *36*, 2607.

[9] a) Otto, M.; Conejero, S.; Canac, Y.; Romanenko, V. D.; Rudzevitch, V.; Bertrand, G. *J. Am. Chem. Soc.* **2004**, *126*, 1016. b) Otto, M. *PhD-Thesis*, Université Paul Sabatier, Toulouse, **2004**.

[10] a) Roesch, L.; Erb, W. *Chem. Ber.* **1979**, *112*, 394. b) Roesch, L.; Starke, U. *Zeitsch. Naturf., Teil B,* **1983,** *38B*, 1292.

[11] Eisch, J. J.; Becker, H. P. *J. Organomet. Chem.* **1979**, *171*, 141.

[12] a) Mitchell, T. N. *J. Organomet. Chem.* **1974**, *71*, 27. b) Mitchell, T. N. *J. Organomet. Chem.* **1974,** *71*, 39.

[13] Seyferth, D. *Acc. Chem. Res.* **1972**, *5*, 65.

[14] Alder, R. W.; Allen, P. R.; Murray, M.; Orpen, A. G. *Angew. Chem. Int. Ed. Engl.* **1996**, *35*, 1121.

[15] a) Sole, S.; Gornitzka, H.; Schoeller, W. W.; Bourissou, D.; Bertrand, G. *Science* **2001**, *292*, 1901. b) Cattoën, X.; Sole, S.; Pradel, C.; Gornitzka, H.; Miqueu, K.; Bourissou, D.; Bertrand, G. *J. Org. Chem.* **2003**, *68*, 911.

[16] Alder, R. W.; Butts, C. P.; Orpen, A. G. *J. Am. Chem. Soc.* **1998**, *120*, 11526.

Electrochemical Synthesis of Functional Organosilanes

B. Loidl, Ch. Grogger,* H. Stüger

Institute of Inorganic Chemistry, University of Technology Graz
Stremayrgasse 16, A-8010 Graz, Austria
Tel.: +43 316 873 8217 — Fax: +43 316 873 8701
E-mail: grogger@anorg.tu-graz.ac.at

B. Pachaly, R. Weidner, T. Kammel, A. Bauer

Consortium für Elektrochemische Industrie GmbH
Zielstattstraße 20, D-81379 München, Germany
Tel.: +49 89 74844 0 — Fax: +49 89 74844 242

Keywords: electrochemical reduction, organofunctional silanes, anionic cleavage, hydrosilanes

Summary: A variety of functional organosilanes were synthesized in two different pathways, both involving an electrochemical step. On the one hand, different hydrodisilanes were prepared by electroreductive coupling of chlorohydrosilanes giving access to alkoxysilanes by treatment with alcohols. The anionic cleavage of *sym*-$(MeO)_2Si_2Me_4$ by NaOMe and addition of functional aromatic halides yielded different organofunctional silanes. The second pathway involved the electroreductive coupling of functional silanes with various functional aromatic, aliphatic, and olefinic halides. All electrochemical reactions were performed in THF using an undivided cell, a sacrificial anode, a constant current density, and $MgCl_2$ and/or LiCl as supporting electrolyte.

Introduction

Organofunctional silanes of the general formula Y–Si(Me_2)R–Y′, with Y and Y′ meaning two functional groups, have special chemical properties. Because of their functionality Y′ bound to the organic spacer R they can undergo chemical bond formation to organic substrates, whereas the functional group Y at the inorganic side of the molecule is a possible linkage to inorganic substrates. As a result of this hybrid character, the applications are widely spread, e.g. in rubber, coatings, thermoplastics, glass fiber, and adhesives [1]. And, because of the rapidly growing market, an easy and cheap synthetic pathway is required by industry.

We have now succeeded in the development of two new methods for the preparation of organofunctional silanes, both involving electroreductive coupling. The first synthesis concept comprises the anionic cleavage of electrochemically formed disilanes and subsequent reaction with organic halides, whereas the second even more general route involves the direct electrochemical Si–C bond formation.

Anionic Cleavage of Functional Disilanes

Watanabe et al. reported that the cleavage of *sym*-dimethoxytetramethyldisilane by a molar equivalent of sodium methoxide in THF leads to methoxydimethylsilylsodium, which reacts *in situ* with bromo- or chlorobenzene to the corresponding methoxydimethylphenylsilane (Scheme 1, **1a**) [2, 3].

Based on this work, we investigated the possibilities of reacting functionalized aromatic halides $BrC_6H_4–Y'$ with the $(MeO)Me_2Si^-$ intermediate, thus yielding organosilyl compounds of the general formula $MeO–SiMe_2C_6H_4–Y'$.

Me Me / MeO–Si–Si–OMe / Me Me → (1. NaOMe; 2. X–C₆H₄–Y') → MeO–SiMe₂–C₆H₄–Y' + MeO–SiMe₂–OMe

	Y′	Yield [%]
1a	H	60 [2]
1b	OMe	60
1c	NMe_2	65
1d	$CH=CH_2$	20

Scheme 1. Isolated products of the cleavage of *sym*-dimethoxytetramethyldisilane and conversion with different aromatic halides.

As can be seen in Scheme 1, *p*-bromoanisole and *p*-bromo-*N*,*N*-dimethylaniline are very suitable for the reaction with methoxydimethylsilyl sodium, thus yielding $MeOPhSiMe_2(OMe)$ (**1b**) and $Me_2NPhSiMe_2(OMe)$ (**1c**) up to 60%. The reaction of *p*-bromostyrene also resulted in the formation of the corresponding organofunctional substrate $CH_2=CHPhSiMe_2(OMe)$ (**1d**), though with lower yields. In contrast to this, the reaction of *p*-bromonitrobenzene as well as of aliphatic and olefinic halides did not give the corresponding product.

In general, yields depend very much on the quality of the sodium methoxide.

Electrochemical Synthesis of Hydrodisilanes

The direct electrochemical synthesis of *sym*-dimethoxytetramethyldisilane by electroreductive coupling of $MeOMe_2SiCl$ is not possible. Nevertheless, the conversion of electrochemically generated *sym*-tetramethyldisilane with methanol in the presence of a catalyst [4] gives access to *sym*-dimethoxytetramethyldisilane, which is demanded for the synthesis of organofunctional silanes by anionic cleavage. Therefore, we investigated the possibilities of generating dihydrodisilanes electrochemically.

Electrochemical Si–Si bond formation has been known since the work of E. Hengge and G. Litscher in 1976, using a divided cell and a mercury anode [5]. Later, T. Shono developed a new undivided cell design employing magnesium as sacrificial anode, which is oxidized and subsequently converted to $MgCl_2$ by the chloride from the chlorosilane [6].

$$2\ H\text{–}SiR_2\text{–}Cl \xrightarrow[Mg]{2e^-} H\text{–}SiR_2\text{–}SiR_2\text{–}H + MgCl_2$$

R = H, Me, Ph

Eq. 1. Electroreductive coupling of chlorohydrosilanes.

We succeeded in the synthesis of different hydrodisilanes by electroreductive coupling of chlorohydrosilanes employing the cell design of T. Shono. All electrolyses were performed in an undivided cell equipped with a magnesium sacrificial anode and a cylindrical stainless steel cathode. The reactions were carried out under constant current conditions (1 mA cm^{-2}) in THF using $MgCl_2$ as supporting electrolyte.

Table 1. Electroreductive synthesis of different hydrodisilanes.

Starting material	Product	By-product	Yield of Product	^{29}Si [ppm]
Me_3SiCl	$Me_3Si\text{–}SiMe_3$	–	90%	–19.2
HMe_2SiCl	$HMe_2Si\text{–}SiMe_2H$	$H(SiMe_2)_3H$	90%	–39.0
HMePhSiCl	HMePhSi–SiPhMeH	–	85%	–36.6(d)
HPh_2SiCl	$HPh_2Si\text{–}SiPh_2H$	$H(SiPh_2)_3H$	20%	–35.4
$H_2PhSiCl$	$H_2PhSi\text{–}SiPhH_2$	$H(SiPhH)_3H$	30%	–61.5

Electroreductive Si–C Bond Formation

The second investigated route to organofunctional silanes consisted in the electrochemical coupling of an organic halide Y′–RCl(Br) with a functionalized chlorosilane Y–Me_2SiCl (Eq. 2).

$$\text{Y}-\underset{\text{Me}}{\overset{\text{Me}}{\underset{|}{\overset{|}{\text{Si}}}}}-\text{X} + \text{X}'-\text{R}-\text{Y}' \xrightarrow{2\,e^-} \text{Y}-\underset{\text{Me}}{\overset{\text{Me}}{\underset{|}{\overset{|}{\text{Si}}}}}-\text{R}-\text{Y}' + \text{X}^- + \text{X}'^-$$

X = Cl, MeO   X′ = Cl, Br

R = alkyl, aryl

Eq. 2. Electroreductive Si–C bond formation.

Shono et al. [7] and Yoshida et al. [8] showed in the early 1980s that electroreductive cross coupling is possible in a divided cell with Pt anodes, but no defined anodic oxidation products were reported. In contrast to this, Bordeau et al. electrolyzed different organic substrates with Me_2SiCl_2 in an undivided cell with sacrificial anodes, thus yielding $MgCl_2$ as by-product [9]. However, to maintain one of the chloro substituents on the silicon a large excess of chlorosilane and the use of an Ni catalyst is unavoidable. Furthermore, HMPA was used as co-solvent.

Applying the same setup as used for the electrolyses of the hydrodisilanes, we now succeeded in the synthesis of various organofunctional silanes. In an undivided cell and with a THF/$MgCl_2$/LiCl electrolyte, all electrolyses were carried out without HMPA or Ni(0) catalysis. Furthermore, we found out that not only is chlorine a suitable leaving group, but so also is the MeO substituent. Thus, dimethylchlorosilane (chloride as a leaving group) as well as dimethoxydimethylsilane (MeO– as a leaving group) were used in the electrolyses of various aromatic, aliphatic, olefinic, and polymeric halides (Cl, Br). After aqueous work-up, the products shown in Scheme 2 were easily recovered and isolated in yields between 60 and 85%.

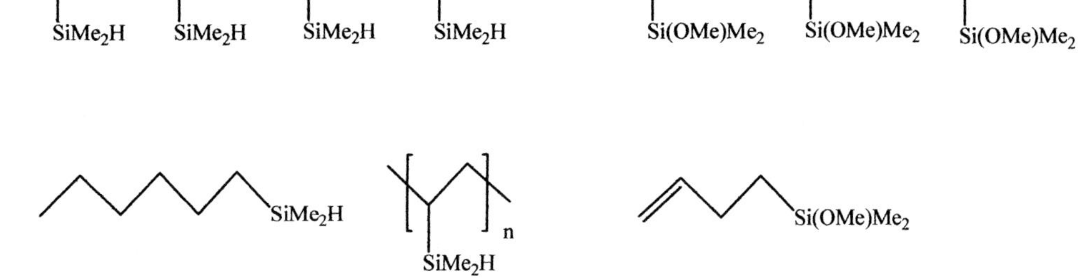

NMe$_2$ / SiMe$_2$H; N(TMS)$_2$ / SiMe$_2$H; OMe / SiMe$_2$H; OTBDMS / SiMe$_2$H; OMe / Si(OMe)Me$_2$; OTBDMS / Si(OMe)Me$_2$; n / Si(OMe)Me$_2$

SiMe$_2$H; n / SiMe$_2$H; Si(OMe)Me$_2$

Scheme 2. Isolated organofunctional silanes synthesized by electroreductive Si–C bond formation.

Direct electrochemical Si–C bond formation appears to be a very simple and elegant method for the synthesis of organofunctional silanes, thus being very interesting both in itself and not least for its potential industrial applications.

Acknowledgments: We want to thank the Consortium für Elektrochemische Industrie GmbH for financial support and the Wacker-Chemie for kindly providing various organosilanes as starting materials.

References

[1] E. P. Plueddemann, *Silane Coupling Agents*, Plenum Press, New York, **1982**.

[2] H. Watanabe, K. Higuchi, M. Kobayashi, M. Hara, Y. Koike, T. Kitahara, Y. Nagai, *J. Chem. Soc., Chem. Commun.* **1977**, 534.

[3] H. Watanabe, K. Higuchi, T. Goto, T. Muraoka, J. Inose, M. Kageyama, Y. Ilzuka, M. Nozaki, Y. Nagai, *J. Organomet. Chem.* **1981**, *218*, 27.

[4] L. Hörner, J. Mathias, *J. Organomet. Chem.* **1985**, *282*, 155.

[5] E. Hengge, G. K. Litscher, *Angew. Chem. Int. Ed. Engl.* **1976**, *15*, 370.

[6] T. Shono, S. Kashimura, M. Ishifune, R. Nishida, *J. Chem. Soc., Chem. Commun.* **1990**, 1160.

[7] T. Shono, H. Ohmizu, S. Kawakami, H. Sugiyama, *Tetrahedron Lett.* **1980**, *21*, 5029.

[8] J. Yoshida, K. Muraki, H. Funahashi, N. Kawabata, *J. Organomet. Chem.* **1985**, *284*, C33.

[9] C. Moreau, F. Serein-Spireau, M. Bordeau, C. Biran, *Organomet.* **2001**, *20*, 1910.

Photoluminescence and Photochemical Behavior of Silacyclobutenes

Duanchao Yan, Andreas A. Hess, Norbert Auner,* Mark Thomson, Michael Backer

Institut für Anorganische Chemie der Johann Wolfgang Goethe-Universität
Marie-Curie-Str. 11, D-60439 Frankfurt, Germany
Tel: +49 69 798 29591 — Fax: +49 69 798 29188
E-mail: Auner@chemie.uni-frankfurt.de

Keywords: silacyclobutene, fluorescence, photochemical reaction, ring-opening reaction, stilbene

Summary: The photoluminescence and photoreactions of 1,1-diorgano-2,3-diphenyl-4-neopentyl-1-silacyclobut-2-enes in organic solvents were studied. It was found that the formation of photolysis products is highly dependent on the reaction conditions. Upon short-wave length (214 nm) irradiation, very complicated product mixtures result, including various ring-opened isomers and decomposition products. Instead, irradiation with moderate wavelength UV light (e.g., 307 nm), gives only simple products: in the presence of trapping agents, such as methanol or water, the ring-opened adducts were isolated, which verify an *in situ* silene formation as a reactive intermediate. In the presence of oxygen "ring-closed" reactions involving the silacyclobutenes' stilbene substructure occurred, giving annelated products that exhibit strong photoluminescence. Without reaction partners no photolysis products were obtained.

Introduction

Silacyclobutenes are compounds of high interest because their carbon/carbon unsaturated strained rings exhibit novel physical and chemical properties. Thus, they are potential precursors for the introduction of very specific optical and electrical properties into silicon-based materials containing silicon–oxygen, silicon–carbon, and silicon–silicon bonds [1].

During the investigation of a series of silacyclobutene compounds it was found that 2,3-diphenyl-3-neopentyl-1-silacyclobut-2-enes (**1**) show an intense blue photoluminescence (PL) upon excitation with UV light in the solid state (Fig. 1) [1].

Here we describe the detailed investigation of a novel PL behavior of compounds **1** in solution and their photochemical reactivity under various conditions.

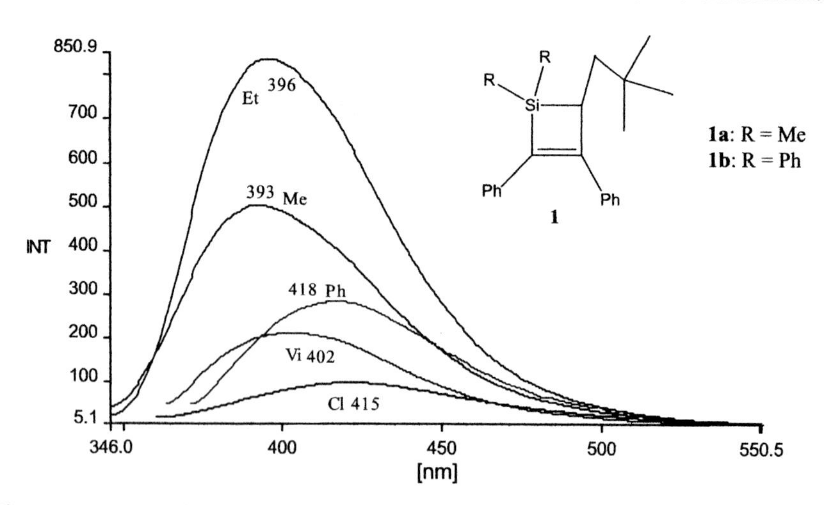

Fig. 1. Fluorescence spectra for silacyclobutenes **1** (R = Cl, Vinyl, Ph, Me, Et) in the solid state (numbers in graph give the maximum-emission wavelength λ_{max}).

Results and Discussion

In contrast to their strong blue PL in the solid state [1], silacyclobutenes **1** exhibit almost no PL in solutions with organic solvents. Surprisingly, the observed very weak PL increases rapidly with time upon irradiation with UV light.

Figure 2 shows the increasing fluorescence of dimethyl-substituted silacyclobutene (**1a**) upon irradiation with 300 nm light in THF solution (1×10^{-5} M). In the beginning of the irradiation, the emission intensity increases very fast, but with time this increase slows down gradually. After 12 h irradiation, the emission intensity reaches a maximum value, which is thousands of times that of the initial solution. The PL intensity decreases slowly after the UV irradiation is stopped, but there is still about 75% intensity PL remaining after three days. If the photolysis solution is kept in the dark, no decrease of PL is observed. These PL measurements were repeated for silacyclobutenes **1** with different substituents at the silicon center (R = Et, Ph, Vinyl, Cl, EtO) and in a wide range of organic solvents (THF, *n*-pentane, benzene, $MeSiCl_3$, etc.), all of them exhibiting similar emission behavior. These results suggest that the increasing PL of compounds **1** in organic solutions originates basically from photoreactions involving the silacyclobutene's four-membered ring substructure and the two phenyl groups at carbon atoms two and three. The increase in emission is only observable in dilute solution. At silacyclobutene concentrations higher than 1×10^{-3} M, the fluorescence becomes very weak. Obviously, under these conditions silacyclobutenes **1** absorb most of the emission light.

In order to understand the photoreactions and to identify reasons that are responsible for the increasing PL, a series of photochemical reactions was designed involving the photolysis of the dimethyl- and diphenyl-substituted silacyclobutenes (**1a**, **1b**) under different reaction conditions such as solvent and reaction partners. The resulting reaction solutions were analyzed by GC-MS.

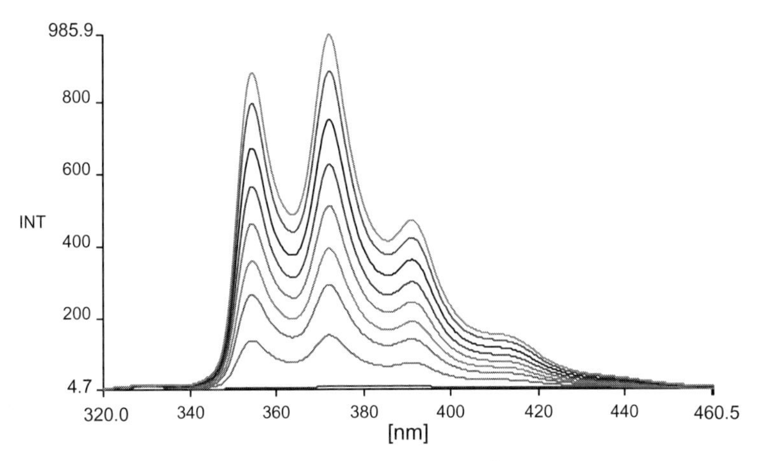

Fig. 2. Time dependence of fluorescence emission spectra for 2×10^{-5} M **1a** in THF solution (λ_{exc} = 300 nm), with 10 min interval time.

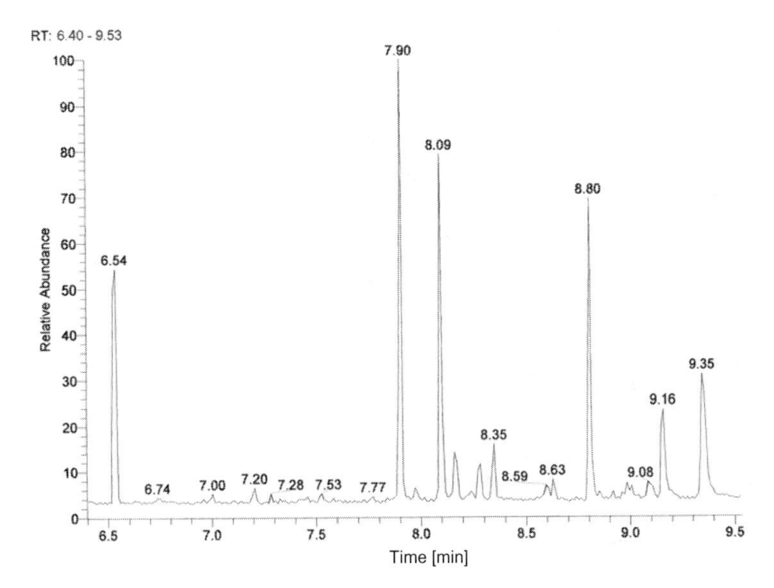

Fig. 3. GC analysis of **1a** in pentane and methanol (1 : 1) irradiated by full spetrum of a Zn lamp.

Figure 3 shows the GC-MS analysis of the photoreaction of silacyclobutene **1a** in a mixture of *n*-pentane and methanol (1:1) exposed to irradiation from a 5 W Zn lamp for 29 h in a quartz cuvette. In the gas chromatogram plot, nine main products and more than ten by-products can be detected in the photolysis solution. The TLC analysis indicated a complex mixture too. This result is not surprising, because the short-wavelength photochemistry of a silacyclobutene follows complicated mechanisms [2].

In sharp contrast, irradiation of silacyclobutenes **1** with filtered 307 nm UV light yields only simple photoproducts: in the presence of trapping agents such as water or methanol in the reaction solution, only one main product was obtained (Fig. 4); in the absence of trapping agents, no photolysis products were formed.

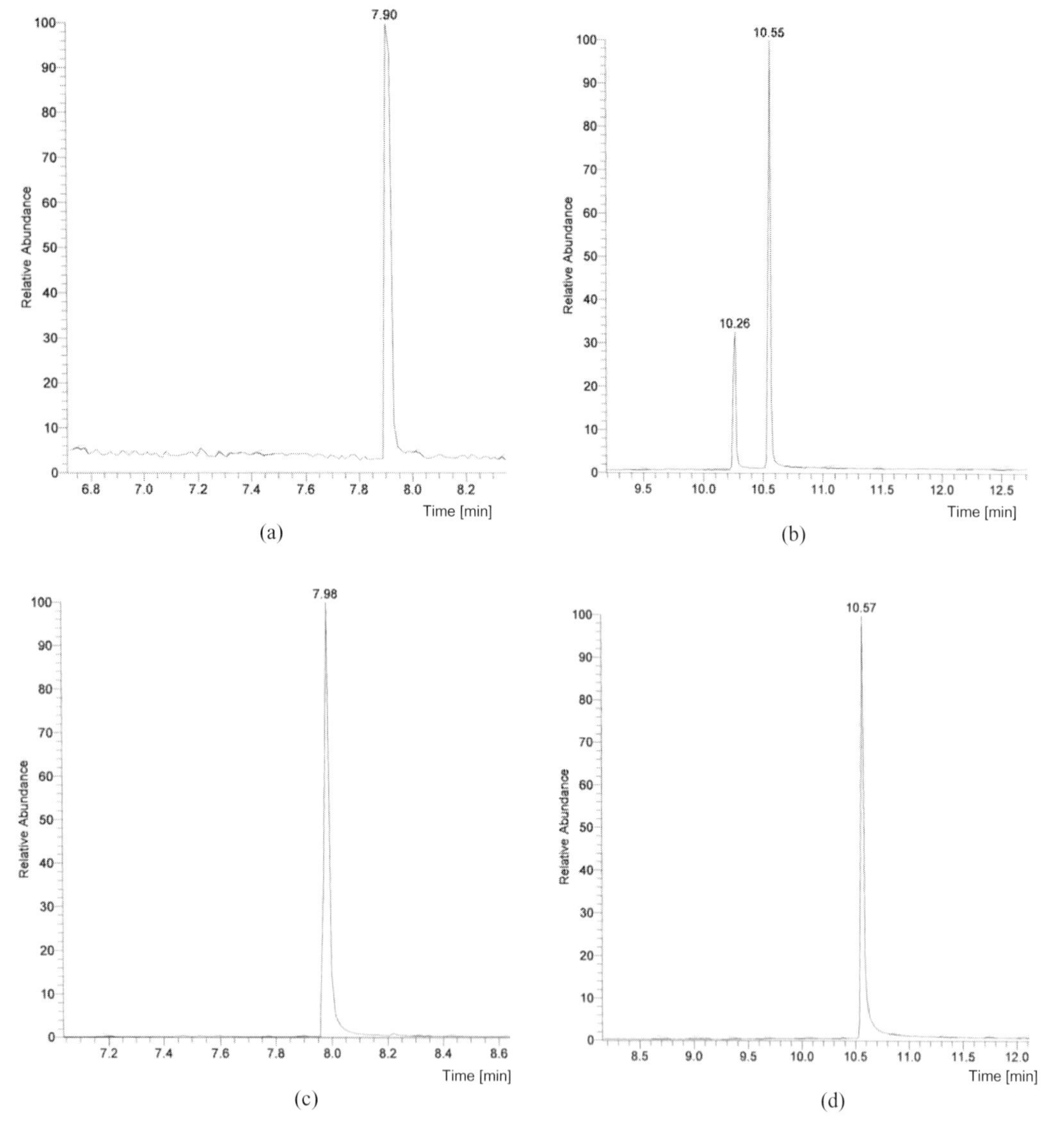

Fig. 4. GC analyses of **1a** or **1b** irradiated with long wavelength (307 nm) UV light for 14 days. (a) **1a** in THF and methanol (2 : 1). (b) **1b** in THF and methanol (2 : 1). (c) **1a** in THF saturated with water. (d) **1b** in THF saturated with water.

The selectively formed photolysis products were identified as ring-opened compounds trapped by the polar reagents used. They were characterized by ^{1}H, ^{13}C, and ^{29}Si NMR spectroscopy

(Scheme 1). ROESY measurements confirm the trans configuration of the molecules **3** and **4** obtained. Surprisingly, PL measurements of these photolysis products in organic solvents show a very weak fluorescence. From that it is concluded that the fluorescence of silacyclobutenes **1** in organic solvents (Fig. 2) does not originate from ring-opened compounds such as **3** and **4**.

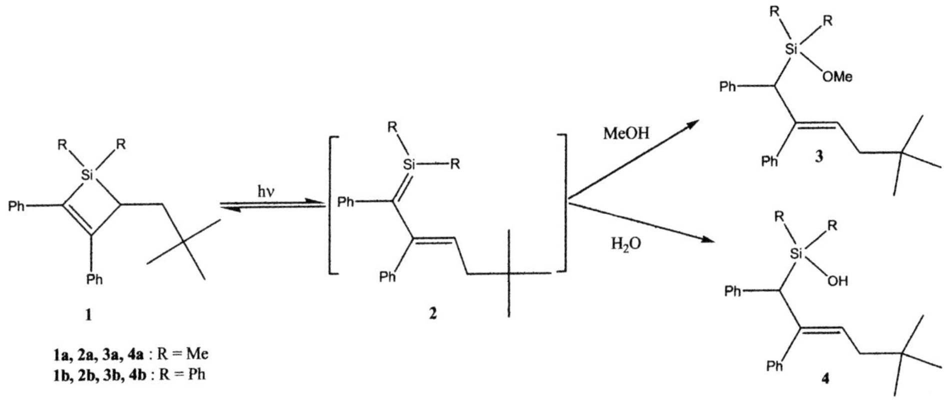

R R Si Ph Ph 1 — hν → [R Si R Ph Ph 2] — MeOH → R R Si OMe Ph Ph 3; — H_2O → R R Si OH Ph Ph 4

1a, 2a, 3a, 4a : R = Me
1b, 2b, 3b, 4b : R = Ph

Scheme 1.

As can be assumed from Scheme 1, the PL might originate from the formation of intermediate silenes **2**. But the observation that irradiated solutions remain fluorescent when they are kept in the dark indicates a relatively high stability of the product formed. It is unreasonable to assume that silenes **2** are stable in organic solutions.

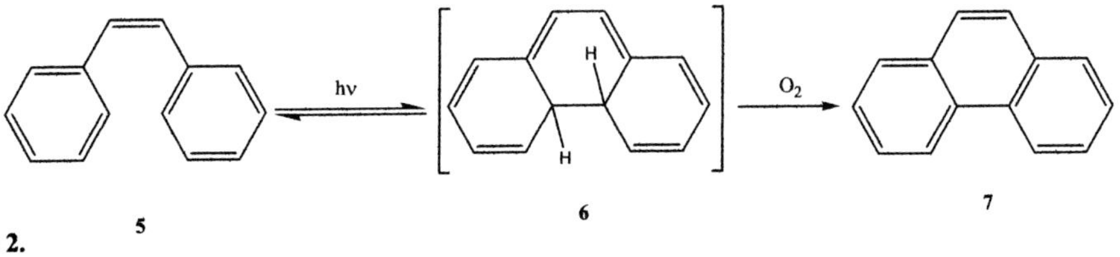

5 — hν → [H H 6] — O_2 → 7

Scheme 2.

R R Si Ph Ph 1 — hν → [R R Si 8] — O_2 → R R Si 9

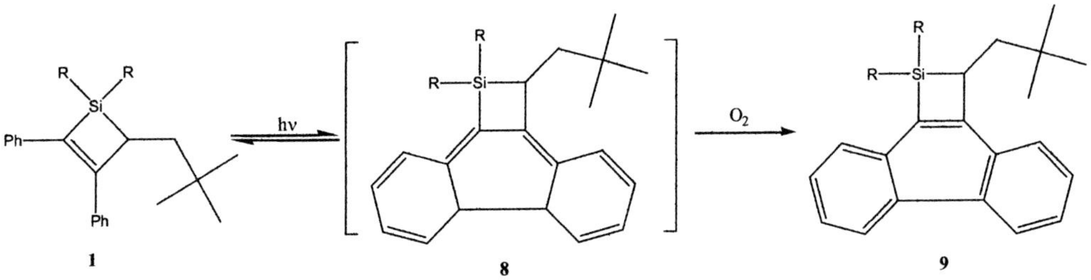

1a, 8a, 9a : R = Me
1b, 8b, 9b: R = Ph

Scheme 3.

From organic literature it is known that UV light transforms *cis*-stilbene into phenanthrene in the presence of air as shown in Scheme 2 [3]. In analogy, the photoreaction of silacyclobutenes **1** may involve the stilbene subunit to give the silicon phenanthrene analog **9** after oxidization with oxygen (Scheme 3).

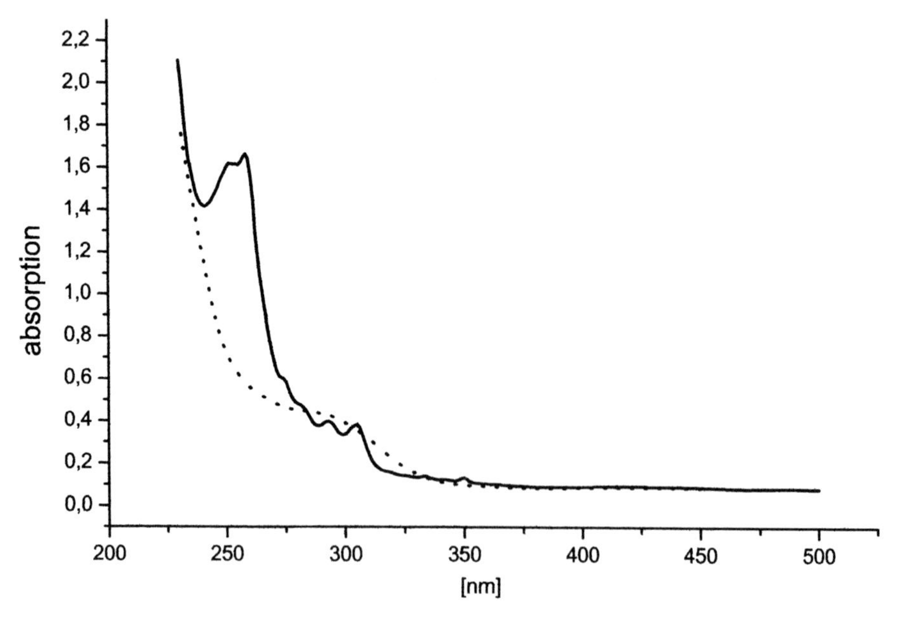

Fig. 5. Absorption spectra for a 1×10^{-5} M pentane solution of **1a** (dotted line) and for the same solution irradiated by 300 nm UV light for 6 h (solid line).

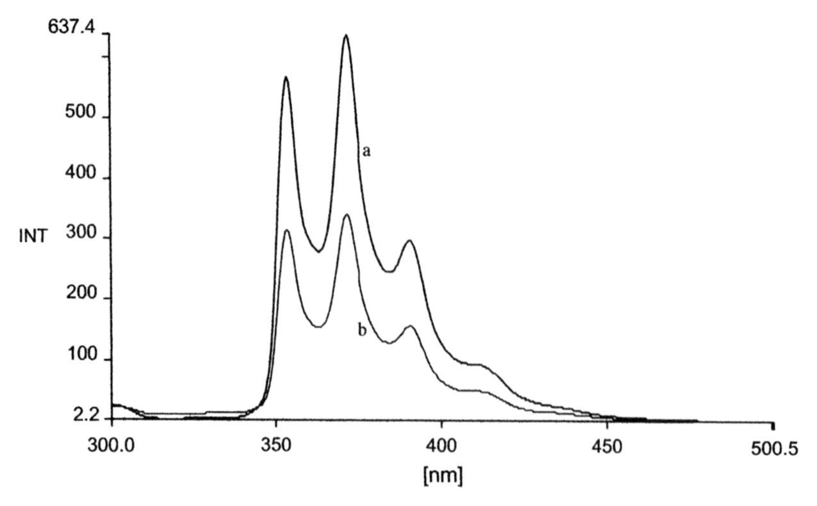

Fig. 6. Fluorescence emission spectra of **1b** and **9b** in THF solution (λ_{exc} = 300 nm). (a) 8×10^{-6} M **9b**; (b) 2×10^{-5} M **1b** irradiated by 300 nm UV light for 4 h.

Moreover, the UV-vis spectrum of **1a** exhibits two new bands at 258 and 300 nm upon irradiation in the presence of air (Fig. 5). These spectral variations are similar to those observed for the photoreaction of stilbene (Scheme 2) [3]. Thus, photoreactions of silacyclobutenes **1** in dry toluene are conducted while oxygen blows through the solution. The resulting photolysis products are separated by chromatography on a silica gel column, and the stable silicon phenanthrene analog **9b** is obtained and characterized by ^{1}H, ^{13}C, and ^{29}Si NMR spectroscopy. Compound **9a** is unstable and gradually decomposes to siloxanes. But the ^{1}H NMR of the photolysis mixture also exhibits the characteristic peaks of the phenanthrene subunit (around 8.55 ppm). Moreover, compound **9b** shows a strong blue PL in organic solutions as shown in Fig. 6. It is obvious that the emission spectrum of **9b**, within experimental accuracy, has the same appearance and position as that of **1b** irradiated by UV light in solution. From that we conclude that the increase in PL of **1** originates from the formation of compounds **9**. Compounds **9** exhibit a high degree of π-conjugation within their organic carbon skeletons. Thus it is not surprising that they show a much stronger PL than the starting silacyclobutene **1** with a lower degree of π-conjugation.

Conclusion

Irradiation of silacyclobutenes **1** in organic solvents with 307 nm light yields simple photoproducts, basically following two different reaction pathways: in the absence of oxygen but in the presence of polar trapping agents, ring opening reactions lead to silabutadiene (silene) formation (Scheme 1), while in the presence of oxygen the silacyclobutene's *cis*-stilbene subunit reacts to give a phenanthrene analog **9** via rearrangement of intermediate **8**. The formation of **9** is responsible for the photoluminescent behavior of silacyclobutenes **1** in organic solutions.

Acknowledgments: We thank the DFG and Dow Corning Corp. for financial support.

References

[1] a) U. Pernisz, N. Auner, Photoluminescence of organically modified cyclosiloxanes, in *Organosilicon Chemistry IV — From Molecules to Materials* (Eds.: N. Auner, J. Weis), VCH, Weinheim **2000**, p. 505; b) U. Pernisz, N. Auner, M. Backer, *Polym. Preprints* **1998**, *39(1)*, 450.

[2] M. G. Steinmetz, B. S. Udayakumar, *Organometallics* **1989**, *8*, 530.

[3] K. A. Muszkat, E. Fischer, *J. Chem. Soc. B* **1967**, 662.

Tris- and Tetrakis-[oligo(phenylenevinylene)]-silanes: Synthesis and Luminescence Behavior

Heiner Detert,* Erli Sugiono

Institut für Organische Chemie, Johannes Gutenberg-Universität Mainz
Duesbergweg 10 – 14, 55099 Mainz, Germany
Tel.: +49 6131 3922111 — Fax: +49 6131 3925396
E-mail: detert@mail.uni-mainz.de

Keywords: silanes, oligo(phenylenevinylene)s, UV-Vis spectroscopy, fluorescence

Summary: The connections of three or four monodisperse oligo(phenylenevinylene)s to a central silicon atom is performed via Wittig-Horner reactions. The terminal rings are substituted with alkoxy side chains. Depending on the ratio of the lengths of the rigid conjugated units and the flexible side chains, transparent films can be obtained from several of these trigonal-pyramidal or tetrahedral molecules. An intense fluorescence in the blue-green region is emitted by molecules of either shape. These compounds are interesting as active materials for electrooptical applications because of their intense fluorescence and improved film-forming capability.

Introduction

Aggregation (or crystallization) is a serious problem for molecular π-conjugated materials which rarely show amorphous glass phases. The boundaries of crystalline regions in emissive or charge-transporting layers creates irregularities or defects, which usually act as traps for mobile electrons or holes. Charges in such traps generate no radiation but dissipate energy as heat or vibrations and thus adversely affect the performance of molecule-based LEDs. Though stilbenoid oligomers exhibit efficient photo- and electroluminescence in the blue to yellow region, the application of these low-molecular-weight molecules is often limited by their aggregation behavior [1]. Oligo(phenylenevinylene)s (OPVs) as side chains [2] or in the main chain of a polymer [3] or dispersed in polymeric matrices [4, 5] are suitable ways to overcome this limitation. An approach to glassy molecular materials was opened by Salbeck [6]: the spiro-connection of two oligo(phenylene) strands via a central spiro-bifluorene unit. This results in three-dimensional molecules of a flattened tetrahedral shape. The less rigid tetraphenylmethane can be used similarly as a central unit; the differences to the original concept are more perfect tetrahedral symmetry, higher flexibility, and, much more important, a broken conjugation due to the central sp^3 carbon. Some examples of such carbon-centered tetrahedra with four oligo(het-)arylene units have recently been described by different groups [7 – 11]. Efficient organic LEDs have been fabricated from these film-forming molecules of intermediate dimensions. Most synthetic strategies use tetraphenylmethane as starting material, which is brominated in the *para*-positions of all four benzene rings. Heck-, Suzuki- or Kumada-coupling reactions are successful routes for the extension of the conjugated system.

The synthesis of tetraarylsilanes via Wurtz cross-coupling of bromoarenes and $SiCl_4$ is a short and efficient route to tetraarylsilanes. Four conjugated units like biphenyl or naphthalene can be attached to the central silicon atom in high yield, but in the case of donor-substituted stilbenoid oligomers, this method gave only poor results [12]. For the preparation of tetrahedral silanes with phenylenevinylene chromophores, the convergent step was not performed on the central silicon atom but on a tetra-*p*-tolylsilane (**1b**) functionalized with four phosphonic esters. The advantage of this approach is the additional extension of the π-system in the final step.

Synthesis and Properties

Tetra-*p*-tolylsilane, prepared via Wurtz-coupling of 4-bromotoluene and $SiCl_4$, was brominated with NBS according to a procedure described by Drefahl [13] yielding a mixture of the three- and four-fold benzylic bromides **1a**, **1b** which can be separated by repeated crystallization from acetone. A more convenient way is to convert the crude mixture in a Michaelis-Arbusow reaction to the phosphonates **2a**, **2b** and to separate these via chromatography (SiO_2/ethyl acetate/ethanol). A second Wohl-Ziegler bromination/Michaelis-Arbusow sequence allows one to transform **2a** into the tetrahedral **2b** in moderate yield.

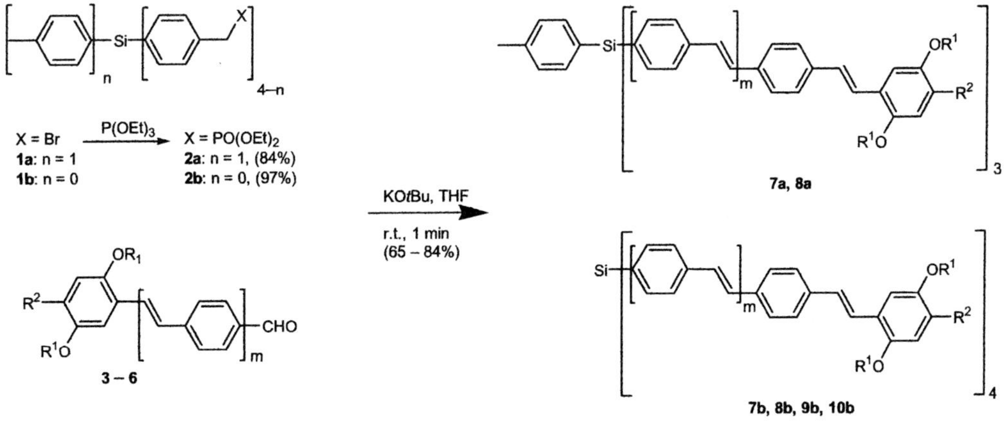

X = Br —$P(OEt)_3$→ X = $PO(OEt)_2$
1a: n = 1 → **2a**: n = 1, (84%)
1b: n = 0 → **2b**: n = 0, (97%)

3 – 6

KO*t*Bu, THF, r.t., 1 min (65 – 84%)

7a, 8a

7b, 8b, 9b, 10b

Scheme 1. Synthesis of tris- and tetrakis-OPV-silanes **7a, 7b, 8a, 8b, 9b, 10b.**

The trigonal-pyramidal **7a, 8a** and tetrahedral molecules **7b**, **8b**, **9b**, and **10b** were prepared via three- or fourfold Horner reactions of the central phosphonate **2a**, **2b** with stilbenoid aldehydes

3 – **6** [12, 14] in reasonable to good yields, provided that the reaction is quenched after about 1 min. The solvent was evaporated *in vacuo*, and the residue redissolved in chloroform, filtered through a pad of basic alumina, and precipitated from ethanol. Prolonged reaction times and excess base caused a nucleophilic attack on the silicon atom resulting in the cleavage of one Si-chromophore bond to yield a tris-OPV-silanol [15]. Particularly **10b** is sensitive to this consequent reaction. The high reactivity of the silicon-connected phosphonate is remarkable, since comparable PO-activated olefinations with analogous tetraarylmethanes require 24 – 48 h [16].

Mass spectroscopy (field desorption technique) reveals an increasing ease of positive charging of these molecules with the extension of their conjugated system. The relative intensities of the molecular ions decrease from 100% (**7a**, **7b**) to 0% (**10b**), whereas ions of higher charge appear with increasing intensity: **8b**, **9b**: M^{2+} = 100%, **10b**: M^{4+} = 100% and M^{5+} = 4%. The molecular ion of **10b** (*m/z* = 3054) could only be detected by Maldi-TOF-spectroscopy.

Table 1. Substitution pattern and properties of tris- and tetrakis-OPV-silanes.

	m	R^1	R^2	m.p. [°C]	Yield	λ_{max}^{abs} (CH_2Cl_2) [nm]	λ_{max}^{em} (CH_2Cl_2) [nm]	λ_{max}^{exc} (film) [nm]	λ_{max}^{em} (film) [nm]
7a	1	C_3H_7	$C(CH_3)_3$	138	84%	382	470	391	473
7b	1	C_3H_7	$C(CH_3)_3$	125	81%	381	470	391	468
8a	2	C_3H_7	H	145	82%	396	469	393	473
8b	2	C_3H_7	H	257	65%	396	470	393	483
9b	2	C_8H_{17}	CH_3	wax	70%	396	481	390	478
10b	3	C_8H_{17}	CH_3	192	67%	412	483	396	502

The film-forming capability depends on the shape and the ratio of the lengths of the conjugated system and the flexible side chains. All compounds are soluble in common solvents and form transparent and fluorescent films on glass. A slow recrystallization was observed for the silanes with three OPV moieties (**7a**, **8a**). As expected, an elongation of the flexible side chains (**8b**, **9b**) results in a strong decrease in the melting point, in contrast to the effect of the extension of the rigid π-system (**9b**, **10b**). The latter does not affect the stability of amorphous films.

The melting points of tetrahedral silanes **7b** and **8b** have been found to be 125 °C and 257 °C. DSC of pristine **7b** (10 °C/min) shows an endothermic transition with an onset temperature of 60 °C and a peak temperature of 78 °C followed by a glass transition at 113 °C. Upon heating to 200 °C, no further transitions were detected; during cooling (10 °C/min 200 to –100 °C), a glass transition in the range 114 – 93 °C occurred. Only this transition (onset 99 °C, T_g 105 °C) was visible during the second heating. The homologous silane **8b** with four distyrylstilbene chromophores starts to melt at 223 °C with a peak temperature at 262 °C. Further heating to 400 °C, cooling to –50 °C, and a second heating scan to 400 °C did not show any visible transition.

Electronic Spectra

Electronic spectra were obtained from solutions in dichloromethane (absorption ca. 10^{-4} M, emission ca. 10^{-6} M) and from spin-coat films on glass substrates. The solid compounds as well as their solutions are light to intense yellow and strongly fluorescent; bathochromic shifts of the absorption band correlate with the elongation of the conjugated system. A connection of the OPV

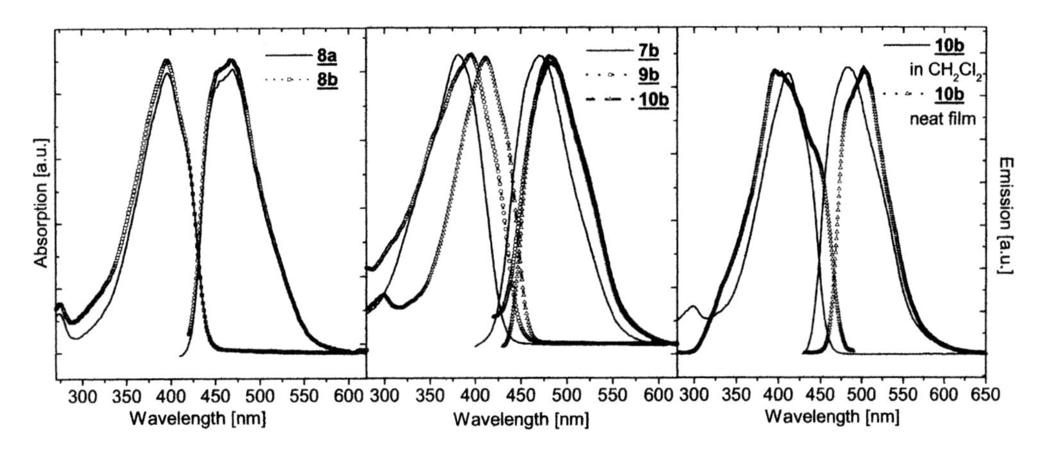

segments via a central silicon atom to three-dimensional scaffolds results in small bathochromic shifts (3 – 7 nm) of the electronic spectra compared with the free chromophore [12]. This can in part be attributed to a small auxochromic effect of the silicon atom [17]; more important is the stabilization of the excited states from inter-chromophore interaction due to the close proximity of the phenyl groups on the central silicon. This effect has also been observed with related carbon-centered tetrahedral [16]. The absorption and emission spectra of tetrahedral and pyramidal compounds are essentially identical (**7a**, **7b**, **8a**, **8b**). An increasing extension of the conjugated system from distyrylbenzene (**7a**, **7b**) to four- and five-ring compounds (**8a**, **8b**, **9b**, **10b**) is reflected in a substantial shift of the absorption band in diluted solutions to longer wavelengths (ca. 15 nm/phenylenevinylene unit, Table 1).

Conclusion

Monodisperse oligo(phenylenevinylene)s are assembled to three-dimensional scaffolds in convergent three- or fourfold Horner reactions of a silane core and stilbenoid aldehydes in moderate to good yields. The central silane is sensitive toward nucleophilic attack. Tetrahedral compounds with short side chains form stable amorphous films but trigonal-pyramidal shaped assemblies tend to recrystallize. The connection of the chromophores to a central silicon atom results in a stabilization of the excited states; an intense blue-green fluorescence is emitted from solution as well as from transparent films.

Acknowledgment: Financial support by the Deutsche Forschungsgemeinschaft is gratefully acknowledged.

References

[1] *Electronic Materials: The Oligomer Approach* (Eds.: K. Müllen, G. Wegner), Wiley, Weinheim, **1998**.

[2] P. Hesemann, H. Vestweber, J. Pommerehne, R. F. Mahrt, A. Greiner, *Adv. Mater.* **1995**, *7*, 388.

[3] Z. Yang, B. Hu, F. E. Karasz, *Macromolecules* **1995**, *28*, 6151.

[4] B. Luther-Davies, M. Samoc, M. Woodruff, *Chem. Mater.* 1996, *8*, 2586.

[5] C. J. Wung, Y. Pang, P. N. Prasad, F. E. Karasz, *Polymer* 1991, *32*, 605.

[6] J. Salbeck, *Ber. Bunsenges. Phys. Chem.* **1996**, *100*, 1667.

[7] C. S. Krämer, T. J. Zimmermann, M. Sailer, T. J. J. Müller, *Synthesis* **2002**, 1163.

[8] T. J. Zimmermann, T. J. J. Müller, *Synthesis* **2002**, 1157.

[9] H.-C. Yeh, R.-H. Lee, L.-H. Chan, T.-Y. J. Lin, C.-T. Chen, E. Balasubramanian, Y.-T. Tao, *Chem. Mater.* **2001**, *13,* 2788.

[10] X.-M. Liu, C. He, J.-W. Xu, *Tetrahedron Lett.* **2004**, *45*, 1593.

[11] M. R. Robinson, S. Wang, G. C. Bazan, Y. Cao, *Adv. Mater.* **2000**, *12*, 1701.

[12] E. Sugiono, *PhD Thesis*, Johannes Gutenberg-Universität, Mainz, **2001**.

[13] G. Drefahl, D. Lorenz, *J. Prakt. Chem.* **1964**, *24 (4)*, 312.

[14] H. Detert, Th. Metzroth, E. Sugiono, *Adv. Synth. Catal.* **2001**, *343*, 351.

[15] All compounds have been characterized by IR, ^{1}H and ^{13}C NMR and mass spectroscopy.

[16] S.-K. Kim, *PhD Thesis*, Johannes Gutenberg-Universität, Mainz, **2000**.

[17] U. Pernisz, N. Auner, Photoluminescence of organically modified cyclosiloxanes, in *Organosilicon Chemistry IV — From Molecules to Materials* (Eds.: N. Auner, J. Weis), VCH, Weinheim, **2000**, p. 505.

Synthesis, Luminescence, and Condensation of Oligo(phenylenevinylene)s with Alkoxysilane End Groups

Heiner Detert, Erli Sugiono*

Institut für Organische Chemie, Johannes Gutenberg-Universität Mainz
Duesbergweg 10 – 14, 55099 Mainz, Germany
Tel.: +49 6131 3922111 — Fax: +49 6131 3925396
E-mail: detert@mail.uni-mainz.de

Keywords: alkoxysilanes, fluorescence, oligo(phenylenevinylene)s

Summary: A synthetic route to highly fluorescent organic semiconductors with rigid connections to curable alkoxysilanes is described. The title compounds are prepared via Heck reactions of monodisperse bromo-OPVs with allyl-alkoxysilanes. A combined extension of the π-system and connection with alkoxysilanes is possible when an alkoxysilyl-styrene is used as a substrate. Hydrolysis and condensation of alkoxysilanes yields linear and cyclic oligo-OPV-siloxanes or three-dimensional networks, thus allowing the transformation of small molecules into luminescent materials with well-defined chromophores.

Introduction

Organic molecules with extended π-systems like poly(phenylenevinylene) (PPV) or polythiophene have become attractive materials for a variety of electrooptical applications [1, 2]. Besides the fully conjugated polymers, oligomers containing only a limited number of repeat units are being intensively studied. These monodisperse molecular species serve as model compounds for a reliable correlation of structure, conjugation length, and properties, and are electronic materials in their own right. Because of their intense fluorescence, stilbenoid oligomers, oligo(phenylenevinylene)s (OPVs) have been used as active layers in light-emitting diodes [3, 4]. A fine-tuning of the solubility, and of the electrical and optical properties of the OPV is possible upon substitution with flexible or bulky side-chain, electron-donating or -accepting groups [5, 6]. A major drawback for the application of these low-molecular-weight materials is their strong tendency toward crystallization. OPVs have been applied as evaporated films [7], in polymeric matrices [8], and as conjugated units in the main chain [9] or as side chains on a flexible polymer [10]. A device-oriented approach uses the immobilization of the functional units to surfaces, often surfaces pretreated with reactive silanes [11, 12], and also siloxane-functionalized chromophores [13, 14].

The synthesis and some properties of monodisperse OPVs connected via rigid spacers to di- and trialkoxysilanes are reported here.

Synthesis via Heck Coupling

The palladium-catalyzed coupling of ethene with bromo- or iodoarenes [15, 16] is a straight forward route for the introduction of terminal vinyl-groups to π-conjugated systems. This procedure has also been applied to the coupling of arenes with vinylsilanes [17, 18]; with more forcing conditions, the vinylalkoxysilane can be used as a synthetic equivalent for ethene [19]. The Heck coupling of donor-substituted iodo-stilbenes with vinylalkoxysilanes proceeds only sluggishly and in the presence of silver salts. Better yields, even with the less reactive bromo-OPVs and in the absence of Ag^+, were obtained in reations with allyltriethoxysilane (**2**). These reactions were studies with a series of substituted distyrylbenzenes (**1**, **12** – **14**). OPVs **1** and **12** – **14** were prepared via consecutive Horner olefinations [20].

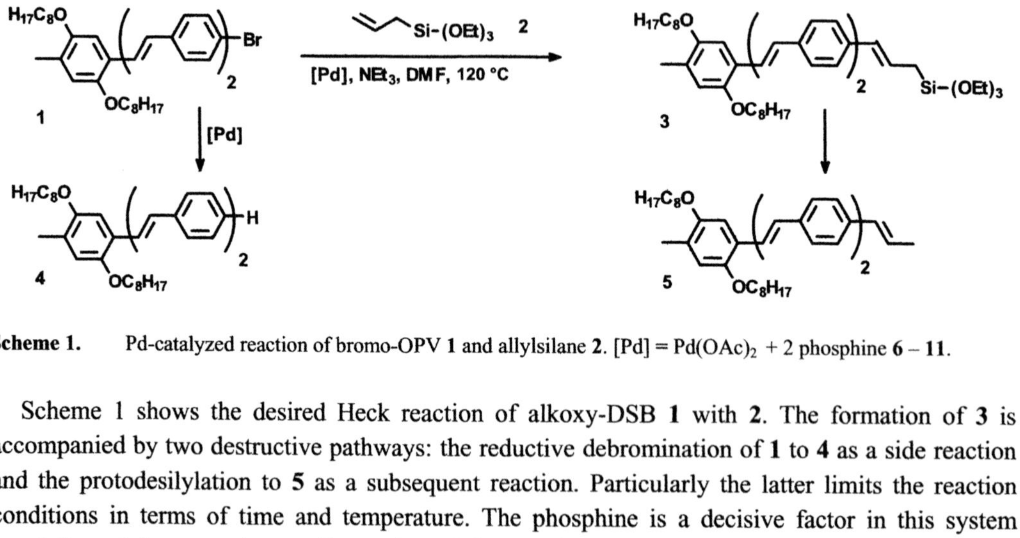

Scheme 1. Pd-catalyzed reaction of bromo-OPV **1** and allylsilane **2**. [Pd] = $Pd(OAc)_2$ + 2 phosphine **6** – **11**.

Scheme 1 shows the desired Heck reaction of alkoxy-DSB **1** with **2**. The formation of **3** is accompanied by two destructive pathways: the reductive debromination of **1** to **4** as a side reaction and the protodesilylation to **5** as a subsequent reaction. Particularly the latter limits the reaction conditions in terms of time and temperature. The phosphine is a decisive factor in this system consisting of three reactions; a fine-tuning of the reaction conditions is possible via electronic and steric effects of the substituents in the phosphine: electron-rich trialkylphosphines **6** and **7** strongly favor the reduction. Fast coupling reactions were observed with tris-*o*-tolylphosphine **8**, the chelating diphosphine dppe **9** being even more efficient in terms of turnover, yield, and suppression of side reactions. Compared with Heck reactions of polycyclic or electron-deficient arenes with **2** [21, 22], the yield of **3** is only moderate. The reactivity of bromo-distyrylbenzenes **1** and **12** – **14** in the coupling reaction is controlled by the substituents on the opposite side of the π-system (Fig. 1, Table 2); a compensation for the electron-donating alkoxy groups by a cyanide (**13**) or exchange of donors with electronically neutral alkyl side chains strongly improves the yields.

Table 1. Influence of phosphine ligands **6** – **12** on the Heck reaction of **1** with **2**.

Ligand	Conversion	3	4	5
Tributylphosphine **6**	36% (3h)	5%	1%	30%
Tricyclohexylphosphine **7**	75% (3h)	1%	22%	50%
Tris-o-tolylphosphine **8**	82% (3h)	59% (51% isolated)	8%	15%
dppe **9**	85% (3h)	68%	11%	7%
dppf **10**	84% (2h)	31%	17%	36%
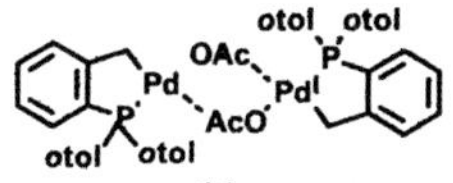 **11**	10% (6h)	5%	2%	4%

[a] Yields given are determined from ^{1}H NMR spectra of the crude reaction mixtures after evaporation of solvent. Comp. **8** isolated yield after chromatography.

Table 2. Influence of substituents on bromo-distyrylbenzene on the Heck reaction with **2**.

	R^1	R^2	R^3	R^4	Si-OPV	Desilylated
1	OC_8H_{17}		CH_3	OC_8H_{17}	59%	15%
12		OC_6H_{13}		OC_6H_{13}	43%	38%
13	OC_8H_{17}		CN	OC_8H_{17}	80%	13%
14	C_6H_{13}			C_6H_{13}	75%	4%

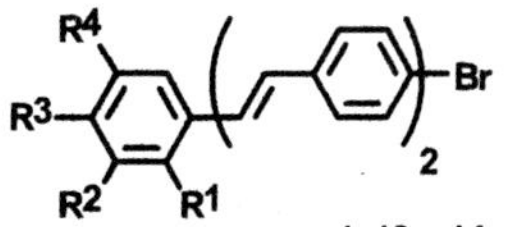

Fig. 1. Bromo-distyrylbenzene.

Styrene is an important olefinic substrate in Heck reactions; a styrene carrying an alkoxysilyl moiety should open up a more convergent approach to the synthesis of the title compounds. Silane **15** was prepared via Barbier-Grignard reaction of *p*-bromostyrene with triethoxymethyl silane [21].

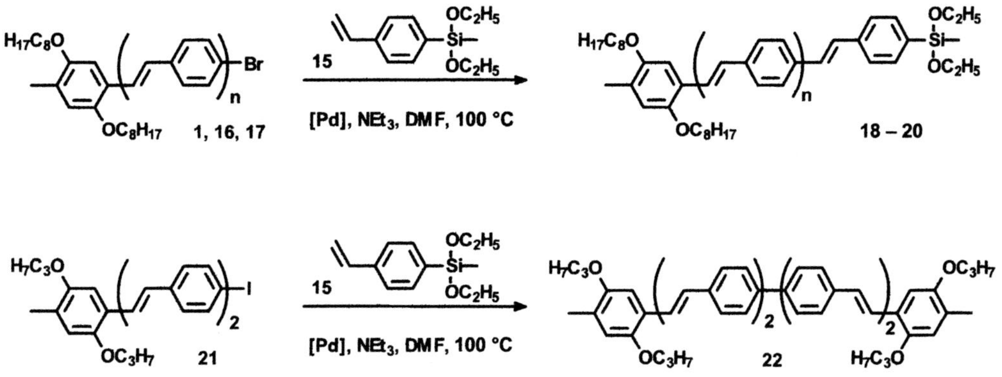

Scheme 2. Heck reactions of bromo- and iodo-OPVs **1**, **16**, **17**, **21**, and alkoxysilyl styrene **15**.

Like the reactions with allylsilane, the Heck reactions with **15** are accompanied by destructive reactions. The reactivity of **15** is sufficient to allow reduced reaction temperatures, most of the starting materials having been consumed within 2 – 3 h. After chromatography, only small amounts (18%) of the silylated distyrylbenzene **18** (n = 1) were isolated, but the yield increased considerably with an elongation of the conjugated system. Yields of 33% of pure ethoxysilyl-OPV **19** with four benzene rings (n = 2) and 39% of the five-ring homolog **20** may be due to the decreasing influence of the donors on the reaction center. An improvement of the reactivity of the halo-OPV by using iodo- instead of bromo-OPVs is a suitable way to get much higher yields of coupling products with styrene. But in the case of **21** and the *p*-ethoxysilyl-styrene **15**, none of the desired silyl-OPVs could be detected. Biphenyl **22**, resulting from a reductive Ullmann-analogous coupling, was the only isolable product (37%); identical results were obtained in the absence of **15**.

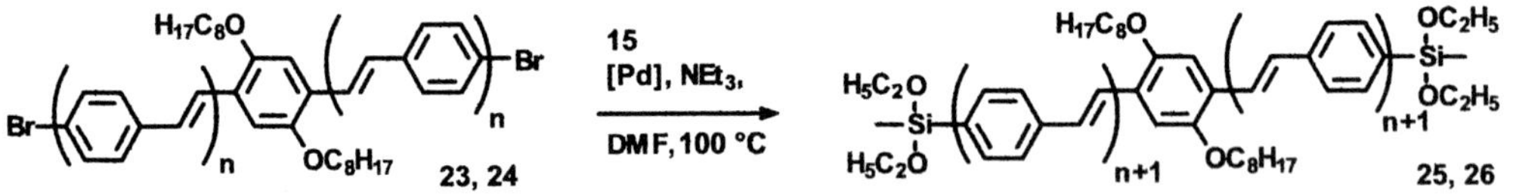

Scheme 3. Twofold Heck reactions of **23**, **24** (n = 0, 1) with **15** forming bis(diethoxysilyl)-OPVs **25**, **26** (n = 1, 2).

The combined extension of the chromophore and introduction of the alkoxysilane using silyl-styrene **15** is also possible with bifunctional bromides; dibromobenzene **23** (n = 0) gave the distyrylbenzene with 2 silane moieties in 16% yield, and the transformation of distyrylbenzene **24** (n = 1) to the 5-ring OPV yielded 33% of **26** (n = 2) after chromatographic separation from 14% of the intermediate with both a bromine and an alkoxysilane.

Hydrolysis of Alkoxysilyl-OPVs

Compounds **3** and **18** – **20** have been prepared to combine luminescent semiconductors with curable units. The transformation of the monomeric compounds to siloxanes of higher molecular weight was performed by hydrolysis of the silicic esters in chloroform/ethanol solution, catalyzed by traces of hydrochloric acid for 2 days (Scheme 4).

Slightly acidic solutions proved to be suitable for the oligomerization; alkaline conditions induced a cleavage of the Si–arene bond. The loss of ethoxy groups and the formation of cyclic oligomers were proven by NMR and Maldi-TOF spectroscopy, e.g., **28** (n = 2) consists of trimers (M = 2138), tetramers (M = 2852) and pentamers (M = 3565) with the relative intensities of 100/30/6. Monomers and oligomers are freely soluble in common solvents. Tin(II)-2-ethyl-hexanoate (ca. 5 mol%) was added to a dioxane solution of **28** (n = 1, m = 1 – 3) and stirred at 95 °C for 1 day. A ring-opening polymerization converted cyclosiloxanes **28** to polymer **30** with rigid π-systems as side chains. GPC analysis of **30** showed the complete loss of cyclosiloxanes and the formation of a polymer with $M_w = 2.4 \times 10^4$ and a polydispersity of 1.2 (THF, referenced to polystyrene standard!). In contrast to the diethoxysilanes **18** – **20**, the hydrolysis of triethoxysilyl-OPV **3** as well as bis(diethoxysilyl)-OPV **26** led via soluble early condensation products to insoluble but still luminescent materials **31, 32**.

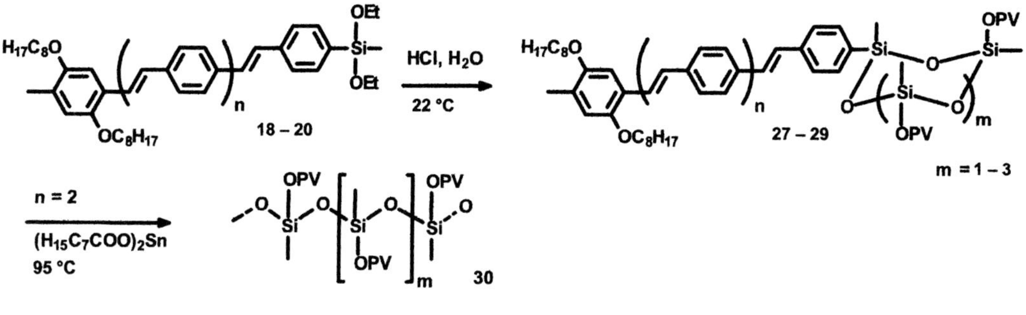

Scheme 4. Hydrolysis of ethoxysilanes and formation of cyclic oligomers **27** – **29**; polymerization of **28**.

Electronic Spectra

The electronic spectra of monomers and oligomers were obtained from solutions in dichloromethane and from films on glass. In dilute solution, the increasing conjugation length of the OPVs is strongly reflected in the bathochromic shifts of the absorption maxima. In the series of the compounds with identical substitution pattern, **3** and **18** – **20**, the absorption maximum is shifted successively to lower energies, as shown by elongation of the π-system of **18** (λ_{max} = 376 nm) with a vinylene segment (λ_{max} = 386 nm), a further phenylene (**19**, λ_{max} = 397 nm), and finally a phenylenevinylene unit (**20**, λ_{max} = 409 nm). An increasing size of the elongating units (2, 6, 8 π-centers) causes nearly equal bathochromic shifts. With the exception of the emission of **3** (λ^F_{max} = 466 nm), the fluorescence spectra are strongly convergent with increasing size of the conjugated systems of **18** – **20**. The first styrene unit **18** → **19** results in a bathochromic shift of 11 nm (**18** λ^F_{max} = 470 nm; **19** λ^F_{max} = 481 nm), the second only 1 nm (**20**, λ^F_{max} = 482 nm), indicating that the convergence limit of the excited state has been reached. The hypsochromic shift of 4 nm **18** → **3**

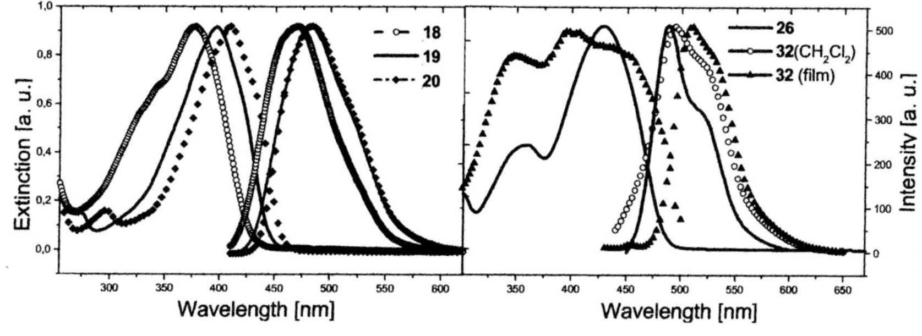

Fig. 2. Spectra of homologous series of alkoxysilyl-OPVs **18** – **20**, of **26**, and of the condensation product **32**.

Table 3. Electronic spectra of alkoxysilyl-OPVs.[a]

OPV	λ_{max}	λ^{F}_{max}	Oligomer	λ_{max}	λ^{F}_{max}
3 (CH_2Cl_2)	386	466	**31** (film)	349	495
18 (CH_2Cl_2)	376	470	**27** (film)	342 (388 sh)	462
19 (CH_2Cl_2)	397	481	**28** (film)	391 (355 sh)	482
20 (CH_2Cl_2)	409	482	**29** (film)	392	507
26 (CH_2Cl_2)	429	488	**32** (film)	394 (433 sh)	510

[a] λ [nm]; solution: (CH_2Cl_2) 10^{-5} mol/L excitation, 10^{-7} mol/L fluorescence, film: cast on glass (**31**, **32** annealed at 110 °C), fluorescence and fluorescence excitation spectra

mostly results from a different shape of the spectra, but the alkoxysilane is an auxochromic unit too [23]. This is best visible with **26**; the spectra of the identical chromophore but with CH_3 instead of silanes [6] are shifted about 5 nm to higher energies (λ_{max} = 424 nm λ^{F}_{max} = 483 nm). The fluorescence of all monomers is efficient in solution (Φ ca. 0.4 – 0.6) and only slightly affected upon oligomerization. Hypsochromic or bathochromic shifts in the solid state result from intermolecular interactions.

Acknowledgments: Generous financial support by the Deutsche Forschungsgemeinschaft is gratefully acknowledged.

References

[1] G. Hadziioannou, *Semiconducting Polymers,* Wiley-VCH, Weinheim **1999**.

[2] K. Müllen, G. Wegner, *Electronic Materials: The Oligomer Approach,* Wiley-VCH, Weinheim, **1998.**

[3] V. Gebhardt, A. Bacher, M. Thelakkat, U. Stalmach, H. Meier, H.-W. Schmidt, D. Haarer, *Adv. Mater.* **1999**, *11*, 119.

[4] F. Meghdadi, G. Leising, W. Fischer, F. Stelzer, *Synth. Met.* **1996**, *76*, 113.

[5] J. L. Segura, N. Martin, M. Hanack, *Eur. J. Org. Chem.* **1999**, 643.

[6] H. Detert, E. Sugiono, *J. Phys. Org. Chem.* **2000**, *13*, 587; *Eur. J. Org. Chem.* **2001**, 2927.

[7] T. P. Nguyen, P. LeRendu, P. Molinié, V. H. Tran, *Synth. Met.* **1997**, *85*, 1357.

[8] U. Stalmach, H. Detert, H. Meier, V. Gebhardt, D. Haarer, A. Bacher, H.-W. Schmidt, *Opt. Mater.* **1998**, *7*, 77.

[9] Z. Yang, I. Sokolik, F. E. Karasz, *Macromolecules* **1993**, *26*, 1188.

[10] P. Hesemann, H. Vestweber, J. Pommerehne, A. Greiner, *Adv. Mater.* **1995**, *7*, 388.

[11] N. Strashnikova, V. Papper, P. Parkhomyuk, G. I. Likhtenshtein, V. Ratner, R. Marks, *J. Photochem Photobiol. A* **1999**, *122*, 133.

[12] S.-C. Ng, W.-L. Yu, A. C. H. Huan, *Adv. Mater.* **1997**, *9*, 887.

[13] D. J. Brondani, R. J. P. Corriu, S. E. Ayoubi, J. J. E. Moreau, M. Wong Chi Man, *J. Organomet. Chem.* **1993**, *451*, C1.
[14] P. Zhu, M. E. van der Boom, G. Evmenenko, P. Dutta, T. J. Marks, *Polym. Prepr.* **2001**, *42(2)*, 579.
[15] H. Detert, E. Sugiono, *J. Prakt. Chem.* **1999**, *341*, 358.
[16] E. Sugiono, Th. Metzroth, H. Detert, *Adv. Synth. Catal.* **2001**, *343*, 351.
[17] K. Karabelas, A. Hallberg, *J. Org. Chem.* **1986**, *51*, 5286.
[18] A. Brethon, P. Hesemann, L. Réjaud, J. J. E. Moreau, M. Wong Chi Man, *J. Organomet. Chem.* **2001**, *627*, 239.
[19] S. Sengupta, S. K. Sadhukhan, *J. Chem. Soc. Perkin* **1999**, *1*, 2235.
[20] E. Sugiono, *PhD-Thesis*, Johannes Gutenberg-Universität Mainz, **2002**.
[21] E. Sugiono, H. Detert, *Synthesis*, **2001**, 893.
[22] C. Carbonneau, R. Frantz, J.-O. Durand, G. F. Lanneau, R. J. P. Corriu, *Tetrahedron Lett.* **1999**, *40*, 5855.
[23] U. Pernisz, N. Auner, Photoluminescence of organically modified cyclosiloxanes, in *Organosilicon Chemistry IV — From Molecules to Materials* (Eds.: N. Auner, J. Weis), VCH, Weinheim, **2000**, p. 505.

How to Make Disilandiyl-Carbon Hybrid Materials: the First ADMET Metathesis Reactions of Organodisilanes

Gabriela Mera, Matthias Driess*

Laboratory of Inorganic Chemistry I, Ruhr-Universität Bochum
Universitätsstrasse 150, 44801, Bochum, Germany
Tel.: +49 234 322 4185 — Fax: +49 234 14378
E-mail:gabriela.mera@ruhr-uni-bochum.de

Keywords: unsaturated disilanes, metathesis reactions, inorganic-organic hybrides

Summary: We report here the first example of an ADMET reaction of unsaturated organodisilanes in the presence of the ruthenium Grubbs catalyst. The structure of the polymer was proven by NMR spectroscopy (^{1}H, ^{13}C, ^{29}Si). Results are also supported by UV-VIS and MALDI-mass spectrometry.

Introduction

Olefin metathesis is a unique carbon skeleton redistribution in which unsaturated carbon-carbon bonds are rearranged in the presence of metal carbene complexes. With the advent of efficient catalysts, this reaction has emerged as a powerful tool for the formation of C-C bonds in chemistry [1]. Olefin metathesis can be utilized in five types of reactions: ring-closing metathesis (RCM), ring-opening metathesis (ROM), respective ring-opening metathesis polymerization (ROMP), cross-metathesis (CM), and acyclic diene metathesis polymerization (ADMET).

The unsaturated organosilicon compounds (see Fig. 1), inorganic materials that are of growing interest and importance, combine electronic and mechanical properties of σ-delocalized polysilanes with those of π-conjugated polyolefins, giving hybrid properties for polymers such as photoluminescence, semiconducting behavior, enhancement of electrical conduction with doping, photoconductivity, non-linear optical properties, and thermochromism. Additionally, their thermal degradation should lead to ceramics (silicon carbide).

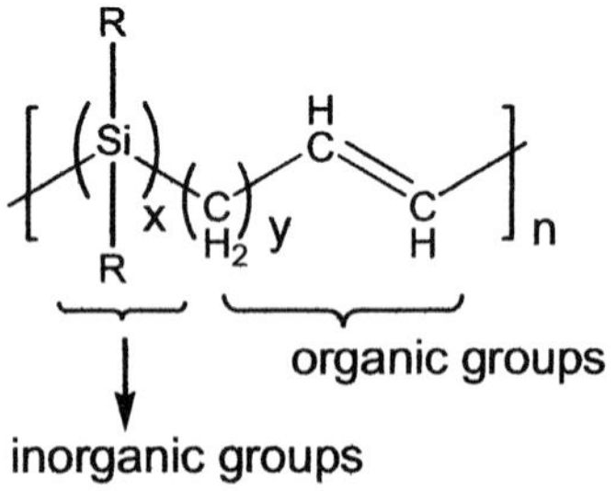

R
Si
R
x
C
H2
y
H
C
C
H
n
organic groups
inorganic groups

Fig. 1. Organosilicon unsaturated hybrid materials.

The examination of linear unsaturated poly(carbodisilanes) has for the most part been neglected because of synthesis difficulties. Until now, the synthesis of these materials has been confined to coupling reactions, thermal cyclopolymerization, and a variety of ring-opening polymerizations including anionic, thermolytic, and catalytic coordination techniques, each with obvious limitations. ADMET polymerization (Fig. 2) avoids many of these limitations and may stimulate the expansion of this class of polymers. This type of polymerization has proven to be a viable synthetic route to unsaturated polymers and copolymers with high molecular weight, including polymers possessing various functionalities [3].

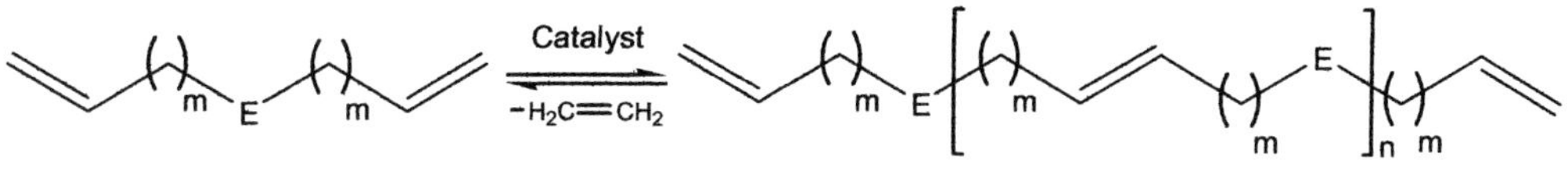

Fig. 2. General representation of ADMET polymerization.

ADMET chemistry is catalyst dependent. Catalyst choice as well as position of the functional group within the monomer makes a significant impact, with respect to both rate of polymerization and molecular weights achieved. Many metal carbenes (Mo, W, and Ru) (Fig. 3) catalyze the metathetical conversion of silicon-containing olefins, but alkenylsilanes, like other unsaturated organosilicon compounds, do not easily participate in olefin metathesis. However, the applications of alkenylsilanes as molecular weight-controlling agents in the polymerization of cyclic olefins and in the modification of existing polymers indicate their metathetic activity.

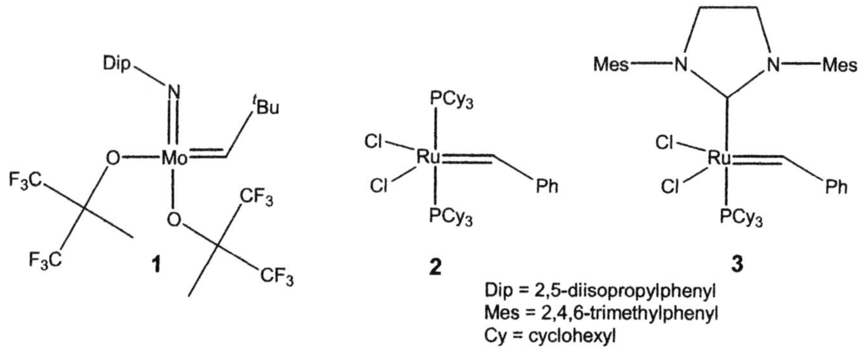

Dip = 2,5-diisopropylphenyl
Mes = 2,4,6-trimethylphenyl
Cy = cyclohexyl

Fig. 3. Three well-defined metathesis catalysts: Schrock's molybdenum alkylidene (**1**) and Grubbs' "first generation" (**2**) and "second generation" (**3**) benzylidene catalysts.

To date there are only a few examples of polymers having a disilandiyl-carbon backbone, but they could not be synthesized by an olefin-metathesis process with Grubbs catalyst [4 – 6]. The aim of our work is to investigate the catalytic activity of the ruthenium-carbene complex $RuCl_2(PCy_3)_2(=CHPh)$ (Grubbs catalyst) in acyclic diene metathesis reactions of different unsaturated organodisilanes.

Synthesis of Novel Unsaturated Monomers

The synthesis of the novel monomers was accomplished using the successful Grignard chemistry for the functionalization of carbosilanes. We present here three types of unsaturated organodisilanes which were prepared by alkenylation of $Me_4Si_2Cl_2$. The alkenyldisilanes **4**, **5**, and **6** are isomers with different positions of the C=C bond (Scheme 1) [7].

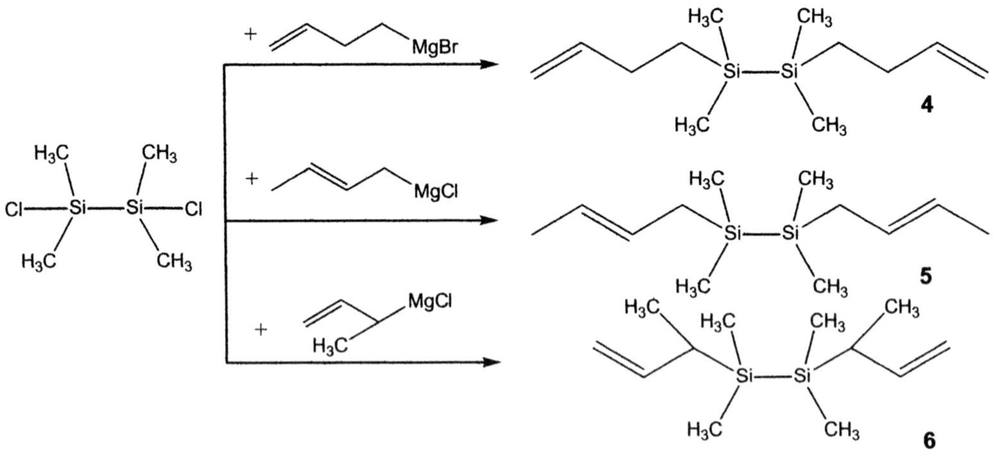

Scheme 1. Synthesis of the novel unsaturated organodisilanes **4** – **6**.

The products **4** – **6** were characterized by means of ^1H -, ^{13}C - (CPD and DEPT 135), ^{29}Si INEPT NMR spectroscopic techniques, UV-VIS spectroscopy, and EI-MS spectrometry (Table 1).

Table 1. Selected spectroscopic data of the unsaturated disilanes (**4** – **6**): ^1H, ^{13}C, ^{29}Si NMR (measured in C_6D_6), UV-VIS (measured in THF solutions) and EI-MS.

	^1H (δ)	^{13}C (δ)	^{29}Si (δ)	UV (λ_{max}) [nm]	EI-MS (M) [%]
4	0.66 (s, 12H, –Si–CH₃), 1.29 (m, 4H, –Si–CH₂–), 2.63 (m, 4H, –CH₂–), 5.58 (m, 4H, =CH₂), 6.40 (m, 2H, =CH–).	–2.96 (Si–CH₃), 15.25 (Si–CH₂–), 29.61 (–CH₂–), 113.79 (=CH₂), 142.00 (=CH–).	–16.79	210	226(5)
5	0.06 (s, 12H, –Si–CH₃), 2.33 (m, 4H, Si–CH₂–), 1.83 (m, 6H, –CH₃), 5.56 (m, 4H, =CH–).	–6.77 (Si–CH₃), 11.80 (Si–CH₂–), 16.00 (–CH₃), 125.50 (=CH–).	–18.60	234	226(10)
6	0.86 (s, 12H, –Si–CH₃), 1.26 (m, 6H, –CH₃), 2.23 (m, 2H, Si–CH–), 5.02 (m, 4H, =CH₂), 6.10 (m, 2H, =CH–).	–6.87 (Si–CH₃), 16.34 (Si–CH–), 19.00 (–CH₃), 114.50 (=CH₂), 137.88 (=CH–).	–17.79	213	226(2)

Catalyst Choice

The number of catalyst systems that initiate ADMET metathesis of unsaturated organosilanes is still very limited. Examples of ADMET polymerization of unsaturated monosilanes exist in the literature, and the metathesis occurs in the presence of multicomponent catalyst systems like Re_2O_7/Al_2O_3 + SnR_4 (or PbR_4) (Finkel'shtein et al. [8 – 11], molybdenum and tungsten alkylidene $[(CF_3)_2MeCO]_2(ArN)M{=}CH(t\text{-}Bu)$, M = Mo, W [12 – 14] and Grubbs' ruthenium catalyst $RuCl_2(=CHPh)(PCy_3)_2$ [15 – 17]. Different unsaturated homopolymers were isolated which are based on carbosilanes and carbosiloxanes.

We have chosen for our metathetical study in the case of unsaturated disilanes "the first generation" of Grubbs catalysts because of the reduced degree of possible isomerization of the products during the time of polymerization [7].

ADMET Polymerization and Characterization of the Polymers

Acyclic diene metathesis (ADMET) polymerization is an equilibrium step condensation during which the production and removal of C_2H_4 (when using terminal olefins) drive the reaction progress [18].

Organic-inorganic hybrid polymers are macromolecules that contain linkages between inorganic groups, such as transition or main group metal atoms and organic groups.

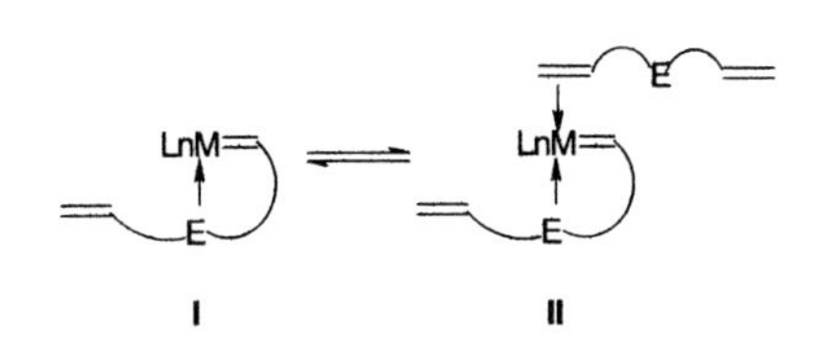

Fig. 4. The "NEGATIVE NEIGHBORING GROUP EFFECT".

The synthesis of the new hybrid systems is realized with the help of a new generation of high-performance, reasonably stable, and, most importantly, exceedingly tolerant catalysts or catalyst precursors such as the ruthenium "first generation" carbene complex $Cl_2(PCy_3)_2Ru{=}CHPh$ introduced by Grubbs and co-workers [2, 7].

The optimum conditions for ADMET chemistry occur by positioning the functional group (Si atoms) at least two methylene units distant from the metathesing olefin in the monomer. If the functional group is in closer proximity, an intramolecular complex can be formed between the donor electrons from the functional group and the electron-deficient metal center ("NEGATIVE NEIGHBORING GROUP EFFECT") [19] (Fig. 4). Terminal dienes are the preferred monomers in the ADMET reactions, for both entropic and steric reasons. Therefore we have chosen the monomer **4** as starting material for ADMET polymerization.

We report here the first ADMET polymerization in the presence of Grubbs catalyst, starting from the new unsaturated disilane **4** (Scheme 2).

Most often, ADMET reactions are conducted in the bulk state (where the catalyst is dissolved in neat monomer) and under reduced pressure (high vacuum) to drive the equilibrium toward polymer formation by maximizing the monomer concentration and removing the ethene by-product.

ADMET polymerizations occur, surprisingly, with identical products using different reaction conditions: (a) without solvent and under reduced pressure; (b) in CH_2Cl_2 or C_6H_6 solutions, Ar atmosphere, room temperature, or elevated temperatures; (c) and even in CH_2Cl_2 or C_6H_6 solutions at room temperature in air [7].

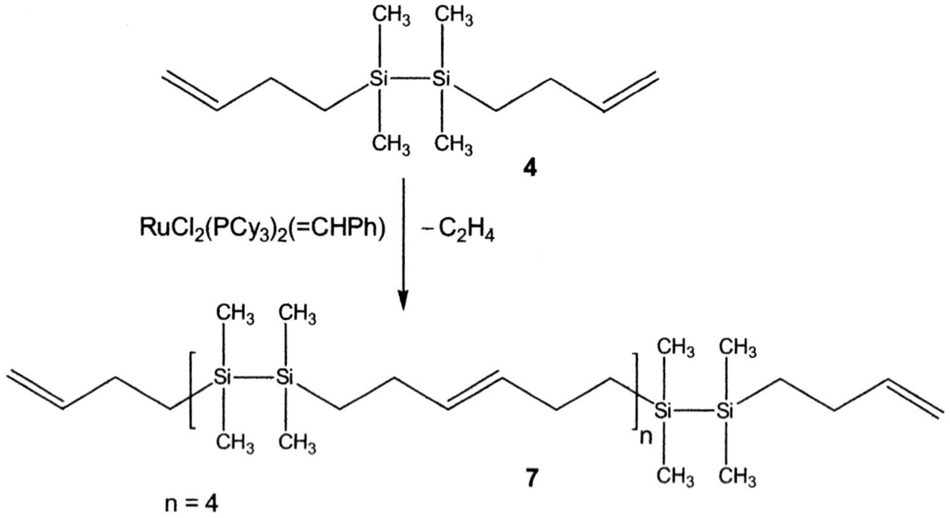

CH_3 CH_3 Si—Si CH_3 CH_3

4

$RuCl_2(PCy_3)_2(=CHPh)$ $-C_2H_4$

CH_3 CH_3 Si—Si CH_3 CH_3 n CH_3 CH_3 Si—Si CH_3 CH_3

7

n = 4

Scheme 2. ADMET polymerization catalysed by Grubbs ruthenium carbene complex.

The oligomeric product **7** was characterized by means of ^{1}H, ^{13}C, ^{29}Si NMR, UV-VIS, (Table 2) TGA-DTA and MALDI-MS (Fig. 6). Figure 5 compares the ^{1}H NMR spectra of the monomer **4** with that of the formed polymeric material. The ^{1}H spectra illustrate the formation of internal unsaturation and the decreased intensities (integral) for the vinyl end-group signals.

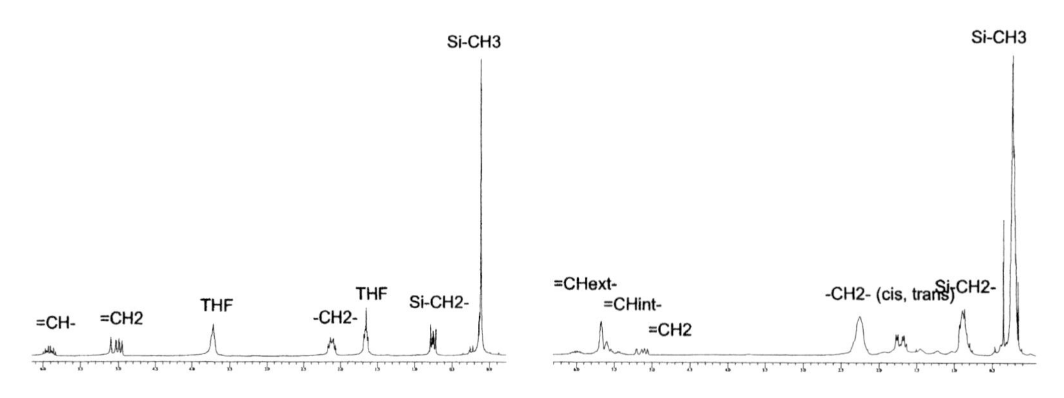

Fig. 5. ^{1}H NMR spectra of of the monomer **4** (left) and polymer **7** (right).

Table 2. Selected spectroscopic data of the polymer **7**: ^{1}H, ^{13}C, ^{29}Si NMR (in C_6D_6) and UV-VIS (solutions in THF – λ_{max} [nm]).

	^{1}H (δ, C_6D_6)	^{13}C (δ)	^{29}Si(δ)	UV [nm]
Polymer	0.08 (s, –Si–CH_3), 0.69 (m, –Si–CH_2–), 2.04 (m, –CH_2–), 4.94 (m, =CH_2), 5.33 (m, =CH-int), 5.87 (m, =CH-ext)	–3.47 (Si–CH_3), 16.01 (Si–CH_2–), 29.05 (–CH_2–), 113.10 (=CH_2), 132.07 (=CH-int), 142.20 (=CH-ext)	–17.33 (br)	253

The presence of the terminal C=C bonds reveals that the degree of oligomerization is relatively low. Indeed, matrix-assisted laser desorbtion/ionization (MALDI) time-of-flight mass spectrometry (TOF/MS) [20] reveals that the polymer mixture contains mostly the pentamer (Fig. 6). The low molecular weight of the polymer is probably because of the decomposition of the catalyst during the process, "negative neighboring group effects", and steric interactions between the methyl groups on the silicon atom adjacent to the C–C double bond and the transition metal catalyst.

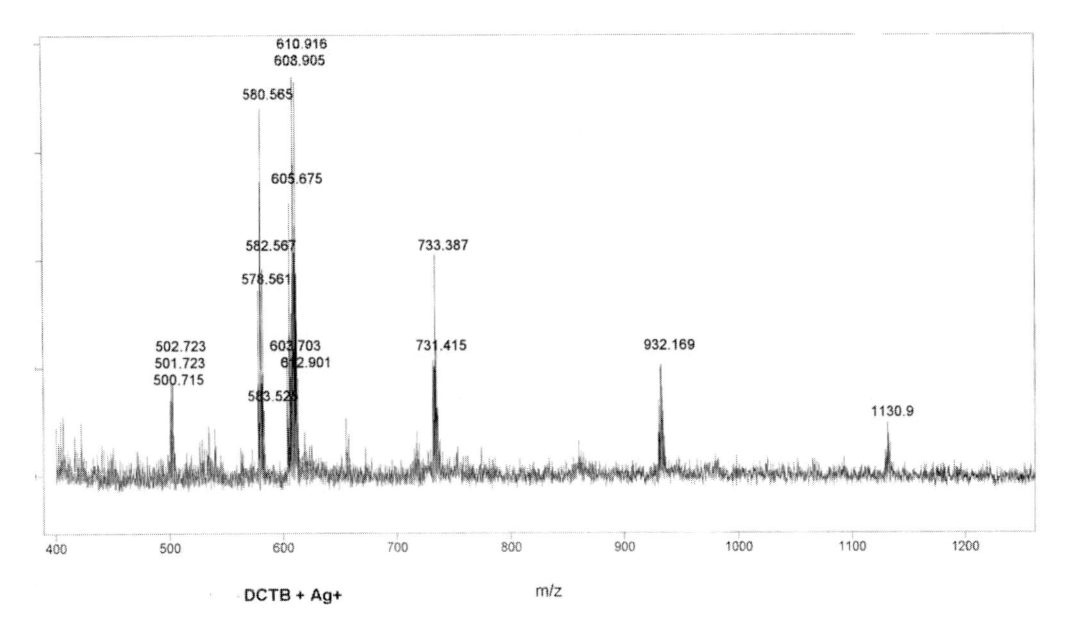

Fig. 6. MALDI mass spectrum of the polymer **7**.

The (disilanylene)-butenylene oligomer shows characteristic strong UV absorption bands (253 – 260 nm), which are significantly red-shifted relative to the absorbtion of 1,1,2,2-tetramethyl di(3-butenyldisilane) (210 nm), because of the higher delocalization of the π electrons through the disilanylene units.

Conclusions

The first unsaturated polycarbodisilane oligomers (mostly pentameric) were synthesized in the presence of Grubbs catalyst starting from the 1,2-dibutenyldisilane (**4**) in an ADMET metathesis reaction. Further work is in progress in order to make higher polymers.

References

[1] a) K. J. Ivin, J. C. Mol, *Olefin Metathesis and Metathesis Polymerization*, Academic Press **1997**; b) A. Fürstner, *Alkene Metathesis in Organic Synthesis*, Springer Verlag, Berlin **1998**.
[2] P. Schwab, R. H. Grubbs, J. W. Ziller, *J. Am. Chem. Soc.* **1996**, *118*, 100.
[3] J. E. Schwendeman, C. A. Church, K. B. Wagener, *Adv. Synth. Catal.* **2002**, *344*, 597.
[4] J. Ohshita, D. Kanaya, M. Ishikawa, *J. Organomet. Chem.* **1989**, *369*, C18.
[5] J. Ohshita, D. Kanaya, T. Watanabe, M. Ishikawa, *J. Organomet. Chem.* **1995**, *489*, 165.
[6] L. Zhang, T. R. Lee, *Polymer Preprints* **1998**, *39*(*1*), 170.
[7] G. Mera, Ruhr-Universität-Bochum, *planned Dissertation*.
[8] E. S. Finkel'shtein, E. B. Portnykh, N. V. Ushakov, V. M. Vdovin, *J. Mol. Catal.* **1992**, *76*, 133.
[9] E. S. Finkel'shtein, B. Marciniec, in *Progress in Organometallic Chemistry* (Eds.: B. Marciniec, J. Chjnowski), Gordon and Breach, London **1995**.
[10] T. Kawai, K. Shiga, T. Suzuki, T. Iyoda, *J. Mol. Catal.* **1999**, *140*, 287.
[11] T. Kawai, K. Shiga, T. Iyoda, *J. Mol. Catal.* **2000**, *160*, 173.
[12] K. B. Wagener, D. W. Smith Jr, *Macromolecules* **1991**, *24*, 6073.
[13] D. W. Smith Jr, K. B. Wagener, *Macromolecules*, **1993**, *26*, 3533.
[14] K. Brzezinska, R. Schiller, K. B. Wagener, *J. Polym. Sci.* **2000**, *38*, 1544.
[15] S. K. Cummings, D. W. Smith Jr, K. B. Wagener, *Macromol. Rapid Commun.* **1995**, *16*, 347.
[16] K. Brzezinska, K. B. Wagener, G. T. Burns, *J. Polym. Sci.* **1999**, *37*, 849.
[17] C. A. Church, J. H. Pawlow, K. B. Wagener, *Polym. Prep.* (Am. Chem. Soc. Div. Polym. Chem.) **2001**, *42* (*1*), 235.
[18] J. E. Schwendeman, C. A. Church, K. B. Wagener, *Adv. Synth. Catal.* **2002**, *344*, 597.
[19] K. B. Wagener, K. Brzezinska, J. D. Anderson, T. R. Younkin, K. Steppe, W. DeBoer, *Macromolecules* **1997**, *30*, 7363.
[20] a) M. W. F. Nielen, *Mass Spectrom. Rev.* **1999**, *18*, 309; b) J. B. Fenn, Nobel Vortrag, *Angew. Chem.* **2003**, *33*; c) K. Tanaka, Nobel Vortrag, *Angew. Chem.* **2003**, *33*.

Application of α,ω-Bis(dimethylvinylsiloxy)alkanes

Piotr Pawluć, Beata Gaczewska, Bogdan Marciniec*

Department of Organometallic Chemistry, Faculty of Chemistry
Adam Mickiewicz University, Grunwaldzka 6, 60-780 Poznan, Poland
Tel:. +48 61 8291366 — Fax: +48 61 8291508
E-mail: marcinb@amu.edu.pl

Keywords: silylative coupling, divinyl silyl ethers, 1,1-bis(silyl)ethenes, siloxylene-alkylene-vinylene oligomers, ruthenium hydride catalyst

Summary: α,ω-Bis(dimethylvinylsiloxy)alkanes of the general formula $CH_2{=}CH(CH_3)_2SiO(CH_2)_nOSi(CH_3)_2CH{=}CH_2$ (where $n = 2 - 4$) in the presence of ruthenium hydride complex $[RuHCl(CO)(PPh)_3]$ undergo effective silylative coupling reaction, yielding under optimum conditions linear oligomers and/or silacyclic products containing an *exo*-methylene bond between silicon atoms. Depending on the number of methylene groups in a molecule of a substrate, either cyclic (when $n = 2$) or oligomeric products (when $n = 4$) can be selectively synthesized with moderate to high yields. When $n = 3$ both cyclic and oligomeric siloxylene-alkylene-vinylenes are formed.

Acyclic diene metathesis (ADMET) polymerization of dialkenyl-substituted organosilicon derivatives occurring in the presence of molybdenum, tungsten, and ruthenium alkylidenes has been commonly used for the synthesis of a wide range of linear unsaturated organosilicon polymers, e.g., polycarbosilanes, polycarbosiloxanes, and related polymers [1 – 8].

On the other hand, organosilicon dienes (except divinyl derivatives) undergo efficient ring-closing metathesis (RCM) under optimum conditions (mostly in the presence of molybdenum and ruthenium alkylidenes) yielding unsaturated silacyclic compounds containing *endo*-cyclic bonds, which have been shown to be interesting intermediates in organic synthesis [1 – 3, 9 – 15].

In contrast to other alkenyl-substituted organosilicon compounds, a metathetical conversion of divinylsilyl derivatives in the ADMET polymerization as well as the RCM process does not occur in the presence of well-defined W, Mo, and Ru carbene complexes because of stereoelectronic effects of the silyl group stimulating non-productive cleavage of the disilylmetallacyclobutane intermediate [16].

On the other hand, divinyl-substituted organosilicon compounds in the presence of ruthenium and rhodium complexes containing or generating M–H and M–Si (M = Ru, Rh) bonds undergo competitive silylative coupling cyclization and polycondensation to give a mixture of oligomers and unsaturated silacyclic compounds [17 – 23], according to Scheme 1.

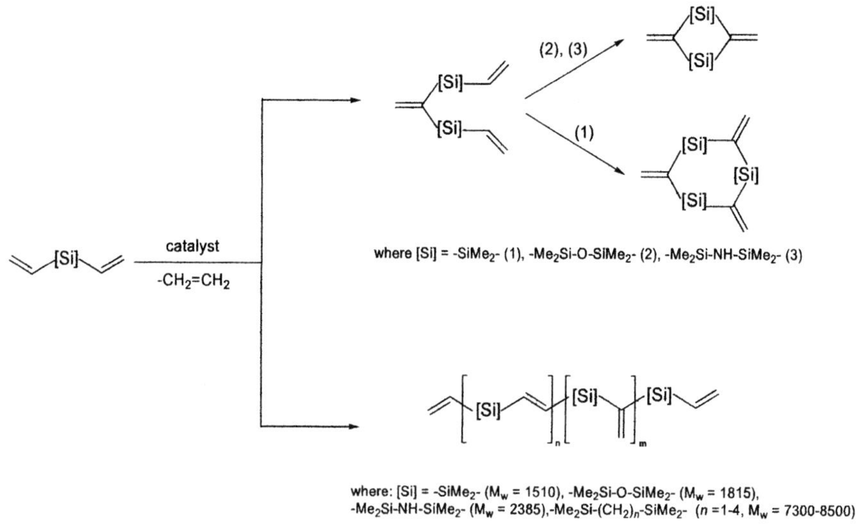

catalyst

$-CH_2=CH_2$

(2), (3)

(1)

where [Si] = $-SiMe_2-$ (1), $-Me_2Si-O-SiMe_2-$ (2), $-Me_2Si-NH-SiMe_2-$ (3)

where: [Si] = $-SiMe_2-$ (M_w = 1510), $-Me_2Si-O-SiMe_2-$ (M_w = 1815), $-Me_2Si-NH-SiMe_2-$ (M_w = 2385), $-Me_2Si-(CH_2)_n-SiMe_2-$ (n =1-4, M_w = 7300-8500)

Scheme 1. Silylative coupling polycondensation.

Although the silylative coupling cyclization is accompanied by linear oligomeric products and the process required a longer reaction time (3 – 4 weeks), this particular method opens up a new route to synthesize silicon-containing *exo*-methylenes, which cannot be prepared via ring-closing diene metathesis.

The silylative coupling reaction of monovinyl-substituted organosilicon compounds proceeds via cleavage of the =C–Si bond of the vinyl-substituted silicon compound and the activation of the =C–H bond of the second vinylsilane molecule. The evidence for this kind of mechanism has been reported previously [24 – 26] and can be generalized for dimerization of divinyl-substituted organosilicon derivatives leading in the following step to competitive oligomerization and silylative coupling ring closure [27].

Although the cyclic and oligomeric products obtained by competitive silylative coupling cyclization and polycondensation represent a new and original class of unsaturated organosilicon derivatives, they usually have rigid structures and cannot be functionalized (except the *exo*-methylene fragment) to produce organosilicon compounds, being important intermediates in organic synthesis.

Therefore, the aim of this work was to synthesize new silacyclic and oligomeric organosilicon compounds containing easily modifiable Si–O–R bonds, via competitive silylative coupling cyclization and polycondensation of divinyl-substituted silyl ethers in the presence of ruthenium hydride complex.

The starting α,ω-bis(dimethylvinylsiloxy)alkanes were easily prepared by the reaction of commercially available chlorodimethylvinylsilane and the respective diol in the presence of triethylamine in moderate to high yield (56 – 90%) as outlined in Scheme 2.

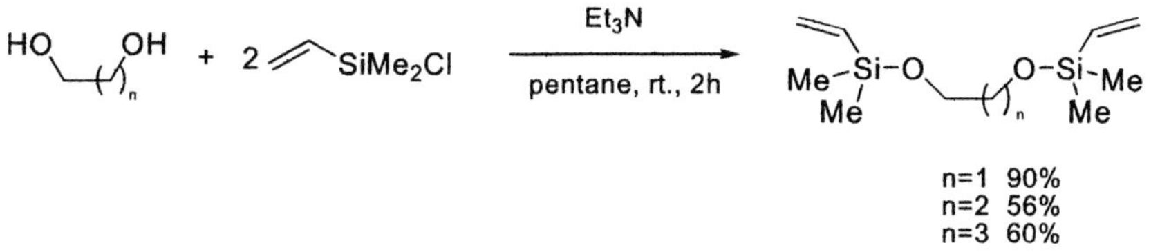

n=1 90%
n=2 56%
n=3 60%

Scheme 2. Synthesis of α,ω-bis(dimethylvinylsiloxy)alkanes.

As has been reported earlier, the silylative coupling reaction of various divinyl-substituted organosilicon monomers catalyzed by ruthenium hydride complex [RuHCl(CO)(PPh_3)$_3$] yielded unsaturated oligomers containing 1,2-bis(silyl)ethene and 1,1-bis(silyl)ethene fragments as well as, under mild conditions, cyclocarbosilanes containing an *exo*-methylene bond between silicon atoms. In order to find the optimum conditions for effective transformation of divinyl silyl ethers in the presence of the above-mentioned ruthenium catalyst, in this work catalytic screenings have been performed to measure the substrate conversion, the distribution, and the yield of cyclic and oligomeric products using GC and GCMS methods.

The high conversion of monomers and the distribution of products in the time of reaction catalyzed by [RuHCl(CO)(PPh_3)$_3$], performed under mild conditions (toluene, 110 °C or 80 °C), show that α,ω-bis(dimethylvinylsiloxy)alkanes undergo silylative coupling cyclization/polycondensation according to Scheme 3.

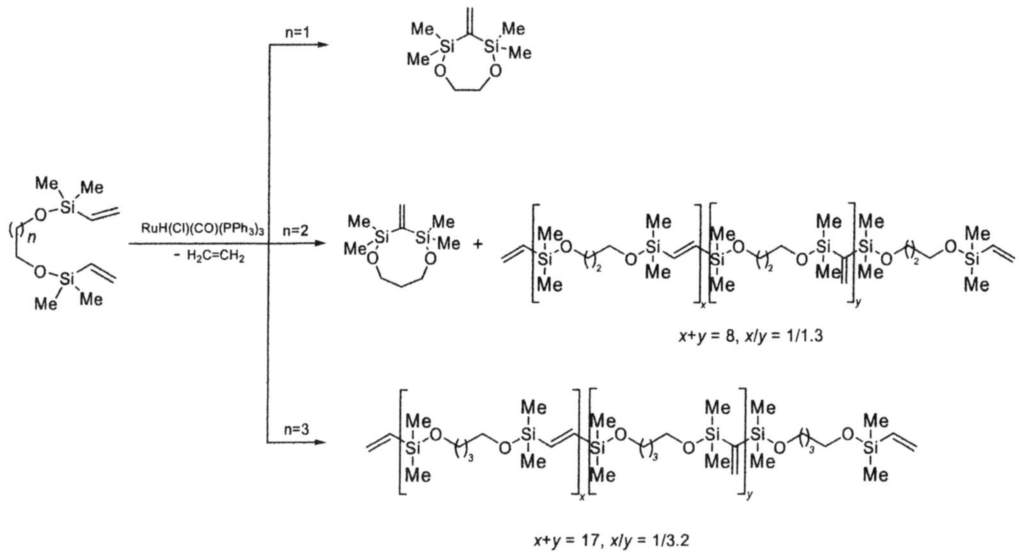

$x+y = 8$, $x/y = 1/1.3$

$x+y = 17$, $x/y = 1/3.2$

Scheme 3. Silylative coupling reaction of α,ω-bis(dimethylvinylsiloxy)alkanes.

The silylative coupling reaction of 1,2-bis(dimethylvinylsiloxy)ethane was effectively catalyzed by 1 mol% of ruthenium hydride catalyst, and the divinyl compound was completely consumed within 1 h at 80 °C. The reaction successfully proceeded without the solvent under air, but toluene could also be employed without affecting either the activity of the catalyst or the selectivity of this process. Application of this catalytic system for silylative coupling cyclization of 1,2-bis(dimethylvinylsiloxy)ethane gave exclusively a cyclic product (isolated yield 85%) with the *exo*-methylene bond between two silicon atoms in the molecule (2,2,4,4-tetramethyl-3-methylene-1,5-dioxa-2,4-disilacycloheptane) accompanied only by trace amounts of oligomers.

In our preliminary report we have shown that the resulting 2,2,4,4-tetramethyl-3-methylene-1,5-dioxa-2,4-disilacycloheptane is a good starting material for synthesizing new organosilicon compounds, because the siloxy functionality can easily undergo transformation to a range of important synthetic intermediates including alkyl-, aryl- and alkenyl-substituted 1,1-bis(silyl)-ethenes and cyclocarbosiloxane [28].

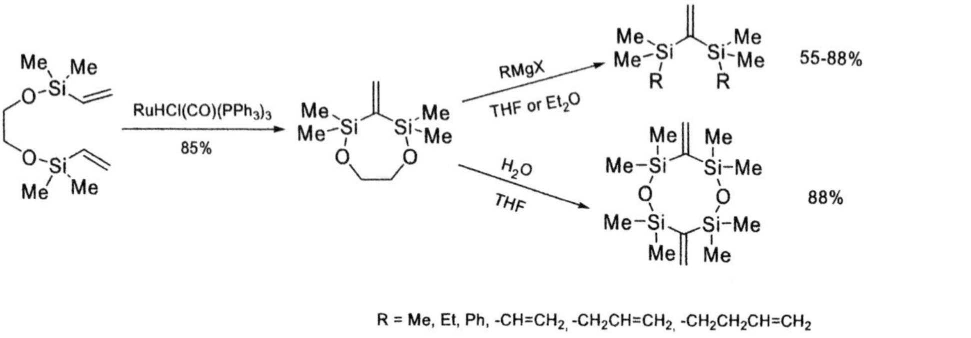

Scheme 4. Application of cyclic silyl ether in organosilicon synthesis.

Reaction of 1,3-bis(dimethylvinylsiloxy)propane with [$RuHCl(CO)(PPh_3)_3$] (1 mol%) in toluene at 110 °C (24 h) gives a mixture of a cyclic product and linear siloxylene-alkylene-vinylene oligomer. The cyclic product 2,2,4,4-tetramethyl-3-methylene-1,5-dioxa-2,4-disilacyclooctane was isolated from the reaction mixture with 49% yield. The oligomer obtained from 1,3-bis(dimethylvinylsiloxy)propane (isolated yield 33%) contains chains with *trans*-1,2-bis(silyl)ethene as well as a 1,1-bis(silyl)ethene fragment, according to Scheme 3. Quantitative analysis of the vinylic region of the ^{1}H NMR spectrum and the ratio of the signals assigned to the $>C=CH_2$ and –CH=CH– groups to those assigned to the terminal vinylic hydrogen atoms enabled a calculation of the molecular weight of the resulting oligomer (M = 1975) and the ratio of both building 1,2-bis(silyl)ethene and 1,1-bis(silyl)ethene fragments in the oligomer chain ($x + y = 8$, x/y (*trans/gem*) = 1/1.3)

In contrast to the above-mentioned monomers, 1,4-bis(dimethylvinylsiloxy)butane in the presence of [$RuHCl(CO)(PPh_3)_3$] (1 mol%) under analogous conditions undergoes silylative coupling polycondensation to yield mainly siloxylene-alkylene-vinylene oligomers (isolated yield 60%). Only very small amounts of intramolecular reaction product (2,2,4,4-tetramethyl-3-

methylene-1,5-dioxa-2,4-disilacyclononane) were found (<4% detected by GCMS), and this compound can be easily separated by vacuum evaporation. Similarly to the oligomers obtained from 1,3-bis(dimethylvinylsiloxy)propane, the chain of the resulting oligomer contains *trans*-1,2-bis(silyl)ethene and 1,1-bis(silyl)ethene units (Scheme 3). The molecular weight of the synthesized oligomer, determined by ^{1}H NMR, was M = 4175 ($x + y$ = 17). The ratio of both building 1,2-bis(silyl)ethene and 1,1-bis(silyl)ethene fragments in the oligomer chain was x/y (*trans/gem*) = 1/3.2.

In conclusion, we have shown that divinyl-substituted silyl ethers of the general formula $CH_2{=}CH(CH_3)_2SiO(CH_2)_nOSi(CH_3)_2CH{=}CH_2$ (where $n = 2-4$) in the presence of ruthenium hydride complex undergo efficient silylative coupling polycondensation to yield selectively linear oligomers (if $n = 4$) or cyclic products (if $n = 2$). Application of 1,3-bis(dimethylvinylsiloxy)-propane led to a mixture of both cyclic and oligomeric products. Facile isolation of the silacyclic compounds makes this method an excellent route for the synthesis of cyclic silyl ethers containing an *exo*-methylene bond between silicon atoms, which cannot be synthesized via ring-closing metathesis. The reaction of the resulting cyclic product with Grignard's reagents is a new effective one-step procedure for the synthesis of 1,1-bis(silyl)ethenes, which are versatile intermediates in organic and organosilicon synthesis.

Acknowledgment: Financial support from the State Committee for Scientific Research (Poland), Grant 3 T09A 145 26, is gratefully acknowledged.

References

[1] K. J. Ivin, J. C. Mol, *Olefin Metathesis and Metathesis Polymerization*, Academic Press, New York, **1997**.
[2] B. Marciniec, C. Pietraszuk, in *Handbook of Metathesis* (Ed.: R. H. Grubbs), Wiley-VCH, Weinheim, **2003**, *Vol 2*, Chapter 13.
[3] B. Marciniec, C. Pietraszuk, *Curr. Org. Chem.* **2003**, *7*, 691.
[4] K. B. Wagener, D. W. Smith, *Macromolecules* **1991**, *24*, 6073.
[5] D. W. Smith, K. B. Wagener, *Macromolecules* **1993**, *26*, 1633.
[6] D. W. Smith, K. B. Wagener, *Macromolecules* **1993**, *26*, 3533.
[7] K. R. Brzezinska, K. B. Wagener, G. T. Burns, *J. Polym. Sci. Part A: Polym. Chem.* **1999**, *37*, 849.
[8] K. R. Brzezinska, R. Schitter, K. B. Wagener, *J. Polym. Sci. Part A: Polym. Chem.* **2000**, *38*, 1544.
[9] *Alkene Metathesis in Organic Synthesis* (Ed.: A. Furstner), Springer, Berlin, **1998**.
[10] S. Chang, R. H. Grubbs, *Tetrahedron Lett.* **1997**, *38*, 4757.
[11] C. Meyer, J. Cossy, *Tetrahedron Lett.* **1997**, *38*, 7861.
[12] P. A. Evans, V. S. Murthy, *J. Org. Chem.* **1998**, *63*, 6768.

[13] T. R. Hoye, M. A. Promo, *Tetrahedron Lett.* **1999**, *40*, 1429.
[14] S. E. Denmark, S. M. Yang, *Org. Lett.* **2001**, *3*, 1749.
[15] A. Okada, T. Ohshima, M. Shibasaki, *Tetrahedron Lett.* **2001**, *42*, 8023.
[16] R. R. Schrock, R. T. DePue, J. Feldman, C. J. Schverin, S. C. Dewan, A. H. Liu, *J. Am. Chem. Soc.* **1988**, *110*, 1423.
[17] B. Marciniec, M. Lewandowski, *J. Polym. Sci. Part A: Polymer Chem.* **1996**, *34*, 1443.
[18] B. Marciniec, M. Lewandowski, *Tetrahedron Lett.* **1997**, *38*, 3777.
[19] T. Mise, Y. Takaguchi, T. Umemiya, S. Shimizu, Y. Wakatsuki, *Chem. Commun.* **1998**, 699.
[20] B. Marciniec, E. Malecka, *Macromol. Rapid. Commun.* **1999**, *20*, 475.
[21] B. Marciniec, M. Lewandowski, E. Bijpost, E. Malecka, M. Kubicki, E. Walczuk-Gusciora, *Organometallics* **1999**, *18*, 3968.
[22] B. Marciniec, E. Malecka, M. Majchrzak, Y. Itami, *Macromol. Symp.* **2001**, *174*, 137.
[23] B. Marciniec, E. Malecka, M. Scibiorek, *Macromolecules* **2002**, *36*, 5545.
[24] Y. Wakatsuki, H. Yamazaki, N. Nakano, Y. Yamamoto, *J. Chem. Soc., Chem. Commun.* **1991**, 703.
[25] B. Marciniec, C. Pietraszuk, *J. Chem. Soc., Chem. Commun.* **1995**, 2003.
[26] B. Marciniec, C. Pietraszuk, *Organometallics* **1997**, *16*, 4320.
[27] B. Marciniec, *Mol. Cryst. Liq. Cryst.* **2000**, *353*, 173.
[28] P. Pawluc, B. Marciniec, G. Hreczycho, B. Gaczewska, Y. Itami, *J. Org. Chem.* **2005**, *70*, 370.

Synthesis and NMR Spectra of Diaryl- and Dihetarylsilacycloalkanes

Luba Ignatovich, Juris Popelis, Edmunds Lukevics

Latvian Institute of Organic Synthesis, Aizkraukles 21, Riga LV-1006, Latvia
E-mail: ign@osi.lv

Keywords: silacyclopentane, silacyclohexane, synthesis, ^{29}Si NMR, mass spectrum

Summary: Two series of 2-furyl-, 2-(4,5-dihydrofuryl)-, 2-thienyl-, and phenyl-silacyclopentanes and silacyclohexanes have been synthesized by the reactions of 1,1-dichloro-1-silacyclopentane and 1,1-dichloro-1-silacyclohexane with corresponding hetaryllithium or phenylmagnesium bromide. The influence of the heterocycle and the nature of the silacycloalkane on the chemical shifts in the ^{1}H, ^{13}C and ^{29}Si NMR spectra has been investigated. NMR data indicate a complex character of the mutual influence of heterocyclic substituents on shielding of the nuclei.

Introduction

Aromatic and heteroaromatic derivatives of Group 14 elements provide a convenient model for the investigation of the interaction of silicon-, germanium-, tin-, or lead-containing substituents with the π-electrons of the ring [1]. It has been found that in the case of (2-furyl)silanes, -germanes, -stannanes, and -plumbanes, the chemical shifts of the protons are determined by the nature of the heteroorganic substituent, the number of furan rings in the molecule, and the character of the interaction of the π-electron system with the heteroatom, as well as the geometric parameters of the π-donor and the heteroatom [2]. In (2-furyl)- and (2-thienyl)alkoxysilanes, competing interactions have been observed: interaction between the central silicon atom with the π-electron system of the furan or thiophene ring on the one hand, and with the p-electrons of oxygen on the other hand [3]. The complex character of the mutual influence of heterocyclic substituents on shielding of the nuclei caused by several effects, including inductive and mesomeric interactions, have been found in 2-thienyl(2′-furyl)- and 2-thienyl[2′-(4′,5′-dihydrofuryl)silanes [4].

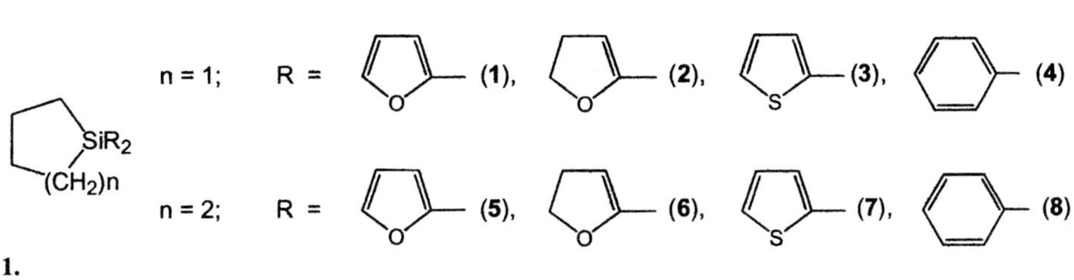

SiR_2 $(CH_2)n$

n = 1; R = (1), (2), (3), (4)

n = 2; R = (5), (6), (7), (8)

O O S O O S

Chart 1.

In order to study the influence of the heteroaromatic ring on the electronic effects in hetarylsilanes as well as the influence of the silicon ring size on the spectroscopic parameters, we have prepared two series of compounds: silacyclopentanes and silacyclohexanes with two aryl or hetaryl substituents at the silicon atom (Chart 1).

Results and Discussion

The complete series of 2-furyl-, 2-(4,5-dihydrofuryl)-, 2-thienyl- and phenyl- silacyclopentanes and silacyclohexanes have been synthesized by the reactions of 1,1-dichloro-1-silacyclopentane and 1,1-dichloro-1-silacyclohexane with corresponding hetaryllithium or phenylmagnesium bromide. Yields, boiling points and mass spectral data of synthesized compounds are presented in Table 1.

Table 1. Analytical and mass spectral data for the compounds **1** – **8**.

Compound	Bp [°C/mmHg]	Yield [%]	*m/z* (rel. abundance) [%]
1	86 – 89/3	51.4	218(M^+,93), 197(14), 186(17), 162(34), 153(27), 147(94), 134(40), 128(20), 118(74), 104(36), 95(78), 89(57), 77(25), 66(25), 53(49), 45(100), 39(44)
2	97 – 99/3	53.0	222(M^+,100), 193(15), 179(13), 166(51), 138(73), 123(64), 110(28), 95(80), 83(51), 67(63), 53(56) ,45(92)
3	123 – 124/4	56.8	250(M^+,100), 222(28), 208(10), 194(100), 182(27), 166(31), 134(71), 111(74), 85(28), 77(12), 51(32), 39(28)
4	145 – 147/4	50.0	238(M^+,50), 210(16), 182(93), 160(39), 132(35), 105(100), 82(28), 53(35), 39(11)
5*	90 – 93/3.5	50.8	232(M^+,55), 176(15), 164(100), 147(15), 136(85), 123(15), 110(24), 103(12), 53(96), 89(14), 77(15), 65(23), 53(28), 45(57)
6	118 – 120/3.5	64.8	236(M^+,100), 208(20), 193(15), 180(23), 166(74), 152(97), 139(63), 124(21), 109(93), 97(82), 81(35), 67(100), 53(59), 45(100)
7	136 – 138/3	49.2	264(M^+,51), 221(17), 208(10), 195(37), 180(100), 152(36), 134(17), 126(10), 111(66), 96(56), 85(18), 65(10), 51(26), 39(25)
8**	155 – 160/3	45.0	252(M^+,27), 209(20), 195(14), 181(52), 174(100), 155(10), 146(19), 132(10), 121(15), 105(85), 96(50), 79(14), 53(18), 41(10)

*Ref. [5], ** Ref. [6]

NMR spectra of solutions of the compounds in $CDCl_3$ were recorded on a Varian Mercury-200 BB instrument with working frequencies (in MHz) 200.04 (^1H), 50.3 (^{13}C), and 39.74 (^{29}Si). The chemical shifts were measured relative to TMS as an internal standard (^1H, ^{13}C) or as an external standard (^{29}Si). Mass spectra were registered on GC-MS HP 6890 (70 eV).

In the ^{1}H spectra of compounds **1** – **3** and **5** – **7** we observed signals of three-spin systems of the AMX type, the chemical shifts and SSCCs of which are characteristic of α-substituted furans or thiophenes, respectively [7] and also multiplets of the AM_2X_2 type, characteristic for protons of the 4,5-dihydrofuryl ring [8]. An unambiguous assignment of signals in the ^{13}C NMR spectra was made on the basis of the ^{13}C-^{1}H SSCCs, measured in spectra without any wideband suppression of protons. The chemical shifts of the ^{1}H, ^{13}C, and ^{29}Si nuclei are listed in Tables 2 and 3.

Table 2. Proton chemical shifts δ [ppm] and coupling constants J [Hz] in R_2Si $(CH_2)_n$.

R	$SiCH_2$	$SiCCH_2$ {$SiCCCH_2$}	$H_3[H_o]$	$H_4[H_m]$	$H_5[H_p]$
			n = 1		
O	1.05 m (6.6)	1.77 m	6.74 dd (3.2; <03)	6.40 dd (3.2; 1.6)	7.69 d (1.6; <0.3)
O	0.84 m (6.6)	1.65m	5.42 t (2.4)	2.61 dt (9.6; 2.4)	4.31 t (9.6)
S	1.14 m (6.8)	1.80 m	7.37 dd (3.4; 0.7)	7.21 dd (4.6; 3.4)	7.66 dd (4.6; 0.7)
	1.10 m	1.78 m	[7.52] m	[7.35] m	[7.33] m
			n = 2		
O	1.10 m (6.8)	1.78 m {1.50 m}	6.74 dd (3.4; 0.9)	6.40 dd (3.4; 1.6)	7.69 dd (1.6; 0.9)
O	0.88 m (6.6)	1.72 m {1.42}	5.37 t (2.6)	2.60 dt (9.8; 2.6)	4.29 t (9.8)
S	1.19 m (6.8)	1.82 m {1.52}	7.33 dd (3.4; 0.8)	7.20 dd (4.6; 3.4)	7.64 dd (4.6; 0.8)
	1.17 m (6.4)	1.75 m {1.50}	[7.52] m	[7.34] m	[7.34] m

The ^{1}H and ^{13}C chemical shifts of the heterocyclic rings indicated that the silacycloalkyl substituents acted as π-acceptors. On the other hand, the hetaryl substituents influenced ^{29}Si chemical shifts as π-donors. The ^{29}Si chemical shifts of the compounds **1** – **8** appeared in much higher fields than those of corresponding 1,1-dimethyl derivatives [9, 10]. In both series they were shifted to the higher fields in the following order of substituents:

< S < O < O

The coupling constants $J(SiC_{(Ar)})$ increased in the same manner.

The ^{29}Si signals in the investigated silacyclohexanes appeared in the higher fields those that in the silacyclopentanes, as already mentioned in the literature for other silacyclanes [10], but the $\Delta\delta^{29}Si$ values were influenced by the nature of the heterocycle and decreased in the following order: 2-thienyl (19.21) > 2-dihydrofuryl (18.79) >2-furyl (18.65)

Table 3. Carbon and silicon chemical shifts δ [ppm] and coupling constants ^{1}J(CSi) [Hz] in $R_2Si(CH_2)_n$.

R	$SiCH_2$	$SiCCH_2$ {SiCCCH}	$C_2(C_i)$	$C_3(C_o)$	$C_4(C_m)$	$C_5(C_p)$	J(CSi)	$J(C_2(C_i)Si)$	^{29}Si
				n = 1					
O	12.30	28.06	156.00	123.44	110.65	148.58	56.4	81.8	−9.72
O	10.92	31.60	158.38	115.31	27.90	71.55	55.6	77.9	−8.87
S	15.40	28.33	136.42	137.33	129.36	132.81	56.1	69.6	−0.07
	13.14	28.72	(137.85)	(135.74)	(128.83)	(130.17)	52.8	(65.5)	9.23
				n = 2					
O	12.04	24.99 {30.46}	156.52	122.95	110.47	148.28	55.28	81.7	−28.37
O	10.66	24.90 {30.35}	158.74	114.87	31.58	71.39	54.23	76.8	−27.66
S	15.00	25.19 {30.56}	136.45	136.81	129.15	132.38	54.58	69.6	−19.28
	12.54	25.33 {30.93}	137.71	135.47	128.84	130.9	51.78	66.0	−12.26

References

[1] E. Lukevics, N. P. Erchak, in *Advances in Furan Chemistry*, (Eds.: E. Lukevics, R. Zinatne), **1978**, p. 216.

[2] M. Magi, E. Lippmaa, E. Lukevics, N. P. Erchak, *Org. Magn. Reson.* **1977**, *9*, 297.

[3] E. Lukevics, O. A. Pudova, Yu. Popelis, N. P. Erchak, *Zh. Obshch. Khim.* **1981**, *51*, 369.

[4] E. Lukevics, L. M. Ignatovich, N. V. Shilina, Yu. Popelis, I. Birgele, *Chem. Heterocycl. Comp.* **1995**, *31*, 1065.

[5] R. A. Benkeser, Y. Nagai, J. L. Noe, R. F. Cunico, P. H. Gund, *J. Am. Chem. Soc.* **1964**, *86*, 2446.

[6] R. A. Widenhoefer, B. Krzyzanowska, G. Webb-Wood, *Organometallics* **1998**, *17*, 5124.

[7] C. Pretsch, S. Seibl, *Tables of Spectral Data for Structure Determination of Organic Compounds*, Springer, NY, **1989**, p. 265.

[8] P. K. Korver, P. J. Van der Haak, H. Steinberg, Th. J. De Boer, *Rec. Trav. Chim. Pays-Bas* **1965**, *84*, 129.

[9] R. L. Scholl, G. E. Maciel, W. K. Musker, *J. Am. Chem. Soc.* **1972**, *94*, 6376l.

[10] M.-L. Filleux-Blanchard, Nguyen-Dinh-An, G. Manuel, *Org. Magn. Reson.* **1978**, *11*, 150.

Synthesis and Biological Activity of Silicon Derivatives of 2-Trifluoroacetylfuran and Their Oximes

Luba Ignatovich, Dzintra Zarina, Irina Shestakova
Skaidrite Germane, Edmunds Lukevics

Latvian Institute of Organic Synthesis, 21 Aizkraukles, Riga, LV-1006, Latvia
E-mail: ign@osi.lv

Keywords: silyl and germyl trifluoroacetylfurans, silyl and germyl trifluoroacetylfuroximes, toxicity, cytotoxicity, neurotropic activity

Summary: A series of silyl-, germyl- and alkyl-substituted trifluoroacetylfurans has been synthesized under Friedel-Crafts electrophilic acylation conditions. Biological investigations have demonstrated that germyl derivatives of trifluoroacetylfuran are more toxic than the silicon analogs. 5-Triethylgermyl-2-trifluoroacetylfuran was the most toxic compound (LD_{50} 11.2 mg/kg, *i.p.* for white mice), i.e. 200 times more toxic than the silicon analog. 5-*t*-Butyl- and 5-trimethylsilyl-2-trifluoroacetylfuran prolong the duration of ethanol anaesthesia by 220 and 140%. The introduction of R_3M (M = C, Si, Ge) substituent in position 5 of the furan ring of 2-trifluoroacetylfuran oxime increased the cytotoxicity against four tumor cell lines MG-22A (mouse hepatoma), HT-1080 (human fibrosarcoma), Neuro 2A (mouse neuroblastoma), and B16 (human melanoma). The oximes were stronger tumor growth inhibitors and $NO^{\cdot}$ inducers than the corresponding ketones.

Introduction

Trifluoromethyl ketones have generated much interest as potent enzyme inhibitors [1]. A series of trimethylsilylated aliphatic and aromatic trifluoromethyl ketones were synthesized and evaluated for anti-acetylcholinesterase activity. One compound in this series — 3-(trimethylsilyl)trifluoroacetylbenzene (MDL-73745, Zifrosilone) — was selected for further investigation for the treatment of Alzheimer's disease [1 – 3].

The introduction of the trifluoroacetyl moiety into aromatics and heteroaromatics has been generally realized by Friedel-Crafts acylation and by reaction of organomagnesium or organolithium reagents with ethyl trifluoroacetate, trifluoroacetic acid, or its salts [4 – 10 and Refs. therein].

The present investigation elucidates the role of the substituent and its influence on neurotropic and antitumor activity of 5-R_3M(M = C, Si, Ge)-substituted 2-trifluoroacetylfurans and their oximes.

Results and Discussion

Trifluoroacetylation of trialkyl(2-furyl)silanes and -germanes and their carbon analog was realized under Friedel-Crafts conditions using trifluroacetic anhydride.

Friedel-Crafts electrophilic acylation of aryl(hetaryl)silanes, -germanes and -stannanes was accompanied by *ipso*-substitution of the trialkylsilyl, -germyl or -stannyl group. Reactions of this kind have gained currency in organic synthesis primarily because they provide a unique possibility for regiospecific insertion of various groups (halogen, nitro, alkyl, acyl, etc.) into the aromatic or heterocyclic ring [11, 12]. In our investigations we succeeded in electrophilic substitution of hydrogen in position 5 of the furan ring without noticeable cleavage of the M–C bond.

As a result, a new series of silicon- and germanium-containing trifluoroacetyl ketones (**2** – **7**) has been prepared and involved in a condensation with hydroxylamine hydrochloride giving the corresponding oximes **8** – **13** (Scheme 1).

R_3M–(2-furyl) $\xrightarrow{(CF_3CO)_2O}$ R_3M–(furan)–C(=O)CF_3 (**1–7**) $\xrightarrow[Na_2CO_3]{NH_2OH \cdot HCl}$ R_3M–(furan)–C(=NOH)CF_3 (**8–13**)

R_3M = Me_3C, Me_3Si, Me_2PhSi, Me_3Ge, Et_3Si, Et_3Ge

Scheme 1.

Biological investigations have demonstrated that the toxicity of trifluoroacetylfuran derivatives **1** – **7** for white mice (*i. p.*) strongly depends on the substituent at position 5 (Table 1). Germyl derivatives **4** and **5** are more toxic than the corresponding silicon derivatives **2** and **3**. 5-Triethylgermyl-2-trifluoro-acetylfuran (**5**) was the most toxic compound (LD_{50} 11.2 mg/kg), 200 times more toxic than the silicon analog. It is interesting to note that trimethyl derivatives of silicon and germanium analogs have comparable toxicity, but substitution of the methyl group by ethyl dramatically changes the toxicity (see Table 1). The substitution of one methyl at the silicon atom by a more bulky *t*-butyl or phenyl group lowers the toxicity of compounds **6** and **7**. The neurotropic activity of 5-trialkylgermyl-2-trifluoroacetylfurans depends on the alkyl substituent at the germanium atom: the 5-triethylgermyl derivative **5** exhibits the highest activity in the hexobarbital anesthesia test and prolongs its duration by 137%. The 5-trimethylgermyl derivative **4** exhibits a stimulating activity in the ethanol anesthesia and completely prevents animals from suffering retrogradal amnesia (RA) (Table 1). The silicon and carbon analogs **2** and **1** prolonged the ethanol anesthesia by 140 and 220%, correspondingly. They had higher anti-Corazol potency but they did not protect animals from retrogradal amnesia. The pharmacological effects of phenamine are diminished by all the trifluroacetyl derivatives except derivative **6** (Table 1).

Table 1. Acute toxicity and neurotropic activity of R_3M–(2,5-furandiyl)–C(=O)CF_3.

R_3M	LD_{50} [mg/kg]	Hypoxia [%]	Corazol induced spasms [%]	Phenamine stereotypy [%]	Anesthesia of control [%] Hexobarbital	Anesthesia of control [%] Ethanol	RA [%]
Me_3C (**1**)	22	120	175	61	133	320	60
Me_3Si (**2**)	112	100	224.7	44	86	240	16.7
Et_3Si (**3**)	2240	128	94	90	89	74	50
Me_3Ge (**4**)	71	113	149	58	96	51	100
Et_3Ge (**5**)	11.2	109	129	84	237	71	80
$Me_2{}^tBuSi$ (**6**)	>1500	117	99	252	105	132	20
Me_2PhSi (**7**)	355	109	106	69	80	115	60

The effect of trifluoroacetylfuran derivatives **1** – **7** on locomotor coordination and muscle tone depends on the substituent and varies to some extent (Table 2). Thus, compounds **1**, **2** and **4** in rotating-rod, tube, and traction tests have ED_{50} in the 4 – 35.5 mg/kg range, but compound **3** in the range 89 – 447 mg/kg. 5-Trimethylsilyl-2-fluoroacetylfuran (**2**) has the highest hypothermic action, but the 5-triethylgermyl derivative **5** has the highest analgesic activity (EC_{50} 0.9 mg/kg) (Table 2).

Table 2. Effects of trifluoroacetylfurans **1** – **7** on locomotor coordination and muscle tone.

R_3M	ED_{50} [mg/kg] Test: Rotating rod	Tube	Traction	Hypothermia	Analgesia
Me_3C (**1**)	4	9	9	10	10
Me_3Si (**2**)	4.1	2.8	35.5	4.5	28.2
Et_3Si (**3**)	447	89	>250	112	>250
Me_3Ge (**4**)	7	14	14	7	14
Et_3Ge (**5**)	> 5	> 5	>5	11.5	0.9
$Me_2{}^tBuSi$ (**6**)	173.9	146.3	145.6	>1000	36.8
Me_2PhSi (**7**)	43.5	225.1	103.8	>1000	23.1

Potential cytotoxic activity of synthesized compounds **1** – **7** was tested *in vitro* on three monolayer tumor cell lines: MG-22A (mouse hepatoma), HT-1080 (human fibrosarcoma), and Neuro 2A (mouse neuroblastoma). Concentrations providing 50% of tumor death effect (IC_{50}) were determined according to the known procedure [13] using 96 well plates.

The compounds **1** – **4** have low activity on HT-1080, MG-22A, and Neuro 2A tumor cell cultures (Table 3). It is interesting to note that substitution of one alkyl substituent by aryl dramatically changes the cytotoxicity of compound from the inactive trimethylsilyl derivative **2** to the highly active compound **7**. 5-Dimethylphenylsilyl-2-trifluoroacetylfuran (**7**) is the most active anti-tumor substance in all the tumor cell lines studied (Table 3). Germyl derivative **5** is a stronger tumor growth inhibitor and NO-inducer than the corresponding silyl analog **3**.

Table 3. Cytotoxicity of R_3M–(furan-2,5-diyl, O)–$C(=O)CF_3$.

Cell line	Method	R_3M						
		Me_3C (1)	Me_3Si (2)	Et_3Si (3)	Me_3Ge (4)	Et_3Ge (5)	$Me_2{}^tBuSi$ (6)	Me_2PhSi (7)
	CV	nce	nce	nce	nce	72*	7	6
HT-1080	MTT	nce	nce	nce	nce	72	22	0.3
	**NO	4	6	5	18	250	350	300
	CV	nce	nce	nce	nce	6	21	0.6
MG-22A	MTT	nce	nce	nce	nce	10	7	0.5
	NO	5	7	15	14	250	150	300
	CV	nce			nce	3.3	40	2.7
Neuro 2A	MTT	nce	nt	nt	nce	4	30	4.5
	NO	5			18	300	67	400

*IC_{50} [μg/mL] providing 50% cell killing effect (CV: coloration; MTT: coloration); crystal violet (CV), 3-(4,5-dimethylthiazol-2-yl)-2,5-diphenyltetrazolinium bromide (MTT). **NO = concentration (%) (CV: coloration); nce = no cytotoxic effect; nt = not tested.

The neurotropic activity of five new oximes **8** – **11** and **13** has been studied (Table 4). Carbon derivative **8** was the most toxic compound (LD_{50} 355 mg/kg). The stimulating effect of phenamine strengthened under the influence of germanium derivative **11** by 266%, but silicon derivative **9** decreased it by 55%. 5-Trimethylgermyl-2-furyltrifluoroacetyloxime (**9**) in comparatively small doses (50 mg/kg) completely prevented retrogradal amnesia caused by electric shock (RA 100%).

The cytotoxic activity of synthesized compounds **8** – **13** was tested *in vitro* on four monolayer tumor cell lines: MG-22A (mouse hepatoma), HT-1080 (human fibrosarcoma), Neuro 2A (mouse neuroblastoma), and B16 (human melanoma). All silicon and germanium derivatives of oximes possess high cytotoxic activity (Table 5). They were more active than 5-unsubstituted oxime (IC_{50} 12 – 19 μg/mL) and also exhibited higher cytotoxicity than the corresponding ketones **1** – **5**. Some selectivity has been observed. For example, the 5-triethylsilyl derivative **10** exhibited high cytotoxic activity on mouse hepatoma (MG-22A) (IC_{50} 0.5 – 0.7 μg/mL), but the germanium analog **12** was more active on mouse neuroblastoma (Neuro 2A) (IC_{50} 0.2 – 0.6 μg/mL) (Table 5). It was also the most active compound against melanoma B16.

Table 4. Acute toxicity and neurotropic activity of R_3M–(2-furyl, 5-position)–$C(=NOH)CF_3$.

R_3M	LD_{50} [mg/kg]	Hypoxia [%]	Phenamine stereotypy [%]	Anaesthesia of control [%]		RA [%]
				Hexobarbital	Ethanol	
Me_3C(**8**)	355	114	40	111	75	60
Me_3Si(**9**)	>1000	126	45	76	91	100
Et_3Si(**10**)	>2000	132	111	146	111	83
Me_3Ge(**11**)	>1000	147	366	100	157	0
Me_2PhSi(**13**)	850	130	82	109	117	40

Table 5. Cytotoxicity of R_3M–(2-furyl, 5-position)–$C(=NOH)CF_3$.

Cell line	Method	R_3M					
		Me_3C (**8**)	Me_3Si (**9**)	Et_3Si (**10**)	Me_3Ge (**11**)	Et_3Ge (**12**)	Me_2PhSi (**13**)
HT-1080	CV	3	0.8	12	2	1.4	1.4
	MTT	11	2.5	19.6	1	2	2.2
	NO	400	650	62	600	600	600
MG-22A	CV	5	0.3	0.5	2	2	3.7
	MTT	9	2.5	0.7	1	2	1.8
	NO	350	750	200	150	700	700
B16	CV	3	29	6	43	0.3	2.3
	MTT	2	38	5	1	1.6	2.1
	NO	850	650	200	200	400	600
Neuro 2A	CV	0,8	38	6	2	0.6	2.9
	MTT	2	31	6	15	0.2	1.4
	NO	600	650	100	150	950	850

Acknowledgment: This work was supported by Latvian Science Foundation Grant No. 180.

References

[1] I. Prous, A. Graul, J. Castaner, *Drugs of the Future* **1994**, *19*, 854.
[2] J. Dow, B. D. Dulery, J.-M. Hornsperger, G. F. Di Francesco, P. Keshary, K. D. Haegele, *Arzneim.-Forsch/Drug. Res.* **1995**, *45(II)*, 1245.
[3] X. Zhu, E. Giacobini, J.-M. Hornsperger, *Eur. J. Pharmacol.* **1995**, *275*, 93.
[4] N. A. Zaitseva, E. M. Panov, K. K. Kocheshkov, *Izv. Akad. Nauk USSR* **1961**, *5*, 831.
[5] V. G. Glukhovtsev, J. V. Ilyin, A. V. Ignatenko, L. Yu. Breznev, *Izv. Akad. Nauk USSR ser. khim.* **1987**, *12*, 2834.
[6] W. S. DiMenna, P. M. Gross, *Tetrahedron Lett.* **1980**, 2129.
[7] J. W. Guiles, *Synlett.* **1995**, 165.
[8] X. Creary, *J. Org. Chem.* **1987**, *52*, 5026.
[9] F. A. J. Kerdesky, A. Basha, *Tetrahedron Lett.* **1991**, 2003.
[10] J.-P. Begue, D. Bonnet-Delpon, *Tetrahedron* **1991**, *47*, 3207.
[11] C. Eaborn, *J. Organomet. Chem.* **1975**, *100*, 43.
[12] W. P. Weber, *Silicon Reagents for Organic Synthesis*, Springer Verlag, Berlin-Heidelberg-New-York, **1983**.
[13] R. J. Riddell, R. H. Clothier, M. F. Balls, *Chem. Toxicol.* **1986**, *24*, 469.

Silicon Diols, Effective Inhibitors of Human Leucocyte Elastase

Graham A. Showell, John G. Montana*

Amedis Pharmaceuticals Ltd., 162 Cambridge Science Park
Milton Road, Cambridge, CB4 0GP, U.K.
Tel.: +44 1223 477910 — Fax: +44 1223 477911
E-mail: graham.showell@amedis-pharma.com

Jenny A. Chadwick, Chris Higgs, Hazel J. Hunt
Robert E. MacKenzie, Steve Price, Tina J. Wilkinson

Argenta Discovery Ltd., 8/9 Spire Green Centre
Flex Meadow, Harlow, Essex, CM19 5TR, U.K.

Keywords: human elastase, protease inhibitors, silicon diol, privileged structure,

Summary: Appropriately substituted silicon diols have been shown to be inhibitors of both aspartyl and metallo proteases. Using the knowledge that the active site of serine proteases consists of two domains, an extended binding site where non-covalent binding interactions occur and a catalytic site where covalent bond-forming and bond-breaking reactions take place, silicon diol compounds were designed and synthesized as potent human leucocyte elastase inhibitors. The peptido-mimetic analog **9** is a potent human elastase inhibitor that provides key interactions at both the prime (C-terminal) and non-prime (N-terminal) sites within the enzyme and demonstrates that the silicon diol presents itself as an important privileged structure within the framework of protease inhibitors.

Introduction

Proteases are well recognized as a significant component of the drug-able genome. The four major classes of protease enzymes (aspartic, metallo, serine, and cysteine) selectively catalyze the hydrolysis of amide bonds in peptides and proteins. They play essential roles in most biological processes and are important therapeutic targets for a multitude of diseases, including cancer, viral, parasitic, fungal and bacterial infections, inflammation and cardiovascular disease. Proteases represent a gene family that is well validated from a drug discovery perspective with numerous drugs targeting these enzymes being approved for clinical use.

The reaction mechanism of proteases is to convert the carbonyl group of the target peptide bond into an enzyme-stabilized tetrahedral diol transition state intermediate. This intermediate further collapses to yield the two-peptide products. The enzyme binds the diol intermediate, the specificity being provided by specific contacts for the amino acids side chains P1, P2, P3, P1′, P2′, P3′ (Fig. 1) and so on, with their respective binding sites on the protein (S1, S2, S3, S1′, S2′, S3′).

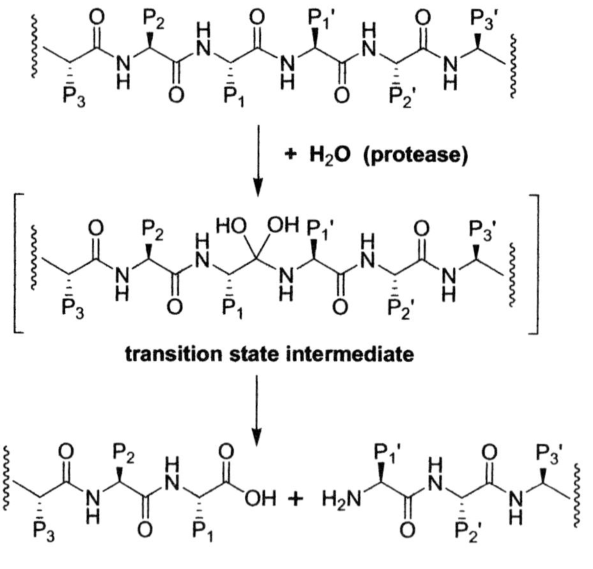

Fig. 1. The transition state intermediate formed during substrate hydrolysis by proteases.

A stable structure that mimics this tetrahedral intermediate would be a transition state analog inhibitor. A silicon diol structure, in which the $C(OH)_2NH$ group of the transition state is replaced by an $Si(OH)_2CH_2$ moiety, mimics that tetrahedral transition state, and it has recently been shown that incorporating silicon diol moieties into peptide mimetics results in potent and selective inhibitors for metallo and aspartyl proteases [1 – 3] (**1** and **2**, Fig. 2).

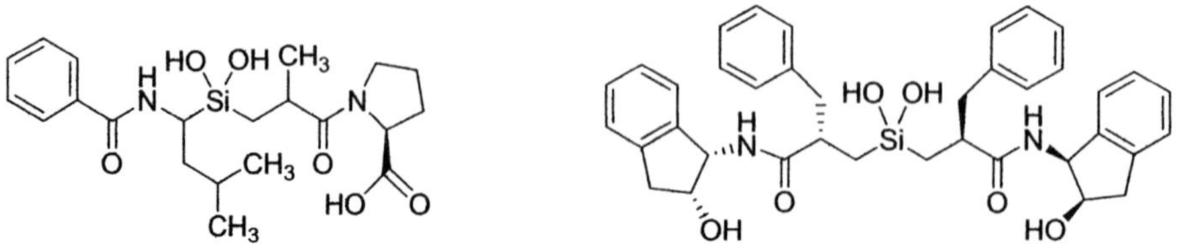

Fig. 2. Silicon diol proteases inhibitors, **1** (angiotensin-converting enzyme (metallo) inhibitor) and **2** (HIV protease (aspartyl) inhibitor).

Design of Serine Protease Inhibitors for Human Leucocyte Elastase

The active site of serine proteases is a catalytic triad of serine, histidine, and aspartyl groups. The substrate binds in the active site forming a Michaelis complex. This exposes the carbonyl group of the amide bond to nucleophilic attack by the active-site serine hydroxyl with the aid of base catalysis of the histidine group. A proton transfer from the histidine then results in expulsion of the amine fragment of the peptide. The acyl-enzyme complex left behind is attacked by water, forming a new tetrahedral intermediate which breaks down to regenerate the serine and the carboxyl fragment of the original substrate (Fig. 1) [4].

Human leucocyte elastase (HE) is released in response to inflammatory stimuli and is responsible for degradation of connective tissues and is implicated in respiratory distress syndrome, rheumatoid arthritis, chronic bronchitis, and pulmonary emphysema. Many inhibitors of HE have been described [5], usually of the electrophilic carbonyl type, and often the key Val-Pro-Val motif has been exploited for potent inhibitory activity (**3** – **5**, Fig. 3). Recently, sivelestat (**6**, HE IC_{50} = 44 nM), a member of an enzyme-acylating series, has been approved in Japan for the treatment of acute lung injury [6].

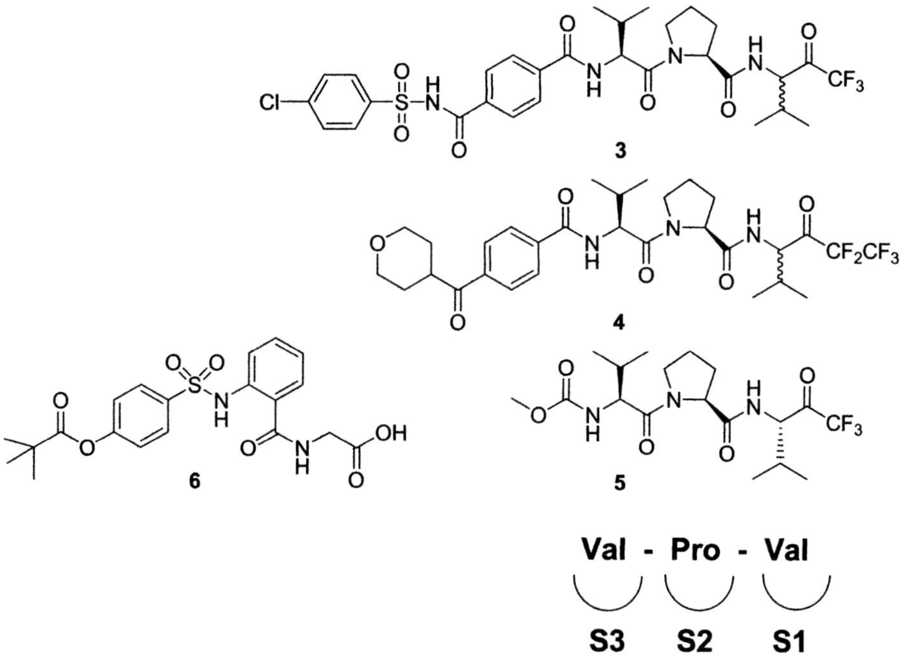

Fig. 3. Val-Pro-Val serine protease inhibitors.

Employing this strategy, a silicon diol "warhead" moiety was incorporated into the Val-Pro-Val motif: however, also realizing that this provided a relatively rigid inhibitor, more flexible analogs were designed around an isoleucine-isobutyl peptide template wherein the isobutyl group would access the hydrophobic S1 pocket and the isoleucine the S2 pocket.

Synthesis of Human Leucocyte Elastase Inhibitors

The synthesis of the Val-Pro-Val analogs was accomplished through the key intermediate (**7**, Scheme 1) using methodology described by Organ et al. [7]. This key building block was then coupled to the proline portion of the molecule using standard peptide-coupling conditions. The capping protecting group was removed by hydrogenolysis and the resulting amine capped with a benzoyl group. The target silicon diol (**8**) was obtained by deprotection of the diphenyl silane group using trifluoroacetic acid (TFA) at 50 °C. The use of TFA was far more convenient to use than triflic acid and gave cleaner products than the boron trifluoride methodology [7]. In addition it was observed that the TFA deprotection progressed more rapidly under microwave conditions.

The synthesis of the flexible isoleucine-isobutyl compound (**9**, Scheme 2) was lengthier, mainly as a consequence of the fact that the compound contains a butyl-carboxamide group capable of picking up binding interactions on the S1′ side of the protein. Synthesis of the key isobutyl-silyl-methylamine compound **10** was achieved over 8 steps. The amine was then coupled in standard fashion with the P2 portion of the molecule and the "prime-side" side chain was modified to the desired amide **11**. Deprotection of the diphenyl-silane using TFA, as described above, provided the target diol **9**.

In vitro Inhibitory Activity of Human Leucocyte Elastase Inhibitors

Silicon diol inhibitors were tested using a standard protocol as follows. The fluorogenic substrate MeO-Suc-Ala-Ala-Pro-Val-AMC was used at 100 μM final concentration with HE. Test compounds **8** and **9** were dissolved in dimethyl sulfoxide (DMSO) and diluted to give a 1% DMSO final concentration in the assay. The compounds were tested at a final concentration range of 100 – 0.005 μM. The reaction was anticipated to be linear for at least 30 min, allowing the calculation of initial rates and percentage inhibitions by comparing with positive control [elastase inhibitor III (MeOSuc-Ala-Ala-Pro-Val-CMK), IC_{50} = 350 nM] values. The IC_{50} values shown in Table 1 represent a mean of 2 determinations. The more rigid analog **8**, based on the Val-Pro-Val motif, is completely void of HE inhibitory activity. Preliminary molecular modeling studies [8] suggest that the P1 isopropyl group cannot access the S1 pocket and no key hydrogen bonds are observed.

In contrast, the more flexible peptido-mimetic analog **9** displays good activity (HE IC_{50} 300 nM). Molecular modeling studies suggest that the S1 and S2 pockets are well accessed by the appropriate substituents along with key hydrogen bond interactions with the backbone of the protein (Val216 and Phe41). In addition the terminal naphthalene group may be involved in π-π stacking interactions within the protein active site.

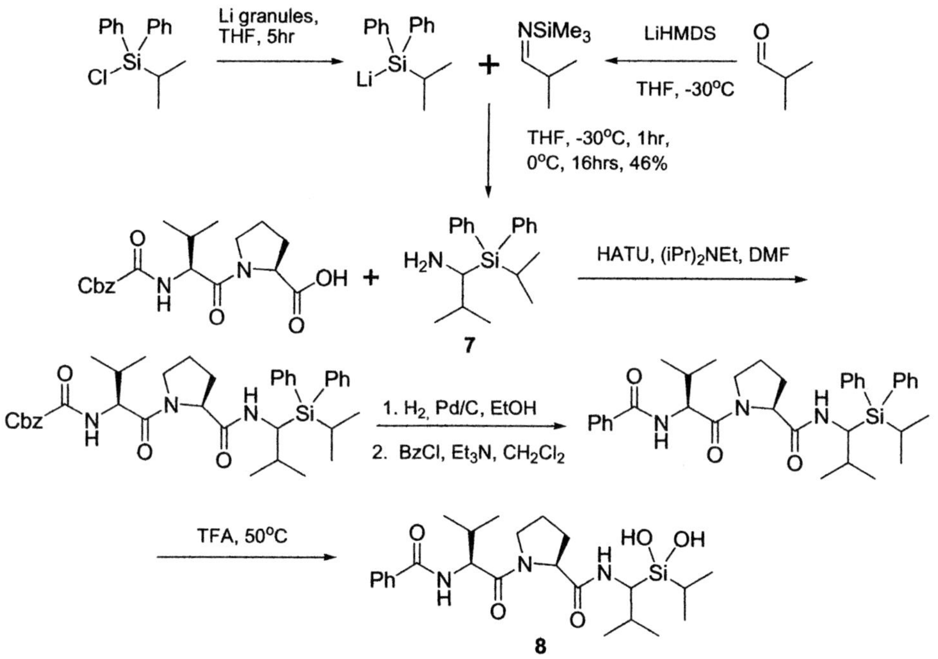

Scheme 1. Synthesis of the silicon diol **8**.

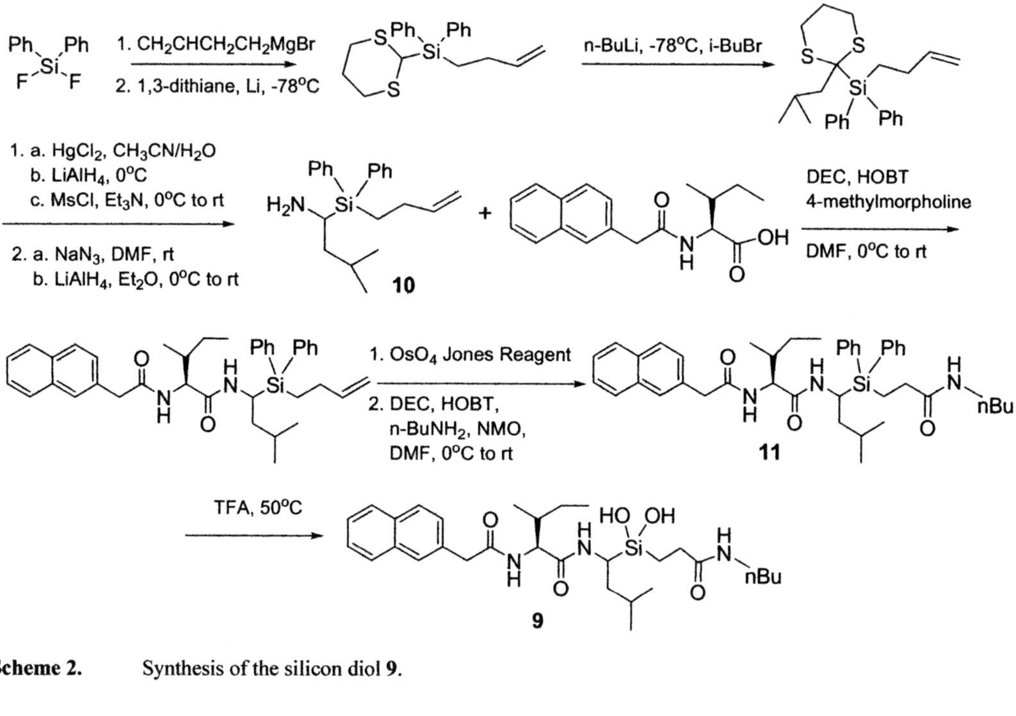

Scheme 2. Synthesis of the silicon diol **9**.

Table 1. Inhibitory activity at human leucocyte elastase for the silicon diol inhibitors **8** and **9**.

Compound	Human leucocyte elastase IC_{50} [nM]
8	>10 000
9	300

Conclusion

The results presented have demonstrated the ability of silicon diols to have serine protease inhibitory activity. The silicon diol **9** is a potent human leucocyte elastase inhibitor. The fact that the silicon diol moiety can be used across the aspartyl, metallo, and serine gene family of proteases shows the power of such a simple privileged structure within the drug design process.

Acknowledgments: The authors would like to acknowledge the contribution to this work from:

- Paul Lickiss and Charles Daka (Imperial College, UK).
- David Otway, Julie Warneck and Hooshang Zavareh (Amedis Pharmaceuticals Ltd., UK).

References

[1] S. McN. Sieburth, T. Nittoli, A. M. Mutahi, L. Guo, *Angew. Chem. Int. Ed.* **1998**, *37*, 812.
[2] C-A. Chen, S. McN. Sieburth, A. Glekas, G. W. Hewitt, G. L. Trainor, S. Erickson-Viitanen, S. S. Garber, B. Cordova, S. Jeffry, R. M. Klabe, *Chem. Biol.* **2001**, *8*, 1161.
[3] M. wa Mutahi, T. Nittoli, L. Guo, S. McN. Sieburth, *J. Am. Chem. Soc.* **2002**, *124*, 7363.
[4] D. Leung, G. Abbenante, D. P. Fairlie, *J. Med. Chem.* **2000**, *43*, 305.
[5] P. D. Edwards, P. R. Bernstein, *Med. Res. Rev.* **1994**, *14*, 127.
[6] H. Ohbayashi, *Expert Opin. Investig. Drugs.* **2002**, *11*, 965.
[7] M. G. Organ, C. Buon, C. P. Decicco, A. P. Combs, *Org. Lett.* **2002**, *4*, 2683.
[8] Molecular modeling studies were performed using the crystal structure of human elastase (pdb code: 1HNE), Sybyl molecular modeling software and a Tripos forcefield. An assumption was made that the silicon diol forms a covalent bond with the catalytic serine residue, Ser195.

σ Ligands of the 1,4′-Silaspiro[tetralin-1,4′-piperidine] Type and the Serotonin/Noradrenaline Reuptake Inhibitor Sila-venlafaxine: Studies on C/Si Bioisosterism

Jürgen O. Daiß, Barbara Müller, Christian Burschka, Reinhold Tacke*

Institut für Anorganische Chemie, Universität Würzburg,
Am Hubland, D-97074 Würzburg, Germany
Tel.: +49 931 888 5250 — Fax: +49 931 888 4609
E-mail: r.tacke@mail.uni-wuerzburg.de

William Bains, Julie Warneck

Amedis Pharmaceuticals Ltd.
162 Cambridge Science Park, Milton Road, Cambridge CB4 0GP, U.K.

Keywords: C/Si bioisosterism, carbon/silicon switch, chirality, σ ligands, serotonin/noradrenaline reuptake inhibitors, silanols, 1,4′-silaspiro[tetralin-1,4′-piperidine]

Summary: Novel potential σ ligands of the 1,4′-silaspiro[tetralin-1,4′-piperidine] type (compounds **1b** – **4b**, isolated as hydrochlorides) and *rac*-sila-venlafaxine (*rac*-**5b**, a silicon analog of the serotonin/noradrenaline reuptake inhibitor *rac*-venlafaxine (*rac*-**5a**)) were synthesized. In addition, the sila-venlafaxine enantiomers (*R*)-**5b** and (*S*)-**5b** and the sila-venlafaxine derivatives *rac*-**6**·HCl and *rac*-**7**·HCl were prepared. Compounds **3b**·HCl and *rac*-**5b**·HCl were structurally characterized by single-crystal X-ray diffraction, and *rac*-**5a**, *rac*-**5b**, *rac*-**6**, and *rac*-**7** were pharmacologically characterized.

Introduction

In a recently published review, silicon chemistry has been demonstrated to be a novel powerful source of chemical diversity in drug design [1]. Sila-substitution of drugs (carbon/silicon switch) is one of the concepts that have been successfully used for the development of new silicon-based drugs (for recent examples, see Ref. [2]).

In context with our systematic studies on sila-substituted drugs [1, 2], we have been interested in the pharmacological properties of novel potential σ ligands of the 1,4′-silaspiro[tetralin-1,4′-piperidine] type, compounds **1b** – **4b** (cf. Ref. [2e] for structurally related σ ligands of the

1,4′-silaspiro[indane-1,4′-piperidine] type). Compounds **1b** – **4b** represent silicon analogs of the σ ligands **1a** – **4a** [3]. In addition, we have been interested in the pharmacological properties of *rac*-sila-venlafaxine (*rac*-**5b**), a sila-analog of the serotonin/noradrenaline reuptake inhibitor *rac*-venlafaxine (*rac*-**5a**) [4] and its derivatives *rac*-**6** and *rac*-**7**. Racemic venlafaxine hydrochloride (Effexor, Wyeth-Ayerst; Efexor, Wyeth, Wyeth-Lederle; Trevilor, Wyeth) is in clinical use as an antidepressant [5]. Sila-substitution of **1a** – **4a** and *rac*-**5a** (→ **1b** – **4b**, *rac*-**5b**) was expected to affect the chemical and physicochemical properties and the structures of these compounds and therefore to alter their biological properties.

We report here on (i) the synthesis of the silicon-based potential σ ligands **1b** – **4b** (isolated as the hydrochlorides), (ii) the synthesis of *rac*-sila-venlafaxine (*rac*-**5b**), (iii) the preparation of the sila-venlafaxine enantiomers (*R*)-**5b** and (*S*)-**5b** via resolution of *rac*-**5b**, (iv) the synthesis of the sila-venlafaxine derivatives *rac*-**6** and *rac*-**7**, (v) the crystal structure analyses of **3b**·HCl and *rac*-**5b**·HCl, and (vi) the pharmacological characterization of *rac*-**5a**, *rac*-**5b**, *rac*-**6**, and *rac*-**7**. Preliminary results of the studies reported here have already been presented elsewhere [6, 7].

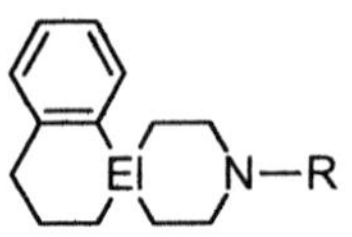

a: El = C
b: El = Si

	R
1	CH_2Ph
2	$4\text{-}CH_2\text{-}C_6H_4\text{-}OMe$
3	$(CH_2)_2Ph$
4	$CH_2CH{=}CMe_2$

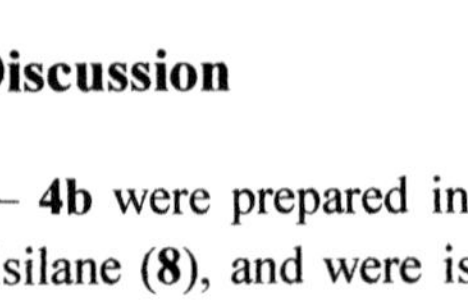

5a, 5b 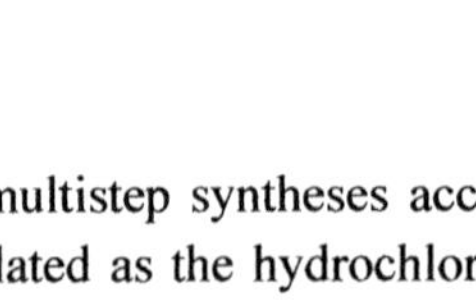 **6** 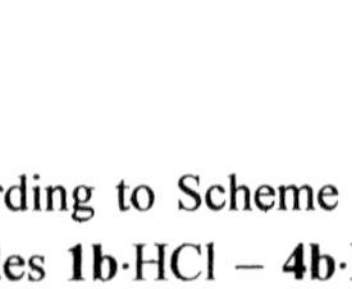 **7**

Results and Discussion

Compounds **1b** – **4b** were prepared in multistep syntheses according to Scheme 1, starting from dichlorodiphenylsilane (**8**), and were isolated as the hydrochlorides **1b**·HCl – **4b**·HCl. In the first step, **8** was reacted with vinylmagnesium chloride to afford diphenyldivinylsilane (**9**), which was treated with trifluoromethanesulfonic acid, followed by addition of triethylammonium chloride, to yield dichlorodivinylsilane (**10**). Reaction of **10** with hydrogen bromide, in the presence of dibenzoyl peroxide, gave bis(2-bromoethyl)dichlorosilane (**11**), which was reacted with 2-bromomagnesio-1-(3-(bromomagnesio)propyl)benzene (obtained by reaction of **15** (**13** → **14** → **15**) with magnesium) to give 1,1-bis(2-bromoethyl)-1-sila-1,2,3,4-tetrahydronaphthalene (**12**). Reaction of **12** with the respective primary amines, in the presence of triethylamine, finally afforded compounds **1b** – **4b** (isolated as the hydrochlorides **1b**·HCl – **4b**·HCl).

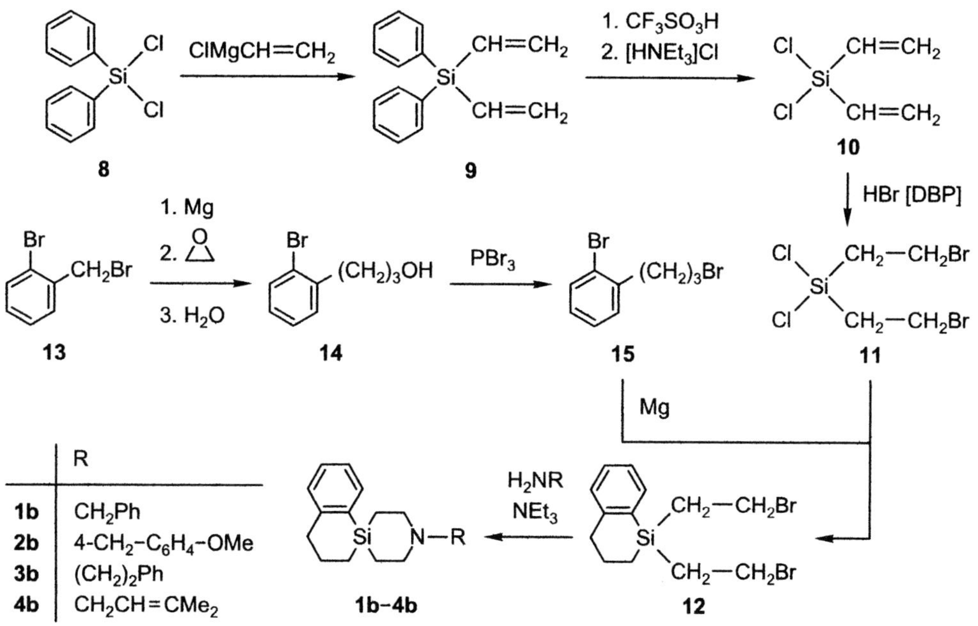

	R
1b	CH_2Ph
2b	$4\text{-}CH_2\text{-}C_6H_4\text{-}OMe$
3b	$(CH_2)_2Ph$
4b	$CH_2CH{=}CMe_2$

Scheme 1. Synthesis of the potential σ ligands **1b** – **4b**.

rac-Sila-venlafaxine (*rac*-**5b**) and its derivatives *rac*-**6** and *rac*-**7** were synthesized in multistep syntheses according to Scheme 2, starting from tetrachlorosilane (**16**) or tetramethoxysilane (**17**). Thus, reaction of **16** with 1,5-bis(bromomagnesio)pentane or 1,4-bis(bromomagnesio)butane afforded 1,1-dichloro-1-silacyclohexane (**18**) and 1,1-dichloro-1-silacyclopentane (**19**), respectively. Methanolysis of **18** and **19**, in the presence of triethylamine, yielded 1,1-dimethoxy-1-silacyclohexane (**20**) and 1,1-dimethoxy-1-silacyclopentane (**21**), respectively. Alternatively, **20** was prepared by reaction of **17** with 1,5-bis(bromomagnesio)pentane. Subsequent treatment of **20** and **21** with (1-phenylvinyl)magnesium bromide (**24**; obtained by reaction of (1-bromovinyl)-benzene (**22**) with magnesium) or (1-(4-methoxyphenyl)vinyl)lithium (**25**; obtained by reaction of 4-methoxyacetophenone 2,4,6-triisopropylbenzenesulfonylhydrazone (**23**) with *n*-butyllithium in the presence of *N,N,N',N'*-tetramethylethylenediamine (TMEDA)) gave the respective 1-(1-arylvinyl)-1-methoxy-1-silacycloalkanes **26** – **28**, which were then reacted with lithium aluminum hydride to yield the 1-(1-arylvinyl)-1-silacycloalkanes **29** – **31**. Reaction of **29** – **31** with dimethylamine, in the presence of lithium dimethyl amide, gave the racemic 1-(1-aryl-2-(dimethylamino)ethyl)-1-(dimethylamino)-1-silacycloalkanes *rac*-**32** – *rac*-**34**, which upon hydrolysis yielded the racemic 1-(1-aryl-2-(dimethylamino)ethyl)-1-silacycloalkan-1-ols *rac*-**5b**, *rac*-**6**, and *rac*-**7** (isolated as the hydrochlorides *rac*-**5b**·HCl, *rac*-**6**·HCl, and *rac*-**7**·HCl).

(*R*)-Sila-venlafaxine ((*R*)-**5b**) was prepared according to Scheme 3 by resolution of *rac*-**5b**, using (+)-10-camphorsulfonic acid ((+)-CSA) as the resolving agent (→ (*R*)-**5b**·(+)-CSA). Treatment of the diastereomerically pure salt (*R*)-**5b**·(+)-CSA (obtained by 3-fold recrystallization from acetone) with an aqueous potassium carbonate solution gave (*R*)-**5b**. The antipode (*S*)-**5b** was prepared analogously, starting from the mother liquor obtained in the preparation of (*R*)-**5b**·(+)-CSA and using (–)-CSA as the resolving agent (→ (*S*)-**5b**·(–)-CSA). The enantiopure hydrochlorides (*R*)-**5b**·HCl and (*S*)-**5b**·HCl were prepared by treatment of (*R*)-**5b** and (*S*)-**5b**, respectively, with an ethereal hydrogen chloride solution. Reaction of (*R*)-**5b** with triphenylphosphonium bromide afforded (*R*)-**5b**·HBr, the crystal structure analysis of which (data not given) allowed the assignment of the absolute configurations of the sila-venlafaxine enantiomers.

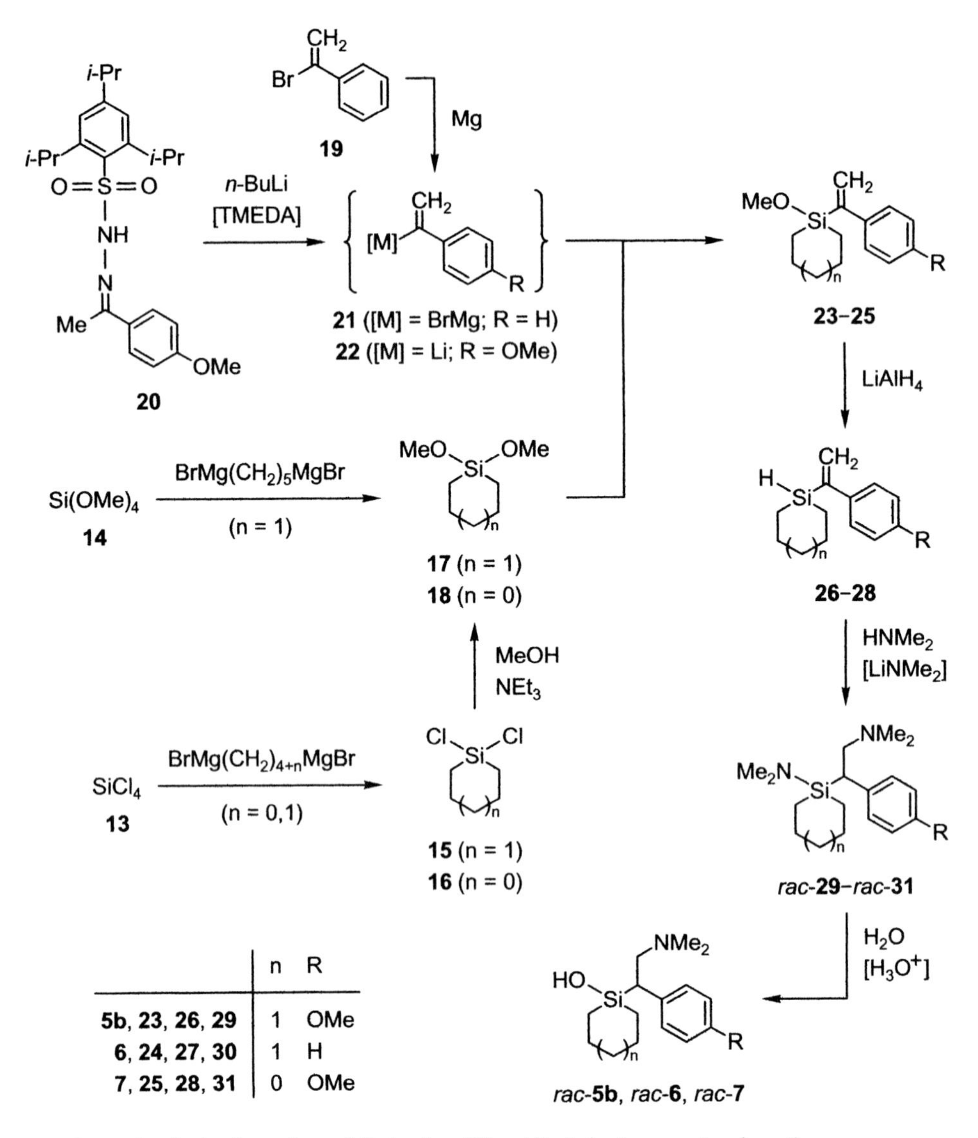

	n	R
5b, **23**, **26**, **29**	1	OMe
6, **24**, **27**, **30**	1	H
7, **25**, **28**, **31**	0	OMe

Scheme 2. Synthesis of *rac*-sila-venlafaxine (*rac*-**5b**) and its derivatives *rac*-**6** and *rac*-**7**.

The identities of compounds **1b** – **4b**, **1b**·HCl – **4b**·HCl, *rac*-**5b**, (*R*)-**5b**, (*S*)-**5b**, *rac*-**5b**·HCl, (*R*)-**5b**·HCl, (*S*)-**5b**·HCl, (*R*)-**5b**·HBr, (*R*)-**5b**·(+)-CSA, (*S*)-**5b**·(–)-CSA, *rac*-**6**, *rac*-**6**·HCl, *rac*-**7**, *rac*-**7**·HCl, **9** – **12**, **14**, **15**, **18** – **21**, **23**, **26** – **31**, and *rac*-**32** – *rac*-**34** were established by elemental analyses and NMR studies (^{1}H, ^{13}C, ^{29}Si). The enantiomeric purities of (*R*)-**5b** and (*S*)-**5b** were determined by ^{1}H NMR spectroscopy in the presence of the chiral solvating agent (*R*)-(–)-1-(9-anthryl)-2,2,2-trifluoroethanol (solvent $CDCl_3$).

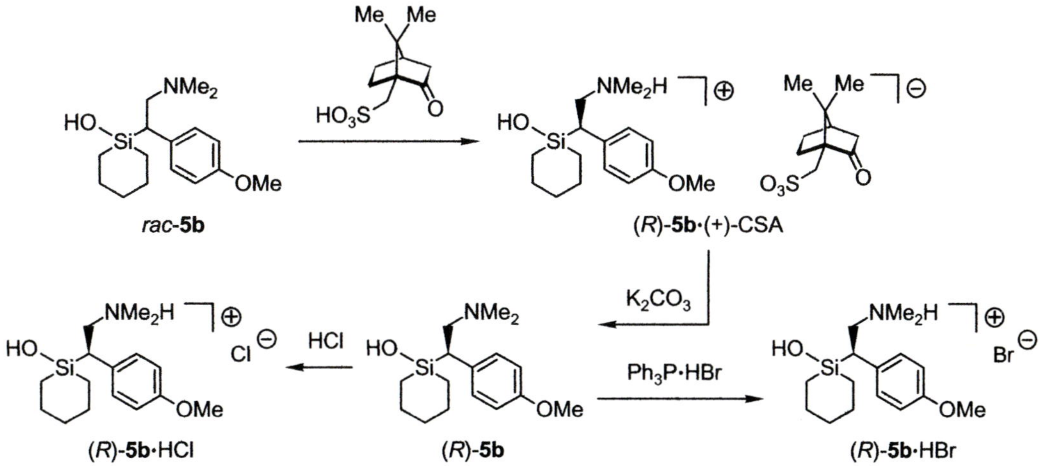

Scheme 3. Preparation of (*R*)-sila-venlafaxine ((*R*)-**5b**), (*R*)-sila-venlafaxine hydrochloride ((*R*)-**5b**·HCl), and (*R*)-sila-venlafaxine hydrobromide ((*R*)-**5b**·HBr).

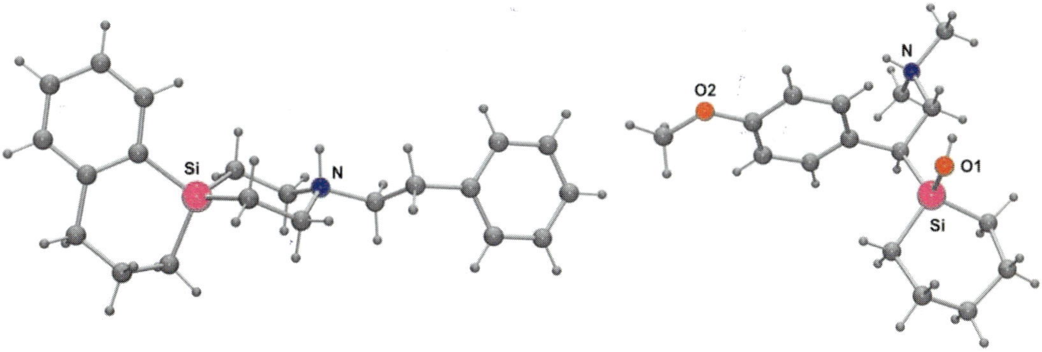

Fig. 1. Structures of the ammonium cations of **3b**·HCl (only one of the four crystallographically independent cations depicted; left) and *rac*-**5b**·HCl (only one enantiomer depicted; right) in the crystal.

Compounds **3b**·HCl and *rac*-**5b**·HCl were structurally characterized by single-crystal X-ray diffraction (T = –100 °C, λ = 0.71073 Å). Suitable single crystals of **3b**·HCl were obtained by cooling of a solution of this compound in trichloromethane from 20 °C to –20 °C (monoclinic space group $P2_1/n$, a = 10.8966(5) Å, b = 22.9911(12) Å, c = 31.8912(16) Å, β = 97.743(6)°, V = 7916.7(7) Å^3, Z = 16). Suitable single crystals of *rac*-**5b**·HCl were obtained by cooling of a boiling saturated solution of this compound in dichloromethane to 4 °C (orthorhombic space group $Pca2_1$, a = 12.8309(16) Å, b = 14.209(2) Å, c = 9.8055(11) Å, V = 1787.7(4) Å^3, Z = 4).

The asymmetric unit of **3b**·HCl contains four ammonium cations, with very similar structures, one of which is depicted in Fig. 1. The structure of one of the two enantiomers of the ammonium cation of *rac*-**5b**·HCl is also shown in Fig. 1.

The carbon/silicon analogs *rac*-**5a** and *rac*-**5b** and the derivatives *rac*-**6** and *rac*-**7** were tested for their *in vitro* efficacy to inhibit the reuptake of serotonin, noradrenaline, and dopamine (all compounds were tested as the hydrochlorides). The results of these studies are depicted in Fig. 2. Pharmacological data for **1b** – **4b**, (*R*)-**5b**, and (*S*)-**5b** will be reported elsewhere.

As can be seen from Fig. 2, sila-substitution of *rac*-**5a** (→ *rac*-**5b**) substantially affects the pharmacological profile with respect to serotonin reuptake inhibition. The other (major) structural changes in the molecular shape of *rac*-**5b** (→ *rac*-**6**, *rac*-**7**) also influenced the pharmacological profile with respect to monoamine selectivity (*rac*-**7**) and/or absolute potency (*rac*-**6**). These results clearly demonstrate that the carbon/silicon switch strategy is a powerful tool for drug design.

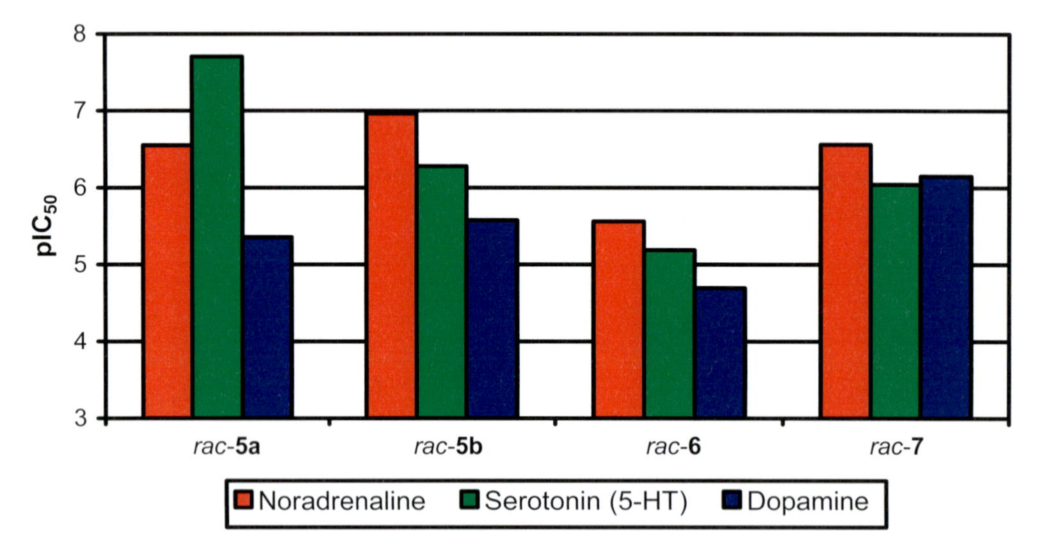

Fig. 2. *In vitro* efficacy of compounds *rac*-**5a**, *rac*-**5b**, *rac*-**6**, and *rac*-**7** regarding serotonin, noradrenaline, and dopamine reuptake inhibition. pIC_{50} denotes the negative decadic logarithm of the half-maximum effect concentration [M]. The monoamine reuptake inhibition profiles of *rac*-**5a**, *rac*-**5b**, *rac*-**6**, and *rac*-**7** were generated via radioligand transporter assays using recombinant human monoamine transporter proteins. The data represent the mean of duplicate analyses.

References

[1] W. Bains, R. Tacke, *Curr. Opin. Drug Discovery Dev.* **2003**, *6*, 526.

[2] (a) M. Merget, K. Günther, M. Bernd, E. Günther, R. Tacke, *J. Organomet. Chem.* **2001**, *628*, 183. (b) R. Tacke, T. Kornek, T. Heinrich, C. Burschka, M. Penka, M. Pülm, C. Keim, E. Mutschler, G. Lambrecht, *J. Organomet. Chem.* **2001**, *640*, 140. (c) R. Tacke, V. I. Handmann, K. Kreutzmann, C. Keim, E. Mutschler, G. Lambrecht, *Organometallics* **2002**, *21*, 3727. (d) R. Tacke, T. Heinrich, *Silicon Chem.* **2002**, *1*, 35. (e) R. Tacke, V. I. Handmann, R. Bertermann, C. Burschka, M. Penka, C. Seyfried, *Organometallics* **2003**, *22*, 916. (f) T. Heinrich, C. Burschka, J. Warneck, R. Tacke, *Organometallics* **2004**, *23*, 361. (g) R. Tacke, T. Heinrich, R. Bertermann, C. Burschka, A. Hamacher, M. U. Kassak, *Organometallics* **2004**, *23*, 4468.

[3] M. S. Chambers, R. Baker, D. C. Billington, A. K. Knight, D. N. Middlemiss, E. H. F. Wong, *J. Med. Chem.* **1992**, *35*, 2033.

[4] (a) G. E. M. Husbands, J. P. Yardley, E. A. Muth (Inventors; American Home Products Corp., USA). Eur. Pat. Appl. EP 0112669 A2 (04.07.1984); *Chem. Abstr.* **1985**, *102*, 5895e. (b) J. P. Yardley, G. E. M. Husbands, G. Stack, J. Butch, J. Bicksler, J. A. Moyer, E. A. Muth, T. Andree, H. Fletcher III, M. N. G. James, A. R. Sielecki, *J. Med. Chem.* **1990**, *33*, 2899.

[5] (a) M. Dierick, A. De Nayer, M. Ansseau, H. D'Haenen, P. Cosyns, W. Verbruggen, A. Seghers, I. Pelc, P. Fossion, G. Stefos, J. Peuskens, M. Malfroid, S. Leyman, A. Mignon, *Curr. Ther. Res.* **2002**, *63*, 475. (b) H. Sauer, S. Huppertz-Helmhold, W. Dierkes, *Pharmacopsychiatry* **2003**, *36*, 169. (c) M. A. Gutierrez, G. L. Stimmel, J. Y. Aiso, *Clin. Ther.* **2003**, *25*, 2138.

[6] Compounds **1b** – **4b**: B. Müller, J. O. Daiß, R. Tacke, *2nd European Organosilicon Days*, Munich, Germany, September 11 – 12, 2003; Abstract P162.

[7] Compounds *rac*-**5b**, (*R*)-**5b**, (*S*)-**5b**, *rac*-**6**, and *rac*-**7**: (a) R. Tacke, J. Daiss (Inventors; Amedis Pharmaceuticals Ltd., U.K.). PCT Int. Pat. Appl. WO 03/037905 A1 (08.05.2003); *Chem. Abstr.* **2003**, *138*, 354097d. (b) J. O. Daiß, R. Tacke, W. Bains, J. Warneck, *2nd European Organosilicon Days*, Munich, Germany, September 11 – 12, **2003**; Abstract A8.

Possible Mechanisms of the Stimulating Effects of Isopropoxygermatran and 1-Ethoxysilatran in Regenerated Liver

M. G. Voronkov, H. N. Muhitdinova, M. K. Nurbekov
M. M. Rasulov, V. M. D'yakov*

Institute of Chemistry, Siberian Branch of Russian Academy of Science, Irkutsk
State Research Center of the Russian Federation, State Research Institute for Chemistry and Technology of Organoelement Compounds, 38, Shosse Entuziastov 111123, Moscow, Russia
Tel./Fax: +7 95 523 0142
E-mail: florasi@online.ru

Keywords: hepatic resection, protein synthesis, aminoacyl-tRNA-synthetases, silatranes, germatranes

Summary: Stimulating effects of isopropoxygermatran (IPG) and 1-etoxysilatran (ES) on liver recovery occur through activation of protein-synthesizing components, aminoacyl-tRNA-synthetases (ARSes) in particular. Increase in the activity of preparations of total ARSes has been shown. Several aspects of the mechanisms of stimulating effects of MA were discussed.

Introduction

Since the 1960s, a new class of biologically active organoelement compounds, silatranes, has been receiving much research attention [1]. Their germanium analogs, germatranes, proved to be not less interesting from the biological activity viewpoint [3]. These two classes possess a whole set of useful properties including promoting liver regeneration [2]. However, the metabolic processes responsible for development of the compensation phenomenon on the whole organ level under the influence of silatrane and germatrane have not yet been identified.

One of the first stages of the chemicals effect on a cell consists in its link with a receptor. Then, the signal about this event may be transferred to other intracellular components in cytoplasm or nucleus. This, in its turn, causes changes in cell activity, namely its biosynthetic activity and the activities of individual enzymes, their complexes, and even whole organelles. In an intricate process of cell response to a signal, a key role belongs to bioprotein and the biosynthesis of its derivatives. Therefore, analysis of ES and IPG effect on mitotic activity and cell sizes of the regenerating liver as well as protein synthesis dynamics is of particular importance.

Investigation Procedure

The experiments were carried out on outbreed male rats of 150 –180 g weight. We resected $^2/_3$ liver mass by the Higgins-Andersen method. The operated rats were divided into two similar groups for control and experiment. Immediately after the operation, ES in a dose of 10 mg/kg intramuscularly or IPG in a dose of 50 mg/kg intraperitoneally were administered only once to the experimental group animals, and 24 h later the injections were repeated. An appropriate amount of standard physiological saline was administered to a control group of rats. Euthanasia was performed 48 h after the operation termination, i.e. 24 h after the second ES administration.

Livers of one experimental group were extracted and fixed in formalin. After fixing, liver pieces were embedded in paraffin and then slices of up to 10 μm thickness were made. The slices were pigmented by hematoxylin-eosin. An average area of a single hepatocyte as well as cell quantity in various mitosis phases were calculated in terms of a thousand elements under microscopical investigation.

Livers of the other experimental group were extracted, reduced to fine particles and frozen in liquid nitrogen for further experiments.

With the purpose of protein fraction isolation of aminoacyl-tRNA-synthetases (ARS), the tissue was triturated from the frozen state, the produced powder was extracted in weak saline neutral buffer in the presence of protease inhibitors [4]. After depositing microsome and membrane fractions, total ARS fraction was obtained by bringing pH to 5.0, protein deposit was dissolved in 0.05 M tris-HCl-buffer, 0.1 mM 2-mercaptoethanol, and 30% glycerin and stored at –20 °C.

Total tRNA and 2% tRNA from yeast and bull's liver were obtained according to a previously described procedure [5]. An enriched fraction of specific tRNA was obtained by preparative electrophoresis in polyacrylamide gel [6]. The enriched tRNA preparations contained from 75 to 150 nmoles of tRNA.

Enzymatic activity of total ARS fraction was registered by inclusion of ^{14}C and ^{3}H labeled amino acids (tryptophan, phenylalanine, and lysine) after enzymatic reaction product (labeled acylated total and/or enriched tRNA fraction) was collected on nitrocellulose supporting filters. Radioactivity was determined by means of an LS-980 counter (Beckman, USA) in a toluene-PPO-POPOP system. ^{14}C-label efficiency amounted to 90% and that of ^{3}H-label 25%.

Student's technique was employed for statistical data manipulation.

Investigation Results

Morphometric analysis demonstrated that the area of hepatocytes in reference group animals varied within the range of 450 ± 20 μm^2, and in control groups 460 ± 25 μm^2. In animals that received ES or IPG, the area of hepatocytes' amounted to 650 ± 40 μm^2 in both cases, i.e. authentically increased ($p < 0.01$) both in regard to reference and control. Total cell quantity in different mitosis phases amounted to 20 ± 3% of total cell quantity in reference rats, 23 ± 4% in control animals, and among rats that received ES or IPG the total amount of fissionable cells increased by 36% on average in comparison with reference ($p < 0.01$) and achieved 32.5% in the case of ES and 38.5% in the case of IPG. These data authentically ($p < 0.05$) exceed almost by 30% and 55%, respectively, the corresponding index in control rats. Unit weight of the cells in metaphase and anaphase in reference rats amounted to 42% of the umber of cells n in the mitotic state, 47% in control animals, and among animals who received ES or IPG it grew in comparison with control and amounted to 57% and 62%, respectively ($p < 0.05$). The mitotic index of hepatocytes in metaphase and anaphase in intact rats was 9.2 ± 1.7, and in control animals 10.5 ± 1.8. In the case of rats who received ES or IPG, this index increased in comparison with control in both cases and reached 17 ± 2.1 at $p < 0.01$.

The results of preliminary kinetic studies intended to determine the optimum concentration of the total ARS preparation, when reaction product accumulation kinetics follows a linear pattern in respect of every used labeled amino acid, are presented in Fig. 1. The latter demonstrates the

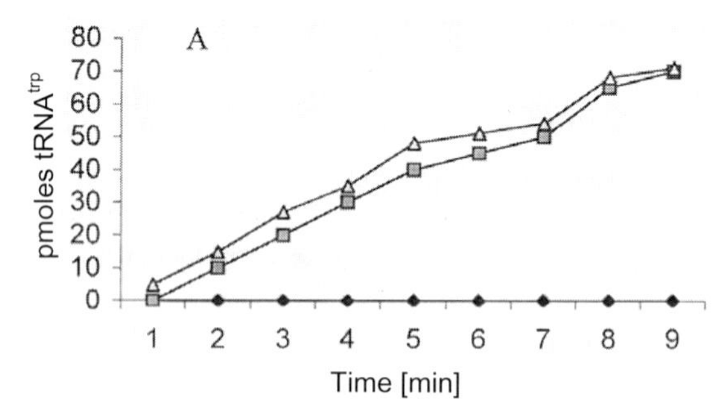

Fig. 1A. $tRNA^{Trp}$ acylation kinetics with total ARS preparations. Relationship between amount (in pmoles) of acylated tRNA produced in the reaction and incubation period with enzyme preparation is demonstrated. (--□--) — control ARS preparation, (--Δ--) — experimental preparation, (--♦--) — internal control over background inclusion (BSA was added instead of enzyme in similar concentration).

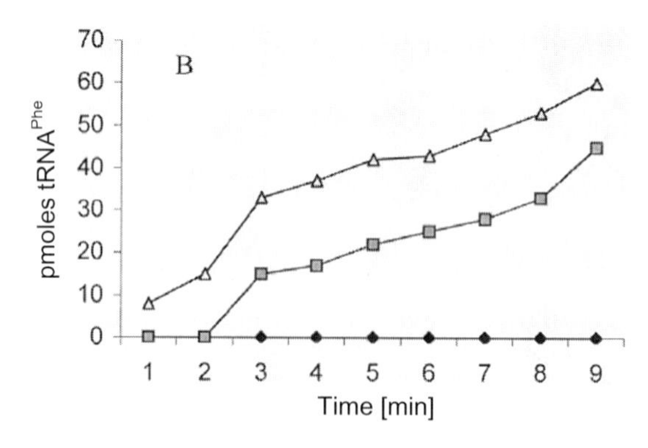

Fig. 1B. $tRNA^{Phe}$ acylation kinetics with total ARS preparations. Relationship between amount (in pmoles) of acylated tRNA produced in the reaction and incubation period with enzyme preparation is demonstrated. (--□--) — control ARS preparation, (--Δ--) — experimental preparation, (--♦--) — internal control over background inclusion (BSA was added instead of enzyme in similar concentration).

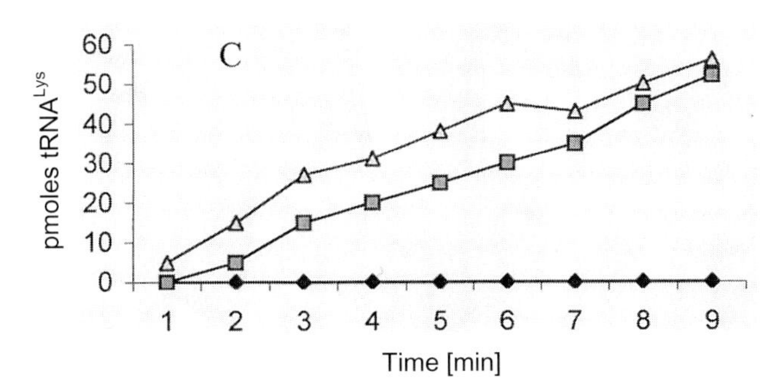

Fig. 1C. tRNA^{Lys} acylation kinetics with total ARS preparations. Relationship between amount (in pmoles) of acylated tRNA produced in the reaction and incubation period with enzyme preparation is demonstrated. (--□--) — control ARS preparation, (--Δ--) — experimental preparation, (--♦--) — internal control over background inclusion (BSA was added instead of enzyme in similar concentration).

relationship between quantity of aminoacyl–tRNA (AA–tRNA) produced in the reaction and in the incubation period. Therefore, in the chosen system, amino acid inclusion change in tRNA testifies to actual processes: either to ARS content increase in a cell or to the change of their specific activities. ES and IPG influence one of the most important protein synthesis reactions. tRNA aminoacylation, largely defining protein synthesis efficiency and velocity, was also analyzed.

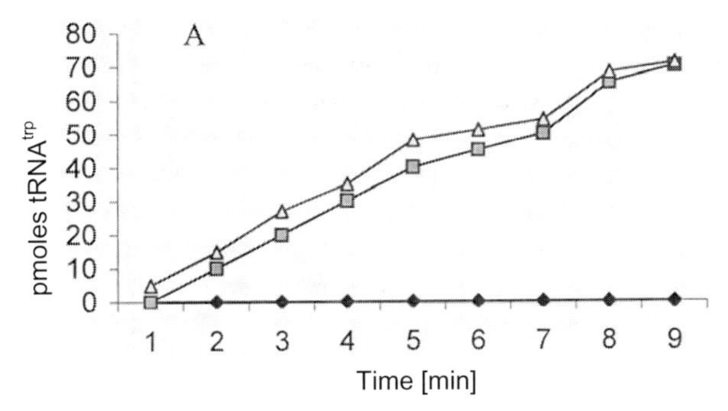

Fig. 2A. tRNA^{Trp} acylation kinetics with total ARS preparations. Relationship between amount (in pmoles) of acylated tRNA produced in the reaction and incubation period with enzyme preparation is demonstrated. (--□--) – control ARS preparation, (--Δ--) – experimental preparation, (--♦--) – internal control over background inclusion (BSA was added instead of enzyme in similar concentration)

When tRNA acylation kinetics was studied on tryptophanyl-ARS-(^{Trp}tRNA) and lysil-ARS-(^{Lys}tRNA) pairs, linear dependence between the aminoacylation reaction and the incubation period was identified just as in the preceding case. Phenylalanine tRNA acylation kinetics also testified to linear dependence. The results of this series of experiments are presented in Fig. 1 (A, B, and C) and Fig. 2 (A, B and C). In Fig. 1, also, are presented ES effects, and in Fig. 2 the influences of IPG on the protein-synthesizing cell mechanism is presented. Figure 2 demonstrates that ES and IPG affect the content of the protein-synthesizing system (enzyme complex components catalyzing the major stage of protein biosynthesis in a cell); tRNA-specific aminoacylation is manifestly increased. The difference between the stimulating effects of ES and IPG is not great; significant increase in the IPG-stimulating effect is observed in the Phe-$tRNA^{Phe}$ pair (Fig. 2B), which correlates with the large IPG effect in morphometric tests.

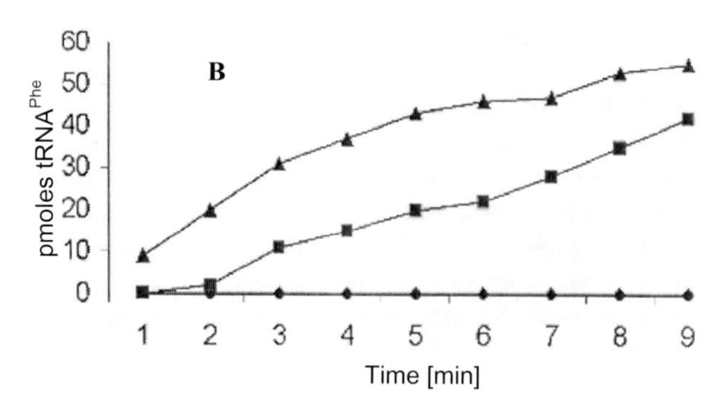

Fig. 2B. $tRNA^{Phe}$ acylation kinetics with total ARS preparations. Relationship between amount (in nmoles) of acylated tRNA produced in the reaction and incubation period with enzyme preparation is demonstrated. (--□--) — control ARS preparation, (--Δ--) — experimental preparation, (--♦--) — internal control over background inclusion BSA was added instead of enzyme in similar concentration).

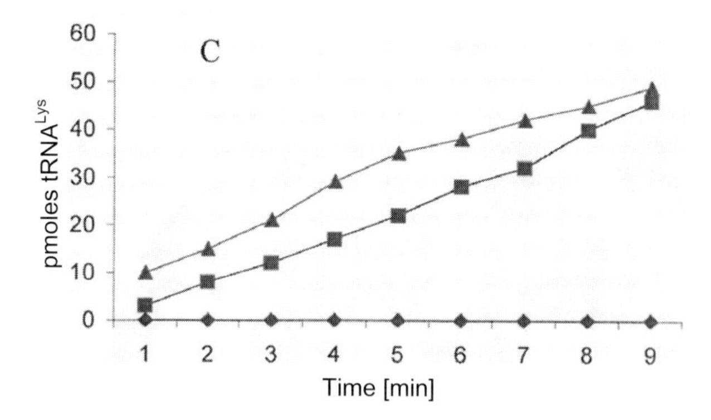

Fig. 2C. $tRNA^{Lys}$ acylation kinetics with total ARS preparations. Relationship between amount (in nmoles) of acylated tRNA produced in the reaction and incubation period with enzyme preparation is demonstrated. (--□--) — control ARS preparation, (--Δ--) — experimental preparation, (--♦--) — internal control over background inclusion BSA was added instead of enzyme in similar concentration).

Results and Discussion

It is known that when xenobiotics containing nitrous heterocycles and organosulfur compounds are administered into an organism, coordinated expression of enzymes, participating in liver detoxication [7, 9], proceeds. This process is sharply promoted by $CdCl_3$. The coordinated expression is explained by common *cis*-regulatory structures in these enzymes' genes. In the authors' opinion, $CdCl_3$ affects a general stage in detoxication enzymes expression [7, 9]. $CdCl_3$ blocks Ca^{2+}-channels of a cell membrane. This suggests that mEH (microsomal epoxide-hydrolase) expression suppression induced by $CdCl_3$ is accomplished through competitive inhibiting of Ca^{2+} transport into a cell [10]. The effects of the $CdCl_3$ ions can also be related to protein kinase C (PKC). GST and mEH genes expression suppression under the $CdCl_3$ effect can be related with PKC activity. Consequently, changes in enzyme expression induced by $CdCl_3$ can be connected with Ca^{2+} intracellular control [6]. As a number of the studied ARSes are metal enzymes, analysis of the interaction of organoelement compounds with metal ions in metal enzymes controlling reparative processes in cells and whole organs of the body is of vital importance.

Thus, IPG and ES administration in doses of 10 mg/kg and 50 mg/kg dramatically facilitates liver regeneration. The analysis of this promoting effect allowed us to observe significant increase in cell protein synthesizing activity on the example of one of its most important components — aminoacyl-tRNA-synthetases. As a number of analyzed ARSes are zinc-containing metalloenzymes [8], organoelement compound interaction with metal ions in metalloenzymes assumes major importance in the provision of reparative process control in organism cells and organs.

References

[1] M. G. Voronkov, V. M. D'yakov, *Silatranes*, Novosibirsk Nauka, **1978**, p. 207.

[2] V. F. Kononenko, L. T. Moskvitina, A. T. Platonova, in *Biologically Active Elements of Group IV B*, Abstracts of All-Russia Symposium, Irkutsk, **1975**, p. 17.

[3] I. G. Kuznetsov, M. M. Rasulov, *Izvestiya AN SSSR*, **1991**, *N3*, 342 – 351.

[4] I. G. Kuznetsov, B. M. Gukasov, M. M. Rasulov, *Khim.-Farm. Zhur.* **1986**, *N9*, 1035 – 1038.

[5] E. Ya. Lukevits, T. K. Gar, M. M. Ignatovich, V. F. Mironov, in *Biologic Activity of Germanium Compounds*, Riga: Zinatne, **1990**, p. 191.

[6] L. A. Mansurova, A. T. Platonova, I. G. Kuznetsov, M. G. Voronkov, In *Biologically Active Compounds of Group IV B*, Abstracts of All-Union Symposium, Irkutsk, **1975**, p. 28.

Wound Healing Effects of some Silocanes and Silatranes

V. M. D'yakov, M. G. Voronkov, V. B. Kazimirovskaya, S. V. Loginov, M. M. Rasulov*

State Research Institute for Chemistry and Technology of Organoelement Compounds,
38, Shosse Entuziastov, 111123, Moscow, Russia
Tel. /Fax: +7 95 523 0017
E-mail: florasi@online.ru

Keywords: connective tissue, regeneration, protein synthesis, RNA, silocanes, silatranes, glycosaminoglycanes

Summary: Stimulating effects of some silatranes and silocanes leading to activation of corneal reparative regeneration after mechanical trauma, protein-synthesizing components, and cicatrizing of experimental burns of rats were studied. Several aspects of the mechanisms of these stimulating effects are discussed.

Introduction

Silatrane physiological activity was discovered by one of the authors in the early 1960s. Most publications deal with the silatrane effect on connective tissue regeneration, including osteosynthesis, hair and wool growth, and healing of wounds and burns. Their monocyclic analogs are of no less practical interest: silocines $R_2Si(OCH_2CH_2)_2NR'$ and silocanes $RR''Si(OCH_2CH_2)_2M$, where M = O, S, Se, CH_2; R = alkyl, aryl, halogenalkyl, acyloxyalkyl et al., R′ = H, CH_2CH_2OH, alkyl, aryl, R″ = alkyl, aryl. Silatranes were previously shown to promote protein synthesis in regenerating rat liver.

We paid particular attention to the study of physiological and therapeutic effects of silatranes and silocanes on one of the thinnest and most complicated varieties of connective tissue — cornea. Silatranes' effect on cornea was studied after mechanical trauma, and after alkaline and acidic burns. We also studied the effect of silocanes in healing tests with wounds and burns.

The results of studying the chloromethylsilatrane effect on eye cornea after mechanical trauma are presented. Among a great number of investigations on tissue regeneration, little attention is paid to cornea recovery. Meanwhile, corneal injuries of different types are a major cause of impairment of eyesight. In most cases, traumas cause formation of coarse scar tissue (leucoma). Our previous investigations let us to suppose that (chloromethyl)silatrane (CMS) should also stimulate reparative processes after injuries to the cornea. Therefore, we present the first results of the study of the effect of CMS on processes of reparative regeneration of the cornea after mechanical trauma.

Investigative Technique

All experiments were carried out on 75 male chinchilla rabbits of mass 1.5 – 2.0 kg. The animals were divided into the following groups:

1st — standard, intact, 5 individuals (10 eyes)

2nd — control, 35 individuals (70 eyes) — animals treated with physiological solution

3rd — 35 individuals (70 eyes) — animals treated with CMS.

In animals of the control and experimental groups after eye treatment with 0.5% dicain solution, a mechanical trauma in the central part of the cornea was produced by a trepan. Wound depth achieved was half of the corneal thickness; diameter was 4 mm.

From the first hour of trauma, 0.8% aqueous solution of CMS (pH 5.5 – 6.0) was administered to eyes of the experimental group 3 times a day for 5 days.

Animals of the control group received physiological salt solution according to the same scheme.

Histological studies of cornea were carried out on the next, fifth, tenth, twenty-fifth, and sixtieth days after trauma. Treatment of sections was conducted by conventional histological methods.

Results

On the next day after trauma to the cornea, necrotic destructive and inflammatory processes developed violently in the control animal given physiological salt solution. The wound base was filled with albumin mucous effusion with different hematogenic cell impurities, segmento-nuclei neutrophiles, mainly, between the exudate and the wound cavity base. Collagenic fascicles of corneal substance itself had chemosis and were homogenized; between them single leucocytes were observed. Part of the keratocytes degenerated. Among the remaining cells, single active prophases with a ball configuration were observed. Cells of the back epithelium from the periphery to the center of the trauma zone degenerated progressively. The fore-epithelium of the cornea near to the lip of the wound gradually became thinner, its cells became flat and stretched, the number of epithelium layers decreased, and finally, thin simple epithelium covered the surface of the side walls of the wound defect under the wound scab like a tongue.

A feature of the traumatized rabbit cornea after receiving CMS was the intact state of the back epithelium even opposite to the wound defect. Intervals between collagen fiber fascicles of the active substance of the cornea in the trauma zone were considerably marked, in spite of the chemosis. In the frontal epithelium at the lips of the side walls, intensive cell growth was observed. Among the keracyte (fibroblasts) in the area of trauma and in the adjacent zone, particularly, cells in mitosis were observed more often than in the control ones.

On the fifth day after trauma in the case of control animals, intensive growth of epithelium regenerated from peripheral sites of the wound defect with immersion into subjacent poorly differentiated connective tissue containing necrotic structures of traumatized cornea was continued. Toward the trauma center, regenerating epithelium became markedly thinner, with pyknosis of its cells and desquamated epithelium layer. Fibroblasts of traumatized sites, in contrast to cells observed over the previous period, were appreciably activated. This was clearly manifested near rim zones with epithelium growth deep down. In the central wound sites, chains of activated fibroblasts and fibrillogenesis symptoms occurred.

A distinctive status of the eye cornea morphology of animals given CMS was epithelization of the whole wound surface. Epithelium growth with immersion into sub-epithelial layer was not observed. Peripheral zones of the wound were covered with a two- to three-layer epithelium wherein a considerable quantity of cells in mitosis was observed. In this zone, collagenic fibers of the cornea itself revealed chemosis, but regulation was observed. Epithelium of adjacent zones comprised four to five layers. In the central sites of the traumatized zone, fibrillogenesis proceeded with higher activity. In contrast to the control and connective tissue, elements at the wound base contained a few necrotized structures, and much newly formed poorly differentiated connective tissue appeared.

On the eleventh day after operation, in the case of the control animals a complete epithelization of the whole traumatized corneal surface took place. However, the thickness of the regenerating bed was not uniform over its entire surface. Thickened sites with epithelium growing deep down alternated with thin sites poorly bonded to subjacent connective tissue of the cornea itself. On the periphery of the cornea itself, the cells were more differentiated; fascicles of friable collagenic fibers were located at random, inter-fascicle glottis was poorly expressed, and a small amount of chemosis of tissue and cell swelling was observed.

The feature of traumatized animal cornea treated by CMS was almost normal recovery of most of the peripheral zone structure, which demonstrated the intensification of regeneration processes. Anisomorphy of epithelial regenerate was evidently expressed in the vertical state; its surface was smooth. Collagenic fibers and their fascicles were arranged in order; cornea itself was without chemosis. Newly formed tissue structure of central trauma sites was less differentiated than that of peripheral ones. In this case, epithelium and regenerated cornea itself (in comparison to control) had a more regulated structure.

On the twenty-third day of the experiment, in the case of control animals all traumatized cornea surface was completely epithelized. However, the thickness of the epithelium layer was not uniform — thin sites alternated with sites of considerable thickness. Epithelium growth with immersion to poorly differentiated regenerating cornea itself was observed. In the cornea itself, intensive fibrillogenesis appeared markedly, and a collection of fibroblasts in mitosis was observed. At these sites, newly formed collagenic fibers and their fascicles were still arranged at random. Connection of epithelium with subjacent conjunctive tissue of cornea in some instances was not very strong. Therefore, epithelium was easily peeled away in the process of preparing formulations. The surface of epithelial regenerate became notched, which means that the recovering processes and restructure of corneal tissue components was continued with sufficient intensity.

Processes of regeneration and restructure of the cornea were practically complete for all animals that received CMS. Only in one case, there was on the cornea a limited site wherein the cornea itself was in a relatively poorly differentiated state. Meanwhile in this case, collagenic fibers and their fascicles occurred in a sufficiently ordered state in comparison to control animals, and regenerated epithelial layer was thin and had a smooth surface.

On the sixtieth day (remote observation time), among control animals, in 50 cases out of 70, in the cornea center leukoma occurred of varying density and dimensions. At the same period, a slight wall in the center of the cornea of rabbits given CMS was observed in 7 cases out of 70 (8.5%).

Obtained data demonstrate that dropping 0.8% aqueous CMS solution (pH 5.5 – 6.0) into eyes after corneal trauma intensifies inflammatory reaction and activates early processes of elimination of necrotic tissue at the point of trauma. CMS provides limitation of pathological process development in the cornea, and everywhere stimulates processes of wound epithelization. Its use ensures histo-characteristics of the epithelium layer, and arranges orderly fibers of the corneal substance itself. This provokes full corneal recovery and prevents leukoma.

We have studied the therapeutic action of monocyclic silatrane analogs using the example of the first obtained silocane derivatives of metacryloxy acid and chloromethylsilocane(CMMS).

We have synthesized previously unknown silocyne, silocane, and silatrane derivatives of acrylic acids according to Scheme 1a in the presence of polymerization inhibitors. Acrylates are also obtained by concurrent synthesis on the basis of previously described halomethylsilocanes via Scheme 1b.

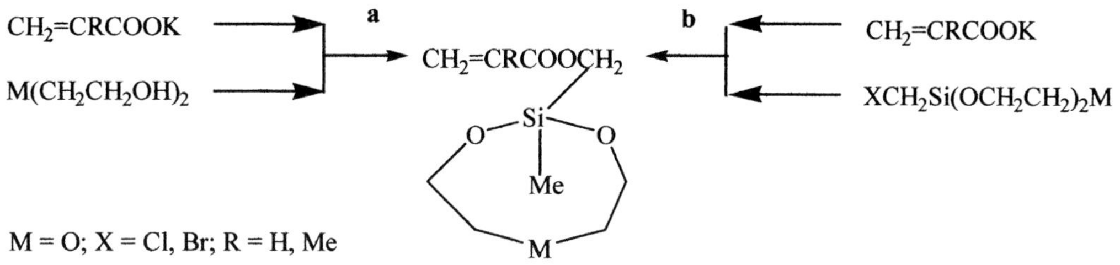

$CH_2{=}CRCOOK$ + $M(CH_2CH_2OH)_2$ —a→ $CH_2{=}CRCOOCH_2$–Si(Me) in ring O–Si–O…M ←b— $CH_2{=}CRCOOK$ + $XCH_2Si(OCH_2CH_2)_2M$

M = O; X = Cl, Br; R = H, Me

Scheme 1.

Silatrane acrylates are obtained in accordance with Scheme 2. In contrast to silocanes their silatrane analogs present the crystalline substances that are not susceptible to easy polymerization. Their toxicity and wound cicatrizing effects have been studied.

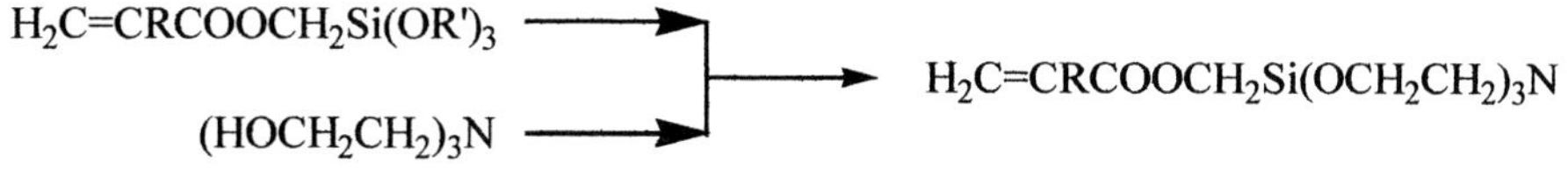

$$H_2C{=}CRCOOCH_2Si(OR')_3 + (HOCH_2CH_2)_3N \longrightarrow H_2C{=}CRCOOCH_2Si(OCH_2CH_2)_3N$$

M = O, NR″; R″ = H, Me, Ph; R′ = Me, Et; R = H, Me

Scheme 2.

All the silocane acrylates even with R = Ph are viscous high-boiling liquids with characteristic odor. They are not toxic ($LD_{50} \geq 2.0$ g/kg).

Their effect on the regeneration of connective tissues at wound defects and deep thermal burns has been studied. In experiments carried out and designed for revealing biological activity of obtained compounds (**A** and **B**), it has been found that they considerably accelerate wound cicatrizing (see Table 1).

CH_2=C(Me)COOCH_2–Si(Me)(OEt)$_2$

Compound A

CH_2=C(Me)COOCH_2–MeSi(OCH_2CH_2)$_2$O

Compound B

Table 1. The effect of alkoxysilane (**A**) and silocane (**B**) on the time required for cicatrizing skin wounds of rats.

Time for primary (I) and secondary (II) scab fall-off [day]							
Control		**Placebo**		**B**		**A**	
I	**II**	**I**	**II**	**I**	**II**	**I**	**II**
9	18	10	19	8	14	8	16
10	19	10	20	8	18	8	16
10	20	10	20	8	18	9	17
11	21	11	21	9	19	9	17
11	21	11	22	9	19	9	18
12	21	12	22	9	19	9	18
12	22	12	23	9	20	9	18
12	23	13	23	10	21	9	18
13	23	14	24	10	21	10	19
13	23	14	27	10	–	10	20
11.3	21.1	11.7	22.1	9.1	18.8	9.0	17.7
±0.84	±1.05	±0.84	±1.68	±0.42	±1.41	±0.21	±0.84

As shown in Table 1, both silane (**A**) and silocane (**B**) activate the cicatrizing of non-complicated skin wounds in experimental animals. It should be noted that for **B** this activation is practically complete before the primary scab falls off; the interval between primary and the secondary scabs remains unchanged. The wound cicatrizing effect of **B** is determined by its influence on granulation-fibroid tissue formation, i.e. the connective tissue component of the cicatrizing process. Silane **A**, along with the same effect on the first process stage, shortens (on the average by one day) the second one — the stage of wound defect epithelization. As a result the total activation effect of **A** is found to be more marked.

Silocanes are no less effective than silatranes in the activation of cicatrizing of burns. Thus, silocanes improve and accelerate scab fall-off (see Table 2).

Table 2. Effect of some silocanes (in comparison with solkoseril)* on cicatrizing of experimental burns of rats.

Time for the primary (I) and secondary (II) scab fall-off [day]							
Control		**B**		**A**		**Solkoseril**	
I	**II**	**I**	**II**	**I**	**II**	**I**	**II**
15	22	13	25	12	24	18	26
15	25	14	25	14	24	19	27
17	25	15	26	14	25	20	29
19	30	16	26	17	26	22	30
21	31	18	28	18	29	23	31
17.4	27.2	15.2	26.0	15.0	25.6	20.4	28.6
±3.06	±4.59	±2.55	±1.53	±3.06	±2.55	±2.55	±2.55

* High efficiency preparation in clinical practice.

From the present results, it follows that both silocane (**B**) and silane (**A**) provide reliable activation of cicatrizing of the deep non-complicated skin thermal burn. The silocane effect on the average time for primary scab fall-off, as completion of granulation-fibroid tissue development at the site of burn necrosis, was more pronounced; the time required is reduced by almost 15%. Somewhat less pronounced (about 6 – 7%) was the activation of secondary scab fall-off, but it is also of practical value, taking into account the fact of non-complicated burn cicatrizing in comparison with solkoseril (a high efficiency preparation in clinical practice), but turned out to be non-effective for a non-complicated burning process in this experiment.

The fact that bioactive silicon in molecules both of silatranes and silocanes intensifies biosynthetic processes (see Table 3) is of great importance.

Table 3. Silocane effect on some biochemical characteristics of granulation-fibroid tissue.

Characteristics	Control	Placebo	B	A
Tissue mass, g	1.51±0.12	1.51±0.22	1.10±0.18	1.13±0.11
Ribonucleic acid*	2.15±0.05	2.26±0.05	2.41±0.01	1.56±0.12
Hydroxyproline*	2.57±0.07	2.71±0.04	2.21±0.07	1.96±0.03
Hydroxylysine*	0.21±0.03	0.23±0.01	0.22±0.01	1.14±0.03
Hexauronic acids*	0.76±0.03	0.76±0.06	0.75±0.02	0.86±0.02
Fractions GAG**				
0.4	45.9	55.2	51.1	50.5
1.2	38.9	27.7	35.9	30.8
2.1	15.2	17.1	13.0	18.7

* g/100 g dried degreased tissue. ** % of total glycosaminoglycans quantity

The average mass of formed granulation-fibroid tissue under the influence of both silocanes decreases. **B** causes decrease in concentration of ribonucleic acids, which demonstrates the increase of the rate of biosynthetic processes in tissue. Collagen concentrations, determined with hydroxyproline, remain nearly at the level of the control series (**A** even causes decrease), but at the same time, a considerable decrease takes place in the concentration of hydroxylysine – an amino acid that is specific for collagens and takes a direct part in cross-bonding of collagenic molecules. It means that in this case collagens differ in lower structure stability (maturity), and this situation may give favorable conditions for morphogenesis of a wound scab, particularly for mesenchymal-epithelial interaction. In addition, a shift in the distribution of glycosaminoglycanes (GAG), determined as their cetylpyridine derivatives, may provide these conditions. A decrease in the fraction dissolved in a 0.4 M solution of sodium chloride comprising mainly hyaluronate is noted; accumulation of the latter is characteristic for non-mature granulation-fibroid tissue. At the same time, fractions of "mature" GAG (predominantly chondroitin-sulfates and heparin) dissolved in 1.2 and 2.1 M solutions of sodium chloride are increased.

The results of our bio-chemical investigations of granulation-fibroid tissue activation are determined by change in the dynamics of metabolic processes in developing a wound scab. Thus, it may be concluded that silocanes under study are able to directly influence the functional state of connective tissue cells participating in the cicatrizing process by activation of their morphogenetic potentials.

Thus, the experimental tests carried out show that (2-methyl-2-metacryloxymethyl-1,3,6,2-thioxasilocane) and (metacryloxymethyl)methyldiethoxysilane are able to activate non-complicated experimental wounds and deep thermal skin burns when these preparations are used as single applications of liniments of high concentration (5%) to the fresh wound surface.

References

[1] M. G. Voronkov, V. M. D'yakov, *Silatranes, Novosibirsk Nauka.* **1978**, p. 206.

[2] V. M. D'yakov, V. B. Kazimirovskaya, M. G. Voronkov, M. M. Rasulov, *Medicina Altera* **2001**, *4*, 15 – 20.

The Development of Methods of Synthesizing Organic Derivatives of Silicon Based on Biogenic Silica

Julia Ubaskina,* Yevgeny Ofitserov

Faculty of Experimental-Chemical Pharmacology, Biochemistry and General Chemistry, Department of Medicine, Ulyanovsk State University, Tolstoi Street 42, 432670, Ulyanovsk, Russia
E-mail: baseou@mail.ru

Keywords: synthesis of silicon organic compounds, biogenic silica, silica solubility, silica soluble form stabilization, vicinal dihydroxy-organic compounds

Summary: The possibility of direct synthesis of silicon organic compounds based on hydrolysis-resistant organic derivatives of silicon by using biogenic silica (from siliceous rocks) as a new raw material is discussed. The complex triethylphosphate with silica, the ammonium salts of tricatechol, and humic acids of monosilicic acids were obtained. Products were identified by proton magnetic resonance spectroscopy, IR spectrophotometry, or elemental analysis.

Silicon-organic compounds (SOC) are increasingly used in all branches of world manufacture. An ecologically pure, economically effective, technologically simple method of SOC synthesis is required. The existing methods of SOC synthesis based on crystal silica require great energy expenditure and the use of chlorinating agents. The wastes from these agents pollute the environment.

Non-chlorine methods that are based on obtaining basic organic compounds of silicon from artificial amorphous silica have many stages and are ineffective, as they use the silica obtained by hydrolysis of tetraalkoxysilane or silicon tetrachloride. So it became necessary to find methods of SOC synthesis based on natural amorphous silica. Such methods are currently being developed by some authors [1 – 3].

According to the literature, such chemical processes exist in natural systems; however, the mechanisms and methods of these syntheses have not yet been found.

Amorphous biogenic silica contained in siliceous rocks is not used widely enough as a raw material for chemical manufacture. So the creation of SOC synthesis methods on a basis of biogenic silica will be able to make these manufacturing methods simpler, to expand the use of siliceous rocks, and to ascertain how the same synthesis happens in nature.

The aim of our work became to explore the possibility of using biogenic silica for SOC synthesis and to develop synthesis methods for basic organic derivatives of silicon from biogenic silica.

In order to fulfill this aim, we set out to study the solubility of biogenic silica, leading to the possibility of its use in SOC synthesis, and factors influencing this solubility, to investigate the methods of stabilization of the obtained soluble forms of silica, and to synthesize basic organic derivatives of silicon that are stable to hydrolysis and oxidation.

The dynamics of the process of dissolution and sorption of biogenic silica from siliceous rocks, diatomite and opoka, has been studied.

It is shown that the solubility of siliceous rocks was 18.6 times higher than the solubility of quartz but 2.3 times lower than the solubility of amorphous silica. The soluble form of silica, monosilicic acid, is unstable. First the concentration of silica in solution increases and reaches a maximum; then it decreases because of processes of sorption and polycondensation of soluble forms.

It has been found that regulators of silica concentration in solution include compounds of Al and Fe.

We found that the solubility of silica decrease with increase in its degree of crystallization and also decreased with increase in the extent of sorption and polycondensation of soluble forms of silica.

The factors increasing the solubility of biogenic silica are: the degree of milling of the rock (the higher the degree, the higher the silica solubility), medium pH in diapason 1 – 4, more than 8, the presence of silicate, elimination of fermentation products or microorganisms producing them, and the presence of organic reagents with vicinal hydroxy-groups, in particular, catechol, tannin, and humic acids.

We determined that, given the presence of vicinal hydroxy-organic compounds, the silica solubility is connected with the formation of basic organic derivatives of penta- and hexacoordinated silicon. We have proposed a scheme of silica dissolution in a saturated solution of catechol on the basis of literature information [1, 4] and our experimental data.

According to this scheme, the first phenoxyate anions are formed.

At the first stage of dissolution of silica in a saturated solution of catechol, the nucleophilic substitution occurs on the silicon atom.

The second stage is analogous to stage I.

At the third stage the dimer catechol ether of monosilicic acid bonded with silica matrix is formed.

At the fourth and fifth stages an intermediate with the pentacoordinated atom of silicon is formed from the tetracoordinated one, and the silicon atom is returned to a stable tetracoordinated state, while splitting of the oxygen bond with silica matrix occurs.

The sixth and seventh, eighth and ninth, and tenth and eleventh stages are analogous to IV and V.

At the eleventh stage the dimer tricatechol ether of monosilicic acid is separated from the dissolving matrix.

It has been shown that the highest yield of the chemically active soluble form of silica is reached with the use of high or medium pH and vicinal dihydroxy organic reagents. The instability of the obtained soluble form of silica illustrates the fact that its stability and accumulation occur over narrow range suitable for direct synthesis of SOC, not only in manufacturing methods, but also in natural systems.

It was found that increase of temperature of monosilicic acid solution caused reduction of its stability in solution.

We determined that vicinal dihydroxy organic reagents stabilize the soluble forms of silica. The stability of monosilicic acid solution is determined by the structure of the stabilizer: ethylene glycol and humic acids do not stabilize soluble forms of silica enough, but glycerin and catechol stabilize the silica solution when they are added at 5 – 7% to the solution. This fact is connected with the formation of hydrogen bonds and stable penta- and hexacoordinated compounds, preventing the processes of polycondensation of silica in solution.

We then began to synthesize organic compounds of silicon stable to oxidation and hydrolysis.

The complex of triethylphosphate with biogenic silica of opoka was obtained by acidification of liquid glass, leached from opoka, and by extraction of the obtained silica gel by triethylphosphate followed by salting out of the organic part by NaCl. The obtained viscous product, of an amber color, which was the complex $(EtO)_3PO \cdot nSiO_2 \cdot mH_2O$, was analyzed by IR and NMR spectroscopic methods.

The formation of the $(EtO)_3PO \cdot nSiO_2 \cdot mH_2O$ complex was confirmed by IR spectrophotometry. The IR spectrum contains stretching vibrations of Si–OH groups with ν(max) 3360 cm^{-1}. The presence of a broad band at the –OH frequency in the Si–OH fragment (3670 cm^{-1}) is evidence for the formation of hydrogen bonds with a phosphoryl group. The presence of hydrogen bonds in the formatting complex corroborates ν(P=O) (1248 and 1200 cm^{-1}).

Additional information was obtained from the ^1H NMR spectrum recorded in CCl_4. There is a triplet from protons of the methyl fragment of the ethoxyl group bonded to an atom of phosphorus with $\delta = 1.52$ ppm, $^3J(HH) = 7.0$ Hz, a quintet from protons of methylene fragments of an ethoxyl group bonded to a P atom with $\delta = 4.25$ ppm and $^3J(PH) = 7.0$ Hz, and a wider than usual band with δ 4.5 ppm from protons of OH groups. The ratio of integral intensities of signals of methylene protons and OH group protons is ~3:2, and we drew the conclusion that there are 4 OH groups for 1 molecule of triethylphosphate. The minimal molecular structure of the complex has to be $(EtO)_3P{=}O \cdot H_4Si_2O_6$.

It was noted that after the formation of the complex two processes occurred: the process of polycondensation in the oligosilicic fragment and the process of reetherification with formation of P–O–Si bonds such that the last process did not proceed significantly.

The formation of the $(EtO)_3PO \cdot nSiO_2 \cdot mH_2O$ complex allows us to conclude that formation of complex silicon compounds with adenosine triphosphate and phosphate-containing organic reagents in natural systems would have such a mechanism.

It is possible to synthesize SOC using $(EtO)_3PO \cdot nSiO_2 \cdot mH_2O$ (Scheme 1):

$$(EtO)_3PO \cdot nSiO_2 \cdot mH_2O \xrightarrow[\text{extraction, reetherification}]{ROH} Si(OR)_4 + 4H_2O \xrightarrow{R'MgX} (RO)_3Si{-}'R$$

Scheme 1.

However the synthesis of ethers of monosilicic acid using such a method presupposes an excess of triethylphosphate, and is characterized by low yield of product and by the necessity for very exact adherence to synthesis conditions. So the practical realization of this method is hindered.

We decided that it would be more effective to use vicinal dihydroxy-organic reagents. The ammonium salt of tricatechol ether of monosilicic acid (formed from biogenic silica from diatomite) was obtained. This salt contains the cation of ammonium, catechol anion and a hexacoordinated atom of silicon as 2:3:1. This salt is a dimer and is formed from the reaction (Eq. 1):

$$6\,C_6H_4(OH)_2 + 4\,NH_4OH + 2Si(OH)_4 \longrightarrow [(C_6H_4O_2)_6Si_2](NH_4)_4 + 12\,H_2O$$

Eq. 1.

Considering the thermodynamic parameters of the reaction, it has been determined that the silicic acid with a hexacoordinated atom of silicon $H_2[Si(OH)_6]$ formed in the process of reaction displaces ammonium ion from phenolates, so that this acid is stronger than catechol. This is in agreement with our experimental data: at the beginning of the reaction process the decrease in the pH of the reaction mixture from 7.16 to 5.85 was observed before the pH became constant.

Therefore, this thermodynamic process can be described in the following way: the first stage is replacing the ammonium ion by catechol from silicates; the second stage is replacing ammonium ion by silicic acid with a hexacoordinated atom of silicon from catecholates.

This method explains the impossibility noted by some authors of the reaction of phenol with silica: at stage I of the process phenol cannot replace the ammonium ion from silicates, because the pK of $Si(OH)_4$ is greater than the pK C_6H_5OH.

The structure of the ammonium salt of the tricatechol ether of monosilicic acid is proved by the data from elementary analysis and IR spectrometry.

The yield of salt on the base of diatomite was 32%; that is 2.4 times lower than the yield of salt on the basis of silica sol. This fact is explained by the quantity of the amorphous part in biogenic silica. Such a yield of the ammonium salt of the tricatechol ether of monosilicic acid on the basis of diatomite is in accordance with yield of liquid glass (40%) leached from native diatomite.

The ammonium salt of the tricatechol ether of monosilicic acid on the basis of biogenic silica of diatomite is the basic organic derivative of silicon.

On the basis of this salt, it is possible to obtain SOC (Scheme 2):

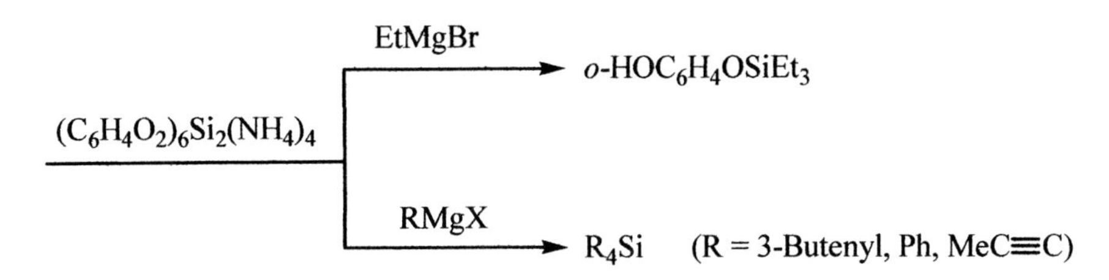

Scheme 2.

So, as we are the first to show, humic acids, being natural polyphenols, actively dissolve the biogenic silica of siliceous rocks, and we used the scheme of obtaining the ammonium salt of the tricatechol ether of monosilicic acid in order to understand the methods of synthesis of biogeochemically active SOC, formed from natural amorphous silica and polyphenols in natural systems.

We obtained the ammonium salts of humic ether of monosilicic acid from biogenic silica of siliceous rock and silicates of clay.

It follows from the literature that the empirical formula of humic acids is $C_{60}H_{52}O_{24}$.

In obtained salts, silicon cannot be present as a silicate, because this would be replaced by salts of stronger acids, such as H_2CO_3, H_3PO_4, and catechol. So we assumed, based on this fact and considering the thermodynamics of formation of $(C_6H_5O_2)_6Si_2(NH_4)_4$, that humic acids with silica form compounds with a hexacoordinated atom of silicon. Ammonium phosphate and water can be present in the product composition as admixtures from the extraction solution.

On the basis of these deductions and elementary analysis data, we deduced a formula of the ammonium salt of the humic ether of monosilicic acid from diatomite: $C_{60}H_{30}O_{24}Si_4(NH_4)_{15}(PO_4)_{10}$. This formula is confirmed by IR analysis of the salt (in comparison with the IR spectra of humic acids and diatomite): the decreasing of ν(OH groups), ν(H–OH) (4000 – 3600 cm^{-1}), displacement of electron density from OH to N (from 3600 to 2850 – 2350 cm^{-1}), amplification of peaks of conjugate rings (1585 cm^{-1}), appearance of Si–O–C bonds (1112 cm^{-1}), formation of atomic ethers (1277 cm^{-1}), decreasing ν(P–OH) and ν(COOH) (decrease of peaks 2750 – 1250 cm^{-1} and disappearance of peak 2726 cm^{-1}).

Using our deductions and elementary analysis of the salt formed from clay, we deduced the formula of the ammonium salt of the humic ether of monosilicic acid: $C_{60}H_{52}O_{24}Si(NH_4)_4(PO_4)_4(H_2O)_4$.

This formula is confirmed by IR spectral analysis of salt (in comparison of IR spectra of humic acids and clay): increasing of ν(OH groups), ν(NH), ν(HOH) (4000 – 3600 cm^{-1}), decreasing ν(P–OH), ν(COOH) (decrease of peaks 2750 – 1750 cm^{-1}, disappearance of peaks 2726 cm^{-1}), displacement of electron density from –OH to N (from 3600 to 2850 – 2350 cm^{-1}), amplification of peaks of conjugate rings (1585 cm^{-1}), formation of atomic ethers (1277 cm^{-1}), appearance of Si–O–C bonds (1097 cm^{-1}), increase of peaks for Si–O–Si bonds (disappearance of peaks 873, 581 cm^{-1}, and decrease of size of peaks in the range from 1041 to 722 cm^{-1} and from 483 to 400 cm^{-1}).

The synthesis of these salts shows the possibility of the realization of direct SOC synthesis from silica in nature.

To conclude, it is necessary to note that the reaction ability of biogenic silica depends on its structure and solubility, determined by various factors. It has been found that soluble forms of silica could be stabilized by 5 – 7% of glycerin and catechol. It has been determined that the amorphous part of the biogenic silica of siliceous rocks formed the complex with triethylphosphate and actively reacts with polyphenols (the simple one, catehol; and complex ones, humic acids), with formation of ethers. The silicon organic derivatives and complexes formed are inert to hydrolysis.

Thus, biogenic silica can be recommended as a raw material for the direct synthesis of simple organic compounds of silicon and the further synthesis of SOC based upon it.

References

[1] A. Weiss, G. Reiff, *Z. Anorg. Allgem. Chem.* **1961**, *311*, 151 – 179.

[2] T. Kemmitt, W. Henderson, *Aus. J. Chem.* **1998**, *51*, 1031 – 1035.

[3] L.-T. Yeh, M. E. Kenney, G. N. Bokerman, J. P. Cannady, O. W. Marko, *Alkoxysilanes and oligomeric alkoxysiloxanes by a silicate-acid route* Assignee: Dow Corning Corporation (Midland, MI); Case Western Reserve University (Cleveland, OH). US 5 183 914: February 2, **1993**. Appl. No.: 692 413 Filed: April 29, 1991.

[4] H. Cheng, R. Tamaki, R. M. Laine, F. Babonneau, Y. Chujo, D. Tredwell, *J. Am. Chem. Soc.* **2000**, *122*, 10063 – 10072.

Organosilicon Chemistry VI

Edited by N. Auner and J. Weis

Further Reading from Wiley-VCH

Jutzi, P., Schubert, U. (Eds.)
Silicon Chemistry
2003. 3-527-30647-1

Auner, N., Weis, J. (Eds.)
Organosilicon Chemistry V
From Molecules to Materials
2003. 3-527-30670-6

Meyer, G., Naumann, D., Wesemann, L., (Eds.)
Inorganic Chemistry in Focus II
2005. 3-527-30811-3

Schubert, U., Hüsing, N.
Synthesis of Inorganic Materials, 2nd Ed.
2005. 3-527-31037-1

Organosilicon Chemistry VI

From Molecules to Materials
Volume 2

Edited by Norbert Auner and
Johann Weis

WILEY-VCH Verlag GmbH & Co. KGaA

Prof. Dr. N. Auner
Department of Inorganic Chemistry
University of Frankfurt
Marie-Curie-Straße 11
60439 Frankfurt am Main
Germany

Prof. Dr. J. Weis
Consortium of Electrochemical Industry GmbH
Zielstattstraße 20
81379 Munich
Germany

Library of Congress Card-No.: applied for

A catalogue record for this book is available from the British Library.

Bibliographic information published by Die Deutsche Bibliothek
Die Deutsche Bibliothek lists this publication in the Deutsche Nationalbibliografie; detailed bibliographic data is available in the Internet at http://dnb.ddb.de

Printed on acid-free paper.

Printing: Strauss GmbH, Mörlenbach
Bookbinding: Litges & Dopf Buchbinderei GmbH, Heppenheim.
Printed in the Federal Republic of Germany

ISBN-13: 978- 3-527-31214-6
ISBN-10: 3-527-31214-5

Preface

In about 1990, the idea of establishing a national conference on organosilicon chemistry was born, bringing together researchers from academia and industry for scientific discussion and exchange of experience and knowledge in order to strengthen the organosilicon-based community, particularly in Germany. When we started with the first Munich Silicon Days in 1992, jointly organized by the *Gesellschaft Deutscher Chemiker* and *Wacker-Chemie GmbH,* we were overwhelmed by the exceptionally large number of participants. The impressive number of participating students and young scientists was especially convincing evidence of the high level of interest in this meeting. The date for the first Munich Silicon Days was not chosen by accident: it marked a very exciting anniversary, the 50th birthday of the *Direct Synthesis*. Because of this celebration, the two pioneers, Prof. Eugene G. Rochow and Prof. Richard Müller, participated in this symposium and were honored with the "Wacker Silicone Award".

Impressed by the scientific quality of the contributions presented and stimulated by the large number of requests for access to information about the symposium from those scientists who could not attend it, we decided to edit the first Volume of *Organosilicon Chemistry — From Molecules to Materials*.

After four successful Munich Silicon Days, in 1992, 1994, 1996, and 1998, the time was ripe to expand the organization from a national to a European level. This happened between 1998 and 2001, and Munich Silicon Days was transformed into European Silicon Days. A European Advisory Board was established, represented equally by academia and industry, which took responsibility to organize future conferences at different places and countries. Meanwhile, the Wacker Silicone Award became a highly regarded honor for outstanding scientists from all over the world. European Silicon Days remained an appropriate venue for the presentation of this prestigious award. At the 2nd European Silicon Days in 2003, the recipient of the Wacker Silicone Award was Prof. Dr. T. Don Tilley of the University of California, Berkeley, who presented highlights of his research in transition metal organosilicon chemistry in a fascinating plenary lecture at this meeting. He continues an impressive list of former awardees.

2003 Prof. Dr. T. Don Tilley
(II. European Silicon Days, September 2003)

2001 Prof. Dr. M. Weidenbruch
(I. European Silicon Days, September 2001)

1998 Prof. Dr. R. Corriu
(IV. Munich Silicon Days, April 1998)

1996 Prof. Dr. H. Schmidbaur
(III. Munich Silicon Days, April 1996)

1994 Prof. Dr. E. Hengge
(II. Munich Silicon Days, August 1994)

1992	Prof. Dr. R. Müller and Prof. Dr. E. G. Rochow (I. Munich Silicon Days, August 1992)
1990	Prof. Dr. H. Sakurai
1989	Prof. Dr. R. West
1988	Prof. Dr. N. Wiberg, Prof. Dr. R. Tacke (junior award)
1987	Prof. Dr. R. Jutzi, Prof. Dr. N. Auner (junior award)

Despite the above-mentioned many changes over the years, we have remained committed to our task of editing the scientific contributions presented during the Symposia. The present edition of *Organosilicon Chemistry VI — From Molecules to Materials* covers the diversity of silicon chemistry as well as the scientific fascination of dealing with this element. This two-volume book is divided into six chapters reflecting the wide reach of organosilicon research. The first volume of *Organosilicon Chemistry VI* contains Chapters I–IV. Chapter I deals with the chemistry of organosilicon-based reactive intermediates, while Chapter II covers different aspects of molecular inorganic silicon chemistry. Chapter III involves the basic chemistry of silicon in the coordination sphere of transition metals, including such topics as silylene complexes, catalysis, and silicon-metal bonds. Finally, Chapter IV contains contributions on the role of silicon in organic and bioorganic chemistry. The second volume of *Organosilicon Chemistry VI* contains Chapters V and VI: Chapter V switches to the often more complex systems of organosilicon compounds in and for industrial applications, whereas Chapter VI covers the contributions on solid silicon-based materials.

The present edition of *Organosilicon Chemistry VI — From Molecules to Materials* will — like all its forerunners — essentially continue to stimulate young researchers to focus on basic silicon science and its transfer into consumer-oriented applications which provide innovative solutions to current problems. In addition, it should serve as a solid basis for learning more about modern developments in a fascinating area of inorganic chemistry which can hardly be surpassed for diversity.

March 2005 Norbert Auner and Johann Weis

Acknowledgments

First of all we thank the authors for their contributions and intense cooperation, which made this overview of current organosilicon chemistry possible. The tremendous work to achieve the attractive layout of this volume was performed by Dr. Yu Yang, and Mrs. Hannelore Bovermann helped to organize the editorial work.

Furthermore we are very grateful to Dr. Sven Holl for his very active assistance to read, compare and correct.

We thank all of them for their admirable engagement!

Prof. Dr. Norbert Auner
Johann Wolfgang Goethe-Universität
Frankfurt

Prof. Dr. Johann Weis
Wacker Chemie GmbH
München

Contents

Chapter I

Organosilicon-Based Reactive Intermediates

Chapter II

Molecular Inorganic Silicon Chemistry

Chapter III

Transition Metals in Organosilicon-Based Chemistry

Chapter IV

Silicon in Organic and Bioorganic Chemistry

Chapter V

Organosilicon Compounds for Industrial Applications

Chapter VI

Silicon-Based Materials

Chapter V

Organosilicon Compounds for Industrial Applications

Silicon Science and Technology — an Industrialist's View of the Future

James W. White

Science and Technology, Dow Corning Corporation
2200 W. Salzburg Road, Mail # C01316, Midland, MI 48686, USA.
E-mail: james.w.white@dowcorning.com

Keywords: industrial silicon chemistry, transportation industry, electronics, biotechnology, energy

Summary: This presentation reflects briefly on changes in industrial silicon chemistry over the last decade and focuses on the outlook for the next 10 years. Set within the context of turbulent global economic conditions, the discussion offers perspectives based on our experience at Dow Corning Corporation, and draws on examples from a variety of industries including transportation, electronics, biotechnology, and energy. There is much to do in silicon-based science and technology over the next decade. Scientific advancement and technological innovations will surely continue, but the old model of supplying specialty silicon-based materials alone will not thrive. We must provide our customers more choices. High quality products together with complete application and service solutions are needed. Fusion of the scientific and commercial disciplines is essential; it will be at the boundaries between the physical, biological, and engineering sciences that we will find the most productive areas. The field of industrial silicon chemistry will provide tremendous opportunities to drive innovation and growth for those who adapt and can exploit these new interfaces. The author concludes that, in these challenging times, there is real cause for optimism in our field.

Turbulent Economic Times

In the decade since the Xth International Symposium on Organosilicon Chemistry in Poznan [1], global economies have experienced boom and bust. The financial markets have fluctuated wildly, fueled by the Internet "dot com" bubble, and, to quote Federal Reserve chairman Alan Greenspan, bouts of "irrational exuberance". Japan has been in the grip of a prolonged recession and the rest of Asia has been on an economic roller coaster ride. In Western Europe and North America, the specialty chemicals business has weathered a period of stagnation, causing many companies to dramatically re-engineer, cutting costs and, of course, levels of employment for scientists. The landscape has changed, too, as companies have been acquired, consolidated, or spun off forming new entities. In the silicon field, new alliances have emerged. All this activity has taken place in a world where we have seen acts of terrorism, a war, and most recently the SARS virus, which brought business in parts of Asia almost to a halt. Together these factors add up to an environment of tremendous uncertainty – an environment that has greatly influenced the availability of R&D funding in all sectors of scientific endeavor. Whether in government, academia, or industry, none of us has been immune. In these circumstances, the *business* of silicon-related science and technology will have to change.

Despite all of this, some companies have done well, even thrived, under these very difficult conditions. There are many examples of companies that have forged ahead – against all odds – to launch innovations that have re-shaped their industries. In our own industry, Shin-Etsu Chemical has reported consistent growth over the period. And my company, Dow Corning Corporation, has emerged stronger than ever: energized, precision-focused, and creating new silicon-based solutions and services for customers around the world. So, before I speculate about the future for silicon, *what of the last 10 years*?

Undoubtedly, the science has advanced and technological innovations continue. I would characterize the period as one of "*steady* technical progress" with "*lack-luster* business growth." Genuine, break-through silicon technology is hard to pinpoint. Some of the new applications I discussed in Poznan have become commercial realities. Novel delivery systems for foam control and drug delivery, gellants for personal care products, and thermoplastic silicon vulcanizates are a few examples. In electronics, Gordon Moore's Law has proven to be remarkably accurate and has driven innovation in silicon-based materials for semi-conductor fabrication. In fact, advances have been made so rapidly that some people argue the technology has outstripped the consumer's demand, preventing industry from being paid for the full value created. Beyond semi-conductors, the market for electronic chemicals continues to expand – especially in passive packaging materials. Some high-temperature silicon-based ceramics are now well established. Yet, there are still question marks about the potential of the more exotic organosilicon-based polymers: the polysilanes, carbosilanes, and semi-conducting polymers. After 20 or 30 years of discovery research, these materials are only just beginning to find utility. Light-emitting diodes and photoresists are examples, but the major limitation, and hence challenge, to the organosilicon community is still centered on practical, efficient, and low-cost methods of production. *So What of the Future?*

To survive as suppliers of specialty silicon-based chemicals, we must adopt new approaches. It is no longer sufficient to simply supply the basic materials. The truth is that while our customers and our partners rely on break-through science and technology, they require it within the context of complete application and service solutions. To be successful today, innovation involves more than just inventing creative new products. It also requires the ability to be fully in touch with the marketplace and a willingness to reassess everything we do through the eyes of our customers. This presents us with a dilemma and a challenge to create this kind of "fusion" – how to continue to add value for our customers even as sources of research funding dry up. Yet, I contend that interest in silicon science has not diminished, and prospects for the specialty chemicals industry in the silicon space are improving. Innovations will continue, but as they do, scientists in our universities and government institutions will need to collaborate evermore closely with industry to bring our work to economic fruition. Simply put, governments and academia have key roles to play, alongside industry, in the future commercial development of the field.

As is often the case in periods of economic downturn, the rate of generation of intellectual property has in fact increased; we are now seeing more than 5 000 US patent filings of interest every year in the *silicones* business. If we broaden the search to include silicon metal and inorganic silicate materials, the figures triple, as shown in Fig. 1.

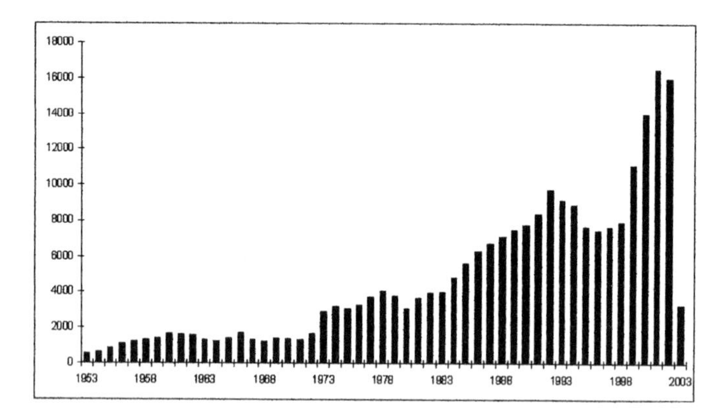

Fig. 1. Organosilicon Patents issued 1953 – 2003. Source: Dow Corning Corporation internal collection (1953 – 1973), Derwent World Patent Index search for Si patent families (1974 – present).

It is in this context, without being limited to *organo*-silicon *chemistry*, that I think we must look for future opportunities in our industry. We must explore *both* the organic and inorganic domains for solutions. The *boundaries* between the physical, biological, and engineering disciplines are blurring, and I believe we will find the most productive areas in this space. It is at the interface of chemistry and other sciences that we can expect to see innovation in silicon-based science and technology – and business growth – in our industry over the next decade. We are already seeing some examples of this happening. For example, Wacker and Shin Etsu have thriving businesses in silicon wafer production, and at Dow Corning, we have invested in areas such as photonics, energy, and biotechnology through a strategic alliance with Genencor International.

Furthermore, I believe we cannot do this using our old models of innovation. Companies recognize today that they can no longer do everything themselves – new alliances are being forged and collaboration between industry and governments is an increasingly important source of innovation. As Harvard Business School Professor Henry Chesbrough has pointed out [2], even though we have enjoyed great success in the past essentially doing everything ourselves, the so-called "virtuous circle of R&D" is breaking. The old model where our own fundamental research flowed easily into new products, generating increased earnings and hence more investment in R&D no longer applies. A new paradigm of "Open Innovation" predominates. Good ideas are widely distributed and no one can claim a monopoly on useful knowledge anymore. Partnerships, spin-off companies, technology in-sourcing, licensing, and collaboration are much more popular models, as companies seek to leverage their R&D spending. Besides, not all of the smart people in the world work for those of us in this industry!

Now I would like to turn to several key industries that I believe will provide business growth and new opportunities for silicon-related science and technology over the next decade:

Transportation

It may seem mundane with many applications in everyday use (the average car today contains > 2 Kg of silicones) but, if you will pardon the pun, I believe this industry will literally "drive" innovation in our field. The automotive industry is a $ 2 trillion global business that is not exempt from the macro trends influencing society. More powerful, cleaner, fuel-efficient, and affordable vehicles are needed. Recycling of materials is gaining momentum, particularly in Japan and here in Europe. The dichotomy presented by the economies of mass production vs the needs for individualism and flexibility for consumer choice is challenging for materials designers. Mass transportation, aerospace, and military applications require ever more sophisticated, stronger, lighter materials that play well to many of the strengths of silicon technology. Here are just a few examples of the ways in which Si-based technology is helping:

- "Green tires" – silane coupling agents and new filler technology impart better mileage through improved rolling resistance, wet traction, and longer life
- Catalytic converters – through catalyst supports and higher performance substrates
- Composites – Si-based resins and ceramic parts for lightweight/flame resistance
- Plastics and thermoplastic elastomers – hoses, cables, Dow Corning's newly launched TPSiV ™ technology
- Sealing, adhering, and barrier technology for sophisticated auto-electronics.

Electronics and Photonics

Even with the recent down-turn in the telecommunications industry, I believe there is still great opportunity for Si-based materials. The lines between electronic and optical function are converging, and siloxane gels, resins, and elastomeric materials will play a key role in OEO devices. Already, major electronics device manufacturers are leading the development of "optical interconnects on multi-chip modules". SiO-based materials have many of the attributes that exactly match the application requirements. They are essentially transparent to the wavelengths employed, and therefore are inherently durable in laser-based applications. Some of our materials last more than 600 h of continuous exposure to high flux laser light at telecom wavelengths. Precision and control of refractive index is also critical; materials that allow its manipulation thermally or by application of an electric field are being designed. Siloxane-based liquid crystals can play a role here. In addition, many of the polymeric Si-based materials can be designed with ease of processing in mind. With these advantages, it will not be long before we see SiO-based polymers used in optical interconnects, polymeric wave-guides, and holographic data storage applications.

Feature size on microprocessors is continually reducing, and the International Technology Roadmap for Semiconductors (ITRS) shows that, if anything, this will continue at an accelerated rate until physical limits of materials are reached within the next 10 years. This will obviously require the introduction of new materials and interfaces that are better able to withstand thermal, electrical, and mechanical challenges. The need for low-K dielectrics is leading to novel porous interlayer and pre-metal dielectrics, while the application of barriers via CVD is growing at the expense of spin-on technology. In the medium term, we anticipate a change from Si low-K dielectrics to doped Si alloy glasses (OSG, organosilicon glasses). Again these trends are driven by growing demand for OEO devices and optical interconnects, with control of refractive index to one part in 10 000 being a critical requirement. Wide band-gap, high-temperature semiconductors based on "slice and dice" Si-C technology are being marketed today. Processing advances to lower costs are much needed.

The need for flexible displays and roll-to-roll processing will also drive materials requirements. Technology advances in photo-definable materials and lithography for ever finer resolutions will require materials that are transparent at < 193 nm and highly etch resistant. Light-emitting diodes, whether organic or inorganic, will require new chemistry and place new demands on barrier materials. In all of this, the understanding and application of basic vapor phase chemistry of silicon-based chemicals will be essential. Precursor design and chemistry is key to CVD processes. For example, we are already seeing the design of SiGe precursors for "strained silicon" semiconductors in applications below 90 nm resolution. At Dow Corning, we believe that precursor chemistry is very important for controlling migration and diffusion in copper damascene systems, and we think that organosilicon glasses will replace silica-based dielectrics.

In a side note, new atmospheric plasma chemistry is enabling silicon chemists to design smart surfaces and do "soft chemistry" in a variety of thin film applications. It is still true today that 70 percent of our business relies in some way on interfacial properties of silicones. Nano or "soft" lithography techniques based on siloxane replication materials are also being employed to "print" circuitry at incredibly high resolution. Looking further ahead, silicon nano particles, wires, and quantum dots will begin to have impact in electronics applications such as computing and sensing. Ultimately, we could see the emergence of molecular electronics – some day.

Biotechnology and the Life Sciences

Much has been written and said about the coming age of biotechnology. In fact, many companies are predicting that the future of the specialty chemicals industry lies in the novel routes, lower costs, and benign processes offered by this technology. The US Chemical Industry report "Technology Vision 2020" [3] calls out biotech as the next major challenge. Companies like Dow Chemical, DuPont, and Monsanto have been investing billions. A study by McKinsey & Co. predicts that about 20 percent of the chemical market will, in some way, be impacted by biotechnology with a potential value creation of \$ 160 billion by 2010. This industry is moving very fast, and the risk takers will reap the most benefit. We will see new products emerge in biopharmaceuticals, biopolymers and materials (the so-called "bioterials"), biosensors, and in the use of biological transformations in chemical manufacturing and remediation.

Dow Corning is already active in biotechnology as we invest in new areas that build on our strengths in technology and innovation. Strange as it may seem, however, we consider Silicon Biotechnology™ as simply a subset of the specialty chemicals industry in this respect. Of course, Dow Corning has been interested in the biological aspects of silicones in human applications and the environment almost since its inception. More recently, we began serious investigation of the biology and biochemical processing of silicon itself. During the mid 1990s, we funded five years of basic research work, which culminated in an alliance with the leading biotechnology company Genencor International in 2001. One journalist called it a "weird science alliance" – maybe he knew enough chemistry to realize that silicon is a metalloid, an inorganic material. Perhaps, in his mind, bringing together the animate (organic, bio) and the inanimate (inorganic, silicon) seemed bit odd. He was probably unaware that Si is the second most abundant element in the earth's crust – about 26 percent vs carbon's 0.08 percent. And of course, carbon is the primary element upon which all life depends, whereas Si has been considered to be nonessential in most living organisms. However, we know that ~15 Giga-tonnes of silica are used every year by diatoms as part of the silica cycle, and many plants depend on silicon for their existence. In fact, one could argue that there is nothing "weird" about this alliance at all; the only question is why it took so long to realize it.

After two years, this effort is beginning to bear fruit in unexpected directions. Enzymatic transformations of useful Si-based chemicals are possible, offering new, low-energy, and proprietary routes to silicon-modified organics, functional siloxanes, and silanes. Highly ordered silicon-containing nano-structures, tubes, and other nano-fabricated materials are accessible via biological routes, and silicon chemistry promises avenues into biosensors and smart detection devices through tethering, MEMS, transducer, and Si-substrate technologies. All this, and we are just beginning to scratch the surface of this fascinating new field of research.

In the next decade, improving health and well-being will be major drivers in the life science businesses. Personal care, household, and laundry markets are continually seeking that "Wow!" or "feel-good factor", while health care offers opportunity for those who choose to manage the associated liability and business risk presented by this market. Encapsulation of actives is providing immediate opportunity for new silicon-based products for the personal care, cleaning, and fabric care markets. Silicon chemistry will play a key role in the future design of new delivery systems, affording improved aesthetics and sensory benefits, and in the active component itself. The ever-growing demand for better personal care products is driving new materials development in siloxane – natural polymer hybrids and advances in silicone formulation science. Improved hair conditioners, new wound and skin care products, and detergents that provide multiple functionality, are a few examples. The use of cosmeceuticals, vitamins, aromatherapy and popular "over the counter" medications is increasing interest in silicone systems. Drug delivery, transdermal patches, silicone matrices, and adhesives that offer lower dosages, consistent and safer delivery, are widely used, and silicone prosthetics are helping improve the quality of life for critically ill people.

The bottom line is that to remain a leader in the specialty chemicals business no one can afford to ignore biotechnology. We believe that is certainly true of the organosilicon field.

Energy

Photovoltaic Cells

The largest single growth segment that I know about is in the production of PV cells. This is a $ 2 billion industry today, which has grown more than 25 percent annually for the last 20 years. Technological advances and efficiency improvements, driven by the global market for alternative energy sources, will see this trend continue. This means we will need to discover and develop improvements in all aspects of the science:

• Single crystal silicon • Polycrystalline silicon	high efficiency 80% of the market	need continuous process improvements
• Thin film deposition • Microcrystalline silicon • "Spheral" silicon and other forms	flexible /lower cost substrates 20% of the market	breakthrough technology required.

Fuel Cells and Batteries

These will provide opportunities for silicon-related materials for niche applications such as

- Sealing and adhering
- Ceramic membranes and bipolar plate materials
- Catalyst supports
- Polymer electrolytes for thin-film batteries.

Alternative Energy

I believe there will be many more opportunities here for silicon-based materials. As sustainable development gains momentum and concern about climate change escalates, alternative energy sources are finding their place in the global market.

Wind power is one of the biggest success stories, especially in Europe. It is becoming more established in North America, and it is already a $ 3 – 4 billion global market. Structures must survive in harsh environments and are being designed with turbine blades up to 100 m diameters and tip speeds of several 100 km/h. Demands on materials are extreme; this is where silicon-based technology can contribute. Composites where the combination of strength, weight, and durability are critical, transmission technology to move the power from source to consumer, and high-performance sealants and adhesives are all needed.

Beyond wind power, hydroelectric, wave, geothermal, and even nuclear energy will all need materials capable of dealing with high temperatures and harsh environments. Growing world demand will most likely mean the continued domination of carbon-based fuels as energy sources, at least for the foreseeable future. But at the same time there will be demand for technology for cleaner, more efficient exploration, production, use, and disposal of those fuels. This means opportunity for traditional solutions in foam control, sealing, and adhering, flow control aids, and insulation, to name a few. More exotic energy forms and better ways to use energy will emerge. Aerogels and xerogels are an interesting class of materials that are finding utility as insulators in white goods and construction applications. Yet another intriguing example, that might create highly efficient magnetic refrigeration systems, is based on novel Gd–Si–Ge alloys.

Conclusions

- All of these applications, and many more, will provide tremendous opportunity for Si-related science and technology where Si is the essential ingredient. So, there is plenty for us to work on.
- We must always listen to the voice of our customers. We need to be fully in touch with their needs – for they have choices.
- To fully realize the potential of the technology, there must be truly effective collaboration between governments, academia and industry. To make the most of scarce resources, these efforts cannot be separated. Old agendas must be thrown out.
- The true innovations will come from a fusion of the traditional disciplines. The interfaces of chemistry, biology, physics, and engineering are where future innovation will occur.
- Inorganic and organosilicon science are equally important in the future development of our field.

I started with the question "Is there a future for Silicon Chemistry?", and clearly I believe we should answer with an emphatic ***yes!*** Certainly, we at Dow Corning are looking forward to the next decade. We look forward to serving our customers with exciting innovations – innovations that are based on our traditional strengths in silicon-related technology fused with novel technologies and new ideas. We have transformed our company to support a new approach – providing complete solutions for our customers to help them solve their problems and seize new opportunities.

References

[1] *Progress in Organosilicon Chemistry* (Eds.: B. Marciniec, J. Chojnowski), Proceedings of The Xth International Symposium on Organosilicon Chemistry, Poznan, Poland, 1993, Gordon and Breach Publishers: Basel, Switzerland, **1995**.

[2] H. W. Chesbrough, *Open Innovation – The New Imperative for Creating and Profiting from Technology*, Harvard Business School: Boston, MA, **2003**.

[3] *Technology Vision 2020*, American Chemical Society: Washington, DC, **1996**.

The Markets for Silicones

M. De Poortere

Centre Européen des Silicones (CES)
Av. E. van Nieuwenhuyzen 4, B-1160 Brussels, Belgium
Tel.: +32 2 676 73 69 — Fax: +32 676 73 01
E-mail: mdp@cefic.be

Keywords: silane, polydimethylsiloxane, end-uses

Summary: Markets for silicones continue to increase faster than average economic growth, driven by the need for higher-performing materials across a huge number of applications.

Introduction

Between 1995 and 1998 the average global growth rate of silicones has been about 6%, with most of the growth coming from developing Asian countries [1]. In 2002, the global market for silicones totaled about € 8 billion, based on a production volume slightly above 2 000 000 tons. The major sales areas are in Europe (33% in 2002) and in North America (34%). The major producers in Europe are members of CES, a sector group of Cefic (the European Chemical Industry Council): Dow Corning, Degussa, GE Bayer Silicones, OSi Specialties (now part of GE Specialty Silicones), Rhodia Silicones, Shin Etsu, and Wacker Chemie.

Silicones are well known for their versatility, which makes them ideally suitable for a variety of applications. The fluids can be used as solvents, as foam-control systems, or as release agents (20% of the total volume). High-molecular-weight silicones are mainly used in rubber applications such as High Temperature Vulcanisable (HTV) and Room Temperature Vulcanisable (RTV) (43%), resins (4%), or specialties (15%). Other applications for silicones are masonry protection (8%), textiles (7%), and paper coatings (3%). Silicones can be uniquely tailored for each application area by substitution by reactive groups, allowing them to be cured by different mechanisms.

The simplest silicones are cyclic or linear polysiloxanes. Their viscosity or volatility characteristics are related to their molecular weight. The higher-molecular-weight siloxanes can carry specific functional groups adapted to their use.

Some highly reactive low-molecular-weight silanes are used as adhesion promoters or coupling agents, and chlorosilanes are used as intermediates. They will be reviewed in turn.

Silanes

Coupling and adhesion promotion agents are used to reinforce fiberglass-to-polymer matrices in composites, and to strongly bond mineral fillers to rubbers and plastics, preventing water absorption and reducing deterioration of electrical insulating properties. They also provide substrate adhesion in adhesives, sealants, and coatings. Other *multifunctional silanes* are increasingly used to crosslink organic polymer molecules. Crosslinking is usually affected after the polymer has been applied or fabricated to a desired shape. The key benefits provided by these crosslinking agents are low-temperature cure, improved thermal stability, creep resistance, hardness, and chemical resistance.

The silicone reactive functions of alkoxy functional silanes form silanols in the presence of water (Eq. 1). The silane triols formed are very reactive and will react very quickly if other hydroxyl groups are present. For example, they will condense or crosslink with other silanols or with the surface of minerals. The hydrophilic character of the silanols will be lost after the reaction has taken place.

$$Y\text{–}CH_2CH_2CH_2\text{–}Si(OR)_3 + 3\,H_2O \longrightarrow Y\text{–}CH_2CH_2CH_2\text{–}Si(OH)_3 + 3\,ROH$$

Eq. 1. Hydrolysis of organofunctional silanes.

The hydrolysis reaction proceeds either through hydroxyl addition in alkaline conditions to the silicon atom, or by proton addition to oxygen in acidic conditions. The base-catalyzed reaction is influenced by the organic group next to the silicon atom: electron-withdrawing groups strongly increase the rate of hydrolysis under alkaline condition. The nature of the organic group has much less influence on acid-catalyzed hydrolysis. Furthermore, the nature of the leaving group –R (generally methyl or ethyl) has a significant influence on both acid- and base-catalyzed hydrolysis rates.

The coupling reaction can be illustrated by the room temperature condensation of two silanols with the hydroxyl groups present on many mineral surfaces (Eq. 2).

The effectiveness of silane coupling agents is greatest with silica, glass, copper, aluminum, or inorganic oxides, while adhesion to carbon black, mineral sulfate fillers, and some other metals is poorest.

The formulator or the chemist can choose among a large variety of functional silanes: with the triethoxysilanol function on one end, the substituents at the other end (Y in Eqs. 1 and 2) can be alkyl, alkylene oxide, isocyanurate, mercapto, polysulfide, amino, ureido, epoxy, isocyanato, vinyl, or methacrylic groups.

One commercially important use of reactive silanes is the modification of polyethylene by vinyltrimethoxysilane in an extruder in the presence of peroxide initiator. The forming of the polymer into cable jacketing, wire insulation, or pipe is usually done in a second step with a moisture-step catalyst added. The resulting pendant trimethoxysilyl groups are activated by moisture or hot water to cause hydrolysis of the silane and condensation to form crosslinks by Si–O–Si bond formation.

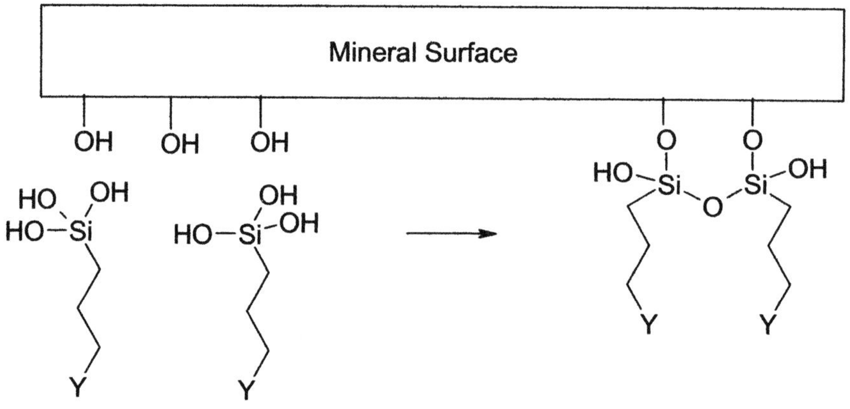

Eq. 2. Coupling of hydrolyzed organofunctional silanes on mineral surfaces.

The tetraalkoxysilanes (tetraalkyl silicate) are a special category of low molecular weight silicones. Masonry products contain tetraalkoxysilanes, which in the presence of air and humidity will form a bond with the masonry and crosslink to a hydrophobic insoluble coating.

Low-Molecular-Weight Siloxanes

The main use of the *cyclic siloxanes*, in particular decamethylcyclopentasiloxane (D5), is in personal care products. D5 helps personal care products spread uniformly and provides a smooth, silky feeling during application. D5 dries quickly without leaving a cooling sensation or residue. A growing use of D5 is in dry cleaning applications in replacement of perchlorethylene.

Polydimethylsiloxanes and Functional Derivatives

Polydimethylsiloxanes (PDMS), sometimes known as dimethicones in the cosmetics and pharmaceutical industries, in their simplest form are straight chain polymers carrying no special substituents and are generally sold according to viscosity or molecular weight. They are being used extensively in premium greases, lubricants, and thermal transfer fluids because of their inertness, their resistance to high temperature oxidation, and their ability to maintain fluidity at low temperatures.

Specific properties can be conferred on PDMS by substitution with functional groups. Some of these functional groups, such as alkoxy, amino, and mercapto will mainly interact with substrates by hydrogen bonding, and are often used for their surfactant properties. Others, such as unsaturated or epoxy groups, can crosslink by covalent bond formation, and will be reviewed in the next section.

PDMS can be modified with polyethers based on either ethylene oxide or propylene oxide. The ratio between the three components will determine the balance of properties according to the triangle in Fig. 1.

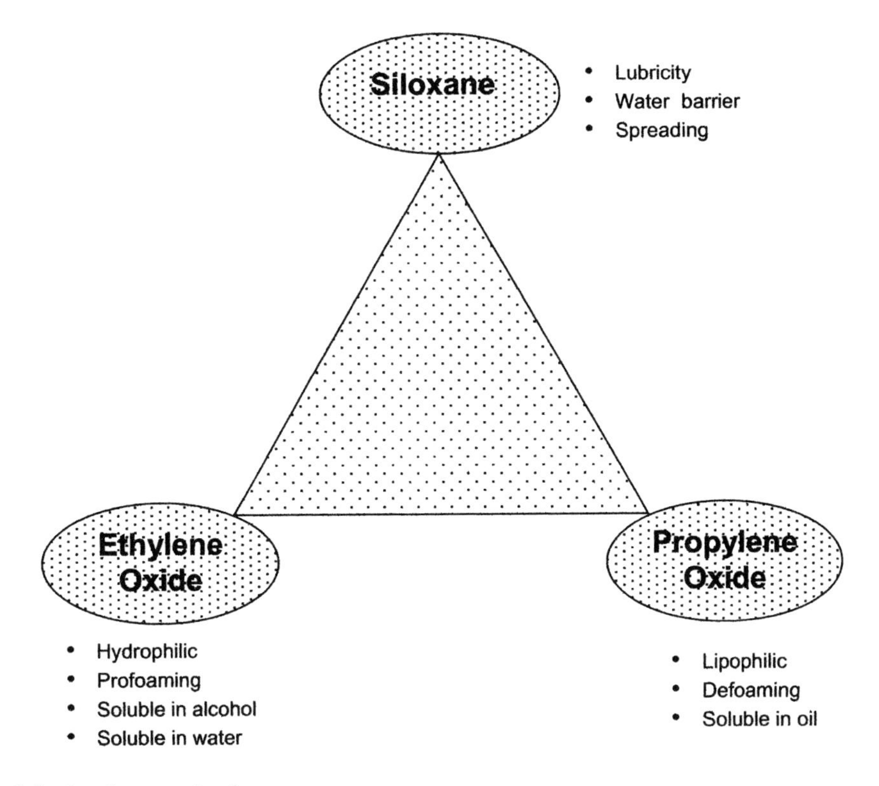

Fig. 1. Polyethersiloxanes triangle.

Furthermore, fine-tuning of the modified alkoxy-substituted PDMS can be done by altering the structure of the alkoxy group: it can be pendant or incorporated into the polymer backbone.

A commercially important category of polyether-modified PDMS is used in polyurethane foam manufacture. Used at a concentration of about 1%, these PDMS derivatives are essential to ensure a uniform distribution of foam properties over the cross section of the foam slab.

The antifoam properties of some of these PDMS are used extensively in industrial applications such as paper mills or domestically in laundry detergents.

Many modifications of polydimethylsiloxanes have been made in response to the needs of *textiles* or of the *fabric care industry*. Amide-functional, amine-functional silicones, and PDMS are key ingredients of fabric softeners and can confer properties such as conditioning, antiwrinkling, ease of ironing, softness, and water absorbency. PDMS and silanol-terminated PDMS confer lubrication, softening, ease of ironing, and medium substantivity. The amino- and amido-functional PDMS are excellent softeners; they have high substantivity and are available in reactive and non-reactive grades. Other formulation parameters can be adjusted to ensure optimal use of the silicones. The particle size of modified polydimethylsiloxanes in emulsion can be adjusted to optimize penetration of the fabric: microemulsions (<150 nm) of modified PDMS are capable of penetrating deeply into fabric yarns to achieve an excellent softening effect by internal lubrication. Macro-emulsions (200 – 1 000 nm), on the other hand, are deposited superficially on the fabric and are therefore designed to facilitate ironing. Several grades of modified PDMS used for fabric care are available depending on the compatibility required with surfactants (non-ionic, anionic or cationic).

PDMS is being used as *medicine* (up to 15 g/day by oral intake) for digestive tract ailments and as direct food additives. The use of "dimethyl polysiloxane" is permitted in jams, jellies, marmalade, pineapple juice, and chewing gum up to 10 mg/kg under the E900 number. PDMS is extensively used in cosmetics, for example, in hand creams. Studies have confirmed that PDMSs have low toxicity.

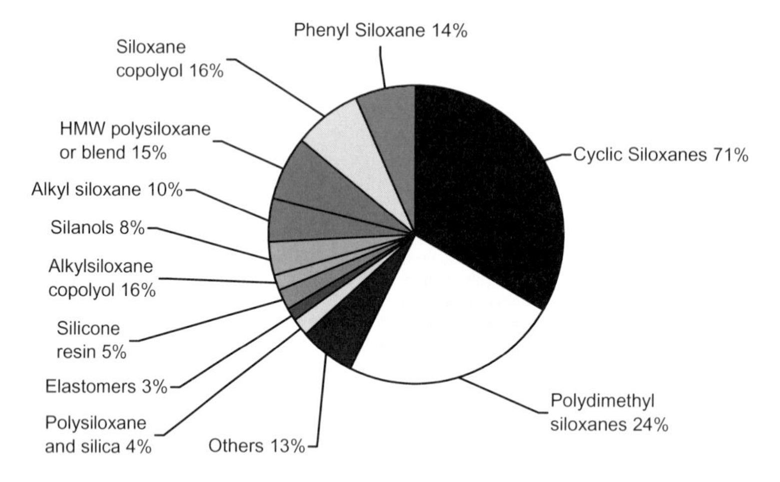

Chart 1. Silicone usage in cosmetics by chemical family (France, 2001), based on a survey of all silicone-containing products (base: 374 products; sum of percentages is greater than 100%).

Personal care and cosmetics are important markets for silicones, and have achieved a high degree of market penetration during the last 25 years. Total sales of silicones in the personal care industry are estimated to be between 21 000 and 23 000 tons/y.

Many different chemical families of silicones, mainly low-molecular-weight siloxanes, PDMS, and functionalized PDMS are used. As seen in Chart 1, which shows the outcome of a survey in France, the largest categories are the cyclic siloxanes (cyclomethicones, as they are named in the personal care industry) and the polydimethylsiloxanes (dimethicones), which have shown strong growth. The sum of the percentages shown in Chart 1 is greater than 100 because several silicones can be used in a single final product. There is a definite trend toward usage of multiple types of silicones in one single formulation that provides different or synergistic benefits.

In Chart 2, the high market penetration of silicones, expressed as a percentage of the number of surveyed brands in personal care products, is evident in a country such as France, where the survey data comes from [2]:

Silicone Usage in Skin Care France — 2001

Hands & Body	Colour	Skin Care Average	Facial Care	Sun Care
54%	55%	61%	62%	88%

Chart 2. Silicone usage in skin care (France 2001), based on 610 products.

A similar market penetration pattern is seen in hair products (shampoo, coloring products, conditioners, fixatives). Silicones are also used in antiperspirants/deodorants, shaving products, shower and bath oils, gels, and liquid soaps.

- Silicone polyethers are used as emulsifiers, resin plasticizers, and light conditioning agents.
- Alkyl methyl siloxanes are noted for their moisturization, and improved compatibility in formulations.
- Blends contain ultra-high-viscosity fluids in low-viscosity fluids to improve spreadability, soft feel, and substantivity.
- Elastomers (see next Section) have a unique silky feel, and are excellent thickeners for silicone- and oil-based formulations.

Reactive Polydimethylsiloxanes

The introduction of different types of reactive groups on the polydimethylsiloxane backbone enables crosslinking reactions to take place. The silicones can be applied as coatings on a variety of substrates, or as sealants. Elastomeric products such as seals, tubing, or other shaped products can be made by extrusion or molding processes.

In consumer markets, keypads, baby bottle nipples, and pacifiers are said to be the applications for molded silicone products with greatest growth potential. Silicone elastomers are particularly well suited for food contact and medical applications.

Many modern architectural realizations would not have been possible if silicone sealants for structural glazing had not minimized the need for mechanical fixing. Each specific application is an illustration of the high-performance properties that can be obtained with silicones.

Silicone emulsion paints for outdoor use based on adhesion promotion technology can offer unique durability benefits, as they provide a unique combination of water repellency, vapor permeability, and low volatiles during application.

The release effect of silicones is put to use in baking paper or baking ware coatings, and pressure-sensitive adhesives release paper coatings.

For textiles, the waterproofing properties of silicones are well known. Friction coatings that reduce slip for stockings tops and socks are examples of an innovative use of silicone properties.

Depending on the functional groups present on the siloxane backbone, silicone sealants can be cured by condensation with crosslinking agents, e.g., acetate, oxime or alkoxy silanes (one-component RTV-1, using ambient humidity to begin the curing, or two-component RTV-2, which do not require humidity to start the curing reaction) or by a radical mechanism with peroxide (HTV rubbers). Liquid silicone rubbers (RTV-2 or LSR) can be cured with platinum catalysts by a hydrosilylation addition mechanism, generating no by-products. Fluorine can be introduced by fluorinated side-chains, imparting high temperature and oil resistance.

As examples of new, fast-growing applications of reactive silicones, *polycarbonate headlight lens coatings* [3] and *microelectronics applications* can be cited.

Polycarbonate automobile headlamp lenses enable a greater freedom of design, lighter weight (and therefore fuel economy), and lower overall manufacturing costs. However, the polycarbonate lens surface is not resistant enough to pass weatherability and abrasion resistance testing equivalent to a 10 year use. Protective hardcoats based on sol-gel-derived hybrid materials have been introduced commercially, and their potential for many other applications is being realized. These hybrid organic-inorganic polymers consist of clear regions or morphologies in which organic structures dominate as well as separate regions in which distinct structures imposed by heteroatoms dominate. They exhibit physical properties which are not a linear or geometric average of these regions [4]. One example of this technology is based on the following reaction scheme (Eqs. 3, and 4):

Silica's surface hydroxyl groups (**1**) will react with a silane (**2**) yielding a silane diol reactive intermediate (**3**).

The silane diol (**3**) will condense during the curing step with another silane diol, forming a crosslinked mineral and organic composite (**4**).

$$\underset{\mathbf{1}}{SiO_2\text{—OH}} + \underset{\mathbf{2}}{RSi(OCH_3)_3} \longrightarrow \underset{\mathbf{3}}{SiO_2\text{—O—Si(OH)}_2\text{R}}$$

Eq. 3. Hydrolysis of a silane in the presence of silica.

$$\underset{\mathbf{3}}{SiO_2\text{—O—Si(OH)}_2\text{R}} \xrightarrow{-H_2O} \underset{\mathbf{4}}{SiO_2\text{—O—Si(R)(OH)—O—Si(R)(OH)—O—}SiO_2}$$

Eq. 4. Condensation of silane diols.

The hardcoat is a silicone-silica composite prepared by hydrolysis of alkoxysilanes in the presence of water and colloidal silica, and can contain an alkoxysilylated UV absorber for protection of the polycarbonate. Adhesion of the topcoat to polycarbonate is provided by a primer, which is a solution of acrylic polymer. The primer is coated on polycarbonate and air dried to a 0.5 μm thick film. The topcoat is applied over the primer layer and air dried before curing at 130 °C for at least 30 min.

The importance of the UV absorber is illustrated by the following photomicrograph (Fig. 2) of uncoated and coated black pigmented polycarbonate exposed to 3 years of Arizona sun:

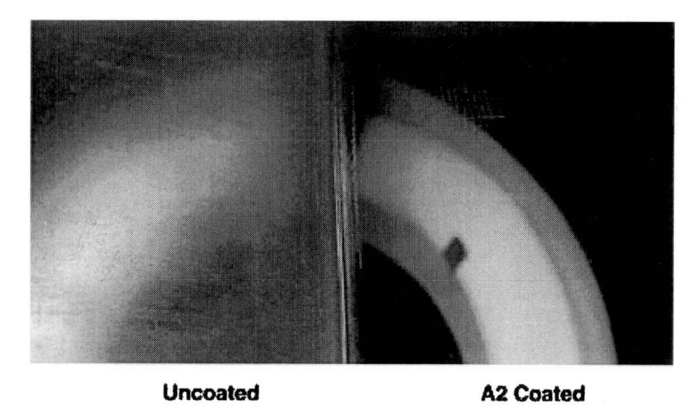

Fig. 2. Effect of UV absorber on polycarbonate hardcoat (*GE Silicones*).

This successful application of silicone hardcoat technology is leading to other applications for polycarbonate parts in automobile manufacturing, such as sunroofs or automobile windows.

In the USA, demand for *electronic polymers* is expected to grown annually 6.4% through 2005 [5]. Growth will be driven by a shift to new electronic packaging technologies such as direct chip attachment as well as higher-cost specialty resins. Silicones can offer several functions to microelectronics:

- Durable dielectric insulation
- Stress-relieving shock and vibration absorption over a very wide temperature and humidity range
- Barriers against environmental contamination
- Thermal management materials.

The following diagram (Fig. 3) illustrates the uses of thermal management materials. Special fillers are used to increase thermal conductivity values. As a consequence, the life and reliability of electronic assemblies are improved.

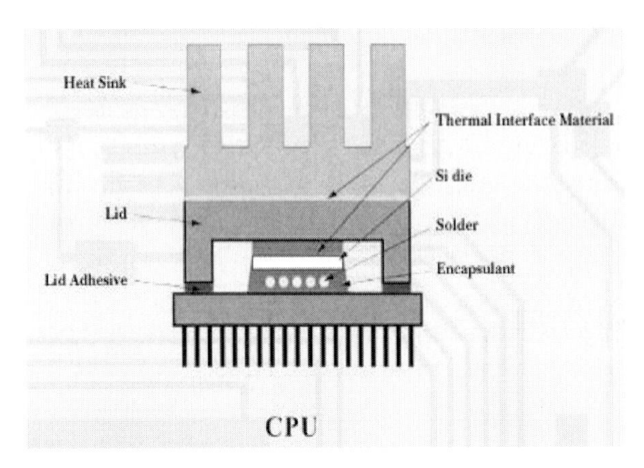

Fig. 3. Heat-conductive silicones in microelectronics *(Dow Corning)*.

Conclusions and Future Prospects

Product innovations continue to expand the useful application range of silicones. Emerging markets, especially in Asia, are providing new opportunities for geographical expansion.

The following markets have been identified as high-growth areas for silicones, driven by the need for longer-lasting and higher-performing materials:

- Personal care products
- High-performance coatings in consumer and industrial applications
- Microelectronics.

Acknowledgment: The author is grateful to CES members for the information and materials provided for this review of silicones markets, which are reproduced with their permission.

References

[1] *Silicones*; CEH Marketing Research Report, SRI **2000**.

[2] *Cosmetic Research — Fiches France*, **2001 – 2002**.

[3] G. Medford, J. Pickett, C. Reynolds, *International Coatings for Plastics Symposium*, Troy (MI) June 4 – 6, **2001**.

[4] B. Arkles, *Commercial Applications of Sol-Gel-Derived Hybrid Materials*, MRS Bulletin, 402 – 407, **2001**.

[5] *Electronic Polymers to 2005*, Research Report No. 1513, Electronics.ca Publications Inc., January **2002**.

Synthesis of Organofunctional Polysiloxanes of Various Topologies

*Julian Chojnowski**

Center of Molecular and Macromolecular Studies, Polish Academy of Sciences
Sienkiewicza 112, 90-363 Łódź, Poland
Tel.: +48 42 6844014 — Fax: +48 42 6847126
E-mail: jchojnow@bilbo.cbmm.lodz.pl

Keywords: polysiloxanes, cyclotrisiloxanes, anionic polymerization siloxane copolymers, block copolymers, dendritic polymers

Summary: The method of the controlled synthesis of polysiloxanes bearing various functional groups in organic radicals pendant to polymer chains and specifically functionalized at a single chain end is discussed. The method is based on quenched anionic ring-opening polymerization of cyclotrisiloxanes having a functional group either in one or in all siloxane units. The copolymerization of the functionalized monomer with the unsubstituted one, hexamethylcyclotrisiloxane, D_3, was also used. The density of functional groups in the macromolecule and their arrangement along the polymer chain was controlled. Polymers were further used for the generation of more complex macromolecular structures, such as block and graft copolymers as well as star-branched, comb-branched, and dendritic branched structures. The method was extended to generate some polysiloxane-inorganic solid hybrids.

Introduction

Polymers of well-defined structures containing functional groups often serve as reactive blocks for the building of more complex macromolecular structures, which are further used for the formation of polymer materials able to perform specific functions [1]. Polysiloxanes are particularly attractive polymers for use as reactive polymer blocks. Their chains are known for unusually high static and dynamic flexibility [2], which gives polysiloxanes many interesting and useful properties, such as unusual hydrodynamic and visco-elastic behavior, high diffusion coefficient, high permeability for gases, high solubility in many solvents, and an easy access to functional groups located on these polymers. In addition, the inorganic backbone gives polysiloxanes a high thermal stability, while amphiphilic character, resulting from the combination of polar Si–O bonds and nonpolar organic substituents, in association with the high flexibility of siloxane chain, is the reason for unusual surface behavior of these polymers.

The anionic ring-opening polymerization (ROP) of cyclotrisiloxanes is a convenient method for the synthesis of well-defined functionalized polysiloxanes [3]. Cyclotrisiloxane bearing a functional group or groups is subjected to anionic polymerization quenched by an additionally functionalized chlorosilane, which leads to polysiloxanes specifically functionalized at the single chain extremity and in groups pendant to the polymer chain. If the functional groups are not tolerated by the anionic polymerization system, a precursor group is introduced to monomer and is transformed into the target functional group after the polymer is formed. In this polymerization, not only are molecular weight and polydispersion controlled, but so also are the topology of polymer and functional groups.

Selectivity of the Anionic ROP of a Functionalized Cyclotrisiloxane

The anionic ROP of cyclotrisiloxane in THF initiated with lithium trialkylsilanolate or butyl lithium is our system of choice for the controlled synthesis of functionalized polysiloxanes. The polymerization proceeds in a selective way. The undesired reactions of back biting, chain transfer and terminal unit exchange may be eliminated. The initiation is fast and complete, and the quenching of polymerization by triorganochlorosilanes is fast and clean. The polymerization may be performed at low temperatures, which increases its selectivity [4, 5]. In addition, the SiOLi/THF system tolerates many functional groups in the monomer.

The powerful tool for examining the selectivity of polymerization is MALDI TOF (Matrix Assisted Laser Desorption Time of Flight) mass spectroscopy, as each macromolecular species in the polymerization system is represented on the MALDI spectrogram by a signal corresponding to its molecular mass. The polysiloxane obtained by anionic ROP of vinyl pentamethylcyclotrisiloxane (VD_2) (Scheme 1) was subjected to the MALDI TOF analysis. The polymerization initiated by Me_3SiOLi [6] or BuLi [5] was performed in THF at –30 °C. It was terminated at a high monomer conversion, above 90%, by Me_3SiCl. The MALDI TOF spectrum of this polysiloxane showed exclusively the signals of macromolecules having undivided monomer units and fragments of initiator and terminator at their chain ends. This result is ample evidence that under these conditions no undesired cleavage of the siloxane chain occurs during this polymerization and initiation and quenching are fast and quantitative.

$$Me_3SiOLi + n\left[(Me_2SiO)_2MeViSiO\right] \xrightarrow[\text{(term. } Me_3SiCl)]{\text{THF, } -30\ ^\circ C} Me_3SiO\left[(Me_2SiO)_2MeViSiO\right]_n SiMe_3$$

$$VD_2$$

Scheme 1. Polymerization of a mixed unit monomer studied by MALDI TOF.

Arrangement of Functional Groups along the Siloxane Copolymer Chain

The MALDI TOF technique gives little information on the microstructure of polysiloxane obtained by the polymerization of a mixed unit monomer, such as VD_2. This monomer may be opened by the lithium silanolate center in one of the three unequivalent sites, leading to different order of siloxane units (Scheme 2) [7]. Information on sequences of units is obtained from a ^{29}Si NMR spectrum taken with quantitative integration. The analysis of the spectrum using Markov statistics showed that the regioselectivity of the reaction performed at room temperature is only 67%. The regioselectivity was significantly higher (89%) if the reaction was performed at –30 °C [5]. In this case the arrangement of vinyl groups is almost regular.

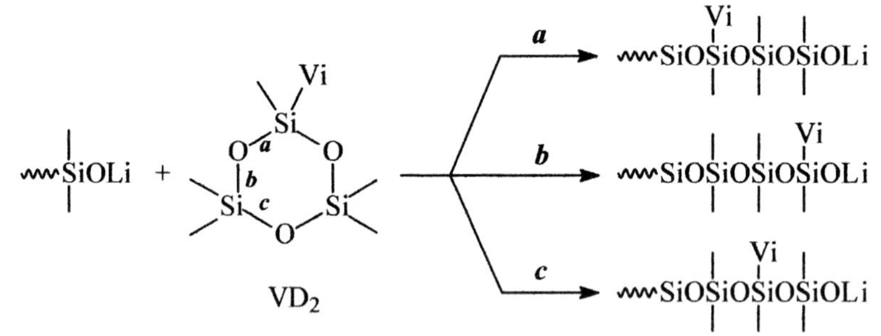

~~~SiOLi + VD_2 → (a) ~~~SiOSi(Vi)OSiOSiOLi ; (b) ~~~SiOSiOSiOSi(Vi)OLi ; (c) ~~~SiOSiOSi(Vi)OSiOLi

Scheme 2. Propagation of monomer with mixed units.

A quite different arrangement is obtained if a vinyl cyclotrisiloxane is copolymerized with hexamethylcyclotrisiloxane (D_3). The reactivity of the former is higher; it enters the polymer chain preferentially, which leads to a gradient distribution of vinyl groups along the chain (Scheme 3) [5, 6]. The knowledge of the monomer reactivity ratios permits one to determine the distribution density function of the vinyl group function along the chain. The gradient distribution of functional groups is between the uniform distribution obtained by polymerization of monomer with mixed units and block distribution, which is formed in sequential copolymerization (see next Section but one).

$$n(\mathrm{ViMeSiO})_3 + n(\mathrm{Me_2SiO})_3 \xrightarrow{\mathrm{BuLi}} \xrightarrow{\mathrm{Me_3SiCl}} \mathrm{BuAAAABAABAABBBABBBBSiMe_3}$$

A (V_3) **B**

Scheme 3. Formation of the gradient copolymer.
~~~

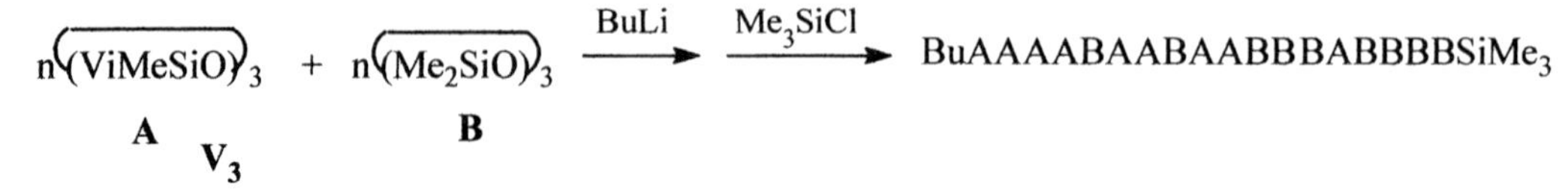

Precursor Functional Groups Pendant to Polymer Chain

Vinyl groups at silicon are versatile precursors, making possible the introduction of a variety of functional groups pendant to the polysiloxane chain by addition reactions such as hydrosilylation [6], hydrophosphination [8], ene-thiol addition [9], or addition of metal complexes [10]. For example, vinyl groups in the block copolymer containing ViMeSiO and Me_2SiO units was subjected to ene-thiol addition of mercaptoacetic acid (Scheme 4). All-siloxane amphiphilic block copolymer with carboxyl groups bonded to the polysiloxane chain via the thioether bridge was generated [11]. Synthesis of such a copolymer by anionic copolymerization of a monomer bearing carboxyl groups would be impossible.

Addition reactions are not the sole processes of the transformation of the vinylsilane group to another functional group. A silylative coupling reaction recently discovered by Marciniec et al. [12], may be used for the introduction of functional groups to polysiloxane, too.

$$BuMe_2SiO(SiMe_2O)_n(\underset{\displaystyle CH{=}CH_2}{SiMeO})_mSiMe_3 + m\,HSCH_2COOH \xrightarrow{AIBN} BuMe_2SiO(SiMe_2O)_n(\underset{\displaystyle CH_2CH_2SCH_2COOH}{SiMeO})_mSiMe_3$$

Scheme 4. The vinyl group as a precursor of functional groups.

Hydrosilylation is the most commonly used reaction for functionalization of silanes in organic radicals [13]. However, the synthesis of cyclotrisiloxanes bearing the silyl hydride group and the controlled polymerization of these monomers is not easy. Paulasaari and Weber [14] succeeded in the synthesis of 2-hydridopentamethylcyclotrisiloxane at a low temperature (–78 °C) and in the polymerization of this monomer on a lithium silanolate center at –78 °C. They obtained the polysiloxane with regularly arranged hydridosilane groups, which was used for synthesis of various functional polysiloxanes by the hydrosilylation process.

Another important precursor is the 3-chloropropylsilane group, which may be converted to many functional groups attached to silicon by the trimethylene bridge using nucleophilic substitution of chlorine [15]. It has been demonstrated that, although chloropropylsilanes are not stable in the presence of alkoxides, the polymerization of cyclotrisiloxanes bearing one or three 3-chloropropyl substituents at silicon proceeds on lithium silanolate in THF in a clean and controlled way, producing 3-chloropropyl-substituted polysiloxane [16]. This polymer was further used for generation of various siloxane copolymers with ionic quaternary salt function [17].

Using of Terminal Silanolate Function for Building More Complex Macromolecular Structures by Macroinitiator and Macromonomer Techniques or by Sequential Copolymerization

The lithium silanolate propagation center is quantitatively converted by the reaction with an additionally functionalized triorganochlorosilane to either a monomer function or to a group able to initiate polymerization. In this way, polysiloxane macromonomers or macroinitiators of atom transfer radical polymerization (ATRP) were produced and used for controlled synthesis of siloxane-organic block and graft copolymers [18, 19]. Polydimethylsiloxanolate was subjected to the reaction with 3-chlorodimethylsilylpropyl bromobutyrate to functionalize the polysiloxane chain end with bromobutyrate. This group was used to initiate the ATRP of polyethylene glycol methylmethacrylate macromonomer to obtain in a controlled way the block-semigraft siloxane-organic copolymer (Scheme 5) [19]. This is an example of the use of combined macroinitiator and macromonomer techniques.

$$Bu(SiO)_{3n-1}SiOLi + ClSi\sim\sim OC(=O)CBr \longrightarrow Bu(SiO)_{3n}Si\sim\sim OC(=O)\text{-}CBr \quad \textbf{I}$$

$$\textbf{I} + CH_2{=}C\text{-}C(=O)\text{-}O(CH_2CH_2O)_mMe \xrightarrow[\text{Cu/(PMDETA)}]{\text{ATRP}} \text{polysiloxane}\sim\sim Si\sim\sim OC(=O)\text{-}C\text{-}(CH_2\text{-}C)_m\text{-}Br \ (\text{PEO})$$

Scheme 5. Synthesis of a block semigraft siloxane-organic copolymer.

All siloxane block copolymers may be synthesized by the sequential copolymerization of two cyclotrisiloxane monomers. For example, unsubstituted hexamethylcyclotrisiloxane, D_3, was first polymerized, and, before full conversion of this monomer, 2,4,6-trivinyltrimethylcyclotrisiloxane, V_3, was added to the polymerization system [11]. The obtained block copolymer showed a high topological purity and a low polydispersion, $M_w/M_n = 1.1$.

Synthesis of Star-Branched and Dendritic-Branched Functionalized Polysiloxanes

Various branched functionalized polysiloxanes may be obtained by the graft-on-graft technique. Living polysiloxane functionalized with pendant vinyl groups is terminated on a low-molecular-weight polyfunctional chlorosilane or on linear polysiloxane bearing the silyl chloride groups, which leads to star-branched or comb-branched polysiloxanes, respectively. The vinyl groups of these polysiloxanes are transformed to the silyl chloride groups by hydrosilylation with $HSiMe_2Cl$, which permits one to use this branched polysiloxane as a reactive core for the grafting of a new generation of branches by the reaction with living functionalized polysiloxane [6].

Vinyl groups were precursors for both functional groups, those used for branching and those giving polymers a specific feature. The density of branching and functional group was controlled by the number of vinyl groups on the monomer. The high density was obtained when V_3 was used, while the polymerization of VD_2 gave a lower density. The copolymerization of VD_2 or V_3 with D_3 was also practiced. The gradient distribution with decreasing density of vinyl groups toward the active propagation center, obtained in this copolymerization, was advantageous. Grafting of these gradient copolymers led to a branched polysiloxane with functional groups located mostly in the external part of the macromolecule.

Living polysiloxane obtained by copolymerization of VD_2 and D_3 was terminated on 1,2-bis(dichloromethylsilyl)ethane, affording tetra arms star-branched polysiloxane with vinyl groups located mostly out of the center of the macromolecule, reaction in **a** Scheme 6 [6]. The coupling reaction occurred to complete conversion at the stoichiometric concentrations of silyl chloride and silanolate groups, giving the star polymer, which showed a high topological purity and fairly low polydispersion ($M_w/M_n = 1.2 - 1.4$).

Grafting of second and third generations of branches (reactions **b** and **c** in Scheme 6) proceeded more slowly because of steric effects. In order to avoid the side reactions, the grafting was stopped well before the full conversion by an excess of Me_3SiOLi, which neutralized remaining silyl chloride groups on the core. In turn, the excess of Me_3SiOLi was neutralized by introduction of Me_3SiCl. A series of dendritic polysiloxanes with vinyl groups located mostly in the outer part of the branched macromolecule was obtained with a fairly high topological purity and $M_w/M_n = 1.4 - 2.0$ [6, 20]. Absolute values of molecular weights $M_n = (0.6 - 5) \times 10^5$ g mol^{-1} were determined by SEC using two detectors, refractive index and multi-angle light scattering, working in tandem in a Wyatt Technology instrumentation system. Polymers had well-defined branching and vinyl group densities.

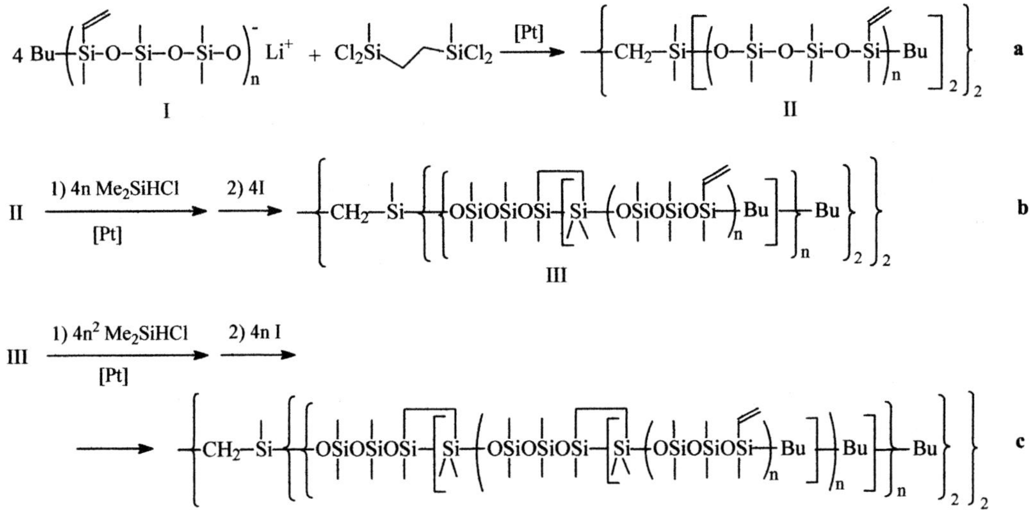

4 Bu–(Si–O–Si–O–Si–O)$_n$ Li^+ + Cl_2Si–CH₂CH₂–$SiCl_2$ →[Pt] {CH₂–Si[(O–Si–O–Si–O–Si)$_n$–Bu]$_2$}$_2$ **a**
I → II

II →(1) 4n Me_2SiHCl, [Pt]; 2) 4I) {CH₂–Si{[OSiOSiOSi–[Si–(OSiOSiOSi)$_n$–Bu]]$_n$–Bu}$_2$}$_2$ **b**
III

III →(1) 4n^2 Me_2SiHCl, [Pt]; 2) 4n I) {CH₂–Si{[OSiOSiOSi–[Si–(OSiOSiOSi–[Si–(OSiOSiOSi)$_n$–Bu])$_n$–Bu]]$_n$–Bu}$_2$}$_2$ **c**

Scheme 6. Synthesis of dendritic-branched, functional polysiloxane.

The vinyl groups on dendritic polysiloxanes were transformed to functional groups. Thus, hydrosilylation was used to introduce various mesogenic groups, and in many cases liquid crystalline dendritic polysiloxanes were obtained [21].

Another method of functionalization was the introduction of functional groups on reactive polysiloxane used as graft in the formation of the last generation of branches. Thus, vinyl groups pendant to a comb-branched polysiloxane were transformed to silyl chloride groups, on which a living polysiloxane bearing a 3-chloropropyl side group was terminated, to generate dendritic polysiloxane functionalized with a 3-chloropropyl group. Those groups were further used to quaternize a tertiary amine to obtain water-soluble dendritic polysiloxane with ionic quaternary ammonium salt groups [20].

Generation of Functionalized Branched Polysiloxane — Silica Hybrids

Two methods were used to obtain functionalized branched polysiloxanes grafted to the surface of silica particles. The first exploits the grafting of the linear living polysiloxane-bearing vinyl groups on the modified silica surface followed by grafting the polysiloxane branches on the polysiloxane attached to SiO_2. The other method is the grafting of preprepared functionalized comb-branched or dendritic-branched polysiloxane on the functionalized SiO_2 surface [22].

Hybrids were analyzed by ^{29}Si and ^{13}C MAS NMR of solids, ^{13}C NMR of hybrid suspension in a polysiloxane solvent, IR, elemental analysis, and thermogravimetry. Hybrids functionalized with vinyl groups and with diphenylphosphine groups bonded to silicon by an ethylene bridge were used as support for metal complex catalysts [10].

Conclusions

Anionic polymerization of functionalized cyclotrisiloxanes is a good method for the synthesis of well-defined functionalized polysiloxane with control of molecular mass, polydispersion, and density and arrangement of functional groups. These polymers may serve as reactive blocks for the building of macromolecular architectures, such as all-siloxane and organic-siloxane block and graft copolymers, star-, comb- and dendritic-branched copolymers and various polysiloxane-inorganic solid hybrids.

Acknowledgments: The research was supported from KBN (State Committee for Scientific Research) Grant PBZ-KBN 15/T09/99/01c and Grant PBZ-KBN/01/CD/2000.

References

[1] K. Matyjaszewski, *Controlled-Living Radical Polymerization*, American Chemical Society, Washington, DC, **2000**.

[2] P. R. Dvornic, in *Silicon-Based Polymers* (Eds.: R. G. Jones, W. Ando, J. Chojnowski), Kluwer Academic Publ., Dordrecht, **2000**, p. 187.
[3] J. Chojnowski, M. Cypryk, W. Fortuniak, K. Kaźmierski, K. Rózga-Wijas, M. Ścibiorek, *ACS Symp. Ser.* **2003**, *838*, 12.
[4] S. Bellas, H. Iatrou, N. Hadjichristidis, *Macromol.* **2000**, *33*, 6993.
[5] J. Chojnowski, M. Cypryk, W. Fortuniak, K. Rózga-Wijas, M. Ścibiorek, *Polym.* **2002**, *43*, 1993.
[6] J. Chojnowski, M. Cypryk, W. Fortuniak, M. Ścibiorek, K. Rózga-Wijas, *Macromol.* **2003**, *36*, 3890.
[7] M. Cypryk, K. Kaźmierski, W. Fortuniak, J. Chojnowski, *Macromol.* **2000**, *33*, 1536.
[8] J. Chojnowski, K. Rózga, *J. Inorg. Organomet. Polym.* **1992**, *2*, 297.
[9] K. Rózga-Wijas, J. Chojnowski, T. Zundel, S. Boileau, *Macromol.* **1996**, *29*, 2711.
[10] Z. Michalska, Ł. Rogalski, J. Chojnowski, W. Fortuniak, M. Ścibiorek, K. Rózga-Wijas, *J. Mol. Catal. A: Chem.* **2004**, *208*, 187.
[11] M. Ścibiorek, N. Gladkova, J. Chojnowski, *Polym. Bull.* **2000**, *44*, 377.
[12] Y. Itami, B. Marciniec, M. Majchrzak, *Organomet.* **2003**, *22*, 1835.
[13] *Comprehensive Handbook of Hydrosilylation* (Ed.: B. Marciniec), Pergamon, Oxford, **1992**.
[14] J. K. Paulasaari, W. P. Weber, *Macromol.* **1999**, *32*, 6574.
[15] U. Deschler, P. Kleinschmit, P. Pauster, *Angew. Chem. Int. Ed. Engl.* **1986**, *25*, 236.
[16] W. Fortuniak, J. Chojnowski, G. Sauvet, *Macromol. Chem. Phys.* **2001**, *202*, 2306.
[17] G. Sauvet, W. Fortuniak, K. Kaźmierski, J. Chojnowski, *J. Polym. Sci, Part A, Polym. Chem.* **2003**, *41*, 2939.
[18] P. J. Miller, K. Matyjaszewski, *Macromol.* **1999**, *32*, 8760.
[19] J. Kurjata, J. Chojnowski, C.-T. Yeoh, N. A. A. Rossi, S. J. Holder, *Polym.* **2004**, *45*, 6111.
[20] J. Chojnowski, W. Fortuniak, M. Ścibiorek, in preparation.
[21] T. Ganicz, T. Pakula, W. Fortuniak, W. A. Stańczyk, E. Białecka-Florjańczyk, *Macromol.*, in press.
[22] K. Rózga-Wijas, J. Chojnowski, W. Fortuniak, M. Ścibiorek, Z. Michalska, Ł. Rogalski, *J. Mater. Chem.* **2003**, *13*, 2301.

A Facile Synthetic Route to Phosphazene Base Catalysts and Their Use in Siloxane Synthesis

Peter C. Hupfield, Avril E. Surgenor,* Richard G. Taylor

Dow Corning Ltd., Cardiff Road, Barry, South Glamorgan, U.K.
Tel.: +44 1446732350 — Fax: +44 1446730495
E-mail: avril.surgenor@dowcorning.com

Keywords: phosphazene base, synthesis, siloxane polymerization

Summary: Phosphazene bases represent a new class of highly active non-ionic catalysts that rapidly polymerize cyclosiloxanes with equilibrium attained in very short reaction times at very low catalyst levels. To date, phosphazene base catalysts have been considered an academic curiosity because of the complicated and hazardous synthetic protocol used to prepare them. A facile synthetic process has been developed, which yields ionic phosphazene bases in three steps with an overall yield of approximately 75%. This is achieved through nucleophilic substitution of ionic phosphonitrilic chloride oligomers with secondary amines, followed by anion exchange. These ionic phosphazenes were found to exhibit similar reactivity in the ring-opening polymerization of cyclosiloxanes to that of the non-ionic phosphazene base.

Introduction

Phosphazene bases represent a class of non-ionic materials which exhibit exceptionally high degrees of basicity, this being dependent upon the structural composition of the molecules [1, 2]. The core structural unit of non-ionic phosphazene bases is derived from a nitrogen basic center double-bonded to a pentavalent phosphorus (P1 unit). This peralkylated triaminoiminophosphorane, through oligomerization, can result in a structurally diverse range of phosphazene bases (Fig. 1).

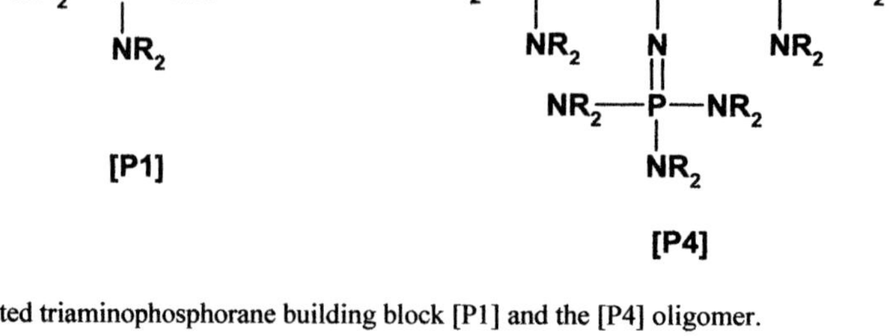

[P1]

[P4]

Fig. 1. Peralkylated triaminophosphorane building block [P1] and the [P4] oligomer.

Phosphazenes derived from oligomerization of the P1 building block yield materials with varying degrees of basicity. For example, the tetrameric phosphazenes (P4) exhibit basicities that are approximately 1×10^{18} times stronger than diazabicycloundecene, DBU, a strongly hindered amine base used extensively in organic reactions [2]. Phosphazene bases, which are derived from the P4 structural unit, readily deprotonate nucleophilic solvents [3 – 7], resulting in ionic phosphonium salts (Fig. 2).

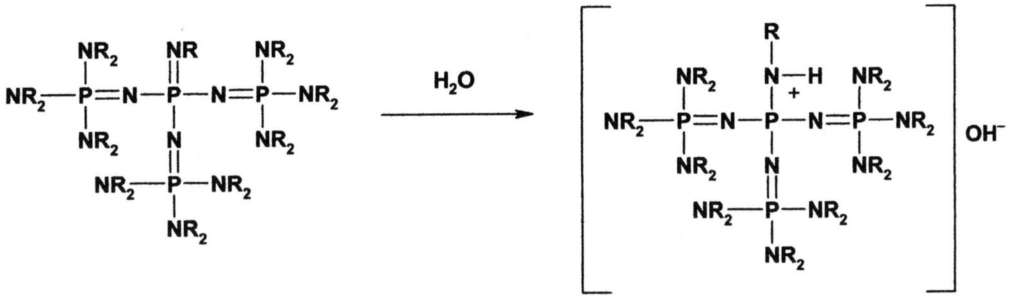

Fig. 2. Phosphonium salt formation by solvent deprotonation.

Through steric hindrance and conjugative effects, these ionic phosphonium salts are very stable to hydrolysis. This, coupled with the lipophilic nature of the cation, results in a very soft, loosely bound ion pair, making materials of this type suitable for use as catalysts in anionic polymerization [8 – 13]. Phosphazene bases have been found to be suitable catalysts for the anionic polymerization of cyclic siloxanes, with very fast polymerization rates observed. In many cases, both thermodynamic and kinetic equilibrium can be achieved in minutes, several orders of magnitude faster than that seen with traditional catalysts used in cyclosiloxane polymerization. Exploiting catalysts of this type on an industrial scale for siloxane polymerization processes has been prevented because of the cost and availability of the phosphazene bases. This paper describes a facile route to materials of this type and their applicability to siloxane synthesis [14].

Results and Discussion

The synthesis of oligomeric peralkylated polyaminophosphazenes based upon the P3 and P4 templates is an elegant (but complicated), hazardous, and expensive multi-step process. To exploit catalysts of this type in silicone synthesis on an industrial scale would require a much simplified and lower-cost synthetic protocol. One of the simplest and lowest-cost routes to a conjugated –P=N– template that forms the framework for the synthesis of phosphazene base materials is via phosphonitrilic chloride oligomers (Fig. 3). These are well-known acidic catalysts used in the silicone industry for the condensation polymerization of silanol-terminated polydimethylsiloxanes. Catalysts of this type are most commonly prepared by the reaction of PCl_5 with NH_4Cl or HMDZ, which yields a mixture of linear oligomeric phosphonitrilic chlorides usually containing the linear oligomeric species P3 through to P5 [15].

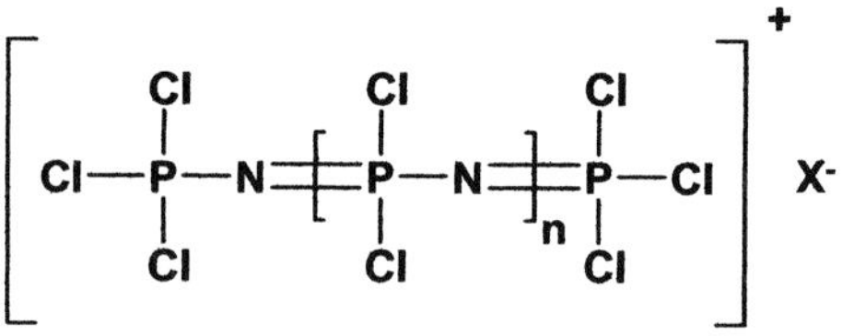

$$\left[Cl-\overset{Cl}{\underset{Cl}{P}}-N{=}\left[\overset{Cl}{\underset{Cl}{P}}-N{=}\right]_n\overset{Cl}{\underset{Cl}{P}}-Cl \right]^+ X^-$$

Fig. 3. Phosphonitrilic chloride oligomer precursor to phosphazene base.

Interestingly, the phosphonitrilic chlorides can be converted into basic phosphonium salts via a two-step process. Firstly, nucleophilic substitution of the phosphonitrilic chloride oligomeric mixture with secondary amines such as dimethylamine or pyrrolidine at elevated temperature yields peralkylated polyaminophosphonium chlorides in high yields, using solvents such as toluene (Fig. 4).

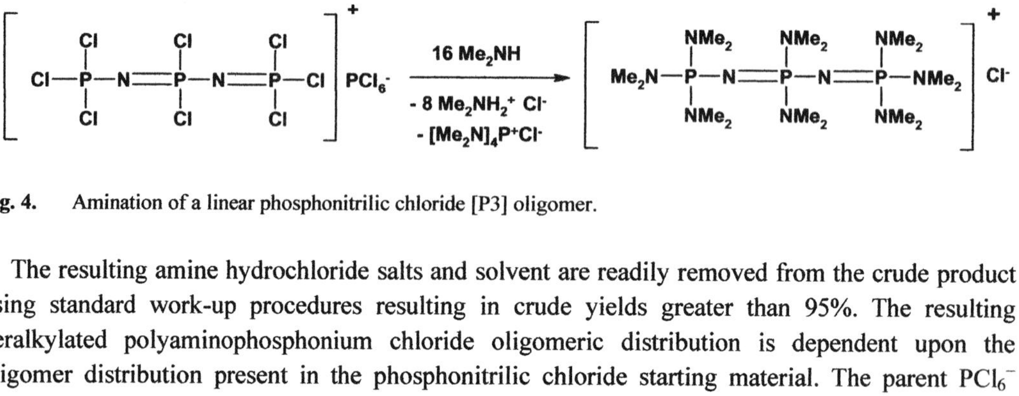

$$\left[Cl-\overset{Cl}{\underset{Cl}{P}}-N{=}\overset{Cl}{\underset{Cl}{P}}-N{=}\overset{Cl}{\underset{Cl}{P}}-Cl \right]^+ PCl_6^- \xrightarrow[\substack{-\,8\,Me_2NH_2^+\,Cl^- \\ -\,[Me_2N]_4P^+Cl^-}]{16\,Me_2NH} \left[Me_2N-\overset{NMe_2}{\underset{NMe_2}{P}}-N{=}\overset{NMe_2}{\underset{NMe_2}{P}}-N{=}\overset{NMe_2}{\underset{NMe_2}{P}}-NMe_2 \right]^+ Cl^-$$

Fig. 4. Amination of a linear phosphonitrilic chloride [P3] oligomer.

The resulting amine hydrochloride salts and solvent are readily removed from the crude product using standard work-up procedures resulting in crude yields greater than 95%. The resulting peralkylated polyaminophosphonium chloride oligomeric distribution is dependent upon the oligomer distribution present in the phosphonitrilic chloride starting material. The parent PCl_6^- anion is converted to the peralkylated tetraaminophosphonium chloride species during the reaction. The resulting products are inactive as polymerization catalysts because of the weakly nucleophilic nature of the chloride anion. In order to obtain an active basic catalyst, the chloride anion is exchanged for a strongly nucleophilic anion such as F^-, OH^-, MeO^-, or Me_3SiO^-. Synthesis of the peralkylated polyaminophosphonium hydroxide is easily achieved with the use of a basic anion exchange resin using methanol or methanol/water as the solvent system. Recovered yields are typically around 80% (Fig. 5).

For anions such as F^-, MeO^- or Me_3SiO^-, anion exchange can be carried out in solvents such as dimethylsulfoxide by reacting the metal salt of the anion with the peralkylated polyaminophosphonium chloride. Yields are typically around 85% after work-up. Using the ion-exchange resin method yields products that contain typically 20 ppm methanol and between 10 and 15% w/w water, which solvate and stabilize the ion-pair. Dehydration of the catalyst results in degradation of the phosphazene backbone. Extending the synthetic study to secondary amines containing β C–H bonds does not result in stable complexes. Whilst the amination reaction proceeds cleanly and in high yield for reagents such as diethyl, dipropyl, and dibutylamine, the anion exchange step results in the formation of neutral complexes identified as phosphine oxides (Fig. 6). This is because of the instability of the phosphonium hydroxide salt that results after basic anion exchange of the peralkylated polyaminophosphonium chloride.

$$\left[(Me_2N)_3P{-}N{=}P(NMe_2)_2{-}N{=}P(NMe_2)_3 \right]^+ Cl^- \xrightarrow[MeOH/H_2O]{\text{Amberlyst } OH^- \text{ Resin}} \left[(Me_2N)_3P{-}N{=}P(NMe_2)_2{-}N{=}P(NMe_2)_3 \right]^+ OH^-$$

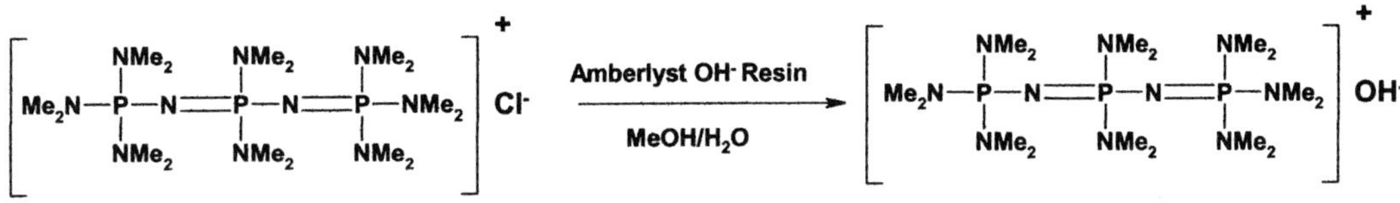

Fig. 5. Basic anion exchange to yield a P3 peralkylated polyaminophosphonium hydroxide.

$$\left[(Et_2N)_3P{-}N{=}P(NEt_2)_3 \right]^+ OH^- \longrightarrow (Et_2N)_3P{=}N{-}P(NEt_2)_2{=}O + Et_2NH$$

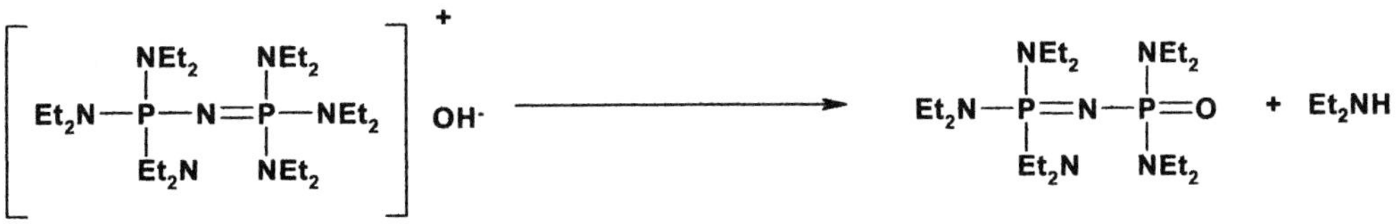

Fig. 6. Formation of peralkylated polyaminophosphine oxide after basic anion exchange.

The synthesis of model compounds from 1,1,1,3,3-pentachloro-3-oxo-1λ^5,3λ^5-diphosphaz-1-ene (Fig. 7) confirmed that this type of rearrangement occurs during anion exchange of the peralkylated polyaminophosphonium chloride oligomers derived from secondary amines containing β C–H bonds.

$$Cl_3P{=}N{-}PCl_2{=}O \xrightarrow[-\,5\,Et_2NH_2^+\,Cl^-]{10\,Et_2NH} (Et_2N)_3P{=}N{-}P(NEt_2)_2{=}O$$

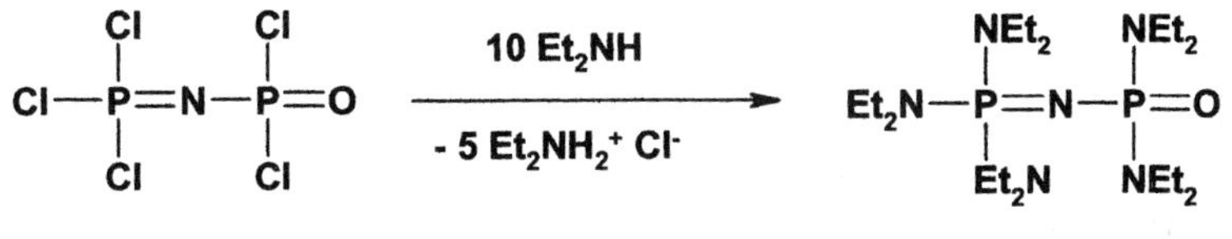

Fig. 7. Synthesis of peralkylated polyaminophosphine oxide model compounds.

Whilst the mechanism has not been proven, it is proposed that this rearrangement is similar to Hoffman rearrangement reactions, which occur with quaternary ammonium compounds containing β C–H bonds [16]. With the peralkylated polyaminophosphonium hydroxides, this rearrangement reaction occurs readily at room temperature, suggesting that in these molecules nitrogen has sufficient electropositive character to facilitate the rearrangement reaction.

1,1,1,3,3,5,5,5-Octakisdimethylamino-1λ^5,3λ^5-triphosphazen-1-ium hydroxide [P3OH] was evaluated as a cyclosiloxane ring-opening polymerization catalyst. Tables 1 and 2 show the results obtained at different reaction times, temperatures, and catalyst concentrations for the synthesis of a polydimethylsiloxane fluid.

Table 1. Properties of pre-devolatilized 30 000 cts polydimethylsiloxane fluid prepared using 1,1,1,3,3,5,5,5-octakisdimethylamino-1λ^5,3λ^5-triphosphazen-1-ium hydroxide [P3OH] as the polymerization catalyst under various reaction conditions.

[P3OH] [ppm]	Reaction time [s]	Reaction temp. [°C]	NVC [%]	Viscosity [cts]	M_w	M_n	M_w/M_n	[SiOH] [ppm]
5	60	170	85.3	19.861	89.000	49.800	1.79	101
5	60	135	86.5	16.407	82.600	47.200	1.75	83
10	40	170	85.7	22.386	84.900	46.400	1.83	64
10	40	135	85.7	17.132	88.365	45.000	1.74	103
15	20	170	86.2	21.517	91.400	49.800	1.84	52
15	20	135	86.8	19.555	98.400	51.800	1.90	89

Table 2. Properties of devolatilized 30 000 cts polydimethylsiloxane fluid prepared using 1,1,1,3,3,5,5,5-octakisdimethylamino-1λ^5,3λ^5-triphosphazen-1-ium hydroxide [P3OH] as the polymerization catalyst under various reaction conditions.

[P3OH] [ppm]	Reaction time [s]	Reaction temp. [°C]	NVC [%]	Viscosity [cts]	M_w	M_n	M_w/M_n	[SiOH] [ppm]
5	60	170	95.7	33.352	96.800	53.300	1.82	74
5	60	135	97.1	26.934	88.700	47.433	1.87	76
10	40	170	96.6	32.468	90.500	49.900	1.81	58
10	40	135	96.8	31.653	92.500	51.300	1.80	61
15	20	170	96.9	30.103	94.000	52.000	1.81	60
15	20	135	97.8	28.254	98.200	55.200	1.78	63

As can be seen from Table 1, 1,1,1,3,3,5,5,5-octakisdimethylamino-1λ^5,3λ^5-triphosphazen-1-ium hydroxide [P3OH] is an effective cyclosiloxane polymerization catalyst with equilibrium attained with very short reaction times across a wide catalyst concentration range. The polymers were prepared using octamethylcyclotetrasiloxane (97.5% w/w) and DC 100cts fluid (2.5% w/w). Based on this ratio of starting materials, the theoretical M_n was calculated to be ~63 000 g/mol. The resulting polymers were found to have an M_n appreciably lower than theoretically expected. However, the raw materials were found to contain approximately 60 ppm water. Assuming that water is rapidly incorporated as silanol groups via endblocking, then the theoretical M_n at 87% NVC (non-volatile content) was calculated to be ~51 000 g/mol, which was in close agreement to that observed experimentally. The silanol levels were higher than that of the unpolymerized materials, which was ~20 ppm, confirming rapid incorporation and conversion of water to silanol.

The polymers were neutralized with approximately 15 ppm bis-silylvinyl phosphate (Fig. 8), and on devolatilization a slight increase in M_n was observed together with a decrease in silanol content, suggesting some post-condensation of silanol.

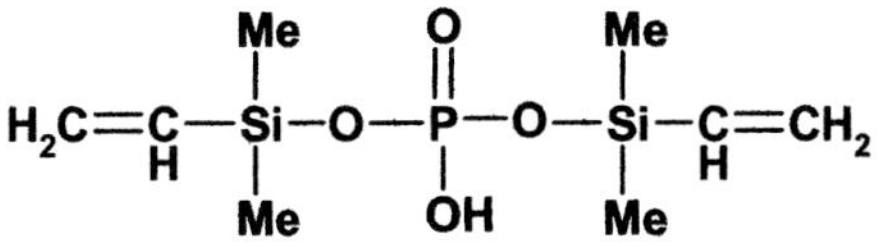

$$H_2C{=}\underset{H}{C}{-}\underset{Me}{\overset{Me}{Si}}{-}O{-}\underset{OH}{\overset{O}{\overset{\|}{P}}}{-}O{-}\underset{Me}{\overset{Me}{Si}}{-}\underset{H}{C}{=}CH_2$$

Fig. 8. Bis-dimethylvinylsilyl phosphate neutralizing agent.

The thermal stability of the neutralized and devolatilized polymers under nitrogen was significantly improved compared with polymers using traditional anionic and cationic polymerization catalysts (Table 3). Decomposition onset temperatures were greater than 500 °C as measured by TGA. This is attributed to the very low levels of catalyst used in the polymerization and also the stability of the neutralized complex that is formed. The appearance of the polymers at these catalyst concentrations and 15 ppm of neutralizing agent was clear without filtration.

Table 3. Thermal stability and appearance properties of polydimethylsiloxanes prepared using 1,1,1,3,3,5,5,5-octakisdimethylamino-1λ^5,3λ^5-triphosphazen-1-ium hydroxide [P3OH] as the catalyst and bis-dimethylvinylsilylphosphate as the neutralizing agent.

[P3OH] [ppm]	Neutralization temp. [°C]	Decomposition onset temp. [°C]	Polymer appearance
5	170	567	Clear
5	135	576	Clear
10	170	582	Clear
10	135	579	Clear
15	170	534	Clear
15	135	543	Clear

Catalysts of this type have the potential to significantly impact the process technology used to manufacture polydimethylsiloxanes. The very short reaction times needed to reach thermodynamic and kinetic equilibrium make these catalysts suitable for use in continuous processes with focus on process intensification. With inherently fast reaction times the reactor stage component can be significantly reduced in size, reducing process hold-up volumes. The enhanced thermal stability under nitrogen can enable much higher devolatilization temperatures to be used in the stripping stage of the process. The ability to produce clear polymers reduces the need for intensive filtration stages in the process.

References

[1] L. A. Paquette, *Encyclopedia of Reagents for Organic Synthesis, Wiley*, **1995**, *6*, 4110.

[2] R. Schwesinger, H. Schlemper,C. Hasenfratz, J. Willaredt, T. Dambacher, T. Breuer, C. Ottaway, M. Fletschinger, J. Boele, *Liebigs Ann.* **1996**, 1055.

[3] H. J. Gais, J. Vollhardt, C. Krueger, *Angew. Chem. Int. Ed. Engl.* **1988**, *27*, 1092.

[4] J. Braun, C. Hasenfratz, R. Schwesinger, H. H. Limbach, *Angew. Chem. Int. Ed. Engl.* **1994**, *33*, 2215.

[5] J.-S. Fruchart, H. Gras-Masse, O. Melnyk, *Tetrahedron Lett.* **2001**, *42*, 9153.

[6] J.-S. Fruchart, G. Lippens, C. Kuhn, H. Gras-Masse, O. Melnyk, *J. Org. Chem.* **2002**, *67*, 526.

[7] A. Solladie-Cavello, *Tetrahedron Lett.* **2002**, *43*, 415.

[8] B. Eβwein, M. Möller, *Angew. Chem. Int. Ed. Engl.* **1996**, *35*, 623.

[9] B. Esswein, J. P. Spatz, M. Möller, *Polym. Prepr. (Am. Chem. Soc. Div. Polym. Chem.)* **1996**, *37*, 647.

[10] S. Förster, S. Kramer, *Macromolecules*, **1999**, *32*, 2783.

[11] P. C. Hupfield, R. G. Taylor, *J. Inorg. Organomet. Polym.* **1999**, *9*, 17.

[12] B. Esswein, N. M. Steidl, M. Möller, *Macromol. Rapid Commun.* **1996**, *17*, 143.

[13] M. E. Van Dyke, S. J. Clarson, *J. Inorg. Organomet. Polym.* **1998**, *8*, 111.

[14] D. Eglin, J. De La Cro Habimana, P. C. Hupfield, A. E. Surgenor, R. Taylor, (Dow Corning Limited, UK), U.S. (**2000**), US 6 448 196.

[15] D. Eglin, J. De La Cro Habimana, S. L. O'Hare, R. Taylor, (Dow Corning Limited, UK), U.S. (**2002**), US 6 350 891; S. Nitzsche, W. Hechtl, E. Wohlfarth, (Wacker-Chemie G.m.b.H.), Ger. Offen. (**1974**), DE 2 229 514.

[16] M. B. Smith, J. March, *Advanced Organic Chemistry* 5th Edition, John Wiley **2001**, p. 1331.

Supramolecular Chemistry and Condensation of Oligosiloxane-α,ω-Diols $HOSiMe_2O(SiPh_2O)_nSiMe_2OH$ (n = 1 – 4)

Jens Beckmann,**[*,†] **Dainis Dakternieks, Andrew Duthie, Richard C. Foitzik

Centre for Chiral and Molecular Technologies, Deakin University
Geelong 3217, Australia

[†] Institut für Anorganische und Analytische Chemie
Freie Universität Berlin, Fabeckstr. 34 – 36, 14195 Berlin, Germany
Fax: +49 30 83852440
E-mail: beckmann@chemie.fu-berlin.de

Keywords: siloxane, silanol, hydrogen bonding, ^{29}Si NMR, X-ray diffraction

Summary: The mild catalytic oxidation of oligosiloxanes, $HSiMe_2SiO(SiPh_2O)_n$-$SiMe_2H$ (**1 – 4**; n = 1 – 4), using Pearlman's catalyst, $Pd(OH)_2/C$, in aqueous THF, is reported to give the corresponding oligosiloxane-α,ω-diols $HOSiMe_2SiO(SiPh_2O)_n$-$SiMe_2OH$ (**5 – 8**; n = 1 – 4) in high yields. The supramolecular structures of **5 – 7** reveal new hydrogen bonding modes, whose influence on the Si–O chain conformations is discussed. The HCl-catalyzed condensation of **5 – 8** is also described, which provided complex mixtures of oligomeric products in the case of **5** and **8**, and high yields of the eight- and ten-membered siloxane rings *cyclo*-$(Me_2SiO)_2(Ph_2SiO)_2$ (**9**) and *cyclo*-$(Me_2SiO)_2(Ph_2SiO)_3$ (**10**) in the case of **6** and **7**.

Introduction

Organosilanols are important key intermediates for the synthesis of a great variety of organometallic materials including siloxane and metallasiloxane oligomers and polymers [1 – 6]. Owing to their inherently high tendency to undergo self-condensation, the preparation of sterically exposed organosilanols requires mild reaction conditions, avoiding heat and traces of acid or base. These requirements are usually fulfilled by *homogeneous* catalytic oxidation reactions of organo-H-silanes in the presence of various transition metal complexes, e.g., based on Cr, Cu, Rh, Re and Ru, but their subsequent removal from the organosilanols is often difficult and cumbersome [7 – 9]. To the best of our knowledge, only the pioneering work of Barnes and Daughenbaugh has utilized immobilized transition metal catalysts (e.g., Pd/C, Pd/Al_2O_3, etc.) for the preparation of organosilanols, thus justifying more work in this area [10]. We have now found that Pearlman's catalyst, $Pd(OH)_2/C$, has superior qualities for the mild *heterogeneous* catalytic oxidation of a variety of organosilanols that are sensitive toward self-condensation [11].

Discussion

The preparation of the oligosiloxanes, $HSiMe_2SiO(SiPh_2O)_nSiMe_2H$ (**1** – **4**; n = 1 – 4), was achieved by the reaction of $HO(SiPh_2O)_nH$ with $HSiMe_2Cl$ in the presence of NEt_3, which acted as an HCl scavenger. The subsequent oxidation of compounds **1** – **4** using a suspension of Pearlman's catalyst $Pd(OH)_2/C$ in aqueous THF produced the corresponding oligosiloxane-α,ω-diols $HOSiMe_2SiO(SiPh_2O)_nSiMe_2OH$ (**5** – **8**; n = 1 – 4) in high yields (Scheme 1). The combination of these two reaction steps effectively provides for an extension of the silanol chains by two additional Si–O linkages.

$$HO(SiPh_2O)_nH + 2\ Me_2SiHCl \xrightarrow[-2\ HNEt_3Cl]{2\ NEt_3} HSiMe_2O(SiPh_2O)_nSiMe_2H \quad \mathbf{1\text{-}4}\ (n = 1\text{-}4)$$

$$\xrightarrow[Pd(OH)_2/C]{2\ H_2O,\ -2\ H_2} HOSiMe_2O(SiPh_2O)_nSiMe_2OH \quad \mathbf{5\text{-}8}\ (n = 1\text{-}4)$$

Scheme 1. Preparation of the oligosiloxane-α,ω-diols $HOSiMe_2SiO(SiPh_2O)_nSiMe_2OH$ (**5** – **8**; n = 1 – 4).

The catalytic oxidation of Si–H groups using Pearlman's catalyst, $Pd(OH)_2/C$, occurred rapidly at ambient temperatures in THF and required no stabilizing phosphate buffer [10]. After the reaction had finished, the catalyst was easily and completely removed by filtration. In contrast to dry palladium catalysts on carbon supports, Pd/C, which possess mostly unreactive Pd(0) sites, Pearlman's catalyst apparently contains more active Pd(II) sites that are surrounded by hydroxy groups and water [12]. The oligosiloxane-α,ω-diols $HOSiMe_2SiO(SiPh_2O)_nSiMe_2OH$ (**5** – **8**; n = 1 – 4) represent crystalline products that can be kept at –18 °C for several months without condensation.

The crystal structure of $HOSiMe_2OSiPh_2OSiMe_2OH$ (**5**) features a polymeric array of intermolecular hydrogen bonds that proceeds along the crystallographic *b*-axis and connects each molecule with four symmetry-related adjacent molecules (Fig. 1). In contrast, the crystal structure of $HOSiPh_2OSiPh_2OSiPh_2OH$ [13] and HOSi*t*-Bu_2OSiMe_2OSi*t*-Bu_2OH [14], the only two other 1,5-hexaorganotrisiloxanediols so far investigated by X-ray crystallography, contain dimers.

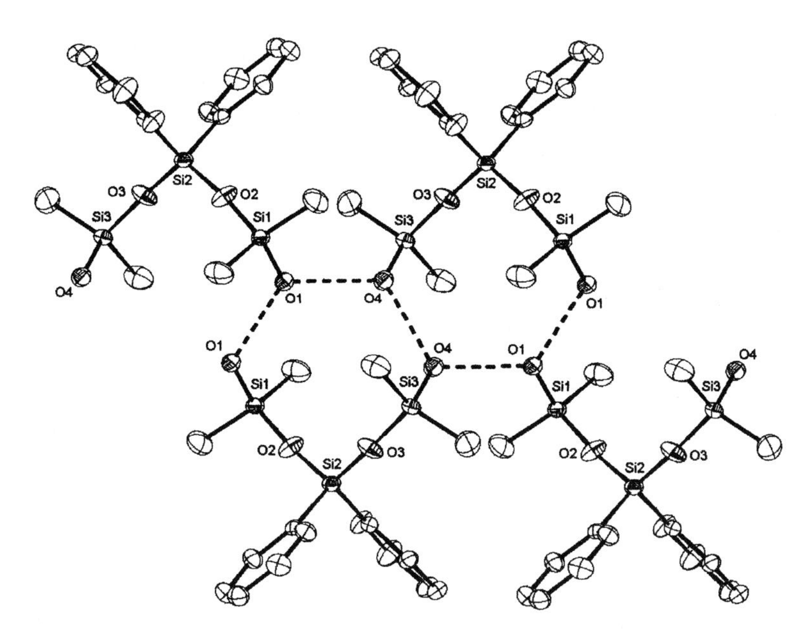

Fig. 1. Crystal and molecular structure of **4** showing 20% probability displacement ellipsoids and the atom numbering scheme. Selected bond parameters [Å, °]: Si–C_{Me}: 1.830(5) – 1.837(4); Si–C_{Ph}: 1.864(3) – 1.866(4); Si–O_{Si}: 1.597(3) – 1.605(3); Si–O_H: 1.621(4) – 1.627(3); Si–O–Si: 168.54(16) – 174.24(16); O1···O1: 2.674(5); O1···O4: 2.698(3); O4···O4: 2.677(4).

The crystal structure of $HOSiMe_2OSiPh_2OSiPh_2OSiMe_2OH$ (**6**) also features a dimer with two independent conformers that have adopted curved-chain conformations (Fig. 2). The hydrogen bonding mode resembles an eight-membered quasi ring of four oxygen and hydrogen atoms. Interestingly, $HOSiPh_2OSiPh_2OSiPh_2OSiPh_2OH$, the only other 1,7-octaorganotetrasiloxanediol yet investigated by X-ray diffraction lacks any O–H···O hydrogen bonds, but shows weak O–H···π bonds and adopts a zigzag chain conformation in the solid state [15]. The different chain conformations apparently imposed by the different hydrogen bond modes are reflected in the average Si–O–Si of 156.0 and 171.7° and demonstrate the influence of crystal packing effects on the molecular geometry.

To the best of our knowledge, $HOSiMe_2O(SiPh_2O)_3SiMe_2OH$ (**7**) comprises the longest α,ω-oligosiloxanediol yet investigated by X-ray crystallography. The crystal structure also features two crystallographically independent conformers, which in this case are connected pairwise into independent, albeit similar dimers (Fig. 3).

The hydrogen bonding modes also resemble eight-membered quasi rings of four oxygen and hydrogen atoms. It is worth noting that for compounds **5** – **7** the number of crystallographically independent Si atoms was generally in good agreement with the number of observed ^{29}Si MAS NMR signals. However, for $HOSiMe_2O(SiPh_2O)_4SiMe_2OH$ (**8**), the only α,ω-oligosiloxanediol for which we failed to obtain suitable single crystals for X-ray crystallography, revealed only four considerably broader signals. This observation was tentatively attributed to the presence of different polymorphs or an even more severe lack of periodicity in the crystal packing.

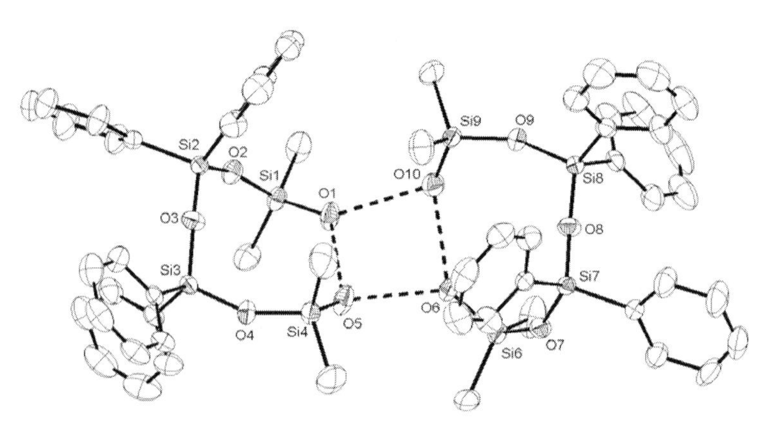

Fig. 2. Crystal and molecular structure of **5** showing 20% probability displacement ellipsoids and the atom numbering scheme. Selected bond parameters [Å, °]: Si–C_{Me}: 1.800(6) – 1.836(7); Si–C_{Ph}: 1.833(5) – 1.857(5); Si–O_{Si}: 1.604(3) – 1.622(4); Si–O_H: 1.612(4) – 1.643(4); Si–O–Si: 143.07(21) – 168.55(21); O1···O5: 2.631(6); O1···O10: 2.711(6); O5···O6: 2.678(5); O6···O10: 2.696(6).

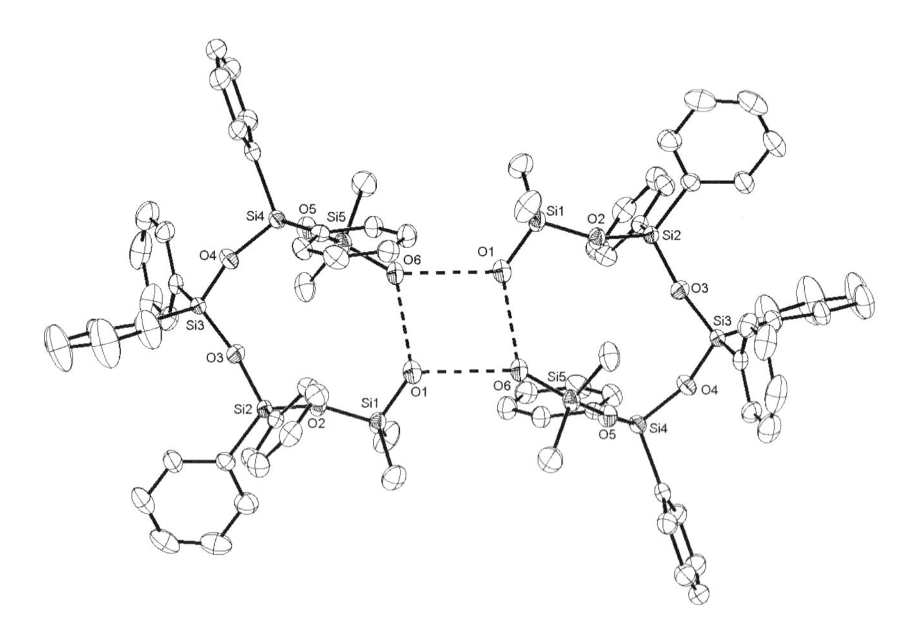

Fig. 3. Crystal and molecular structure of **6** showing 20% probability displacement ellipsoids and the atom numbering scheme (the second independent conformer is not shown). Selected bond parameters [Å, °]: Si–C_{Me}: 1.815(4) – 1.835(6); Si–C_{Ph}: 1.831(5) – 1.868(3); Si–O_{Si}: 1.597(3) – 1.627(3); Si–O_H: 1.609(3) – 1.638(2); Si–O–Si: 145.37(16) – 170.31(16); O1···O6: 2.760(4), 2.770(4); O7···O12: 2.680(4), 2.684(4).

Condensation of $HOSiMe_2SiO(SiPh_2O)_nSiMe_2OH$ (5 – 8; n = 1 – 4)

Like most sterically exposed organosilanols, $HOSiMe_2SiO(SiPh_2O)_nSiMe_2OH$ (**5** – **8**; n = 1 – 4) are sensitive to self-condensation, especially at elevated temperatures and/or in the presence of acid or base. In an effort to investigate whether compounds **5** – **8** are suitable precursors for the preparation of defined dimethylsiloxane – diphenylsiloxane block copolymers, we have studied their HCl-catalyzed condensation in a small amount of THF at 60 °C for 48 h. The crude condensation products were studied by ^{29}Si NMR spectroscopy, which revealed in the case of **5** and **8** the formation of complex mixtures of low-molecular-weight oligomers and evidence for incompletely condensed $-Me_2SiOH$ and $-Ph_2SiOH$ end groups. In the case of **6** and **7**, the ^{29}Si NMR spectra were surprisingly simple and consistent with the (almost) exclusive formation of the eight- and ten-membered tetra- and pentasiloxane rings *cyclo*-$(Me_2SiO)_2(Ph_2SiO)_2$ (**9**) and *cyclo*-$(Me_2SiO)_2(Ph_2SiO)_3$ (**10**) respectively (Eq. 1). Compounds **9** and **10** were isolated as low-melting solids in very high yields.

$$\underset{\mathbf{6},\ \mathbf{7}\ (n = 2, 3)}{HOSiMe_2O(SiPh_2O)_nSiMe_2OH} \xrightarrow[-\ H_2O]{HCl\ cat.} \underset{\mathbf{9},\ \mathbf{10}\ (n = 2, 3)}{cyclo\text{-}(Me_2SiO)_2(Ph_2SiO)_n}$$

Eq. 1. Selective HCl-catalyzed condensation of $HOSiMe_2SiO(SiPh_2O)_nSiMe_2OH$ (**7**, **8**; n = 2, 3).

Compound **10** is apparently the first pentasiloxane ring investigated by X-ray crystallography. The molecular structure features strain-free Si–O–Si bond angles that fall in the range from 149 to 165° and a heavily puckered ring conformation (Fig. 4).

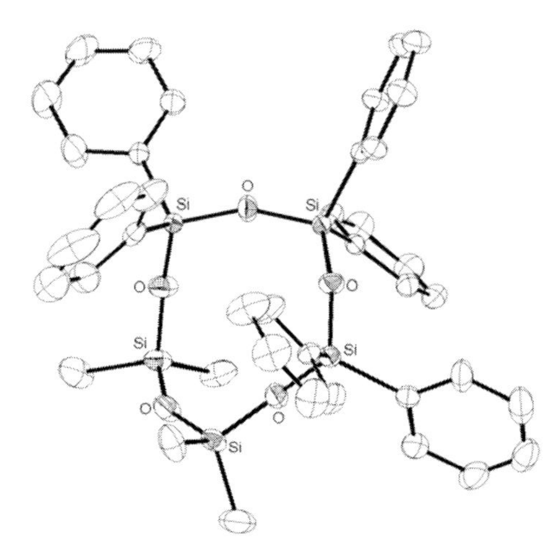

Fig. 4. Molecular structure of **10** showing 20% probability displacement ellipsoids and the atom numbering scheme. Selected bond parameters [Å, °]: Si–O: 1.594(3) – 1.622(4); $Si–C_{Me}$: 1.820(7) – 1.847(7); $Si–C_{Ph}$: 1.843(4) – 1.860(4); C–Si–C: 111.08(19) – 112.65(30); C–Si–O: 107.23(17) – 111.87(21); O–Si–O: 108.01(19) – 110.73(19); Si–O–Si: 149.05(23) – 164.49(24).

Acknowledgments: The Australian Research Council (ARC) is thanked for financial support. Dr. Jonathan White (The University of Melbourne) is gratefully acknowledged for the X-ray crystallography data collection.

References

[1] W. Noll, *Chemistry and Technology of the Silicones*; Academic Press, New York, **1968**.
[2] J. A. Semlyen, S. J. Clarson, *Siloxane Polymers*, Prentice Hall, Englewood Cliffs, **1991**.
[3] R. Murugavel, A. Voigt, M. G. Walawalkar, H. W. Roesky, *Chem. Rev.* **1996**, *96*, 2205.
[4] L. King, A. C. Sullivan, *Coord. Chem. Rev.* **1999**, *189*, 19.
[5] V. Lorenz, A. Fischer, S. Giessmann, J. W. Gilje, Y. Gun'ko, K. Jacob, F. T. Edelmann, *Coord. Chem. Rev.* **2000**, *206 – 207*, 321.
[6] J. Beckmann, K. Jurkschat, *Coord. Chem. Rev.* **2001**, *215*, 267.
[7] P. D. Lickiss, *Adv. Inorg. Chem.* **1995**, *42*, 147.
[8] P. D. Lickiss, The Synthesis and structures of silanols, in *Tailor-Made Silicon-Oxygen Compounds* (Eds.: R. Corriu, P. Jutzi), Vieweg, Braunschweig, **1996**, p. 47.
[9] P. D. Lickiss, 2001 Polysilanols in *Chemistry of Organic Silicon Compounds, Vol. 3* (Eds.: Z. Rappoport, Y. Apeloig), John Wiley & Sons Ltd, Chichester, **2001**, p. 695.
[10] G. H. Barnes, N. E. Daughenbaugh, *J. Org. Chem.* **1966**, *31*, 885.
[11] J. Beckmann, D. Dakternieks, A. Duthie, R. C. Foitzik, *Silicon Chem.* **2003**, *2*, 27.
[12] J. Tsuji, *Palladium Reagents and Catalysts: Innovations in Organic Synthesis*, John Wiley & Sons Ltd., Chichester, **1995**.
[13] H. Behbehani, B. J. Brisdon, M. F. Mahon, K. C. Molloy, M. Mazhar, *J. Organomet. Chem.* **1993**, *463*, 41.
[14] Graalmann, U. Klingebiel, W. Clegg, M. Haase, G. M. Sheldrick, *Chem. Ber.* **1984**, *117*, 2988.
[15] J. Beckmann, K. Jurkschat, D. Müller, S. Rabe, M. Schürmann, *Organometallics* **1999**, *18*, 2326.

The Reactivity of Carbofunctional Aminoalkoxysilanes in Hydrolytic and Reetherification Reactions

V. A. Kovyazin, V. M. Kopylov, A. V. Nikitin

State Research Center of the Russian Federation, State Research Institute for Chemistry and Technology of Organoelement Compounds, 38, Shosse Entuziastov, Moscow, 111123, Russia
Tel.: +7 095 2737161 — Fax: +7 095 2737206
E-mail: vlkov@aport.ru

S. P. Knyazev, E. A. Chernyshev

M. V. Lomonosov Academy for Fine Chemical Technology, 86, Prospekt Vernadskogo, 113127, Moscow, Russia

Keywords: carbofunctional organosilicon amines, hydrolysis, reetherification

Summary: The alkoxy group reactivity in carbofunctional organosilicon amines $H_2NR^1Si(OR)_3$ in hydrolytic and reetherification reactions was studied. Modeling calculations of electronic and molecular parameters and thermodynamic functions of organosilicon amines were performed by computer chemistry methods. The obtained calculated parameters of the molecules agree with experimental kinetic data in terms of alkoxy group reactivity of carbofunctional aminoalkylalkoxysilanes.

The alkoxy group reactivity in carbofunctional organosilicon amines of general formula $H_2NR^1Si(OR)_3$, where R = Et, $R^1 = (CH_2)_3$ (**Ia**), R = Me, $R^1 = (CH_2)_3$ (**Ib**), R = Et, $R^1 = CH_2(CH_3)CH$ (**Ic**), R = Et, $R^1 = (CH_2)_6NHCH_2$ (**Id**) was analyzed in hydrolytic (Eq. 1) and reetherification (Eq. 2) reactions that proceeded without a catalyst:

$$H_2NR^1Si(OR)_3 + 1.5\,H_2O \longrightarrow [H_2NR^1SiO_{1.5}]_n + 3\,ROH$$

Eq. 1.

$$NH_2RSi(OR^1)_3 + 3\,R^2OH \longrightarrow NH_2RSi(OR^2)_3 + 3\,R^1OH$$

$R^2 = Me_3Si, EtOCH_2CH_2$.

Eq. 2.

Hydrolysis with water was carried out in an appropriate alcohol at various temperatures and ratios. The reaction was controlled by GLC in terms of initial amine concentration.

On the basis of the relationship between conversion of compounds **Ia–d** and time ($[Am]_0$ = 1 mol/L, $[H_2O]_0$ = 1.5 mol/L), ratios of initial hydrolysis rates were found to be: $w_{0(\mathbf{Id})} : w_{0(\mathbf{Ib})} : w_{0(\mathbf{Ic})} : w_{0(\mathbf{Ia})} = 200 : 89 : 50 : 1$.

A similar order of alkoxy group reactivities is maintained for the first constant in the reetherification reaction by trimethylsilanol or the monoethyl ether of ethylene glycol.

Modeling calculations of electronic and molecular parameters and thermodynamic functions of 1-aminopropyltriethoxysilane (**II**), 2-aminopropyltriethoxysilane (**III**), and 3-aminopropyltriethoxysilane (**Ia**), which had a different position of the N atom in respect to Si, were performed by computer chemistry methods (Gaussian 98, HyperChem 6.0 software).

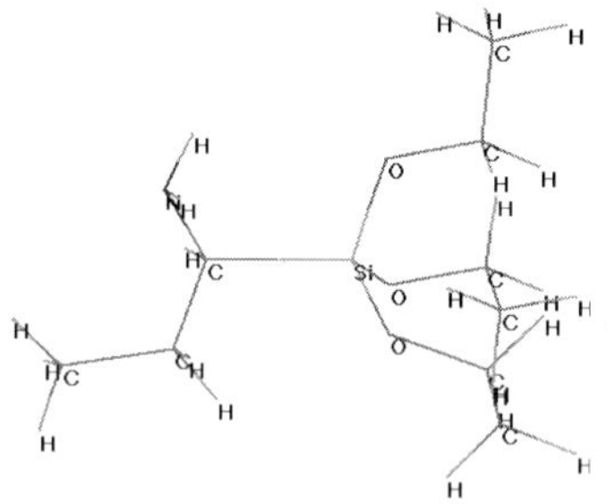

(II) 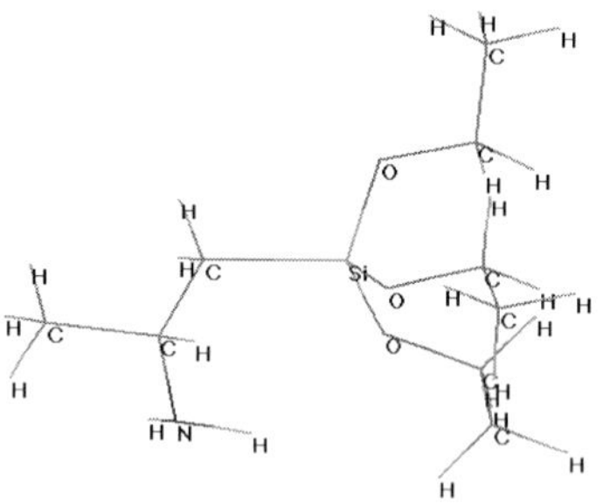 (III) 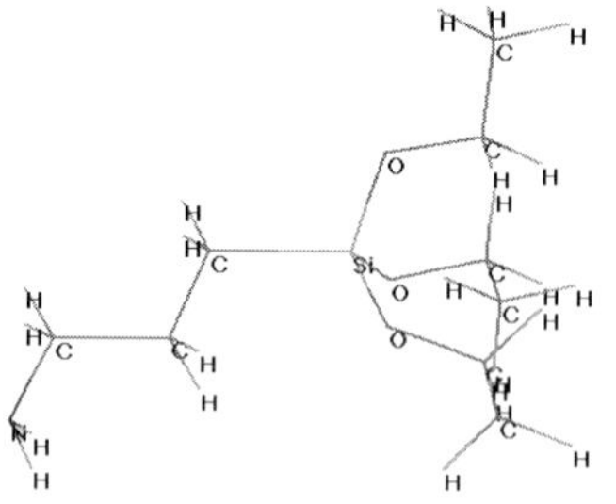 (Ia)

These calculations were compared with the analogous ones for propyltriethoxysilane (**IV**).

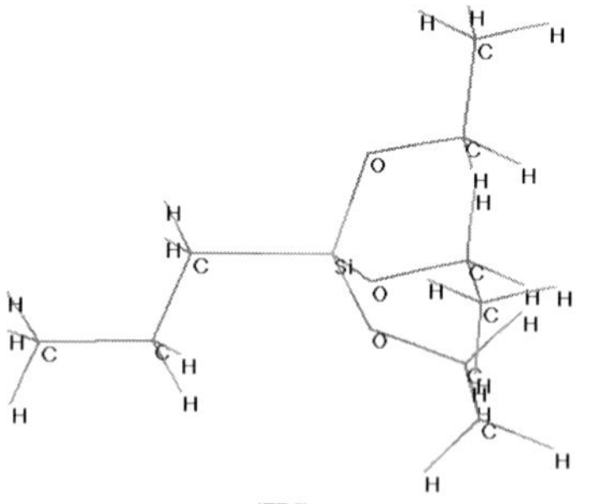

(IV)

The main targets of the investigation were: the determination of particular qualities of the electronic structure of aminopropyltriethoxysilane isomers and their trimethylsilyl derivatives by quantum chemistry methods, and estimation of their thermodynamic parameters and regularity of reactivity.

The construction, modeling, and preliminary optimizing of molecule structure were performed with the use of software HyperChem and approximation by molecular mechanic methods with use of force field MM+. The main purpose of this stage was the determination of starting geometric parameters of separate molecules, which were necessary for the calculations with assistance of software Gaussian 98. Optimizing of the structure for all molecules was carried out by minimizing of potential energy with gradient accuracy not less then 10^{-5} kcal/Å·mol.

To solve the problem put by the Hartree-Fock method with *ab initio* approximation (non-empirical calculations) was chosen. All calculations have been done with the assistance of Gaussian 98/A7 software and use the extended valence-splitting basis, which included diffusive and polarized d- and p-functions — 6-31G(d,p). The correlation amendments were performed with use of Density Functional Theory (DFT) in B3LYP approximation.

Thermodynamic parameters (H, S, G) were calculated at 25 °C and 1 atmosphere (standard conditions).

Optimizing the structure and thermodynamic calculations were carried out for a separate molecule, which was considered in the main singlet state, with assistance of Gaussian 98. To increase the calculation accuracy, "Tight" option was added to "Opt" command. The calculations of normal vibration and thermodynamic parameters for all systems were done by "Freq" command. To exclude the symmetry error, which resulted in the appearance of false frequencies, the command "NoSymm" was used during the calculations of molecules of reagents and reaction products.

The analysis, which was done after calculation, showed the absence of false frequencies for all investigated molecules. It confirms that points corresponding to minimal potential energy have been found. The analogous calculations with the use of HyperChem 6.0 and semi empirical approximation PM3 have been carried out for the comparison, with the above-mentioned results. Optimizing of the molecular structure was carried with use of the "Polak-Ribiere" algorithm until the minimal potential energy was reached. The accuracy of calculation was not less then 10^{-5} kcal/Å·mol.

All structural, electronic, and thermodynamic molecular parameters necessary for analysis were picked up from *out*-files (Gaussian 98) or *log*-files (HyperChem) and treated by MS Excel. All results obtained are presented in the Table 1. These are:

- values of general energy and thermodynamic functions H, S and G and the relative values ΔH, ΔS and ΔG
- parameters of two upper occupied and lower vacant MO and values of energetic gap between them
- values of charges (Q) on Si, N and O atoms.

Analysis of obtained results has shown:

- The general energy of molecule (E), enthalpy (H), and free energy (G) all increase in the order 2-aminopropyltriethoxysilane, 3-aminopropyltriethoxysilane, and 1-aminopropyltriethoxysilane; entropy increases in different way: 2-aminopropyltriethoxysilane, 1-aminopropyltriethoxysilane, and 3-aminopropyltriethoxysilane.

- The energy of the lower vacant orbital, which is localized mainly on the Si atom, naturally increases in the order 1-aminopropyltriethoxysilane, 2-aminopropyltriethoxysilane, and 3-aminopropyltriethoxysilane.
- The energy of the upper occupied orbital, which is localized mainly on the N atom of the amino group, remarkably decreases in the order 1-aminopropyltriethoxysilane, 2-aminopropyltriethoxysilane, and 3-aminopropyltriethoxysilane.
- The energy of the upper occupied orbital, which is localized mainly on the O atom, increases in the same order 1-aminopropyltriethoxysilane, 2-aminopropyltriethoxysilane, and 3-aminopropyltriethoxysilane.
- The value of the energetic gap between the lower vacant and upper occupied orbitals increases in the same manner.
- The value of the charge on the Si, N, and O atoms almost does not depend on the particular molecule investigated.

Conclusions

a) Frontier orbital parameters regular changes were found in a series of the studied molecules:

- one can expect aminopropylethoxysilane reactivity to increase in respect of nucleophiles in the order: (propyltriethoxysilane), 3-aminopropyl-, 2-aminopropyl-, 1-aminopropyltriethoxysilane;
- nucleophile activity of amine groups may increase in the same order (3-aminopropyl-, 2-aminopropyl-, 1-aminopropyltriethoxysilane);
- Si–O–C group nucleophilic activity of the oxygen atom increases in inverse order: 1-aminopropyl-, 2-aminopropyl-, 3-aminopropyltriethoxysilane.

b) Parameters of the charge distribution over the analyzed molecule atoms do not carry significant information in regard of electronic structure peculiarities and reactivity regularities.

The obtained calculated parameters of the molecules agree well with experimental kinetic data in terms the reactivity of alkoxy groups of carbofunctional aminoalkoxysilanes.

Table 1. Calculated data of thermodynamic functions and parameters 1-NH2-, 2-NH2-, and 3-NH2-PrSi(OEt)3 (HyperChem-PM3 and Gaussian98-B3LYP/6-31G**).

Parameter	Unit	1-NH$_2$-PrSi(OEt)$_3$		2-NH$_2$-PrSi(OEt)$_3$		3-NH$_2$-PrSi(OEt)$_3$	
		PM3	6-31G**	PM3	6-31G**	PM3	6-31G**
E	a.e.		–926.8882159		–926.8965125		–926.8914874
	kcal/mol	–58633.47161	–581631.1609	–58633.87361	–581636.3671	–58633.32097	–581633.2138
$\Delta E = E - E_{min}$	kcal/mol	0.4020039	5.206195318	0	0	0.5526424	3.153297988
S	cal/mol ·K	–	158.007	–	155.323	–	158.961
$\Delta S = S - S_{min}$	cal/mol ·K	–	2.684	–	0	–	3.638
H	a.e.		–926.540492		–926.549018		–926.543578
	kcal/mol	–261.8718082	–581412.9609	–262.2738121	–581418.311	–261.7211697	–581414.8974
$\Delta H = H - H_{min}$	kcal/mol	0.4020039	5.350145997	0	0	0.5526424	3.41365168
G	a.e.	–	–926.615566	–	–926.622817	–	–926.619105
	kcal/mol	–	–581460.0705	–	–581464.6206	–	–581462.2913
$\Delta G = G - G_{min}$	kcal/mol	–	4.550071385	–	0	–	2.329315264
E_{LVO}	a.e.		0.05897		0.062		0.06423
	eV.	0.67094	1.6045737	0.782	1.68702	0.85334	1.7476983
E_{UOO}	a.e.		–0.21057		–0.21862		–0.22252
	eV	–8.41694	–5.7296097	–9.10438	–5.9486502	–9.24943	–6.0547692
E_{UOO-1}	a.e.		–0.27528		–0.27028		–0.26849
	eV.	–10.16168	–7.4903688	–10.08722	–7.3543188	–10.10754	–7.3056129
$\Delta E = E_{LVO} - E_{UOO}$	eV	9.08788	7.3341834	9.88638	7.6356702	10.10277	7.8024675
Q Si		0.954616	1.04999	1.000556	1.057807	0.997326	1.067787
Q N		–0.011801	–0.607879	–0.053059	–0.599285	–0.037682	–0.594887
Q O		–0.451286cp	–0.577811average	–0.452179cp	–0.574382cp	–0.444711cp	–0.572327cp

Study of Octyltriethoxysilane Hydrolytic Polycondensation

N. S. Plekhanova, V. V. Kireev

D. Mendeleev University of Chemical Technology of Russia
9, Miusskaya Square, 125047, Moscow, Russia
Tel.: +7 95 273 63 64
E-mail: nadp@nm.ru

V. V. Ivanov, V. M. Kopylov

State Research Institute for Chemistry and Technology of Organoelement Compounds,
38, Shosse Entuziastov, 111123, Moscow, Russia

Keywords: octyltriethoxysilane, hydrolytic polycondensation, siloxanes

Summary: Hydrolytic polycondensation of octyltriethoxysilane (OTES) was studied by ^{1}H and ^{29}Si NMR spectroscopy, GLC, and Fischer titration. Hydrolysis rate in terms of ethoxy groups was found to exceed polycondensation rate during the reaction in ethyl alcohol with hydrochloric acid as a catalyst. The molecular mass of hydrolytic polycondensation oligomers reaches a maximum value at molar ratio water/OTES = 1.5.

Introduction

Organoalkoxysilanes are widely used for coatings production [1], as electrolytes [2], adhesives [3], crosslinking agents [4, 5], water-repellent components for glass, building materials, and textiles [6 – 8].

Organoalkoxysilanes are used because of their ability to interact with hydroxyl groups of various materials and undergo hydrolytic polycondensation (HPC) in the presence of water, transforming to polyorganosiloxanes. HPC of organoalkoxysiloxane may proceed without a catalyst, but, as a rule, nucleophilic or electrophilic catalysts are employed for reaction promoting. Amines, alkali metal hydroxides, and silanolates or siloxanolates can be used as nucleophilic catalysts, and Lewis acids and organic and inorganic acids as electrophilic ones. Organoalkoxysilane HPC rate decreases with the growth of substituent size including alkoxy groups at silicon [9 – 11]. In the case of acid catalysis, with the increase of electron-accepting properties of the substituent at silicon, HPC rate is inhibited, and, in the case of basic catalysis, it increases [9, 12].

Organoalkoxysilane HPC was studied by means of ^{29}Si NMR spectrometry [13, 14] and gas chromatography [15]. Tetraethoxysilane (TEOS), methyltriethoxysilane (MTES), phenyltriethoxysilane (PhTES) HPC was looked at in [16], and MTES hydrolysis rate was found to be much higher than that of TEOS and PhTES. TEOS and MTES hydrolytic product polycondensation rates are comparable, but that of PhTES is much lower. OTES HPC kinetics at the water-decane interphase was analyzed in [17] by interfacial tension method. OTES polycondensation and hydrolytic reaction rate constants were determined.

OTES is currently the most promising organoalkoxysilane for use as a water repellent [18 – 20]. For the determination of optimum conditions for the use of OTES, its HPC processes have to be investigated under various conditions. However, there is a lack of literary data concerning the processes of OTES condensation and hydrolysis. Therefore, this paper deals with the investigation of OTES HPC for electrophilic catalysis.

Experimental

Initial OTES: B.p. = 98 °C/2 mm, n_d^{20} = 1.416 – 1.418, d_4^{20} = 0.88, product of Penta-91 Ltd.

OTES HPC was conducted in 93% ethanol environment at a temperature of 30 °C with hydrochloric acid addition (HCl, 36.5%) as a catalyst.

OTES HPC is controlled by the GLC method with *n*-xylol as a reference; LKhM-80 chromatograph, column 2000 × 4 mm, stationary phase SE-30 on silanized Chromosorb-W, helium carrier gas.

Water in the reaction mixture was determined by coulometric titration with Fischer reagent (GOST 24614-81) using a "Titricoulon" ROKBA device. In order to decrease the side-reaction effect (Si–OH interaction with methanol), a graphic extrapolation technique was applied to the titration curves.

The reaction mixture was studied by ^{1}H and ^{29}Si NMR spectroscopy on a Bruker AM-360 spectrometer with a working frequency of 360.13 MHz.

MWD parameters of HPC products were determined using a "Knauer" gel chromatograph by the HPC method, "Shodex" styrogel columns, refractometry detector, toluene solvent, and scaling by polystyrene.

HPC Procedure

Mix 50 g OTES and 5 g *n*-xylol in a round-bottomed three-necked flask equipped with a stirrer and thermometer. Add 15 mL 93% ethanol to the mixture. Keep the reaction mixture at 30 °C with stirring. Then add the specified quantity of water 0.4 – 3.0 mol per mol OTES (taking into account the water present in the ethyl alcohol) and the calculated quantity of concentrated hydrochloric acid to provide a concentration of 0.03 mol/L. Sampling was performed at defined intervals and analysis was carried by means of ^{1}H and ^{29}Si NMR spectroscopy, GLC, and Fischer methods.

HPC and reaction products analysis was performed as described above, but 0.8 mol water was introduced per mol OTES (taking into account the water present in the ethyl alcohol), and various quantities of hydrochloric acid were introduced in order to provide a concentration within the range of 0.001 – 0.003 mol/L.

Results and Discussions

The OTES HPC process involves hydrolytic stages with silanol formation (Eq. 1) and polycondensation with the production of siloxanes (Eqs. 2 and 3):

$$\equiv SiOC_2H_5 + H_2O \longrightarrow \equiv SiOH + C_2H_5OH$$

Eq. 1.

$$\equiv SiOH + \equiv SiOH \longrightarrow \equiv SiOSi\equiv + H_2O$$

Eq. 2.

$$\equiv SiOH + \equiv SiOC_2H_5 \longrightarrow \equiv SiOSi\equiv + C_2H_5OH$$

Eq. 3.

General scheme of OTES HPC can be represented by Eq. 4:

$$m\, C_8H_{17}Si(OC_2H_5)_3 + n\, H_2O \xrightarrow[-C_2H_5OH]{} [C_8H_{17}Si(OC_2H_5)_a(OH)_b(O)_c]_m$$

Eq. 4.

Equation 1 shows that maximum consumption of three water molecules per OTES molecule is observed at hydrolysis of three ethoxy groups. Because of consumption of silanol groups according to Eq. 2, water consumption decreases to 1.5 mol per mol OTES. Silanol group excess may be formed when hydrolytic reaction proceeds with water content above 1.5 mol per mol OTES. In this case, water content may drop at the beginning because of hydrolysis of ethoxy groups and then increase because of condensation of silanol groups according to Eq. 3.

Study of OTES HPC at specified molar ratios water/OTES = 1.5 and lower HCl concentration equal to 0.03 mol/L demonstrated a sequential drop of water content in the time domain (Fig. 1). When the water/OTES ratio is 2.0 or 3.0, the water content quickly decreases at first and then quickly increases (Fig. 1).

Thus, for example, for molar ratio of water/OTES = 0.8, water conversion amounts to 94% during the first 15 min and 97.9% in 24 h. At the same time, according to ^1H NMR data, water consumption on cleavage of ethoxy groups amounts to 70% during the first 15 min and 97.9% in 24 h. This testifies to the fact that the hydrolytic reaction according to Eq.1 proceeds faster than polycondensation reactions according to Eqs. 2 and 3.

For molar ratio of water/OTES = 3.0 during the first 15 min, water content amounts to 42.6% of the starting quantity, in 24 h 48.3%, and in 14 days ~50%. High residual water content at water/OTES ratio equal to 2.0 and 3.0 shows that under specified conditions polycondensation processes are of deep types.

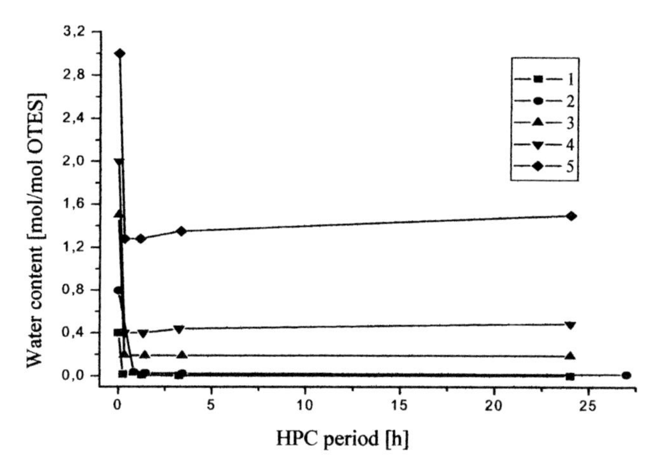

Fig. 1. Water content change during HPC of OTES. HCl = 0.03 mol/L, molar ratios of water/OTES 0.4 (1), 0.8 (2) 1.5 (3), 2.0 (4), 3.0 (5).

OTES content determination by the GLC method proved that the main portion of OTES, just like water, reacts during the first 15 min (Fig. 2a). The remaining OTES portion is gradually consumed by participation in hydrolytic and polycondensation reactions with formation of oligosiloxanes. OTES vanishes in reaction products when the water/OTES ratio ≥ 0.8. At lower ratios OTES remains, and its quantity increases as the water/OTES ratios decrease (Fig. 2a).

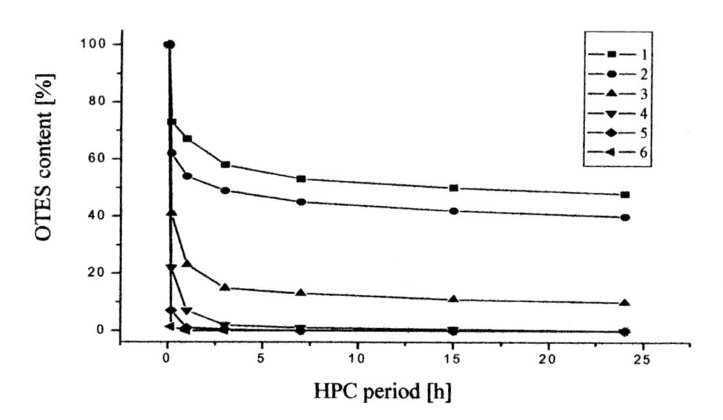

Fig. 2a. OTES content change at HPC. HCl = 0.03 mol/L, molar ratios of water/OTES 0.32 (1), 0.4 (2), 0.6 (3), 0.8(4) 1.0 (5), 1.2 (6).

Analysis of ^{29}Si NMR spectra of HPC products in 26 days at molar ratios water/OTES = 0.32 and 0.4 demonstrated that the reaction mixture includes OTES with chemical shifts of silicon atoms

δ = 45.0 ppm, 1,1,3,3–tetraethoxy-1,3-dioctyldisiloxane (dimer) with chemical shift of silicon atoms δ = 52.9 ppm and 1,1,3,5,5–pentaethoxy-1,3,5-trioctyltrisiloxane (trimer) with chemical shift of end silicon atom δ = 53.2 ppm and middle silicon atom δ = –61.2 ppm. At water/OTES ratio = 0.6 along with the above-mentioned products multiplet signals of silicon atoms in more complex HPC products with chemical shift δ = –50.3 to –50.8 ppm and δ = –59.1 to –60.0 ppm are registered. A still more complicated picture is observed in spectra on further increase of water/OTES molar ratio.

Analysis of HPC products by the GLC method proved the presence of dimer, identified by the mass spectrometry technique. Dimer content dependence on time at various molar ratios of water/OTES was determined by GLC, silicon signal integral intensities in spectra of ^{29}Si NMR, and mass spectroscopy (Fig. 2b).

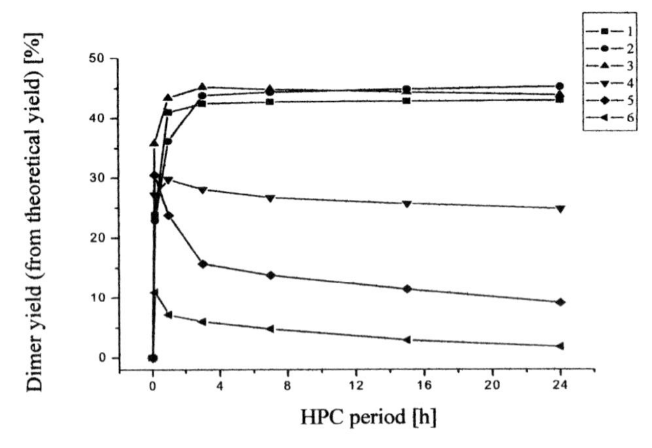

Fig. 2b. Siloxane dimer content change during HPC. HCl = 0.03 mol/L, molar ratios of water/OTES 0.32 (1), 0.4 (2), 0.6 (3), 0.8 (4) 1.0 (5), 1.2 (6).

Figure 2b and results of HPC product composition determination show that in 26 days (Table 1), dimer content in all ratios of water/OTES reaches its peak. The highest dimer content in the reaction mixture is observed when 0.3 – 0.6 mol water are used for one mol OTES. Its quantity reaches a maximum in 3 h. When the water/OTES ratio increases to 0.8 – 1.2 mol per mol OTES, dimer product yield reaches its maximum during the first few hours and then gradually decreases until full disappearance. The dimer content decrease with time testifies to the participation of dimer products in the polycondensation reaction along with OTES. When water is used in an amount exceeding 1.2 mol per mol OTES, the dimer is absent at the initial hydrolytic stage.

Analysis of HPC products by ^{1}H NMR spectra showed the presence of proton signals in the spectrum in the form of two triplets with 0.73 ppm and 0.46 ppm centers, corresponding to chemical shifts of $-CH_3$ groups in ethanol and ethoxy groups as well as two quartets in the 3.7 ppm and 3.5 ppm areas corresponding to chemical shifts of $-CH_2$ groups in ethanol and ethoxy groups, and multiplet signals of protons in the 1.0 – 1.2 ppm area corresponding to chemical shift of a $-C_8H_{17}$ group. Proton signals of internal standard *n*-xylol are in the 6.9 ppm area corresponding to a chemical shift of aromatic ring protons, and 2.1 ppm corresponding to methyl group protons. On the basis of the proton signal, integral intensities of reaction products and the standard ethoxy group content was determined in HPC products. The results are presented in Table 2.

On the basis of water and ethoxy group contents in the reaction products, the hydroxyl group content was calculated and included in Table 2.

Table 1. HPC products–composition in 26 days.

Water/OTES [mol/mol]	Content OTES [%]	Dimer content [%]	Trimer content [%]
0.32	40.0	42.9	17.1
0.40	25.0	38.2	21.8
0.60	6.0	29.9	37.4

Table 2. Ethoxy and silanol groups content in products of OTES HPC and residual water in the reaction mixture.

Water/OTES	Content in hydrolyzate [mol/mol OTES]		
[mol/mol]	Residual water	Ethoxy groups	Si–OH groups
	HPC period 3 h		
0.4	0.01	2.39	0.18
0.8	0.02	1.60	0.16
1.5	0.19	0.75	0.37
2.0	0.44	0.22	0.34
3.0	1.30	0.06	0.46
	HPC period 24 h		
0.4	0	2.21	–
0.8	0.02	1.40	–
1.5	0.19	0.35	–
2.0	0.49	0.10	–
3.0	1.45	0.05	–

Table 2 demonstrates that at the initial stage of hydrolysis silanol groups are present in the reaction mixture at any water/OTES ratios. Use of large amounts of water for ethoxy group hydrolysis results in silanol group content increase. However, because of polycondensation, concentration of silanol groups decreases and in 24 h is insignificant.

The influence of catalyst concentration and duration of HPC on the change in the content of –OEt and Si–OH groups at molar ratio water/OTES = 0.8, depending on hydrochloric acid concentration (0.001 – 0.030 mol/L) is shown in Figs. 3a–c. One can see that during the first 15 min 88 – 95% of the water and 65 – 85% of the ethoxy groups react.

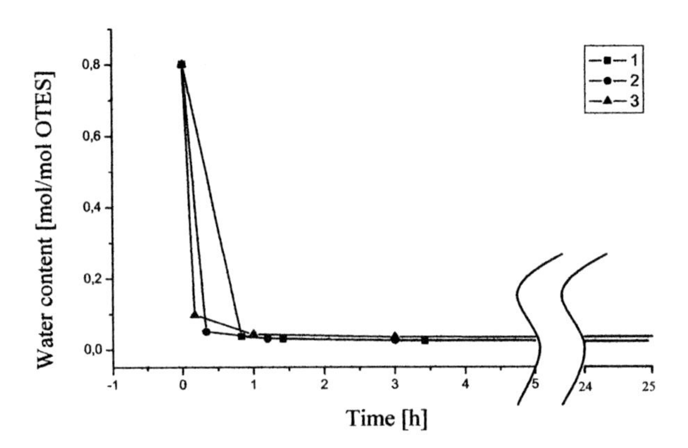

Fig. 3a. Water content change during HPC of OTES. Molar ratio of water/OTES = 0.8, hydrochloric acid concentration 0.03 (1), 0.003 (2), 0.001 (3).

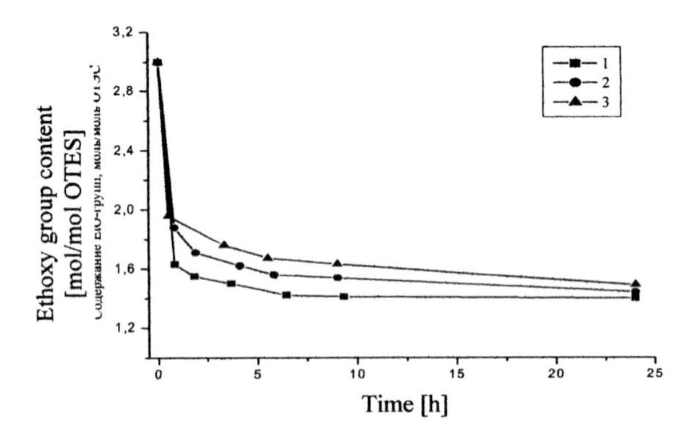

Fig. 3b. Ethoxy group content change during HPC of OTES. Molar ratio of water/OTES = 0.8, hydrochloric acid concentration 0.03 (1), 0.003 (2), 0.001 (3).

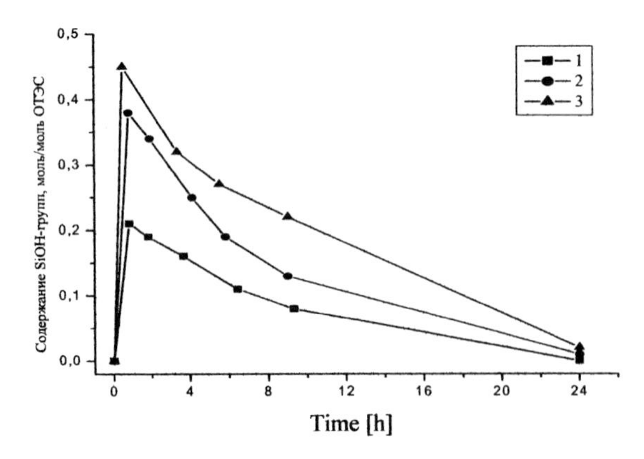

Fig. 3c. Silanol groups content change during HPC of OTES. Molar ratio of water/OTES = 0.8, hydrochloric acid concentration 0.03 (1), 0.003 (2), 0.001 (3).

Maximum silanol group content (Fig. 3c) is observed at low hydrochloric acid concentrations. The silanol groups are gradually consumed during 24 h, indicating completion of the condensation process.

On OTES HPC, when the water/OTES ratio is equal to 0.8, clear homogeneous siloxane solutions are produced. Further increase of water/OTES ratio is accompanied by the formation of two immiscible phases.

Gel permeation chromatography was used for the analysis of products formed during OTES HPC. The obtained data are presented in Table 3.

Table 3. MWD parameters of products of OTES HPC.

Water/OTES [mol/mol]	Fraction of HPC products	M_n	M_w	M_w/M_n
0.4	–	570	700	1.23
0.8	–	900	1020	1.13
	Combined*	800	1100	1.36
1.5	From the top layer	560	770	1.38
	From the bottom layer	950	1250	1.32

*Molecular weights were found after dissolving both fractions in toluene.

With increase in the water/OTES molar ratio, the molecular weight of formed siloxanes increases and achieves a maximum at a water/OTES ratio equal to 1.5. Investigation of two-phase HPC product composition showed that the top layer is a water-alcohol solution of a low-molecular-mass siloxane fraction, and the bottom layer is a relatively higher siloxane fraction. In general, the products are siloxanes of relatively low-molecular mass (M_w = 650 – 1250), with ethoxy and hydroxyl groups on the silicon.

From the data obtained, the compositions of the products obtained were calculated. The results are presented in Table 4.

Table 4. Compositions of the products formed by OTES HPC.

Water/OTES [mol/mol]	**Time of HPC [h]**		**Average degree of polymerization of end products [$\overline{n}$]****
	3	**24**	
0.4	$[C_8H_{17}Si(OEt)_{2.39}(OH)_{0.18}(O)_{0.21}]_n$	$[C_8H_{17}Si(OEt)_{2.21}(O)_{0.4}]_n$	2.84
0.8	$[C_8H_{17}Si(OEt)_{1.6}(OH)_{0.16}(O)_{0.62}]_n$	$[C_8H_{17}Si(OEt)_{1.4}(O)_{0.8}]_n$	4.70
1.5*	$C_8H_{17}Si(OEt)_{0.75}(OH)_{0.37}(O)_{0.94}]_n$	$[C_8H_{17}Si(OEt)_{0.35}(O)_{1.33}]_n$	6.18

* For combined fraction. ** $\overline{n}$ value was obtained by dividing oligomer's M_w (Table 3) by the molecular mass of the corresponding product's empirical formula after 24 h of HPC.

References

[1] H. Schmidt, B. Seiferling, *Mater. Res. Soc. Symp. Proc.* **1989**, *73*, 739.
[2] D. Ravaine, A. Seminel, Y. Charbouillot, M. Vincent, *J. Non-Cryst. Solids* **1986**, *82,* 210.
[3] H. H. Huang, G. L. Wilkes, J. G. Carlson, *Polymer.* **1989**, *30,* 2001.
[4] K. P. Hoh, H. Ishida, J. L. Koenig, *Polym. Compos.* **1990**, *11,* 121.
[5] A. Serier, J. P. Paskault, T. M. Lam, *J. Polym. Sci. Chem.* **1991**, *29*, 1225.
[6] Y. Akamatsu, K. Makita, H. Inaba, T. Minami, *Thin Solid Films* **2001**, *389*, 138 – 145.
[7] H. Marina, Pat. 0 552 874 EP **1993**.
[8] L. A. Stark-Kasley, P. J. Popa, T. M. Gentle, D. E. Hauenstein, L. D. Kennan, Pat. 5 421 866 USA **1995**.
[9] T. Mizuno, J. Phalipoou, J. Zarzucki, *Glass Technol.* **1985**, *26*, 39.
[10] R. C. Mehrotra, *Non-Cryst. Solids* **1988**, *100*, 1.
[11] K. A. Andrianov, *Organic Silicon Compounds*, (US Dept. of Commerce, Washington, DC, Trans, 59–11239) State Scientific Publishing House for Chemical Literature. Moscow, **1995**.
[12] F. Surivet, T. M. Lam, J. P. Pascault, Q. T. Pham, *Macromolecules* **1992**, *25*, 4309.
[13] A. Vainrub, F. Devreux, J. P. Boilet, F. Chaput, M. Sarkar, *Mater. Sci. Eng. B.* **1996**, *37*, 197.
[14] D. M. Heenan, S. E. Friberg, J. Sjoblom, G. C. Farrington, submitted to *J. Phys Chem.*
[15] K. Izumi, H. Tanaka, Y. Uchida, N. Tohge, T. Minami, *J. Mater. Sci. Lett.* **1993**, *12*, 724.
[16] Y. Sugahara, T. Inoue, K. Kuroda, *J. Mater. Chem.* **1997**, *7*(*1*), 3 – 59.
[17] R. Lindberg, J. Sjoblom, *J. Colloids Surf. A: Physicochem. Eng. Aspects* **1998**, *135*, 53 – 58.
[18] A. Be, D. T. Liles, F. G. P. Wilhelmi, Pat. 5 919 296 USA **1999**.
[19] M. J. Chen, A. Chaves, Pat. 5 393 330 USA **1995**.
[20] E. Schamberg, G. Koerner, H. Fritsch, M. Grasse, R. Sucker, Pat. 5 091 002 USA **1992**.

A Study of the Dependence of Silicone Compositions on the Initial Structure and Composition of Oligoorganosiloxanes

S. R. Nanushyan, E. I. Alekseeva,* A. N. Polivanov

State Research Center of the Russian Federation, State Research Institute for Chemistry and Technology of Organoelement Compounds, 38, Shosse Entuziastov, 111123, Moscow, Russia

Tel.: +07 95273 7180 — Fax: +07 95273 1327

E-mail: nanush@eos.incotrade.ru

Keywords: polyaddition, thermal stability, oligomer

Summary: The backgrounds for elastomer and glassy silicone compositions cured by a polyaddition mechanism have been developed. Their thermal stability as a function of basic siloxane chain framework and change in some physical–chemical and physical–mechanical characteristics over a wide temperature range are studied.

Oligoorganosiloxanes of the general formula given in Scheme 1 were produced through catalytic rearrangement of corresponding organocyclosiloxanes or by hydrolytic co-polycondensation of organochloro-(ethoxy-, acetoxy)silanes.

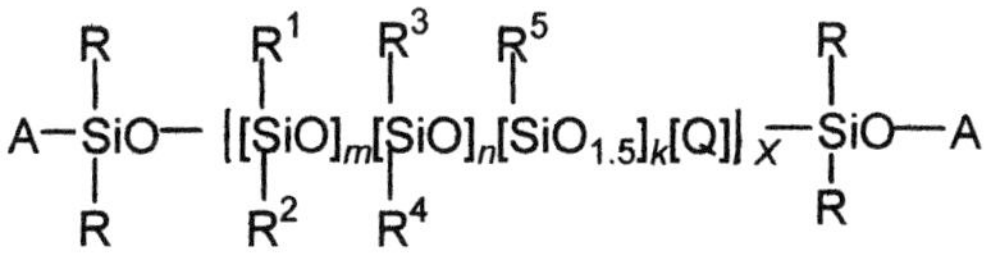

$$A-\underset{R}{\overset{R}{|}}SiO-\{[\underset{R^2}{\overset{R^1}{|}}SiO]_m[\underset{R^4}{\overset{R^3}{|}}SiO]_n[\overset{R^5}{|}SiO_{1.5}]_k[Q]\}_x-\underset{R}{\overset{R}{|}}SiO-A$$

where: A= CH_2=CH, SH, H, CH_3; $m + n + k + l = 1$; $x = 10–100$;

R= CH_3, CH_2=CH; R^1 =R^2 =CH_3; R^1=R^2 =C_6H_5; R^1= CH_3, R^2 = C_6H_5;

R^3= CH_3; R^4 = H; R^3 = CH_3, R^4 =$CF_3CH_2CH_2$; R^3 = CH_3, R^4 = CH_2=CH;

R^5 = CH_3, C_6H_5

Q = [structure: biphenyl-bridged Si–O unit] or [structure: biphenyl-bridged –Si–O– ring unit]

Scheme 1. Total composition of silicone oligomers with hetero siloxy units.

Composites with preset curing rates and temperature, suitable physical and mechanical performance, and high optical properties can be obtained by modifying the type and ratio of the synthesized oligoorganosiloxanes by the employment of various complex catalysts based on platinum adducts with polyunsaturated siloxanes, additional crosslinking agents, fillers, etc.

Elastomers (general name — SIEL) are the compositions based on linear α,ω-bis-trivinyl-(dimethylvinyl)siloxyoligoorganosiloxanes with molecular weight (M_w) 50 000 – 100 000 and α,ω-bis-trimethyl(dimethylhydro)(dimethylmercapto)siloxyoligoorganosiloxanes with M_w 800 – 1500.

Refractive index fluctuates in the range 1.37 – 1.57 depending on the basic siloxane chain framework (Fig. 1). The influence of the siloxane chain framework on the refractive index was studied over a wide temperature range, and hysteresis was found on the curve of refractive index vs temperature (Fig. 2).

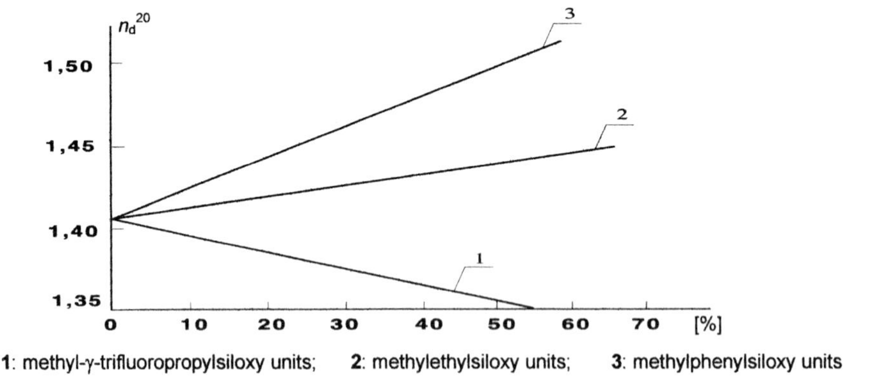

Fig. 1. Regulation of the refractive index by introduction of hetero units into polydimethylsiloxane.

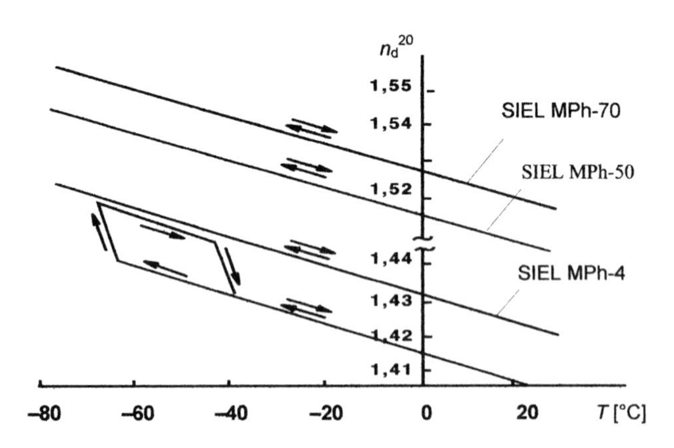

Fig. 2. Temperature dependence of the refractive index of cured SIEL MPh compounds for different contents of methylphenylsiloxy units.

Glassy high strength, providing the possibility of mechanical treatment materials (general name — STYK), are resins based on linear and branching oligovinyl- and oligohydridesiloxanes with viscosity ~800 – 1200 MPa s. Compositions of some resins are shown in Table 1.

Table 1. Composition of STYK compounds.

	Content of organosiloxy units [mol part]						
Sample	**$(CH_3)_3SiO_{0.5}$**	**$C_6H_5SiO_{1.5}$**	**$(CH_3)HSiO$**	**CH_3CH_2CHSiO**	**$CH_3C_6H_5SiO$**		
I	–	0.20	0.20	0.20	0.40	–	–
I	–	0.20	0.20	0.20	0.20	0.20	–
III	–	0.20	0.20	0.20	–	0.40	–
IV	0.05	–	0.225	0.225	–	0.50	–
V	0.05	–	0.175	0.175	–	0.60	–
VI	–	0.25	0.25	0.25	–	–	0.25
VII	–	0.20	0.20	0.20	–	–	0.40

The properties of polyaddition materials SIEL and STYK were analyzed by methods of long-term isothermal aging, DSK, TGA, and DTA (for example DTA and TGA curves of SIEL with methylphenylsiloxy units are presented) and radiography within a wide temperature range (Figs. 3–5).

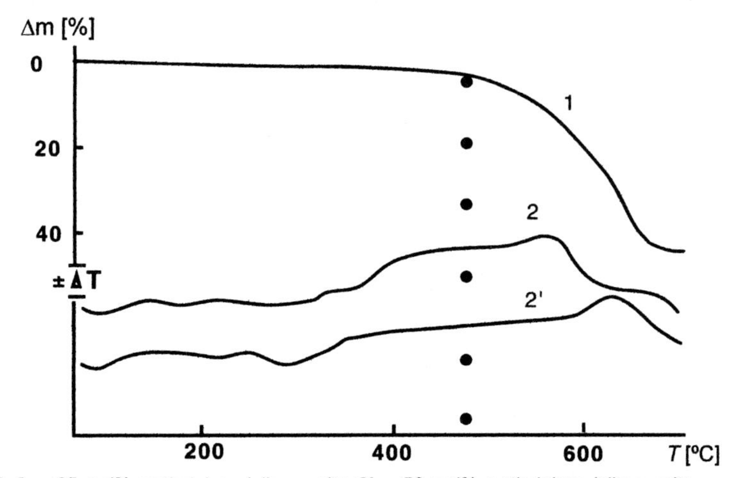

1, **2** — 25 mol% methylphenylsiloxy units; **2'** — 50 mol% methylphenylsiloxy units.
Speed of heating: 2.5 deg/min. Weight of sample: 100 mg.

Fig. 3. The TGA and DTA curves of SIEL MPh compounds.

Total composition operating range was found to be wide enough: from –90 to 350 °C. At a similar degree of crosslinking of the composite samples of the elastomers, thermal stabilization increases according to basic siloxane chain framework in the series: M_2; MF; MPh; Ph_2, where M_2

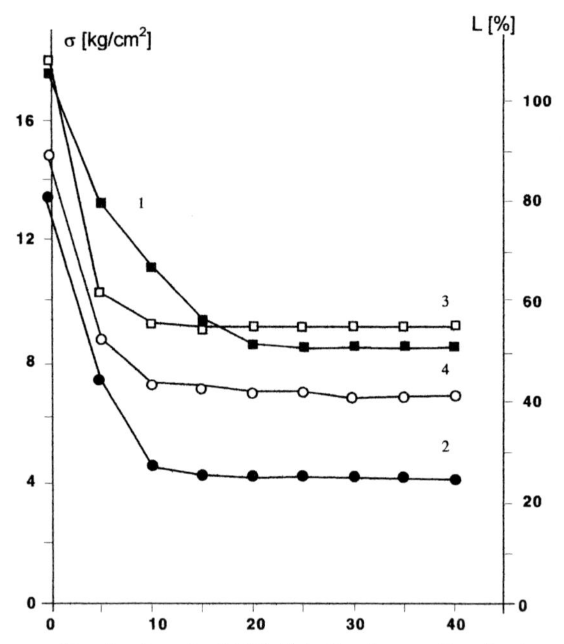

1 – change of tensile strength of SIEL Mph; **2** – change of tensile strength of SIEL Ph_2;
3 – change of elongation of SIEL Mph; **4** – change of elongation of SIEL Ph_2.

Fig. 4. Change of physical-mechanical properties on thermal aging (250 °C) of compounds SIEL MPh and SIEL Ph_2.

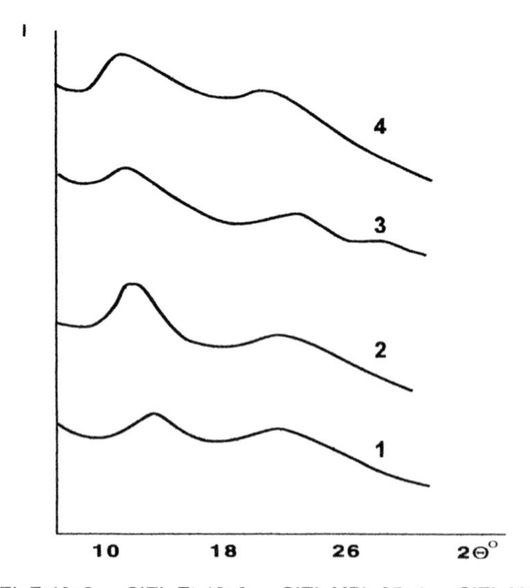

1 — SIEL F-10; **2** — SIEL Et-12; **3** — SIEL MPh-25; **4** — SIEL Mph-50.
Temperature range: –120 °C – 20 °C

Fig. 5. Diffractograms of SIEL compounds with different structures of siloxane chains.

= dimethylsiloxane unit; MF = methyl-γ-trifluoropropylsiloxane unit; MPh = methylphenylsiloxane unit, and Ph_2 = diphenylsiloxane unit. Significant increase in thermal stability on transfer from the foregoing series to fluorine and, particularly, phenanthrene-type substituents (from 220 – 250 °C to 350 – 400 °C) was determined for glassy materials on simultaneous increase of samples hardness and strength (Table 2 and Fig. 6).

Table 2. Thermal stability of STYK resins.

Sample	Temperature [°C]					The final mass loss [%]
	The beginning of destruction	5% mass loss	10% mass loss	25% mass loss	The end of destruction	
I	180	360	430	600	840	46.0
II	260	470	540	680	880	50.2
III	260	470	570	710	920	54.0
IV	240	470	570	590	–	–
V	270	530	610	730	–	–
VI	220	400	485	625	820	50.0
VII	260	435	525	650	820	50.0

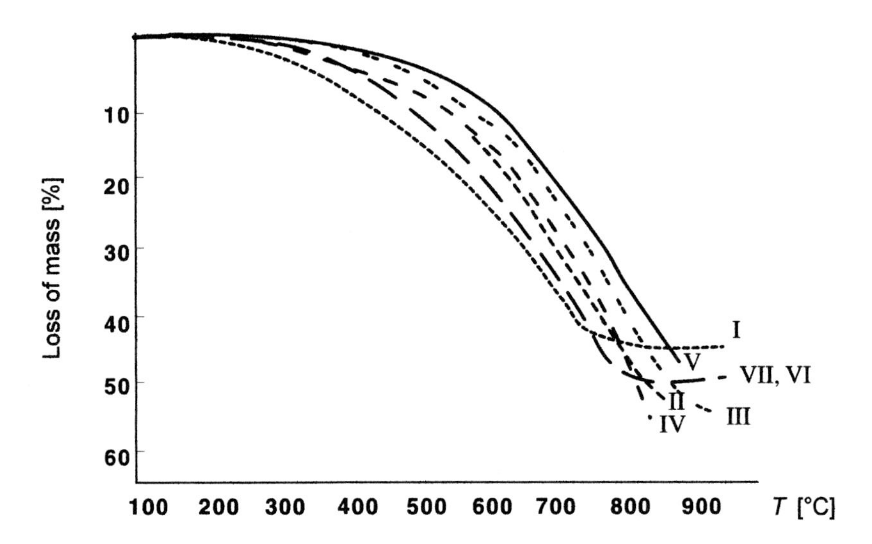

Fig. 6. TGA curves of STYK compounds.

As seen from Table 2, the sample **V** containing 60% phenanthrenesiloxy units has the best thermal stability. The influence of the siloxane chain framework on adhesive properties of compositions STYK is very great. So, the adhesion (τ) does not decrease, thanks to the introduction of phenyl-containing units, up to a temperature of 300 °C (Fig. 7).

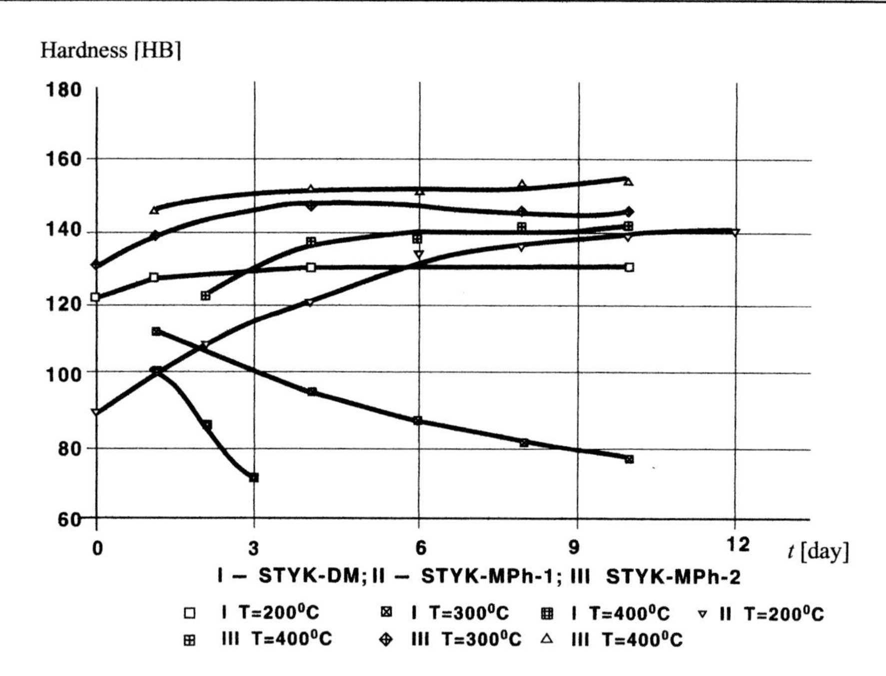

Fig. 7. Study of STYK compounds aged at different temperatures.

Temperature and irradiation effects on functional properties of synthesized compositions were studied in regard to the nature of the substituent in the basic siloxane chain as well as the degree of crosslinking (polymerization). Substantial differences in adhesion properties of SIEL and STYK compounds were found under similar operating conditions, which can be explained by their structural peculiarities. Test results showed a high enough radiation resistance (RR) of SIEL and STYK materials. The RR of compounds tends to increase in a series similarly to the change of thermal stability. When total irradiation dose is below $10^4 - 10^5$ rad, the material performance does not actually change (Table 3). When the dose is above 10^7 rad, the properties of the first members of the series begin to degrade (up to partial material destruction), while last members of the series are quite stable within the analyzed range, and STYK compounds containing fluorine and, particularly, phenanthrene chains can be used up to 10^8 rad.

Table 3. Investigation of irradiation effect on characteristics of SIEL compounds.

	Physical mechanical characteristics					
	Tensile strength [kg/cm^2]			Elongation [%]		
Composition	Before irradiation	After dose 1^x	After dose 2^{xx}	Before irradiation	After dose 1	After dose 2
SIEL MPh	22.6	26.3	24.4	100	105	97
SIEL Ph_2	11.4	16.3	16.2	109	113	92

^{x} Dose 1 – 5.2×10^4 rad; ^{xx} Dose 2 – 1.06×10^7 rad.

Study of Rheological Properties of Oligoethylsiloxane-Based Compositions

A. O. Gureev, T. V. Koroleva, M. B. Lotarev, I. I. Skorokhodov, E. A. Chernyshev*

State Research Institute for Chemistry and Technology of Organoelement Compounds, 38, Shosse Entuziastov, 111123, Moscow, Russia
Tel.: +7 95 2737291 — Fax: +7 95 2737982
E-mail: aleksgur@mail.ru

Keywords: oligoethylsiloxane, compositions, rheology, free volume theory

Summary: Rheological properties of compositions containing oligoethylsiloxane, poly-α-olefin, and dioctylsebacinate were studied. Both energetic and structural factors were shown to affect the fluidity of compositions.

Introduction

Oligoethylsiloxane-based compositions with various lubricity-promoting additives are currently attracting much attention. Lubricity and behavior at low temperatures of such compositions have been thoroughly investigated [1].

Rheological properties of the oligoethylsiloxane-based compositions with additives are of vital importance for the selection of maintenance conditions of various items, but they have not been studied yet. No data on such composition–fluidity relationships (unconfined fluctuating free volume and viscous flow activation energy values) are available.

Therefore, the study of rheological properties, the determination of the activation energy of viscous flow, composition–fluidity relationships, and key factor effects of the lubricity-promoting additive on the fluidity of oligoethylsiloxane-based compositions were of interest.

Experimental Methods

The compositions based on oligoethylsiloxane fluid and organic additive were the main subject of this research. The oligoethylsiloxane fluid consisted of siloxane with a predominantly linear structure, dynamic viscosity η^{20} = 44 mPa s, and average molecular mass M_n = 900 g/mol (hereinafter called the "base"). The organic part was a mixture of poly-α-olefins (catalytic oligomerization product of higher α-olefins of C_{10} fraction with medium degree of oligomerization 7 – 8 and molecular mass M_n = 1070 g/mol) and dioctylsebacinate in a ratio of 4:1, and its dynamic viscosity amounted to η^{20} = 198 mPa s (hereinafter called the "additive"). The compositions were produced by means of mixing base and additive in the following ratios: 85:15, 80:20, 75:25, 65:35 (hereinafter called "compositions").

The measurements were conducted on a rotary viscometer "Viscoster UT550" of HAAKE. "Applications software" was used for treatment of the results. A "Cone Board" instrument was used. Radius = 25 mm; Aperture angle = 3.49×10^{-2} rad; kegel PK5* 2.0 GRAD cone; running clearance = 100 μm; sample volume = 1.2 cm^3. The measurements were conducted at shear rate $30 - 2.2 \times 10^3$ s^{-1} and within a temperature range from –20 to 50 °C. Sample holding time at preset temperature before measurements was 20 min.

Results and Discussions

Dynamic viscosity values (η) were obtained within the defined temperature range for the compositions as well as for initial base and additive (see Table 1).

Table 1. Initial component and composition dynamic viscosity (η) dependence on temperature.

Composition No.	Base : Additive ratio	Viscosity (η) [mPa s], at temperature (T) [K] 253	263	273	283	293	303	313	323
–	Base	132	106	78	58	44	33	26	–
1	85 : 15	203	127	96	67	48	40	28	–
2	80 : 20	249	157	104	71	51	38	29	22
3	75 : 25	237	178	117	79	56	40	31	–
4	65 : 35	295	187	–	77	56	42	31	–
–	Additive	4725	1575	700	–	198	111	77	–

The analyzed fluids at shear rates $30 - 2.2 \times 10^3$ s^{-1} showed the properties of a Newtonian fluid.

The comparison of dynamic viscosity experimental values (η_{exp}) obtained by us with the values calculated by the additive law showed significant negative deviation of experimental dynamic viscosity values from the design parameters, reaching 75% at 20 °C and 440% at –20 °C (see Table 2).

Table 2. Experimental and predictive viscosity dependence on component ratio.

Composition No.	η^{20}_{exp} [mPa s]	η^{20}_{pred}* [mPa s]	Deviation [%]	η^{-20}_{pred} [mPa s]	η^{-20}_{pred}* [mPa s]	Deviation [%]
1	48	67	40	203	821	304
2	51	75	47	249	1051	322
3	56	83	48	237	1280	440
4	56	98	75	295	1740	490

* $\eta^{20}_{pred} = \sum x_i \cdot \eta^{20}_i$, where x_i – weight part of i-th component.

Figure 1 shows that the viscosity–temperature relationship revealed by plotting $\ln\eta$ against $1/T$ is linear for the base (**1**) and compositions 1 (**2**) and 2 (**3**) with additive content up to 20 wt%; for other analyzed fluids (**4**), (**5**), and (**6**) this relationship is not linear. Therefore, we may conclude that the flow of some analyzed fluids does not obey the activation mechanism described by the Arrhenius-Frenkel-Eiring equation within the studied temperature range.

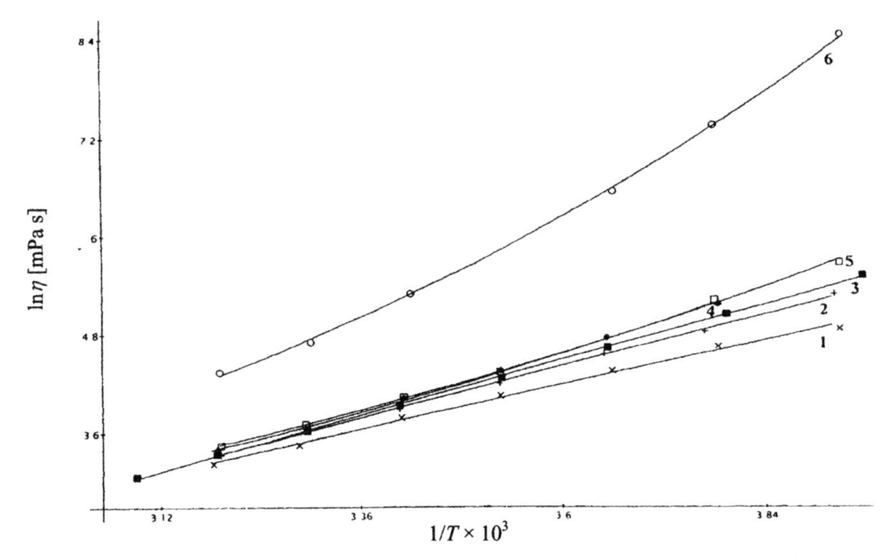

Fig. 1. Viscosity–temperature relationship in Arrhenius coordinates. Curve numbers correspond to liquid numbers of Table 3.

Table 3. Predicted constant values in Falcher-Tamman equation of initial components and compositions.

Composition	ln*A*	*B*	T_0	B_T* [K]	E_η* [kJ/mol]
Base (**1**-Fig. 1–3)	–5.35	3100	25	2277	18.9
Composition 1 (**2**-Fig. 1–3)	–5.91	3130	49	2651	22.0
Composition 2 (**3**-Fig. 1–3)	–3.24	1608	69	2755	22.9
Composition 3 (**4**-Fig. 1–3)	–1.27	830	135	2868	23.8
Composition 4 (**5**-Fig. 1–3)	–0.84	735	142	2970	24.7
Additive (**6**-Fig. 1–3)	–1.57	839	170	4781	39.7

* at temperature 20 °C.

Figure 2 demonstrates that the viscosity–temperature relationship revealed by plotting $\ln\eta$ against $1/(T–T_0)$ is linear for all compositions studied as well as for oligoethylsiloxane and additive. The calculations show that all the analyzed fluid flows within the analyzed temperature range with maximum relative error 2%, obeying the Falcher-Tamman equation (free volume concept) [3]:

$$\eta = A \exp [B/(T–T_0)]$$

Eq. 1.

where T = temperature in Kelvin; A, B and T_0 = constants.

Constants of the Falcher-Tamman equation were calculated for the analyzed fluids (see Table 3).

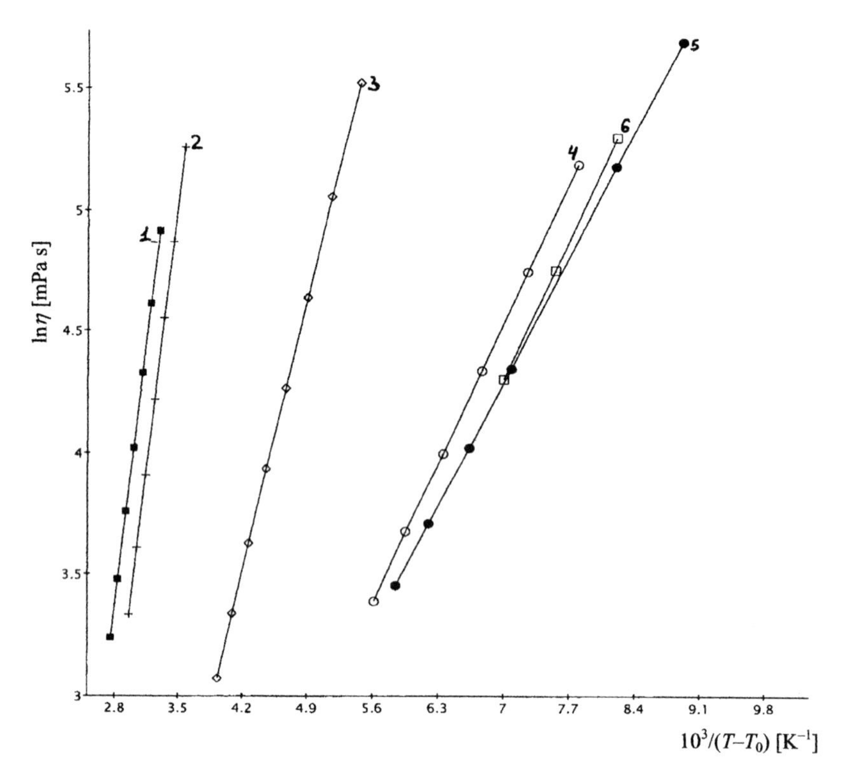

Fig. 2. Dynamic viscosity dependence on temperature in Falcher-Tamman equation. Curve numbers correspond to liquid numbers of Table 3.

The free fluctuation volume portion, calculated from Eq. 2, is the value characterizing liquid bulk structure.

$$f_{v(\eta)} = (T–T_0)/B$$

Eq. 2.

Free fluctuation volume portions of the analyzed compositions depend on temperature, this dependence being linear. The higher the additive content (from 15 to 35 wt%), the stronger is the free fluctuation volume portion dependence on temperature (see Fig. 3).

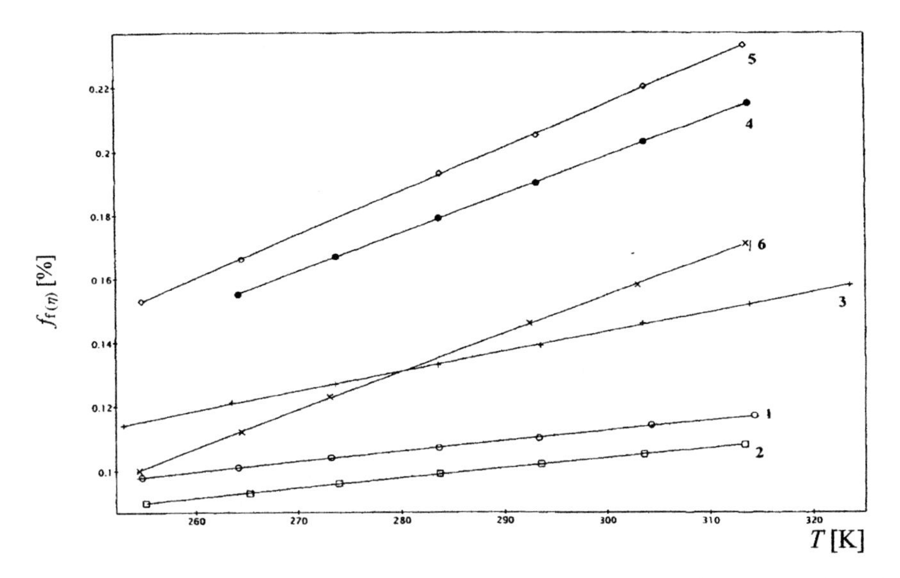

Fig. 3. Dependence of free fluctuation volume portion on temperature for various compositions. Curve numbers correspond to composition numbers of Table 3.

If we assume that on mixing fluids not only the occupied but also the free volume is of the additive type (no effect of composition volume change at mixing) [3], total free volume of the mixture can be found by the equation:

$$f_c = \varphi_1 f_1 + \varphi_2 f_2$$

Eq. 3.

where f_1 and f_2 represent free volume portions of the components before mixing, and φ_1 and φ_2 the weight percent of components in the composition.

Thus, calculated free fluctuation volume of the compositions at a temperature of 293 K increases monotonously from 0.120 to 0.132 (see Fig. 4a) with the additive content increase from 15 to 35 wt%.

According to the calculations, on the basis of the experimental data, the free volume portion of composition 2, involving base and additive in 85:15 ratio at a temperature of 293 K amounts to 0.102. This value is not only less than the free volume portion calculated for the mixture by the additive law, but it is less than the free volume portion of the starting components of the composition (see Fig. 4a). The additive content increase in the composition from 20 to 35 wt% results in a free fluctuation volume increase at a temperature of 293 K from 0.139 to 0.205; the two compositions thus have a larger free volume than that of the initial components of the compositions (see Fig. 3).

The free fluctuation volume dependence on composition does not change with temperature increase or decrease (see Figs. 3 and 4b).

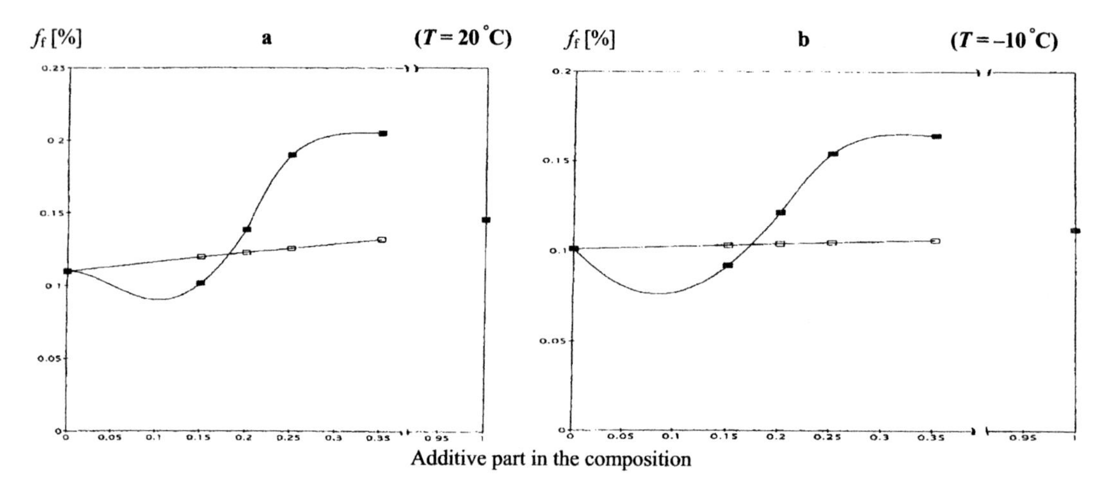

Fig. 4. Dependence of free fluctuation volume of compositions on additive content. Free fluctuation volume values obtained by the additive law are designated by white squares and those calculated by the experimental data by black squares.

Another key feature of the fluid is the design parameter B_T (see Table 3), obtained by differentiation of the Falcher-Tamman equation [3], which is proportional to the viscous flow activation energy:

$$E_\upsilon = B_T R$$

Eq. 4.

where R = the universal gas constant.

Additive content increase in the compositions results in a regular increase in the value of B_T and the values of the viscous flow activation energy at 20 °C (see Table 3).

The foregoing data concerning the applicability of the Falcher-Tamman equation to the description of the viscosity–temperature relationship of the compositions and the nonapplicability of the Arrhenius-Frenkel-Eiring equation testify to the fact that the viscous flow of the compositions influences not only energetic but also structural factors that are considered by free volume theory.

Free fluctuation volume defines the spatial packing density of fluid molecules. With the increase of additive content, the free fluctuation volume at first decreases, which may testify to free volume compression [4]. Further increase of additive content in the composition results in significant loosening of molecule spatial packing; the higher the temperature, the stronger is the dependence of composition molecule packing loosening on additive content. According to A. I. Bachinskii's equation [3]:

$$\eta^{-1} \approx \upsilon_f$$

Eq. 5.

where υ_f = free fluctuation volume, significant increase in the dependence of experimental free fluctuation volume values on composition in comparison with the dependence of additive values of the free fluctuation volume on composition results in the foregoing insignificant change of composition viscosity in comparison with that calculated by the additive law (see Figs. 4a and 4b).

Intermolecular packing loosening in compositions is evidently determined by intermolecular component interaction in compositions. This is confirmed by the above-mentioned fact: with the increase of additive content in the compositions, viscous flow activation energy values increase, testifying to the increase in intermolecular interaction values. Viscous flow activation energy increases from 18.9 to 22.0 kJ/mol (see Table. 3) with the introduction of 15% of the additive (whose viscous flow activation energy value amounts to 39.7 kJ/mol) to oligoethylsiloxane. Further increase of additive content to 35% raises viscous flow activation energy value insignificantly (up to 24.7 kJ/mol). Therefore, the dependence of viscosity on temperature only slightly increases with additive content increase (see Fig. 1).

Thus, the conducted research showed that absolute reaction rate theory is not applicable to the explanation of the composition's viscosity–temperature relationship. It was found that the free volume theory allows us to describe the viscosity–temperature relationship with satisfactory accuracy within the studied temperature range from minus –20 to 50 °C. Parts of the free fluctuation volume and viscous flow activation energy values determining fluids properties were calculated.

Acknowledgments: The authors wish to express gratitude to Doctor A. A. Mosine for the provided software for the application of Falcher-Tamman equation parameters and to M. S. Starshov for his assistance in performaning measurements.

References

[1] V. V. Zverev, T. V. Koroleva, A. O. Gureev (GNIIChTEOS), RF 2 194 741, **2002**.

[2] I. A. Lavygin, I. I. Skorokhodov, L. V. Sobolevskay, *Plasticheskiye Massy*. **1984**, *12*, 14.

[3] G. V. Vinogradov, A. Ya. Malkin, *Polymer Rheology*, Moscow, Khimiya, **1977**, p. 438.

[4] A. J. Hill, M. D. Zipper, G. P. Simon, MACROAKRON`94: 35th IUPAC Int. Union Pure and Appl. Chem. Int. Symp. Macromol.: Abstr. — Akron (Ohio), **1994**, p. 868.

Silacyclobutene-PDMS Copolymers — Siloxanes with Unusual Thermal Behavior

Michael W. Backer,* Jonathan P. Hannington, Peter R. Davies

New Ventures R & D, Polymer and Organics Technology Platform, Dow Corning Limited, Cardiff Road, Barry, Vale of Glamorgan, CF63 2YL, United Kingdom
Tel.: +44 1446 723712 — Fax: +44 1446 730495
E-mail: michael.w.backer@dowcorning.com

Norbert Auner

Institut für Anorganische Chemie der Johann Wolfgang Goethe-Universität
Marie-Curie-Str. 11, D-60439 Frankfurt, Germany

Keywords: polydimethylsiloxane, silacyclobutene, differential scanning calorimetry

Summary: A series of copolymers of 1,1-dichloro-2,3-diphenyl-4-neopentyl-1-silacyclobut-2-ene and *oligo*-dimethylsiloxane-α,ω-diols having different chain lengths has been prepared and investigated by Modulated Differential Scanning Calorimetry (MDSC) as well as Thermogravimetric Analysis (TGA). With respect to PDMS homopolymers, the insertion of low mole percentages of the bulky silacyclobutene unit demonstrates a significant influence on the sub-ambient temperature morphology, e.g., shift of the glass transition temperature or complete inhibition of cold crystallization and melting transition points. On the other hand, the copolymers exhibit thermal stability up to temperatures of 400 – 500 degrees Celsius. Comparisons to diphenyl-siloxy-modified PDMS are discussed.

The insertion of diphenylsiloxy units into polydimethylsiloxanes drastically affects the sub-ambient temperature behavior of the siloxanes by shifting the glass transition point to higher temperatures or inhibiting the crystallization [1]. Cyclosilanes having bulky aromatic substituents attached to the rigid ring core are suggested to show a similar influence on the thermal properties of PDMS.

1,1-Dichloro-2,3-diphenyl-4-neopentyl-1-silacyclobut-2-ene is generated by reaction of trichlorovinylsilane / *tert*-butyl lithium with diphenylacetylene in nonpolar solvents (see Eq. 1). The dichloro-functionality can be used for all types of substitution reactions, hydrolysis, and copolymerisation reactions [2 – 4]. The stilbene subunit also exhibits strong fluorescence behavior [5 – 7].

Equivalent amounts of 1,1-dichloro-2,3-diphenyl-4-neopentyl-1-silacyclobutene (**1**) and oligo-dimethylsiloxane-α,ω-diols $HO(Me_2SiO)_nH$ (**2** – **5**), with n = 2 (**2**), 4.3 (**3**), 12.4 (**4**), and 38.3 (**5**)

— average chain lengths were determined by ^{29}Si NMR analysis — were reacted in warm toluene at 50 °C in the presence of 2.1 equivalents of triethylamine and 0.05 equivalents of dimethylaminopyridine DMAP, as shown in Eq. 2. Polymers formed were precipitated by the addition of methanol. After removal of the solvent, the copolymers **7** – **16** were obtained as colorless, viscous oils. In the case of tetramethyldisiloxanediol (**2**), only a mixture of very short chains and small cyclosiloxanes was formed. Since the addition of methanol to the reaction mixture did not initiate a polymer precipitation, the solvents were evaporated, and a yellow, very high viscous oil (**6**) was obtained.

$Cl_3Si{-}CH{=}CH_2$ + *t*-BuLi $\xrightarrow{\text{Pentane}}$ [intermediate: Cl_3Si–CH(Li)–CH_2–$C(CH_3)_3$] $\xrightarrow[\text{- LiCl}]{\text{PhC≡CPh}}$ **1**

Eq. 1. Preparation of 1,1-dichloro-2,3-diphenyl-4-neopentyl-1-silacyclobut-2-ene (**1**).

1 + $H{-}O{-}[Si(Me)_2{-}O]_n{-}H$ (**2** – **5**) $\xrightarrow[\text{- }R_3NHCl]{NR_3\text{ / DMAP, toluene}}$ $H{-}O{-}\big[[Si(Me)_2{-}O]_n{-}Si(\text{silacyclobutene})\text{–}O{-}[Si(Me)_2{-}O]_n\big]_p{-}H$ (**6** – **16**)

Eq. 2. Co-condensation reaction of silacyclobutene **1** with α,ω-dimethyl-oligo-siloxanediols **2** – **5**.

Experimental and analytical data for the preparation of copolymers **6** – **16** are listed in Table 1. It is obvious that the yield increases with the chain length of the starting oligomer **3** – **5**. The degree of polymerization (DP) was calculated by determination of the integral ratios of the signals of siloxy-units versus silanol endgroups in ^{29}Si NMR analysis.

The sub-ambient thermal properties of the starting oligo-siloxanediols and the silacylobutene copolymers (10 –13 mg sample range) were analyzed in hermetic aluminum pans using a TA 2920 DSC instrument. Under a helium purge rate of 25 ml min^{-1}, a cooling rate of 5 °C min^{-1} down to –155 °C has been applied followed by a heating rate of 10 °C min^{-1}.

The graph of the differential scanning calorimetry analysis of oligo-dimethylsiloxanediol **5** is displayed in Fig. 1. In the heating cycle, the glass transition is detected at a temperature of –124.4 °C; however, no cold crystallization or melting transition is observed. This is surprising with respect to trimethylsiloxy endcapped siloxanes of similar chain lengths (DP 25) that normally exhibit a cold crystallization in the region of –93 °C and a melting transition at –61 °C. Thus, the type of endgroup seems to influence the crystallization behavior as well.

Table 1. Copolymerization reaction of silacyclobutene **1** with siloxanediols **2** – **5**.

No	Amount of 1 [mmol]	Ratio of siloxanediols [mmol] 2	3	4	5	Yield [%]	Average chain length (DP)	Ratio Si_{Me} : Si_{cycl}
6	8.38	8.38	–	–	–	80.4	–	2.4
7	10.00	–	10.00	–	–	29.8	92.8	7.1
8	10.00	–	–	10.00	–	64.3	176.7	15.0
9	5.00	–	–	–	5.00	78.2	293.5	63.5
10	10.00	–	5.00	5.00		50.5	165.6	12.0
11	10.00	–	7.50	–	2.50	72.7	646.9	22.0
12	5.00	–	2.50	–	2.50	72.5	242.1	41.4
13	5.00	–	1.25	–	3.75	79.1	240.6	62.8
14	10.00	–	–	7.50	2.50	70.7	301.0	43.3
15	5.00	–	–	2.50	2.50	69.0	350.0	49.9
16	5.00	–	–	1.25	3.75	84.5	293.5	63.5

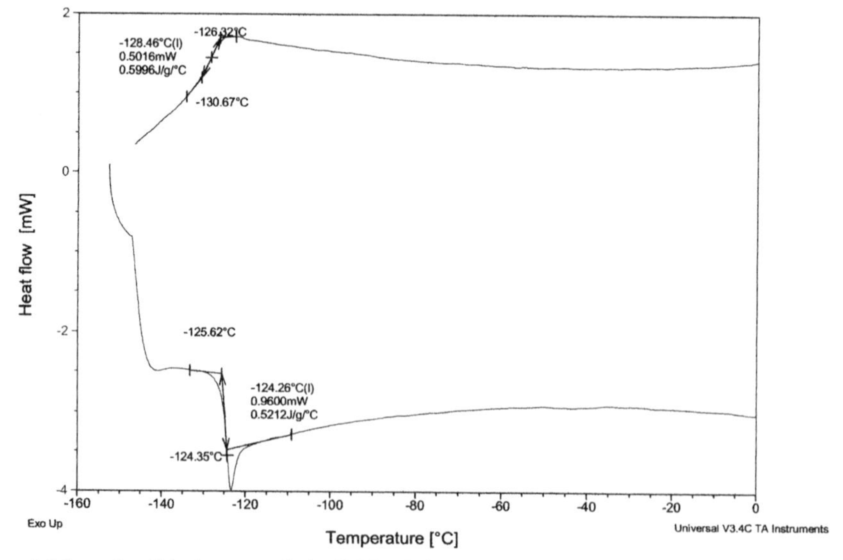

Fig. 1. Differential Scanning Calorimetry analysis (DSC) of oligo-dimethylsiloxanediol (**5**).

A typical graph of differential scanning calorimetry analysis of a dimethylsiloxane polymer with higher viscosity (350 cSt) is shown in Fig. 2. The glass transition is observed at –127.9 °C, the cold crystallization at –96.5 °C, and melting transitions at –51.7 and –38.3 °C, respectively.

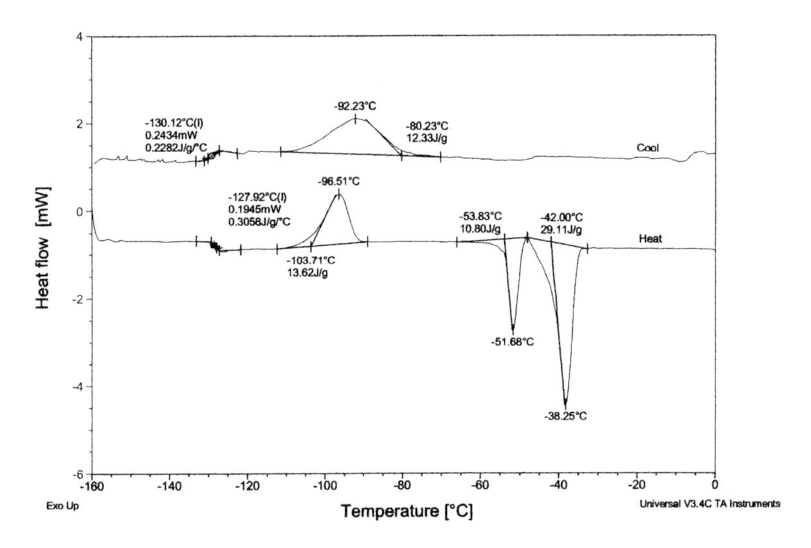

Fig. 2. Differential Scanning Calorimetry analysis (DSC) of a trimethylsiloxy endcapped PDMS fluid, 350 cSt.

The insertion of the bulky silacyclobutene subunit into polymers in a high ratio not only shifts the glass transition to significantly higher temperatures, but also inhibits the cold crystallization as demonstrated in Fig. 3, which shows the DSC analysis of copolymer **7**. The copolymer has a $Si_{cyclobutene}$ content of about 19%. The sharp signal in the cooling cycle can be assigned to traces of remaining solvent.

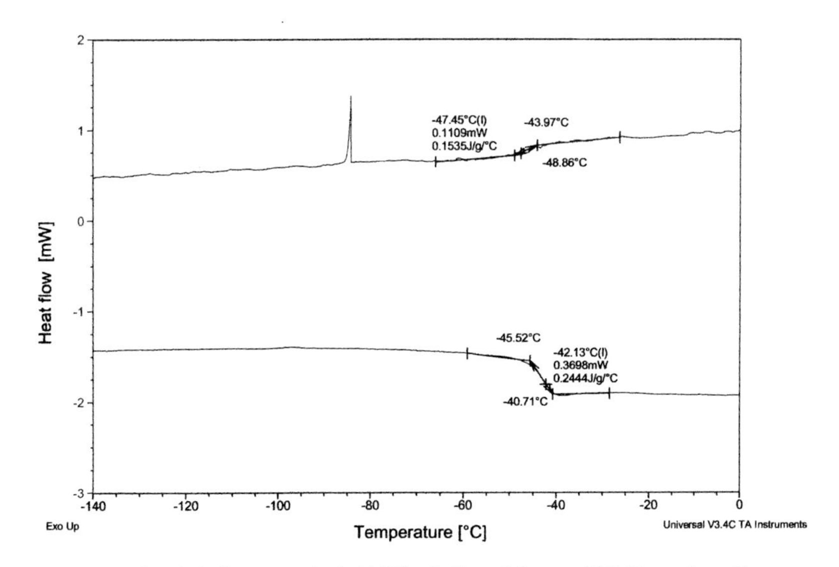

Fig. 3. Differential Scanning Calorimetry analysis (DSC) of silacyclobutene-PDMS copolmer **7**.

This observation is supported by the data in Table 2, which lists the DSC results of the siloxanediols **3** – **5** as well as of the silacyclobutene containing copolymers **6** – **16**. It has been found that the shift of the glass transition temperature is related to an increase of the silacyclobutene ratio. In the same way, the heat capacity change at the glass transition increases with increasing $Si_{cyclobutene}$ ratio.

Table 2. Differential Scanning Calorimetry data of siloxane-diols **2** – **5** and copolymers **6** – **16**.

No	T_g (cooling) [°C]	T_g (heating) [°C]	ΔC_p (heating) [J g^{-1} K^{-1}]
3	–108.9	–106.4	0.44
4	–	–117.4	0.45
5	–128.5	–124.4	0.52
6	–10.1	–4.7	0.28
7	–42.1	–47.5	0.24
8	–98.3	–95.3	0.33
9	–121.5	–119.5	0.46
10	–81.7	–85.1	0.34
11	–117.0	–114.9	0.34
12	–119.8	–117.6	0.42
13	–120.9	–119.3	0.51
14	–113.1	–112.7	0.43
15	–118.8	–117.4	0.44
16	–120.5	–118.2	0.45

In comparison to copolymer **9**, the insertion of about 3% of diphenylsiloxy units into a PDMS copolymer having a molecular weight of 9300 leads to a glass transition at –120.8 °C. However, the copolymer was still showing a cold crystallization at 75 °C and a melting transition at –45.1 °C [8]. Only copolymers with higher ratios of diphenylsiloxy units showed inhibition of the crystallization.

The thermal stability of the copolymers **6** – **16** was studied using a TA 2950 TGA instrument in a nitrogen atmosphere. Figure 4 shows the thermogravimetric analysis of copolymer **9**. The copolymer exhibits thermal stability up to a temperature of over 400 °C before the major weight loss is observed. A smaller weight loss of 2% has been detected at about 345 °C. The amount of organic residue at 800 °C is about 4.4%.

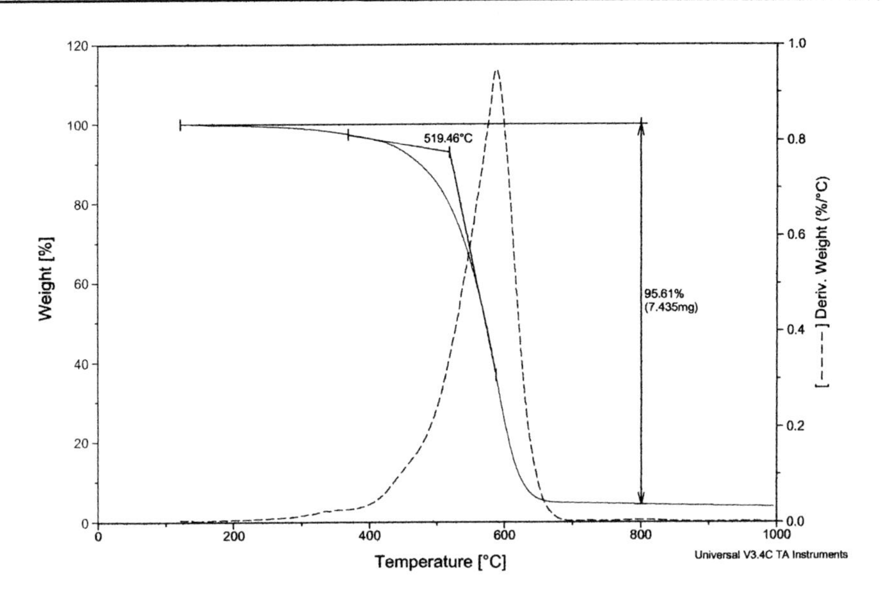

Fig. 4. Thermogravimetric analysis (TGA) of copolymer **9**.

The thermogravimetric analysis data of the copolymers **6 – 16** are listed in Table 3. The temperature at 2% weight loss, and the weight loss at 400 °C and 800 °C are given.

It has been found that an increasing level of silacyclobutene units results in a reduction of the temperature of the first onset of weight loss temperature. It also leads to an increase in the residual organic mass at 800 °C.

Table 3. Thermogravimetric Analysis data of copolymers **6 – 16**.

No	Temperature at 2% weight loss [°C]	Weight loss at 400 °C [%]	Weight loss at 800 °C [%]
6	151.7	48.7	92.0
7	266.4	6.5	67.6
8	300.9	4.3	83.5
9	345.3	3.6	95.6
10	305.1	5.2	67.5
11	310.2	4.6	81.7
12	323.5	4.2	81.9
13	341.7	3.9	86.5
14	319.8	4.7	79.2
15	346.2	4.3	91.5
16	350.4	3.6	93.5

Conclusion

Co-condensation reactions of a bulky dichloro-silacyclobutene with several oligo-dimethylsiloxanediols have generated copolymers with $Si_{cyclobutene}$ content of 1.5% to about 30%, having variable and mixed dimethylsiloxy blocks. The thermal analysis of oligosiloxanediols and silacyclobutene copolymers thereof has shown that the observation of cold crystallization and melting is strongly dependent on the type of endgroup ($-OSiMe_3$ vs $-OH$). On the other hand, the insertion of the bulky silacyclobutene units totally inhibits the crystallization even in longer polymers. The glass transition is shifted to higher temperatures and the associated heat capacity change is increased with increasing silacyclobutene ratio. Therefore, the silacyclobutene could be an alternative to diphenylsiloxy units for preventing crystallization in low-temperature applications. The copolymers exhibit a high thermal stability at temperatures above 400 °C; however, the increase in the silacyclobutene ratio lowers the temperature of onset of thermal loss, but leads to a higher amount of organic residue.

Acknowledgments: MWB thanks Sian B. Rees (Dow Corning Limited), John P. Cannady (Dow Corning Corporation), and Ian Moss (Dow Corning Limited) for the scientific discussions and technical support of the project throughout his postdoctoral placement.

References

[1] R. Bischoff (Dow Corning Limited), unpublished results.

[2] N. Auner, C. Seidenschwarz, E. Herdtweck, *Angew. Chem.* **1991**, *103*, 1172; *Angew. Chem. Int. Ed. Engl.* **1991**, *30*, 1151.

[3] M. Backer, M. Grasmann, W. Ziche, N. Auner, C. Wagner, E. Herdtweck, W. Hiller, M. Heckel, Silacyclobutenes — synthesis and reactivity, in *Organosilicon Chemistry II — From Molecules to Materials* (Eds.: N. Auner, J. Weis), VCH-Verlagsgesellschaft, Weinheim, **1996**, p. 41.

[4] M. Backer, *PhD-Thesis*, Humboldt-Universität zu Berlin, **1999**.

[5] U. Pernisz, N. Auner, M. Backer, *Polym. Prepr.* **1998**, *39(1)*, 450.

[6] U. Pernisz, N. Auner, Photoluminescence of organically modified cyclosiloxanes, in *Organosilicon Chemistry IV — From Molecules to Materials* (Eds.: N. Auner. J. Weis), Wiley-VCH, Weinheim/New York, **2000**, p. 505.

[7] U. Pernisz, N. Auner, M. Backer, Photoluminescence of phenyl- and methyl-substituted cyclosiloxanes, in *Silicones and Silicone-Modified Materials* (Eds.: S. J. Clarson, J. J. Fitzgerald, M. J. Owen, S. D. Smith) *ACS Symposium Series 729*, Am. Chem. Soc., Washington DC, **2000**, *Chap. 7*, p. 115.

[8] N. Auner, R. Bischoff, J. Hannington, S. Rees (Dow Corning Limited), GB2 326 417, **1998**.

Electrochemically Initiated “Silanone Route” for Functionalization of Siloxanes

Robert Keyrouz, Viatcheslav Jouikov*

UMR 6510, Laboratory of Molecular and Macromolecular Electrochemistry,
University of Rennes I, Av. de General Leclerc
Campus de Beaulieu, 35042 Rennes, France
Tel.: +33 22 323 6293 — Fax: +33 22 323 6292
E-mail: vjouikov@univ-rennes1.fr

Keywords: silanones, electrosynthesis

Summary: The reaction of organyl dichloro- or dialkoxysilanes $R^1R^2SiX_2$ (R^1, R^2 = Alk, Ph, Vi, OAlk; X = Cl, OAlk) with electrochemically generated superoxide anion provides very potent reactive intermediates which insert into linear or cyclic permethyl cyclosiloxanes to give siloxane products with one additional R^1R^2SiO fragment. Relative and absolute rate constants of silanone generation from different precursors were determined showing that the process is faster with X = Cl and with less bulky R^1 and R^2 groups. For the first time the use of ionic liquids in the “silanone route” of siloxane functionalization was demonstrated.

The generation of silicon analogs of ketones, diorganylsilanones $R_2Si{=}O$, and the control of their reactions remain challenging topics of silicon chemistry. The transient formation of silanones was postulated in a number of reactions such as thermal transformations of oxygen-containing silicon compounds [1], reactions of silenes with carbonyl compounds [2], the oxidation of silylenes by DMSO [3] and epoxides [4], and the oxidation of dichlorosilanes by O-donor reagents under relatively mild conditions [1, 5] etc. Although the silanones as such have never been observed in solutions at room temperature and there are still no convenient methods of their generation, they are believed to be very reactive intermediates in above cited and other processes of silicon organic compounds.

We recently proposed a method of electrochemical activation of oxygen to render it more nucleophilic and reactive toward dichloro- and alkoxysilanes [6, 7]. Since corresponding silanones were thought to be intermediates in this reaction, this process aimed to explore a very attractive idea of using molecular or even atmospheric oxygen for generating silanones from simple and accessible starting material, like chlorosilanes, thus providing a “silanone route” for functionalization of organosilicon compounds.

Several control experiments showed that dichloro- and dialkoxysilanes would react with nucleophilic forms of O_2. First, it was found that under the usual conditions these compounds do not react with O_2 itself. After bubbling oxygen for 4 h through 10^{-2} mol L^{-1} solutions of $Ph_2Si(OMe)_2$, Ph_2SiCl_2 or of hexamethyldisiloxane in DMF containing 0.1 M Bu_4NPF_6, no new products were obtained. Applying a negative potential of –1.8 V vs SCE to an electrode (diorganyl dichlorosilanes and dialkoxysilanes are electrochemically inactive at these potentials [8]), plunged into the same solutions without passing oxygen, did not result in the formation of any products either.

Only the reduction of oxygen in the presence of dichloro- or dimethoxysilanes results in their transformation into the mixture of cyclic and linear siloxanes. These siloxane products arise from autocyclization and oligomerization of electrogenerated transient silanones and their insertion into Si–O bonds of primarily formed siloxanes.

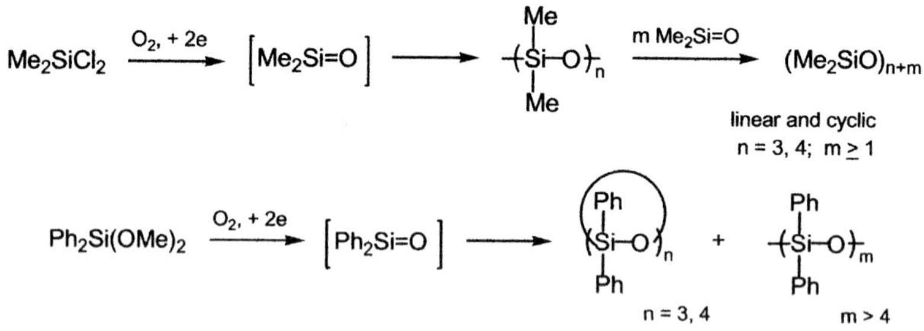

$$Me_2SiCl_2 \xrightarrow{O_2,\ +2e} [Me_2Si{=}O] \longrightarrow (SiMe_2{-}O)_n \xrightarrow{m\ Me_2Si{=}O} (Me_2SiO)_{n+m}$$

linear and cyclic
n = 3, 4; m ≥ 1

$$Ph_2Si(OMe)_2 \xrightarrow{O_2,\ +2e} [Ph_2Si{=}O] \longrightarrow (SiPh_2{-}O)_n\ (\text{cyclic}) + (SiPh_2{-}O)_m$$

n = 3, 4 m > 4

Scheme 1. Autocyclization and oligomerization of silanones electrogenerated from dichloro- and dialkoxy silanes.

Besides olygomerization, one of the typical reactions of silanones is the insertion into Si–O or other polar bonds [1]. So the cathodic generation of diphenylsilanone was performed using hexamethyldisiloxane (HMDS) as a silanone trap. The reduction of O_2 dissolved to saturation in a 0.1 M solution of Bu_4NPF_6 in DMF or THF in the presence of $Ph_2Si(OMe)_2$ and HMDS resulted in formation of an insertion product — 1,1,1,5,5,5-hexamethyl-3,3-diphenyltrisiloxane (Scheme 2). With permethylcyclosiloxanes D_n (n = 3, 4), used as trapping molecules instead of HMDS, corresponding cyclic products with extended cycles, $D_nD^{Ph_2}$, were obtained. The results of siloxane backbone extension with different silanone precursors and traps are exemplified in Table 1.

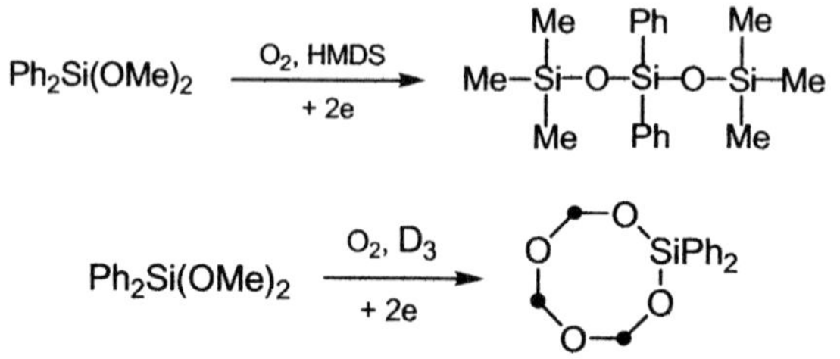

$$Ph_2Si(OMe)_2 \xrightarrow[+2e]{O_2,\ HMDS} Me_3Si{-}O{-}SiPh_2{-}O{-}SiMe_3$$

$$Ph_2Si(OMe)_2 \xrightarrow[+2e]{O_2,\ D_3} \text{cyclo-}(\bullet O)_3 SiPh_2 O$$

Scheme 2. The reaction of linear and cyclic siloxanes with diphenylsilanone generated from diphenyl dialkoxysilane. • stands for Me_2Si group.

Table 1. Insertion of R^1R^2SiO into siloxanes by electrochemical oxygenation of $R^1R^2SiX_2$ (DMF/0.1 M Bu_4NPF_6, T = 45°C).

R^1	R^2	X	Trap	Q, F [a]	Products [b] (Yield, [%]) [c]
Me	Vi	Cl	D_3	2.2	D_3D^{Vi} (29.6), D_4D^{Vi} (2.3)
Me	Oct	Cl	D_3	2.3	D_3D^{Oct} (37), D_4D^{Oct} (6)
Me	Me	Cl	D_3	2.0	D_4 (46), D_5 (4.3)
Me	$(CH_2)_3CF_3$	MeO	D_3	2.2	$D_3D^{(CH_2)_3CF_3}$ (34), $D_4D^{(CH_2)_3CF_3}$ (5.2)
Me	$(CH_2)_3NH_2$	EtO	D_3	2.2	$D_3D^{(CH_2)_3NH_2}$ (32)
Me	$(CH_2)_3Cl$	Cl	D_3	1.9	$D_3D^{(CH_2)_3Cl}$ (39), $D_4D^{(CH_2)_3Cl}$ (4.4)
Me	$(CH_2)_3CN$	Cl	D_3	2.0	$D_3D^{(CH_2)_3CN}$ (34)
Ph	Ph	MeO	D_3	2.0	D_3D^{Ph2} (33), D_4D^{Ph2} (7.6)
Ph	MeO	MeO	D_3	1.9	$D_3D^{Ph,MeO}$ (32), $D_4D^{Ph,MeO}$ (5.2)
Ph	Ph	EtO	D_3	1.9	D_3D^{Ph2} (25.8), D_4D^{Ph2} (7.8)
Ph	Vi	MeO	D_3	2.1	$D_3D^{Ph,Vi}$ (37)
Ph	Ph	EtO	HMDS	2.1	$MD^{Ph2}M$ (48)
Ph	Ph	MeO	HMDS	1.7	$MD^{Ph2}M$ (46)
Ph	Ph	Cl	Et_3SiH	2.0	$Et_3SiOSiPh_2H$

[a] Related to 1 mol of $R^1R^2SiX_2$; [b] In General Electric nomenclature: M = Me_3SiO, D = Me_2SiO, D^R = MeRSiO, $D^{R^1R^2}$ = R^1R^2SiO; [c] Isolated yields after column chromatography on SiO_2.

Methoxytrimethylsilane in this process reacted with diphenylsilanone to give $Me_3SiOSi(Ph)_2OMe$. On the other hand, when trimethoxysilanes were taken, $RSi(OMe)_3$, only two OMe groups reacted to form intermediate silanone whereas a third one was introduced and conserved in the molecule of the ring expansion product (Scheme 3).

$$PhSi(OMe)_3 \xrightarrow[+\ 2e]{O_2,\ D_3}$$

OMe Si Ph O O O O

Scheme 3. Phenyltrimethoxysilane as silanone precursor; the insertion into D_3.

When $MeHSi(OMe)_2$ was used as the silanone precursor, not only the expected D_3D^H and D_4D^H products but also higher derivatives D_5D^H and even D_6D^H were formed (Scheme 4). There was also found a trace amount of a product whose composition corresponds either to the double insertion into the cycle of the silanone trap, $D_3D^HD^H$, or to the silanone insertion into the Si–H bond, of a primarily formed mono-functionalized product, $D_3D^{OSiMeH_2}$. Triethylsilane in this process formed triethylsiloxydiphenylsilane as the only product.

The "silanone route" of functionalization of siloxanes turned out to have a high functional group tolerance – not only alkoxy and vinyl group can thus be inserted into siloxane molecules, but also NH_2, Cl, CF_3 and CN (Table 1). In the following, these groups can be used for additional specific functionalization, grafting, or special cure of these siloxanes.

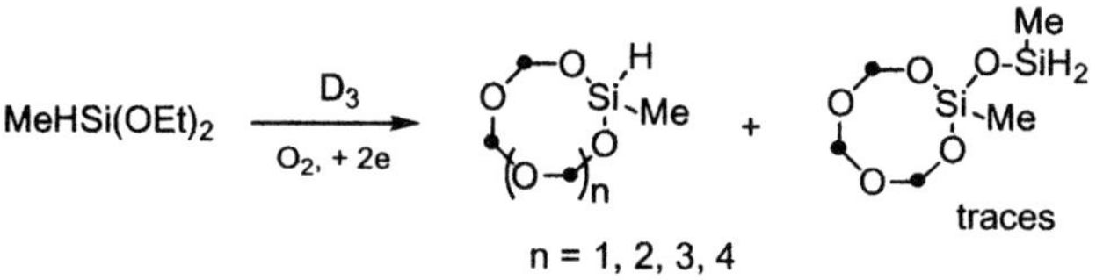

Scheme 4. The reaction of a Si–H containing precursor with superoxide ion in the presence of D_3.

From the kinetic point of view, the fact that silanones were never observed in solution could be explained by that the rates of their formation being usually slower than the rates of their ensuing reactions. This observation seems to be true at least for ordinary substituents at Si like Me, Vi, Et, Ph. A very small amount of autocyclization products in the electrolyses with silanone traps suggests kinetic boundary conditions $k_1 << k_1'$ and $k_2 << k_2'$ (Eq. 1), and therefore that k_1 and k_2 can be considered as rate-determining steps in this process. Now, a competitive reaction scheme can be applied for the determination of the relative efficiency of dialkoxysilanes vs dichlorosilanes for electrogeneration of diorganylsilanones. The three kinetic conditions (similar mechanism of both parallel processes; $C_{Ph_2Si(OMe)_2} = C_{PhViSiCl_2} >> C_{O_2^{-\cdot}}$ and low conversion, $\alpha \approx 5\%$ [9]) allow us to determine the ratio $k_1 : k_2$ simply as the ratio of the yields of corresponding products, $Y_1 : Y_2$ (Eq. 1). From the experiments with couples of parent compounds, $Ph_2SiCl_2 : ViPhSiCl_2$ and $Ph_2Si(OMe)_2 : ViPhSi(OMe)_2$, the ratio $k_{Ph_2} : k_{PhVi} = 0.31$ was obtained, showing that steric and electronic effects from the substituent at Si are very important factors in the formation of silanones.

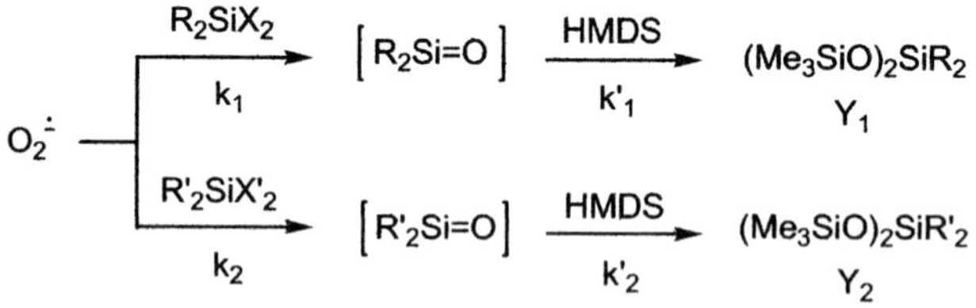

Eq. 1. Competitive reactions method with one electrogenerated reagent and two substrates, R_2SiX_2 and $R'_2SiX'_2$.

The nature of the leaving groups is also critical: the ratio $k^{OMe} = 0.56 \times k^{Cl}$ suggests that dichlorosilanes are about two times more reactive than dialkoxysilanes and this is supposedly the reason why, other conditions being equal, the process with alkoxysilanes is visibly more selective [7].

Cyclic voltammetry allowed to shed some light to the nature of active form of oxygen and to determine the rates of its reaction with different silanone precursors. Normally, any distinct steps of the reduction of oxygen are not seen at a macroelectrode so it is not clear which form, $O_2^{\bullet -}$ or O_2^{2-}, is actually being generated. However at a Pt or a glassy carbon (GC) microelectrode in anhydrous DMF or THF, one can clearly observe two reduction peaks of oxygen ($E_p^1 = -0.75$, $E_p^2 \cong -2.0$ V vs SCE), whose shape depends on the polarity of the solvent (Fig. 1).

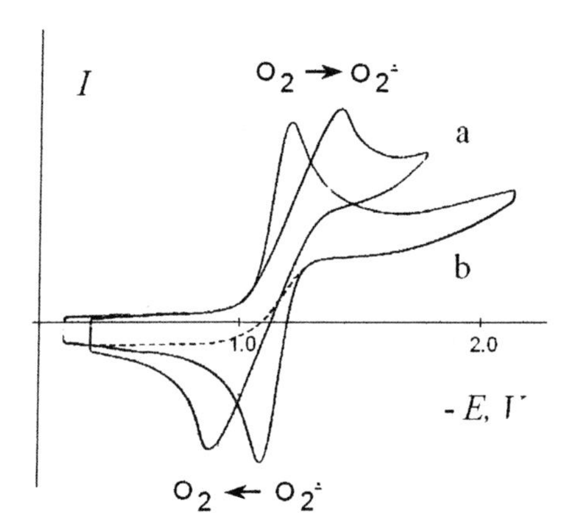

Fig. 1. Voltammograms of a saturated solution of O_2 in (a) THF, (b) DMF. Supporting electrolyte 0.1 M n-Bu_4NPF_6. 0.8 mm GC disk electrode, scan rate v = 200 mV s^{-1}. Dashed line corresponds to the back scan after addition of 25 mmol of $Ph_2Si(OMe)_2$.

The first reduction step provides superoxide anion, which was shown to react with added silanone precursors as is seen from the decrease up to a total disappearance of the oxidation peak of $O_2^{\bullet -}$. Using the ratio of these signals, $i_p(O_2^{\bullet -}) : i_p(O_2)$, and the kinetic treatment proposed in [10] for reversible electron transfer followed by an irreversible reaction of ion-radicals and modifying it to pseudo-first order reactions, we determined absolute rate constants of nucleophilic addition of electrogenerated superoxide anion on several silanone precursors (Scheme 5).

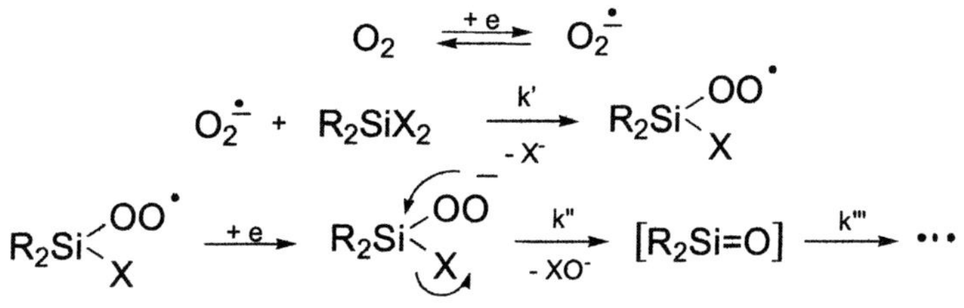

$$O_2 \underset{}{\overset{+e}{\rightleftharpoons}} O_2^{\bullet -}$$

$$O_2^{\bullet -} + R_2SiX_2 \xrightarrow[-X^-]{k'} R_2Si(OO^{\bullet})X$$

$$R_2Si(OO^{\bullet})X \xrightarrow{+e} R_2Si(OO^{-})X \xrightarrow[-XO^-]{k''} [R_2Si{=}O] \xrightarrow{k'''} \cdots$$

Scheme 5. Electrochemically detectable reactions of superoxide ion in the first steps of silanone formation. The bimolecular reaction with k' is rate determining.

In agreement with the competitive kinetic measurements, dichlorosilanes react faster than dialkoxysilanes. Bulkier substituents slow the process compared to less encumbering substituents at Si (Table 2). For triethylsilane, used as silanone trap, it was found that Si–H bond is also reactive toward $O_2^{\bullet-}$ but the rate constant of this process is about two orders smaller than that of the reaction of superoxide with dichloro- or dialkoxysilanes. Therefore hydrosilanes can act as efficient silanone traps in this process.

Table 2. Rate constants of the reaction of electrogenerated $O_2^{\bullet-}$ with silylated substrates (v = 200 mV s^{-1}).

Silane	τ, [s][a]	k [L mol^{-1} s^{-1}]
$Ph_2Si(OMe)_2$	2.50	5.11
$Ph_2Si(O^nPr)_2$	2.50	1.04
$Ph_2Si(O^tBu)_2$	2.50	0.62
Ph_2SiF_2	2.50	24.8
$SiCl_4$		hydrolysis
Et_3SiH	4.15	4.1×10^{-2}

[a] τ is the time elapsed between E_p and E_{max}, $\tau = (E_p - E_{max})/v$; v is the scan rate [10].

We examined the possibility to perform the silanone electrolyses in ionic liquids (IL) — new quasi-permanent media which attract more and more attention as solvents for "green chemistry" with very promising technological properties [11]. ILs are often immiscible with water and many organic solvents, thus reducing product separation to a simple extraction. The ionic conductivity and good electrochemical stability of ILs present a very appealing electrochemical medium, combining solvent and supporting electrolyte in one system. To circumvent the main drawback of these systems in electrochemistry — low currents due to small diffusion rates — and to intensify the electrolysis, we have chosen pyridinium-based ILs with a fluorinated alkyl chain, $[C_5H_5N^+C_8F_{18}]NTf_2$, which revealed higher O_2 solubility. Unlike N,N'-dialkylimidazolium salts, this liquid has no acid protons, so the protonation of electrogenerated superoxide is thus avoided. In addition, the viscosity of this IL considerably decreases upon dissolving silanone precursors and silanone trap.

$$NC(CH_2)_3MeSiCl_2 \xrightarrow[O_2,\ +2e]{Me_3Si\text{-}O\text{-}SiMe_3} (Me_3Si\text{-}O)_2Si(Me)CH_2CH_2CH_2CN$$

Scheme 6. Functionalized silanone insertion into hexamethyldisiloxane in $[C_5H_5NC_8F_{18}]NTf_2$ ionic liquid.

Table 3. The "silanone" functionalization of HMDS in IL (P = 1 bar, T = 25 °C).

Silanone precursor	Media	Q [F mol^{-1}]	Product
$NC(CH_2)_3(Me)SiCl_2$	$[C_5H_5N^+C_8H_{18}]NTf_2$	2.2	$MD^{(CH_2)_3CN}M$
$H_2N(CH_2)_3(Me)Si(OEt)_2$	$[C_5H_5N^+C_8F_{18}]NTf_2$ [a]	2.1	$MD^{(CH_2)_3NH_2}M$

[a] 10 : 1 mixture $[C_5H_5N^+C_8F_{18}]NTf_2$: $[C_5H_5N^+C_8H_{18}]NTf_2$.

When $H_2N(CH_2)_3(Me)Si(OEt)_2$ and $CN(CH_2)_3(Me)SiCl_2$ were used in this process, corresponding amino- and cyano-derivatives were obtained (Scheme 6, Table 3).

A PAR-362 potentiostat in three-electrode mode was used. The potentials were controlled vs Ag/0.1 M $AgNO_3$ in CH_3CN reference electrode. A graphite cylinder was used to diffuse O_2 into the solution under cathodic polarization, and a glassy carbon felt was used as an anode in electrolyses.

Organylsilanes, cyclosiloxanes and HMDS were purchased from ABCR and used as received. The ionic liquids were prepared according to procedures described elsewhere [12] and dried overnight in vacuum (3 mm Hg at 80 °C) before using for electrolysis. Residual water in solvents was controlled by Karl Fisher titration. Oxygen was dried by passing it through a cartridge filled with P_2O_5 on glass wool.

The pre-electrolysis of 1 mmol of Me_3SiCl for about 1.5 h preceded the main electrolysis to eliminate H^+ formed from hydrolyses of the chlorosilane by residual water. The silanone precursor (50 mmol) and a trap (55 mmol) were then injected into the cell using a syringe, and the electrolysis was continued with constant flow of O_2 through the solution (5 mL min^{-1}) and the current density $j = 5 - 7$ mA cm^{-2}, the cathode potential not to exceed –2 V. From THF and DMF solutions, the products were isolated as described earlier [6, 7]; when the process was carried out in IL, thorough extraction with diethyl ether, evaporation of the latter and distillation of the organic residue were effected to isolate the products.

Acknowledgment: The authors gratefully acknowledge the financial support of this project by Rennes Metropole.

References

[1] M. G. Voronkov, *J. Organomet. Chem.* **1998**, 557, 143.
[2] L. E. Gusel'nikov, Z. A. Kerzina, Y. P. Polyakov, N. S. Nametkin, *Zh. Obsch. Khim.* **1982**, *52*, 467.
[3] H. S. D. Soysa, H. Okinoshima, W. P. Weber, *J. Organomet. Chem.* **1977**, *133*, C17.
[4] W. F. Goure, T. J. Barton, *J. Organomet. Chem.* **1980**, *199*, 33.
[5] M. G. Voronkov, S. V. Basenko, *J. Organomet. Chem.* **1995**, *500*, 325.
[6] D. S. Fattakhova, V. V. Jouikov, M. G. Voronkov, *J. Organomet. Chem.* **2000**, *513*, 170.
[7] R. Keyrouz, V. Jouikov, *New J. Chem.* **2003**, *27*, 902.
[8] V. Jouikov, *Rus. Chem. Rev.* **1997**, *66*, 509.
[9] V. V. Jouikov, L. A. Grigorieva, *Electrochim. Acta* **1996**, *15*, 2489.
[10] R. S. Nicholson, I. Shain, *Anal. Chem.,* **1964**, *4*, 706.
[11] P. Wasserscheid, W. Keim, *Angew. Chem., Int. Ed.* **2000**, *39*, 3772.
[12] P. Lozano, T. De Diego, D. Carrié, M. Vaultier, J. L. Iborra, *Biotechnol. Lett.* **2001**, *23*, 1529.

Monofunctional Silicone Fluids and Silicone Organic Copolymers

*Wolfgang Keller**

Wacker-Chemie GmbH, Johannes-Hess-Str. 24, 84489 Burghausen, Germany
Tel.: +49 8677 837419 — Fax: +49 8677 8867419
E-mail: wolfgang.keller@wacker.com

Keywords: monofunctional silicone fluids, silicone copolymers

Summary: Monofunctional silicone fluids and silicone organic copolymers are specialty products which offer the ideal supplement to today's standard silicones and polymers. By use of these materials, desirable silicone properties like certain surface effects are conferred onto organic systems, in most cases without compromising the convenient handling of the base formulations.

Introduction

Almost all functional silicone fluids of today's industrial production are either of a cyclic nature, containing the appropriate residues, or are linear oils bearing reactive functionalities at both ends or in the chain. The chemical nature of silicone synthesis done by equilibration and condensation is prohibitive for formation of asymmetrical silicones, in contrast to organic molecules like oleic acid or even ethanol. Currently there is only one way of preparing monofunctional silicone fluids, which is through kinetic anionic ring opening polymerization of the cyclic silicone monomer hexamethylcyclotrisiloxane (D3).

Monofunctional Silicone Fluids

Synthesis of Monofunctional Fluids

Synthesis of monofunctional fluids usually proceeds via anionic ring opening of D3 by butyl lithium in an aprotic solvent (for example THF) and subsequent polymerization to a lithium polydimethylsiloxanolate. This may be neutralized to a silanol functional fluid by acids or terminated with chlorosilanes to the corresponding functional fluids.

With its new proprietary process, Wacker Silicones is now positioned to produce monofunctional silicone fluids from D3. Pilot plant product samples of different molecular weights and functionalities are available for customers on request (Fig. 1).

$$n\,[SiMe_2O]_3 + R'M \xrightarrow[\text{catalyst}]{\text{solvent}} \xrightarrow[-MX]{+R_3SiX} R'[SiMe_2O]_{3n}SiR_3$$

known structure: M = Li, R = functional or alkyl group, R' = Bu, X = Cl
new structures: M = H, R = functional or alkyl group, R' = R"O, X = R"O,
R" = (branched) alkyl, e.g. Me, Et, ^{i}Pr, Bu, ^{t}Bu, 2-Bu, ...

Fig. 1. Synthesis of monofunctional siloxanes prepared from D3 following the traditional and the new route.

Product Range and Quality

The current product range of developmental samples covers monofunctional fluids of aminoalkyl, chloroalkyl, vinyl, and (meth)acryloyl types. Molecular mass is 1700 and 4000 Da, and product viscosities reach from 15 to about 80 mPas. The appearance of the fluids is optically clear and colorless to slightly amber. Technologically accessible future functionalities include carbinol, epoxide, substituted phenyl, hydrosilyl, carboxy, sulfhydryl, polyglycol, and other groups (see literature attached). In Fig. 2, currently available products are sketched.

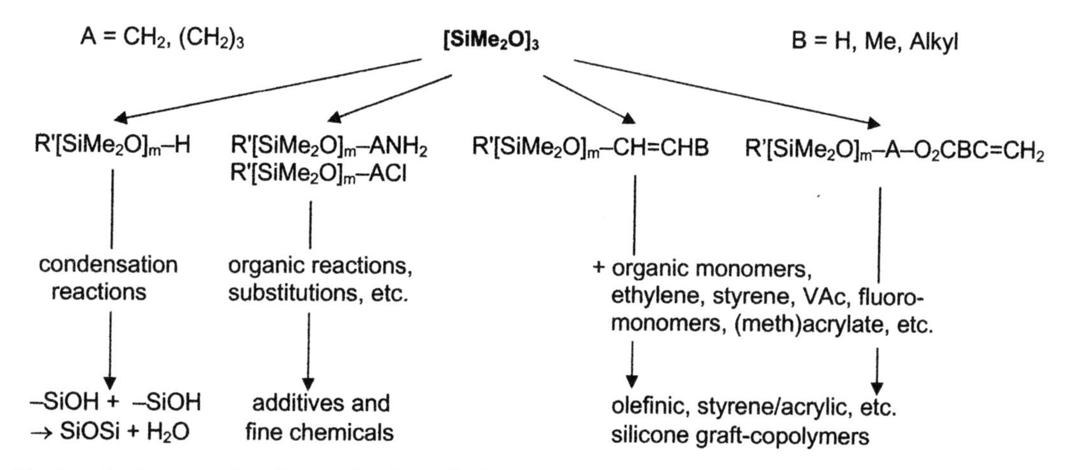

Fig. 2. Product examples of new molecules available from D3, their reactions and downstream products.

Applications of Monofunctional Silicones

Since the principal chemical structures of these new products are almost identical to the already known monofunctional silicone fluids, the former can be used in all applications already described for the latter, be it in literature, science, or industry.

Because of their monofunctionality, the title compounds may be used as a silicone offset for many organic molecules and additives. While the monosilanol oil is prone to spontaneous condensation reactions and has limited pot life, the amino- and chlorofunctional fluids will confer the silicone haptics, surface tension, weatherability, and other sought-after siloxane properties onto the organic formulation.

Also, because of their non-crosslinking behavior in many polymer analogous reactions and radical polymerizations, the macromonomers are especially useful in incorporating silicone performance into organic systems, resulting in silicone organic copolymers and polymer additives.

Silicone-Organic Copolymers

Fig. 3. Appearance of silicone organic copolymers: left — poly[silicone-co-(butylacrylate/styrene)], right — poly[silicone-co-styrene]

General Aspects

In addition to the monofunctional fluids, Wacker has recently developed block- and graft-copolymers of these silicone macromonomers with organic monomers.

A large variety of monomers and their mixtures were successfully polymerized with functional silicones to the corresponding copolymers, e.g., vinyl acetate, styrene, methyl methacrylate, butyl acrylate, acryl amide, acrylic acid, and other unsaturated monomers with special residues.

By applying and combining classical and innovative polymerization concepts, polysiloxane copolymers with different degrees of phase segregation morphologies can be prepared from the raw materials. Samples of standard and customized copolymers are available on request.

Properties and Applications of Silicone Copolymers

By combining typical properties of silicones with those of organic polymers, specialty copolymers of the desired product performance are obtained.

Melt viscositiy, reactivity, solubility, and other chemical and/or physical parameters like hydrophobicity or compatibility with specific organic systems can be widely varied. Also, the silicone portion gives these copolymers the typical siloxane feel and surface structure.

Silicone copolymers containing Wacker macromonomers may be used beneficially in all applications and industries where other hybrid materials from silicones and organic monomers are known to perform, especially those using silicone macromonomers described in the literature.

Silicone Copolymer Products and Product Formulations

Wacker Silicone Copolymers are available as

- Solutions, solid resins, elastomers (formulated as beads, powder, or granulate)
  Emulsions and dispersions (neutral, anionically or cationically stabilized)
  Thermoplastics of various *Tg* values (from below 0 °C to about 100 °C)
- Hyperbranched and crosslinked copolymers
  Polymers of different phase separation levels
- Wide range of silicone contents (from 1 to over 40 wt% realized).

Application Examples and Literature on Silicone Macromonomers and Copolymers, Their Solutions and Dispersions

The following list – though not comprehensive – gives an overview of potential applications for monofunctional fluids and silicone copolymers.

General silicone surface effects (various applications)

- Silicone haptics; enhancement of slip and slip angle; release effects; low friction; soil, dirt, and water repellency; weatherability; UV stability

Special applications and effects

- Raw materials in chemical reaction types for which the contained functional group is known (for example amino functional silicones)
- Non-bleeding softener or pot life regulator in silicone elastomers and gels, production of vibration-absorbing rubber (less crosslinked than regular HTV rubber)
- Antistructuring agent and rheology additive in silicones
- Additives and comonomers in thermoset, moisture, UV, and catalytically curing resins or coatings (e.g., acrylic, epoxy, melamine, and unsaturated polyester systems)
- Comonomers in industrial polymers (PMMA, PS, PVC, PE, PP etc.) and copolymers (ABS, EPDM, SBR, SAN, TPOs, special elastomers, etc.)
- Surface modification of fillers, powders, and silica; film-forming additive (all industries) for inducing the surface effects mentioned above
- Plastic, rubber, and non-iron metal processing (extrusion aid and release additive for plastics and rubber; e.g., PUR, thermosets, casting resins, high melting point mold release agents and additives)
- Plastic modification (impact modification, low/high temperature stability, flame retardant, PU foam stabilizer, blend compatibilizer)
- Plastic additivation (vapor permeability, surface effects), additive for slip agents and (cooler) lubricants

- (Injection) molded/cast articles with silicone surface effects mentioned above
- Dental masses, oxygen/gas-permeable contact lenses and membranes, semi-permanent bladder coating
- Medicinal bandages and adhesives (sprayable, elastic, easily removable)
- Additive for liquid and powder coating formulations (less brittle, self-levelling, high/low T stability, scratch and rub resistance, transparency, surface effects above)
- Automotive coatings, clearcoats, laquers, etc.
- Binder for various multi-component and hybrid coatings (abrasives, powder, fillers) with silicone surface properties mentioned above
- Additive for polymer binders and adhesives (architectural paints and coatings, asphalt and bitumen enhancement, road marking)
- Use as impregnating agents or additivation thereof (inorganic and porous surfaces, building materials, anti-graffiti)
- Additive in cosmetics and consumer care for induction of surface effects: decorative cosmetics, hair care (fixative, conditioning), cremes, rheology modification, SPF booster, nail care; polish formulations for wood, floor, metal, car, shoe, coating and household care applications
- Textile and leather treatment and care, tissue impregnating and binder additive, cork release coatings
- Additive for paper making and coating, release coating and modification, toner treatment (surface effects), additive in PSA formulations.
- Process aid (antifoam and deaeration agent in, e.g., paper and textile industry, in agrochemicals, dispersion production and formulation, adhesives)

Patent Literature (selection)

- Cosmetic: US 6383502; US 6248316; US 02/131948
- Coatings: US 6310169; EP 1251151; EP 1134241
- Plastics: US 5204438; US 6331589; DE 2717227
- Others: US 6420480; US 5039761; US 5057578; US 02/0005933; DE 4234898; EP 0839869; EP 0852609; EP 1140428

Conclusion

Because of their exceptional behavior and process ability, silicone-based copolymers will take their place as specialties within the silicone polymer class. With these silicone hybrid copolymers, new possibilities open up for adding value and performance to polymer and additive formulations.

Acknowledgments: A world of thanks to everyone who contributed to the work described and the successful publishing of this paper – especially to my dear colleagues at Wacker-Chemie GmbH.

Tailoring New Silicone Oil for Aluminum Demolding

Philippe Olier, Luc Delchet, Stefan Breunig*

Rhodia Silicones SAS, 1 et 55, rue des Frères Perret
BP 22, 69191 Saint-Fons Cedex, France
Tel.: +33 4 72 73 71 72 — Fax: +33 4 72 73 74 40
E-mail: stefan.breunig@eu.rhodia.com

Keywords: aluminum, foundry, release

Summary: Standard organomodified silicone oils are currently used in water-based formulations for foundry release applications. They are generally used in combination with organic waxes and oils. The use of silicone oil limits mold fouling. The use of the right type of modified silicone oil assures excellent release properties, low gas formation, and the paintability of demolded articles. Because of the application parameters, only a small amount of demolding agent is transferred to the mold surface; a high amount of demolding agent is lost because of splashing and bouncing. By tailoring new silicone oil, a new concept was developed. Our target was to optimize product efficiency by increased product transfer to the mold surface, even at elevated temperatures, so reducing product loss. At the same time an efficient cooling of the mold is assured and mold fouling remains low.

Introduction

The use of aluminum in the automobile industry is steadily increasing. The replacement of steel permits a decrease in fuel consumption of the vehicles due to weight reduction. In the car industry, aluminum is mainly used in body parts, starters, generators, compressors, admission pipes, oil vats, gearbox housings, cylinder heads, and even engine blocks.

The parts are obtained by injection of molten aluminum into steel molds. The molds weigh some tons and are expensive to produce. Maximization of their lifetime is an important factor.

The melting point of pure aluminum is 660 °C. The temperature of the molten aluminum alloys used in the foundries is at around 720 – 750 °C. The injection is done under pressure so that mold filling is as homogeneous as possible. Under these conditions the aluminum will stick to the steel surface of the mold. Demolding would be impossible.

Release agents are used to protect the steel mold from the molten aluminum. They provide lubrication and protection of the mold surface even under pressure. Efficient demolding of the cooling and solidifying aluminum part is ensured.

At 700 °C the organic structure of the release agent is partially destroyed. Among other products, silica, carbon black, water, and carbon dioxide are formed. It is important to ensure low gas (CO_2) formation during aluminum injection, so that the parts do not contain bubbles, inclusions, or surface faults that compromise their structural stability.

Any thermal decomposition of the release agent must lead to material that does not cause mold fouling and generates only low amounts of waste gases. Specifically modified silicone oils are best suited.

This kind of product is well known in the literature. Generally, fully modified PDMS backbones are employed in which on each repeating unit one methyl group is replaced by a lubricating function. These are generally introduced by hydrosilylation reaction on methylhydrogenosiloxanes. In most cases C6 to C18 alkyl chains and/or α-methylstyrene are grafted onto the silicone backbone. This kind of product was been described in the literature since the 1970s [1] and is in the public domain.

What remains problematic is the optimization of the product efficiency. There are two aspects to be looked at:

- the transfer efficiency of the active components of the emulsion to the mold surface
- the cooling power of the emulsion.

The more product that is transferred to the mold surface the more economic its use becomes. The better the emulsion cools the surface the less product is needed, or the more the emulsion may be diluted.

The aim is to achieve good wetting of the hot mold surface by the emulsion. But good wetting is not sufficient. The projected emulsion droplets that touch the surface will first spread (wet), then partially splash and partially retract and finally rebound. So for optimum efficiency the droplet should have an excellent wetting power but at the same time a limited tendency for retraction and rebound.

Formulators tend principally to optimize the composition of the emulsion, so as to get optimum wetting with limited retraction and rebound.

They have to evaluate their formulations at elevated temperatures, as the Leidenfrost effect hinders the initial wetting.

The Leidenfrost effect describes the "hovering" of a droplet above a hot surface. In the case of release agents for aluminum demolding, the emulsion droplet may not touch the metallic mold surface because of the rapid evaporation of the water contained in the droplet. This phenomenon occurs above a certain temperature, called the Leidenfrost temperature. The formulations are optimized for a maximum Leidenfrost temperature, so that the product efficiency is maximized even at elevated temperatures.

A prolonged and optimized direct contact of the droplet with the mold surface permits an increased transfer of the active ingredients, and the evaporating water can now ensure efficient cooling of the mold at the same time. So the longer the emulsion droplets stay in contact with the mold surface the more product may be transferred and the more heat can be drained by the evaporating water.

Analysis of the literature shows that there is more than one approach.

Most concern the use of specific surfactants or surfactant mixtures that permit the lowering of the Leidenfrost temperature [2]. Others concern the increase of the oil viscosity contained in the emulsion [3 – 5] and/or the use of specific organic waxes [6]. Another approach concerns the thickening of the emulsion or the increase of its elongational viscosity [7].

All have limitations. Thickeners cause mold fouling or increased waste gas formation. A highly increased silicone oil viscosity may lead to spreading problems and thereby to sticking. Increased amounts of organic waxes lead to mold fouling. The surfactant route is limited at very high temperatures.

The Objective

The performance optimization of demolding agents was mostly left in the hands of specialized formulators. They optimized the composition of the aqueous demolding agent to achieve good performance. But they are limited to the use of the available organomodified silicone oils.

So we made it our objective to offer them a better performing organomodified silicone oil. During our studies we realized that the highest increase in efficiency can be obtained by optimizing the transfer capability of an emulsion. We therefore looked for a way to optimize the transfer efficiency of the organomodified silicone oil emulsion by optimizing the structure of the modified silicone oil but without optimizing the overall emulsion formulation.

In this way we offer the key to formulators for optimum performance by still leaving them the liberty to optimize their formulations depending on the specific end use.

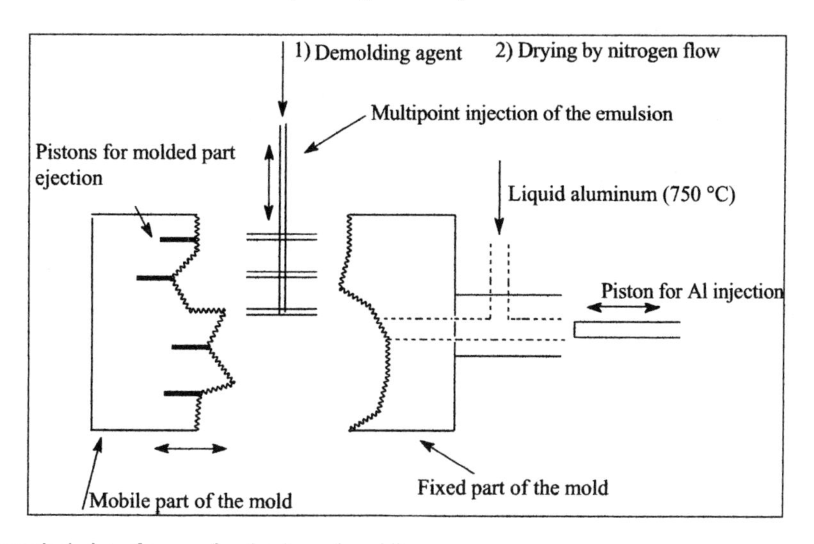

Fig. 1. Schematic design of a press for aluminum demolding.

Product Evaluation

So as to be able to evaluate the different candidates, we needed laboratory tests that simulate the application.

On an industrial aluminum foundry press the release agent is sprayed onto the hot vertical mold surface after the release of every aluminum workpiece and at the beginning of every demolding cycle (see Fig. 1).

This spray application of the demolding agent can be simulated by letting a drop fall onto a hot 45° inclined surface [8, 9]. We designed a test accordingly.

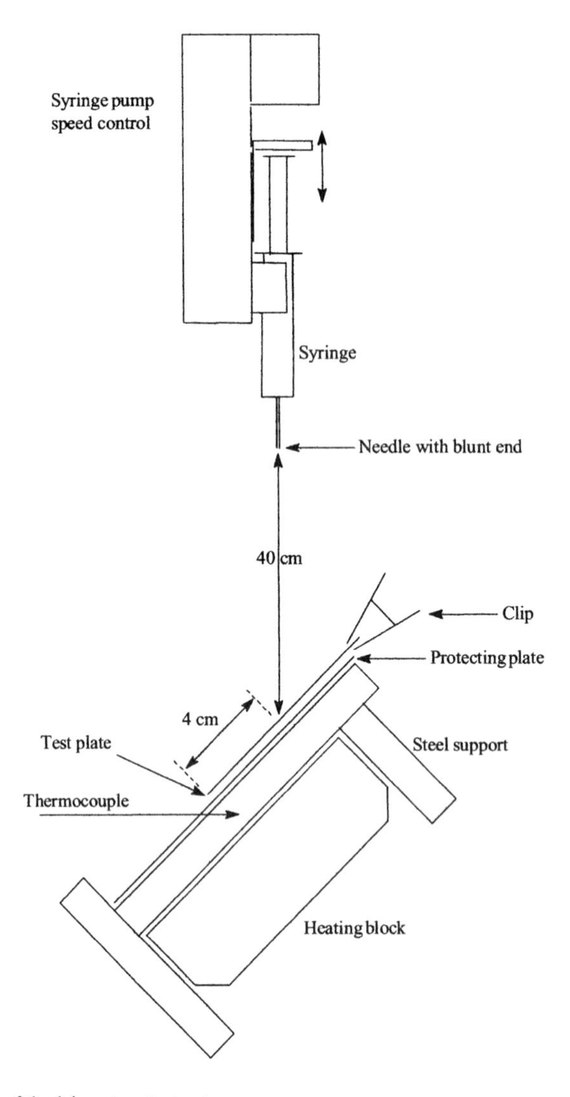

Fig. 2. Schematic design of the laboratory test setup.

Figure 2 shows the test setup. The emulsion drops fall onto a 1 × 7 cm metallic plate. The plate is heated to the desired temperature.

The temperature of the plate can be measured by IR camera, and the temperature evolution during product application can be followed when taking into account the emissivity change due to product deposit. An ultra-rapid camera can survey the drop impact at the same time. The amount of product transferred can be determined by weight difference of the small metal plate before and after the application.

Dependence on Oil Viscosity

We first tried to evaluate the effect of increasing oil viscosity in an otherwise equivalent emulsion environment. We checked this by using standard PDMS oils of 1000, 5000, 60 000 and 125 000 cps in emulsion. These tests were conducted with emulsions diluted to 10% solids.

We found an important increase in the amount of product transferred when passing from the usually used 1000 cps viscosity oils to high viscosity versions (Graph 1).

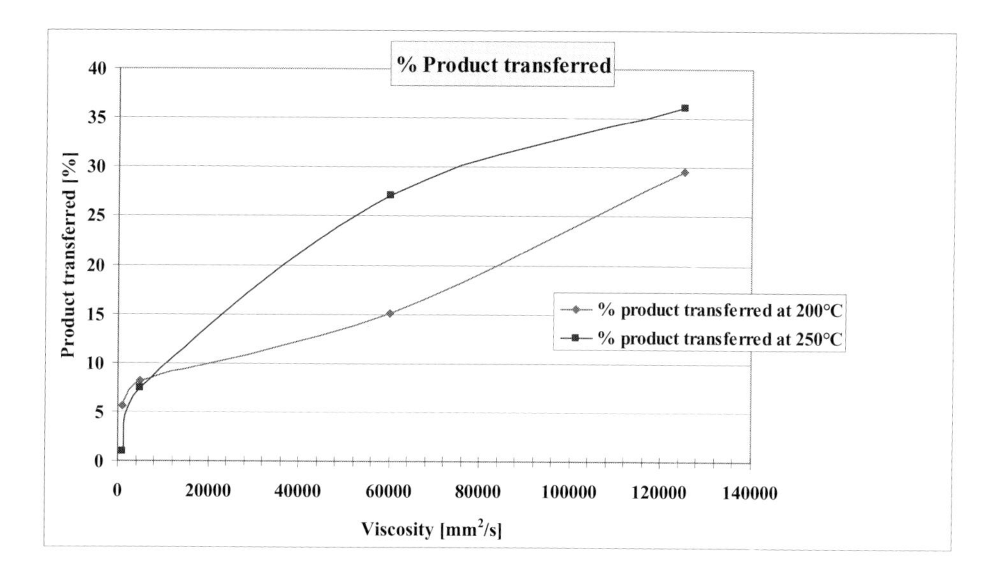

Graph 1. Oil viscosity dependence of transfer efficiency.

Viscosity is an important factor. Standard PDMS oils spread well and form an efficient release film on the mold surface, but they do not perform in this application when paintability of the demolded articles is required.

Therefore, we tried highly viscous organomodified silicone oils, like styrene modified PDMS. The transfer performance was excellent but the spreading of the product seemed insufficient as the release performance in foundry tests was poor. The molded aluminum parts were sticking spot-wise to the mold surface.

The idea was to integrate lubricating functions and surface wetting functions into the same silicone oil, so that efficient transfer to the mold surface is assured at the same time as excellent wetting and therefore release performance.

Product Synthesis

A large number of various surface-wetting functions were introduced into the silicone chain so as to optimize the efficiency.

The product synthesis starts from a standard linear SiH-functional silicone oil: M-D'$_{50}$-M, where M = $(CH_3)_3Si\text{-}O_{1/2}$ and D' = $H(CH_3)SiO_{2/2}$. Approximately 94% of the available SiH functions are grafted by hydrosilylation in a one-pot reaction with α-methylstyrene or 1-dodecene moieties. Even in the presence of an excess of this first moiety the hydrosilylation reaction slows down rapidly once 94% of the SiH functions have reacted. This is probably because of steric hindrance.

The residual ~6% SiH functions are then available for grafting of a second function, the surface active moiety. The simple addition of this sterically less demanding moiety to the reaction mixture, after obtention of 94% grafting of the first function, allows the hydrosilylation reaction to go to completion.

Thus, in a one-pot reaction, the lubricating moiety and the surface wetting moiety are grafted onto the silicone backbone (Fig. 3). A numerical average 94:6 distribution of those functions on all polymer chains is assured.

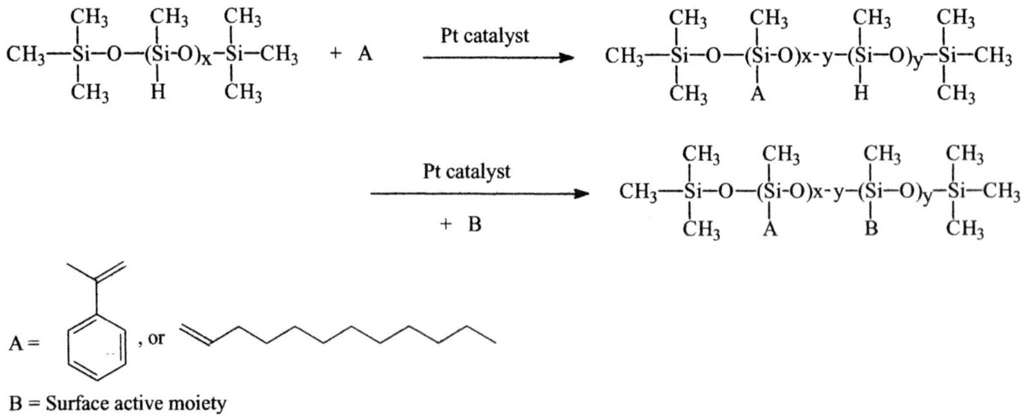

$$CH_3\text{-}Si(CH_3)_2\text{-}O\text{-}(Si(CH_3)(H)\text{-}O)_x\text{-}Si(CH_3)_2\text{-}CH_3 + A \xrightarrow{\text{Pt catalyst}} CH_3\text{-}Si(CH_3)_2\text{-}O\text{-}(Si(CH_3)(A)\text{-}O)_{x-y}\text{-}(Si(CH_3)(H)\text{-}O)_y\text{-}Si(CH_3)_2\text{-}CH_3$$

$$\xrightarrow[\text{+ B}]{\text{Pt catalyst}} CH_3\text{-}Si(CH_3)_2\text{-}O\text{-}(Si(CH_3)(A)\text{-}O)_{x-y}\text{-}(Si(CH_3)(B)\text{-}O)_y\text{-}Si(CH_3)_2\text{-}CH_3$$

A = [α-methylstyrene structure], or [1-dodecene structure]

B = Surface active moiety

Fig. 3. Stepwise one-pot synthesis.

The result is a new modified silicone oil (H 1680) containing both lubricating and metal surface wetting functions. The latter help also in the preparation of an emulsion of this oil and they have an anti-rust effect.

Preparation of a Standard Emulsion

So as to be able to compare the new oil's performance with that of market standards, we have chosen to use a standard emulsion formulation for all modified silicone oils.

Typically the modified silicone oils are mixed with a standard alkylpolyethyleneglycolether surfactant and a small amount of water. The mixture is stirred under high shear and then stepwise diluted with water to approximately 56% solids.

Product Evaluation

Product Evaluation at Comparable Solids Content

Transfer

A variety of organomodified silicone oils designed for foundry demolding application were evaluated in the product transfer test. See Fig. 2 for test setup details.

The emulsified product (50 mL), diluted to 0.2% solids, fell dropwise onto a small (1 × 7 cm) heated metal plate. The plate was degreased, sand blasted, and weighed before use. The chosen temperatures were 200 °C and 250 °C, as these are the most representative for the application.

The plate was then weighed after application of the emulsion. The weight difference of the metal plate in relation to the amount of emulsion applied and its solids content gives the amount of product transferred (Eq. 1).

$$\text{Solids transferred} = (\text{MPA} - \text{MPB}) \times 10\,000/(\text{WE} \times \text{SC})$$

where : MPA = weight of metal plate after application [g]
MPB = weight of metal plate before application [g]
WE = weight of emulsion applied [g]
SC = solids content of the emulsion applied [%]

Eq. 1. Calculation of the transfer performance.

By this test we evaluated our standard oil for metal demolding, Rhodorsil H1658, the new development Rhodorsil H1680 PEX, and three modified silicone oils of competitors, each in the standard emulsion formulation.

We also compared these results with those of two emulsions which are market references for aluminum demolding (MR = market reference). See Table 1 for results.

As the data shows, the new H1680 performs a lot better than standard organomodified silicone oils. At 250 °C more than seven times the amount of product is transferred. At 200 °C it is still about four times more efficient than the competitive oils, based on the same type of emulsion system (see also Graph 2).

Table 1. Comparison of the transfer performance.

Emulsion type	Oil	Solids transferred at 200 °C [%]	Solids transferred at 250 °C [%]
Standard	H1658	1.42	0.33
Standard	H1680	5.35	2.54
Standard	Competition 1	0.82	0.05
Standard	Competition 2	1.06	0.26
Standard	Competition 3	0.29	0.06
Formulation 1 (MR)	Competition 3	1.83	0.82
Formulation 2 (MR)	Competition 3	0.51	1.20

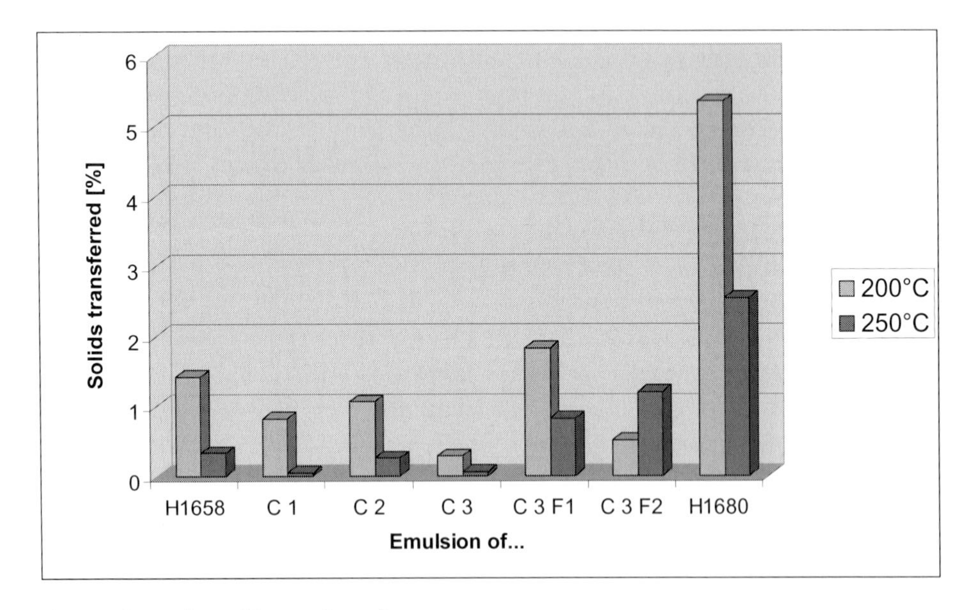

Graph 2. Comparison of the transfer performance.

An efficient formulation (MR) may partially compensate for the lack of efficiency of standard organomodified silicone oils. The formulation may only achieve optimum performance over a certain temperature range, however. Some perform best in the low-temperature range: others are excellent at very high temperatures.

The overall performance of the standard emulsion of H1680 is excellent, even in comparison with the market references at 200 °C and 250 °C. The emulsion of H1680 is three times more efficient at 200 °C and 250 °C than the excellent Formulation 1 without having been optimized.

There is still a good margin for possible performance improvement by formulators.

Ultra-Rapid Camera Observation

So as to optically document the difference in cooling and product transfer efficiency of H1680 in emulsion we set up a test (refer to Fig. 2) without using a small weighed metal plate. We let drops of the emulsion to be evaluated fall directly onto a hot stainless steel metal plate. The new plates were degreased before use.

We observed the wetting, splashing, and bouncing behavior of the droplets coming into contact with the hot metal plate by a high-speed camera. We could thus observe the behavior of the emulsions when coming into contact with the hot metal surface. This helps to explain the product efficiency differences observed.

At 250 °C we applied emulsions of a solids content of 0.2%, which is a concentration typically, used in foundries.

We checked the performance of the laboratory standard emulsion of H1680 versus an excellent market reference (Formulation 1).

The following pictures were taken at 12 ms after drop impact (Figs. 4 and 5).

The first two correspond to the first drop impacting on the hot metal surface (Fig. 4). Already it is obvious that more product is transferred to the hot metal plate in the case of H1680 emulsion. The transferred solids content darkens the surface. Splashing behavior on the other hand still looks similar.

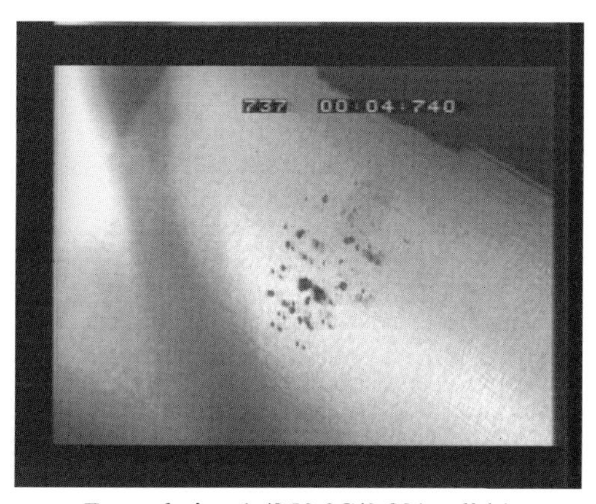

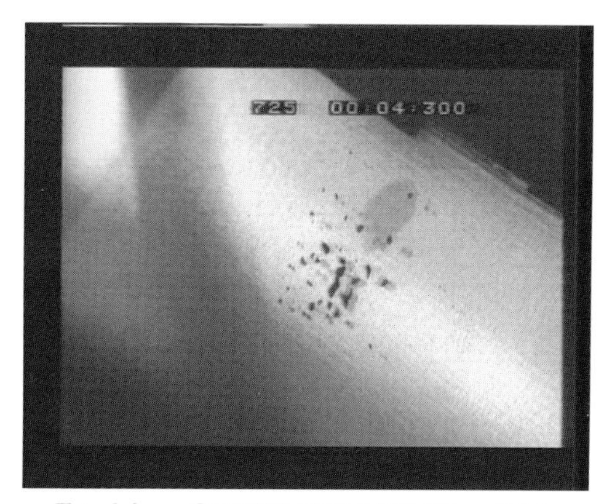

Formulation 1 (250 °C/0.2% solids) Emulsion of H1680 (250 °C/0.2% solids)

Fig. 4. Impact behavior of the first drop on the hot metal surface after 12 ms.

After 300 drops have fallen onto the hot surface (Fig. 5) there is still more product transferred in the case of the emulsion of H1680. But the splashing and bouncing are now reduced, which implies that the surface has been cooled efficiently. The time during which the droplet stays in contact with the hot metal surface increases. Product transfer and cooling therefore also increase.

With the reference emulsion the splashing and bouncing occur earlier. Less product is transferred and the cooling is less efficient.

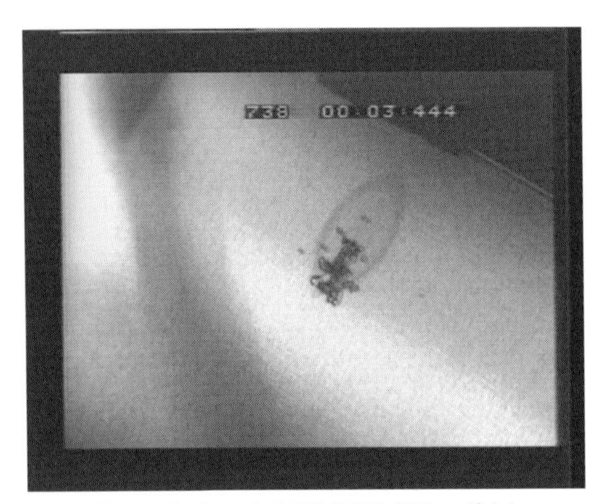

Formulation 1 (250 °C/0.2% solids)

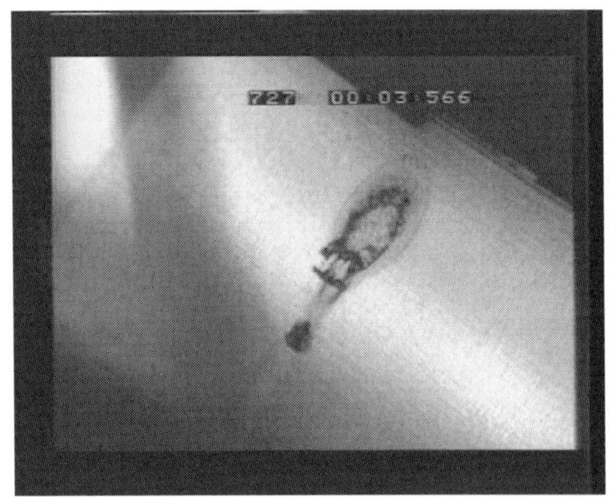

Emulsion of H1680 (250 °C/0.2% solids)

Fig. 5. Impact behavior at 12 ms after 300 drops.

IR Camera Evaluation

The above setup was also equipped with an infrared camera, so that we could follow the temperature development of the impact point on the hot plate during emulsion application. Emissivity modification due to surface modification as product is deposited on the hot metal surface has to be taken into account.

The emulsion of H1680 was found to have a better cooling effect than the reference emulsion (Formulation 1).

After the impact of 300 droplets at a plate temperature of 250 °C the reference emulsion was able to lower the temperature of the impact point to 214 °C.

The emulsion of H1680 allowed the plate to cool down to 135 °C under the same conditions. The cooling efficiency of the latter is therefore much greater.

Table 2. Comparison of the cooling effect at 0.2% solids.

Emulsion type	Oil	Solids content [%]	Initial temperature [°C]	Temperature after 300 drops [°C]
Standard	H1680	0.2	250	135
Formulation 1 (MR)	Competition 3	0.2	250	214

H1680 Emulsion Advantage at Reduced Concentrations

Transfer

As the performance of the standard emulsion of H1680 exceeded even that of a specially formulated product at 0.2% solids, we decided to evaluate the new product at a reduced concentration (Table 3).

Table 3. Transfer performance advantage of H1680.

Emulsion type	Oil	Solids	Solids transferred at 200 °C [%]	Solids transferred at 250 °C [%]
Standard	H1680	0.05	2.48	1.18
Formulation 1 (MR)	Competition 3	0.2	1.83	0.82

The results show that even at a 4 times lower concentration there is still 1.3 to 1.4 times more product transferred to a hot metal surface with the emulsion of H1680 (see Graph 3).

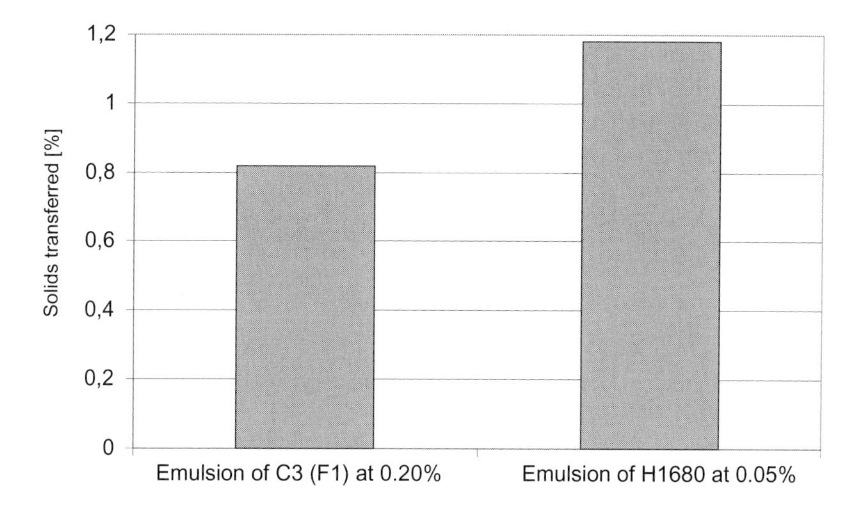

Graph 3. Transfer performance advantage of H1680.

IR Camera Observation

The result of the product transfer evaluation is supported by the temperature evolution measurement by IR camera during the product application (Table 4).

The temperature is lowered to 207 °C with the standard emulsion of H1680, whereas the formulated emulsion of a standard oil only cools the metal surface to 214 °C.

Table 4. Cooling effect advantage of H1680.

Emulsion type	Oil	Solids content [%]	Initial temperature [°C]	Temperature after 300 drops [°C]
Standard	H1680	0.05	250	207
Formulation 1 (MR)	Competition 3	0.2	250	214

Ultra-Rapid Camera Observation

When optically observing the amount of product transferred to a hot metal plate it becomes obvious that the H1680 emulsion at 0.05% solids still gives rise to a higher product transfer than the market reference at 0.2% solids (Fig. 6). The transferred solid content darkens the surface.

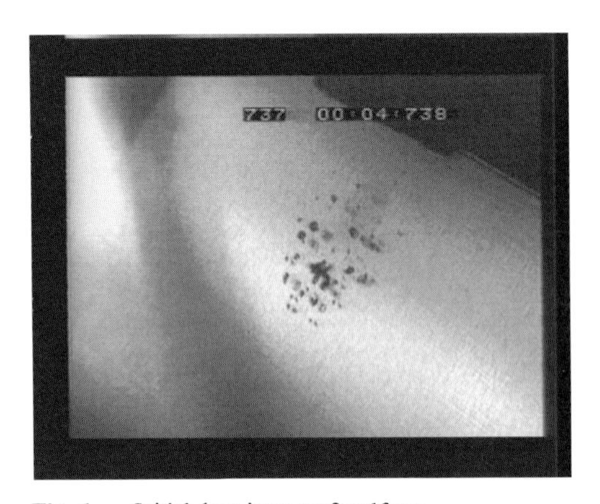

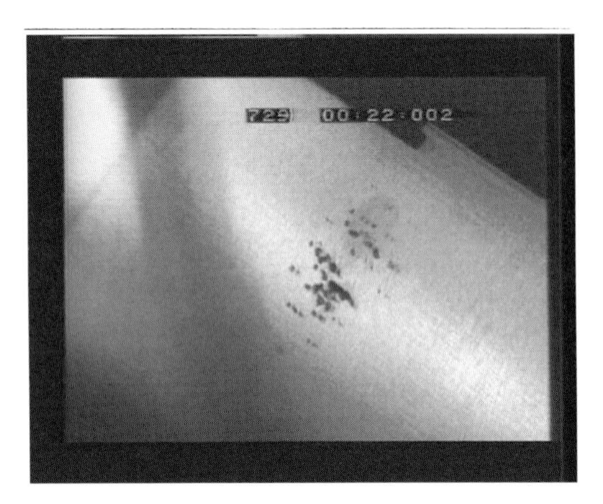

Fig. 6. Initial drop impact after 10 ms.

On checking the wetting behavior of the two emulsions it becomes obvious that the H1680 emulsion at 0.05% solids ensures a more homogeneous wetting of the metal surface than the market reference at 0.2% solids. The latter shows a wetting gradient with a distinct fringe at the edge (Fig. 7).

The H1680 emulsion deposit shows no irisation, so the layer thickness is greater than the visible light wavelengths, in contrast to the market reference.

Furthermore, bouncing occurs later in the case of the H1680 emulsion in spite of the 4 times lower solid content.

Thus, the efficiency of the standard emulsion of H1680 in a dilution to 0.05% solids is still better than that of the reference emulsion at 0.2% solids.

This means that with less active product a higher amount of release agent is available on the hot metal mold surface. This means an increased efficiency, less product loss, and also less waste product to be treated.

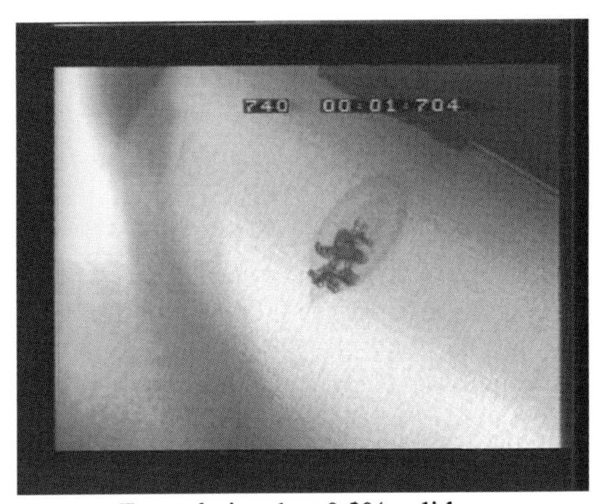

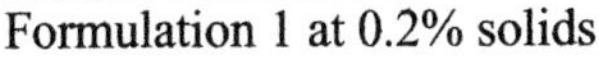
Formulation 1 at 0.2% solids

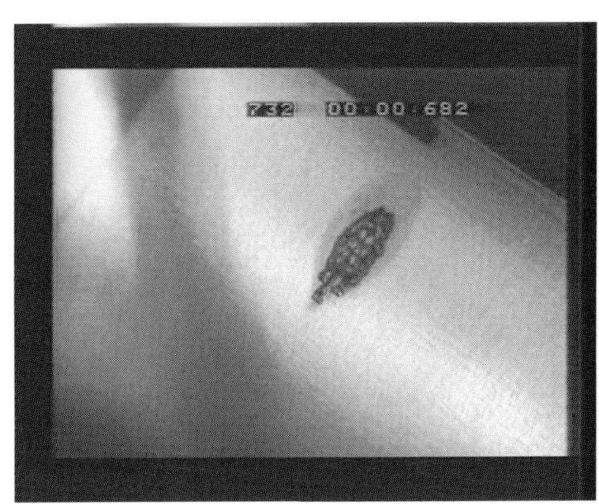

Emulsion of H1680 at 0.05% solids

Fig. 7. Impact behavior at 10 ms after 600 drops.

Conclusion

We have shown that by specially tailoring a new silicone oil for foundry applications, product efficiency can be at least quadrupled and dry waste material can be reduced by a factor of four.

This helps formulators to offer better performing products at reduced cost. This increased efficiency has a second benefit, which is a reduced amount of waste sludge, as less active product is lost by bouncing and splashing off the mold surface.

Acknowledgment: The authors would like to thank Jean-Yves Martin from the Rhodia Research center Lyon for his help in taking ultra-rapid camera pictures and making IR camera evaluations.

References

[1] Dow Corning, US 212 727, **1972**.
[2] Dow Corning Toray, US 5 401 801, **1995**.
[3] Dow Corning Toray, EP 0 825 004, **1996**.
[4] Takemoto Oil & Fat, JP 08 003 578-A, **1996**.
[5] Dow Corning Toray, JP 10 212 411, **1997**.
[6] Acheson, EP 0 550 028, **1992**.
[7] Toyota, JP 11 077 233, **1997**.
[8] L. Wachters, N. Westerling, *Chem. Eng. Sci.* **1966**, *21*, p. 1047 – 1056.
[9] B. Prunet-Foch, F. Legay, M. Vignes-Adler, C. Delmotte, *J. Coll. Int. Sci.* **1998**, *199*, 151.

A New Generation of Silicone Antifoams with Improved Persistence

John Huggins,* Kevin Chugg, Christopher Roos, Sabine Nienstedt

GE Bayer Silicones GmbH, Bayerwerk, D-51368 Leverkusen, Germany
Tel.: +49 214 30 57631 — Fax: +49 214 30 72183
E-mail: John.Huggins@gesm.ge.com

Keywords: antifoams, defoaming, black liquor

Summary: In this study, the knock-down and persistence performance of conventional antifoams were compared with a new generation of silicone antifoams with optimized higher viscosity polymer properties at different temperatures. A series of three new antifoam compounds TP 3860, TP 3861, and TP 3862, with viscosities ranging from 30 000 to 100 000 mPa.s, were studied. The defoaming performance was tested in a 0.375% solution of Mersolat H30 with a pH of 6.5. Conventional antifoams are highly effective at defoaming this ionic aqueous system at room temperature. By comparison, the room temperature defoaming effectiveness of the compounds TP 3860-3862 significantly decreases as the viscosity increases. Not only is the knock-down effectiveness greatly decreased, but also the defoaming persistence at room temperature is adversely affected if the viscosity greatly exceeds 30 000 mPa.s. At 70 °C in Mersolat this picture changes entirely. At this higher temperature, the new higher-viscosity compounds show equivalent knock-down behavior and superior persistence. The persistence increases in parallel with the compound viscosity throughout the series, and the highest-viscosity compound TP 3862 exhibits superior defoaming persistence, clearly outperforming all other compounds at 70 °C. The overall comparison ranks TP 3860 as the best all-round defoamer for applications with a wide temperature range, performing equally well at room temperature and above. For applications such as paper manufacture that require a maximum of persistence at higher temperatures the new compound TP 3862 is the clear favorite.

Introduction

Foaming problems occur widely in aqueous industrial processes, wherever inherent or added surfactants are present and agitation can lead to aeration. Unwanted foams can also cause product defects in a variety of aqueous and nonaqueous coating applications.

A broad range of defoaming products, or antifoams, have been developed to meet these diverse needs. Some of the most common and versatile antifoams are based upon polydimethylsiloxanes and hydrophobic silica particles. These antifoam compounds are the basis of a wide variety of emulsion antifoam formulations.

The goals of these investigations were to examine the influence of the polymer structure and viscosity on the antifoam compound performance. Of particular interest was the influence of the compound viscosity on the antifoam performance at the higher temperatures typical of many foaming situations.

A number of testing methods for antifoams are known and used. In this study we used a recirculation pump rig as shown in Fig. 1.

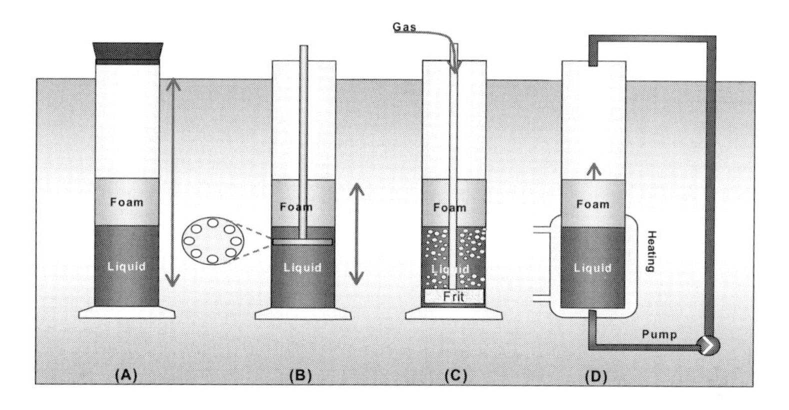

Fig. 1. Antifoam testing methods. A: Measuring Cylinder; B: Stroke Rig; C: Aeration Rig; D: Pump Recirculation Rig.

The antifoam performance is measured in terms of the knock-down level and the persistence (Fig. 2). The knock-down level is the minimum level of foam that is obtained directly after adding the antifoam while continuously recirculating the foaming liquid. The persistence is measured as the time required for the foam level to reach a defined height.

Three new silicone antifoam compounds with optimized polymer structures and viscosities ranging from 26 000 to 100 000 mPa.s were compared with conventional silicone antifoams.

Table 1. Antifoam compounds tested.

Compound	Viscosity
AF 9000	1 000
AF 100 %	2 400
CF	2 000
TP 3860	26 000
TP 3861	43 000
TP 3862	100 000

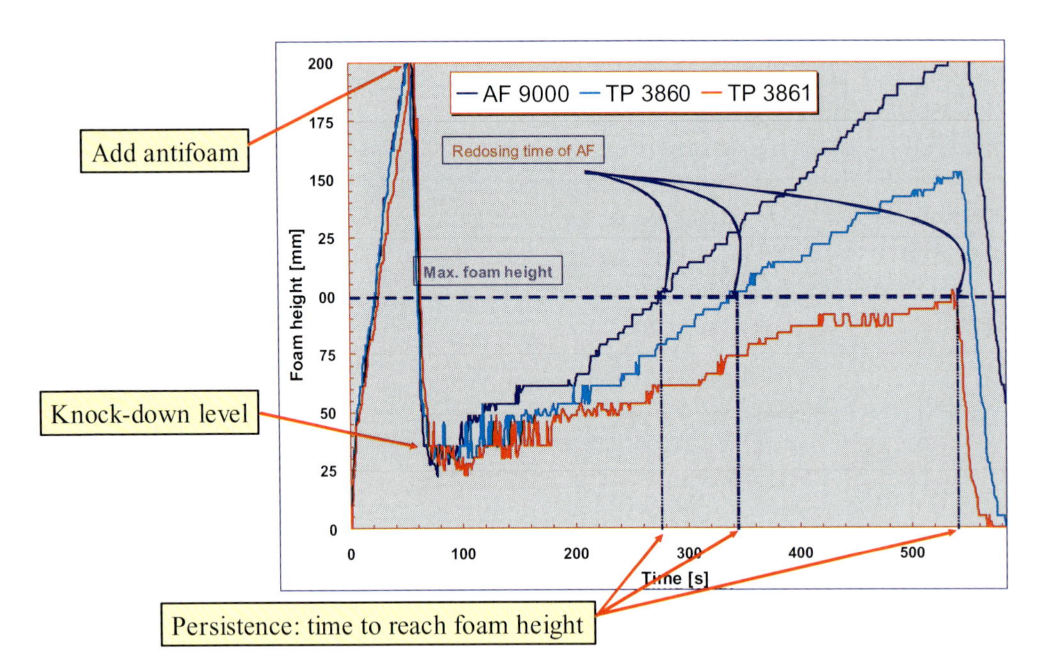

Fig. 2. Measuring performance in pump rig.

The antifoam compounds were tested in 0.375% Mersolat H30, a sodium alkyl sulfonate with pH 6.5. The Persistence Value was calculated according to the formula: Persistence Value = [mL KD/(seconds to 125 mL – seconds to KD)] × 150. The results at 23 °C and 70 °C are given in Tables 2 and 3, respectively.

At 23 °C, antifoam CF and AF 9000 both have a superior knock-down effect, but TP 3860 has the best persistence and best overall antifoam performance. In the series TP 3860/3861/3862, increasing antifoam viscosity decreases knock-down and persistence performance significantly.

Table 2. Defoaming in Mersolat H30 at 23 °C.

Compound	Knock-down [mL]	Knock-down speed	Persistence value
AF 100 %	35	Fast	39
CF	25	Fast	24
AF 9000	30	Fast	15
TP 3860	35	Fast	15
TP 3861	45	Slow	36.5
TP 3862	80	Slow	41

The series TP 3860/3861/3862 give increasing persistence at 70 °C as the viscosity of the antifoam increases. TP 3862 gives the superior overall performance at 70 °C, whereas TP 3860 is the best universal antifoam at all temperatures.

Table 3. Defoaming in Mersolat H30 at 70 °C.

Compound	Knock-down [mL]	Knock-down speed	Persistence value
AF 100 %	25	Fast	23
CF	20	Fast	13
TP 3860	20	Fast	8
TP 3861	25	Fast	7
TP 3862	35	Fast	5.5

Antifoaming Black Liquor

After formulating into corresponding 30% emulsions, TP 3862 gives clearly superior antifoam performance at 70 °C in paper manufacture black liquor, Fig. 3.

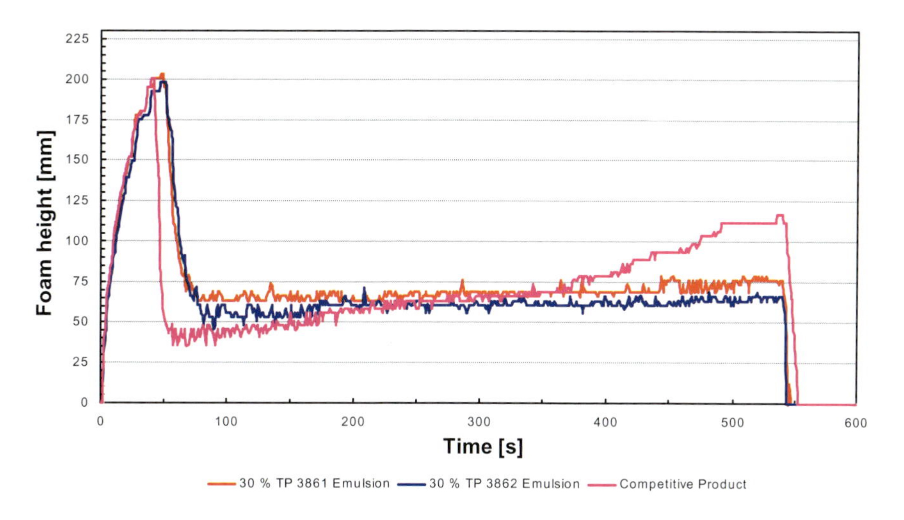

Fig. 3. Defoaming in black liquor at 70 °C (20 ppm AF).

Silicone Mist Supressors for Fast Paper Coating Processes

Jos Delis, John Kilgour*

GE Bayer Silicones GmbH & Co KG
Street, Postal Code, Bergen op Zoom, Netherlands
Tel.: +31 164 29 3253 — Fax: +31 164 29 3477
E-mail: Jos.Delis@gesm.ge.com

Keywords: silicones, label systems, release coatings

Summary: Silicone coated paper and films are used for manufacturing labels, decorative laminates, transfer tapes, etc. Manufacturing lines for making these products have increased line speeds for productivity reasons. Silicone suppliers are challenged to develop silicone systems which cure at these higher line speeds while maintaining other coating properties like coverage and stable release. Novel tailor-made silicones consisting of rigid and elastomeric moieties can be used as rheology modifiers in order to avoid misting, which is a processing problem appearing at higher line speeds. We report about a new anti-mist additive system which significantly reduces or eliminates silicone mist on high-speed coating without negatively impacting on outstanding coating performance.

Silicone-coated papers and films find their application in many areas where release of an adhesive is needed. By far the biggest application can be found in the label industry. Other applications can be found in graphic art and in hygiene applications, e.g. in diaper tapes. Another big area where these products can be found is that of bakery and food, in which silicone-coated paper can give release against bakery products or act as grease-proof wrapping paper.

Label stock is the combination of a release liner and a label. The former is the silicone-coated paper or film while the latter is the face stock with the adhesive. The silicone coating on the paper or film normally has a weight of around 1 gram per square meter (g/m^2) while that of the adhesive is normally around 30 (g/m^2).

The state-of-the-art silicone systems used in label stock application are normally solventless and thermal curing. Base polymers for these systems are vinyl-functionalized polydimethylsiloxanes having viscosities of around 200 – 600 mPa.s. Cross-linkers normally are hydride-functionalized polydimethylsiloxanes with a viscosity of around 25 mPa.s. These two components are cross-linked by a platinum catalyst, which can be the Karstedt catalyst. Additionally an inhibitor is added to the silicone mixture to prevent curing before it is applied on the substrate. These inhibitors are low-boiling compounds, which evaporate upon heating in the curing oven. Subsequently, curing of the silicone polymers will occur (see Fig. 1).

Silicone coatings on paper and film have to give good and stable release against adhesives. The most important requirement is that the silicone is fully cured. The amount of silicone-extractables in the coating should be lower than 3 – 5%. Furthermore, the coating should give a sufficient coverage of the substrate at a coat-weight of 1 g/m². Additionally, the liquid silicone coating should be very easily processable on the application system.

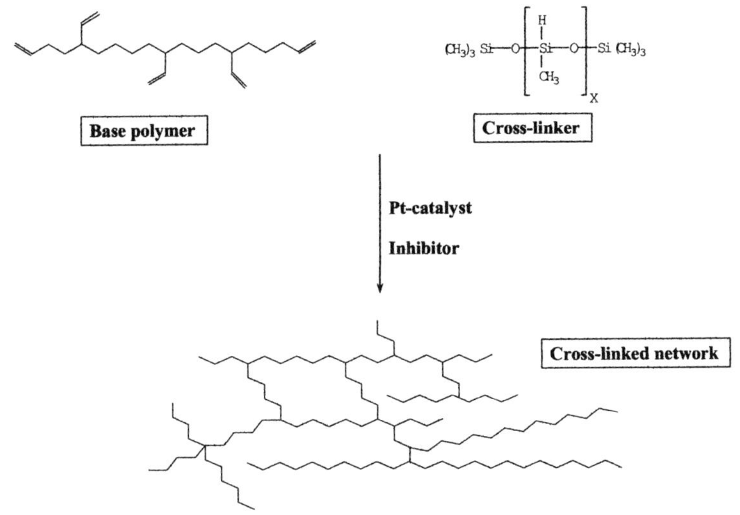

Base polymer

$(CH_3)_3Si-O-[SiH(CH_3)-O]_x-Si(CH_3)_3$

Cross-linker

Pt-catalyst

Inhibitor

Cross-linked network

Fig. 1. Cross-linking reaction of solventless thermal curing siloxanes.

In Fig. 2 a schematic drawing of a coating machine is shown. The unwinding station of the substrate is shown in the middle of the machine at the bottom. After unwinding, the substrate is transported to the 5-roll application system on the left, which applies the liquid silicone mixture onto the substrate. Subsequently, the substrate with the liquid silicone goes through the oven, which heats up the substrate to temperatures of around 150 °C. The substrate is rolled up on the winding station after curing of the silicone or the coated substrate goes through the next operation, which can be application of the adhesive, drying of this compound if necessary, and application of the face stock. In this way a full laminate structure is formed.

A detail of a state-of-the-art 5-roll application system is shown in Fig. 3. The liquid silicone bath is situated between the first two rolls. The liquid silicone is subsequently transferred to the next steel and rubber roll, and the latter transfers the silicone on the substrate. The steel backing roller presses the paper against the rubber applicator roller.

Like many industries, the industry using silicones is also facing high cost pressures. This means that the industry is looking for lower raw material prices, lower coat weights, and thinner substrates.

To gain productivity, higher line speeds are practiced. Silicone suppliers have to deliver silicone systems which cure faster while providing the same high coverage standards and process properties. The consequence of higher coating speeds is the formation of mist.

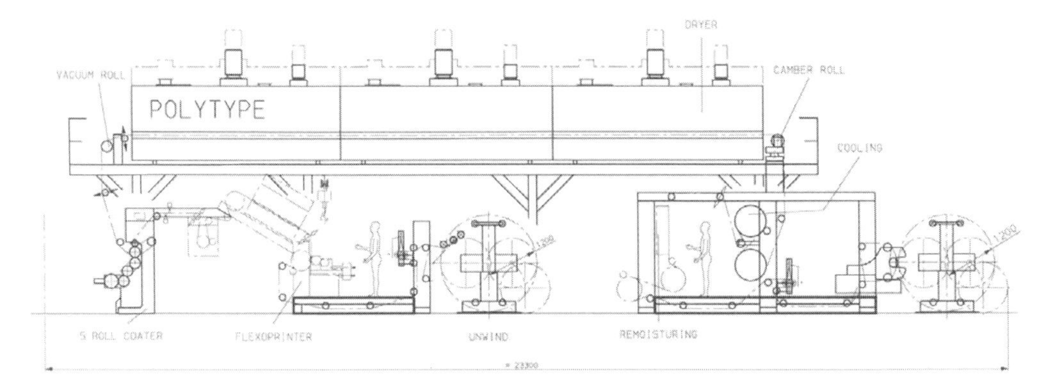

Fig. 2. Schematic picture of a coating machine for silicone release liners.

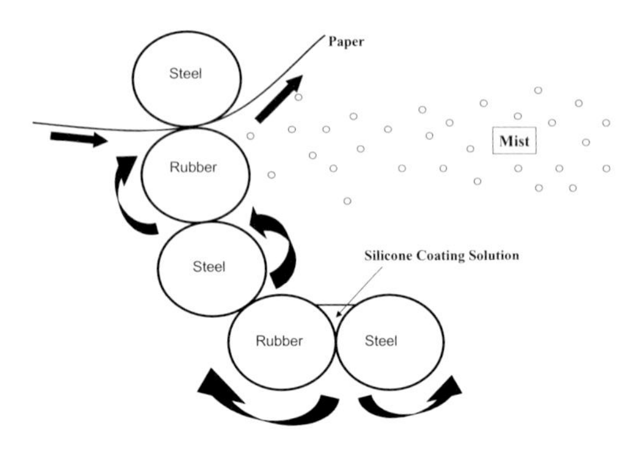

Fig. 3. Schematic detail of the 5-roll application system.

Misting is the generation of small droplets of silicone capable of being shot into the environment (Fig. 3). Misting is an issue from an economical and an environmental point of view. Systems must comply with all clean air regulations and offer the best possible method of application.

There are several factors which have influence on the mist formation. One can think of the diameter and setting of the rollers of the application system, the type of the rollers, etc. However, the most important factors are the speed of the machine and hence the speed of the rollers and the viscosity of the silicone-based polymer.

Speeds of the silicone coating machines have increased dramatically over the past decades. While state-of-the-art machines running at 200 m/min in 1975, machines are running at 1000 m/min nowadays. Pilot machines running at 1500 m/min have even been built by machine manufacturers. Severe mist formation can normally be observed at speeds of 600 m/min and above.

The effects of the machine speed and the viscosity of the silicone polymer on misting are shown in Fig. 4. The amount of mist is measured in mg/m^3 by, e.g., a Dust Track, which measures the amount of mist by light scattering.

Older silicone systems having a viscosity of around 500 mPa.s showed high mist formation when line speeds increased over time. To decrease the amount of mist it was decided to switch to silicone systems of lower viscosity, which resulted in systems of viscosities around 250 mPa.s. However, since line speeds are still increasing, one has to work with additives to suppress mist formation.

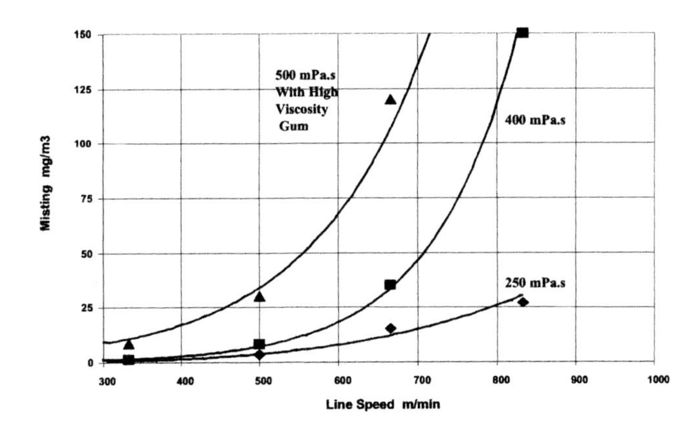

Fig. 4. Influence of viscosity of the silicone and of the line speed on mist formation.

Our theory behind mist formation is schematically illustrated in Fig. 5. This shows the applicator roller, which is in contact with the paper, thereby transferring the liquid silicone to the paper, generating film split. There exists high pressure at the spot where the applicator roller is in contact with the paper surface. This pressure is released when this contact is lost, which causes cavitation. Bubbles and fibrils are formed. Mist particles will be formed if the fibrils break at several places. If one can make sure that the fibrils break at one spot only, a clean film split takes place and no mist will be formed.

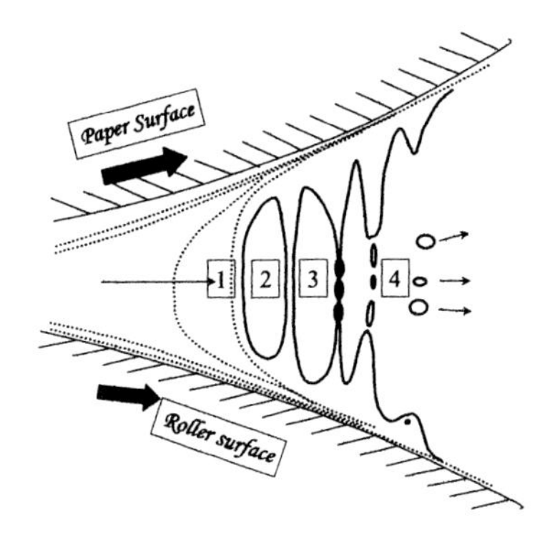

Fig. 5. Schematic drawing of mist formation during film split.

One of the approaches to reach this goal was to add small solid particles to the siloxane mixture in such a quantity that there is on average one particle per fibril, which will cause a weak spot in the fibril and thus an early break of the fibril at only one place.

Indeed, particles will help to change the break characteristics. The technique works at speeds up to 800 m/min, while at higher speeds misting takes over again.The next approach was to change the rheology of the silicone in such a way that misting is diminished. The siloxane mixture was given more elastic properties, since siloxanes of around 250 mPa.s only have Newton-like viscous properties.

Figures 6 and 7 show that this approach works.

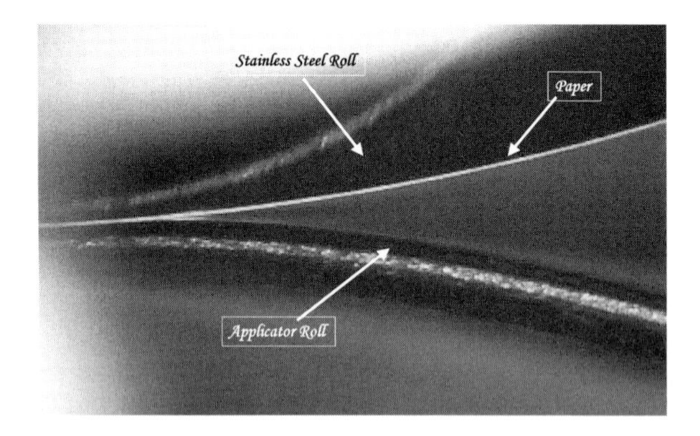

Fig. 6. High-speed camera picture of silicone system without anti-mist additive.

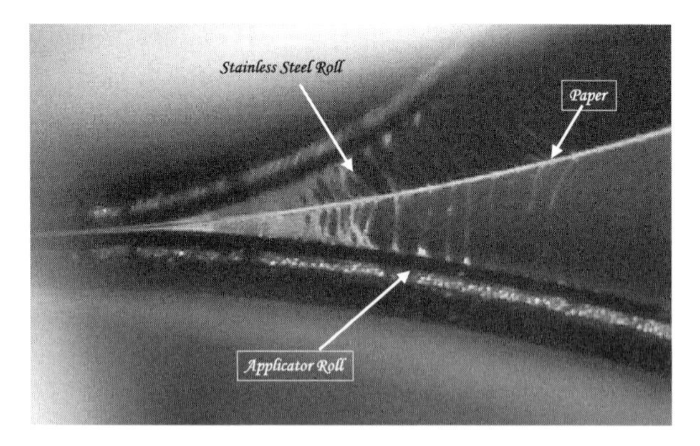

Fig. 7. High-speed camera picture of silicone system with anti-mist additive SLAM2000.

Figure 6 shows a high-speed camera picture of the applicator roller and paper substrate onto which the liquid silicone is transferred. This silicone mixture does not contain the anti-mist additive. The film split occurs very quickly after contact between applicator roller and substrate is lost. The high amount of mist can be observed in the area close to the paper and applicator roller.

Figure 7 shows the analogous picture of a silicone system containing anti-mist additive SLAM2000 (the commercial name). Very surprisingly, the film split occurs much later in the process and the fibrils of the silicone system can be easily observed. This is because of the higher elasticity of the silicone system. Breakage of these fibrils, however, does not cause mist formation, since, if a fibril breaks, the remaining silicone is returned to the paper or the applicator roller because of the elasticity of the silicone system.

Figure 8 shows the effect of the SLAM2000 anti-mist additive on the quantitative amount of mist in the commercially available silicone system SL6425. An amount of 2% of SLAM2000 reduces the amount of mist from around 125 mg/m^3 to 1.5 mg/m^3 at a machine speed of 800 m/min.

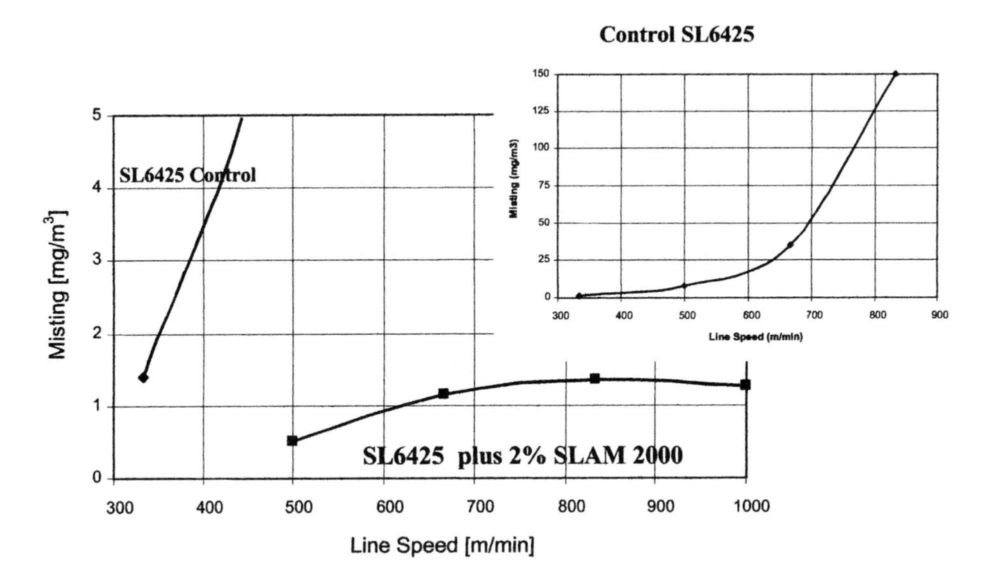

Fig. 8. Mist measurements of systems with and without anti-mist additive SLAM2000.

We have shown that with the SLAM2000 anti-mist additive we can tackle one of the most challenging processing problems of the silicone coating industry nowadays. With the increasing demands of this industry, new challenges and interesting solutions will surely be revealed in the future.

Silicone Copolymers for Coatings, Cosmetics, and Textile and Fabric Care

Kurt Stark*

Wacker Polymer Systems GmbH & Co. KG
Johannes-Hess-Strasse 24, D-84489, Burghausen, Germany
Tel.: +49 8677 83 7945 — Fax: +49 8677 886 7945
E-mail: kurt.stark@wacker.com

Keywords: silicone macromers, polymerization, polyvinyl acetate, polyvinyl alcohol, polyvinyl acetal

Summary: Wacker Specialties succeeded in developing proprietary processes to manufacture silicone copolymers. The problems of combining silicone and vinyl polymers were overcome. So, for example, silicone-modified polyvinyl acetate, silicone-modified polyvinyl alcohol, and silicone-modified polyvinyl acetal were obtained on a laboratory or pilot plant scale. These products, having very interesting properties, could be advantageously used for various applications.

Copolymers of Vinyl Acetate and Silicone [1]

Wacker Specialties have succeeded in developing a proprietary process to manufacture a copolymer of vinyl acetate and silicone.

The copolymerization is performed in solution with radical-forming initiators at temperatures of about 70 °C. The silicone macromer contains polymerizable vinyl groups on both chain ends. Such silicone macromers are produced by Wacker GmbH and are available with different chain lengths. The copolymerization with vinyl acetate is very facile and take place statistically [2]; after polymerization, no double bonds can be found. To avoid crosslinking, the molecular weight has to be adjusted by the use of chain transfer agents or by the solvent and its concentration.

The choice of the solvent is important, too. By choosing an appropriate solvent or a special mixture of solvents, a transparent solid resin results. Thus, the phase separation between polyvinyl acetate part and the silicone part of the copolymer, two normally incompatible polymers, can be reduced or even be completely avoided.

Figure 1 schematically shows the reaction between vinyl acetate and vinyl-functionalized silicone macromer.

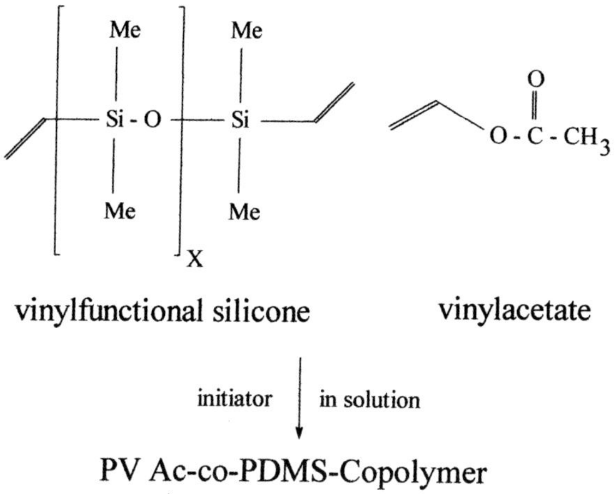

vinylfunctional silicone vinylacetate

initiator ↓ in solution

PV Ac-co-PDMS-Copolymer

Fig. 1. Reaction scheme to produce copolymers of vinyl acetate and vinyl-functionalized silicone macromers.

The silicone content of the polyvinyl acetate-silicone copolymers can be varied by up to 50% or more. The glass transition temperature T_g is usually in the range between 10 and 35 °C. Typical molecular weights M_w are between 10 000 and approximately 80 000 g/mol.

Other vinyl esters and other monomers (e.g., acrylates and methacrylates, or ethylene and vinyl silanes) can also be copolymerized with vinyl-functional silicones. The introduction of chemical functions (e.g., carboxyl groups) and charged groups is possible in order to further improve the properties.

Figure 2 shows the rheological behavior in the molten state of a typical copolymer of vinyl acetate and silicone macromer with 20 wt% silicone (oscillating rheological measurement at a frequency of 1 Hz with the Bohlin rheometer CVO 120 HR, measured with a plate/plate system). As can be seen, the loss modulus G'' is always higher than the storage modulus G' in the whole temperature range. The phase angle δ is always above 45°. This means that the copolymer exhibits marked thermoplastic properties (the phase angle δ is defined as: $\tan \delta = G''/G'$).

Wacker Specialties' copolymers of vinyl acetate and silicone have the following properties:

- consistency: hard/brittle to waxy material (function of T_g)
- transparent solid resin, no phase separation
- thermoplastic behavior, low melt viscosity
- resistance to blocking; release properties
- good binding power (→ PVAc content)
- emolliency and "good feeling" (→ silicone content)
- use as additive leads to very smooth surfaces with a very high slip effect
- readily soluble in organic solvents like methanol, THF, ethyl acetate, MEK, toluene, or a solvent mixture of 50% water, 50% ethanol
- temperature stability up to 200 °C.

With these properties, copolymers of vinyl acetate and silicone can be used for coatings, generally as additives (e.g., for hydrophobic modification, impact modification, flexibilization, lubrication, surface modification), in cosmetic formulations (e.g., hair styling agents like hair sprays or in skin care, imparting good feeling), or as a fabric care agent.

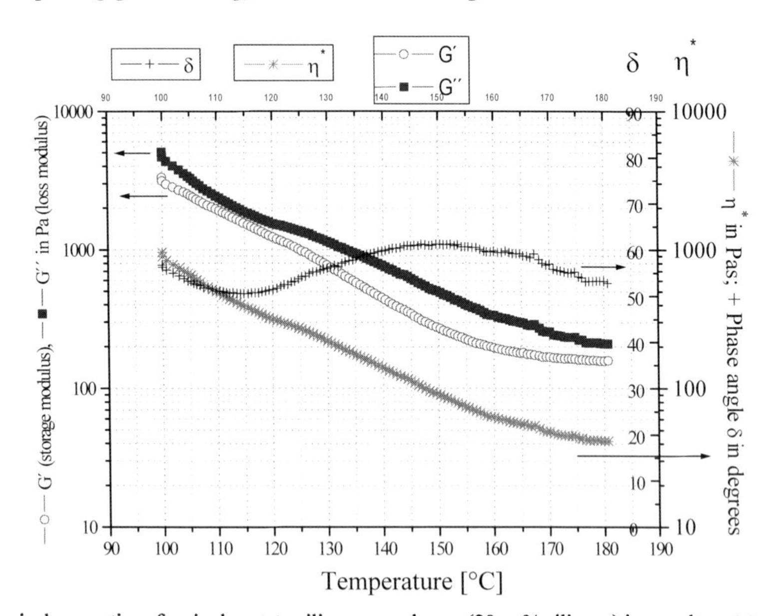

Fig. 2. Rheological properties of a vinyl acetate-silicone copolymer (20 wt% silicone) in a molten state.

Saponified Polyvinyl Acetate-Silicone Copolymers (Silicone-Modified Polyvinyl Alcohols) [1]

The saponification (transesterification) of a copolymer of vinyl acetate and silicone is performed in an alcoholic solvent (like methanol) with, for example, sodium hydroxide as catalyst. This leads to a "silicone-modified" polyvinyl alcohol. Depending on the saponification conditions, a new copolymer of vinyl alcohol and silicone ($z = 0$) or a new terpolymer of vinyl alcohol, vinyl acetate, and silicone ($z > 0$) is obtained. A schematic structural formula is given in Fig. 3.

Typical compositions are in the range PVAc (Z): 0 – 30 wt%; PVOH (Y): 50 – 95 wt%; silicone: 5 – 40 wt%.

The higher the silicone content and the lower the degree of saponification (z = high), the poorer the solubility in water. The saponification reaction with alkaline catalysts can break the silicone chain. Nevertheless, more than 50 wt% of silicone remains covalently bound to the polyvinyl alcohol part after saponification. With a special saponification process found at Wacker Specialties it is even possible to leave more than 60 wt% and up to 99 wt% of the total silicone content covalently bound to the polyvinyl alcohol part after saponification. The unbound silicone can be removed, if desired.

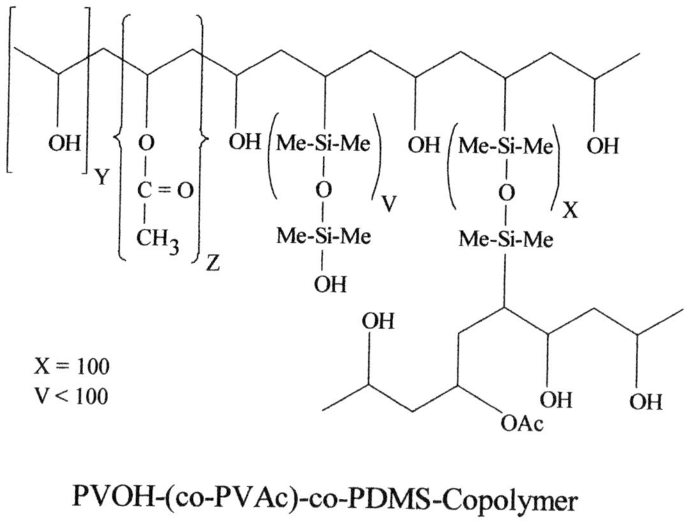

X = 100
V < 100

PVOH-(co-PVAc)-co-PDMS-Copolymer

Fig. 3. Schematic structural formula of silicone-modified polyvinyl alcohol.

Wacker Specialties' saponified polyvinyl acetate-silicone copolymers (silicone-modified polyvinyl alcohols) have the following properties:

- available as powder or aqueous solution; soluble or dispersible in water: a colloidal solution is obtained (more or less turbid, depending on silicone content; typical particle mean diameter: about 200 – 400 nm)
- static surface tension at saturation: about 35 mN/m
- good emulsifying properties, perhaps compatibilizing effect
- emolliency and creaminess on skin.

With these properties, saponified polyvinyl acetate-silicone copolymers can be used in cosmetic formulations (e.g., for hair and skin care) such as shampoos, lotions, and cremes or as a fabric care agent.

Silicone-Co-Polyvinyl Acetals (Silicone-Modified Polyvinyl Acetals) [3]

Wacker Specialties have also succeeded in the synthesis of new copolymers on the polyvinyl acetal basis containing long silicone chains. So, on the laboratory scale, polyvinyl acetal-co-polydimethylsiloxane-copolymers are obtained.

Starting from a copolymer of vinyl acetate and silicone made by radical polymerization in solution, the saponification leads to a "silicone-modified" polyvinyl alcohol. This modified polyvinyl alcohol is dissolved in water and then acetalized with aldehydes (butyr- or acetaldehyde) under acidic conditions (e.g., with hydrochloric acid as catalyst). This gives a silicone-modified polyvinyl acetal.

The silicone chains are bound covalently to the polyvinyl acetal chain, and there is only a very low amount of free silicone. In this way, all disadvantages resulting from free silicone, like phase separation/inhomogenity or migration, are eliminated.

A schematic structural formula is given in Fig. 4.

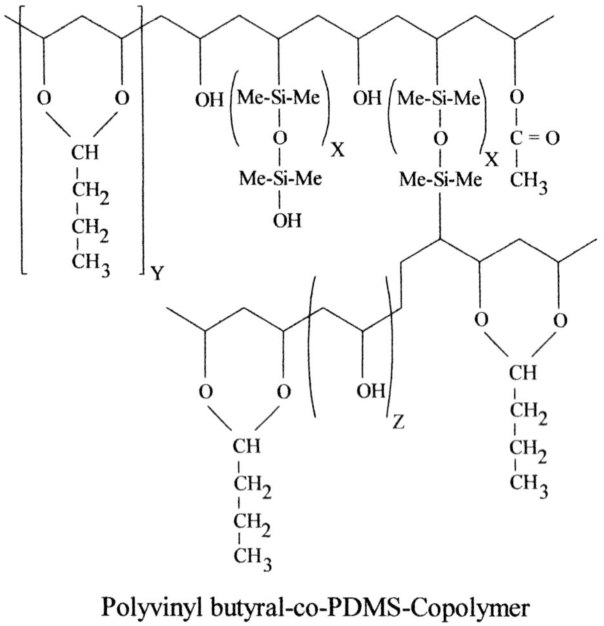

Polyvinyl butyral-co-PDMS-Copolymer

Fig. 4. Schematic structural formula of silicone-modified polyvinyl butyral.

Typical compositions of polyvinyl acetal-co-PDMS-copolymers are in the range:

Acetate groups:	1 – 25 wt%
Acetal groups (Y):	45 – 60 wt%
OH groups (Z):	10 – 20 wt%
Polydimethylsiloxane (PDMS):	5 – 35 wt% (6 – 40 mol%).

Wacker Specialties' silicone-modified polyvinyl acetals have the following properties:

- white powder (when dried immediately)
- glass transition temperature (T_g):
  - polyvinyl butyral: 46 – 52 °C
  - mixed polyvinyl acetal: 63 – 65 °C

- readily soluble in organic solvents like ethanol: a colloidal solution is obtained (more or less turbid, depending on silicone content; typical particle mean diameter: 100 – 500 nm)
- no phase separation in solution
- "amphiphilic behaviour" in solution due to different solubilities of silicone and organic polymer → compatibilizing effect
- viscosity (10% in ethanol) extremely low: < 10 mPas
- good adhesion to substrates due to polyvinyl acetal part
- excellent pigment compatibility and pigment dispersion in printing inks
- excellent rheological behavior in printing inks
- water repellency when used as coating
- excellent release properties when used as coating
- very smooth surfaces with an extremely high slip effect when used as coating
- low migration, no bleeding out of silicone.

With these properties, silicone-modified polyvinyl acetals could be used as

- binder
- primer
- coating (release, water repellent, polish, lacquers, powder coating)
- printing inks
- sheets for safety glass laminates.

References

[1] K. Stark, R. Singer (Wacker Polymer Systems), DE-A 10 215 962, **2002**.
[2] Y. Tezuka, A. Fukushima, K. Imai, *Makromol. Chem.* **1985**, *186*, 685.
[3] K. Stark (Wacker Polymer Systems), DE-A 10 338 479, **2003**.

Silicone-Based Copolymers for Textile Finishing Purposes — General Structure Concepts, Application Aspects, and Behavior on Fiber Surfaces

*H. Lange, R. Wagner**

GE Bayer Silicones
Building R20, 51368 Leverkusen, Germany
Tel.: +49 214 30 67715
E-mail: roland.wagner@gesm.ge.com

A. Hesse, H. Thoss, H. Höcker

Deutsches Wollforschungsinstitut an der RWTH Aachen e.V.
Veltmanplatz 8, 52062 Aachen, Germany

Keywords: softener, textile, durability

Summary: A structure concept for durable hydrophilic silicone softeners is outlined. ABC-triblock copolymers forming polyloop polyquaternary structures were found to be very substantive on cotton fibers. A study on the adsorption kinetics suggests a two-step process: (1) rapid coverage of the fiber surface, and (2) slow diffusion/spreading into the interior of the fiber. Washing experiments and subsequent evaluations of the soft hand prove the durability of the finish. XPS data indicate that after five washing cycles 60 to 70% of the new durable hydrophilic softeners remain on the fibers. For nondurable hydrophilic softeners a maximum substantive portion ranging from 20 to 40% was determined.

Introduction

Aminosilicone softeners are widely used for textile finishing purposes. The superior soft hand of these silicone finishes is based on the unique flexibility of the Si–O–Si bond system. A major disadvantage of conventional aminosilicone finishes is the pronounced hydrophobic character. The textile's ability to adsorb water and sweat is considerably reduced.

There is a growing demand for high end finishes combining the typical silicone soft hand and the hydrophilic character of cotton fibers. Silicone-aminopolyether block copolymers and comb-like polyether modified aminosilicones have been developed for these purposes. However, it was found that this hydrophilic derivatization of aminosilicones causes a significant substantivity reduction.

Therefore, commercially available hydrophilic silicone finishes are nondurable. The hand changes considerably after one washing cycle, and the typical silky silicone character of the textile is lost. The additional use of ester quat-based rinse additives does not solve this problem. Ester quats provide a nondurable hydrophobic character.

In order to overcome this uncomfortable situation, strongly substantive hydrophilic silicones have to be developed.

Chemical Concept

Quaternary ammonium moieties are known to be more substantive than aminostructures. Therefore, silicones containing a sufficient number of quat moieties should be promising candidates for durable softeners.

Silicone quats with an increasing number of quat moieties have been synthesized, formulated, applied on knitted cotton, and tested in washing experiments.

Fig. 1. Silicone monoquats, internal and terminal substitution.

Initially, silicones bearing a single internal or terminal quat moiety were evaluated (Fig. 1) [1]. It was found that these materials do not provide a durable hydrophilic silicone soft hand.

Fig. 2. Silicone diquats, internal and terminal substitution.

The same is true for silicone-based diquats (Fig. 2). Neither internal nor terminal derivatives showed an acceptable substantivity in presence of anionic detergents.

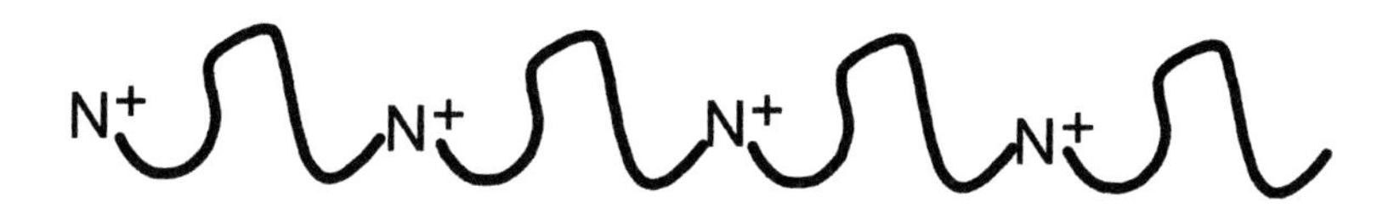

Fig. 3. Silicone-based polyquaternized polyloop structures; ABC-triblock copolymers.

A further increase in the number of substantive groups within one molecule yields polyquaternized polyloop structures (Fig. 3).

Certain silicone-based ABC-triblock copolymers have been identified which provide an excellent soft hand, keep the textiles hydrophilic, and are sufficiently substantive in the presence of anionic washing formulations. The soft hand is noticeable by the customer over repeated washing cycles.

Adsorption Kinetics

An in-depth understanding of the adsorption mechanism is the key for the optimization of silicone softener structures, the subsequent softener formulation, and the adjustment of process parameters.

XPS has been used to analyze the time dependence of the fiber surface coverage. It was found that a maximum surface coverage is reached almost immediately after the start of the finishing procedure (Scheme 1) [2]. The surface silicon level reached after 30 s remains constant over the whole adsorption time.

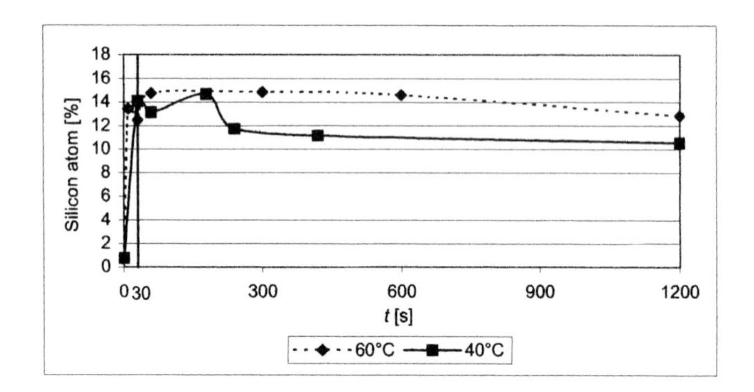

Scheme 1. Silicon content on a cotton fiber; treatment with silicone softener at 40 and 60 °C, 0 to 1200 s, XPS measurement.

Fluorescently labeled quat materials have been used to determine the total silicone uptake of the fibers indirectly. The decline of the fluorescence signal of the finishing liquor is equivalent to the amount of silicone adsorbed on the fiber surface and diffused into the interior of the fiber (Scheme 2).

The kinetic data in Scheme 2 indicate that surface coverage and diffusion into the fiber interior are complete after approx. 300 s. Obviously, the diffusion proceeds significantly more slowly than the fiber surface coverage.

Based on these kinetic data a two-step process can be proposed: rapid maximum surface coverage and subsequent slow diffusion into the fiber interior.

Fluorescence micrographs support this assumption. Initially, labeled polyquats are predominantly located in thin layers at the fiber surface (Fig 4. left). Substantial softener quantities have reached the inner fiber layers and the fiber core after 600 s (Fig 4. right). Here, the broad fluorescence ring structures indicate the diffusion into the interior of the fiber.

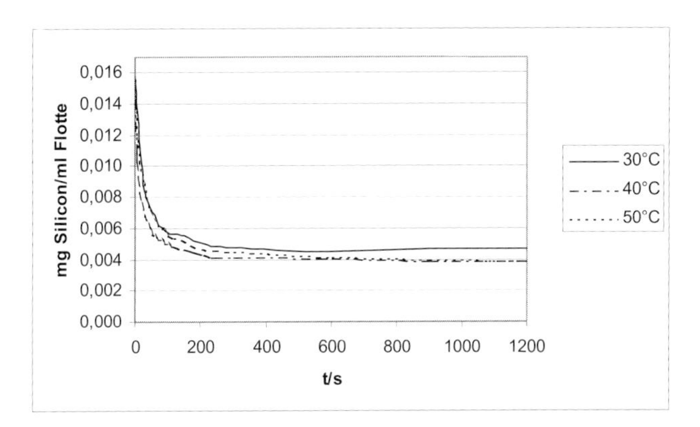

Scheme 2. Concentration of fluorescently labeled silicone softener in aqueous phase; treatment at 30, 40 and 60 °C, 0 to 1200 s.

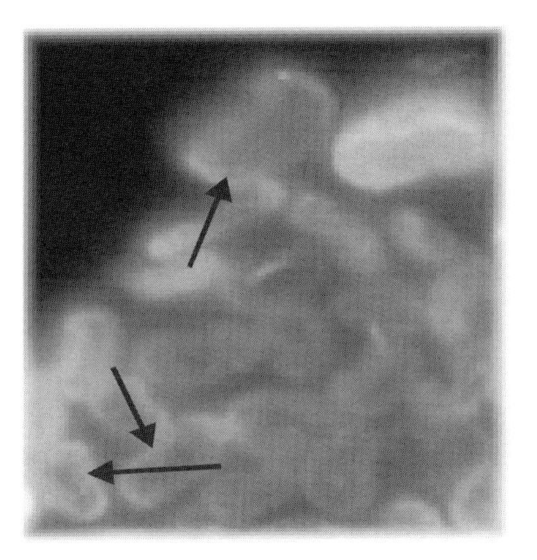

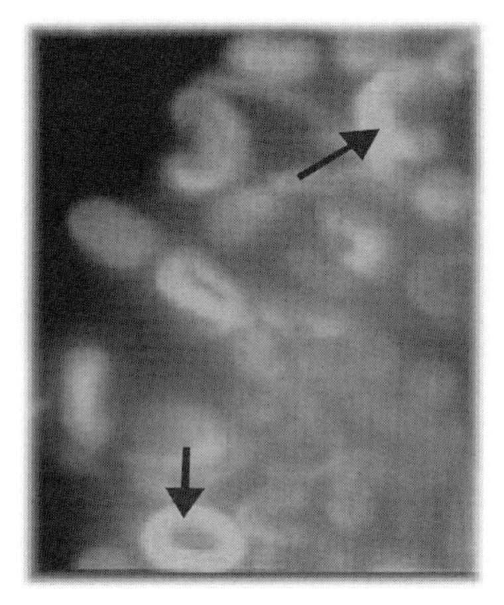

Fig. 4. Cross section of cotton fibers treated with fluorescently labeled silicone softener; left: after 30 s treatment, right: after 600 s treatment, 40x magnification.

Durability and Soft Hand

The durability of the adsorbed polyquat softener layers is the necessary precondition for a permanent silicone soft hand.

This durability of the new silicone softeners on cotton fibers has been quantified by means of XPS immediately after the finishing process and after repeated washing cycles in the presence of anionic detergents (Scheme 3).

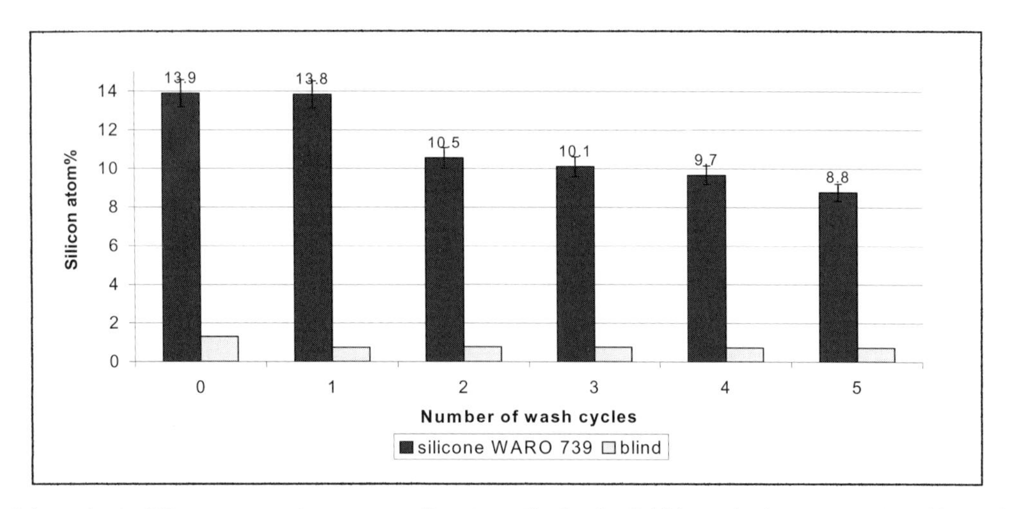

Scheme 3. Silicon concentration on cotton fibers immediately after finishing and subsequent 1 – 5 washing cycles, 40 °C, anionic detergent, XPS measurement.

The data for WARO 739 describe the typical behavior of a hydrophilic softener which gives the desired soft hand after five washing cycles. The initial silicon amount is reduced to 63% after five washing cycles [3].

The XPS analysis of hydrophilic softeners with insufficient soft hand after five washing cycles leads to the conclusion that in those cases not more than 20 – 40% of the materials remained on the fibers. The soft hand of cotton textiles after several washing cycles is clearly a function of the substantive softener portion.

Extended evaluations proved the superior soft hand of textiles finished with the new materials over commercially available nondurable hydrophilic softeners (Table 1).

Table 1. Knitted cotton, soft hand evaluation after five washing cycles at 40 °C, anionic detergent, 9 test persons (TP), WARO 720 and 739 durable hydrophilic softeners, commercial benchmark nondurable hydrophilic softener of polyether modified aminosiloxane type.

	TP1	TP2	TP3	TP4	TP5	TP6	TP7	TP8	TP9	Ø
WARO 720	1	1	1	1	1	2	2	2	1	1.33
WARO 739	2	2	2	2	2	1	1	1	2	1.66
commercial benchmark	3	3	3	3	3	3	3	3	3	3.00

Fate of Silicone Quats and Ecological Considerations

Finishing process parameters and XPS data can be used to quantify the fate of nondurable and durable hydrophilic softeners (Table 2) and the ecological consequences.

Table 2. Calculated silicone quantities on finished cotton and after up to five washing cycles (from XPS data); washes at 40 °C, anionic detergent, ester quat additions calculated according to supplier recommendations.

	Silicone after finishing (g/kg textile)	Silicone after 5 washes durable (g/kg textile)	Silicone after 5 washes drain (g/kg textile)	Ester quat add. over 5 washes drain (g/kg textile)
WARO 393 (nondurable)	2.19	0.88	1.31	17.5
WARO 739 (durable)	2.19	1.36	0.83	0

60% of the nondurable hydrophilic softener WARO 393 is removed from the fiber surface during five washing cycles. Because of the resulting poor soft hand, standard rinse additives containing ester quats (5 × 3.5 g) have to be added. These ester quats are nondurable.

Following the same routine, 38% of the durable hydrophilic softener WARO 739 is removed over five washing cycles. The hydrophilic silicone soft hand is maintained over all cycles despite the silicone loss. An addition of ester quat-based rinse additives is not necessary.

Two immediate conclusions can be drawn from the data in Table 2. The application of durable hydrophilic softeners reduces the silicone concentration in wastewaters considerably. Further, a broad implementation of durable hydrophilic silicone finishes makes the use of ecologically critical rinse additives largely redundant.

Acknowledgment: The project "Polysiloxane-Based Durable Hydrophilic Softeners" was supported by the Bundesministerium für Bildung und Forschung (project no. 0339779/0).

References

[1] R. Wagner, H. Lange, A. Hesse, H. Höcker, *Melliand Textilberichte* **2002**, 469.
[2] H. Thoss, A. Hesse, H. Höcker, R. Wagner, H. Lange, *Melliand Textilberichte* **2003**, 314.
[3] A. Hesse, H. Höcker, R. Wagner, H. Lange, *Melliand Textilberichte* **2003**, 547.

Silanes as Efficient Additives for Resins

H. W. G. van Herwijnen, S. Kowatsch, R. A. Wagner**

Dynea Austria GmbH, Hafenstraße 77, A-3500, Krems, Austria
Tel.: +43 2732 8990 — Fax: +43 2732 899 1329
E-mail: Hendrikus.van.Herwijnen@dynea.com, Robert.Wagner@dynea.com

Keywords: resin, insulation, impregnation

Summary: Silanes are used as additives in numerous examples of resin chemistry. From these progresses in two applications, mineral wool production and paper impregnation, are highlighted. In the first application, silane acts as a typical coupling agent to improve the binding between resin and fiber, whereas in the latter, silane functions as a modifier to improve surface properties of the final laminate.

Mineral Wool Application

Introduction

Binders are at the heart of mineral and glass wool [1]. The best binders are based on resole resins prepared by the polycondensation reaction of phenol with formaldehyde in the presence of a basic catalyst [2]. To such resoles, urea is added to improve fire resistance and scavenge free formaldehyde.

In a typical mineral wool factory, stone (or glass) is melted to produce mineral fibers, which are collected by depositing on a moving conveyer belt. The aqueous binder is sprayed onto these fibers as they drop onto this collecting conveyer belt. The layer of fiber with binder is then compressed and shaped into the form of the desired insulation and passed through an oven. In here, the wool is heated until the resin is completely cured, the end product being insulation mats of inorganic fibers bonded together in an organic matrix.

However, when such insulation mats are exposed to moisture, the bonding strength is reduced, which can affect the quality of the insulation material. To prevent this problem, on the one hand, a hydrophobic agent is added to make the mineral wool more water repellent, and on the other hand silane is included in the binder mixture to enhance the bond between the inorganic fibers and the organic binder.

The silane most commonly used for this purpose is γ-aminopropyl triethoxy silane. However, also other aminosilanes and glycidyl trialkoxy silanes may also be applied [3].

The amount of silane can be as little as 0.001 wt% or as much as 2.0 wt% [3] based on the weight of the resin solids. When employing more than 2.0 wt%, problems with precipitation of silane may start to occur. According to our experience, the amount of silane that preferably should be employed in practice is 0.2 wt%. More silane might still improve the product, but as silane is relatively expensive, the extra costs do not justify the additional gain in strength.

Laboratory Tests

It is very hard to simulate the production of mineral wool in the laboratory since this involves the melting of stone or glass. Nevertheless, a method, the so-called "dog bone" test, does exist. One should realize, however, that this model only gives indicative values and that it is always necessary to do final tests on-line in production in order to obtain reliable values [4].

In the test, sticks called "dog bones" are made of sand bound together with resin. Then, the flexural strength to break these sticks is measured. In order to determine the wet strength, the sticks are boiled twice in water with subsequent drying before being broken.

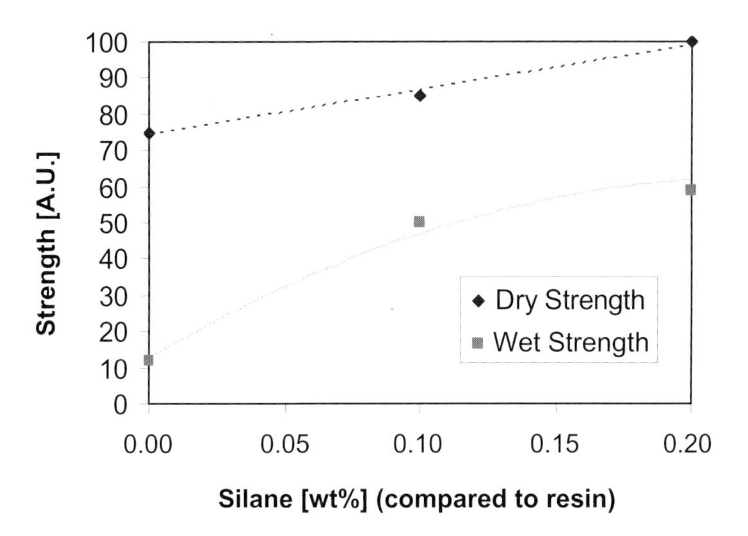

Fig. 1. Influence of silane bonded with Dynea resin Prefere 72 5526M.

Figure 1 shows the dry and wet strength of sticks bonded with the Dynea insulation resin "Prefere 72 5526M" with different amounts of γ-aminopropyl triethoxy silane. The amount of silane (wt%) is expressed as a percentage of the solid content of the resin. It can be clearly seen that silane greatly enhances the wet strength, whereas the influence on the dry strength remains small.

Chemistry

The supposed chemical reactions responsible for the function of γ-aminopropyl triethoxy silane are depicted in Scheme 1. Silanes can be hydrolyzed in the presence of water under basic or acidic conditions. Aminosilanes, however, do not require pH adjustments. The basic amino group acts as a catalyst for hydrolysis, and the resulting aminosilanol is stable (1A). Then, the silanol bonds condense with reactive oxide or hydroxide sites of the mineral fiber, e.g., with silanol surface groups. First, hydrogen bonds between the silanol groups are formed, and then reaction can occur with one or more surface silanol groups resulting in covalent binding (1B). Remaining silanol groups may connect to form polysiloxanes (1C). Further, the reactive amino groups react with the methylol groups of the resin (1D). Upon curing, the network between organic binder and mineral fiber is completed (1E).

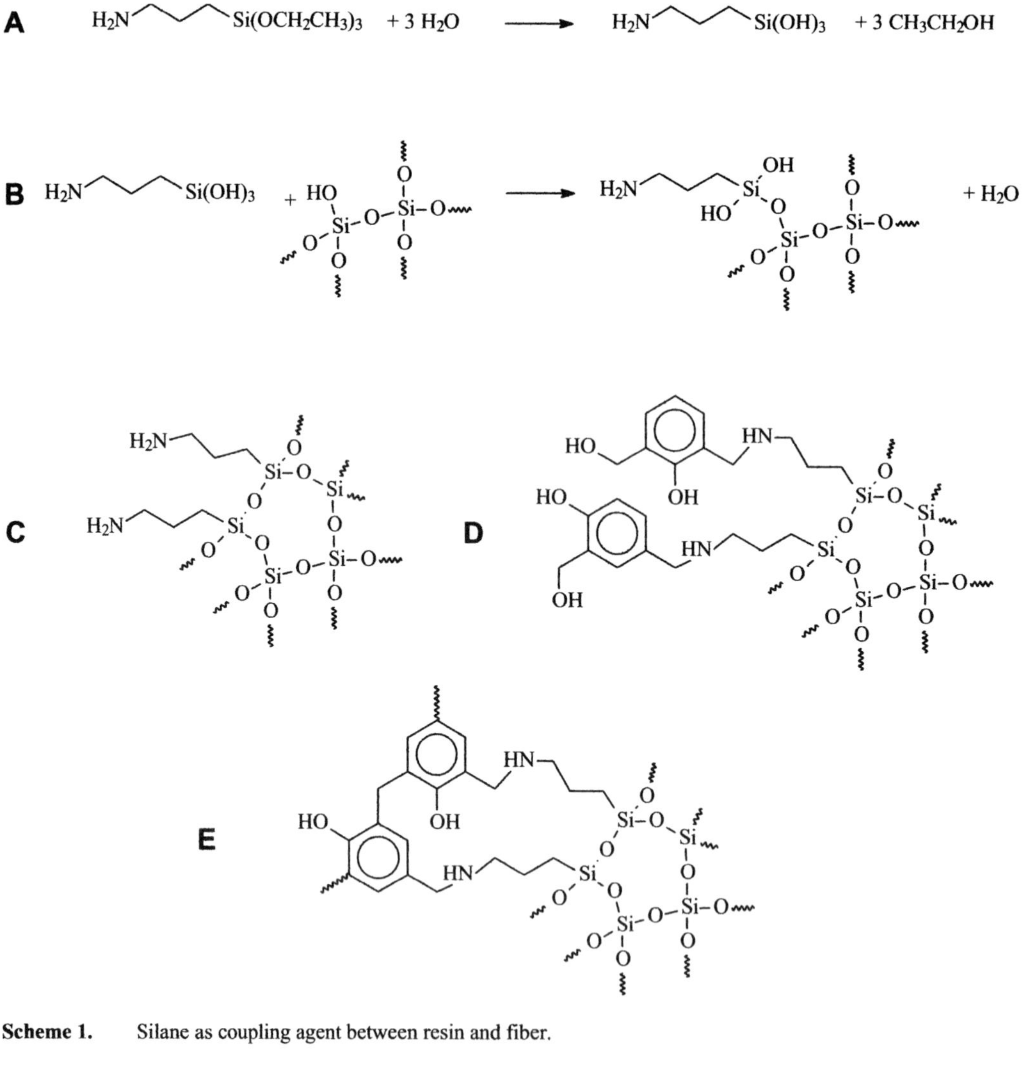

A H_2N–(CH$_2$)$_3$–$Si(OCH_2CH_3)_3$ + 3 H_2O ⟶ H_2N–(CH$_2$)$_3$–$Si(OH)_3$ + 3 CH_3CH_2OH

B H_2N–(CH$_2$)$_3$–$Si(OH)_3$ + HO–Si(surface) ⟶ H_2N–(CH$_2$)$_3$–Si(OH)$_2$–O–Si(surface) + H_2O

C

D

E

Scheme 1. Silane as coupling agent between resin and fiber.

As mentioned before, after the binder (including the silane) is sprayed onto the fibers, the wool travels through a collection chamber toward the curing oven. Silane diffuses from the resin toward the resin/fiber interface [5]. The best result is obtained when all the silane has the time and opportunity to reach the mineral fiber surface before the resin cures. Silane that remains "captured" within the cured resin matrix cannot act as a coupling agent and, therefore, does not improve the final product.

In order to get all silane molecules onto the surface, it is possible to treat the fibers with the coupling agent before the binder is supplied [6]. In this case, the silane is added to the stream of air advancing along the peripheral surface of the spinning wheel. In this manner, the coupling agent is contacted with the fibers during or shortly after their formation. The binder can be supplied afterwards in a conventional matter. Thus, better wet strengths can be obtained using the same amount of silane, the reason being that more silane is present on the interface between fiber and resin.

Paper Impregnation Application

Introduction

The modern household furnishing, flooring, automotive, and electronic industries are hardly conceivable without impregnation resins [7].

These low-molecular-weight, thermosettable polycondensation products of formaldehyde with urea, melamine, or mixtures of both belong to the family of aminoplasts [8]. All of them give duroplastic films after final processing [9]. Products of this industry are omnipresent in almost every private and public building [10].

Melamine laminates distinguish themselves from other surface materials such as coatings and thermoplastic foils by the unique combination of surface hardness, colorability, designs, low price, and resistance against various chemicals. Melamine-formaldehyde (MF) resins are essential parts of these laminates.

In the last few years the *intrinsic* resistance of melamine films against water or other sources of soiling (e.g., acrylic sprays) was found not to fulfill the requirements for some applications.

Resin Production and Lamination Process

Since the early days of MF condensation chemistry [11], the way of synthesis has been and is still actively researched by industrial and academic groups [12]. In a typical synthesis, a conventional MF paper impregnation resin is obtained in a polycondensation reaction of melamine with formaldehyde under basic catalysis and heat. In an initial methylolation or hydroxymethylation step, mainly three different species are generated: monomethylolmelamine (mmm), dimethylolmelamine (dmm) and trimethylolmelamine (tmm).

After about 20 min, the formation of di- and trimers begins to yield the so-called "A stage" resin in which melamine is linked via methyleneether and methylene bridges (see Scheme 2). This A stage resin is then impregnated into décor paper by means of an impregnation line.

To reach this goal, the resin has to be formulated with additives in order to achieve the desired properties of the impregnate. Moreover, a catalyst has to be added to the dip to enable a further condensation, which is accelerated by heat. In this step the resin is transformed into the so-called "B stage", in which low-molecular-weight oligomers are formed.

In a further step toward the well-known products, impregnated paper is laminated (pressed) onto different substrates (e.g., plywood, MDF, or laminates of PF-impregnated Kraft paper). Regarding the molecular structure of an MF resin during that process, a change from B stage to "C stage" (molecules of high to infinite molecular weight; crosslinked, insoluble, infusible) takes place. At the end of the press process, 90 – 95% of the molecules are crosslinked to give a three-dimensional network with infinite molecular weight.

dimethylolmelamine = dmm + monomethylolmelamine = mmm → ($-H_2O$) methyleneether bridged dimelamine

mmm + m → ($-H_2O$) methylene bridged dimelamine

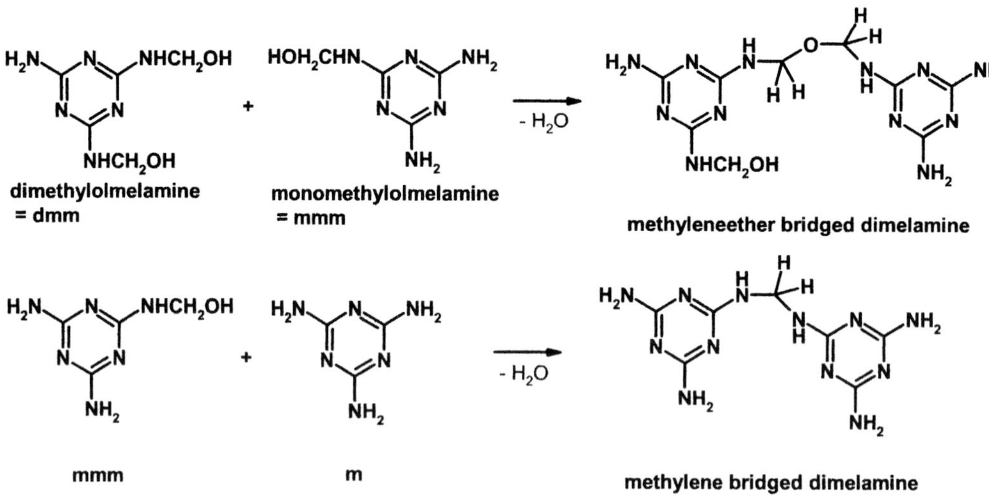

Scheme 2. Formation of A-stage resin.

Role of Silanes

In order achieve the demands of expanded applications, new additives for surface modifications were developed. These additives are often fluoro compounds [13], sometimes fluoroalkylsilanes [14], and in many cases alkylsilanes. When these additives are incorporated into a polymer matrix, the extremely valuable properties of these chemicals like chemical inertness, thermo-oxidative stability, and resistance against water can be transferred to the whole polymer system. This incorporation can either be achieved by chemical bonding to the resin matrix or by formation of an interpenetrating polymer network.

Chemistry [15]

In many cases different trialkoxy silanes, as depicted in the assumed chemical reaction (Scheme 3), are applied. These compounds hydrolyze stepwise in water under both basic and acidic catalysis to give the corresponding silanetriols.

$$R{-}Si(OAlk)_3 \xrightarrow{H_2O} R{-}Si(OH)_3 + 3\ AlkOH$$

$R = C_6F_{13}, C_4H_9, C_8H_{17}, C_{12}H_{17}F_9$

Alk = Me, Et

Scheme 3. Hydrolysis of silanes.

Under optimized conditions, the hydrolysis is very fast, while the undesired condensation reaction to siloxanes is much slower.

After hydrolysis, the silanol groups are able to condense with the reactive methylol groups in hydroxymethylated di- and trimelamine as well as with unbridged methylolated melamine species (Scheme 4) to yield the rigid and infusible three-dimensional networks between resin and silane after final curing. Of course, reactions between silanol groups and hydroxyl groups of the cellulose in the paper would seem to be possible, but this has never been proven without doubt.

Less important is the condensation of silanols with free amine groups in different melamine species. This is unlikely in terms of bond strength considerations of Si–O vs Si–N. However, an undesired side reaction like the self-condensation of silanols to polysiloxanes can become important if the amount of silane in the dip exceeds 2 wt%.

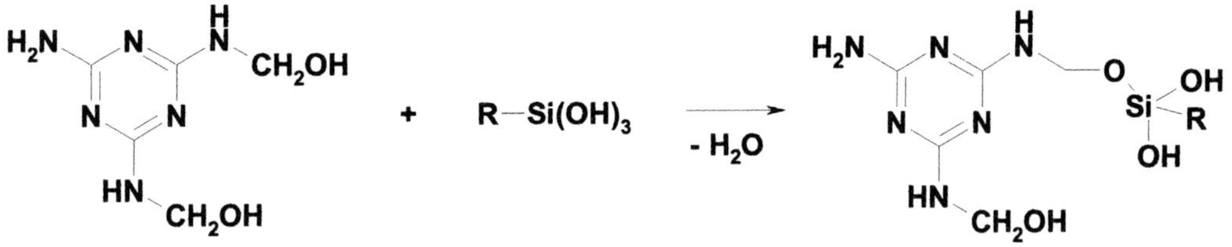

$$\text{dmm} + R{-}Si(OH)_3 \xrightarrow[-\,H_2O]{} \text{dmm–}CH_2{-}O{-}Si(OH)_2R$$

Scheme 4. Reaction of dmm with silanol.

Laboratory and Application Tests

First, the stability of the formulation containing resin, additives, and catalysts is judged. Second, the reactivity of the dip has to be in a certain range. Third, on ready-made laminates (either hpl or plywood based) the surface tension of water droplets is determined. In this test it can easily be seen that paper with hydrophobic additives is more repellent toward liquids than ordinary paper (Fig. 2).

Fig. 2. Droplet on untreated and treated paper.

Clearly it can be seen that the repulsion properties are derived from lowering the critical surface energy of the hydrophobic surface below that of the wetting liquid, thus creating a chemical barrier against the penetration by the liquid.

Acknowledgment: The authors thank Peter Czerny, Martin Emsenhuber, Clemens Gartner, Michael Meneder, Karin Peterschofsky, and Kristina Steiner (Dynea Austria GmbH), Alexander Tseitlin (Dynea Canada Ltd.), and Andreas Zaumseil (Dynea Erkner GmbH) for practical work and helpful discussions.

References

[1] R. Bruyea, *Dynea Technology in Focus*, April **2002**, 10.

[2] A. Knop, L. A. Pilato, *Phenolic Resins*, Springer Verlag, Berlin, **1985**.

[3] Monsanto Company, GB 1,063,654, **1967**.

[4] This test is also in use for testing resins for coated abrasives. In this case, the method corresponds well to reality.

[5] B. W. Lipinski, *Seifen-Öle-Fette-Wachse*, **1972**, *98*, 230.

[6] E. L. Hansen, (Rockwool International), EP 0 367 194, **1984**.

[7] H. Steindl, in *Proceedings of the European Laminates Conference, Supplement*, Vienna, **2002**.

[8] A. Einhorn, A. Hamburger, *Ber. Dtsch. chem. Ges.* **1908**, *41*, 24; *Liebigs Ann. Chem.* **1908**, *361*, 122; H. Staudinger, *Makromolekul. Chem.* **1953**, *11*, 81; G. Zigeuner, *Monatsh.* **1951**, *82*, 847; **1955**, *86*, 517, 523.

[9] J. Binder, K. Lepedat, R. A. Wagner, in *Proceedings of the Asia Pacific Laminates Workshop*, Shanghai, **2003**; J. Binder, R. A. Wagner, in *Proceedings of the European Students Conference on Physical Organic and Polymer Chemistry*, Vienna, **2003**.

[10] W. Puwein, in *Proceedings of the European Laminates Conference*, Vienna, **2002**; C. O'Carroll, in *Proceedings of the European Laminates Conference*, Vienna, **2002**.

[11] CIBA, US Patent 2 197 357, **1937**; O. Okano, *J. Am. Chem. Soc.* **1952**, *74*, 5728.

[12] R. A. Wagner, J. Binder, in *Proceedings of the Xth Austrian Chemical Days*, Linz, **2002**.

[13] A. Cigognini, in *Proceedings of the European Laminates Conference*, Vienna, **2002**.

[14] K. Weißenbach, B. Standke, in *Proceedings of the 1st European Silicon Days*, Munich, **2001**.

[15] E. P. Plueddemann, *Silane Coupling Agents*, Plenum Press, New York, **1991**.

New UV-Curable Alkoxysiloxanes Modified with Tris(trimethylsilyl)methyl Derivatives

Anna Kowalewska, Włodzimierz A. Stańczyk*

Center of Molecular and Macromolecular Studies, Polish Academy of Sciences
Sienkiewicza 112, 90-363, Lodz, Poland
Tel.: +48 42 6818952 ext 203 — Fax: +48 42 6847126
E-mail: anko@cbmm.lodz.pl

Keywords: tris(trimethylsilyl)methyl, UV curing, alkoxysilane

Summary: New tris(trimethylsilyl)methyl-modified siloxane systems bearing alkoxysilyl groups were prepared. Photo-acid-catalyzed hydrolysis and condensation of reactive ethoxysilyl moieties gave interesting novel materials which can be of value as organic/inorganic coatings or silica aerogels.

Introduction

Siloxane materials functionalized with reactive groups have been of long-standing interest for many years. The Si–O–Si skeleton provides excellent flexibility at low temperatures as well as low surface tension and high thermal stability, whereas side functional groups may serve as potential cross-linking centers [1]. Photo-initiated cationic polymerization carried out in the presence of aryl iodonium or sulfonium salts seems to be a versatile tool among other techniques of cross-linking [2]. It provides a fast and complete cure, as well as convenient *in situ* generation of the catalyst. Its insensitivity toward oxygen makes the process especially interesting for industrial application (release coatings, adhesives, photoresists, abrasion-resistant coatings, etc.). The mechanism of cationic photo-curing utilizing iodonium salts has been well established [3]. The initiating species are protons of strong acids, which are generated during the photolysis of onium salts.

All cationically polymerizable monomers can be potentially used in this process; however, the main study has been focused so far on the most reactive oxirane and vinyl ethers [4]. Alkoxysilane derivatives – the most common acid-sensitive monomers for the synthesis of siloxane materials through the use of sol-gel methods – were not used extensively. Only a few examples of their application in photo-activated cross-linking can be noted, mainly in *co*-reaction with oxirane sites [5]. Typically, alkoxysilanes are subjected to an acid- or base-catalyzed process involving hydrolysis of an ≡SiOR group and then condensation of the formed silanol with another molecule bearing an ≡SiOH or ≡SiOR function to give a siloxane linkage [6]. It was of interest to combine the properties of cross-linked silicone materials with the ones provided by sterically overloaded tris(trimethylsilyl)methane (Tsi) derivatives [7]. This report presents preliminary results on bifunctional Tsi-modified oligosiloxanes, cross-linked by means of hydrolysis and condensation of ethoxysilyl moieties in the presence of photo-acids. The materials may display improved adhesion of the coating to the support because of silanols (adhesion promoters) and increased thermal stability due to the presence of sterically shielding Tsi functions.

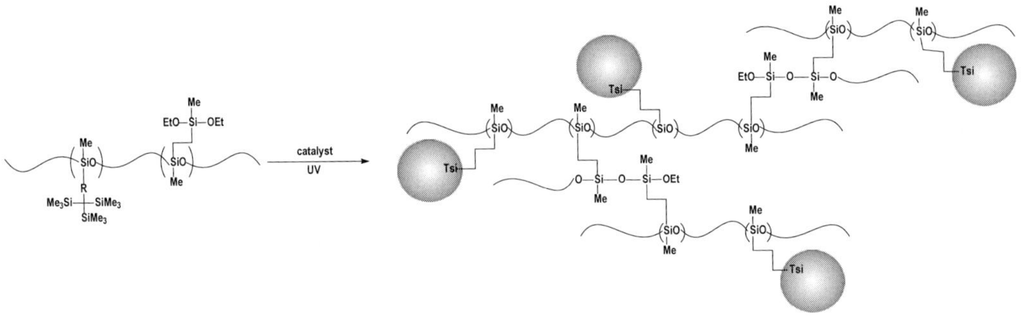

Scheme 1. Formation of Tsi-siloxane network by photo-activated, acid-catalyzed hydrolysis of alkoxysilyl groups.

Results and Discussion

The tris(trimethylsilyl)methyl modified oligosiloxanes bearing also reactive alkoxy side substituents can be made in a simple one-pot/two step hydrosilylation reaction procedure in the presence of a platinum divinyltetramethyldisiloxane complex.

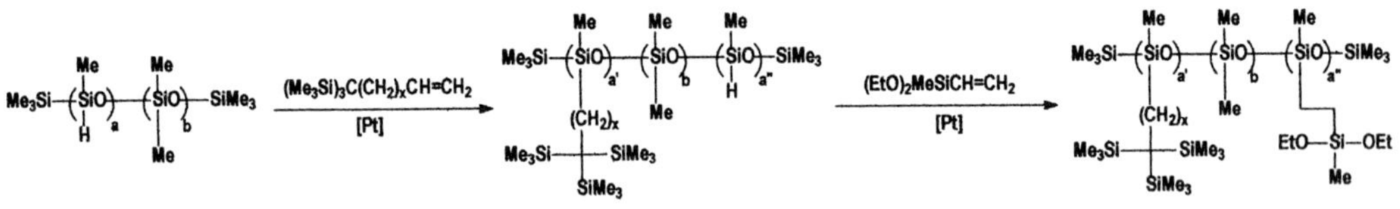

Scheme 2. Preparation of Tsi/alkoxysilane-functionalized oligosiloxanes.

The Tsi/EtO-functionalized polysiloxanes were cross-linked in the presence of trifluoromethanesulfonic acid generated in an UV-initiated decomposition of $Ph_2IOSO_2CF_3$. The process is based on acid-catalyzed hydrolysis of ≡SiOEt groups and formation of ≡SiOH functions and their subsequent condensation. We have found that the use of $Ph_2IOSO_2CF_3$ gave much better results than Ph_2IBF_4, widely used in photo-initiated polymerization of epoxides [8].

Traces of water from air are sufficient to promote the hydrolysis of ≡SiOEt groups in a thin polymer film and to give cross-linked material. Alkoxysilane groups hydrolyze rapidly under UV radiation with cationic photoinitiator, which was proved by ^{13}C NMR in the solid state (disappearance of resonances corresponding to OEt groups in cured materials). However, because of the low ratio of water to alkoxysilane, the condensation was relatively slow and might have not been complete. The acid present in the reaction medium may also catalyze reverse reactions of siloxane hydrolysis or ethanolysis. Therefore only some of the silanol groups were further reacted to form Si–O–Si linkages, and the cross-linked oligomers required post-UV cure baking.

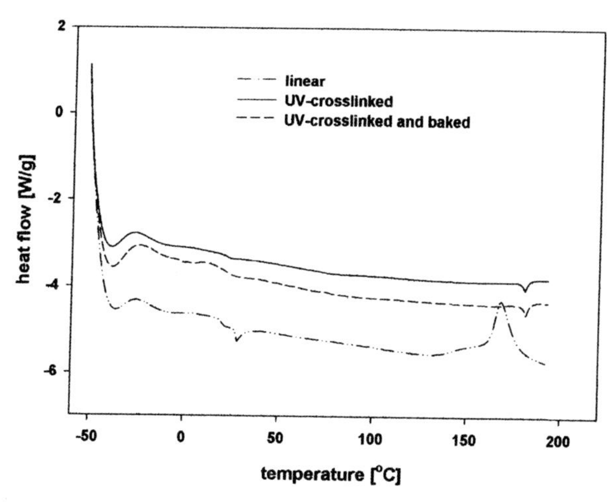

Fig. 1. Thermal condensation of silanols (DSC measurements: heating rate: 20 °C/min, temperature range: 1st heating –50 °C to 200 °C, 2nd heating –50 °C to 300 °C).

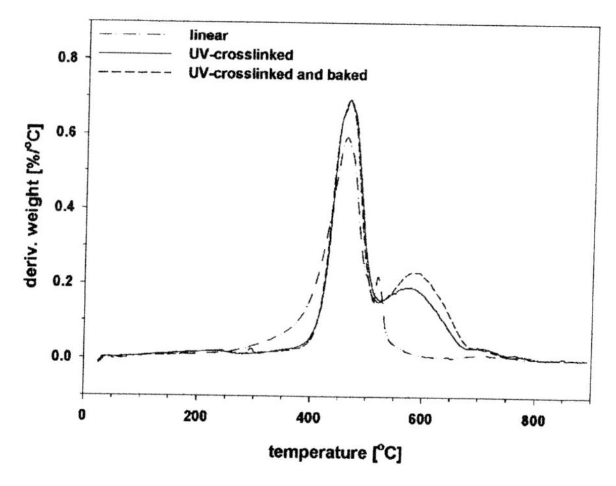

Fig. 2. Thermal decomposition of Tsi-siloxanes with various degrees of cross-linking (TGA measurements – heating rate: 20 °C/min).

Direct observation of the condensation of silanols was possible by the DSC method. DSC measurements performed for starting oligomers and the products at various stages of curing indicate thermal condensation of reactive ≡SiOH functional groups. The peaks that appeared at 181 °C with heating rate 20 °C/min correlate with temperatures of condensation and molecular weight increase observed for silanol-terminated PDMS heated under isothermal conditions [9]. The peak observed during the second heating run is substantially smaller. Still, even the cured material baked at 140 °C for 5 min displayed similar peaks, pointing to an incomplete condensation of silanol groups.

TGA analyses were performed for polymer samples having different degrees of cross-linking. The decomposition of the linear oligomer starts at about 200 °C. Once cured and baked, the formed siloxane network is more thermally stable, and the decomposition begins at temperatures higher by 100–150 °C. The results are similar to those reported for analogous Tsi-modified siloxanes cross-linked by means of photo-initiated cationic polymerization of epoxides [8].

The density of cross-linking can be adjusted by *co*-condensation of multialkoxy-silane monomers, such as $[Me(EtO)_2(CH_2)_2SiMe_2]O$ or $[Me(EtO)_2(CH_2)_2SiMe_2O]_4$. DSC and TGA analyses did not show much difference in the amount of unreacted ≡SiOH or reduction of thermal resistance.

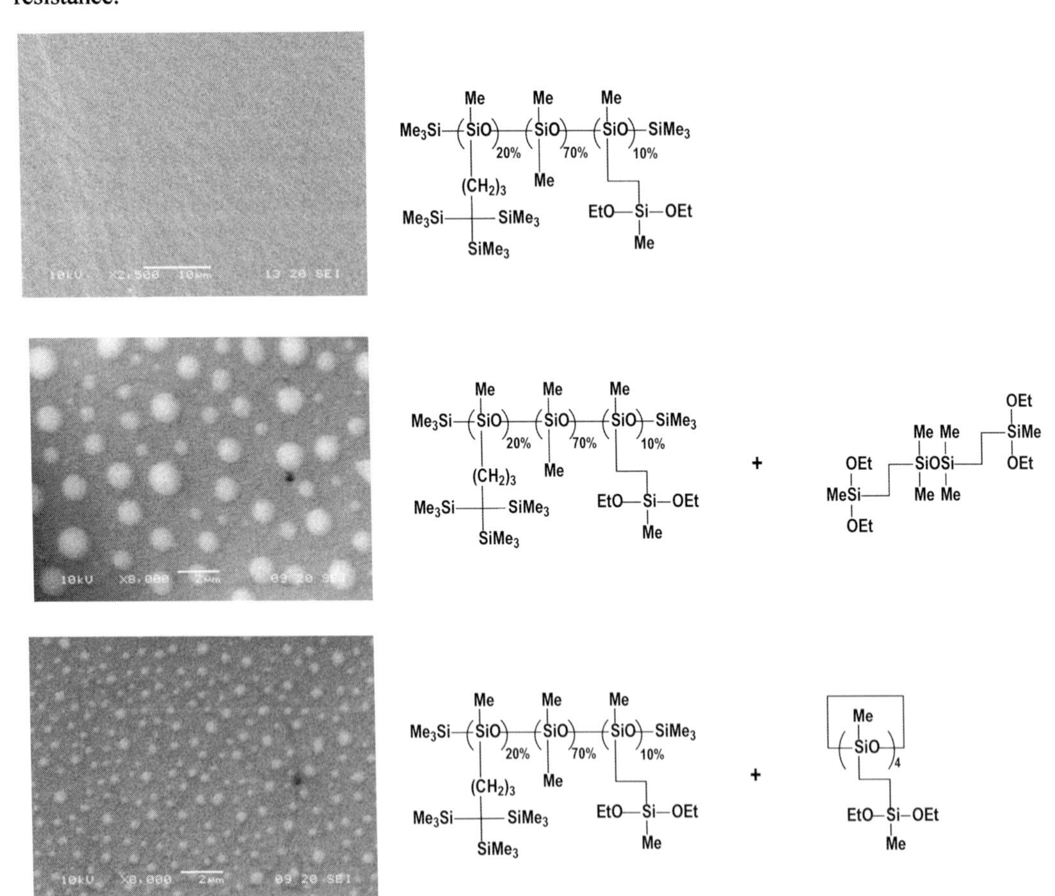

Scheme 3. SEM pictures of Tsi/alkoxysilane-modified oligosiloxanes cured in the presence of *co*-alkoxysilane monomers.

References

[1] T. C. Kendrick, B. Parbhoo, J. W. White, Siloxane polymers and copolymers in *The Chemistry of Organic Silicon Compounds* (Eds.: S. Patai, Z. Rappaport), Wiley, **1989**, p. 1347.

[2] L. Abdellah, B. Boutevin, B. Youssef, *Progress in Organic Coatings* **1994**, *23*, 201.

[3] J. V. Crivello, *J. Polym. Sci., Part A* **1999**, *37*, 4241 and references therein.

[4] J. V. Crivello, D. Bi, *J. Polym. Sci., Part A* **1993**, *31*, 3109;
J. V. Crivello, J. L. Lee, D. A. Conlon, *J. Radiat. Curing* **1983**, *1*, 6;
J. V. Crivello, G. Löhden, *Chem. Mater.* **1996**, *8*, 209.

[5] J. Chen, M. D. Soucek, *Eur. Polym. J.* **2003**, *39*, 505; S. Wu, M.T. Sears, M. D. Soucek, W. J. Simonsick, *Polymer* **1999**, *40*, 5675; J. V. Crivello, K. Y. Song, R. Ghoshal, *Chem. Mater.* **2001**, *13*, 1932; J. V. Crivello, Z. Mao, *Chem. Mater.* **1997**, *9*, 1554; J. V. Crivello, D. Bi, *J. Polym. Sci., Part A* **1993**, *31*, 3121.

[6] C. J. Brinker, G. W. Sherer, *Sol-Gel Science*, Academic Press, New York, **1990**.

[7] A. Kowalewska, W. A. Stańczyk, S. Boileau, L. Lestel, J. D. Smith, *Polymer*, **1999**, *40*, 813.

[8] A. Kowalewska, W. A. Stańczyk, *Chem. Mater.* **2003**, *15*, 2991.

[9] T. C. Kendrick, B. Parbhoo, J. W. White, Siloxane polymers and copolymers in *The Chemistry of Organic Silicon Compounds* (Eds.: S. Patai, Z. Rappaport), Wiley, **1989**, p. 1320.

New Methacrylic Silanes: Versatile Polymer Building Blocks and Surface Modifiers

Jürgen Pfeiffer

Wacker-Chemie GmbH, Johannes-Hess-Straße 24, 84480 Burghausen, Germany
Tel.: +49 8677 83 8277 — Fax: +49 8677 83 7202
E-mail: juergen.pfeiffer@wacker.com

Keywords: organofunctional silanes, methacryloyloxyalkyl silanes, hydrolysis, kinetics, NMR spectroscopy, copolymerization, coatings, scratch resistance

Summary: Methacryloyloxymethylsilanes have been obtained via a convenient synthetic procedure suitable for large-scale production in excellent yield via a three-step process starting from methylchlorosilanes. The thus obtained methacryloyloxymethylsilanes show exceptionally high reactivity with respect to hydrolysis and condensation reactions. In some cases, hydrolysis rates up to 40 times higher than in case of the parent 3-methacryloyloxypropyltrimethoxysilane were observed. In addition, it was found that the high reactivity is retained on changing from the methacrylic functionality-bearing silanes to saturated analogs, which is not the case if vinyltrimethoxy silane is compared with propyltrimethoxysilane. Copolymerization of the methacryloyloxymethylsilanes with common (meth)acrylic monomers leads to silyl-functional polymers. Coatings obtained after acid-catalyzed curing exhibit an enhanced scratch resistance compared to an analogous polymer containing 3-methacryloyloxypropyltrimethoxysilane, presumably because of an increased crosslinking density.

Introduction

Although unsaturated silanes like vinyl or 3-methacryloyloxypropyl alkoxysilanes have been established in various industrial applications for several decades now, there is still a demand for alternative substances. Special interest is focused on lower amounts of VOCs liberated during hydrolysis, adjustable reactivity, and availability. Manufacturing processes currently in use for the established compounds are based on hydrosilation and/or nucleophilic substitution from precursors like allylchloride or allyl methacrylate.

In order to solve the problems mentioned above and to widen the product range of unsaturated alkoxysilanes, various methacryloyloxymethylsilanes have been synthesized, and their reactivity profile has been investigated.

Synthesis of Methylene-Bridged Silanes

The methacryloyloxymethyl alkoxysilanes are easily available starting from methyl trichlorosilane, dimethyl dichlorosilane, or trimethyl chlorosilane (**1**), respectively (Scheme 1). Photochlorination of these precursor silanes almost exclusively leads to the chloromethyl silanes. Alkoxylation with methanol or ethanol leads to the chloromethyl(methyl)alkoxysilanes (**2**). Nucleophilic substitution under phase transfer catalysis conditions and distillative isolation finally yields the methacryloyloxymethylsilanes (**3**), making the production independent of expensive hydrosilation technology.

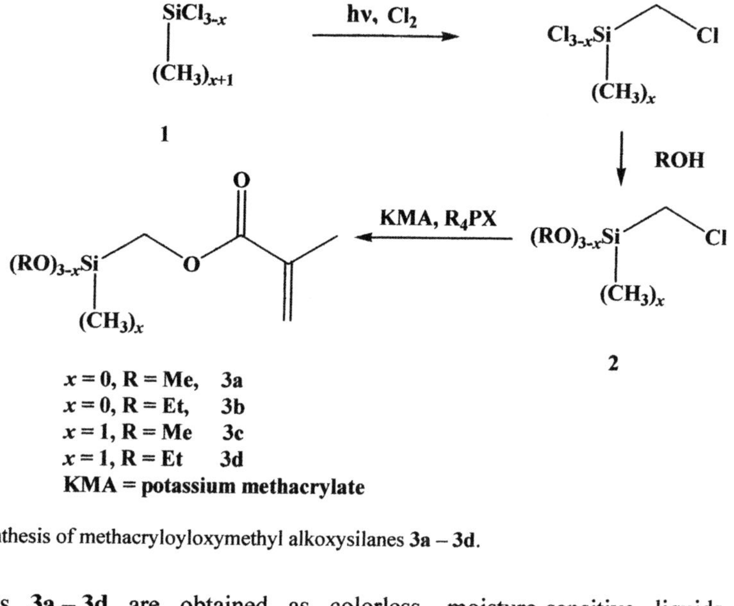

$SiCl_{3-x}(CH_3)_{x+1}$ (**1**) $\xrightarrow{h\nu,\ Cl_2}$ $Cl_{3-x}Si(CH_3)_x$–CH_2Cl $\xrightarrow{ROH}$ $(RO)_{3-x}Si(CH_3)_x$–CH_2Cl (**2**) $\xrightarrow{KMA,\ R_4PX}$ $(RO)_{3-x}Si(CH_3)_x$–CH_2–O–C(O)–C(CH_3)=CH_2

x = 0, R = Me, 3a
x = 0, R = Et, 3b
x = 1, R = Me 3c
x = 1, R = Et 3d
KMA = potassium methacrylate

Scheme 1. Synthesis of methacryloyloxymethyl alkoxysilanes **3a – 3d**.

The products **3a – 3d** are obtained as colorless, moisture-sensitive liquids. A radical polymerization inhibitor (e.g., HQME) is added and the products are stored under light exclusion in order to prevent undesired polymerization.

Hydrolysis and Condensation Reactivity

In order to elaborate the reactivity profile of compounds **3a – 3d**, the rate constants of the first hydrolysis step were determined by ^{1}H NMR spectroscopy (pH = 4 or 9, 30-fold molar excess H_2O). In addition to the methylene-bridged silanes **3a – 3d**, 3-methacryloyloxypropyl-trimethoxysilane **3e**, propyl trimethoxy silane **4a**, vinyl trimethoxy silane **4b**, as well as butyric acid (trimethoxysilyl)methyl ester **5** (as saturated analog for **3a**) shown in Scheme 2 were investigated for comparison.

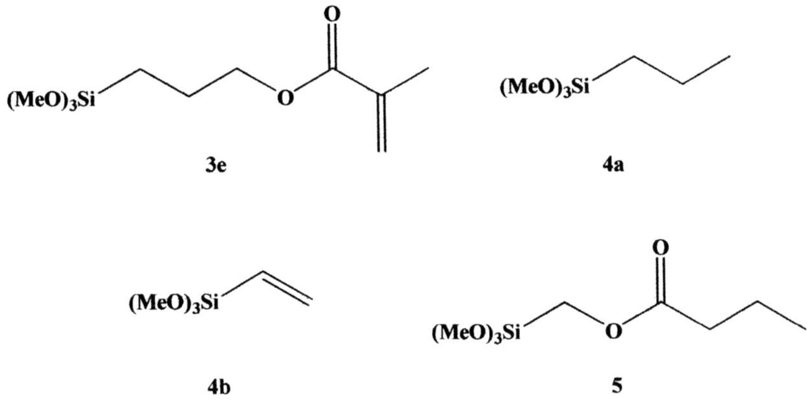

Scheme 2. Compounds **3e** – **5** subjected to hydrolysis for comparison.

The results of these experiments can be summarized as follows [1] (Fig. 1):

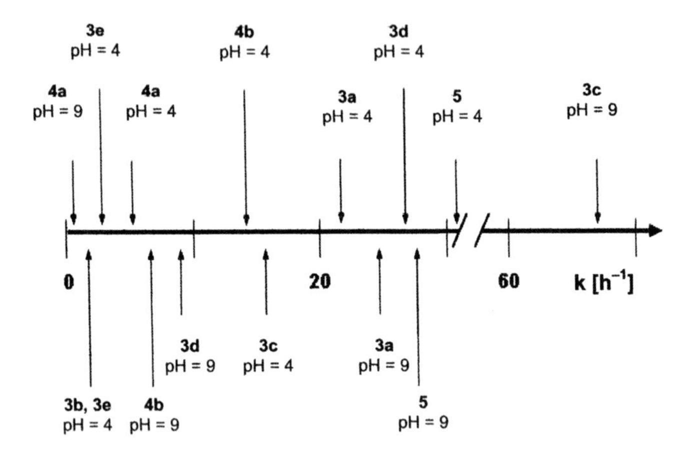

Fig. 1. Hydrolysis rate constants for compounds **3a** – **3e**, **4a** – **4b**, and at pH = 4 and 9.

- A broad range of hydrolysis rates are found, providing the possibility to select a suitable methacryloyloxyalkyl-functional silane for special applications. All methylene-bridged methacrylic silanes are more reactive than the long known silane **3e**, surprisingly even if they bear ethoxy substituents at the silicon atom.
- The dialkoxy compounds **3c** and **3d** both show hydrolysis rates comparable to their trifunctional counterparts **3a** and **3b**. Comparing the rate constants for 3-methacryloyloxy-propyltrimethoxy- and the difunctional 3-methacryloyloxypropyl(dimethoxy)methylsilane

3f, the latter displays a much lower reactivity.

- The ethoxysilanes **3b** and **3d** are almost as reactive as the methoxy-substituted analogous silanes **3a** and **3c**. Usually, ethoxysilanes show a very much lower hydrolysis reaction rate than their methoxy counterparts.
- Propyl trimethoxysilane **4a** exhibits a much lower reaction rate than that of trimethoxy vinyl silane **4b**. This is the result of a reduced electron density at the silicon atom in **4a** compared to **4b** bearing the vinyl group. In contrast, the high reactivity in the case of **5**, the saturated counterpart of **3a**, is not only retained, but enhanced. This can be explained by a more effective electron transfer from the ester moiety to silicon in the case of **5** compared to **3a** due to the missing electron delocalization. The high hydrolysis speed found for compound **5** indicates that the high reactivity of **3a** can be transferred into a polymer in the case of a copolymer synthesis.

In order to obtain data for the hydrolysis as well as condensation rates of the monomers **3a** and **3e** incorporated in polymers, copolymers of 0.5 mol% silane with methyl methacrylate have been synthesized in methyl ethyl ketone (30% solids content). Catalyzing the crosslinking reaction with an aq. Me_4NOH-solution, it turned out that PMMA-**3a** copolymer needed only 13 min to solidify, whereas 95 min were necessary for a PMMA-**3e** copolymer (0.5 mol% **3e**) to become a gel.

The surprisingly high reactivity of the methacryloyloxymethylsilanes **3a–d** may be explained in terms of an electronic influence of the ester group on the electron density at the silicon atom. In principle, three different possibilities may be assumed, namely

- intermolecular interactions between the silicon atom of a first molecule with the ester group of a second molecule (Scheme 3 A),
- intramolecular activation of the silicon atom by interaction with the carbonyl group by forming a silatrane analog five-membered ring (Scheme 3 B), and
- intramolecular activation of the silicon moiety by electron transfer from the ester group via the methylene group either through space, via the σ-bonds, or by hyperconjugation (Scheme 3 C).

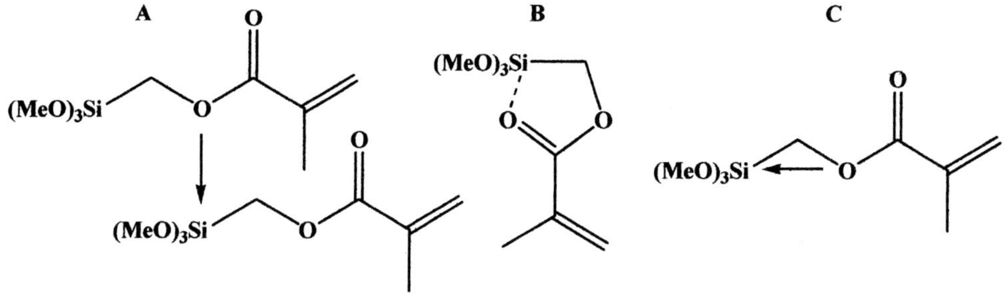

A B C

$(MeO)_3Si$ O O $(MeO)_3Si$ O O $(MeO)_3Si$ O O $(MeO)_3Si$ O O

Scheme 3. Model structures to explain the enhanced hydrolysis reactivity of methacryloyloxymethyl-silanes **3a–d** compared to **3e**.

The first possibility is unlikely, because this effect should also play a significant role in the case of **3e**, bearing a propylene spacer. Evidence for one of the latter two models can be drawn from the comparison of the ^{29}Si NMR data of compounds **3a–d** with **3e** and 3-methacryloyloxypropyl-(dimethoxy)methylsilane **3f**, depicted in Table 1.

Table 1. ^{29}Si NMR data for the methacryloyloxyalkyl alkoxysilanes **3a–f**.

Compound	δ [ppm]	Compound	δ [ppm]
3a	–53.3	**3d**	–14.9
3b	–56.6	**3e**	–42.8
3c	–11.4	**3f**	–3.1

This data reveal that the methacryloyloxymethylsilanes **3a–d** show a high-field shift of 8–14 ppm compared to their counterparts **3e–f** bearing a propylene spacer. This indicates a partial electron transfer from one of the ester oxygen atoms to silicon.

A comparable effect was also found, for example, in the case of substituted diphenylsilanes. The chemical shift of diphenylmethylchlorosilane was found to be 10.6 ppm. Introduction of a methoxymethyl substituent at the *ortho* position of the phenyl group was accompanied by a high-field shift of about 7 ppm leading to a chemical shift of 3.4 ppm by intramolecular coordination [2]. A decision whether the influence on the electronic structure of silicon is due to intramolecular ring formation (Scheme 3 B) or hyperconjugation (Scheme 3 C) cannot be made based on the present data.

Potential and Applications

The reactivity profile of compounds **3a–d** opens a vast field of potential applications. In general, their utilization can allow

- replacement of methoxy by ethoxysilanes, avoiding the liberation of toxic methanol during hydrolysis.
- reduction of the VOC amount liberated in the reaction course by 1/3 by engaging the dialkoxysilanes **3c** or **3d** instead of the trialkoxysilanes **3a** or **3b**
- realization of highly reactive moisture-curing polymer systems
- formation of highly crosslinked systems because of the high reactivity.

Because of the reactivity pattern described, silanes **3a–d** are of high interest in various material treatment application fields like glass fiber sizing, filler surface modification, or sol-gel processes. In addition, **3a–d** are of special interest for the production of novel silylfunctional organic polymers via copolymerization or grafting. These polymers can be moisture cured to give coatings or crosslinked thermoplastics with improved mechanical and chemical properties.

In a first attempt to evaluate the possibility to utilize the high hydrolysis and condensation reactivity of compounds **3a–d** compared to **3e**, copolymers consisting of methacryl silanes **3a–e**, styrene, and isobutyl methacrylate have been synthesized using 2,2′-azobis(2-methylbutyronitrile) as a radical polymerization initiator. Clear coat formulations containing these polymers were prepared, using *p*-toluenesulfonic acid (*p*TSA) as a crosslinking catalyst. After applying the coating formulation (no additional crosslinking mechanism was engaged) onto glass plates and curing, the scratch resistance was measured using the Peters-Dahn test method [3]. Results are given in Fig. 2.

The data obtained clearly demonstrate that the scratch resistance of polyacrylate coating materials can be strongly enhanced by incorporation of methacryloyloxymethylsilanes (the reasons for the low value of gloss retention observed for **3c** are not yet fully understood). As an additional benefit, the cured clearcoats revealed an extremely high chemical resistance: no change in the films were observed after a 500 h exposure to 30% aq. H_2SO_4 test or after 100 cycles in the MEK rub test [4].

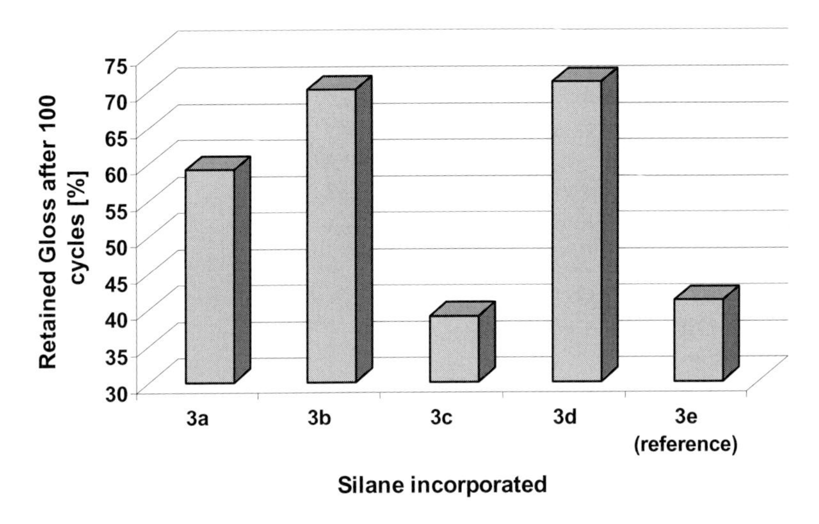

Fig. 2. Gloss retention of cured films of polyacrylates modified with methacryloyloxyalkylsilanes **3a – 3e** after 100 scratch cycles in a Peters-Dahn tester. The resins contained 30 mol% silane; curing conditions were 30 min at 130 °C using 0.5% *p*TSA as catalyst.

Concluding Remarks

A variety of methacryloyloxymethyl alkoxysilanes have been synthesized. The compounds show exceptionally high reactivity rates with regard to hydrolysis and condensation compared to the parent 3-methacryloyloxypropyltrimethoxysilane **3e**. This high reactivity renders compounds **3a–d** highly interesting for the modification of inorganic surfaces and makes them very attractive as versatile building blocks for the synthesis of silane-modified moisture-crosslinkable organic polymers.

References

[1] S. Altmann, J. Pfeiffer, *Monatsh. Chem.* **2003**, *134*, 1081.

[2] U. H. Berlekamp, A. Mix, P. Jutzi, H. G. Stammler, B. Neumann, Oxygen, phosphorus or sulfur donor ligands in higher-coordinated organosilyl chlorides and triflates, in *Organosilicon Chemistry IV — From Molecules to Materials* (Eds.: N. Auner, J. Weis), VCH, Weinheim, **2000**, Wiley-VCH, Weinheim, p. 489.

[3] In the Peters-Dahn test method, a defined type of Scott Brite steel wool is moved over the test substrate back and forth under a defined weight (500 g in the case of this study). The gloss retention is measured after 100 scratch cycles.

[4] In the H_2SO_4 test, the cured film is treated with 38% aq. H_2SO_4 at 25 °C and 50% humidity. The MEK rub test consists of rubbing a cloth saturated with MEK over the coating under a defined weight (300 g in the present study).

Hydrolysis Studies of Silane Crosslinkers in Latexes

Jeffrey A. Cooke, Weizhen Cai, Alain Lejeune

GE Specialty Materials - OSi Specialties
7, rue du pré Bouvier
CH 1217 Meyrin, Geneva, Switzerland
Tel.: +41 22 989 23 48 — Fax: +41 22 785 11 40
E-mail: alain.lejeune@gesm.ge.com

Keywords: silane, crosslinking, hydrolysis, coating latex, adhesion promoter

Summary: Silanes have long been used as crosslinkers and adhesion promoters in solventborne systems. However, as silanes are inherently water reactive, it is often more technically demanding to use them successfully in waterborne formulations. The present study introduces a method for following the rate of hydrolysis and subsequent crosslinking by silanol condensation of copolymerized silanes directly in fully formulated latexes. Low-temperature separation of the volatiles from latex solids is the basis of these methods. Analysis of the volatiles for alcohol of hydrolysis can provide information on the degree to which the alkoxy groups have reacted with water; the solid polymer can then be analyzed by ^{29}Si NMR spectroscopy to determine crosslinking. In addition, crosslinking can be followed in the cured film. The implications of the findings for formulation stability are discussed, as are recommendations for the most efficient use of silane monomers in waterborne systems.

Introduction

The use of silanes in the solventborne coatings industry as adhesion promoters and crosslinkers is well known and widely practiced. The advantages of silanes in coatings systems have been known for quite some time; they were first introduced almost a half century ago [1, 2]. As additives to formulations or as monomers in polymer synthesis, these products impart many desirable performance characteristics to a cured film. Coatings prepared with silanes can show improved mar and abrasion resistance, improved solvent resistance, improved acid and alkali resistance, and often remarkable adhesion to various substrates [3]. The use of silanes in coatings is not limited to interaction between the organic polymer and the substrate, however. Pigments pre-treated with silanes show increased dispersibility and binding, resulting in lower viscosities and better hiding power. Thus, silanes have steadily grown to become an indispensable component in many aspects of the coatings industry.

Organosilane coupling agents consist, in general, of an organic functionality and one or more silicon atoms, each with one, two or three hydrolyzable alkoxy ligands. The organic moiety provides a "handle" that can react with functionality on a polymer backbone. The hydrolyzable groups can then react with water, forming reactive silanol groups, and then bond to a filler or substrate, or self-condense to form siloxane crosslinks. The reactivity of alkoxy substituents on silanes is sensitive to many factors, including steric hindrance, temperature, moisture content, and pH. Many studies on the reactions and reactivity of alkoxysilanes have appeared in the literature [4, 5].

Silanes can be polymerized into the backbone of a polymer during its synthesis [6]. Emulsion polymerization of methacryloxyalkyl or vinyl functional organosilanes has been shown to be a particularly useful method of incorporating a crosslinking silane into a waterborne system. Subsequent to reaction of the methacryl or vinyl functionality, the alkoxy groups are left available either to react with a substrate or filler, or to crosslink upon film formation.

In the coatings industry, emulsion polymerization, as a synthetic tool for the preparation of acrylic and vinyl polymers, allows the manufacturer access to a wide array of polymeric systems, simply based on the types and ratios of the monomers used [7]. However, coatings derived from these emulsion polymers often are deficient in performance because of lack of an internal crosslinking mechanism. Silanes can remedy this deficiency.

While methoxy-substituted vinyl and methacryloxypropyl silanes have long been used in the coatings industry to prepare silylated polymers through emulsion polymerization, the inherent reactivity of these monomers only allowed incorporation of the crosslinking monomer at low levels, typically on the order of a quarter to a half mole percent based on monomers. This limitation on level of incorporation of methoxy-substituted monomers limits the final crosslink density and ultimately the performance of the silylated polymer. Through the use of vinyl and methacryloxyalkyl silanes substituted with bulky alkoxy groups, namely *iso*-propoxy, silanes with sufficient stability to survive emulsion polymerization conditions have been prepared, and are useful in much higher concentrations than "traditional" silane monomers [8]. The shelf lives of acrylic and vinyl acrylic polymers containing up to 3 – 5 mol percent *iso*-propoxy-substituted silane can exceed one year, retaining full crosslinking potential.

In order to improve silane technology for waterborne coatings, knowledge and control of silane hydrolysis and condensation during synthesis and application of silylated latexes is most critical. However, because of the low concentration of the silane monomers and the intractability of the latex systems, it has historically been very difficult to analyze quantitatively the degree to which silanes were hydrolyzed in these "real life" systems. Gas chromatographic determination, for example, is prone to sampling errors and to hydrolysis occurring in the injection port of the gas chromatograph. As mentioned previously, there have been many studies on the hydrolysis rates of silanes in many different model systems. However, interactions with the resin system, surfactants, fillers and other additives may provide for significant differences in the rate of hydrolysis of a silane when compared to a homogeneous model system.

For application as a functional crosslinker or coupling agent in a one-component waterborne coating, a silane has to meet two important demands: it must (1) be stable in the formulation for the required shelf life, and (2) be reactive after the coating formulation is applied to the substrate. Both of these characteristics are imperative in order to provide a well-crosslinked coating film with superior performance. Hydrolysis and condensation of the silane are factors that control both the stability and performance of the silane in the formulation. Therefore, understanding the extent of hydrolysis at a given point in the life of the latex would be extremely useful.

To determine the nature of the silicon moieties in a polymer, clearly the easiest method would be a technique that provides a direct observation of the silicon atom and meaningful, interpretable information on the atom. Nuclear magnetic resonance spectroscopy tuned to the ^{29}Si isotope (^{29}Si NMR) is a tool of this nature; it can directly probe the state of the silicon atom, and with it one can often readily determine the extent to which Si-O-Si crosslinks (from silanol condensation), have formed. One can observe spectra of silicon-containing compounds either dissolved in a solvent or in the solid state. Liquid-state ^{29}Si NMR, while the most sensitive, cannot be used quantitatively on heterogeneous systems such a latex formulations. Therefore, one must separate the liquid and solid portions of the latex (without heat, which would promote hydrolysis and condensation) and use the solid residue for the ^{29}Si NMR experiments.

Indirect methods for probing the silicon center of molecules can be just as powerful, provided they can be shown to be quantitative and reproducible, and thus meaningful. In alkoxysilane hydrolysis, by definition one alcohol molecule is generated for each alkoxy group hydrolyzed. Thus, if one can measure the free alcohol in a system containing alkoxysilanes, such as a silylated-latex-based coatings formulation, then one can measure the degree to which the silanes have undergone hydrolysis. Of course, this approach is especially useful when the only source of alcohol is silane hydrolysis.

In this study we describe a method used to measure the reactions that silylated latexes can undergo in coatings formulations. ^{29}Si NMR is shown to reveal crosslinking by silanol condensation occurring in silylated latex synthesis and cured films. In addition, a complementary method for quantitative determination of degree of alkoxysilane hydrolysis is described. Low-temperature separation of the latex solids from the volatile components followed by gas chromatographic analysis of the distillate can provide accurate and reproducible measurement of the alcohol generated by the hydrolysis of the alkoxysilanes used in the formulation.

These methods offer an opportunity to measure the phenomena that cause instability of silane-containing formulations, and are valuable to the continued understanding of the complex factors that should be considered when formulating with silanes in real-world waterborne systems.

Experimental

Butyl acrylate/methyl methacrylate/methacrylic acid copolymers with silylated latexes were prepared essentially according to a general procedure suitable to real-world systems.

A trap-to-trap (T-T) distillation apparatus was employed to separate the latex solids from the volatile components. The distillation procedure allows freeze drying of the latexes with condensation of all volatiles that are quantitatively condensed.

GC analysis was then performed in order to determine quantitatively the content of alcohol in liquid distillate samples after the distillation was complete.

In parallel, the dry latexes, obtained by the freeze drying procedure, and coated latexes were analyzed by solid-state ^{29}Si NMR in order to determine the extent of condensation of silanols (SiOH) into siloxane bonds (SiOSi).

Results and Discussion

NMR Studies

Silicon centers with two or three crosslinked groups were successfully resolved in our experiments, but a silicon with only one crosslink is difficult to resolve from a silicon that has not undergone hydrolysis. Additionally, silicon atoms substituted with SiO–H and SiO–C bonds resonate at very similar frequencies, so it is not possible to directly examine the extent of hydrolysis itself. One can at best calculate the extent of hydrolysis of the alkoxy groups by examining the extent of crosslinking.

One advantage of the solid-state NMR investigation is that one is able to study films at different stages of curing (provided the films are cast on a low-surface-energy substrate from which they can easily be removed) to determine the changes in degree of crosslinking over time or with the application of heat.

Figure 1 is a spectrum of the nonvolatile portion of a silylated latex containing three mole percent MTIPS [9].

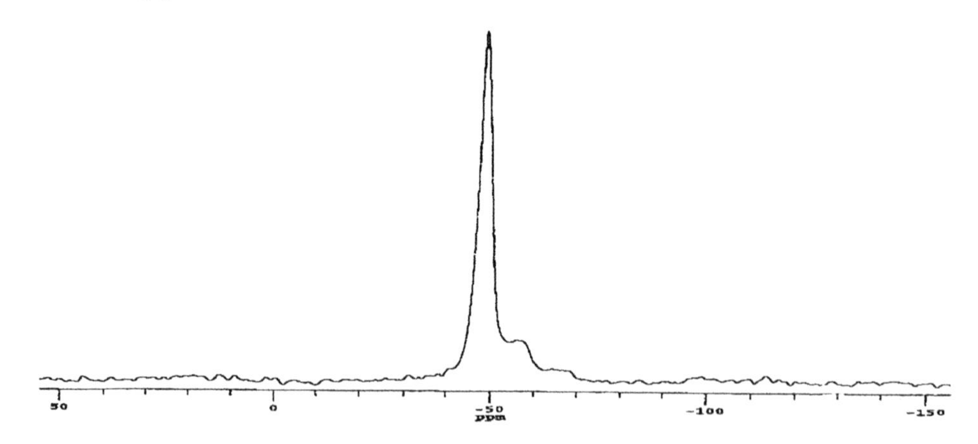

Fig. 1. ^{29}Si NMR spectrum of solids from three mole percent MTIPS (Methacryl Tri *Iso*propoxy Silane) latex synthesis.

The large signal at ca. –50 ppm is silicon in unhydrolyzed silane, and also silicon that has one Si–O–Si linkage (i.e., of the three oxygen atoms connected to the silicon, only one is bonded to another silicon). The two smaller shoulders upfield of –50 ppm correspond to silicon atoms that have two and three Si–O–Si linkages, respectively. It is clear that some modest amount of crosslinking of the latex has taken place during the synthesis and/or isolation and drying of the polymer. This is supported by the insolubility of the dried polymer in organic solvents compared to the high solubility of the unsilylated latex control.

Figure 2 is a spectrum of the a cured film drawn down from the original latex used above, but after a 120 °C, 20 min cure cycle.

It is clear that the shoulders visible in Fig. 2 have markedly increased in intensity relative to the large peak. This indicates that there has been significant further silane crosslinking during the cure of the drawn-down film. Again, physical evidence of crosslinking such as increased gel content and solvent double rub resistance are also seen (see Appendix for typical performance characteristics).

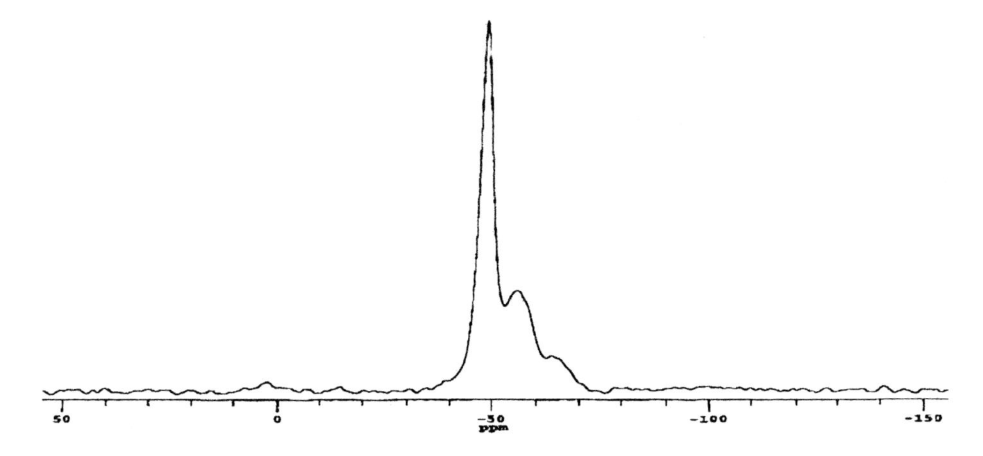

Fig. 2. ^{29}Si NMR spectrum of cured film from three mole percent MTIPS latex synthesis.

Figure 3 is a spectrum of the nonvolatile components of a latex synthesized as described above using three mole percent VTIPS (Vinyl Tri *Iso*Propoxy Silane). As previously, the large peak at approximately –50 ppm is unhydrolyzed and mono-crosslinked silicon. The medium peak at approximately –80 ppm, however, is not seen in the methacryloxy silane latex. This peak interestingly corresponds to silicon which is substituted with an *intact, unpolymerized* vinyl group.

Additionally, the chemical shift of the peak at ca. –80 ppm indicates that the silicon also has condensed three times, forming three crosslinks. As there is no detectable unreacted silane (or organic) monomer remaining in the latex, the conclusion that must be reached is that the silane monomer has condensed either with other silane monomers or with silane incorporated into the backbone of the growing polymer chain *before* the vinyl group polymerized, and that being bonded to the polymer chain protected some of the vinyl groups in the oligomer from reacting. It should be noted that with all of the other monomers in this system, vinyl silanes have lower reactivity ratios, indicating that they will have lower polymerization toward acrylic monomers through their vinyl groups under these conditions. It is possible that slower addition of the silane monomer may allow more of the silane to be incorporated through reaction of the vinyl group, possibly increasing the overall performance of the cured film.

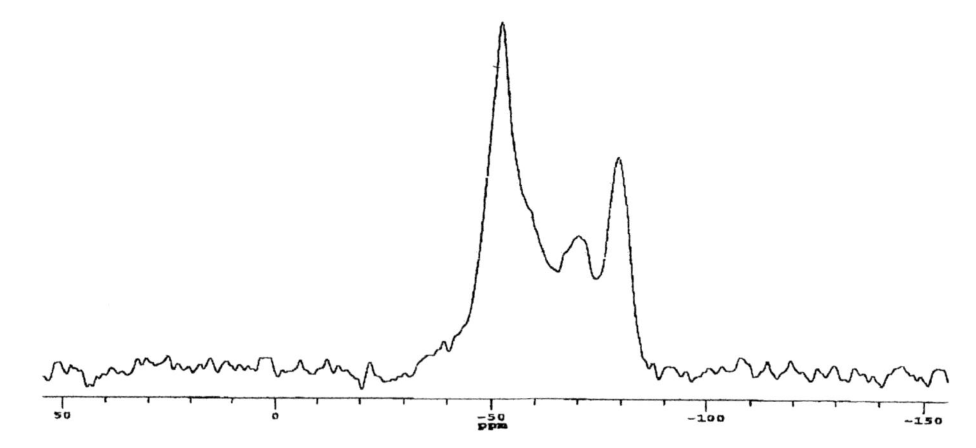

Fig. 3. ^{29}Si NMR spectrum of solids from three mole percent VTIPS latex synthesis.

Figure 4 is a spectrum of the same polymer, but after the cure step. While not as pronounced as in the previous case, it is clear that the intensity of the resonances of the Si-O-Si crosslinks in the approximate range –55 to –70 ppm has increased.

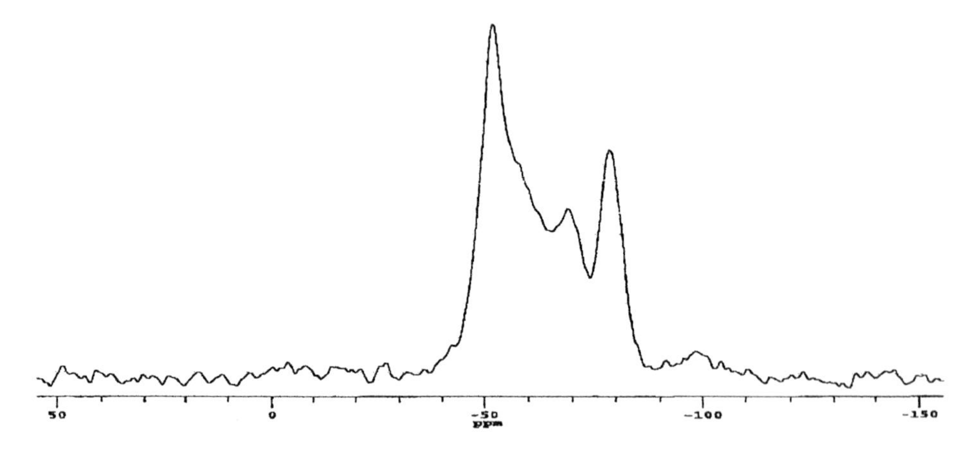

Fig. 4. ^{29}Si NMR spectrum of cured film from three mole percent VTIPS latex synthesis.

Because the silicon of the vinyl silane-containing acrylic polymers is attached directly to the backbone of the polymer, it is expected that the silanols will be less able to condense and crosslink than the less sterically encumbered silanols of the MTIPS latexes. The vinyl silane signal did not change in intensity during cure.

The vinyl silane reacts more slowly than the methacrylate in free-radical polymerizations. In addition, vinyl silanes are known to hydrolyze relatively quickly compared to alkyl-modified silanes [4]. These two facts make it reasonable to assume that the cause of the oligomerization seen in the vinyl silane polymers and not the methacryloxypropyl analogs is that the vinyl silanes hydrolyze and condense to a greater degree before they are polymerized. As vinyl silanes rarely homopolymerize, once the other monomers have reacted, there would not be further free-radical polymerization of any vinyl silane monomers.

Analytical Determination of Alcohol of Hydrolysis

In addition to ^{29}Si NMR analysis, one can measure the alcohol of hydrolysis in order to probe the reactivity of the silane moiety. While one cannot determine extent of crosslinking by this method because it does not measure condensation, one should see a good correlation between rate of silane hydrolysis and shelf life of the formulation.

Before the trap-to-trap distillation technique could be used to quantify the degree of hydrolysis of the silanes in the latex samples, the identity of the recovered compounds was established and the reproducibility and accuracy of both the distillation and the response of the gas chromatograph were determined.

With confidence in the method assured, silylated latexes were subjected to the method. The polymerization of the silylated latexes takes approximately 4.5 h to complete. Samples of the reaction mixtures were removed at various points after addition of monomer and initiator, and also at the end of the reaction, just prior to neutralization of the latex with ammonia.

Figure 5 shows the percentage of alkoxy groups that had been hydrolyzed in the system during the reaction.

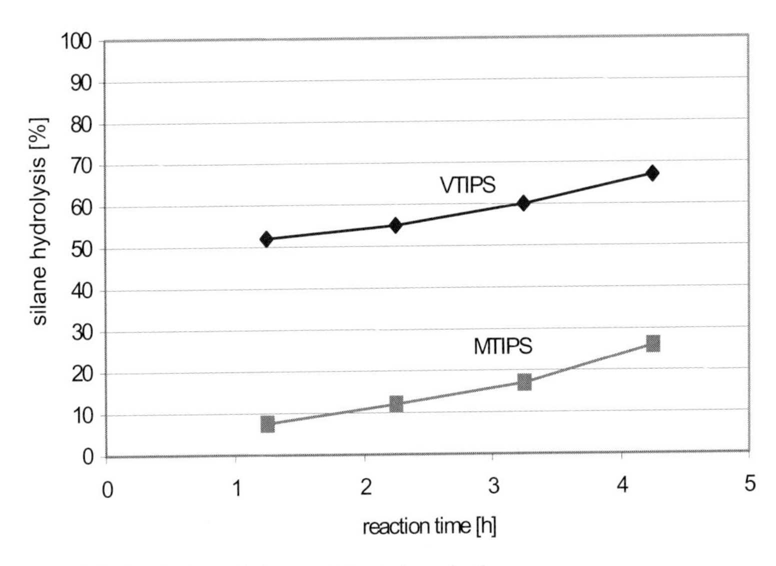

Fig. 5. Hydrolysis of hindered silanes during emulsion polymerization.

Both silanes show significant hydrolysis even early in the reaction. The relative amount of hydrolyzed silane in the mixtures is fairly constant throughout the reaction, only rising slightly in both cases. As expected, the vinyl silanes hydrolyzed to a much greater extent than the methacryl silanes.

Presumably, any silane is much more prone to hydrolysis prior to polymerization, when it is not nearly as sterically encumbered and "protected" from water as it would be after incorporation into the growing polymer. The data support this. As the monomers are added by a "starved-feed" method, common when monomers of different reactivities are used, one can presume that at some point a steady state concentration of monomer is established. If the monomeric silane, and not the polymerized silane, were the species that predominantly underwent hydrolysis, then we would expect to see a relatively constant level of hydrolysis throughout the polymerization. If there were significant hydrolysis occurring at the silane centers on the polymer backbone, then one would expect to see the percent of hydrolyzed silane steadily increasing throughout the course of the reaction. While it does appear that hydrolysis is occurring both before and after polymerization, the amount of "post-polymerized" hydrolysis seems to be greater in the methacryl example. One might expect this due to the relatively more accessible silicon center, separated from the bulky polymer by five atoms instead of zero for the vinylsilane. This argument only becomes stronger when one considers that, in this redox-initiated polymerization without buffer, the pH in both cases gradually drops from near neutrality at the onset to approximately 2 – 3 at the conclusion. The rate of hydrolysis increases as the pH shifts away from neutrality (to either acidic or basic conditions) [4].

Compared to the bulky VTIPS and MTIPS, the analogous trimethoxy-substituted silanes would be expected to hydrolyze to a much greater extent under the same reaction conditions. This is borne out by experiment; the following table shows the percent of alkoxy groups that hydrolyze during the synthesis of silylated latexes containing 0.25 mol percent methacryloxypropyl trimethoxysilane (MTMS) and vinyl trimethoxysilane (VTMS), respectively.

Silane	Hydrolysis [%]
MTMS	67
VTMS	92

In addition, VTMS hydrolyzes to a much greater extent than MTMS, an expected pattern similar to the *iso*propoxy examples. Clearly the bulk of the *iso*propoxy groups on the hindered silanes is very beneficial in protecting the silane from premature reaction during latex polymerization, when compared to the more "traditional" methoxy substituted silane monomers. This has implications for the ultimate properties of the final films, as the stability of the *iso*propoxy silanes allows levels of incorporation, and thus final performance, that are simply not available when using methoxy-substituted silane monomers.

This indirect measurement of the degree of hydrolysis of the silicon species during the emulsion polymerization is in close agreement with the conclusions from the ^{29}Si NMR spectra. Both indicate that there is a significant fraction of alkoxy groups that are not hydrolyzed, and both support the conclusion that the vinyl silane hydrolyzed to a much greater extent during the course of the reaction.

Conclusions

Overall, it is clear that significant hydrolysis of silanes in waterborne formulations occurs when the hindered silanes are used as monomers in emulsion polymerization. ^{29}Si NMR spectroscopy on latex solids also reveals that a significant amount of hindered vinyl silane monomer forms condensates with itself, so that not all of the vinyl groups are incorporated into the backbone of the growing polymer through free-radical polymerization.

One might surmise that this alkoxysilane hydrolysis might be detrimental to the shelf life of the silylated latex formulations. On the contrary, the remarkable stability to storage of these silylated latexes indicates that the silanol is quite well protected from damaging premature crosslinking by condensation. In fact, some hydrolysis during the emulsion polymerization of the bulky *iso*propoxy groups may be necessary to sufficiently "activate" the silylated latex to crosslinking upon film cure. Similar conclusions have been reached by other authors [10].

The bulk of the *iso*propoxy groups on the vinyl and methacryl silanes is unique among silane monomers in that, through their enhanced stability, levels of incorporation of silane an order of magnitude greater than "standard" methoxy-substituted monomers is possible. Thus, much higher performance can be achieved using these silane monomers.

References

[1] E. P. Plueddeman, *Silane Coupling Agents*, 2nd Edition. Plenum Press, NY, **1991**.

[2] G. L. Witucki, *J. Coatings Tech.* **1993**, *65*, 57.

[3] M. J. Chen, A. Chaves, F. D. Osterholtz, E. R. Pohl, W. B. Herdle, *Surf. Coat. Int.* **1996**, *79*, 539.

[4] F. D. Osterholtz, E. R. Pohl, *J. Adhesion Sci. Tech.* **1992**, *6*, 127.

[5] E. R. Pohl, F. D. Osterholtz, in *Molecular Characterization of Composite Interfaces* (Ed.: G. Kumar), Plenum Publishing, New York, **1985**, 157.

[6] a) J. R. Grawe, B. G. Bufkin, *J. Coatings Tech.* **1978**, *50*, 67; b) T. R. Bourne, B. G. Bufkin, G. C. Wildman, J. R. Grawe, *J. Coatings Tech.* **1982**, *54*, 69.

[7] For comprehensive coverage of emulsion polymerizaton see: *Emulsion Polymerization and Emulsion Polymers* (Eds.: P. A. Lovell, M. S. El-Aasser), Wiley, New York, **1997**.

[8] M. J. Chen, F. D. Osterholtz, US Patent 5 827 922.

[9] Details of the acquisition of the NMR spectra herein are available from the author.

[10] D. T. Liles, D. L. Murray, US Patent 5 932 651. The authors suggest that the silane monomers may exist primarily as silanol after polymerization.

Mastering Crosslinking in Silicone Sealants

Jean-Marc Pujol,* Joseph-Marc François, Bernard Dalbe

Rhodia Silicones, 55 rue des frères Perret, B. P. 22, 69191, Saint-Fons Cedex, France
Tel.: +33 4 72 73 76 74 — Fax: +33 4 72 73 74 50
E-mail: jean-marc.pujol@eu.rhodia.com

Keywords: crosslinking, model, silicone, sealant

Summary: Silicone sealants are used in the construction industry to make waterproof seals and joints and to bond window frames to buildings for structural glazing. Upon curing through air moisture diffusion, these paste-like products give high-performance elastomers. Heterogeneous properties have been observed. Crosslink density gradients have been observed in cured sealants. A mathematical model simulating moisture diffusion curing has been developed to follow the crosslink density evolution during cure. The model explains different behaviors observed for different curing chemistries such as fast-reacting acetoxysilane systems or slower-reacting alkoxysilane systems. It is possible to understand why highly reactive silanes can sometimes lead to tacky sealants with a surface not fully cured. Alkoxysilane-based systems can still present gradients of properties while fully cured. This is related to the migration of aminoalkylsilanes. They are not fully soluble in the silicone system and they migrate toward the surface leading to a higher degree of crosslinking.

Introduction

Silicone elastomers can be obtained through condensation of functional polysiloxanes with air moisture. During the curing process, several molecules diffuse in the elastomer: atmospheric moisture which is the reagent leading to crosslinking, the reaction product (acetic acid, methanol...), and aminofunctional silanes used as adhesion promoters, which can migrate to the surface. The diffusion of these species influences the properties of the cured elastomer. The crosslink density changes with the elastomer sample thickness. Properties on the face exposed to air during the cure may be different from the properties of the sealant at the substrate surface. Gradients exist in the elastomer during and after the cure.

Mathematical models simulating the crosslinking of a sealant versus thickness and time have been described [1, 4]. They relate the crosslink density to the water consumed. The objective of this work was to develop a new model including the concentration in reactive functional groups to better describe the curing elastomer.

Crosslinking Model

The chemical reactions involved in the crosslinking are reported in Eq. 1. The functional polysiloxane is hydrolyzed releasing a by-product, and then a condensation reaction builds a tridimensional network. K_1 and K_2 are the kinetic constants.

$$\mathrm{Me{-}Si(OR)_2O{-}\left[Si(Me)_2O\right]_n{-}Si(OR)_2{-}Me} + 2\,\mathrm{H_2O} \xrightarrow{K_1} \mathrm{Me{-}Si(OR)(OH)O{-}\left[Si(Me)_2O\right]_n{-}Si(OR)(OH){-}Me} + 2\,\mathrm{ROH}$$

Hydrolysis

$$2\ \mathrm{Me{-}Si(OR)(OH)O{-}\left[Si(Me)_2O\right]_n{-}Si(OR)(OH){-}Me} \xrightarrow{K_2} \mathrm{Me{-}Si(OR)O{-}\left[Si(Me)_2O\right]_n{-}Si(OR){-}Me} \ (\text{linked by two } {-}O{-} \text{ bridges to}) \ \mathrm{Me{-}Si(OH)O{-}\left[Si(Me)_2O\right]_n{-}Si(OH){-}Me} + 2\,\mathrm{ROH}$$

Condensation

Eq. 1. Hydrolysis and condensation reactions.

The curing of silicone elastomers is initiated by water, which diffuses from the surface exposed to air. Silicones are very permeable to gases. The water enters the sealant following simple diffusion. The chemical reaction is assumed to be first order with respect to the water concentration and to the hydrolyzable group concentration. The rate of water concentration increase in the sealant $\frac{\partial[H_2O]}{\partial t}$ is determined by the rate of water diffusion minus the rate of the water consumption reaction as stated in Fig. 1 and Eq. 2.

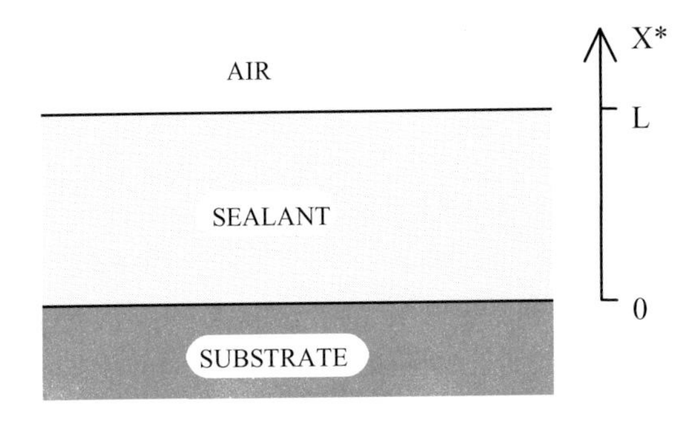

Fig. 1. Sealant representation.

$$\frac{\partial[H_2O]}{\partial t^*} = D\frac{\partial^2[H_2O]}{\partial x^{*2}} - K_1[H_2O][OR]$$

diffusion hydrolysis

$$\frac{\partial[OR]}{\partial t^*} = -K_1[H_2O][OR] - K_2[OH][OR]$$

$$\frac{\partial[OH]}{\partial t^*} = K_1[H_2O][OR] - K_2[OH][OR]$$

Eq. 2. Equation system.

The system is a non-linear coupled partial derivatives equation system. It has been solved by a numerical method. The resolution subroutine is provided by the NAG (Numerical Algorithm Group) library. The model developed has been used to simulate the curing of silicone sealants. For K_1 and K_2, literature values vary from 0.01 to 0.0001 L/mol s [5 – 11]. The concentration in reactive groups was set to 0.5 mol/L, which corresponds to the concentration in curing agent typically used in a silicone sealant.

The crosslink density is related to the condensation reaction. The model estimates the crosslink density versus time and thickness as seen on Fig. 2 for an example with $K_1 = K_2 = 0.005$ L/mol s. The crosslink density is high at the surface where the water is present at the beginning. It decreases with the thickness while the water is diffusing and increases with time, up to full cure. The water concentration is presented on Fig. 3, where a progressive diffusion is observed. Figures 4 and 5 present the OR and OH concentrations. Reactive groups disappear with time, quickly at the surface, more slowly in the mass where water has to diffuse. The OH concentration is zero at the beginning and increases quickly to a low value; however it does not decrease with time, meaning that silanols remain in the cured sealant.

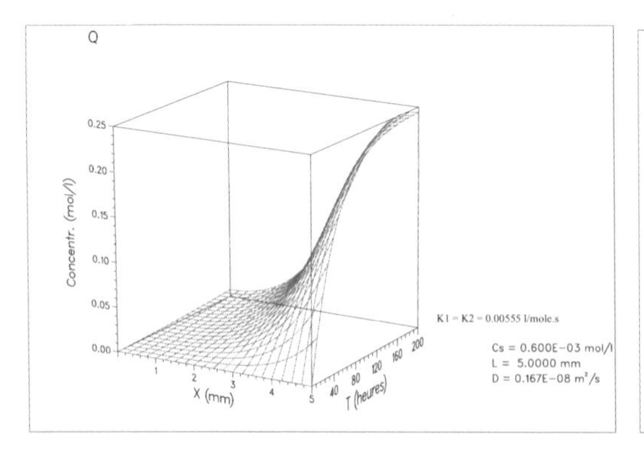

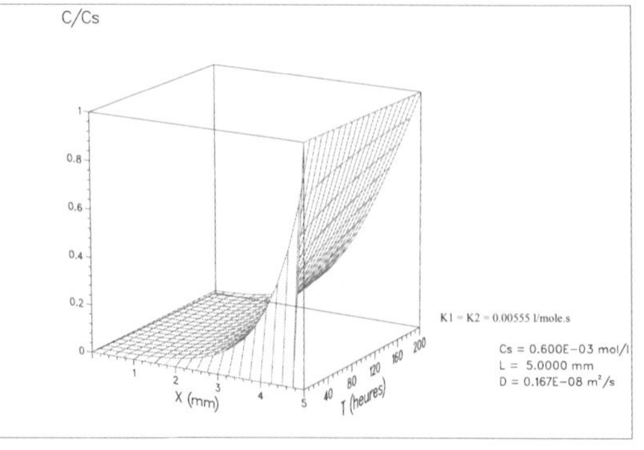

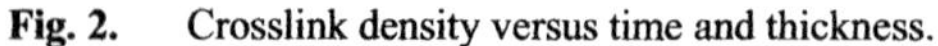

Fig. 2. Crosslink density versus time and thickness.

Fig. 3. Water concentration versus time and thickness.

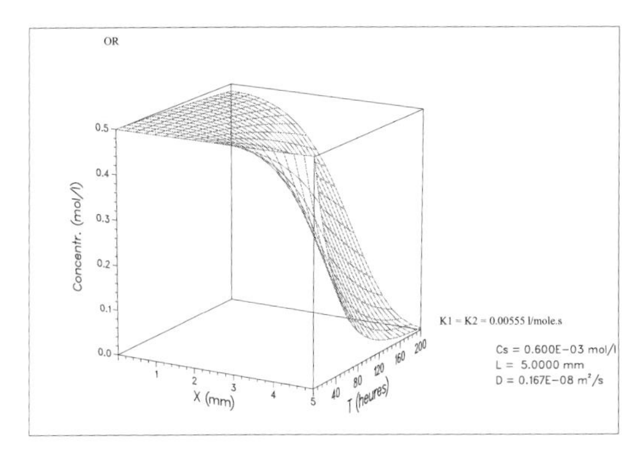

Fig. 4. SiOR concentration versus time and thickness.

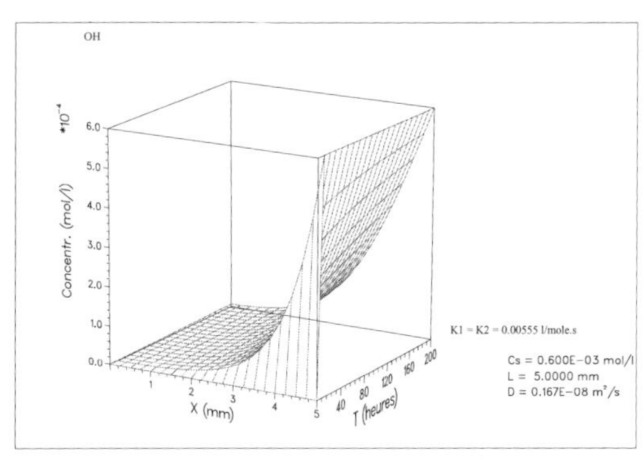

Fig. 5. SiOH concentration versus time and thickness.

It is interesting to observe variations due to different reaction rates. Fast hydrolysis and condensation are reported on Fig. 6, slow hydrolysis and condensation are reported on Fig. 7. When the chemical reactions are fast compared to the diffusion of water, there is a very fast surface curing. The crosslinking is controlled by the water diffusion. As soon as the water diffuses in the uncured sealant, it reacts and is consumed by the hydrolysis reaction. It does not diffuse further until the sealant is fully cured. At the surface the full cure is quickly reached. A full cure at 5 mm is obtained in 100 h, this is close to the experimental data observed for an acetic type sealant (Fig. 8).

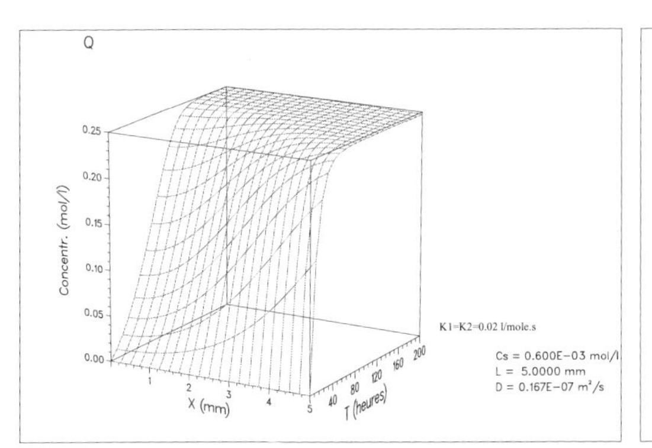

Fig. 6. Crosslink density versus time and thickness for fast hydrolysis and condensation.

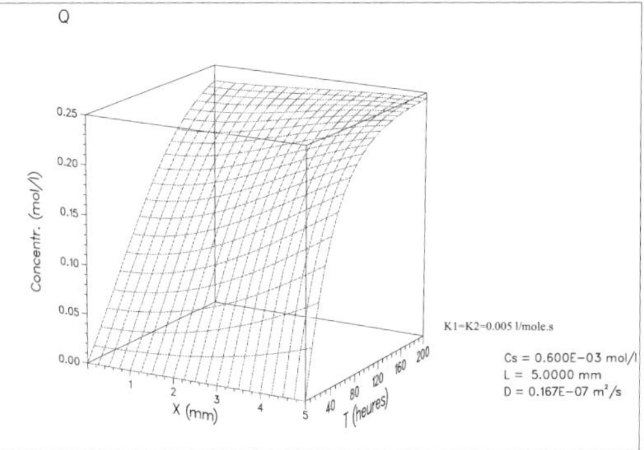

Fig. 7. Crosslink density versus time and thickness for slow hydrolysis and condensation.

On the other hand, when the hydrolysis and condensation reactions are slow compared to diffusion, there is a reaction control, i.e. the water molecules diffuse before reacting. This is a progressive curing. The differences between the surface and the unexposed face are not so strong. This type of behavior is closer to Alkoxy type sealants, which harden progressively with time even after full diffusion of water.

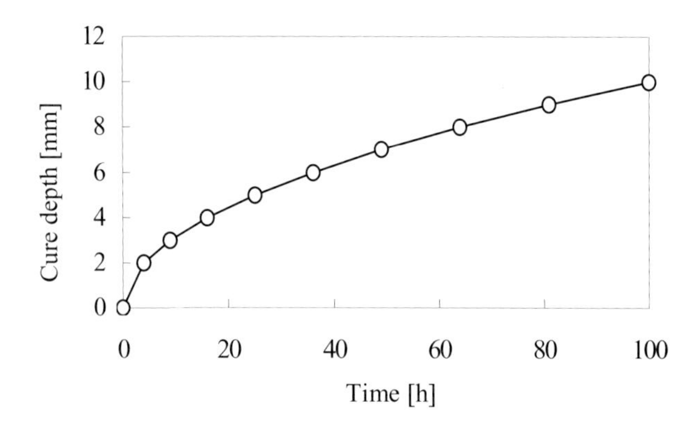

Fig. 8. Cure depth versus time for an acetic silicone sealant.

It is also interesting to model the case with a fast hydrolysis and a slow condensation. Figure 9 shows a crosslink density higher in the sealant than at the surface. During the cure, the crosslink density goes through a maximum between the surface not fully cured and the bottom yet uncured before water diffusion. This situation is related to a fast hydrolysis at the surface with a high concentration in available water. The Si–OR groups hydrolyze quickly and produce a large amount of Si–OH groups as seen in Fig. 10. They cannot condense with Si–OR groups, which have disappeared, and they condense only very slowly together. This undercure is sometimes observed for acetic silicone sealant, where a tacky surface due to the surface excess of low reactive Si–OH groups can be obtained.

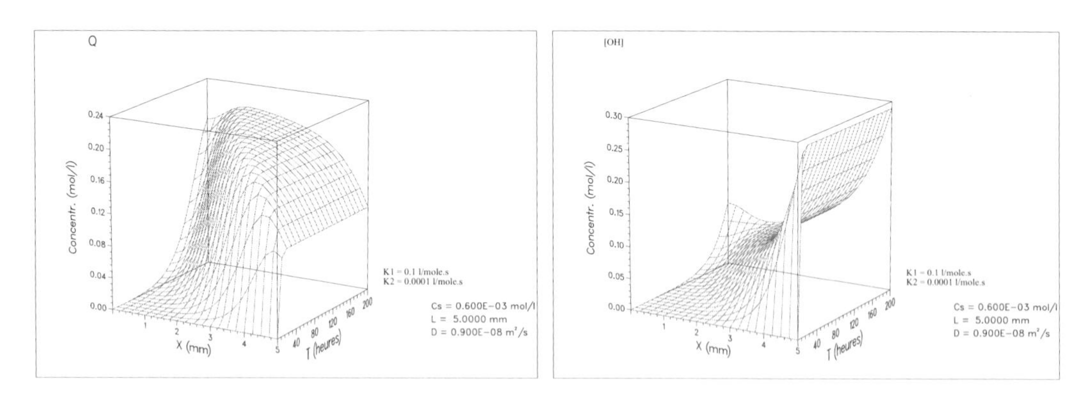

Fig. 9. Crosslink density versus time and thickness for fast hydrolysis and slow condensation.

Fig. 10. SiOH concentration versus time and thickness for fast hydrolysis and slow condensation.

Application

As stated, the model can explain different behaviors in relation to the kinetic constants. The hydrolysis and condensation kinetics can be adjusted by modification of the hydrolyzable group or of the silicon atom substitutive groups. The use of different organosilanes to endcap silanol-terminated polydimethylsiloxane has been described [12] with the use of organometallic derivatives of tin or titanium, which catalyze the cure.

The R group methyl, ethyl, propyl, butyl, pentyl, vinyl, or phenyl (Scheme 1) influences the hydrolysis rate of the silane endcapping group. The size and the inductive effect of the substituents R′, acetyl, methyl, ethyl, or imine modify the reactivity as well.

endcapping group **by-product**

R' = Me, Et, Pr, Bu, Pe, Vi, Ph

R = Ac, Me, Et, (Me)(Et)C=N

~~~~Si(OR)(OR)–R' ⟶ ROH

Scheme 1. Functional substitutive groups.

With these tools, it has been possible to develop a whole range of silicone sealants with cure kinetics designed for many different requirements (Fig. 11). The acetoxy sealants are fast curing, and thus well adapted to glazing applications. The oxime sealants are better suited for adhesion on plastics. For applications where a longer tooling time is required, such as expansion seals or perimeter joints, Alkoxy sealants have been developed with a longer skin-over time.

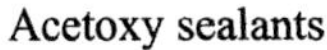

Acetoxy sealants Oxime sealants Alkoxy sealants

Fig. 11. Range of silicone sealants.

Conclusion

A model for the cure of silicone sealants has been developed. Crosslink density evolution has been studied. The model leads to a better understanding of behavior differences between fast curing acetic type silicone sealants and Alkoxy type silicone sealants with a slower cure. The model explains problems such as tacky surfaces. The data obtained in this study help in designing formulations well-tailored to specific application requirements.

References

[1] J. Falender, *ACS Polymer Preprints* **1979**, *20(2)*, 467.
[2] W. Schoenherr, J. Falender, *ACS Polymer Preprints* **1981**, *22(1)*, 190.
[3] J. M. Pujol, *Rubber World* **2000**, *222(3)*, 43.
[4] De Buyl, J. Comyn, N. E. Shepard, N. P. Subramaniam, in *Third International Symposium on Silanes and other Coupling Agents*, **2001**, Newark, New Jersey.
[5] G. Helary, G. Sauvet, *Eur. Polym. J.* **1992**, *28(1)*, 37.
[6] M. Guibegia-Pierron, G. Sauvet, *Eur. Polym. J.* **1992**, *28(1)*, 29.
[7] F. D. Blum, W. Meesiri, H. J. Kang, J. E. Gambogi, *J. Adhesion Sci. Technol.* **1991**, *5(6)*, 479.
[8] E. R. Pohl, F. D. Osterholtz, *Polym. Sci. Technol.* **1985**, *27*, 157.
[9] G. J. Kallos, J. C. Tou, R. M. Malczewski, W. F. Boley, *Am. Ind. Hyg. Assoc.* **1991**, *52(7)*, 259.
[10] J. Chojnowski, M. Cypryk, K. Kazmierski, K. Rozga, *J. Non-Cryst. Solids* **1990**, *125*, 40.
[11] J. Sanchez, A. McCormick, *J. Phys. Chem.* **1992**, *96*, 8973.
[12] J. M. Pujol, C. Prébet, *J. Adhesion Sci. Technol.* **2003**, *17(2)*, 261.

Selecting the Right Aminosilane Adhesion Promoter for Hybrid Sealants

Helmut Mack*

Degussa. DRTT,
Dr.-Albert-Frank-Straße 32, 83308 Trostberg, Germany
Tel.: +49 8621 86 3673 — Mob: +49 170 7634275
E-mail: helmut.mack@degussa.com

Keywords: silane, adhesion promoter, sealant

Summary: MS Polymer® pre-polymers offer a wide range of physical properties and can be formulated into a variety of moisture-curable hybrid sealants. Silanes greatly influence the performance of hybrid sealants and play a key role in the success of these products. This paper provides an overview of recent developments with novel and non-sensitizing diaminosilanes used as adhesion promoters to make highly demanding and low modulus construction sealants. Monomeric diaminosilanes are among the most widely used adhesion promoters for hybrid sealants. Novel oligomeric diaminosilane DYNASYLAN® 1146 brings the additional benefit of not being skin sensitizing. As a consequence sealants made with oligomeric diaminosilane DYNASYLAN® 1146 do not require special allergenic health labeling (directives 1999/45/EC and 67/548/EEC). In addition to this very important benefit, the oligomeric diaminosilane DYNASYLAN® 1146 brings additional features to hybrid sealants such as reduced modulus, excellent primerless adhesion on polar substrates, negligible yellowing, and low odor during curing. Oligomeric diaminosilane DYNASYLAN® 1146 can help reduce sealant costs. The preparation of hybrid sealants and the impact of diaminosilane adhesion promoters on the sealant's shelf life, cure rate, physical properties, and adhesion on various substrates is reviewed.

Silanes

Silanes have been used in sealants since the early 1960s when they were introduced successfully in polysulfide formulations. It was found that only small amounts of silane were needed in the formulation to improve adhesion to surfaces such as glass, metals, and, in some cases, concrete without using a primer. More recently, hybrid sealants (made of an organic polymer backbone and a silane crosslinker) have been widely commercialized. Many of these hybrid sealants have superior properties and will bond to a variety of substrates.

DYNASYLAN® silanes are bifunctional molecules that possess dual reactivity and act as adhesion promoters, crosslinkers, and moisture scavengers in many different sealant products. Silanes can be used as additives, primers, and co-monomers in sealant applications. Silane adhesion promoters act as "molecular bridges" between two chemically different materials. Hybrid sealants are based on organic polymers, while the substrates they hold together are often inorganic in nature, for example, glass, metal, or concrete. Silanes improve adhesion by chemically bonding to both types of materials simultaneously, forming a "molecular bridge" that is strong, durable, and resistant to the negative effects of moisture and temperature. Silanes open the door to such performance benefits as

- superior adhesion, wet and dry
- improved heat and moisture resistance
- flexibility in formulation
- improved wetting
- improved mechanical properties
- improved filler dispersion
- improved solvent resistance
- improved weatherability

Silanes are versatile products that provide improved product performance and quality. The properties and effects of silanes are defined by their molecular structure:

$$\mathbf{Y}\text{–}(CH_2)_n\text{–}Si(\mathbf{OX})_3$$

Y = organofunctional group
OX = silicon-functional group
n = typically 0 or 3

The organofunctional group **Y** links with the polymer. This group must be chosen carefully to ensure maximum compatibility with the resin. The silicon-functional groups **OX**, usually alkoxy groups, must be hydrolyzed to the silanol (Si–OH) first before they can bond to the substrate or crosslink.

Silanes for Adhesion

Adhesion is a complex phenomenon based on a number of different mechanisms. In contrast to simple mechanical adhesion, silanes enable a sealant to bond chemically to a substrate, resulting in more durable adhesion. The bond is less susceptible to the negative effects of moisture and temperature. The nature of the substrate surface also plays an important role in achieving good adhesion. The more chemically active sites (preferably hydroxyl groups) the substrate has, the better the adhesion will be. A further improvement in adhesion can be achieved by applying a silane primer prior to sealant application. To achieve optimal adhesion, the functionalities of the silane must be carefully matched to the type of polymer matrix used and the physical properties desired. While the exact amount of silane required for property enhancements must be determined empirically for an individual formulation, one percent (1%) based on sealant weight is a good starting point.

Unlike traditional polyurethanes, diaminosilanes are typically used as adhesion promoters in hybrid sealants to obtain good adhesion to various substrates.

Silanes for Crosslinking

Silanes have long been used to endcap polymers such as silicones, polyethers, and polyurethanes to tailor modulus, tensile strength, and elongation. In addition to endcapping a polymer, silanes can be incorporated in the polymer chain by free-radical polymerization or by reaction with active groups on the polymer chain. Crosslinking occurs in the presence of ambient moisture, and the rate can be varied by adjusting the pH of the system. MS Polymer® is endcapped with methyldimethoxysilane. In the presence of moisture the methoxy groups can be hydrolyzed to silanols. The use of an appropriate catalyst leads to the formation of a durable siloxane network. Silane crosslinking technology offers fast, room-temperature cure. The resulting siloxane (Si–O–Si) crosslink is very strong, imparting mechanical, chemical, and weather resistance to the polymer. The final polymer can be thought of as a polyether cured through siloxane groups.

Silanes for Moisture Scavenging

Because of their faster hydrolysis rates vinylsilanes are typically used as water scavengers in moisture-cured hybrid sealants. Vinylsilanes capture excess moisture even in the presence of other silanes added to the formulation for other purposes. Addition of vinylsilanes to a sealant formulation results in improved package stability and prevents premature cure during compounding (i.e. skin formation).

Hybrid Sealants

The excellent adhesion of MS Polymer® sealants to conventional construction substrates such as glass, metal, and plastics (e.g., PVC) is derived from the widely used silane adhesion promoter *N*-(β-aminoethyl)-3-aminopropyltrimethoxysilane, DYNASYLAN® DAMO-T. Another aspect of using a diaminosilane is the intrinsic change of pH that results in fast hydrolysis and condensation reactions of the MS Polymer®. To guarantee a certain shelf life, MS Polymer® sealants are generally formulated with vinyltrimethoxysilane, DYNASYLAN® VTMO as a water scavenger.

The goal of the research reported here was to determine the effects of non-sensitizing diaminosilane adhesion promoter DYNASYLAN® 1146 on the mechanical properties and the adhesion of a standard MS Polymer® sealant. DYNASYLAN® DAMO-T is used as a reference.

Experimental

Sealant samples were prepared by employing KANEKA's MS Polymer® S203H and S303H as a binder. MS Polymer® S203H is a linear polyether with a methyldimethoxysilane group at each end. It is used to produce very soft sealants with low modulus and low hardness. MS Polymer® S303H is a higher functionality polyether that is used for firmer, but still soft, sealants with somewhat higher modulus, lower tack, and hardness. The viscosity of the two MS Polymer® pre-polymers is around 10 000 mPa s. The plasticizer selected for this study was supplied by EXXONMOBIL. Jayflex® DIUP (di-*iso*undecyl phthalate) is commonly used in sealants and mastics because of its permanence and product stability. The filler was obtained from IMERYS and is a fine stearate-coated ground calcium carbonate Carbital® 110S (specific surface area 5 m^2/g; mean particle size 2.0 µm). KERR-McGEE CHEMICALS titanium dioxide Tronox® R-FK-2 is a micronized rutile pigment with hydrophobic properties. Dispersibility in nonpolar polymers is easy and plasticizer absorption is low. The stabilizers selected for this study are CIBA SPECIALTY CHEMICALS Tinuvin® 327 and Tinuvin® 770. Tinuvin® 327 is a long-wavelength UV radiation absorber of the hydroxyphenyl-benzotriazole class, which imparts good light stability to organic polymers. Tinuvin® 770 is a hindered amine light stabilizer (HALS) for applications demanding light stability for thick sections and products with high surface area. The crosslinking catalyst Metatin® 740 was supplied by ROHM AND HAAS and is di-(*n*-butyl)tin bisketonate. This tin compound works best as a crosslinking catalyst and is widely accepted and used to cure MS Polymer®. The CRAY VALLEY rheology modifier employed is Crayvallac® Super and is a micronized amide wax. When activated at rather low temperatures it imparts excellent dispersion, aiding both extrusion and application.

The two key points of producing a single component MS Polymer® sealant are to eliminate moisture from fillers and pigments (e.g., $CaCO_3$, TiO_2) and to avoid moisture contact. Therefore the filler and pigment were pre-dried physically in an air-ventilated oven at 120 °C for 12 h prior to compounding. Pre-drying may also be achieved chemically by DYNASYLAN® VTMO during compounding. The following sealant formulation was used to evaluate the performance characteristics of a MS Polymer® sealant:

Component	phr	wt%
- Binder blend		24.5
MS Polymer® S203H	60	
MS Polymer® S303H	40	
- Filler		49.0
Carbital® 110S	200	
- Pigment		2.5
Tronox® R-FK-2	10	
- Plasticizer		17.2
Jayflex® DIUP	70	
- Rheology modifier		4.9
Crayvallac® Super	20	
- Crosslinking catalyst		0.2
Metatin® 740	1	
- Stabilizers		0.5
Tinuvin® 327	1	
Tinuvin® 770	1	
- Drying agent		0.7
DYNASYLAN® VTMO	3	
- Adhesion promoter		0.5
DYNASYLAN® DAMO-T	2	
or DYNASYLAN® 1146		
Total:	408	100.0

Sealants were prepared in a MOLTENI Labmax using a one-liter, double-planetary mixer equipped with an oil-heated jacket mixing vessel. Typically, the plasticizer was charged at room temperature (approx. 20 °C), and an equal amount of filler and pigment were added slowly with intense stirring to facilitate thorough dispersion. When the desired level of dispersion was obtained, the MS Polymer® binder, two thirds of the DYNASYLAN® VTMO drying agent, the rheology modifier, the stabilizers, and the rest of the filler and pigment were added. The resulting suspension was mixed and stripped under full vacuum (< 5 mm) at 90 °C for 120 min to achieve full rheological activation and dispersion (pasty consistency). The sealant was then cooled to 50° C. In the absence of vacuum, the last third of the DYNASYLAN® VTMO drying agent, the diaminosilane DYNASYLAN® DAMO-T or DYNASYLAN® 1146 adhesion promoter, and the crosslinking catalyst were added and the mixture was stirred for an additional 30 min at 50 °C. With all the components thoroughly mixed, the sealant was de-aerated by applying vacuum (< 5 mm) for 5 min. The finished sealant was removed and packaged in cartridges.

Physical properties were evaluated using standard DIN or ASTM specifications. The sealants were filled into Teflon® molds to form homogeneous test pieces of comparable thickness. The specimens were then moisture cured and conditioned at 25 °C and 50% relative humidity for 14 days before mechanical property testing. The hardness of the cured sealant samples was measured by Shore A. Shelf life at 50 °C was determined for a maximum of 21 days. Tack-free times were determined by finger touch under ambient conditions. For adhesion testing the substrates were first wiped with either methyl ethyl ketone (aluminum, steel, glass, concrete, wood) or methanol (PVC, PMMA, ABS, polystyrene), then washed with detergent, rinsed with distilled water, and allowed to air dry prior to preparation of the test specimens. Specimens were cured for 14 days at ambient conditions.

DYNASYLAN® 1146 Adhesion Promoter

Monomeric diaminofunctional silanes such as DYNASYLAN® DAMO-T and DYNASYLAN® 1411 are among the most widely used adhesion promoters for MS Polymer® sealants. However, when formulated at dosages of 0.1 to 1.0 wt.% in a sealant, special allergenic health labeling is required. Oligomeric diaminosilane DYNASYLAN® 1146 is not skin sensitizing. As a consequence, sealants made with oligomeric diaminosilane DYNASYLAN® 1146 do not require special allergenic health labeling. In addition to this, oligomeric diaminosilane DYNASYLAN® 1146 adhesion promoter has the following beneficial characteristics:

- chemical multifunctionality
- low volatility and viscosity
- high flash point
- greatly reduced volatile by-product
- improved film-forming properties on substrates
- polymer by OECD definition.

Results

Summarized below are the effects of two diaminosilane adhesion promoters (DYNASYLAN® DAMO-T and DYNASYLAN® 1146) on MS Polymer® sealant properties based on the above given formulation:

Silane		
DYNASYLAN® VTMO [phr]	3	3
DYNASYLAN® DAMO-T [phr]	2	
DYNASYLAN® 1146 [phr]		2
Performance		
Skin formation [h]	1–2	1–2
Tensile strength [N mm^2]	1.3	1.6
Elongation at break [%]	610	860
100% modulus [MPa]	0.4	0.3
Cure [mm after 1 day]	2.3	2.8
Cure [mm after 7 days]	5.4	6.4
Shore A	32	33
Shelf life at 50 °C [d]	21	21
Adhesion to		
- anodized aluminum	CF	CF
- steel	CF	CF
- stainless steel	CF	CF
- teak wood	CF	CF
- glass	CF	CF
- PVC	CF	CF
- PMMA	C3A7	C2A8
- concrete	C3A7	C8A2
- polystyrene	AF	AF
- ABS	AF	C2A8

CF: cohesive failure (100%)

AF: adhesive failure (100%)

Sealants incorporating a typical loading (two parts per hundred parts resin) of DYNASYLAN® DAMO-T adhesion promoter are known to offer good adhesion to glass, aluminum, various polar plastics, and concrete. Decreasing the silane concentration results in poor adhesion. It is evident that the type and amount of diaminosilane adhesion promoter has a significant effect on the crosslink density of the MS Polymer® sealant. The replacement of DYNASYLAN® DAMO-T by oligomeric diaminosilane DYNASYLAN® 1146 results in improvements in elongation at break performance. Oligomeric diaminosilane DYNASYLAN® 1146 provides more flexible crosslink sites, resulting in a more flexible sealant.

Because of its unique oligomeric structure, diaminosilane DYNASYLAN® 1146 exhibits good wetting on substrates and achieves weather- and moisture-resistant bonds to substrates with difficult adhesion properties. Clearly, when added as adhesion promoter, oligomeric diaminosilane DYNASYLAN® 1146 represents the best choice when trying to formulate a low-modulus MS Polymer® sealant having good overall flexibility and wet adhesion performance. In addition, no special allergenic health labeling is required.

Our laboratory testing has confirmed that oligomeric diaminosilane DYNASYLAN® 1146 offers an enhanced performance when used in hybrid sealants. Laboratory testing, however, cannot account for all the variations of application and cure that will be experienced in the field. Therefore, we always recommend that users validate their product's performance under the appropriate conditions of service. Depending on the final application, formulators need to choose the "right" diaminosilane or silane blend to achieve the desired characteristics of the sealant.

Isocyanatomethyl-Dimethylmonomethoxy Silane: A Building Block for RTV-2 Systems

*Wolfgang Ziche**

Wacker-Chemie GmbH, Friedrich-von-Heyden-Platz 1, 01612 Nünchritz, Germany
Tel.: +49 35265 73180 — Fax: +49 35265 743180
E-mail: wolfgang.ziche@wacker.com

Keywords: RTV-2, isocyanatomethyl, silane, permeability

Summary: Two-component room temperature-vulcanizing, condensation-curing systems (RTV-2) are well known in silicone chemistry. Even silicone-based materials cannot fulfill all requirements in diverse applications. It is therefore desirable to combine the curing properties of silicone-based systems with those of other polymer backbones. The use of isocyanatomethyl-dimethylmonomethoxysilane allows the straightforward derivatization of, e.g., hydroxyl-terminated polymers, which yield mono-silanol-terminated polymers upon hydrolysis.

Introduction

Two-component room temperature-vulcanizing, condensation-curing systems (RTV-2) are well known in silicone chemistry (e.g., Wacker ELASTOSIL® M and RT product range, ELASTOSIL® SG 500 and IG 25). Their applications are for potting microelectronic components, fully automatic sealing, mold making, or bonding. They are generally composed of silanol-terminated polydimethylsiloxanes as one component and a curing agent as a second component which contains alkoxy silane crosslinkers and a condensation catalyst. In contrast to one-component room temperature-vulcanizing, condensation-curing systems (RTV-1), they have the advantage of fast and deep-section cure independent of the supply of external moisture. An additional advantage is that by varying the type of curing agent and amount used, the pot life and demolding times can be increased or decreased.

Even silicone-based materials cannot fulfill all requirements in diverse applications. One property of silicone materials is their high gas and moisture permeability, which is disadvantageous in, e.g., edge sealing of insulation glass, where a barrier for water and gas has to be provided. Hydrocarbon polymer-based materials like poly-isobutene possess these desirable properties. However, they lack the ready curability and adhesion of RTV silicone materials. Preferably the advantages of the hydrocarbon base polymer and the silane condensation reaction should be combined.

One-component room temperature-vulcanizing, condensation-curing materials (RTV-1) based on di- or trialkoxysilyl terminated polymers cannot be used, as the vulcanizate formed on the surface of the applied material is water vapor impermeable. Further diffusion of moisture is prevented, and thus deep section cure does not take place (Scheme 2). The solution is to use mono-silanol-terminated polymers.

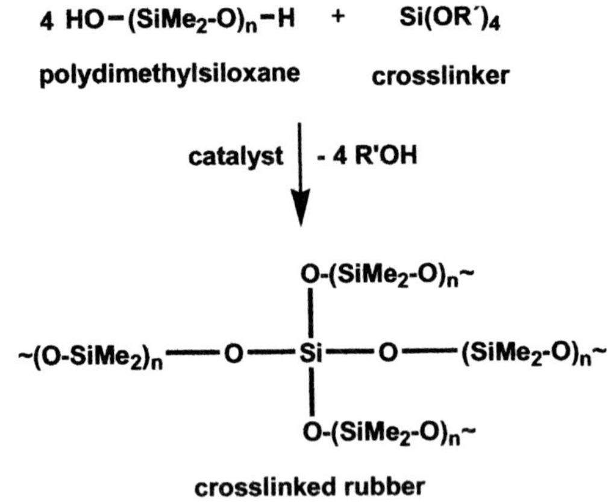

$$4\ HO{-}(SiMe_2{-}O)_n{-}H \quad + \quad Si(OR')_4$$

polydimethylsiloxane crosslinker

catalyst ↓ $-\ 4\ R'OH$

$$\sim(O{-}SiMe_2)_n{-}O{-}Si(O{-}(SiMe_2{-}O)_n\sim)_2{-}O{-}(SiMe_2{-}O)_n\sim$$

crosslinked rubber

Scheme 1. Basic RTV-2 reaction.

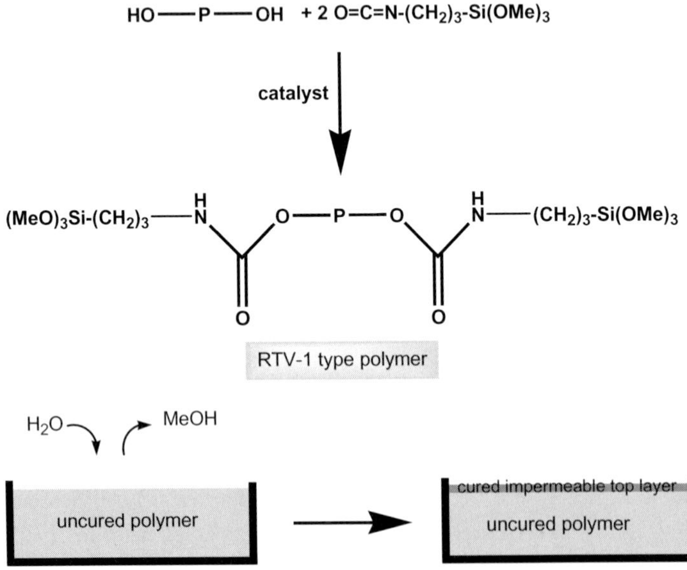

$$HO{-}P{-}OH \ + 2\ O{=}C{=}N{-}(CH_2)_3{-}Si(OMe)_3$$

catalyst ↓

$$(MeO)_3Si{-}(CH_2)_3{-}NH{-}C(=O){-}O{-}P{-}O{-}C(=O){-}NH{-}(CH_2)_3{-}Si(OMe)_3$$

RTV-1 type polymer

H_2O → ← MeOH

uncured polymer ⟶ cured impermeable top layer / uncured polymer

Scheme 2. Why one-component moisture-curing systems are not feasible when the polymeric backbone (denoted by P) is not permeable to moisture.

Synthesis of Silanol-Terminated Polymers

Synthesis of mono-silanol-terminated polymers is possible by the use of the title compound isocyanatomethyl-dimethylmonomethoxy silane. The isocyanato group readily reacts with polymers having terminal groups containing active hydrogen.

Examples of commercially available polymers are the Poly bd® R45-HTLO Resin of Sartomer (hydroxyl-terminated homopolymer of polybutadiene) or Kraton™ Liquid L-2203 (hydroxyl-terminated ethylene/butylene polymer) of Kraton Polymers.

The mono-methoxy-terminated polymers are then hydrolyzed to yield the wanted mono-silanol-terminated base polymers for RTV-2 materials.

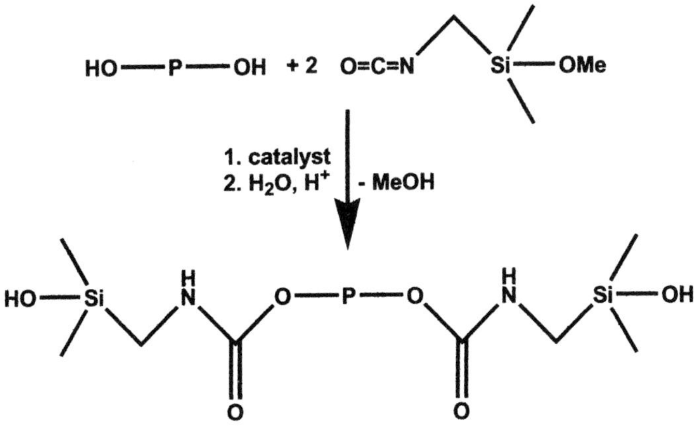

HO—P—OH + 2 O=C=N Si—OMe

1. catalyst
2. H_2O, H^+
- MeOH

HO—Si N O—P—O N Si—OH

Scheme 3. Synthesis of a modified silanol-endcapped polymer.

Furthermore, it is conceivable to use standard chain elongation techniques, e.g., with diisocyanates known from polyurethane chemistry, to adjust the viscosity of the HO-P-OH polymers prior to the derivatization with the isocyanatomethyl-dimethylmonomethoxy silane. This of course also influences the mechanical properties of the vulcanizates.

More elegantly, the chain elongation may also be achieved by condensation of the silanol groups to disiloxane units. In this way the limited availability of dihydroxyl functionalized polymers can easily be overcome, and no extra chain extenders with different reactivity are necessary.

Properties

Provided that the condensation of the silanol end groups is well controlled, the polymers obtained are all flowable. A ^{29}Si NMR spectrum of a typical, partially condensed polymer is shown in Fig. 1.

The polymers can be compounded with appropriate fillers, plasticizers, and other additives.

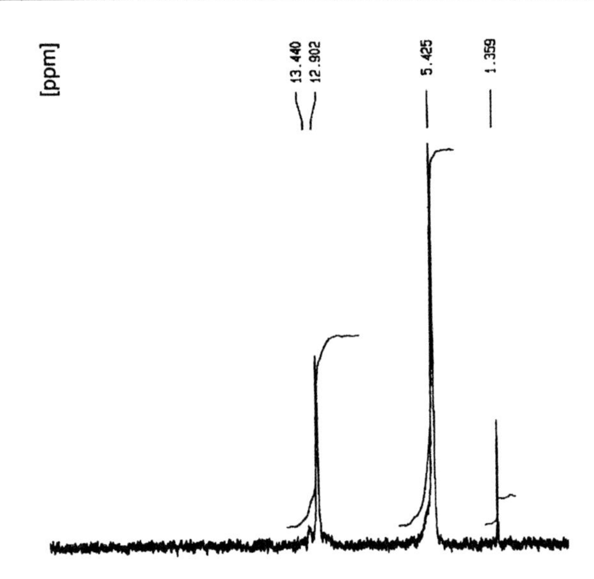

Fig. 1. ^{29}Si NMR of a partially condensed polymer (Si–OH at 13 ppm and Si–O–Si at 5.4 ppm).

Water Vapor Permeability of Vulcanizates

The concept of lower water vapor permeability has been tested in comparison to a silicone based RTV-2 system (Elastosil® SG 500 from Wacker Silicones) according to DIN 53122-1 (samples were 2 mm thick). The Poly bd® RT 45 HTLO-based sample also contained ground calcium carbonate as a filler. It can be clearly seen that the overall permeability is significantly lower for the novel RTV-2 system.

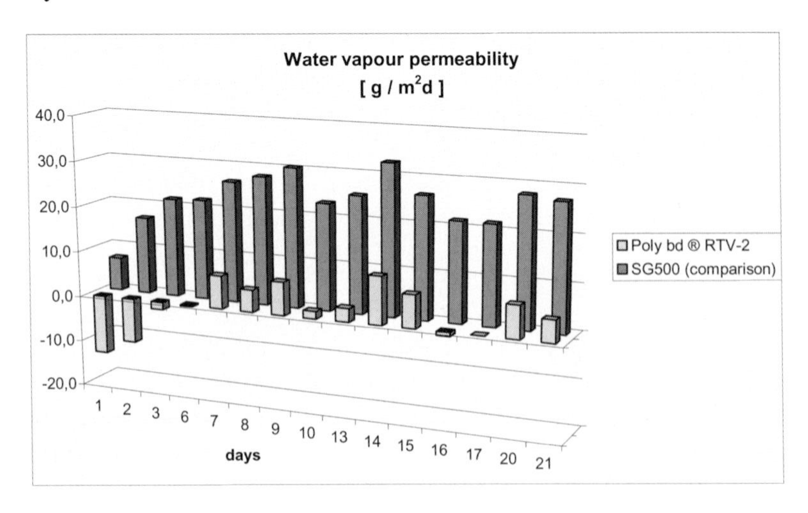

Fig. 2. Water vapor permeability.

Summary and Outlook

The use of the building block isocyanatomethyl-dimethylmonomethoxy silane allows the ready synthesis of mono-silanol-terminated hydrocarbon polymers. These then can be used in the same way as their polysiloxane congeners in RTV-2 applications. The advantages that the vulcanizates then have mainly depend on the performance properties which the base polymer HO-P-OH introduces into the RTV-2 system.

The well-defined end group functionality also allows the derivatization and manipulation of the mono-silanol functional polymers by the same methods that are suitable for silanol-terminated polysiloxanes (e.g., equilibration with polysiloxanes and NR_4OH or "$PNCl_2$" catalysis).

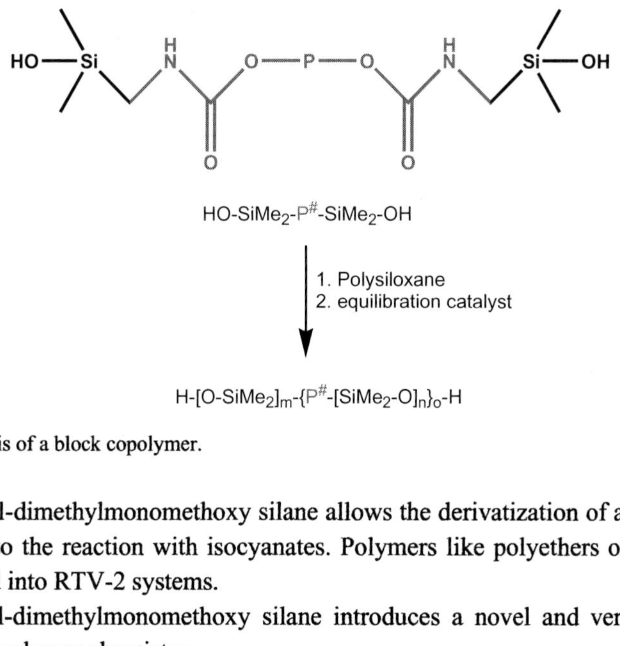

HO—Si ... N(H) ... O—P—O ... N(H) ... Si—OH

$HO\text{-}SiMe_2\text{-}P^{\#}\text{-}SiMe_2\text{-}OH$

1. Polysiloxane
2. equilibration catalyst

$H\text{-}[O\text{-}SiMe_2]_m\text{-}\{P^{\#}\text{-}[SiMe_2\text{-}O]_n\}_o\text{-}H$

Scheme 4. Synthesis of a block copolymer.

Isocyanatomethyl-dimethylmonomethoxy silane allows the derivatization of any polymer having groups susceptible to the reaction with isocyanates. Polymers like polyethers or polyesters can be used and formulated into RTV-2 systems.

Isocyanatomethyl-dimethylmonomethoxy silane introduces a novel and versatile concept into silyl-functionalized polymer chemistry.

The Influence of Different Stresses on the Hydrophobicity and the Electrical Behavior of Silicone Rubber Surfaces

Roland Bärsch*

Hochschule Zittau/Görlitz (FH), FB Elektro- und Informationstechnik
Theodor-Körner-Allee 16, D-02763 Zittau, Germany
Tel.: +49 3583 61 1235 — Fax: +49 3583 61 1213
E-mail: R.Baersch@hs-zigr.de

Heiko Jahn

Siemens AG, Power Transmission and Distribution, High-voltage
PTD H 372, D-13623 Berlin, Germany
Tel.: +49 30 386 36745 — Fax: +49 30 386 26683
E-mail: Heiko.Jahn@siemens.com

Helmut Steinberger

GE Bayer Silicones GmbH&Co. KG, Geb. R20, D-51368 Leverkusen, Germany
Tel.: +49 241 30 65267
E-mail: helmut.steinberger@gesm.ge.com

Robert Friebe

Bayer AG, Abt. BIS-SUA-BM-SM, Geb. Q 18, D-51368 Leverkusen, Germany
Tel. +49 214 30 71450
E-mail: robert.friebe.rf@bayerindustry.de

Keywords: silicone rubber, surface, hydrophobicity, electrical behavior, PDMS

Summary: Based on current knowledge of silicone rubber surfaces and on the methods of evaluating the material-specific properties, influences of different stresses on the hydrophobicity performance and the electrical behavior are presented. The contact angle serves as a diagnosis quantity for evaluating the hydrophobicity. To find correlation to the electrical behavior of insulating surfaces, leakage current measurements were carried out on well-defined pre-stressed LSR-elastomer surfaces with droplet layers. The measurement results are compared to the dynamic contact angle of the surface. Furthermore, the influence of low-molecular-weight (LMW) fluids on the lifetime of the hydrophobicity was investigated.

Introduction

For a long time silicone rubber has been widely used as the housing material of composite insulators for high-voltage outdoor applications. These materials have proved their hydrophobicity and their high resistance to tracking compared to other polymeric insulating materials. In contrast to conventional insulators made of porcelain or glass, the wettability of silicone rubber surfaces is very low. This so-called hydrophobicity causes a low level of leakage currents under real-life conditions. The measured values are far below the critical values for thermally induced material deterioration of organic insulation materials. Hydrophobicity is a dynamic property that can be influenced by exterior stresses, e.g., diffusion of water (rain, fog), electrical surface discharges, or the pollution of the insulating surface with dust. The hydrophobicity can be lost and can be recovered. It can also be transferred through a pollution layer on the silicone rubber surface. Figure 1 shows examples of droplet layers on a silicone rubber surface at different levels of hydrophobicity.

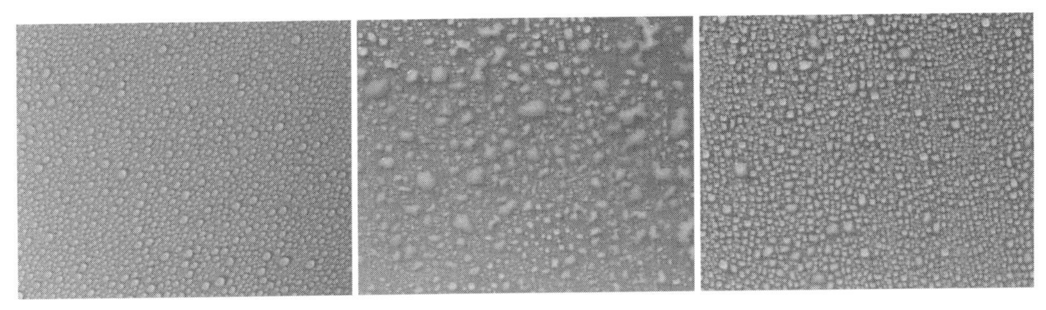

Fig. 1. Droplet configurations on a silicone rubber surface after spraying with a water bottle (distilled water); new sample (left), sample after 2 min stress with electrical discharges (middle), stressed sample after 2 h recovery (right).

Actually there are defined no minimum requirements for the stability or the dynamics of the hydrophobicity of silicone elastomers. The presently existing knowledge has been achieved from long-time field and laboratory tests. Because of the complexity of the stress factors, the results are not always comparable. Thus, systematic laboratory experiments have to be performed to be able to validate the influence of different material parameters. This paper deals with experiments on the influence of single stress parameters on the hydrophobicity. The results are compared with the electrical behavior of droplet layers on the silicone rubber surfaces.

Evaluation of the Hydrophobicity

The hydrophobicity can be evaluated by measuring the contact angle resulting from the equilibrium of the surface tensions according to Young's equation. It was shown in [1] that the equilibrium contact angle Θ_S is a theoretical value valid only for ideal, smooth, and chemical homogenous surfaces. Further, the history of the droplet can influence the measured value [1, 2]. In service, the hydrophobicity of polymeric insulators is often evaluated by the STRI method. This means that distilled water is sprayed on the surface, and the shape and the distribution of the resulting droplets are evaluated [3]. This method is not very exact. Therefore, for measurements the dynamic contact angles were used. They can be easily understood by considering a droplet rolling down an inclined plane (see Fig. 2). If the plane has an angle such that the droplet just starts to move, then the angle on the front side is the advancing angle Θ_A (AA) and the angle on the back side is the receding angle Θ_R (RA). In the laboratory, the angles are measured by increasing (AA) or decreasing (RA) the volume of the measuring droplet (see Fig. 2). For statistical reasons, the angles are measured at 10 different locations distributed on the specimens surface.

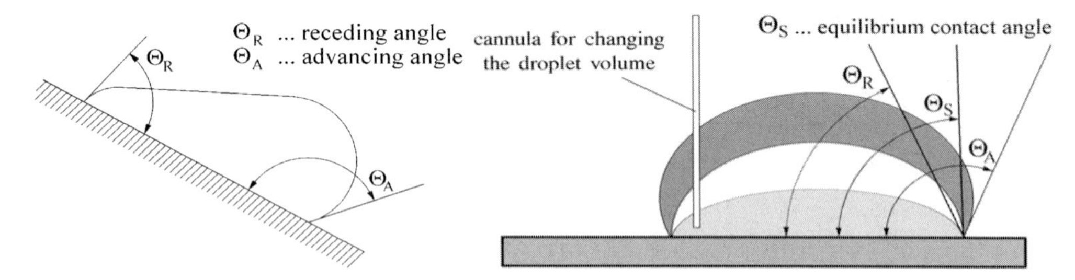

Fig. 2. Definition of the dynamic contact angle (left) and real-life measuring set-up for the dynamic contact angles (right).

Influence of Single Stress Factors on Silicone Rubber Specimens

For the evaluation of the influence of different stresses on the hydrophobicity of silicone rubber surfaces, the following methods were used:

- Water storage of specimens
- Stress with partial discharges (so-called corona discharges)
- Pollution of specimens with an artificial pollution layer made of fumed silica.

The influence of these procedures was documented by measuring the dynamic contact angles.

Figure 3 shows, left, that the receding angle decreases with increasing storage time. Compared to the value measured on the surface of a new specimen, the angle can reach 0% if the storage time is long enough. The value of the advancing angle decreases only a few percent. From the viewpoint of high-voltage insulators, this behavior can lead to completely wetted paths when a droplet rolls down a shed. This region is then electrically short-circuited, and surface partial discharges with critical current values (some mA) can be initiated. For the effect of water see also [9].

The different curves for the receding angles represent different amounts of freely diffusible low-molecular-weight silicone components (LMW). Typical silicone rubbers for high-voltage insulators have a content of LMW between 2 and 3%. It was found that a lower amount of LMW causes a faster loss of surface hydrophobicity. The curve with partially removed LMW represents a material that was thermally treated by a post-curing process. During this process about 1% of the LMW content is lost. The curve marked "LMW removed" represents a sample that was stored in an oven at 180 °C until the mass was constant. This material shows the fastest loss of hydrophobicity during storage in water. This can be explained by a theory of transport processes in polymers [4].

Based on the different values of receding angle after water storage, the samples were dried to find out whether the angles reach the value measured on a new sample. It was found that the angles stabilize at a value of not more than 60% of the start value (Fig. 3, right). Not all of the water seems to be removed after drying. The effect of the amount of LMW content is also visible.

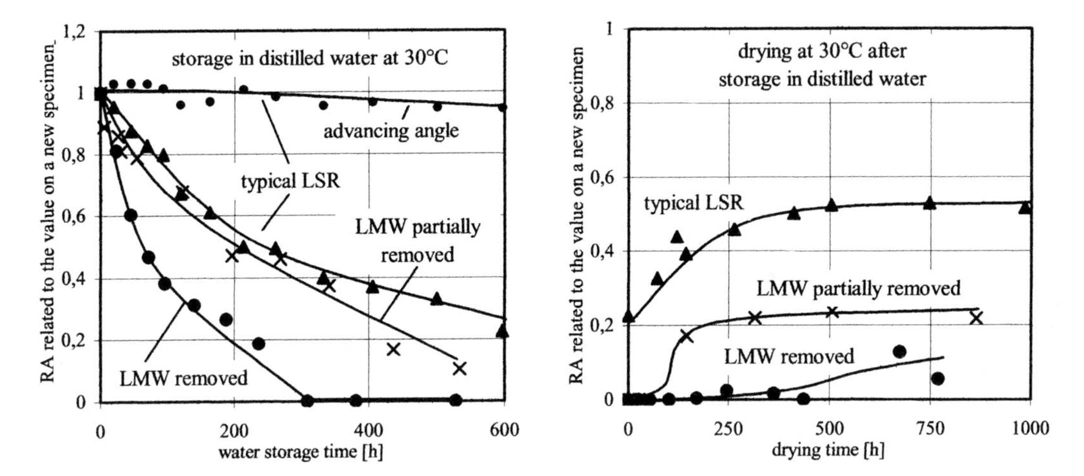

Fig. 3. Dynamic contact angles as a function of water storage time (left) or drying time respectively (right).

The loss of hydrophobicity under the influence of corona discharges occurs much faster than is the case during water storage. To investigate this behavior, a test set-up as described in [5] was used. Figure 4 shows (left) that a nearly complete loss of hydrophobicity occurs within some 10 min. The necessary time for recovery is much longer. After up to some 100 h the receding angle of new samples is reached (Fig. 4, right). It can be observed that also more than 100% of the value for new samples can be measured. The reason is that corona discharges also influence the surface quality. An increased surface roughness was found after stress with corona discharges. Besides this effect, chemical reactions of the silicone rubber surface can also be observed [6].

In contrast to a water-stored specimen, a clear influence of additional LMW could not be found, either during the stress with corona discharges or during the recovery process. It is possible that such an influence would occur when the LMW was completely extracted before the corona treatment.

The transfer of hydrophobicity onto solid pollution layers is possible by the diffusion of LMW that is contained in the silicone elastomers. This means that LMW diffuses into this layer and is lost when the layer is removed. In service this can be caused by wind or rain. To enable comparison of different materials, it was found useful to evaluate the dynamics of this process by measuring the dynamic contact angles on the surface of such pollution layers. For the tests, the procedure with a pollution layer of fumed silica described in [9] was used. With respect to a lifetime of high-voltage outdoor insulators of at least 25 years, we have to ask at the same time for the long-time effects and the repeatability of the transfer effect.

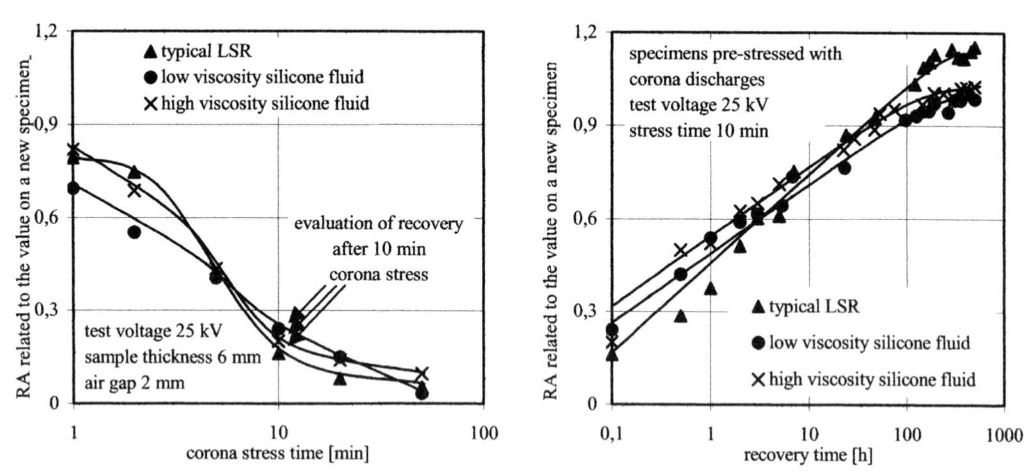

Fig. 4. Receding angle as a function of the duration of corona discharges (left) and recovery of receding angle after the end of the stress (right).

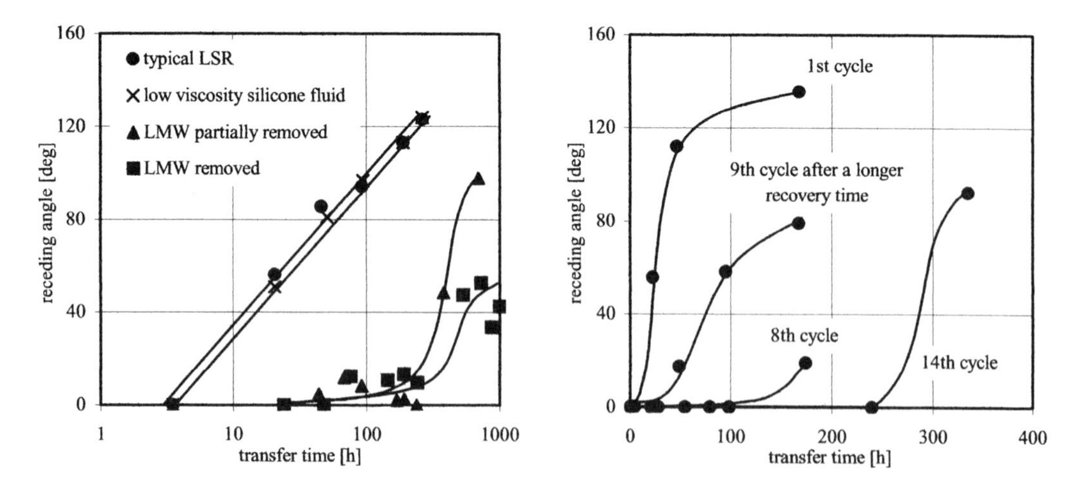

Fig. 5. Receding angle on a pollution layer of fumed silica applied to a silicone rubber sample, measuring new samples with different LMW contents (left) and repeated measuring during cycles of pollution and cleaning (right).

Figure 5 shows (left) that for a fast hydrophobicity transfer, LMW is essentially. Materials with a decreased amount of LMW also show an increase in receding angles but need a significantly longer time. It can be considered that new LMW can be generated within the silicone polymer bulk as result of a chain scission [7]. It was shown that additional LMW does not further accelerate the transfer process.

If weekly cycles of polluting and cleaning of specimens are performed, some of the LMW is taken away with the pollution layer. It is shown in Fig. 5 (right), that with the number of performed cycles the transfer process becomes slower (compare 1st and 8th cycles). After the 8th cycle a recovery period of 4 weeks was performed. It was found that the transfer of LMW into the pollution layer occurs significantly faster in the 9th cycle, but not as fast as in the 1st cycle. This shows that LMW from the bulk needs time to diffuse to the surface of the specimen. Even after 14 cycles a transfer effect can be proved. A lifetime for this process cannot be estimated from these curves. This should be the object of further investigations (see also Ref. [10]).

Electrical Measurements

For high-voltage insulators in service the contact angle is an abstract property that is not easy to measure and has to be interpreted. To compare the influence of transfer of LMW into an artificial pollution layer on the leakage current, a high-voltage electrode arrangement as shown in Fig. 6 was used.

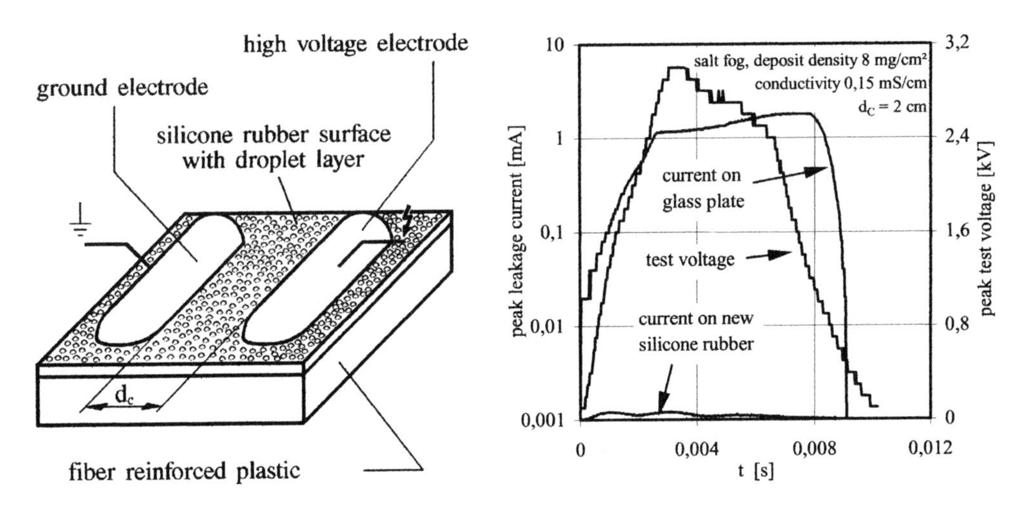

Fig. 6. Test set-up for evaluation of the electrical behavior of droplet layers on silicone rubber samples (left) and examples of measured leakage currents (right).

The test sample was placed in a small fog chamber. A salt fog with a conductivity of 0.15 mS/cm was generated with an ultrasonic dispenser and was transported to the chamber. On the sample, the configuration of a droplet layer can be observed. The specific deposit density was permanently increased and measured by weighing a reference plate.

The electrode configuration was stressed with a voltage impulse of nearly a sine-half-wave shape. This waveform was used to avoid thermal influence on the pollution layer that would occur on a permanently stressed sample. The leakage current was measured by a digital storage oscilloscope.

The both extremes of possible leakage current waveform are shown in Fig. 6. The waveform measured on a clean silicone rubber surface is rather low, in a range of some μA. The droplets are ball-shaped and insulated from each other. The current is mostly influenced by the impedance of the test configuration and of the droplet layer. A good example of a hydrophilic layer is a clean glass plate. Droplets will spread on this surface and a homogeneous layer will be configured that becomes thicker with increased deposit density. The current through this layer can nearly be calculated and is in the range of some mA. This current is used as reference in Fig. 7 and it is shown that calculated and measured values are nearly identical.

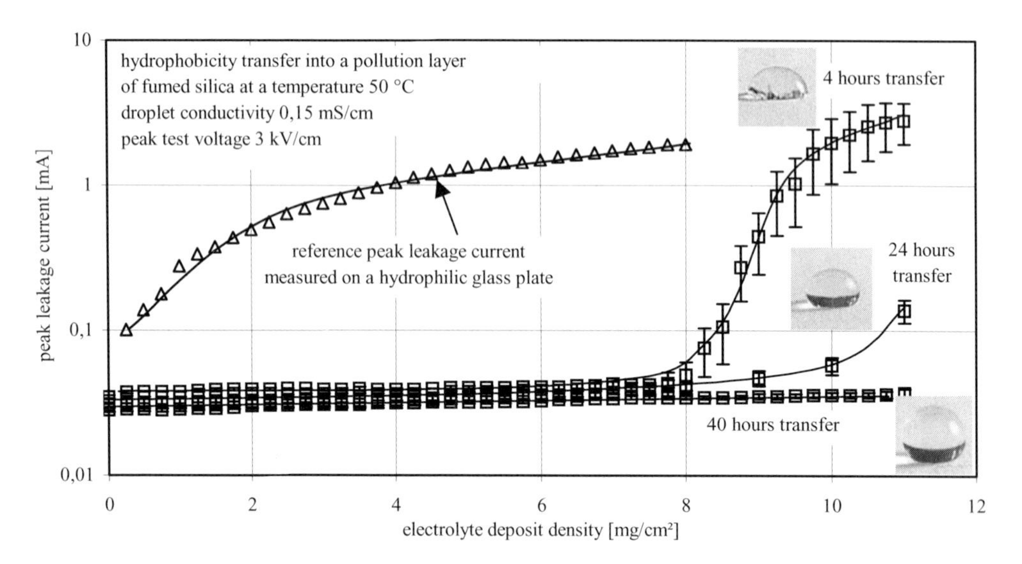

Fig. 7. Leakage current measurements on a pollution layer of fumed silica after different times for recovering the hydrophobicity.

At the beginning of the transfer process the pollution layer is hydrophilic and the measured current is identical to that for a glass plate. Even after 4 h the leakage current is suppressed until a deposit density of 8 mg/cm² is reached (Fig 7). Then the leakage current increases very fast and the values are in the range for hydrophilic layers. After 24 h the leakage current is significantly lower, and after 40 h the leakage current is identical to that for a clean surface without pollution layer. On a pollution layer, very high contact angles can be observed. The droplets are ball shaped and the effect seems to be comparable to the so-called Lotus-effect, described for some plants [8].

The described measurements can be repeated with different pre-stressed samples. Then correlation can be found to the contact angles, as shown in Fig. 8. Over a wide range of receding angles, the leakage current is rather low. If the receding angle reaches about 20°, the leakage current increases rapidly. For the same receding angle, the leakage current increases with decreasing advancing angles. This can be explained by the configuration of droplets. If the surface tension of the sample is too low, then droplets spread more easily during the voltage impulse and cannot reconfigure to a round shape after voltage stress. So they can run into one another, and water paths occur between the electrodes. This is also observed during multi-factor stress tests as described in Ref. [9].

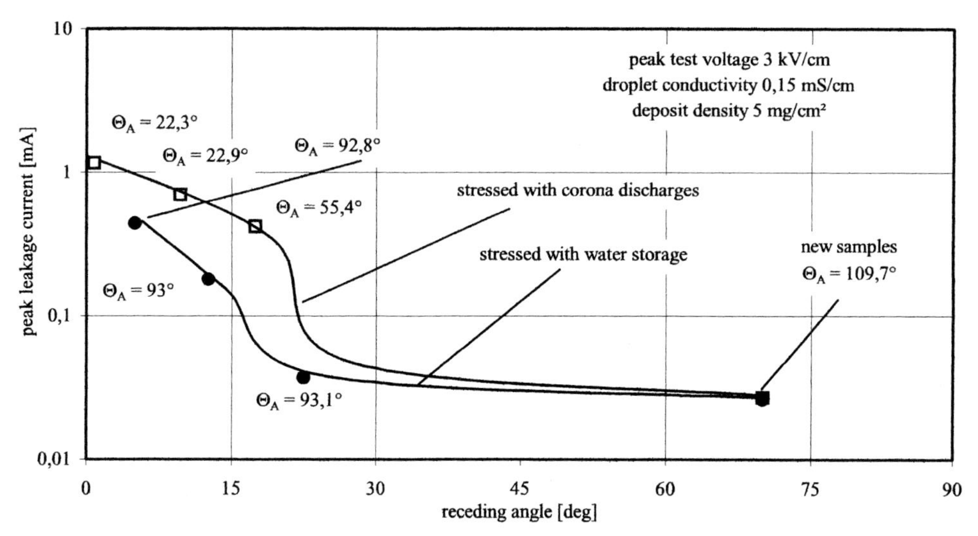

Fig. 8. Peak leakage current of pre-stressed samples as a function of the receding angle, with advancing angle as parameter.

Summary

The influence of single stress factors on the hydrophobicity of silicone rubber surfaces was investigated. It was found that the hydrophobicity decreases during water storage. The receding angle is strongly influenced by water diffusion. It does not recover completely after drying at elevated temperature. Corona discharges have also a very strong influence on the hydrophobicity of silicone rubber surfaces. Both the receding angle and the advancing angle decrease very fast until water droplets completely spread on the surface. The contact angles recover to the level of new samples.

Diffusible low-molecular-weight components have an important influence on the hydrophobicity of silicone rubber samples. They are not only responsible for the transfer of hydrophobicity to adsorbed pollution layers but also for the hydrophobicity loss as result of water storage. Removal of low-molecular-weight components decreases the hydrophobicity performance significantly. Addition of fluids to the silicone rubber bulk does not show significant advantages.

Cycles of pollution and cleaning of silicone rubber samples show that hydrophobicity transfer is a dynamic process that depends on the amount of freely diffusible fluid components. It seems probable that the transfer effect is present during the whole lifetime of a silicone rubber for high-voltage insulators.

Measurements of the leakage current and comparison with dynamic contact angles have shown a correlation between the wetting behavior and the electrical performance of insulating surfaces. A change of both the advancing and the receding angles can influence the leakage current.

Further investigations should be performed to improve comprehension of the dynamic processes that determine hydrophobicity performance of silicone rubber surfaces. A deeper knowledge of the role of low-molecular-weight fluids is still needed. This is also the key to understanding the long-time behavior and will enable the lifetimes of these materials to be estimated.

References

[1] D. Y. Kwok, T. Gietzelt, K. Grundke, H.-J. Jacobasch, A. W. Neumann, *Langmuir* **1997**, *13* 2880.

[2] H. Janssen, U. Stietzel Contact angle measurement on clean and polluted high-voltage polymer insulators in *10th ISH, Conf. Proc. Vol. 3*, p. 149 – 152, Montreal, August **1997**.

[3] STRI Guide 92/1: *Hydrophobicity Classification Guide* Swedish Transmission Research Institute, **1992**.

[4] F. Müller-Plathe, O. Biermann, Simulation von transportprozessen in polymeren, in *Spektrum der Wissenschaft*, Dossier 2/1999: Software, p. 51 – 55, April **1999**.

[5] H. Jahn, *Zur Bewertung stofflicher und herstellungsbedingter Einflußgrößen auf das Hydrophobie- und Erosionsverhalten von Silikonelastomeroberflächen* Dissertation, TU Dresden, Shaker Verlag GmbH, **2003**.

[6] H. C. Hillborg, *Loss and Recovery of Hydrophobicity of Polydimethylsiloxane after Exposure to Electrical Discharges* PhD Thesis, KTH Stockholm, **2001**.

[7] H. Janssen, A. Herden, H. C. Kärner, LMW components in silicone rubbers and epoxy resins, in *11th ISH, Conf. Proc. Vol. 4*, p. 4.18.S17 – 4.21.S17, London, August **1999**.

[8] W. Barthlott, C. Neinhuis, *Planta* **1997**, *202*, 1.

[9] J. Lambrecht, *Über Verfahren zur Bewertung der Hydrophobieeigenschaften von Silikonelastomer-Formstoffen*, Dissertation, TU Dresden, Shaker Verlag GmbH, **2001**.

[10] R. Bärsch, U. Gustke, J. Lambrecht, H. P. Wolf, On the reproducibility and the long-term stability of the hydrophobicity transfer ability of silicones, in *13th ISH, Conf. Proc.*, p. 183 ff, Delft, August **2003**.

Silicone Magnetoelastic Composite

G. V. Stepanov, E. I. Alekseeva, A. I. Gorbunov*

State Research Institute for Chemistry and Technology of Organoelement Compounds
38, Shosse Entuziastov, 111123, Moscow, Russia
Tel.: +7 95 273 7180 — Fax: +7 95 273 1327
E-mail: alena@eos.incotrade.ru

L. V. Nikitin

Moscow State University, Physical Department

Keywords: silicone matrix, magnetic filler, elasticity

Summary: This chapter describes research into magnetic silicone composite properties. The main components of this composite are silicone matrix and magnetic fillers (fine iron and ferrite powders). Composites with changing elasticity can be obtained by modifying matrix composition and the content of fillers. These products are "smart materials", and it can be expected that they will have wide application in electronics and electrical engineering

The different types of silicone elastomers (polycondensation, polyaddition, and radical vulcanization polymers) were tested as matrices for magnetoelastic composites [1, 2]. Magneto-elastic is a new magnetic composite capable of changing its properties under the action of a magnetic field. The investigated materials represent a new type of composite consisting of small magnetic particles dispersed within a highly elastic polymeric matrix. They are named "smart materials" [3]. These materials are unique in their capacity to increase elongation [3 – 5] and elasticity [6, 7] in a magnetic field by up to 100%.

It was found that the silicone polyaddition polymers are more suitable as a matrix because of the absence of by-products from the process of vulcanization and the formation of ethane bridges (Scheme 1) in the polymer, which significantly strengthens the end product, while high elasticity is retained.

The main object of this research was a study of highly elastic polymer matrix formation processes for magnetelastic composites. An elastic polymer consisting of hydride- and vinyl-containing oligomers was synthesized by hydrolytic copolycondensation of corresponding organochlorosilanes and catalytic polymerization of organocyclosiloxanes with vinyl-containing disiloxanes. The matrix, with tailor-made physical and mechanical performance, was produced by two techniques:

- Varying vinyl and hydride groups ratio in starting components;
- Plasticization of high-strength polymers through introduction of polymethylsiloxane fluid (PMS) into the matrix (Fig. 1).

$$\sim\!\sim Si\text{–}CH{=}CH_2 + H\text{–}Si\sim\!\sim \longrightarrow \sim\!\sim Si\text{–}CH_2\text{–}CH_2\text{–}Si\sim\!\sim$$

$$Si\text{–}H + H_2O \longrightarrow n\, Si\text{–}OH + H_2\uparrow$$

$$Si\text{–}H + HO\text{–}Si \longrightarrow Si\text{–}O\text{–}Si + H_2\uparrow$$

$$2\, SiH + O_2 \longrightarrow 2\, Si\text{–}OH$$

$$2\, Si\text{–}OH \longrightarrow Si\text{–}O\text{–}Si + H_2O$$

Scheme 1. Polyaddition polymerization.

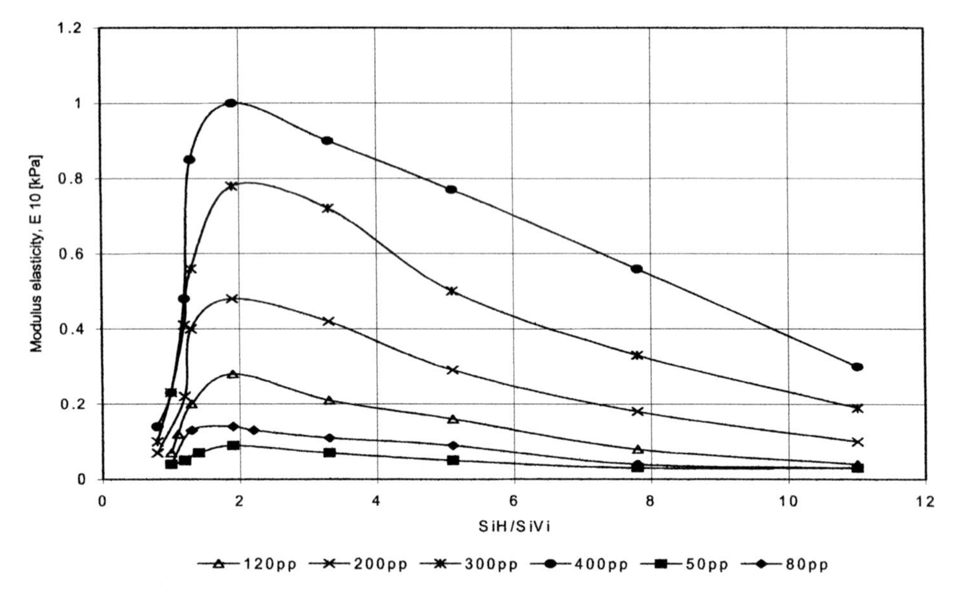

Fig. 1. Modulus of elasticity as a function of ratio SiH/SiVi at different contents of PMS-200.

Magnetoelastic (ME) parameters — elasticity or Young's Modulus (E), strength (P), elongation at failure (L) were evaluated. Strength decrease accompanied by elasticity increase for both synthetic methods of highly elastic matrix was observed, i.e. P/E or E/P ratio was constant for a specified silicone composition. ME was produced by means of mixing silicone matrix with magnet powders. Then the composition was polymerized at high temperature (120 – 150 °C) for 1 – 3 h.

For the purpose of composition optimization of the silicone matrix, two types (I, II) of polyaddition compounds were researched. Special additional agents (QM) were synthesized for increasing tensile strength.

Composition I consists of hydride-containing oligomer (M_W ~800, content of active hydrogen

0.4 – 0.5%) and oligomer with terminal dimethylvinylsiloxy units (M_W ~75 000).

Composition II consists of the same hydride-containing oligomer and oligomer with terminal trivinylsiloxy units (M_W ~70 000).

The physical mechanical properties for initial and filled materials, with compositions, are presented in Tables 1 and 2.

Table 1. Influence of reinforcing agent on physical–mechanical parameters of composition I.

NN	Content of QM [%]	Initial composition			Filled composition		
		P [kPa]	*L* [%]	*P/L*	*P* [kPa]	*L* [%]	*P/L*
1	–	70	179	0.39	400	230	1.7
2	0.2	130	188	0.70	870	282	3.1
3	0.5	270	305	0.90	1050	296	3.5
4	0.8	240	192	1.25	1130	245	4.6
5	1.0	3.90	201	1.95	1270	215	5.9

Table 2. Influence of reinforcing agent on physical–mechanical parameters of composition II.

NN	Content of QM [%]	Initial composition				Filled composition			
		P [kPa]	*L* [%]	*E* [kPa]	*P/E*	*P* [kPa]	*L* [%]	*E* [kPa]	*P/E*
1	–	450	130	400	1.1	1650	244	560	2.9
2	0.2	760	191	370	2.05	2000	223	324	6.1
3	0.5	1130	239	280	4.03	1550	205	397	3.9
4	0.8	870	224	210	4.14	1130	165	306	3.7
5	1.0	590	210	150	3.9	1030	193	170	6.1

It is seen from Tables 1 and 2 that the introduction of QM increases the tensile strength of polymer matrix and ME while preserving the high elongation (parameter *P/L* increases too). It is very interesting that the introduction of magnetic filler increases the composite tensile strength threefold with some increasing of elongation.

As shown in Tables 1 and 2, other conditions being equal, compositions of series II have higher tensile strength than analogous compositions of series I. This is explained by additional ethane bridges in the polymer net thanks to three terminal vinyl groups.

As seen from Fig. 1 and Table 3, the introduction of plasticizer significantly increases the elongation and decreases the elasticity and tensile strength.

The silicone matrix used for composite production was most favorable, as it allowed simple modifying adjustment of composite viscous and elastic properties according to composition, temperature, and time of polymerization. The presence of filler in the cured composition may influence mechanical properties in two ways. On the one hand, filled composition strength outperforms the initial strength; on the other hand, active filler surface may partly inhibit polymerization, up to its complete blocking. All samples (Table 3) were prepared on the base composition III (compound SIEL 159 – 254 is the trade mark). So, Fe^{2+} ions can exist on the iron metallic filler and magnetite surface. The initial surface is oxidized and covered by oxide film in which Fe is fully oxidized and is in the Fe^{3+} condition. However, along with the increase of temperature up to 150 °C, exchange processes begin in the filler. The surface oxygen of the oxide film diffuses into filler particles, and Fe^{2+} ions are formed on the surface. These ions are very active inhibitors of polyaddition reaction. The studies showed that the rate of polymerization was decreasing when the filler concentration was increasing; at the same time, the elasticity was decreasing. If the magnetic powder was fresh or old powder was ground and introduced into the polymer, the process of polymerization was practically blocked. Polymerization was not observed at high filling of polymer (~10 vol%).

The study of the influence of plasticizer on physical-mechanical properties of ME is presented in Table 3. Introduction of plasticizer causes elongation increase (up to 1100 %) and reduction of elasticity (up to 1.7 kPa) at the same time.

Table 3. Influence of filler and plasticizer concentration on physical mechanical parameters of composition III.

NN	Filler [%]	Content of plasticizer in polymer [%]	*P* [kPa]	*L* [%]	*E* [kPa]	*P*/*E*
1	50	0	1650	–	400	4.1
2	60	0	1720	–	510	3.3
3	70	0	1300	–	220	5.9
4	60	50	28	1130	1.76	16*
5	75	50	173	309	9.6	18*
6	60	0	1800	200	540	3.3
7	60	50	370	410	31	12
8	60	75	53	550	5.5	9.6

* In this case the hydride-containing oligomer was added at the rate of 15 wt%, and the stoichiometric quantity is 10 wt% (Scheme 1, reaction 1). As a result, ratio *P*/*E* was increased.

Our investigations showed that composite elasticity increase in a magnetic field. The increase in elasticity in a magnetic field is higher when the starting composite elasticity is lower. ME elasticity varied within the range 1 – 500 kPa. In Fig. 2, the dependence of elasticity on magnetic field is shown. For samples with initial elasticity 2 kPa placed in a magnetic field (the value of the magnetic field is 4 kOe), elasticity increases by up to 800%. When initial composite elasticity increases up to 200 kPa, no elasticity increase in the magnetic field occurs.

Detailed study of ME properties demonstrated that in the case of ME synthesis in the magnetic field, significant anisotropy of ME elasticity was observed. Elasticity measured for an ME sample along the magnetic filler orientation axis is a few times higher than the sample elasticity in the perpendicular direction (Fig. 3).

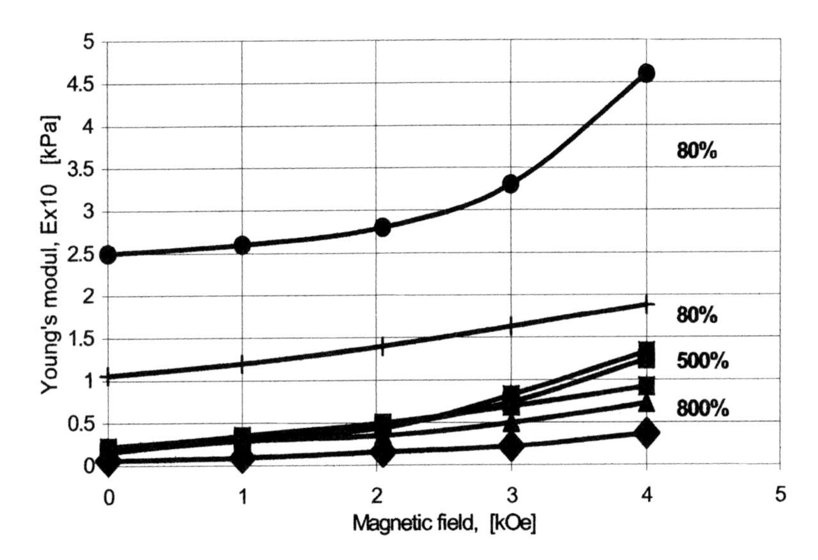

Fig. 2. Dependence magnetoelastic elasticity on size of magnetic field. The percent (%) is ratio of elasticity in magnetic field to initial elasticity.

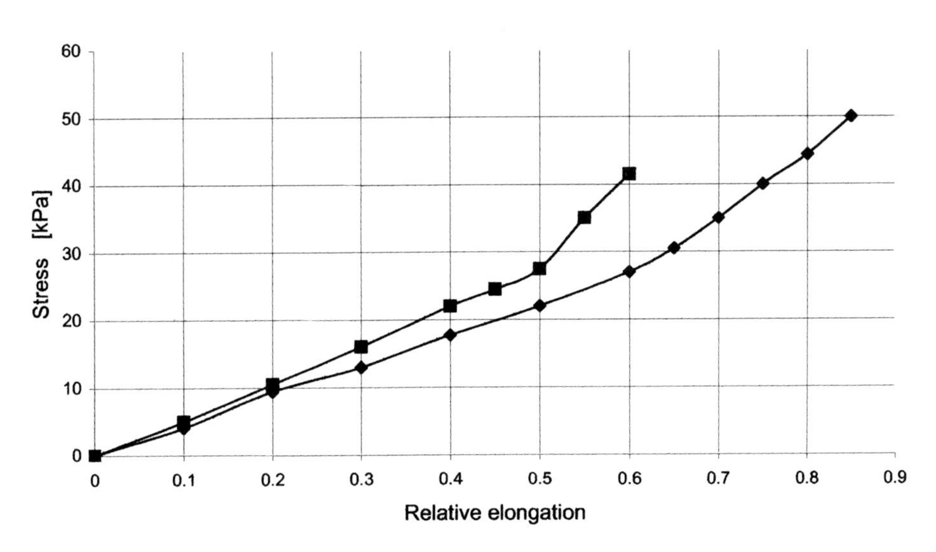

Fig. 3. Dependence of stress in a sample on its relative deformation.

Conclusion

A polymeric matrix based on a silicone polyaddition composition is promising for the creation of magnetic control material. This type of polymer allows us to regulate the elastic properties of the ME in the range 1 – 500 kPa and elongation up to 1100%. In the magnetic field the elasticity increased by 800 %.

Acknowledgment: This work is supported by NATO Grant SfP 977 998.

References

[1] S. Nanushyan, E. Alekseeva, A. Polees, *J. Chemicheskaya Prom.* **1995**, *N11*, p. 691.
[2] M. A. Brook, *Silicon in Organic and Polymer Chemistry*, John Wiley & Sons Inc., **1999**.
[3] D. Kouzoudis, C. A. Grimes, *Smart Mater. Structure* **2000**, *9*, 885 – 889.
[4] L. Lanotte, G. Ausanio, V. Iannott, *Phys. Rev. B*, **2001**, *6*, 63.
[5] M. Zrinyi, D. L. Szabo, H.-G. Kilian, *Poly. Gels Networks* **1999**, *6*, 441 – 454.
[6] L. V. Nikitin, L. S. Mironova, G. V. Stepanov, A. N. Samus, *Poly. Sci. A* **2001**, *43(4)*, 443.
[7] J. D. Carlson, M. R. Jolly, *Mechanotronics* **2000**, *10*, 555 – 569.

Modifiers for Compounded Rubbers Based on Fluoro- and Phenylsiloxane Rubbers

*O. G. Ryzhova, T. N. Korolkova, S. N. Kholod**

State Research Institute for Chemistry and Technology of Organoelement Compounds,
38 Shosse Entuziastov, 111123, Moscow, Russia
Tel.: +7 95 273 7189, 750 3185 — Fax: +7 95 750 3185, +7 95 913 2538
E-mail: gamm@ostrov.net, eos@eos.incotrade.ru

Keywords: rubbers, modifiers, siloxanes

Summary: Comprehensive investigation of the influence of nine modifiers on the properties of fluoro- and phenylsiloxane rubber was carried out. The synthesized modifiers were analyzed by DTA, TGA, and ^{29}Si NMR methods. Modifiers' polydispersivity was evaluated in terms of molecular-mass distribution (MMD). The modifiers were found to produce a selective effect on the performance of phenyl- and fluorosiloxane-based rubbers, and at their optimum content an improvement in most rubber properties was observed. The optimum amount of the introduced modifier depends on its structure.

As can be seen from the literature [1], the properties of natural rubber may be widely varied by modification with different oligomeric additives to rubber mixtures. Oligomers with various chemical structures and modifiers with polyfunctional action are predominantly used. In the presence of the latter, change in properties of rubber compounds in a desired manner is achieved with more efficiency. The chemical structures of the modifiers and their concentration in the oligomeric composition of the raw rubber have a substantial effect on rubber properties (elasticity, frost and heat resistance, strength, dynamic and fatigue characteristics, hardness) and thus rubber processing facilitates tackiness, cohesive strength, and vulcanization parameters. Therefore this article deals mainly with the development of synthesis methods for polyfunctional modifiers as well as with the determination of the relationship between the structure and properties of the obtained fluoro- and phenylsiloxane rubber.

Modifiers were synthesized by the method described in [2] with optimization of certain process stages (see Eq. 1).

$$3n\ Ph_2Si(OH)_2 + n\ [Me_2SiNH]_3 \longrightarrow HO\{[Ph_2SiO]_a\text{–}[SiMe_2O]_b\}_nH + 3n\ NH_3$$

Eq. 1.

For the investigations, the following compounds were synthesized:

(a) Linear oligosiloxanes with terminal hydroxyl groups — α,ω-dihydroxydimethyldiphenyl-polysiloxanes (DMDPhS) of general formula:

$$HO\{[Ph_2SiO]_a\text{–}[SiMe_2O]_b\}_nH, \quad \text{where } a \approx b$$

Modifiers of group (a) were synthesized according to Eq. 1. Characteristics of obtained modifiers with a molecular mass* of 3 200 up to 22 000 are presented in Table 1.

(b) Linear oligosiloxanes with terminal hydroxyl groups – α,ω-dihydroxydimethyldiphenyl-oligosiloxanes (DMDPhS^M) of general formula:

$$HO\{[Ph_2SiO]_a\text{–}[Me_2SiO]_b\text{–}[SiRO]_c\}_nH$$

where R = —OAlO— (with O below) (V–VII); Si (silacyclopentene) (VIII); Si, Si (dibenzodisilin) (IX)

Monomers containing the corresponding fragment R are introduced into the reaction at the stage of heterofunctional condensation.

(c) Modifiers of cyclolinear structure containing three functional fragments that provide additional spatial cross-linkers (MVPhS) of general formula:

$$[Me_3SiO]_a[Me_2SiO]_b[MeVinSiO]_c[Ph_2SiO]_d[PhSiO_{1.5}]_f[PhSi(OH)O]_g$$

Table 1. Physical-chemical characteristics of modifiers of group (a) (DMDPhS).

Modifiers		η*** [cst]	M_M	n_D^{20}
DMDPhS	I	1.92	3200	1.559
DMDPhS	II	2.30	5000	1.563
DMDPhS	III	5.24	13800	–
DMDPhS	IV	7.06	22000	1.567
DMDPhS^M	V	2.47	5000	–

*** kinematic viscosity of 20% solution in toluene.

* Calculation of reagents introduced into the reaction for determination of molecular mass was conducted by the formula: $m = 3n/m$; $n = (M_m - 219)/272$, where m = number of diphenylsilandiol (DPhSDO) moles; n = degree of polymerization; M_m = required molecular mass; 219 = molecular mass of hexamethyldisilazane, 272 = molecular mass of DPhSDO.

Modifiers of group (c) were synthesized according to methods [3, 4] by hydrolysis of a mixture of organochlorosilanes with further catalytic rearrangement.

Identification of all structural fragments in the obtained modifiers was performed by ^{29}Si NMR spectroscopy, and their ratio was determined. The ^{29}Si NMR spectrum of group (a) oligomers consists in signals of two groups: D (18 – 21 ppm) and D^{ph}_2 (45 – 47 ppm). The content of OH groups accounted for less then 2% and is not determined in the spectrum. Observed fission (Fig. 1) for signals of D and D^{ph}_2 fragments demonstrates the disturbance of fragments in oligomeric chains at polycondensation.

For oligomers of group (a) chain parameters of microstructures were estimated in accordance with procedure described in Ref. 5; chain structure was indicated by irregularity coefficient B. A Value of $B > 1$ indicates the trend to alternative chain structure. The ^{29}Si NMR spectrum of group (b) modifiers is similar to the spectrum in Fig. 1; however, the fission level of signals is considerably higher. Modifiers of this type represent a block with alternating D and D^{ph}_2 fragments connected by fragment R.

The Ratio of the MVPhS structural fragments determined by ^{29}Si NMR spectroscopy is as follows:

$[Me_3SiO] : [Me_2SiO] : [MeVinSiO] : [Ph_2SiO] : [PhSiO_{1.5}] : [PhSi(OH)O] = a : b : c : d : f : g = 7.65 : 16.6 : 23.4 : 31.3 : 13.7 : 7.35.$

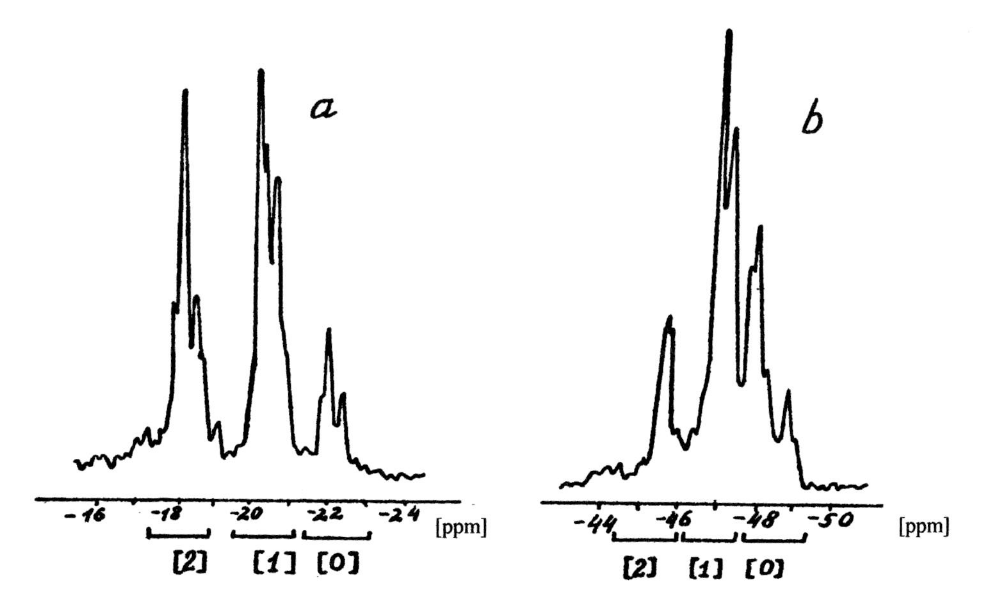

Fig. 1. ^{29}Si NMR spectrum of α,ω-dihydroxydimethyldiphenyloligosiloxane in d_6-acetone. *a*: silicon atoms signal of D fragment; *b*: silicon atoms signal of D^{Ph}_2 fragment.

As seen from characteristics given in Table 2 for mass-molecular distribution (MMD) of D^{ph}_2, all the modifiers are characterized by high homogeneity excluding DMDPhS^{m}, testifying to all introduced monomers in the chain construction. In modifier DMDPhS (VI), a low molecular fraction is present (13%).

Table 2. Molecular mass characteristics of modifiers.

Modifiers		M_n	M_w	M_z	M_w/M_n	M_z/M_n	Q**
DMDPhS	**I**	1320	3020	6400	2.28	2.11	–
DMDPhS	**II**	1600	3600	6890	2.25	1.91	–
DMDPhS^{M}	**V**	1700	4100	8000	2.41	1.95	0.205
DMDPhS^{M}	**VI**	1500	3800	7300	2.53	1.92	0.088
DMDPhS^{M}	**VII**	1800	4300	8400	2.40	1.95	0.086
DMDPhS^{M}	**VIII**	1740 370	3300 390	6200 410	1.90 1.05	1.88 (87%) 1.05 (13%)	0.19
DMDPhS^{M}	**IX**	900	1380	3000	1.53		0.068
MVPhS	**X**	2700	8910	23160	3.30	2.60	–

** Ratio of R-containing monomer to sum of initial reagents of Eq. 1.

All synthesized oligomers, as shown by DTA and TGA data, are characterized by high thermal stability. Thermal oxidative destruction occurs at temperature as high as 500 °C.

Table 3. Effect of modifier's molecular mass on fluorosiloxane rubber properties.

Properties	Modifiers DMDPhS, concentration in mass parts											
	DMDPhS II				**DMDPhS III**				**DMDPhS IV**			
	0*	**4**	**6**	**8**	**0***	**5**	**7**	**10**	**0***	**5**	**7**	**10**
Tensile strength [MPa]	8.0	7.0	7.4	6.3	7.1	6.7	7.2	6.2	7.1	6.7	7.2	6.2
Relative elongation [%]	400	380	365	310	325	375	430	425	325	400	350	425
Relative elongation [%]	0	0	0	0	0	0	0	0	0	0	0	0
Tearing strength [kP/m]	18	19	23	18	25	30	25	31	25	25	30	21
Shore hardness	73	75	74	72	70	63	66	68	70	67	68	68
Thermal treatment of rubber at 250 °C for 24 h												
Tensile strength [MPa]					2.7	3.1	2.9	3.0	2.7	2.8	3.4	3.5
Relative elongation [%]					225	350	290	265	225	250	265	280
Relative elongation [%]					225	350	290	265	225	250	265	280
Shore hardness					70	68	71	70	70	69	73	72

* Modifier-free control specimen.

The effect of synthesized modifiers on fluorosiloxane rubbers is determined on the basis of fluorosiloxane rubber CKTFT-100 P[†]. As seen from data presented in Table 3, characteristics of fluorosiloxane rubber depend on quantity of introduced modifier and its molecular mass (M_m). Properties of rubber compounds are highly dependent on modifier concentrations. Optimal modifier content amounts to 6 – 7%. Increase in molecular mass of modifier on transition from modifier II to modifier VI leads to increase in tear strength and relative elongation in comparison to the control specimen (modifier concentration = 0). Also, after thermal treatment at 250 °C for 24 h in the presence of modifier VI, improvement of all fluorosiloxane rubber compound properties is observed.

Modifiers of group (b) have a more pronounced influence on fluorosiloxane rubber compound characteristics (Table 4). Optimal content of each modifier is individual and varies from 3.5% for modifier VI up to 7% for modifier VIII. Thus, in the presence of modifier VI, increase of strength to 9.4 MPa (instead 6.4 MPa for the control specimen) may be achieved, and Shore hardness is increased to 60.

Achieved improvements in parameters are retained after thermal treatment (200 °C for 24 h) and Shore hardness is increased to 65.

Table 4. Effect of synthesized modifiers on the behavior of fluorosiloxane rubber compounds.

Properties	**Modifiers, concentration in mass parts**									
	DMDPhS^{M} VII			**DMDPhS^{M} VI**		**DMDPhS^{M} IX**			**DMDPhS^{M} VIII**	
	0*	**5**	**7**	**0***	**3.5**	**0***	**5**	**7**	**0***	**4**
Tensile strength [MPa]	7.1	7.5	6.2	6.4	9.4	7.1	6.5	8.4	8.0	6.4
Relative elongation [%]	325	415	500	410	360	325	515	350	400	290
Relative elongation [%]	0	10	10	20	10	0	25	0	0	0
Tearing strength [kP/m]	13	20	14	–	–	13	30	26	18	22.5
Shore hardness	70	65	55	55	60	70	55	55	73	83
Thermal treatment of rubber at 200 °C for 150 h				**Thermal treatment of rubber at 200 °C for 24 h**						
Tensile strength [MPa]	5.5	5.4	4.7	7.0	8.6	5.5	4.5	5.6		
Relative elongation [%]	250	300	350	400	350	250	400	330		
Shore hardness	65	55	55	60	65	65	60	60		

* Modifier-free control specimen.

The influence of synthesizing modifiers on phenylsiloxane rubber compounds was studied on the basis of siloxane rubber CKTF–2103[‡]. Data given in Table 5 show that the effect of modifiers of

[†] Composition of rubbers in all cases remains constant: CKTF-100P – 100 mass parts, peroxide – 1.5 mass parts, Aerosic A-300 – 40 mass parts (dispersing agent and iron oxide also remain constant).

[‡] Composition of rubbers in all cases remains constant: CKTF rubber – 100 mass parts, peroxide –1.5 mass parts, Aerosic A-300 – 40 mass parts (dispersing agent and iron oxide also remain constant).

group (a) on phenylsiloxane rubber compound is similar to their effect on fluorosiloxane rubber compounds: characteristics of rubber compounds improve with increase in M_m. Thus, on increase to 5000 M_m all parameters are increased by 1.4 times, excluding hardness, which stays unchanged. The character of the modifier has also a marked effect on phenylsiloxane rubber compounds. In the presence of modifiers VII, even in amounts of 3%, hardness increases by 1.5 times, tearing strength by 1.1 times, residual elongation by 1.3 times, and moderate changes in other parameters were observed.

Table 5. Effect of modifier on the character of phenylsiloxane rubber compounds.

Properties	Modifiers [concentration in mass parts]													
	DMDPhS II			DMDPhS^{M} VIII				DMDPhS^{M} V		DMDPhS^{M} VI		MVPhS X		
	0*	5	6	0*	6	0*	3	0*	5	0*	6	0*	5	10
Tensile strength [MPa]	6.3	7.7	9.0	6.3	7.5	7.5	7.2	7.5	7.2	5.4	6.5	7.5	7.4	6.2
Relative elongation [%]	370	540	530	370	420	375	490	375	375	350	450	375	375	400
Relative elongation [%]	–	–	–	–	–	0	0	0	0	0	5	0	0	0
Tearing strength [kP/m]	17	27	24	17	18.5	15	23	15	18	14.4	17.8	15	17	24
Shore hardness	56	57	56	56	62	54	57	54	60	–	45	54	65	65

* Modifier-free control specimen.

As modifier concentration increases to 6%, all parameters increase by 1.1 times. Effects of modifiers V and IX are similar. Only increase in tearing strength and hardness of rubbers are observed, with slight changes in the rest of the parameters.

Analysis of data presented in the tables shows that the presence of modifiers has selective effects on fluoro- and phenylsiloxane rubber. Some parameters are considerably improved; others remain unchanged or slightly degrade. The most effective modifiers for fluorosiloxane rubber are types VI and VIII. For phenylsiloxane rubbers, a more effective modifier appears to be of type II. Improvement of rubber hardness is observed in the presence of modifiers containing a reactive peroxide group, the vinyl compound in MVPhS, for example.

Effects of modifiers are more pronounced in the case of lower parameters of initial raw rubber. Introduction of modifiers to rubber compositions likely allows us to eliminate variability of properties of the raw rubbers produced.

References

[1] A. A. Dontsov, A. A. Kanaysova, T. V. Litvinova, *Rubber – Oligomeric Composition in Production of Rubber Articles*, Moscow/Chimia, **1986.**

[2] Ru Patent, No 1 587 881.

[3] Authorship Certificate, No. 4 864 157, (25/06/**1996**).

[4] Ru Patent, No. 1 678 018 {07/07/**1993**}

[5] O. G. Ryzhova, B. V. Molchanov, E. V. Denisova, *Vysokomolek. Soed. A* **1988**, *30*(*6*), 1294.

Polyphenylsilsesquioxane–Polydiorganosiloxane Block Copolymers

Natal'ya Yr. Semenkova, Sergei R. Nanushyan, Pavel A. Storozenko

State Research Center of the Russian Federation, State Research Institute for Chemistry and Technology of Organoelement Compounds, 38, Shosse Entuziastov, 111123, Moscow, Russia

Tel.: +7 95 273 1327 — Fax: +7 095 273 1327

E-mail: nanush @ eos.incotrade.ru

Keywords: polyorganosiloxanes, polyphenylsilsesquioxane–polydiorganosiloxane block copolymers, heat-resistant composites.

Summary: Comparative analysis is carried out of some physical–chemical and physical–mechanical properties of polyorganosiloxane block copolymers of linear and linear-ladder structure. Advantages of polyorganosiloxane block copolymers are demonstrated in terms of thermal stability and physical-mechanical performance. There is an opportunity for the development of new heat-resistant composites for various branches of industry.

Introduction

The history of siloxane rubber manufacture is related to the search for materials combining high heat- and frost-resistance with good physical–chemical performance.

The most significant progress in this area results from the implementation of an idea for micro heterogeneous two-phase block copolymer development, including flexible dimethylsiloxane and rigid polyphenylsilsesquioxane blocks (PPhSO–PDOS block copolymers) in a series of pure polyorganosiloxanes. There is also a possibility to obtain composite elastomers retaining their strength, flexibility, and adhesion at elevated (>300 °C) temperatures.

The thermal stability of composites is mainly determined by the thermal stability of the polymer support. High-molecular-weight organosilicon compounds are the largest group of thermally stable polymers.

In the process of heating, polyorganosiloxane rubbers undergo a series of chemical transformations where, along with destruction and depolymerization reactions, intramolecular conversions, polymerization-type, exchange, and crosslinking reactions proceed.

The analysis and systematization of all factors affecting the above-mentioned processes significantly define the development history of the synthesis of new heat-resistant polymers. Nominally, this process can be divided into the following stages:

- Modification of the organic framework of the linear polydimethylsiloxane molecule (Scheme 1)
- Introduction of volume groupings having bonds different from siloxane bonds into the siloxane chain
- Replacement of a few atoms with Si heteroatoms in the siloxane chain
- Introduction of heterocyclic siloxane fragments into the siloxane chain.

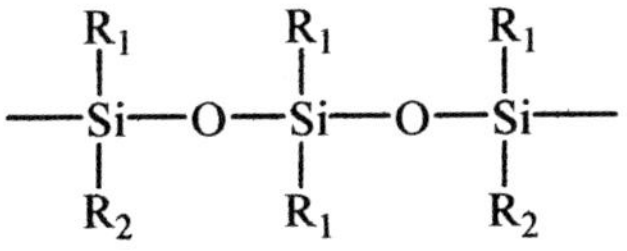

$$-\underset{R_2}{\overset{R_1}{Si}}-O-\underset{R_1}{\overset{R_1}{Si}}-O-\underset{R_2}{\overset{R_1}{Si}}- \qquad R_1, R_2 = CH_3\,, C_6H_5\,, CH_2{=}CH, C_2H_5\,, \text{et al.}$$

Scheme 1. Modification of linear polydimethylsiloxane.

Polyorganosiloxane block copolymers of linear-ladder structure have the general formula shown in Scheme 2.

$$\{[C_6H_5SiO_{1.5}]_a[C_6H_5(OH)SiO]_b[R_2SiO]_c\}_n,$$

$(a = 0.15 - 0.30;\ b = 0.01 - 0.03;\ c = 0.60 - 0.80;\ n = 130 - 220;$ R = aliphatic or aromatic radicals)

Scheme 2. General formula of polyorganosiloxane block copolymers of linear-ladder structure.

Polyorganosiloxane block copolymers of linear-ladder structure are the most promising compounds for the development of new composites because of their unusual structure. The advantages of block copolymers in comparison with α,ω-dihydroxypolydimethylsiloxane become obvious on comparative analysis of their physical–chemical and physical–mechanical properties.

The investigated substances are shown schematically in Scheme 3.

Comparative Thermogravimetric Analysis (TGA)

Let us look at the behavioral peculiarities of α,ω-DHPS and PPhSO oligomer during heating together in air at the rate of 5 °C/min to 900 °C (TGA).

The curves obtained by thermal gravimetrical analysis (TGA), differential thermal analysis (DTA) and differential thermal gravimetrical analysis (DTG) (Fig. 1, Table 1) are compared.

TGA data (Fig. 1, Table 1) analysis testifies to the fact that the temperature of the onset of destructive processes is shifted to a higher range in comparison with linear α,ω-DHPS.

Protection of the linear polydimethylsiloxane chain with the bulky ladder polyphenylsilsesquioxane (PPhSO) block copolymer molecule inhibits its cyclic depolymerization and retains the organic framework to a large extent.

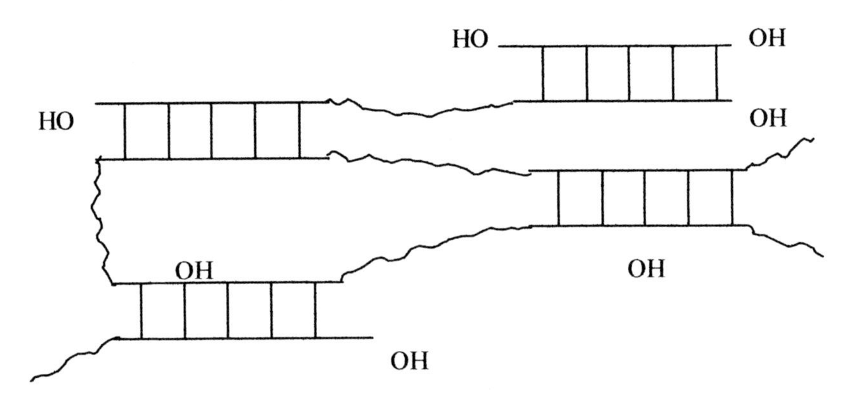

a) Polyphenylsilsesquioxane–polydiorganosiloxane block copolymer (PPhSO–PDOS)

b) Polyphenylsilsesquioxane ladder fragment (PPhSO)

R_1
—Si—O—
R_2

c) Polydiorganosiloxane linear fragment ($R_1 = R_2 = CH_3$, C_6H_5; $R_1 = CH_3$, $R_2 = C_6H_5$)

HO OH

d) α,ω-dihydroxypolydiorganosiloxane (α,ω-DHPS)

Scheme 3. The substances investigated.

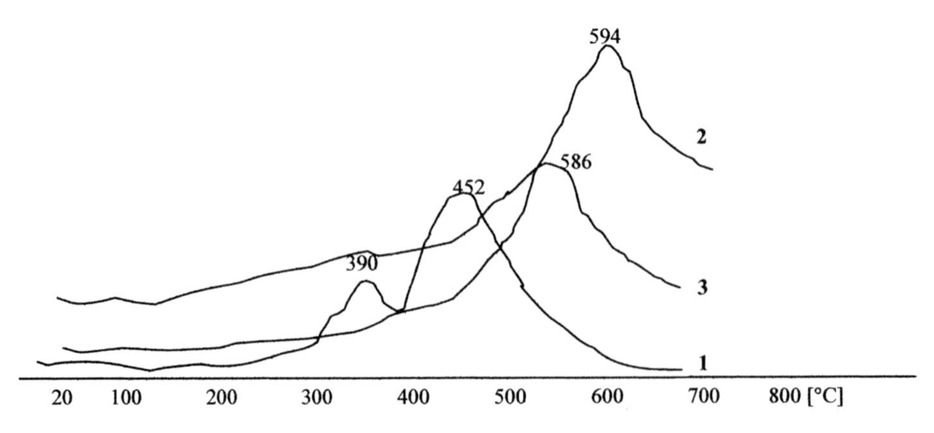

Fig. 1. DTA in air environment, heating rate 5 °C/min (**1**: α,ω-DHPS, **2**: PPhSO-oligomer, **3**: PPhSO–PDOS block copolymer).

Table 1. Comparative thermogravimetric analysis (TGA).

Polymer sample	Temperature of weight loss [°C]		Temperature of volatiles extraction maximum (and areas) [°C] (DTG)		Weight loss at 900 °C [%]
	10%	20%			
α,ω-DHPS	328	366	370 and 448	(300 – 470)	82
PPhSO-oligomer	539	572	390 and 500	(488 – 631)	55
PPhSO–PDOS Block copolymer	395	462	563	(360 – 550)	63

Some Peculiarities of the Physical–Mechanical Properties of Unfilled and Filled Vulcanizates Based on PPhSO-PDOS Block Copolymers

The significant peculiarity of α,ω-DHPS and PPhSO–PDOS block copolymers consists in the physical–mechanical properties of unfilled vulcanizates.

All the unfilled vulcanizates based on linear α,ω-DHPS are distinguished by a tensile strength maximum of 1 kg/cm and a percentage elongation at rupture of 80 – 120%.

The molecules of PPhSO–PDOS block copolymers have a micro heterogeneous structure.

PPhSO fragments, whose glass transition temperature is high (> 390 °C), exist in the solid crystalline state, and soft PDOS fragments with a glass transition temperature of –70 °C (crystallization starts at –35 °C) are in the liquid state.

PPhSO and PDOS fragments are linked to the block copolymer molecule, actually behaving like a filled system, thus providing high mechanical performance of the unfilled vulcanizates.

On the basis of combining linear-ladder PPhSO–PDOS block copolymers, a whole set of useful properties have been developed (Table 2).

The new composites are used in various branches of industry as coatings, sealants, adhesives, and compounds serviceable within a wide temperature range.

Table 2. Properties of PPhSO–PDOS block copolymer-based composites.

Performance	Characteristic
Serviceability within a temperature range	From –50 to 450 °C (short-term from –150 to 550 – 600 °C)
Tensile strength of unfilled vulcanizates after exposure in air at 23 ± 2 °C/3days	40 – 50 kg/cm
Percentage elongation of unfilled vulcanizates after exposure in air at 23 ± 2 °C/3days	300 – 600%
Adhesion strength	Is retained within the whole temperature range
Insulating properties: electric strength	Up 55 kV/mm (within temperature range of 25 – 350 °C). ≥ 30 kV/mm (under the conditions of 95% relative humidity of the environment).
Possibility of vulcanizate filling for various functional materials development	Provides high degree of filling (up to 1000 parts per 100 parts of the block copolymer).

Thermoplastic Silicone Elastomers

O. Schäfer

Wacker-Chemie GmbH, Johannes-Hess-Str. 24, 84489, Burghausen, Germany
Tel.: +49 8677 83 2436 — Fax: +49 8677 83 2536
E-mail: oliver.schaefer@wacker.com

J. Weis, S. Delica, F. Csellich, A. Kneißl

Consortium für eletrochemische Industrie GmbH
Zielstattstr. 20, 81379, München, Germany

Keywords: elastomers, silicone, copolymers, urea

Summary: Today, silicone elastomers are some of the most important substances in the silicones business and have received widespread attention. They are principally based on high-molecular-weight polydimethylsiloxane (PDMS) chains, crosslinked chemically and blended with active fillers (e.g. fumed silica) in order to obtain useful mechanical properties. The crosslinking step is often carried out with catalysts at high temperatures (high-temperature vulcanizing), resulting in a thermoset material. Processing of such materials typically requires rubber equipment specially designed for thermoset materials. To avoid the additional crosslinking step in producing silicone rubbers, we modified the PDMS backbone with urea segments, which tend to crystallize at room temperature or at least tend to be separated from the siloxane matrix. The segments are able to melt reversibly at elevated temperatures and behave as non-permanent crosslinking sites in the modified rubbers. By carefully choosing the appropriate organic segment, it is possible to obtain silicone rubbers with melting or softening points from 80 up to 170 °C. By modifying the amount of organic segments it is also possible to produce very soft or very hard materials. PDMS-containing silicone-urea copolymers display a combination of very interesting properties. These include very low glass transition temperatures (–120 °C), high oxidative stability, low surface energy, hydrophobicity, high gas permeability, good electrical properties, and biocompatibility. These materials display tensile strengths up to 14 MPa and do not need additional fillers.

Introduction

Today, silicone elastomers are some of the most important substances in the silicones business and they have received widespread attention. They are principally based on high-molecular-weight polydimethylsiloxane (PDMS) chains, crosslinked chemically and blended with active fillers (e.g. fumed silica) in order to obtain useful mechanical properties.

The crosslinking step is often carried out with catalysts at high temperatures (high-temperature vulcanizing), resulting in a thermoset material. To obtain reasonable curing rates even at elevated temperature additional catalysts have to be compounded into the polymer base. Processing of such materials typically requires rubber equipment specially designed for thermoset materials.

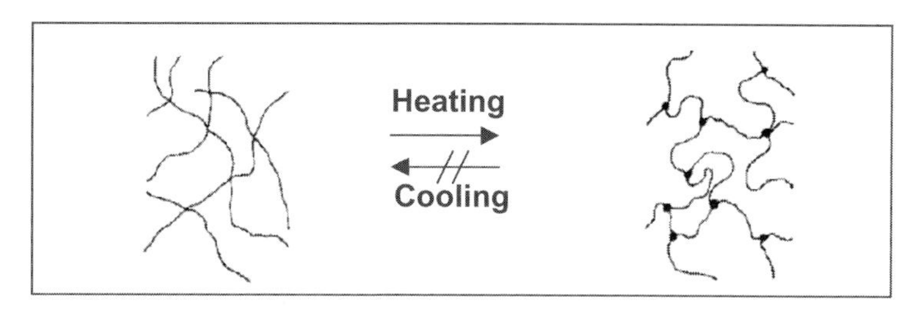

Fig. 1. Crosslinking of conventional thermoset silicone rubbers.

Thermoplastic elastomers, e.g., polyurethane elastomers, have an entirely different structure that is heterogenic. They have an elastomeric matrix and phase-separated "hard" blocks, which act as embedded physical crosslinking sites. The hard blocks can be softened at elevated temperatures to obtain a single-phase melt that is easily processed. Upon cooling, the two-phase nature is recovered and the material becomes a solid again.

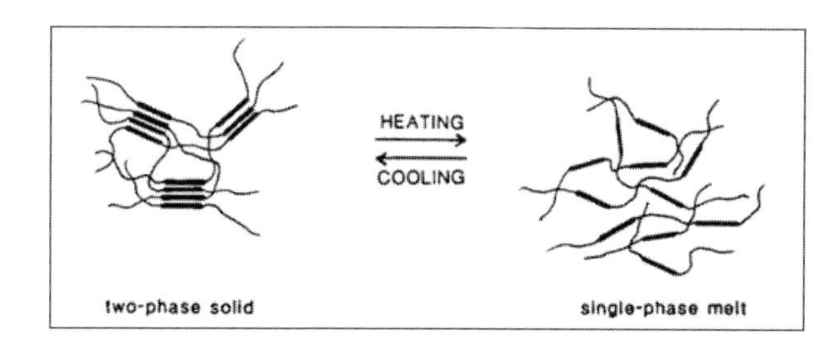

Fig. 2. Thermal behavior of block copolymers.

Thermoplastic Siloxane-Urea Block Copolymers

Since 1980, siloxane-urea block copolymers prepared by the reaction of bisaminoalkyl-terminated silicones and diisocyanates [1] have been well known in the literature. Silicone is responsible for the elastomeric behavior and urea for the thermal behavior. Nevertheless, to obtain high-molecular-weights in the polymerization of such copolymers, the starting compounds have to be sufficiently pure [2]. Therefore, one key step in the synthesis of such block copolymers is the preparation of appropriate starting products. Suitable organofunctional silicone fluids for the synthesis of thermoplastic siloxanes include α,ω-aminopropyl-terminated siloxanes.

Synthesis of Siloxane-Urea Block Copolymers

The usual way of preparing such aminopropyl-terminated siloxanes is the synthesis via a base-catalyzed equilibration reaction [1, 2]. A more convenient way is the termination reaction of silanol-terminated siloxanes with special organofunctional silanes [3, 4].

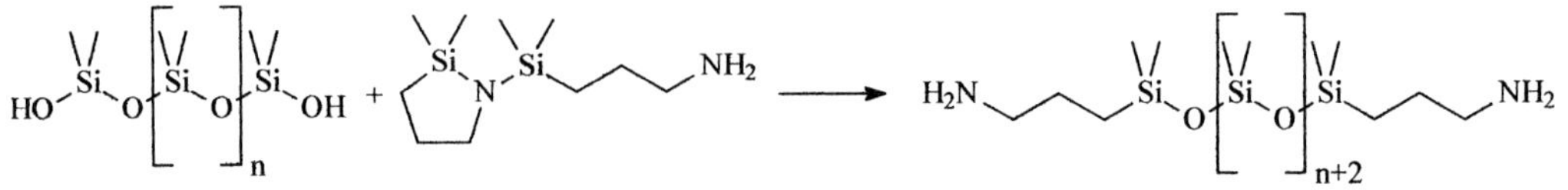

Scheme 1. Reaction between cyclic silazane and α,ω-silanol-terminated siloxanes.

This reaction yields the desired silicone fluids with high purities in a quantitative way [4]. By choosing appropriate silanol fluids, amino-functional fluids with different molecular weights can be obtained [5].

Starting from these aminofunctional siloxanes, PDMS-urea block copolymer can be obtained easily [6]. The synthesis is performed by polyaddition of aminoalkyl-terminated polydimethylsiloxanes with diisocyanates to yield thermoplastic silicone elastomers (TPSE). The reaction between the amino and the isocyanato group gives highly polar urea groups, which solidify at room temperature (Scheme 1).

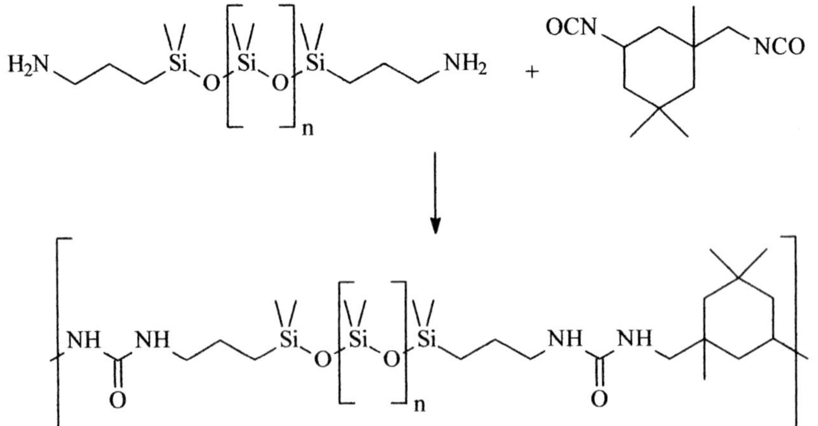

Scheme 2. Reaction between isophoronediisocyanate and α,ω-aminopropyl-terminated siloxanes.

The highly polar urea groups are linked together by strong hydrogen bonding. This "bond" can be split at elevated temperatures in a reversible process which gives the material a thermoplastic nature.

The synthesis of siloxane-urea block copolymers can be performed either in solution or in the melt at elevated temperatures up to 200 °C. We favor the use of dynamic mixers or extruders to ensure intensive mixing of the components. Surprisingly, adding aminosiloxane and isocyanate to get the block copolymers (2K process) is not the only possibility. We can also add the silanol fluid, the cyclic aminosilane, and the diisocyanate consecutively to the extruder for a successful solventless preparation of siloxane-urea block copolymers (3K process).

Properties of Thermoplastic Siloxane-Urea Block Copolymers

The properties of the PDMS-urea block copolymer can be modified by changing either the diisocyanate or the molecular weight of the aminopropyl-functional silicone fluids.

The variation of the diisocyanate affects the softening temperature, the optical properties, and the mechanical properties of the copolymer. By choosing aliphatic diisocyanates, transparent materials with lower softening temperatures are obtained. Aromatic diisocyanates yield mechanically strong elastomers with higher softening temperatures.

Altering the molecular weight of the silicone compound affects the siloxane content, the mechanical properties, and the softening temperatures.

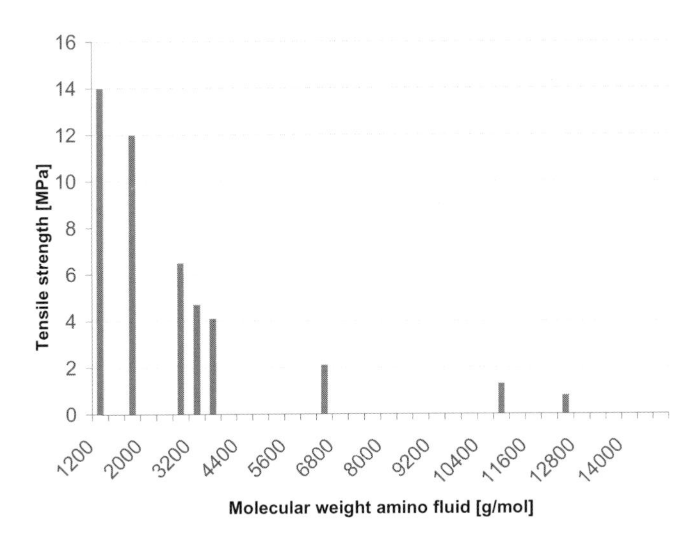

Fig. 3. Mechanical properties of PDMS-urea block copolymers based on 4,4'-methylenebis(cyclohexylisocyanate) and various amino-functional fluids.

The plastic behavior changes drastically with increasing silicone content. Because of the intermolecular forces, a siloxane-urea copolymer with a silicone content up to 70% can be treated like an impact resistant plastic. Beyond 70% silicone content, the character of the material changes to a thermoplastic elastomer. The elastomeric behavior is reflected by max. elongations of up to 2000%. Nevertheless, the ultimate tensile strength is also clearly affected by the molecular weights of the siloxane diamines used.

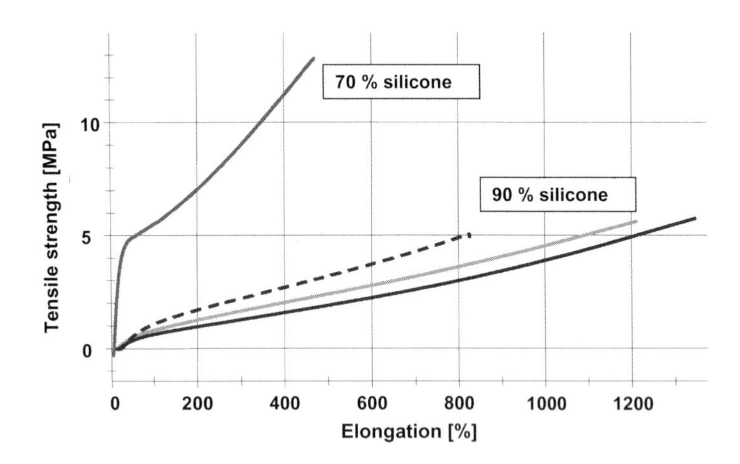

Fig. 4. Different stress-strain characteristics of siloxane-urea block copolymers with different silicone contents.

Naturally, the Shore A hardness is also altered by varying the molecular weight of the silicone fluids. By using the same diisocyanate, it is possible to obtain materials in the range of 80 to 20 Shore A. In contrast to covalently crosslinked silicone rubbers, these block copolymers are highly transparent, still soluble in various organic solvents, and processable with common thermoplastic equipment. This is reflected in the various uses, e.g., gelling agents in cosmetics [7], anti-fouling coatings, additives for processing plastics [8, 9], preparation of PSA [2], various biomaterial applications, optical applications, and making release films [10]. These thermoplastic silicones are available under the name GENIOMER® [11] from WACKER SILICONES, Munich.

Summary

We used a new silane which readily permits quantitative conversion of silanol-terminated fluids into aminopropyl-terminated fluids. The reaction between aminopropyl-terminated fluids and diisocyanates proceeds smoothly within a few minutes, either in solution or in the melt. The preparation of siloxane-urea block copolymers is performed in either a two- or a three-component process. By carefully choosing the inorganic segment defined by the corresponding silicone fluid, it is possible to obtain silicone rubbers with different material characteristics. The mechanical properties can be tuned from very soft to very hard. Those materials display tensile strengths up to 14 MPa without requiring additional fillers and can be used for diverse applications.

Acknowledgment: The authors of this paper would like to thank Prof. I. Yilgör for various helpful discussions.

References

[1] I. Yilgör, A. K. Sha'aban, W. P. Steckle, D. Tyagi, G. L. Wilkes, J. E. McGrath, *Polymer* **1982**, 1802.
[2] C. M. Leir, J. J. Hoffman, L. A. Tushaus, G. T. Weiderholt, EP 250 248, Minnesota Mining (3M).
[3] O. Schaefer, A. Bauer, L. Brader, B. Pachaly, V. Frey, EP 1 195 379, Consortium für elektrochemische Industrie GmbH.
[4] O. Schaefer, A. Bauer, B. Pachaly, V. Frey, EP 1 201 699, Consortium für elektrochemische Industrie GmbH.
[5] Available from WACKER SILICONES, Munich, Germany.
[6] DE 3 143 994 A1, Bayer AG, US 3 963 679, Allied Signal.
[7] G. Riess, M. Mendolia, H. W. Schmidt, *Macromol. Symp.* **2002**, *181*, 123 – 134.
[8] I. Yilgör, E. Yilgör, S. Suzer, *J. Appl. Pol. Sci.* **2002**, *83*, 1652 – 1634.
[9] O. Kulikov, Macro 2004, Paris, to be published.
[10] A. A. Sherman, W. R. Romanko, M. H. Mazurek, K. C. Melancon, C. J. Nelson, J. Seth, WO 96/34029, Minnesota Mining (3M).
[11] GENIOMER® is a registered trademark of Wacker-Chemie GmbH.

Silicone Hybrid Copolymers — Structure, Properties, and Characterization

*Wolfgang Hiller**

Technische Universität München, Lichtenbergstr. 4, 85747, Garching, Germany
Tel.: +49 89 289 13133 — Fax: +49 89 289 14512
E-mail: wolfgang.hiller@lrz.tum.de

Wolfgang Keller

Wacker-Chemie GmbH, Johannes-Hess-Str. 24, 84489, Burghausen, Germany
Tel.: +49 8677 83 7419 — Fax: +49 8677 88 67419
E-mail: wolfgang.keller@wacker.com

Keywords: silicone copolymers, surface analytics, atomic force microscopy

Summary: Silicone copolymers are hybrid materials with varying phase separation levels. The polymers themselves as well as surfaces formed by these copolymers were analyzed by a variety of polymer and surface characterization methods. Atomic force microscopy was found to be especially suitable for analysis of thin polymer films. Both surface and bulk properties are dominated by the domain size and the silicone content.

Silicone Organic Hybrids — Introduction and Classification

Silicone hybrid materials include all macromolecular assemblies or structures which are built of pure silicone polymers as one part and organic or inorganic materials as the other.

Within the scope of this article, silicone hybrids are defined as copolymers containing structural units of both silicone polymers and classical organic polymers.

These hybrid materials from organosilicon and organic polymers can be classified in a number of ways, for example according to their

- chemistry (composition) or substitution pattern
- field of application or target market
- preparation or synthetic route.

In the following, the analytical work conducted on several types of silicone-organo-copolymers made from silicone fluids and organic monomers by radical polymerization will be discussed.

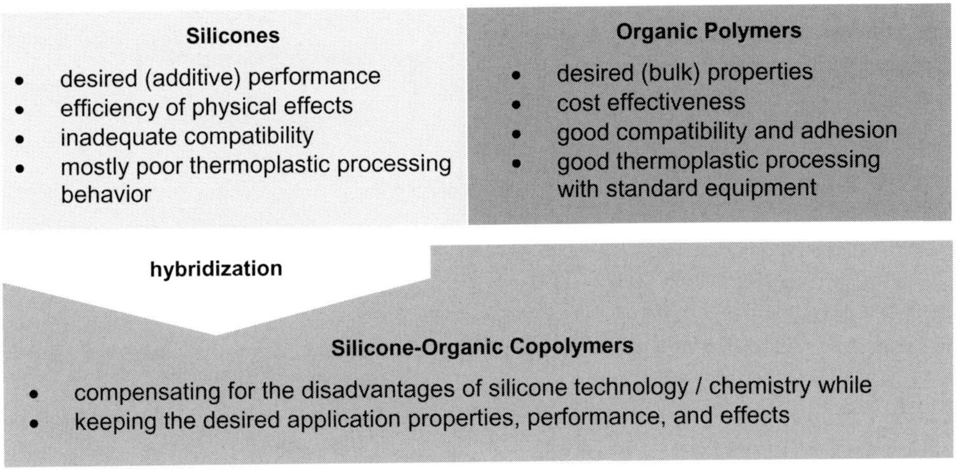

Silicones

- desired (additive) performance
- efficiency of physical effects
- inadequate compatibility
- mostly poor thermoplastic processing behavior

Organic Polymers

- desired (bulk) properties
- cost effectiveness
- good compatibility and adhesion
- good thermoplastic processing with standard equipment

hybridization

Silicone-Organic Copolymers

- compensating for the disadvantages of silicone technology / chemistry while
- keeping the desired application properties, performance, and effects

Synthetic Approach

Monofunctional siloxanes are readily prepared by anionic polymerization of hexamethylcyclotrisiloxane (D3) following the route illustrated in Fig. 1 [1].

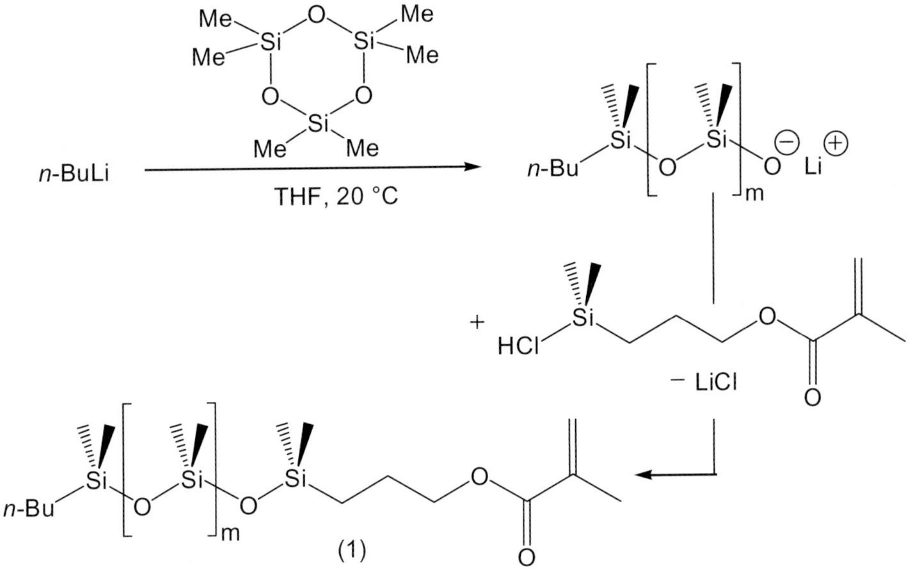

n-BuLi

THF, 20 °C

– LiCl

(1)

Fig. 1. Synthesis of monofunctional siloxanes prepared from D3.

The resulting methacrylic siloxanes are then polymerized through radical polymerization to the corresponding copolymer structures (Fig. 2) [2 – 4].

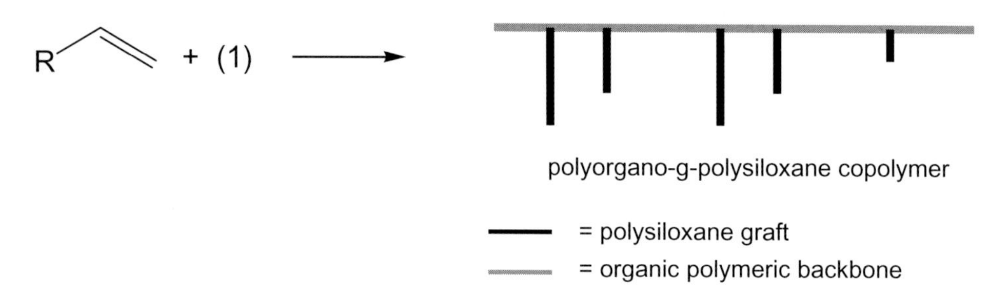

Fig. 2. Copolymerization reaction of monofunctional silicone macromonomers with organic monomers.

Copolymer Characterization

Characterization of the copolymers containing 10 – 30 wt% of silicone macromonomers was done by GPC, DSC, SEM, TEM, surface analysis (SIMS, XPS, contact angle), and especially atomic force microscopy (both Tapping Mode™ and force modulation mode).

DSC Measurements

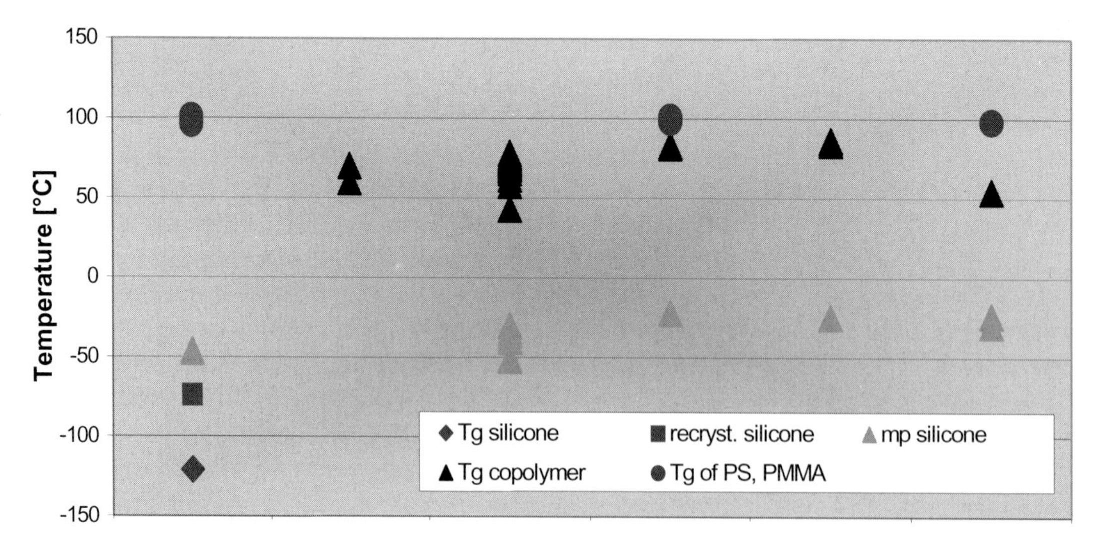

Fig. 3. Glass transition temperatures shown by copolymers of different phase separation levels. High phase separation and larger domain sizes lead to two glass transition temperatures (polymer + silicone, 1st column), while statistical copolymers indicate only one Tg for the specific copolymer.

TEM and SEM Pictures

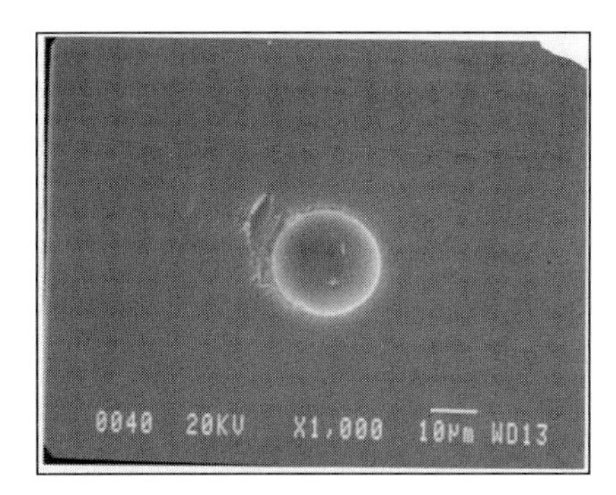

Fig. 4. TEM (left) and SEM (right) pictures of a 10 wt% silicone (statistically polymerized polymer sample). No phase separation is shown at this concentration level.

Atomic Force Microscopy

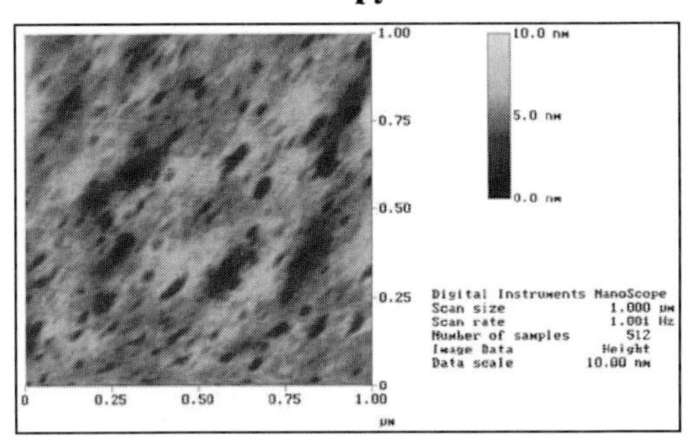

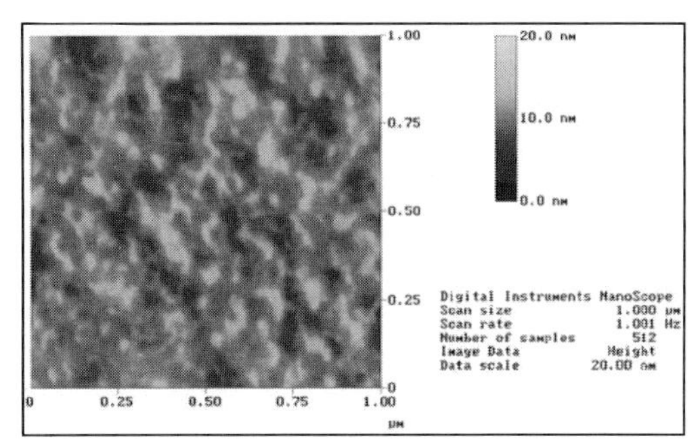

Fig. 5. Tapping Mode™ pictures of a thin solvent-cast copolymer film with 20 (left) and 30 (right) wt% of a siloxane macromonomer. Clearly seen is the transition from a segregated domain structure to an interpenetrating sponge morphology.

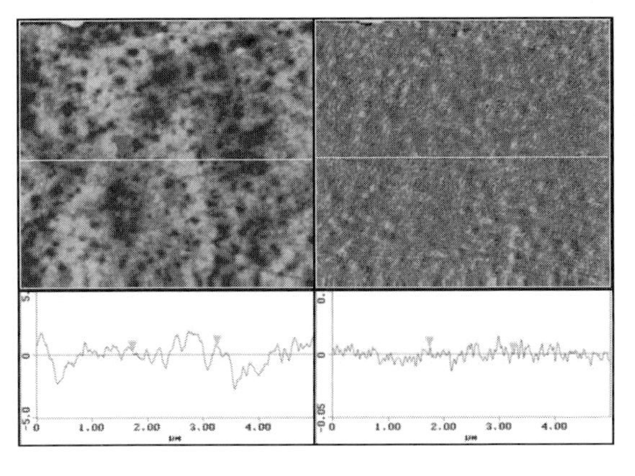

Fig. 6. Tapping Mode™ (left) versus force modulation mode (right) picture with the corresponding height and elasticity modulus profiles of a solution-cast 20 wt% silicone thin copolymer film.

Though the elasticity picture is a better indicator of the phase segregation morphology, even the height-sensitive Tapping Mode™ clearly shows artifacts of the domains. The section cuts are proof of the non-identity of the pictures as well as of their quality.

Surface Analysis by SIMS and XPS

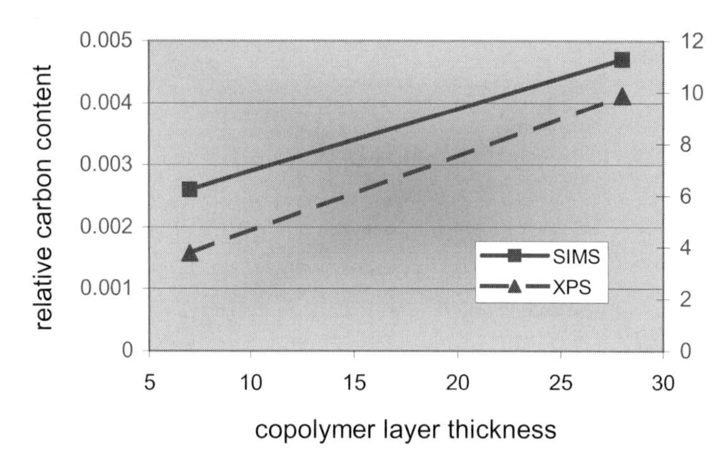

Fig. 7. Surface analysis (XPS, SIMS) of silicone copolymer films as a function of their thickness. Surprisingly, thinner films show a higher relative siloxane concentration at their surface than thicker ones. As expected, the silicon metal concentration measured by secondary ion mass spectrometry is higher because of the lower depth profiled by the method.

Conclusion

Analytical characterization of hybrid silicone copolymer systems shows distinct phase separation patterns depending on the level of silicone concentration in the copolymer. Copolymers with a siloxane content of 10 wt% or lower do not show any domains.

References

[1] Y. Yamashita, *Polymer Bulletin* **1982**, *7*, 289 – 294.

[2] Y. Yamashita, *Polymer Bulletin* **1983**, *10*, 368 – 372.

[3] Y. Yamashita, *Makromol. Chem.* **1984**, *185*, 9 – 18.

[4] Y. Yamashita, *Macromolecules* **1985**, *18*, 580 – 582.

Hydrolytic Stability of Silicone Polyether Copolymers

M. G. Pigeon, A. M. Czech, S. J. Landon*

GE Silicones – OSi Specialties,
769 Old Saw Mill River Rd, 10591, Tarrytown, NY, USA
Tel.: +1 914 784 4963 — Fax: +1 914 784 4922
E-mail: Michael.Pigeon@gesm.ge.com

Keywords: silicone, polyether, water stability, pH stability

Summary: Silicone polyether copolymers are widely used in formulations as aesthetic ingredients and as excellent dispersants, emulsifiers, and foam modifiers. Unlike silicone homopolymers, most silicone polyethers are compatible with aqueous systems. Since many formulations vary in pH, long-term stability is often questioned. The study was undertaken to define a "functional" pH range for silicone polyether copolymers and provide formulators with practical guidelines for working with these ingredients. This was achieved by investigating the effects of varying structures and properties on the long-term stability of silicone polyether copolymers in aqueous media. Selected copolymers were evaluated across wide pH and concentration ranges using reverse-phase HPLC.

Introduction

It is known that silicone polyether copolymers can have limited stability in aqueous media. The stability will depend on several factors, which include pH, concentration, structure of the copolymer, and the molecular weight of the copolymer. In this work the stability of a wide range of copolymers was investigated in aqueous media at various pH values and concentrations. The goal was to provide formulators and users with a tool to predict the long-term stability of silicone polyether surfactants in their own applications and formulations.

Experimental

To determine the stability of silicone polyether copolymers, solutions were prepared by dispersing the copolymers in standard buffer solutions ranging in pH from 2 to 12 and concentrations ranging from 0.5 to 2.0 weight percent. High-performance liquid chromatography (HPLC) was used to monitor the stability of the polymers over time. Reverse-phase HPLC is an ideal method for separating silicone polyether copolymers from the other components in the buffer solutions. A typical HPLC chromatogram of a silicone polyether copolymer is shown in Fig. 1.

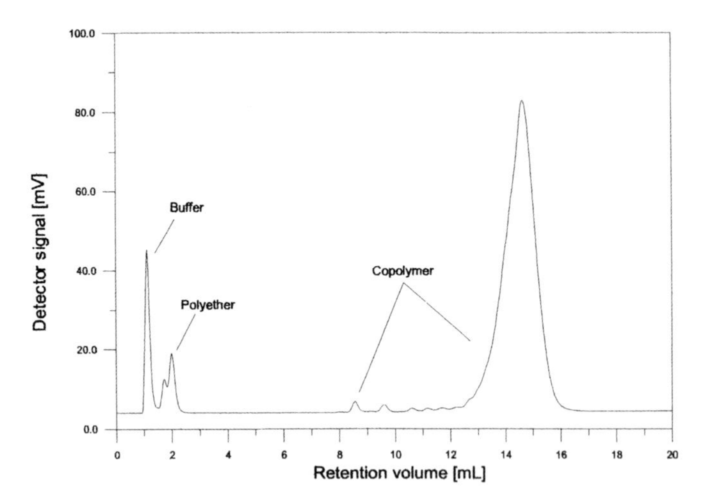

Fig. 1. Typical reverse-phase HPLC chromatogram of a silicone polyether copolymer.

To more conveniently classify the stability of a copolymer, the percentage of unhydrolyzed material was plotted against time. Two points on the curves were used to characterize the stability of the copolymer. This is represented in Fig. 2.

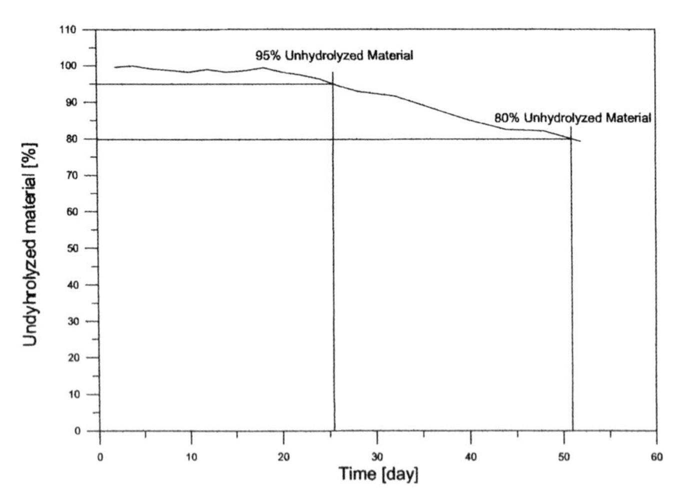

Fig. 2. Percent of unhydrolyzed material curve for a silicone polyether copolymer. 95% Unhydrolyzed — The amount of time over which there is no hydrolysis. 80% Unhydrolyzed — The amount of time to establish a hydrolysis trend.

To fully analyze a silicone polyether copolymer, solutions ranging in pH from 2 to 12 were studied. A typical set of hydrolysis curves for a copolymer is shown in Fig. 3.

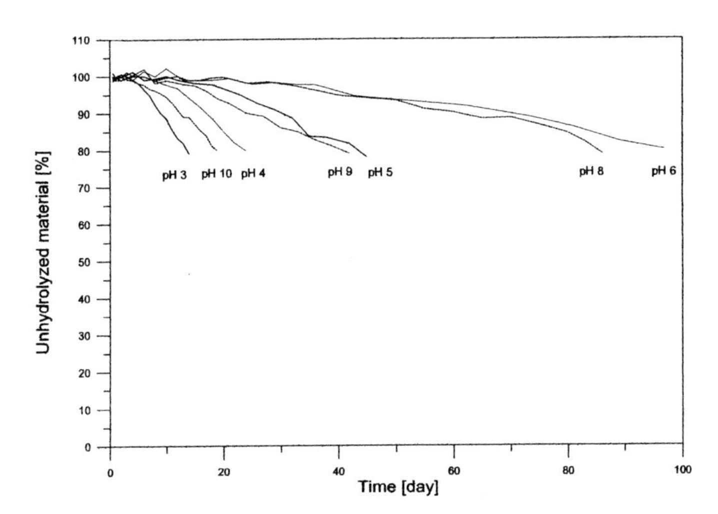

Fig. 3. Percentage of unhydrolyzed material versus time at ph values from 2 to 12.

Methodology and Equipment

Waters Alliance® 2690 HPLC chromatograph
Alltech® 500E evaporative light-scattering detector with low temperature adapter
Column – Phenomenex® LUNA C18 end cap, 5 micron particle size, 75 × 4.6 mm.
Gradient – Water/Methanol/Isopropanol at 1.0 mL/min.

All solutions were analyzed without dilution using 10 microlitre injection volumes.

Results and Discussion

There is a wide range of commercially available silicone polyether copolymers. They fall into three classes: conventional, hydrolyzable and trisiloxane. These classes are described below.

Conventional Copolymers

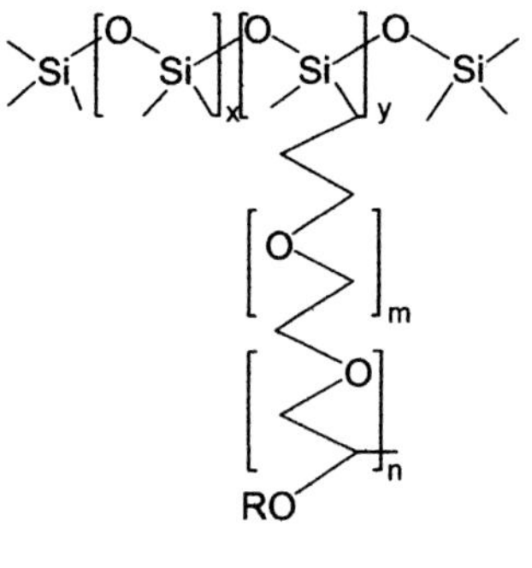

Conventional silicone polyether copolymers have excellent stability as the polyether groups are attached to the siloxane backbone through a series of hydrolytically stable Si–C bonds. Analysis of materials with some hydrolysis revealed that these levels of hydrolysis had no effect on performance properties.

Hydrolyzable Copolymers

Hydrolyzable silicone polyether copolymers are made by condensation chemistry and contain Si–O–C bonds between the silicone chain and polyether chains. This linkage offers limited resistance to hydrolysis under neutral and slightly alkaline conditions but breaks down quickly in acidic media.

$$(\mathrm{Me{-}Si})_{y-2}-\left[(\mathrm{O}\overset{\mathrm{Me}}{\underset{\mathrm{Me}}{\mathrm{Si}}})_{x/y}-\mathrm{O}\left(\diagup\!\!\diagdown\mathrm{O}\right)_n\left(\diagdown\!\!\diagup\mathrm{O}\right)_m\right]_y$$

Trisiloxane Copolymers

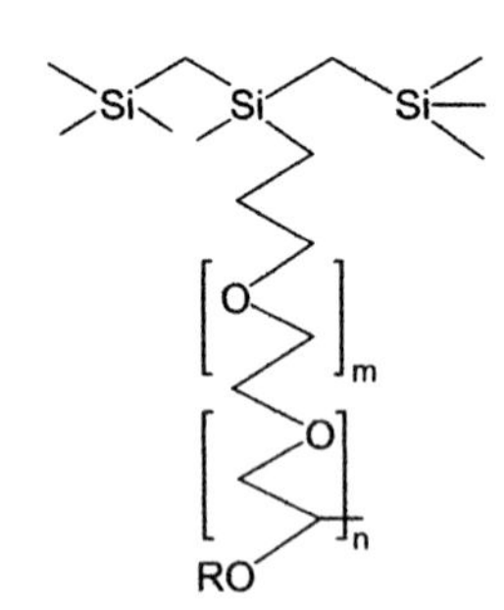

Trisiloxanes represent a unique subset within conventional silicone polyether copolymers. Their well-defined compact siloxane backbone affords distinctive application benefits (e.g., spreading and wetting). However, they tend to be more prone to hydrolysis.

Summary of Results

A number of silicone polyether copolymers were studied. These are detailed in Table 1 and are listed in order of increasing stability. To provide a tool for prediction of stability, 80% unhydrolyzed material was chosen as the stability limit. A summary of these times is shown in Table 2 and Fig. 4.

Table 1. Molecular weight, solubility, and architecture of silicone polyether copolymers studied.

Designation	Polyether	Molecular Weight	Solubility in Water	Architecture
Copolymer I	PEG5/PPG3	< 1 000	Dispersible	Trisiloxane
Copolymer II	PEG20/PPG15	> 10 000	Soluble	Hydrolyzable
Copolymer III	PEG8	< 5 000	Soluble	Pendant
Copolymer IV	PEG8	< 5 000	Dispersible	Linear
Copolymer V	PEG8	< 5 000	Dispersible	Pendant
Copolymer VI	PEG20/PPG23	> 10 000	Soluble	Pendant
Copolymer VII	PEG8	> 10 000	Insoluble*	Pendant
Copolymer VIII	PPG12	< 5 000	Insoluble*	Linear
Copolymer IX	PPG12	< 5 000	Insoluble*	Pendant

* Hydrolysis experiments run in water-isopropanol solutions to produce stable dispersions.

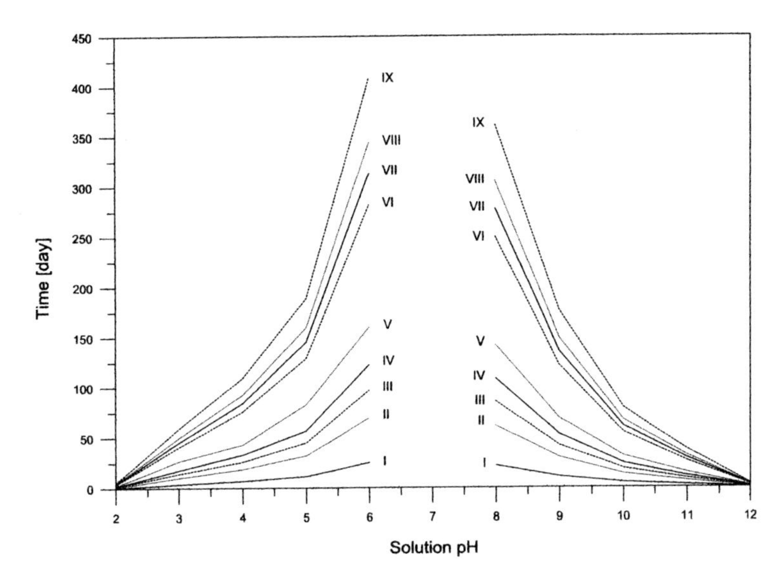

Fig. 4. Time to reach 80% unhydrolyzed material for all silicone polyether copolymers in acidic and basic media at 1% concentration.

Table 2. Time to reach 80% unhydrolyzed material for all copolymers at all pH levels (1% concentration).

	Silicone Polyether Copolymer (Time in Days)								
pH	**I**	**II**	**III**	**IV**	**V**	**VI**	**VII**	**VIII**	**IX**
2	0.33	0.92	1.29	1.63	2.13	3.75	4.17	4.58	5.42
3	3.58	9.98	14.00	17.69	26.41	40.69	45.24	49.69	58.81
4	6.65	18.55	26.00	32.86	42.94	75.60	84.07	92.33	109.27
5	11.51	32.09	45.00	56.85	82.85	128.93	145.45	159.75	189.05
6	24.81	69.17	97.00	122.56	160.15	281.96	313.54	344.37	407.53

	Silicone Polyether Copolymer (Time in Days)								
pH	**I**	**II**	**III**	**IV**	**V**	**VI**	**VII**	**VIII**	**IX**
8	22.00	61.34	86.00	108.67	142.01	250.01	278.01	305.35	361.35
9	10.74	29.96	42.00	53.07	69.35	122.10	135.78	149.12	176.48
10	4.86	13.55	19.00	24.01	31.37	55.24	61.42	67.46	79.84
11	2.30	6.42	9.00	11.38	14.87	26.18	29.11	31.97	37.83
12	0.21	0.59	0.82	1.04	1.36	2.40	2.67	2.93	3.47

Conclusions

The rate of hydrolysis of conventional silicone polyether copolymers decreases as:

- Solubility in water decreases
- Molecular weight increases (>10 000)
- Concentration of the copolymer increases (from 1% to 5%).

The recommended pH range for maintaining long-term stability of conventional silicone polyether copolymers is 5.5 to 8.5.

Hydrolyzable silicone polyether copolymers have limited stability in aqueous media due to the presence of Si–O–C linkages. The recommended pH range for maintaining optimum stability of hydrolyzable silicone polyether copolymers is 6.0 to 8.0.

Trisiloxane copolymers are susceptible to rapid hydrolysis under both acidic and basic conditions. The recommended pH range for maintaining stability of trisiloxane copolymers is 6.5 to 7.5. Solutions maintained at these pH levels have been shown to be stable for more than two years. A summary of these results is shown in Table 3.

Table 3. Optimum pH ranges for stability of silicone polyether copolymer types.

Silicone Polyether Copolymer Type	pH Range for Optimum Stability
Conventional Copolymers	5.5 – 8.5
Hydrolyzable Copolymers	6.0 – 8.0
Trisiloxane Copolymers	6.5 – 7.5

Silane-Crosslinking High-Performance Spray Foams

Barbara Poggenklas, Heinrich Sommer, Volker Stanjek,* Richard Weidner, Johann Weis

Consortium für elektrochemische Industrie GmbH,
Zielstattstr. 20, 81379 Munich, Germany
Tel.: +49 89 74844 232 — Fax: +49 89 74844 242
E-mail: Volker.Stanjek@wacker.com

Keywords: spray foam, isocyanate free, silane terminated, prepolymers

Summary: As all conventional spray foams contain monomeric isocyanates, they can present toxicological hazards. Therefore, an isocyanate-free substitute for these foams has been developed. This isocyanate-free foam is cured by a crosslinking reaction of silane-terminated prepolymers (STPs). This silane crosslinking system exhibits some new properties, and several problems occur that are completely unknown with ordinary isocyanate spray foams. In particular, crack formation during foam curing turns out to be a crucial problem. Therefore it was necessary to develop modified STPs and to use special STP-propellant mixtures.

Introduction — Conventional Spray Foams

All conventional spray foams are based on polyurethane (PU) systems [1]. They provide excellent mechanical properties and outstanding thermal insulation. Thus, they are widely used especially in construction, i.e. pipe or building insulation and interior applications.

Because of their easy handling, 1-component spray foams are mainly in use in Europe. These foams are cured by a reaction of a mixture containing both isocyanate-terminated prepolymers and monomeric isocyanates with moisture.

All spray foams contain monomeric isocyanates, which can present toxicological and environmental hazards [2]. Isocyanates like TDI (toluenediisocyanate) or MDI (methylendiphenyl-diisocyanate) are even suspected to cause cancer [2, 3].

A New Curing Reaction for Spray Foams

Conventional (1-component) polyurethane spray foams are cured by a reaction of NCO-prepolymers with moisture to generate an urea unit and CO_2 [1]. The latter serves as a propellant or

– more often – as a co-propellant along with a physical blowing agent.

$$\sim N{=}C{=}O \xrightarrow[-\,CO_2]{+\,H_2O\ (\text{moisture})} \sim NH_2 \xrightarrow{+\,\sim N{=}C{=}O} \sim NH{-}CO{-}NH\sim$$

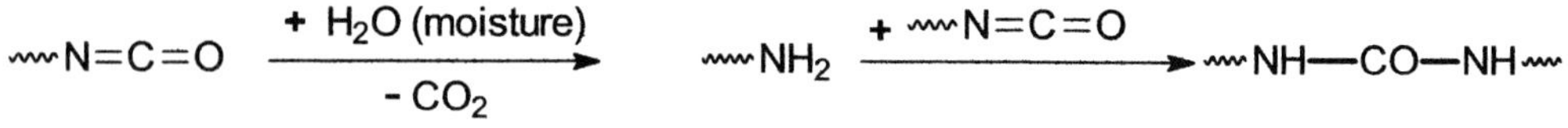

Scheme 1. Curing reaction of conventional spray foams.

To develop toxicologically safe spray foams, it is necessary to find a curing reaction that does not involve isocyanates but is able to generate rigid foams from low-viscosity prepolymers. These conditions can be met by silane-terminated prepolymers (STPs).

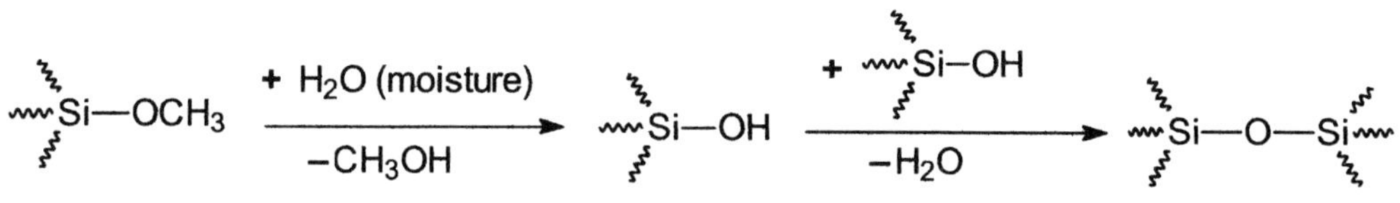

$$\sim Si{-}OCH_3 \xrightarrow[-CH_3OH]{+\,H_2O\ (\text{moisture})} \sim Si{-}OH \xrightarrow[-H_2O]{+\,\sim Si{-}OH} \sim Si{-}O{-}Si\sim$$

Scheme 2. Curing reaction of silane crosslinking foams (STP-foams).

As this new curing reaction generates no gaseous by-products, the foam must be blown by a physical blowing agent exclusively.

STP Preparation

The STPs are prepared by a reaction of NCO-terminated prepolymers and aminosilanes. In particular, STPs prepared from the anilinosilane **1** exhibit superior properties. The STP preparation can be previously performed inside the spray can. As a very small excess of the aminosilane is employed, an NCO-free product is obtained.

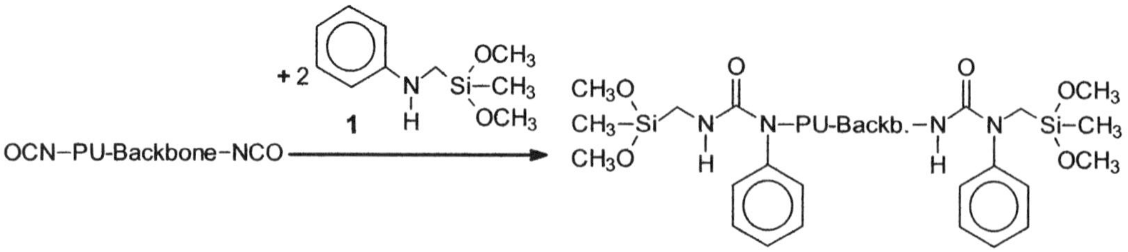

OCN–PU-Backbone–NCO + 2 **1** (PhNH–CH₂–Si(CH₃)(OCH₃)₂) → (CH₃O)₂(CH₃)Si–CH₂–N(Ph)–CO–NH–PU-Backb.–NH–CO–N(Ph)–CH₂–Si(CH₃)(OCH₃)₂

Scheme 3. STP-Preparation from a NCO-terminated prepolymer and anilinosilane **1**.

The NCO-terminated prepolymer can be synthesized from a diisocyanate (e.g., TDI, MDI) and a polyol. The anilinosilane **1** can be prepared from the corresponding chloromethylsilane and aniline, i.e. from low cost components, too.

A foamable compound is produced by compounding the obtained prepolymer mixture with a physical blowing agent, a foam stabilizer, and further additives (e.g., adhesion promoters, catalysts, etc.).

Adjustment of Reactivity and Viscosity

Conventional STPs only exhibit a moderate reactivity towards humidity, even when quite large amounts of (tin) catalysts are employed [4]. Additionally, their viscosity is far too high [5].

These problems can be surmounted by the use of an α-STP in which the silane moieties are connected to a urea group by a methylene spacer. The resulting prepolymers have an extremely high but adjustable reactivity towards moisture [6, 7].

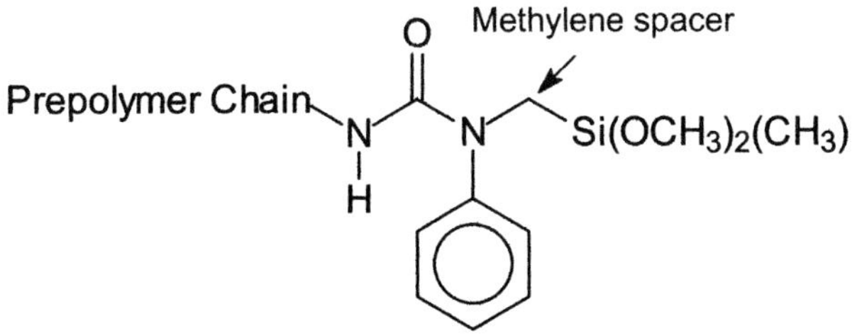

Methylene spacer

Prepolymer Chain–NH–C(=O)–N(Ph)–CH_2–$Si(OCH_3)_2(CH_3)$

Scheme 4. α-STP that exhibits a very high reactivity toward moisture and dramatically enhanced curing rates.

As the α-STPs are much more reactive than conventional γ-silanes, a reactive diluent (vinyl- or methyltrimethoxysilane) can be added to adjust the viscosity of the prepolymer without any decrease in the curing speed [6, 8]. During foam curing, the reactive diluent is incorporated into the emerging polymer. Therefore the content of this compound has only a little effect on the hardness of the cured polymer.

Prevention of Crack Formation

Under certain circumstances, STP foams tend to give a distinct crack formation during foam curing.

The main cause of the crack formation is the diffusion of the propellant. Being fairly polar, the propellant can diffuse rather fast through the likewise polar foam lamellas. Thus, inside the foam an underpressure results.

Consequently, the cracks can be avoided by using a non-polar blowing agent like propane/butane. However, unfortunately these non-polar propellants are completely insoluble in the STP. Employing propane/butane, it is impossible, to obtain foamable and stable mixtures.

This crucial problem can be solved by modifying the STP itself. A small amount of long alkyl chains is inserted into the STP to improve the solubility of the propane/butane. These alkyl end groups are generated by adding the corresponding alcohols (fatty alcohols) to the NCO-terminated prepolymer. It is sufficient to terminate 5 – 10% of all end groups with alkyl chains. All remaining end groups of the STP are silane terminated.

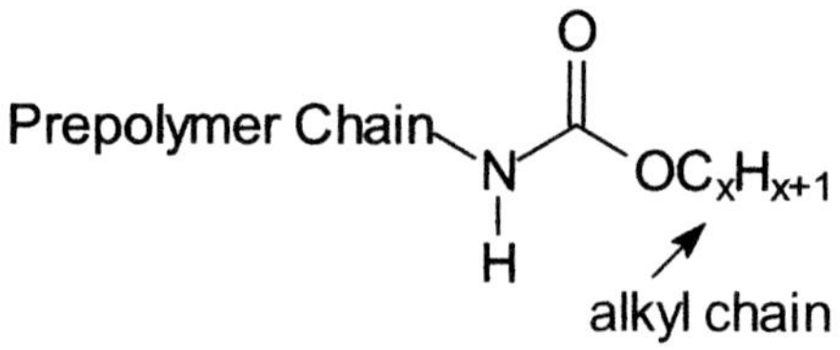

Prepolymer Chain–NH–C(=O)–OC_xH_{x+1}

alkyl chain

Scheme 5. End group of an alkyl-modified STP that can be used with propane/butane as propellant.

Summary

An isocyanate-free silane-crosslinking spray foam has been developed. The cured foam exhibits good mechanical properties. In particular, it possesses high hardness and very good elasticity. The compressive strength and foam densities are comparable to common PU-spray foams, too. Because of the high reactivity of the α-STP, high or even extremely high curing rates can be achieved without tin catalysts. Using various catalysts (e.g., *tert*-amines, phosphorous acid esters) the tack-free time can be adjusted to between 1 and 30 min.

No shrinking of the foam during curing could be observed. As shown in the following picture, the resulting foam provides a good structure

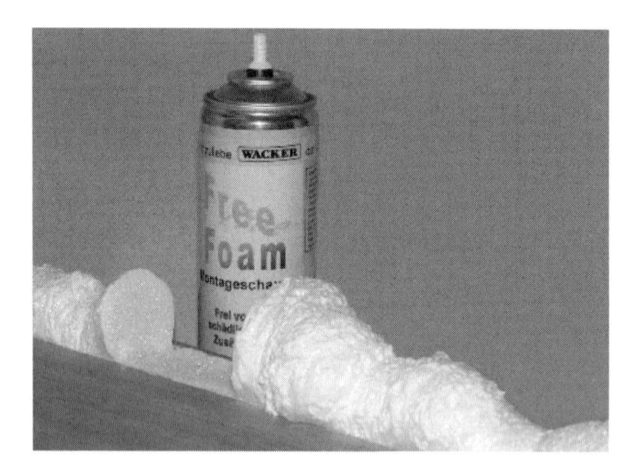

Fig. 1. Photo of a silane-crosslinking foam.

The properties of the spray foam have been tested completely. The foam meets all demands of the market.

References

[1] M. Szycher, *Szychers' Handbook of Polyurethanes*, London, **1999**.

[2] U.S. Environmental Protection Agency, Toxicological Review of Methylene Diphenyl Diisocyanate (MDI), **1999**.

[3] TDI already has to be labeled as carcinogenic: (carcinogenic Category 3).

[4] H. Sommer, M. Pauls, EP 1 098 920, **1998**.

[5] F. D. Osterholtz, E. R. Pohl, *J. Adnesion Sci. Technol.* **1992**, *6*(*1*), 127.

[6] V. Stanjek, W. Schindler, B. Pachaly, B. Bauer, WO02/66 532, **2001**.

[7] W. Schindler, V. Stanjek, B. Pachaly, WO03/18 658, **2001**.

[8] W. Schindler, B. Bauer, V. Stanjek, B. Pachaly, WO03/14 226, **2001**.

Phase Behavior of Short-Chain PDMS-*b*-PEO Diblock Copolymers and Their Use as Templates in the Preparation of Lamellar Silicate Materials

Guido Kickelbick,* Nicola Hüsing

Institute of Materials Chemistry, Vienna University of Technology
Getreidemarkt 9, A1060, Wien, Austria
Tel.: +43 1 58801 15321 — Fax: +43 1 58801 15399
E-mail: guido.kickelbick@tuwien.ac.at

Keywords: polysiloxane, block copolymers, mesoordered silicate materials

Summary: PDMS-*b*-PEO short-chain diblock copolymers were prepared via anionic ring-opening polymerization of cyclosiloxanes. Applying this method, various well-defined block copolymers with different compositions were synthesized and their phase behavior was investigated. The polymers predominantly showed lamellar phases in aqueous solutions. At small surfactant concentrations, vesicles were formed, as observed via cryogenic TEM. The aggregates of the diblock copolymers were used for the formation of lamellar thin films, applying the evaporation-induced self-assembly approach.

Introduction

Polysiloxanes have attracted much interest since their discovery because of their extraordinary properties such as the high flexibility of the backbone expressed by their low glass transition point, their optical transparency, biocompatibility, etc. More recently, the remarkable surface-active properties of polysiloxanes have been shown by the development of so-called "superspreaders" based on trisiloxanes [1]. The low cohesive energy of siloxanes is believed to be responsible for the unusual properties of this type of surfactants [2]. This makes them ideal segments of surface-active block copolymers, which are, for example, regularly used as structure-directing agents in the synthesis of mesostructured materials [3]. Usually, pure organic surfactants are applied for this purpose such as the non-ionic commercially available amphiphilic block copolymers consisting typically of either polyether (Pluronic®; Brij®) or polyhydroxyl(glyceroglycolipid) sugar derivatives as the polar group. These polymers are only available with specific chain lengths; sometimes they are inhomogeneous and show high polydispersities.

Therefore, we searched for amphiphilic polymers that are available, through preparative methods, in various compositions via controlled polymerization techniques. In this report we present the synthesis and the phase behavior of short-chain amphiphilic diblock copolymers composed of polydimethylsiloxane (PDMS) as the hydrophobic and poly(ethylene oxide) (PEO) as the hydrophilic component.

Phase Behavior of PDMS-*b*-PEO Diblock Copolymers

The synthesis of the diblock copolymers was carried out via a coupling of the two preformed segments (Scheme 1) as described earlier [4].

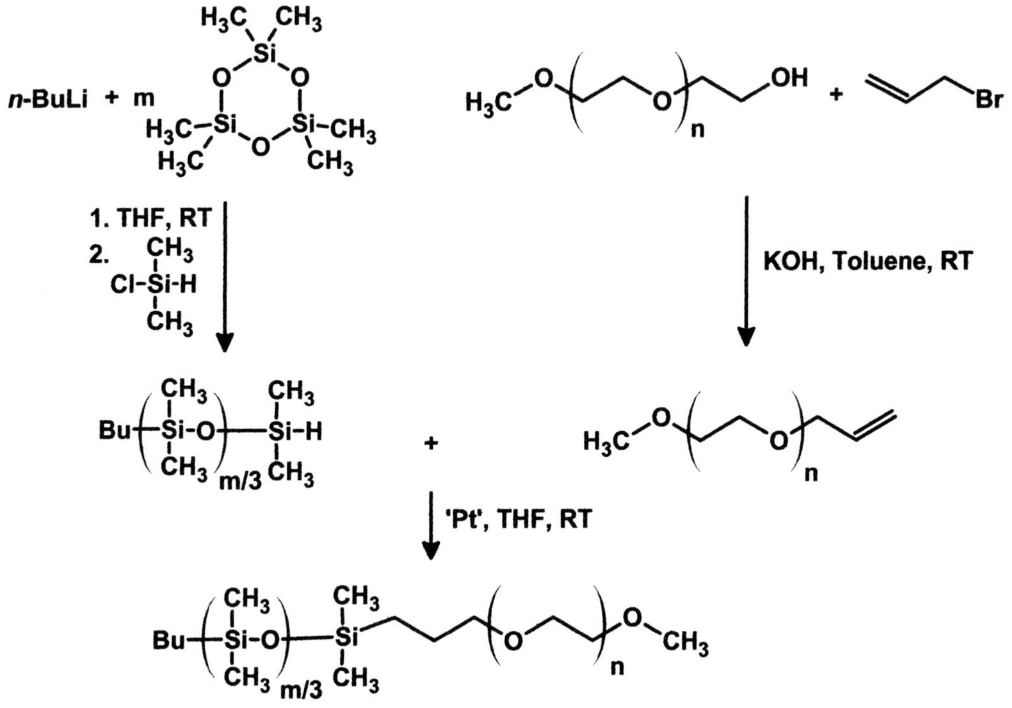

n-BuLi + m [hexamethylcyclotrisiloxane] → 1. THF, RT; 2. Cl-Si(CH$_3$)$_2$-H → Bu-[Si(CH$_3$)$_2$-O]$_{m/3}$-Si(CH$_3$)$_2$-H

H_3C-O-(CH$_2$CH$_2$O)$_n$-CH$_2$CH$_2$-OH + allyl bromide → KOH, Toluene, RT → H_3C-O-(CH$_2$CH$_2$O)$_n$-CH$_2$CH=CH$_2$

+ → 'Pt', THF, RT → Bu-[Si(CH$_3$)$_2$-O]$_{m/3}$-Si(CH$_3$)$_2$-(CH$_2$)$_3$-(OCH$_2$CH$_2$)$_n$-O-CH$_3$

Scheme 1.

The aggregation behavior of selected diblock copolymers with various compositions was investigated applying several techniques, such as polarized optical microscopy (POM), tensiometry measurements, fluorescence studies, deuterium NMR spectroscopy, SAXS measurements, and cryogenic TEM [4, 5]. In systematic studies we particularly focused on the effect of an increase in the dimethylsiloxane chain length on the aggregation behavior of the investigated surfactants.

The variation of the surface tension with the surfactant concentration was determined for several surfactant molecules, and the critical aggregation concentration (CAC) was calculated from the obtained results. The length of the dimethylsiloxane block was varied from DMS_4-*b*-EO_{12} to DMS_{18}-*b*-EO_{12}, and it was shown that (a) the CAC values increase with increasing size of the hydrophobic segment until a maximum is obtained at 10 DMS units, and that (b) the values for the surfactants with short hydrophobic chains are unrealistically low. This behavior can probably be explained by the extremely hydrophobic PDMS segment inducing tight packing of the hydrophobic segments, which, in turn, may force the poly(ethylene oxide) chain to become more extended than normal.

The aggregation of three surfactants with constant PEO chain length, i.e. DMS_4-*b*-EO_{12}, DMS_6-*b*-EO_{12}, DMS_{10}-*b*-EO_{12}, was also investigated by steady-state fluorescence using the emission spectrum of pyrene. These measurements showed similar CAC values to those obtained by tensiometry.

It is noteworthy and interesting that both the tensiometry and the fluorescence measurements give an order of CAC values for the series of copolymers with 4, 6 and 10 DMS units that is opposite to the expected order. The usual trend for surfactants is that the aggregation, which is usually micelle formation, starts at lower concentration the longer the hydrophobic tail. With the dimethylsiloxane surfactants studied in this work, the trend is the opposite. We believe the reason for this is that we observe the formation of vesicles instead of micelles. (Vesicles are dispersions of a lamellar phase in water.)

While the formation of micelles in the solution was expected, cryo-TEM studies proved that the short-chain PDMS-*b*-PEO diblock copolymers spontaneously form vesicles at low concentrations in aqueous solutions [4]. Cryo-TEM images of the studied systems are shown in Fig. 1.

A 0.023 wt% aqueous solution of DMS_4-*b*-EO_{12} shows a dispersion of mostly globular unilamellar vesicles. The vesicles are polydisperse, ranging from 20 nm up to several hundreds of nm in diameter. The hydrophobic layers of the vesicles exhibit a thickness of ~3 – 4 nm. For comparison, the extended length of the hydrophobic block of the diblock copolymer was calculated to be ~2.1 nm including the butyl end group and the propyl spacer, and the whole diblock copolymer has an extended length of ~7.0 nm. Hence, the measured thickness of the hydrophobic layers may be explained by a parallel arrangement of the extended hydrophobic blocks or by loose packing in a bilayer structure with coiled hydrophobic segments.

When the concentration of DMS_4-*b*-EO_{12} was increased to 5 wt%, the cryo-TEM image revealed that the vesicles start to arrange into partially well-ordered multilayers over several hundreds of nanometers, which may mark the beginning of the formation of a well-defined lamellar structure. The thickness of the hydrophobic walls in these multilayers measures ~5 nm, which approximates to twice the extended length of the hydrophobic moiety of the polymer.

A diblock copolymer with an extended hydrophobic siloxane part was examined to compare the influence of the length of the hydrophobic block on the aggregation behavior in water. The extended length of the hydrophobic part of DMS_{10}-*b*-EO_{12} was calculated to be 3.9 nm, and the extended length of the complete diblock copolymer equals ~8.8 nm. Similarly to the DMS_4-*b*-EO_{12}/water mixture, no micelles could be observed in a 0.12 wt% sample of DMS_{10}-*b*-EO_{12} in water (3.3 times the CAC). Multilamellar vesicles of different sizes were present in coexistence with unilamellar vesicles of more than 1 μm in diameter. The thickness of the hydrophobic parts of the vesicles is very uniform and varies around a mean value of ~7 nm. The latter value equals approximately twice the theoretical extended length of the hydrophobic domain of the surfactant. This observation seems to support the conclusion that the hydrophobicity of the siloxane part of the surfactant leads to a complete repulsion of water molecules from the hydrophobic bilayer. On the other hand, the length of the siloxane block obviously results in the formation of a nearly perfect bilayer structure with siloxane chains ordered parallel to each other. The small variation of the thickness of the hydrophobic domains may be explained by the polydispersity of the siloxane block precursor.

Cryo-TEM images of solutions with increasing surfactant concentration show ordered areas, similar to those in the image obtained for DMS_4-*b*-EO_{12}. Increasing the length of the siloxane block leads to an increased tendency to form multilamellar aggregates. The ordered lamellae span over several μm and most of them are closed. The thickness of the hydrophobic layers varies around a value of ~6 nm. Compared to the value of 7 nm measured before, the diameter of the lipophilic moieties is somewhat reduced. The hydrophilic interlayer in the lamellar areas has a thickness of ~5 nm, which is smaller than the value found for the multilamellar vesicles.

The binary phase behavior of three selected PDMS-*b*-PEO diblock copolymers was investigated in more detail. POM revealed the lyotropic behavior of the PDMS-*b*-PEO diblock copolymers via the observation of birefringent patterns between two glass slides (Fig. 2). This method allowed us a fast determination of the phase diagrams depending on the water-to-surfactant ratio and the temperature. The phase diagrams of the investigated surfactants were dominated by large areas of lamellar phases, which were confirmed via deuterium NMR spectroscopy and SAXS measurements [6].

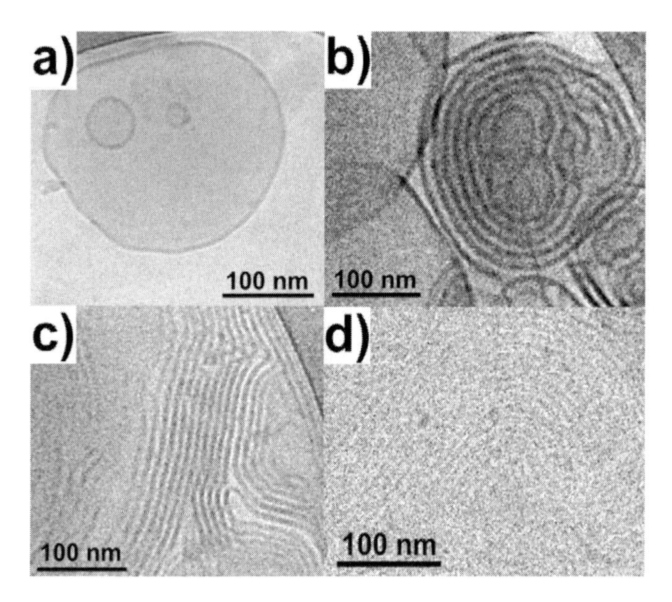

Fig. 1. Cryo-TEM images of DMS_4-*b*-EO_{12} diblock copolymer in water: a) 0.023 wt% solution, b) 5 wt% solution compared to DMS_{10}-*b*-EO_{12} diblock copolymer, c) 0.12 wt% solution, and d) 5 wt% solution.

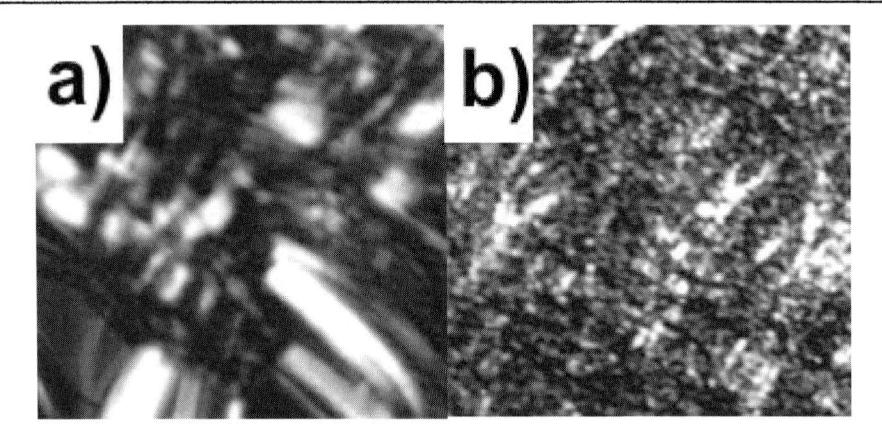

Fig. 2. Polarizing light microscopy (POM) image of aqueous samples at 25 °C (magnification 100) of a) 55 wt% DMS_6-*b*-EO_{12} and b) 80 wt% DMS_6-*b*-EO_{12}.

Application of PDMS-*b*-PEO Diblock Copolymers as Templates for Mesoordered Silica Materials

We were interested in the potential of the synthesized diblock copolymers as templates in the synthesis of mesostructured silicate materials. Our focus was the preparation of mesoordered thin films via the evaporation-induced self-assembly process [7]. For this purpose PDMS-*b*-PEO were mixed with a precursor solution containing the network-building species, and films were produced via a dip-coating procedure [8]. XRD diffraction patterns confirmed the formation of lamellar materials which were stable upon calcination at 430 °C/3 h.

While the films formed by using common templates such as Brij56® show a strong contraction of the structure upon calcination because of the thermally driven network condensation, loss of the organic template, and/or phase transitions, the PDMS-*b*-PEO-templated films revealed no contraction of the structure with the *d*-spacing for the calcined films remaining unchanged.

Studies of the diffraction patterns for calcined PDMS-*b*-PEO block copolymer samples prepared with (a) different block copolymer concentrations and (b) different block lengths showed that the higher the concentration (up to 15%) of the surfactant the more pronounced is the order in the final material. In addition, variation of the block length has also a profound influence on the obtained structures. The TEM image of a calcined DMS_{18}-*b*-EO_{12}/silica film with 15 wt% of surfactant with a corresponding layered structure is shown in Fig. 3. The layer spacing obtained from the TEM image (about 5 nm) is consistent with the long-range order parameter obtained from the XRD pattern.

However, the question whether the mesostructure obtained after heat treatment is a hexagonal or a lamellar one remains unclear. ^{29}Si and ^{13}C CP-MAS NMR studies have been performed to investigate the thermal degradation of the block copolymer within the mesostructured silica upon heat treatment in air at different temperatures ranging from 20 to 600 °C. The ^{29}Si NMR spectrum of the as-synthesized sample (20 °C) presents peaks at 11 and –20 ppm characteristic of the end and the chain atoms of the PDMS polymer (11 ppm $(CH_3)_3SiO$, –20 ppm $(CH_3)_2Si(OSi)_2$) and in the range from –85 to –109 ppm for the Q units of the condensed silicate species (Q^2 –85 ppm $(RO)_2Si(OSi)_2$, Q^3 –98 ppm $(RO)Si(OSi)_3$, Q^4 –109 ppm $Si(OSi)_4$ with R = H or CH_3CH_2). The number of Q^4 units increases with temperature, as expected for a higher degree of condensation. At 600 °C, complete degradation of the organic moieties results in pure silica. Above 300 °C the PDMS units start to decompose, and with increasing calcination temperature a new peak at about –60 ppm appears which can be assigned to T-units (R-Si(O_3)), probably arising from the presence of methyl silicon moieties or methylene spacers between two silicon atoms.

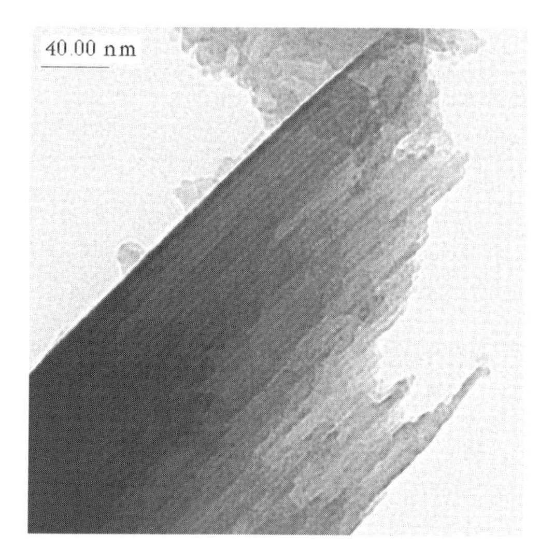

Fig. 3. TEM image of a calcined DMS_{18}-*b*-EO_{12}/silica film.

The ^{13}C CP-MAS NMR spectra are in agreement with the ^{29}Si CP-MAS NMR results. They show that the organic part of the surfactant can be easily removed by calcination at relatively low temperatures (200 – 300 °C).

Conclusions

Amphiphilic inorganic-organic block copolymers (PDMS-*b*-PEO) with variable block lengths were successfully synthesized. The obtained polymers revealed predominantly lamellar phases in their aqueous solutions. At low concentrations the spontaneous formation of vesicles was observed. The polymers can be applied as structure-directing agents in the synthesis of mesostructured silica-based thin films.

Acknowledgments: We would like to thank the PhD students Beatrice Launay and Josef Bauer for their intensive and fruitful work on the project. We also thank Martin Andersson, Prof. Anders Palmqvist, and Prof. Krister Holmberg from the Department of Applied Surface Chemistry, Chalmers University of Technology, Göteborg, Sweden for their help in the investigation of the phase behavior. Furthermore, we thank the Fonds zur Förderung der wissenschaftlichen Forschung, Austria, the Swedish Foundation for Strategic Research, and the European Cooperation in the Field of Scientific and Technical Research (COST) Action D19 for their financial support of this work. Finally we thank Wacker-Chemie GmbH for providing many chemicals.

References

[1] R. M. Hill, *Curr. Opin. Coll. Interface Sci.* **1998**, *3*, 247.

[2] M. J. Owen, *Ind. Eng. Chem., Prod. Res. Dev.* **1980**, *19*, 97.

[3] G. J. J. de A. A. Soler-Illia, E. L. Crepaldi, D. Grosso, C. Sanchez, *Curr. Opin. Colloid Interface Sci.* **2003**, *8*, 109.

[4] G. Kickelbick, J. Bauer, N. Hüsing, M. Andersson, A. Palmqvist, *Langmuir* **2003**, *19*, 3198.

[5] G. Kickelbick, J. Bauer, N. Hüsing, M. Andersson, K. Holmberg, *Langmuir* **2003**, *19*, 10073.

[6] G. Kickelbick, J. Bauer, N. Huesing, M. Andersson, K. Holmberg, *Colloids and Surfaces, A* **2005**, *254*, 37.

[7] C. J. Brinker, Y. Lu, A. Sellinger, H. Fan, *Adv. Mater.* **1999**, *11*, 579.

[8] N. Hüsing, B. Launay, J. Bauer, G. Kickelbick, D. Doshi, *J. Sol-Gel Sci. Technol.* **2003**, *76*, 609.

Nanoscale Networks for Masonry Protection

A. Lork, F. Sandmeyer*

Wacker-Chemie GmbH, Wacker Silicones
Johannes-Hess-Straße 24, D-84489 Burghausen, Germany
Tel.: +49 8677 83 7671 — Fax: +49 8677 83 5765
E-mail: anette.lork@wacker.com

J. Köhler, J. Weis

Consortium für elektrochemische Industrie GmbH
(Corporate Research Facility of Wacker-Chemie)
Zielstattstrasse 20, D-81378 Munich, Germany

Dedicated to Prof. Dr. Rochus Blaschke on the occasion of his 75th birthday

Keywords: silicone resin network, masonry protection, building material, natural stone impregnation, silicone coating of mineral substrates, structure-effect principle of trifunctional silicones

Summary: Suitable silicones offer excellent distribution in the building materials applications illustrated. The nanometer-thin silicone resin network structures, which previously eluded observation, can nowadays be rendered visible with the possibilities afforded by high-resolution field-emission scanning electron microscopy (FE-SEM). Complete silicone coverage indicates good attachment to the mineral substrates studied. Silicone resin networks cover quartz and calcite substrates quite well. In contrast to silanes, oligomers and higher-molecular-weight siloxanes generate exceptionally good hydrophobic properties even on calcite. By virtue of the perfect encapsulation, microstructure reinforcement, and hydrophobic properties of nanoscale networks, complex damage processes in microstructures of building materials, arising from weathering influences, can be minimized. The organosilicon compounds can be rendered extremely efficient by admixing additives in quantities of 0.1 – 0.5 wt% to the building material (integral water-repellent treatment) or filler coating.

Organosilicones enable major demands imposed on masonry protection to be fulfilled:

- Very low water uptake, by reduction in the polarity of mineral fillers in paints and hydrophilic building materials.
- High water vapor diffusion, because there is very little reduction in the cross-

section of capillary pores. Moisture coming from within the masonry will evaporate.

- Reinforcement of the microstructure when used in excess as a paint binder without clogging any pores. Even additive-size application quantities can have a slight reinforcing effect on the microstructure of building materials.
- Long-term stability under exposed conditions, implying a permanent anchoring of the hard silicone resin network (organo-modified quartz structure) to the substrate.

As a tool for visualizing nanoscale silicone structures within building materials, high-resolution imaging techniques are indispensable. It has been conclusively shown that the high-resolution microscopic studies are not only useful for visualizing the failure of unsuitable silicones for building stone impregnation; they also provide access to a fundamentally superior understanding of the structure of building materials.

A model (arbitrary assembly) of a network structure of a methyl silicone resin was built. Further molecular modeling of siloxane derivatives on quartz and calcite revealed that a complete coverage of the surface can best be obtained by cyclic Si_3-siloxanes with methyl groups on quartz, because of similarity in characteristic bond length of siloxanes and quartz and the small size of the methyl groups. In contrast, the crystal structure of calcite has different distances and can only be covered partially. Also, larger side-chains like *iso*-octyl are rather space demanding and prevent optimal covalent bonding.

Introduction

Organosilicon compounds are particularly suitable for use in masonry protection. By virtue of their organo-inorganic hybrid nature [1, 2], they bring about a noticeable reduction in the level of water absorbed by coating systems and by mineral building materials (e.g., natural building stone, lime sandstone, bricks, plaster, gypsum plasterboard), which are usually highly porous. Aside from their pronounced hydrophobic nature, two important properties of polysiloxanes need to be mentioned that virtually predestine them for use in water-repellent treatment of hydrophilic building materials. These are their good spreading power and the formation of more or less strong interactions with different mineral substrates.

It has been known from ancient times [3 – 6] that moisture which gains ingress into building materials gives a major boost to weathering mechanisms. However, water itself is very often not what leads to weathering in the form of loosening of the forces holding together the grains of the mineral components in the microstructure of the paint and building materials. Rather, the cause is alternating damp and dry spells that trigger physicomechanical interfacial processes in the substrate surfaces [5 – 9].

Consequently, the requirements that are imposed on hydrophobic coatings and water-repellent agents for treating façades are high. This is particularly true in the preservation of monuments, as it is the moral duty of every generation to protect invaluable cultural jewels from the past for its successors [10, 11]. For this reason, the "modern" organosilicon compounds used in façade protection must be expected to offer not only good water repellency but also, and especially, lasting durability in the building material itself [4, 12 – 15].

Paints and impregnating treatments primarily utilize network-forming, trifunctional methylsilicone resins and alkylalkoxy silanes of general composition $RSi(OR)_3$ (R = alkyl group, $-CH_3$ to $-C_8H_{17}$; OR = alkoxy group) [16 – 18]. The monomeric precursors of the silicone resins bear three alkoxy functions, which hydrolyze in ambient moisture to form silanol groups and which both form macromolecular condensation products and interact via OH groups with the ionic, mineral substrate. The resultant polysiloxane layer (silicone resin network) envelops mineral substrates because of the high polarity of the Si–O– bond. The always outwardly directed, nonpolar alkyl groups of the nanoscale silicone resin network [1, 2] reduce wetting by liquid water.

Effective stabilization and the protection afforded by fully condensed silicone resin networks in the microstructure of paint and building materials are manifested by largely uninterrupted polymeric coatings on all mineral substrates and nanoscale grain bridges that do not restrict gas diffusion. The aim of this paper is therefore to study whether modern, high-resolution field-emission scanning electron microscopy (HR-FEM) can serve to unequivocally identify thin silicone resin coatings in the range 1 – 100 nm and whether it is at all possible to detect the distribution of this protective layer in the microstructures of different building materials.

It is well known that organosilicon compounds do not become equally well attached to all mineral substrates. While silicates always readily lend themselves to coating with silane and polysiloxane [19 – 21], the same cannot always be said of calcium carbonate [19, 20]. Calcium carbonate (calcite), a widely used filler, is generally considered difficult to cover with silanes. Given the proven, good attachment of silicone resin to calcium carbonate fillers in silicone resin emulsion paints [2, 22], the question arises as to whether only higher polymeric siloxanes are able to form hydrophobic protective coatings on calcium carbonate.

Use of Organosilicon Compounds in Masonry Protection

Durability of Organosilicon Compounds in Different Building Materials

Silicone-based binders for paints and impregnating agents exhibit long-term water repellency and very good durability. Although it seems plausible that silicones' good weathering resistance derives from their high interfacial tension with water and their lack of reactive groups/double bonds, comprehensive exposure tests of the long-term effect of water-repellent treatment of natural building stone/building materials have occasionally produced contradictory results [4].

Silicone Resin Emulsion Paint

The fact that "mineral-paint-like" silicone resin emulsion paints, which have very open pores yet extremely high weathering resistance, suggests *inter alia* that the silicone resin binder employed has good durability. This has been documented in numerous reference objects whose coatings were up to 30 years old [12, 13]. Furthermore, comparative exposure tests on various paint systems have revealed that hydrophobic silicone resin emulsion paint was the only one to offer adequate protection for natural stone at risk of severe weathering [23, 24]. Aside from the seemingly undegradable hydrophobicity exhibited by the reference buildings, which were examined at regular intervals [13], the outstanding properties of surface-dry (hydrophobic) silicone resin emulsion paint that need to be mentioned are the high water vapor diffusion ability and the low tendency to be colonized by algae [12, 23].

Impregnation of Building Stones and Building Materials

Organosilicon compounds have been successfully used to impregnate building materials and natural building stone for approximately 35 years [25 – 27]. However, when the effectiveness of diverse treatments on both buildings and exposed samples was evaluated, the results suggested varying degrees of protection [4, 5, 14, 28 – 30].

Although long-term weathering experiments on impregnated building/natural stone materials leave no doubt about the good durability of organosilicon compounds [4, 31 – 34], there are authors who postulate that water-repellent agents lose effectiveness after a few years [4, 5, 28, 30, 35]. However, there is agreement that it will take a reliable body of data to adequately prove degradation of silicones' effectiveness and that each case of water-repellent treatment where inadequate protection has occurred needs to be checked.

One recourse is to use low-destructive test methods that can quantify the efficacy of protective agents [27] and provide evidence of improper use or application of the silicone [4, 15, 27, 28, 32, 36, 37]. High-resolution microscopy would offer the advantage of visualizing the reasons for failure of hydrophobic treatments just after application, *before* they became visible very much later when the weathered materials started to fail [38, 39]. Weak areas such as inadequate uptake of the treatment, poor penetration depth, and incompleteness of coating could be documented just after the masonry treatment had been applied.

There is reason to believe that the molecular structure of crosslinked trifunctional silanes/siloxanes may also influence the durability of the resulting silicone resin network [14, 40 – 42]. Allowance must be made here for the fact that such considerations are models in nature and also that we still do not have adequate knowledge of the molecular principles of fully condensed silicone resin networks [15].

The Right Organosilicon Compound for Every Substrate

The *binders used for silicone resin emulsion paints* [16] are almost exclusively methylsilicone resins, $R–SiO_{3/2}$ ($R = CH_3$), in combination with an organic polymer binder. Aqueous silicone resin emulsions predominate. As a solvent-free solution cannot be produced from a solid resin, high-viscosity silanes/siloxanes/silicone resin mixtures are used that differ initially from the solid resin in the degree of polymerization.

In the case of highly hydrophilic *gypsum building materials* [43 – 45], additive admixtures of polymethylhydrogensiloxane (H-siloxane) or trimethoxymethylsilane (MTMO) as active ingredients are enough to reduce the water absorption of the gypsum significantly to a few percent [44, 46]. The use of H-siloxane is state of the art. Not only is the water repellency very good, but neither the mechanical properties nor the setting time of the plaster are impaired.

For *treating façades*, low-molecular-weight compounds based on monomeric alkyltrialkoxysilanes and derivative oligomeric siloxanes are used to improve penetration. Since the methyl groups have less stability toward alkalis, water-repellent agents nowadays usually also contain a mixture of methyl and *iso*-octyl silicones. Silane and siloxane can be combined to produce tailor-made water repellents for diverse substrates. Silicone microemulsion concentrates (SMK) are highly effective on a large number of substrates [25]. These solvent-free silanes/siloxanes/silicone mixtures yield ready-to-use silicone microemulsions when diluted with water.

Alkali siliconates are obtained by highly alkaline condensation of methyltrimethoxysilane with potassium hydroxide. In the field of masonry protection, potassium methyl siliconates of general formula $CH_3Si(OH)_2O^- K^+$ are used predominantly for factory impregnation of building materials, but hardly ever nowadays for façade treatment [16, 25, 27].

Mechanisms of Silicone Resin Network Attachment to a Mineral Substrate

On account of the large difference in the electronegativities of Si and O (Δ = 1.7), the Si–O–Si bonds of the polysiloxanes are highly polar (see Fig. 1), and this explains the strong affinity of the polar siloxane backbone for the equally polar, mineral substrate. By comparison, the C – H bond has a low electronegativity difference (Δ = 0.4), and a substantially lower surface tension. Also worth mentioning is the high bond energy of the Si–O–Si bond (444 kJ mol^{-1}) [47], which plays a crucial role in the high durability of the silicone resin network. The energy of the Si–C bond is much lower at 306 kJ mol^{-1} (C–C = 345 kJ mol^{-1}). The low barrier to rotation of the Si–O–Si bond facilitates attachment to mineral substrates.

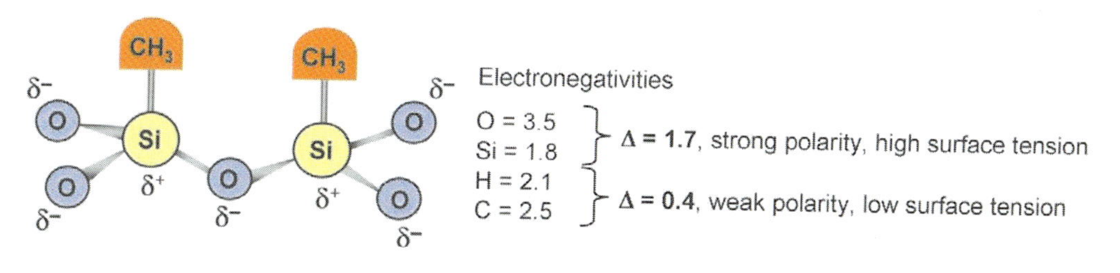

Fig. 1. Electronegativity values for the bonds in a siloxane molecule (after Pauling).

The Silicone Resin Network in Paints and Impregnated Building Materials

The microstructure of highly porous paints/building materials/natural building stone necessitates preventive measures against water ingress and thus against chemical and microbial attack. For an understanding of the mode of action of silicone resin binders and water repellents in the gypsum and building stone under consideration, it is important to know about the distribution and attachment of the substrate and the degree to which it is enveloped by the organosilicon compounds.

Sample Preparation and Analytical Methods

The *paint samples* were prepared from a high-grade silicone resin emulsion paint [48] with a binder ratio of 10 wt% styrene acrylate to 10 wt% silicone resin emulsion (both 50% solids). The coatings were applied to a mineral substrate and dried at room temperature. Part of each sample was etched for a few minutes in 10% nitric acid so that the organic polymer could be visualized better.

As is usual practice, the *gypsum samples* were bulk impregnated. As the gypsum hemihydrate (plaster of Paris) was being mixed with water, polymethylhydrogensiloxane (H-siloxane) was added in the usual amount of < 0.5% and also, for test purposes, in excessive doses of 2%. Experience has yielded good results with a water-to-gypsum ratio WGR (weight of water/weight of calcium sulfate hemihydrate) of 0.5 to 0.8, which produces a pore volume of less than 50%.

The *stone specimens were impregnated* by fully immersing them for two minutes in silicone microemulsion concentrate (diluted 1:10) and potassium siliconate (diluted 1:4). There is a general correlation between absorption of the treatments and the inner surface of the various stone types. Treatment under optimum conditions in the laboratory always produces better impregnation than when it is performed on an actual building.

Coating of problematic mineral substrates with silicone (calcite, gypsum hemihydrate) was tested by the dry method. To this end, 100 g samples of filler were treated with either 0.1 or 0.5 wt% silicone active ingredient and then mixed at high speed for at least 3 min so that the silicone would be adequately distributed.

All paint, gypsum, and natural stone samples were broken perpendicularly to the surface so as to produce rough surfaces. To avoid artifacts, most analyses were performed either on ultra-thin carbon-coated samples for the purpose of improving the conductivity or, if possible, on non-coated samples. Ultra-thin carbon-coated powder preparations were made from silicone-coated fillers.

High-resolution *field-emission scanning electron microscopy (HRFE-SEM)* combined with techniques for rapidly examining building surfaces at low magnification is suitable for unequivocal visualization and interpretation of the nanoscale network structures expected in building materials. A prerequisite for surface imaging at maximum resolution is in-lens detection in the low kV mode (< 1 kV).

The Microstructure of Silicone Resin Coatings

Silicone resin emulsion paints always contain two binders, i.e. some of the organic polymer binder in the emulsion paint is replaced by silicone resin. Silicone resin emulsion paints are predominantly formulated at high PVC (pigment volume concentration > 60%), as a result of which the overall binder content is low. The PVC is a theoretical description of the volume fraction of the pigments/fillers relative to the overall volume of the dry coat.

Theory Behind Silicone Resin Emulsion Paint

The principle behind highly porous silicone resin emulsion paints is shown schematically in Fig. 2. On account of the high polarity of the Si–O–Si bond of the silicone resin molecule to the polar mineral substrate, the silicone resin should envelop all fillers. Because of the binder deficit, not enough film-forming, organic polymer binder is present to seal the pores between the fillers and hence prevent water ingress. Instead, only narrow adhesive bridges can form, which bond individual filler particles and pigments but are unable to seal the pores. Such a high-porosity paint containing approx. 45 wt% hydrophilic fillers/pigments can only act as a good water repellent and reinforce the microstructure throughout the network if all fillers have been almost completely enveloped by the hydrophobic, protective polymeric layer.

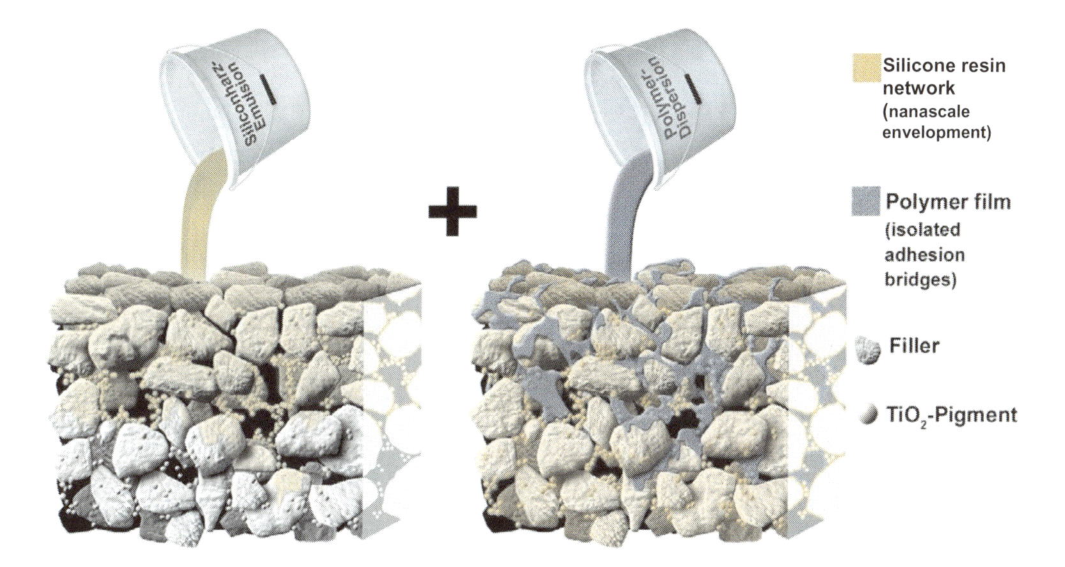

Fig. 2. Highly simplified model of the microstructure of a silicone resin emulsion paint.

The Silicone Resin Network and the Polymer Binder in the Paint Microstructure

The microstructure of a silicone resin emulsion paint is shown in Fig. 3. The overview even at low magnification (Fig. 3a) indicates that one binder forms thick films (black arrows) into which sinks some of the approx. 300-nm-sized white pigment (titanium dioxide). This is the organic polymeric binder. Because of the deficit of organic polymer, pore voids are open and voids between fillers are only partly filled (Fig. 3c), i.e., only small adhesive bridges are present (black arrows in Figs. 3a,c), which determine the binding power.

Exposed filler surfaces are always coated by a nanoscale silicone resin network, although this is only apparent at much higher magnification, as can be seen from the detail (white arrow) in Fig. 3b.

The silicone resin network (polymethylsilicic acid) coats the surfaces of both large calcite grains and small titanium dioxide particles with an ultra-thin veil-like film. Clear evidence of an existing network coating is the "apparent" rounding and void filling just a few nanometers thick (Figs. 3b,d) between the mineral phases (white arrows).

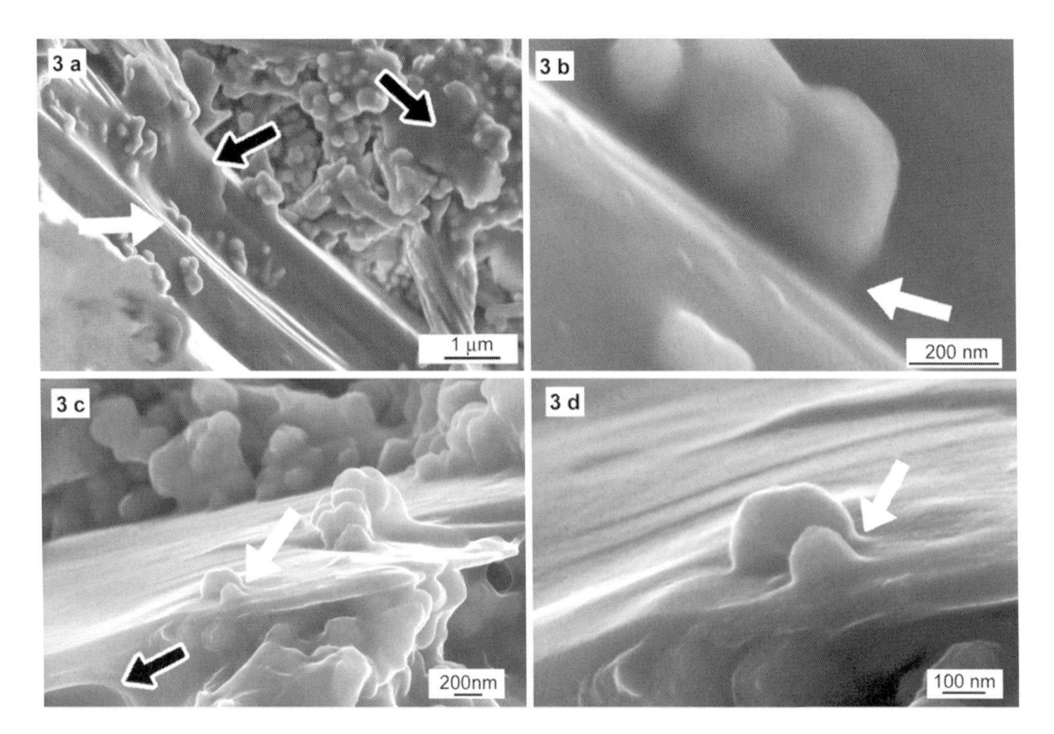

Fig. 3. FE-SEM secondary electron (SE) images of representative microstructures in a silicone resin emulsion paint. **a, c** Overview: localized adhesive bridges consisting of thick polymeric binder films (black arrows), which only partially wet the calcitic filler (**a**) and a clay mineral (**c**). **b, d** Detailed views: ultra-thin silicone resin network envelops nanoscale titanium dioxide (white arrows) and its substrate. The arched meniscus-like rim (**d**) toward the unfilled pore void (see arrow) is a characteristic feature.

On account of the minimum size of its primary polymer particles (approx. 100 nm in the case of silicone resin emulsion paints), the organic polymer film must differ from the silicone resin network by being significantly thicker (Fig. 3a). Primary particles of an arbitrary, coarse-particle polymer dispersion are shown in Figs. 4a,b. Despite their having undergone accelerated drying, polyhedral deformation of the – ideally – circular primary particles (Fig. 4b) indicates incipient film formation resulting from increased capillary forces [49, 50]. The material in the polymer voids (see arrow) consists of auxiliaries that, after coalescence, are no longer visible as hydrophilic components in interstitial phases in the undisturbed polymer film.

The distribution of the polymer dispersion in building materials is sometimes better visualized by etching methods (see "Sample preparation and analytical methods" above), which partially dissolve calcitic fillers or even etch them away. This approach offers a surprising further advantage: whereas un-etched polymer films always have a smooth surface (Figs. 3a,c), etched polymer films always have a roughened surface (Fig. 4c), and this always permits reliable identification of the organic binder. The film seems to retain a memory of its particulate history [49], and the round structures result from corrosive acid etching of the hydrophilic interstitial phases between the primary particles [22, 48]. The polymer film (Fig. 4d), broken perpendicularly, impressively illustrates the precise double-layer nature of perfectly coalesced polymer particles, which undergo appreciable deformation during film formation, and shrink.

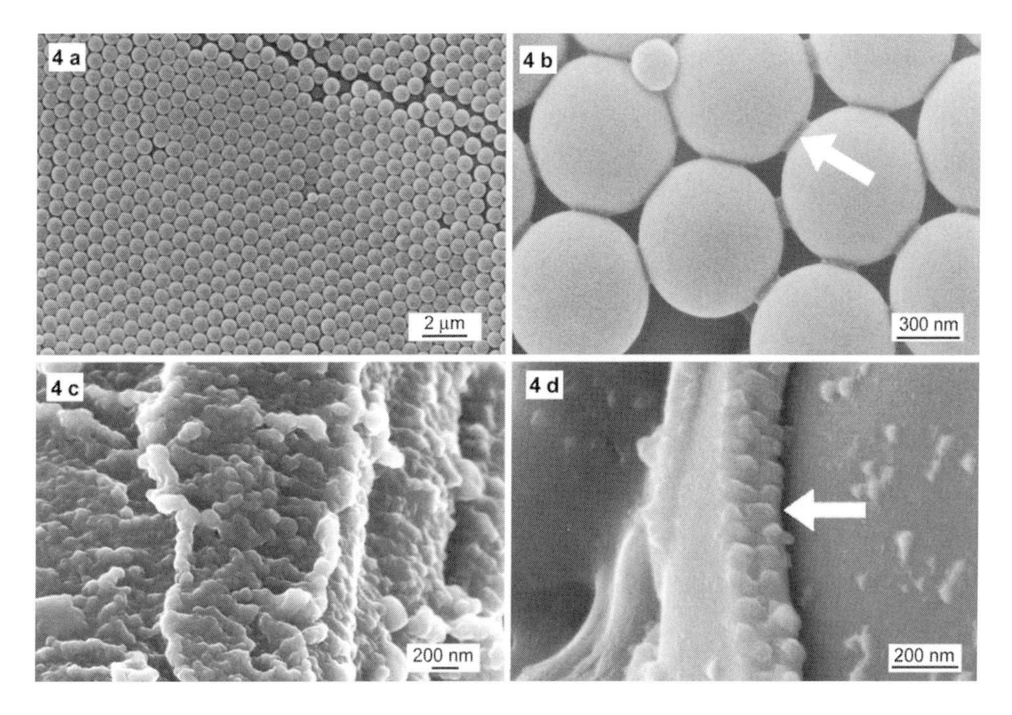

Fig. 4. SE images of organic polymer binders (**a, b**) in the primary form as dispersion and (**c, d**) coalesced in silicone resin coatings. **a, b** Typical particle shape of the primary latex particles after rapid drying in a vacuum. **c, d** Plaster/paint samples etched with 10% HNO_3. **c** Plan view of a coalesced polymer binder of a silicone resin plaster and **d** cross-section of of a polymer film partly covering a clay mineral particle within a silicone resin emulsion paint.

Strength of the Silicone Resin Network in the High-Porosity Coating Microstructure

Silicone resin forms uninterrupted nanoscale coatings on fillers and pigments, i.e., all pores/capillaries are enveloped hydrophobically. Additionally, the silicone resin network reinforces the microstructure, the visible features of this effective stabilization being polymeric grain bridges and filled voids. The high durability and lack of hygroplasticity and thermoplasticity are salient properties of silicone resins, since dry and less tacky façades are less prone to microbial attack and dirt pick-up. Elasticity and good binding power render the organic polymer binder the ideal complement to the hard and brittle silicone resin binder.

The Silicone Resin Network in Building Material Microstructure and Building Stone

The microstructure of building materials and building stone is riddled with pores and capillaries that are interconnected. Clearly, given pore volumes of 40 – 50% in gypsum building materials [44 – 46] and 10 – 20% in the natural stone examined [51], the overall porosity, especially the pore radius distribution, influences the material properties. These determine how quickly water and water vapor penetrate into a microstructure, whether they are stored or rapidly emitted again, and what, for example, occurs during freezing.

Nanoscale Networks in Bulk-Impregnated Gypsum Building Materials

The microstructure of the gypsum sample illustrated in Figs. 5a,b is highly porous and consists of

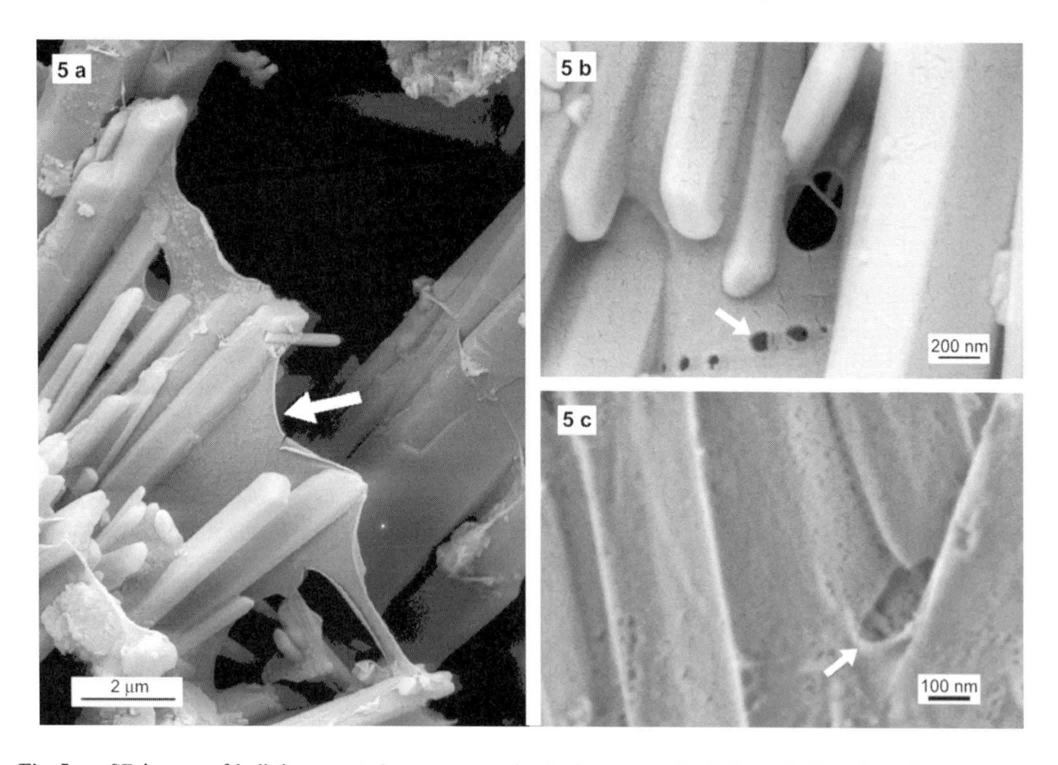

Fig. 5. SE images of bulk-impregnated gypsum samples broken perpendicularly. **a**, **b** Overview: Gypsum microstructure of samples with high water-gypsum ratio (WGR = 1.0), overdosing with H-siloxane at 2 wt%. The loosely packed gypsum microstructure, which depends on the water-to-gypsum ratio, exhibits free polymethylsilicic acid films in crystal interstices only when silicone is present in excess. **c** The usual market H-siloxane dosage of < 0.5 wt% is found to produce ultra-thin silicone films on gypsum crystals, as shown in non-carbon-coated samples: the presence of silicone is indicated by menisci (see arrow).

loosely intergrown, thin, prismatic to tabular crystals (see also "Sample preparation and analytical methods"). To render the silicone resin network more easily identifiable in the gypsum microstructure, gypsum samples were prepared with a high WGR and, uncommonly for practical use, high H-siloxane doses. The loosely packed gypsum crystals are connected by thin film webs of polymethylsilicic acid – condensed silicone resin network (see arrow), with the typical arc-like menisci toward the free pore void (Fig. 5a). Also only to be found at extreme overdosing are pitted nanolayers (Fig. 5b) in narrow crystal interstices. Suspected network coats on the gypsum crystals cannot be identified at the magnifications employed here.

At low H-siloxane doses, very high magnification is necessary for unequivocal identification of networks. Figure 5c shows evidence of a nano-permeable, apparently fully enveloping silicone film on a gypsum crystal. Further clues supporting the presence of the silicone resin network in addition to crystal edge-rounding is the pittedness that is atypical of gypsum. A reflection of the much lower silicone dose here is the faint presence of menisci (see arrow) in narrow crystal voids (Fig. 5c).

Factors Underpinning the Success of H-Siloxane in Water-Repellent Treatment of Gypsum

Although the maximum pore radius distribution in conventional gypsum microstructures usually lies within the range of capillary suction action (100 nm to 1 mm) [44, 52], the roughly 50% water absorption that occurs in untreated gypsum is often reduced to less than 5 wt% [44, 45] at standard market application rates of just 0.3 – 0.5 wt% H-siloxane. The optimum amount to use depends upon the gypsum raw material, the fineness of the grain in the plaster of Paris, its specific surface area, and the reaction temperature.

Hydrophobic lining of the pores therefore produces a statistical reduction in water transport within the gypsum microstructure [44]. High-resolution microscopy confirms the very good distribution of H-siloxane in the gypsum microstructure, which necessitates excellent spreading power on the part of the silicone. Even unintended overdosing with H-siloxane during gypsum production [53, 44] has an insignificant effect on the gypsum crystal habit, and impairment of compressive and flexural tensile strength of the gypsum building materials need not be feared. Tests on buildings have confirmed the very good durability of gypsum building materials treated with silicone water repellent [54]. Even additive-size application quantities can increase compressive, tensile, and bending strength in microstructures [43, 45, 46].

Opinions differ as to how the H-siloxane works [44 – 46, 53]. Whereas one may consider hydrolysis to be the all-important active principle, in which surface-controlled processes involve the interaction of hydroxyl groups with the rehydrated gypsum hemihydrate [46, 45], another mechanism is based not on hydrolysis, but on the optimum distribution of inert methyl groups. Rapid condensation (with release of hydrogen) to amorphous polymethylsilicic acid substantially lowers the degree of hydrophobicity [44].

Nanoscale Networks in Impregnated Red Sandstone and Shell Limestone

The stone composition and physical characteristics of the two sedimentary stones examined from the Lower Triassic Buntsandstein Series (Wüstenzeller Buntsandstein) and a shell limestone from the Middle Triassic Muschelkalk Series have been reported in detail [51].

Very simply, a sandstone has a grain composition of stone detritus (quartz, feldspar, detritic mica/clays) and a binder that may be clayey, quartzitic, or carbonatic (or mixtures thereof) [51, 55]. The mineral composition of limestone is mostly calcium carbonate, often derived from precipitated calcium carbonate and fossil shell remnants from marine organisms and a calcitic binder matrix [51, 56, 57].

Basically, stone samples impregnated under optimum conditions in the laboratory usually have a better distribution and higher penetration depth of impregnating agent than identical stone materials on actual façades [55, 27]. The dilution series was discussed in "Sample preparation and analytical methods".

Microstructure of an Impregnated Red Sandstone (Wüstenzeller Sandstein)

One difficulty associated with studying the distribution of silicone resin coats with thicknesses in the double-digit *nanometer* range [12, 22, 23, 48] in stone microstructures lies in the separation between phases belonging to the substrate and the polymeric network. Since certain nanostructures of altered feldspar crystal surfaces (overview photograph, see Fig. 7b) or growth forms of quartz in the untreated state can give the deceptive appearance of an overlapping silicone resin network, a detailed knowledge of the growth and alteration forms of relevant stone-forming minerals is necessary.

In *non-impregnated stone samples* (Figs. 6a,b), high resolution of layer silicate structures (mica, clay minerals) fails to show signs of crystal edge rounding or "bonding" of phyllosilicate layers or nanoscale crystals by the substrate's own phases. Individual, ultra-thin fibrous illite crystals (Fig. 6a) can be seen without difficulty. At high magnification, the mica layer surfaces (Fig. 6b) are largely smooth on account of their perfect cleavability along (001), and the prism surfaces, i.e. the crystal edges (see arrow), are demarcated by sharp edges.

In the *impregnated red sandstone microstructure* (Figs. 6c–f), layer silicates of ultra-thin and edge-rounding silicone resin films are coated with a characteristic rough surface structure. The detailed view (Fig. 6d) reveals the outlines of nanoscale clay mineral crystals that disappear under a structured veil of polymethylsilicic acid that is foreign to the substrate. The section magnification (Fig. 6f) of the clay mineral monocrystal in Fig. 6e (see white arrow) shows a fully intact silicone resin envelope that clearly encloses the prism surface (crystal edge). Only at maximum magnification can structures of the silicone resin network in the double-digit nanometer range be visualized on the clay mineral (Fig. 6f).

An example of the use of a silicone that is unsuitable for natural stone impregnation because of non-intact substrate wetting is shown in Fig. 7. Wüstenzeller Buntsandstein was treated with a potassium siliconate that is chiefly used to very great effect in building materials.

In the red sandstone microstructure, the poor wetting manifests itself as silicone coats of different thicknesses on quartz (Fig. 7a); locally, the substrate is more or less exposed (see arrows). The very finely divided, almost homogeneously distributed, pale particles on the polymer coats are a typical siliconate precipitate of potassium carbonate.

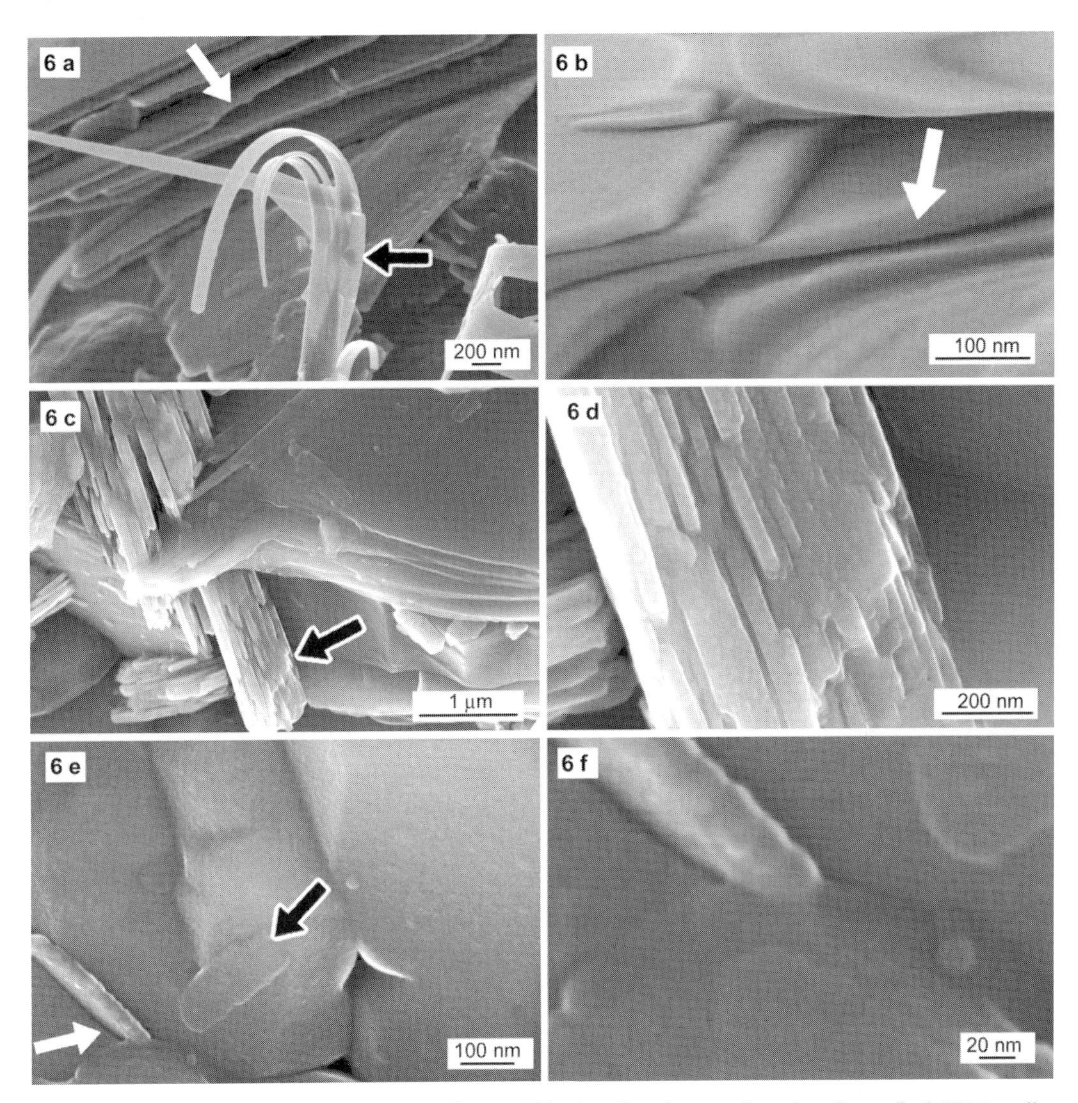

Fig. 6. Comparative SE images of an untreated Lower Triassic red sandstone and a red sandstone (both Wüstenzeller Buntsandstein) impregnated with a silicone microemulsion concentrate (SMK). **a**, **b** Untreated: Mica layer surfaces (white arrows) and clay mineral fibers (illite, black arrow) without signs of polymeric encapsulation. **c** – **f** Impregnated with SMK: **d** A detailed view from Fig. 6c: clay mineral-mica intergrowths reveal structured silicone films on the prism plane of the clay mineral. **e** Clay mineral intergrowth with quartz, coated with a veil of silicone resin: (white arrow) plan view of crystallographic prism plane (110) or (010); (black arrow) plan view of sheet plane (001). **f** Detailed view from Fig. 6e: clay mineral perfectly encapsulated by structured silicone resin network.

The feldspar crystal (Fig. 7b) with typical alteration forms (see black arrow) is also enveloped with a non-intact coat. The fracture edge reflects the alternating thickness of the silicone network coat (see white arrows). Adhesion to the substrate is not so good, and, on a large scale, there is a lack of network envelopment on the right half of the crystal (black arrow). Pale potassium carbonate crystals accumulate on crystallographically preferred directions.

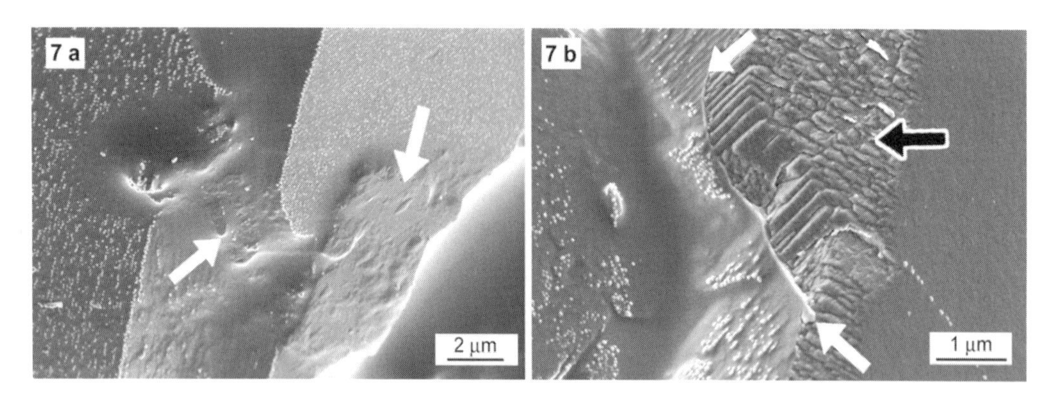

Fig. 7. SE images of a red sandstone (Wüstenzeller Buntsandstein) impregnated with potassium siliconate for comparison with the stone specimens from Fig. 6 that are treated with the more suitable silicones SMK. **a** Partly very thick silicone layer with poor local wetting (see white arrows) and homogeneously distributed precipitate of K_2CO_3. **b** Non-intact silicone coat on altered feldspar with fracture edge (see white arrows); white K_2CO_3 crystallites are heterogeneously distributed.

Microstructure of an Impregnated Shell Limestone (Krensheimer Muschelkalk)

Since growth structures and alteration forms of certain crystallographic planes can differ significantly from calcite crystals in limestone, an equally detailed knowledge of primary surface structures is necessary in order that polymeric wetting may be clearly identified.

Unlike the case for the untreated limestone (Fig. 8a), extremely fine nanometer-sized crystal voids between the calcite crystals of treated samples are largely filled with silicone resin (Figs. 8c,d,f). As with impregnated red sandstone (Figs. 6d,e,f), a nanostructured network surface is present at very high magnification. The envelopment of extremely fine structures is shown by a radial structured crystallite intergrowth of presumably aragonite ($CaCO_3$, modified crystal structure); the outlines of the fine crystallites partially disappear in the network coating (Fig. 8f). Crystal tips "project" partly into the polymeric nanolayer of the enveloped calcite substrate (see black arrows). Fine nanopores are perfectly lined. Problematic limestone [58, 59, 60] is thus definitely coatable with suitable silicones.

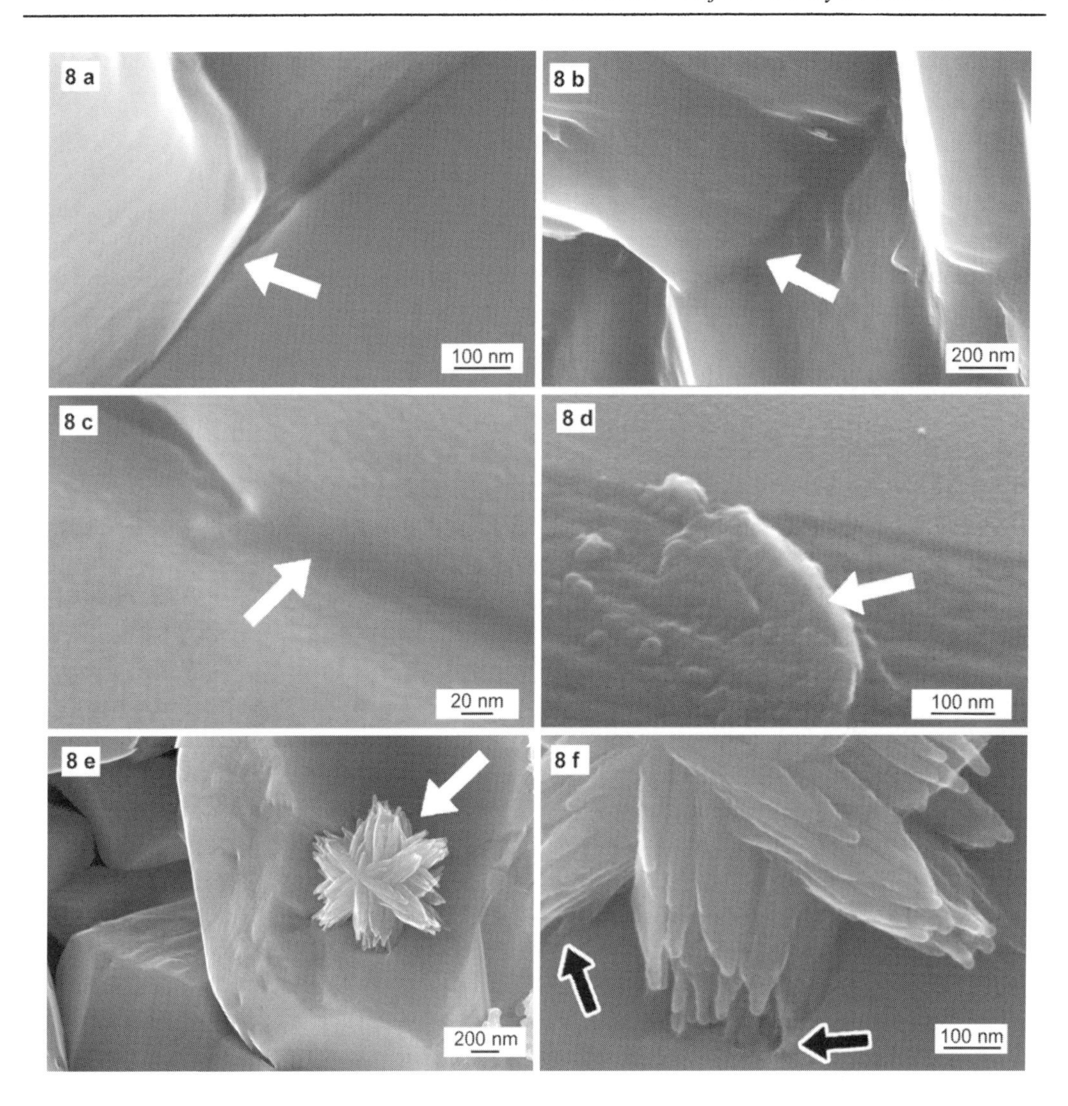

Fig. 8. Comparative SE images of an untreated shell limestone (Krensheimer Muschelkalk) and a shell limestone impregnated with a silicone microemulsion concentrate (SMK): **a** Untreated: smooth calcite crystal surfaces, no filling of grain voids (see arrow). **b** – **f** Impregnated with SMK: **b** Overview: sealing of grain-to-grain bridges (see arrow) as evidence of network coat. **c**, **d** At high resolution, nanostructured and complete covering coats are visible: **c** Interstitial filling and meniscus formation (see arrow). **d** Crystal edge rounding and "veiling" of crystallites (see arrow). **e**, **f** Presumably aragonite accretion (see white arrow) on calcite. **f** Radial structured mineral intergrowth with ultra-thin silicone resin network film; interfaces with calcitic crystal are lined (black arrows).

Structural Studies as an Aid to Checking the Success of Impregnation

It was shown for two impregnated, problematic types of stone (clay-rich, carbonatic) that ultra-thin networks are formed on all mineral and matrix components (except in Fig. 7). High-resolution microstructural analysis usually permits a separation between substrate-own mineral phases and a polymeric network with thicknesses in the double-digit nanometer range [12, 22, 35, 61]. Capillary pores become lined, and moisture-sensitive mineral components, such as clay minerals, become completely enveloped, without visible restriction of efficient moisture-transport processes. Assuming proper application of organosilicon impregnating agents, excellent durability of the silicone resin network is to be expected on account of selective changes to the polarity of the mineral and matrix surfaces in the natural stones. In contrast to received expert opinion [29, 62], it was shown that readily spreadable silicones can reach capillaries/pores of < 100 nm and can envelop structures.

The nanostructures of the networks shown are characteristic of the fully condensed silicone microemulsion employed, whose resin phases in the emulsion are already of a size ranging from 5 nm to approx. 50 nm. Impregnation of sandstones rich in clay minerals usually reduces the water absorption by approx. 40 wt% relative to the untreated material, but hygric dilatation decreases only by 20% [5]. Hygric swelling (hygric dilatation) of building stone rich in clay minerals is considered to be a key mechanism of damage in the weathering of natural stone [5, 7, 30, 62, 63]. In contrast to concrete literature references [62], *no* non-intact silicone coating of clay minerals was observed in the red sandstone studied. Sealing of the layer structures along the prism crystallographic planes (010) and (110) limits the occlusion of water and thus swelling by hydratable cations at the inter-crystalline base surfaces of clay minerals. Cryo-FEM studies on watered, impregnated sandstones rich in clay minerals [64] confirm this.

Real-life impregnation of exposed façades with silicone-based water-repellent agents does not always lead to the desired effect. To minimize mistakes during impregnation in the future, timely microscopy studies of effectiveness and a quality assessment of the impregnation measures are recommended. Factors that can later lead to low durability of the hydrophobic treatment [4], such as unsuitable silicone composition (Fig. 7) or preparation of the silicone active ingredient [28] with incomplete substrate coverage or a non-optimal application method with too low a penetration depth [15, 27, 37], can be identified immediately. Should a decline in efficacy of the water-repellent treatment nevertheless reveal itself at a later point, it would be possible to correlate the susceptibility to damage with the already documented application errors or natural stone properties [7, 28, 62, 64, 65].

What is the Reason for the Difference in the Extent of Attachment of Organosilicon Compounds to Diverse Mineral Substrates?

It is known that organosilicon compounds attach themselves outstandingly to silicate substrates, but not always to calcium carbonate [19, 20]. A surprising finding was that *polymeric silicones* attach themselves very well to calcitic fillers in silicone resin emulsion paints (see "The microstructure of silicone resin coatings" above). Coating tests designed to replicate the manufacturing process of silicone resin emulsion paints using aqueous millbases documented for the first time the complete coating of calcitic fillers and white pigments (titanium dioxide) by silicone [2, 12, 22]. It was also demonstrated that silicone resin coats polar mineral substrate surfaces *immediately* and is *irreversibly* anchored there. A further example of effective siloxane coating (H-siloxane) of calcite is known [66].

Nevertheless, with regard to filler treatment and limestone impregnation, it is still considered difficult to coat calcium carbonate with silicone. As shown in Table 1, the commercial *monomeric silanes* that are used for coating the surfaces of fillers nowadays are not effective for coating calcite [19, 20, 21]. The idea that the production of a silicone-coated surface is difficult may result from the misconception that attachment of high silicone polymers to calcium carbonate is equally problematic. The issue of attachment of silicone to the mineral surfaces has still not been finally resolved [62].

Table 1. Effectiveness of coating mineral substrates with silanes.

Silane effectiveness	Substrates
Excellent	Quartz
	Glass
	Aluminum, Copper
Good	Alumino-Silicates (e.g. Feldspar, Mica, Clay: Talc)
	Inorganic Oxides (e.g. Fe_2O_3, TiO_2, Cr_2O_3)
	Steel, Iron
	Asbestos
Slight	Zinc
	Lead
⇧	Calcite ($CaCO_3$)
	Gypsum ($CaSO_4 \cdot 2H_2O$)
	Baryte ($BaSO_4$)
Poor	Graphite, Carbon Black

Formation of Nanoscale Networks on Mineral Substrates Used in Building

Simple dry-coating trials on mineral substrates afford a way of checking the ability of silicones to become attached to polar surfaces (see "Sample preparation and analytical methods" above). Calcium carbonate is so important here because it is the most commonly employed filler in façade-coating systems. The organosilicon compounds employed were a monomer (MTMO), a partially characterized MTMO oligomer, and the high polymer H-siloxane used in special commercial fillers for fire-fighting (Table 2).

Table 2. Measured contact angles of water droplets (5μL) on hydrophobically treated mineral substrate powders of different particle size fractions. Silicone active ingredient added at a rate 0.1 – 0.5 wt% on solids.

	Calcite (100 mm)	Calcite (5 μm)	Precipitated calcite (PCC, 300 nm)	Gypsum hemi-hydrate
Monomer (0.5 wt%) MTMO	–	Droplet absorbed	Droplet absorbed	Droplet absorbed
Oligomer (0.5 wt%) Partly condensed MTMO	–	151 ± < 1	–	146 ± < 1
Polymer (0.5 wt%) H-Siloxane	140 ± 1	147 ± < 1	148 ± 2	142 ± 1
Polymer (0.1 wt%) H-Siloxane	142 ± 1	141 ± 1	Droplet absorbed	Droplet absorbed

OH functionality < 3 wt%

The choice of the calcitic fillers took into account the fact that the specific surface plays a critical role in the quality of the silicone coating. For example, a precipitated calcium carbonate (approx. 300 nm particle size) with a specific surface of 7 $m^2 g^{-1}$ (BET) is harder to coat than a coarse-grained calcite some 100 μm larger and having a specific surface of just 0.3 $m^2 g^{-1}$. Of no relevance to building practice is the surface treatment of highly reactive plaster of Paris (gypsum hemihydrate: $CaSO_4 \cdot ½ H_2O$).

Measuring the contact angle [67] of water droplets on scattered and slightly compacted silicone-coated filler powders affords a good means of quantifying the wetting behavior and the resultant hydrophobic effect (Table 2). It can be seen that the monomeric methyltrimethoxysilane (MTMO) has practically no effect in filler coating (a fact which is not due exclusively to the reduced quantity of active ingredient as a result of the high methoxy fraction), whereas short chain oligomers of higher silicone polymers generate exceptionally good hydrophobic properties on hydrophilic, mineral substrates. The measured contact angles of 140 – 151° are strikingly high when compared with the contact angles of normal silicone resin emulsion paints (120°) or super-hydrophobic paints (144°).

Microscopy Provides the Evidence

The low water-wettability of silicone-treated calcite powder (Table 2) must be the outcome of good envelopment by a hydrophobic protective layer. The surface structures of H-siloxane-treated calcium carbonate (Fig. 9d,e) with their tenth-of-a-nanometer-sized, veil-like, silicone films and missing grain interstices between the nanoscale dust particles (see arrow) clearly point to this. Precipitated calcium carbonate (PCC) illustrates, with its apparently uniform coating, the typical meniscus-shaped interstitial fillings (Fig. 9e) of the scalenohedric crystals (see arrow) toward the free pore space. An approx. 5 mm water droplet on *silicone-covered calcium carbonate* (Fig. 9b) confirms, despite deformation (due to its own weight), a low degree of wetting, and the drops only evaporate after spending many hours on the powder surface.

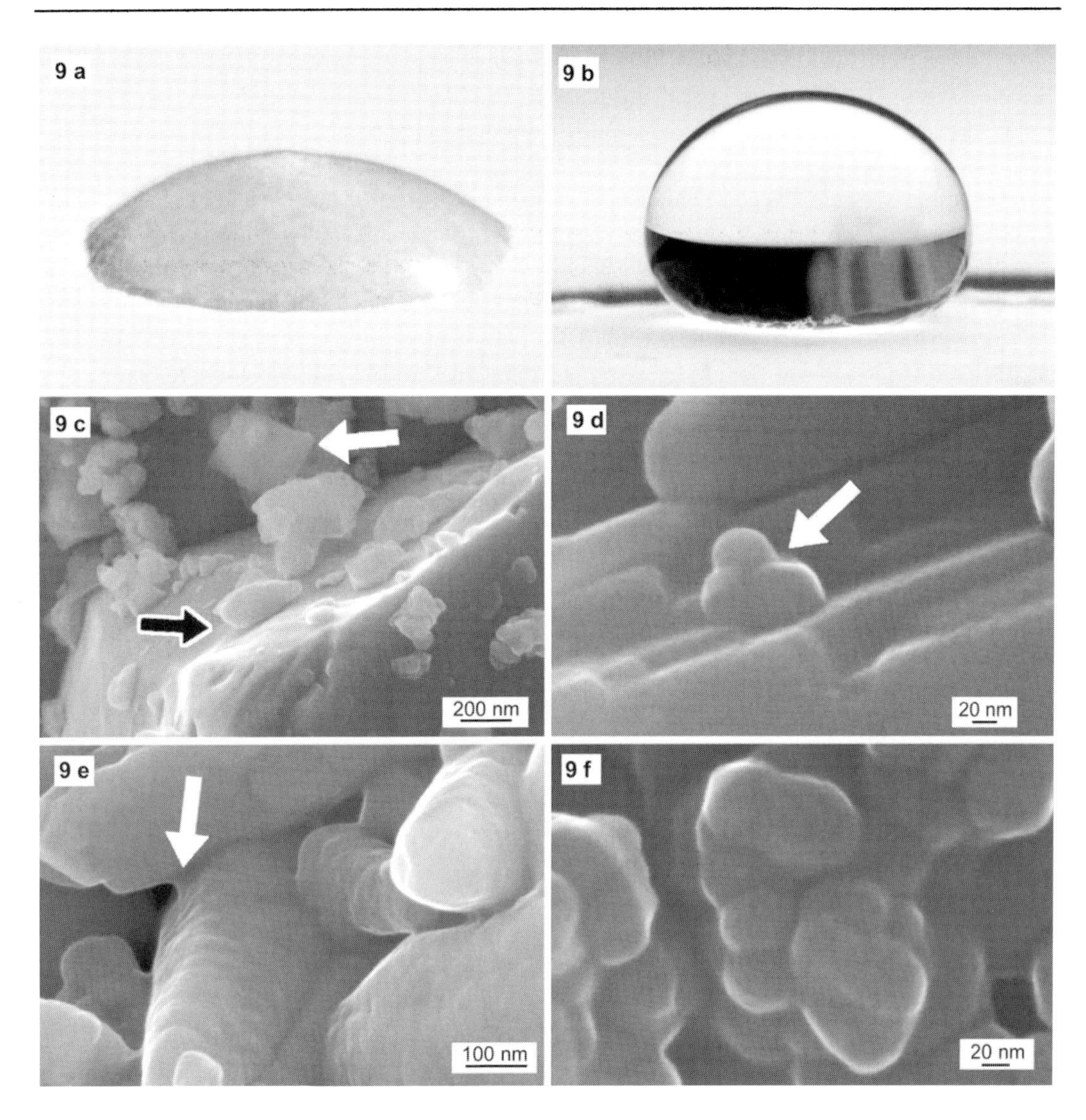

Fig. 9. FE-SEM SE micrographs of untreated cristobalite compared with calcite and phthalocyanine blue coated with polymethylhydrogensiloxane (H-siloxane) added in a concentration of 0.5 wt%. **a**, **b** Photographs of large water drops (approx. 5 mm in size) on powder substrate: wetting of (**a**) untreated silicate substrate (cristobalite) and (**b**) silicone-treated calcium carbonate (calcite, Cc). **c** Untreated cristobalite: sharp fracture surfaces. **d** – **f** Silicone-coated substrates: **d** crystal-edge/fracture-surface rounding of Cc; interstitial filling indicative of nanoscale silicone network envelopment, **e** precipitated scalenohedrally formed calcite (PCC) with meniscus-like silicone bridges in the grain interstices. Despite a high specific surface (7 m^2 g^{-1}), there is no evidence of an incomplete coverage of the network shown in micrograph 9e. **f** Silicone-treated phthalocyanine blue.

For comparison, *an untreated filler* (SiO_2 in the modification of cristobalite, Fig. 9c) shows sharp fracture shapes (white arrow), which differ markedly from those of silicone-enveloped fillers. Furthermore, missing interstitial fillings toward an overlying ground dust (black arrow) indicate that cristobalite was not treated with silicone. On contact with water, the slightly compressed cristobalite powder is wetted immediately (Fig. 9a), and the water drop is absorbed within a few seconds.

On account of their primary surface coating, visualization of the silicone coating of nonpolar, organic pigments (Fig. 9f) is problematic. For comparison purposes, adequate studies are needed on each untreated material.

Significance of the Crystal Chemistry of Mineral Surfaces for the Attachment of Silicones

An understanding of the mechanisms by which organosilicon compounds become attached to mineral substrates requires a knowledge of the crystal structure of the substrates examined. There is no doubting the presence of copious functional silanol groups (Fig. 10) on silicate material (quartz, feldspar, mica, clay) [2, 19 – 21], whose surface concentration on quartz fracture planes is

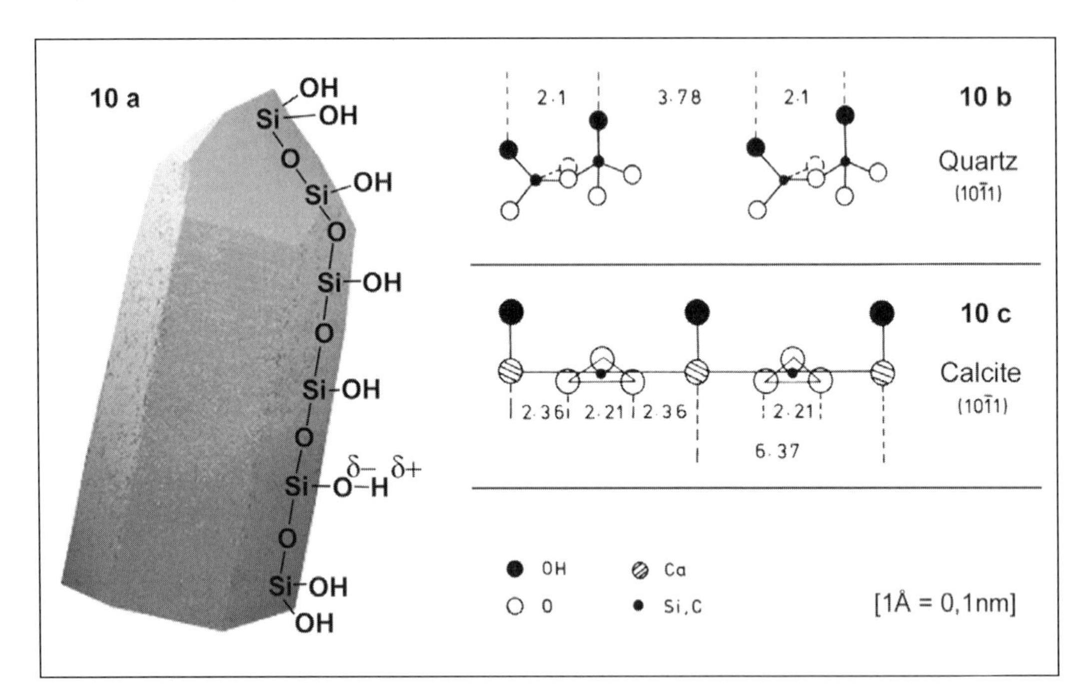

Fig. 10. Distances between silanol groups (Si–OH) on quartz and calcite surfaces in Å (1 Å = 0.1 nm). **a** Highly simplified schematic showing arbitrary arrangement of OH groups on a quartz crystal (adapted from Ref. [19]). **b** Distances between silanol groups on quartz surfaces (10$\bar{1}$1 crystallographic plane). **c** Distances between hydroxylated calcium ions on calcite surfaces (10$\bar{1}$1 crystallographic plane), drawn from Refs. [68, 57], **b** and **c** taken from [20]).

estimated at 8.2 μmol m^{-2} for the $(10\bar{1}1)$ crystal face [68]. Obviously, the presence of hydroxyl groups on mineral surfaces influences the adsorptive behavior toward polar sorbents. The adhesive property of the polar Si–O–Si backbone of silicone is likely to be enhanced if more silanol groups are present on the siloxane or are formed by ongoing hydrolysis of the alkoxy groups.

In contrast, calcium carbonate is still widely classified today as inert (free of functional groups) [19, 21]. This is contradicted by crystallographic structural evidence [20, 57]: calcite fracture planes are defect structures where Ca^{2+} ions are trying to retain their hexavalent coordination to the crystal surface through hydroxyl groups. The appreciable rise in pH on "disturbed" calcite surfaces [20, 69] shows that copious amounts of $Ca(OH)_2$ must be present.

The distances shown in Fig. 10c enable adequate numbers of functional OH groups to be identified on calcite surfaces. The distances between the reactive silanol groups on quartz (Fig. 10b) and calcite rule out, from geometrical considerations, the possibility of poorer or better attachment of the silicone resin network in either of the two substrates that might explain differences in the adhesion mechanism, e.g., to quartz [20]. Adhesion to silicate substrates proceeds by chemical bonds (high bond energy of the Si–O bond). In the case of calcite, the attachment mechanisms are different, despite adequate numbers of OH groups, because the interfacial reactivities are different from those of silicates [20, 19]. Adhesion proceeds via Ca–O–Si bonds with lower bond energies that are similar to those of hydrogen bonds.

Importance of Nanoscale Networks in Nanotechnology: Outlook

Images of special networks are shown in Figs. 11 and 12: these are exposed networks, *without* their original substrate. How is it possible to obtain a faithful reproduction of the nanoscale network envelope of a pre-existent carrier material?

Only an organic substrate (template) is needed! When a piece of cotton cloth (Figs. 11b,c) or some cellulose fibers (Fig. 11a) are immersed for just a few seconds in 7% or 15% silicone resin solution and then ashed at 450 °C, the result is a copy of the structural motif of the organic starting material bearing the physicochemical properties of an inorganic material, namely SiO_2. Either ashing at 450 °C or cold ashing in plasma burns away the organic methyl groups of the silicone resin to leave behind pure silicate replicas. Figure 11 compares a "nanoscopically" copied cloth napkin after hot ashing (Fig. 11b) and cold ashing (Fig. 11c). The higher shrinkage of the hot-ashed cotton napkin may be clearly seen (Fig. 11b). Shrinkage of up to 40% is standard during calcination at 500 °C [70].

The detailed view of individual thread structures of the silicate replica (Figs. 12a,b) of a cotton napkin reveals a major difference in the surface morphology. As expected, greater shrinkage in the sample ashed at 450 °C (Fig. 11b) is reflected in a more greatly deformed, contracted structure (Fig. 12a). Less shrinkage of the cold-ashed sample (Fig. 11c) is indicated by the hollow SiO_2 backbone of a primary cotton thread (Fig. 12b) with a heavily pitted surface texture. It is questionable whether this is due to a different degradation mechanism for the methyl group during cold ashing or simply to a thin covering layer from silicone resin impregnation.

Fig. 11. SE micrographs of nanoscale, silicate networks, which faithfully reproduce the original structure of their organic substrates following immersion in 7 – 15% silicone resin solution. **a** Cellulose from fiber-fill after calcination at 450 °C. **b**, **c** Cotton napkin after ashing, **b** by calcination at 450 °C, and **c** after cold ashing.

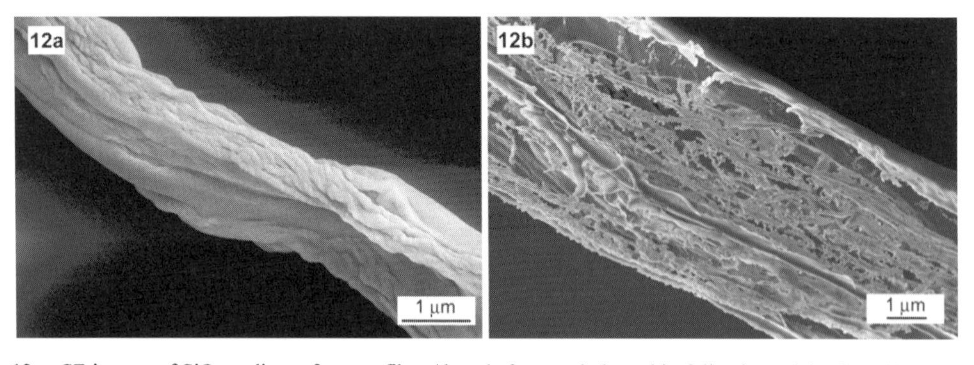

Fig. 12. SE images of SiO_2 replicas of cotton fibers/threads from a cloth napkin following calcinations (450 °C) and cold ashing: **a** Pronounced shrinkage following calcination at high temperatures, **b** Hollow shape of the cotton thread after cold ashing.

Nanoscale Networks as Building Blocks for Further Innovative Technologies of the Future

The possibility of exactly copying nanoscale, organic structures (template) in a casting process by coating with a silicate layer [70] or nano-tubes to precise blueprints with organic polymers [71] is currently a major interest in application and research. Microscopically accurate negative prints of bio-templates or microfibers still retain in some cases the functionality of their originals when they are based on the structure. At the same time, they possess the properties of another, inorganic material – pure SiO_2. But it is not only silicate replicas which are of interest. Composite materials, e.g., the high-performance silk of spiders' webs, can also be rendered brittle by coating with silane. Silicones also open up new paths to special functionalities [72]. Basically, however, it is accepted that the application potential of replicated nanoscale structures has not yet been discovered to any appreciable extent, let alone tapped.

Model Ideas about the Silicone Resin Network and Substrate Attachment

Finally, the question arises as to the molecular structure of the polymeric, nanoscale silicone resin network, with one partial aspect still largely unknown, namely a possible connection between molecular structure and durability in practice [14, 15, 40, 41].

It is suspected that the molecular structure of fully condensed trifunctional silanes/silicones with different alkyl radicals determines their durability [14, 15, 40, 41]. As the alkyl chain length increases, the efficacy of fully condensed silicones on natural stone frequently declines (see, e.g., Refs [41, 55, 73]). Monomers, such as pure *iso*-octylsilanes are suspected of having a limited capability [65, 15, 74] to form networks. It was also found that silanes may even boost hygric dilatation relative to untreated substrates [4, 5, 62]. SIMS analyses of substrates impregnated with $RSi(OR)_3$ ($R = C_3H_7$ and C_8H_{17}) mixtures only provide clues of C_3-substituted silsesquioxane, a fact which might suggest preferential polycondensation of the silanes with short-chain alkyl radicals [65]. However, the possibility that this is due to constraints of the method used cannot be excluded. Substrate surfaces or addition of catalyst can exert a positive influence on network formation [15, 41, 65].

Furthermore, little is known about the molecular basics of the fully condensed network [15, 65]. Whether modern spectroscopic analytical techniques, such as DRIFT and XPS combined with TOF-SIMS [15, 42, 65], reproduce the molecular structure of the polycondensed silicone resin network with enough accuracy will be revealed by future studies [65].

Opinions as to how a silicone resin network is attached to mineral surfaces (mostly quartz), all gleaned from this study and from the literature [19, 20, 22, 66, 68, 69], may be summarized as follows.

- On all mineral fillers examined, there are enough OH groups available for the silanol groups of the siloxanes to crosslink with.
- There is concrete evidence from experimental practice that organosilicon water repellents bearing long alkyl groups hydrolyze in preference to low-molecular condensates [73]. There is some controversy as to whether, for example, silanes bearing exclusively alkyl groups (e.g., C_8) for impregnation purposes are capable at all of producing satisfactory

network formation under certain conditions or whether C_1/C_8 siloxane mixtures should be used. It is known that there is a higher spatial requirement of aliphatic side-groups like the sterically more demanding C_8 groups of arbitrary silane/siloxane molecules. There is unanimity, however, that full condensation to the three-dimensionally linked silicone network is significantly influenced by the alkyl group, the degree of polymerization, and the possible mixtures of short- and long-chain organic groups (e.g., C_1/C_8).

- On a wide range of mineral surfaces (SiO_2, calcite, sheet silicates), functional OH groups have totally different distances. From purely geometrical considerations, the result is fewer or more linking sites for crosslinking with the resin, a fact that leads to more or less punctiform crosslinkage. Nevertheless (and this is one outcome of the microstructural determinations presented in this paper), a methyl silicone resin network exhibits comparably good attachment to quartz, to clay, and to calcite (not only as building materials, but also as comparable powdery substrates). This may be attributable to a few punctiform chemical links (e.g. to quartz) and/or to purely physical adhesion (such as that suspected in the case of calcite). For this reason, it is fundamentally not possible to infer that methyl siloxanes confer better hydrophobicity on certain mineral surfaces. However, there are conceivably differences as regards durability, because Si–O–Si bonds between the silicone resin network and the quartz substrate have to be different from the Ca–O–Si interactions with a calcite.

Molecular modeling was used in an attempt to answer the questions of structure and bonding of the silicone resin network that are associated with the three viewpoints above. In other words, this method was used to investigate the relationship between the molecular architecture of the silicone resin and the ability to bond or to attach to mineral surfaces. To compare those abilities on a molecular level is not trivial, since the steric properties of such resins can only be described sufficiently in statistical terms.

Modeling of Methyl and *iso*-Octyl Siloxane Derivatives on Quartz and Calcite Surfaces

To compare the bonding behavior of silicone resins bearing either methyl or *iso*-octyl side-chains to quartz and calcite crystal surfaces, the investigations were performed on small cyclic Si_3-siloxane molecules, Si_8-cage-type silsesquioxanes, and ladder-type siloxanes [75].

The X-ray structures of quartz and calcite were obtained from the FIZ Karlsruhe ICSD database [76]. The chosen model structures were constructed in MATERIALS STUDIO [77] under periodic boundary conditions and were then energy-minimized using the program DISCOVER with COMPASS force field until a "maximum derivative" better than 0.01 kcal $mol^{-1}Å^{-1}$ was achieved.

Network Structure of a Methyl Silicone Resin (Arbitrary Assembly)

A molecular model of a network formed by methyl silicone resin could look like the molecular fragment in Fig. 13. To build this arbitrary assembly, a selection of typical siloxane molecules, found in the literature, served as methyl derivatives [14, 28, 42, 65, 71, 75, 78 – 81]. Cage molecules, ladder molecules, and siloxane chains were energy-minimized separately and then crosslinked manually to form a polysiloxane network. The bond lengths, bond valences, bond angles, constitution, and proportion of the model shown are therefore realistic. Of course, only an arbitrary assembly can be shown with all the features described below in a compiled view, its main characteristic being that there are only two covalent bonds between the siloxane and the quartz surface.

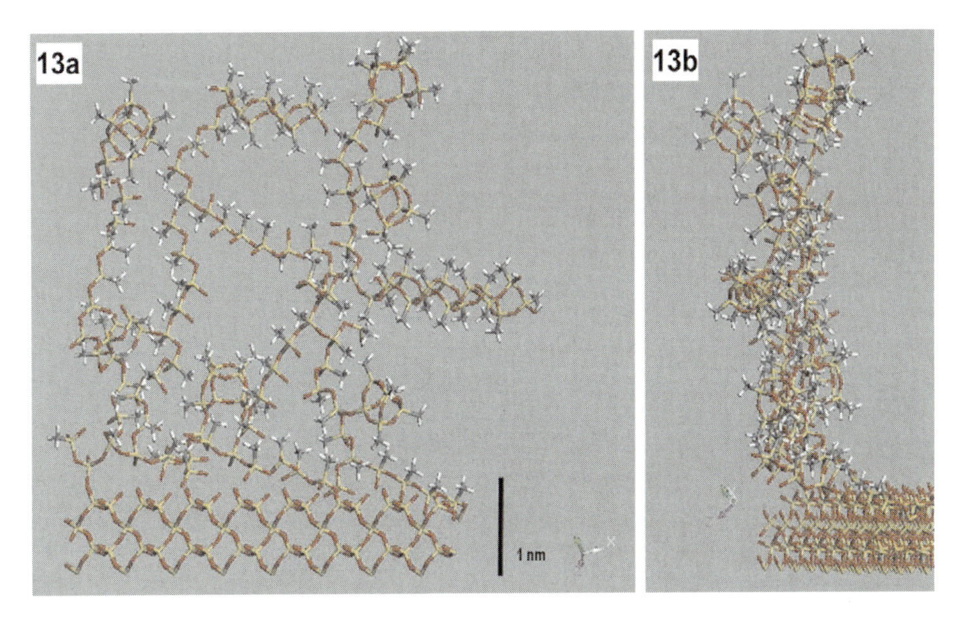

Fig. 13. Modeling of a water-repellent silicone resin network on quartz (arbitrary assembly), viewed from the front (**a**) and from the side (**b**). The three-dimensional network is built up of two-dimensional monolayers (methyl derivative) for optical reasons (this leads to seemingly unsaturated oxygen atoms, which form the oxygen bridges in the three-dimensional case). A simplified silicone resin network on quartz could be composed of covalently bound polysiloxane chains, incompletely crosslinked polysiloxane, and embedded silsesquioxanes or homosilsesquioxanes (cages and ladders), going from the bottom (quartz) to the top [28, 42, 75, 78, 79]. The network has been optimized from graphical aspects. Total height: approx. 50 Å.

Comparison of Methyl and *iso*-Octyl Side-Chains

The effect of changing the side-chains from methyl to *iso*-octyl in siloxanes in order to obtain higher and stronger hydrophobic layers has only been investigated on a quartz surface. All methyl derivatives were stable, with a ΔE ranking of

ladder type $\geq$ cage type >> cyclic Si_3-siloxane (Figs. 14 a,c,e).

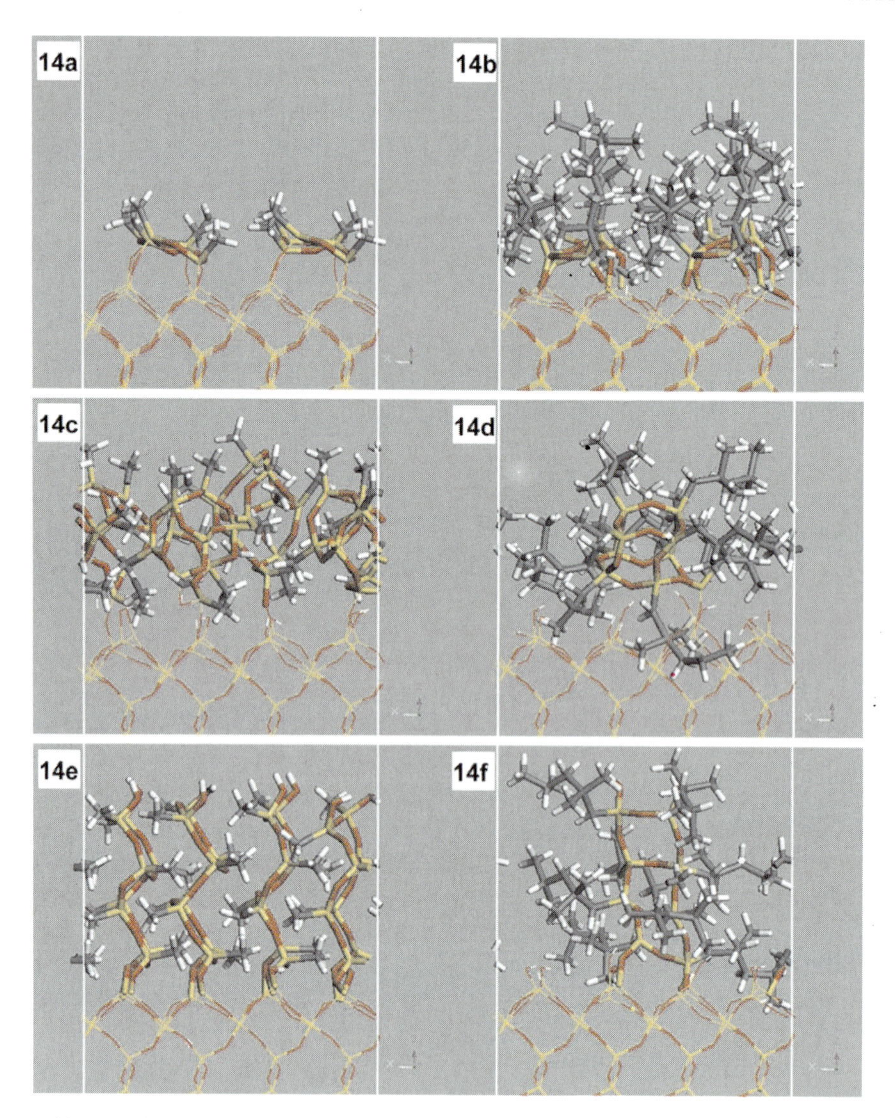

Fig. 14. A side view along the *y* axis of bonding of cyclic Si_3-siloxane molecules: cage type and ladder type siloxane molecules at maximum possible occupancy on an α-quartz (001) surface (3 × 4 unit cell) after manual matching on the X-ray structure, followed by energy minimization with DISCOVER/COMPASS at "periodic-boundary" conditions, with cut-off radius 9.5 Å (white vertical lines mark the computational box). **a**, **c**, **e**: methyl derivatives, **b**, **d**, **f**: *iso*-octyl derivatives, **a**, **b**: cyclic Si_3-siloxanes, **c**, **d**: cage molecules (type VII according to Ref. [75]), **e**, **f**: ladder molecules. Specifications of occupancy (per 3 × 4 unit cell), where N = number of bound siloxane molecules, R = number of aliphatic side-chains, F = number of remaining free OH groups on the quartz surface, ΔE = bond enthalpy (kcal mol^{-1} unit cell). **a**: $N = 4$, $R = 12$, $F = 0$, $\Delta E = -319.5$, **b**: $N = 4$, $R = 12$, $F = 0$, $\Delta E = +650.9$, **c**: $N = 4$, $R = 36$, $F = 8$, $\Delta E = -1950$, **d**: $N = 1$, $R = 9$, $F = 11$, $\Delta E = +357.5$, **e**: $N = 4$, $R = 32$, $F = 4$, $\Delta E = -2005$, **f**: $N = 1$, $R = 8$, $F = 10$, $\Delta E = -173.9$.

In general, the quartz surface was covered regularly by the methyl derivatives, but not by the *iso*-octyl derivatives. It turned out that the latter had the same ΔE ranking, but that all *iso*-octyl condensation products were much less stable than their methyl counterparts. These properties were easily explained by the high spatial requirements of the *iso*-octyl side-chains (Figs. 14b,d,f). Therefore, the ladder type (*iso*-octyl), with its vertical arrangement, was the poorest candidate in this respect. Additionally, there is another effect: only in the case of the cyclic Si_3-siloxane, which is energetically the worst candidate, did no free (hydrophilic) OH residues remain on the quartz surface after systematic condensation. Ladder and cage types reacted only with one (cage) or two (ladder) OH residues: the remaining OH groups of twelve possible ones in the 3×4 unit cell were merely covered with aliphatic side-chains.

In summary (assuming maximum-possible surface coverage), all methyl siloxane derivatives have balanced coverage possibilities, whereas the three *iso*-octyl systems exhibit too much repellency, which would increase tremendously if they were extended to infinity. Thus, in reality, such pure *iso*-octyl systems would not be formed at all. This tallies with results from hydrophobization experiments on dominating *iso*-octyl derivatives that failed to perform adequately in special cases [73]. However, it was found, that catalytic effects may increase polycondensation [65].

Comparison of Quartz and Calcite Surfaces

Closest packing having been assumed on the surface, manual matching of the model structures revealed a major difference between quartz and calcite. The distance between the reactive hydroxyl groups in cyclic Si_3-siloxane molecules was 4.91 Å. This matched perfectly with the distances for a quartz surface, with the result that it was easily covered in full. In contrast, many of the distances between the hydroxyl residues on the calcite surface, which were represented by the Ca–Ca distances [20] (coordination number of the Ca ion should remain 6) were not suitable for bonding cyclic Si_3-siloxane molecules or other oligovalent structural elements of siloxane resins (Fig. 15g). Only on the (101) surface was a suitable Ca–Ca distance of about 4.9 Å found. But even this bonding of Si_3-siloxane to calcite was only realized by two of the three possible hydroxy functions of this siloxane. The third group failed to bond for steric reasons. This became very obvious when an energy minimization was performed: the close contacts between the methyl groups of vicinal siloxane molecules after manual modeling (Fig. 15d) were resolved during the energy minimization. Flipping of the siloxane rings from equatorial to about 30° diminished the extent to which the calcite surface was completely covered (Figs. 15f, 15h).

In summary, condensing these siloxane molecules on calcite was less effective when compared to a quartz surface (only two bonds per molecule, only one possible surface (101) and the flipping of the siloxane rings). *The final conclusion* is that a mixture of high-valence structural elements (e.g., trivalent cyclic Si_3-siloxanes) with short side-chains to saturate the OH groups and low-valence components with voluminous side-chains was best at building up a sufficiently hydrophobic layer on top of a quartz (or, less effectively, calcite) surface. Better defined model systems with reduced coverage (less than 100%) or a mixture of methyl and *iso*-octyl side-chains

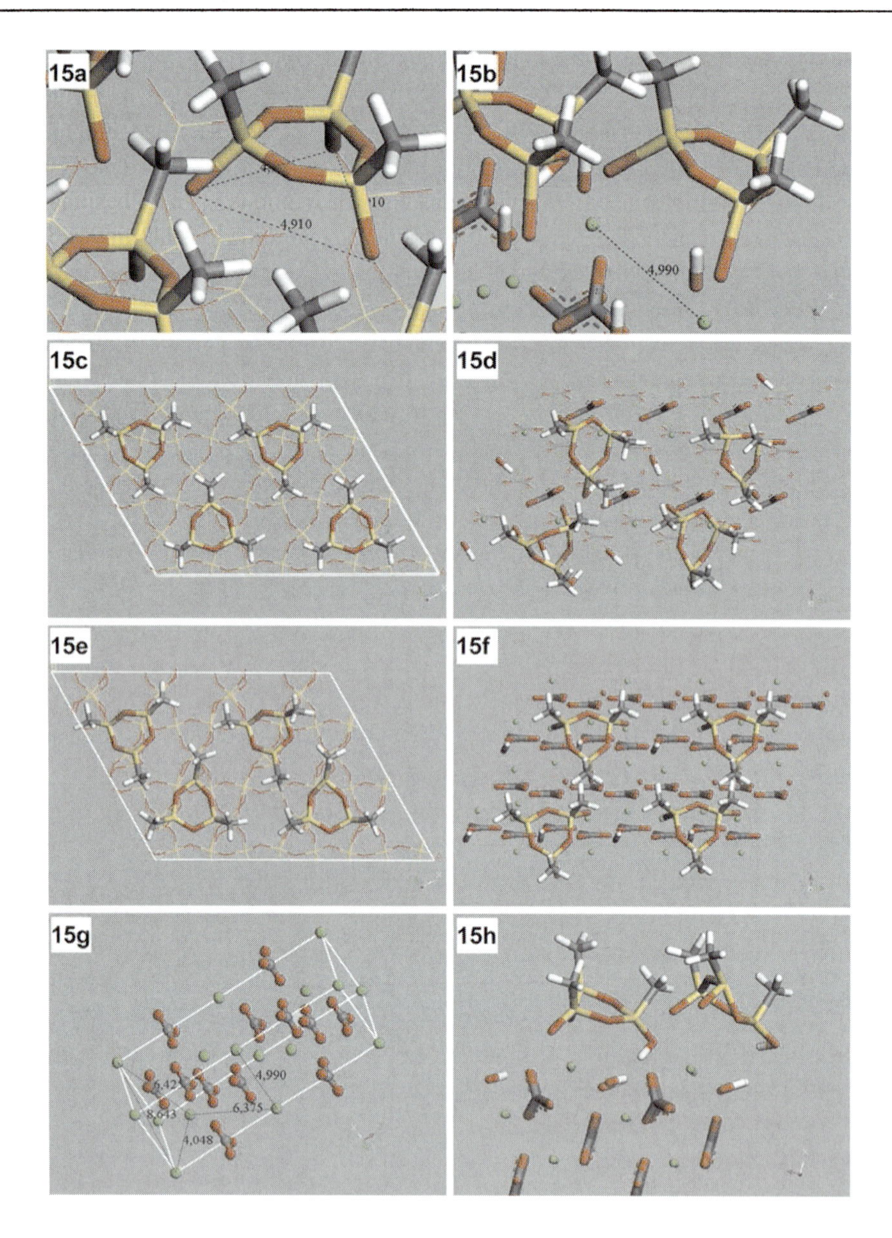

Fig. 15. Bonding of a cyclic Si_3-siloxane molecule on an α-quartz (001) surface (**a**, **c**, **e**) and bonding of a cyclic Si_3-siloxane molecule on a rhombohedral calcite (101) surface (**b**, **d**, **f**). **a**: detail of Fig. 15c, **c**: manual matching on the X-ray structure, **e**: after energy minimization with DISCOVER/COMPASS (white lines mark the computational boxes), **b**: detail of Fig. 15d, **d**: manual matching on the X-ray structure, **f**: energy minimization with DISCOVER/COMPASS, **g**: calcite with characteristic Ca–Ca distances (green spheres) and calcium carbonate anions (C atoms gray, O atoms red), **h**: as Fig. 15f, side view along the *y* axis (two rings removed for free view).

should be investigated with molecular dynamics simulations [82, 82] to find optimal mixtures.

Conclusion

Building materials exposed to weathering are corroded by the action of atmospheric influences, especially water; destruction is unavoidable in the long term. Water-repellent treatments can certainly not totally stop these harmful processes, but, given adequate envelopment and attachment to the substrate, nanoscale silicone resin networks can retard material decomposition because of their high durability. Since damage caused by hydrophobic measures can be virtually ruled out if the treatments are properly applied, the organosilicon compounds used in masonry protection will become more and more widely used.

To avoid mistakes in conservation methods involving silicones on structures of architectural and historical importance, the routine use of high-resolution microscopy makes sense, because microstructural analysis can provide indirect evidence of damage mechanisms before damage is even macroscopically visible under subsequent conditions of exposure [38, 39]. Such a demanding approach, however, necessitates a scientific understanding of mineral substrates, their physicomechanical interactions, and the symptoms of weathering that can occur in them under adverse conditions. In this regard it has been shown that high-resolution microscopic studies are not only useful at diagnosing damage, they also may play an increasingly important role in research and product development.

Methyl silicone resin networks have been shown to cover all substrates equally well. However, the way in which the network is attached is fundamentally different in each case, which could have repercussions on the durability. As is known from the literature on quartz and calcite surfaces, reactive hydroxyl groups exist that will react with siloxanes. This means that for a high-molecular resin with a reasonable degree of conformational freedom a sufficient number of reactive hydroxy groups at suitable distances can be found on the mineral surfaces. In contrast, the quality of binding is different in quartz (covalent Si–O–Si bridge) and calcite (Ca–O–Si interaction), which will influence long-term stability.

We are nowhere close to knowing the potential benefits that nanoscale silicone resin networks, with their special structure-effect relationship, will be able to offer innovative technologies of the future (nanotechnology). By virtue of their organic-inorganic hybrid nature, silicones open up new paths to special functionalities and enhanced properties of materials.

Acknowledgments: The authors would like to thank Leonhard Gollwitzer, Wacker-Chemie GmbH, Munich for providing the impregnated stone test specimens and Dr. Petra Herrmann, BASF Coatings, Münster and Dr. Herbert Juling, Amtliche Materialprüfanstalt Bremen for permission to use Figs. 4a and 4b. Many thanks are also due to Dr. Hans-Joachim Mergner, Hygiene Institut des Ruhrgebietes, Gelsenkirchen for cold ashing the silicone resin-treated cotton fabric. Special thanks are due to Dr. Jörg Schäffer for performing the molecular modeling and for his committed support in all questions of its interpretation. We also thank Dr. Richard Weidner, Wacker-Chemie GmbH, for his helpful comments on silicone structures.

References

[1] J. Pfeiffer, J. Weis, CLB Chem. in Lab. und Biotech., in *Silicon-Chemie* (*2*), **2002**, *53*(*4*), 128 – 135.

[2] A. Lork, I. König-Lumer, H. Mayer, *European Coatings Journal* **2003**, *4*, 132.

[3] L. G. W. Verhoef, *Proceedings of Hydrophobe III, 3rd International Conference on Surface Technology with Water Repellent Agents*, **2001**, p. 21 – 36.

[4] A. E. Charola, *Proceedings of Hydrophobe III, 3rd International Conference on Surface Technology with Water Repellent Agents,* **2001**, p. 3 – 19.

[5] E. Wendler, in *Saving our Architectural Heritage, the Conservation of Historic Stone Structures,* Dahlem Workshop Report (Eds.: N. S. Baer, R. Snethlage), Wiley-VCH, **1997**, p. 181 – 196.

[6] H. M. Künzel, K. Kießl, M. Krus, *Int. Zeitschr. für Bauinstandsetzung* **1995**, 267.

[7] E. Wendler, in *Jahresberichte Steinzerfall – Steinkonservierung* (Ed.: R. Snethlage), Vol. 1, Ernst & Sohn, Berlin, **1991**, p. 71 – 76.

[8] R. Snethlage, Denkmalpflege und Naturwissenschaft im Gespräch, in *Sonderheft der Publikationsreihe des Verbundforschungsprojektes Steinzerfall und Steinkonservierung* (Ed.: N. Gerner), Workshop Fulda, **1990**, p. 18 – 22.

[9] A. Lork, M. Steiger, Th. Grodthen, (in preparation).

[10] Wacker-Chemie, *Conserving a Cultural Heritage* (Ed.: Wacker-Chemie GmbH), Munich, **1997**, p. 1 – 180.

[11] L. Küsten, *Forschung für den Denkmalschutz* (Ed.: Bundesministerium für Forschung und Technologie BMFT), Hermann Daniel GmbH, **1994**, p. 1 – 122.

[12] A. Lork, in *Future Trends in Coatings Technology* (Reprint of Congressional Paper of the 17th SLF Congress), September 7 – 9, Stockholm, **2003**, p 375 – 386.

[13] L. Gollwitzer, in *Munich Tour* (Special Marketing Paper), Wacker-Chemie GmbH, München, **2001**, p. 1 – 31, (unpublished).

[14] Ch. Bruchertseifer, S. Brüggerhoff, K. Stoppek-Langner, J. Grobe, M. Jursch, H.-J. Götze, in *Oranosilicon Chemistry III — From Molecules to Materials* (Eds.: N. Auner, J. Weis), VCH, Weinheim, **1998**, p. 531 – 537.

[15] Ch. Bruchertseifer, H.-J. Götze, R. Störger, P. Albers, in *4th Intern. Kolloquium Werkstoffwissenschaften und Bauinstandsetzen*, TAE, December 17 – 19, Esslingen, **1996**, p. 257 – 271.

[16] H. Mayer, in *Wässrige Siliconharz-Beschichtungssysteme für Fassaden* (Ed.: W. Schulze), Renningen-Malmsheim, expert Verlag, Vol. 522, **1997**, p. 63 – 80.

[17] J. Pfeiffer, J. Weis, CLB Chem. in Lab. und Biotech., in *Silicon-Chemie* (*1*), **2002**, *53*(*3*), 84 – 89.

[18] W. Noll, *Chemie und Technologie der Silicone*, 2nd edn., Weinheim, Verlag-Chemie, **1968**.
[19] G. Wypych, *Handbook of Fillers*, 2nd edn., Plastics Design Library, Toronto, **1999**.
[20] R. Snethlage, Steinkonservierung 1979 – 1983, in *Arbeitshefte des Bayerischen Landesamtes für Denkmalpflege, München*, **1984**, *22*, p. 101 – 108.
[21] J. E. Moreland, in *Handbook of Fillers and Reinforcements for Plastics* (Eds.: H. S. Katz, J. V. Milewski), Van Nostrand Reinhold, New York, **1978**, p. 136 – 159.
[22] A. Lork, *Surface Coatings International Part B: Coatings Transactions, OCCA*, **2004**, *87*, p. 41 – 46.
[23] Ch. Herm, Th. Warscheid, in *Jahresberichte Steinzerfall — Steinkonservierung* (Ed.: R. Snethlage), Vol. 5, Ernst & Sohn, Berlin, **1995**, p. 159 – 173.
[24] Ch. Herm, *Doctoral Dissertation,* Ludwig-Maximilians-Universität, München, **1997**.
[25] H. Mayer, M. Roth, in *SD Wacker-Chemie, München*, **1995**, p. 1 – 5.
[26] H. Leisen, *Die Geowiss.* **1992**, *10* (*9*, *10*), 277 .
[27] E. Wendler, L. Sattler, *Bautensch. Bausan.*, BMFT Statusseminar, Wuppertal, **1989**, p. 71 – 75.
[28] Ch. Bruchertseifer, *Doctoral Dissertation*, Universität Bochum, dissertation.de, Berlin, **2000**.
[29] E. Wendler, *Die Geowiss*, **1992**, *10* (*9*, *10*), 290 – 292.
[30] E. Wendler, in *Proceedings of the 1st International Symposium of the Surface Treatment of Building Materials with Water Repellent Agents* (Eds.: F. H. Wittmann, A. J. M. Siemens, L. G. W. Verhoef), Delft, Netherlands, **1995**, 25/1 – 25/11.
[31] A. Boué, in *4th Intern. Kolloquium Werkstoffwissenschaften und Bauinstandsetze*, TAE, December 17 – 19, Esslingen, **1996**, 245 – 256.
[32] R. P. J. Van Hees, J. A. G. Koek, H. De Clercq, E. De Witte, L. Binda, E. D. Ferrieri, E. Carraro, Protection and conservation of the cultural heritage, Res. Rep. No. 7, in *Evaluation of the Performance of Surface Treatments for the Conservation of Historic Brick Masonry* (Ed.: R. P. J. van Hees), Luxembourg, **1998**, p. 33 – 53.
[33] H. Schoonbrood, *Wacker Technical Information*, Melbourne, Australia, **1999**, p. 1 – 16.
[34] E. Wendler, D. D. Klemm, R. Snethlage, in *Durability of Building Materials and Components* (Eds.: J. M. Baker, J. P. Nixon, A. J. Majumbar, H. Davies), London, **1991**, p. 203 – 212.
[35] E. De Witte, H. De Clercq, R. De Bruyn, A. Pien, in *Proc. 1st Int. Symp. on Surface Treatment of Building Materials with Water Repellent Agents* (Eds.: F. H. Wittmann, A. J. M. Siemens, L. G. W. Verhoef), Delft, Netherlands, **1995**, p. 5/1 – 5/10.
[36] A. Gerdes, F. H. Wittmann, in *Proceedings of Hydrophobe III, 3rd Int. Conf. on Surface Technology with Water Repellent Agents*, Aedificatio Publishers, Zürich, **2001**, p. 111 – 122.
[37] L. J. A. R. van der Klugt, J. A. G. Koek, in *Proc. 1st Int. Symp. of the Surface Treatment of Building Materials with Water Repellent Agents* (Eds.: F. H. Wittmann, A. J. M. Siemens, L. G. W. Verhoef), Delft, Netherlands, **1995**, p. 19/1 – 19/14.
[38] R. Blaschke, *Beton-Instandsetz.*, BMI 1/97, **1997**, p. 183 – 187.
[39] H. Juling, F. Schlütter, A. Lork, J. Schad, R. Blaschke, Tagungsbericht Bd. 3, 12. ibausil, P 11, Weimar, **1994**, p. 95 – 102.
[40] J. Grobe, K. Stoppeck-Langner, A. Benninghoven, B. Hagenhoff, W. Müller-Warmuth, S.

Thomas, Spectroscopic investigations on silylated inorganic substrates, in *Organosilicon Chemistry I — From Molecules to Materials* (Eds.: N. Auner, J. Weis), Wiley-VCH, Weinheim, **1994**, p. 325 – 329.

[41] K. Stoppek-Langner, K. Meyer, A. Benninghoven, in *Organosilicon Chemistry III — From Molecules to Materials* (Eds.: N. Auner, J. Weis), Wiley-VCH, Weinheim, **1998**, p. 520 – 525.

[42] K. Stoppeck-Langner, J. Grobe, in *Jahresberichte Steinzerfall – Steinkonservierung 1994 – 1996,* (Ed. R. Snethlage), Fraunhofer IRB Verlag, Stuttgart, **1998**, p. 11 – 22.

[43] A. Lork, K. Haneburger, D. Gerhardinger, *Gypsum brochure,* Wacker-Chemie GmbH, München, **2002**, p. 1 – 19.

[44] L. Jakobsmeier, *Doctoral Dissertation*, Tech. Univ. München, **2000**, München.

[45] W. Keller, in *Wacker Special Marketing Report,* **1996**, p. 1 – 25, München (unpublished).

[46] W. Keller, *Dipl. Dissertation*, Tech. Univ. München, **1996**, München.

[47] J. Greene, D. Haeussler, B. Berglund, *PCI*, Sept., **2002**, p. 30 – 36.

[48] A. Lork, I. König-Lumer, H. Mayer, *ECJ*, Dec. **2002**, Vincentz, Hannover, p. 14 – 21.

[49] M. Schwartz, R. Baumstark, *Water Based Acrylates for Decorative Coatings*, 1st edn., Vincentz, Hannover, **2001**, p. 286.

[50] O. Wagner, in *Wässrige Siliconharzbeschichtungssysteme für Fassaden* (Ed.: W. Schulze), Renningen-Malmsheim, expert Verlag, Vol. 522, **1997**, p. 113 – 164.

[51] W. D. Grimm, in *Bayerisches Landesamt für Denkmalpflege*, Ah. 50, **1990**, München, p. 1 – 655.

[52] B. Meng, *Doctoral Dissertation*, Rheinisch-Westfälische Technische Hochschule (RWTH) Aachen (ibac), ABBF Bd. 3, **1993**, Aachen.

[53] A. Lork, *Wacker Lab. Notiz 2178*, Burghausen, **2001**, p. 1 – 27. (unpublished).

[54] M. Götze, R. Hempel, K.-H. Tausch, *ZKG Internat.* **1995**, *48*(*10*), 555.

[55] D. J. Honsinger, *Doctoral Dissertation*, RWTH Aachen, **1990**, Aachen.

[56] H. Füchtbauer, *Sedimente und Sedimentgesteine, Sediment-Petrologie II,* 4th Edn., E. Schweizerbart'sche Verlagsbuchhandlung, Stuttgart, **1988**.

[57] F. Lippmann, *Sedimentary Carbonate Minerals*, 1st edn., Springer-Verlag, Berlin, **1973**, p. 1 – 228.

[58] H. De Clercq, E. De Witte, in *Proc. Hydrophobe II*, Zürich, **1998**, p. 21 – 34.

[59] V. Verges-Belmin, D. Garnier, G. Orial, A. Bouineau, R. Coignard, in *2nd Internat. Symp. Conservation of Monuments in the Mediterranean Basin*, Geneva, Nov. **1991**, p. 421 – 437.

[60] V. Antonelli, in *Proceedings of 3rd International Congress on Deterioration and Preservation of Stones*, October 14 – 27, C. R. E. O., Venezia, **1979**, p. 629 – 644.

[61] D. Honsinger, in *Jahresberichte Steinzerfall – Steinkonservierung, Denkmalpflege und Naturwissenschaft im Gespräch,* Workshop Fulda, March 6 – 7, (Ed.: R. Snethlage), Fraunhofer IRB Verlag, Stuttgart, **1990**, p. 30 – 31.

[62] R. Snethlage, E. Wendler, D. D. Klemm, in *Denkmalpflege und Naturwissenschaft – Natursteinkonservierung I* (Ed. R. Snethlage), Verlag Ernst & Sohn, Berlin, **1995**, p. 127 – 146.

[63] G. Hilbert, E. Wendler, *Bautensch. Bausan.*, *18. Jg.*, **3/1995**, p. 60 – 64.

[64] W. G. Burchard, G. Clooth, J. D. Neisel in *Denkmalpfl. und Naturwiss. im Gespräch*, Workshop Fulda (Eds.: A. Boué, K. Gertis, H. Höcker, H. D. Koehne, H. R. Sasse, R. Snethlage), cre art®, Fulda, **1990**, p. 116 – 118.

[65] Ch. Bruchertseifer, S. Brüggerhoff, J. Grobe, K. Stoppek-Langner, in *Proceedings of the 1st International Symposium of the Surface Treatment of Building Materials with Water Repellent Agents* (Eds.: F. H. Wittmann, A. J. M. Siemens, L. G. W. Verhoef), Delft, Netherlands, **1995**, p. 27/1 – 27/11.

[66] UK Patent Application GB 2 355 453 A, Preparing hydrophobic calcium carbonate by surface treating with a siloxane, Dow Corning Corporation, **2001**, Michigan, USA.

[67] U. Zilles, *Welt der Farben* **2001**, *6*, 18.

[68] W. Stöber, *Kolloid Z.* **1956**, *145* (*H1*), 117.

[69] S. Yariv, H. Cross, in *Geochemistry of Colloid Systems for Earth Scientists*, Springer-Verlag, Berlin, **1983**.

[70] Ch. Göltner-Spickermann, *Nachr. Chem.* **2003**, *51*, 1035.

[71] A. Greiner, J. H. Wendorff, M. Steinhart, *Nachr. Chem.* **2004**, *52*, 426.

[72] N. Auner, B. Ziemer, B. Herrschaft, W. Ziche, P. John, J. Weis, *Eur. J. Inorg. Chem.* **1999**, 1087.

[73] J. Grobe, K. Stoppeck-Langner, W. Müller-Warmuth, S. Thomas, A. Benninghoven, B. Hagenhoff, *Nachr. Chem. Tech. Lab.* **1993**, *41*(*11*), 1233.

[74] B. Hagenhoff, *Doctoral Dissertation*, University of Münster, **1993**.

[75] M. G. Voronkov, V. I. Lavrent'yev, *Top. Curr. Chem.* **1982**, *102*, 199.

[76] FIZ-Karlsruhe Inorganic Crystal Structure Database ICSD (2004), Programm Version 1.3.3.

[77] Fa. Accelrys, Materials Studio with DISCOVER 3.01 and COMPASS – force field version 3.0.

[78] O. Schneider, in *Proc. 9th Ing. Congr. Therm. Analys.*, August 21 – 26, **1988**, Jerusalem.

[79] O. Schneider, private communication, **2004**.

[80] H. Jancke, G. Engelhardt, *J. Organomet. Chem.* **1983**, *247*, 1329.

[81] F. J. Feher, D. A. Newman, J. F. Walzer, *J. Am. Chem. Soc.* **1988**, *111*, 1741.

[82] J. Köhler, M. Hohla, R. Söllner, H. J. Eberle, *Supramolecular Science* **1998**, *5*, 101.

[83] J. Köhler, M. Hohla, R. Söllner, M. Amman, *Supramolecular Science* **1998**, *5*, 117.

Chapter VI

Silicon-Based Materials

Mesostructured Silica and Organically Functionalized Silica — Status and Perspectives

Ferdi Schüth, Yanqin Wang, Chia-Min Yang, Bodo Zibrowius*

Max-Planck-Institut für Kohlenforschung,
Kaiser-Wilhelm-Platz 1, 45470 Mülheim, Germany
Tel.: +49 208 3062372 — Fax: +49 208 3062995
E-mail: schueth@kofo.mpg.de

Keywords: MCM-41, mesoporous, organosilica

Summary: Ordered mesoporous silica has now been known for something over ten years, and tremendous progress has been made to produce various structures. However, even more diverse than the structures are the framework compositions which have been synthesized. In the context of this volume, framework compositions with organic groups either attached to the surface or fully incorporated into the framework are the most relevant ones. In this contribution, the state of the art of this class of materials will be reviewed, including recent developments from the group of the authors.

Introduction

Ordered mesoporous materials, especially silicas, were discovered in the early 1990s [1, 2] and are of tremendous current interest, both from a scientific and an applications point of view. The materials consist of a periodic arrangement of silica walls (the atomic structure of which strongly resembles amorphous silica) separated by pores. Many review papers are available, covering different aspects of this type of material [3]. Since the synthesis of the original MCM-41 material with hexagonally packed pores and the cubic MCM-48, many other structures have been obtained, which often, but not always, have their counterparts in surfactant chemistry. Judicious choice of the templates, synthesis conditions, additives, and precursor species allows fine tuning of the resulting materials, by which control of structure, pore sizes, wall thickness, framework composition, and other properties is possible. Figure 1 summarizes, for the best-known material, MCM-41, the most salient features, which are (i) X-ray diffraction patterns with reflections in the low-angle range, indicating periodicity on the length scale of several nanometers, (ii) a TEM in which this periodicity can also be observed as contrast between the walls and the pores, and (iii) a sorption isotherm with a steep step in the intermediate range of pressures, corresponding to a narrow pore size distribution in the mesopore range.

The key in the synthesis of these ordered mesoporous materials is the use of surfactants, which can be ionic isolated molecules, neutral molecules, block copolymers, and others. Removal of the surfactants to create an accessible pore system is possible in the simplest case by calcination in air, but also different leaching procedures have been developed. If the template is removed, highly porous materials result, having surface areas of typically around 1000 m^2/g and pore volumes on the order of 1 mL/g. The surface properties of the silica-based materials correspond roughly to those of conventional amorphous silica, and thus all the techniques to anchor organic groups to the surface which have been developed for silica are applicable, as well as co-condensation procedures, involving, for instance, tetraalkoxysilanes and alkyltrialkoxysilanes to produce hybrid organic/inorganic materials. Since the title of this book series is "Organosilicon Chemistry", the following discussion is exclusively focused on materials in which Si–C bonds are present, which are introduced into the solids either by surface modification or directly during the condensation of the silica. For information on the synthesis, properties and applications of the pure silica materials and those with frameworks based on other elements, the reader is referred to the review articles already mentioned, which provide easy access to the original literature. But even if this paper is restricted to the class discussed above, the vast literature available does not allow a complete coverage, given the limited space available, since the field has grown tremendously in only a few years. We therefore attempt to sketch the major lines of development and to cite review papers wherever possible.

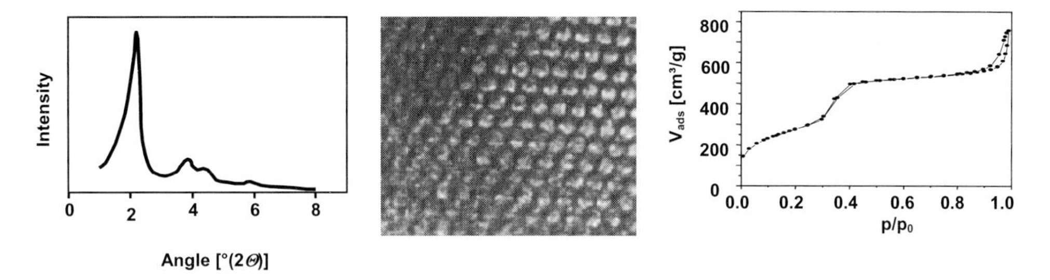

Fig. 1. Typical XRD pattern of an MCM-41 type material (left), typical TEM of an MCM-41 type material (middle), and typical sorption isotherm of an MCM-41 type material (right).

Introduction of Organosilicon Species via Surface Modification

Surface modification is possible by essentially all the reactions which have been developed over the past decades for amorphous silica. Excellent compilations of the different reagents and modifications achieved for ordered mesoporous silica are given in reviews by Anwander [4] and by Sayari and Hamoudi [5]. In almost all cases, trialkoxysilanes, trichlorosilanes, or disilazanes are used, which bear the organic functional group that is to be grafted to the surface of the silica. Also, several other review papers on surface modification with organosilicon groups are available [6 – 8].

The surface silanol groups are the sites which provide the anchor point of the functional silanes to the surface. For conventional hydroxylated silica, the density of silanol groups is remarkably constant, irrespective of the origin of the silica, at about 4–6 Si–OH/nm^2 [9]. This, however, can be different for ordered mesoporous materials. Concentrations of between 1.4 and 1.9 groups/nm^2 have been determined for the cubic MCM-48 [10]. From the data given by other authors, a concentration between 2 and 3 groups/nm^2 can be estimated to be present in MCM-41 as well as MCM-48 (values are estimated because data in the literature were determined normalized to weight and no specific surface areas were given) [11]. However, it is often not clear to what extent the surface of these materials is dehydroxylated by the thermal pretreatment, and thus a direct comparison with fully hydroxylated silica is difficult. Also, pore size has an effect on silanol density. For SBA-15, with its larger pores pretreated at 550 °C, a concentration of 1.07 OH/nm^2 was reported, which roughly corresponds to the value expected after such a pretreatment for conventional silica also [12]. In addition, there is evidence that the silanol groups are not distributed homogeneously over the surface, but that more hydrophobic and more hydrophilic patches exist in these materials and the concentration changes with surface area [13]. For post-synthesis modification, this implies a non-homogeneous loading with the modifying group. But even if the silanol groups are homogeneously distributed, the lower density compared to conventional hydroxylated silica may correspond to a lower podality (on average) of the anchored silicon species and thus possibly to higher susceptibility to hydrolytic bond cleavage.

The first surface modification by trimethylsilylation was reported in one of the seminal papers of the Mobil Oil group [14]. Already in this publication the decrease of pore size caused by the additional space occupied by the trimethylsilyl groups was reported. Such early studies were mainly of a fundamental nature, aimed at investigating the accessibility of the silanol groups, the degree of silylation which can be achieved, and the changes of pore sizes. Jaroniec et al., for instance, observed by modification with different silylating agents that surface coverages of around 2.5 μmol/m^2 could be achieved [15], which corresponds to the conversion of a major fraction of the available silanol groups. This was corroborated by work of Zhao and Lu, who found that the coverage was between 35 and 80%, depending on the conditions of the pretreatment of the silica [16]. Higher pretreatment temperatures led to higher coverages, because more hydrogen-bonded silanol groups were converted to reactive terminal silanols.

However, more interesting from an application point of view are silylation reactions which introduce new functions into the materials. These can be created either directly or in subsequent further steps after silylation. Most simple is the direct conversion of the silica to a basic material by reaction, for instance, with 3-aminopropyltriethoxysilane [17]. Also two-step processes have been employed to synthesize basic materials, where first chloropropyl groups are anchored to the surface with subsequent conversion of the chloro group into an amine. In order to remove the residual, unreacted silanol groups, a second silylation with hexamethyldisilazane can be used. Such materials were found to be reasonably active in different base-catalyzed reactions, such as Knoevenagel condensations and Michael additions. A survey of the catalyzed reactions and the types of modification used can be found in Ref. [5].

Functionalization to create acidic functions is mostly done in two steps to create sulfonic surface sites. First a thiol is attached to the surface, for instance by reaction of the silica with 3-mercapto-propyltriethoxysilane. This is then oxidized with hydrogen peroxide and acidified with sulfuric acid

[18]. However, the non-oxidized thiol modified materials are also interesting, since they can be used to very efficiently remove trace levels of heavy metal contaminants from water [19].

Introduction of Pendant Organosilicon Species via Co-Condensation

Co-condensation of tetraalkoxysilanes and organotrialkoxysilanes provides a very convenient and simple one-step procedure to create silica with pendant organic groups. For silica xerogels, such processes have long been known and are extensively covered in the literature [20, 21]. The first transfer to ordered mesoporous materials was reported at about the same time by Burkett et al. [22] and by Macquarrie [23]. Since then, various different organoalkoxysilane precursors have been used to produce a wide variety of organically functionalized silicas, also reviewed in [5]. While the direct synthesis by co-condensation is certainly simpler than post-functionalization of a preformed ordered silica, there are some disadvantages associated with the one-pot synthesis, which are (i) the type of functional group to be introduced has to be compatible with the synthesis conditions for the ordered mesoporous material, (ii) the surfactant template can no longer be removed by calcination, and thus extraction processes have to be developed which often lead to incomplete template removal, (iii) the presence of silanes with different hydrophobicity changes the conditions for the formation of the surfactant-silica liquid crystal phase, thus possibly making it difficult to synthesize a desired phase, and (iv) the different hydrolysis rates of the different silanes may lead to inhomogeneity of the resulting product.

However, there are also advantages. Post-synthesis modification has been reported to result in an inhomogeneous distribution of the organic groups in the silica, with groups enriched on the external surface and close to the pore mouths, while they are depleted in the interior of the materials [24]. According to the same authors, introducing the same functional group via co-condensation leads to more homogeneous loading of the silica. Another advantage is certainly the simplicity of the preparation. This can be illustrated by the synthesis of an SBA-15 [25] material bearing carboxylic acid groups, which has recently been developed by us [26]. Carboxylic acid groups had been introduced into small pore HMS-type silica before, by co-condensation of tetraethoxysilane and 2-cyanoethyltriethoxysilane and subsequent hydrolysis of the cyano-group. A similar strategy was adopted by us, however, involving an ethyleneoxide-propyleneoxide-ethyleneoxide triblock copolymer as template (Pluronic P-123, $EO_{20}PO_{70}EO_{20}$). These triblock copolymer templates have the advantage that they result in materials with substantially increased pore size. However, solvent extraction of the template is not straightforward, since the latter is partly embedded in the silica walls.

For the synthesis, 2-cyanoethyltriethoxysilane was prehydrolyzed in an HCl-containing solution of the polymer template, and after 30 min tetraethoxysilane was added. The solid product, which was recovered by filtration, was obtained after 20 h reaction at 40 °C and subsequent aging for one day at 90 °C. The problematic step of template removal was made possible by a process developed earlier by us, involving ether cleavage of the block copolymer by ca. 50% sulfuric acid [27]. This process extracts the block copolymer and incidentally hydrolyzes the cyano group to result directly in the formation of the porous carboxylate-containing silica. Figure 2 demonstrates the success of this process, which was monitored by NMR spectroscopy.

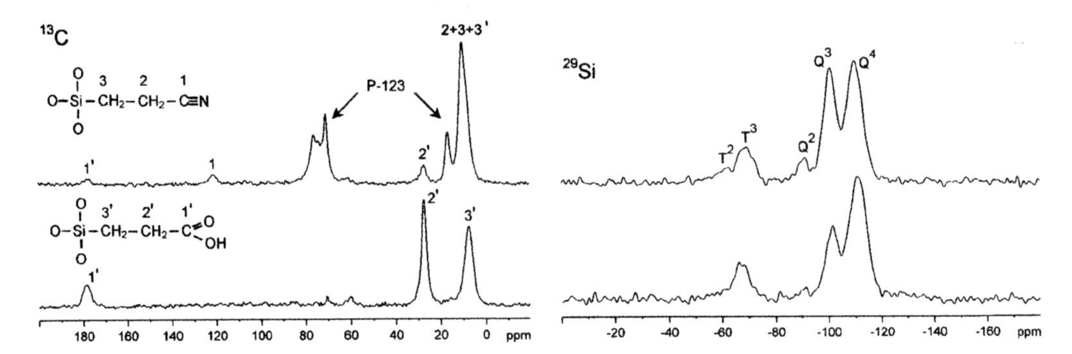

Fig. 2. ^{13}C NMR and ^{29}Si NMR spectra of cyano-functionalized SBA-15 (top traces) and carboxylate-functionalized material obtained after extraction with 48 wt% H_2SO_4 (bottom traces).

One can clearly see that the signals corresponding to the surfactant have almost completely vanished, and also the signal corresponding to the cyano-carbon is absent in the lower trace. The silicon spectrum demonstrates that the additional acid treatment leads to increased condensation of the framework, as evidenced by the higher proportion of T^3 and Q^4 species.

The strong influence of organotrialkoxysilanes on the phase formation in ordered mesoporous materials can be observed in the synthesis of modified, block copolymer-templated silica. In this case, we have investigated the synthesis of vinyl-modified materials. It has been rather difficult to synthesize the cubic members of the family of large-pore ordered mesoporous silica. However, the hydrophobic/hydrophilic proportion of the silicon species, which is coassembled with the surfactant, influences the curvature and thus the nature of the phase formed in a synthesis. While hydrophilic silicon species will favor high curvatures, i.e. direct the synthesis toward the formation of hexagonal materials, more hydrophobic silicon species will rather favour cubic or lamellar phases with lower curvatures. This can indeed be observed, if one co-condenses tetraethoxysilane and triethoxyvinylsilane [28]. Keeping all other parameters constant, hexagonally ordered vinylsilica is obtained up to concentrations of 10 – 15% vinylsilane. At higher concentrations, however, the higher hydrophobicity of the synthesis mixture induces the formation of the cubic Ia3d-structure, having a lower curvature. Figure 3 (left) shows the diffraction patterns for three samples with increasing vinylsilane concentration in the synthesis mixture, Fig. 2 (right) a TEM of the cubic phase obtained at the highest vinylsilane concentration of 20%.

As for the carboxylate-modified materials, surfactant extraction is easy, using the ether cleavage procedure with sulfuric acid. This creates porous materials with surface areas around 500 m^2/g and pore sizes in the region of 6 nm. ^{13}C and ^{29}Si NMR spectroscopy demonstrate the almost complete removal of the surfactant. The Si–C bond is not affected by the surfactant removal. However, slight changes in the signals in the vinyl range could indicate that to some extent coupling reactions between neighboring vinyl groups may take place.

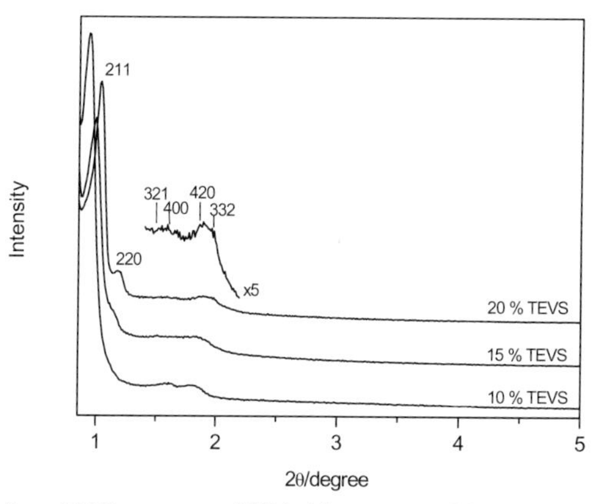

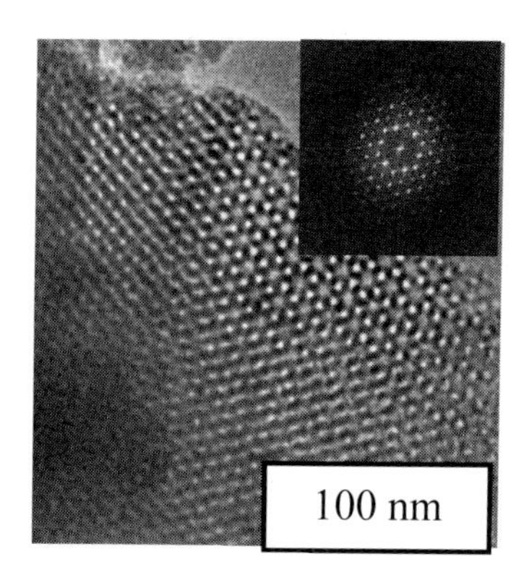

Fig. 3. XRD patterns of SBA-15-type materials synthesized with different amounts of triethoxyvinylsilane (TEVS) after surfactant extraction with sulfuric acid (left) and TEM image of material synthesized with 20% TEVS (right).

Organosilicas by Incorporation of Bridging Organic Groups in the Walls

The most exciting development in the field of organically modified ordered mesoporous materials of recent years is certainly the synthesis of true "organosilicas", where each silicon atom in the walls of the ordered porous material is connected via three oxo-bridges and one organic group to the surrounding silicon atoms. The synthesis of such materials was independently discovered by three different groups [29 – 31], and relies on a combination of the concepts which had been developed earlier for disordered organosilicas [20] and those developed for the synthesis of ordered mesoporous materials by surfactant templating. The basis for these approaches is the replacement of simple alkoxysilanes as silica precursors by organically bridged silsesquioxanes of the type $(RO)_3Si–R'–Si(OR)_3$. Even if 100% of the bridged precursor species are used, highly ordered materials can still be obtained.

Various different bridging groups have been used in subsequent years. In the first papers cited above, the relatively simple ethane or ethylene bridges have been used. An interesting feature observed with the ethylene bridge was the expansion of the lattice upon bromination, which would be consistent with transformation from sp^2 hybridization to sp^3 hybridization of the carbon species in the walls [30]. In subsequent publications, the range of organic groups incorporated in the framework was substantially expanded, now including methylene, acetylene, C_3-C_8 alkyl chains, butane, benzene, thiophene, bithiophene, ferrocene (all reviewed in [5]), and biphenyl [32]. Additionally, 1,3,5-tris(triethoxysilyl)benzene has been used as a precursor for an organosilica, providing an organic moiety which does not connect two, but three silicon atoms [33]. These groups, however, are thermally not very stable and are converted to bisiloxybenzene and monosiloxybenzene materials upon thermal treatment.

A related thermal effect has been reported for methylene-bridged organosilica [34]. The methylene-bridged materials are thermally extraordinarily stable, the methylene groups being retained even at temperatures exceeding 500 °C in air. However, they are converted from bridging groups between two silicon atoms to pendant groups between 400 and 700 °C. At the latter temperature, all organic groups had been removed from the sample. This process is not only interesting because of the thermal stability and the interesting transformations occurring, but also since it may allow the synthesis of materials with a very high concentration of pendant groups. Normally the concentration is restricted to about 20% in co-condensation procedures, because about 75% of tetraethoxysilane are necessary to provide sufficient cross-linking for structural integrity.

An exciting recent development also relied on the preparation of a 100% organosilica, in this case benzenesilica. The regularly ordered mesoporous materials have order on the mesoscale, but on the atomic scale, i.e. within the walls, the structure very closely resembles that of amorphous silica. From the first reports on, there have been attempts to create structure in the walls also on the atomic scale, but only with the synthesis of a benzenesilica was this quest finally successfully completed [35]. Both in the XRD and in the TEM, periodicity on a length scale of about 0.75 nm was observed perpendicular to the channel axis, which can be interpreted as alternating layers of benzene and silicate lining the walls of the mesopores. Since this provides a surface patterning on the molecular scale in materials with very high surface areas, one may envisage unprecedented control of reactivity toward small molecules in such solids.

Also, for organosilicas with organic groups fully incorporated in the walls, most investigations were carried out with small molecular surfactants as used in the synthesis of MCM-41 and related materials. Expansion of the pore sizes, however, would be attractive to widen the range of possible applications. We therefore tried to combine the synthesis of SBA-15 with that of the organosilicas to produce such samples. This proved to be possible when P-123 was used as surfactant. A hexagonally ordered ethenesilica with a pore size of about 6 nm was obtained. Figure 4 shows an XRD pattern and a corresponding TEM, demonstrating that also with triblock copolymer surfactants well-ordered periodic organosilicas can be synthesized.

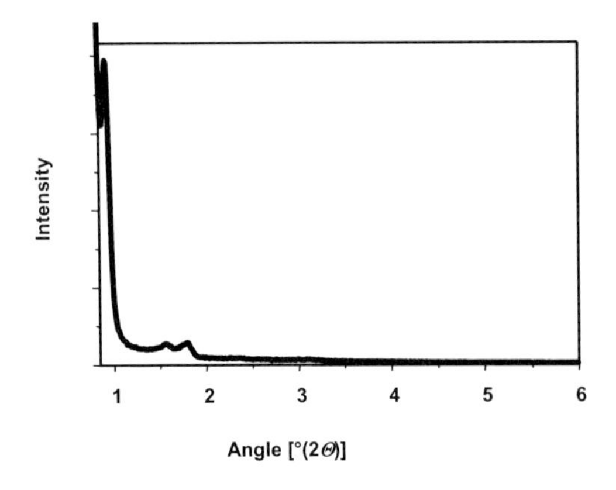

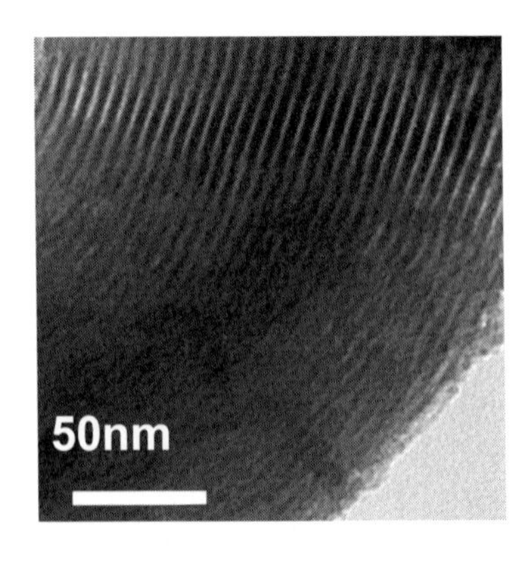

Fig. 4. XRD pattern of an SBA-15-type ethylenesilica synthesized with P123 as surfactant (left) and corresponding TEM (right). Pore size calculated from nitrogen sorption is around 6 nm.

Perspectives

Organically modified periodic mesoporous silicas have been more thoroughly investigated for only about five years. Already in this short period of time, tremendous progress has been made with respect to the type of functional groups which can be incorporated in the materials. This is certainly to a large extent because of the much longer experience with similar approaches to modify disordered silica, and much more can probably be learned from work in this field.

After the initial phase on ordered organosilicas has been used to develop primarily the synthesis strategies and tools for characterization, the future will show what kind of functions can be integrated via organosilicon groups. First attempts to incorporate different metal centers in the wall on the one hand and in the channels on the other have already proven to be successful [36]. Since organic groups in the framework are accessible for reactions, which may change their spatial requirements, interesting actuator properties and mechanical functions can be expected. Diazobenzene moieties, for instance, are photoswitchable; if they were incorporated, irradiation could lead to changes in the dimensions of the samples. Also, interesting optical properties can be envisioned, for instance, if transparent monolithic samples are loaded with molecules with high hyperpolarizability, such as para-nitroaniline, into the materials. Oriented channels could produce oriented molecules, which would give rise to macroscopically observable non-linear optical properties. For efficient performance, phase matching is often necessary, which means that refractive indices have to be adjusted. This may easily be possible by a controlled modification of the composition of the wall. The interesting aspect associated with using ordered mesoporous materials instead of sol-gel silica is in fact the order of the pore. If one considers, for instance, a possible photoswitchable material, the switching effect may average over the whole sample if the groups are randomly oriented, while a macroscopic effect could result for ordered materials. Whether such visions or other developments which so far do not exist can indeed be realized, the years to come will show.

Acknowledgments: Work of our group in this field was over the years supported by the DFG, the FCI and the EU, in addition to the generous basic funding by the Max-Planck-Society.

References

[1] C. T. Kresge, M. E. Leonowicz, W. J. Roth, J. C. Vartuli, J. S. Beck, *Nature* **1992**, *359*, 710.

[2] T. Yanagisawa, T. Shimizu, K. Kuroda, C. Kato, *Bull. Chem. Soc. Jpn.* **1990**, *63*, 988.

[3] U. Ciesla, F. Schüth, *Microporous Mesoporous Mater.* **1999**, *27*, 131; F. di Renzo, A. Galarneau, P. Trens, F. Fajula, Micelle templated materials in *Handbook of Porous Solids* (Eds.: F. Schüth, K. S. W. Sing, J. Weitkamp) Wiley-VCH, Weinheim, **2002**, p. 1311; F. Schüth, *Chem. Mater.* **2001**, *13*, 3184; J. Y. Ying, C. P. Mehnert, M. S. Wong, *Angew. Chem. Int. Ed.* **1999**, *38*, 56.

[4] R. Anwander, *Chem. Mater.* **2001**, *13*, 4419.

[5] A. Sayari, S. Hamoudi, *Chem. Mater.* **2001**, *13*, 3151.

[6] D. Brunel, *Microporous Mesoporous Mater.* **1999**, *27*, 329.
[7] A. Stein, B. J. Melde, R. C. Schroden, *Adv. Mater.* **2000**, *12*, 1403.
[8] M. J. MacLachlan, T. Asefa, G. A. Ozin, *Chem. Eur. J.* **2000**, *6*, 2507.
[9] L. T. Zhuravlev, *Langmuir* **1987**, *3*, 316.
[10] M. Widenmeyer, R. Anwander, *Chem. Mater.* **2002**, *14*, 1827.
[11] H. Landmesser, H. Kosslick, W. Storek, R. Fricke, *Solid State Ionics* **1997**, *101*, 271.
[12] C. Nozaki, C. G. Lugmair, A. T. Bell, T. D. Tilley, *J. Am. Chem. Soc.* **2002**, *124*, 13194.
[13] M. F. Ottaviani, A. Galarneau, D. Desplantiers-Giscard, F. Di Renzo, F. Fajula, *Microporous Mesoporous Mater.* **2001**, *44*, 1.
[14] J. S. Beck, J. C. Vartuli, W. J. Roth, M. E. Leonowicz, C. T. Kresge, K. D. Schmitt, C. T.-W. Chu, D. H. Olson, E. W. Sheppard, S. B. McCullen, J. B. Higgins, J. L. Schlenker, *J. Am. Chem. Soc.* **1992**, *114*, 10834.
[15] C. P. Jaroniec, M. Kruk, M. Jaroniec, A. Sayari, *J. Phys. Chem. B* **1998**, *102*, 5503.
[16] X. S. Zhao, G. Q. Lu, *J. Phys. Chem. B* **1998**, *102*, 1556.
[17] A. Cauvel, G. Renard, D. Brunel, *J. Org. Chem.* **1997**, *62*, 749.
[18] W. D. Bossaert, D. E. De Vos, W. M. Van Rhijn, J. Bullen, P. J. Grobet, P. A. Jacobs, *J. Catal.* **1999**, *182*, 156.
[19] X. Feng, G. E. Fryxell, L. Q. Wang, A. Y. Kim, J. Liu, K. M. Kemner, *Science* **1997**, *276*, 923.
[20] J. Wen, J. L. Wilkes, *Chem. Mater.* **1996**, *8*, 1667.
[21] D. A. Loy, K. J. Shea, *Chem. Rev.* **1995**, *95*, 1431.
[22] S. L. Burkett, S. D. Sims, S. Mann, *Chem. Commun.* **1996**, 1367.
[23] D. J. Macquarrie, *J. Chem. Soc., Chem.Commun.* **1996**, 1961.
[24] M. H. Lim, A. Stein, *Chem. Mater.* **1999**, *11*, 3285.
[25] D. Zhao, J. Feng, Q. Huo, N. W. Melosh, G. H. Frederickson, B. F. Chmelka, G. D. Stucky, *Science* **1998**, *279*, 548.
[26] C.-M. Yang, B. Zibrowius, F. Schüth, *Chem. Commun.* **2003**, 1772.
[27] C.-M. Yang, B. Zibrowius, W. Schmidt, F. Schüth, *Chem. Mater.* **2004**, *16*, 2918.
[28] Y. Wang, C.-M. Yang, B. Zibrowius, B. Spliethoff, M. Lindén, F. Schüth, *Chem. Mater.*, submitted.
[29] S. Inagaki, S. Guan, Y. Fukushima, T. Ohsuna, T. Terasaki, *J. Am. Chem. Soc.* **1999**, *121*, 9611.
[30] B. J. Melde, B. T. Holland, C. G. Blanford, A. Stein, *Chem. Mater.* **1999**, *11*, 3308.
[31] T. Asefa, M. J. MacLachlan, N. Coombs, G. A. Ozin, *Nature* **1999**, *402*, 867.
[32] M. P. Kapoor, Q. Yang, S. Inagaki, *J. Am. Chem. Soc.* **2002**, *124*, 15176.
[33] M. Kuroki, T. Asefa, M. Whitnal, M. Kruk, C. Yoshina-Ishii, M. Jaroniec, G. A. Ozin, *J. Am. Chem. Soc.* **2002**, *124*, 13886.
[34] T. Asefa, M. J. MacLachlan, M. Grondey, N. Coombs, G. A. Ozin, *Angew. Chem. Int. Ed.* **2000**, *39*, 1808.
[35] S. Inagaki, S. Guan, T. Ohsuna, O. Terasaki, *Nature* **2002**, *416*, 304.
[36] R. J. P. Corriu, A. Mehdi, C. Reyé, C. Thieuleux, *Chem. Commun.* **2003**, 1564.

Physical-Chemical Features of Synthetic Amorphous Silicas and Related Hazard and Risk Assessment

M. Heinemann, A. Bosch*

Wacker-Chemie GmbH, Johannes-Hess-Strasse 24, 84489 Burghausen, Germany
Tel.: +49 86 77 83 40 22 — Fax: +49 86 77 83 68 14
E-mail: mario.heinemann@wacker.com

M. Stintz

Lehrstuhl für Mechanische Verfahrenstechnik, TU Dresden

W. Vogelsberger

Institute of Physical Chemistry, University of Jena

Keywords: synthetic amorphous silicas, particle sizes, dissolution kinetics, hazard and risk assessment

Summary: Synthetic Amorphous Silica (SAS) is highly pure, crystalline-free, silicon dioxide which may be produced as pyrogenic silica, precipitated silica, and silica gel. Amorphous and crystalline polymorphs of silicon dioxide ("silica") differ in their hazard to human health. Dissolution kinetics and particle size data of SAS have been determined in order to get a better differentiation between the polymorphs. The dissolution rate coefficients have a range of 1×10^{-4} to 5×10^{-4} mg cm² day^{-1}. The clearance rates of SAS from the lung by dissolution are sufficiently high to make the contributions of macrophage-mediated particulate clearance rates incommensurable and negligible. The relevant particle size fraction of SAS as marketed, handled, or used covers the range from 100 to 2000 μm (geometric). The inhalable and respirable fractions according to EN/DIN481 have been calculated. Result: <1 wt%. SAS as commercial products do not fall in the class of nanomaterials. With this background, a better understanding of inhalation toxicology of SAS is possible, affording basis for credible health risk assessments.

Introduction

Synthetic Amorphous Silica (SAS) is highly pure, crystalline-free, silicon dioxide, which may be produced as pyrogenic silica, precipitated silica, and silica gel. Various types of SAS may be differentiated in relation to their process technologies – "wet-route" processes (precipitated silica and silica gel) and the thermal route (pyrogenic silica). SAS can be surface treated to render it hydrophobic. The produced amounts are low compared to the untreated SAS; the surface treatment does not change the solid properties of the inorganic polymeric silicon dioxide. This paper looks at two major physical-chemical properties of SAS – dissolution and particle size – in an attempt to better understand SAS, to differentiate more clearly between the polymorphs of silicon dioxide, and to evaluate health hazards.

Silica Polymorphs

Silica, independently of its form and method of preparation, is found under RNCAS 7631-86-9. In September 1996, based on a review of existing literature, one international agency, the International Agency for Research on Cancer (IARC), changed its carcinogen rating for the crystalline polymorphs of silica – quartz and cristobalite – to a group 1 classification (carcinogenic to humans). The amorphous polymorphs remain as a group 3 classification (not classifiable as carcinogenic to humans) [1]. Furthermore, natural forms of amorphous silica often contain crystalline impurities (e.g., up to 65% cristobalite in the case of calcination). Amorphous silicas as by-products from electrometallurgical processes differ notably from SAS in morphology and composition, e.g., "heavy metal". Therefore, it is essential to distinguish carefully between the different polymorphs of silica. Figure 1 covers the different polymorphs of silica and their related CAS Reg. Numbers.

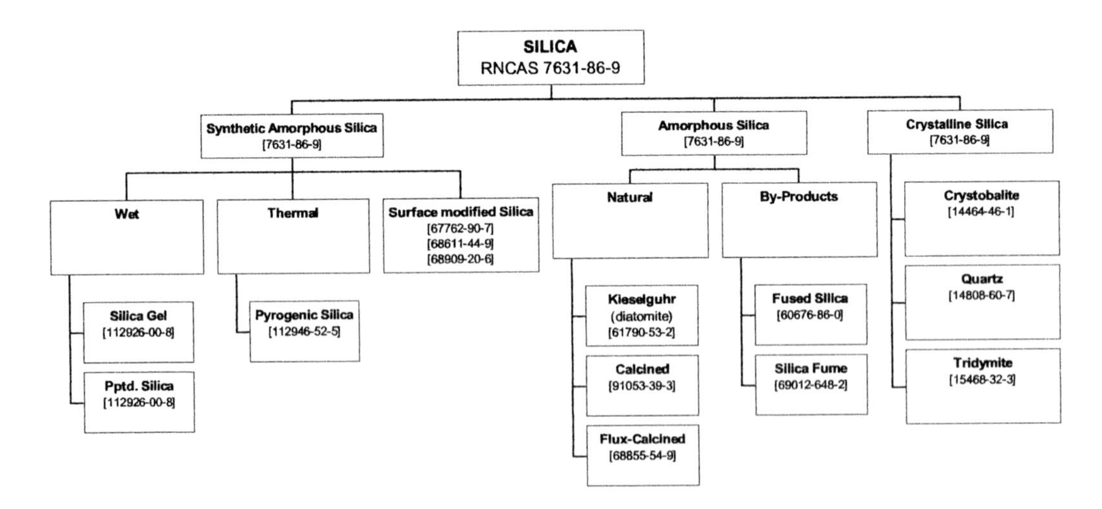

Fig. 1. Silica polymorphs and related CAS Reg. numbers.

Dissolution Kinetics

Quartz has an extremely low dissolution rate in water [2]. In contrast, experiments revealed fast dissolution of SAS in water [3]. Experiments in biological media with SAS confirmed these findings [4, 5].

Experimental

- pH 7.4 (37 °C); buffer composition and total ion content equal to "body liquid"
- *L*-α-Dipalmitonylphoshatidylcholine as lung surfactant (DPPC)
- Dissolved monomer determination: Photometric determination of silicomolybdic acid with ammonium molybdate according to Motomizu
- Test materials: SAS, typical pyrogenic silica: 200 m^2/g – Cab-O-Sil M5; 400 m^2/g – HDK T40
- The concentration-time curves are shown in Fig. 2.
- K[mg·cm^2/d]; calculated $dM/dt = -kA$ [6]

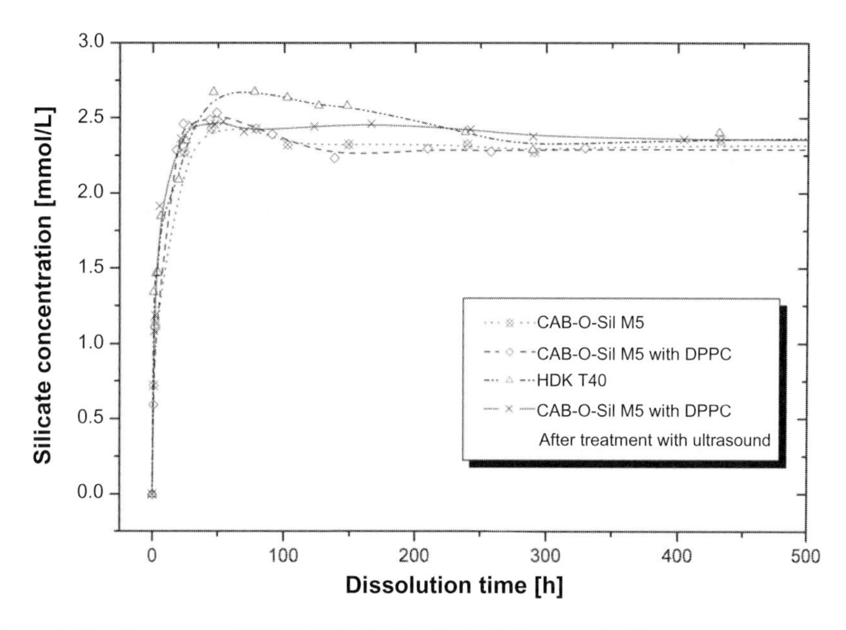

Fig. 2. Concentration-time curves of SAS in biological media, typical pyrogenic silicas.

Calculated dissolution rate coefficients have a range of 1×10^{-4} to 5×10^{-4} mg cm^2 day^{-1} depending on the SAS type [4]. Related to the low to high value of the dissolution rate, SAS will be partly retained for no more than 30 to 6 days in the lung. Based on these findings, a predictive theoretical clearance and retention model for soluble particle has been developed [6]. The clearance rates of SAS by dissolution are sufficiently high to make the contributions of macrophage-mediated particulate clearance rates incommensurable and negligible. Thus, clearance by dissolution is at least 20 times faster than particulate clearance. Credible health risk assessments need reliable knowledge of effective relative doses of concerned particulates. With this background, a better understanding of inhalation toxicology of SAS is possible.

Particle Sizes

The basic principles of SAS particle formation are shown for pyrogenic silica (Fig. 3). Hydrolysis of chlorosilane leads to SiO_2 molecules. Nucleation, condensation, and coagulation lead to primary particles of about 5 – 50 nm. By varying process parameters, the mean particle size, the particle size distribution, and the degree of aggregation / agglomeration can be varied over relatively broad ranges. However, primary particles under the conditions of the reaction zone – about 1600 °C in the inner zone – collide and firmly stick together building stable SiO_2 aggregates. Primary particles do not exist outside the reaction zone. Aggregates subsequently form agglomerates of SiO_2. Because of the strong physical attraction forces particle growth goes even further; "flocks" are formed.

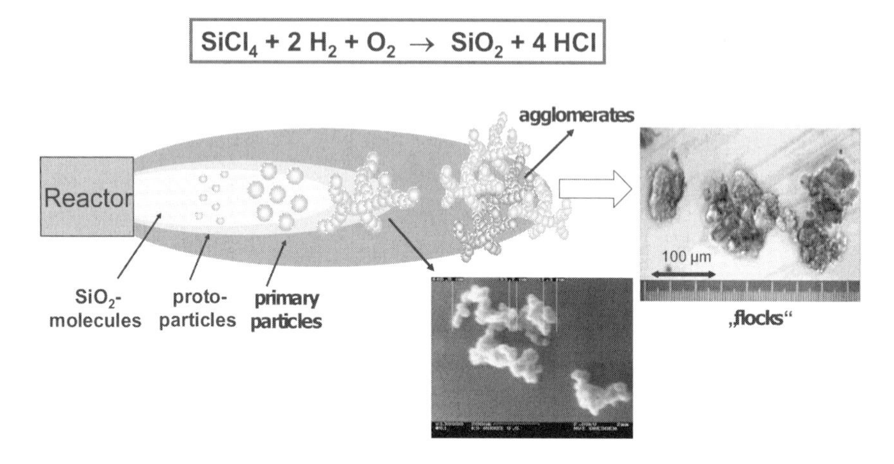

Fig. 3. Pyrogenic silica particle formation.

Many methods can be used to measure the particle sizes of particulates. However, each physical measurement principle provides in general a different distribution curve. Recently, standard test methods for dry particles in air, such as dry sieving (DIN 66165), laser diffraction (LD, ISO 13320), particle image velocimetry (PIV), time-of-flight (TOF), cascade impactor (CI), scanning electron microscopy (SEM), and light microscopy have been applied to determine the particle sizes of SAS. Some of the methods proved to be destructive to the sample; other require extreme "dilution" of the sample or need the density of the samples [7]. The density of 2.2 g/cm³ of the primary particle (literature) does not reflect the shape and the porosity of the "real" particle, the agglomerate. The effective density of such a "fractal" structure as an agglomerate has been determined to be 0.075 g/cm³; this fits well with the macroscopic bulk density of about 0.05 g/cm³ [8]. In the discussion of particle sizes of SAS, these findings have to be taken into account to give reliable data.

Particle Sizes of SAS as Marketed, Handled, and Used

The condition of the SAS can be characterized as a dry powder state, high solids concentration, and little or no dispersion. Laser diffraction with a sedimentation shaft for sample feeding (low shearing) and classical sieving (DIN 66 165) proved to be useful tools to determine the particles sizes properly [7].

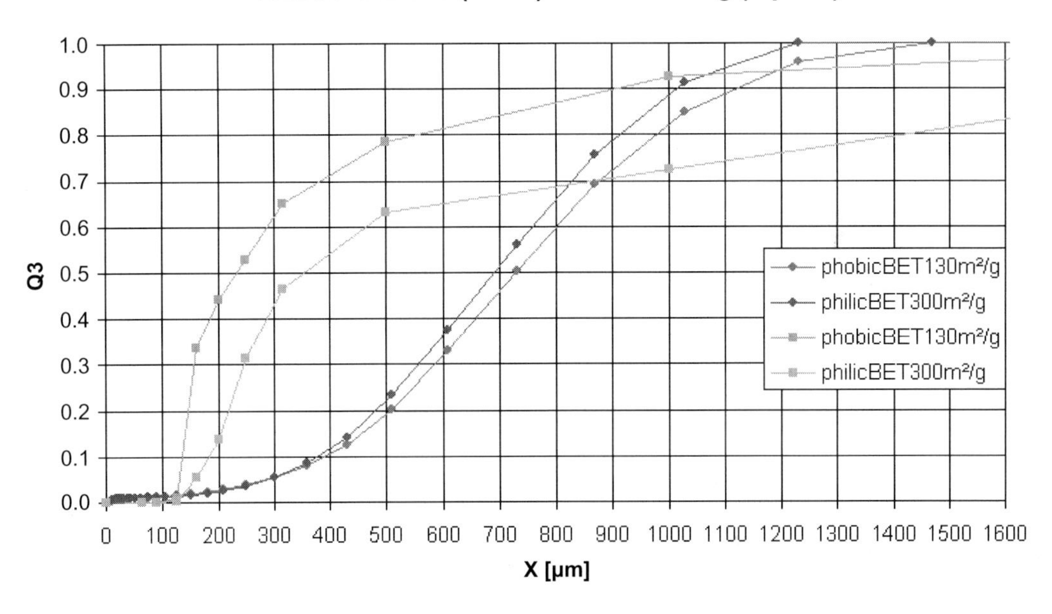

Fig. 4. Geometric particle size distributions of different pyrogenic silicas. Correlated aerodynamic diameter can be calculated from LD by using an effective density (agglomerate) of 0.075 g/cm³ [8].

The relevant particle size fraction for the products as marketed, handled, or used covers the range from 100 to 2000 µm (geometric). The inhalable and respirable fractions have been calculated according to EN/DIN481. Result: <1 wt% [7].

SAS as commercial products do not fall in the class of nanomaterials [7 – 9].

Hazard Definition and Risk Assessment of SAS

Dissolution and particle size data are important factors that affect toxicity resulting from exposure to SAS particles. These data enable:

- a better understanding of the mode of action of the different silica polymorphs in the human organism
- a differentiated risk assessment of SAS in comparison to quartz and/or ultrafine particles
- a differentiated assessment of animal inhalation test results for hazard definition.

References

[1] *IARC Monographs*, Vol. 68, IARCPress, **1997**.

[2] R. K. Iler, *The Chemistry of Silica*, John Wiley & Sons, New York, **1979**.

[3] W. Vogelsberger, T. Mittelbach, A. Seidel, *Ber. Bunsenges, Phys. Chem.* **1996**, *100*(*7*), 1118.

[4] W. Vogelsberger, Some results of dissolution experiments carried out with different kinds of amorphous silica, *CEFIC Internal Report* **2002**.

[5] F. Roelofs, *Master Thesis*, Friedrich-Schiller-University, Jena, **2002**.

[6] W. Stöber, B. Wong, W. Koch, H. Windt, A simple pulmonaryretention model for inhaled soluble particles of limited biologicla residence time, in *CEFIC Internal Report* **2000**.

[7] M. Stintz, M. Heinemann, Particle analysis of pyrogenic (fumed) silicas at technical concentrations and under technical handling conditions, in *Wacker Internal Report* **2001**.

[8] H. Barthel, M. Heinemann, M. Stintz, B. Wessely, *Chem. Eng. Technol.* **1998**, *21*, 9.

[9] M. Stintz, M. Heinemann, S. Ripperger, *Chemie Ing. Tech.* **2002**, *74*, 546.

Particle Size of Fumed Silica: a Virtual Model to Describe Fractal Aggregates[†]

Christoph Batz-Sohn*

Degussa AG, BU Aerosil & Silanes, Rodenbacher Chaussee 4, 63457 Hanau, Germany
Tel.: +49 6181 594623 — Fax: +49 6181 5974623
E-mail: Christoph.batz-sohn@degussa.com

Keywords: dynamic light scattering, fractal aggregates, fumed silica, particle size, simulation

Summary: Fumed oxides that are produced in gas-phase processes, such as silica and aluminum oxide, consist of clusters of aggregated primary particles. The aggregate [2] size of these particles is an important variable in many applications. However, current procedures for measuring particle sizes all assume that the particles have a spherical shape and are thus not truly capable of determining aggregate size. The results of such particle size measurements are consequently called "equivalent spherical diameter" (ESD) [3], but these results vary from method to method. This paper shows that it is feasible to use the number of primary particles per aggregate, rather than the ESD, as a measure of the particle size of clusters of this type. The method is based on dynamic light scattering (photon correlation spectroscopy, PCS), which has proven itself in the analysis of fumed oxides [4, 5]. A numerical simulation based on random, computer-generated model aggregates is used to modify the well-known Stokes-Einstein equation so that the number of primary particles can be determined.

Computer Generated Model Aggregates

Apart from their morphology, the aggregates of spherical primary particles can be ideally described by three basic parameters: r is the radius of the primary particle, x the number of primary particles per aggregate, and γ the coefficient of penetration of the particles to each other.

Many basic properties of such aggregates can be directly calculated (e.g., volume, specific surface area, $\cdots$) [1]. Some properties related to the particle size and of further interest like envelope radius or the area of projection cannot be calculated directly. Thus, a computer algorithm was implemented which randomly creates aggregates meeting the selected starting conditions of x, r and γ (see the references for a detailed description).

Implementation of the algorithm for generating aggregates also includes, as a major function, a

[†] A more detailed version of this paper has been published in Ref. [1].

numerical method for determining the projected area (shade) of the aggregate that has just been generated.

The *x*, *y*, *z* data that this algorithm generates represent virtual models for fractal [6] aggregates that are composed of primary particles (see Fig. 4). The algorithm makes no attempt to simulate a particular creation process for aggregates (such as in a flame) and is consequently incapable of producing a correlation to flame parameters or the like. In contrast, it is an independent (virtual) formation process for fractal aggregates, which will in turn possess particular characteristic properties for this very reason. For the comparison with other simulation results, the radius of gyration, the fractal dimension D_f and the fractal prefactor k_f are given [6].

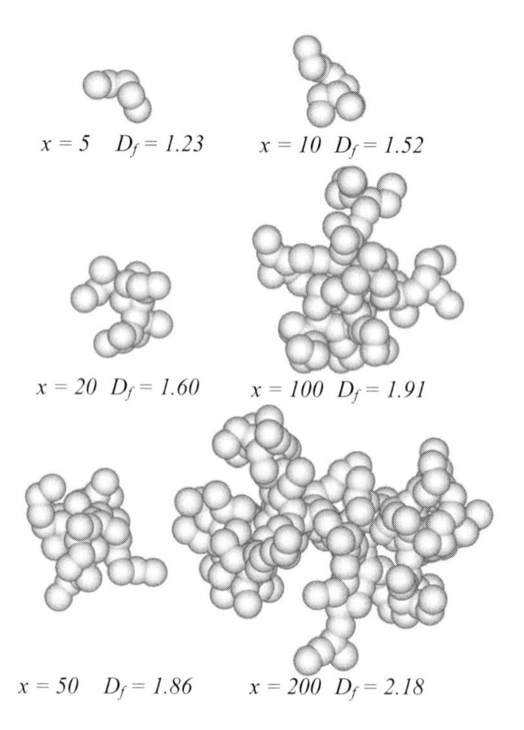

Fig. 1. Examples of aggregate models generated by the algorithm.

Simulation Results

The model system is used to determine the dependence of the projected area and the envelope radius on the given parameters. It is easy to show and see that both scale with the radius of the primary particles. It is therefore possible to transfer the results obtained for a given primary-particle radius to other radii. In the calculations presented here, the value of the particle radius was 10 nm. Since the algorithm generates aggregates randomly, there exists substantial fluctuation in the values determined for the projected area and the envelope radius. As in the case of real experiments, it is consequently also necessary to resort to statistics in this “virtual” experiment. A total of 287 model aggregates, consisting of 8, 20, 55, and 150 primary particles were used, the coefficient of penetration of each primary particle in turn assuming the values of 0.1, 0.25, and 0.4. The following table presents the results of the calculations.

Table 1. Results from the simulation calculations. The value $r = 10$ nm was used as the radius of the primary particles. The table states the mean and standard deviation for each variable.

No. of primary particles	Penetration coefficient	Projected area (calculated acc. Numerical derivation)	Envelop diameter (calculated acc. Numerical derivation)	Radius of gyration	Fractal dimension	Fractal prefactor
x	γ	A_f [nm^2]	$2w_r$ [nm]	R_g [nm]	D_f	k_f
8	0.10	1856 ± 235 (1841)	89 ± 10 (85)	22.7 ± 1.6	1.51 ± 0.15	2.30 ± 0.07
8	0.25	1606 ± 225 (15001)	76 ± 10 (74)	18.8 ± 1.4	1.52 ± 0.14	3.10 ± 0.12
8	0.40	1347 ± 202 (1159)	69 ± 9 (63)	15.0 ± 1.1	1.53 ± 0.12	4.28 ± 0.28
20	0.10	4097 ± 462 (4168)	125 ± 17 (132)	36.9 ± 1.4	1.65 ± 0.06	2.20 ± 0.15
20	0.25	3496 ± 486 (3395)	110 ± 16 (113)	30.2 ± 1.6	1.68 ± 0.10	2.98 ± 0.16
20	0.40	2796 ± 357 (2623)	93 ± 12 (95)	24.2 ± 1.6	1.68 ± 0.09	4.33 ± 0.30
55	0.10	9865 ± 1148 (10140)	186 ± 29 (193)	59.1 ± 2.4	1.84 ± 0.08	1.96 ± 0.25
55	0.25	8148 ± 878 (8260)	155 ± 19 (164)	49.6 ± 2.6	1.84 ± 0.08	2.77 ± 0.26
55	0.40	6371 ± 761 (6380)	132 ± 13 (136)	38.8 ± 1.7	1.88 ± 0.09	3.98 ± 0.46
150	0.10	24056 ± 2302 (24078)	267 ± 33 (262)	89.6 ± 4.0	2.05 ± 0.07	1.55 ± 0.19
150	0.25	19890 ± 1938 (19614)	232 ± 21 (222)	74.0 ± 2.2	2.06 ± 0.07	2.27 ± 0.34
150	0.40	14916 ± 1633 (15150)	191 ± 21 (182)	59.2 ± 1.8	2.05 ± 0.07	3.63 ± 0.35

Numerical Derivations of A_f and w_r

We can now divide the computed projected area by the sum of the projected areas of all primary particles to obtain a percentage that we can use to normalize the values. Plotting these data against each other, it can be seen that this percentage drops exponentially as the number of primary particles per aggregate increases. We can confirm that the drop is indeed exponential by plotting the percentage against the natural logarithm of the number of primary particles (see Fig. 2). The result is nearly parallel fitted lines, each depending on the coefficient of penetration. By fitting to all data of Table 1 we can thus derive the equation as shown in Fig. 2.

It should be noted that the empirical factors 0.085 and 1.1 were not derived directly from the slopes and intercepts of the fitted lines of Fig. 5. They were obtained by fitting the entire family of parameters from the formula to the results exhibited in Table 1. Such a fit takes into consideration the interacting terms between the two parameters, which a direct evaluation of the fitted lines would not accomplish.

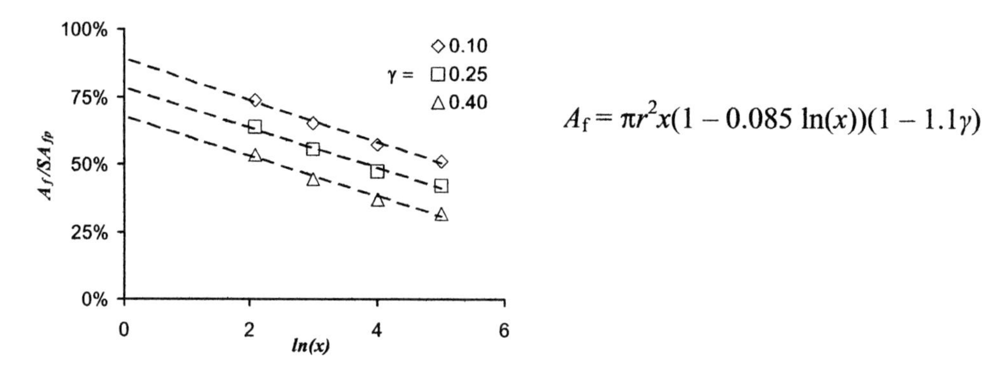

Fig. 2. Numerical derivation of A_f.

A similar approach is succesful for the derivation of a formula for the envelope radius. It can be seen that the envelope diameter increases steadily as the aggregates grow, but that the increase is "damped." The function $\ln(x)^{3/2}$ describes this dependence well (see Fig. 3). The corresponding evaluation and the fit to the values in Table 1 produces the following

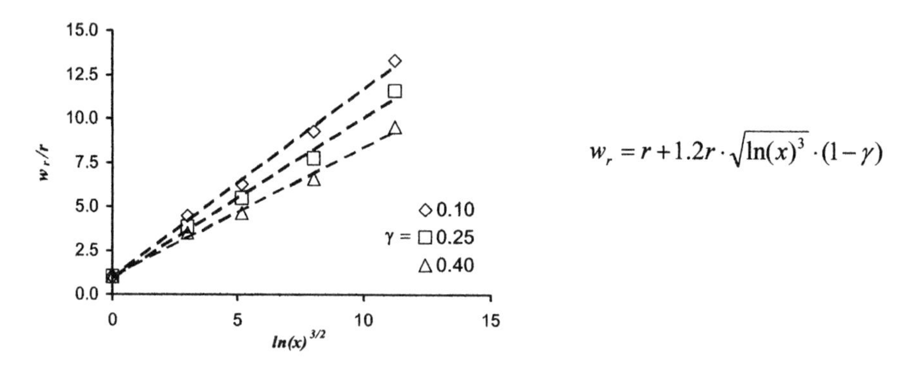

Fig. 3. Numerical derivation of w_r.

Modifying the Stokes-Einstein Equation for Fractal Aggregates

Our understanding of the equations above now enables us to derive an alternative to the Stokes-Einstein equation which, starting from the diffusion coefficients, allows us to determine a measure that is more meaningful for fractal aggregates than the equivalent spherical diameter.

The literature in the field of fluid mechanics [8] provides us with an expression as the definition of the frictional drag coefficient ζ. For spheres, disks, cylinders, and similar bodies, the diameter is used as the characteristic length l. Similarly, $2w_r$ is used as the envelope diameter for further derivations. (Abbreviations used: ρ_f: density of the fluid, η: viscosity of the fluid, v: velocity of the particles, k: Boltzmann's constant, T: absolute temperature)

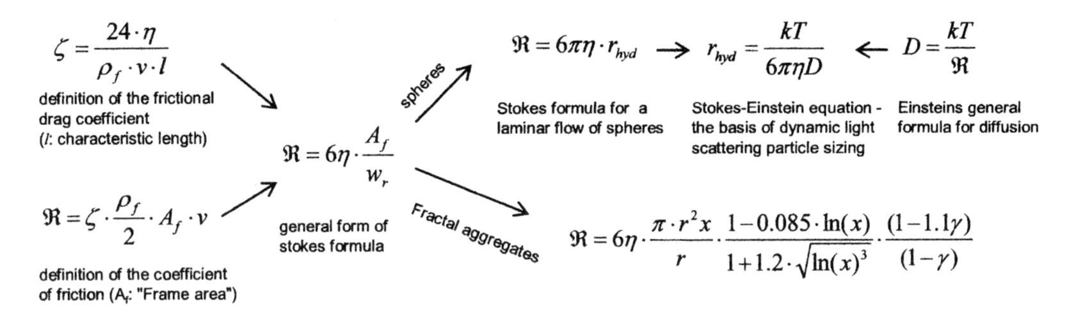

Scheme 1. Derivation of a modified Stokes-Einstein equation for fractal aggregates from basic fluid mechanics and the numerical derived expressions for A_f and w_r.

The definition of the coefficient of friction leads to a general form of Stokes formula, which enables us to use our expressions for the "Frame area" and the "characteristic length" instead of using the expression for spheres.

If we knew the primary particle size and coefficient of penetration (both of which can be determined from TEM images), we would then obtain the number of primary particles as a direct measure of the size of the aggregate.

$$6\eta \cdot \pi \cdot r_{hyd} = 6\eta \cdot \frac{\pi \cdot r^2 x}{r} \cdot \frac{1 - 0.085 \cdot \ln(x)}{1 + 1.2 \cdot \sqrt{\ln(x)^3}} \cdot \frac{(1 - 1.1\gamma)}{(1 - \gamma)} \longrightarrow \frac{r_{hyd}}{r} = x \cdot \frac{1 - 0.085 \cdot \ln(x)}{1 + 1.2 \cdot \sqrt{\ln(x)^3}} \cdot \frac{(1 - 1.1\gamma)}{(1 - \gamma)}$$

Scheme 2. Since it is not easy to modify the evaluation software in commercially available PCS measuring instruments, one can always select the hydrodynamic or Stokes radius (equivalent spherical radius) rather than using the roundabout approach. It is possible to eliminate several parameters by setting both terms, Stokes formula and its new version, equal.

Application

Figure 4 shows the impact of the number of primary particles per aggregate and the coefficient of penetration on the result of a dynamic light-scattering particle sizing of those aggregates. The impact of γ is very low. Therefore, it is now very easy to give the number of primary particles as a result of a PSD measurement rather than the equivalent spherical diameter $2\ r_{hyd}$.

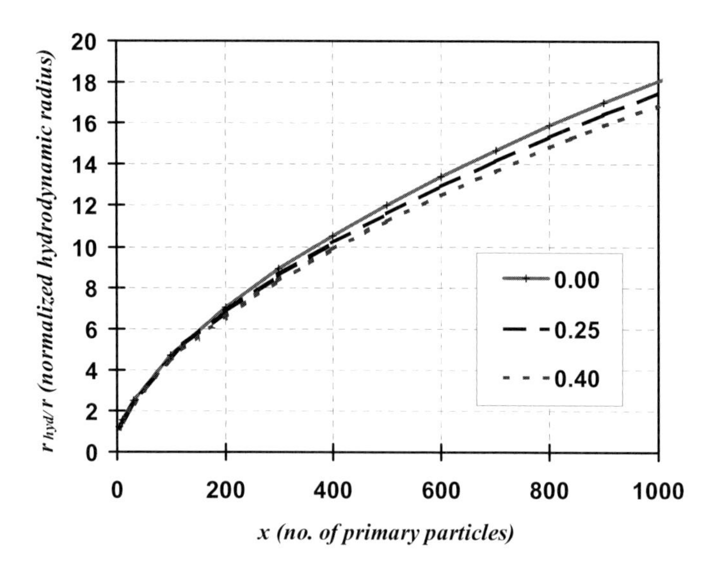

Fig. 4. Plotting of the formula of Scheme 2 shows the dependence of the hydrodynamics radius on the number of primary particles per aggregate. In contrast, the impact of the coefficient of penetration is very low.

As an example, Fig. 5 visualizes the results of particle size distribution curves on two different Alumina C dispersions.

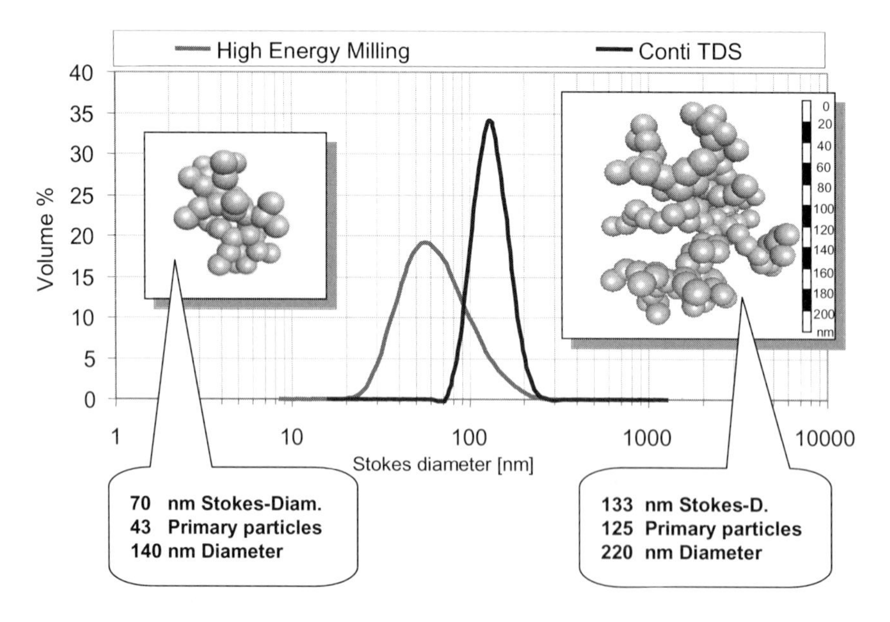

Fig. 5. Particle size measurements of Alumina C dispersions and possible new visualization.

References

[1] C. Batz-Sohn, *Part. Syst. Charact.* **2003**, *20*, 370.
[2] The term "aggregate" is used in the sense of DIN 53206 for a collection of primary particles that have tightly intergrown together. "Agglomerates," in contrast, are only loose combinations of aggregates and/or primary particles. The terms aggregate and agglomerate are used in the reverse sense in English-speaking countries.
[3] B. R. Jennings, K. Parslow, *Proc. R. Soc. Lond. A* **1988**, *419*, 137.
[4] H. Barthel, M. Heinemann, M. Stinz, B. Wessely, *Chem. Eng. Technol.* **1998**, *21*, 745.
[5] T. Wriedt, *Part. Part. Syst. Charact.* **1998**, *15*, 76.
[6] B. B. Mandelbrot, *The Fractal Geometry of Nature*, W. H. Freeman, New York, **1983**.
[7] A. M. Brasil, T. L. Farias, M. G. Carvalho, *Aerosol Sci. Technol.* **2000**, *33*, 440.
[8] P. Grassmann, *Der umströmte Einzelkörper*, in: *Physikalische Grundlagen der Chemie-Ingenieur-Technik* (Eds.: H. Mahler, O. Fux.), H. R. Sauerländer & Co., Frankfurt/Main, **1961**, Volume 1, pp. 706ff.

Characterization of Size and Structure of Fumed Silica Particles in Suspensions

Frank Babick,* Michael Stintz, Herbert Barthel

Lehrstuhl für Mechanische Verfahrenstechnik, TU Dresden, 01062 Dresden, Germany
Tel.: +49 351 463 33724 — Fax: +49 351 463 37058
E-mail: Frank.Babick@mailbox.tu-dresden.de

Mario Heinemann

Wacker-Chemie GmbH, Werk Burghausen, 84480 Burghausen, Germany

Keywords: fumed silica, laser diffraction, ultrasonic spectroscopy

Summary: Suspensions of fumed (silica) powders consist of particle structures in different size ranges (from nanosized primary particles to agglomerates in the micrometer range). The complexity of these particle structures determines the macroscopic properties of the suspensions. Hence, there is a need for characterization methods that reflect the state of dispersion. The paper shows, using the example of laser diffraction spectroscopy and ultrasonic extinction spectroscopy, that the use of a single measurement technique cannot describe the state of dispersion comprehensively and that the appropriate characterization method is determined by those properties responsible for the product quality. While laser diffraction is very sensitive to agglomerates in the micrometer range, it does not detect aggregates in the submicron range. Ultrasonic spectroscopy shows rather the internal structure of aggregates but not the large agglomerates.

Introduction

The macroscopic properties of liquid suspensions of fumed powders of silica, alumina etc. are not only affected by the size and structure of primary particles and aggregates, which are determined by the particle synthesis, but as well by the size and structure of agglomerates or mesoscopic clusters, which are determined by the particle-particle interactions, hence by a variety of product- and process-specific factors like the suspending medium, solutes, the solid concentration, or the employed mechanical stress. However, it is still unclear how these secondary and tertiary particle structures can be adequately characterized, and we are a long way from calculating product properties from them [1, 2].

This paper focuses on the application of two particle sizing methods, laser diffraction spectroscopy and ultrasonic attenuation spectroscopy, to the characterization of suspensions of fumed powders.

Comparison of Methods

In the case of fumed powders, the results of particle size analysis depend very strongly on the characterization method. Each method measures a different particle property, from which sphere equivalent diameters are calculated. The underlying models assume homogeneous, spherical particles, which does not apply to the porous aggregates and agglomerates of these materials.

Figure 1 shows the SEM image of aggregates of fumed alumina which were filtered on a membrane and the particle size distributions, which where obtained by different characterization methods. Obviously, the analysis of TEM images provides us with the size of the primary particles, while all other methods measure larger particle structures. Before discussing the differences between the methods, it should be noted that the measurements were conducted at different particle concentrations (PCS and LD at highly diluted samples, $\phi < 0.1$ vol%; USS at the undiluted sample, $\phi = 3$ vol%), while the sample preparation before dilution was identical (stirring for 8 h, 8 min ultrasonic homogenization).

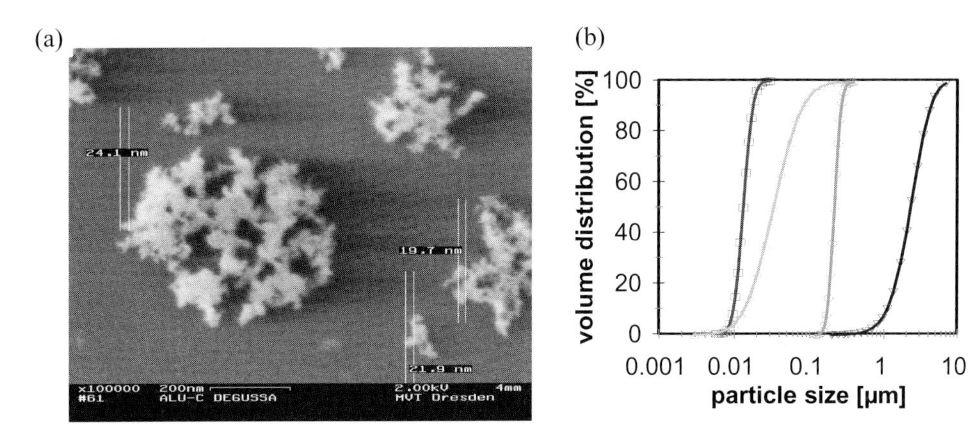

Fig. 1. Fumed alumina: SEM image (a) size distributions from TEM images, ultrasonic spectroscopy (USS), dynamic light scattering (DLS), and laser diffraction (LD) (b).

By comparison with the SEM image, one can conclude that ultrasonic spectroscopy measures the size of particle structures in the range between the primary particles and the stable aggregates, which are formed immediately after the particle synthesis. In contrast, laser diffraction apparently ignores all particle structures in the submicron range but shows particles in the micrometer range.

The first question we will examine in this paper is: do these micrometer particles really exist? Laser diffraction spectroscopy analyses the angular intensity spectrum of diffracted laser light and is actually restricted to the size range from 1 μm to 1000 μm. It may be possible that the analysis software will get somewhat "confused" by the nanosized particles.

Remembering that ultrasonic spectroscopy measures the frequency spectrum of sound attenuation of suspensions from 0.01 μm to 1000 μm [3], the more difficult question (that cannot be answered here) will be: how are the diffraction pattern and the attenuation spectrum related to size parameters of fumed powders?

Sensitivity to Agglomerates

In order to check whether micrometer particles, suggested by laser diffraction, really exist, we filtered suspensions of fumed silica through microsieves with mesh sizes at 50 μm and 20 μm. Indeed, laser diffraction reflected the filtering steps (Fig. 2a). However, the volume fractions of the respective oversize material obtained by laser diffraction (ca. 50 vol%) are considerably larger than their weight fractions, which were determined by weighing the sieves before and after filtering (ca. 0.5 wt%).

This discrepancy results firstly from the high porosity of agglomerates of fumed silica (> 90% [1]), and secondly from the inadequacy of the Fraunhofer diffraction theory in the case of highly porous materials. If one assumes a relative refractive index of the particle of 1.01, the Mie theory predicts approximately the same forward scattering intensity as Fraunhofer's diffraction model for 20 μm particles, while for 1 μm particles Fraunhofer's model gives a 300-fold higher intensity than Mie. This means that, according to Fraunhofer's model, 1 μm particles cause considerably less measured diffraction intensity than really exists. Hence, laser diffraction spectroscopy dramatically overestimates the large agglomerates (and probably ignores all aggregates below 0.5 μm, because of their uniform scattering profile).

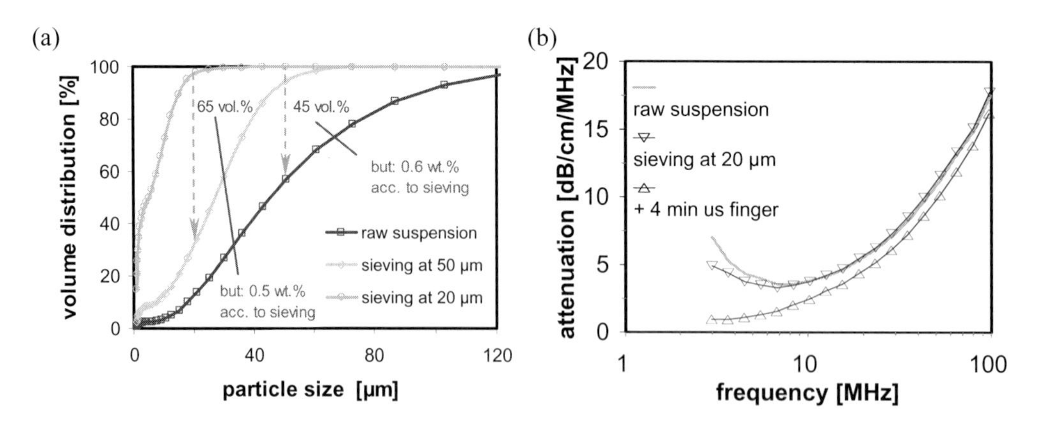

Fig. 2. Effect of filtering a fumed silica suspension through microsieves on the results of laser diffraction (a) and ultrasonic spectroscopy (b).

In contrast to laser diffraction, the removal of large agglomerates from the suspension by sieving is hardly reflected in the attenuation spectra (Fig. 2b). This does not mean that it is insensitive to agglomerates (here: weight fraction < 1 %), but that it is considerably less sensitive than laser diffraction.

Degree of Dispersion

According to the previous section, laser diffraction should reflect the disagglomeration of large agglomerates significantly better than ultrasonic spectroscopy. This is confirmed by Fig. 3 (cf. Fig. 2b). While ultrasonic spectroscopy does not show any significant change after an initial disagglomeration step, laser diffraction still measures significant size reduction. Besides, we found that, additionally to size reduction, the optical concentration is dramatically reduced when disagglomeration is conducted with the sample inside the laser diffractometer. Therefore we propose a slight modification for the disagglomeration control with laser diffraction.

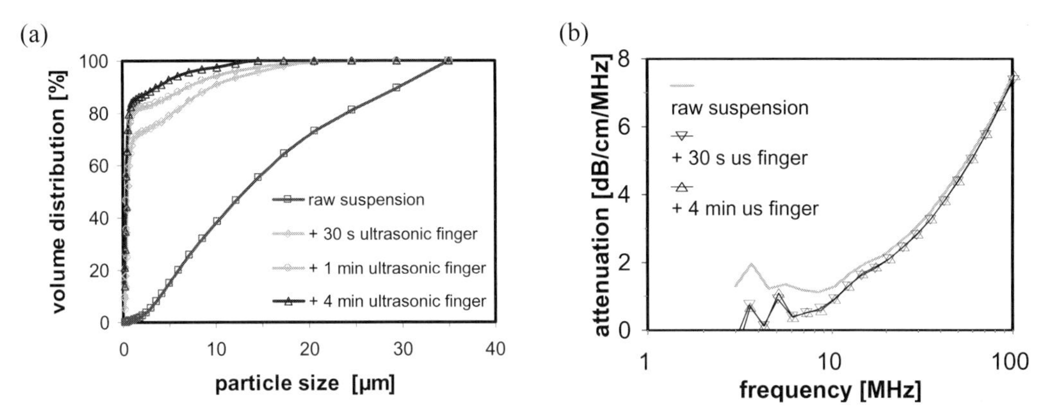

Fig. 3. Effect of disagglomeration of a fumed silica suspension on the results of laser diffraction (a) and ultrasonic spectroscopy (b).

First, a defined amount of compact, well-stabilized particles ($x \approx 100$ µm) are measured with the laser diffractometer. Afterwards, a defined amount of the suspension in question is added to the reference system. The amount of agglomerates is reflected by the mixing ratio found by the analysis. In the case of compact particles, the mixing ratio will not be affected by disagglomeration, but for fumed powders it will. The method is reproducible and detects quality differences between particles.

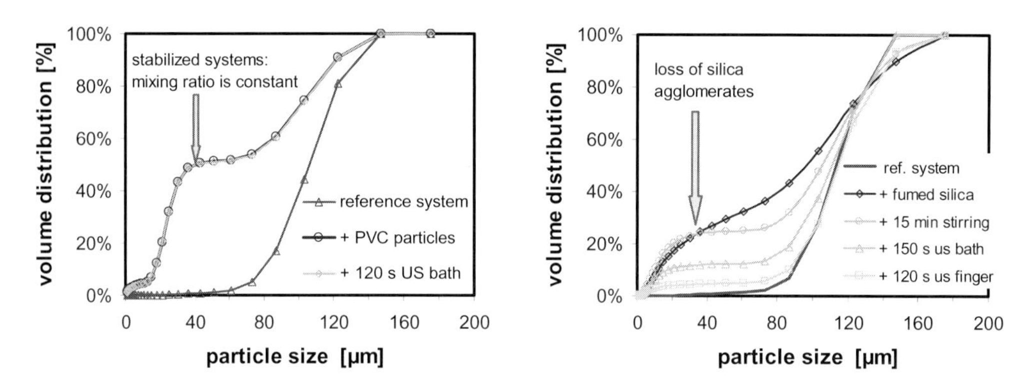

Fig. 4. Laser diffraction results of disagglomeration processes after adding compact particles to a stabilized dispersed system (a) and after adding fumed silica suspension (b).

Detection of Coarse Particles

For some applications the product quality is spoiled less by coarse agglomerates than by coarse compact particles. Again, laser diffraction overestimates these particles (Fig. 5).

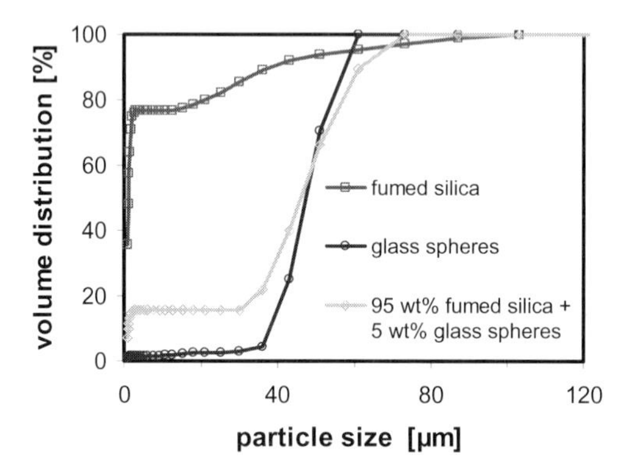

Fig. 5. Sensitivity of laser diffraction to coarse, compact particles (40 µm to 50 µm).

However, laser diffraction can only be used as an off-line measurement technique, as it requires relatively low particle concentrations. In contrast to it, ultrasonic spectroscopy can be applied as an on-line technique, but the sensitivity towards coarse particles is not as high as for laser diffraction (Fig. 6).

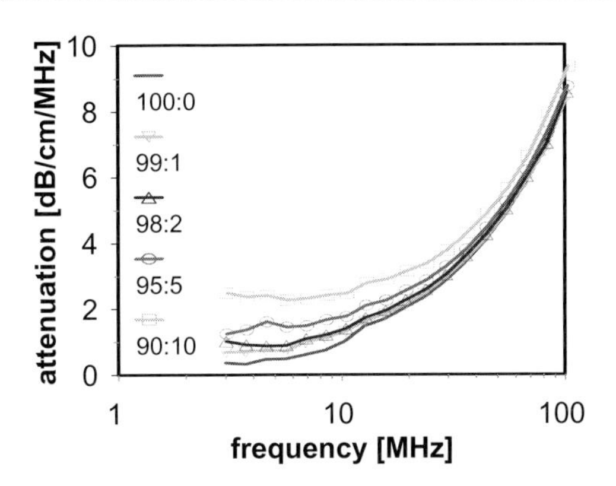

Fig. 6. Sensitivity of ultrasonic spectroscopy to coarse, compact particles (1.5 μm).

References

[1] H. Barthel, M. Heinemann, M. Stintz and B. Wessely, *Part. Part. Syst. Charact.* **1999**, 169 – 176.

[2] M. Stintz, H. Barthel, M. Heinemann, S. Ripperger, in *World Congress on Particle Technology* 4, 21 – 25.07.**2002**, Sydney, Proceedings, Int. Number 298.

[3] F. Babick, F. Hinze, M. Stintz, S. Ripperger, *Part. Part. Syst. Charact.* **1998**, 230 – 236.

Adsorption of Water on Fumed Silica

E. Brendlé, F. Ozil, H. Balard

IGCLab SARL, Carreau Rodolphe, 68840, Pulversheim, France
Tel.: +33 389480042 — Fax: 33 389480963
E-mail: contact@igclab.com

H. Barthel*

Wacker-Chemie GmbH, Werk Burghausen, D-84489, Burghausen, Germany
Tel.: +49 8677 834144 — Fax: +49 8677 833644
E-mail: herbert.barthel@wacker.com

Keywords: pyrogenic silica, water adsorption, inverse gas chromatography, hydrophilic surface

Summary: The specific surface areas of a series of silica samples were calculated starting from water adsorption isotherms and using the BET method. The seven examined pyrogenic silica samples exhibit surface area values, determined using water, that are much lower than, but proportional to, those determined using nitrogen.

Introduction

Water is present in the atmosphere, interacting with its environment and adsorbing on or condensing on most surfaces. Water adsorption on silica significantly changes its surface properties, depending on the amount of adsorbed water. Hence, from academic but also from industrial points of view, the knowledge of the amount of adsorbed water, in relation with the partial water pressure (adsorption isotherms), is of importance.

Several methods allowing the determination of adsorption isotherms exist. The most "economical" one is the use of saturated salt solutions, generating known water partial pressures, followed by the measurement of the silica weight uptake. However, the time needed to obtain a suitable adsorption isotherm is excessively long (several weeks), and experimental precautions need to be taken to achieve meaningful results. Specialized equipments, quite expensive ones, have been developed for the water adsorption isotherm determination, based on highly sensitive microbalances.

Inverse Gas Chromatography at finite concentration conditions (IGC-FC) offers another possibility to perform such determinations. Furthermore, IGC readily provides the data required for the calculation of adsorption energy distribution functions. The aim of the present study was to compare the water adsorption behavior of seven pyrogenic silica samples produced by Wacker Chemie.

Inverse Gas Chromatography at Finite Concentration (IGC-FC)

A large part of the adsorption isotherm may easily be computed from one single GC peak [1] in the way illustrated in Fig. 1, for relative pressures (P/P_0) ranging from near zero to 0.3. (Above this relative pressure domain, point-by-point determination of the isotherm is required: each point needing an additional probe injection).

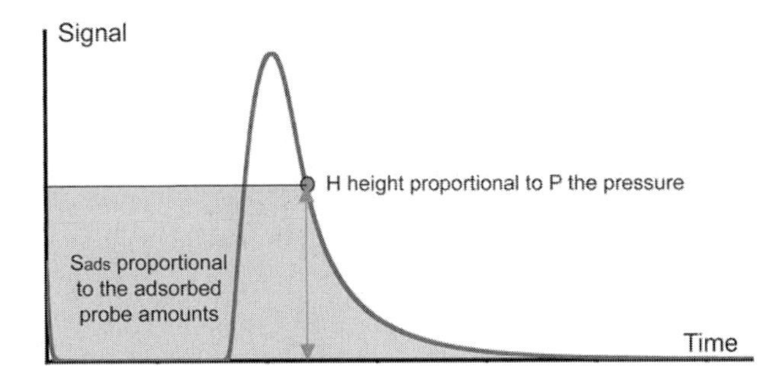

Fig. 1. Principle of the determination of a point of the adsorption isotherm at relative pressures below 0.3.

From the isotherms, the following information is obtained:

- The specific surface area of silica samples: the part of the isotherm corresponding to relative pressures between 0.05 and 0.3 allows the evaluation of the specific surface area and of a constant (C_{BET}) that is indicative of the adsorption energy.
- The energetic heterogeneity of silica surfaces

From an energetic point of view, the surface of silica is heterogeneous: the surface is best described by assuming domains corresponding to discrete values of the adsorption energy (defining "local" isotherms). The sum of the local isotherms reconstructs the observed isotherm.

A major advantage of the determination of adsorption isotherms by IGC is the fact that the first derivative of the isotherm is readily calculable by application of the following relationship:

$$\left(\frac{\partial N}{\partial P}\right)_{L,\, t_R} = \frac{1}{RT}\frac{D_c}{m}(t_R - t_0)$$

where : L is the length of the column, N is the number of molecules at a given point on the GC peak, P is the adsorption pressure, t_R is the injected probe retention time, and t_0, the retention time of a poorly adsorbing solute such as air, which allows us to calculate the dead volume. Dc is the corrected gas flow and m the mass of solid inside the column.

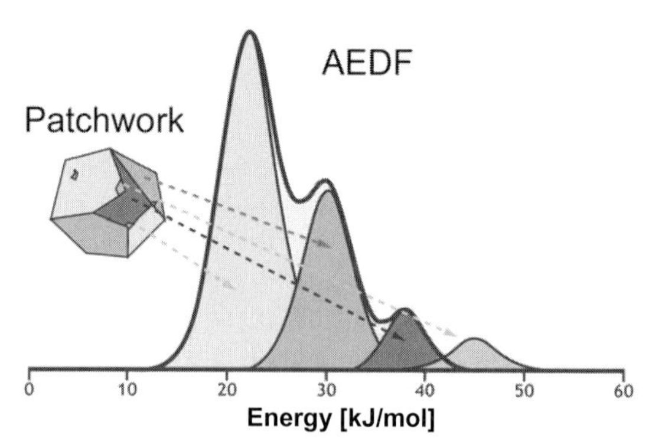

Fig. 2. AEDF: Number of adsorption sites vs adsorption energy. (Patchwork distribution of surface sites).

From the values of the derivatives of the isotherm, one gets access to the adsorption energy distribution functions (AEDF) that differentiate the various adsorption sites families (each family being characterized by a mean value of the adsorption energy). Figure 2 schematically illustrates this possibly in the case of a solid surface having 4 families of adsorption sites, in a patchwork distribution. Such patterns are typical of an energetic heterogeneous surface.

Experimental

The characteristics of HDK pyrogenic silica samples, from Wacker-Chemie, are given in Table 1. Inverse gas chromatography at finite concentration conditions requires samples acting like a chromatographic phase. Hence, the very fine silica samples were first agglomerated in order to obtain particles sizes between 250 and 400 μm. Those materials were then used to fill 30 cm long stainless columns of 2 mm internal diameter. Before measurements, these columns were outgassed one night at 140 °C under a helium carrier gas flow. The measurements were carried out at 40 °C using a Shimadzu GC-8A equipment fitted with a TCD detector. The water probe molecules were injected using micro-syringes.

Table 1. Denomination and surface area of the silica samples.

Name	Code	$S_{BET}N_2$ [m²/g]
HDK D05	D05	50
HDK C10P	C10P	98
HDK S13	S13	130
HDK V15	V15	150
HDK N20	N20	190
HDK T30	T30	300
HDK T40	T40	400

Results

The peaks resulting from the water injections are exploited using the ECP procedure (Elution at the Characteristic Point) [1] for the computation of the isotherms. The observed curves are made up of two domains:

- $0 < P/P_0 < (0.15 \text{ or } 0.25)$
- $(0.15 \text{ or } 0.25) < P/P_0 < 0.85$

The first domain is used to determine the specific surface area ($S_{BET}(H_2O)$) and the BET constant values. Moreover, the large number of available experimental points allows the calculation of the corresponding adsorption energy distribution function of water on silica.

Figure 3 displays the obtained isotherms. From their shape, these isotherms are comparable to those displayed in Gun'ko's paper [2]. The results are shown in Table 2. As expected, the water uptake increases with increasing silica surface areas $S_{BET}(N_2)$. The computation of the "water weight ratio" (W%) gives values between 0.9 and 3.0% of water per gram of silica These values are in good agreement with the Cw, 105(wt%) given in Gun'ko's paper [2].

The BET constant values ($C_{BET}(H_2O)$) are between 3.5 and 5, indicating that the isotherms are type II isotherms of the BET classification. Yet these values are rather low, indicating that the water molecules do not interact strongly with the silica surface.

The other interesting observation concerns the surface area determined using the H_2O monolayer capacity. We notice that the surface area values are far away from the surface area values determined from the nitrogen adsorption measurements.

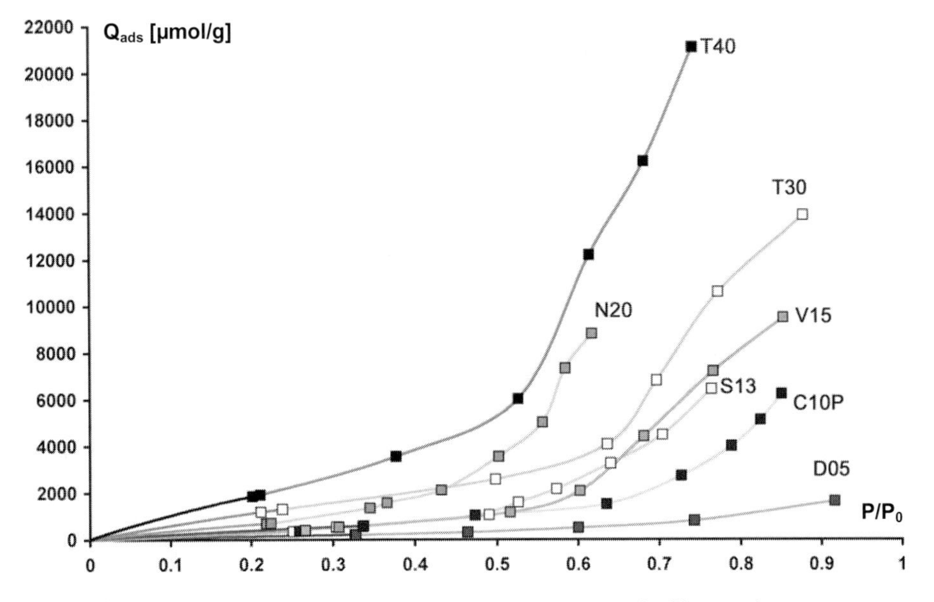

Fig. 3. Water adsorption isotherms determined, at 40 °C, on Wacker's pyrogenic silica samples.

Table 2. Results of the exploitation of the water desorption isotherms obtained at 40 °C.

SAMPLES	H_2O monolayer [µmole/g]	S_{BET} (H_2O)* [m^2/g]	C_{BET} (H_2O) [n.u.]	Surf. Ratio [%]	*W*% [%]
HDK D05	227 ± 6	16 ± 1	5.1 ± 0.4	32	0.4
HDK C10P	500 ± 15	36 ± 1	4.2 ± 0.1	37	0.9
HDK S13	588 ± 13	42 ± 1	4.7 ± 0.1	32	1.1
HDK V15	586 ± 22	42 ± 1	4.5 ± 0.2	28	1.1
HDK N20	941 ± 22	68 ± 3	5.0 ± 0.2	36	1.7
HDK T30	1650 ± 139	118 ± 10	3.8 ± 0.6	39	3.0
HDK T40	2703 ± 163	194 ± 11	4.6 ± 0.2	49	4.9

* Assuming a water cross sectional area of 11.9 Å^2.

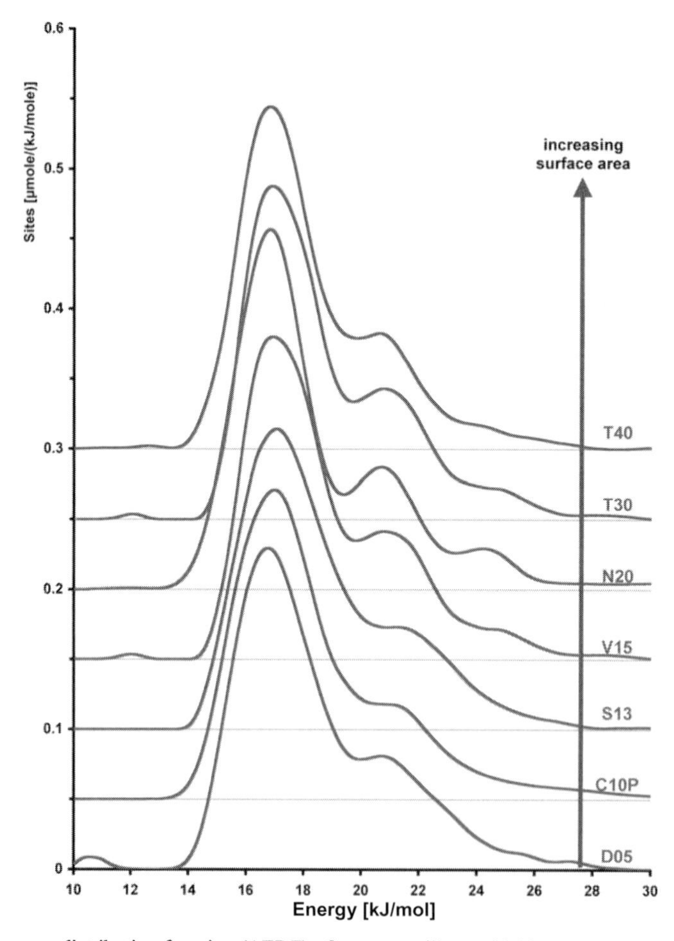

Fig. 4. Adsorption energy distribution function (AEDF) of water on silica at 40 °C.

Most of the surface area values determined with water correspond to 30 – 40% of the surface area values calculated from the nitrogen adsorption measurements. Furthermore, the evolution of this ratio seems to be related to the nitrogen S_{BET} value of silica. Indeed for the samples having $S_{BET}(N_2)$ values equal or higher than 200 m^2/g, the water measured surface area corresponds to 40 – 50% of the latter.

The last point concerns the water adsorption energy distribution functions that give a "fingerprint" of the surface energetic heterogeneity of the examined solids [3]. The adsorption energy distribution functions (AEDF) of water on the silica are displayed in Fig. 4.

We note that the AEDF gives comparable shapes for all examined solids. They are composed by two components: the first, located around 17 kJ/mole, is the most important. The second is situated around 21 – 22 kJ/mol.

Discussion

The results of the present study give an interesting picture of the behavior of pyrogenic silicas toward water.

Indeed, strong interactions were expected between water and the surface silanol groups considering the ability of water and silanol to form hydrogen bonds: water "irreversible" physical adsorption should occur. After verification, this seems not to be the case.

Davydov [4] describes similar behavior on Aerosil. He specifies that all physically adsorbed water may be removed by vacuum evacuation even at ambient temperature. This confirms that the water adsorption on silica is not as strong as initially supposed. This is also corroborated by the low BET constant values.

The other surprising results are the low surface area values computed from the water adsorption measurements. These may look surprising, but our results are in agreement with those of Gun'ko [2] and Davydov [4]. In our opinion, these results show the silica's surface heterogeneity from a "hydrophilic-hydrophobic" point of view.

According to Davydov "*Thus surfaces of silica carrying mainly siloxane bonds may be considered as "hydrophobic". Such a surface is unstable and, in contact with water vapor, is gradually hydrolyzed. But it is important that siloxane bonds existing on hydroxylated silica surfaces may be considered as hydrophobic parts of hydroxylated silica surfaces*". Hence, water adsorption is supposed to occur mainly on the hydroxylated silica surface part. Thus, at a first approximation, we may consider that the surface area determined from water adsorption isotherms corresponds only to the hydrophilic or hydroxylated part of the silica surface.

It is also noteworthy that the hydrophilic part of the silica surfaces remains quasi constant (30 – 40%) for surface area values lower than 200 m^2/g. The hydrophilic part of the surface increases as the surface area values do exceed 200 m^2/g.

Conclusion

This study demonstrates the ability of IGC, at finite concentration conditions, to determine quickly (within one or two days depending on the desired accuracy) water adsorption isotherms, with relative pressures ranging from 0 to 0.85. Moreover, IGC provides isotherms made up of several hundreds of experimental points. This permits the computation of meaningful adsorption energy distribution functions.

Furthermore, our results indicate that water interacts less strongly with the silica surface than postulated. Moreover, only part of the silica surface may be considered as hydrophilic.

References

[1] J. R. Conder, C. L. Young, *Physico-Chemical Measurement by Gas Chromatography*, Wiley, New York, **1979**.

[2] V. M. Gun'ko, I. F. Mironyuk, V. I. Zarko, V. V. Turov, E. F. Voronin, E. M. Pakhlov, E. V. Goncharuk, R. Leboda, J. Skubiszewska-Zieba, W. Janusz, S. Chibowski, Yu. N. Levchuk, A. V. Klyueva, *J. Colloid Interface Sc.* **2001**, *242*, 90.

[3] H. Balard, E. Brendlé, *Kautsch. Gummi Kunstst.* **2002**, *55*, 464.

[4] V. Ya. Davydov, Adsorption on silica surfaces, in *Adsorption on Silica Surfaces* (Ed. E. Papirer), Marcel Dekker, Inc. New York, **2000**, p. 63 – 118.

Methylene Chloride Adsorption on Pyrogenic Silica Surfaces

E. Brendlé, F. Ozil, H. Balard*

IGCLab SARL, Carreau Rodolphe, 68840, Pulversheim, France
Tel.: +33 389480042 — Fax: 33 389480963
E-mail: contact@igclab.com

H. Barthel

Wacker-Chemie GmbH, Werk Burghausen, D-84489, Burghausen, Germany
Tel.: +49 8677 834144 — Fax: +49 8677 833644
E-mail: herbert.barthel@wacker.com

Keywords: pyrogenic silica, methylene chloride adsorption, Inverse Gas Chromatography, hydrophilic surface

Summary: The specific surface areas of a series of silica samples were calculated starting from CH_2Cl_2 adsorption isotherms and using the BET method. The seven examined pyrogenic silica samples present surface area values, determined using methylene chloride, that are lower than, but proportional to, those determined using nitrogen. It appears that the sum $S_{BET}(CH_2Cl_2) + S_{BET}(H_2O)$ is almost equal to $S_{BET}(N_2)$.

Introduction

The presence of water on solid surfaces and in particular on the silica surface is at the origin of known problems. A related concern is the determination of the hydrophilicity/hydrophobicity character of silica surfaces.

Water adsorption studies are obviously indicated for the evaluation of the hydrophilic character of silica. Besides, according to our earlier studies, methylene chloride (CH_2Cl_2), used as a probe for Inverse Gas Chromatography (IGC) measurements at finite concentration conditions, appears to be an alternative choice for determining the hydrophobicity of silica surfaces. Indeed, earlier indications suggest that this probe, when used under appropriate conditions, is not interacting with the hydrophilic silanol groups. The present work is complementary to the study of water adsorption on fumed silica samples, described in this book [1].

Inverse Gas Chromatography at Finite Concentration (IGC-FC)

Relevant information may be obtained by determining CH_2Cl_2 adsorption isotherms using inverse gas chromatography at finite concentration conditions. The adsorption isotherms, in fact desorption isotherms, are readily acquired when applying IGC. The principle of this procedure is given in this book [1].

Experimental

The characteristics of HDK pyrogenic silica samples, from Wacker-Chemie, are given in Table 1.

The measurements were carried out at 30 °C using a Fisons HRGC Mega 2 8533 apparatus fitted with dual FID detectors. The methylene chloride probe molecules were injected using micro syringes.

Table 1. Denomination and surface area of the silica samples.

Name	Code	$S_{BET}N_2$ [m²/g]
HDK D05	D05	50
HDK C10P	C10P	98
HDK S13	S13	130
HDK V15	V15	150
HDK N20	N20	190
HDK T30	T30	300
HDK T40	T40	400

Table 2. Applied treatments and conditioning conditions.

Code	Temperature	Carrier gas	Measurements
D030	30 °C	0% RH	30 °C, 0% RH
D100	100 °C	0% RH	30 °C, 0% RH
D160	160 °C	0% RH	30 °C, 0% RH
D300	300 °C	0% RH	30 °C, 0% RH
D400	400 °C	0% RH	30 °C, 0% RH
RH45	30 °C	45% RH	30 °C, 45% RH
RH56	30 °C	56% RH	30 °C, 56% RH
RH75	30 °C	75% RH	30 °C, 75% RH

In order to compare the effect of the surface hydroxyl groups and water contents on the methylene chloride adsorption process, various conditioning conditions were applied. The latter are summarized in Table 2. Of course, the D_{xxx} conditioning conditions were chosen in order to modify the surface hydroxyl group concentration of silica, whereas the conditioning under a controlled humidity gas stream was selected in order to modify the surface water contents.

The desired relative humidity value is obtained by passing the carrier gas through a saturator. The desired relative water pressure is achieved by controlling both the temperature of the chromatograph oven and that of the saturator.

Results

Using Dry Carrier Gas

The samples were conditioned following the D160 protocol (conditioning at 160 °C, under dry carrier gas stream). The methylene chloride adsorption isotherms were determined on each sample and used to compute the BET surface area ($S_{BET}(CH_2Cl_2)$), the corresponding BET constant (C_{BET}), and the adsorption energy distribution functions (AEDF).

The results obtained from the adsorption isotherms exploitation are shown in Table 3. These results correspond to the mean value and relative variation determined starting from at least 3 chromatograms.

Table 3. S_{BET} and C_{BET} values determined from CH_2Cl_2 adsorption isotherms (outgassing at 160 °C).

Samples	Monolayer [μmol/g]	$S_{BET}(CH_2Cl_2)$ [m²/g]	Δ [%]	$C_{BET}(CH_2Cl_2)$ [n.u]
HDK D05	169 ± 5	31 + 1	61	5.6 ± 0.2
HDK C10P	362 ± 4	65 ± 1	66	6.6 ± 0.1
HDK S13	453 ± 14	82 ± 2	63	6.9 ± 0.2
HDK V15	496 ± 9	90 ± 2	60	6.8 ± 0.2
HDK N20	936 ± 36	169 ± 1	89	6.1 ± 0.1
HDK T30	1219 ± 39	220 ± 7	73	8.5 ± 0.4
HDK T40	1282 ± 27	232 ± 5	61	10.9 ± 0.2

Δ: Ratio between surface area values measured with N_2 and CH_2Cl_2.

The C_{BET} values corresponding to the methylene chloride adsorption are between 5.6 and 10.9, demonstrating a relatively low interactivity of those probe molecules with the silica surfaces. However, they are large enough to make them correspond to typical type II isotherms, allowing the application of the BET method. The general tendency observed is a slight increase of the BET constant values with increasing surface areas.

The $S_{BET}(CH_2Cl_2)$ values were computed from the monolayer capacity assuming a cross-sectional area of 30 Å² for the adsorbed methylene chloride molecule. When comparing the S_{BET}

values determined using CH_2Cl_2 to those obtained with nitrogen at 77K, we notice that those obtained using methylene chloride are systematically lower. Moreover, a constant relative variation is observed (60 – 70%, mean value of 67%) between the present values and surface areas computed from the nitrogen adsorption data.

The surface energetic heterogeneity determination constitutes an additional aspect of the present study. This was performed by means of the methylene chloride adsorption energy distribution functions (AEDF) computation, relating the number of interactive surface sites to the desorption energy of each individual site. The latter are displayed in Fig. 1.

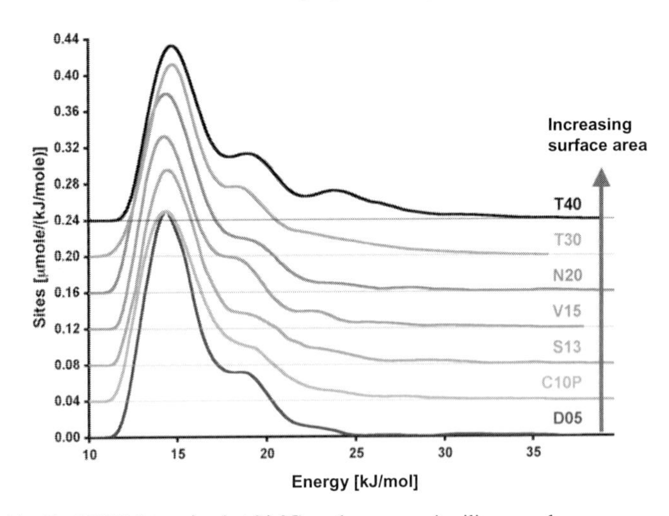

Fig. 1. Methylene chloride AEDF determined at 30 °C on the pyrogenic silica samples.

The methylene chloride AEDF value, determined at 30 °C on silica samples, are very similar. They present two components or site populations: the first is located around 14.7 kJ/mol and the second around 19 kJ/mol. Moreover, the number of high-energy adsorption sites increases with the S_{BET} (N_2) values. Again, this observation is consistent with the observed C_{BET} value increase.

Using Heat Treatments of Silica Samples

The number of silica hydroxyl groups decreases after heat treatments. Hence, the study of the influence of such treatment on the $S_{BET}(CH_2Cl_2)$ values and on the corresponding AEDF is of interest.

Those investigations were carried out on HDK C10P. The different applied outgassing conditions between 30 and 300 °C, described in Table 2, are identified as: D030, D100, D160, D300, and D400. The results obtained from the adsorption isotherms exploitation are gathered in Table 4. A slight decrease, with the applied heat treatment temperature, of the $S_{BET}(CH_2Cl_2)$ values (BET surface areas determined with methylene chloride), is observed. The C_{BET} values show the same trend.

Table 4. S_{BET} and C_{BET} values determined from CH_2Cl_2 adsorption isotherms on HDK C10P: effect of outgassing temperature.

Outgassing	Monolayer [µmol/g]	$S_{BET}(CH_2Cl_2)$ [m^2/g]	Δ [%]	C_{BET} (CH_2Cl_2) [n.u]
D030	408 ± 9	74 ± 2	76	7.7 ± 0.3
D100	401 ± 7	72 ± 1	73	7.4 ± 0.1
D160	362 ± 4	65 ± 1	66	6.6 ± 0.1
D300	377 ± 6	67 ± 1	68	6.7 ± 0.1
D400	353 ± 5	64 ± 1	65	5.4 ± 0.1

Δ: Ratio between surface area values measured with N_2 and CH_2Cl_2.

The examination of the AEDF curves of Fig. 2 points out that the location of the highest adsorption energy peak follows the same evolution as the C_{BET} values. Indeed, the latter location is slightly shifted toward the lower energies with increasing conditioning temperatures. Besides, the AEDF remain very similar to those obtained after conditioning at 160 °C as described above.

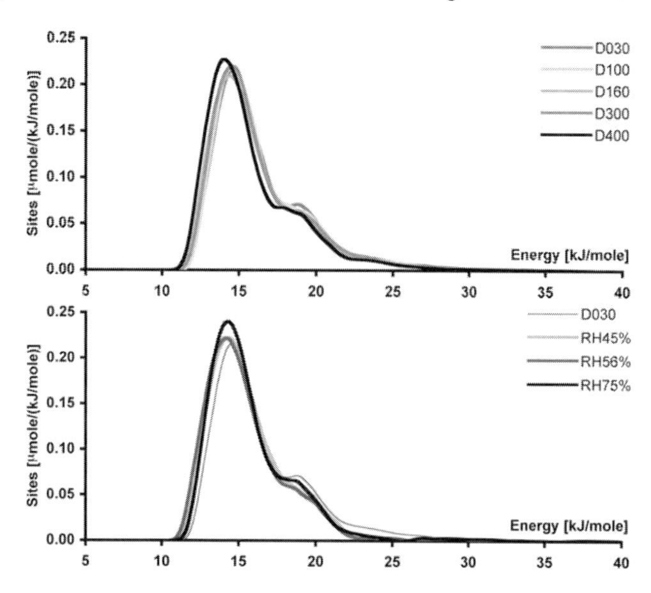

Fig. 2. Methylene chloride AEDF on HDK C10P: effect of outgassing temperature and relative humidity.

Using Wet Carrier Gas

The influence of physically absorbed water molecules on the methylene chloride adsorption and interaction capacity of silica will now be investigated. We generated, at 30 °C, three carrier gases having controlled percentages of moisture (RH%). The silica samples were equilibrated for several hours with these wet gases in order to achieve a constant yield of physically absorbed water. The methylene chloride IGC-FC measurements were carried out maintaining these conditions.

The methylene chloride adsorption isotherms were determined at 30 °C for each selected relative humidity. The BET surface area ($S_{BET}(CH_2Cl_2)$), the corresponding BET constant (C_{BET}), and the adsorption energy distribution functions (AEDF) were then computed. The results obtained from the evaluation of the adsorption isotherms are shown in Table 5.

Table 5. S_{BET} and C_{BET} values determined from CH_2Cl_2 adsorption isotherms on HDK C10P: effect of relative humidity.

RH%	Monolayer [μmol/g]	$S_{BET}(CH_2Cl_2)$ [m^2/g]	Δ [%]	C_{BET} (CH_2Cl_2) [n.u]
D030	408 ± 9	74 ± 2	76	7.7 ± 0.3
RH45	378 ± 14	68 ± 3	77	5.3 ± 0.1
RH56	378 ± 6	68 ± 1	71	4.8 ± 0.1
RH75	315 ± 4	57 ± 1	72	5.1 ± 0.1

Δ: Ratio between surface area values measured with N_2 and CH_2Cl_2.

The results of Table 5 indicate that the surface areas of the silica samples $S_{BET}(CH_2Cl_2)$ decrease when the relative humidity increases, i.e., when the amount of physically adsorbed water increases. On the other hand, the BET constant values also decreases with increasing adsorbed water amounts. However, this decrease is not smooth like the one observed when the samples are heat-treated. Another interesting observation concerns the fact that in spite of the presence of water, the BET constant values remain above 4.5. In other words, the isotherms are still of type II.

Figure 2 shows the effects of the adsorbed water amounts on the methylene chloride AEDF. Again the shape of these functions remains unchanged. Finally, as mentioned before, the evolution of C_{BET} values and those of the AEDF principal peak location seem to be related.

Discussion

In the light of these experiments, how may we answer the following questions:

- Why is $S_{BET}(CH_2Cl_2) < S_{BET}(N_2)$?
- Is this indicative of particular silica surface characteristics?

An answer to the first question would be to suppose a wrong choice of the cross-sectional area of the methylene chloride molecule. In the case of nitrogen, a standard molecule for the determination of the specific surface area of solids, various values were also postulated (16.2, 12.9 or 17.7Å^2) depending on the nature of the adsorbent and the orientation of the adsorbed molecule.

This may also be the case for CH_2Cl_2. A unique value of the surface area occupied by the adsorbed molecule is not found in the literature. Most often, a value close to 30 Å^2 is recommended. However, underestimated methylene chloride surface areas are also reported for other materials, for example, for carbon blacks and microporous carbons [2]. The authors do not give any sound explanation, but point out the need of further experimentation in order to reach a better understanding of the adsorption mechanism of that molecule.

Why such a phenomena?

Firstly, taking into account the polarity and the hydrophobicity of dichloromethane, we may assume that dichloromethane does not interact with the surface hydroxyl groups. Hence, the $S_{BET}(CH_2Cl_2)$ surface area value should only correspond to the "hydrophobic" fraction of the silica surface. Hence, the $S_{BET}(H_2O)$ value should only correspond to the "hydrophilic" fraction of the surface.

Consequently, by sum of the ($S_{BET}(CH_2Cl_2)$ and $S_{BET}(H_2O)$ values should be close to the $S_{BET}(N_2)$ value. This could actually be verified.

Conclusion

The results of this study look very promising: in particular, it now seems possible, by IGC, to readily determine the hydrophobic and hydrophilic fractions of silica surfaces.

References

[1] E. Brendlé, F. Ozil, H. Balard, H. Barthel, *Adsorption of water on fumed silica*, in *Organosilicon Chemistry VI — From Molecules to Materials*, **2003**.

[2] P. J. M. Carrott, M. M. L. Ribeiro Carrott, I. P. P. Cansado, *Carbon* **2001**, *39*, 465.

Pyrogenic Silica — Mechanisms of Rheology Control in Polar Resins

Torsten Gottschalk-Gaudig, Herbert Barthel*

Wacker-Chemie GmbH, 84480 Burghausen, Germany

Tel.: +49 8677 83 2648 — Fax: +49 8677 83 2536

E-mail: torsten.gottschalk-gaudig@wacker.com

Keywords: fumed silica, particle interaction, rheology

Summary: Fumed silica is a widely used industrial rheology control additive for composites, adhesives, sealants, and coatings. Despite the broad range of applications, the mechanisms which control the rheological properties of fumed silica dispersions are not fully understood. By applying different rheological techniques and theoretical calculations it could be shown that for hydrophilic and hydrophobic silica grades different colloidal interaction forces are active. The influence of these interparticulate interactions on the rheology of fumed silica dispersions in reactive resins is also discussed.

Introduction

Pyrogenic silica is an amorphous synthetic colloidal silicon dioxide formed in a hydrothermal reaction of chlorosilanes in an oxygen-hydrogen flame and consists of nano-sized primary particles which are chemically fused to sinter aggregates with a mean hydrodynamic equivalent diameter of approx. 200 – 500 nm [1]. Strong van der Waals forces between the aggregates finally result in the formation of larger space-filling structures such as micrometer-sized agglomerates or flocs. The colloidal nature of pyrogenic silica is demonstrated by a large surface area of 50 – 450 m^2/g. The surface is smooth without micropores and covered by silanol groups which can be chemically modified in order to render the initially hydrophilic silica hydrophobic.

Pyrogenic silica is commonly used in industry as a powerful rheological additive to impart shear thinning, thixotropy, or yield points to liquid media such as paints or adhesives. Several intrinsic properties of pyrogenic silica have been identified to be responsible for its outstanding performance as rheological additive: (a) a space-filling fluffy structure which is able to immobilize large quantities of liquids. Using the approach of fractality to describe these structures a mass fractal dimension $d_f < 2.7$ has been determined for pyrogenic silica [2]; (b) a non-porous smooth surface of the primary particles resulting in a large specific surface area which provides the basis for effective particle-particle and particle-matrix interactions, giving rise to the formation of clusters or a percolating network [3]

However, additional factors must be considered in order to understand and finally predict the rheological behavior of silica-resin mixtures. In particular, the type and strength of particle interactions and the influence of adsorption processes on these interactions are key factors which govern the rheology of liquid media containing pyrogenic silica for rheology control.

In general the rheological behavior of silica dispersions can be described according to Fig. 1.

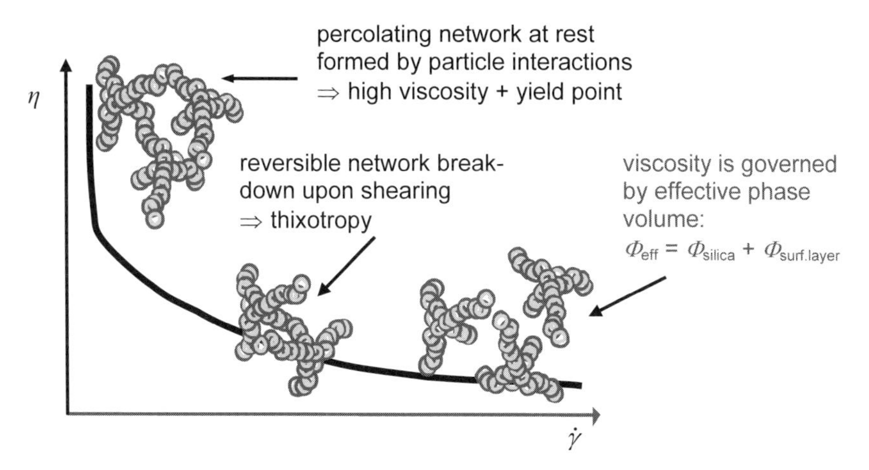

Fig. 1. Relationship between shear rate, network structure, and viscosity.

Silica dispersions exhibit a gel structure at rest, provided that there is a sufficiently high silica volume fraction. Upon shearing, this structure is degraded with increasing shear stress / shear rate until at very high shear stresses the viscosity reaches a plateau. This high-shear end viscosity is solely governed by hydrodynamic interactions between the fluid medium and the silica particles, with an effective phase volume $\Phi_{eff} = \Phi_{silica} + \Phi_{surf.layer}$. This process of structure degradation combined with a decrease in viscosity is reversible, giving rise to thixotropy [4].

It is evident that for a profound understanding of the rheological properties of fumed silica dispersion at rest and under shear conditions, detailed information about the nature and strength of the interparticulate interaction forces is indispensable.

The aim of this paper is to demonstrate that, depending on the type of silica (hydrophilic or hydrophobic), different kinds of colloidal interaction forces are effective, to show how these forces are influenced by the composition of the liquid matrix, and finally to elucidate the influence of the colloidal forces on the rheological properties of silica-resin dispersions. For this purpose, experimental results obtained from steady-shear and dynamic rheological experiments together with theoretical calculations will be presented.

Experimental Details

Two grades of fumed silica with different degrees of surface treatment and different polarity have been studied: a non-treated, hydrophilic Wacker HDK® N20 (BET surface area 200 m^2/g, 1.8 $SiOH/nm^2$ equivalent to 100% residual SiOH) and a fully silylated hydrophobic Wacker HDK® H18 (carbon content 4.5%, 15% residual SiOH), the latter being covered by a chemically grafted PDMS layer. Both silicas are products from Wacker-Chemie GmbH, Germany. The resins used in this study are an unsaturated polyester resin (UP resin), a co-condensate of a diol, maleic acid, and orthophthalic acid, and a vinyl ester resin (VE resin), a co-condensate of glycidine, methacrylic acid, and bisphenol A bearing pendant OH groups. Both resins have a styrene content of 35% and were provided by DSM, NL. Additionally, the rheology of both silica types in pure styrene was studied.

For the evaluation of the rheology of the silica dispersions, different test methods were applied: (a) a shear rate-controlled relaxation experiment at $\dot{\gamma}$ = 0.5 s^{-1} (conditioning), 500 s^{-1} (shear thinning), and 0.5 s^{-1} (relaxation) to evaluate the apparent viscosity, the relaxation behavior, and thixotropy; (b) shear yield-stress measurements using a vane technique introduced by Nguyen and Boger [5]; (c) low deformation dynamic tests at a constant frequency of 1.6 s^{-1} in a stress range of ca. 0.5 – 100 Pa. All samples contained 3 wt% of fumed silica.

Results and Discussion

Commonly, the thickening of liquids by hydrophilic silica is explained by the formation of H-bonds between the silanol groups of silica particles [6]. According to this model, the stability of silica gels in styrene and toluene, two fluids with comparable dielectrical properties, should be more or less identical. Figure 2 depicts the shear yield-stress experiments using the vane geometry of HDK N20 in styrene and toluene.

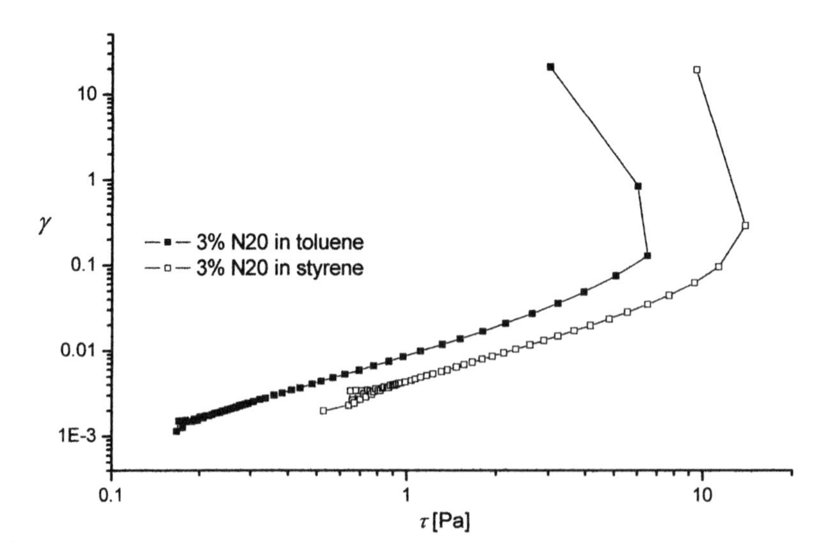

Fig. 2. Shear yield-stress experiments using the vane geometry of HDK N20 in styrene and toluene.

The critical shear stress, where the gel structure of the dispersion collapses, is for the system N20/styrene significantly higher than that for an N20/toluene dispersion. The critical shear stress is correlated to the strength of the particle-particle interactions. Hence, the attractive forces between silica particles in styrene are supposed to be stronger than those in toluene. However, this result cannot be explained by particle-particle interactions solely based on H-bonds. Their strength in styrene and toluene should be comparable because of the similar dielectrical properties of both liquids and should therefore result in comparable network stability.

A straightforward explanation for the differences in gel stability is that the attractive particle-particle interactions are not mainly driven by interparticulate H-bond but by van der Waals forces. This idea is supported by pair potential calculations using an approximate expression given by Israelachvili based on the Lifshitz theory of van der Waals forces [7].

The results of these calculations are shown in Fig. 3.

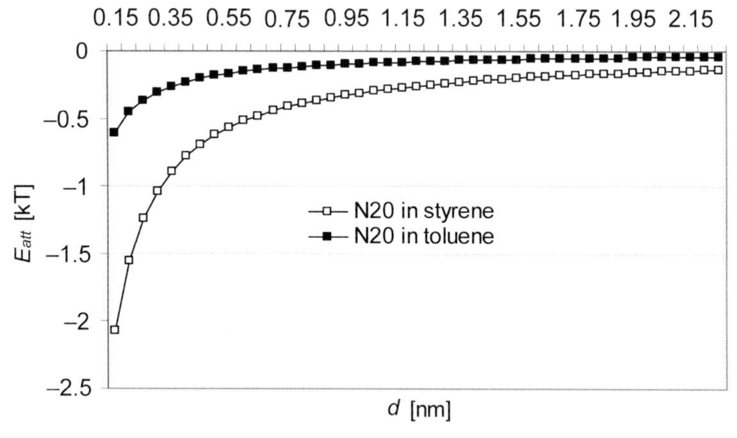

Fig. 3. Pair potential calculations of SiO_2 in styrene and toluene using an approximate expression for the Lifshitz theory of van der Waals forces.

The calculations for silica primary particles with a diameter of 12 nm reveal at least qualitatively that the attractive interaction potential E_{att} for silica dispersed in styrene is stronger by a factor of ca. 4 than that for silica dispersions in toluene. This result demonstrates that the common concept of explaining the rheological properties of hydrophilic silica dispersions by means of interparticulate H-bonds is questionable. In silica-resin mixtures also, adsorption of resin oligomers due to the high surface energy of the silica particles has to be taken into account [8]. Adsorption of polymers/oligomers will sterically hamper the formation of interparticulate H-bonds.

In Fig. 4, the influence of the styrene content on the relative viscosity of UP and VE dispersions containing hydrophilic and hydrophobic silica using a shear rate-controlled relaxation experiment is shown.

In general, the increase in the styrene content from 35 wt% to 60 wt% results in an increase in the relative viscosity of the N20 and H18 dispersions. This effect is more pronounced for dispersions containing hydrophilic silica N20 than for H18. Even more interestingly, the relaxation rate of N20 dispersions after applying high shear forces strongly depends on the styrene content. Dispersions containing low volumes of styrene exhibit a retarded relaxation, which can be in the range of several minutes up to hours for N20 dispersed in VE resin. However, addition of styrene increases the relaxation rate in such a way that the low-shear end viscosity is almost instantaneously approached after finishing the high-shear phase.

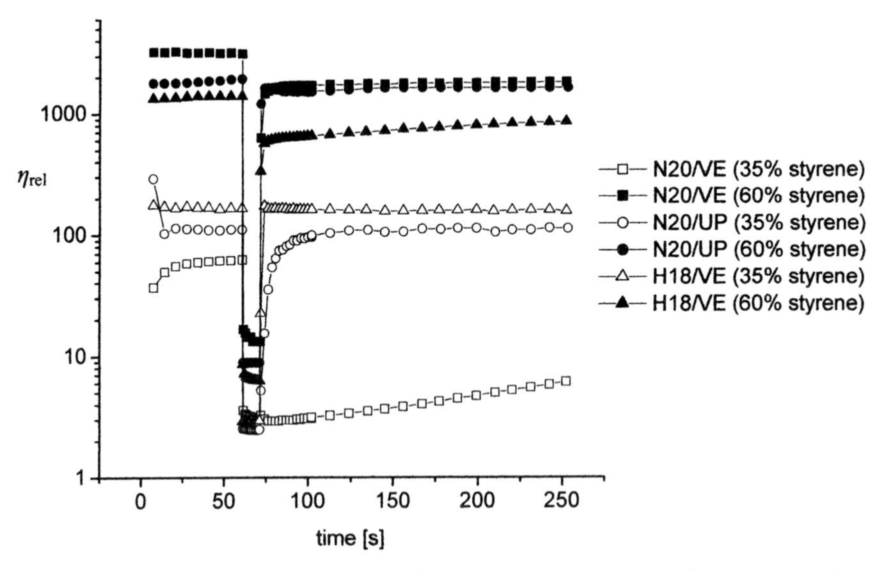

Fig. 4. Shear rate-controlled relaxation experiment at $\dot{\gamma} = 0.5\ s^{-1}$ (conditioning), 500 s^{-1} (shear thinning), and 0.5 s^{-1} (relaxation) of 3% N20 and H18 dispersed in VE and UP resins.

In order to evaluate the rest structure of the silica-resin dispersions, small-amplitude dynamic measurements have been performed. Figure 5 depicts the stress sweeps of HDK N20 and H18 dispersed in UP and VE resins, respectively.

The storage modulus of H18 dispersions in VE resin with low styrene content exhibits high plateau values in the linear viscoelastic regime with $G' > G''$ (G'' omitted for clarity). This is indicative of a viscoelastic solid with percolating gel structure [9]. However, addition of styrene results in a pronounced decrease of the storage modulus.

For N20 dispersions in UP and VE resins a different behavior is observed. N20-VE resin dispersions containing 65% of resin show liquid behavior with low storage modulus and $G' < G''$. Addition of styrene increases the storage modulus with $G' > G''$. Hence, a viscoelastic solid is obtained. N20-UP resin dispersions with a resin content of 65% already show the properties of a viscoelastic solid ($G' > G''$). However, addition of styrene only slightly increases the storage modulus.

The rheological properties of highly hydrophobic fumed silica HDK H18 as outlined in this contribution can be readily explained by treating the particles as a composite material of a SiO_2 core covered by a thin grafted PDMS shell [10]. The thickness of the PDMS layer strongly depends on the solvent quality of the liquid medium. Liquids with a good solvency for PDMS (better than THETA conditions) will cause a swelling of the PDMS layer. This results in weaker attractive particle-particle interactions because of the barrier function of the PDMS layer. Additionally, the swelling of the PDMS layer by the diffusion of solvent molecules into the layer will cause a better matching of the dielectrical and optical properties of the liquid medium and the adlayer. This results in similar Hamaker constants of the homogenous phase and the layer. Hence, according to Eq. 1, which gives an approximation for the effective Hamaker constant, the attractive net interaction force based on van der Waals interactions will be reduced [7].

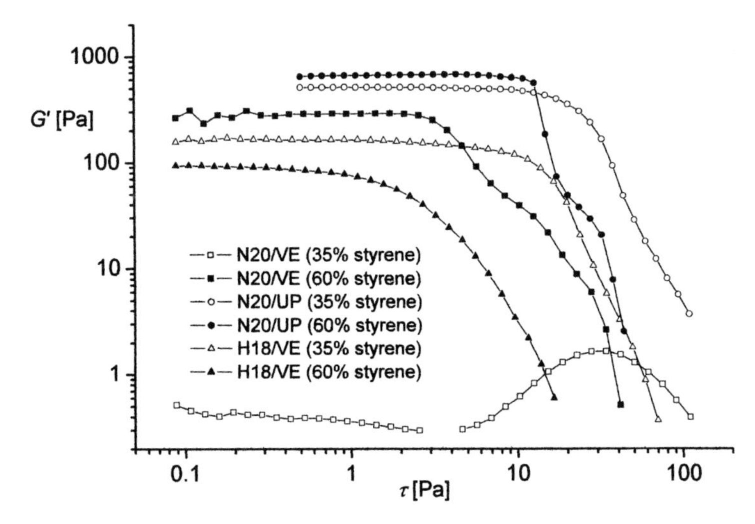

Fig. 5. Stress sweep of HDK N20 and H18 dispersed in UP and VE resins recorded at a constant frequency of $\omega = 10$ rad/s.

$$A_{131} \approx ((A_{11})^{0.5} - (A_{33})^{0.5})^2$$

Eq. 1

This means, for the system H18/VE resin, that addition of styrene, which shows reasonable solvency for PDMS, will cause swelling of the grafted PDMS surface layer and therefore reduce the net interaction force between the dispersed particles. On the other hand, swelling of the surface layer will result in an increase in the effective phase volume $\Phi_{eff} = \Phi_{silica} + \Phi_{surf.layer}$ of the silica particles, which gives rise to a higher relative viscosity. These effects are demonstrated by the dynamic and steady-shear rheological tests, which reveal a decrease in the storage modulus upon addition of styrene but an increase in the relative viscosity.

The effect of PDMS surface layer swelling is visualized in Fig. 6.

However, for hydrophilic silica N20, an explanation of the rheological effects by means of particle interactions is less straightforward. The high surface energy of hydrophilic silica will result in complex adsorption phenomena. Hence, the constitution, thickness, and segment density gradient in the surface adsorption layer is difficult to predict.

In the case of low-molecular-weight polar resins such as VE resins, relatively thin and dense adsorption layers can be assumed. This should result in low viscosities due to low effective phase volumes of the dispersed phase and weak interparticulate interactions forces according to steric stabilization. However, addition of a solvent like styrene will influence the Hamaker constant of the liquid medium and of the adlayer and the structure of the adlayer in terms of swelling and/or multilayer formation. In particular, any multilayer formation could result in surface layer entanglement depending on the solvency of the liquid medium expressed in terms of the Flory-Huggins parameter χ [11]. These effects should dramatically influence the viscosity and rest structure of the dispersion, as seen in the experiments.

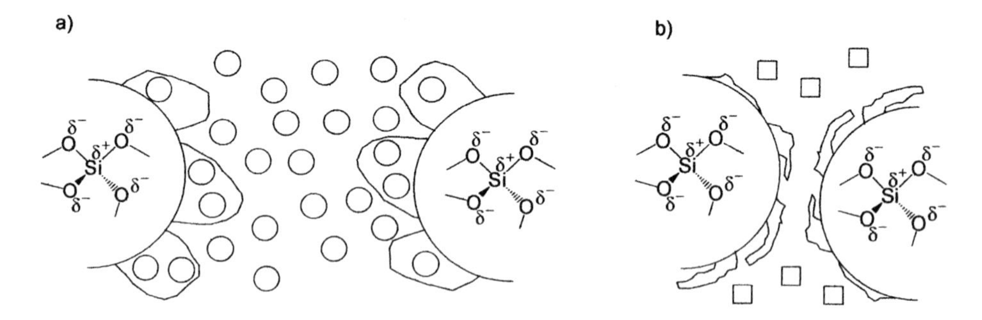

Fig. 6. Effect of the solvent quality on the PDMS surface layer thickness and particle-particle interaction force: a) PDMS surface layer swelling due to good solvency of the liquid medium; b) compression of PDMS layer due to low solvency of the liquid medium.

For long-chained resin oligomers the situation is even more complex. In addition to the adsorption phenomena resulting in steric stabilization, polymer bridging has also to be taken into account. In general, polymer bridging should be more relevant in diluted systems, because of a less dense adlayer, which might facilitate the simultaneous adsorption of polymer coils at two different particles [12]. This model is at least in general supported by our results of the rheological tests.

In order to illustrate the discussion above, Fig. 7 depicts results of more detailed calculations of the attractive pair potentials E_{att} between silica particles according to an approach suggested by Vincent based on the original work of Vold [13].

The calculations for hydrophobic silica reveal that with increasing styrene content of the liquid phase the strength of the pair potentials decrease. The magnitudes of the potentials are reasonable and can be assigned to a weakly flocculated gel.

The particle interactions based on van der Waals forces between hydrophilic particles covered by a resin layer are almost completely screened. This implies that, for a discussion of the rheologically relevant interaction forces, non-DLVO effects such as polymer bridging have also to be taken into account as discussed above. For a theoretical treatment of these effects, approximations given by Napper and Vincent, respectively, can be applied [11, 14].

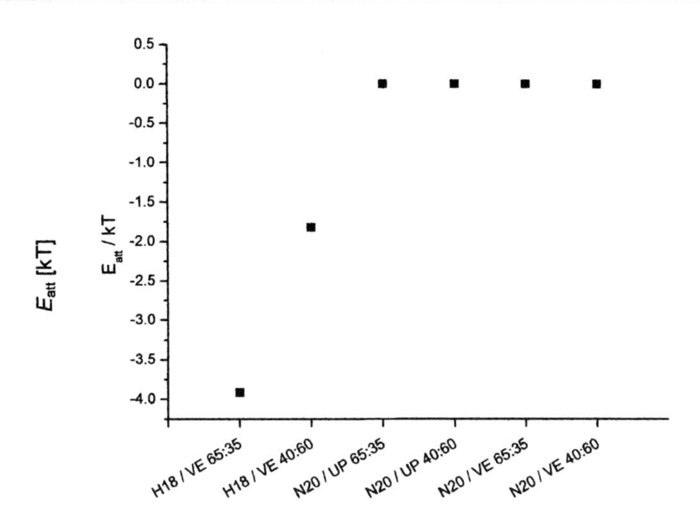

Fig. 7. Attractive interaction pair potential at a surface-surface separation of 0.15 nm, particle radius of 6.5 nm, surface layer thickness of 2.0 nm, and assumed variable layer composition: H18 / VE (65:35): 0.9 PDMS, 0.1 styrene; H18 / VE (40:60): 0.6 PDMS, 0.4 styrene; N20 / UP (65:35): 0.9 UP, 0.1 styrene; N20 / UP (40:60): 0.8 UP, 0.2 styrene; N20 / VE (65:35): 0.9 VE, 0.1 styrene; N20 / VE (40:60): 0.8 VE, 0.2 styrene.

References

[1] V. Khavryutchenko, H. Barthel, E. Nikitina, *Macromol. Symp.* **2001**, *169*, 7.

[2] H. Barthel, *Colloids Surf. A: Physicochemical and Engineering Aspects* **1995**, *101*, 217.

[3] H. Barthel, F. Achenbach, H. Maginot, *Proc. Int. Symp. on Mineral and Organic Functional Fillers in Polymers (MOFFIS 93)* **1993**, 301.

[4] R. G. Larson, *The Structure and Rheology of Complex Fluids*, Oxford University Press, New York, **1999**.

[5] Q. D. Nguyen, D. Boger, *J. Rheol.* **1985**, *29*, 335.

[6] G. Michael, H. Ferch, *Schriftenreihe Pigmente, Vol. 11*, Degussa AG.

[7] J. Israelachvili, *Intermolecular and Surface Forces*, 2nd. edn., Academic Press, London, **1991**.

[8] H. Barthel, M. Dreyer, T. Gottschalk-Gaudig, V. Litvinov, E. Nikitina, in *Organosilicon Chemistry V — From Molecules to Materials* (Eds.: N. Auner, J. Weis), VCH Weinheim, **2001**, p. 752.

[9] T. F. Tadros, *Adv. Colloid Interface Sci.* **1996**, *68*, 97.

[10] M. A. Bevan, S. N. Petris, D. Y. C. Chan, *Langmuir* **2002**, *18*, 7845.

[11] B. Vincent, J. Edwards, S. Emmett, A. Jones, *Colloids Surf.* **1986**, *18*, 261.

[12] Y. Otsubo, *Adv. Colloid Interface Sci.* **1994**, *53*, 1.

[13] D. W. J. Osmond, B. Vincent, F. A. Waite, *J. Colloid Interface Sci.* **1973**, *42*, 262.

[14] D. H. Napper, *J. Colloid Interface Sci.* **1977**, *58*, 390.

Silica Adhesion on Toner Surfaces Studied by Scanning Force Microscopy (SFM)

Mario Heinemann,* Ute Voelkel, Herbert Barthel

Wacker-Chemie GmbH, Johannes-Hess-Str. 24, D-84489 Burghausen, Germany
Tel.: +49 8677 83 4022 — Fax: +49 86 77 83 68 14
E-mail: mario.heinemann@wacker.com

Sabine Hild

Experimental Physics, University of Ulm, D-89069 Ulm, Germany

Keywords: pyrogenic silica, toner, toner-silica adhesion, scanning force microscopy (SFM)

Summary: SFM is an ideal tool to determine the morphology of toner and toner-silica interfaces. Domains with different elasticity and adhesion have been found. SFM determines the real silica particle dimensions, aggregates of 100 – 200 nm after blending with the toner particles. 0.5 wt% loading leads to a monolayer with 300 m^2/g silicas. 50 m^2/g silica show lower adhesion than 300 m^2/g silica. PDMS-treated silica reveals in the phase image a dark halo surrounding the silica particles. The dark halo indicates a softer layer – an indication of an increased adhesive interaction between the silylation layer of silica particles and the toner substrates. The chemically prepared toner showed the strongest adhesive interaction with PDMS-treated silicas. After soft blending, bright spots of silica are visible; after hard blending, the silica particles are more embedded in the toner surface.

Introduction

Pyrogenic silicas have long been used as surface additives on toner particles for electrophotographic applications, like photocopiers or laser printers. Pyrogenic silicas, while altering toner particle flow and charging ability, also have a strong influence on toner particle adhesion [1, 2]. The control of toner adhesion to various surfaces within a photocopier is of the highest importance for proper operation of the copier as well as for maintaining image quality on the output document. The focus of this paper is on the characterization of toner surface morphology and the determination of adhesion of pyrogenic silica particles on a variety of toner surfaces by using dynamic SFM techniques. Tapping Mode (TM) and Pulse Force Mode (PFM) were utilized for imaging and modeling of toner surfaces and toner-silica particles. Toner-making technology differentiates between the pulverizing route and the chemical route (CPT) [10 – 12]. Products from both technologies have been integrated in our study. SFM enables us to image surface topography features and surface properties at the same time. Basic principles of SFM techniques are described in Refs. [3 – 6].

SEM images of a typical pulverized toner and a CPT are shown in Fig. 1. An estimate of global toner volume is about 160 000 mt (2001) with an average of >10% CAGR.

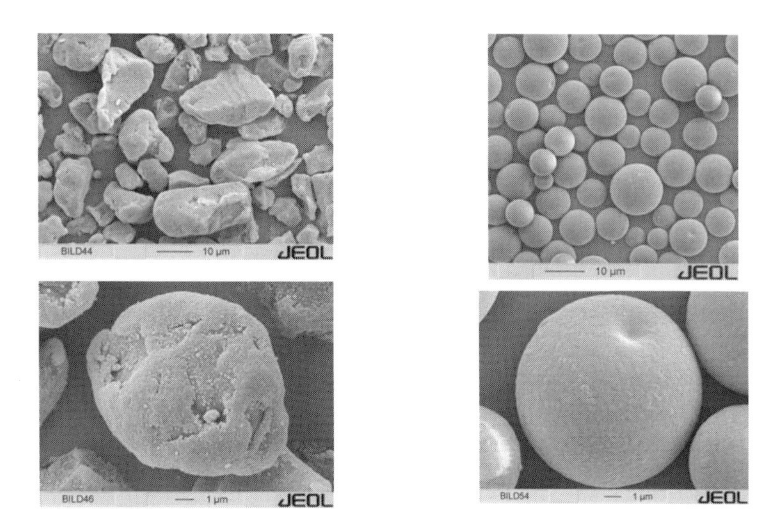

Fig. 1. SEM pictures of pulverized (conventional) toner vs CPT.

Experimental

Materials

- Toner: nonmagnetic; color (magenta); resin: polyester-epoxy; d_{50} = 12 µm; T_{soft} = 95 °C
- Toner: nonmagnetic; b/w: CPT – technology; d_{50} = 8 µm
- Toner: monocomponent magnetic; resin: styrene-acrylic; d_{50} = 13 µm; T_{soft} = 100 °C
- Pyrogenic (fumed) silica HDK (Wacker-Chemie GmbH):
  H05TX (BET hydrophilic base silica: 50 m²/g),
  H30TX (BET hydrophilic base silica: 300 m²/g) surface treatment: polydimethylsiloxane/ hexamethyldisilazane (PDMS/HMDS)
  H05TD (BET hydrophilic base silica: 50 m²/g),
  H30TD (BET hydrophilic base silica: 300 m²/g) surface treatment: polydimethylsiloxane PDMS
  H05TM (BET hydrophilic base silica: 50 m²/g),
  H30TM (BET hydrophilic base silica: 300 m²/g) surface treatment: hexamethyldisilazane HMDS

Impact of Blending Conditions

To evaluate the impact of the blending conditions, a low-shear blender (Turbula) and a high-shear blender (Henschel-type) were used to blend toner particles with silica. Typical loading rates of silicas onto toner varied between 0.2 and 2.0 wt%. For our study the loading was 0.5 wt% silica.

SFM Measurements

Silica-toner systems have been investigated by this technique very recently [7, 9]. A Nanoscope D3100 Scanning Force Microscope (Veeco Instruments, Santa Barbara, CA) was used to acquire the images. Pictures were taken under ambient conditions (30% humidity; 278 K) in TM using a standard microfabricated silicon cantilever with aluminum reflex coating (Olympus, AC160TS). In our studies, the free amplitude was set to 2 V and the amplitude damping r_{SP} was adjusted to 0.65 – 0.7 (1.325 – 1.35 V), which is a suitable and moderate force for imaging. Under these conditions, a high phase shift corresponds to high stiffness regions of the sample. In this case, higher phase shift, noticeable as brighter regions of the samples, can be attributed to harder areas of the samples. In TM, the topography is measured as the constant distance of the tip normal to the surface by keeping the amplitude of the oscillating tip constant. The level of the force applied to the surface is given by so-called amplitude damping r_{SP}. r_{SP} is defined as the magnitude of the engaged amplitude to the free-air amplitude (A_0) of the oscillating tip ($r_{SP} = A/A_0$). Assuming that the oscillation of the cantilever is damped similarly at all sample locations, height data are complemented by simultaneously measured phase shift data [8]. The phase image outlines domains of varying material contrast at or near the surface. Depending on the chosen r_{SP}, the phase contrast can either be dominated by the adhesive properties of the surface or by the stiffness due to the chosen experimental parameters. Adjusting the set point ratio below 0.8 results in stiffness-dominated contrast in the phase image.

SFM Sample Preparation

To avoid displacement of the particles during SFM imaging, they have to adhere properly on the supporting surface. One possibility is to disperse them by air flow on sticky tapes (Tesa, double-sided, Beiersdorf). Because the adhesive polymer matrix of the tape covers the particles after one day of storage, images have to be taken only on freshly prepared samples.

Results and Discussions

Pulverized Toner, Styrene-Acrylic: Silica Free

The pulverized styrene-acrylic toner shows a rock-like structure leading to a surface morphology with step edges up to several hundred nanometers in height and plateaus with up to 2 µm lateral extension. On the plateaus, only small variations in height can be found. The phase shift image of such a plateau region shows an inhomogeneous morphology. The main part of these areas appears dark with an averaged phase shift of about 30, indicating the softer area. This area contains bright spots of about 50 nm diameter with a phase shift of about 40° showing hard domains. This indicates a phase-separated structure of softer methacrylic areas (dark) and harder polystyrene domains.

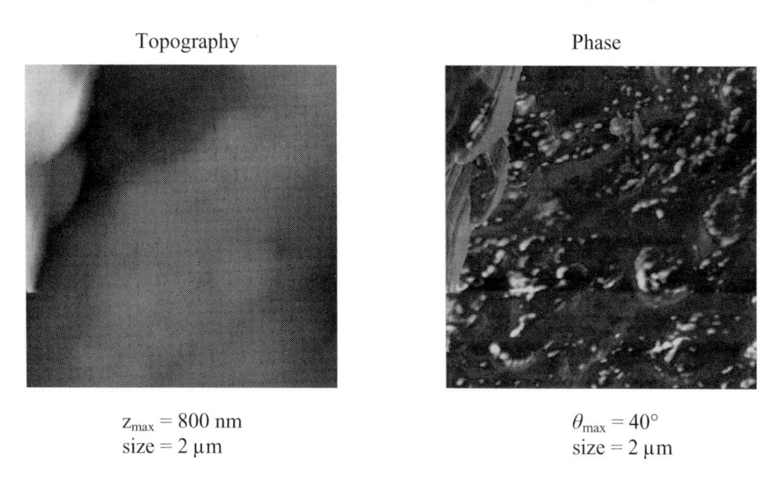

Fig. 2. Topography and phase of a polystyrene/polyacrylic toner: silica free.

Pulverized Toner: Polyester, Silica Free

The overall shape of this toner is more round and the surface is less rough than the styrene-acrylic one without sharp edges. The phase images show maximum phase shift of about 35°. Only small variations in phase shift can be detected, pointing to a mainly homogeneous surface.

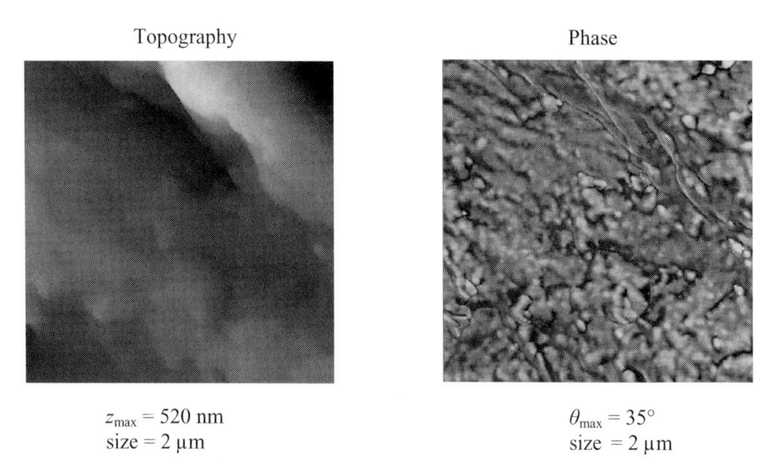

Fig. 3. Topography and phase of polyester toner: silica free.

CPT: Silica Free

The CPT show a nearly spherical shape with a smooth surface. The maximum height is at 100 nm.

The maximum phase shift of about 26° is smaller than both of the other investigated toners. This indicates a soft surface.

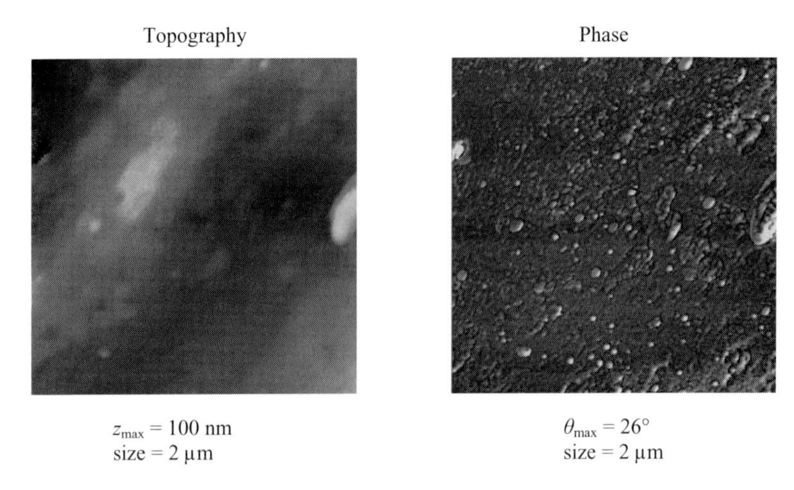

Fig. 4. Topography and phase of CPT: silica free.

Silica-Loaded Toner Surfaces: Pulverized Toners

After blending on all toner surfaces, pyrogenic silica aggregates (definition according DIN53206) can be detected from the significantly higher phase shifts. The sizes of these aggregates on the toner surface are in the range of about 100 – 500 nm, depending on the silica type used.

Pulverized Toner: Polyester; Silica 0.5 wt%; BET 50 m²/g; HMDS-Treated

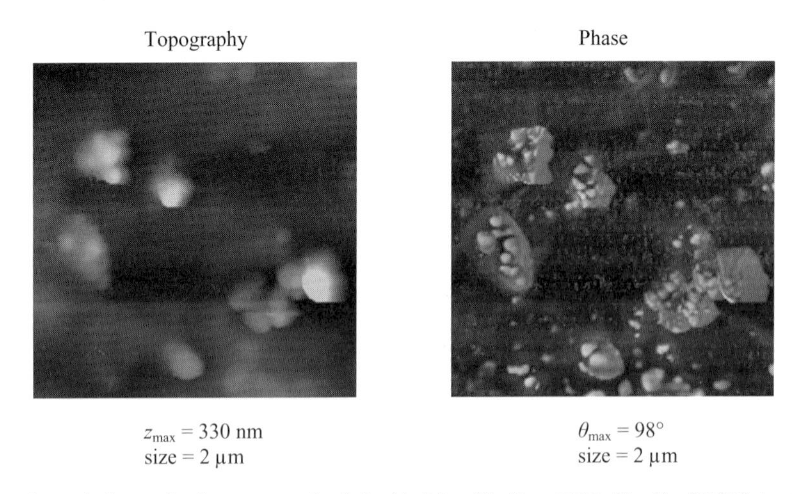

Fig. 5a. Topography and phase of polyester toner loaded with 0.5 wt% silica; BET: 50 m²/g; HMDS-treated; HDK H05TM.

HMDS-modified silica aggregates with a diameter of about 300 nm can be detected on the polyester toner surface. Even in the topography image the raspberry-like substructure of the particles can be detected. These bigger silica aggregates have a lowered adhesion to the toner, and partial removal of the aggregates during the TM scan has been observed. In the phase image, the particles exhibit a high phase shift of about 98°. These particles are much harder than the toner resin.

Pulverized Toner: Polyester; Silica 0.5 wt%; BET 50 m²/g; PDMS-Treated

PDMS-treated silica particles are less pronounced in the topography image, indicating softer particles which can be compressed. This assumption is supported by the fact that the maximum phase shift is smaller than that for the HMDS-modified silica particles. In the phase image, silica aggregates of about 150 nm can be found. The silica aggregates are surrounded by a larger, dark area. A silylation layer based on a short-chain PDMS is formed after the reaction of the corresponding silica with PDMS (indicating softening around the particles), probably because of the interaction of the PDMS-based silylation layer and the resin.

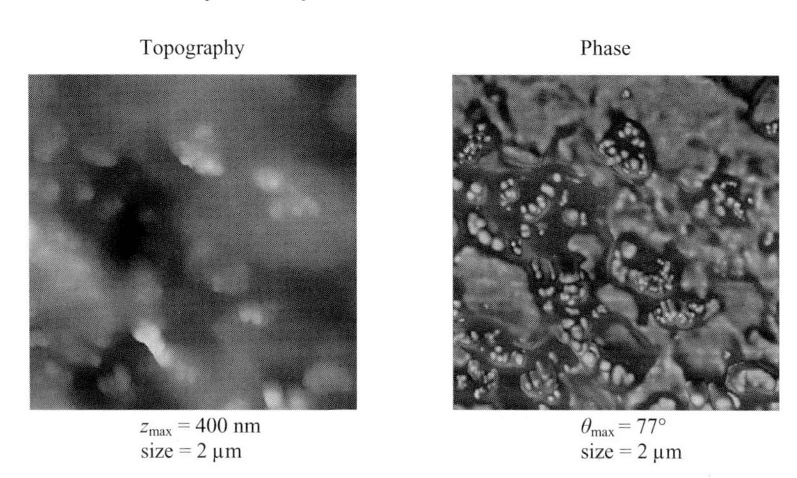

Fig. 5b. Topography and phase of polyester toner: 0.5 wt% silica; BET: 50 m²/g; PDMS-treated; HDK H05TD.

Pulverized Toner: Polyester; Silica 0.5 wt%; BET 300 m²/g; PDMS-Treated

However, PDMS-modified silica aggregates seem to coat nearly the entire toner particle. Their topographical diameter is doubled compared to the HMDS-coated particles, but the diameter estimated from the phase image is very similar to the one found for the HMDS-modified ones. The maximum phase shift is reduced compared to the samples above. This indicates that the silica aggregates are covered by a soft material – the silylation layer. This soft material covers not only the particles but also the resin surface in between them. This can be seen by the decreased phase shift in between the silica particles compared to the pure resin. As in sample (Fig. 5b), probably the PDMS silylation layer interacts with the toner resin surface, increasing the overall adhesion.

Fig. 5c. Topography and phase of polyester toner loaded with 0.5 wt% silica; BET: 300 m²/g; PDMS-treated; HDK H30TD.

CPT: 0.5wt% Silica; BET 50 m²/g; HMDS/PDMS

Silica aggregates are visible on the surface of the CPT as clusters with a diameter up to 500 nm. In the phase image these aggregates show the highest phase shift, indicating that they are the hardest part of the surface. Around the particles an area with very low phase shift can be found. The thickness of this soft area extends to 100 nm around the particles. Similarly to the PDMS-modified silica particles on the polyester toner, the silylated surface interacts with the toner surface. The large extension of the soft area indicates that the interaction is stronger than for the polyester resin.

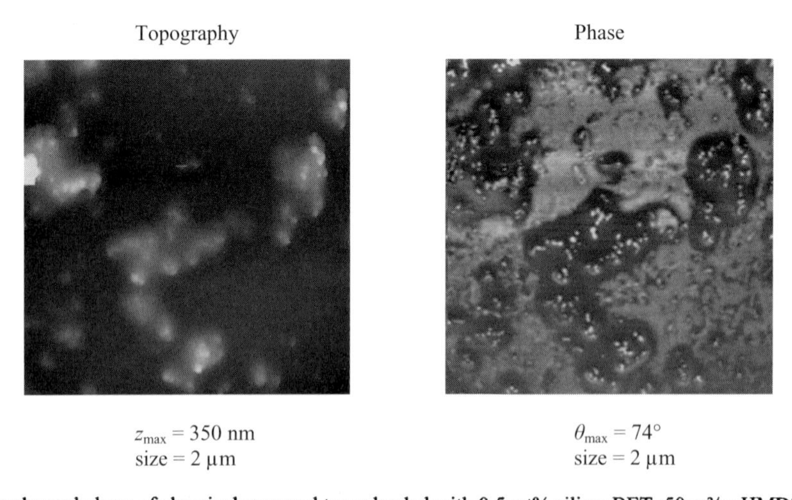

Fig. 6a. Topography and phase of chemical prepared toner loaded with 0.5 wt% silica; BET: 50 m²/g; HMDS/PDMS-treated; HDK H05TX.

CPT: 0.5 wt% Silica; BET 300 m²/g; HMDS/PDMS

On the surface of the toner a large number of aggregates can be observed with a diameter between 100 and 200 nm. As seen in Fig. 6a, the particles can be correlated to bright (hard) parts of the surface. The particles are homogeneously spread over the surface. In between the particles, again a dark, soft phase can be observed. In the lower right hand edge of the picture a small amount of the pure toner surface can be detected. Comparing the phase shift of this area to the phase shift in between the particles, the pure polymer shows a higher phase shift. This indicates a softening due to interaction between silylated silica particles and the toner surface.

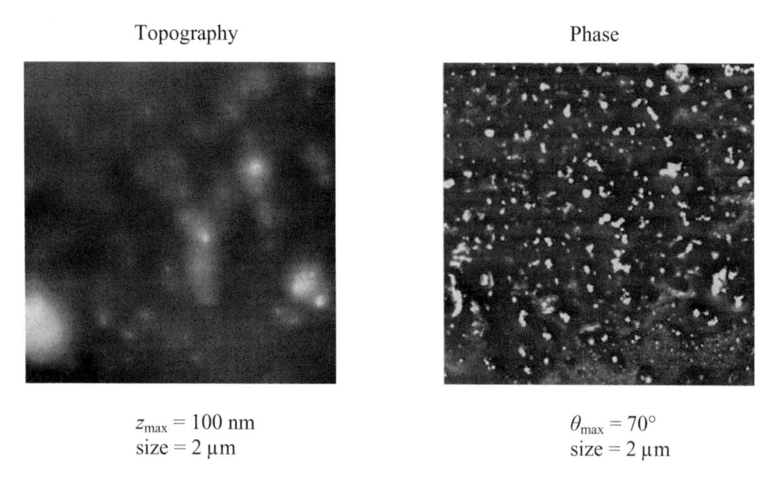

Fig. 6b. Topography and phase of chemically prepared toner loaded with 0.5 wt% silica; BET: 300 m²/g; HMDS/PDMS-treated; HDK H30TX.

Comparing PDMS/HMDS- and HMDS-modified silica particles, a lower phase shift of up to 70° has been observed for the PDMS/HMDS silicas, and for the HMDS silicas a higher phase shift of up to 90° has been found. Differences in phase shift values indicate the impact of surface modification on the local hardness of the silica particles. "PDMS" modification leads to a softer, polymer-like grafting, whereas pure HMDS modification only increases the hydrophobicity by a hard monolayer formation of trimethylsiloxy groups. HMDS-treated silicas seemed to show a weaker interaction with the toner resin surfaces. In contrast, PDMS/HMDS-treated silicas show stronger adhesion to the toner resin surfaces, so they can easily be imaged at high resolution.

The Impact of Blending

Low-shear blending leads to isolated bright particles. In the phase image only darker spots can be found. The dimension of the dark spots is in the same dimension than the silica particles. One possible interpretation of this effect is that under low-shear the silylated particles adhere to the surface. This leads to a softening of the contact area. But the adhesion is not strong and the particles can easily be removed. PDMS which is not chemically bound to the silica particle remains on the surface, leading to soft spheres at the surface.

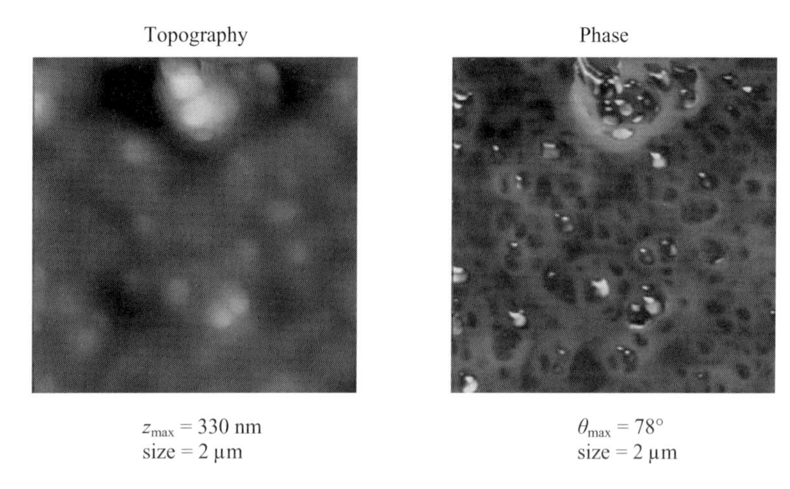

Fig. 7a. Topography and phase of polystyrene-acrylic toner loaded with 0.5 wt% silica; low-shear conditions; BET: 300m²/g; PDMS-treated; HDK H30TD.

After blending under high-shear conditions, the surface seems to be as smooth as the pure styrene-acrylic toner. In the topography no particles can be detected, but in the phase shift they can be detected by a high phase shift. It seems that the particles are squeezed into the surface. The silylation layer of the silica will interact with the toner surface forming a softer surface area that can be observed by a lower phase shift than the pure resin.

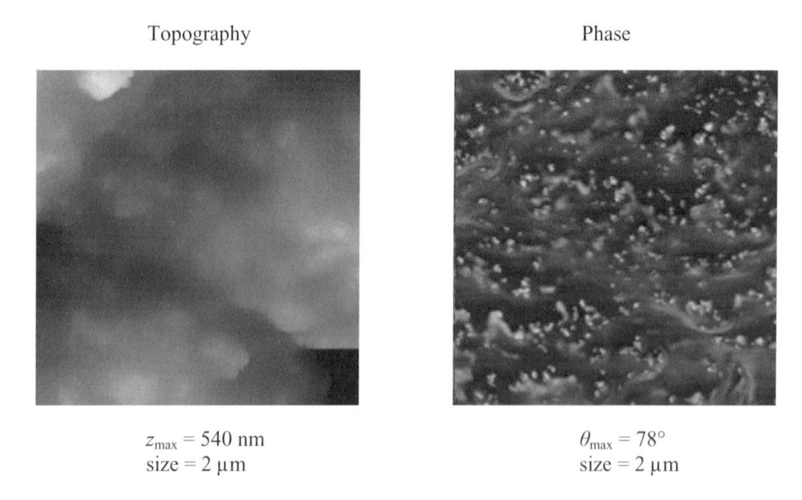

Fig. 7b. Topography and phase of polystyrene-acrylic toner loaded with 0.5 wt% silica; high-shear conditions; BET: 300 m²/g; PDMS-treated; HDK H30TD.

Conclusion

- Pure toner shows differences in surface topography from rock-like to spherical depending on the preparation. Phase imaging gives a first insight into surface properties. The chemically prepared toner is softer than the other two. Because phase shift is also sensitive to chemical surface properties, the application of complementary experiments using Pulsed Force Mode are necessary. Micro-hardness measurements using force vs distance curves are also possible.
- Larger particles show lower adhesion due to smaller contact area.
- Surface modification by PDMS cause a polymeric surface layer on the silica particles leading to higher adhesion due to softening of the toner surface and increase in contact area.
- Chemically prepared toner shows higher interaction than the other resins.
- Low-shear attaches the particles to the surface but high shearing squeezes the particles into the surface.

References

[1] H. Barthel, M. Heinemann, PPIC/JH '98, **1998**, p. 428 – 431.

[2] H. Barthel, M. Heinemann, *Electrophotography* (The Society of the Electrophotography of Japan) **1995**, *34*(*4*), 401.

[3] H. A. Mizes, K.-G. Loh, R. J. D. Miller, S. K. Ahuji, E. F. Grabowski, *Appl. Phys. Lett*, **1991**, *59*, 2901.

[4] G. L. Binning, C. F. Quate, Ch. Gerber, *Phys. Rev. Lett.* **1986**, 56, 930.

[5] Th. Stifter; E. Weilandt, O. Marti, S. Hild, *Appl. Phys. A* **1998**, *66*, S597.

[6] D. Sarid, V. Elings, *J. Vac. Sci. Technol. B* **1991**, 9(2).

[7] M. Heinemann, U. Voelkel, H. Barthel, S. Hild, Morphology of toner-silica interfaces, IS&T's NIP17: *International Conference on Digital Printing Technologies*, **2001**, 845.

[8] S. N. Magonov, V. Elings, M.-H. Whangbo, *Surf. Sci.* **1997**, 375, L385 – L391.

[9] M. Heinemann, U. Voelkel, H. Barthel, S. Hild, S. Koenig, S.-C. Imhof, Silica adhesion on toner surfaces studied by scanning force microscopy, IS&T's NIP18: *International Conference on Digital Printing Technologies*, **2002**, 651.

[10] M. Heinemann, U. Voelkel, H. Barthel, *5th Annual Toner, Ink Jet Ink & Imaging Conference*, IMI, **2002**.

[11] D. Tyagi, DPP2003: *International Conference on Digital Production Printing, Industrial Applications*, **2003**, p. 207 – 210.

[12] G. E. Kmiecik-Lawrynowicz, DPP2003: *International Conference on Digital Production Printing, Industrial Applications*, **2003**, p. 211 – 213.

Characterization of Silica-Polymer Interactions on the Microscopic Scale Using Scanning Force Microscopy

*Sigrid König, Sabine Hild**

Experimental Physics, University of Ulm
Albert-Einstein-Allee 11, D-89081, Ulm, Germany
Tel.: +49 731 502 3016 — Fax: +49 731 502 3036
E-mail: sabine.hild@physik.uni-ulm.de

Keywords: silica particles, adhesion, Scanning Force Microscopy

Summary: Pyrogenic silica particles are used to enhance flow and support triboelectric charging. Since is has been shown that surface chemistry and surface energy have an impact on the system pyrogenic silica-toner, investigations of silica-polymer interactions on the nanoscopic scale are of importance. Scanning Force Microscopy (SFM) is an effective tool to describe surface features and the mechanical and chemical properties of small particles. Thus, dynamic SFM techniques like Tapping Mode (TM) and Pulsed Force Mode (PFM) are used to image toner surfaces and study the toner-pyrogenic silica interaction. The effect of different surface chemistries, providing harder and softer silylation layers, on the adhesion to a conventional toner based on polystyrene (PS) and acrylates has been investigated to develop models of actual pyrogenic silica-toner adhesion. One aim is to model and investigate the interaction between the silica particles and the polymeric surfaces. Therefore pull-off forces between a silylated tip and a polystyrene/polymethylmethacrylate blend are measured.

Introduction

Pyrogenic silicas are key ingredients in a wide range of applications such as additives in electrographic toners and developers. They enhance flow and support triboelectric charging. The long-term stability of the toner performance in terms of charge stability and high flow even at high temperature and humidity are the most challenging requirements, which are based to a significant extent on the adhesion of pyrogenic silica particles to toner particles [1 – 2]. Although is has been shown that the shape, surface chemistry, and surface energy of the toner particles have an impact on the system pyrogenic silica-toner, little is known about the silica-polymer interaction on the nanoscopic scale. It is crucial to be able to visualize the size distribution of pyrogenic silica particles and their distribution on toner particle surfaces. Beside this, the characterization of the adhesion between pyrogenic silicas and polymeric surfaces on a microscopic scale is important.

Scanning Force Microscopy (SFM) is an effective tool, which helps to answer these unsolved questions. It enables us to image surface topography on the nanometer scale; basic principles of these techniques are described in Ref. [3, 4]. Dynamic techniques like Tapping Mode (TM), where the tip is periodically brought into contact with the surface during the imaging process [4], enable us to investigate delicate polymer samples. Assuming that the oscillation of the cantilever is damped similarly at all sample locations, height data are complemented by simultaneously measured phase shift data [5]. The phase image outlines domains of varying stiffness at or near the surface when the applied force is in the repulsive force regime [6]. Most recently, silica toner surfaces have been investigated by this technique [7, 8].

SFM also enables us to measure specific interaction forces between a small silicon tip and the surface. The pull-off forces between the tip and the surface estimated from Force vs Distance Curves (FDC) can be correlated to the adhesive interactions between tip and surface [9]. Recording of FDCs line-by-line allows us to image surface topography and adhesive surface properties simultaneously [10]. This technique has some disadvantages, like the requirement for a large amount of data acquisition and analysis, which have been alleviated by the invention of the Pulsed Force Mode (PFM). The PFM simplifies and accelerates the measurements of adhesive properties with high lateral resolution [11, 12].

Besides imaging the toner-silica surfaces with respect to the silica modification, one aim of this study is to model and investigate the interaction between the silica particles and polymeric surfaces. Under ambient conditions, silicon cantilevers are covered with SiO_2. The radius of curvature of the silicon tip is about 10 nm, which is of the same order of magnitude as that of primary particles of diameter of about 100 nm in a pyrogenic silica aggregate. This enables us to use the tip as a model for silica particles. Since the pyrogenic silicas used in polymeric applications are so-called surface-modified silicas, a surface-modified tip would provide a more realistic picture of the polymer-silica interactions. Thus, silica-polymer interactions can be studied by adhesive force measurements in Pulsed Force Mode using silane-modified tips. Since conventional toners are often based on polystyrene/acrylate blends, we chose polystyrene/polymethyl methacrylate (PS/PMMA) films as a model system for the investigation of the silica particle-polymer surface interactions. These immiscible polymers will form a phase-separated structure, which enables us to investigate the interactions between the SFM tip and hydrophilic/hydrophobic surfaces at the same time.

Experimental

Samples

A mono-component, magnetic styrene-acrylic resin toner with an average particle size of d_{50} = 13 μm and a softening temperature T_{soft} = 100 °C was used for the following experiments. The toner was loaded with 5 wt% pyrogenic (fumed) silica WACKER HDK of BET = 300 m^2/g. All pyrogenic silicas showed a high negative tribocharge. To study the influence of the surface chemistry, the silicas were modified by (A) hexamethyldisilazane (HMDS), leading to a silane monolayer, (B) dimethyldichlorosilane, forming a polydimethylsiloxane-like layer (PDMS), and (C) dimethyldichlorosilane/hexamethyldisilazane (PDMS/HMDS).

Dispersing particles directly on a support like sticky tape will provide the surface of silica-loaded toner without additional external changes. To avoid displacement of the particles during SFM imaging, they have to adhere properly to the supporting surface. One possibility is to disperse them by air flow on sticky tapes (Tesa, double-sided, Beiersdorf). Because the adhesive polymer matrix of the tape covers the particles after one day of storage, images have to be taken only on freshly prepared samples.

Spin-coated PS/PMMA films were used as samples to study silica-polymer interactions. For this purpose, solutions of 10 mg polymer in 1 mL toluene were prepared from each polymer. The PMMA and PS solutions were mixed in a ratio of 2:1. When the mixed solution was spin-coated on a freshly cleaned silicon surface a phase-separated film of 100 nm thickness was obtained. Cantilevers were modified either by deposition in the saturated atmosphere of hexamethyldisilizane or immersion in a 2% solution of dimethyldichlorosilane in toluene. Typically, measuring the contact angle of the sample after modification gives proof of hydrophobization. However, because of the small size of the tips the contact angle cannot be estimated. Thus, at the same time as the cantilever modification, small pieces of silicon wafers were modified to enable contact angle measurements to be made.

Scanning Force Microscopy Measurements

A Dimension 3100 Scanning Force Microscope (Veeco Instruments, Santa Barbara, CA) was applied to acquire the images of the silica-loaded toners particles. Pictures are taken under ambient conditions (30% humidity; 278 K) in TM using a standard micro-fabricated silicon cantilever with aluminum reflex coating (Olympus, AC160TS). In our studies, the free amplitude was set to 2 V, and the amplitude damping r_{SP} was adjusted to 0.65 – 0.7 (1.325 – 1.35 V), which is a suitable moderate force for imaging in the repulsive regime. Under these conditions a high phase shift, noticeable as brighter regions of the sample, corresponds to stiffer regions of the sample. In the case of toner particles, higher phase shift can be attributed to harder areas of the samples with respect to the silica particles. To receive a better impression of the surface morphology, the images are subjected to a second-order surface-flattening process. After this procedure, the primary spherical topography appears flat. Details of, e.g., attached silica particles can be better identified.

The Pulsed Force Mode allows investigations of the tip-sample interactions on a more quantitative level. For PFM measurements, the Dimension 3100 SFM was equipped with a Pulsed Force Mode box (WITec Instruments, Ulm, Germany). For adhesive force measurements, a standard micro-fabricated silicon cantilever with aluminum reflex coating (Olympus, AC240TS) with spring constants of about 2 N/m was used. To choose cantilevers with similar spring constants, their resonance frequencies were measured [13]. All measurements are performed under ambient conditions using the same set point of 0.5 V.

Results and Discussion

Toner-Silica Interaction

The conventional styrene-acrylic resin (Fig. 1A, upper image) show a rock-like structure leading to a surface morphology with step edges up to several 100 nm in height and plateaus of up to 2 μm lateral extension. On the plateaus, only small variations in height can be found. The phase shift image of such a plateau region reveals an inhomogeneous morphology (Fig. 1A, lower image). The main part of these areas appears dark with an averaged phase shift of about 20°, indicating a softer area. Inside this area bright spots of about 50 – 100 nm diameter with a phase shift of about 30° can be found, indicating harder domains. These observations can be explained by the formation of a phase-separated structure of softer acrylic areas (dark) and harder polystyrene domains.

When the toner is loaded with pyrogenic silica particles, these appear as isolated protrusions at the surface (Fig. 1B–D). Depending on the modification, the diameters of the protrusions range between 50 and 200 nm. When HMDS is used as the silylation reagent, a silane monolayer is formed. In this case the diameter of the particles is ca. 100 nm. Such particles are adsorbed mainly as isolated aggregates to the toner surface (Fig. 1B). PDMS-modified particles seem to coat almost the entire toner particle (Fig. 1C). Their topographical diameter is doubled compared to that of the HMDS-coated particles. This can be explained by the formation of a polymer-like PDMS layer. HMDS/PDMS-coated silicas appear only as isolated protrusions with a diameter up to 250 nm (Fig. 1D).

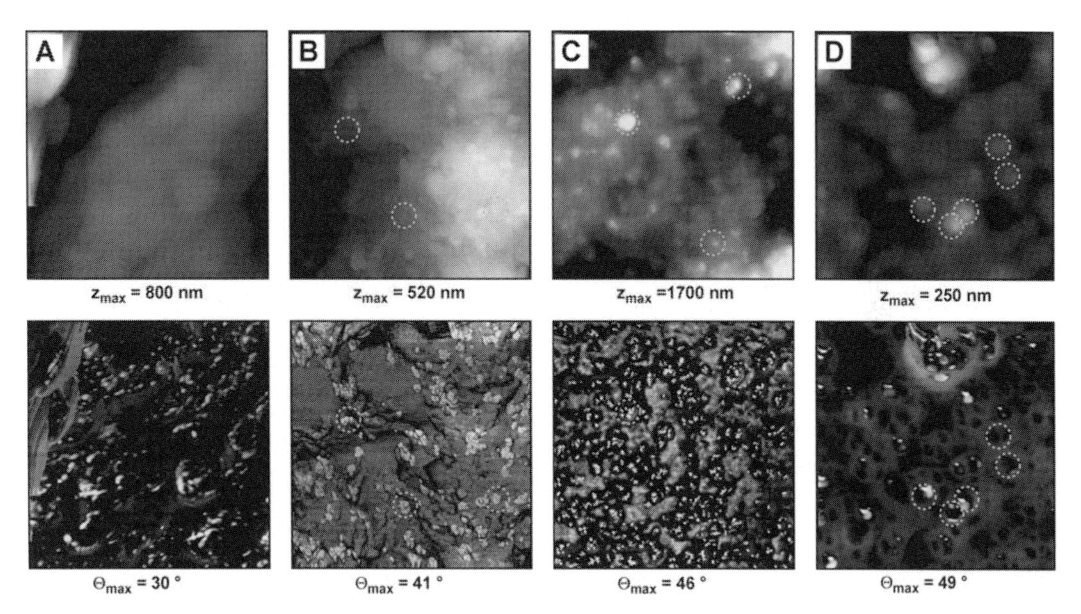

Fig. 1. Tapping Mode images of toner surface without (A) and with chemically modified silicas (B: HMDS; C: PDMS, D: HMDA/PDMS), revealing both the topography (upper row) of the particles and the changes in material contrast (phase images, lower row). Scan range: 2 μm.

The lower row of Fig. 1B–D reveals the phase shift images of silica-loaded toner. For all samples, silica particles can be detected by the significantly higher phase shifts, up to 45°. The diameter of such bright spots ranges between 50 and 100 nm. For HMDS-modified silicas, the diameter estimated from the phase shift image is similar to the one found for the protruding particles (Fig. 1B), confirming the formation of just a monolayer. The phase shift image of the PDMS-modified samples also show bright particles, but these particles show a dark collar of about 50 to 100 nm thickness, indicating a softer region (Fig. 1C). Because of the formation of this soft shell, the silicas appear larger in the topography image. The soft material also covers the resin surface. Probably, the PDMS interacts with the toner resin surface. HMDS/PDMS-modified particles also consist of a hard core and a softer shell, but they are somewhat larger. In the phase shift image (Fig. 1D), isolated darker, softer domains appear, only a few of which contain brighter particles. This observation can be explained as follows. In the case of a mixed modification the PDMS layer is not chemically bonded to the silica particles. Thus, after the PDMS interacts with the toner resin, the hard silica particles can be removed.

Model System

To simulate the toner-silica interactions, PS/PMMA blends were used as a model for polystyrene/acrylic toners. Silicon tips modified either with HMDS or with PDMS were applied to model surface-treated silica particles. The Pulsed Force Mode images of PS/PMMA films displayed

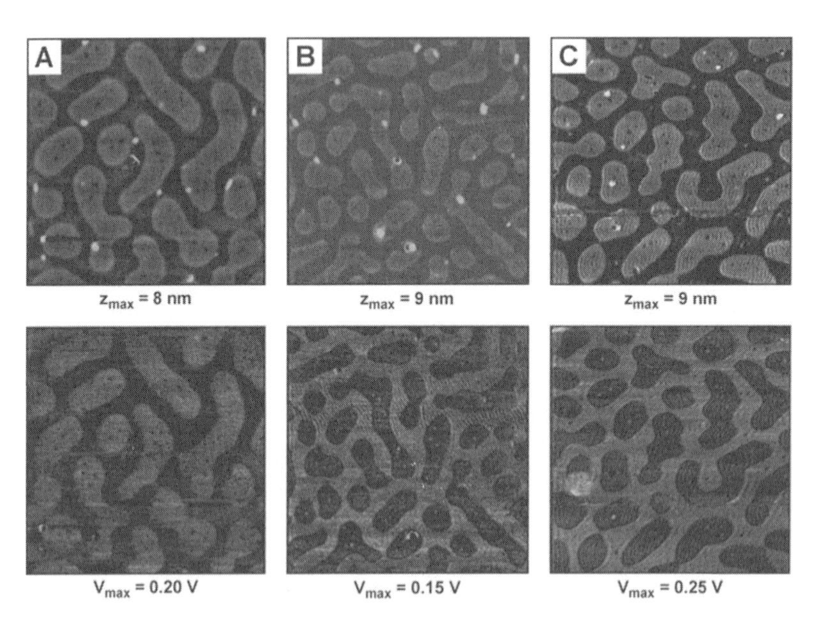

Fig. 2. Pulsed Force Mode images of PS/PMMA blend show no distinct changes in the topography (upper row) when silylated tips are used but a clear dependence on the adhesive contrast (lower row). With a pure silicon tip (A), the elevated areas show higher adhesion; after silylation, contrast is reversed. The HMDS-modified tip (B) shows a lower maximum force than the PDMS-treated tip (C).

in Fig. 2 show no distinct changes in the topography (upper row) using silylated tips. In the topography images, islands of one polymer deposited on the other can be seen. These elevated islands can be attributed to PMMA domains; the lower regions are PS areas [12]. The pull-off force images displayed in the lower row reveal a clear dependence on the adhesive contrast.

When the blend is imaged with the non-modified tip, the elevated PMMA islands appear brighter than the PS domains (Fig. 2A). This indicates stronger interaction between a hydrophilic silicon oxide tip and PMMA. After silylation, the adhesive force contrast is reversed (Fig. 2B,C). This indicates that the interactions between a hydrophobic tip and PS are stronger than the adhesive interaction with PS. Comparison of the maximum adhesive forces obtained for an HMDS-modified tip (Fig. 2B) with those obtained for a PDMS-treated tip (Fig. 2C) reveals higher forces for the latter. Since both tips are hydrophobic, the difference cannot be explained by a variation in surface energy. More likely it is caused by differences in the mechanical properties of the silylated tips; HMDS monolayers are harder than PDMS polymeric layers. The softer tip is tackier than the harder one.

Conclusion

A comparison of HMDS with PDMS indicates the impact of surface modification on the local hardness of the silica particles. Pure HMDS modification only increases the hydrophobicity by formation of a hard monolayer of trimethylsiloxy groups. Thus, HMDS-treated silicas seemed to give weaker interaction with the toner resin surfaces, and a smaller amount of particles can be found on the toner surface. The PDMS modification leads to a softer, polymer-like grafting. This grafted layer can interact with the toner resin surface, leading to stronger adhesion of the particles. PDMS/HMDS-treated silicas are covered by a PDMS layer which is not chemically bonded. Thus, silica particles can be removed while PDMS stay on the toner surface.

PFM experiments performed with surface-modified tips allow steady imaging of polymer blends with respect to the pull-off forces. Comparing non-modified, hydrophilic tips with modified hydrophobic tips reveals the inversion of the difference between the adhesive forces, indicating even an inversion of the strength of interactions between hydrophilic and hydrophobic polymer surfaces and the SFM tip or silica particles. It could also be shown that the hardness of the silane layer influences the measured pull-off forces. The harder HMDS modification leads to lower adhesive forces, like the softer PDMS modification, confirming the results obtained for toner-silica particles.

Acknowledgment: We would like to thank Wacker-Chemie GmbH, Werk Burghausen for their financial support. H. Barthel, M. Heinemann, and U. Voelkel are also thanked for providing the samples and fruitful discussions.

References

[1] H. Barthel, M. Heinemann, *PPIC/JH '98*, **1998**, 428.
[2] W. H. Barthel, M. G. Heinemann, *Electrophotography* **1995**, *34(4)*, 401.
[3] G. Binning, C. F. Quate, Ch. Gerber, *Phys. Rev. Lett.* **1986**, *56*, 930.
[4] D. Sarid, V. Elings, *J. Vac. Sci. Technol.* **1991**, *B9(2)*, 1287.
[5] S. N. Magonov, V. Elings, M.-H. Whangbo, *Surf. Sci.* **1997**, *375*, L385.
[6] G. Bar, Y. Thomann, R. Brandsch, H.-J. Cantow, M.-H. Whangbo, *Langmuir* **1997**, *13*, 3807.
[7] M. Heinemann, U. Voelkel, H. Barthel, S. Hild, Morphology of toner-silica interfaces, *IS&T's NIP17: International Conference on Digital Printing Technologies*, **2001**, p. 845.
[8] M. Heinemann, U. Voelkel, H. Barthel, S. Hild, S. Koenig, C.-S. Imhof, Silica adhesion on toner surfaces studied by scanning force microscopy in *IS&T's NIP18: International Conference on Digital Printing Technologies*, **2002**, p. 651.
[9] H. A. Mizes, K. H. Loh, R. J. D. Miller, S. K. Ahuji, E. F. Grabowski, *J. Appl. Phys. Lett.* **1987**, 59, 2901.
[10] K. O. Van der Werf, C. A. J. Putman, B. G. Degrooth, Greve, *J. Appl. Phys. Lett.* **1994**, *65*, 1195.
[11] Rosa-Zeiser, E. Weilandt, S. Hild, O. Marti, *Meas. Sci. Technol.* **1997**, *8*, 1333.
[12] H.-U. Krotil, E. Weilandt, Th. Stifter, O. Marti, S. Hild, *Surf. Interface Anal.* **1999**, *27*, 341.
[13] N. A. Burnham, X. Chien, C. S. Hodgas, G. A. Matei, E. J. Thoreson, C. J. Roberts, M. C. Davies, S. J. B. Tendler, *Nanotechnology* **2003**, *14*, 1.

Advanced Hydrophobic Precipitated Silicas for Silicone Elastomers

Kenichi Kawamoto, Christian Panz*

Degussa AG, Advanced Fillers & Pigments, Brühlerstr. 1, 50389 Wesseling, Germany
Tel.: +49 2236 763063 — Fax: +49 2236 762039
E-mail: christian.panz@degussa.com

Keywords: hydrophobic, silica, silicone elastomer, silicone rubber

Silicone elastomers ("silicone rubber") filled with silica have outstanding properties, and many commonly used products are made of these materials. Silicone rubbers filled with silica exhibit high mechanical strength and high chemical stability, and can be used at temperatures ranging from –50 °C to +300 °C [1].

Unlike other elastomeres, unfilled silicone polymers achieve only very low mechanical strength when cured. Adequate strength is only obtained by incorporating reinforcing fillers. Highly dispersible silicas with high surface areas are used almost exclusively for this purpose [2]. The silica improves mechanical properties like tensile strength, elongation, and tear resistance. The tensile strength of silicone elastomers reinforced with these silicas can be as much as 50 times that of unfilled systems.

Figure 1 shows a silicone rubber sample without silica. Mechanical properties are so poor that it can be crumbled between the fingers.

Figure 2 demonstrates a silicone sample reinforced with hydrophobic precipitated silica. It is nearly impossible to tear the silicone sample by hand.

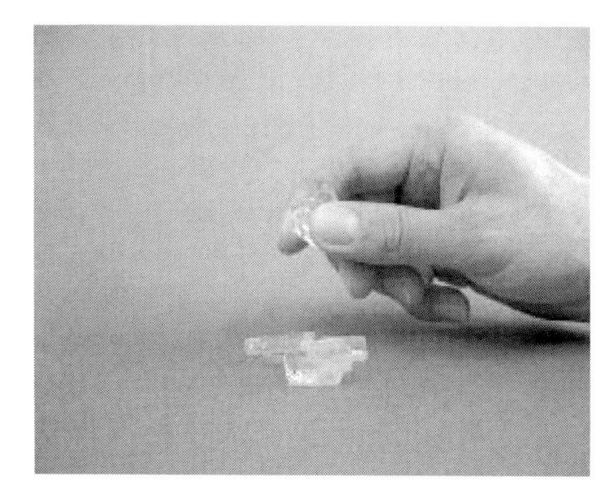

Fig. 1. Silicone rubber without silica.

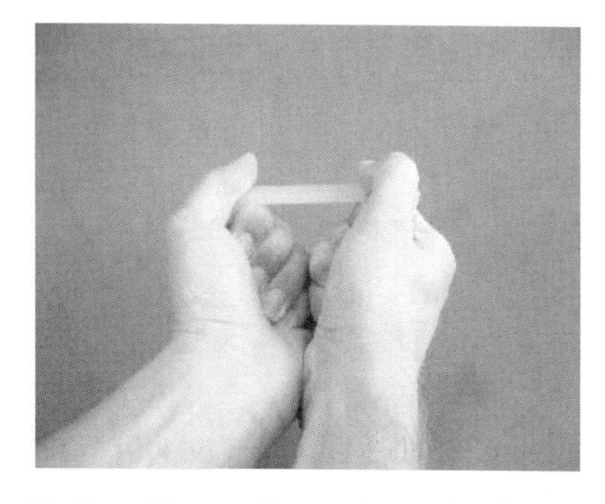

Fig. 2. Silicone rubber reinforced with silica.

However, adding silica can induce other difficulties concerning rheology (during processing) [3], moisture uptake [4], lower transparency, or sometimes even discoloration.

To meet the requirements, it is necessary to design silicas with high purity and suitable structures and surface properties.

Figure 3 shows mixtures of precipitated silicas in polydimethylsiloxane with different structures, surface areas, pore volumes, and pore size distributions. The silica on the right exhibits high transparency in dimethylpolysiloxanes.

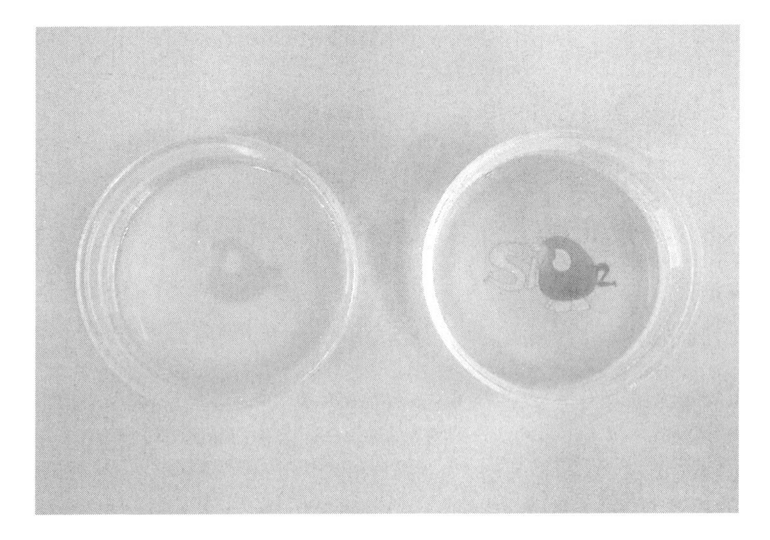

Fig. 3. Transparency of precipitated silicas in polydimethylsiloxane (5% silica).

The basic silica must have an adequate surface area (> 50 m^2/g), the right pore size distribution (so that the polymer can reach the silica surface) and an adapted silanol group density, and the silica must be highly dispersible.

To control the surface properties of silica (reduced filler–filler interactions, advanced filler-polymer interactions, minimized moisture uptake of the silica), it is necessary to modify the silica surface with silanes. Thickening behavior of silica, for example, is related to structure and surface properties of the silica.

Figure 4 shows a hydrophilic and the corresponding hydrophobic precipitated silica mixed with dimethylpolysiloxane. The mixture with the hydrophilic silica gives an almost solid compound. In contrast, the well hydrophobized silica leads to a liquid "silica-in-oil" dispersion.

Figure 5 shows silicas with different amounts of silanol groups, which cause very different moisture uptake from humid air. The advanced hydrophobic silica on the right adsorbs only a little water from the air. This silica is made using a hydrophilic silica with a very low silanol group density, a hydrophobization agent which fits perfectly to the basic silica, and a hydrophobization process which is very well adapted to the silica and the hydrophobization agent. Low moisture uptake is an important property for several applications, e.g., insulators, cable applications, and silicone systems which are cured by atmospheric moisture, e.g., silicone sealants. Introduction of too much water can lead to poor electrical properties of silicone insulators and bad shelf life in moisture-curing silicone systems.

Fig. 4. Hydrophilic and hydrophobic silica in polydimethylsiloxane (20% silica).

Fig. 5. Water content of different precipitated silicas at 50% air humidity.

Through the combination of special basic silicas, suitable silicone elastomers, and adequate hydrophobization conditions, it is possible to generate high-performance silicas with new properties at competitive prices.

References

[1] A. Tomanek, *Silicone und Technik*, Carl Hanser-Verlag, München, **1990**.

[2] R. Bode, H. Ferch, H. Fratzscher, *Gummi Asbest Kunstst.* **1967**, *20*(*12*), 699.

[3] Y. Todani, A. Ueda, *Nippon Gomu Kyokaishi* **1977**, *50*(*6*), 379.

[4] U. Schachtely, R. Nowak, *Schriftenreihe Pigmente*, Degussa AG Nr. *63*, 6.

Iodine Insertion into Pure Silica Hosts with Large Pores

R. Nechifor, P. Behrens*

Institut für Anorganische Chemie, Universität Hannover
Callinstrasse 9, 30167, Hannover, Germany
Tel.: +49 511 762 5188 — Fax: +49 511 762 3006
E-mail: Peter.Behrens@mbox.acb.uni-hannover.de

Keywords: insertion compounds, zeosils, zeolite analogues, iodine

Summary: Iodine was inserted into the voids of crystalline microporous materials built from pure SiO_2 (zeosils), namely UTD-1 (DON), SSZ-24 (AFI), CIT-5 (CFI), and ITQ-4 (IFR), by vapor phase loading. All these hosts possess large pores. Their iodine insertion compounds exhibit characteristic colors. The properties of these compounds were further studied by powder X-ray diffraction, UV-Vis and Raman spectroscopy. It turned out that the insertion compounds of the large-pore zeosils are unstable when removed from a saturated iodine atmosphere, a property in which they differ from the iodine insertion compounds of medium-pore zeosils. This instability hampered the characterization of these substances.

Introduction

Zeosils are microporous solids with tetrahedral frameworks, which are similar to those of aluminosilicate zeolites, but which are built from pure SiO_2 [1, 2]. With their neutral frameworks, zeosils do not show the typical properties of zeolites such as ion exchange, hydrophilicity, and catalytic activity; instead, these materials are hydrophobic and non-reactive. Zeosils find their main applications as highly selective adsorbents for sorbing nonpolar molecules from wet gas streams or aqueous solutions.

In basic research, zeosils can be used as geometrically structuring media: the host pore system can be filled with guest molecules. Because of the electroneutrality of the zeosils, the host-guest interactions are very weak. The dimensions (channel diameters, interconnection of channels) of the host pore system then regulate the arrangement of the inserted molecules in a purely geometric fashion; especially, the geometry of the pore system may allow for or eliminate interactions between the guest species.

This idea was tested by the insertion of several different inorganic species into zeosils. For example, the insertion of sulfur [3] or selenium [4] into the zeosil host decadodecasil 3R (DDR)

with cage-like voids leads to compounds containing S_7 and Se_6 rings, respectively, instead of the more stable chalcogen rings with eight atoms (which would be too large to fit into the cages). For the insertion of mercury halides HgX_2 (X: Cl, Br, I) [5, 6] and halogens (Br_2 and I_2), a larger variety of zeosil hosts with ten-membered ring channels (channel diameter ca. 5.5 Å) was employed. The results were especially interesting for the insertion compounds of iodine, where it was possible to control the guest-guest interactions by the different dimensionalities of the pore systems of the zeosil hosts [7, 8].

Here, we extend the investigations on iodine insertion compounds of zeosils to host compounds with larger pore diameters (see Fig. 1): ITQ-4, CIT-5, SSZ-24 and UTD-1, with twelve- or fourteen-membered ring windows and channel diameters between 6 and 10 Å. All the pore systems consist of uni-dimensional channels. In ITQ-4, the channel walls are corrugated, and a channel may be described as a series of large cages.

Experimental

The synthesis of the insertion compounds is a multi-step procedure involving first the preparation of the host material using organic molecules as structure-directing agents (SDAs). This is followed by the removal of the organic part occluded in the pores (channels) and finally the insertion of the guest material (Scheme 1). After the removal of the template, the host materials maintain a high crystallinity, as judged from X-ray diffraction.

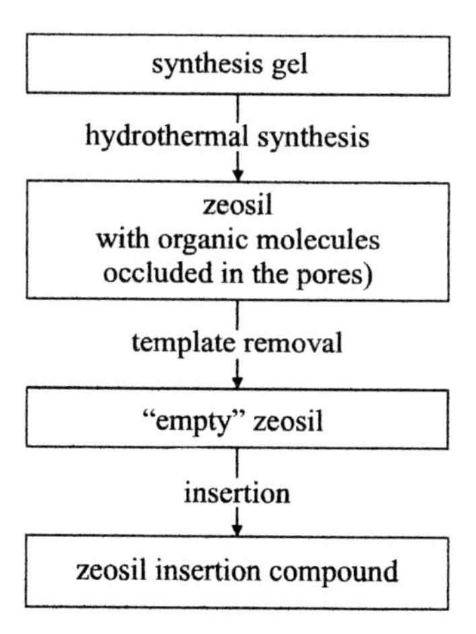

Scheme 1. Synthetic procedure for the insertion of zeosils.

Synthesis of the Host Compounds

The hydrothermal synthesis of the host compounds takes place in Teflon-lined steel autoclaves with 10 mL free volume. The reaction mixture and parameters were optimized in order to obtain large crystals with high yields [9]. The calcination process takes place in air and the temperature is chosen in order to ensure the complete combustion of the SDA, while maintaining a high crystallinity of the host material (Table 1). After calcination, the zeosils are colorless, except for UTD-1, which has a gray color due to cobalt oxide species blocking the pores. These are removed by washing with diluted hydrochloric acid.

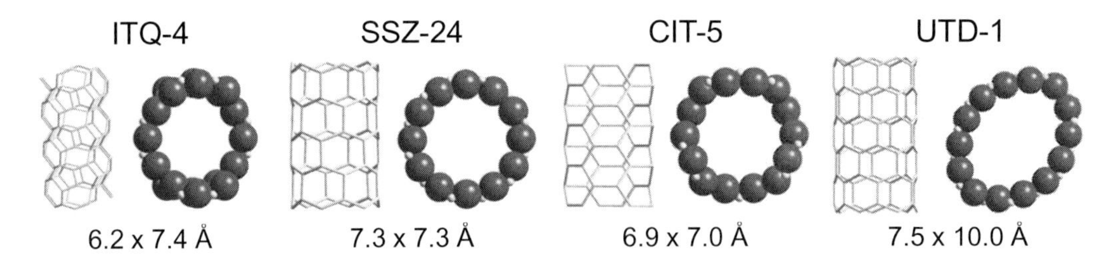

Fig. 1. Structures and pore dimensions of zeosil frameworks: ITQ-4 [10], SSZ-24 [11], CIT-5 [12] and UTD-1 [13]. In the wire models, Si is at the nodes and oxygen atoms are omitted for clarity. In the space-filling presentation, small light balls are Si, large dark balls are O.

Table 1. Parameters for the synthesis of the zeosils and the unit cell compositions of their iodine insertion compounds as determined by thermogravimetric investigations.

Zeosil	SDAs	reaction mixture SiO_2:SDA:H_2O	Reaction temp. [°C]	Reaction time [d]	Calcination temp. [°C]	Calcination time [d]	Unit cell composition I_2:SiO_2
ITQ-4	N-benzyl-chinuclidinium	1:0.5: 3–12	150	13	600	4	1.75:32
SSZ-24	N-methyl-sparteinium	1:0.5:8–14	175	5	650	7	1.45:24
CIT-5	N-methyl-sparteinium	1:0.5:8.5–20	175	6	650	6	1.13:32
UTD-1	Decamethyl-cobalticinium	1:0.13:61	150	14	550	6	1.61:64

Insertion Experiments

For the insertion, the zeosil and solid iodine were put in an evacuated glass tube, separated from each other by glass-wool, and heated in an oven to 160 °C for approximately five days. After this period, the ampoules were slowly cooled down to ambient temperature. The unit cell compositions of the insertion compounds as determined by thermogravimetry are given in Table 1.

Results and Discussions

All zeosils readily form insertion compounds with iodine, as shown by thermogravimetry and by the fact that the materials exhibit characteristic colors after insertion. As was shown in Ref. [7] for the iodine insertion compounds of zeosils with medium-sized pores (4 – 5.5 Å), the color gives indications about the degree of interaction between the iodine molecules in the voids of zeosils. Iodine-UTD-1 and iodine-ITQ-4 have a dark violet-to-brownish color, which resembles that of iodine-MFI. This color is similar to that of liquid iodine and indicates rather strong interactions between the iodine molecules in the large channels of UTD-1 and in the cage-like corrugations of the channels of ITQ-4 (as well as in the cage-like cross-sections of the channels of MFI [7, 14]). Iodine-CIT-5 possesses a violet color, iodine-SSZ-24 a red-violet, indicative of isolated iodine molecules. This is similar to I_2 in the vapor phase and to the insertion compound iodine-DDR [7, 14]. The pore system of the zeosil DDR consists of small cages, and it is difficult to correlate the violet color with the rather large channels of SSZ-24 and CIT-5. The color of the compounds is an especially important indicator for the properties of the insertion compounds of large-pore zeosils, because it turned out that these are much less stable than the compounds based on medium-pore zeosils. After opening the ampoule, they lose iodine rapidly, as can be observed by the color changing from brown or violet to white. This fact hampers the application of further characterization methods such as powder X-ray diffraction, thermogravimetric analysis, UV-Vis and Raman spectroscopy.

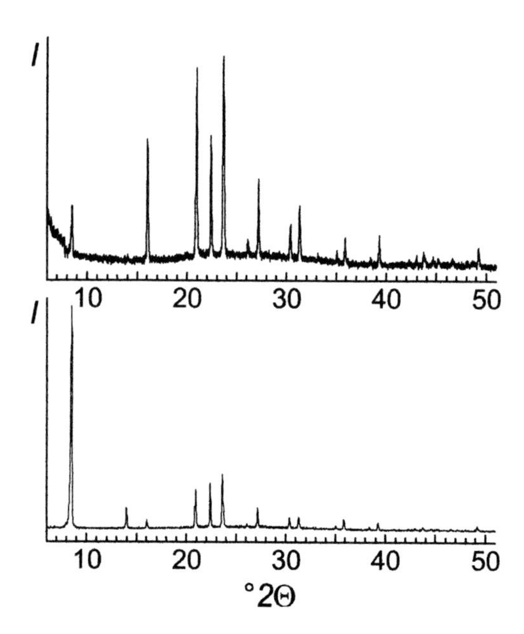

Fig. 2. Powder X-ray diffraction patterns of calcined SSZ-24 (lower trace) and of its insertion compound with iodine (upper trace).

The powder X-ray diffraction pattern of the as-synthesized, the calcined, and the loaded zeosils were recorded at room temperature. The diffraction data show significant differences in the reflection intensities of the loaded samples as compared to the calcined zeosils. Generally, the diffraction peaks in the low *2Θ* Theta region are suppressed upon insertion [14]. This can be explained by an increased electron density in the channels after insertion (Fig. 2).

The UV-Vis spectra of the insertion compounds, shown in Fig. 3, reveal the general problem of their characterization. All spectra look very similar, especially in the visible region. This finding is in contrast to the different colors of the compounds that they exhibit when they are still in the ampoules used for the preparation. In fact, all spectra correspond closely to that of iodine-DDR [7, 14], which in turn resembles that of gaseous iodine. Possibly, the insertion compounds have become partially deinserted during the measurement or the sample preparation. With lowering the concentration of the iodine molecules within the channels, the interactions between them will be reduced. Correspondingly, spectral characteristics of isolated molecules are observed.

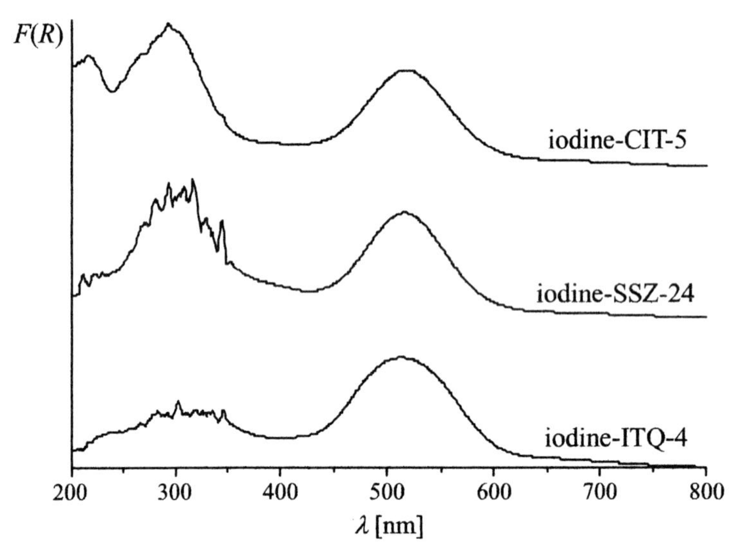

Fig. 3. UV-vis spectra of iodine insertion compounds of zeosils.

A very sensitive method to investigate the interactions between the occluded I_2 molecules is Raman spectroscopy [7, 14]. Intermolecular interactions between the inserted iodine molecules cause a weakening of the intramolecular bonds and lead to a shift of the stretching vibrations to lower frequencies as compared to the isolated molecule in the gas phase. Raman spectra are shown for iodine-ITQ-4 and iodine-SSZ-24 in Fig. 4. Iodine-ITQ-4 shows a single band at 207 cm^{-1}. This is characteristic of isolated I_2 molecules (for iodine-DDR this band is observed at 208 cm^{-1} [7, 14], for gaseous iodine at 213 cm^{-1} [15]) and may correspond to a partially deinserted sample. Iodine-SSZ-24 appears to be more stable. In addition to the peak at 207 cm^{-1}, there is a signal at 202 cm^{-1}, characteristic of iodine molecules interacting in one dimension [7, 14], and a broad emission band located around 180 cm^{-1}. This band is similar to the Raman spectrum of liquid iodine [16, 17] and thus characteristic of interacting iodine molecules.

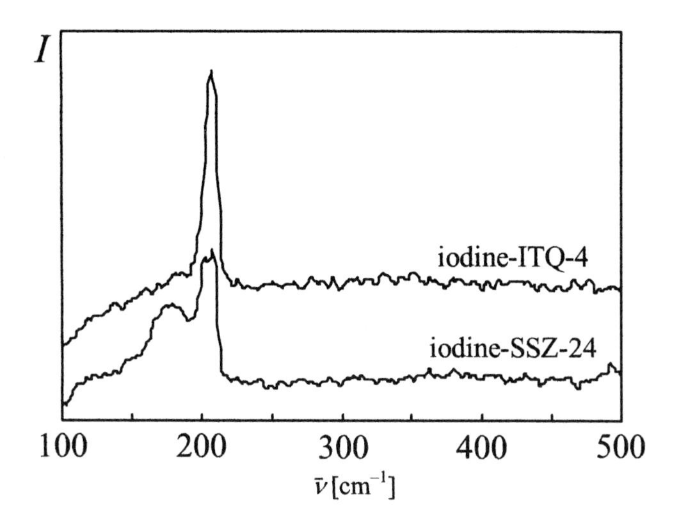

Fig. 4. Raman spectra of iodine insertion compounds of zeosils.

Conclusions

Iodine forms insertion compounds with large-pore zeosils, as evidenced by X-ray diffraction and by the characteristic colors of the substances. These compounds are, however, much less stable than those of medium-pore zeosils. This hampers their further characterization. The colors indicate that, in most of the insertion compounds, interactions between the iodine molecules are present. The Raman spectrum of SSZ-24 further verifies the presence of such interactions.

Acknowledgments: This work was supported by the "Fonds der Chemischen Industrie". R.N. thanks the Graduate Program "New Materials with Tailored Properties", which is sponsored by the State of Niedersachsen, for a Georg Christoph Lichtenberg fellowship. We thank PD Dr. Claus H. Rüscher from the Institut für Mineralogie of the Universität Hannover for recording the Raman spectra.

References

[1] H. Gies, B. Marler, U. Werthmann, in *Molecular Sieves, Vol. 1: Synthesis* (Eds.: G. Karge, J. Weitkamp), Springer, Berlin, Heidelberg, New York, **1998**, p. 35.

[2] P. Behrens, A. M. Schneider, in *Extended Abstracts of Silica '98*, Mulhouse, **1998**, p. 25.

[3] G. Wirnsberger, H. P. Fritzer, H. Koller, P. Behrens, A. Popitsch, *J. Molec. Struct.* **1999**, *480 – 481*, 699 – 704.

[4] G. Wirnsberger, H. P. Fritzer, R. Zink, A. Popitsch, B. Pillep, P. Behrens, *J. Phys. Chem. B*

1999, *103*, 5797 – 5801.

[5] P. Behrens, M. Hartl, G. Wirnsberger, A. Popitsch, B. Pillep, *Stud. Surf. Sci. Catal.* **2001**, *135*, 10.

[6] G. Wirnsberger, B. Pillep, A. Popitsch, P. Knoll, P. Behrens, *Chem. Eur. J.* **2002**, *8*, 3927.

[7] G. Wirnsberger, A. Popitsch, H. P. Fritzer, G. van de Goor, P. Behrens, *Angew. Chem. Int. Ed. Engl.* **1996**, *35*, 2777 – 2779.

[8] G. Flachenecker, P. Behrens, G. Knopp, M. Schmitt, T. Siebert, A. Vierheilig, G. Wirnsberger, A. Materny, *J. Phys. Chem. A* **1999**, *103*, 3854 – 3863.

[9] R. Jäger, *PhD Thesis,* Universität Hannover, **2002**.

[10] P. A. Barrett, M. A. Gamblor, A. Corma, R. H. Jones, L. A. Villaescusa, *Chem. Mater.* **1997**, *9*, 1713 – 1715.

[11] R. Bialek, W. M. Meier, M. Davis, M. J. Annen, *Zeolites* **1991**, *11*, 438 – 442.

[12] P Wagner, M. Yoshikawa, M. Lovallo, K. Tsuji, M. Tsapatsis, M. E. Davis, *Chem. Commun.* **1997**, *22*, 2179 – 2180.

[13] C. C. Freyhardt, M. Tsapatsis, R. F. Lobo, K. J. Balkus Jr., M. E. Davis, *Nature* **1996**, *381*, 295.

[14] G. Wirnsberger, *Doctoral Dissertation,* Graz University of Technology, **1998**.

[15] W. Holzer, W. F. Murphy, H. J. Bernstein, *J. Chem. Phys.* **1970**, *52*, 399.

[16] R. J. Magana, J. S. Lannin, *Phys. Rev. B* **1985**, *32*, 3819.

[17] M. Yao, N. Nakamura, H. Endo, Z. *Phys. Chem.* **1998**, *157*, 569.

Metal-Doped Silica Nano- and Microsized Tubular Structures

*M. Milbradt**

Inorganic and Analytical Chemistry, University of Paderborn
Warburger Str. 100, 33098 Paderborn, Germany
Tel.: +49 5251 602572
E-mail: marc.milbradt@gmx.de

H. C. Marsmann, S. Greulich-Weber

Experimental Physics, University of Paderborn
Warburger Str. 100, 33098 Paderborn, Germany

Keywords: sol–gel process, nanotubes, metal-doped silica gel

Summary: In this paper, the synthesis of silica-based nano- and microsized tubular structures is described. The nanotubes were obtained by a sol–gel process and investigated by ^{29}Si MAS NMR spectroscopy and electron microscopy. These tubular structures can be doped with several metals by adding metal ion solutions at the sol–gel synthesis.

Silica-based nano- and microsized tubular structures have been known since the mid-1990s [1]. The preparation using the sol–gel process is a low-temperature process at room temperature and offers scope for manipulation of, e.g., the size and shape of these tubes. Silica-based tubular structures have many advantages, such as easy accessibility, stability, and the possibility of surface functionalization. They can be used for catalysis, separation, reinforcing materials, and fillers for plastics and ceramics.

The SiO_2 nanotubes were obtained using the sol–gel method at room temperature in a solution of DL-tartaric acid, ethanol, and tetraethylorthosilicate (TEOS) via the addition of ammonium hydroxide [1]. They had a square shape. Scheme 1 shows the mechanism of the sol–gel process.

In the ^{29}Si MAS NMR spectrum (Fig. 1) two signals are observed at the chemical shifts –112 ppm (Q^4) and –104 ppm (Q^3). The signal at –93 ppm shows the Q^2 groups at a very small intensity. For identifying the silicon atoms the Q notation is used. Q^n is a silicon atom bonded to four oxygen atoms in a tetrahedral environment. The superscript denotes that n oxygen atoms were connected to further silicon atoms (Scheme 2).

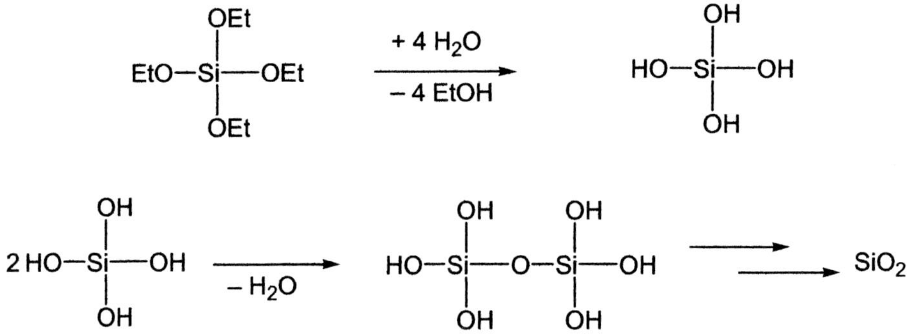

$$Si(OEt)_4 \xrightarrow[-4\,EtOH]{+4\,H_2O} Si(OH)_4$$

$$2\,Si(OH)_4 \xrightarrow[-H_2O]{} (HO)_3Si{-}O{-}Si(OH)_3 \longrightarrow\longrightarrow SiO_2$$

Scheme 1. Mechanism of the sol–gel process.

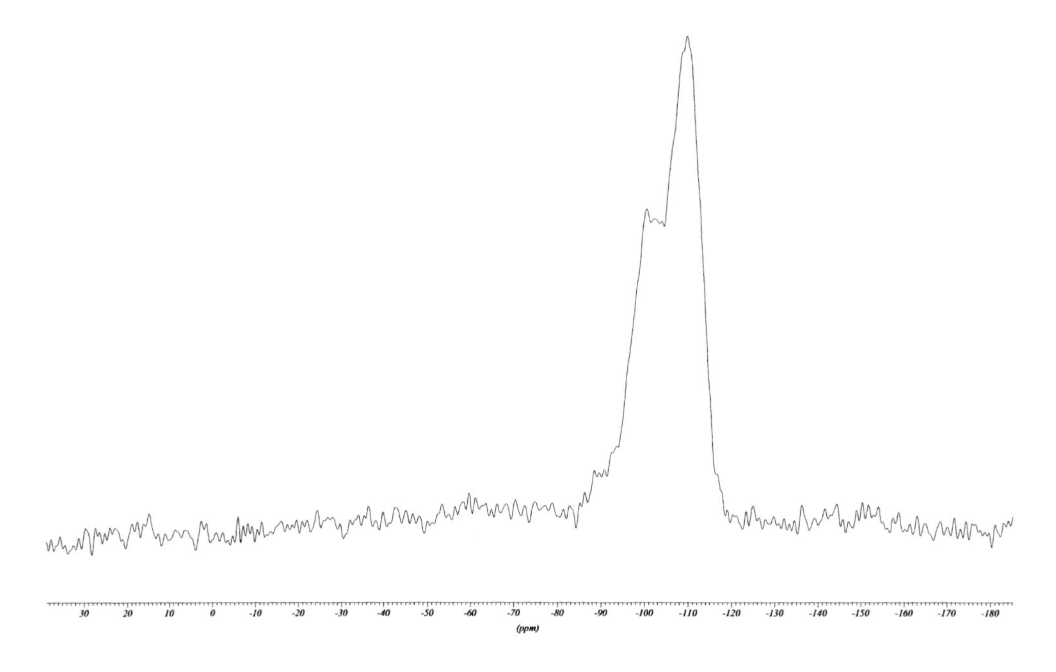

Fig. 1. Typical ^{29}Si MAS NMR spectrum of silica-based nanotubes.

The concentrations of Q^2, Q^3 and Q^4 can be taken from the ^{29}Si MAS NMR spectrum: 3.3% for Q^2, 58.5% for Q^3 and 38.2% for Q^4. The Q^2 and Q^3 groups were from incompletely hydrolyzed/condensed TEOS, while Q^3 groups were mainly on the surface of the nanotubes.

Metal-doped silica gels exhibit a wide range of optical properties which allow them to be used for optical and optoelectronic applications. The preparation is done via the sol–gel process. Emission and absorption maxima as well as quantum yield can be adjusted by built-in functionalized silanes [2] or adsorbed semiconductor or metal colloids, respectively [3, 4].

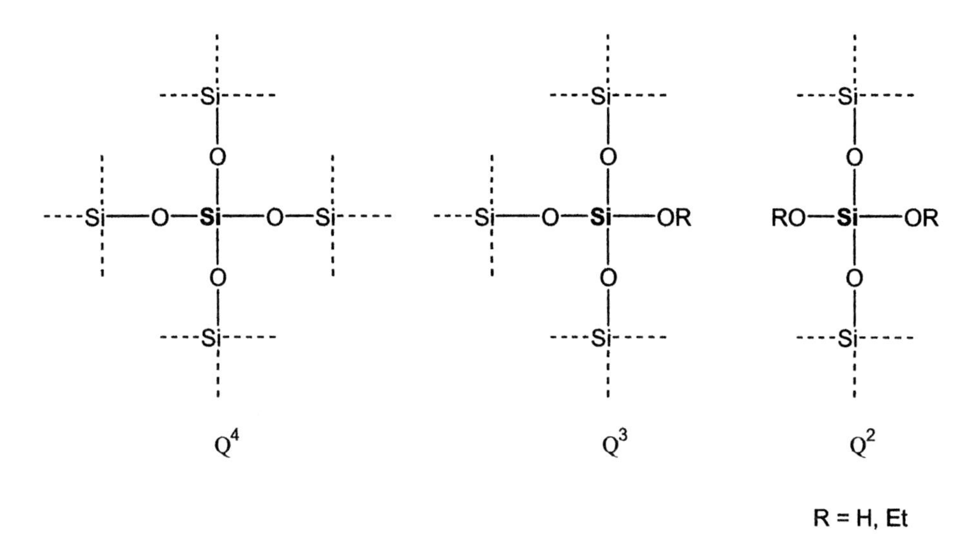

Scheme 2. Q^n notation.

Fig. 2. SEM image of silica tubular structures.

It is our aim to combine the outstanding properties of nano- microtubes and metal-doped silica gel. There are many kinds of dopands, e.g., transition metal and rare-earth ions. A further advantage is the high heat resistance of these tubular structures up to 250 °C with organically functionalization, and up to 800 °C with only metal functionalization. Metal ions can be built directly in the silica backbone or adsorbed on the surface. Functionalized silanes can also be used to coordinate metal ions and build them into the tubes.

The metal-doped tubular structures are investigated by various spectroscopic methods (e.g., ESR).

The signal in Fig. 3 shows an antiferromagnetically coupled Cr^{3+}–Cr^{3+} pair or a Cr^{5+} in a tetrahedral environment.

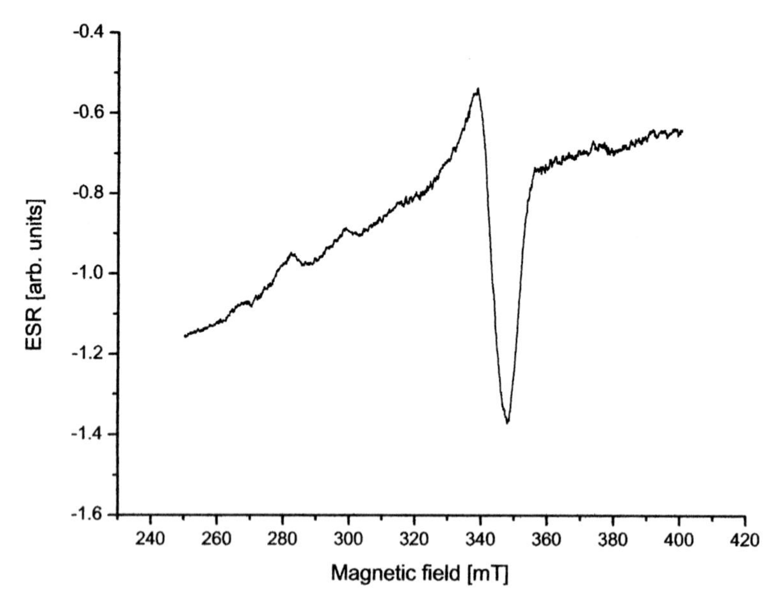

Fig. 3. ESR spectrum of Cr-doped nanotubes.

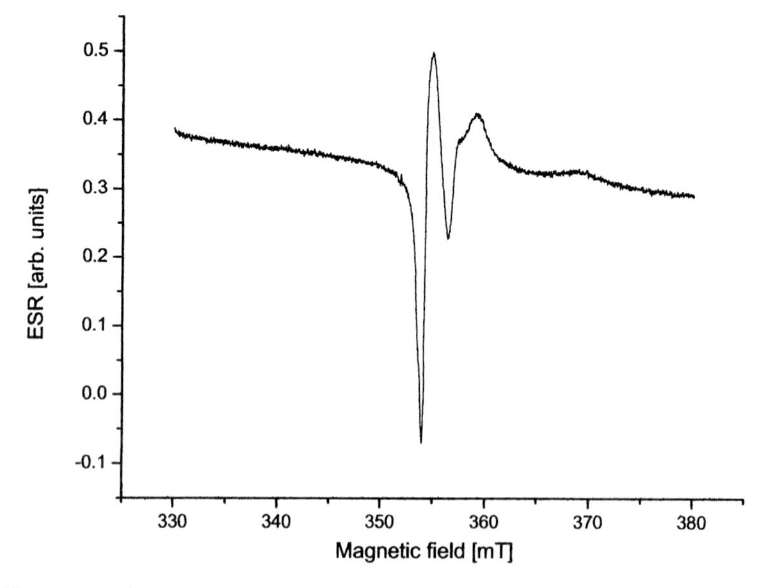

Fig. 4. ESR spectrum of Cu-doped nanotubes.

The signal in Fig. 4 shows Cu^{2+} coordinated with N in an octahedral environment. The Cu-complex may be adsorbed on the surface of the nanotubes.

Both ESR spectra were measured at 9.83 GHz/10K.

The SEM image in Fig. 5 shows a larger amount of non-tubular structures in the presence of metal ions.

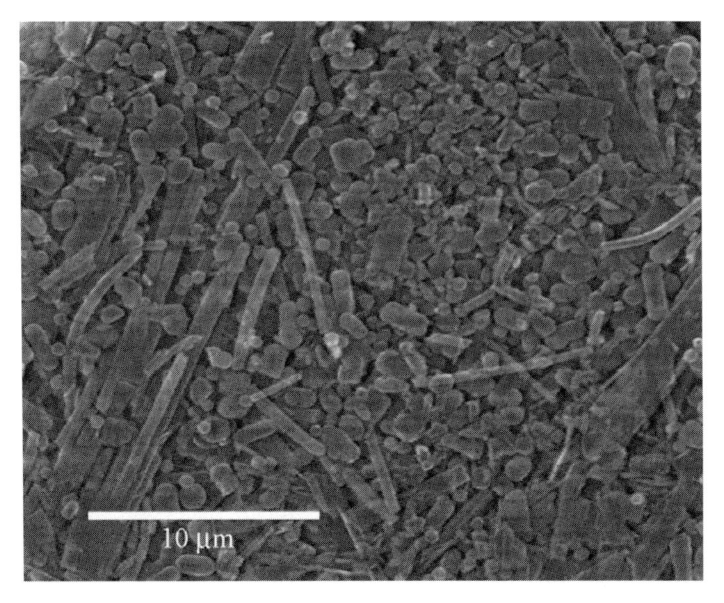

Fig. 5. SEM image of Cu-doped tubular structures.

References

[1] H. Nakamura, Y. Matsui, *J. Am. Chem. Soc.* **1995**, *117*, 2651.

[2] M. Iwasaki, J. Kuraki, S. Ho, *J. Sol–Gel Science and Technology* **1998**, *13*, 587.

[3] S. T. Selvan, T. Hayakawa, M. Nogami, *J. Non-Cryst. Solids* **2001**, *291*, 137.

[4] S. T. Selvan, T. Hayakawa, M. Nogami, *J. Phys. Chem B* **1999**, *103*, 7064.

Branched Functionalized Polysiloxane–Silica Hybrids for Immobilization of Catalyst

K. Rózga-Wijas, J. Chojnowski, W. Fortuniak, M. Ścibiorek*

Centre of Molecular and Macromolecular Studies,
Polish Academy of Sciences, 112 Sienkiewicza, 90-363 Łódź, Poland
Tel.: +48 426818952 — Fax: +48 426847126
E-mail: krysia@bilbo.cbmm.lodz.pl

Keywords: silica hybrid, modified silica, dendritic polysiloxane, functionalized polysiloxane, supported catalyst, platinum, hydrosilylation

Summary: Two methods are used for synthesis of branched functionalized polysiloxane–SiO_2 hybrids. In the first approach, the branched polymer is generated on the functionalized silica surface by graft-on-graft techniques. The graft is living linear polysiloxane functionalized in side groups synthesized by controlled polymerization or copolymerization of cyclotrisiloxanes, while the reactive core contains silyl chloride groups. The other approach uses the grafting of the branched polysiloxane-containing vinyl groups on the functionalized silica surface by hydrosilylation. Hybrids had vinyl or diphenylalkylphosphine functional groups on the polysiloxane moiety. They were destined for the immobilization of metal complex catalysts.

Introduction

In the present study, we describe the methods of preparing the silica hybrids of linear and branched functional polysiloxanes which could be used as a support for metal complex catalysts. The way in which the catalyst operates when it is attached to the polysiloxane moiety of the hybrid suspended in a polysiloxane solvent should be similar to the way it operates when in solution. Thus, its high catalytic activity is expected. On the other hand, it is easily separated from the reaction products and may be recycled or used in the continuous process. A high catalytic activity and specificity may be achieved if a polymer with a highly branched structure is used for the immobilization of catalysts [1 – 3]. Considerable amounts of catalytic groups may be placed in the external part of the branched macromolecule.

High static and dynamic flexibility of the polysiloxane chain, associated with a very low energy barrier to rotation around their skeletal bonds and a low energy of deformation of the SiOSi bond angle, make the polymer soluble in many solvents. The catalyst attached to such a mobile polymer chain, which can adopt many conformations, is available for the interaction with reactants in a catalyzed system.

Modification of Silica Surface

Silica gel with a relatively large particle size (0.06 – 0.5 mm) was used to ensure its sedimentation. It showed relatively high values of specific surface area (300 – 550 m^2g^{-1}) and medium pore size (60–100 Å).

The surface modification was a two-step process including introduction of $-SiCH=CH_2$ functions and their transformation to $-SiCH_2CH_2SiCl$. In the first step, the silica surface was subjected to silylation using a silicone coupling agent: vinyltrichlorosilane, vinyltriethoxysilane, or 3-mercaptopropyltrimethoxysilane [4]. The modified silica contained 1.0×10^{-3} to 1.8×10^{-3} mol g^{-1} of attached vinyl or thiol functional groups. These groups were further transformed to silyl chloride or silyl hydride functions by hydrosilylation or ene-thiol addition.

Full experimental details are described in Ref. [5].

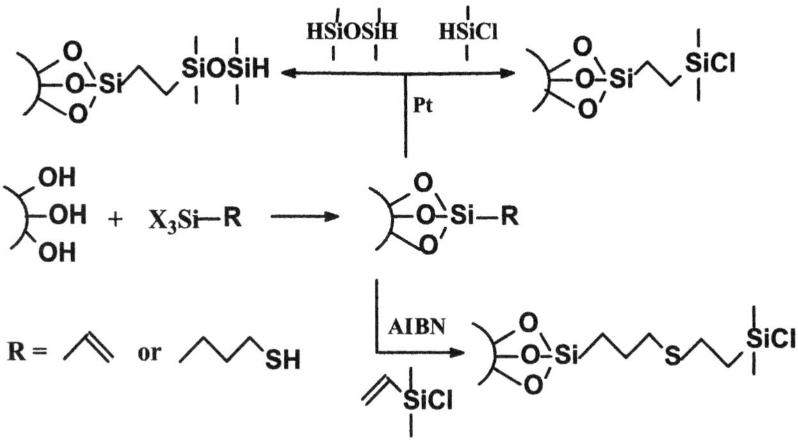

Scheme 1.

Grafting of Linear Functionalized Living Polysiloxane onto the SiO_2-Particle Surface and onto Silica Hybrids

The hybrids were prepared by a coupling reaction between the living polysiloxane-bearing vinyl or diphenylphosphinoethyl groups bonded to silicon and having the lithium silanolate group at one chain end and the chloride group at the silica surface.

The living polysiloxanes having various densities of vinyl groups were obtained by anionic living ring-opening polymerization of 2,4,6-trivinyl-2,4,6-trimethylcyclotrisiloxane (V_3) or by copolymerization of this monomer with hexamethylcyclotrisiloxane (D_3) or with 2-vinyl-2,4,4,6,6-

pentamethylcyclotrisiloxane (VD_2) [6, 7]. The living siloxane-bearing phosphine groups are obtained by polymerization of 2-diphenylphosphinoethyl-2,4,4,6,6-pentamethylcyclotrisiloxane in THF initiated by butyllithium. It leads to a polymer with phosphine groups pendant to the polysiloxane chain, uniformly spread along the polymer chain [8].

The vinyl groups on grafted polysiloxanes were transformed into the silyl chloride groups (Scheme 2). Living polysiloxanes were terminated on these groups, grafting new generations of branches. The yield of grafting on the linear polysiloxane is small (13%). A somewhat larger yield was observed for grafting on branched polymer (19%). Thus, the functional groups on the branched polymer, which does not penetrate so deeply into the pores in the silica, are more available for the external reactant as compared with those on linear polymer [5].

The functionalization of the polysiloxane-silica hybrid with tertiary phosphine groups bonded to silicon by the ethylene bridge was accomplished by grafting the living polysiloxane bearing these groups to the silica surface functionalized with silyl chloride groups. Phosphines are regarded as excellent ligands for transition metal catalysts.

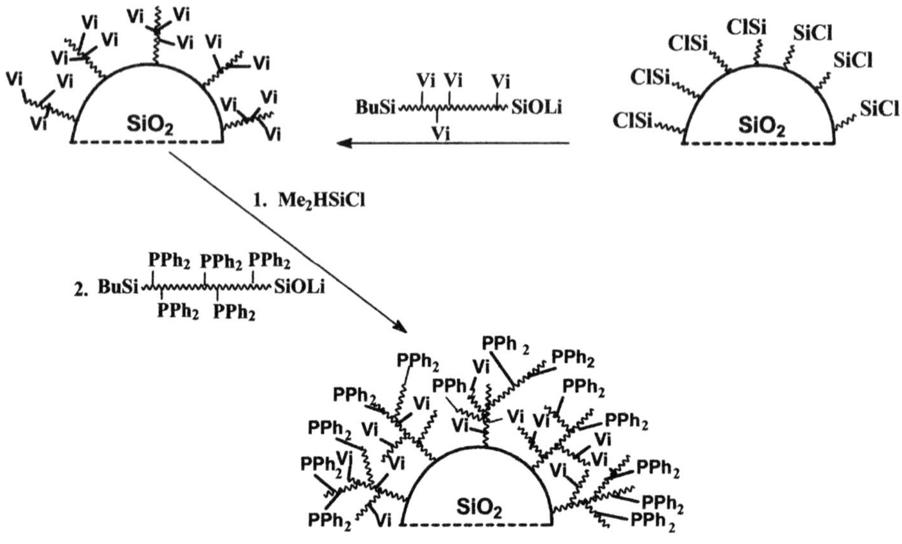

Scheme 2.

The grafting of the living vinylated polysiloxanes to the reactive groups (Si-Cl) on the silica surface occurs with a yield from 12 to 20%. In the same conditions, the polysiloxane bearing the phosphine groups is attached to the surface with a still low yield, about 14%. The characteristics of selected hybrids are presented in Table 1. Full details are described in Ref. [5].

The polysiloxane-silica hybrids were characterized by elemental analysis, ^{13}C NMR of the particle suspension, ^{13}C and ^{29}Si solid-state NMR spectra, FT IR spectroscopy and thermogravimetry. The examples of ^{29}Si NMR solid-state and suspension ^{13}C NMR spectra are shown in Fig. 1.

Table 1. Grafting of linear polysiloxanes onto the silica surface.

No	Grafting	Graft[a] M_n,	N_{vi},	D/V	Sb	Modified silica D_{fg}[c] 10^3 [mol g^{-1}]	Y [%]	Silica hybrid C_{vinyl} 10^3 [mol g^{-1}]	W_c/SiO_2 [g/g]
1		5000[b]	6.54	8.9	A	0.21	20.0	0.27	0.26[d]
2	)~SiCl + LiD_mV_nBu	2500	17.51	0.68	B	0.43	19.5	1.47	0.26[d]
3		10070	42.50	2	B	0.46	12.2	1.02	0.31
4	)~SiCl + LiV_nBu	2350	25.81	0	C	0.54	20.0	2.78	0.33[d]
5	)~SiCl + $Li(D_2P)_mBu$	2450	5.59	2.0	B	0.43	14.2	0.34	0.18[d]
6		3000	6.83	2.0	B	–	–	0.45	0.25
7	)~SiOSiH + $BuV_nD_mSiMe_3$	3090	10.14	2.78	A	0.67	11.2	0.68	0.30[d]

D = [Me_2SiO], V = [$(CH_2{=}CH)MeSiO$], P = [$(PPh_2CH_2CH_2)MeSiO$], N_{Vi}: number of vinyl groups (from ^{1}H NMR), D/V: ratio of the number of dimethylsiloxane units to the number of vinylmethylsiloxane units, Sb: symbol, D_{fg}: density of function groups, Y: yield of grafting, C_{vinyl}: Content of vinyl groups, W_c: weight content of polymers, a: data from analysis of unreacted polysiloxanes isolated after grafting, b: data from stoichiometry, c: per gram of modified silica, d: from Ref. [5].

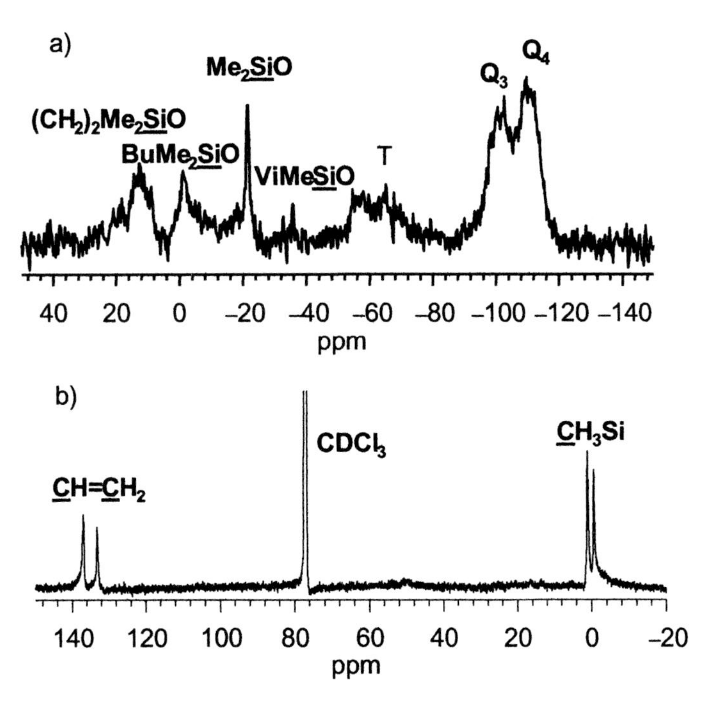

Fig. 1. MAS ^{29}Si NMR (a) and suspension ^{13}C NMR (b) spectra of the hybrid bearing grafted poly-[vinylmethylsiloxane-co-dimethylsiloxane] (M_n = 2500, 0.26 g polymer per 1 g SiO_2). T = C–Si(O)$_3$, Q_3 = Si(OSi)$_3$(O), Q_4 = Si(OSi)$_4$, (Table 1, entry 2).

Grafting of Branched Functionalized Polysiloxanes onto the Silica Surface

Two highly branched polysiloxanes, bearing pendant vinyl groups, were grafted onto the modified silica surface. Synthesis of the hybrid support was performed in two subsequent hydrosilylation reactions catalyzed by Pt(0) Karstedt catalyst. In the first reaction, vinylated silica reacted with 1,1,3,3-tetramethyldisiloxane to give Si–H sites on the silica surface. In the next stage these sites reacted with the vinyl groups of the highly branched dendritic (M_n = 140 000, 760 vinyl groups in macromolecule) or comb-branched polysiloxanes (M_n = 33 400, 270 vinyl groups in macromolecule) generated separately by the method described earlier [6, 7]. The yield of grafting both polymers is low (0.30 – 0.35 g of polymer per 1 g SiO_2). This may be because of the limited ability of the high-molecular-weight branched polymer to penetrate into the pores of the silica gel particles [5].

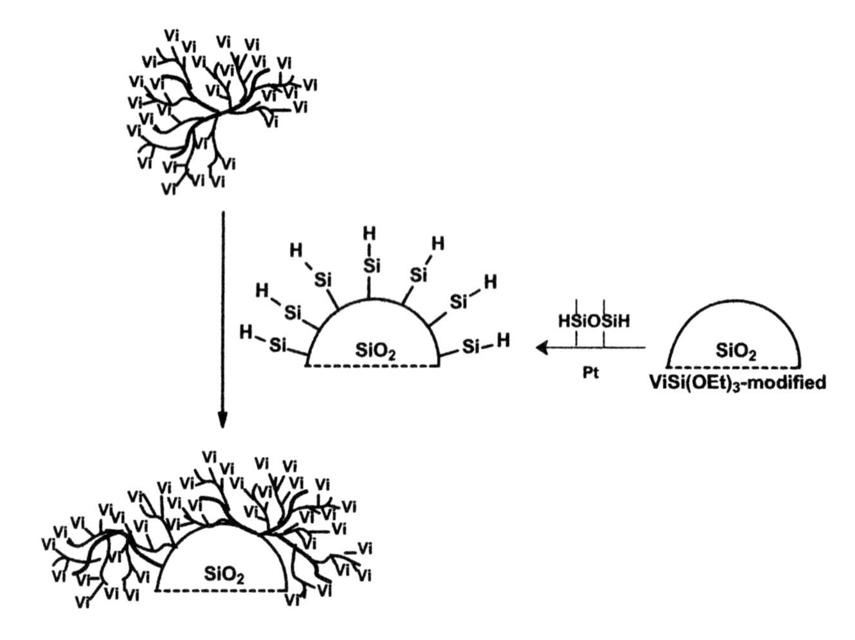

Scheme 3.

Catalytic Tests for a Catalyst Immobilized on Hybrids

The polysiloxane - SiO_2 hybrids were tested for their capability for heterogenizing homogeneous catalysts. The platinum(II) complex was obtained by replacement of ligands in the parent metal complex $PtCl_2(PhCN)_2$ by Vi groups.

The catalytic activity of the immobilized complex catalyst in hydrosilylation of 1-hexene with dimethylphenylsilane in toluene was found to be high. The catalytic tests were performed by Prof. Z. Michalska in the Institute of Polymers, Technical University, and are described in [10].

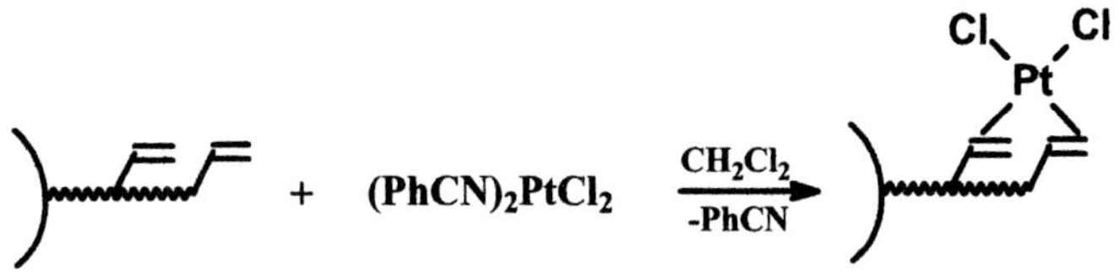

$$\text{polymer-diene} + (PhCN)_2PtCl_2 \xrightarrow[-PhCN]{CH_2Cl_2} \text{polymer-diene-}PtCl_2$$

Scheme 4.

$$Me_2PhSiH + CH_2{=}CH(CH_2)_3CH_3 \longrightarrow Me_3PhSi(CH_2)_5CH_3$$

Scheme 5.

Acknowledgment: The research was supported by the National Research Council (KBN) Grant No: PB2/15/T09/99/01c.

References

[1] S. C. Bourque, H. Alper, L. E. Manzer, P. Arga, *J. Am. Chem. Soc.* **2000**, *122*, 956.

[2] L. Rapartz, D. F. Foster, R. E. Morris, A. M. Z. Slavin, D. J. Cole-Hamilton, *J. Chem. Soc., Dalton Trans.* **2002**, 1997.

[3] M. T. Reetz, D. Giebel, *Angew. Chem. Int. Ed.* **2000**, *39*, 2498.

[4] J. Brinker, G. W. Scherer in *Sol–Gel Science*, Academic Press, San Diego **1990.**

[5] K. Rózga-Wijas, J. Chojnowski, W. Fortuniak, M. Ścibiorek, Z. M. Michalska, Ł. Rogalski, *J. Mater. Chem.* **2003**, *13*, 2301.

[6] J. Chojnowski, M. Cypryk, W. Fortuniak, K. Rózga-Wijas, M. Ścibiorek, *Polymer* **2002**, *43*, 1993.

[7] J. Chojnowski, M. Cypryk, W. Fortuniak, M. Ścibiorek, K. Rózga-Wijas, *Macromolecules* **2003**, 36, 3890.

[8] J. Chojnowski, K. Rózga-Wijas, *J. Inorg. Organometal Polym.* **1992**, *2*, 297.

[9] J. Chojnowski, M. Cypryk, W. Fortuniak, K. Kaźmierski, K. Rózga-Wijas, M. Ścibiorek in *Synthesis and Properties of Silicones and Silicone-modified Materiale* (Eds.: S. J. Clarson, J. Fitzgerald, M. J. Owen, S. D. Smith, M. E.Van Dyke) ACS Washington D.C. **2003**, Ch. 2, p 12.

[10] Z. M. Michalska, Ł. Rogaliki, K. Rózga-Wijas, J. Chojnowski, W. Fortuniak, M. Ścibiorek, *J. Mol. Cat.* **2004**, *208,* 187.

Investigations into the Kinetics of the Polyamine–Silica System and Its Relevance to Biomineralization

Petra Bärnreuther,* Michael Jahns, Ilka Krueger, Peter Behrens

Institut für Anorganische Chemie, Universität Hannover
Callinstraße 9, 30167 Hannover, Germany
Tel.: +49 511 762 3660 — Fax: +49 511 762 3006
E-mail: Peter.Behrens@mbox.acb.uni-hannover.de

Sandra Horstmann, Henning Menzel

Institut für Technische Chemie, Abt. TC Makromolekularer Stoffe
Technische Universität Braunschweig
Hans-Sommer-Str. 10, 38106 Braunschweig, Germany
Tel.: +49 531 391 5361 — Fax: +49 531 391 5357
E-mail: h.menzel@tu-bs.de

Keywords: silica condensation, polyamines, biomineralization, silica

Summary: Linear polypropylene imines with low molecular mass were found in the shells of diatoms and are thought to influence the biomineralization process in these algae [1]. In order to elucidate the chemistry in polyamine–silica model systems, we have investigated the kinetics of silica condensation in the presence of various polyethylene and polypropylene imines using the molybdate method. All amines accelerate the silica condensation, with the degree of acceleration depending on the chemical nature of the polyamines (ethylene or propylene linkages, methylated or non-methylated), their degree of polymerization, and their architecture (linear or branched). The degree of acceleration seems to scale with the increase in pH which the amines cause in the reaction solution. Therefore, it appears that the main influence of the polyamine on the first steps of the silica condensation reaction (to which the molybdate method is sensitive) is by the adjustment of the pH value. Interestingly, nature has chosen amines which give rise to only a moderate degree of acceleration and not those which provide the strongest acceleration. In the natural biosilicification systems, this may allow for more time to form the intricate macrostructures of the biominerals.

Introduction

Biominerals are widespread in nature [2, 3]. They occur, e.g., as calcium phosphates in bones and teeth or as magnetite in magnetotactic bacteria [4]. Biominerals are composite materials consisting of the inorganic mineral and an organic matrix. They are formed by an organism at ambient temperature and pressure and at a physiological pH value in a reasonable period of time. The organic matrix influences the formation and the properties of the biomineral in a variety of ways.

Silica in amorphous form is generated by a wide variety of species, including horsetail plants, marine sponges, and diatoms [5]. In recent years, some constituents of the organic matrix of biosilicifying organisms have been identified, for example, the proteins of the silicatein family from marine sponges [6] and silaffins from diatoms [7, 8]. Silaffins are short-chain peptides which carry unusual modifications, namely linear chains of polypropylene imine (PPI), with most of the imino groups methylated. Such PPI polymers occur also in a free, unbound state in all diatoms investigated, with chain length P_n varying between 8 and 22 monomer units; in some diatoms, especially those which show rather simple morphologies of their silica shells, only PPI and no silaffins were found [1]. For silaffins as well as for polyamines, first experiments using material which had been extracted from diatom shells have shown that these substances influence the precipitation of silica [9]. Sumper has proposed a phase separation model, according to which the formation of simple silica patterns in diatoms is caused by self-organization processes in polyamine–silica solutions, thus assigning a central role to the polyamines [10].

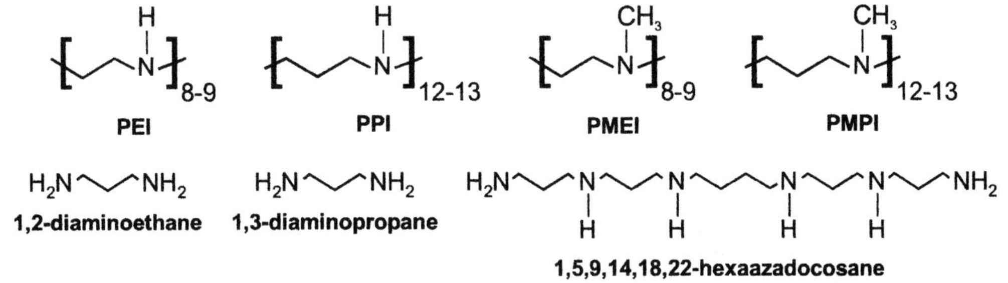

PEI (n = 8-9) PPI (n = 12-13) PMEI (n = 8-9) PMPI (n = 12-13)

1,2-diaminoethane 1,3-diaminopropane 1,5,9,14,18,22-hexaazadocosane

Fig. 1. Polyamines used in this study.

In contrast to extracts obtained from nature, which usually consist of a mixture of substances and are available only in very limited amounts, synthetic model systems have the advantages of controlled purity and good availability. Such systems enable the scientist to vary reaction parameters in a systematic fashion and to adjust individual functions to analyze their influence on the system. However, model systems should be as close as possible to the natural system. Therefore, we have undertaken a study to elucidate the chemical behavior of polyamine–silica systems, including not only polyamines which can be purchased, but also polyamines specifically synthesized, with structures as close as possible to the naturally occurring ones [11]. Here, we report on investigations on the kinetics of silica condensation in the presence of various polyamines

(Fig. 1). These include linear polyethylene and polypropylene imines of varying degree of polymerization (P_n = 10 to 20) in non-methylated (PEI and PPI) and methylated (PMEI and PMPI) forms, which were specifically synthesized. Hexaazadocosane, which may be regarded as a model for a PPI with P_n = 6, was prepared by Michael addition of acrylonitrile to spermidine [12]. In addition, a commercially available PEI with high molecular mass and with a branched polymer architecture was employed.

Experimental

Preparation of the Amines

According to Saegusa et al. [13], linear polyamines (PEI and PPI) were synthesized starting from oxazolines or 1,3-oxazines with subsequent hydrolysis or methylation, respectively [11]. The degree of polymerization was varied between 10 and 20. Here we present results of investigations on PEI with P_n= 8 – 9 and on PPI with P_n= 12 – 13.

Kinetic Measurements

The kinetics of the silicic acid condensation was investigated using the molybdate method introduced by Alexander [14]. Upon addition of a molybdate solution to a silicate solution, monomeric and dimeric silicic acid react to form the yellow molybdosilicate $[SiMo_{12}O_{40}]^{8-}$ [5]. The change in the absorption of this solution reflects the decrease in monomeric and dimeric silicic acid due to the condensation reaction. Because of the reaction equilibrium between silicic acid and the molybdate as well as between monomeric, dimeric, and oligomeric silicic acids, it is important to adhere to a well-defined, constant, and reproducible time protocol for the formation of the molybdosilicate.

The polyamine solutions were prepared by dissolving the amine in 10^{-5} M hydrochloric acid so that the amine concentrations were 1 mg/L, 10 mg/L, 100 mg/L and 1000 mg/L, respectively. The silicic acid solution was prepared by hydrolysis of tetramethoxysilane in the stoichiometric amount of 0.01 M hydrochloric acid. To the freshly prepared silicic acid solution, the amine solution was added so that the initial silicic acid concentration was 0.16 mol/L. The acid concentration was 1.13 mmol/L. The progress of the condensation was followed by taking samples from the reacting mixture, diluting the sample to obtain a silicic acid concentration appropriate for the UV-vis measurement, and adding this diluted solution to the acidic molybdate solution. After exactly 10 min reaction time, the absorbance of the yellow solution was measured at a wavelength of 400 nm.

Results and Discussion

As shown in Fig. 2a, linear PEI (P_n = 8 – 9) shows a significant acceleration of the reaction rate compared to the condensation reaction under similar conditions but without the addition of an amine. In contrast, branched PEI with its high molecular mass shows nearly no influence on the condensation rate. This implies that the molecular mass and the architecture of the polymer (linear or branched) have a definite influence on the reaction rate. The commercially available branched PEI is therefore not a good model [15] for the naturally occurring polyamines. The methylated analogon of the linear PEI, namely PMEI (polymethylethyleneimine), has a significantly less marked effect on the condensation rate, i.e., the degree of methylation also influences the accelerating effect of the polyamine.

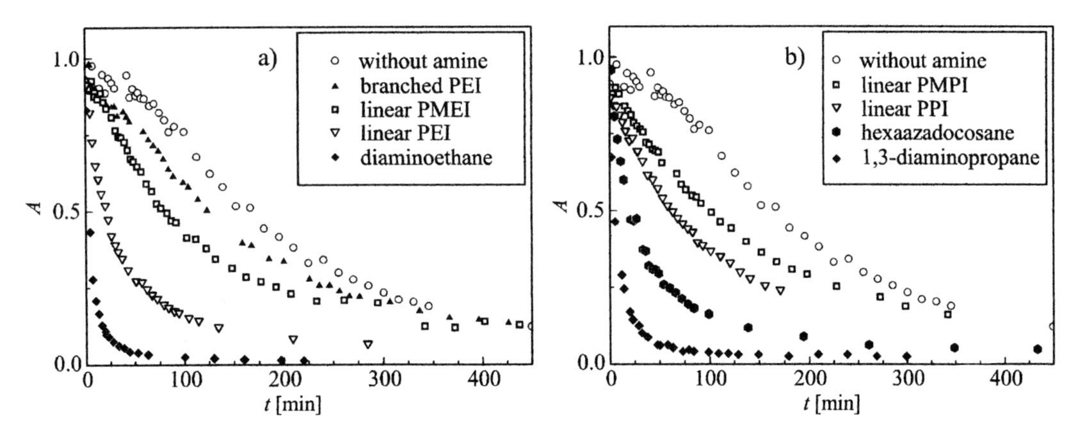

Fig. 2. Plot of the absorbance of the molybdosilicate versus the reaction time: (a) polyethylene amines, (b) polypropylene amines. Amine concentration: 10 mg/L.

Similar behavior is found for the polypropyleneimine derivatives, as shown in Fig. 2b. Again the linear PPI ($P_n = 12 - 13$) shows a significant acceleration of the condensation rate, whereas the methylated variant PMPI shows a less marked effect. A comparison of the hexaazadocosane with the linear PPI shows that the hexaazadocosane with a smaller molecular mass (corresponding to a P_n of 6) shows a stronger acceleration of the condensation. However, hexaazadocosane has a ratio of secondary to primary amine groups of 2 : 1. As can be seen by the examples of diaminoethane (Fig. 2a) and 1,3-diaminopropane (Fig. 2b), both of which contain only primary amino groups, primary amino groups generally appear to exert a stronger acceleration effect.

The concentration of the polyamines in the natural process of diatom biosilicification is not known. Figure 3 summarizes the influence of the concentration of our synthetic polyamines on the condensation reaction. As expected, a higher concentration gives rise to a stronger acceleration.

Kinetic plots of the different condensation reactions assuming rate laws of first, second, and third order are shown in an exemplary way for linear PPI in Fig. 4. The occurrence of these reaction orders has been observed and explained before [5, 16]. When linear regions appear in these plots, the reaction can be assumed to be of the corresponding order. In most of the cases, a single order of reaction is not sufficient to explain a condensation reaction over the whole course of reaction time. However, different reaction orders can be assigned to different phases of the reaction [5, 16].

For a slow reaction like the condensation without amines, there is an "initial phase" where the molybdosilicate concentration is constant, i.e., apparently no condensation occurs. In this period the dimerization of monosilicic acid takes place [5]. Because of the nature of the molybdate method, both monomeric and dimeric silicic acid form the molybdosilicate, so that during this period the absorption is constant.

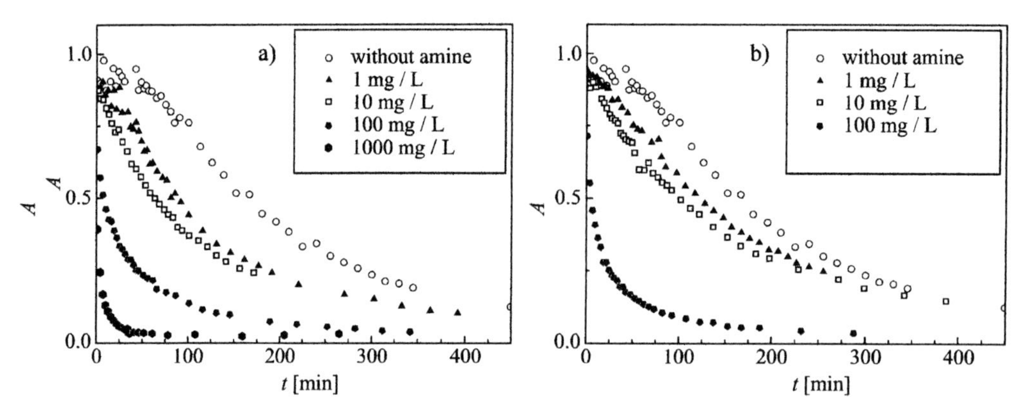

Fig. 3. Plot of the absorbance of the molybdosilicate versus the reaction time for different amine concentrations; (a) linear PPI, (b) linear PMPI.

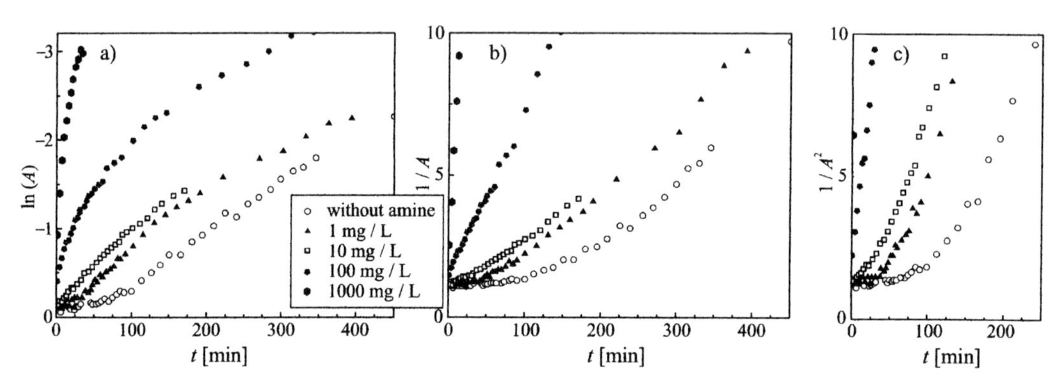

Fig. 4. Kinetic plots for the condensation reactions with linear PPI: (a) assuming a first order reaction, (b) assuming a second order reaction, (c) assuming a third order reaction.

For fast condensation reactions the kinetic plots yield a reaction order of two following the initial phase. For slow condensation reactions, a reaction order of one is found in this phase (e.g., at PPI concentrations of 1 and 10 mg/L). However, it cannot be excluded that for these reactions there is also an early phase with a reaction order of three, as was reported by Perry et al. [5]. According to this work, a reaction order of two, when observed for fast condensation reactions, can be explained by the overlap of reaction phases with reaction orders of one and three.

Table 1 presents the pH values measured during the condensation reactions. For all condensation reactions where an amine was added there is a small shift in the pH value within the first 60 min of the condensation.

Table 1. Variation in the pH values (initial, final, and average value) of the condensation mixtures.

Polyamine	Concentration [mg/L]	pH	Average pH
without amine	–	4.4 → 4.3	4.3
branched PEI	10	4.5 → 4.3	4.4
PMEI	10	4.8 → 4.4	4.6
linear PEI	10	5.6 → 4.7	5.1
hexaazadocosane	10	5.4 → 4.8	5.1
PMPI	10	4.9 → 4.2	4.5
linear PPI	1	4.5 → 4.3	4.4
linear PPI	10	5.1 → 4.4	4.7
linear PPI	100	5.2 → 4.6	4.6
linear PPI	1000	5.6 → 5.9	5.7

The pH value of the reaction mixture shows a relation to the condensation rate: the higher the pH value, the faster the condensation reaction. Of course, this relation is well known from simpler silica chemical systems. It thus appears that polyamines do not directly influence the first steps of the condensation reaction, for example by catalytic action or by coordinative interaction between the amine and the silicic acid anions, but that the accelerating effect the polyamines exert is related to the increase of the pH value, i.e. to their basicity. As the molybdate method probes only the first steps of the condensation reaction, significant interactions may, however, occur between larger silicate species and the amines.

The basicity of polyamines depends on their degree of polymerization, their architecture (linear or branched), and their chemical make-up (methylated or non-methylated) [17, 18]. It is then obvious that these properties should also influence the acceleration effect. It is remarkable that nature has not chosen the fastest polyamine–silica system for the biosilicification in diatoms, although a rapid construction of novel shells is needed during cell division. On the other hand, however, the biomineralization process also requires time in order to form the intricate shapes and special morphologies characteristic of diatom shells.

Acknowledgments: This work was supported by the Schwerpunktsprogramm 1117 "Prinzipien der Biomineralisation" of the DFG and by the Fonds der Chemischen Industrie.

References

[1] N. Kröger, R. Deutzmann, C. Bergsdorf, M. Sumper, *Proc. Nat. Acad. Sci. USA* **2000**, *97*, 14133.

[2] E. Bäuerlein, *Biomineralization*, Wiley-VCH, Weinheim, **2000**.
[3] S. Mann, *Biomineralization*, Oxford University Press, Oxford, **2002**.
[4] E. Bäuerlein, *Angew. Chem. Int. Ed.* **2003**, *42*, 614.
[5] C. C. Perry, T. Keeling-Tucker, *J. Biol. Inorg. Chem.* **2000**, *5*, 537.
[6] K. Shimuzu, J. Cha, G. D. Stucky, D. E. Morse, *Proc. Natl. Acad. Sci. USA* **1998**, *95*, 6234.
[7] N. Kröger, R. Deutzmann, M. Sumper, *Science* **1999**, *286*, 1129.
[8] N. Kröger, S. Lorenz, E. Brunner, *Science* **2002**, *298*, 584.
[9] N. Kröger, R. Deutzmann, Ch. Bergsdorf, M. Sumper, *Biochemistry* **2000**, *97*, 14133.
[10] M. Sumper, *Science* **2002**, *295*, 2430.
[11] H. Menzel, S. Horstmann, P. Behrens, P. Bärnreuther, I. Krueger, M. Jahns, C*hem. Commun.* **2003**, *24*, 2994.
[12] B. Dietrich, M. W. Hosseini, J.-M. Lehn, R. B. Sessions, *Helv. Chim. Acta* **1983**, *66*, 1262.
[13] S. Kobayashi, T. Saegusa, *Ring-opening Polymerization Volume 2*, Elsevier Applied Science Publishers, London, New York, **1984**, p. 761.
[14] B. G. Alexander, *J. Am. Chem. Soc.* **1953**, *75*, 2887.
[15] E. G. Vrieling, T. P. M. Beelen, R. A. van Santen, W. W. C. Gieskes, *Angew. Chem. Int. Ed.* **2002**, *41*, 1543.
[16] C. Harrison, N. Loton, *J. Chem. Soc. Faraday Trans.* **1995**, *91*, 4287.
[17] G. J. M. Koper, M. H. P. van Genderen, C. Elissen-Roman, M. W. P. L. Baars, E. W. Meijer, M. Borkovec, *J. Am. Chem. Soc.* **1997**, *119*, 6512.
[18] M. Borkovec, G. J. M. Koper, *Macromolecules* **1997**, *30*, 2151.

Polyol-Modified Silanes as Precursors for Mesostructured Silica Monoliths

Nicola Hüsing,* Doris Brandhuber, Viktoria Torma, Christina Raab

Institute of Materials Chemistry, Vienna University of Technology
Getreidemarkt 9, A-1060 Vienna, Austria
Tel.: +43 1 5880115322 — Fax: +43 1 5880115399
E-mail: nicola.huesing@tuwien.ac.at

Herwig Peterlik

Institute of Materials Physics, University of Vienna
Boltzmanngasse 5, A-1090 Vienna, Austria

Keywords: sol–gel chemistry, mesostructured monoliths, glycol-modified silanes

Summary: Large silica monoliths with a hierarchical network structure comprising periodic mesopores in a macroporous architecture are prepared using polyol-modified silanes as silica precursors and lyotropic liquid crystalline phases of surfactants in water as structure-directing agents. Tetrakis(2-hydroxyethyl) orthosilicate, tetrakis(2-hydroxypropyl) orthosilicate and tetrakis(2,3-dihydroxypropyl) orthosilicate were obtained by transesterification reaction from tetraethyl-orthosilicate and the corresponding multifunctional alcohols. These silanes allow for a true liquid crystal templating of surfactant phases, since they are soluble in water and show good compatibility with surfactant–water phases. The monolithic materials show a unique but tunable macro- and mesostructure, which is characterized by N_2-sorption measurements, small-angle X-ray scattering (SAXS), transmission electron microscopy (TEM), and scanning electron microscopy (SEM).

Introduction

Porous materials with a hierarchical organization of the network are desired for a broad variety of applications, including chromatography and catalysis, because of the obvious multiple benefits that arise from each of the pore size regimes, e.g., micro/mesopores for size- or shape-selective applications and macropores for reduction of diffusion limitations to the active sites [1]. Despite the drastic progress in the synthesis of porous inorganic materials with control of pore sizes from nanometers to micrometers, the preparation of materials with a simultaneous tailoring of morphology (monoliths, fibers, films, powders) and pore structures still remains a challenging task.

One of the key steps in the synthesis of these materials continues to be the use of organic templates to spatially pattern the deposition of inorganic solids. The concept of using templates or structure-directing agents for a specific design of inorganic network structures is adapted from nature, where archetypes of synthetic strategies for the construction of organized materials across a range of length scales can be found, e.g., the mineralization of the beautiful silica exoskeletons of diatoms or radiolaria in the presence of organic biopolymers [2]. The synthetic chemist applies these concepts from biomineralization processes to different types of materials. In the case of microporous crystalline compounds, the organic additives used as templates are single molecules and lead to zeolitic structures [3]. In contrast, syntheses of ordered mesoporous oxides employ supramolecular templates such as liquid-crystal aggregates of surfactants, and micromolding methods using emulsion droplets, latex spheres, or bacterial threads have been used to prepare inorganic solids with pores in the micrometer range [3 – 5].

However, the preparation of materials with structures on different length scales typically involves more complicated synthetic procedures, e.g., for porous materials two or more pore-building processes have to be synchronized. The implementation of these techniques toward the production of *large monolithic systems exhibiting porous structures on different length scales* is only just beginning to emerge [1, 6].

Polyol-Modified Silanes

In our approach, modified silanes have been used as precursor molecules for the formation of an inorganic network such as tetrakis(2-hydroxyethyl) orthosilicate (EGMS), tetrakis(2-hydroxy-propyl) orthosilicate (PGMS), and tetrakis(2,3-dihydroxypropyl) orthosilicate (GLMS) (Scheme 1). To simplify the following text, the term "glycol" means that the compound has more than one alcohol group, thus including GLMS.

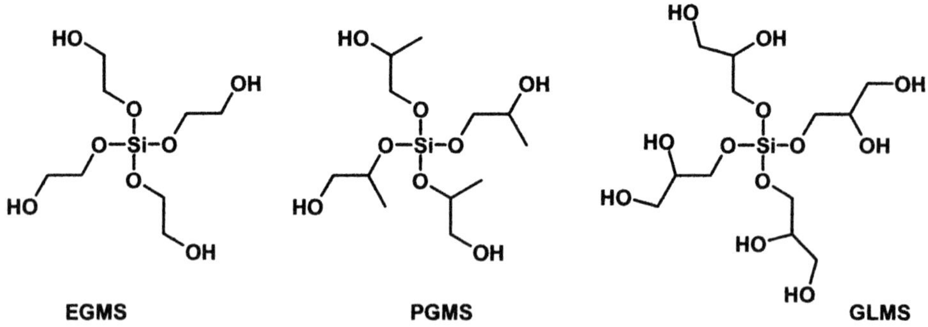

EGMS PGMS GLMS

Scheme 1. The different precursors, schematically shown in their monomeric form.

The product of the transesterification reaction is not a well-defined molecular species, even with the silicon-to-alcoholate ratio adjusted to about 1 : 4 in all cases. This is probably because diols and other multifunctional alcohols allow the formation of bridged and chelating silane species in equilibrium with non-covalently bound alcohols, as in the case of the ethylene glycol-modified silane shown in Scheme 2.

For the highly viscous precursor, complete alcohol exchange was proven by ^{1}H and ^{13}C NMR studies, and ^{29}Si NMR investigations showed a single peak at about –83 ppm for all different silanes. No higher Q-species than Q^0 are present in the precursor mixture, suggesting once more that the oligomers are formed because of strong intermolecular hydrogen bonding.

Modifications of alkoxysilanes with diols or polyols such as ethylene glycol, glycerol, etc. have been well known since the middle of the last century; however, their application was hampered by their hydrolytic instability [7, 8]. Only in recent years is interest in this type of glycol-modified silanes increasing again because of some obvious advantages:

- A high water solubility (no co-solvent is required to homogenize typical sol–gel solutions).
- The hydrolysis and condensation reactions can be initiated without acid or base catalysis just at neutral pH by addition of water.
- Hydrolysis results in the release of a biocompatible alcohol.

The latter point was the one first recognized as an advantage in the synthesis of bio-silica gels as seen in the application of glycerol-modified silanes for the incorporation of otherwise incompatible biomolecules [9, 10].

The extraordinarily good compatibility of ethylene glycol-modified silanes with lyotropic liquid-crystal phases of surfactant molecules was demonstrated by Hoffmann in 1998 [11, 12].

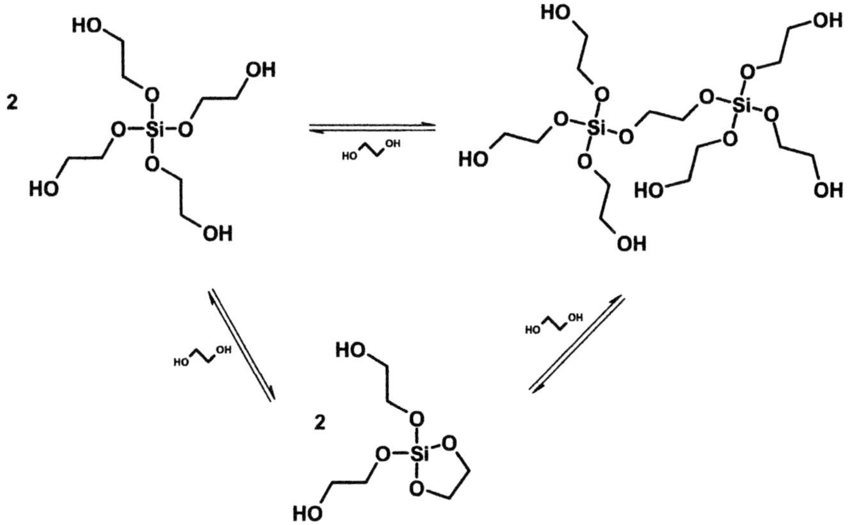

Scheme 2. Formation of bridged and chelating species exemplarily shown for tetrakis(2-hydroxyethyl)-orthosilicate.

Therefore, these silanes are ideal candidates for true liquid-crystal templating approaches in which the silane is mixed with a highly concentrated surfactant–water mixture. In our studies, EGMS, PGMS, or GLMS were added to a homogeneous mixture of the surfactant Pluronic P123 and water or aqueous HCl (10^{-2} M), resulting in a composition (by mass percentage) of SiO_2/P123/H_2O or HCl = 10/30/70. We chose the block copolymer P123 as the surfactant, not only because of its well-investigated phase behavior in water, but also because of its chemical analogy (non-ionic polyoxyethylene moieties as the hydrophilic part of the surfactant) to the glycols which are released during sol–gel processing of the glycol-modified silanes.

Interestingly, little is known about the hydrolysis and condensation behavior of glycol-modified silanes. In our studies, the pH was varied from an acidic medium to neutral (using pure water), and the gelation times of the silane processed in the presence of the surfactant (Pluronic P123) were followed. It can immediately be seen (Fig. 1) that the modified silanes show an extraordinary condensation behavior compared to other commercially available tetraalkoxysilanes such as tetraethyl orthosilicate, with gel times being very fast in neutral conditions, increasing towards pH = 2 and decreasing again in a more acidic medium. The reasons for this kind of behavior can be manifold; however, the charges involved during the condensation reactions definitely play a decisive role, switching from negatively charged silanolates above pH = 3, to positively charged silanol moieties, thus crossing the point of zero charge (PZC) for silica at pH = 2 – 3. Close to the PZC, the condensation reaction is the slowest.

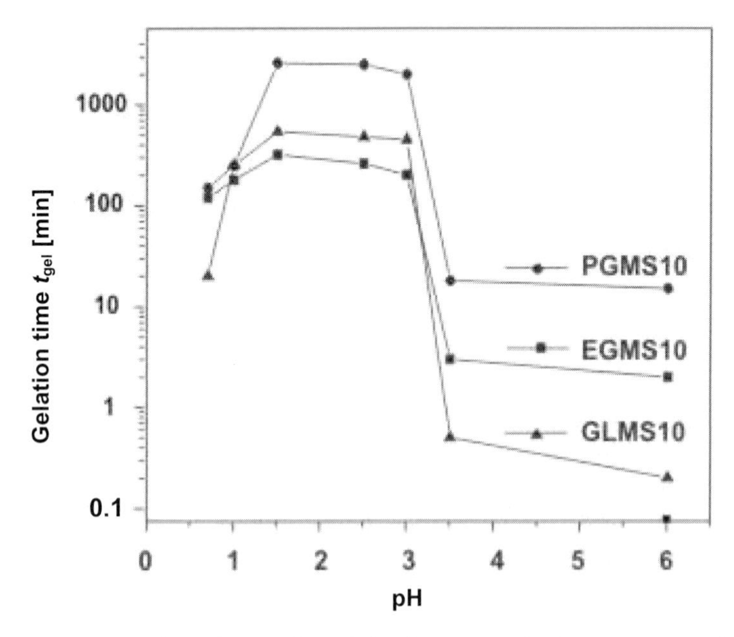

Fig. 1. Gel times as a function of pH of the starting solution for EGMS, PGMS, and GLMS modified silanes in the presence of surfactant (Pluronic P123); (note that the gelation time is presented in a logarithmic scale).

White, monolithic wet gels were obtained from the various starting mixtures. Drying of the wet gel monoliths was performed by silanization of the whole monolith body with a solution of 10 wt% trimethylchlorosilane in petroleum ether for 24 h, leading to an immediate, visible extraction of the surfactant and aqueous pore liquid. After washing with petroleum ether (three times within 24 h) and ethanol (three times within 24 h), the wet gel bodies were dried by heating them to 200 °C (heating rate of 1 °C/h) at ambient pressure.

Table 1 gives some representative synthesis conditions, gel times, and architectural features of the resulting gels determined by small-angle X-ray scattering and nitrogen sorption measurements.

Table 1. Synthesis conditions, gelation times, and physicochemical properties of gel samples from different silanes. t_{gel} corresponds to the gel times. Specific surface areas, S^{BET}, the total pore volume, V^t, and pore size distributions with D^{BJH} representing the maximum were determined from N_2 adsorption–desorption experiments. t^{wall} is the wall thickness of the material and was calculated from the sorption experiments and small-angle X-ray diffraction data as $[(2d_{100}/\sqrt{3})-D^{BJH}]$.

			SAXS	N_2 Adsorption-desorption			
Samples	c_{HCl} [M]	t_{gel} [min]	d_{100} [nm]	S^{BET} [m^2 g^{-1}]	V^t [cm^3 g^{-1}]	D^{BJH} [nm]	t_{wall} [nm]
EGMS	0	2	11.4	1010	1.9	7.8	5.4
PGMS	0	10	10.8	820	2.7	–	–
GLMS	0	<1	–	–	–	–	–
EGMS	10^{-2}	260	11.1	960	1.6	6.4	6.4
PGMS	10^{-2}	2500	10.1	890	0.8	3.7	7.4
GLMS	10^{-2}	480	–	800	0.7	–	–

Monolithic dry gels with a hierarchical network structure and an interconnected, multilevel pore system were obtained after drying (Fig. 2). Even within this small pH regime from pH = 6 to 2, the macroscopic gel morphology can to a large extent be controlled. In addition, the choice of glycol has a decisive influence on the architeture of the network.

The use of tetrakis(2-hydroxypropyl) orthosilicate (PGMS) results in gels with a particulate appearance and a very distinct periodicity of the mesophase with a repeating unit distance of about 10 to 11 nm and a corresponding monomodal pore diameter of about 4 nm. Tetrakis(2-hydroxy-ethyl) orthosilicate (EGMS) gives a more cellular architecture of the network with single strands 1 – 3 μm in length and 0.5 μm in diameter, with a highly ordered 2D hexagonal honeycomb mesostructure with a repeating unit distance a little above 11 nm. The pores seem to be larger than for PGMS gels (6 to almost 8 nm in diameter). Gels prepared from the tetrakis(2,3-dihydroxy-propyl) orthosilicate (GLMS) did not show any long-range ordering under the given synthetic conditions, and exhibit particulate structures typical of silica-based sol–gel materials. For each system, the optimal synthesis parameters such as silane/P123 ratio and acid concentration have to be fine tuned to get optimal results with respect to the long-range periodicity of the mesostructure (see also Fig. 2). We found that with increasing gel times, thus slow condensation rates, the system obviously has more time to equilibrate, resulting in a higher degree of ordering of the mesophase [13].

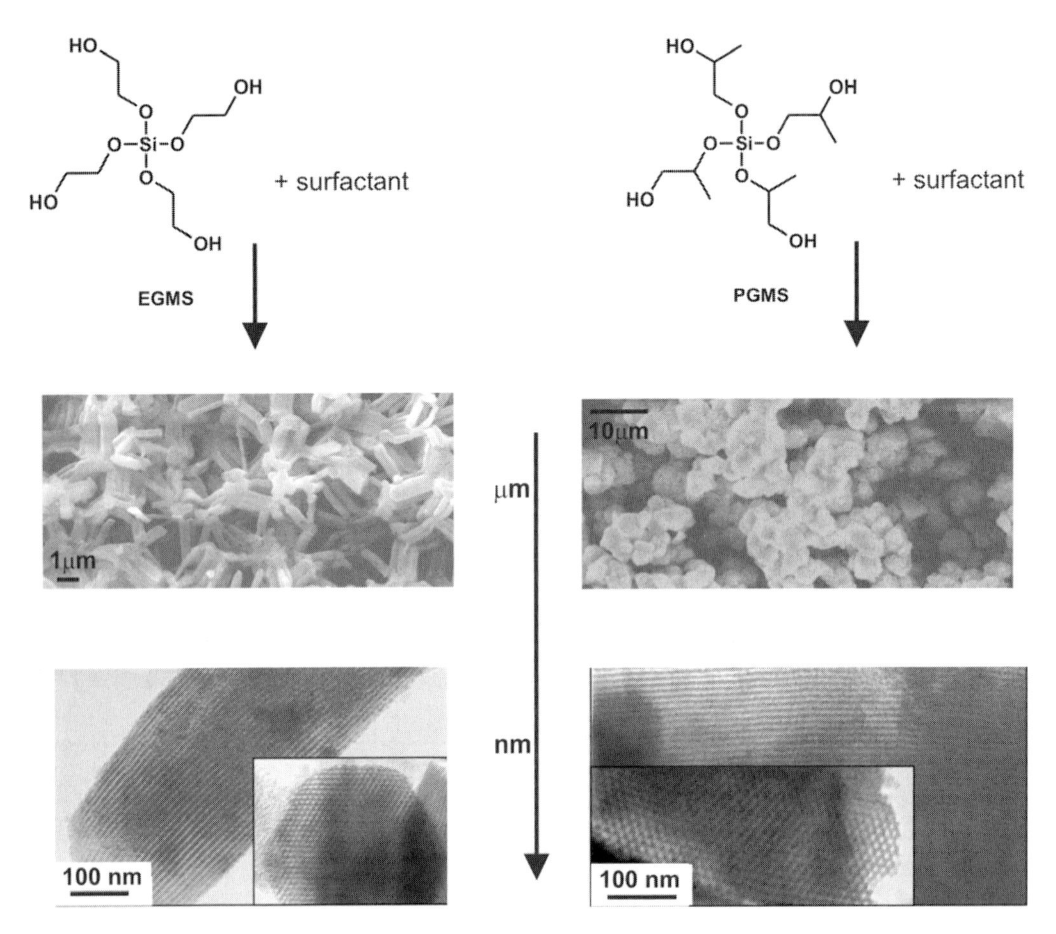

Fig. 2. Schematic representation of the precursors (EGMS, left and PGMS, right) and the structure of the gels obtained with 10^{-2} M HCl in the micrometer range (SEM images), which is in both cases composed of mesoscopically arranged pores (TEM images).

Acknowledgment: The authors would like to thank Wacker Chemie GmbH and BASF for their kind donation of chemicals. V.T. acknowledges the financial contribution of the FWF within the scope of a Lise-Meitner fellowship. We also express our thanks to Dr. M. Puchberger for his assistance with the various NMR measurements.

References

[1] J.-H. Smatt, S. Schunk, M. Linden, *Chem. Mater.* **2003**, *15*, 2354.

[2] S. Mann, *Biomineralization: Principles and Concepts in Bioinorganic Materials Chemistry*, Oxford University Press, **2001**.

[3] *Handbook of Porous Solids* (Eds.: F. Schüth, K. S. W. Sing, J. Weitkamp), VCH, Weinheim, **2002**.

[4] C. T. Kresge, M. E. Leonowicz, W. J. Roth, J. C. Vartuli, J. S. Beck, *Nature* **1992**, *359*, 710.

[5] B. T. Holland, C. F. Blanford, A. Stein, *Science* **1998**, *281*, 538.

[6] K. Nakanishi, *J. Porous. Mater.* **1997**, *4*, 67 – 112 and references therein; K. Nakanishi, R. Takahashi, T. Nagakane, K. Kitayama, N. Koheija, H. Shikata, N. Soga *J. Sol–Gel Sci. Technol.* **2000**, *17*, 191.

[7] A. Goneberg, A. Verheyden, Belgian Patent 510 419 (to Union chimique belge Soc.), **1952**; H. Krimm, H. Schnell, German Patent 1 136 114 (to Farbenfabrik Bayer), **1962**; H. A. Vaughn, British Patent 989 379, (to General Electric Co.), **1965**; E. P. Goldberg, E. J. J. Powers, *Polym. Sci. Polym. Phys. Ed.* **1964**, *2*, 835.

[8] R. C. Mehrotra, R. P. Narain, *Indian J. Chem.* **1966**, *5*, 444.

[9] I. Gill, A. Ballesteros, *J. Am. Chem. Soc.* **1998**, *120*, 8587.

[10] M. A. Brook, J. D. Brennan, Y. Chen, WO 03/102 001, (to McMaster University, Ontario) **2003**.

[11] K. Sattler, M. Gradzielski, K. Mortensen, H. Hoffmann, *Ber. Bunsenges. Phys. Chem.* **1998**, *102*, 1544.

[12] K. Sattler, H. Hoffmann, *Progr. Colloid Polym. Sci.* **1999**, *112*, 40.

[13] N. Hüsing, C. Raab, V. Torma, H. Peterlik, A. Roig, *Chem. Mater.* **2003**, *14*, 2690.

Self-Organized Bridged Silsesquioxanes

J. J. E. Moreau, L. Vellutini, M. Wong Chi Man, C. Bied*

Laboratoire Hétérochimie Moléculaire et Macromoléculaire (UMR-CNRS 5076)
ENSCM – 8, rue de l'école normale, 34296 Montpellier CEDEX 5, France
Tel.: +33 4 67 14 72 19 — Fax: +33 4 67 14 43 53
E-mail: mwong@cit.enscm.fr

Keywords: bridged silsesquioxanes, chirality, self-assembly

Summary: An innovative and general method for the synthesis of shape-controlled bridged silsesquioxanes by the hydrolysis of urea-derived organo-bis(trialkoxy)silanes has been developed. This method relies on the ability of the hydrogen bonds of the urea groups to self-assemble the molecules in a supramolecular architecture. The controlled hydrolysis of the latter efficiently leads to new hybrid silsesquioxanes with sought-after morphologies. Following the reaction conditions (acid- or base-catalyzed hydrolysis) and also depending on the solvent used (purely aqueous or a mixture of water and ethanol), different forms of hybrid materials were obtained from the hydrolysis of chiral diureido derivatives of *trans*-diaminocyclohexane. Right- and left-handed helices were formed under acid-catalyzed hydrolysis of the enantiopure (R,R) and (S,S) compound respectively. Upon base-catalyzed hydrolysis, tubular or ball-like structures were obtained from the pure enantiomers and the racemic mixture, respectively.

Introduction

Growing interest is being paid to the preparation of nano-structured materials with sought-after properties. The hydrolysis-condensation of organo-alkoxysilanes represents a mild process to achieve hybrid silicas with tunable properties owing to the wide variety of organic components one can incorporate in these solids [1, 2].

In this expanding field, organo-bridged bis(trialkoxysilanes) [3, 4] are good candidates for the synthesis of bridged silsesquioxanes with targeted properties [5, 6]. Until recently, these hybrids have always been reported as amorphous solids. Materials chemists are now spending much effort in trying to organize the organic moieties in the silicate framework in order to improve their efficiency.

We have recently developed a new and general method for the preparation of self-organized bridged silsesquioxanes [7 – 9]. The introduction of urea groups into the bridging organic units creates a self-assembly of the molecular precursors owing to the hydrogen bonding, affording peculiar supramolecular architectures. In this work, we describe the synthesis of a chiral organo-bridged bis(triethoxysilane). According to the hydrolysis reaction conditions, helical, tubular, and spherical solids were obtained.

Results

The molecular precursors were prepared by the reaction of 2 molar equivalents of 3-isocyanatopropyl(triethoxy)silane with (R,R)-, (S,S)- or racemic *trans*-diaminocyclohexane. These molecular precursors were then separately hydrolyzed in acidic or basic conditions to form the corresponding hybrid silicas (Scheme 1).

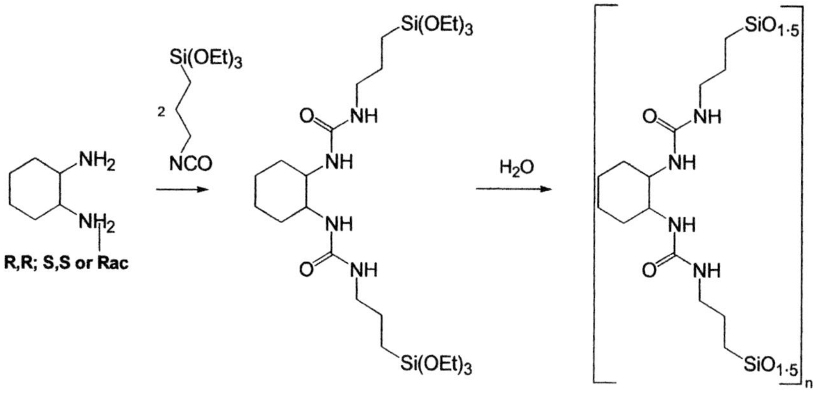

R,R; S,S or Rac

Scheme 1.

In all cases, the solid-state NMR (^{29}Si and ^{13}C) spectra showed the presence of the organic fragment whatever the catalyst used (acid or base). Only T^2, C–Si(OSi)$_2$(OR) and T^3, C–Si(OSi)$_3$ are observed from ^{29}Si NMR, proving the preserved Si–C bonds in the hybrid network after the hydrolysis-condensation process.

Acid-Catalyzed Hydrolysis

The hydrolysis was performed in an aqueous solution of HCl with a molar ratio of precursor: H_2O:HCl of 1:600:0.2. The mixture was stirred in an oil bath (80 °C) for 2 h and was left standing for 2 days at the same temperature. The white precipitate was washed with water to neutral pH and was then dried, leading to a white powder.

The Scanning Electronic Microscopy (SEM) images of the material formed from the racemic precursor showed featureless granular solids, while rope-like structures were observed for the hybrid materials formed from the enantiopure precursors. Under high magnification, bundles of helical fibers (width ranging from 0.5 to 2 μm and length up to 15 μm) could be distinguished. Interestingly, the helices of the (R,R)-derived hybrid were right-handed (Fig. 1a), whereas those observed from the (S,S)-derived hybrid were left-handed (Fig. 1b). These results represent the first chirality transcription from the molecular precursors to the hybrid solid silicas.

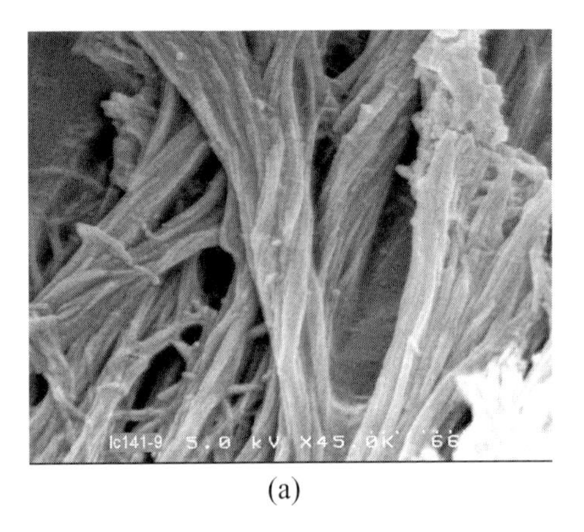

(a)

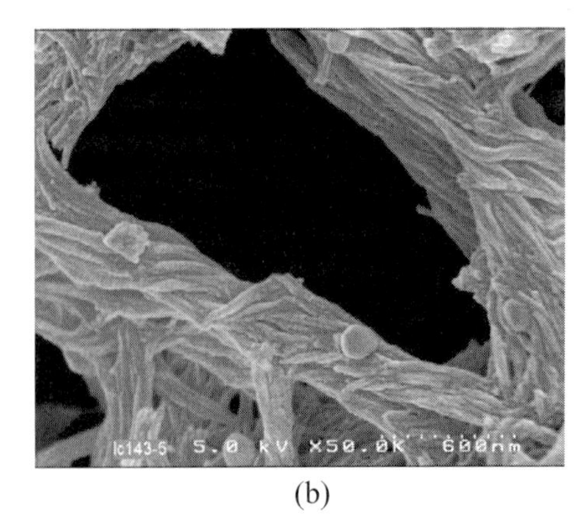

(b)

Fig. 1.

Base-Catalyzed Hydrolysis

The reaction was done in a mixture of ethanol and water (V/V: 2/5) at the same temperature as that above. The precursor (pure enantiomers or racemic) was first dissolved in ethanol, and on addition of water a white precipitate appeared. An aqueous NaOH solution was then added until pH 12 was attained, and the mixture was heated in an oil bath at 80 °C for 6 days. In the case of the racemic mixture, the precipitate completely dissolved after 10 min, and the formation of a white precipitate was observed after 0.5 h. This dissolution was never observed for the pure enantiomers, the reaction occurring in a heterogeneous solid/liquid state. Usual work-up (washing with water until neutral pH and drying at 110 °C) resulted in white powders in all cases. The features obtained from the SEM images of the enantiopure-derived materials were completely different from those of the racemic material (Fig. 2).

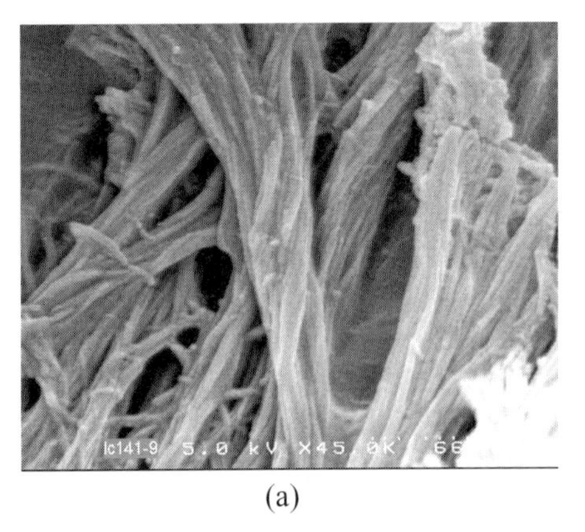

(a)

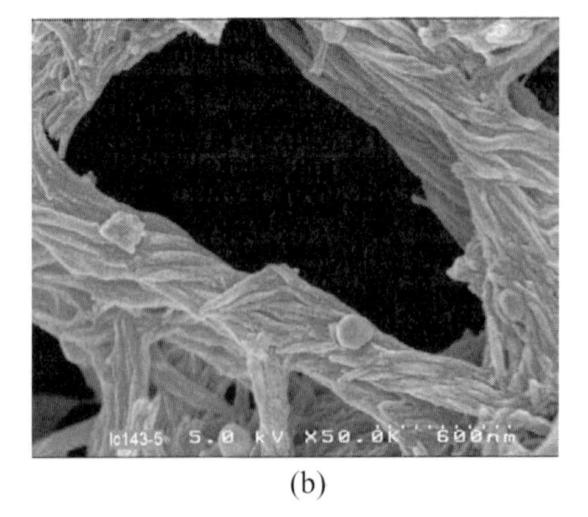

(b)

Fig. 2.

On the one side, hollow tubular hybrids (length up to 15 μm) with surprisingly rectangular cavities (size ranging from 0.3 to 1.6 μm) formed from the (R,R) and the (S,S) precursors (Fig. 2a). On the other hand, only spherical ball-like structures (0.3 – 2 μm) were obtained from the racemic compound (Fig. 2b).

To better understand the mechanistic formation of these peculiar shapes, we analyzed the precipitate formed by addition of water to the ethanolic solution of the precursors. The precipitate, which was still quite soluble in chloroform, could be analyzed by liquid NMR (^{1}H, ^{13}C and ^{29}Si) and was proved to correspond to the starting molecular precursor. This was also confirmed by melting point measurements, which gave exactly the same values as for the starting compounds (210 °C for the pure enantiomers and 155 °C for the racemic mixture). Hence, at this stage, no hydrolysis had occurred yet. Interestingly, SEM pictures of the precipitate (not shown here) for both the enantiomerically pure and the racemic precursors showed rigid rectangular rods (length up to 12 μm and width ranging from 0.3 to 1.5 μm). The striking similarity of the size and the shape of the rods with those of the channel of the tubular hybrid suggests that the latter could result from the former's shape. Since, in the case of the (R,R)- and the (S,S)-enantiomers, the reaction seemed to occur in a heterogeneous manner, it looks plausible that at the early stage (soon after the addition of aqueous NaOH), hydrolysis-condensation took place at the surface of the rectangular rods, leading to the formation of an insoluble film around these crystalline rods (moulding effect). Ethanol was released during the reaction process, and, as a good solvent, it dissolved the precursor at the inner side of the rods, creating the hollow tubes. The resulting aggregates then condensed at the outer surface of the tubes. This crystal-templating effect has already been reported for the formation of tubular silicas [10 – 11]. In our case, a self-templating system is observed. Concerning the racemic material formed, the reaction was homogeneous at the early stage. The formation of the corresponding spherical balls is explained by a nucleation growth process.

Conclusion

This work gives a new and general method for the preparation of shape-controlled hybrid organo-silicas. It is based on the supramolecular interactions between the molecular precursors. Upon acid-catalyzed hydrolysis, the chirality of the enantiopure molecular precursor was transcribed to the solid materials, leading to rope-like helices with controlled handedness. The base-catalyzed hydrolysis of the pure enantiomers led to hollow tubular hybrids, whereas ball-like structures were formed from the racemic mixture. A self-templating of the organic crystals of the pure enantiomers accounts for the formation of the tubes, whereas the spherical balls were formed according to a classical nucleation growth phenomenon. This challenging field represents an amazing bottom-up approach for the synthesis of new multifunctional shape-controlled materials.

Acknowledgments: The authors are grateful to the CNRS and the "Ministère de la Recherche et de l'Education de France" for financial support.

References

[1] C. J. Brinker, G. W. Scherer, *Sol–Gel Science: the Physics and Chemistry of Sol–Gel Processing*, Academic Press, San Diego, **1990**.

[2] C. Sanchez, F. Ribot, *New J. Chem.* **1994**, *18*, 1007.

[3] K. J. Shea, D. A. Loy, O. W. Webster, *Chem. Mater.* **1989**, *1*, 512.

[4] R. J. P. Corriu, J. J. E. Moreau, P. Thépot, M. Wong Chi Man, *Chem. Mater.* **1992**, *4*, 1217.

[5] U. Schubert, *New J. Chem.* **1994**, *18*, 1049.

[6] R. Abu-Reziq, D. Avnir, J. Blum, *Angew. Chem. Int. Ed.* **2002**, *41*, 4132.

[7] J. J. E. Moreau, L. Vellutini, M. Wong Chi Man, C. Bied, *J. Am. Chem. Soc.* **2001**, *123*, 1509.

[8] J. J. E. Moreau, L. Vellutini, M. Wong Chi Man, C. Bied, J.-L. Bantignies, P. Dieudonné, J.-L. Sauvajol, *J. Am. Chem. Soc.* **2001**, *123*, 7957.

[9] J. J. E. Moreau, L. Vellutini, M. Wong Chi Man, C. Bied, *Chem. Eur. J.* **2003**, *9*, 1594.

[10] H. Nakamura, Y. Matsui, *J. Am. Chem. Soc.* **1995**, *117*, 2651.

[11] F. Miyaji, S. A. Davis, J. P. H. Charmant, S. Mann, *Chem. Mater.* **1999**, *11*, 3021.

Carbamatosil Nanocomposites with Ionic Liquid: Redox Electrolytes for Electrooptic Devices

Vasko Jovanovski, Boris Orel, Angela Šurca Vuk*

National Institute of Chemistry, Hajdrihova 19, SI – 1000 Ljubljana, Slovenia
Tel.: +386 1 4760290 — Fax: +386 1 4760300
E-mail: boris.orel@ki.si

Keywords: ionic liquid, redox electrolytes, sol–gel nanocomposites, hybrid electrochromic cells

Summary: Short single end-capped carbamatosil alkoxysilane precursors were synthesized from mono-ether-terminated polyethyleneglycols and 3-isocyanato-triethoxysilane. A redox gel electrolyte was made from alkoxysilane precursors with oxalic acid (OxA) as a gelation promoter, while inorganic (LiI, NaI, KI, NH4I) or organic (methylpropylimidazoliumiodide) iodide and I_2 were used as the source of redox species, and non-volatile triethyleneglycol was used as a co-solvent. Hybrid electrochromic cells (WO_3/carbamatosilalkoxysilane/inorganic or organic iodide + I_2/ OxA/TEG/TEOS/Pt) were constructed and their electrochromic stability tested by cyclic voltammetry.

Introduction

The hybrid electrochromic (HEC) cell represents an alternative to the better known three-layered battery-type and liquid EC devices with incorporated luminophores [1]. While the battery-type EC cells have not yet attained popularity as "smart" windows for buildings, the liquid EC devices have already become widely used as rear-view mirrors for cars, preventing glare by their ability to control electrically the level of the reflected light. HEC cells, in contrast to three-layered battery-type EC cells, consist of the active electrochromic material (usually TiO_2 with attached viologen dye [2]) deposited on a transparent conducting oxide electrode facing a counter electrode with a deposited catalytically active thin layer of platinum. The space between them is filled with a liquid electrolyte, usually an organic solvent containing I^-/I_3^- or a ferrocene/ferrocenium redox couple. Because only two active layers are needed for the operation of the HEC cells, this type of EC cell represents a considerable step forward in simplicity of design compared with the three-layered battery-type EC devices, reducing the cost of fabrication and the number of possible modes of failure. In this work we used — instead of a nanocrystalline TiO_2 [2] sensitized with viologen derivatives and a liquid electrolyte — the non-hydrated WO_3 film prepared via the peroxo sol–gel route and a semi-solid electrolyte [3].

Despite the advantages of the HEC cells, there are some drawbacks. Unlike the battery-type EC devices, which exhibit a memory effect, the HEC cells show self-erasing properties. This means that to maintain the fully colored state the potential must be continuously applied on the cell. This drawback is compensated for by much faster coloring-bleaching kinetics, attaining fractions of a second for certain devices [3]. The fast kinetics is achieved because of the absence of an ion storage counter-electrode, having limited diffusion of the intercalating ions shuttled from the active electrode across the ionic conductor to the counter-electrode.

In HEC cells, liquid electrolytes are usually employed containing KI and I_2 dissolved in a suitable organic solvent (ethylenecarbonate, propylenecarbonate, acetonitrile, etc.). The liquid electrolyte wets well the nanocrystalline structure of the active electrode, while a high concentration of redox species (I^-/I_3^-) assures a high ionic conductivity that contributes to the fast coloring-bleaching speeds. However, since the electrolyte contains highly reactive I^-/I_3^- species, risks of leaking exist. To avoid leakages, polymer (gel) electrolytes [4] and electrolytes gelled with nanocrystalline silica particles have already been developed, mainly to be used for a dye-sensitized photoelectrochemical cell (DSPEC) [5, 6].

Recently, we reported about the silica-reinforced nanocomposite redox electrolytes having a semi-solid consistency and tested them in the HEC cells employing WO_3 film [7, 8]. The redox electrolyte was made using the sol–gel route from bis end-capped triethoxysilane chemically bonded via the urea groups to a long (M_W = 4000) poly(propyleneglycol) chain (ICS-PPG4000 for short). Redox conductivity was attained by incorporating KI and I_2 in the ethanolic solution of ICS-PPG4000 precursor. The precursor solution gelled by the addition of the acetic acid (AcOH). The resulting hybrid silica gels contained a complex mixture of I^-/I_3^- species, EtOH, and ethylacetate as a reaction product. The latter compound appeared as a consequence of the solvolysis reactions caused by the AcOH catalyst. This resulted in non-hydrated gels with a low silanol content. Although the HEC cells exhibited a highly reversible and persistent electrochromic effect (>1800 repetitive coloring-bleaching cycles), after few months they ceased to work; EtOH evaporated (cells were not sealed to increase the severity of testing) and KI crystallized in the electrolyte entrapped in the cells. Although the evaporation of EtOH could be prevented by appropriate sealing, resulting in higher longevity of the HEC cell, we decided rather to avoid evaporation of EtOH, replacing it with a suitable co-solvent. The following criteria were taken into account in choosing the co-solvent: (i) it dissolves KI and I_2, (ii) it favors the formation of I_3^- and I^- from the added KI + I_2, (iii) it should be compatible with a carboxylic acid catalyst and sol–gel precursor, (iv) it should support diffusion of charged species by increasing the ionic conductivity of the gel electrolyte, and (v) the boiling point should be above 250 °C to give a negligible vapor pressure at the operating conditions of the HEC (~80 °C) cell. Sulfolane (bp ~280 °C), chosen first, fulfilled most of the mentioned criteria (i–v), and the HEC cells composed of ICS-PPG 4000/ KI + I_2/ sulfolane/ AcOH [7] performed more than 3000 repetitive coloring-bleaching cycles before they deteriorated. However, AcOH catalyst and the corresponding reaction products of the solvolysis reactions (i.e. ethylacetate) are both relatively low-boiling-point liquids, and unsealed HEC cells failed to work after few months.

In this paper we report construction of redox electrolytes based on a new type of sol–gel precursors (2-methoxyethyl-3-(triethoxy-λ^4-silyl)propylcarbamate (ICS-2ME) and 2-(2-methoxy-ethoxy)ethyl-3-(triethoxy-λ^4-silyl)propylcarbamate (ICS-DEM)) compatible with high-boiling point co-solvent (TEG). A novel ionic liquid (i.e. ethylpropylimidazoliumiodide) similar to that reported by Bonhote et al. [9] was synthesized and used as a source of iodide ions. We chose this type of ionic liquid to avoid the addition of MI salts (M = Li, K, Na); however, to ensure the presence of cations needed for insertion into WO_3 film, the weak AcOH (pK = 4.27) was replaced with the stronger oxalic acid (OxA) (pK = 1.23). Since the single end-capped carbamatosils are not easy to gel, tetraethoxysilane (TEOS) was added as a network former. Here we focus on reporting the synthetic routes of the sol–gel precursors and ionic liquid, while the role of OxA catalyst for performing the solvolysis reactions of a complex precursors' mixture will be published elsewhere. The suitability of the gels as redox electrolytes for HEC cells was demonstrated.

Experimental

Synthesis of Alkoxysilane Precursors and Ionic Liquid

The synthesis of single or bis end-capped carbamatosil precursors is straightforward, although it takes some time (a few days). In all cases we used 3-isocyanatopropyltriethoxysilane (ICS) in combination with various polyethyleneglycols or mono-ether-terminated polyethyleneglycols that are commercially available. The synthesis of two single end-capped carbamatosil precursors is described below.

Synthesis of ICS-2ME (*3*)

11.76 g (0.154 mol) of 2-methoxyethanol (**2**) was dissolved in 40 mL of tetrahydrofurane (THF). To this solution, 38.24 g (0.154 mol) of ICS (**1**) was added dropwise (Scheme 1). The solution was then heated slightly below reflux temperature (64 °C) for 48 h with constant stirring. The solvent was removed by distillation under reduced pressure. 48.1 g of product **3** was obtained with a high yield (96%).

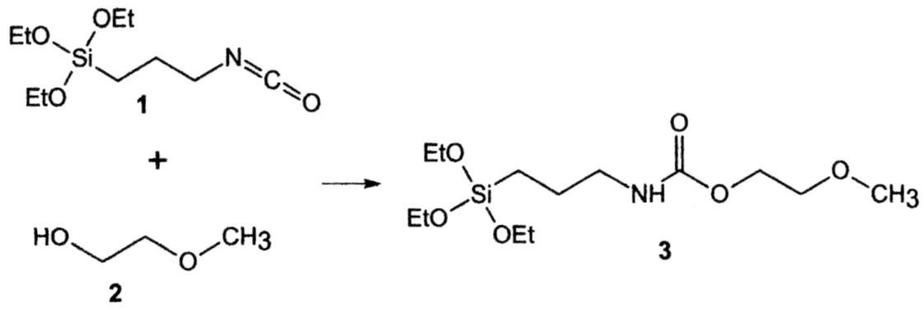

EtO OEt Si N C O 1 + HO O CH3 2 → EtO Si EtO OEt NH O O O CH3 3

Scheme 1. Synthesis of ICS-2ME precursor.

Synthesis of ICS-DEM (**5**)

The procedure was the same as above. Instead of 2-methoxyethanol, 16.35 g (0.136 mol) of diethyleneglycol-monomethylether (**4**) was dissolved in THF and 33.65g (0.136 mol) of ICS (**1**) was added (Scheme 2). 47.7 g (95%) of the product **5** was obtained after isolation.

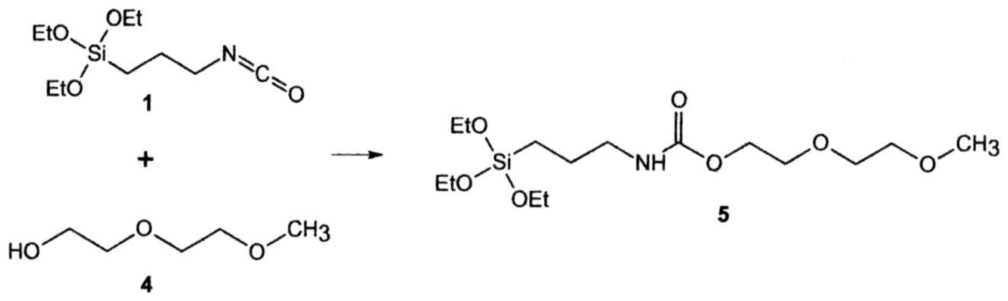

Scheme 2. Synthesis of ICS-DEM precursor.

Synthesis of 1-Ethyl-3-propylimidazolium Iodide Ionic Liquid ($EtPrIm^+I^-$)

Synthesis of 1,3-dialkylimidazolium ionic liquids does not require very much effort. One can start from commercially available 1-methylimidazole for a one-step or imidazole for a two-step synthesis (Scheme 3).

(i) 1-Ethylimidazole was synthesized according to Bonhote [9]. From 50 g (0.734 mol) of imidazole, 55 g (0.81 mol) of sodium ethoxide, and 88 g (0.81 mol) of bromoethane, 48.5 g 1-ethylimidazole was obtained with a 69% yield.

(ii) Under vigorous stirring 88 g (0.52 mol) of propyliodide (*Fluka*) was added dropwise over 1 h to a solution of 41.25 g (0.43 mol) of 1-ethylimidazole in 200 mL of 1,1,1-trichloroethane (*Fluka*). The mixture was then refluxed for 3 h. Ionic liquid was decanted from the hot solution in a separating funnel, washed twice with 100 mL of 1,1,1-trichloroethane, and dried under reduced pressure. 100.6 g (88%) of 1-ethyl-3-propyl-imidazolium iodide was obtained.

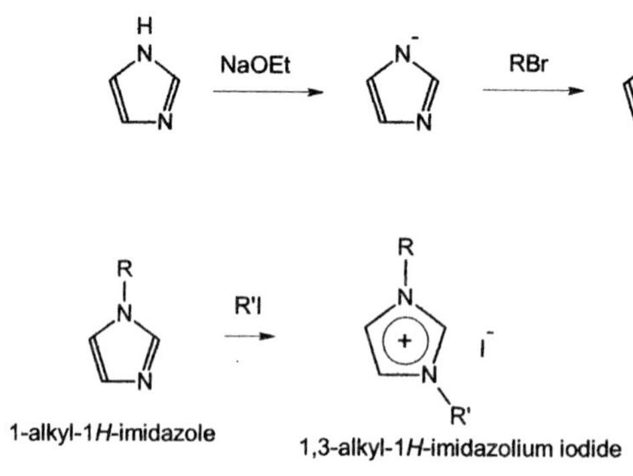

Scheme 3. Synthesis of 1-ethyl-3-propylimidazolium iodide ionic liquid.

Preparation of a Gel Electrolyte for a Hybrid Electrochromic Device Application

The preparation of a precursor solution followed by the construction and characterization of HEC is presented in Scheme 4. To a solution of OxA in TEG (1:1 wt ratio), the inorganic iodide salt (MI: M = Li, K, Na, NH_4) or ionic liquid ($EtPrIm^+I^-$) and I_2 were added ($I^-:I_2$ = 10:1 mol ratio). Both silica precursors (ICS-2ME or ICS-DEM) and TEOS (1:1 wt ratio) were then admixed to form a sol, which was used for the construction of the HEC. The combinations of silica precursors with different inorganic iodides and $EtPrIm^+I^-$ are shown in Table 1.

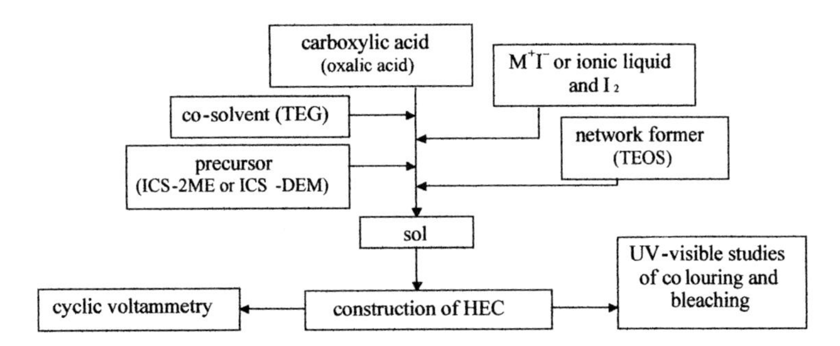

Scheme 4. Preparation of a procedure of a redox electrolyte for HEC cell.

Table 1. Redox electrolytes tested in HEC cells. ΔOD_{1c}, ΔOD_{100c} — optical density at 634 nm for the 1st and 100th cycle [$\Delta OD = \log(T_b/T_c)$].

	M^+	ΔOD_{1c}	ΔOD_{100c}
ICS-2ME/TEOS/TEG/OxA/M^+I^-	Li^+	1.4	1.3
	Na^+	1.3	1.0
	K^+	0.6	0.6
	NH_4^+	1.0	0.5
	$EtPrIm^+$	0.3	0.2
ICS-DEM/TEOS/TEG/OxA/M^+I^-	Li^+	1.0	0.2
	Na^+	1.4	1.1
	K^+	1.4	1.3
	NH_4^+	1.6	1.2
	$EtPrIm^+$	0.8	0.5

The HEC cells were constructed from nanocrystalline WO_3 films prepared using the peroxo sol–gel route [3] and Pt-coated SnO_2:F-glass substrate served as a counter elelectrode. The HEC cells were assembled in such a way that a drop of the sol (redox electrolyte) was placed on the WO_3 film and immediately covered with a Pt counter electrode [10]. No sealing of the HEC cells was used; the electrolyte gelled inside the HEC.

Instrumental

In situ UV-visible spectroelectrochemical measurements of the HEC cells were made using an HP 8453 diode-array spectrophotometer in combination with an EG&G PAR model 273 potentiostat-galvanostat. The cyclovoltammetric measurements were scanned from 0 V to –2 V, and reversed to 2 V and then back to 0 V at scan rates of 20 and 50 mV/s. Reported potentials correspond to the potential applied to the WO_3 film (working electrode) with respect to the Pt-coated SnO_2:F-glass counter-electrode. The electrochromic effect of HEC cells was expressed by the variation of optical density (ΔOD) while their stability was estimated considering the fading of ΔOD after 100 consecutive colouring/bleaching cycles.

Results

Before the electrochromic properties of the HEC cells, schematically shown in [10], employing various redox electrolytes are given, the operating principles of the cell are briefly discussed to show the role of the redox electrolyte. When a cathodic potential is applied to the WO_3 film, the inserted charge needs compensation to stabilize the W^{5+} oxidation state. That is achieved by the insertion of M^+ (Table 1) ions coming from the redox electrolyte with I^-/I^-_3 couple (Eq. 1):

$$WO_3 + x\,M^+ + x\,e^- \longleftrightarrow M_xWO_3$$

Eq. 1. Intercalation of cations and electrons in WO_3 film.

Upon application of cathodic potentials, the electrolyte is disturbed, changing the concentration of I^- species at the WO_3|electrolyte interface. Because of the potential gradient originating from the applied potential, I^- ions should diffuse across the redox electrolyte to bring charge to the positively charged Pt counter-electrode, where the following redox reaction is performed:

$$3\,I^- \longleftrightarrow I_3^- + 2\,e^-$$

Eq. 2. Electrochemical transformation of I^-/I_3^- redox pair.

However, the diffusion of I^- takes place in limited extent [11] due to the addition of iodine which, as a Lewis acid acceptor, reacts with Lewis base donor I^- to form I^-_3 ion. The latter is referred to the building block which catenates forming various polyiodides. The increased content of iodine leads the increase in the connectivity between I^-_3 and iodine. Thus, the distance between iodine and triiodide is reduced and the transport of charge carriers became easier. The charge transport is established via the Grotthus (a relay type) mechanism, enabling the transport of the charge without net transport of mass, assuring the electrical contact between WO_3 and Pt counter-electrode inside the HEC cell [12].

The reversible redox reaction (Eq. 2) also takes place at the WO_3|electrolyte interface, where it competes for electrons needed for performing reduction of W^{6+} to W^{5+}. At stationary conditions, when potential is applied across the cell, there are enough electrons to provide W^{5+} ions, which assure the colored state of HEC cell and the reaction $I_3^- + 2e^-$ (WO_3|electrolyte interface) $\leftrightarrow 3I^-$ is prevailing. Therefore, the charge is transported across the cell during the application of a potential (not shown here). However, when potential is switched off, the reduction reaction (Eq. 2) of I^-_3 at the WO_3|electrolyte interface becomes dominant that causes self-erasing of the cell.

The insertion of M^+ ions was inferred from the *in situ* UV-VIS spectroelectrochemical measurements performed on the HEC cell in the course of potential sweeping; transmittance dropped at cathodic potentials and was restored during the anodic scan (Fig. 1). Insertion of M^+ ions and consequent colouration was fast and took place only in a few seconds, in contrast to the longer time for coloring-bleaching changes in battery-type EC cells, where the coloring-bleaching kinetics is governed by the properties of both intercalation electrodes. The electrochromic kinetics of HEC cells is therefore mostly limited, with the electric conductivity of the redox electrolyte not exceeding $10^{-4} - 10^{-3}$ S/cm.

The functioning of the HEC cell with $WO_3/(I^-/I_3^-)$ redox electrolyte is facilitated by the properly adjusted energy levels of the conduction band (CB) of WO_3 (0.53 V vs NHE) and by the redox potential of the I^-/I_3^- couple lying nearly at the same potential (~0.5 V vs NHE) [13]. In DSPEC cells, the efficiency of the cells depends largely on the recombination losses of photoejected electrons with oxidized dye molecules or redox couple, which is in contact with the large area TiO_2 film. The I^-/I_3^- couple could in that case act as an electron scavenger, reducing the number of holes (i.e. I_3^- ions) at the photoanode (i.e. TiO_2)/electrolyte interface), and consequently decreasing the number of I_3^- needed at the Pt/electrolyte interface to accept electrons coming through the external circuit to the Pt counter-electrode. DSPEC cells function well because the CB of TiO_2 is lies close to the 1 V vs NHE electrode that prevents high recombination losses. The recombination losses do not represent a drawback for the HEC cell's operation because potential is constantly applied across the cell.

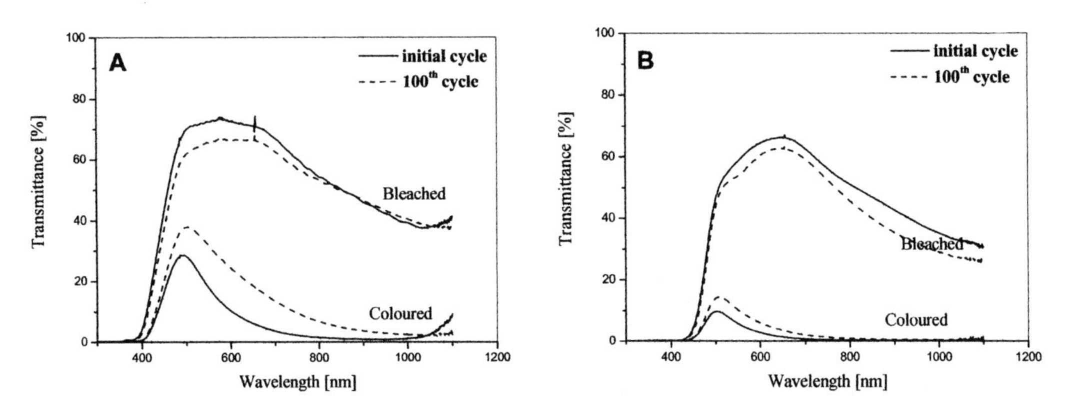

Fig. 1. *In situ* UV-visible spectroelectrochemical measurements of HEC cells employing: A) ICS-2ME/TEOS/TEG/OxA/$NH_4I + I_2$ and B) ICS-DEM/TEOS/TEG/OxA/$NH_4I + I_2$ redox electrolytes.

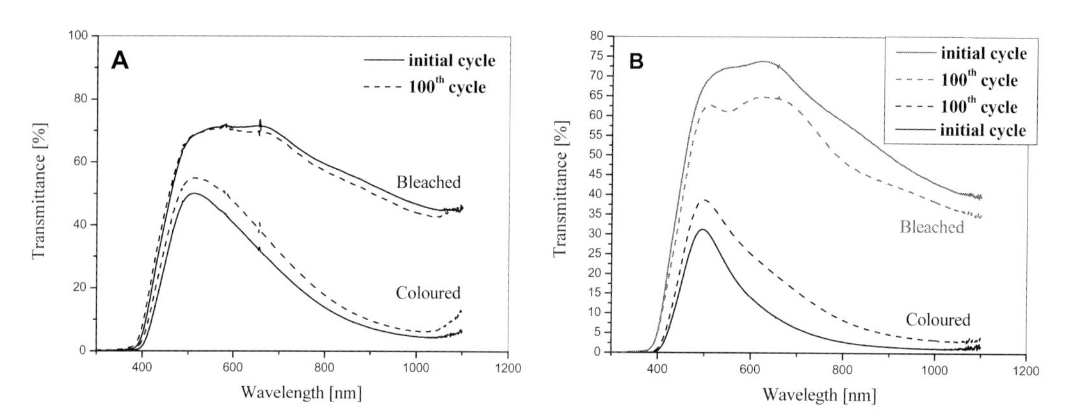

Fig. 2. *In situ* UV-visible spectroelectrochemical measurements of HEC cells employing: A) ICS-2ME/TEOS/TEG/OxA/EtPrIm$^+$I$^-$ and ICS-DEM/TEOS/TEG/OxA/EtPrIm$^+$I$^-$ redox electrolytes.

The coloring-bleaching changes of the HEC cells (four examples presented in Figs. 1 and 2) employing redox electrolytes (Table 1) with composition ICS-DEM/TEOS/TEG/ OxA/M$^+$I$^-$ (or EtPrIm$^+$I$^-$) + I_2 and ICS-2ME/TEOS/TEG/OxA/M$^+$I$^-$ (or EtPrIm$^+$I$^-$) + I_2, where $M^+ = NH_4^+$, K^+, Na^+ and Li^+, showed that all HEC exhibited reversible EC effect up to the 100th cycle except those where LiI was used as a source of the intercalating species. This we related to the affinity of LiI to water. The variations in the EC effect were partially due to the non-uniformity in the prepared WO_3 film and the thickness of the electrolyte, which varied, as no spacers were used for the assembly of the HEC cells. For HEC cells where ionic liquid was used, the intercalating species were protons, while other cell's coloring actually relied on the alkali ions and not protons, which could be generated by the presence of oxalic acid catalyst.

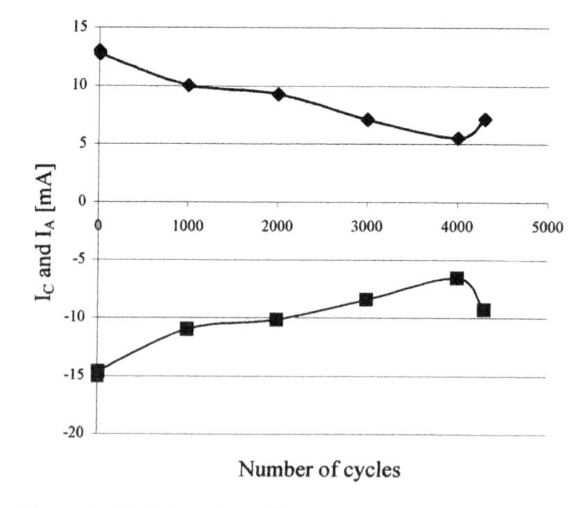

Fig. 3. Dependence of the cathodic (I_C) and anodic (I_A) current on the number of coloring-bleaching cycles for the HEC employing ICS-2ME/TEOS/TEG/OxA/$NH_4I + I_2$ redox electrolyte (active electrode surface area ~4 cm^2).

One of the HEC cells having the composition ICS-2ME/TEOS/TEG/OxA/NH4I + I_2 was continuously cycled more than 5000 times. Current densities continuously dropped in the course of cycling, as shown in Fig. 3. Examination of the electrolyte after the cell was dismantled under SEM and optical microscope did not reveal crystallites as formed with other electrolytes prepared with either sulfolane or EtOH in combination with ICS-PPG4000/KI + I_2 and AcOH or valeric acid catalyst [7, 8, 10]. This demonstrates the advantages of the electrolyte's composition based on the combination of precursors with a network former (TEOS) and suitable co-solvents (TEG). Further studies are under way to assess the advantages of the synthesized precursors and ionic liquids for HEC cells, and the application of these redox electrolytes in the DSPEC cells is to be tested.

Conclusions

Some novel carbamatosil-type precursors were synthesized to be compatible with other components of the redox electrolytes: co-solvents, i.e. TEG and $(EtPrIm)^+I^-$ ionic liquid, OxA catalyst used to perform solvolysis of carbamatosil precursors, and inorganic salts (M^+I^-) serving as a source of the intercalating species (NH_4^+, Li^+, Na^+ and K^+) into a nanocrystalline non-hydrated WO_3 film. The stability of the electrochromic effect of a hybrid electrochromic cell with a composition WO_3/redox electrolyte/Pt counter-electrode was assessed, showing the advantages of using co-solvents having low vapor pressure for attaining long-term electrochromic stability of the redox electrolyte in a hybrid electrochromic cell.

Acknowledgments: Authors wish to thank to Dr. Urška Lavrenčič Štangar for the help in preparing the manuscript. One of us (V.S) is greatly acknowledged to Ministry for Science, Education and Sports (MŠZŠ) of Slovenia for PhD grant.

References

[1] R. D. Rauh, *Electrochim. Acta* **1999**, *44*, 3165.
[2] R. Cinnsealach, G. Boschloo, S. N. Rao and D. Fitzmaurice, *Sol. Energy Mater. Sol. Cells* **1998**, *55*, 215.
[3] U. Opara Krašovec, R. Ješe, B. Orel, J. Grdadolnik, G. Dražič, *Monatsh. Chem.* **2002**, **133,** 1115.
[4] K. Tennakone, G. K. R. Senadeera, V. P. S. Perera, I. R. M. Kottegoda, L. A. A DeSilva, *Chem. Mater.* **1999**, *11*, 2474.
[5] E. Stathatos, P. Lianos, C. Krontiras, *J. Phys. Chem. B* **2001**, *105*, 3486.
[6] E. Stathatos, P. Lianos, U. Lavrenčič Štangar, B. Orel, *Adv. Mater.* **2002**, *14*, 354.
[7] B. Orel, A. Šurca Vuk, R. Ješe, P. Lianos, E. Stathatos, P. Judeinstein, Ph. Colomban, *Solid State Ionics*, in press.

[8] B. Orel, U. Lavrenčič Štangar, A. Šurca Vuk, P. Lianos, Ph. Colomban, *MRS Proc.* **2002**, *725*.

[9] P. Bonhote, A.-P. Dias, N. Papageorgiou, K. Kalyanasundaram, M. Graetzel, *Inorg. Chem.* **1996**, *35*, 1168.

[10] A. Šurca Vuk, M. Gabrscek, B. Prel, Ph. Colomban, *J. Electrochem. Soc.* **2004**, *151,* E150.

[11] N. Papageorgiou, Y. Anatassov, M. Armand, P. Bohnote, H. Pettersen, A. Azam, M. Graetzel, *J. Electrochem. Soc.* **1996**, *143*, 3099.

[12] H. Stegemann, A. Rohde, A. Reiche, A.Schnittke, H. Fuellbier, *Electrochim. Acta* **1992**, *37*, 379.

[13] I. Shiyanovskaya, M. Hepel, *J. Electrochem. Soc.* **1998**, *145*, 33981.

Silicone Nanospheres for Polymer and Coating Applications

Jochen Ebenhoch,* Helmut Oswaldbauer

Wacker-Chemie GmbH, D-84480 Burghausen, Germany
Tel.: +49 8677 832436 — Fax: +49 08677 836962
E-mail: jochen.ebenhoch@wacker.com

Keywords: silicone nanospheres, impact resistance, scratch resistance

Summary: Based on patented Wacker technology, elastomeric and highly crosslinked silicone and silicone-organopolymer core-shell materials can be synthesized and manufactured on a large scale. These novel tailor-made materials consist of monodisperse primary particles which can be homogeneously dispersed in thermosetting and thermoplastic resins to improve mechanical properties, e.g., impact resistance, leading to a variety of possibilities for polymer and coating applications. Mechanical studies in several systems exhibited significant effects even with very low particle loading levels.

Introduction

Silicone nanospheres with different particle diameters, crosslinking density, and chemical functionalization are accessible by aqueous hydrolysis-condensation sequences of silane and siloxane precursors [1–3] and subsequent isolation. Grafting of functionalized particles with organopolymers [1] or surface modification [2] results in nanosized silicone domains which are readily dispersible in monomeric and polymeric systems. A variety of these versatile, tailor-made products will soon be launched by Wacker on a commercial scale.

Polymer Modification

The properties of thermosetting and thermoplastic resin systems are continually improved to meet increasing performance requirements of end users. One way to enhance material properties is to incorporate nano-modifiers, based on elastomeric silicone particles, which are optionally grafted with other (acrylic) polymers to control dispersibility, viscosity, and other parameters. As an example, epoxy resin formulations have been modified with silicone nanospheres to improve low-stress behavior. Table 1 shows the outstanding fracture toughness improvement of silicone core-shell nanospheres, even at very low particle loading levels.

Table 1. Fracture toughness of modified epoxy resins. [a]

Modifier	0%	3%	5%	10%
K_{Ic}	0.60	1.39	1.75	2.02

[a] Bisphenol-A diglycidylether, thermally cured with DDS at 160 °C.

Some benefits of Wacker silicone nanospheres for thermosets are

- Excellent dispersibility
- Agglomerate-free distribution
- Maintenance of matrix T_g
- Good adhesion
- Low viscosity

In the case of acrylic molding sheets and semi-finished products, Wacker have elaborated a product concept based on silicone core-shell technology, which enables the manufacture of highly transparent particle-modified materials with improved RT and low-temperature (–40 °C) impact resistance and high weatherability [4]. Other transparent applications (e.g., epoxy resins, polycarbonates) are showing encouraging results, too.

More Polymer Application Fields

(a) Toughening of thermosetting resins

- Epoxy molding compounds,
- Liquid encapsulation and potting compounds, e.g., for automotive parts
- Thermal and RT-curing adhesives, underfiller materials
- Composites, fiber-reinforced, e.g., for transportation or in printed wiring boards
- PCB build-up materials

(b) Improvement of other polymers (toughening and mar resistance)

- Acrylic sheets and casting compounds, e.g., for sanitary use and outdoor applications
- Acrylic, cyanoacrylic, and other adhesive systems
- Polystyrenes (e.g., ABS, SAN), -acetals, -amines, polycarbonates and other thermoplastic resins
- High-performance polymers, e.g., PPS

Coating Applications

There are not many industrial fields where key requirements are of the same complexity as for coating applications. Improvement of scratch resistance is a challenge for today's liquid automotive topcoat and powder coating systems. On the other hand, impact resistance of waterborne and powder coats is currently not sufficient and has to be improved to meet today's customer requirements. Wacker silicone nanospheres offer a solution for both problems – significantly higher scratch resistance by using functionalized and/or grafted highly crosslinked silicone particles with diameters of ≤ 20 nm – and better impact resistance by incorporation of core-shell particles with elastomeric silicone rubber cores. Other properties, e.g., UV and chemical resistance, adhesion, gloss, and transparency are maintained or even improved at the same time.

As an example, Table 2 shows how Wacker core-shell particles are able to enhance the impact resistance at RT and –27 °C of an acrylic powder coating system by a factor of 2 – 3.

Table 2. Impact resistance of powder coating composition.[a]

Modifier	0%	5%	10%
at RT [Ibs]	24	48	69
at –27°C [Ibs]	24	38	48

[a] Acrylic GMA with 0.5% carbon black, cured with DDDA at 150 °C.

Some other benefits of Wacker silicone nanospheres for powder coatings are

- Good surface appearance
- No gloss reduction
- No silicones bleeding
- Chemical resistance
- Excellent weatherability
- Reduction of overspray

More Coating Application Fields (Toughening and Scratch Resistance)

- Polyester and other non-acrylic powder coatings
- Waterborne and "high-solid" coatings
- Liquid solvent-borne coatings, e.g., for automotive topcoats and industrial coatings
- Permanent PCB coatings (photoresist ink)
- Others, e.g., flooring (parquet)

Other Application Fields

- Cosmetics and hair-care (shampoo and conditioner)
- Dental applications (composites, artificial teeth and dentures)
- Textile applications (fiber treatment and modification)

References

[1] M. Geck (Wacker-Chemie GmbH), EP 0 744 432.
[2] K. Mautner (Wacker-Chemie GmbH) EP 0 492 376.
[3] N. Jungmann, M. Schmidt, J. Weis, J. Ebenhoch, *Macromol.* **2002**, *35*, 6851.
[4] Patent filed.

New Approaches and Characterization Methods of Functional Silicon-Based Non-Oxidic Ceramics

*J. Haberecht, F. Krumeich, K. Hametner, D. Günther, R. Nesper**

ETH Zurich, ETH Hönggerberg-HCI H139
W.-Pauli-Str. 10, CH-8093 Zurich, Switzerland
Tel.: +41 1 6323069 — Fax: +41 1 6321149
E-mail: nesper@inorg.chem.ethz.ch

Keywords: Si-ceramic, precursor, TEM, laserablation-ICP-MS

Summary: Si–B–N–C materials derived from different silyl-functionalized ethynylborazines were investigated. Pt-catalyzed hydrosilylation of *B,B',B''*-triethynyl-borazine with $HSiCl_3$ leads to B-tris(trichlorosilylvinyl)borazine (**2**) in quantitative yield with a selectivity of 80% β-substituted product. Subsequently, hydrogenation of the trichlorosilyl groups of **2** leads to B-tris(silylvinyl)borazine **4**. Starting from **4** or polymer **3P**, a highly durable Si–B–N–C ceramic is obtained after pyrolysis under inert atmosphere. The composition of the ceramic material corresponds exactly to the backbone of the precursor molecules **2**, **4**, and $SiBNC_2$. From this synthetic route, very compact materials are formed. The ceramic yield of approximately 94% (from **1**) represents a new class via this synthetic route. As a new method, Laser ablation-ICP-MS was used for the characterization of the materials. The results demonstrate the high potential of this direct solid sampling technique for the characterization of such samples.

Introduction

The synthesis of quaternary silicon-based non-oxidic ceramics from molecular precursors with outstanding chemical and physical properties directly from molecular precursors has been investigated intensively since the early 90s [1]. Nevertheless, polycarbosilanes and related polymers are still the most important among non-oxide polymer precursors [2]. Here we discuss two different strategies which have both been investigated in order to form the desired ceramic material. A major advantage of our reaction pathway is the extreme stability of the borazine precursor molecules, which can easily be functionalized. Because of the large amount of multiple bonds in the precursors, a high degree of crosslinking should also be present in the ceramic material. In contrast to established procedures [1], the direct hydrogenation of precursor molecules and their pyrolysis have been investigated for the first time. With our synthetic method, shrinking effects can be reduced and more compact materials produced. Until now, this concept was discussed only for polymer precursors [3].

Molecular Precursors

Hydrosilylation of **1** with trichlorosilane has been performed with Pt on charcoal (1% by weight) in quantitative yields (Scheme 1). The B-tris(trichlorosilylvinyl)borazine (**2**) was obtained with a high regioselectivity of approximately 80% β-*trans* hydrosilylation product [4]. Pure **2** can be obtained by fractional crystallization of the synthesized product from hexane. For further synthesis, both α- and β-hydrosilylation products can be used. No further hydrosilylation was observed in this case. In order to interconnect the single-source precursor molecule **2** to a pre-ceramic polymer, methylamine was added to the solution of **2** in hexane, and a high viscosity, colorless oil was formed. By changing the reaction parameters (excess of CH_3NH_2, temperature), the viscosity of the polymer can be varied [5]. The obtained polymer (**3P**) is pure after evaporation of the solvent, which is checked by NMR. Other solvents like thf or toluene are also possible for the reaction, as well as for dissolution of the polymer. Furthermore, ethylamine leads to similar results in the formation of the polymer.

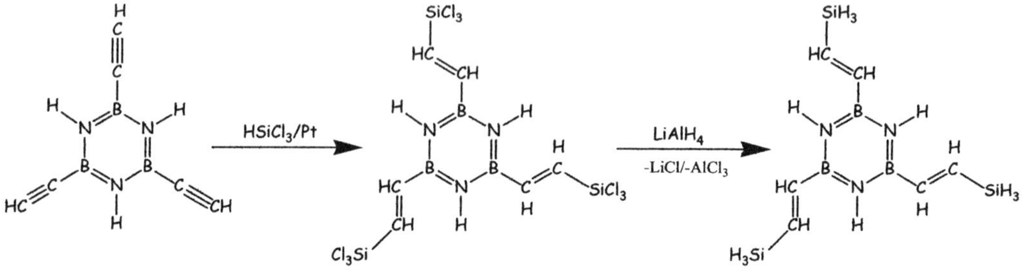

HSiCl₃/Pt → LiAlH₄, -LiCl/-AlCl₃

1 **2** **4**

Scheme 1. Synthesis of silylvinyl-substituted borazines **2** and **4** as molecular precursors for the ceramic materials.

To increase the ceramic yield during the pyrolysis, a completely new synthetic strategy was carried out for the borazine molecules. Instead of an interconnection of the single-source molecules via sol-gel ammonolysis, the precursor molecule **2** was selectively hydrogenated (Scheme 1) [6]. As pointed out before, the target molecule B-tris(silylvinyl)borazine (**4**) contains no leaving group except hydrogen for the pyrolytic conversion. The hydrogenation reaction can be carried out with different reagents, like $LiAlH_4$, $NaAlH_4$ or the so-called superhydride $LiH(BEt_3)$. From the reaction with a small excess of $LiAlH_4$ in thf, **4** was obtained after careful purification in ~50% yield. Hydrogenation was complete after one day as monitored with ^1H and ^{29}Si NMR spectroscopy. The reaction has to be carried out at a maximum starting temperature of –20 °C and is selective for the –$SiCl_3$ functional groups. No reactions with the multiple bonds of the vinyl-borazines have been observed. B-tris(silylvinyl)borazine (**4**) is a white solid, soluble in any common solvent, but difficult to deliver from residual solvents. The Si–H functional groups are characterized by two strong vibration modes at 2155 cm^{-1} (stretching) and 913 cm^{-1} (bending). A ^{1}H-coupled ^{29}Si NMR spectrum of B-tris(silylvinyl)borazine (**4**) (Fig. 1) shows all expected Si–H couplings. The ^{1}J(Si–H) coupling constant was measured with 199 Hz, the ^{2}J(HSi–CH) with 13.2 Hz, and the ^{3}J(HSi–C=CH) with 10.6 Hz. Isotropic chemical shift values of the described compounds are given in Table 1.

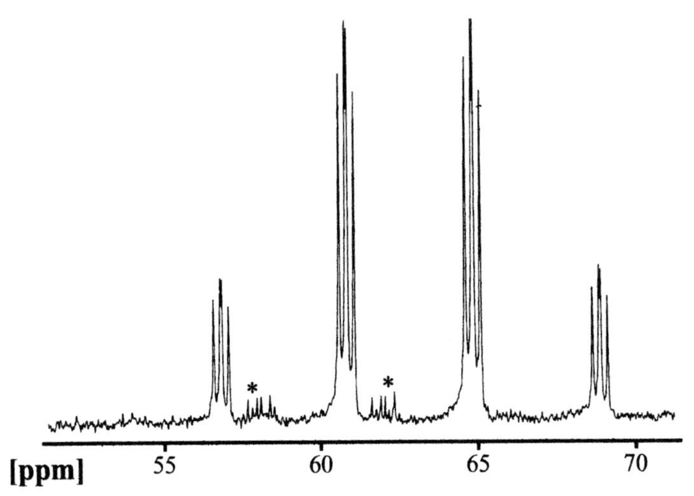

Fig. 1. ^{1}H-coupled ^{29}Si NMR spectrum of tris(E-silylvinyl)borazine (**4**). * denotes signals from residual α-hydrosilylation product.

Table 1. Isotropic ^{29}Si chemical shift values of **2**, **3P** ($CDCl_3$), and **4** (C_6D_6) in ppm.

	Si–C=C–B	C=C(B)–Si
2	–3.3	0.2
3P	–31.4	–27.9
4	–62.9	–60.1
Ceramic	–43[a]	

[a] $w_{½}$ = 1400 Hz

Pyrolysis and Material Characterization

Thermal treatment of the polymeric precursor (**3P**) and molecular precursors (**4**) result in an amorphous material (Scheme 2). After the precursor-to-ceramic conversion, a black, dense ceramic was obtained. The density of the material was determined to be 1.6 – 1.7 g/cm^3. One of our main interests was to reduce the accompanying formation of gaseous by-products during pyrolysis. As shown in Fig. 2, TG measurements up to 1500 °C reveal a weight loss of ca. 37% in the case of **3P** and only ca. 6% for **4** (see also Scheme 2). Thus, tris(silylvinyl)borazine (**4**) is the first molecular precursor for Si–B–N–C ceramics with the potential to give ceramic yields similar to those derived from polymeric precursors like polysilazane and its derivatives [3]. The evaporation of hydrogen can be observed mainly in the temperature range from 600 °C to 1500 °C for **4**. The TG analysis of **3P** suggests, according to literature, three steps of thermal degradation [1]. The first of these three clearly divided steps can be assigned to completion of the poly-condensation (250 – 400 °C) by the separation of methylamine and small amounts of HCN (weight loss ~20%). Furthermore, a fragmentation in the range of 400 – 600 °C (weight loss ~14%, mainly methane and also HCN) and at last the elimination of residual hydrogen (>1000 °C, weight loss ~3%) can be observed.

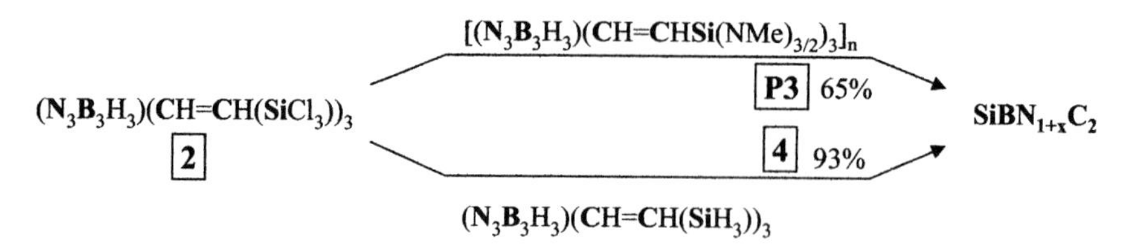

Scheme 2. Formation (pyrolysis) of the ceramic material from B-tris(trichlorosilylvinyl)borazine (**2**) via different synthetic pathways.

X-ray powder investigations after thermal treatment up to 1650 °C reveal the amorphous state of the ceramic material. Additionally, HRTEM investigations reveal the completely amorphous state at atomic scale of the samples heated up to 1650 °C. Furthermore, elemental maps of Si, B, N, and C recorded by electron spectroscopic imaging confirm a homogeneous distribution of the elements [5]. The ^{29}Si-NMR spectra of the ceramic materials show one signal at –43 ppm ($w_{½}$ = 1400 Hz) for a typical four-fold coordination of the silicon.

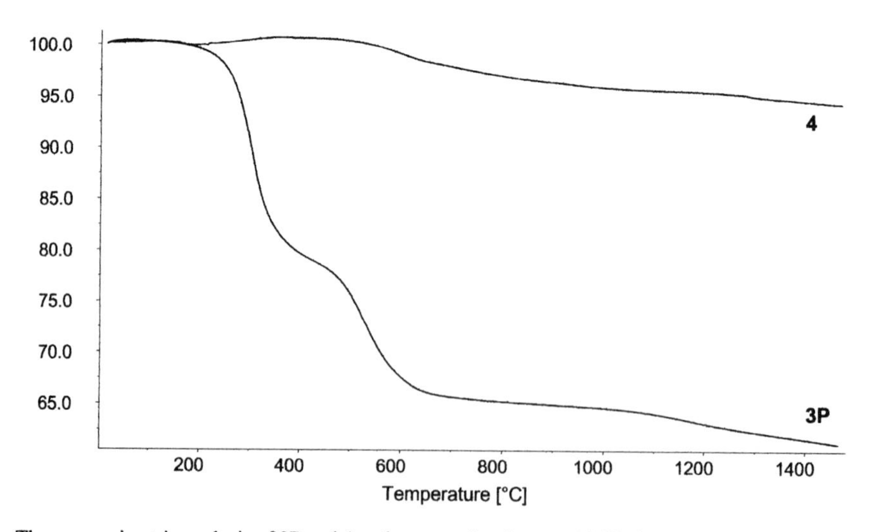

Fig. 2. Thermogravimetric analysis of **3P** and **4** under argon, heating rate 10 K/min.

For further characterization of the material, a high-temperature DTA/TG has been carried out. The ceramic material, obtained from **3P** and pyrolyzed at 1000 °C, shows only a weight loss of 2% up to 2000 °C. Powder X-ray diffraction analysis as well as HRTEM investigations of the sample indicates the crystallization of Si_3N_4 and SiC at the mentioned temperatures.

Laser-Ablation ICP-MS

Because of difficulties and high costs/complexity for digestions of ceramic materials, Laserablation-ICP-MS was used as a new method for the characterization of our material [7]. The sampling on the ceramic surface was carried out using a 193 nm Excimer laser ablation system (GeoLas, MicroLas, Göttingen, Germany). An aerosol produced from the laser sample interaction was transported using helium as carrier gas to an inductively coupled plasma mass spectrometer (ICP-MS, ELAN 6100, Perkin-Elmer, Norwalk, USA), where the aerosol was vaporized, atomized, and ionized. The NIST 610 reference glass material was used as an external standard for calibrating the instrument. Every sample was analyzed at different positions using a spatial resolution of 80 μm (see Table 2). The measured Si/B ratio was 2.55 for **3P** and 2.64 for **4**, with standard deviations of 0.04. The expected Si/B ratio from the borazine precursors (2.6) was found in the ceramic material with a completely homogeneous distribution, confirming the result of the ESI investigations [5]. No residual chlorine or other metallic impurities were detected in the product. From additional analytical results (C/N analyses), the composition of the produced ceramic material can be described as $SiBN_{1+x}C_2$ (for **4**: x = 0; for **3P**: x = 1.5) within experimental error.

Table 2. LA-ICP-MS analysis of the ceramic material. *left*: hole after ablation; *right*: detailed analysis values of selected isotopes. Values are in ppm.

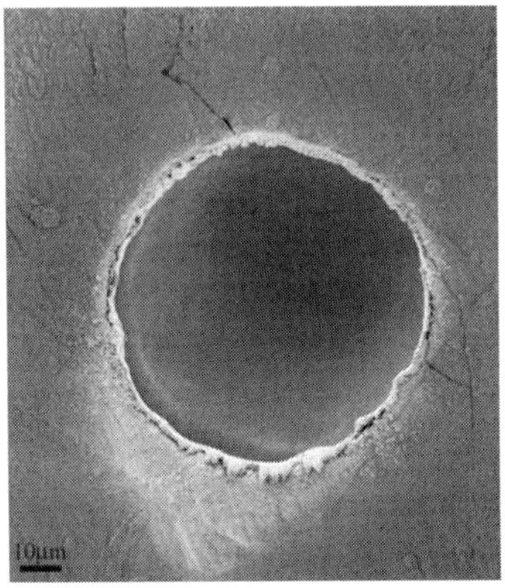

Precursor	^{11}B	^{29}Si [a]	^{30}Si	Si/B ratio[b]
3P	140351	364900	365687	2.55
3P	139930	364900	365939	2.56
3P	140405	364900	360577	2.54
3P	137164	364900	363640	2.59
4	133816	364900	368464	2.67
4	135214	364900	365650	2.64
4	133087	364900	360926	2.68
4	134247	364900	362985	2.65

[a] reference isotope; [b] standard deviation from analysis of ^{10}B, ^{11}B, ^{28}Si, ^{29}Si, and ^{30}Si: 0.04 .

Physical Properties

In view of the search for novel applications, conductivity measurements of our ceramic material show promising results. The amorphous ceramics show a semiconductor behavior in the measured temperature range of 300 – 600 °C. At room temperature, the value is approximately 10^2 $(\Omega m)^{-1}$, comparable to typical III/V-semiconductors. It will take further investigations, testing and tuning this behavior.

Conclusion

Our synthetic route opens a new and wide range for synthesis of further precursor molecules for Si–B–N–C ceramics. Starting with hydrosilylation of **1** with dichloroalkylsilanes HCl_2SiR and subsequent polymerization or hydrogenation, the chemical composition of the ceramic material can be varied like a construction kit by changing the silyl-substituents R (Cl, alkyl, phenyl). Because of many multiple bonds in the borazine molecules, a very high thermal stability was achieved. The backbone of the precursor molecules remains intact during pyrolysis. Furthermore, it was possible to increase the ceramic yield to 94% by using tris(silylvinyl)borazine, the first molecule which could be directly pyrolyzed.

Acknowledgments: Thanks to Dr. T. Jäschke (MPI FKF Stuttgart) for HT-DTA/TG, and ETH Zürich for financial support.

References

[1] M. Jansen, T. Jaeschke, B. Jaeschke, in *High Performance Non-oxide Ceramics I, Vol. 101* (Ed.: M. Jansen), Springer-Verlag, Berlin, Heidelberg, **2002**, pp. 137.

[2] E. Kroke, Y. Li, C. Konetschny, E. Lecomte, C. Fasel, R. Riedel, *Mater. Sci. Eng.* **2002**, *26*, 97.

[3] M. Weinmann, J. Schuhmacher, H. Kummer, S. Prinz, J. Q. Peng, H. J. Seifert, M. Christ, K. Muller, J. Bill, F. Aldinger, *Chem. Mater.* **2000**, *12*, 623.

[4] J. Haberecht, A. Krummland, F. Breher, B. Gebhardt, H. Rüegger, R. Nesper, H. Grützmacher, *Dalton Trans.* **2003**, 2126.

[5] J. Haberecht, F. Krumeich, H. Grützmacher, R. Nesper, *Chem. Mater.* **2004**, *16*, 418.

[6] R. Nesper, J. Haberecht, H. Grützmacher, PCT Int. Appl. WO 2 004 069 768, **2003**.

[7] D. Günther, I. Horn, B. Hattendorf, *Fresenius J. Anal. Chem.* **2000**, *368*, 4.

Preceramic Polymers for High-Temperature Si–B–C–N Ceramics

Markus Weinmann, Markus Hörz, Anita Müller, Fritz Aldinger

Max-Planck-Institut für Metallforschung
Heisenbergstraße 3, D-70569 Stuttgart, Germany
Tel.: +49 711 6893127 — Fax: +49 711 6893138
E-mail: weinmann@mf.mpg.de

Keywords: precursor, thermolysis, Si–B–C–N

Summary: Precursor-derived quaternary Si–B–C–N ceramics frequently possess an enhanced thermal stability compared to SiC, Si_3N_4 or Si–C–N ceramics. The stability of the materials towards crystallization and/or decomposition is directly connected to the molecular structure and the elemental composition of the polymeric precursors. This paper highlights recent investigations on the synthesis of boron-modified polysilazanes and polysilylcarbodiimides. Hydroboration of polyvinylsilazanes and dehydrocoupling reactions of boron-modified silanes with ammonia or amines as well as cyanamide are described. It is shown that simple organosilicon chemistry provides a means to efficiently optimize ceramic yields and tune elemental composition as well as thermal properties of the polymer-derived ceramics.

Introduction

Since the publications of Jansen [1] and Riedel [2] in the early 1990s, there has been an increasing interest in precursor-derived Si–B–C–N ceramics. They obtained Si–B–C–N precursors by metathesis reactions of methylamine with $Cl_3SiNHBCl_2$ or ammonolysis of $B(C_2H_4Si(CH_3)Cl_2)_3$ (releasing $[B(C_2H_4Si(CH_3)\text{-}NH)_3]_n$, in the literature referred to as T2-1 [2]). Crucial subjects were a time-intensive processing (separation of the precipitates from the polymer solutions) and low ceramic yields (ca. 50%). Alternative syntheses which avoided formation of solid by-products were published by Seyferth and Sneddon. They reported on dehydrocoupling of borane dimethylsulfide, $BH_3{\cdot}SMe_2$ [3], or borazine derivatives [4] with polysilazanes. However, materials derived from such precursors possessed unsatisfactory high temperature stability.

Hydroboration of Poly(vinylsilazanes)

Precursors obtained by ammonolysis of $B(C_2H_4Si(R)Cl_2)_3$ (R = Cl, H; Scheme 1) [5] have much higher ceramic yields than derived from $B(C_2H_4Si(CH_3)Cl_2)_3$ [2]. The reason for the improved ceramic yield in $[B(C_2H_4Si(NH)_{1.5})_3]_n$ (**2*M***, 84%), which was derived from $B(C_2H_4SiCl_3)_3$, is an increased cross-linking density. Si–H and N–H groups in $[B(C_2H_4SiH\text{-}NH)_3]_n$ (**1*M***, ceramic yield 88%), in contrast, provide latent reactivity. They undergo dehydrocoupling reactions during thermolysis, which *in situ* increase the cross-linking density and thus avoid depolymerization and volatilization of low-molecular-weight species. Because of the low solubility of **1*M*** and the insolubility of **2*M*** in organic solvents, the processing is difficult and product yields are unsatisfactory.

Both polymer yields and processing can be improved by modifying the synthetic approach. Hydroboration of vinyl-substituted poly- or oligosilazanes $[(H_2C{=}CH)Si(R)\text{-}NH]_n$, (R = H, $(NH)_{0.5}$, CH_3, Scheme 1) using $BH_3{\cdot}SMe_2$ delivers the corresponding boron-modified polysilazanes $[B(C_2H_4Si(R)\text{-}NH)_3]_n$, **1*P*** – **3*P*** in 100% yield [5].

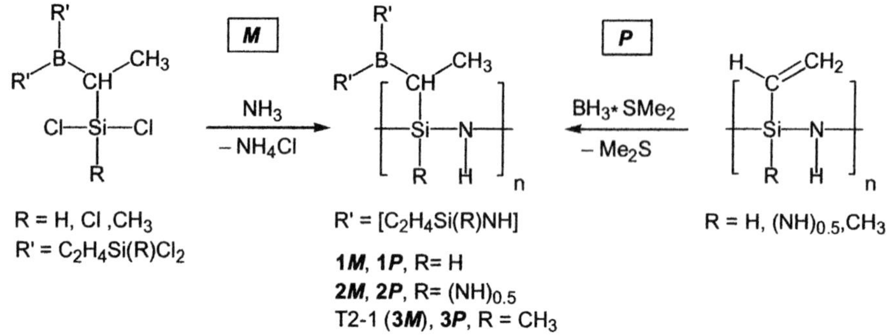

Scheme 1. Synthesis of boron-modified polysilazanes by ammonolysis of $B(C_2H_4Si(R)Cl_2)_3$ (monomer route → index ***M***) or hydroboration of vinyl-substituted polysilazanes (polymer route → index ***P***) [2, 5].

Remarkably, ceramic yields were not influenced by the reaction pathway applied. They are mainly a function of the molecular structure of the precursors, i.e. the nature of the silicon-bonded substituents R. The methyl group in **3*M*** (T2-1 [2]) and **3*P*** is responsible for low ceramic yields (ca. 50%). It does not contribute to cross-linking reactions and is split off at ~500 °C. In contrast, **2*M*** and **2*P*** (ca. 84% ceramic yield) are highly cross-linked; consequently depolymerization reactions are inhibited. Ceramic yields are highest in **1*M*** and **1*P***. This is because of the possibility to cross-link during thermolysis by dehydrocoupling of Si–H and N–H units, as mentioned above.

Dehydrocoupling of $B(C_2H_4Si(R)H_2)_3$ with NH_3 or H_3CNH_2

An alternative access to boron-modified polysilazanes is a dehydrocoupling of ammonia or alkyl amines with tris(hydridosilylethyl)boranes, $B(C_2H_4Si(R)H_2)_3$ (R = H, CH_3; C_2H_4 = CH_2CH_2, $CHCH_3$; Scheme 2) [6]. This avoids the formation of solid by-products and allows for the synthesis of highly cross-linked precursors. The starting compounds are best obtained from chloro vinylsilanes $(H_2C{=}CH)Si(R)Cl_2$ (R = Cl, CH_3), which are initially reacted with $LiAlH_4$ in diethyl ether. The hydrido vinylsilanes $(H_2C{=}CH)Si(R)H_2$ that form are difficult to handle as neat products and therefore reacted *in situ* by distilling the volatile components of the reaction mixture into a solution of $BH_3{\cdot}SMe_2$ in toluene [7].

It was observed that ammonolysis of $B(C_2H_4Si(R)H_2)_3$ (Scheme 2, route A) requires basic catalysts such as *n*-butyl lithium. The reaction is performed in analogy to the potassium hydride-catalyzed cross-linking of cyclic silazanes described by Seyferth et al. [8]. Most probably, *n*-BuLi initially deprotonates the weak nucleophile ammonia with the formation of lithium amide and evaporation of *n*-butane. The stronger nucleophilic amide then replaces a silicon-bonded hydride, which subsequently deprotonates ammonia, leading to the evolution of molecular hydrogen. The silylamines that arise are not stable under the reaction conditions applied (refluxing solvent), and by fast condensation of ammonia the polymeric precursors form [6].

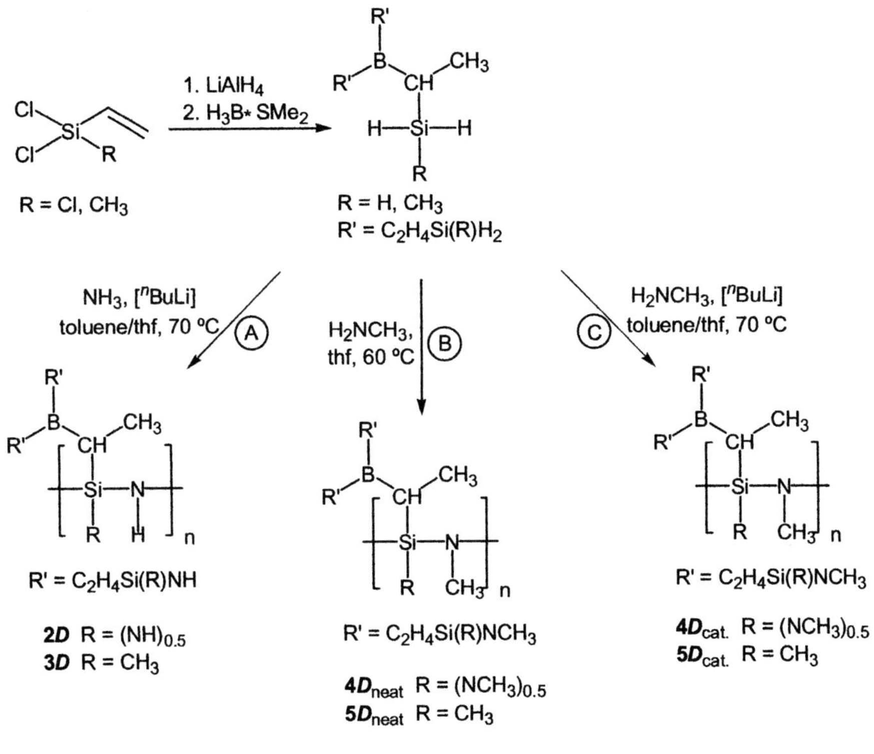

Scheme 2. Dehydrocoupling reactions (→ index ***D***) of tris(hydridosilylethyl)boranes, $B(C_2H_4Si(R)H_2)_3$ (R = H, CH_3) with ammonia or methylamine [6].

Since methylamine is a stronger nucleophile than ammonia, aminolysis of $B(C_2H_4Si(R)H_2)_3$ can be performed without catalyst (Scheme 2, route B). Nevertheless, aminolysis was also carried out in the presence of catalytic amounts of *n*-BuLi (Scheme 1, route C) according to route A. IR and NMR spectroscopy as well as elemental analysis showed that the nitrogen contents in the polymers (and also those of the derived ceramic materials) were higher if the reactions were performed with catalyst [6].

High-temperature thermogravimetric analysis (TGA) of the ceramic materials and their chemical compositions are given in Fig. 1. TGA investigations reveal a significant difference in the mass stability of the materials. Ceramics derived from ***2D***, ***4D*****cat.** and ***5D*****cat.** decompose at 1450 – 1550 °C. This is the typical temperature range where ternary Si–C–N materials decompose [9]. Against this, ceramics obtained from ***3D***, ***4D*****neat** and ***5D*****neat** do not decompose below 1900 °C. The thermal stability is directly connected with the nitrogen content in the materials. Thermally stable materials possess nitrogen contents of below 21 wt%. If the nitrogen contents exceed 25 wt%, ceramics decompose around 1500 °C. Attempts to systematically adjust various Si : B : N ratios by a dehydrocoupling of $B(C_2H_4Si(R)H_2)_3$ and ammonia or methylamine failed because of the volatility of the latter compounds and their partial loss during synthesis.

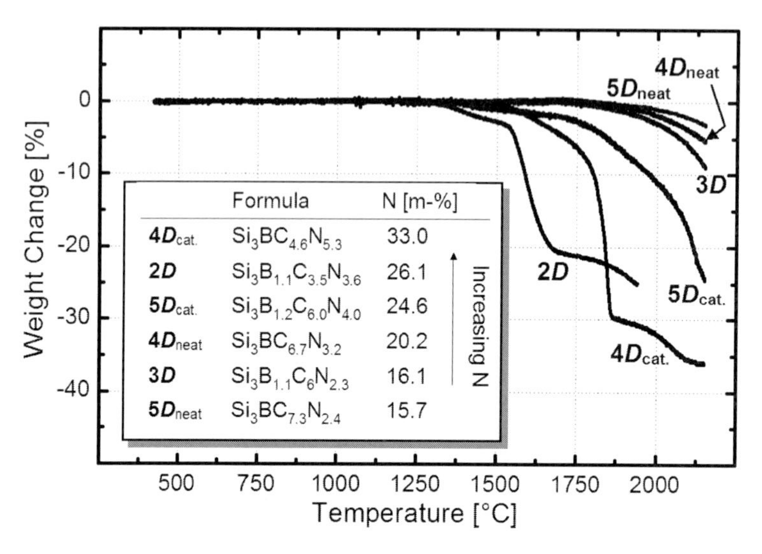

	Formula	N [m-%]
4Dcat.	$Si_3BC_{4.6}N_{5.3}$	33.0
2D	$Si_3B_{1.1}C_{3.5}N_{3.6}$	26.1
5Dcat.	$Si_3B_{1.2}C_{6.0}N_{4.0}$	24.6
4Dneat	$Si_3BC_{6.7}N_{3.2}$	20.2
3D	$Si_3B_{1.1}C_6N_{2.3}$	16.1
5Dneat	$Si_3BC_{7.3}N_{2.4}$	15.7

Fig. 1. High-temperature TGA of ceramics obtained from ***2D*** – ***5D***. Heating rate $T < 1400$ °C: 5 °C/min, $T > 1400$ °C: 2 °C/min; argon atmosphere. The inserted table gives the experimentally determined chemical compositions of the materials [6].

Dehydrocoupling of $B(C_2H_4SiH_3)_3$ with $H_2N–CN$

Cyanamide was chosen as an alternative nitrogen source. In contrast to NH_3 or H_3CNH_2 it is a solid. It is not volatile and it is sufficiently soluble in polar solvents. According to Scheme 3 various polymer compositions could be realized.

Polymerization reactions were performed by dissolving $B(C_2H_4SiH_3)_3$ with different amounts of cyanamide in tetrahydrofuran [10]. A catalyst as required for the dehydrocoupling of $B(C_2H_4Si(R)H_2)_3$ with NH_3 or H_3CNH_2 was not used. The solutions were refluxed, whereby strong hydrogen evolution and precipitation of the precursors were observed. After 12 h the mixtures were cooled to room temperature, and all volatile components were removed at 25 °C/10^{-1} mbar. The colorless residues were rinsed with diethyl ether and dried (130 °C/10^{-2} mbar).

Thermolysis delivered ceramics in 75 – 84% yield, depending on the molecular structure. The polymers with the lowest degree of cross-linking **H-N2**, **H-N3**, and **H-N3.5**, had the highest ceramic yields of approx. 83 – 84%. The more highly cross-linked precursors **H-N4** and **H-N5** gave ceramics in 81% and 80% yield, respectively, whereas the most highly cross-linked polymer **H-N6** gave ceramics in 75% yield. These findings again show that latent reactive sites in the precursor are essential for obtaining high polymer-to-ceramic conversion yields.

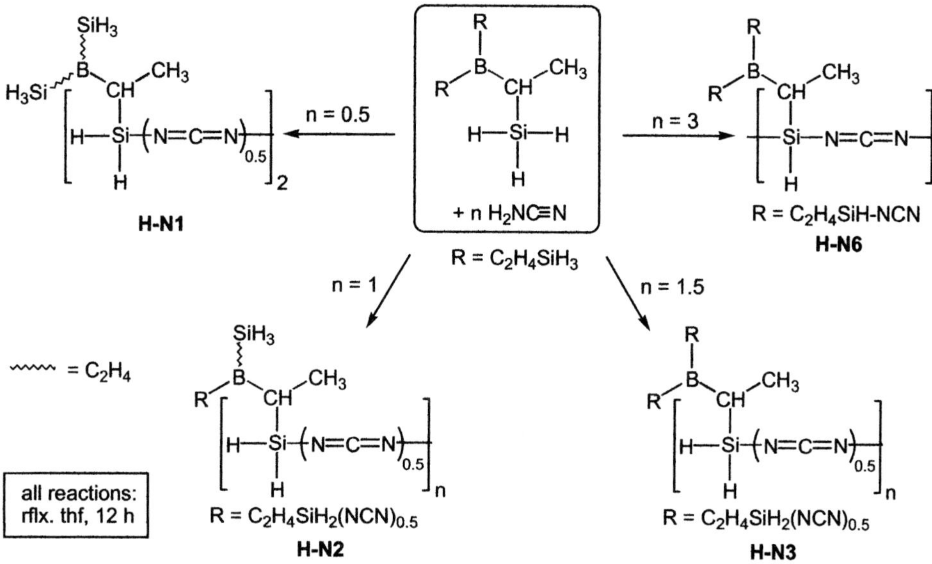

Scheme 3. Synthesis of boron-modified poly(silylcarbodiimides) by a dehydrogenative coupling of $B(C_2H_4SiH_3)_3$ with cyanamide, $H_2N–CN$. Molar ratios of 1 : 0.5 to 1 : 3 were chosen. The indexes **N1** – **N6** give the numbers of nitrogen atoms per boron atom. Chemical compositions range from ($C_{6.5}H_{20}NSi_3B$) (**H-N1**) to ($C_9H_{15}N_6Si_3B$) (**H-N6**). Likewise, boron-modified poly(silylcarbodiimides) were obtained from $B(C_2H_4Si(CH_3)H_2)_3$ [10].

High Temperature Investigations

As a consequence of the high ceramic yields and similar thermolysis behavior, the Si : B : N ratio in **H-N2** – **H-N6** was transferred into the ceramic materials [11, 12]. High-temperature TGA investigations (Fig. 2) show a clear trend. Thermal stability of the ceramics increases with decreasing N content. Whereas nitrogen-rich **H-N5** and **H-N6** ceramics start to decompose at ca. 1650 °C, those derived from **H-N2**, **H-N3**, and **H-N3.5** are stable up to at least 1900 °C. Degradation of **H-N4** ceramics begins at 1720 °C. In comparison to that of **H-N5** and **H-N6** it appears retarded. Obviously, this material possesses a critical amount of nitrogen with respect to thermal stability.

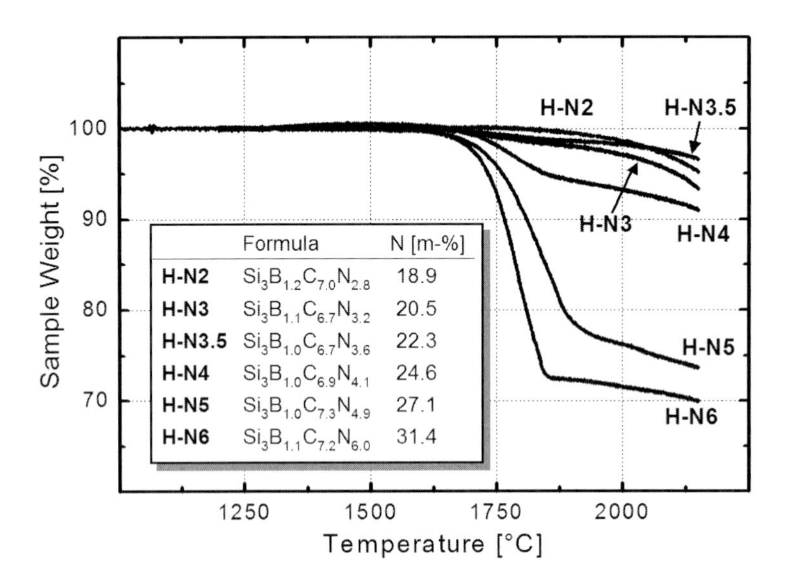

Fig. 2. High-temperature TGA of ceramics obtained from **H-N2** – **H-N6**. Heating rate $T < 1400$ °C: 5 °C/min, $T >$ 1400 °C: 2 °C/min; argon atmosphere. The inserted table gives the experimentally determined chemical compositions of the materials [11, 12].

These observations are reflected in the phase evolution of the materials, which was studied in the 1400 – 2000 °C range by X-ray diffraction (XRD, Fig. 3a). Ceramics obtained from **H-N4**, **H-N5**, and **H-N6** crystallize at around 1600 °C and only SiC reflections appear, whereas ceramics derived from **H-N2**, **H-N3**, and **H-N3.5** do not crystallize below 1800 °C. At higher temperature, crystallization of α/β-SiC and β-Si_3N_4 takes place. Remarkably, the intensity of β-Si_3N_4 reflections does not decrease even after annealing the sample at 2000 °C [11, 12].

The microstructure of **H-N2** after annealing at 1900 °C is shown in Fig. 3b. It consists of SiC and Si_3N_4 nanocrystals. They are embedded in an amorphous matrix consisting of turbostratic B–C–N. It is supposed that the matrix encapsulates Si_3N_4 crystals (resulting in a pressure stabilization of Si_3N_4) and that it decreases the carbon activity [13]. Both effects can act concurrently, and, as a consequence, the degradation of Si_3N_4 and thus the decomposition of the Si–B–C–N ceramic is shifted to higher temperatures than expected.

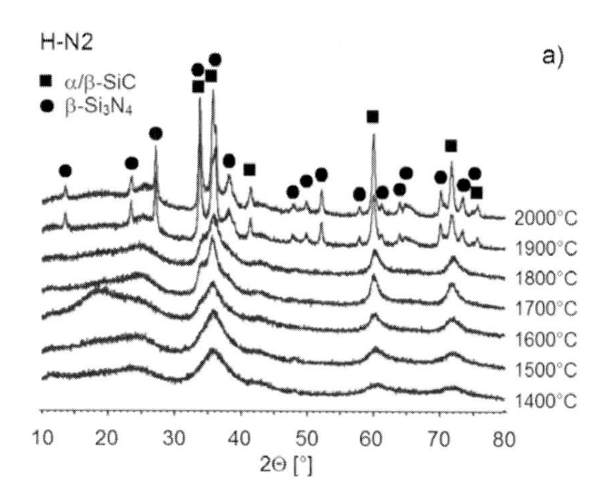

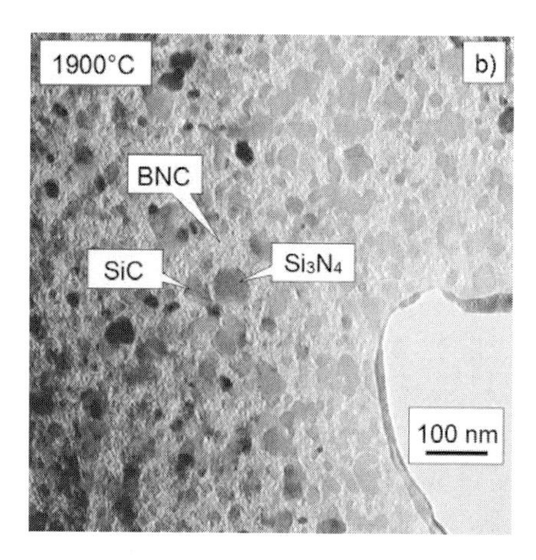

Fig. 3. Phase evolution of **H-N2**-derived ceramics investigated by (a) XRD and (b) TEM. XRD patterns were obtained after annealing **H-N2** ceramics at 1400 – 2000 °C (1 bar N_2, 3 h). The as-obtained material is amorphous. α/β-SiC and β-Si_3N_4 crystallization occurs after annealing at 1900 °C. The TEM image (b) of the 1900 °C sample shows SiC (30 – 50 nm) and Si_3N_4 crystals (40 – 60 nm), which are embedded in an amorphous matrix consisting of boron, nitrogen, and carbon [11, 12].

References

[1] (a) M. Jansen, H. P. Baldus, DE 4 107 108 A1, **1992**. (b) H. P. Baldus, O. Wagner, M. Jansen, *Mater. Res. Soc. Symp. Proc.* **1992**, *271*, 821.

[2] R. Riedel, A. Kienzle, W. Dressler, L. Ruwisch, J. Bill, F. Aldinger, *Nature* **1996**, *382*, 796.

[3] D. Seyferth, H. Plenio, *J. Am. Ceram. Soc.* **1990**, *73*, 2131.

[4] K. Su, E. E. Remsen, G. A. Zank, L. G. Sneddon, *Chem. Mater.* **1993**, *5*, 547.

[5] M. Weinmann, J. Schuhmacher, H. Kummer, S. Prinz, J. Peng, H. J. Seifert, M. Christ, K. Müller, J. Bill, F. Aldinger, *Chem. Mater.* **2000**, *12*, 623.

[6] M. Weinmann, S. Nast, F. Berger, G. Kaiser, K. Müller, *Appl. Organomet. Chem.* **2001**, *15*, 867.

[7] M. Weinmann, T. W. Kamphowe, P. Fischer, F. Aldinger, *J. Organomet. Chem.* **1999**, *592*, 115.

[8] D. Seyferth, G. H. Wiseman, *J. Am. Ceram. Soc.* **1984**, *67*, C-132.

[9] E. Kroke Y.-L. Li, C. Konetschny, E. Lecomte, C. Fasel, R. Riedel, *Mat. Sci. Eng.* **2000**, *R26*, 97 and literature cited therein.

[10] M. Weinmann, M. Hörz, F. Berger, A. Müller, K. Müller, F. Aldinger, *J. Organomet. Chem.* **2002**, *659*, 29.

[11] M. Weinmann, Habilitation-Thesis, Universität Stuttgart, **2003**.

[12] M. Weinmann, in preparation.

[13] H. J. Seifert, F. Aldinger, *Structure and Bonding* **2002**, *101*, 1.

Heterochain Polycarbosilane Elastomers as Promising Membrane Materials

N. V. Ushakov, E. Sh. Finkelshtein*

A. V. Topchiev Institute of Petrochemical Synthesis RAS,
29, Leninskii prospect, 119991, Moscow, Russia

*E. G. Krasheninnikov**

RRC Kurchatov Institute, 1, Acad. Kurchatov Square, 123182 Moscow, Russia

Keywords: silacyclobutanes, polymerization, polysiltrimethylenes, gas separation, membranes

Summary: This work is devoted to the systematic study of hydrocarbon gas separation parameters of polydimethylsilmethylene (**I**) and polysiltrimethylenes bearing various groups at the Si atom (R^1 = R^2 = Me (**II**), $CH_2Si(Me_2)Ph$ (**III**); R^1 = Me, R^2 = CH_2SiMe_3 (**IV**), $(CH_2)_3SiMe_3$ (**V**), *m*-Tol (**VI**)) deposited on polyamide hollow fibers.

Earlier, S. A. Stern showed that poly(dimethyl)siltrimethylene elastomer $–Si(Me_2)CH_2–$ has separation properties with respect to gas mixtures containing CH_4, CO_2, and C_3H_8 [1]. We have published comparative gas separation data for poly(dimethyl)silmethylene and polysiltrimethylenes having bulky groups such as α-naphthyl and cyclohexyl at Si atoms [2]. In this work, we realized the first measurement and systematic study of gas separation parameters of polysiltrimethylenes bearing various substituents at Si atoms ($–SiR_2CH_2CH_2CH_2–$). Most of the studied polycarbosilanes have been prepared first by us according to the ring-opening polymerization (ROP) (Scheme 1) using mono- and disilacyclobutanes as monomers. Unknown earlier, **IIIm** was synthesized by interaction of 2.5 equivalents of $Ph(Me_2)SiCH_2MgCl$ with 1.0 equivalent of 1,1-dichloro-1-silacyclobutane in Et_2O or THF followed by heating the reaction mixture at 55 – 60 °C for 3 h. Distillation *in vacuo* gave a wide fraction containing **IIIm** (130 – 155 °C/0.1 mm Hg). The latter was purified by means of column liquid chromatography on neutral alumina using hexane as a solvent (R_f = 0.63 ± 0.01). Distillation *in vacuo* gave **IIIm** with 72 % yield (b.p. 149 – 153 °C/0.01 mmHg; n_D^{20}1.5503; d_4^{20}0.9649); ^1H NMR (Bruker MSL-300, $CDCl_3$, δ(ppm): 0.40 (s, 4H, $SiCH_2Si$), 0.52 (s, 12H, $SiCH_3$), 1.16 (t, 4H, α-CH_2 cycl), 2.18 (m, 2H, β-CH_2 cycl), 7.50 (m, 6H, *m*- and *p*-protons arom.), 7.67 (m, 4H, *o*-protons arom.). Analysis Calcd. for $Si_3C_{21}H_{32}$ (%): Si 22.85, C 68.40, H 8.75. Found (%): Si 23.11, C 68.32, H 8.49.

ROP of **I** and **II** has been realized by two methods – thermally [3] and catalytically in the presence of Pt complexes [4]. The best way to prepare polysilalkylenes of high molecular weight ($\geq 10^6$) and narrow MMD (1.02 – 1.7) is by thermoinitiated ROP of highly pure monomers. Both types of ROP have been realized according to techniques published in [5], resulting in 75 – 94% polymer yields. Polymer structures were confirmed by ^{1}H NMR spectra. For **V**, ^{1}H NMR spectrum ($CDCl_3$) contains δ(ppm) –0.078 s and –0.027 s (CH_3Si and $(CH_3)_3Si$ corresp.), 0.549 t and 0.573 t ($SiCH_2C$ main and side chains) and ~1.32 m ($SiCCH_2C$ main and side chains).

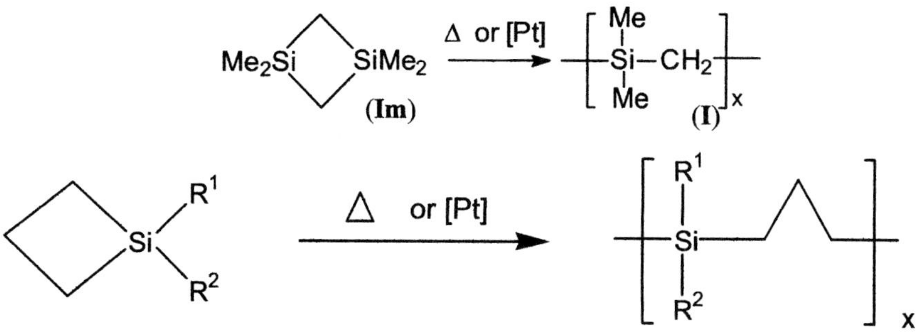

$R^1 = R^2$ = Me (**IIm**), CH_2SiMe_2Ph (**IIIm**);
R^1 = Me, R^2 = CH_2SiMe_3 (**IVm**),
$(CH_2)_3SiMe_3$ (**Vm**), *m*-Tol (**VIm**)

$R^1 = R^2$ = Me (**II**), CH_2SiMe_2Ph (**III**);
R^1 = Me, R^2 = CH_2SiMe_3 (**IV**),
$(CH_2)_3SiMe_3$ (**V**), *m*-Tol (**VI**)

Scheme 1.

These are the ways to control polymer molecular weight on thermoinitiated polymerization:

- The direct relationship between molecular weight of polysilalkylenes ring-opening polymerization and temperature allows us to obtain polymers of desired molecular weight and narrow M_{WD} (1.01 – 1.3) by precisely controlling reaction temperature. This relationship for polymerization of methyl derivatives of mono- and disilacyclobutanes is displayed in Fig. 1.
- Preparation of low-molecular-weight polysilalkylenes can be by the use of stop reagents. Figure 2 shows the relationship between polydimethylsiltrimethylenes η and quantity of a stop reagent (here *n*-hexylalkohol) added to the starting monomer.

Polymers **I** – **VI** were deposited on polysulfone (PSF) and polyamide (PA) hollow fibers according to the technique [6]. Gas transport parameters of the prepared composite membranes for the pair C_4H_{10}/CH_4 have been measured. The results are presented in Table 1.

The polymer **I** deposited on PA hollow fibers has also been investigated over a wide range of upstream pressures (0.2 – 1.6 atm). The results are reported in Fig. 3.

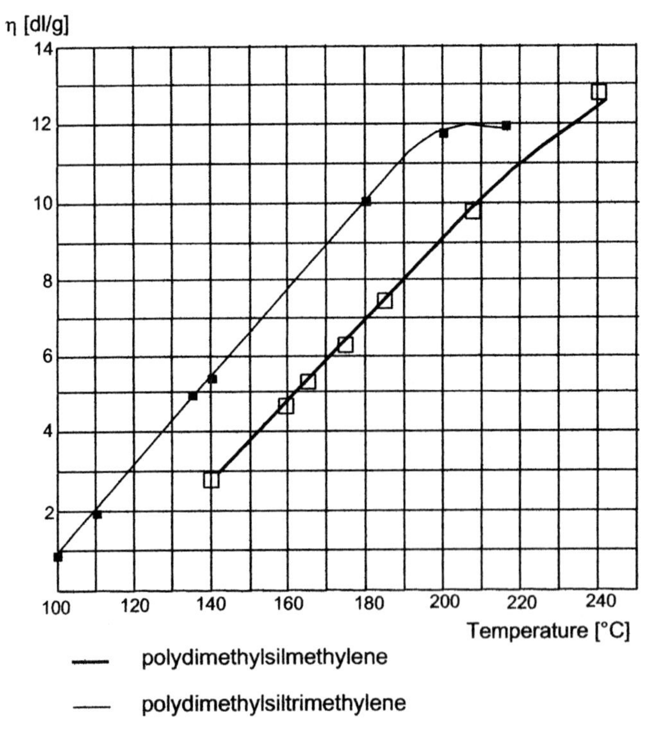

Fig. 1. Relationship between polysilalkylenes η and polymerization temperature.

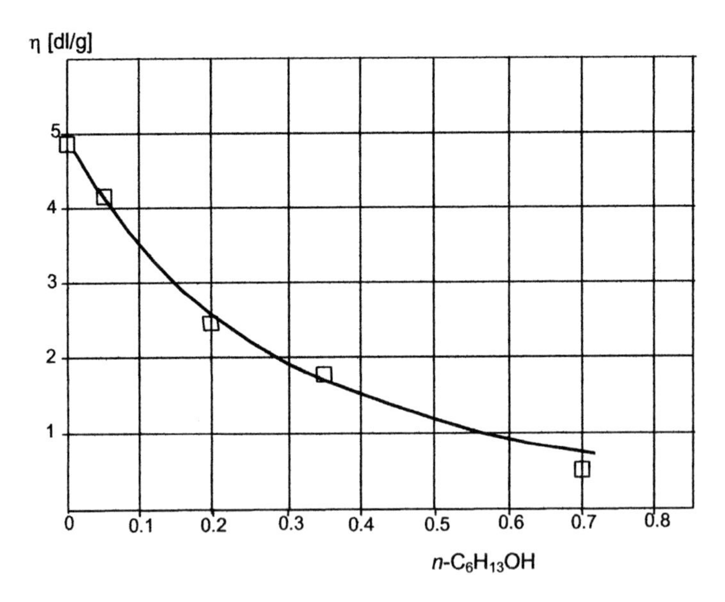

Fig. 2. Relationship between M_W of polydimethylsiltrimethylene and quantity of M_W moderator n-$C_6H_{13}OH$ added to the starting monomer.

Table 1 shows that the best permeation and separation parameters are in polysilmethylene (**I**) and polysiltrimethylene (**II**) (α C_4H_{10}/CH_4 = 27 – 39). Polysiltrimethylene (**IV**) having Me and Me_3SiCH_2 substituents also demonstrated a high separation factor α C_4H_{10}/CH_4 = 28.1, but substantially lower permeability. Summing up all the results, it is possible to conclude that increase in substitute volume leads to increase in T_g and decrease in gas separation parameters.

Table 1. Permeability of polymers and ideal separation factors for composite hollow fiber membranes (20 °C).

Polymer	Mol. *W.* (M_w/M_n)	T_g, [°C]	Hollow Fiber	Permeability [$Lm^{-2}h^{-1}$]		α O_2/N_2	Permeability [$Lm^{-2}h^{-1}$]		α C_4H_{10}/CH_4
				O_2	N_2		C_4H_{10}	CH_4	
I	357 000 (3.43)	–92	PSF	79.6	39.8	2.0	3650	94.0	38.8
I	1 600 000 (1.13)	–92	PA	55.1	20.5	2.68	2219	61.4	36.1
II	1 400 000 (1.16)	–75	PA	14.1	5.79	2.44	457	16.8	27.2
III	1 200 000 (2.4)	–36	PA	71.9	29.9	2.40	1294	85.1	15.2
IV	1 650 000 (1.14)	–51	PA	5.04	2.0	2.52	194	6.91	28.1
V	1 420 000 (1.2)	–46	PA	8.26	3.56	2.32	118	4.5	26.2
VI	760 000 (1.44)	–25	PA	4.15	1.85	2.24	76.4	4.36	17.5

PSF: membrane supported by polysulfone hollow fibers. PA: membrane supported by polyamide hollow fibers.

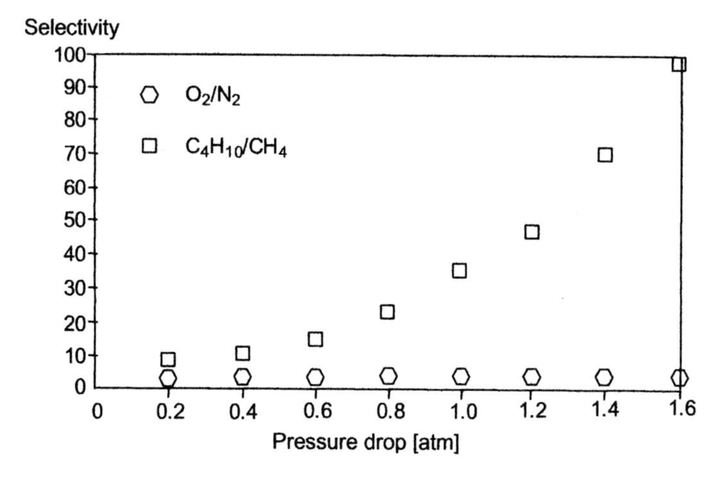

Fig. 3. Ideal separation factors as a function of pressure drop in the **I**/PA membrane.

Figure 3 shows that increase in gas pressure leads to a rise in C_4H_{10} permeability and, therefore, increase in α C_4H_{10}/CH_4 value.

Acknowledgment: This work has been supported by the NATO Science for Peace Programme (Project No. 972 638).

References

[1] S. A. Stern, V. M. Shah, B. J. Hardy, *J. Pol. Sci., Part B.* **1987**, *25*, 1263.

[2] E. Sh. Finkelshtein, N. V. Ushakov, E. G. Krasheninnikov, Yu. P. Yampolskii, *Organosilicon 33Symposium,* April 6–8, **2000**, Saginaw, Michigan, USA. Program and Abstracts, PB.24.

[3] N. S. Nametkin, V. M. Vdovin, V. I. Zavialov, *Izv. Akad. Nauk SSSR, Ser. Chim.* **1965**, 1448.

[4] D. R. Weyenberg, L. E. Nelson, *J. Org. Chem.* **1965**, *30*, 2618.

[5] N. V. Ushakov, E. Sh. Finkelshtein, E. D. Babich, *J. Polym. Sci.* **1995**, *37*, 320.

[6] The Method of coating was developed in collaboration with the RRS Kurchativ Institute.

Chemical Functionalization of Titanium Surfaces

D. Cossement, Z. Mekhalif, J. Delhalle, L. Hevesi*

University of Namur, Department of Chemistry
61, rue de Bruxelles, B-5000 Namur, Belgium
Tel.: +32 81 724538 — Fax: +32 81 725451
E-mail: laszlo.hevesi@fundp.ac.be

Keywords: titanium, silanization, radical initiator, polymer film, surface chemistry

Summary: AIBN-type radical polymerization initiators have been grafted onto poly-crystalline titanium surfaces allowing synthesis of polymer films covalently bound to the surfaces. Vinylic monomers such as styrene, methyl methacrylate, and 4-chloromethyl-styrene have been used; the pendant benzyl chloride moiety present at the outer surface of the polymer film obtained from the latter monomer has allowed further functionalization of the system. In the case of polystyrene films on Ti, molecular weights of the polymer have been estimated to be $M_w \approx 25\ 000$; $M_n \approx 10\ 000$ (Pd ≈ 2.5).

Most metal surfaces interact with organic molecules through physisorption or chemical reactions of their tiny external oxide layer. It has been shown that, under suitable conditions, organosilane coupling agents [1] are able to form Metal–O–Si bonds that are strong enough to ensure long term stability of the organic layer on the metal surface sufficient for various applications.

In the case of titanium, a number of earlier [2] and more recent [3] studies revealed that Ti–O–Si bonds can be formed using alkoxysilanes [2, 3], chlorosilanes [4], or hydrosilanes [5]. In the latter case, Ti–O–Si bonds resulted from the reaction of surface Ti–OH groups with the reagent's Si–H groups accompanied by molecular hydrogen production. This type of reaction was also shown to be operative between hydrosilanes and Si–OH-containing surfaces [6].

We now report our results on the *in situ* formation of thin polymer films covalently bound to polycrystalline titanium surfaces. As compared to the existing strategies [7], the novelty of our approach resides in the preparation of self-assembled monolayers (SAM) of azo-bis(*iso*butyronitrile), AIBN-type radical chain initiators bearing trichlorosilyl ($-SiCl_3$) end groups allowing for easy and efficient chemical grafting onto the TiO_2-covered titanium surfaces [8]. Dipping the obtained titanium plates into solutions of vinyl monomers at high enough temperature to start homolysis of the grafted initiators led to the formation of various chemically grafted polymer films. A very similar methodology has been described for the preparation of chemically grafted vinyl polymer films on silica (SiO_2) surfaces [9].

Results

Two AIBN-type radical polymerization initiators have been synthesized starting from commercial 4,4′-azo-bis(4-cyano)valeric acid (**1**) (Scheme 1): ACTP {3″-trichlorosilylpropyl-4,4′-azo-bis(4-cyano)valerate} (**4a**) and ACTU {11″-trichlorosilylundecyl-4,4′-azo-bis(4-cyano) valerate} (**4b**). After thorough drying of the wet diacid **1**, it was transformed into its chloride (**2**) using an excess of phosphorus pentachloride in dichloromethane. Reaction of **2** with the unsaturated alcohols allyl alcohol and 10-undecen-1-ol respectively gave the esters **3** in high yields which in turn led to ACTP (**4a**) and ACTU (**4b**) on Pt-catalyzed hydrosilylation.

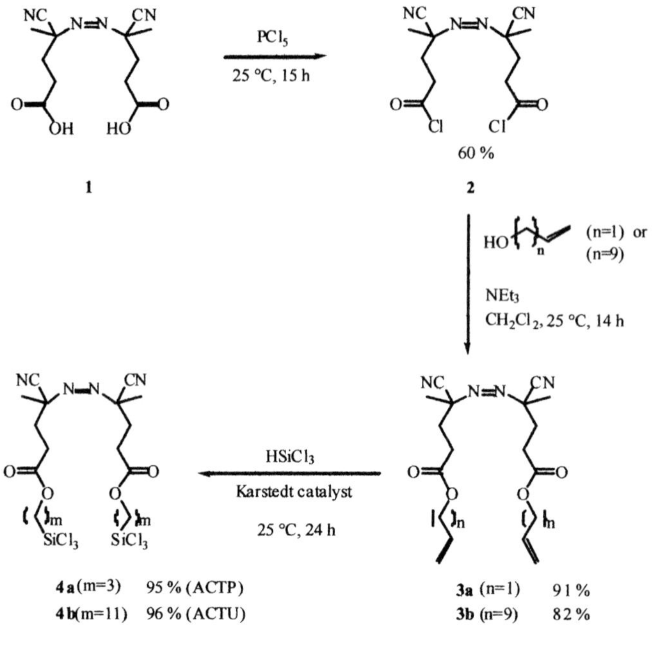

PCl_5, 25 °C, 15 h

60 %

(n=1) or (n=9)

NEt_3, CH_2Cl_2, 25 °C, 14 h

$HSiCl_3$, Karstedt catalyst, 25 °C, 24 h

4a(m=3) 95 % (ACTP)
4b(m=11) 96 % (ACTU)

3a (n=1) 91 %
3b (n=9) 82 %

Scheme 1.

Detailed study of the effect of various sample preparation conditions using octadecyltrichlorosilane as coupling agent led us to conclude that careful polishing of the commercial titanium plates (followed by dipping into a 10^{-3} molar toluene solution of the coupling agent for 2 h) gave good quality densely-packed self-assembled monolayers (SAM).

The two initiators ACTP and ACTU were grafted onto polished titanium plates according to the procedure worked out for octadecyltrichlorosilane. Water drop contact angles measured for the polished titanium (Ti_{pol}), Ti-ACTP, and Ti-ACTU samples were the following: 62°, 75°, and 85°, respectively. These values are in good agreement with the value of 73° measured for a SAM-covered glass surface obtained from 11-cyanoundecyltrichlorosilane [10]. The somewhat higher value of 85° measured for Ti-ACTU may indicate a better organized monolayer.

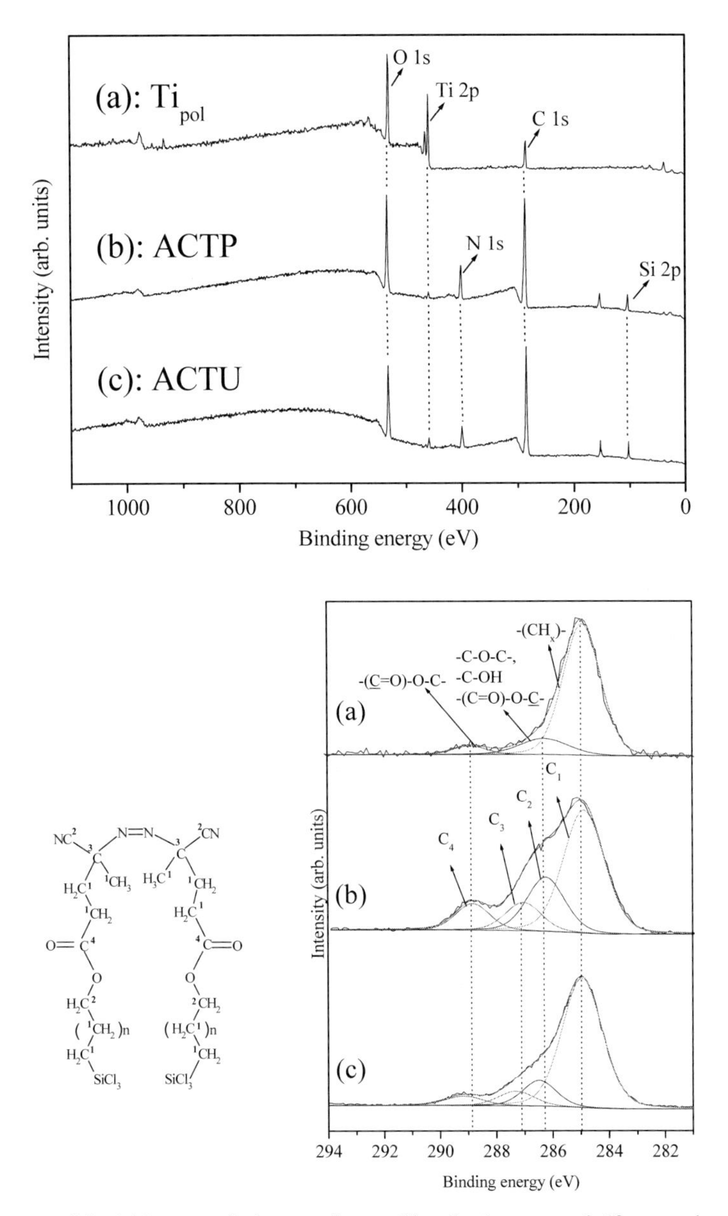

Fig. 1. XPS spectra of the initiators grafted onto polycrystalline titanium; upper half: general spectra, lower half: core level C1s spectra before (a) and after grafting of ACTP (b) and ACTU (c).

Figure 1 shows the XPS spectra corresponding to these titanium samples together with details of the C1s regions. It is important to note concerning the (Ti_{pol}) sample that the XPS spectra show the presence of a significant amount of carbonaceous contaminants. However, in the course of our preliminary study with octadecyltrichlorosilane, we have shown that those contaminants almost completely disappear from the surface during grafting of the silane coupling agent. Therefore, we are quite confident about the efficient grafting of the two radical polymerization initiators also supported by the N1s portion of the spectra (not shown), exhibiting the expected two components in a 1:1 ratio and located at 399.7 and 401.2 eV (–CN and N=N nitrogens, respectively).

In the next step, we carried out the polymerization experiments using three vinylic monomers, i.e. styrene, methyl methacrylate, and 4-chloromethyl-styrene, by dipping the appropriate titanium plates into deoxygenated monomer/toluene solutions (1:2, v/v) at 90 °C for 24 h. Interestingly, grafted initiators ACTP and ACTU gave very comparable polymer films from the two styrene monomers, whereas polymerization of methyl methacrylate was found to be significantly better on Ti-ACTP than on Ti-ACTU.

As can be seen in Fig. 2A, extensive polymerization of 4-chloromethyl-styrene has taken place under the conditions used, since after polymerization (Fig. 2A(b)) only carbon and chlorine signals can be observed in the XPS spectra. Moreover, the detailed C1s spectrum (Fig. 2B) exhibits the characteristic peaks of the expected *poly*-(4-chloromethyl-styrene).

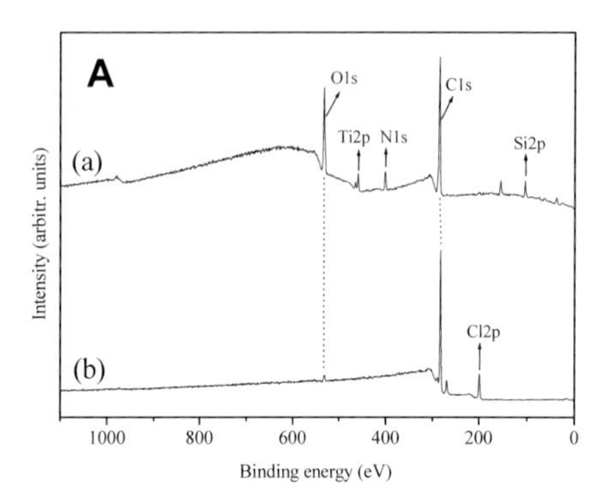

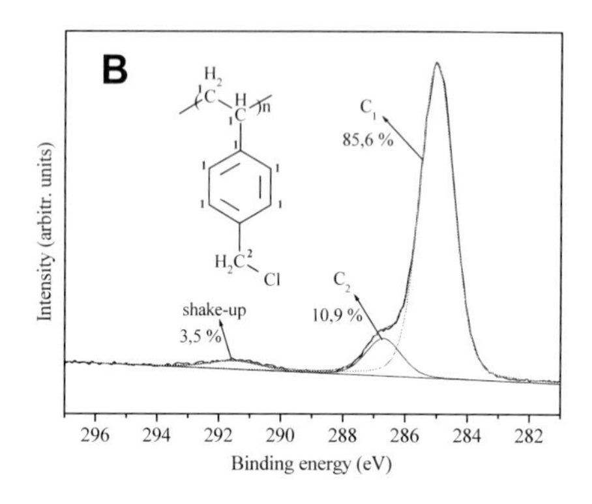

Fig. 2. **A**: XPS spectra of (a) ACTU-grafted titanium plate and (b) the same after polymerization (90 °C, 24 h) of 4-chloromethyl-styrene; **B**: C1s spectrum of *poly*-(4-chloromethyl-styrene) film obtained in (b).

At this stage we examined the efficiency of further functionalizations at the surface of *poly*-(4-chloromethyl-styrene) films bound to the titanium plates. For this purpose the plates were immersed in pure butylamine (24 h, reflux) and in a DMSO solution of sodium azide (10^{-3} M, 24 h, 65 °C), respectively.

Figure 3 shows XPS spectra relative to the former experiment: the chlorine signal (spectrum (a)) has been replaced by that of nitrogen (spectrum (b)). The latter signal (not shown) has two components [11]: one assigned to the free amine appearing at lower energy (400 eV), and the other corresponding to the ammonium form appearing at higher energy (402 eV). In addition, a C/N ratio of 12.9 was observed, which compares well with the theoretical value of 13.

These spectroscopic data are fully supported by the measured contact angles: 100° for the starting hydrophobic *poly*-(4-chloromethyl-styrene) film and 81° for the more hydrophilic *poly*-(4-butylaminomethyl-styrene) film.

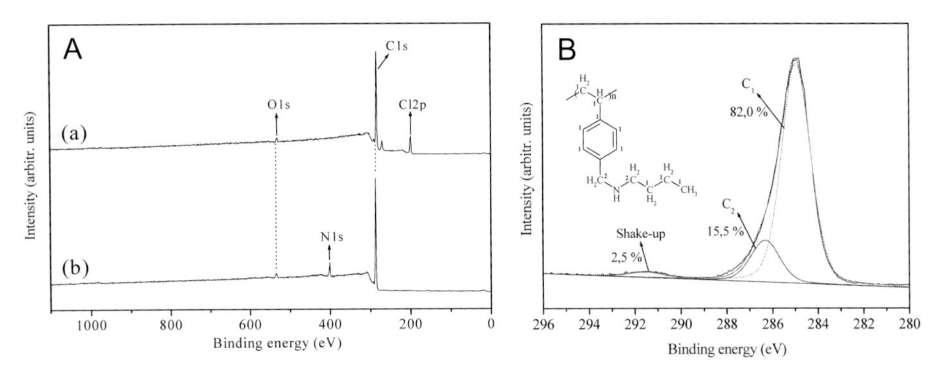

Fig. 3. **A**: XPS spectra of (a) *poly*-(4-chloromethyl-styrene)-covered titanium plate and (b) the same after reaction with *n*-butylamine (24 h, reflux); **B**: C1s spectrum of *poly*-(4-butylaminomethyl-styrene) film obtained in (b).

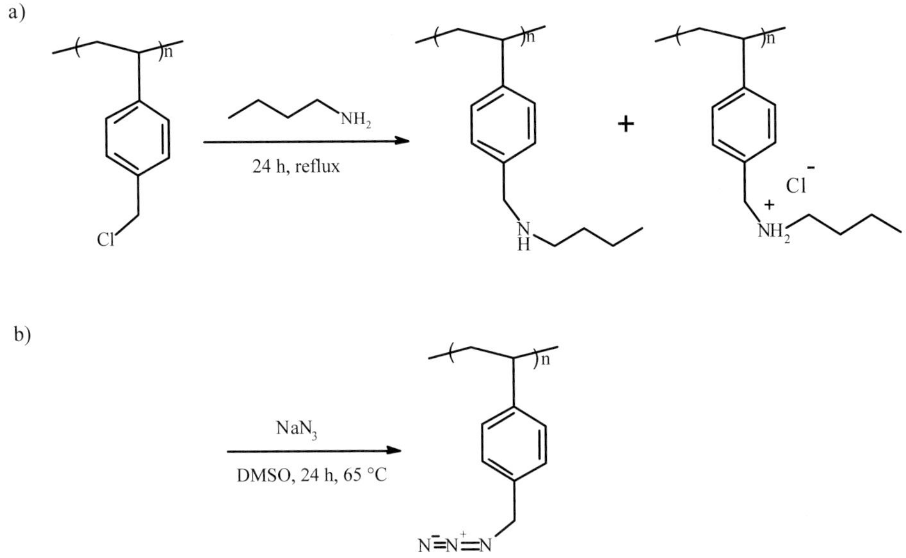

a)

NH_2

24 h, reflux

b)

NaN_3

DMSO, 24 h, 65 °C

Scheme 2.

In a similar fashion, all the data (contact angle and XPS) were in agreement with an efficient substitution reaction of the benzylic chloride atoms by azide ions (Scheme 2b), except for the observed too high C/N ratio of 4.9 instead of the theoretical value of 3.0. It has been reported [12], however, that X-ray exposure can cause loss of azide groups, thereby increasing the C/N ratio.

For the purpose of molecular weight determination 10 g titanium dioxide (rutile) was grafted with ACTU in the same way as were the plates. Polymerization of styrene was carried out as previously; the polymer-covered rutile was thoroughly extracted with toluene in a Soxhlet apparatus. Polystyrene was then detached from the TiO_2 particles as shown in Scheme 3 and subjected to molecular weight measurements by size exclusion chromatography (SEC), which gave the results: $M_w \approx 25\,000$; $M_n \approx 10\,000$ (Pd ≈ 2.5).

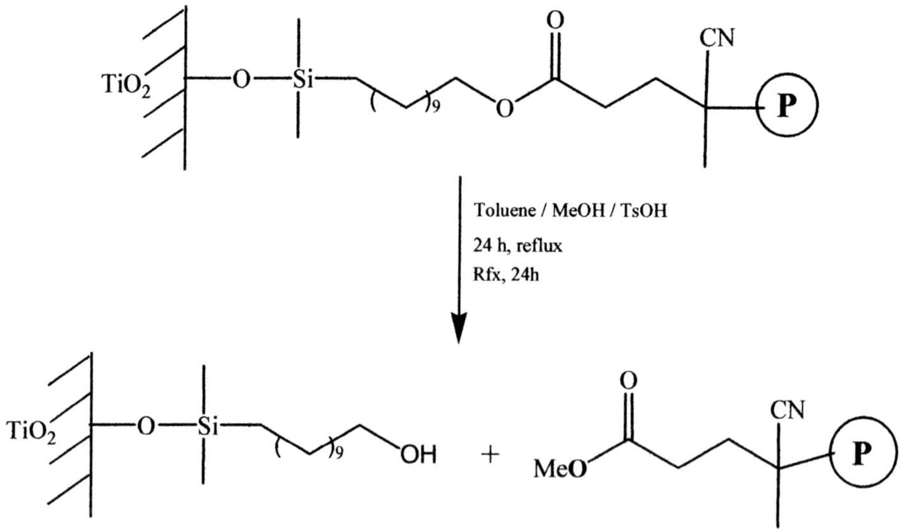
TiO2
O
Si
9
O
O
CN
P
Toluene / MeOH / TsOH
24 h, reflux
Rfx, 24h
TiO2
O
Si
9
OH
+
O
MeO
CN
P

Scheme 3.

Thus, we can conclude that the method reported here is a convenient one for grafting chemically thin polymer films onto oxidized metal surfaces. The grafted polymer films can then be further functionalized for various technological or biomedical applications.

Acknowledgment: One of us (D. C.) gratefully acknowledges the *Fonds pour la Recherche dans l'Industrie et l'Agriculture* (FRIA) for predoctoral fellowships. The authors also thank Professors R. Legras and J. Devaux (Laboratoire de Physique et de Chimie des Hauts Polymers, Université Catholique de Louvain, Louvain-la-Neuve) for their help with the molecular weight determinations.

References

[1] Plueddemann, E. P. *Silane Coupling Agents*; Plenum Press, New York, **1982**.

[2] a) T. Osa, M. Fujihara, *Nature* **1976**, *264*, 349 – 350; b) P. R. Moses, R. W. Murray, *J. Am. Chem. Soc.* **1976**, *98*, 7435 – 7436 ; c) M. Fujihara, N. Ohishi, T. Osa, *Nature* **1976**, *268*, 226 – 228; d) P. R. Moses, L. M. Wier, J. C. Lennox, H. O. Finklea, J. R. Lenhard, R. W. Murray, *Anal. Chem.* **1978**, *50*, 576 – 585 ; e) H. O. Finklea, R. W. Murray, *J. Phys. Chem.* **1979**, *83*, 353 – 358.

[3] a) L. Gamble, M. B. Hugenschmidt, C. T. Campbell, T. A. Jurgens, J. W. Jr. Rogers, *J. Am. Chem. Soc.* **1993**, *115*, 12096 – 12105; b) T. A. Jurgens, J. W. Jr. Rogers, *J. Phys. Chem.*

1995, *99*, 731 – 743; c) L. Gamble, L. S. Jung, C. T. Campbell, *Langmuir* **1995**, *11*, 4505 – 4514; d) L. Gamble, M. A. Henderson, C. T. Campbell, *J. Phys. Chem. B* **1998**, *102*, 4536 – 4543.

[4] a) D. Cossement, Y. Delrue, Z. Mekhalif, J. Delhalle, L. Hevesi, *Surf. Interface Anal.* **2000**, *30*, 56 – 60; b) D. Cossement, C. Pierard, J. Delhalle, J.-J. Pireaux, L. Hevesi, Z. Mekhalif, *Surf. Interface Anal.* **2001**, *31*, 18 – 22; c) D. Cossement, F. Plumier, J. Delhalle, L. Hevesi, Z. Mekhalif, *Synth. Met.* **2003**, *138*, 529 – 536.

[5] A. Y. Fadeev, T. J. McCarthy, *J. Am. Chem. Soc.* **1999**, *121*, 12184 – 12185.

[6] a) J. N. Greeley, L. M. Meeuwenberg, M. M. B. Holl, *J. Am. Chem. Soc.* **1998**, *120*, 7776 – 7782; b) J. Moineau, *Doctoral Dissertation*, Université de Montpellier II, **2002**.

[7] See for example: G. J. Fleer, M. A. Cohen Stuart, J. M. H. M. Scheutjens, T. Cosgrove, B. Vincent, *Polymers at Interfaces*, Chapman & Hall, London, **1993** and references therein.

[8] D. Cossement, *Doctoral Dissertation*, University of Namur, **2002**.

[9] a) O. Prucker, J. Rühe, *Macromolecules* **1998**, *31*, 592 – 601; b) O. Prucker, J. Rühe, *Macromolecules* **1998**, *31*, 602 – 613; c) O. Prucker, J. Rühe, *Langmuir* **1998**, *14*, 6893 – 6898; d) J. Habicht, M. Schmidt, J. Rühe, D. Johannsmann, *Langmuir* **1999**, *15*, 2460 – 2465; e) H.-J. Butt, M. Kappl, H. Mueller, R. Raiteri, W. Meyer, J. Rühe, *Langmuir* **1999**, *15*, 2559 – 2565; f) M. Ruths, D. Johannsmann, J. Rühe, W. Knoll, *Macromolecules* **2000**, *33*, 3860 – 3870.

[10] L. Jeanmart, *PhD Thesis*, University of Namur, **2001**.

[11] G. Beamson, D. Briggs, *High Resolution XPS Spectra of Organic Polymers,* The Scienta ESCA A300 Database, John Wiley and Sons, Chichester, **1992**.

[12] H. Heise, M. Stamm, M. Rauscher, H. Duschner, H. Menzel, *Thin Solid Films* **1998**, *199*, 327.

Documentation of Silicones for Chemistry Education and Public Understanding

Michael W. Tausch

University of Duisburg-Essen, Institute of Chemistry Duisburg
Lotharstrasse 1, 47057 Duisburg
Tel.: +49 203 379 2207 — Fax: +49 203 379 1729
E-mail: *M.Tausch@uni-duisburg.de*

Keywords: silicones, chemical education, public understanding

Summary: As silicones represent compounds situated at the border between organic and inorganic chemistry, they constitute excellent examples for introducing and communicating core principles of chemistry. On the basis of laboratory experiments and the dialogue between chemists from the industry and from the university a CD-ROM has been developed and linked via "www.theochem.uni-duisburg.de/DC" into the world wide web. The CD-ROM contains information on the structure, properties, preparation, and applications of silicones as well as experiments, details on the chemical background, teaching suggestions, and interactive modules. The main goal of this product is to achieve a contribution to the modernization of chemistry curricula in schools and universities. Furthermore, it is suitable for improvement of the public understanding of science, technology, and industrial products.

Features of the CD-ROM

- the CD-ROM runs in an auto starting modus
- it is equipped with a clearly arranged introductory module as well as with an appealing module "search for the silicone" ("Finden Sie das Silicon")
- the navigation is comfortable and self-explaining
- both German and English versions are available via WACKER-Chemie GmbH Munich
- a Chinese version will appear in 2005

Contents of the CD-ROM

- information on the preparation, properties, and applications of silicones
- 20 experiments (procedures, discussions, and teaching suggestions)
- 43 worksheets

- 8 videos (Windows-Media or Quicktime)
- 2 interactive Flash-animations
- 38 slides (MS-Powerpoint, MS-Word or Acrobat Reader)
- for further information see "www.theochem.uni-duisburg.de/DC"

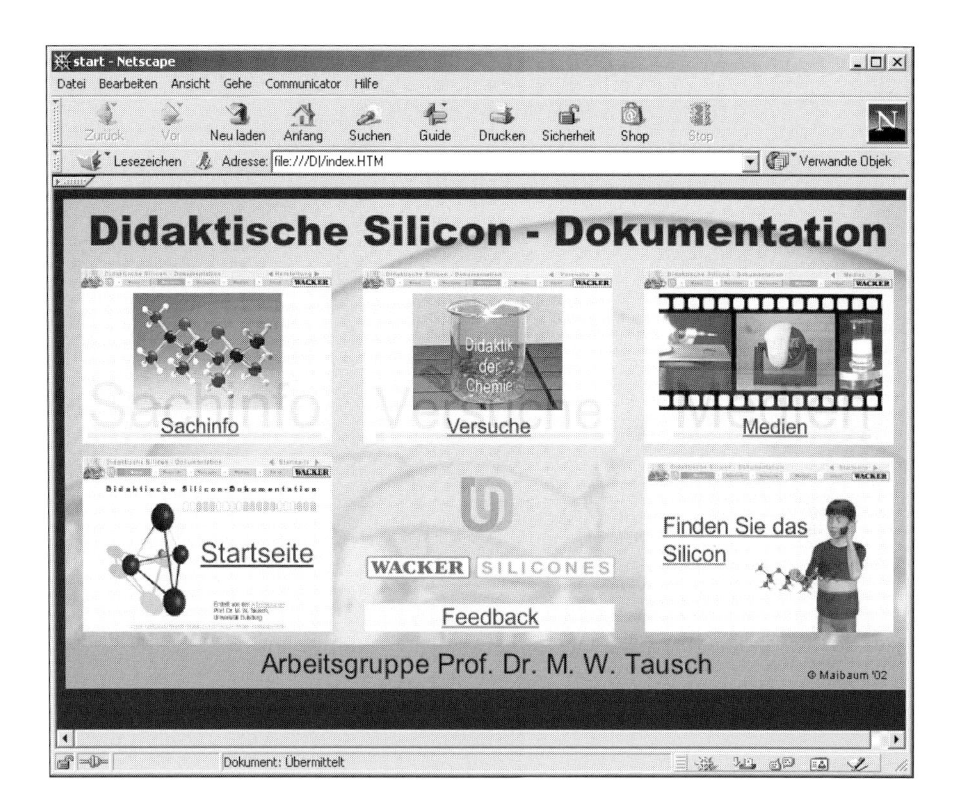

References

[1] M. W. Tausch, *Chemie S II Stoff-Formel-Umwelt* (Textbook for high schools and colleges), C. C. Buchner, Bamberg (**1993**, **1999**, **2001**).

[2] M. W. Tausch, W. Held, J. Weis, Silicone, Special edition of the journal *Praxis der Naturwissenschaften – Chemie in der Schule*, **2002**, *51*(*7*).

Author Index

Subject Index

T